Chemistry

for advanced level

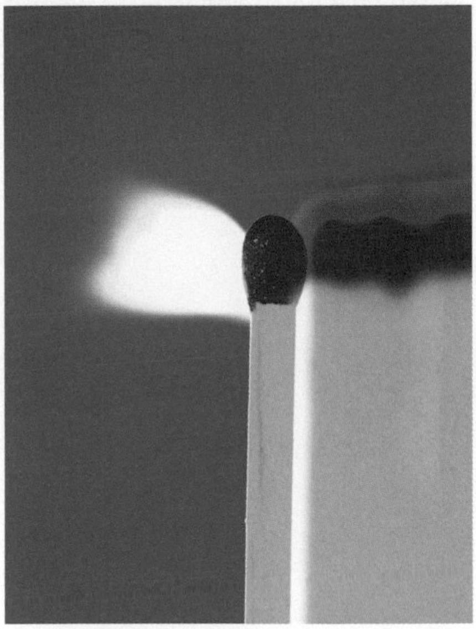

Peter Cann
Peter Hughes

JOHN MURRAY

Cover:
The reaction occurring in a match head during ignition is an example of an exothermic redox reaction (see Chapter 7) with a low activation energy (see Chapter 8).

The match head consists of a mixture of the oxidant potassium chlorate (V) (see Chapter 17) and the reducing agents sulphur and tetraphosphorus trisulphide, embedded in a solid glue matrix. The frictional heating caused by striking the match on a rough surface gives the reactants enough energy to overcome the activation barrier, and the exothermic reaction proceeds with ever-increasing rapidity, setting light to the wood of the match stick.

$$6KClO_3 + S + P_4S_3 \longrightarrow 6KCl + 4SO_2 + P_4O_{10}$$

First published in 2002
by John Murray (Publishers) Ltd
50 Albemarle Street
London W1S 4BD

Layouts by Eric Drewery
Artwork by Barking Dog Art
Cover design by John Townson/Creation

Typeset in 10/12pt Goudy by Servis Filmsetting Ltd, Manchester
Printed and bound in Spain by Bookprint, S.L., Barcelona

A catalogue entry for this title is available from the British Library

ISBN 0 7195 8602 X

Contents

PHYSICAL CHEMISTRY

INORGANIC CHEMISTRY

ORGANIC CHEMISTRY

Introduction

Chemists are inquisitive creatures. So are students. In our teaching experience, spanning over three decades, we have found that students who choose to study chemistry at the post-16 stage of their education are doubly blessed with this desire to understand what makes the world tick. They possess an enquiring attitude that demands reasoned, logical answers. Given the right stimulus and encouragement, this can often result in a lively interaction between student and teacher, and between student and student.

Chemistry is first and foremost an empirical science. It is best practised and understood in the manner in which it started, at the laboratory bench. Virtually all our present-day chemical knowledge of the properties of elements and compounds, and how they react together to form new materials, has been discovered by skilled experimentalists, often with a degree of serendipity. Over the past century and a half, however, chemists have become increasingly interested in developing theories to explain their laboratory-based observations, and to predict new lines of discovery. These range from the simple, yet fundamentally important, ideas of van't Hoff and others in the 1870s, about the fixed three-dimensional shapes of molecules, to the more abstruse quantum-mechanical explanations of bonding and reactivity that were developed in the 1920s. These had to await the arrival of the powerful computing facilities of the latter decades of the twentieth century to realise their full predictive potential.

Chemistry today is an exciting mix of the practical and the theoretical. Theory plays an increasing part in helping the experimentalist to design and carry out investigations. Conversely, theoretical chemists relish the challenge of explaining the new and unexpected discoveries of the experimentalist, such as high-temperature superconductors, or the chemical properties of the new allotropes of carbon called the fullerenes. The integrated nature of the subject is clearly illustrated by recent advances in the chemical applications of spectroscopy. Whereas in the first half of the twentieth century it may have taken an organic chemist a year or so to work out the molecular structure of a complex naturally-occurring compound, the same problem can be solved today in a matter of hours using modern physical techniques.

Chemistry is the central science, at the crossroads of biology and its associated disciplines on the one hand, and physics on the other. Chemistry relies on physics for its understanding of the fundamental building blocks of matter, and biology relies on chemistry for an understanding of the structures of living organisms, and the processes that go on inside them that we call life. Standing at this crossroads, the chemist is uniquely positioned to understand, and make significant contributions to, many interdisciplinary areas of current and future importance. The chemistry-based sciences of biochemistry, genetic engineering, pharmacology, and polymer and material science will all make increasing contributions to our physical and material well-being in the future. Chemists are also playing a key role in the fight against industrial society's pollution of our environment.

About this book

This book attempts to go some way towards satisfying the curiosity of the able student, and to answer the questions of the inquisitive. Although based firmly on the AS and A2 specifications of the various Examination Boards, it extends these topics in areas where an application, or a more fundamental explanation, is deemed to be appropriate. These extensions are clearly delimited from the main text in panels, and can be bypassed on first reading.These panels are also used to describe topics (mostly at A2 level) that are included in only one of the Examination Boards' specifications. The reader can readily establish the relevance of these panels to their course by referring to the Specification matrix on page viii.

It would be wrong, however, to consider this a book just for the high flier. The majority of students starting an AS course in chemistry come from a background of GCSE Science: Double Award, and the initial chapters start at a level and a pace that is more suited to all such students. Because of the changes in the emphasis of school mathematics courses over the past few decades, the mathematical concepts required for chemistry, although simple in principle and few in number, are not practised as much by 13–16-year-old students as they used to be. Some of these students come to AS chemistry with the belief that they will find the mathematics difficult. We hope to demonstrate that, as long as the processes are understood, rather than learned by rote, the mathematics in both the AS and A2 chemistry courses is well within the grasp of those who have gained a grade C at GCSE.

Students also sometimes consider that chemistry is a subject full of difficult concepts. This is not true. Most of chemistry is based on the very simplest idea of electrostatics: like charges repel; unlike charges attract. When the subtle ramifications of this generalisation are studied during the AS/A2 course, students should constantly remind themselves of the inherent simplicity of this relationship.

Apart from the extension materials mentioned above, the reader will also find within the text worked examples, illustrating the application of the material just read; and further practice questions, which allow students to try out their newly found understanding. Numerical answers to these questions can be found at the end of the book.

Each chapter starts with an outline of its contents, and concludes with a summary of the key facts and ideas that have been introduced. At the end of the book the reader will find questions taken from the recent past papers of the Examination Boards.

Whilst the AS specification is quite closely prescribed, and the various Boards' specifications contain a large degree of overlap, their A2 specifications vary widely. To help the reader decide which chapters or topics are relevant to his or her particular course of study, we have produced a matrix showing the relationship of the contents of this book to the specification of each Board.

AS and A level specification/syllabus matrix

Board	AQA					Edexcel				OCR						WJEC				CCEA				CIE		
Module	1	2	3	4	5	1	2	4	5	2811	2812	2813	2814	2815	2816	1	2	4	5	1	2	4	5	Physical	Inorganic	Organic
1	✔					✔				✔						✔	✔	✔		✔				✔		
2	✔					✔				✔						✔				✔				✔		
3	✔					✔				✔						✔				✔				✔		
4	✔					✔				✔				✔		✔				✔				✔		
5		✔					✔					✔					✔			✔				✔		
6				✔	✔			✔				✔					✔			✔	✔	✔	✔			
7		✔				✔			✔	✔				✔		✔			✔	✔		✔	✔	✔		
8		✔					✔					✔					✔				✔			✔		
9		✔					✔					✔					✔				✔			✔		
10					✔			✔						✔					✔	✔		✔		✔		
11				✔					✔						✔				✔		✔	✔	✔	✔		
12				✔				✔							✔		✔		✔			✔		✔		
13					✔				✔								✔		✔	✔		✔		✔		
14													✔	✔				✔		✔			✔			
15	✔				✔	✔		✔		✔				✔		✔			✔	✔		✔			✔	
16	✔					✔				✔				✔		✔			✔		✔				✔	
17		✔				✔				✔						✔			✔						✔	
18								✔											✔						✔	
19					✔				✔					✔					✔				✔		✔	
20		✔					✔					✔					✔					✔			✔	
21																			✔	✔	✔					
22			✔	✔			✔	✔	✔		✔		✔				✔	✔			✔	✔				✔
23			✔				✔				✔						✔				✔					✔
24			✔	✔			✔		✔		✔		✔				✔				✔	✔				✔
25			✔				✔	✔	✔		✔							✔			✔		✔			✔
26			✔				✔				✔							✔			✔					✔
27				✔					✔				✔				✔	✔					✔			✔
28				✔				✔			✔		✔					✔				✔				✔
29								✔	✔		✔		✔					✔				✔	✔			✔
30				✔				✔	✔				✔					✔					✔			✔
31				✔					✔				✔					✔					✔			

Modules in the tinted columns make up the A2 part of the course.
For a more detailed analysis, including the CIE Singapore syllabus, please contact the publisher.

Acknowledgements

Authors' acknowledgements

We should like to thank the editorial staff at John Murray for their invaluable help and freely-given advice, willingly provided both during the preparation of the manuscript and its subsequent transformation into this finished volume.

Our thanks are also due to our wives Margaret and Iris, for their support, fortitude and patience during the book's lengthy gestation period.

Examination boards

Examination questions have been reproduced with kind permission from the following examination boards:

AQA (Assessment and Qualifications Alliance)
CCEA (Northern Ireland Council for the Curriculum, Examinations and Assessment)
CIE: University of Cambridge Local Examinations Syndicate (UCLES)
Edexcel
OCR (Oxford, Cambridge and RSA Examinations)
WJEC (Welsh Joint Education Committee)

Artwork figure acknowledgement

Figures 23.3, 23.4, 23.6, 23.12, 23.13 Adapted from Chris Conoley and Phil Hills, *Collins Advanced Science: Chemistry*, 1998. Collins Educational.

Photo credits

The publisher is grateful to the following, who have kindly permitted the reproduction of copyright photographs:

Cover Tek Image/Science Photo Library; **p.1** Science Photo Library/James King Holmes; **p.2** Mary Evans Picture Library; **p.3** *tl* John Townson/Creation, *tr* John Townson/Creation, *c* Ace Photo Library, *bl* Environmental Images, *br* Colorsport; **p.5** Science Photo Library/Eye of Science; **p.7** *all* John Townson/Creation; **p.10** *all* John Townson/Creation; **p.16** Milepost 92½; **p.18** *all* John Townson/Creation; **p.22** *l* Science Photo Library, *r* Science Photo Library/A Barrington Brown; **p.24** Science Photo Library/Geoff Tompkinson; **p.26** Science Photo Library; **p.29** Science Photo Library/Martin Dohrn; **p.30** Science Photo Library/James King Holmes; **p.32** Science Photo Library/Geco UK; **p.39** *all* John Townson/Creation; **p.77** *all* John Townson/Creation; **p.87** Bruce Colman; **p.95** *tl, tr, cl, cr* John Townson/Creation, *bl, br* Science Photo Library; **p.100** *all* John Townson/Creation; **p.101** *l* John Townson/Creation, *r* Robert Harding Photo Library; **p.103** *all* John Townson/Creation; **p.115** Science Photo Library; **p.116** Rex Features; **p.117** Cussons; **p.120** Science Photo Library/NASA; **p.136** *all* John Townson/Creation; **p.137** *all* John Townson/Creation; **p.138** *all* John Townson/Creation; **p.153** *all* John Townson/Creation; **p.155** *all* John Townson/Creation; **p.161** *all* Andrew Lambert; **p.163** *all* Andrew Lambert; **p.167** *all* Science Photo Library; **p.168** Science Photo Library/Malcolm Fielding/Johnson Matthey; **p.171** Rex Features; **p.172** Science Photo Library; **p.192** *all* John Townson/Creation except *bl* Science Photo Library/NASA; **p.197** Science Photo Library; **p.201** *all* Science Photo Library; **p.202** *tl* Science Photo Library/G Brad Lewis, *tr* Science Photo Library/Charles D Winters, *b* Science Photo Library/Dr Kari Lounatmaa; **p.203** Science Photo Library/Perlstein, Jerrican; **p.204** *t* Science Photo Library/J C Revy, *c* Ace Photo Library, *b* Science Photo Library/George Haling; **p.209** Andrew Lambert; **p.235** Andrew Lambert; **p.239** *l* Science Photo Library, *r* John Townson/Creation; **p.249** John Townson/Creation; **p.255** John Townson/Creation; **p.259** John Townson/Creation; **p.261** Science Photo Library; **p.264** Mary Evans Picture Library; **p.265** John Townson/Creation; **p.266** John Townson/Creation; **p.270** Science Photo Library/Philippe Plailly; **p.271** *t* Science Photo Library/I Andersson, Oxford Molecular Biophysics Lab, *bl* Science Photo Library, *br* Science Photo Library/Alfred Pasieka; **p.274** *t* Courtesy of Hitachi Europe, *bl, br* John Townson/Creation; **p.276** John Townson/Creation; **p.277** Science Photo Library/Dept of Physics, Imperial College; **p.280** John Townson/Creation; **p.281** Science Photo Library/James King Holmes; **p.284** *t* Science Photo Library/Geoff Thompkinson, *b* Science Photo Library/James King Holmes; **p.285** Robert Harding; **p.286** Science Photo Library; **p.293** John Townson/Creation; **p.296** *t* Science Photo Library, *b, r* Oxford Scientific Films; **p.302** *tl, tc, tr, br* John Townson/Creation, *bl* Science Photo Library; **p.306** Andrew Lambert; **p.308** *all* John Townson/Creation; **p.311** Environmental Images; **p.315** Robert Harding; **p.322** *all* John Townson/Creation; **p.323** Science Photo Library/John Paul Kay, Peter Arnold Inc; **p.327** *all* John Townson/Creation; **p.329** Andre Lambert; **p.332** *all* John Townson/Creation; **p.336** Science Photo Library/Sinclair Stammers; **p.337** Science Photo Library/Astrid & Hans Frieder

Michler; **p.339** *all* John Townson/Creation; **p.340** *all* John Townson/Creation; **p.341** John Townson/Creation; **p.343** John Townson/Creation; **p.356** John Townson/Creation; **p.357** John Townson/Creation; **p.361** Ace Photo Library; **p.362** *all* John Townson/Creation; **p.364** *all* John Townson/Creation; **p.365** *t* John Townson/Creation, *b* Elizabeth Whiting; **p.369** Science Photo Library; **p.371** *all* Science Photo Library; **p.373** Science Photo Library; **p.375** Robert Harding; **p.376** Science Photo Library; **p.377** Skyscan/Peter Smith Photography; **p.378** Science Photo Library; **p.379** Science Photo Library; **p.380** *tl, tr, c* Science Photo Library, *b* Ace Photo Library; **p.381** John Townson/Creation; **p.386** *all* John Townson/Creation; **p.387** *all* John Townson/Creation; **p.388** Andrew Lambert; **p.393** Robert Harding; **p.399** *all* John Townson/Creation; **p.400** *all* John Townson/Creation; **p.407** Bellingham & Stanley; **p.408** *all* John Townson/Creation; **p.411** *all* John Townson/Creation; **p.421** *all* John Townson/Creation; **p.424** Ace Photo Library; **p.426** Science Photo Library/Maxmillian Stock; **p.429** Science Photo Library/David Frazier; **p.441** *all* John Townson/Creation; **p.444** John Townson/Creation; **p.450** John Townson/Creation; **p.456** Robert Harding; **p.457** *l* Kwikfit, *all* John Townson/Creation; **p.459** Science Photo Library; **p.464** *all* John Townson/Creation; **p.474** *l* Science Photo Library/NASA, *r* Colorsport; **p.479** *all* John Townson/Creation; **p.486** John Townson/Creation; **p.489** *tl* Science Photo Library/Andrew Syred, *bl* Science Photo Library/David Hall, *r* Science Photo Library/Peter Manzel; **p.490** Robert Harding; **p.495** *tl* Science Photo Library; *bl,br* John Townson/Creation; **p.514** Science Photo Library/Sinclair Stammers; **p.520** Science Photo Library; **p.533** *all* Andrew Lambert; **p.535** *all* John Townson/Creation; **p.554** *all* John Townson/Creation; **p.569** *l* Science Photo Library/Don Fawcett, *r* South American Pictures; **p.572** John Townson/Creation; **p.576** *l* Science Photo Library/National Institutes of Health, *r* Science Photo Library/Laboratory of Molecular Biology, MRC; **p.579** Science Photo Library/J. C. Revy; **p.581** *l* Robert Harding, *tr* Colorsport, *br* John Townson/Creation; **p.584** Science Photo Library/J. C. Revy; **p.585** Roger Scruton; **p.588** *all* John Townson/Creation; **p.589** *all* John Townson/Creation; **p.590** *all* John Townson/Creation; **p.607** Science Photo Library/Tek Image.

l = left, *r* = right, *t* = top, *b* = bottom, *c* = centre

The publisher has made every effort to contact copyright holders. If any have been inadvertently overlooked they will be pleased to make the necessary arrangements at the earliest opportunity.

PHYSICAL CHEMISTRY

1 Chemical formulae and moles

In this chapter we introduce chemical formulae, and show how these can be worked out using the valencies (or combining powers) of atoms and ions. We also introduce the chemists' fundamental counting unit, the mole, and show you how it can be used to calculate both empirical formulae and the amounts of substances that react or are formed during chemical reactions.

The chapter is divided into the following sections.

As you read through the chapter, work through the Examples, and then try the Further practice questions that follow them.

1.1 Introduction

What is chemistry?

Chemistry is the study of the properties of matter. By **matter**, we mean the substances that we can see, feel, touch, taste and smell – the stuff that makes up the material world. Passive observation forms only a small part of a chemist's interest in the world. Chemists are actively inquisitive scientists. They seek to understand why matter has the properties it does, and how to modify these properties by changing one substance into another through chemical reactions.

Chemistry as a modern science began a few hundred years ago, when chemists started to relate the observations they made about the substances they were investigating to theories of the structure of matter. One of the most important of these theories was the Atomic Theory. It is less than 200 years since John Dalton put forward his idea that all matter was composed of **atoms**. His theory stated that:

- the atoms of different elements were different from each other
- the atoms of a particular element were identical to each other
- all atoms were unchangeable over time and could be neither created nor destroyed
- all matter was made up from a relatively small number of elements (Dalton thought about 50) combined in various ways.

■ **Figure 1.1** John Dalton, who first suggested the modern Atomic Theory.

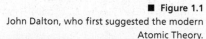

Although Dalton's theory has had to be modified slightly, it is still a useful starting point for the study of chemistry.

Since that time chemists have uncovered and explained many of the world's mysteries, from working out how elements are formed within stars to discovering how our genes replicate. On the way they have discovered thousands of new methods of converting one substance into another, and have made millions of new compounds, many of which are of great economic and medical benefit to the human race.

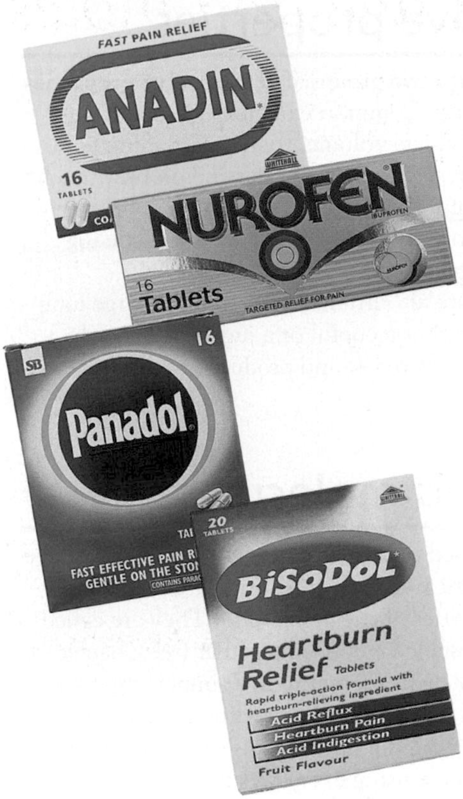

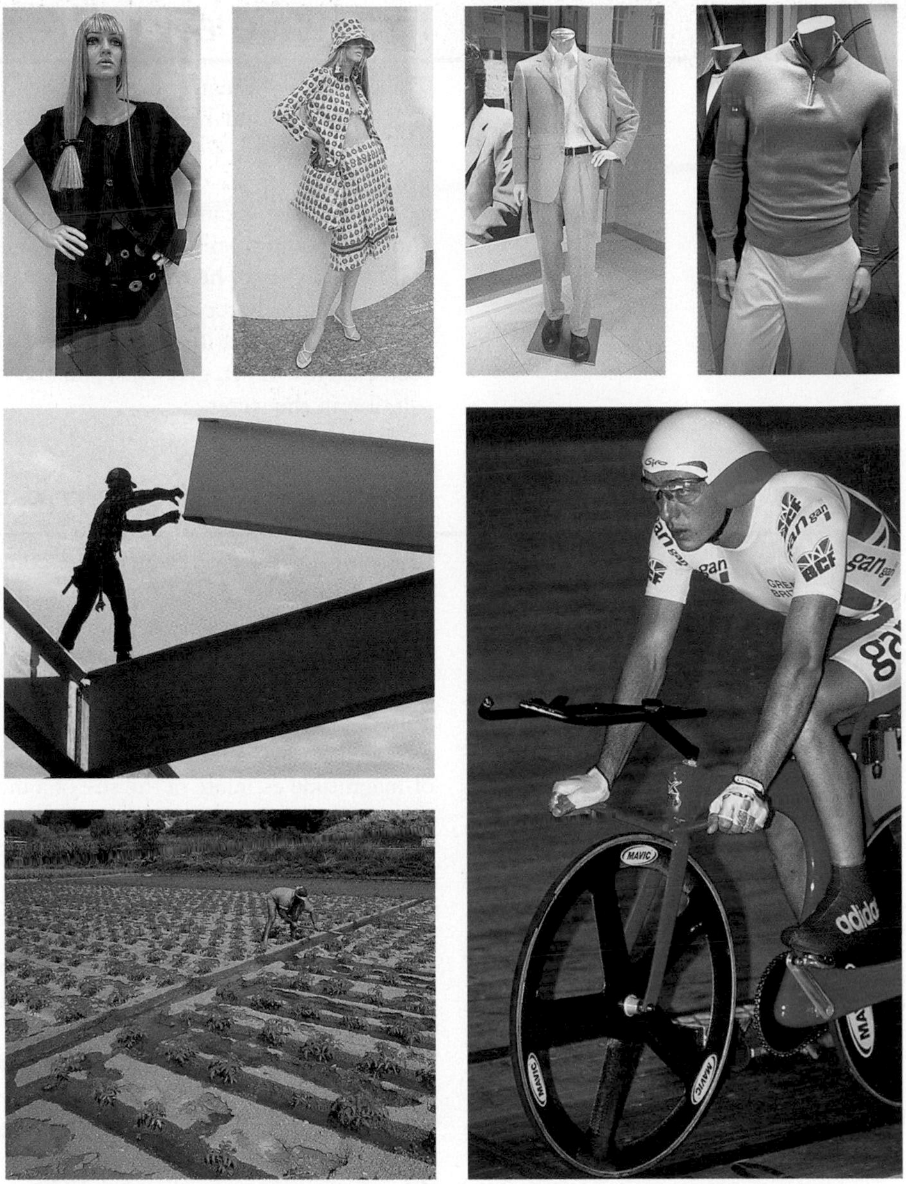

■ **Figure 1.2**
Some examples of economic, medical and agricultural benefits of chemistry.

Classifying matter – elements, compounds and mixtures

Chemists classify matter into one of three categories.

- **Elements** contain just one sort of atom. Although the atoms of a particular element may differ slightly in mass (see Chapter 2, page 23), they all have identical chemical reactions. Examples of elements include hydrogen gas, copper metal and diamond crystals (which are carbon).
- **Compounds** are made up from the atoms of two or more different elements, bonded together chemically. The ratio of elements within a particular compound is fixed, and is given by its chemical formula (see page 9). The physical and chemical properties of a compound are always different from those of the elements that make it up. Examples include sodium chloride, water (containing hydrogen and oxygen atoms) and penicillin (containing hydrogen, carbon, nitrogen, oxygen and sulphur atoms).
- **Mixtures** consist of more than one compound or element, mixed but chemically uncombined. The components can be mixed in any proportion, and the properties of a mixture are often the sum of, or the average of, the properties of the individual components. Examples include air, sea water and alloys such as brass.

1.2 Intensive and extensive properties

The properties of matter may be divided into two groups. The **extensive properties** depend on how much matter we are studying. Common examples are mass and volume – a cupful of water has less mass, and less volume, than a swimming pool. The other group are the **intensive properties**, which do not depend on how much matter we have. Examples include temperature, colour and density. A copper coin and a copper jug can both have the same intensive properties, although the jug will be many times heavier (and larger) than the coin.

The chemical properties of a substance are also **intensive**. A small or large lump of sodium will react in the same way with either a cupful or a jugful of water. In each case it will fizz, give off steam and hydrogen gas, and produce an alkaline solution in the water.

1.3 The sizes of atoms and molecules

Just as we believe that elements are composed of identical atoms, so we also believe that compounds are made up of many identical units. These units are the smallest entities that still retain the chemical properties of the compound. They are called **molecules** or **ions**, depending on how the substance is bonded together (see Chapter 4).

Molecules are extremely small – but how small? Sometimes, a simple experiment, a short calculation and a little thought can lead to quite a startling conclusion. The well-known oil drop experiment is an example. It allows us to obtain an order-of-magnitude estimate of the size of a molecule using everyday apparatus.

■ *Experiment* The oil drop experiment

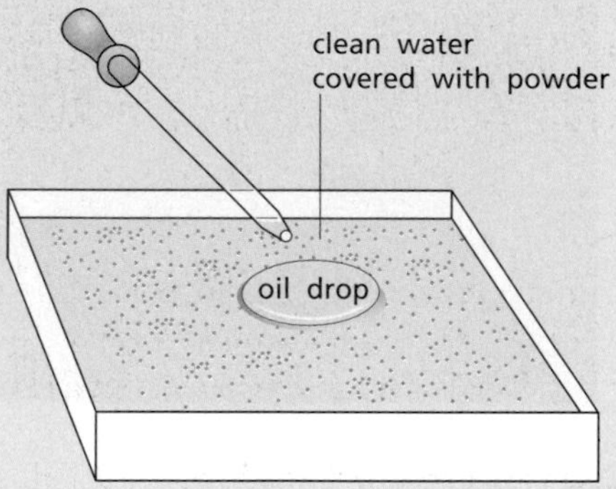

■ **Figure 1.3** The oil drop experiment.

A bowl is filled with clean water, and some fine powder such as talcum powder is sprinkled over the surface. A small drop of oil is placed on the surface of the water, as shown in Figure 1.3. The oil spreads out. As it does so, its pushes the powder back, so that there is an approximately circular area clear of powder.

We can measure the volume of one drop by counting how many drops it takes to fill a micro measuring cylinder. (If we know the oil's density, an easier method would be to find the mass of, say, 20 drops.) We can calculate the area of the surface film by measuring its diameter. Assuming the volume of oil does not change when the drop spreads out, we can thus find the thickness of the film. This cannot be smaller than the length of an oil molecule (though it may be bigger, if the film is several molecules thick – there is no way of telling). The following are typical results:

$$
\begin{aligned}
\text{volume of drop} &= 1.0\times10^{-4}\,\text{cm}^3 \\
\text{diameter of oil film} &= 30\,\text{cm} \\
\text{hence, radius of film} &= 15\,\text{cm} \\
\text{area of film} &= \pi r^2 \\
&= 3.14\times15^2 \\
&= \mathbf{707\,cm^2} \\
\text{volume} &= \text{area}\times\text{thickness} \\
\therefore\ \text{thickness} &= \frac{\text{volume}}{\text{area}} \\
&= \frac{1.0\times10^{-4}\,\text{cm}^3}{707\,\text{cm}^2} \\
&= \mathbf{1.4\times10^{-7}\,cm}\ (1.4\times10^{-9}\,\text{m})
\end{aligned}
$$

Distances this small are usually expressed in units of **nanometres**:

$$1\ \text{nanometre (nm)} = 1\times10^{-9}\ \text{metres (m)}$$

So the film is 1.4 nm thick. Oil molecules cannot be larger than this.

Since molecules are made up of atoms, atoms must be even smaller than the oil molecule. We can measure the sizes of atoms by various techniques, including X-ray crystallography. A carbon atom is found to have a diameter of 0.15 nm. That means it takes 6 million carbon atoms touching each other to reach a length of only 1 mm!

■ **Figure 1.4** Coloured scanning tunnelling electron microscope of carbon nanotubes, comprised of rolled sheets of carbon atoms. Individual atoms are seen as raised bumps on the surface of the tubes.

Standard form

The numbers that chemists deal with can often be very large, or very small. To make these more manageable, and to avoid having to write long lines of zeros (with the accompanying danger of miscounting them), we often express numbers in **standard form**.

A number in standard form consist of two parts, the first of which is a number between 1 and 10, and the second is the number 10 raised to a positive or negative power. Some examples with their fully written-out equivalents are given in Table 1.1.

■ **Table 1.1** Standard form.

Standard form	Fully written-out equivalent
6×10^2	600
7.142×10^7	71 420 000
2×10^{-6}	0.000 002
3.8521×10^{-4}	0.000 385 21

If the 10 is raised to a positive power, the superscript tells us how many digits to the right the decimal point moves. As in the examples in the table, we often have to add zeros to allow this to take place. If the 10 is raised to a negative power, the superscript number tells us how many digits to the left the decimal point moves. Here again, we often have to add zeros, but this time to the left of the original number.

Significant figures

In mathematics, numbers are exact quantities. In contrast, the numbers used in chemistry usually represent physical quantities which a chemist measures. The accuracy of the measurement is shown by the number of significant figures to which the quantity is quoted.

If we weigh a 1p piece on a digital kitchen scale, the machine will tell us it has a mass of 4 g. A 1-decimal place balance will give its mass as 3.5 g, whereas on a 2-decimal place balance its mass will be read as 3.50 g (or more, if it is old and tarnished).

We should interpret the reading on the kitchen scale as meaning that the mass of the penny lies between 3.5 g and 4.5 g. If it were just a little lighter than 3.5 g, the scale would have told us its mass was 3 g. If it were a little heavier than 4.5 g, the read-out would have been 5 g. The 1-decimal place balance narrows the range, telling us the mass of the penny is between 3.45 g and 3.55 g. The 2-decimal place balance narrows it still further, to between 3.495 g and 3.505 g.

In this way the number of significant figures (1, 2 or 3 in the above examples) tells us the accuracy with which the quantity has been measured.

The same is true of volumes. Using a $100\,cm^3$ measuring cylinder, we can measure a volume of $25\,cm^3$ to an accuracy of $\pm 0.5\,cm^3$, so we would quote the volume as $25\,cm^3$. Using a pipette or burette, however, we can measure volumes to an accuracy of $\pm 0.05\,cm^3$, and so we would quote the same volume as $25.0\,cm^3$ (that is, somewhere between $24.95\,cm^3$ and $25.05\,cm^3$).

Most chemical balances and volumetric equipment will measure quantities to 3 or 4 significant figures. Allowing for the accumulation of errors when values are calculated using several measured quantities, we tend to quote values to 2 or 3 significant figures.

In the examples in Table 1.1, 6×10^2 has one significant figure, and 7.142×10^7 has four significant figures.

1.4 The masses of atoms and molecules

Being so small, atoms are also very light. Their masses range from 1×10^{-24} g to 1×10^{-22} g. It is impossible to weigh them out individually, but we can measure accurately their **relative masses**, or how heavy one atom is compared with another. The most accurate way of doing this is by using a mass spectrometer (see Chapter 2, page 24).

Originally, the atomic masses of all the elements were compared with the mass of an atom of hydrogen:

$$\text{relative atomic mass of element E} = \frac{\text{mass of one atom of E}}{\text{mass of one atom of hydrogen}}$$

This is because hydrogen is the lightest element, so the relative atomic masses of all other elements are, conveniently, greater than 1.

Because of the existence of isotopes, and the central importance of carbon in the masses of organic compounds, the modern definition uses the isotope carbon-12, ^{12}C (see Chapter 2) as the standard of reference:

$$\textbf{relative atomic mass}\text{ of element E} = \frac{\text{average mass of one atom of E}}{\frac{1}{12}\text{ the mass of one atom of }^{12}C}$$

The difference between the two definitions is small, since a carbon-12 atom has almost exactly 12 times the average mass of a hydrogen atom (the actual ratio is 11.91:1). Relative atomic mass is given the symbol $\mathbf{A_r}$. Since it is the ratio of two masses, it is a dimensionless quantity – it has no units.

The masses of atoms and subatomic particles (see Chapter 2) are often expressed in atomic mass units. An **atomic mass unit** (amu) is defined as $\frac{1}{12}$ the mass of one atom of carbon-12. It has the value of 1.6606×10^{-24} g.

Although we cannot use a laboratory balance to weigh out individual atoms, we can use it to weigh out known ratios of atoms of various elements, as long as we know their relative atomic masses. For example, if we know that the relative atomic masses of carbon and magnesium are 12 and 24 respectively, we can be sure that 12 g of carbon will contain the same number of atoms as 24 g of magnesium. What is more, 24 g (12 × 2) of carbon will contain twice the number of atoms as 24 g of magnesium. Indeed, we can be certain that any mass of carbon will contain twice the number of atoms as the same mass of magnesium, since the mass of each carbon atom is only half the mass of a magnesium atom.

Similarly, if we know that the relative atomic mass of helium is 4 (which is one-third the relative atomic mass of carbon) we can deduce that identical masses of helium and carbon will always contain three times as many helium atoms as carbon atoms.

1.5 The mole

Chemists deal with real, measured quantities of substances. Rather than counting atoms individually, they prefer to count them in units that are easily measurable. The gram is a conveniently sized unit of mass to use for weighing out matter (a teaspoon of sugar, for example, weighs about 5 g). The chemist's unit of amount, the **mole** (symbol **mol**), is defined in terms of grams:

> 1 mole of an element is the amount that contains the same number of atoms as there are in 12.000 g of carbon-12.

■ Figure 1.5 One-tenth of a mole of each of the elements aluminium, sulphur, bromine and lead.

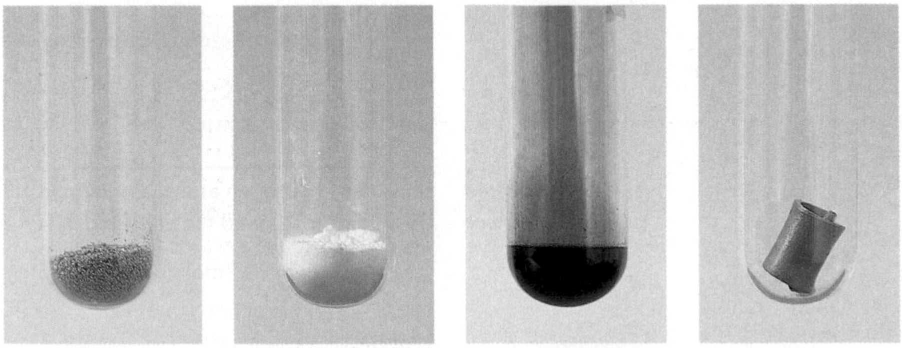

a Al 2.7 g **b** S 3.2 g ($\frac{1}{80}$ mol S_8) **c** Br 8.0 g ($\frac{1}{20}$ mol Br_2) **d** Pb 20.7 g

The mass of one mole of an element is called its **molar mass** (symbol **M**). It is numerically equal to its relative atomic mass, A_r, but is given in grams per mole:

relative atomic mass of carbon = 12
molar mass of carbon = 12 g mol^{-1}
relative atomic mass of magnesium = 24
molar mass of magnesium = 24 g mol^{-1}

It follows from the above definition that there is a clear relationship between the mass (m) of a sample of an element and the number of moles (n) it contains:

$$\text{amount (in moles)} = \frac{\text{mass}}{\text{molar mass}}$$

or

$$n = \frac{m}{M} \text{ mol} \qquad (1)$$

Example What is the amount (in moles) of carbon in 30 g of carbon?

Answer Use the value A_r (carbon) = 12, to write its molar mass, and use equation (1) above:

$$m = 30\,\text{g}$$

$$M = 12\,\text{g mol}^{-1}$$

$$\therefore \quad n = \frac{30\,\text{g}}{12\,\text{g mol}^{-1}}$$

$$= 2.5\,\text{mol}$$

Further practice Using the following A_r values: O = 16, Mg = 24, S = 32, calculate the amount of substance (in moles) in each of the following samples:

1 24 g of oxygen
2 24 g of sulphur
3 16 g of magnesium

As we saw on page 6, the actual masses of atoms are very small. We would therefore expect the number of atoms in a mole of an element to be very large. This is indeed the case. One mole of an element contains a staggering 6.022×10^{23} atoms (six hundred and two thousand two hundred million million million atoms). This value is called the **Avogadro constant**, symbol ***L***.

$$L = 6.022 \times 10^{23}\,\text{mol}^{-1}$$

The approximate value of $L = 6.0 \times 10^{23}\,\text{mol}^{-1}$ is often adequate, and will be used in calculations in this book.

The relationship between the number of moles in a sample of an element and the number of atoms it contains is as follows:

number of atoms = $L \times$ number of moles

or

$$N = Ln \qquad (2)$$

Example How many hydrogen atoms are there in 1.5 mol of hydrogen atoms?

Answer Use equation (2), and the value of L given above:

$$L = 6.0 \times 10^{23}\,\text{mol}^{-1}$$

$$n = 1.5\,\text{mol}$$

$$N = 6.0 \times 10^{23}\,\text{mol}^{-1} \times 1.5\,\text{mol}$$
$$= 9.0 \times 10^{23}$$

1.6 Atomic symbols and formulae

Each element has a unique **symbol**. Symbols consist of either one or two letters. The first is always a capital letter and the second, if present, is always a lower case letter. This rule avoids confusions and ambiguities when the symbols are combined to make the formulae of compounds. For example:

- the symbol for hydrogen is H
- the symbol for helium is He (not HE or hE)
- the symbol for cobalt is Co (not CO – this is the **formula** of carbon monoxide, which contains two atoms in its molecule, one of carbon and one of oxygen).

Symbols are combined to make up the **formulae** of compounds. If more than one atom of a particular element is present, its symbol is followed by a subscript giving the number of atoms of that element contained in one formula unit of the compound. For example:

- the formula of copper oxide is CuO (one atom of copper combined with one atom of oxygen)
- the formula of water is H_2O (two atoms of hydrogen combined with one atom of oxygen)
- the formula of phosphoric(V) acid is H_3PO_4 (three atoms of hydrogen combined with one of phosphorus and four of oxygen).

Sometimes, especially when the compound consists of ions rather than molecules (see Chapter 4), groups of atoms in a formula are kept together by the use of brackets. If more than one of a particular group is present, the closing bracket is followed by a subscript giving the number of groups present. This practice makes the connections between similar compounds clearer. For example:

- the formula of sodium nitrate is $NaNO_3$ (one sodium ion, Na^+, combined with one nitrate ion, NO_3^-, which consists of one nitrogen atom combined with three oxygen atoms)
- the formula of calcium nitrate is $Ca(NO_3)_2$ (one calcium ion, Ca^{2+}, combined with two nitrate ions).

Note that in calcium nitrate, the formula unit consists of one calcium, two nitrogens and six oxygens, but it is not written as CaN_2O_6. This formula would not make clear the connection between $Ca(NO_3)_2$ and $NaNO_3$. Both compounds are nitrates, and both undergo similar reactions of the nitrate ions.

The formulae of many ionic compounds can be predicted if the valencies of the ions are known. (The valency of an ion is the charge on the ion.) Similarly, the formulae of several of the simpler covalent (molecular) compounds can be predicted if the covalencies of the constituent atoms are known. (The covalency of an atom is the number of covalent bonds that the atom can form with adjacent atoms in a molecule.) Lists of covalencies and valencies, and examples of how to use them, are included on pages 58 and 93.

Example How many atoms of each element are present in one formula unit of:

a $Al(OH)_3$

b $(NH_4)_2SO_4$?

Answer **a** The subscript after the closing bracket multiplies all the content of the brackets by three. There are therefore three OH (hydroxide) groups, each containing one oxygen and one hydrogen atom, making a total of three oxygen atoms and three hydrogen atoms, together with one aluminium atom.

b Here there are two ammonium groups, each containing one nitrogen atom and four hydrogen atoms, and one sulphate group, containing one sulphur atom and four oxygen atoms. In total, therefore, there are:

- two nitrogen atoms
- eight hydrogen atoms
- one sulphur atom
- four oxygen atoms.

Further practice How many atoms in total are present in the formula units of each of the following compounds?

1 NH_4NO_3

2 $Na_2Cr_2O_7$

3 $KCr(SO_4)_2$

4 $C_6H_{12}O_6$

5 $Na_3Fe(C_2O_4)_3$

1.7 Moles and compounds

Relative molecular mass and relative formula mass

Just as we can weigh out a mole of carbon (12 g), so we can weigh out a mole of a compound such as ethanol (alcohol). We first need to calculate its **relative molecular mass, M_r**.

To calculate the relative molecular mass (M_r) of a compound we add together the relative atomic masses (A_r) of all the elements present in the compound (remembering to multiply the A_r values by the correct number if more than one atom of a particular element is present). So for ethanol, C_2H_6O, we have:

$$\begin{aligned} M_r &= 2A_r(C) + 6A_r(H) + A_r(O) \\ &= 2(12) + 6(1) + 16 \\ &= 46 \end{aligned}$$

Just as with relative atomic mass, values of relative molecular mass are ratios of masses, and have no units. The molar mass of ethanol is $46\,g\,mol^{-1}$.

For ionic and giant covalent compounds (see Chapter 4), we cannot, strictly, refer to their relative molecular masses, as they do not consist of individual molecules. For these compounds, we add together the relative atomic masses of all the elements present in the simplest (empirical) formula. The result is called the **relative formula mass**, but is given the same symbol as relative molecular mass, M_r. Just as with molecules, the mass of one formula unit is called the molar mass, symbol M. For example, the relative formula mass of sodium chloride, NaCl, is calculated as follows:

$$\begin{aligned} M_r &= A_r(Na) + A_r(Cl) \\ &= 23 + 35.5 \\ &= 58.5 \end{aligned}$$

The molar mass of sodium chloride is $58.5\,g\,mol^{-1}$.

a H_2O 1.8 g   b $K_2Cr_2O_7$ 29.4 g   c $CuSO_4.5H_2O$ 25.0 g

■ **Figure 1.6** One-tenth of a mole of each of the compounds water, potassium dichromate(VI) ($K_2Cr_2O_7$) and copper(II) sulphate-5-water ($CuSO_4.5H_2O$).

We can apply equation (1) (page 7) to compounds as well as to elements. Once the molar mass has been calculated, we can relate the mass of a sample of a compound to the number of moles it contains.

Example 1 Calculate the relative molecular mass of glucose, $C_6H_{12}O_6$.

Answer

$$\begin{aligned} M_r &= 6A_r(C) + 12A_r(H) + 6A_r(O) \\ &= 6(12) + 12(1) + 6(16) \\ &= 72 + 12 + 96 \\ &= 180 \end{aligned}$$

Example 2 How many moles are there in 60 g of glucose?

Answer Convert this relative molecular mass to the molar mass, *M*, and use the formula in equation (1):

$$n = \frac{m}{M}$$

$$m = 60\,\text{g}$$
$$M = 180\,\text{g mol}^{-1}$$

$$\therefore\; n = \frac{60\,\text{g}}{180\,\text{g mol}^{-1}}$$
$$= \mathbf{0.33\,mol}$$

Further practice

1 Calculate the relative molecular masses of the following compounds. (Use the list of A_r values on page 671.)
   a iron(II) sulphate, $FeSO_4$
   b calcium hydrogencarbonate, $Ca(HCO_3)_2$
   c ethanoic acid, $C_2H_4O_2$
   d ammonium sulphate, $(NH_4)_2SO_4$
   e the complex with the formula $Na_3Fe(C_2O_4)_3$

2 How many moles of substance are there in each of the following samples?
   a 20 g of magnesium oxide, MgO
   b 40 g of methane, CH_4
   c 60 g of calcium carbonate, $CaCO_3$
   d 80 g of cyclopropene, C_3H_4
   e 100 g of sodium dichromate(VI), $Na_2Cr_2O_7$

3 What is the mass of each of the following samples?
   a 1.5 moles of magnesium sulphate, $MgSO_4$
   b 0.333 moles of aluminium chloride, $AlCl_3$

A mole of what?

When dealing with compounds, we need to define clearly what the word 'mole' refers to. A mole of water contains 6×10^{23} molecules of H_2O. But because each molecule contains two hydrogen atoms, a mole of H_2O molecules will contain two moles of hydrogen atoms, that is, 12×10^{23} H atoms. Likewise, a mole of sulphuric acid, H_2SO_4, will contain two moles of hydrogen atoms, one mole of sulphur atoms and four moles of oxygen atoms. A mole of calcium chloride, $CaCl_2$, contains twice the number of chloride ions as does a mole of sodium chloride, NaCl.

Sometimes this also applies to elements. The phrase 'one mole of chlorine' is ambiguous. One mole of chlorine molecules contains 6×10^{23} Cl_2 units, and thus contains 12×10^{23} chlorine atoms (2 mol of Cl).

1.8 Empirical formulae

The **empirical formula** is the simplest formula that shows the relative number of atoms of each element present in a compound.

If we know the percentage composition by mass of a compound, or the masses of the various elements that make it up, we can determine the ratios of atoms.

The steps in the calculation are as follows:

1 Divide the percentage (or mass) of each element by the element's relative atomic mass.
2 Divide each of the figures obtained in step 1 by the smallest of those figures.
3 If the results of the calculations do not approximate to whole numbers, multiply them all by 2 (or exceptionally 3) to obtain whole numbers.

Example Calculate the empirical formula of an oxide of iron that contains 70% Fe by mass.

Answer The oxide contains iron and oxygen only, so the percentage of oxygen is 100 − 70 = 30%.

Following the steps above:

1 Fe: $\frac{70}{56} = 1.25$ O: $\frac{30}{16} = 1.875$

2 Fe: $\frac{1.25}{1.25} = \mathbf{1.00}$ O: $\frac{1.875}{1.25} = \mathbf{1.50}$

3 Multiply both numbers by 2: Fe = 2, O = 3.

Therefore empirical formula = $\mathbf{Fe_2O_3}$.

Further practice Calculate the empirical formula of each of the following compounds:

1 a sulphide of copper containing 4.00 g of copper and 1.00 g of sulphur

2 a hydrocarbon containing 81.8% carbon and 18.2% hydrogen

3 a mixed oxide of iron and calcium which contains 51.9% iron and 18.5% calcium by mass (the rest being oxygen).

1.9 Equations

Mass is conserved

A chemical equation represents what happens during a chemical reaction. A key feature of chemical reactions is that they proceed with no measurable alteration in mass whatsoever. Many obvious events can often be seen taking place – the evolution of heat, flashes of light, changes of colour, noise and evolution of gases. But despite these sometimes dramatic signs that a reaction is happening, the sum of the masses of all the various products is always found to be equal to the sum of the masses of the reactants.

This was one of the first quantitative laws of chemistry, and is known as the **Law of Conservation of Mass**. It can be illustrated simply but effectively by the following experiment.

■ *Experiment* The conservation of mass

A small test tube has a length of cotton thread tied round its neck, and is half filled with lead(II) nitrate solution. It is carefully lowered into a conical flask containing potassium iodide solution, taking care not to spill its contents. A bung is placed in the neck of the conical flask, so that the cotton thread is trapped by its side, as shown in Figure 1.7. The whole apparatus is then weighed.

The conical flask is now shaken vigorously to mix the contents. A reaction takes place, and the bright yellow solid lead(II) iodide is formed. On re-weighing the conical flask with its contents, the mass is found to be identical to the initial mass.

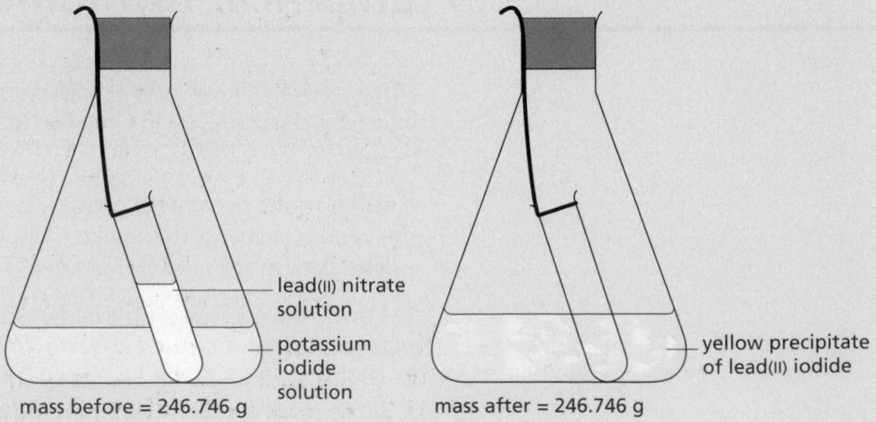

■ **Figure 1.7** During the formation of lead(II) iodide, mass is conserved.

Balanced equations

The reason why the mass does not change during a chemical reaction is because no atoms are ever created or destroyed. The same number of atoms of each element are present at the end as were there at the beginning. All that has happened is that they have changed their chemical environment. In the above example, the change can be represented in words as:

lead(II) nitrate solution + potassium iodide solution → solid lead(II) iodide + potassium nitrate solution

■ **Figure 1.8**
The formation of lead(II) iodide.

There are several steps we must carry out to convert this **word equation** into a **balanced chemical equation**:

1 Work out and write down the formula of each of the compounds in turn, and describe its physical state using the correct one of the following four **state symbols**:

(g) = gas (l) = liquid (s) = solid (aq) = aqueous solution (dissolved in water).

For the above reaction we have:

lead(II) nitrate solution is $Pb(NO_3)_2(aq)$
potassium iodide solution is $KI(aq)$
solid lead(II) iodide is $PbI_2(s)$
potassium nitrate solution is $KNO_3(aq)$.

The equation now becomes:

$$Pb(NO_3)_2(aq) + KI(aq) \rightarrow PbI_2(s) + KNO_3(aq)$$

2 The next step is to balance the equation. That is, we must ensure that we have the same number of atoms of each element on the right-hand side as there are on the left-hand side.
   a Looking at the equation in step **1** above, we notice that there are two iodine atoms on the right, in PbI_2, but only one on the left, in KI. Also, there are two nitrate groups on the left, in $Pb(NO_3)_2$, but only one on the right, in KNO_3.
   b We can balance the iodine atoms by having two formula units of KI on the left, that is 2KI. (Note that we cannot change the formula to KI_2 – that would not correctly represent potassium iodide, which always contains equal numbers of potassium and iodide ions.)
   c We can balance the nitrates by having two formula units of KNO_3 on the right, that is $2KNO_3$. This also balances up the potassium atoms, which, although originally the same on both sides, became unbalanced when we changed KI to 2KI in step **b**.

The fully balanced equation is now:

$$Pb(NO_3)_2(aq) + 2KI(aq) \rightarrow PbI_2(s) + 2KNO_3(aq)$$

It is clear that we have neither lost nor gained any atoms, but that they have swapped partners – the iodine was originally combined with potassium, but has ended up being combined with lead; the nitrate groups have changed their partner from lead to potassium.

Example Write the balanced chemical equation for the following reaction:

zinc metal + hydrochloric acid → zinc chloride solution + hydrogen gas

Answer Following the steps given above:

1 zinc metal is $Zn(s)$
hydrochloric acid is $HCl(aq)$
zinc chloride solution is $ZnCl_2(aq)$
hydrogen gas is $H_2(g)$ (hydrogen, like many non-metallic elements, exists in molecules made up of two atoms).
The equation now becomes:

$$Zn(s) + HCl(aq) \rightarrow ZnCl_2(aq) + H_2(g)$$

2 a There are two hydrogen atoms and two chlorine atoms on the right, but only one of each of these on the left.
b We can balance both of them by just one change – having two formula units of HCl on the left.

The fully balanced equation is now:

$$Zn(s) + 2HCl(aq) \rightarrow ZnCl_2(aq) + H_2(g)$$

Further practice 1 Copy the following equations and balance them.

a $H_2(g) + O_2(g) \rightarrow H_2O(l)$
b $I_2(s) + Cl_2(g) \rightarrow ICl_3(s)$
c $NaOH(aq) + Al(OH)_3(s) \rightarrow NaAlO_2(aq) + H_2O(l)$
d $H_2S(g) + SO_2(g) \rightarrow S(s) + H_2O(l)$
e $NH_3(g) + O_2(g) \rightarrow N_2(g) + H_2O(l)$

2 Write balanced symbol equations for the following reactions.

a magnesium carbonate → magnesium oxide + carbon dioxide
b lead + silver nitrate solution → lead nitrate solution + silver
c sodium oxide + water → sodium hydroxide solution
d iron(II) chloride + chlorine (Cl_2) → iron(III) chloride
e iron(III) sulphate + sodium hydroxide → iron(III) hydroxide + sodium sulphate

1.10 Using the mole in mass calculations

We are now in a position to look at how the masses of the individual substances in a chemical equation are related. As an example, take the reaction between marble chips (calcium carbonate) and hydrochloric acid:

$$CaCO_3(s) + 2HCl(aq) \rightarrow CaCl_2(aq) + H_2O(l) + CO_2(g)$$

When this reaction is carried out in an open conical flask on a top-pan balance, its mass is observed to decrease. (Note that this is not due to the destruction of matter – as was mentioned on page 12, the overall number of atoms does not change during a chemical reaction. Rather, it is due to the fact that the gaseous carbon dioxide produced escapes into the air.) We can use the knowledge gained in this chapter to calculate the answer to the following question:

- By how much would the mass decrease if 50 g of marble chips were completely reacted with an excess of hydrochloric acid?

We use the following steps:

1 We can use equation (1) (page 7) to calculate the number of moles of calcium carbonate in 50 g of marble chips:

$$n = \frac{m}{M} \qquad M_r(CaCO_3) = 40.1 + 12 + 3(16) = 100.1$$

$$\therefore \quad M = 100.1\,g\,mol^{-1}$$

$$n = \frac{50\,g}{100.1\,g\,mol^{-1}}$$

$$= \mathbf{0.50\,mol\ of\ CaCO_3}$$

2 From the balanced equation above, we see that one mole of calcium carbonate produces one mole of carbon dioxide. Therefore the number of moles of carbon dioxide produced is the same as the number of moles of calcium carbonate we started with, namely **0.50 mol of carbon dioxide**.

3 Lastly, we can use a rearranged form of equation (1) to calculate what mass of carbon dioxide this corresponds to.

$$n=\frac{m}{M} \quad \text{so} \quad m=n\times M \quad M_r(CO_2)=12+2(16)=44$$

$$\therefore \quad M=44\,g\,mol^{-1}$$
$$\text{also } n=0.50\,mol$$

$$\therefore \quad m=0.50\,mol\times 44\,g\,mol^{-1}$$
$$=\mathbf{22\,g}$$

The loss in mass (due to the carbon dioxide being evolved) is **22 g**.

The three steps can be summarised as shown in Figure 1.9.

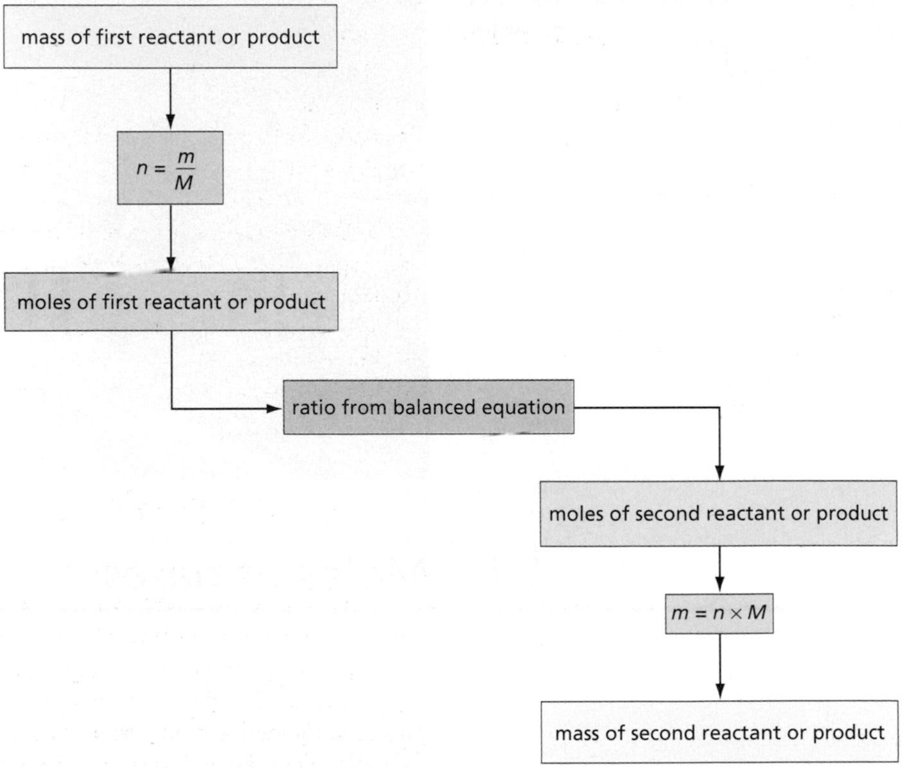

■ **Figure 1.9**
Finding the mass of a product from the mass of reactant, or vice versa.

Example The highly exothermic **thermit reaction** (see Figure 1.10, page 16) is used to weld together steel railway lines. It involves the reduction of iron(III) oxide to iron by aluminium. Use the chart in Figure 1.9 to calculate what mass of aluminium is needed to react completely with 10.0 g of iron(III) oxide according to the equation:

$$2Al_{(s)} + Fe_2O_{3(s)} \rightarrow Al_2O_{3(s)} + 2Fe_{(s)}$$

Answer **1** $M_r(Fe_2O_3)=2(55.8)+3(16)=111.6+48=159.6$

$$\therefore \quad M=159.6\,g\,mol^{-1}$$

$$\text{number of moles of iron(III) oxide } (n) = \frac{m}{M}$$
$$=\frac{10.0\,g}{159.6\,g\,mol^{-1}}$$
$$=\mathbf{0.0627\,mol}$$

2 From the balanced equation, one mole of iron(III) oxide reacts with two moles of aluminium, therefore:

number of moles of aluminium (n) $= 0.0627 \times 2 =$ **0.125 mol**

3 $A_r(\text{Al}) = 27 \quad \therefore \quad M = 27\,\text{g mol}^{-1}$

mass of aluminium $= n \times M$

$= 0.125\,\text{mol} \times 27\,\text{g mol}^{-1}$

$=$ **3.38 g**

Further practice

1 What mass of silver will be precipitated when 5.0 g of copper are reacted with an excess of silver nitrate solution?

$$Cu(s) + 2AgNO_3(aq) \rightarrow Cu(NO_3)_2(aq) + 2Ag(s)$$

2 What mass of ammonia will be formed when 50.0 g of nitrogen are passed through the Haber process (assume 100% conversion)?

$$N_2(g) + 3H_2(g) \rightarrow 2NH_3(g)$$

■ **Figure 1.10**
The thermit process is used to weld steel railway lines together.

1.11 Moles of gases

The molar masses of most compounds are different. The molar volumes of most solid and liquid compounds are also different. But the molar volumes of gases (when measured at the same temperature and pressure) are all the same. This strange coincidence results from the fact that most of a gas is in fact empty space – the molecules take up less than a thousandth of its volume at normal temperatures (Chapter 4 looks at gases in more detail). The volume of the molecules is negligible compared with the total volume, and so any variation in their individual size will not affect the overall volume. At room temperature (25 °C, 298 K) and normal pressure (1 atm, 1.01×10^5 Pa)

the molar volume of any gas = 24.0 dm^3 mol^{-1}

So we can say that:

$$\text{amount of gas (in moles)} = \frac{\text{volume (in dm}^3)}{\text{molar volume}}$$

$$n = \frac{V}{24.0}$$

or

$$\text{volume of gas (in dm}^3) = \text{molar volume} \times \text{moles of gas}$$

$$V = 24.0 \times n$$

Example What volume of hydrogen (at room temperature and pressure) will be produced when 7.0 g of iron are reacted with an excess of sulphuric acid?

Answer The equation for the reaction is as follows:

$$Fe(s) + H_2SO_4(aq) \rightarrow FeSO_4(aq) + H_2(g)$$

1 $A_r(Fe) = 55.8$

$\therefore \quad M = 55.8\,g\,mol^{-1}$

$$\text{amount (in moles) of iron} = \frac{m}{M} = \frac{7.0\,g}{55.8\,g\,mol^{-1}} = \mathbf{0.125\,mol}$$

2 From the balanced equation, one mole of iron produces one mole of hydrogen molecules, therefore:

$$\text{number of moles of } H_2 = \mathbf{0.125\,mol}$$

3 volume of H_2 in dm^3 = molar volume × moles of H_2

$= 24.0\,dm^3\,mol^{-1} \times 0.125\,mol$

$= \mathbf{3.0\,dm^3}$

1.12 Moles and concentrations

Many chemical reactions are carried out in solution. Often it is convenient to dissolve a reactant in a solvent in advance, and to use portions as and when needed. A common example of this is the dilute sulphuric acid you find on the side shelves in a laboratory. This has been made up in bulk, at a certain concentration. Solutions are most easily measured out by volume, using measuring cylinders, pipettes or burettes. Suppose we need a certain amount of sulphuric acid (that is, a certain number of moles) for a particular experiment. If we know how many moles of sulphuric acid each 1 cm^3 of solution contains, we can obtain the required number of moles by measuring out the correct volume. For example, if our sulphuric acid contains 0.001 mol of H_2SO_4 per 1 cm^3, and we need 0.005 mol, we would measure out 5 cm^3 of solution.

In chemistry the concentrations of solutions are normally stated in units of **moles per cubic decimetre** (= moles per litre). The customary abbreviation is **mol dm^{-3}**. Occasionally the older, and even shorter, abbreviation **M** is used.

If 1 dm^3 of a solution contains 1.0 mol of solute, the solution's concentration is 1.0 mol dm^{-3} (or 1.0 M, verbally described as 'a one molar solution').

If more moles are dissolved in the same volume of solution the solution is more concentrated. Likewise, if the same number of moles is dissolved in a smaller volume of solution, the solution is also more concentrated. For example, we can produce a 2.0 mol dm^{-3} solution (2.0 M, 'two molar') by:

- either dissolving 2 mol of solute in 1 dm^3 of solution
- or dissolving 1 mol of solute in 0.5 dm^3 of solution.

$$\text{concentration} = \frac{\text{amount (in moles)}}{\text{volume of solution (in dm}^3\text{)}}$$

$$c = \frac{n}{V}$$

or

$$\text{amount (in moles)} = \text{concentration} \times \text{volume (in dm}^3\text{)}$$

$$n = c \times V$$

Unlike the mass of a solid, or the volume of a gas, which are extensive properties of a substance (see page 4), the concentration of a solution is an intensive property. It does not depend on how much of the solution we have. Properties that depend on the concentration of a solution are also intensive. For example, the rate of reaction between sulphuric acid and magnesium ribbon, the colour of aqueous potassium manganate(VII), the sourness of vinegar and the density of a sugar solution depend not on how much solution we have, but only on how much solute is dissolved in a given volume.

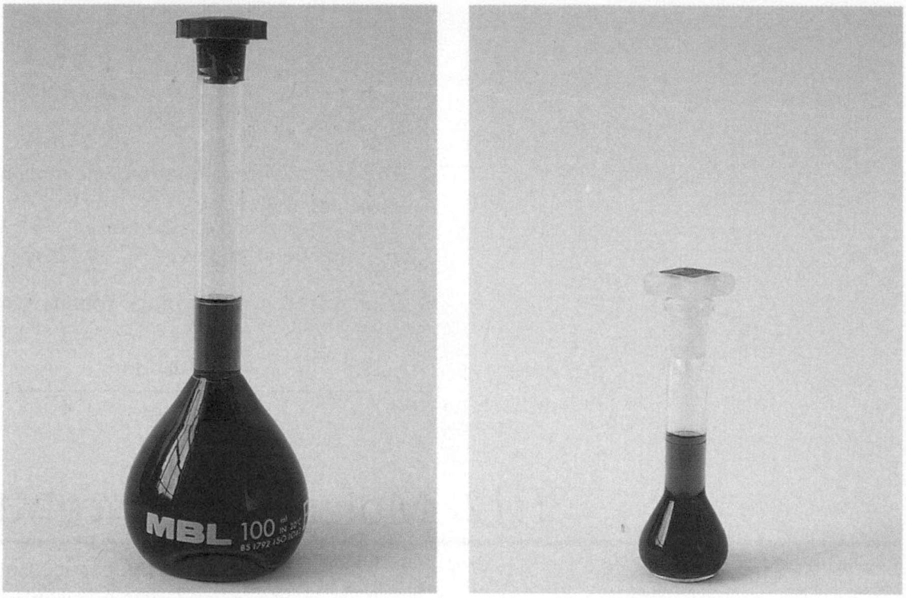

■ **Figure 1.11** Colour is related to concentration, an intensive property that does not depend on how much solution is there.

Example 1 What is the concentration of a solution, $200\,cm^3$ of which contains 0.40 mol of sugar?

Answer Use the first equation in the panel above:

$$c = \frac{n}{V}$$ remember, $1000\,cm^3 = 1.0\,dm^3$ so $200\,cm^3 = 0.20\,dm^3$

$$c = \frac{0.40\text{ mol}}{0.20\text{ dm}^3}$$

$$= \mathbf{2.0\,mol\,dm^{-3}}$$

Example 2 How many grams of salt, NaCl, need to be dissolved in $0.50\,dm^3$ of solution to make a $0.20\,mol\,dm^{-3}$ solution?

Answer Use the second equation in the panel above:

$$n = c \times V \qquad c = 0.20\,mol\,dm^{-3} \qquad V = 0.50\,dm^3$$

$$n = 0.20\,mol\,dm^{-3} \times 0.5\,dm^3$$

$$= \mathbf{0.10\,mol} \text{ of NaCl}$$

Now use equation (1) (page 7) to convert moles into mass:

$$n = \frac{m}{M} \quad \text{so} \quad m = n \times M$$

$M_r(NaCl) = 23 + 35.5$
$= 58.5$
$\therefore\ M = 58.5\,g\,mol^{-1}$
$n = 0.10\,mol$

$$m = 0.10\,mol \times 58.5\,g\,mol^{-1}$$

$$= \mathbf{5.85\,g} \text{ of salt, NaCl}$$

Further practice 1 What are the concentrations of solute in the following solutions?

a 2.0 mol of ethanol in $750\,cm^3$ of solution

b 5.3 g of sodium carbonate, Na_2CO_3, in $2.0\,dm^3$ of solution

c 40 g of ethanoic acid, $C_2H_4O_2$, in $800\,cm^3$ of solution

2 How many moles of solute are in the following solutions?
  a 0.50 dm^3 of a 1.5 mol dm^{-3} solution of sulphuric acid.
  b 22 cm^3 of a 2.0 mol dm^{-3} sodium hydroxide solution.
  c 50 cm^3 of a solution containing 20 g of potassium hydrogencarbonate, $KHCO_3$, per dm^3.

We shall come across the concentrations of solutions again in Chapter 6, where we look at the technique of titration.

1.13 Calculations using a combination of methods

At the heart of most chemistry calculations is the balanced chemical equation. This shows us the ratios in which the reactants react to give the products, and the ratios in which the products are formed. This is called the stoichiometry of the reaction. Most calculations involving reactions can be broken down into a similar set of three steps as was described for mass calculations on page 15 (see Figure 1.12).

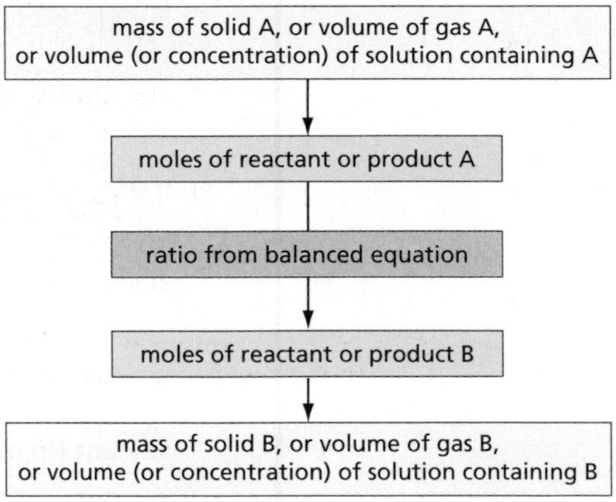

■ **Figure 1.12** Calculations involving the stoichiometry of a reaction.

Further practice

Use the chart above and the A_r values on page 711 to answer the following questions.

1 What volume of hydrogen gas will be given off when 2.3 g of sodium metal react with water?

$$2Na(s) + 2H_2O(l) \rightarrow 2NaOH(aq) + H_2(g)$$

2 The equation for the complete combustion of methane, CH_4, in oxygen is:

$$CH_4(g) + 2O_2(g) \rightarrow CO_2(g) + 2H_2O(g)$$

Calculate the volume of oxygen needed to burn 4.0 g of methane.

3 What volume of 0.50 mol dm^{-3} sulphuric acid, H_2SO_4, is needed to react exactly with 5.0 g of magnesium, and what volume of hydrogen will be evolved?

$$Mg(s) + H_2SO_4(aq) \rightarrow MgSO_4(aq) + H_2(g)$$

4 What mass of sulphur will be precipitated when an excess of hydrochloric acid is added to 100 cm^3 of 0.20 mol dm^{-3} sodium thiosulphate solution?

$$Na_2S_2O_3(aq) + 2HCl(aq) \rightarrow 2NaCl(aq) + SO_2(g) + H_2O(l) + S(s)$$

5 What would be the concentration of the hydrochloric acid produced if all the hydrogen chloride gas from the reaction between 50 g of pure sulphuric acid and an excess of sodium chloride was collected in water, and the solution made up to a volume of 400 cm^3 with water?

$$NaCl(s) + H_2SO_4(l) \rightarrow NaHSO_4(s) + HCl(g)$$

Summary

- **Atoms** and **molecules** are small and light – about 1×10^{-9} m in size, and about 1×10^{-22} g in mass.
- **Relative atomic mass** and **relative molecular mass** are defined in terms of $\frac{1}{12}$ the mass of an atom of carbon-12.
- **One mole** is the amount of substance that has the same number of particles (atoms, molecules, etc.) as there are atoms in 12.000 g of carbon-12.
- The relative molecular mass, M_r, of a compound is found by summing the relative atomic masses, A_r, of all the atoms present.
- The **empirical formula** of a compound is the simplest formula that shows the relative number of atoms of each element present in the compound.
- Chemical reactions take place with no change in mass, and no change in the total number of atoms present.
- Chemical equations reflect this – when balanced they contain the same numbers of atoms of each element on their left-hand and right-hand sides.
- The following equations allow us to calculate the number of moles present in a sample:

$$n = \frac{m}{M}$$

$$\text{amount (in moles)} = \frac{\text{mass}}{\text{molar mass}}$$

$$n = \frac{V}{24.0}$$

$$\text{amount (in moles)} = \frac{\text{volume of gas (in dm}^3\text{)}}{24.0}$$

$$n = c \times V \text{ or } n = c \times \frac{v}{1000} \quad (v \text{ is in cm}^3)$$

$$\text{amount (in moles)} = \text{concentration of solution (in mol dm}^{-3}\text{)} \times \text{volume (in dm}^3\text{)}$$

2 The structure of the atom

In this chapter we introduce the three subatomic particles – the electron, the proton and the neutron. We look at their properties, and how they are arranged inside the atom. We explain why some elements form isotopes, and how their relative abundances can be measured using the mass spectrometer. We review the radioactivity of some unstable isotopes, and how they can be of use to the human race. We outline the types of energy associated with the particles in chemistry. Against this background we look at how the electrons are arranged around the nucleus, and how this arrangement explains the positions of elements within the periodic table, their ionisation energies and the sizes of their atoms.

The chapter is divided into the following sections.

2.1 The discovery of the subatomic particles

Our understanding of atoms is very much a nineteenth- and twentieth-century story. In 1808 John Dalton published his atomic theory. This revived the ancient Greek idea of atoms as being indivisible constituents of matter (the name 'atom' is derived from the Greek word meaning 'cannot be cut'). Dalton suggested that all the atoms of a given element were identical to each other, but differed from the atoms of every other element. His atoms were the smallest parts of an element that could exist. They could not be broken down nor destroyed, and were themselves without structure.

For most of the nineteenth century the idea of atoms being indivisible fitted in well with chemists' ideas of chemical reactions, and was readily accepted. Even today chemists believe that atoms are never destroyed during a chemical reaction, but merely change their partners. However, in 1897 J.J. Thomson discovered the first **subatomic particle**, that is, a particle that is smaller than an atom. It was the **electron**. He found that it was much lighter than the lightest atom, and had a negative electrical charge. What is more, he found that under the conditions of his experiment, atoms of different elements produced the same electron particles.

Since atoms are electrically neutral objects, if they contain negatively charged electrons they must also contain particles with a positive charge. An important experiment carried out by Ernest Rutherford, Hans Geiger and Ernest Marsden in 1911 showed that the positive charge in the atom is concentrated into an incredibly small **nucleus** right in the middle of it. They estimated that the diameter of the nucleus could not be greater than 0.00001 times that of the atom itself. Eventually, Rutherford was able to chip away from this nucleus small positively charged particles. He showed that these were also identical to each other, no matter which element they came from. He called this positive particle the **proton**. It was much heavier than the electron, having nearly the mass of the hydrogen atom.

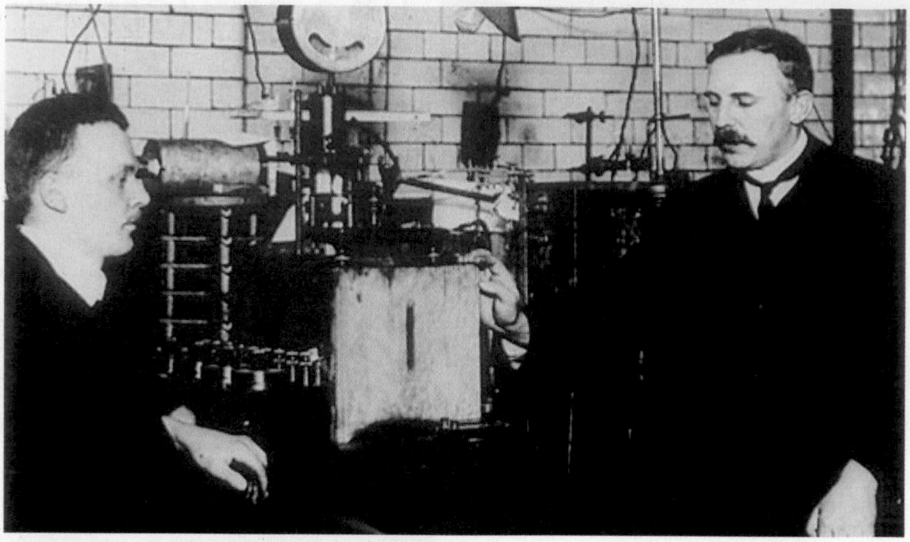

■ **Figure 2.1**
Ernest Rutherford (right) and Hans Geiger in their laboratory at Manchester University in about 1908. They are seen with the instrumentation they used to detect and count α-particles, discussed on pages 26–7.

■ **Figure 2.2**
James Chadwick discovered the neutron in 1932, and was awarded a Nobel Prize in 1935 for his work.

It was another 20 years before the last of the three subatomic particles, the **neutron**, was discovered. Although its existence was first suspected in 1919, it was not until 1932 that James Chadwick eventually pinned it down. As its name suggests, the neutron is electrically neutral, but it is relatively heavy, having about the same mass as a proton.

The nineteenth-century picture of the atom had therefore to be changed. In a sense it had become more complicated, showing that atoms had an internal structure, and were made up of other, smaller particles. But looked at in another way the picture had become simpler – the 90 or so different types of atom needed to make up the different elements had been replaced by just three subatomic particles. It was presumed that these, in different amounts, made up the atoms of all the different elements.

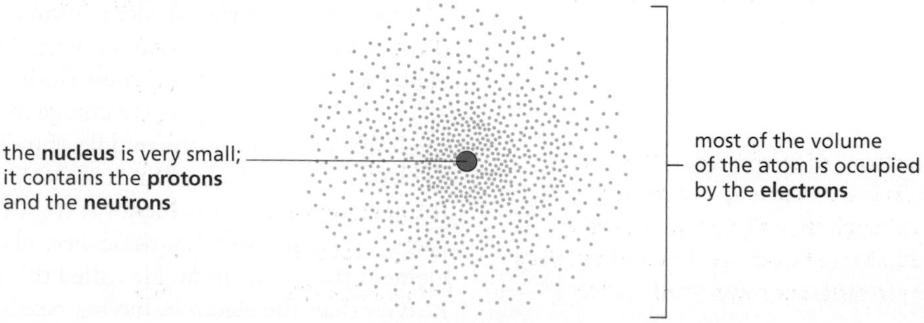

the nucleus is very small; it contains the protons and the neutrons
most of the volume of the atom is occupied by the electrons

■ **Figure 2.3**
The picture of the atom in the early twentieth century.

2.2 The properties of the three subatomic particles

Table 2.1 lists some of the properties of the three subatomic particles.

■ **Table 2.1**
The properties of the subatomic particles. Note that the masses given in the last row are given relative to $\frac{1}{12}$ the mass of an atom of carbon-12. These masses are often quoted relative to the mass of the proton instead, when the mass of the electron is $\frac{1}{1836}$, and the masses of the proton and neutron are both 1.

Property	Electron	Proton	Neutron
Electrical charge/coulombs	-1.6×10^{-19}	$+1.6\times10^{-19}$	0
Charge (relative to that of the proton)	-1	$+1$	0
Mass/g	9.11×10^{-28}	1.673×10^{-24}	1.675×10^{-24}
Mass/amu (see page 7)	5.485×10^{-4} or $\frac{1}{1823}$	1.007	1.009

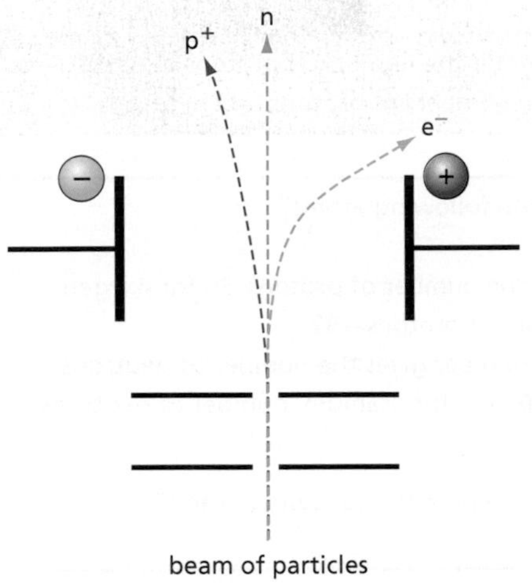

■ **Figure 2.4**
The behaviour of protons, neutrons and electrons in an electric field.

Because of their relative masses and charges, the three particles behave differently in an electric field, as shown in Figure 2.4. Neutrons are undeflected, being electrically neutral. Protons are attracted towards the negative pole, and electrons towards the positive pole. Electrons are deflected to a greater extent than protons because they are much lighter.

The picture of the atom assembled from these observations is as follows.

- Atoms are small, spherical structures with diameters ranging from 1×10^{-10} m to 3×10^{-10} m.
- The massive particles (protons and neutrons) are contained within a very small central nucleus having a diameter of about 1×10^{-15} m.
- The electrons occupy the region around the nucleus. They are to be found in the space inside the atom outside the nucleus, which is virtually the whole of the atom.
- All the atoms of a particular element contain the same number of protons. This also equals the number of electrons within those atoms.
- The atoms of all elements except hydrogen also contain neutrons. These are housed in the nucleus along with the protons. Virtually the only effect they have on the properties of the atom is to increase its mass.

2.3 Isotopes

At the same time as Rutherford and his team were finding out about the structure of the nucleus, it was also discovered that some elements contained atoms that had different masses, but identical chemical properties. These atoms were given the name **isotopes**, since they occupy the same (*iso*) place (*topos*) in the periodic table. The first isotopes to be discovered were those of the unstable radioactive element thorium. (Thorium is element number 90 in the periodic table.) In 1913, however, Thomson was able to show that a sample of neon obtained from liquid air contained atoms with a relative atomic mass of 22 as well as those with the usual relative atomic mass of 20. These heavier neon atoms were stable, unlike the thorium isotopes. Many other elements contain isotopes, some of which are illustrated in Table 2.2.

■ **Table 2.2**
The relative abundances of some isotopes.

Isotope	Mass relative to hydrogen	Relative abundance
Boron-10	10.0	20%
Boron-11	11.0	80%
Neon-20	20.0	91%
Neon-22	22.0	9%
Magnesium-24	24.0	79%
Magnesium-25	25.0	10%
Magnesium-26	26.0	11%

Isotopes of an element differ in their composition in only one respect – although they all contain the same numbers of electrons and protons, they have different numbers of neutrons.

2.4 Extending atomic symbols to include isotopes

For most chemical purposes, the atomic symbols introduced in Chapter 1 are adequate. If, however, we wish to refer to a particular isotope of an element, we need to specify its **mass number**, which is the number of protons and neutrons in the nucleus. We write this as a superscript before the atomic symbol. We often add the atomic number as a subscript before the symbol. So for carbon:

- ${}^{12}_{6}C$ is the symbol for carbon-12, the most common isotope, containing 6 protons and 6 neutrons, with a mass number of 12
- ${}^{14}_{6}C$ is the symbol for carbon-14, which contains 6 protons and 8 neutrons, with a mass number of 14.

The **atomic number (proton number)** of an atom is the number of protons in its nucleus.
The **mass number** of an atom is the sum of the numbers of protons and neutrons.

Example How many protons and neutrons are there in the following atoms?
a ${}^{18}_{8}O$ **b** ${}^{235}_{92}U$

Answer The subscript gives the atomic number, that is, the number of protons. So for oxygen, number of protons = **8**, and for uranium, number of protons = **92**.
Subtracting the atomic number from the mass number gives the number of neutrons. So for oxygen, number of neutrons = 18 − 8 = **10**, and for uranium, number of neutrons = 235 − 92 = **143**.

Further practice How many protons, electrons and neutrons are there in the following atoms?
1 ${}^{23}_{11}Na$ **2** ${}^{127}_{53}I$

2.5 The mass spectrometer

The masses and the relative abundances of individual isotopes are easily measured in a **mass spectrometer**. This is a machine in which ionised atoms are accelerated to a high velocity, and their trajectories (paths) are then deflected from a straight line by passing them through a magnetic field. A magnetic field has a similar bending effect on moving charged particles as the electric field we met on page 23.

■ **Figure 2.5** (right) Schematic diagram of a mass spectrometer.

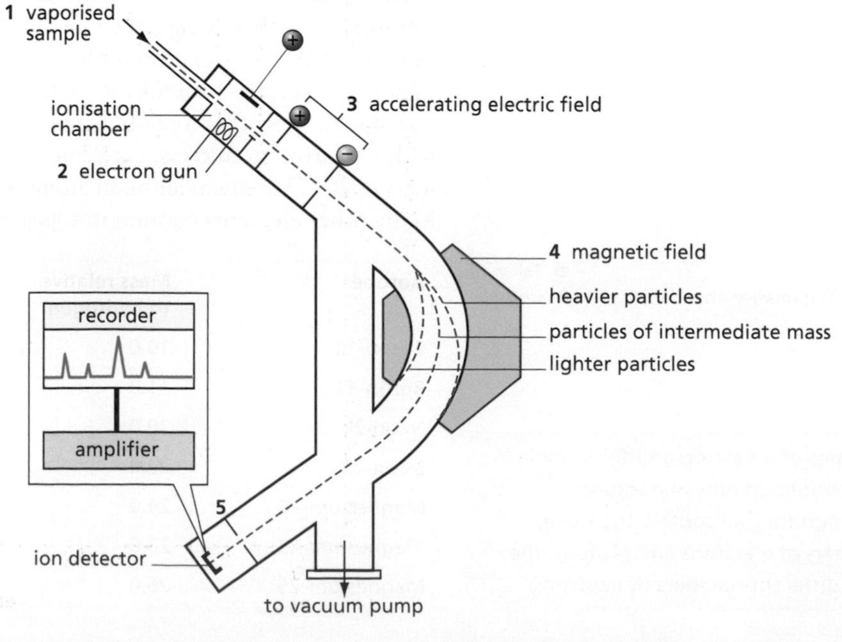

■ **Figure 2.6** (below) A modern mass spectrometer.

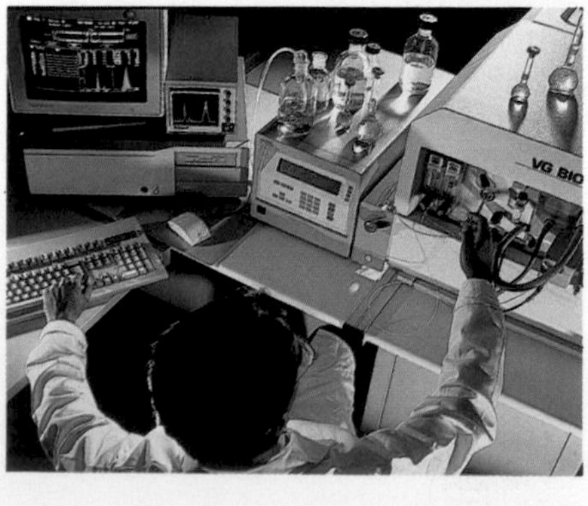

Five processes occur in a mass spectrometer:

1 If not already a gas, the element is vaporised in an oven.
2 Electrons are fired at the gaseous atoms. These knock off other electrons from the atoms:

$$M + e^- \rightarrow M^+ + 2e^-$$

3 The gaseous ions are accelerated by passing them through an electric field (at a voltage of 5–10 kV).
4 The fast-moving ions are deflected by an electromagnet. The larger the charge on the ion, the larger the deflection. On the other hand, the heavier the ion, the smaller the deflection. Overall, the deflections are proportional to the ions' charge to mass ratios. If all ions have a +1 charge (which is usually the case), the extents of deflection will be inversely proportional to their masses.
5 The deflected ions pass through a narrow slit and are collected on a metallic plate connected to an amplifier. For a given strength of magnetic field, only ions of a certain mass pass through the slit and hit the collector plate. As the (positive) ions hit the plate, they cause a current to flow through the amplifier. The more ions there are, the larger the current.

The ions may travel a metre or so through the spectrometer. In order for them to do this without hitting too many air molecules (which would deflect them from their course), the inside of the spectrometer is evacuated to a very low pressure. When the situation is such that a steady stream of ions is being produced, the current through the electromagnet is changed at a steady rate. This causes the magnetic field to change in strength, and hence allows ions of different masses to pass successively through the slit. A **mass spectrum** is produced, which plots ion current against electromagnetic current. This is equivalent to relative abundance against mass number. Figure 2.7 shows the mass spectrum of krypton.

A mass spectrum enables us to analyse the proportions of the various isotopes in an element. However, nowadays by far the major use of the mass spectrometer is in analysing the formulae and structures of organic and inorganic molecules. We shall return to this application in Chapter 31.

Figure 2.7 Mass spectrum of krypton.

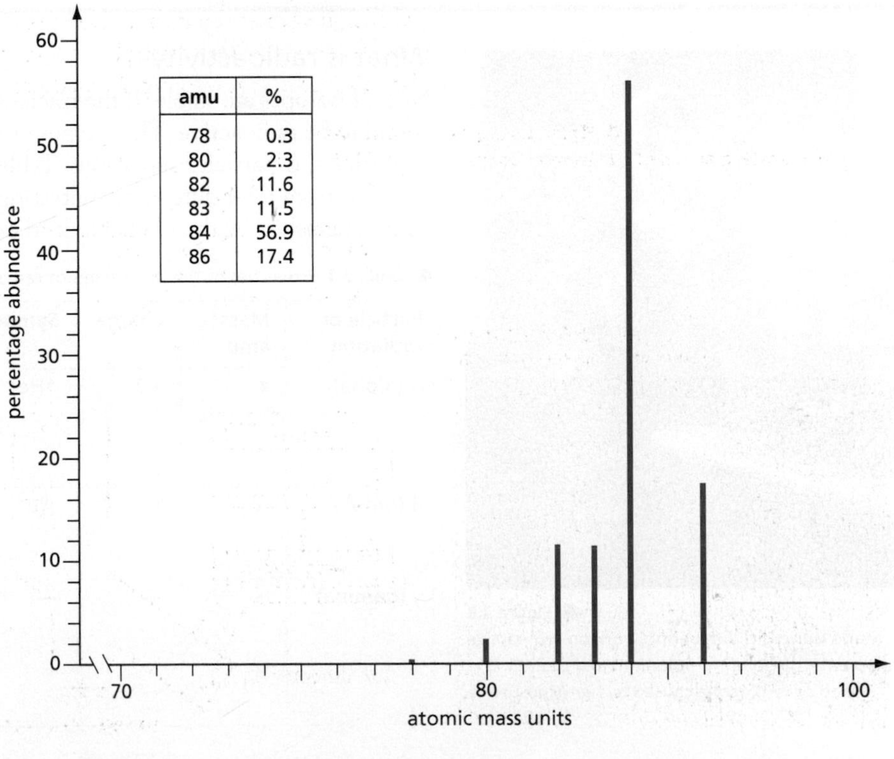

amu	%
78	0.3
80	2.3
82	11.6
83	11.5
84	56.9
86	17.4

Example 1 Chlorine consists of two isotopes, with mass numbers 35 and 37, and with relative abundances 76% and 24% respectively. Calculate the average relative atomic mass of chlorine.

Answer The percentages tell us that if we took 100 chlorine atoms at random, 76 of them would have a mass of 35 units, and 24 of them would have a mass of 37 units.

$$\text{total mass of the 100 random atoms} = (35 \times 76) + (37 \times 24) = 3548\,\text{amu}$$

$$\text{so average mass of one atom} = \frac{3548}{100} = 35.5\,\text{amu}$$

that is, $A_r = \mathbf{35.5}$

Example 2 Calculate the average relative atomic mass of krypton from the table in Figure 2.7.

Answer We can extend the 100-random-atom idea in the above example to include fractions of atoms. Thus the average mass of one atom of krypton

$$= \frac{(78 \times 0.3) + (80 \times 2.3) + (82 \times 11.6) + (83 \times 11.5) + (84 \times 56.9) + (86 \times 17.4)}{100}$$

$= \mathbf{83.9}$

Further practice

1 Chromium has four stable isotopes, with mass numbers 50, 52, 53 and 54, and relative abundances 4.3%, 83.8%, 9.5% and 2.4% respectively. Calculate the average relative atomic mass of chromium.
2 Use the figures in Table 2.2 (page 23) to calculate the average relative atomic masses of boron, neon and magnesium to 1 decimal place.
3 (Harder) Iridium has two isotopes of mass numbers 191 and 193, and its average relative atomic mass is 192.23. Calculate the relative abundances of the two isotopes.

2.6 Radioactivity

■ **Figure 2.8** Marie Curie did pioneering work on radioactive elements, discovering polonium and radium. She died of leukaemia caused by her exposure to radiation.

What is radioactivity?

Not all isotopes are stable. If the nucleus of an unstable isotope emits radiation, it is said to be **radioactive**. Three types of radiation have been discovered. They are α-particles, β-particles and γ-rays. Table 2.3 describes their nature.

Apart from their different penetrating powers, the three types of radiation can be distinguished by their behaviour in an electric field (see Figure 2.9).

■ **Table 2.3** Properties of the three types of radiation.

Particle or radiation	Mass/ amu	Charge	Symbol	Description	Penetrating power
α (alpha)	4	+2	$^{4}_{2}He^{2+}$	The nucleus of a helium atom: 2 protons + 2 neutrons	Poor – stopped by 0.001 cm of metal, or a sheet of paper
β (beta)	$\frac{1}{1823}$	1	$^{0}_{-1}e^{-}$	An electron travelling at very high speed	Medium – stopped by 2 cm of aluminium
γ (gamma)	—	—	—	High energy electromagnetic radiation, like X-rays	Good – passes through 10 cm of aluminium, but is stopped by 10 cm of lead

Figure 2.9
The behaviour of α-particles, β-particles and γ-rays in an electric field.

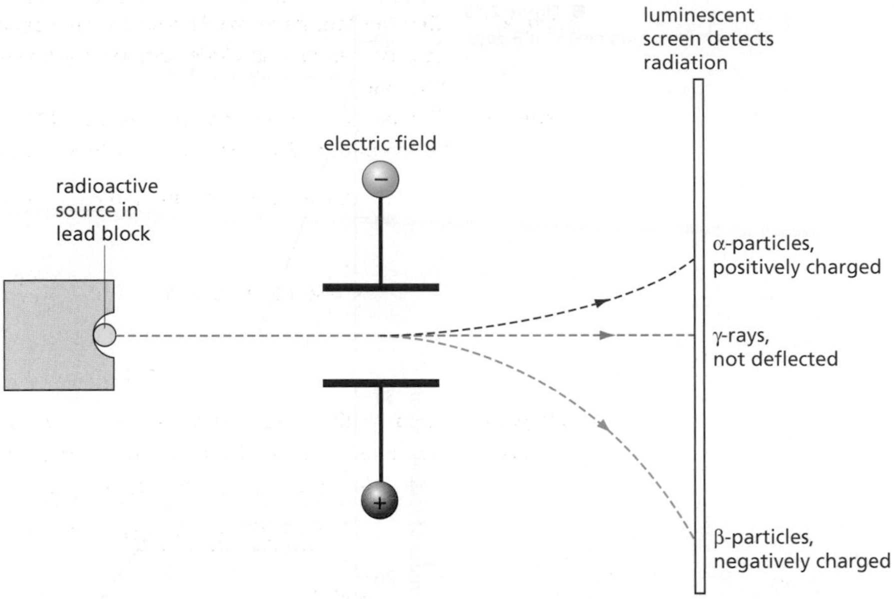

The emission of a γ-ray has little effect on a nucleus other than to decrease its energy, allowing it to end up in a more stable state. (But sometimes γ-ray emission is coupled with the fission of the nucleus into smaller pieces.) The emission of an α-particle or a β-particle, however, causes the nucleus to change its atomic number. This **transmutation** results in the production of a new element. The identity of the new element can be deduced from the following summary.

- The loss of an α-particle results in the atomic number decreasing by 2, and the mass number decreasing by 4.
- The loss of a β-particle has no effect on the mass number, but results in the atomic number increasing by 1.

Example What isotopes will result from the following nuclear decay processes?

a the loss of an α-particle from the $^{238}_{92}U$ nucleus

b the loss of a β-particle from the $^{14}_{6}C$ nucleus

Answer **a** The mass number will decrease by 4, and the atomic number will decrease by 2. The element whose atomic number is (92 – 2) = 90 is thorium:

$$^{238}_{92}U - ^{4}_{2}He \rightarrow ^{234}_{90}Th$$

b The mass number will not change, but the atomic number will increase by 1, from 6 to 7:

$$^{14}_{6}C - ^{0}_{-1}e \rightarrow ^{14}_{7}N$$

Further practice Predict the isotopes that will result from the following decay processes:

1 $^{239}_{92}U - ^{0}_{-1}e \rightarrow ?$

2 $^{218}_{84}Po - ^{4}_{2}He \rightarrow ?$

Half-life

Radioactive decay is a first-order process (see Chapter 11). Its rate is proportional to the amount of decaying isotope present. The decrease in the amount of isotope with time is an **exponential** one, and its **half-life** is constant (see Figure 2.10, overleaf). The half-life is the time taken for half of a radioactive sample to decay. The slower the rate of decay, the longer is the half-life. As can be seen in Figure 2.10, since the rate of decay is slowing down in proportion to the amount of isotope left, the point when all the isotope will have decayed will never be reached – there will always be some left, no matter how little.

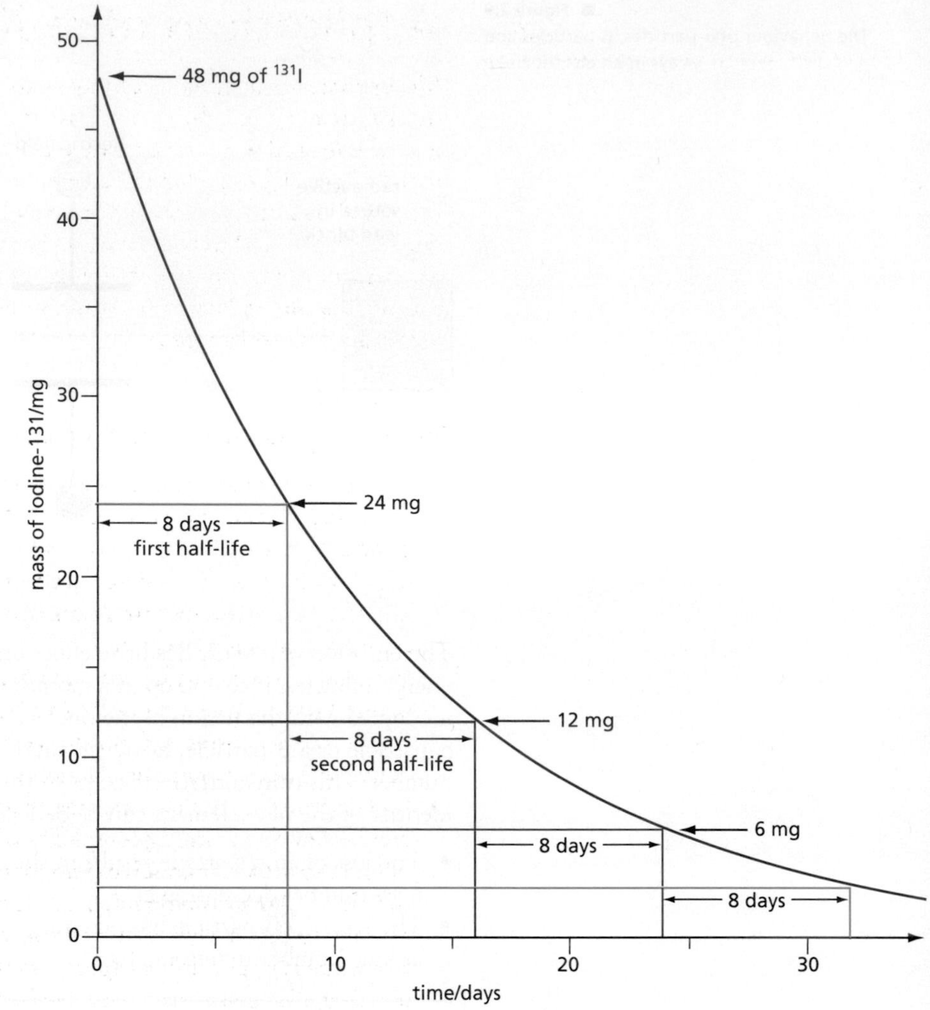

■ **Figure 2.10** The half-life of iodine-131 is 8 days.

Table 2.4 lists some common radioactive decay processes, with their half-lives.

■ **Table 2.4** Radioactive decay processes.

Element	Isotope	Decay route	Half-life
Carbon	^{14}C	β-emission	5730 years
Cobalt	^{60}Co	β-emission	5.3 years
Strontium	^{90}Sr	β-emission	29 years
Iodine	^{131}I	β-emission	8 days
Uranium	^{238}U	α-emission	4.5×10^9 years
Americium	^{241}Am	α-emission	230 years

Example Strontium-90 contaminated the atmosphere in fairly large quantities during the testing of atomic weapons in the 1950s. It entered the food chain through being washed by rain onto the pastures where milking cows grazed. Being similar to calcium in its chemical reactions, it was incorporated into the bones of children growing at the time.

If a child absorbed 40 mg of strontium-90 during his major growing period in 1956, how much will be left in his bones by the year 2014?

Answer The interval between 1956 and 2014 is 58 years, which is two half-lives of 29 years (see Table 2.4). After one half-life, the amount would have reduced to $\frac{40}{2} = 20$ mg, and after another half-life the 20 mg would have reduced to $\frac{20}{2} =$ **10 mg**.

Further practice Take the age of the Earth as 4.2×10^9 years, and the half-life of uranium-235 as 7×10^8 years. If there are 1×10^{12} tonnes of uranium-235 in the Earth's crust at the present time, how much was there when the Earth began?

2.7 Applications of radioactivity

Most people think of radioactivity as an artificial human invention which is dangerous and a bad thing, of no use to anyone. It is certainly true that in too large a dose, radioactivity can cause illness or death. But so too can many other things, such as heat, light, medicines, cigarettes and alcohol. There are in reality many beneficial uses of radioactivity, for example, in medicine, in food processing, in quality control and in protecting us from fire. As for being an artificially invented phenomenon, the amount of ionising radiation we receive each year from cosmic rays and natural radioactivity is almost ten times as much as that caused by artifically created isotopes.

Medicine

In medicine, radioactive isotopes find several major uses.

- The isotope cobalt-60 decays by γ-ray emission with a half-life of 5 years. The radiation is sufficiently damaging to destroy living cells. This is useful in the treatment of localised cancers. The patient is exposed to several controlled doses of radiation, each dose being aimed at the cancerous tissue from a different direction. By this means damage to healthy cells surrounding the cancerous ones is minimised.
- Ingested iodine is readily concentrated by the body into the thyroid gland. Large doses of radioactive iodine-131 can be used to treat thyroid cancers. Much smaller doses are sometimes used to monitor thyroid activity, in order to diagnose over- or under-production of the iodine-containing hormone thyroxin.
- Radioactive technetium finds many uses as a tracer. Being a transition element, technetium forms strong complexes with many ligands, which in turn can be tailored to attach themselves temporarily to various organs or components of the blood. An **autoradiograph** can then be taken. The patient stands in front of a radiation-sensitive screen, and the γ-rays emitted from the technetium-99 atoms cause bright dots to appear on the screen. Thus the effectiveness of the blood supply to various organs can be measured.

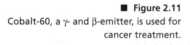

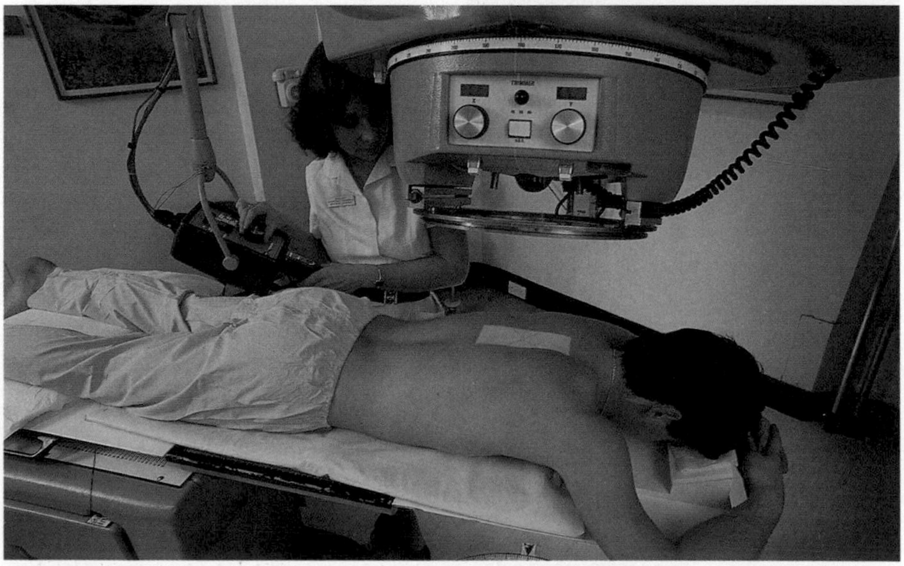

■ **Figure 2.11** Cobalt-60, a γ- and β-emitter, is used for cancer treatment.

Food

The irradiation of foodstuffs with γ-rays kills all microorganisms in the food without using heat or adding harmful or taste-affecting chemicals. Public concerns have not allowed this irradiation to be used very much in the UK, but in the USA irradiation has been used without ill effect for many years. The sterilisation of hospital and surgery equipment is regularly carried out using γ-ray sources.

Smoke alarms

An artificial radioisotope we unknowingly come across everyday is americium-241. Smoke alarms contain a small amount of this α-emitter. The α-particles cause the air inside the detector to become ionised, and when a potential difference is applied between two plates, the ionised air conducts and a small current flows. Smoke particles interrupt this current, and cause a current-sensing device to trigger the alarm circuit. As you can see from Table 2.4 (page 28), the half-life of americium-241 is long enough for it to last at least one human lifetime!

Quality control

β-emitters are often used for quality control in steel-rolling mills or when sheets of plastic are being rolled. The thicker the sheet, the more β-particles are absorbed, and so fewer particles reach the detector on the other side. The change in signal from the detector can be amplified and used to adjust automatically the pressure on the rollers to ensure a constant thickness (see Figure 2.12).

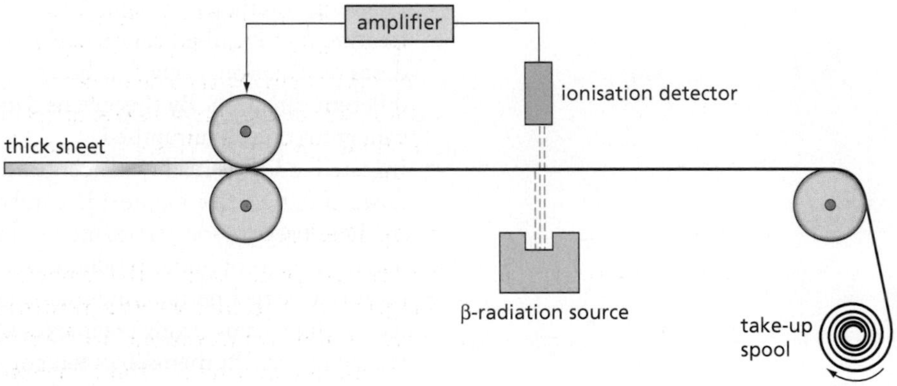

■ **Figure 2.12**
β-radiation is used in a thickness detector.

Carbon dating

Radiocarbon dating uses the naturally occurring isotope carbon-14 to determine the age of a carbon-containing object. A small but constant amount of carbon-14 is produced in the upper atmosphere by the bombardment of nitrogen-14 atoms by neutrons (which in turn are formed from cosmic rays hitting other atoms).

$$^{14}_{7}N + ^{1}_{0}n \rightarrow ^{14}_{6}C + ^{1}_{1}p$$

Over the years, the rate of production has been equalled by the rate of radioactive decay, so the proportion of carbon-14 in the carbon dioxide in the atmosphere is constant, at about one atom of carbon-14 to 8×10^{11} atoms of carbon-12. This also equals the proportion in the biosphere, because all carbon atoms incorporated into the structures of plants and animals come directly or indirectly from atmospheric carbon dioxide. When a living plant or animal dies, however, its carbon atoms are 'fossilised'. No new carbon atoms are incorporated from the atmosphere, but those carbon-14 atoms that are present in the object continue to decay. The $^{14}C:^{12}C$ ratio slowly reduces, in accord with the half-life of carbon-14, which is 5730 years. So by measuring the $^{14}C:^{12}C$ ratio, the age of the object can be estimated.

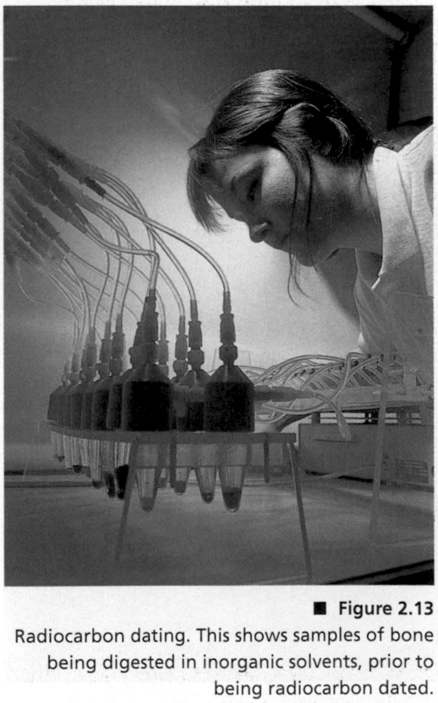

■ **Figure 2.13**
Radiocarbon dating. This shows samples of bone being digested in inorganic solvents, prior to being radiocarbon dated.

Harmful effects of radiation

As mentioned above, radiation is harmful to living tissue. Not only can it cause physical and irreparable damage to cells, but it can also cause chemical damage to the DNA within the nucleus. This can lead to mutations and cancer. α-radiation is the most harmful, because it is the most energetic. But it is also the most easily absorbed, so it is fairly easily protected against. If ingested in the form of dust, however, α-emitters can cause serious damage to internal organs. γ-rays also cause harm, the degree of which depends on their energy. Many low-energy γ-rays pass through our bodies every day from cosmic radiation, without damaging our tissues.

The danger of a radioactive isotope depends on several factors: how much there is of it; what the ionising power of its radiation is; where in the environment it is (for example, the radioactive noble gas radon is harmless in the open atmosphere, but in an enclosed space such as a house it can build up and cause an unacceptably high dose of radiation); and how long its half-life is.

Isotopes used as tracers in medicine usually have short half-lives. After their use in diagnosis they quickly decay, leaving small quantities which are harmless. Isotopes with long half-lives are generally more damaging to the environment, because they cannot be eliminated easily. The plutonium-239 produced in nuclear reactors has a half-life of 24360 years, so it will be with us for many generations. No one has been able to influence the rate of radioactive decay by temperature or any other means.

2.8 Chemical energy

The concept of energy is central to our understanding of how changes come about in the physical world. Energy therefore underpins our study of chemical reactions.

Chemical energy is made up of two components – **kinetic energy**, which is a measure of the motion of atoms, molecules and ions in a chemical substance, and **potential energy**, which is a measure of how strongly these particles attract each other.

Kinetic energy

Kinetic energy increases as the temperature increases. Chemists use a scale of temperature called the **absolute temperature scale**, and on this scale the kinetic energy is directly proportional to the temperature (see Chapter 4, page 104).

Kinetic energy can be of three different types. The simplest is energy due to **translation**, that is, movement from place to place. For monatomic gases (gases made up of single atoms, for example helium and the other noble gases), all the kinetic energy is in the form of translational kinetic energy. For molecules containing two or more atoms, however, there is the possibility of **vibration** and **rotation** as well (see Figure 2.14). Both these forms of energy involve the movement of atoms, even though the molecule as a whole may stay still. The principal form of kinetic energy in diatomic gases is translation, but in more complex molecules such as ethane, vibration and rotation become the more important factors.

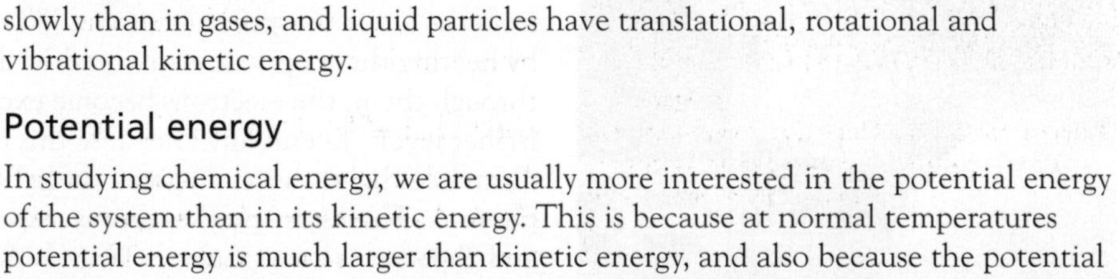

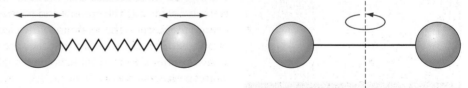

■ **Figure 2.14** The two atoms in a diatomic gas molecule, such as nitrogen, behave as though they are joined by a spring, which lets them vibrate in and out. The two atoms can also rotate about the centre of the bond.

In solids, the particles are fixed in position, and the only form of kinetic energy is vibration. In liquids, the particles can move from place to place, though more slowly than in gases, and liquid particles have translational, rotational and vibrational kinetic energy.

Potential energy

In studying chemical energy, we are usually more interested in the potential energy of the system than in its kinetic energy. This is because at normal temperatures potential energy is much larger than kinetic energy, and also because the potential energy gives us important quantitative information about the strengths of chemical bonds.

Potential energy arises because atoms, ions and molecules attract and repel each other. These attractions and repulsions follow from the basic principle of electrostatics, that unlike charges attract and like charges repel. Ionic compounds contain particles with clear positive and negative charges on them. Two positively charged ions repel each other, as do two negatively charged ions. A positive ion and a negative ion attract each other. Atoms and molecules, which have no overall charge on them, also attract each other, as we shall see in Chapter 3, pages 79–82.

It requires energy to pull apart a sodium ion, Na^+, from a chloride ion, Cl^-. The potential energy of the system increases because we need to move a force F (equal to the force of attraction between the two ions) a distance d apart (see Figure 2.15). In a similar way we increase the potential energy of a book if we pick the book up off the floor and put it on a desk (see Figure 2.16).

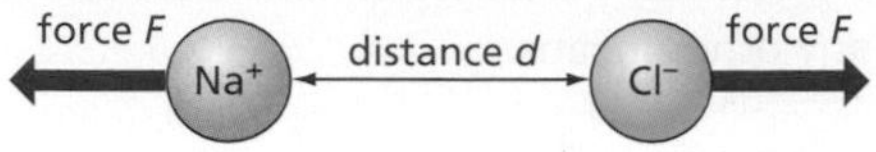

■ **Figure 2.15** To separate oppositely charged ions through a distance *d*, a force *F* is required.

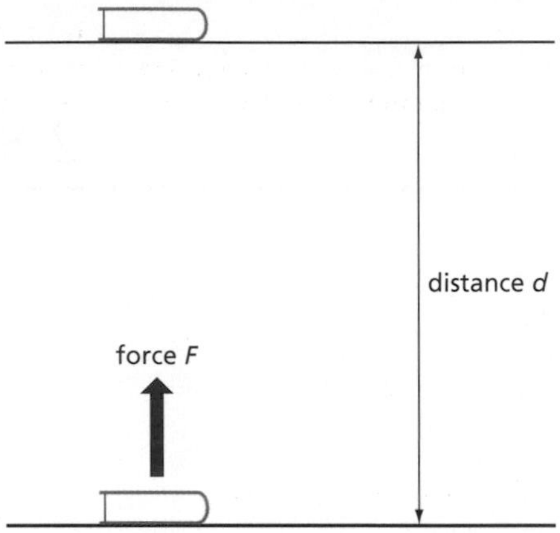

■ **Figure 2.16** Energy is needed to lift a book from the floor and put it on the desk.

In contrast, if we start with a sodium ion and a chloride ion separated from each other and then bring them together, the potential energy decreases. We also decrease the potential energy of a book if we allow it to fall from the desk to the floor.

2.9 The arrangement of the electrons – energy levels and orbitals

On page 22 we concluded that the atom was made up of a very dense, very small nucleus containing the protons and neutrons, and a much larger region of space around the nucleus that contained the electrons. We now turn our attention to these electrons. We shall see that they are not distributed randomly in this region of space. They occupy specific volume regions, called **orbitals**, which have specific energies associated with them.

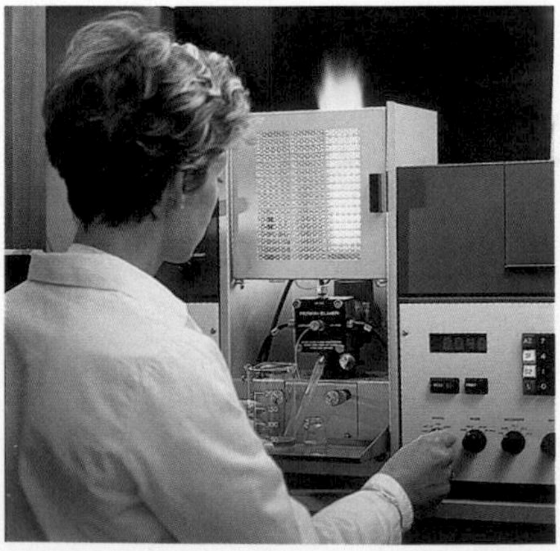

■ **Figure 2.17** An atomic absorption spectrometer.

Energy levels and emission spectra

Evidence for these different 'energy levels' of electrons within atoms came initially from the emission spectra of atoms. When gaseous atoms are given energy, either by heating them up to several hundred degrees, or by passing an electric current through them, the electrons become **excited** and move from lower energy levels to higher levels. Eventually, they lose this energy again by radiating visible or ultraviolet light. An atom gives out light in specific amounts as 'packets' called **photons**. This is the process that is responsible for the flame colours of the Group 1 and 2 elements (see Chapters 14 and 16). We can analyse the radiated light with a device called a spectroscope, shown in Figure 2.18. This shows us that atoms do not emit light with an equal intensity throughout the spectrum. Only a few very specific frequencies are emitted, and these are unique to an individual element. All the atoms of a particular element radiate at the same set of frequencies, which are usually different from those of all other elements (see Figure 2.19).

■ **Figure 2.18**
Outline diagram showing how a spectroscope works.

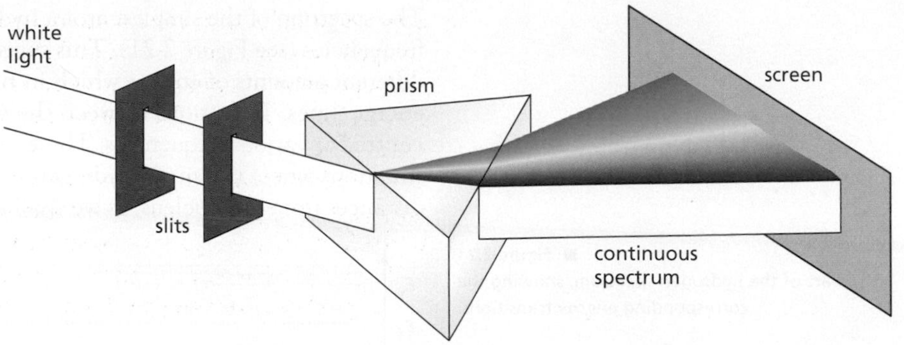

■ **Figure 2.19**
a White light spectrum. **b** Sodium emission spectrum. **c** Cadmium emission spectrum.

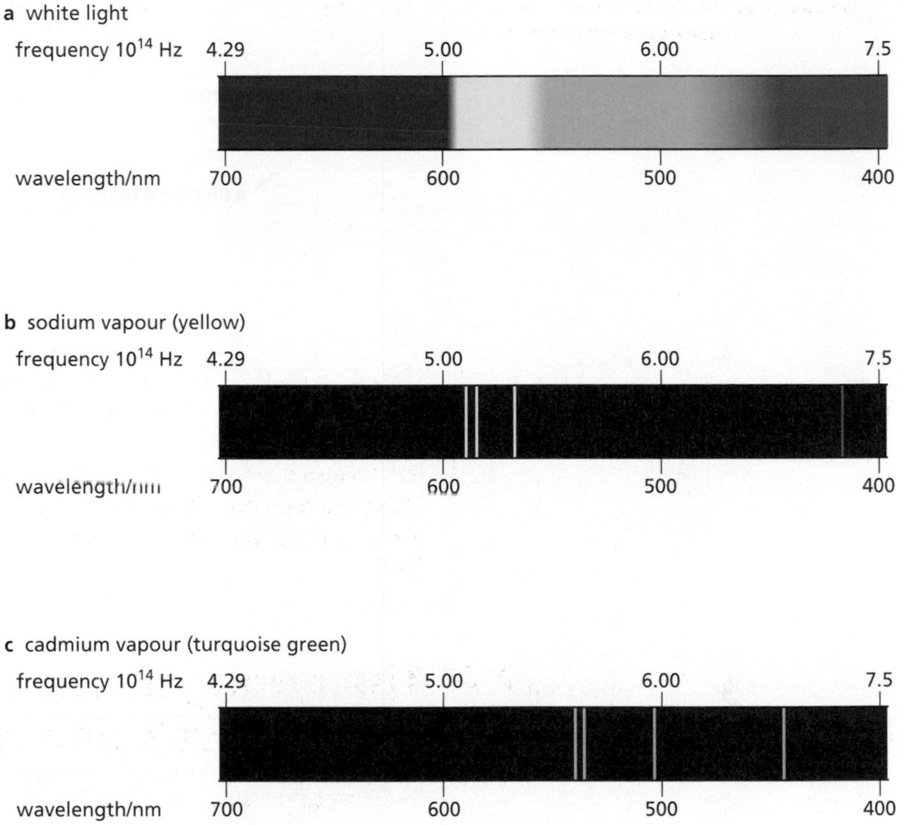

The energy E of a photon of light is related to its frequency, f, by Planck's equation:

$$E = hf$$

where h is the Planck constant.

If photons of a particular frequency are being emitted by an atom, this means that the atom is losing a particular amount of energy. This energy represents the difference between two states of the atom, one more energetic than the other (see Figure 2.20).

■ **Figure 2.20**
A photon is emitted as the atom moves between state E_1 and state E_2.

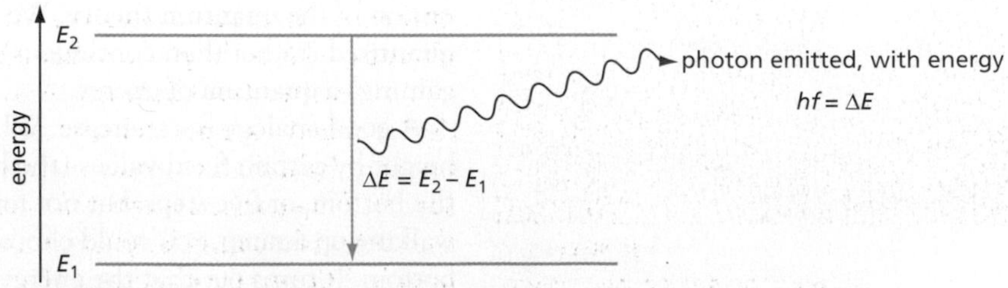

The spectrum of the simplest atom, hydrogen, shows a series of lines at different frequencies (see Figure 2.21). This suggests that the hydrogen atom can lose different amounts of energy, which in turn suggests that it can exist in a range of energy states. Transitions between the various energy states cause photons to be emitted at various frequencies. These energy states can be identified with situations where the single hydrogen electron is in certain orbitals, at specific distances from the nucleus, as we shall see.

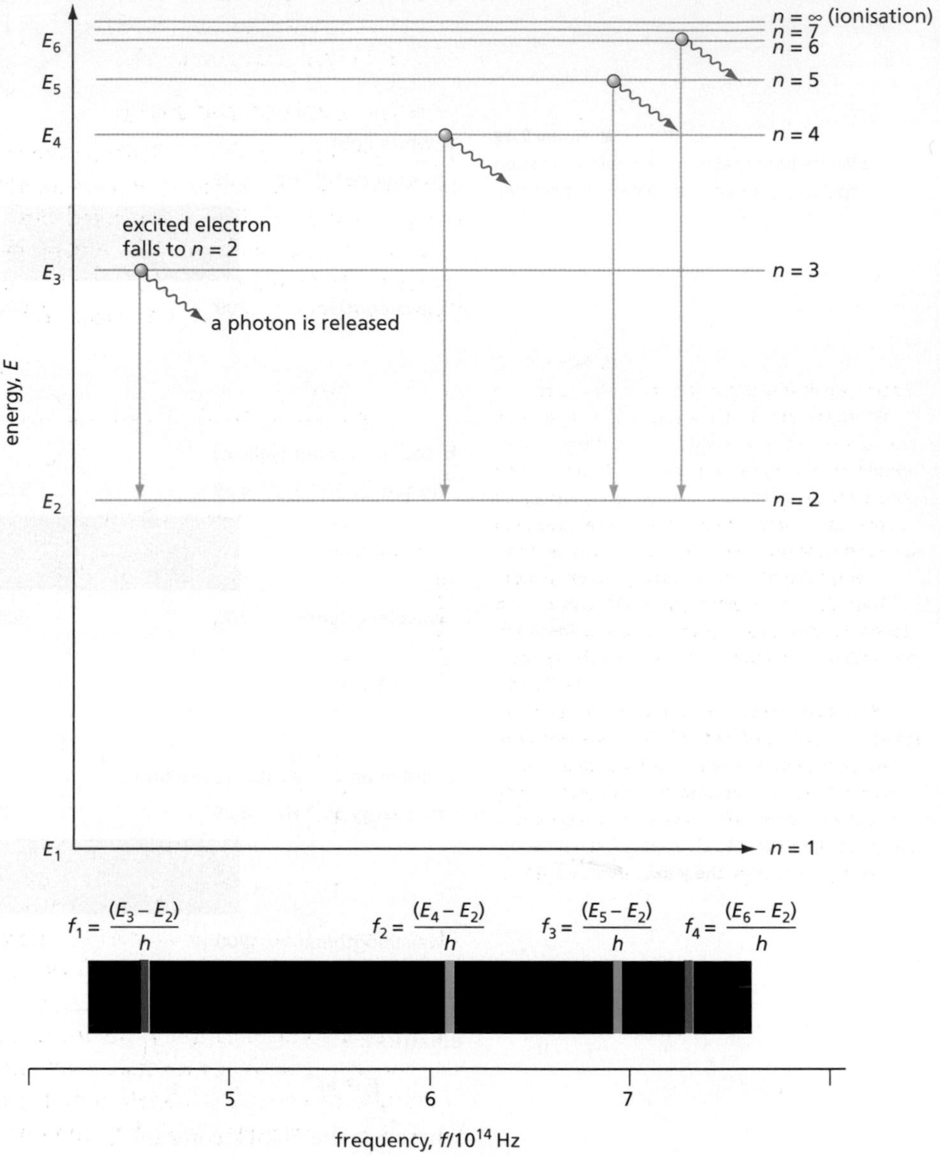

■ **Figure 2.21** Part of the hydrogen spectrum, showing the corresponding energy transitions.

Quantum theory

The fact that only certain frequencies of light are emitted, rather than a continuous spectrum, is compelling evidence that the energy of the hydrogen atom can take only certain values, not a continuous range of values. This is the basic notion of the **quantum theory**. We say that the energy of the hydrogen atom is **quantised** (rather than continuous). It can lose (or gain) energy only by losing (or gaining) a **quantum** of energy.

A good analogy is a staircase. When climbing a staircase, you increase your height by certain fixed values (the height of each step). You can be four steps from the bottom, or five steps, but not four and a half steps up. By contrast, if you were walking up a ramp, you could choose to be at any height you liked from the bottom. It turns out that the energy of all objects is distributed in staircases rather than in ramps, but if the object is large enough, the height of each step is vanishingly small. The energy values then seem to be virtually continuous, rather than stepped. It is only when we look at very small objects like atoms and molecules that the height of each step becomes significant.

The size of an energy quantum (the height of each step) is not fixed, however. It depends on the type of energy (electronic, vibrational, translational, etc. – see Chapter 5) we are considering. We shall return to this point in Chapter 14 when we study spectroscopy. We can use the methods of quantum mechanics to calculate not only the energies but also the probability distributions of orbitals (see Figure 2.22 and Figure 2.23, page 37).

Energy levels in the atom

Hydrogen is a very small spherical atom with only two particles, a proton and an electron. Apart from energy of movement (translational kinetic energy), the only energy it can have, is that associated with the electrostatic attraction between its two particles (see page 31). The different energy levels are therefore due to different electrostatic potential energy states, where the electron is at different distances from the proton (see Figure 2.22).

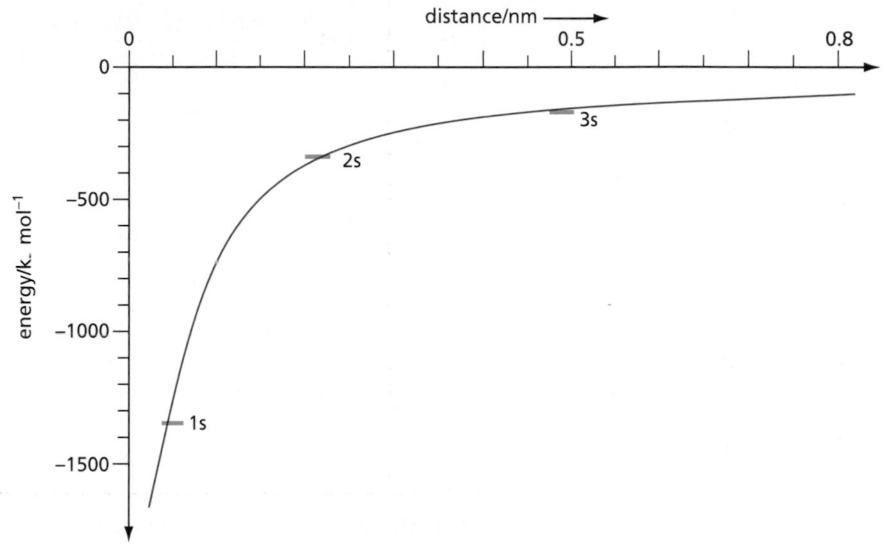

■ **Figure 2.22**
Energy against electron–proton distance for the hydrogen atom. The orange curve represents how the potential energy of the hydrogen atom would vary with electron–proton distance if the energy could take on any value, as predicted by simple electrostatic theory. But, as we have seen above, the potential energy is quantised, and can only take on certain values, shown in blue. Therefore the electron can only be at certain (average) distances from the nucleus. These are the electronic orbitals, and are given the symbols 1s, 2s, etc.
Note that the energy of the proton–electron system is usually defined as being zero when the two particles are an infinite distance apart. As soon as they start getting closer together, they attract each other and this causes the potential energy of the system to decrease. This is why the energy values on the *y*-axis are all negative.

These energy levels, associated with different distances of the electron from the nucleus, are called **orbitals**, by analogy with the orbits of the planets at different distances from the Sun.

The orbitals are arranged in **shells**. Each shell contains orbitals of roughly the same energy. The shell with the lowest energy, the one closest to the nucleus, contains only one orbital. Shells with higher energies, further out from the nucleus, contain increasingly large numbers of orbitals, according to the formula:

$$\text{number of orbitals in shell } n = n^2$$

Table 2.5 shows the number of orbitals in each shell, according to this equation.

■ **Table 2.5**
The number of orbitals in each shell.

Principal shell number	**Number of orbitals**
1	1
2	4
3	9
4	16
5	25
6	36

Quantum mechanics

During the 1920s, various experimental results suggested that electrons (and other small particles) do not always behave in the way that we would predict particles to behave. In particular, Clinton Davisson and Lester Germer, in 1925, showed that a beam of electrons could be diffracted by a crystal lattice, just as a beam of light could be diffracted by a grating. This showed that electrons have wave properties, similar to those of light.

Louis de Broglie suggested that the wavelength of the wave associated with a particle could be related to the particle's momentum by the simple equation:

$$\text{wavelength, } \lambda \times \text{momentum, } p = h$$

(The numerical value for h, the Planck constant, is $6.62 \times 10^{-34}\,\text{J s}$, or $6.62 \times 10^{-34}\,\text{kg m}^2\,\text{s}^{-1}$.) The equation applies to all particles, no matter what their size. If, however, the particle is large and heavy, its momentum (= mass × velocity) is also large, so its wavelength will be correspondingly small. Its wave properties can therefore be ignored. For small particles like the electron, on the other hand, the wavelength becomes much larger than the particle itself.

Two examples will make this clear:

1 A snooker ball of mass 0.14 kg travelling at $5.0\,\text{m s}^{-1}$:

$$\text{wavelength, } \lambda = \frac{h}{p} = \frac{6.62 \times 10^{-34}\,\text{kg m}^2\,\text{s}^{-1}}{(0.14\,\text{kg}) \times (5.0\,\text{m s}^{-1})} \qquad [p = m \times v]$$

$$\approx \mathbf{1 \times 10^{-33}\,m}$$

2 An electron of mass $9 \times 10^{-31}\,\text{kg}$ travelling at $6 \times 10^6\,\text{m s}^{-1}$:

$$\text{wavelength, } \lambda = \frac{h}{p} = \frac{6.62 \times 10^{-34}\,\text{kg m}^2\,\text{s}^{-1}}{(9 \times 10^{-31}\,\text{kg}) \times (6 \times 10^6\,\text{m s}^{-1})}$$

$$\approx \mathbf{1 \times 10^{-10}\,m}$$

In the first case, the wavelength of the snooker ball is insignificant compared with either its size, or the size of the snooker table it is travelling on. The wavelength of the electron, however, is about the same as the diameter of the atomic 'snooker table' that it bounces around in. Its wavelength is over 1 million times as large as its own diameter!

Clearly, therefore, the electron cannot be treated like a particle with a well-defined extent. A new method was needed to allow chemists to calculate the dynamics of subatomic systems. The tool was provided by Erwin Schrödinger in 1926, when he suggested using a differential equation that related the energy of an electron to an expression known as a wave function (given the symbol ψ, psi). The wave function, in turn, could be used to calculate the speed and the most likely positions of the electron. Schrödinger's equation can only be solved if the energy of the electron has one of a limited number of values, so it automatically predicts the quantisation of energy within the atom. Three sets of quantum numbers restrict the solutions to the equation. These correspond to the different types of orbital that we know occur within the atom, and allow us to calculate the shapes and sizes of these orbitals.

These three quantum numbers are given the symbols n, l and m_l, and can take the following values:

- the **principal quantum number**, n, determines the shell the orbital is in ($n = 1, 2, 3, \ldots$)
- the **orbital quantum number**, l, determines the shape of orbital ($l = 0, 1, 2,$ up to $n - 1$)
- the **orientation quantum number**, m_l, determines the direction the orbital points in ($m_l = -l, \ldots 0, \ldots +l$).

We shall see in the next section how these are used to describe the orbitals within an atom.

One of the most useful consequences of the wave nature of the electron is that, like any other waves, electrons on different atoms can interact. This allows them to occupy orbitals between atoms, called molecular orbitals, which arise from the overlapping and mixing of the atomic orbitals on the adjacent atoms. The formation of covalent bonds is thus readily explained (see Chapter 3).

2.10 Subshells and the shapes of orbitals

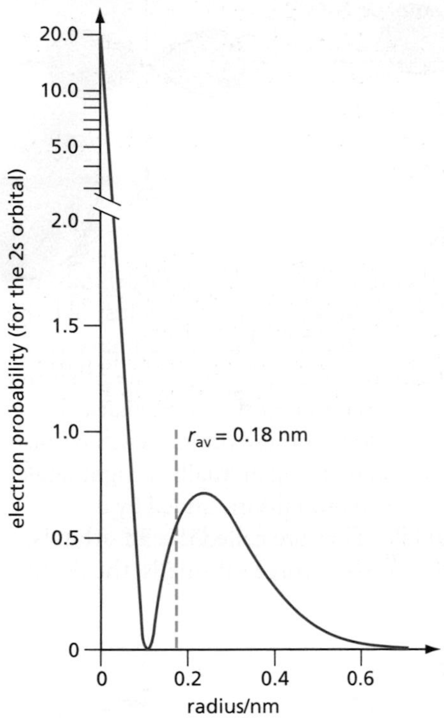

■ **Figure 2.23** (above) The electron probability distribution for a 2s orbital.

■ **Figure 2.24** (right) The probability distribution for the Earth–Sun distance.

Electron probability and distance from the nucleus

Unlike a planet in a circular orbit, whose distance from the Sun does not change, an electron in an orbital does not remain at a fixed distance from the nucleus. Although we can calculate, and in some cases measure, the average electron–nuclear distance, if we were to take an instantaneous snapshot of the atom, we are quite likely to find the electron either further away from or closer to the nucleus than this average distance. The graph of probability of finding the electron against its distance from the nucleus for a typical orbital is shown in Figure 2.23, and by contrast a similar one for the Earth–Sun distance is shown in Figure 2.24. Notice that, because of the gradual falling off of the electron probability, there is a finite (but very small) chance of finding an electron a very long way from the centre of the atom.

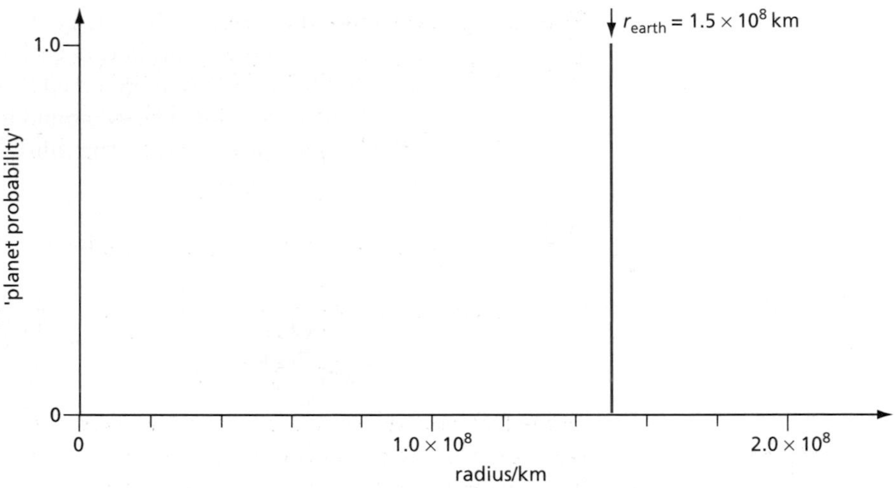

The first shell – s orbital

The single orbital in the first shell is spherically symmetrical. This means that the probability of finding the electron at a given distance from the nucleus is the same no matter what direction from the nucleus is chosen. This orbital is called the **1s orbital**.

The second shell – s and p orbitals

Of the four orbitals in the second shell, one is spherically symmetrical, like the orbital in the first shell. This is called the **2s orbital**. Its probability curve is shown in Figure 2.23. The other three second-shell orbitals point along the three mutually perpendicular *x*-, *y*- and *z*-axes. These are the **2p orbitals**. They are called, respectively, the $2p_x$, $2p_y$ and $2p_z$ orbitals. They do not overlap with each other. For example, an electron in the $2p_x$ orbital has a high probability of being found on or near to the *x*-axis, but a zero probability of being found on either the *y*-axis or the *z*-axis. These two different types of orbital, the 2s and the 2p, are of slightly different energies. They make up the two **subshells** in the second shell of orbitals.

When an electron is located in an s orbital, there is a fair chance of finding it right at the centre of the atom, at the nucleus. But the distribution curve for the 2p orbital in Figure 2.25 shows that there is a zero probability of finding a p electron in the centre of the atom. This is general for all p orbitals. Electrons in these orbitals tend to occupy the outer reaches of atoms.

In three dimensions, an s orbital can be likened to a soft sponge ball, whereas a p orbital is like a long spongy solid cylinder, constricted around its centre to form two lobes, as shown in Figure 2.26 (overleaf).

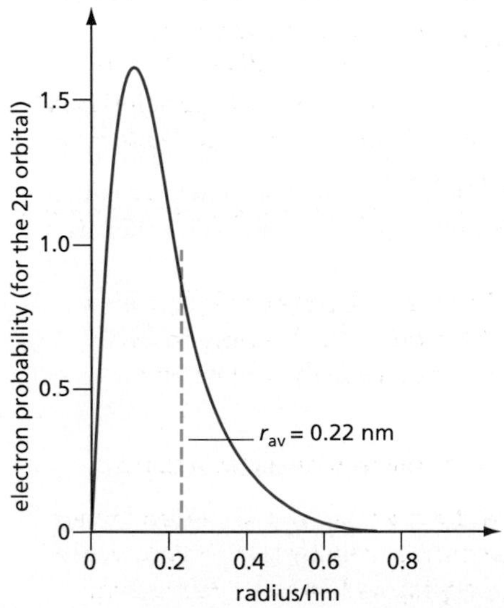

■ **Figure 2.25** The electron probability distribution for a 2p orbital.

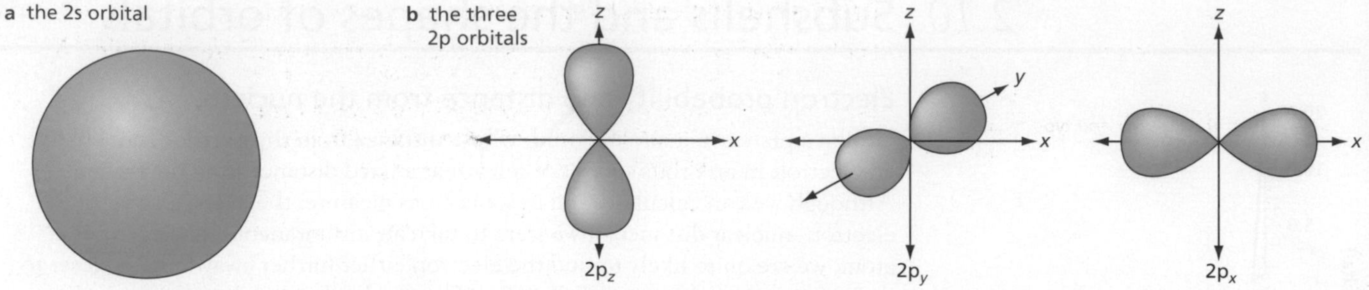

■ **Figure 2.26** The shapes of **a** the 2s orbital and **b** the three 2p orbitals.

The third shell – s, p and d orbitals

From the n^2 formula on page 35 we can predict that there will be nine orbitals in the third shell. One of these, the 3s orbital, is spherically symmetrical, just like the 2s and 1s orbitals. Three **3p orbitals**, the $3p_x$, $3p_y$ and $3p_z$ orbitals, each have two lobes, pointing along the axes in a similar fashion to the 2p orbitals. The other five have a different shape. The most common interpretation of the mathematical equations that describe their shape suggests that four of these orbitals each consist of four lobes in the same plane as each other, and pointing mutually at right angles, whereas the fifth is best represented as a two-lobed orbital surrounded by a 'doughnut' of electron density around its middle. They are called the **3d orbitals**, illustrated in Figure 2.27. So the third shell consists of three sub-shells, the 3s, the 3p and the 3d subshells.

■ **Figure 2.27** The shapes of the five 3d orbitals. Four of them have the same shape, but in different orientations. The fifth has a different shape, as shown in **e**.

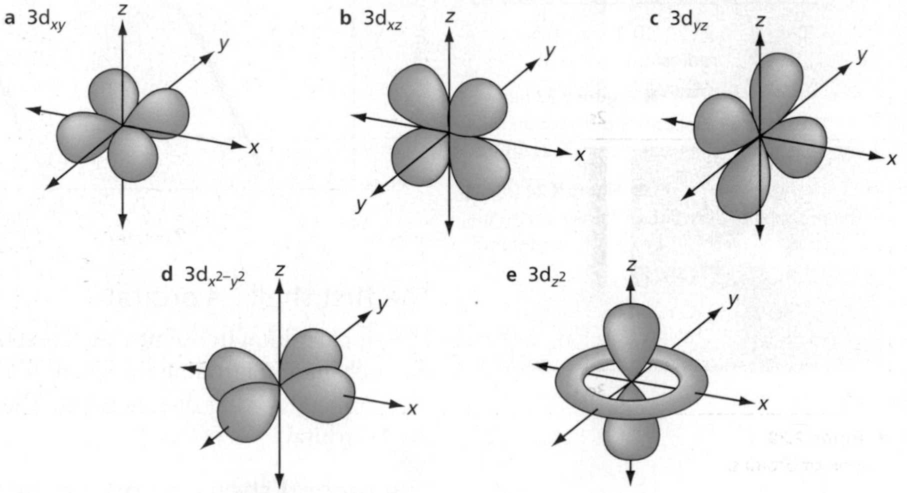

The fourth shell – s, p, d and f orbitals

The process can be continued. In the fourth shell, we predict that there will be $4^2 = 16$ orbitals. One of these will be the 4s orbital, three will be the 4p orbitals, five will be the 4d orbitals, leaving seven orbitals of a new type. They are called the **4f orbitals**. Each consists of many lobes pointing away from each other, as shown in Figure 2.28.

■ **Figure 2.28** The shape of one 4f orbital. There are eight lobes pointing out from the centre.

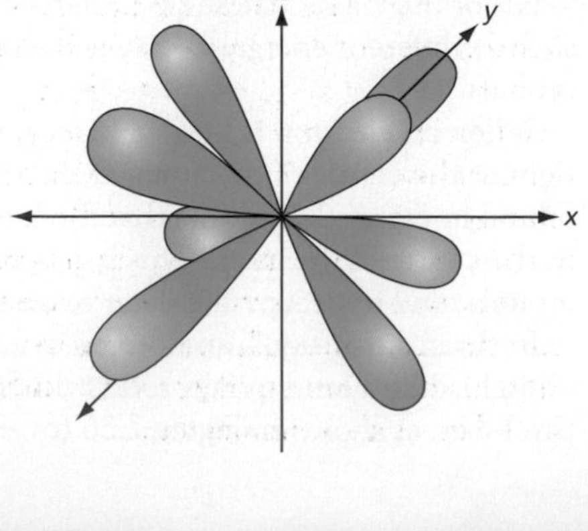

Shells and orbitals in summary

The number and type of orbitals within each shell is summarised in Table 2.6.

■ Table 2.6
Summary of the number and type of orbitals in each shell.

Shell number, n	Number of orbitals within the shell, n^2	Number of orbitals of each type (i.e. number of orbitals in each subshell)			
		s	p	d	f
1	1	1			
2	4	1	3		
3	9	1	3	5	
4	16	1	3	5	7

Example What is the total number of orbitals in the fifth shell? How many of these are d orbitals?

Answer Total number of orbitals $= 5^2 = \mathbf{25}$. Of these, **5** are d orbitals (there are 5 d orbitals in every shell above the third shell).

Further practice How many orbitals will be in: **a** the 5p subshell, **b** the 5f subshell?

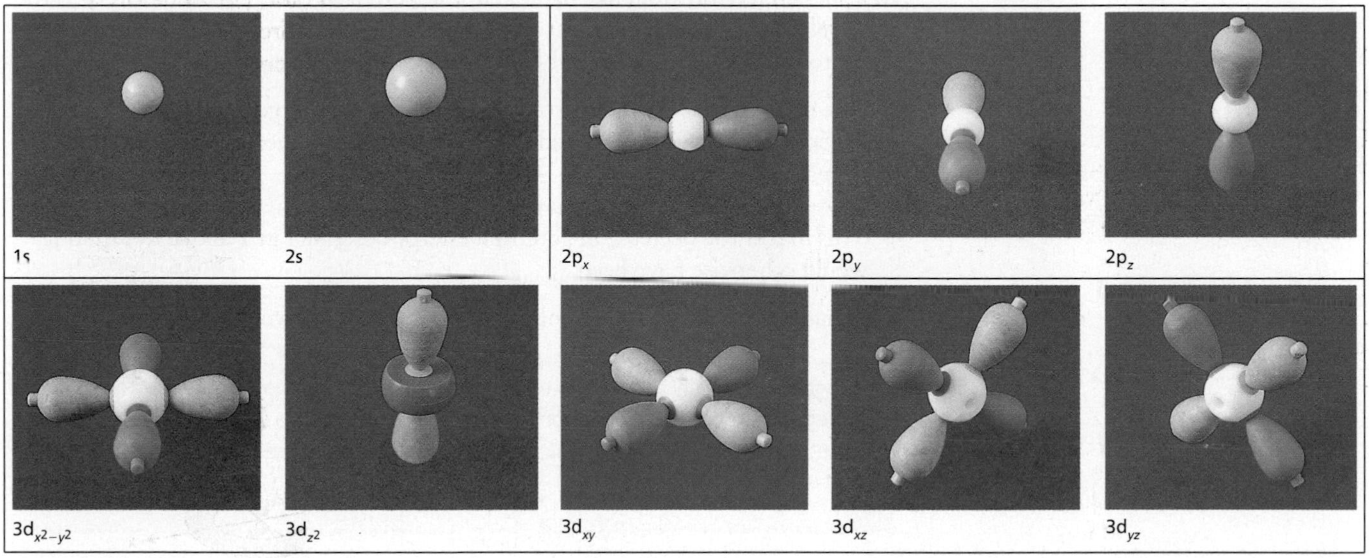

■ Figure 2.29
Models of orbitals.

2.11 Putting electrons into the orbitals

The atoms of different elements contain different numbers of electrons. A hydrogen atom contains just one electron. An atom of uranium contains 92 electrons. How these electrons are arranged in the various orbitals is called the atom's **electronic configuration**. This is a key feature in determining the chemical reactions of an element.

The electrons around the nucleus of an atom are most likely to be found in the situation of lowest possible energy. That is, they will occupy orbitals as close to the nucleus as possible. The single electron in hydrogen will therefore be in the 1s orbital. The electronic configuration of hydrogen is written as $1s^1$ (and spoken as 'one ess one'). The next element, helium, has two electrons, so its electronic configuration is $1s^2$ ('one ess two').

For reasons that we shall look at later, we find that each orbital, no matter what shell or subshell it is in, cannot accommodate more than two electrons. The third electron in lithium, therefore, has to occupy the orbital of next lowest energy, the 2s. The electronic configuration of lithium is $1s^2\,2s^1$. Beryllium, $1s^2\,2s^2$, and boron, $1s^2\,2s^2\,2p^1$, follow predictably.

When we come to carbon, $1s^2 2s^2 2p^2$, we need to differentiate between the three 2p orbitals. Because electrons are all negatively charged, they repel each other electrostatically. The three 2p orbitals are all of the same energy. Therefore we would expect the two p electrons in carbon to occupy different 2p orbitals (for example, $2p_x$ and $2p_y$), as far away from each other as possible. This is in fact what happens. Likewise the seven electrons in the nitrogen atom arrange themselves $1s^2 2s^2 2p_x^{\,1} 2p_y^{\,1} 2p_z^{\,1}$. Only when we arrive at oxygen, $1s^2 2s^2 2p_x^{\,2} 2p_y^{\,1} 2p_z^{\,1}$, do the 2p orbitals start to become doubly occupied. For many purposes, however, there is no need to distinguish between the three 2p orbitals, and we can abbreviate the electronic configuration of oxygen to $1s^2 2s^2 2p^4$.

This process continues until element number 18, argon. With argon, the 3p subshell is filled, and we would anticipate the next electron starting to occupy the 3d subshell. It does not, however. The nineteenth electron in the next element, potassium, occupies the 4s subshell. We shall now explain why this is the case.

2.12 Shielding by inner shells

In a single-electron atom like hydrogen, the potential energy is due entirely to the single electrostatic attraction between the electron and the proton. When we move to the two-electron atom helium, two changes have occurred:

1 The nucleus now contains two protons, and has a charge of +2. It will therefore attract the electrons more, and reduce the potential energy of the system, that is, make it more stable.
2 However, the second electron in the atom will repel the first (and vice versa). This makes the decrease in potential energy described in **1** above less than it would otherwise have been.

This has a clear effect on the ionisation energies of the two elements.

The **ionisation energy** of an atom (or ion) is defined as the energy required to remove completely a mole of electrons from a mole of gaseous atoms (or ions). That is, the ionisation energy is the energy change for the following process:

$$X_{(g)} \rightarrow X^+{}_{(g)} + e^-$$

$$H(g) \rightarrow H^+(g) + e^- \qquad \Delta H = 1312\ kJ\,mol^{-1}$$
$$He(g) \rightarrow He^+(g) + e^- \qquad \Delta H = 2372\ kJ\,mol^{-1}$$

(ΔH, the enthalpy change for the ionisation, is equal to the ionisation energy for this process. See Chapter 5, page 110, for an explanation of this terminology.)

It takes more energy to remove an electron from a helium atom than from a hydrogen atom. This shows that, compared to the energy when the electron is at infinity (see Figure 2.22, page 35), the energy of the 1s orbital has become more negative in helium (see Figure 2.30).

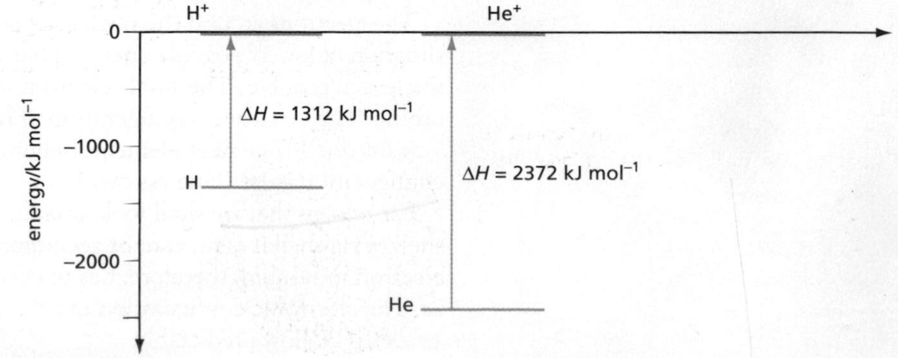

■ **Figure 2.30**
The energy needed to remove an electron from a hydrogen atom and a helium atom.

■ **Figure 2.31**
Variation of the energy of the 1s orbital with atomic number.

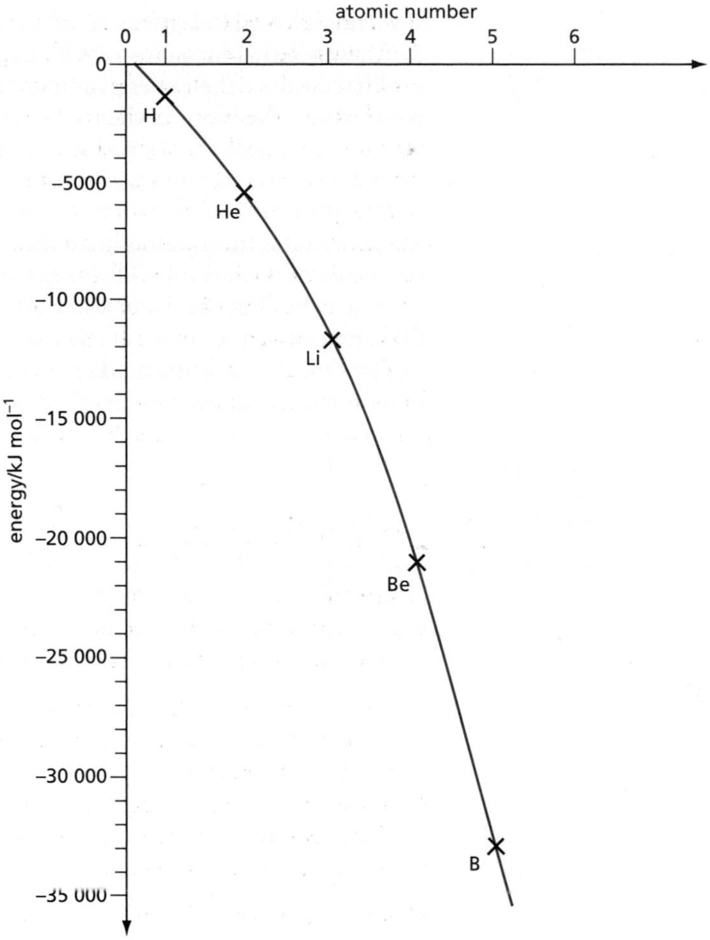

This decrease in energy for the 1s orbital continues as the atomic number increases (see Figure 2.31). The decrease in energy is also true for other orbitals. The reason is that as the number of protons in the nucleus increases, the electrons in a particular orbital are attracted to it more. The decrease in energy is not regular, however, and is not the same for all orbitals. This is due to two factors:

1 The average distance from the nucleus of a p orbital is larger than that of the corresponding s orbital in the same shell (see Figures 2.23 and 2.25, page 37). Electrons in p orbitals therefore experience less of the stabilising effect of increasing nuclear charge than electrons in s orbitals. For a similar reason, electrons in d orbitals experience even less of the increasing nuclear charge than electrons in p orbitals.
2 All electrons in outer shells are to some extent shielded from the nuclear charge by the electrons in inner shells. This shielding has the effect of decreasing the effective nuclear charge. In the case of lithium, for example, the outermost electron in the 2s orbital does not experience the full nuclear charge of +3 (see Figure 2.32). The two electrons in the filled 1s orbital mask a good deal of this charge. Overall, the **effective nuclear charge** experienced by the 2s electron in lithium is calculated to be +1.3, which is considerably less than the actual nuclear charge of +3.

■ **Figure 2.32**
The lithium nucleus and inner shell.

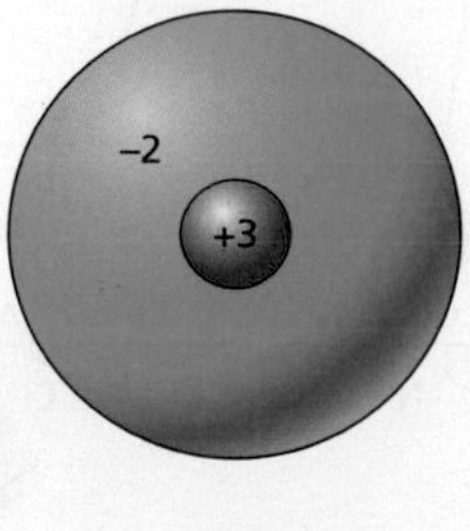

If we now plot orbital energy against atomic number for several orbitals (see Figure 2.33), we see that for a given shell, because of the above factors, electrons in its s orbital decrease their energy, as more electrons are added, faster than those in its p orbitals do. Electrons in the p orbitals in turn decrease their energy faster than electrons in the d orbitals, and so on. In particular, electrons in the 4s and 4p orbitals decrease their energy faster than electrons in the 3d orbitals. So much so that by the time 18 electrons have been added (it so happens), the next most stable orbital is the 4s rather than the 3d. The orbital filling continues after potassium in the order 4s–3d–4p–5s–4d–5p–6s–4f–5d–6p–7s, reflecting the 'lagging behind' of the d and f orbitals. This is due to the effective shielding, by filled inner electron shells, of these orbitals from the increasing nuclear charge.

The simple mnemonic diagram in Figure 2.34 will help you to remember the order in which the orbitals are filled.

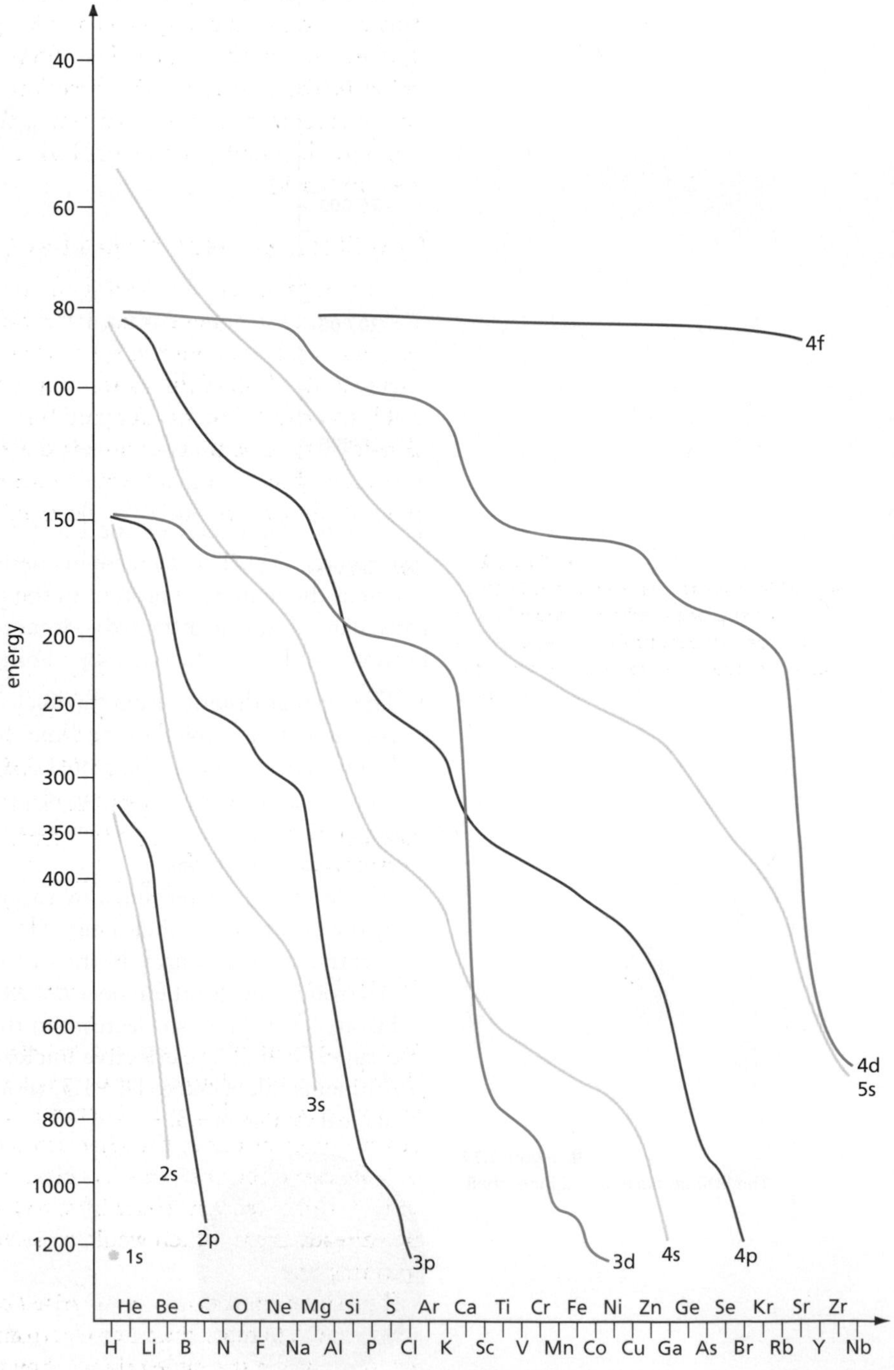

■ **Figure 2.33**
Graph of orbital energy against atomic number for the principal quantum numbers 1–5.

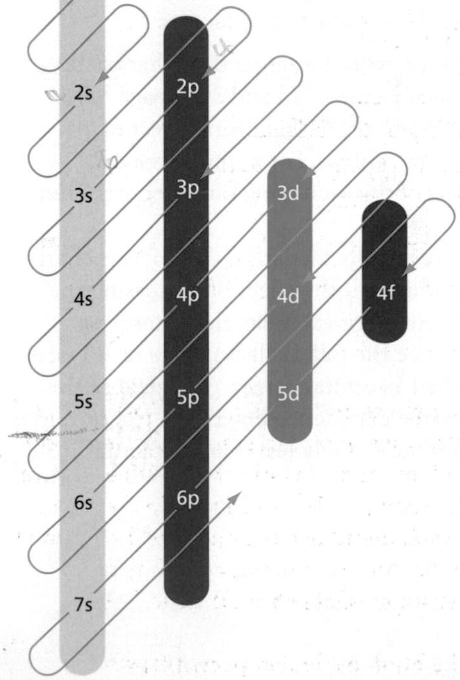

■ **Figure 2.34**
Mnemonic for the order of filling orbitals.

2.13 Electron spin and the Pauli principle

Electrons are all identical. The only way of distinguishing them is by describing how their energies and spatial distributions differ. Thus an electron in a 1s orbital is different from an electron in a 2s orbital because it occupies a different region of space closer to the nucleus, causing it to have less potential energy. An electron in a $2p_x$ orbital differs from an electron in a $2p_y$ orbital because although they have exactly the same potential energy, they occupy different regions of space.

Two electrons with the same energy and occupying the same orbital must be distinguishable in some way, or else they would, in fact, be one and the same particle.

Experiments by Otto Stern and Walther Gerlach in the 1920s (see the panel below) showed that an electron has a magnetic dipole moment. A spinning electrically charged sphere is predicted to produce a magnetic dipole – it acts like a tiny magnet, with a north and a south pole. Therefore the most common explanation for the results of the Stern–Gerlach experiment is that the electron is spinning on its axis, and the direction of spin can be either clockwise (let us say) or anticlockwise. These two directions of spin produce magnetic moments in opposite directions, often described as 'up' (given the symbol ↑) and 'down' (given the symbol ↓).

The Stern–Gerlach experiment

In 1921 Otto Stern and Walther Gerlach looked at the effect of an inhomogeneous (uneven) magnetic field on beams of individual metal atoms. A beam of silver atoms was split into two as it travelled through the field (see Figure 2.35). Half of the atoms were attracted to the stronger part of the magnetic field, and half to the weaker part. Silver atoms have the electronic configuration $[Kr]4d^{10}5s^1$, so contain an unpaired electron. The odd electron gave the atom an overall magnetic moment, which either reinforced the magnetic field, or opposed it, depending on whether the electron's spin was 'up' or 'down'.

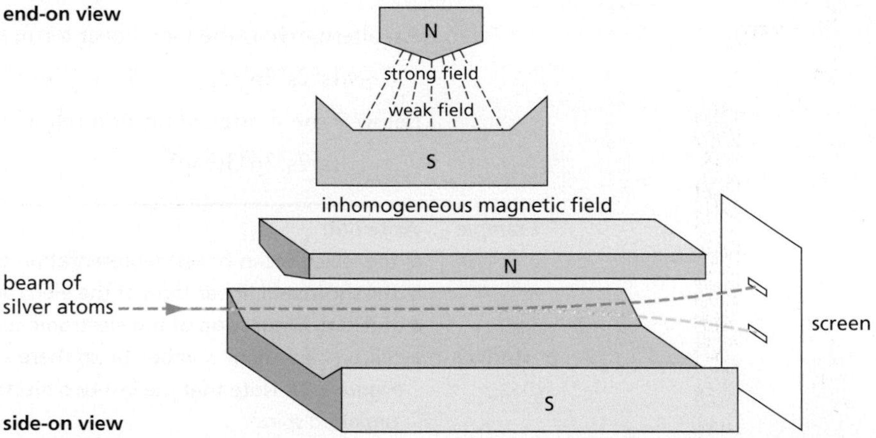

■ **Figure 2.35** In the Stern–Gerlach apparatus, the beam of silver atoms is split into two different components because of the two different spin orientations of the odd electron in the silver atom.

We could therefore distinguish between two electrons in exactly the same orbital if they had different directions of spin. All electrons spin at the same rate, and there are only two possible spin directions. Therefore there are only two possible ways of describing electrons in the same orbital (for example, 1s↑ and 1s↓). So there can only be two electrons in each orbital, and each must have the opposite direction of spin. A third electron would have to have the same spin direction as one of the two already there, which would make it indistinguishable from the similarly spinning one.

The situation is neatly summarised by the **Pauli exclusion principle**, which states that 'No more than two electrons can occupy the same orbital, and if two electrons are in the same orbital, they must have opposite spins'.

2.14 Filling the orbitals

We are now in a position to formulate the rules to use in order to predict the electronic configuration of the atoms of the elements, and also the ions derived from them. Collecting together the conclusions of sections 2.9–2.12, we arrive at the following:

1 Work out from the element's atomic number (and the charge, if an ion is being considered) the total number of electrons to be accommodated in the orbitals.
2 Taking the orbitals in order of their energies (see Figure 2.34, page 42), fill them from the lowest energy (1s) upwards, making sure that no orbital contains more than two electrons.
3 For subshells that contain more than one orbital with the same energy (the p, d and f subshells), place the electrons into different orbitals, until all are singly occupied. Only then should further electrons in that subshell start doubly occupying orbitals. For example, place one electron each into $2p_x$, $2p_y$ and $2p_z$ before putting two electrons in any p orbital.
4 Two electrons sharing the same orbital must have opposite spins.

This procedure is known as the **Aufbau** or 'building-up' principle.

The electronic configurations of atoms and ions can be represented in a variety of ways. These are illustrated here by using sulphur (atomic number 16) as an example. Figure 2.36 shows the 'electrons-in-boxes' diagram.

■ **Figure 2.36** Electrons-in-boxes diagram showing the electronic configuration of sulphur. Note that the boxes can be arranged in one row, with no 'steps' to indicate energy levels (as below).

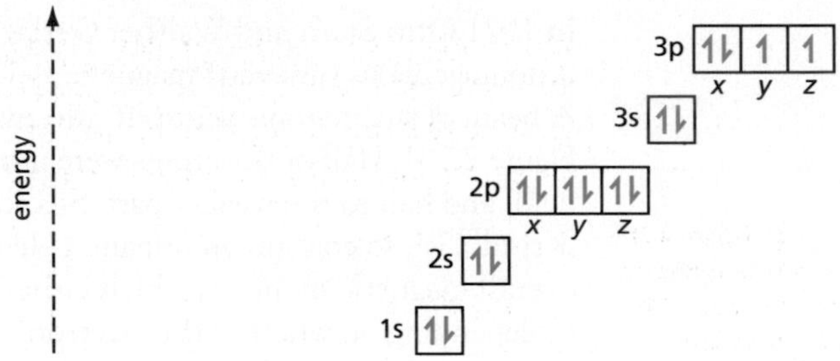

An alternative is the long linear form, specifying individual p orbitals.

$$1s^2 2s^2 2p_x^{\,2} 2p_y^{\,2} 2p_z^{\,2} 3s^2 3p_x^{\,2} 3p_y^{\,1} 3p_z^{\,1}$$

Below is the shortened linear form, which is the most usual representation:

$$1s^2 2s^2 2p^6 3s^2 3p^4$$

Example Write out:

a the 'electrons-in-boxes' representation of the silicon atom
b the shortened linear form of the electronic configuration of the magnesium atom
c the long linear form of the electronic configuration of the fluoride ion.

Answer a Silicon has atomic number 14, so there are 14 electrons to accommodate, as shown in Figure 2.37. Note that the last two electrons go into the $3p_x$ and $3p_y$ orbitals, with unpaired spins.

■ **Figure 2.37**

b Magnesium has atomic number 12. The 12 electrons doubly occupy the lowest six orbitals in energy: $1s^2 2s^2 2p^6 3s^2$.

c Fluorine has atomic number 9. The F^- ion will therefore have $(9 + 1) = 10$ electrons: $1s^2 2s^2 2p_x^{\,2} 2p_y^{\,2} 2p_z^{\,2}$.

Further practice Write out the long linear form of the electronic configuration of the following atoms or ions:
1 N **2** Ca **3** Al^{3+}.

2.15 Experimental evidence for the electronic configurations of atoms – ionisation energies

A major difference between electrons in different types of orbital is their energy. We can investigate the electronic configurations of atoms by measuring experimentally the energies of the electrons within them. This can be done by measuring **ionisation energies** (see page 40).

Ionisation energies are used to probe electronic configurations in two ways:

- successive ionisation energies for the same atom
- first ionisation energies for different atoms.

We shall look at each in turn.

Successive ionisation energies

We can look at an atom of a particular element, and measure the energy required to remove each of its electrons, one by one:

$$X(g) \rightarrow X^{+}(g) + e^{-} \qquad \Delta H = IE_1$$
$$X^{+}(g) \rightarrow X^{2+}(g) + e^{-} \qquad \Delta H = IE_2$$
$$X^{2+}(g) \rightarrow X^{3+}(g) + e^{-} \qquad \Delta H = IE_3 \quad \text{etc.}$$

These **successive ionisation energies** show clearly the arrangement of electrons in shells around the nucleus. If we take the magnesium atom as an example, and measure the energy required to remove successively the first electron, the second, the third, and so on, we obtain the plot shown in Figure 2.38.

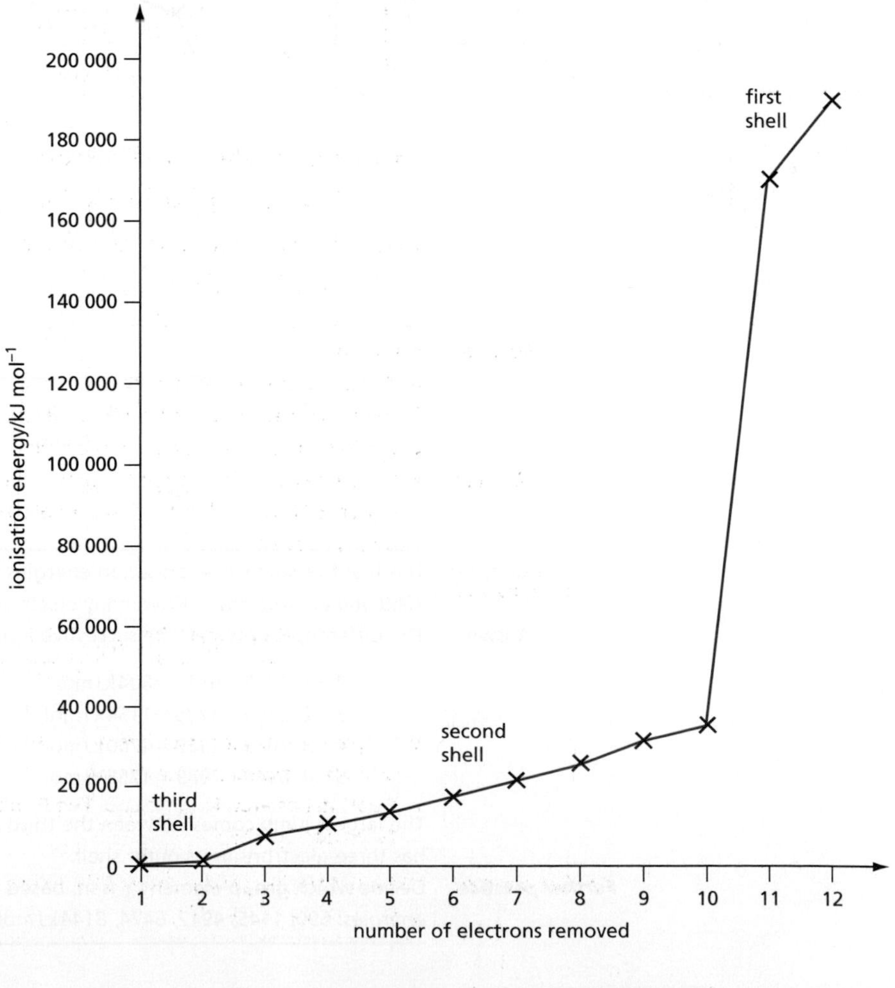

■ **Figure 2.38** Graph of the twelve ionisation energies of magnesium against electron number. The electronic configuration of magnesium is $1s^2\ 2s^2\ 2p^6\ 3s^2$.

Successive ionisation energies are bound to increase because the remaining electrons are closer to, and less shielded from, the nucleus. But a larger increase occurs when the third electron is removed. This is because once the two electrons in the outer (third) shell have been removed, the next has to be stripped from a shell that is very much nearer to the nucleus (the second shell). A similar, but much more enormous, jump in ionisation energy occurs when the eleventh electron is removed. This has to come from the first, innermost shell, right next to the nucleus. These two large jumps in the series of successive ionisation energies are very good evidence that the electrons in the magnesium atom exist in three different shells.

The jumps in successive ionisation energies are more apparent if we plot the logarithm of the ionisation energy against atomic number, as in Figure 2.39.

Figure 2.39 Graph of logarithms of the twelve ionisation energies of magnesium against electron number.

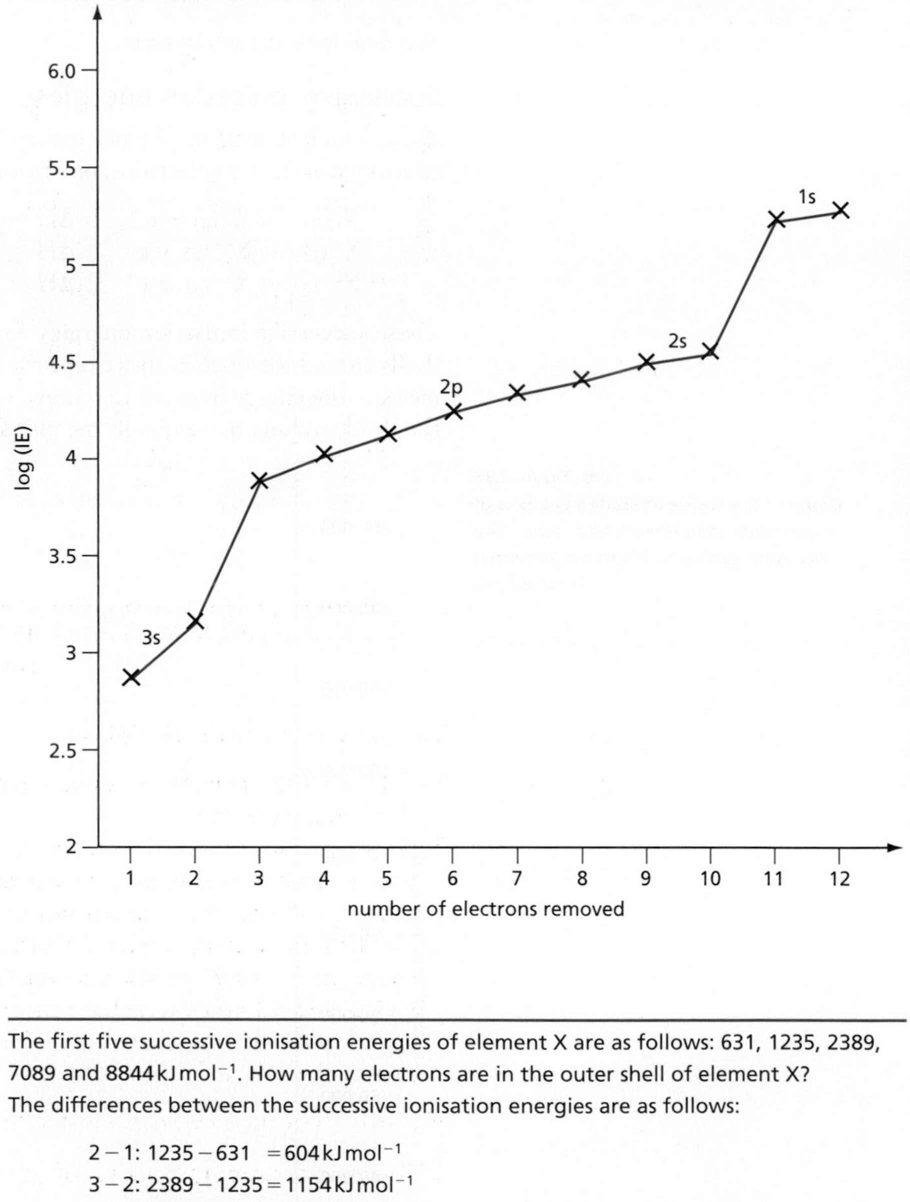

Example The first five successive ionisation energies of element X are as follows: 631, 1235, 2389, 7089 and 8844 kJ mol^{-1}. How many electrons are in the outer shell of element X?

Answer The differences between the successive ionisation energies are as follows:

2 − 1: 1235 − 631 = 604 kJ mol^{-1}

3 − 2: 2389 − 1235 = 1154 kJ mol^{-1}

4 − 3: 7089 − 2389 = 4700 kJ mol^{-1}

5 − 4: 8844 − 7089 = 1755 kJ mol^{-1}

The largest jump comes between the third and the fourth ionisation energy, therefore X has three electrons in its outer shell.

Further practice Decide which group element Y is in, based on the following successive ionisation energies: 590, 1145, 4912, 6474, 8144 kJ mol^{-1}.

First ionisation energies

The second way that ionisation energies show us the details of electronic configuration is to look at how the first ionisation energies of elements vary with atomic number. Figure 2.40 shows a plot for the first 40 elements.

■ **Figure 2.40**
First ionisation energies for the first 40 elements.

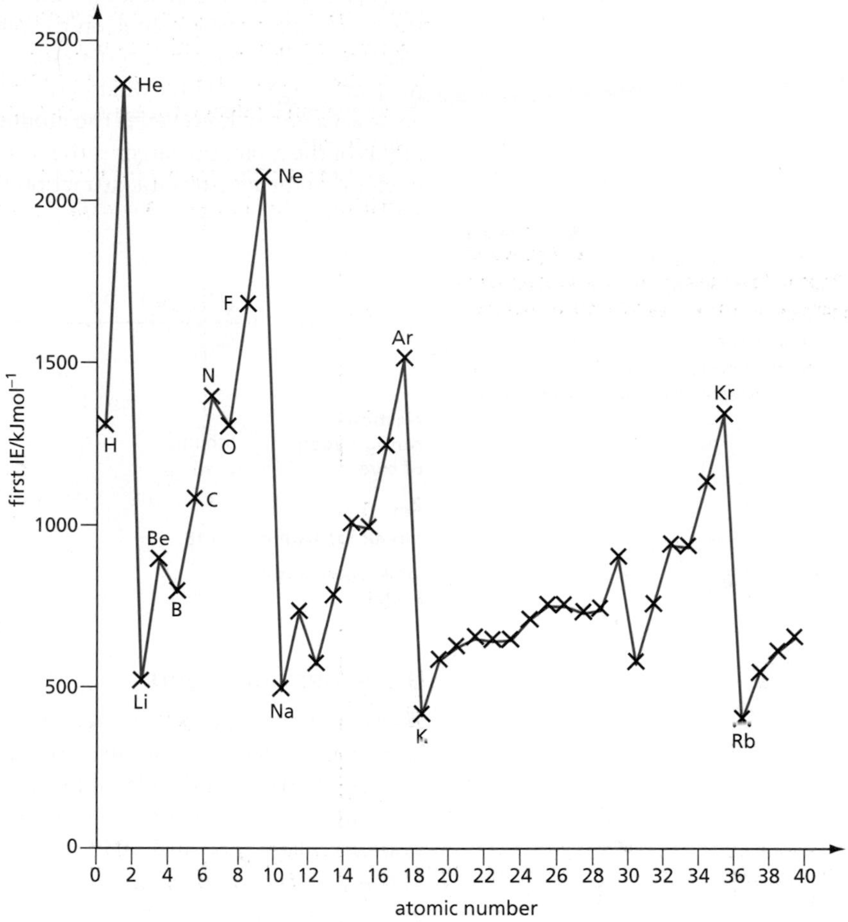

This graph shows us the following:

1 All ionisation energies are strongly endothermic – it takes energy to separate an electron from an atom.
2 As we go down a particular group, for example, from helium to neon to argon, or from lithium to sodium to potassium, ionisation energies decrease. The larger the atom, the easier it is to separate an electron from it.
3 The ionisation energies generally increase on going across a period. The Group 1 elements, the alkali metals, have the lowest ionisation energy within each period, and the noble gases have the highest.
4 This general increase across a period has two exceptions. For the first two periods these occur between Groups 2 and 3 and between Groups 5 and 6.

We shall comment on each of these in turn.

1 The endothermic nature of ionisation energies

This is due to the electrostatic attraction between each electron in an atom (even the outermost one, which is always the easiest to be ionised) and the positive nucleus. It is worth remembering that this even applies to alkali metals like sodium, which we usually think of as 'wanting' to form ions. We must bear in mind, however, that ionisation energies as plotted in Figure 2.40 apply to the ionisation of isolated atoms in the gas phase. Ions are much more stable when in solid lattices or in solution. We shall be looking at this in detail in Chapter 10.

2 The group trend

The nuclear charge experienced by an outer electron is the **effective nuclear charge**, Z_{eff}. It can roughly be equated to the number of protons in the nucleus, *P*, minus the number of electrons in the inner shells, *E*. As we explained in the comparison of hydrogen and lithium on page 41, inner shells shield the effect of the increasing nuclear charge on the outer electrons. Because of this, the outer electrons of elements within the same group experience roughly the same effective nuclear charge no matter what period the element is in. What does change as we go down a group, however, is the atomic radius (see Figure 2.41). The larger the radius of the atom, the larger is the distance between the outer electron and the nucleus, so the electrostatic attraction between them is smaller.

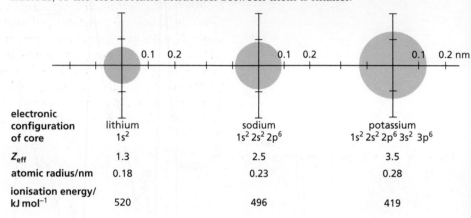

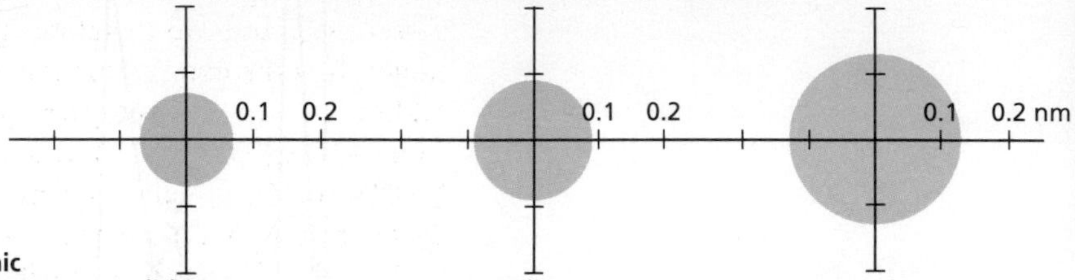

electronic configuration of core	lithium $1s^2$	sodium $1s^2\,2s^2\,2p^6$	potassium $1s^2\,2s^2\,2p^6\,3s^2\,3p^6$
Z_{eff}	1.3	2.5	3.5
atomic radius/nm	0.18	0.23	0.28
ionisation energy/ kJ mol^{-1}	520	496	419

■ **Figure 2.41**
The effective nuclear charge, Z_{eff}, and sizes of the cores of three alkali metals. The core comprises the nucleus plus all the inner shells of electrons. Despite the increase in Z_{eff}, the ionisation energies decrease from Li to K due to the increased electron–nucleus distance.

3 The periodic trend

As we go across a period, we are, for each element, adding a proton to the nucleus, and an electron to the outermost shell. The extra proton will, of course, cause the nucleus to attract all the electrons more strongly. Electrons in the same shell are (roughly) at the same distance from the nucleus as each other. They are therefore not particularly good at shielding each other from the nuclear charge. As a result of this the effective nuclear charge increases. This causes the electrostatic attraction between the ionising electron and the nucleus to increase also. Table 2.7 illustrates this for the second period.

■ **Table 2.7**
Comparing the values of $(P - E)$ and the effective nuclear charge, Z_{eff}, for Period 2. Z_{eff} is not exactly equal to $(P - E)$ for two reasons:
1 Electrons in s orbitals penetrate the inner orbitals to a certain extent, and are therefore less shielded by them than one might have predicted.
2 Electrons in the same shell do, to a certain extent, shield each other from the nucleus. This effect becomes larger as the outer shell becomes more full of electrons, and so the discrepancy between Z_{eff} and $(P - E)$ increases as we cross a period.

Element	Number of protons, *P*	Number of inner shell electrons, *E*	$(P - E)$	Effective nuclear charge, Z_{eff}
Li	3	2	1	1.3
Be	4	2	2	1.9
B	5	2	3	2.4
C	6	2	4	3.1
N	7	2	5	3.8
O	8	2	6	4.5
F	9	2	7	5.1
Ne	10	2	8	5.8

4 The exceptions

The two exceptions to the general increase in ionisation energy across a period (Figure 2.40) arise from different causes. In boron ($1s^2\,2s^2\,2p^1$) the outermost electron is in a 2p orbital. The average distance from the nucleus of a 2p orbital is larger than that of a 2s orbital (see Figures 2.23 and 2.25). We would therefore expect the outermost electron in boron to experience less electrostatic attraction than the outermost 2s electron in beryllium. So the ionisation energy of boron is less.

The other exception is the decrease in ionisation energy from nitrogen to oxygen. This occurs when the fourth 2p electron is added, and is related to the fact that there are just three 2p orbitals. As we have seen before, because they are of the same electrical charge, electrons repel each other. The three successive electrons added to the series of atoms boron–carbon–nitrogen are therefore most likely to go into the three orbitals $2p_x$, $2p_y$ and $2p_z$. These orbitals are of equal energy and at right angles to each other, so allowing the electrons to be as far apart as possible. They will therefore experience the least electrostatic repulsion from each other, and so, overall, the atomic system will be the most stable. In oxygen, however, the fourth 2p electron has to be accommodated in an orbital that already contains an electron. These two electrons will be sharing the same region of space (by the Pauli principle, of course, their spins will have to be in opposite directions), and will therefore repel each other quite strongly. This repulsion is larger than the extra attraction experienced by the new electron from the additional proton in oxygen's nucleus. So the energy needed to ionise the electron from the oxygen atom is less than the ionisation energy of the nitrogen atom. Similar repulsions are experienced by the new electrons in the fluorine and neon atoms, so the ionisation energy of each of these elements is, like oxygen's, about $430\,kJ\,mol^{-1}$ less than one might have expected (see Figure 2.42).

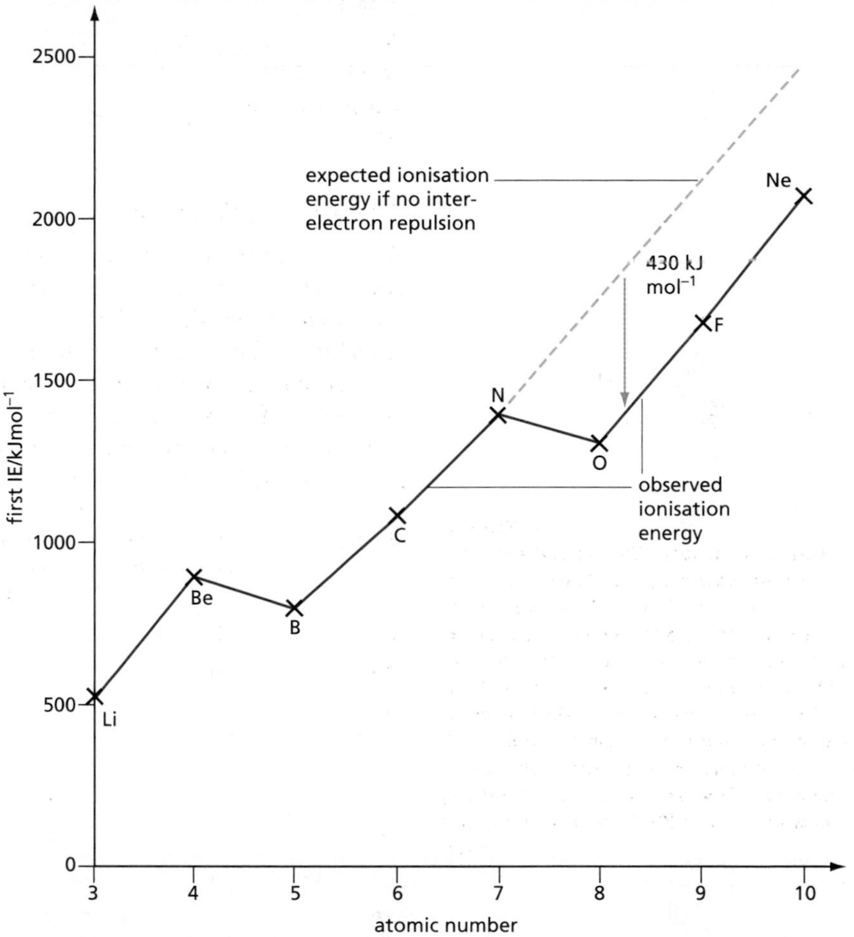

■ Figure 2.42 First ionisation energy against atomic number for Period 2.

Once the second shell has been filled, at neon, the next new electron (in sodium) has to occupy the 3s orbital, and starts to fill the third shell. This is further out from the nucleus than the 2p electron in neon (remember, electrostatic attraction decreases with distance). It is also more shielded from the nucleus (by two inner shells, rather than the one in neon; hence Z_{eff} for sodium is only 2.5). On both accounts, we would expect a large decrease in ionisation energy from neon to sodium, which we indeed observe (see Figure 2.40).

2.16 The effect of electronic configuration on atomic radius

At first sight, the radius of an atom might seem an easy quantity to visualise. But as we saw in Figure 2.23, page 37, the outer reaches of atoms have an ever-decreasing electron probability, which is still greater than zero even at large distances. A filled orbital is a pretty elusive affair that can easily be squashed or polarised. We must therefore be prepared to accept that the value of the atomic radius will not be a fixed quantity, but will depend on the atom's environment.

With this proviso, however, we can use the theories developed above to predict how the atomic radius might change with atomic number. We have seen that we might expect two major influences on the size of an atom. One of these will be the number of shells – the more shells, the bigger the atom should be. We should see this effect as we go down a group. The other influence will be the effective nuclear charge – the larger the charge, the more the orbitals are pulled in towards the nucleus, and so the smaller the atom should be. We should see this effect as we go across a period. These two factors combine to produce a predictable pattern in the plot of atomic radius against atomic number, which is borne out by experimental observations (see Figure 2.43).

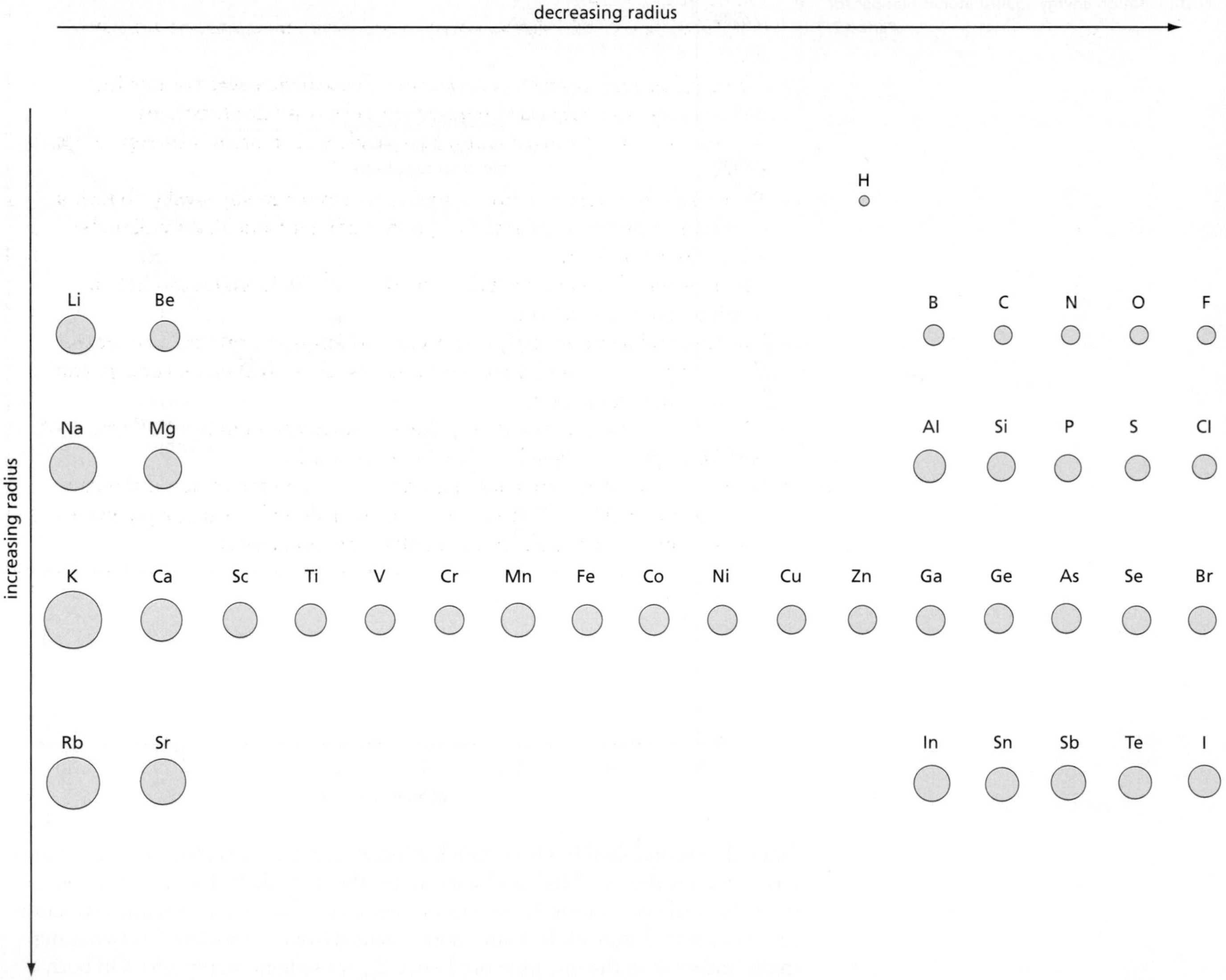

■ **Figure 2.43**
Atoms get larger down a group, as the number of shells increases. They get smaller across a period, as the effective nuclear charge pulls the electrons closer to the nucleus.

Summary

- All atoms are made of the same three **subatomic particles** – the **electron**, the **proton** and the **neutron**.
- Their relative electrical charges are, respectively, -1, $+1$ and 0.
- Their masses (relative to that of the proton) are, respectively, $\frac{1}{1836}$, 1 and 1.
- The **atomic number** is the number of protons contained in the nucleus of an element's atoms. It tells us the order of the element in the periodic table.
- **Isotopes** are atoms of the same element (and therefore with the same proton number) but with different numbers of neutrons.
- The **mass number** of an atom is the sum of the numbers of protons and neutrons it contains.
- The full symbol for an atom shows its mass number as a superscript and its atomic (proton) number as a subscript. For example, $^{12}_{6}C$ shows a carbon atom with mass number 12 and atomic number 6.
- Masses and relative abundances of isotopes can be measured using the **mass spectrometer**.
- Unstable isotopes decay radioactively by the emission of α-particles, β-particles or γ-rays.
- Although they can be harmful to life, radioactive isotopes find many uses in medicine and industry.
- The **half-life** of a radioactive isotope is constant, no matter how much is present.
- **Chemical energy** has two components – the **kinetic energy** of moving particles, and the **potential energy** due to electrostatic attractions.
- The electrons are arranged around the nucleus of an atom in energy levels, or **orbitals**.
- When an electron moves from a higher to a lower energy level (orbital), a quantum of energy is released as a photon of light (sometimes visible, but often ultraviolet light).
- The number of possible orbitals in shell n is n^2. Each orbital can hold a maximum of two electrons.
- The first shell contains only one orbital, which is an s orbital. The second shell contains one s and three p orbitals, the third shell one s, three p, and five d orbitals, and so on.
- The electrons in an atom occupy the lowest energy orbitals first. Orbitals of equal energy are occupied singly whenever possible.
- Successive ionisation energies of a single atom, and the trends in the first ionisation energies of the element across periods and down groups, give us information about the electronic configurations of atoms.

3 Chemical bonding in simple molecules

This chapter is the first of two which look at how atoms bond together to form molecules and compounds, and how particles arrange themselves into larger structures to form all the matter we see around us. In this chapter we look at the various types of covalent bond. We also explain some of the properties of simple covalent molecules.

The chapter is divided into the following sections.

3.1 Introduction

Of the total number of individual chemically pure substances known, several million are **compounds**, formed when two or more elements are chemically bonded together, and less than 100 are **elements**. Only six of these elements consist of free, unbonded atoms at room temperature. These are the noble gases. All other elements exist as individual or giant molecules, or metallic lattices, in which their atoms are chemically bonded to each other.

This shows us that the natural state of most atoms, the state in which they have the lowest energy, is the bonded state. Atoms prefer to be bonded to each other. They give out energy when they form bonds. On the other hand, it always requires an input of energy to break a chemical bond – bond breaking is an endothermic process (see Figure 3.1).

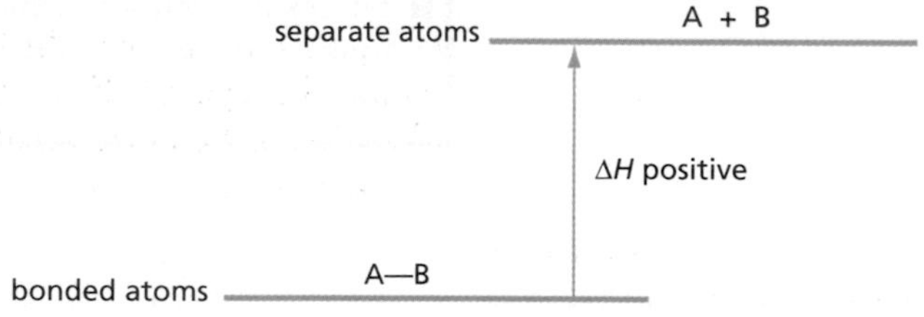

■ **Figure 3.1** Bond breaking is endothermic.

The bond strength can be identified with ΔH in Figure 3.1, that is, the enthalpy change that occurs when 1 mol of bonds in a gaseous compound is broken, forming gas phase atoms. (See Chapter 5, page 125.)

We saw in Chapter 2 that the negatively charged electrons in an atom are attracted to the positively charged nucleus. An electron in an atom has less potential energy than an electron on its own outside an atom. It therefore requires energy to remove an electron from an atom. In this chapter we shall discover a similar reason why atoms bond together to form compounds – because the electrons on one atom are attracted to the nucleus of another, causing the system to have less potential energy when bonded than in its unbonded state. (The three major types of bonding, covalent, ionic and metallic, differ only in how far this attraction to another nucleus overcomes the attraction of the electron to its own nucleus.)

3.2 Covalent bonding – the hydrogen molecule

We shall look first at the simplest possible bond, that between the two hydrogen atoms in a hydrogen molecule. Imagine two hydrogen atoms, initially a large distance apart, approaching each other. As they get closer together, the first effect will be that the electron of one atom will experience a repulsion from the electron on the other atom. But this will be compensated by the attraction it will experience towards the other atom's nucleus (in addition to the attraction it always experiences from its own nucleus). Remember that the electron in an atom spends some of its time at quite a large distance from the nucleus (see Chapter 2, page 37). As the hydrogen atoms get closer still, the two electrons will encounter an even greater attraction to the opposite nucleus, but will also continue to repel each other. Eventually, when the two nuclei become very close together, they in turn will start to repel each other, since they both have the same (positive) charge. The most stable situation will occur when the attractions of the two electrons to the two nuclei is just balanced by the electron–electron and nucleus–nucleus repulsions. A **covalent bond** has formed. The decrease in energy at this point from the unbonded state is called the **bond energy**, and the nuclear–nuclear distance is known as the **bond length** (see Figure 3.2).

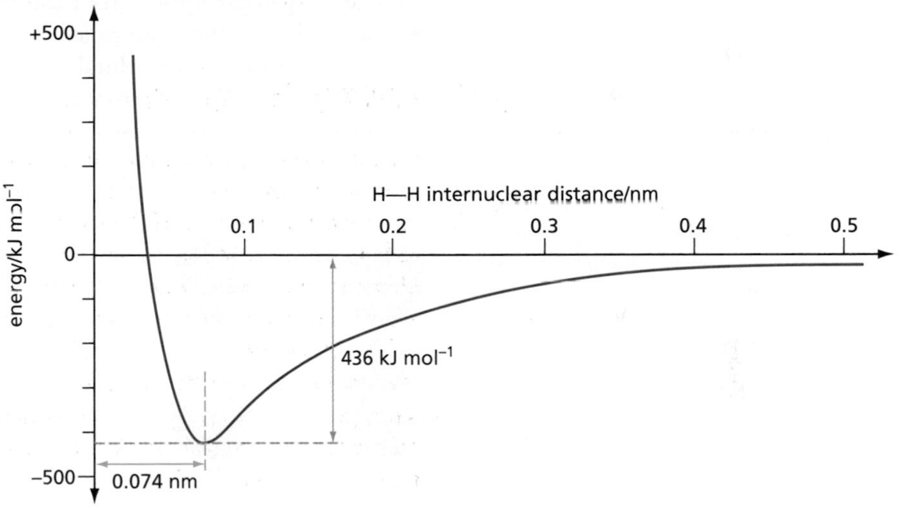

■ **Figure 3.2** Potential energy against internuclear distance for the hydrogen molecule.

Being attracted to both the hydrogen nuclei, the two electrons spend most of their time in the region half-way between them, and on the axis that joins them. This is where the highest **electron density** (or **electron probability**) occurs in a single covalent bond (see Figure 3.3).

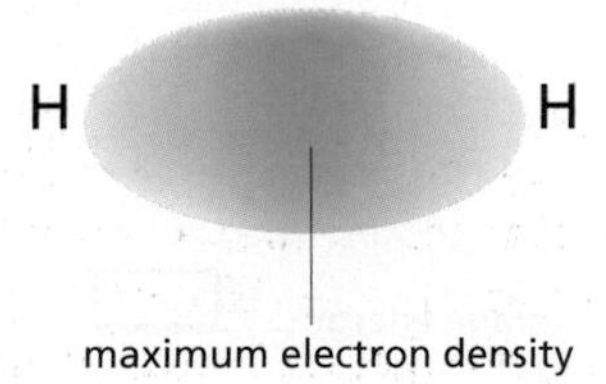

■ **Figure 3.3** A molecular orbital is a region between two bonded atoms where the electron density is concentrated.

The two electrons are therefore **shared** between the two adjacent atoms. As we mentioned in Chapter 2, page 35, we use the word (atomic) **orbital** to describe the region of space around the nucleus of an atom that is occupied by a particular electron. In a similar way, we can use the term **molecular orbital** to describe the region of space within a molecule where a particular electron is to be found.

The concept of creating molecular orbitals by overlapping atomic orbitals is developed on page 54, and in other panels in this chapter.

Molecular orbital theory (1): the hydrogen molecule

The **molecular orbital (MO) theory** was developed to answer the following questions: where in space do the electrons go when a covalent bond is formed, and how does their potential energy change as they become part of the bond? The theory takes the atomic orbitals of the atoms concerned, and mathematically combines the quantum mechanical equations (see the panel on page 36) that represent these orbitals. This combination produces new equations, which are related to new orbitals (called **molecular orbitals**) that depict the regions of space occupied by the bonding electrons. The number of new orbitals formed is the same as the number of atomic orbitals combined. Just as for atomic orbitals (see Chapter 2, page 35), the equations that represent the new molecular orbitals can be used to calculate the energy of an electron in those orbitals. By this means, bond energies can be calculated.

The theory can be illustrated by the formation of the hydrogen molecule as follows.

- Two hydrogen atoms, H_A and H_B, each with a 1s orbital, come together.
- The two 1s orbitals overlap, mix, and produce two new molecular orbitals.
- The two new orbitals are called the σ (sigma) and the σ* (sigma-star) **orbitals**.
- The new equations can be solved to find out the shapes and energies of these two molecular orbitals.
- It turns out that the σ orbital has a lower energy than the 1s orbital in the isolated hydrogen atom, whilst the σ* orbital has a higher energy.
- The σ orbital is therefore called the **bonding orbital**, and the σ* orbital is called the **antibonding orbital**.
- The equations also show that the σ orbital occupies the region of space between the two hydrogen atoms, but the σ* orbital is on the outside of the molecule (see Figure 3.4).

Figure 3.4 Two 1s orbitals can overlap to form two molecular orbitals, σ and σ*.

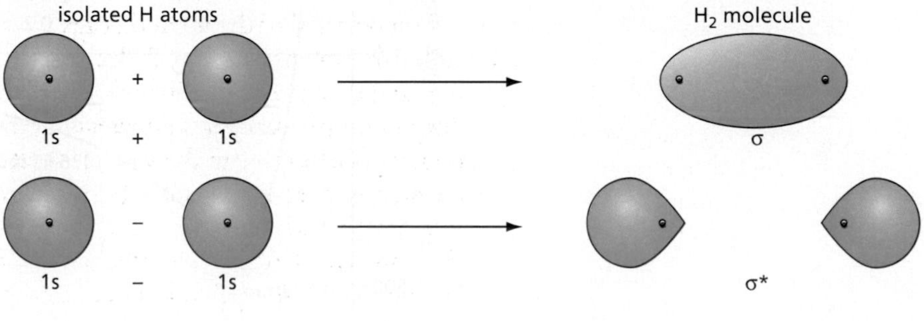

Figure 3.5 The energy levels of the molecular orbitals σ and σ*.

How the energies of the two new orbitals relate to the original 1s orbital energy is shown in Figure 3.5, which also shows that the two 1s electrons end up in the molecular orbital of lowest energy, the σ orbital.

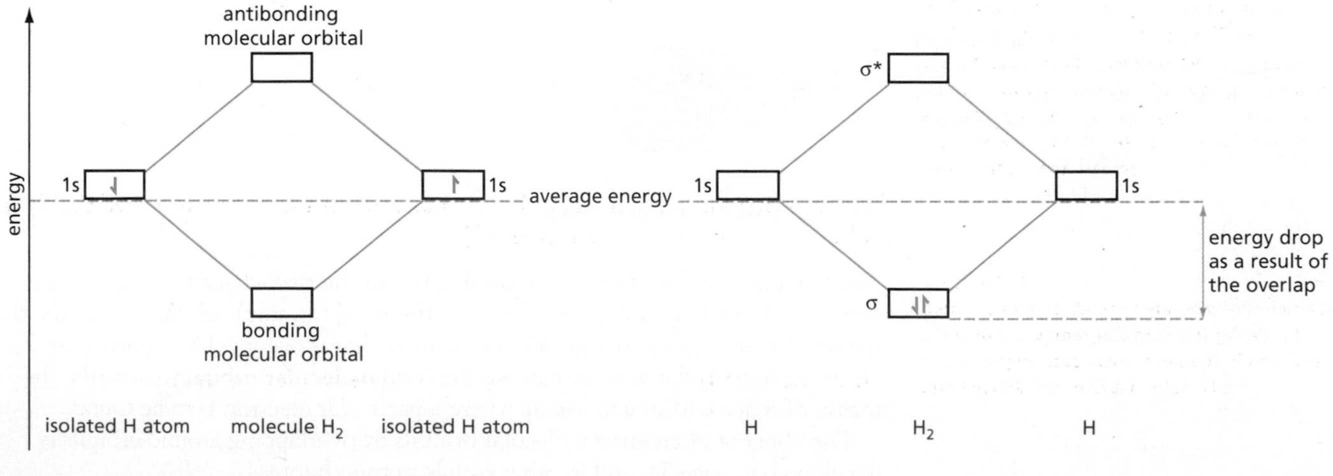

3.3 Representing covalent bonds

Depending on the context, and the information we want to put over, we can use various ways to represent a covalent bond, as listed below.

- The dot-and-cross diagram:

  H• + ×H → H $\overset{\bullet}{\times}$ H

- The dot-and-cross diagram including Venn diagram boundaries – this is preferred for more complicated molecules, as it allows an easy check to be made of exactly which bonds the electrons are in:

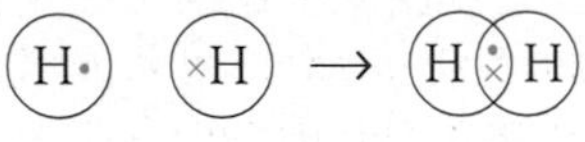

- The line diagram, which uses one line for each bond – this is not very informative for small simple molecules, but it is often less confusing than drawing out individual electrons for larger more complex molecules:

  H• + •H → H – H

3.4 Covalent bonding with second-row elements

Bonding and valence-shell electrons

When an atom bonds with others, it is normally only the electrons in the outermost shell of the atom that take part in bonding. This outermost shell is called the **valence shell**. All electrons in the valence shell can be considered together as a group – when we look in simple terms at bonding, we can ignore the distinction between the various subshells (s, p, d, etc.) in the valence shell. The number of electrons in the valence shell is the same as the group number of the element in the periodic table (see Chapter 15).

If an atom has more than one electron in its valence shell, it can form more than one covalent bond to other atoms. For example, the beryllium atom in beryllium hydride, BeH_2, has two 2s electrons which are paired. When forming a compound, these can unpair themselves and form two bonds with two other atoms. The boron atom ($2s^2 2p^1$) can form three bonds. The bonding in beryllium hydride and boron hydride is shown in Figure 3.6.

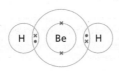

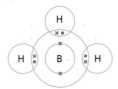

■ Figure 3.6 Dot-and-cross diagrams showing beryllium hydride, BeH_2, and boron hydride, BH_3. (It is worth remembering that all electrons are identical, no matter which atom they came from. Our representing them as dots and crosses is merely for our own benefit, to make it clear where they came from.)

The carbon atom in methane, CH_4, has four electrons in its valence shell ($2s^2 2p^2$), and so forms four bonds (see Figure 3.7).

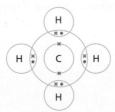

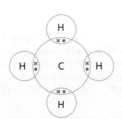

■ Figure 3.7 Dot-and-cross diagram for methane, CH_4. Often, for clarity, bonding diagrams show only the electrons in the outer shell, omitting the inner shells. Both diagrams are shown here.

The general rule is that:

Elements with one to four electrons in their valence shells form the same number of covalent bonds as the number of valence-shell electrons.

Carbon also forms strong bonds to itself (as do some other elements). An example of this bonding is shown in Figure 3.8.

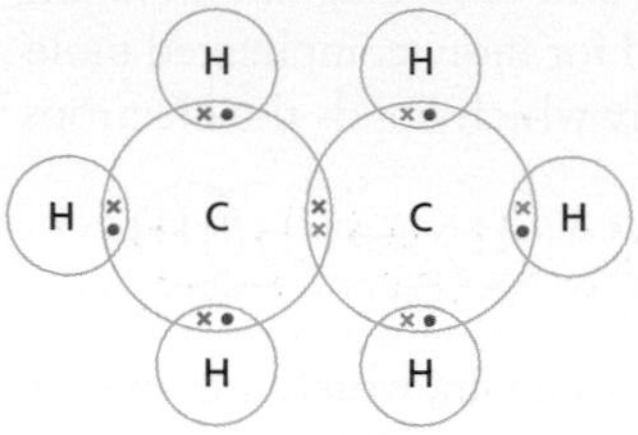

■ **Figure 3.8**
Dot-and-cross diagram for ethane, C_2H_6.

More than four valence-shell electrons

When the number of valence-shell electrons is greater than four, the maximum possible number of bonds is not always formed. This is because elements of the second row of the periodic table have only four orbitals in their valence shells (2s, $2p_x$, $2p_y$, $2p_z$) and so cannot accommodate more than four pairs of electrons. The number of bonds formed is restricted by this overall maximum of eight electrons, since every new bond brings another electron into the valence shell. For example, nitrogen ($2s^2 2p^3$) has five valence-shell electrons. In a molecule of ammonia, NH_3, there are three N—H bonds. These involve the sharing of three nitrogen electrons with three from the hydrogen atoms. These three additional electrons bring the valence shell total to eight. This is as many as the four orbitals in the second shell can hold. The outer shell is therefore filled, with a **full octet** of electrons. No further bonds can form, and the remaining two of the five electrons in nitrogen's valence shell remain unbonded, as a lone pair, occupying an orbital associated only with the nitrogen atom (see Figure 3.9).

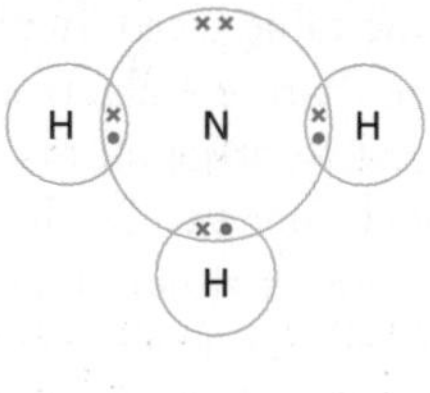

■ **Figure 3.9**
Dot-and-cross diagram for ammonia, NH_3.

The general rule is that:

Elements of the second period with more than four electrons in their outer shells form (8 − *n*) covalent bonds, where *n* = the number of valence-shell electrons.

Similarly oxygen ($2s^2 2p^4$), with six electrons in its valence shell, can form only two covalent bonds, with two lone pairs of electrons remaining. Fluorine ($2s^2 2p^5$), with seven valence-shell electrons, can form only one bond, leaving three lone pairs around the fluorine atom (see Figure 3.10).

■ **Figure 3.10**
Dot-and-cross diagrams for water, H_2O, and hydrogen fluoride, HF. Oxygen can form two bonds (8 – 6) and fluorine just one (8 – 7).

Example Draw diagrams showing the bonding in nitrogen trifluoride, NF_3, and hydrogen peroxide, H_2O_2.

Answer

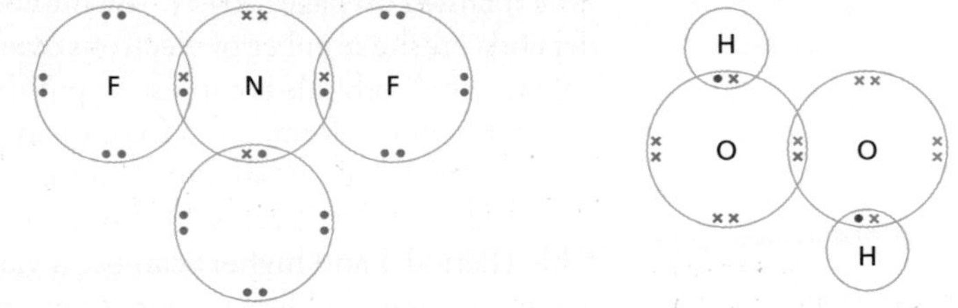

NF_3 H_2O_2

■ **Figure 3.11**

Further practice Draw diagrams showing the bonding in the following molecules:
1 BF_3 **2** N_2H_4 **3** CH_3OH **4** CH_2F_2.

3.5 Covalent bonding with third-row elements

Unlike elements in the second row of the periodic table, those in the third and subsequent rows can utilise their d orbitals in bonding, as well as their s and p orbitals. They can therefore form more than four covalent bonds to other atoms. Like nitrogen, phosphorus ($1s^2 2s^2 2p^6 3s^2 3p^3$) has five electrons in its valence shell. But because it can make use of five orbitals (one 3s + three 3p + one 3d) it can utilise all five of its valence-shell electrons in bonding with fluorine. It therefore forms phosphorus pentafluoride, PF_5, as well as phosphorus trifluoride PF_3 (see Figure 3.12).

■ **Figure 3.12**
Dot-and-cross diagrams for phosphorus trifluoride, PF_3, and phosphorus pentafluoride, PF_5.

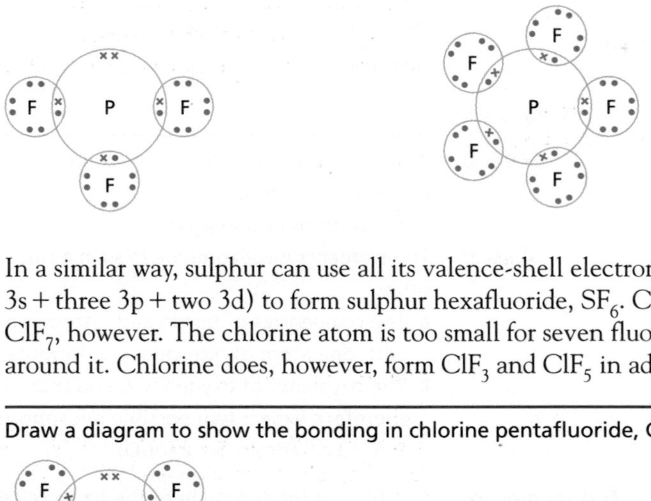

In a similar way, sulphur can use all its valence-shell electrons in six orbitals (one 3s + three 3p + two 3d) to form sulphur hexafluoride, SF_6. Chlorine does not form ClF_7, however. The chlorine atom is too small for seven fluorine atoms to assemble around it. Chlorine does, however, form ClF_3 and ClF_5 in addition to ClF.

Example Draw a diagram to show the bonding in chlorine pentafluoride, ClF_5.

Answer

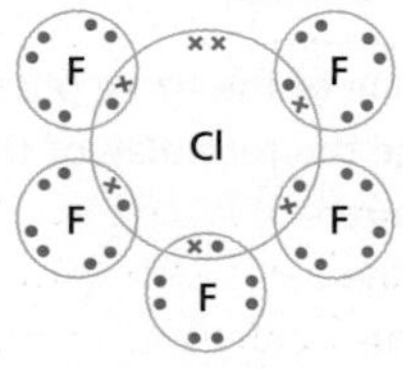

■ **Figure 3.13**

Further practice
1 Sulphur forms a tetrafluoride, SF_4. Draw a diagram to show its bonding.
2 a How many valence-shell electrons does chlorine use for bonding in chlorine trifluoride, ClF_3?
b So how many electrons are left in the valence shell?
c So how many lone pairs of electrons are there on the chlorine atom in chlorine trifluoride?

3.6 The covalency table

As explained on pages 55–57, the number of covalent bonds formed by an atom depends on the number of electrons available for bonding, and the number of valence-shell orbitals it can use to put the electrons into. For many elements the number of bonds formed is a fixed quantity and is termed the **covalency** of the element. It is often related to its group number in the periodic table. As shown above, elements in Groups 5–7 in the third and subsequent rows of the periodic table (Period 3 and higher) can use a variable number of electrons in bonding. They can therefore display more than one covalency. Table 3.1 shows the most usual covalencies of some common elements.

Table 3.1
Covalencies of some common elements.

Element	Symbol	Group number	Covalency
Hydrogen	H	1	1
Lithium	Li	1	1
Beryllium	Be	2	2
Boron	B	3	3
Carbon	C	4	4
Nitrogen	N	5	3
Oxygen	O	6	2
Fluorine	F	7	1
Aluminium	Al	3	3
Silicon	Si	4	4
Phosphorus	P	5	3 or 5
Sulphur	S	6	2 or 4 or 6
Chlorine	Cl	7	1 or 3 or 5 or 7
Bromine	Br	7	1 or 3 or 5 or 7
Iodine	I	7	1 or 3 or 5 or 7

Example What could be the formulae of compounds formed from the following pairs of elements?

a boron and nitrogen
b phosphorus and oxygen

Answer The elements must combine in such a ratio that their total covalencies are equal to each other.

a The covalencies of boron and nitrogen are both 3, so one atom of boron will combine with one atom of nitrogen. The formula is therefore BN (3 for B = 3 for N).
b The covalency of oxygen is 2, and that of phosphorus can be either 3 or 5. We would therefore expect two possible phosphorus oxides: P_2O_3 (2×3 for P $= 3 \times 2$ for O) and P_2O_5 (2×5 for P $= 5 \times 2$ for O).

Further practice

1 What are the formulae of the three possible oxides of sulphur?
2 Use Table 3.1 to write the formulae of the simplest compounds formed between:
  a carbon and hydrogen
  b oxygen and fluorine
  c boron and chlorine
  d nitrogen and bromine
  e carbon and oxygen.

3.7 Dative bonding

So far, we have looked at covalent bonds in which each atom provides one electron to form the bond. It is possible, however, for just one of the atoms to provide both bonding electrons. This atom is called the **donor atom**, and the two electrons it provides comes from a lone pair on that atom. The other atom in the bond is called the **acceptor atom**. It must contain an empty orbital in its valence shell. This kind of bonding is called **dative bonding** or **coordinate bonding**. The apparent covalency of each atom can increase as a result of dative bonding.

For example, when gaseous ammonia and gaseous boron trifluoride react together, a white solid with the formula NH_3BF_3 is formed. The nitrogen atom in ammonia has a lone pair. The boron atom in boron trifluoride has an empty 2p orbital, with which the nitrogen's lone pair overlaps, as shown in Figure 3.14.

Figure 3.14 Forming a dative (coordinate) bond in NH_3BF_3.

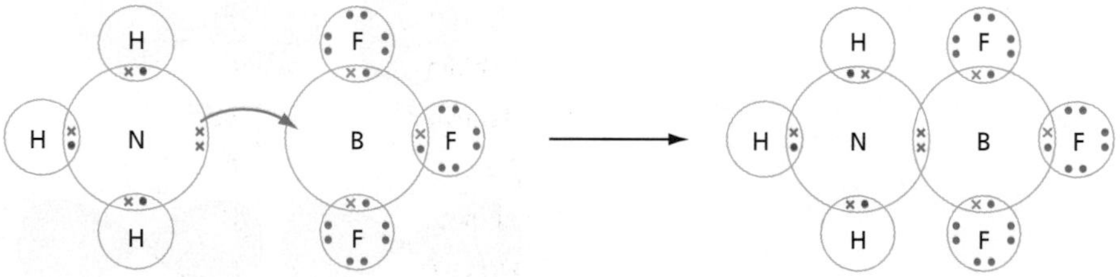

The dative bond, once formed, is no different from any other covalent bond. For example, when gaseous ammonia and gaseous hydrogen chloride react, the white solid ammonium chloride is formed. The lone pair of electrons on the nitrogen atom of ammonia has formed a dative bond with the hydrogen atom of the hydrogen chloride molecule (see Figure 3.15). (At the same time, the electrons in the H Cl bond form a fourth lone pair on the chlorine atom. With 18 electrons and only 17 protons, this now becomes the negatively charged chloride ion. Ionic bonding is covered in detail in Chapter 4.)

Figure 3.15 Forming an ionic bond in ammonium chloride, NH_4Cl.

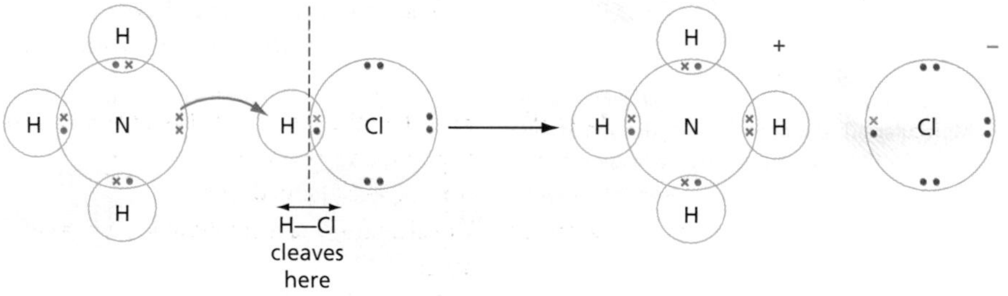

The shape of the ammonium ion is a regular tetrahedron, as is that of the methane molecule (see Figure 3.16). All four N—H bonds are exactly the same. It is not possible to tell which one was formed in a dative way.

Figure 3.16 Ammonium and methane are both tetrahedral.

3.8 Molecular orbitals from the overlap of atomic orbitals

We can gain a fairly good idea of the shape of a molecular orbital (and hence where the electrons in a bond are to be found) by seeing where the maximum **orbital overlap** occurs when we bring two isolated atoms close together. Figure 3.17 shows three examples.

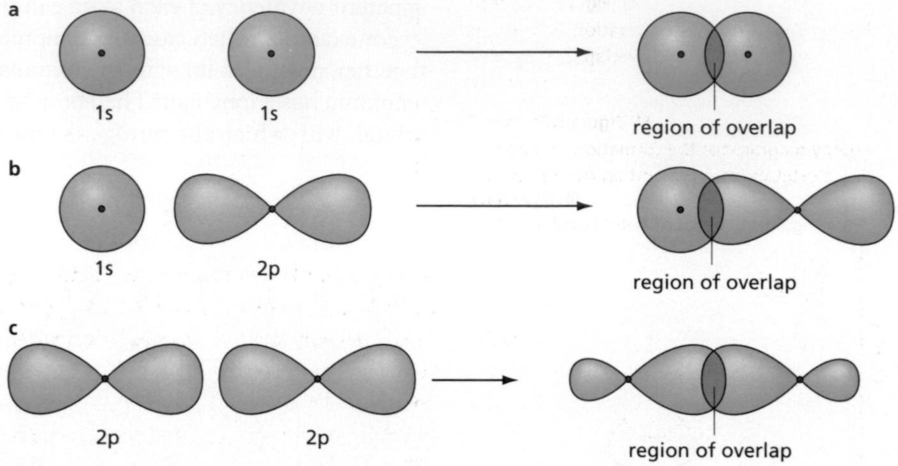

■ **Figure 3.17** The shapes of molecular orbitals formed by overlapping atomic orbitals.

The three combinations of orbitals in Figure 3.17 occur in many situations.

- The H—H bond in the hydrogen moledule arises through the overlap of two 1s orbitals on adjacent hydrogen atoms (see Figure 3.18**a**).
- The C—H bonds in alkanes arise through the overlap of the 1s orbital on hydrogen and a 2p orbital on carbon (see Figure 3.18**b**).
- In a similar way, the C—C bonds in alkanes arise through the 'end-on' overlap of two 2p orbitals of adjacent carbon atoms (see Figure 3.18**c**).
- Similar overlap, either between themselves or with the 1s orbital of a hydrogen atom, occurs with the 2p orbitals of nitrogen, oxygen and fluorine.

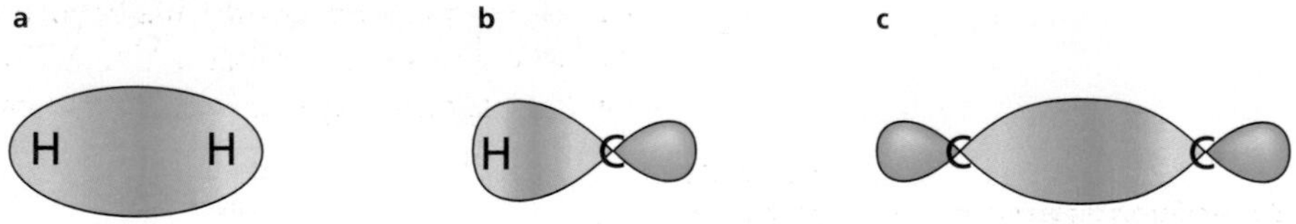

■ **Figure 3.18** Molecular orbitals between **a** hydrogen atoms **b** carbon and hydrogen atoms **c** carbon atoms.

Orbitals formed in the ways described above are called σ (sigma) **orbitals**.

- Two p orbitals can also undergo sideways overlap, to produce a π (pi) **orbital** (see Figure 3.19). These π orbitals occur in compounds containing double or triple bonds (see page 61).

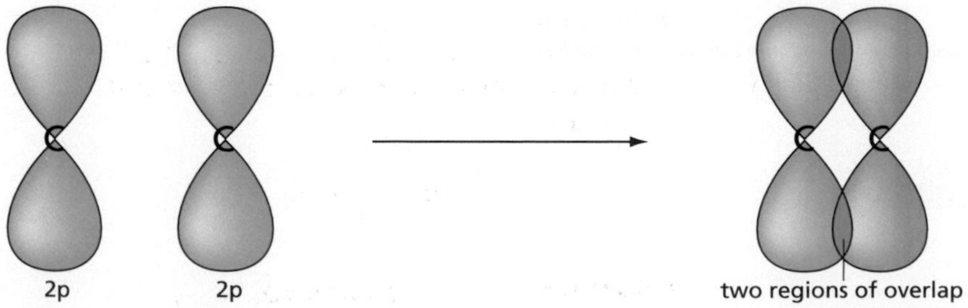

■ **Figure 3.19** Sideways overlap of two p orbitals to form a π orbital.

The panel opposite describes in outline the molecular orbital theory treatment of these processes.

Molecular orbital theory (2): overlapping different orbitals

In the panel on page 54, we saw how the overlap of two 1s orbitals of two different hydrogen atoms produces two new orbitals, the σ (bonding) and the σ* (antibonding) molecular orbitals. We can develop this principle to the overlap of other orbitals with each other. On page 60 we saw how s and p orbitals can overlap in a variety of ways. This gives σ, σ*, π and π* orbitals.

The energy diagrams for σ and σ* orbitals are shown in Figures 3.20 and 3.21.

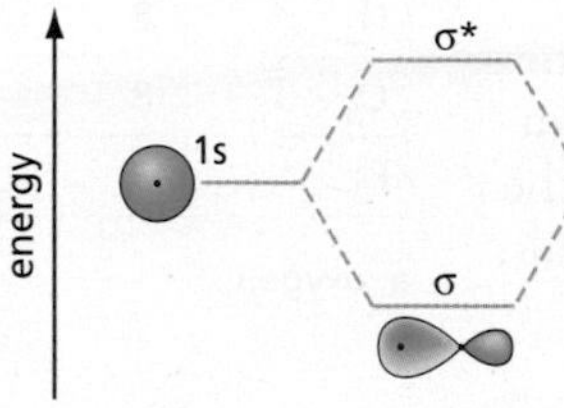

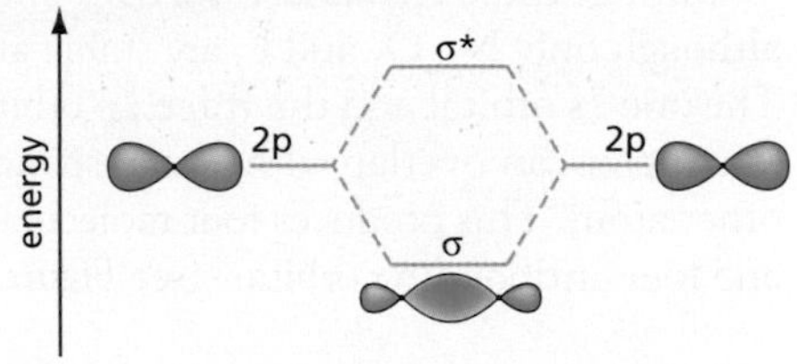

■ **Figure 3.20** (left)
Energy diagram for the formation of σ and σ* orbitals from the overlap of a 1s and a 2p orbital.

■ **Figure 3.21** (right)
Energy diagram for the formation of σ and σ* orbitals from the end-on overlap of two 2p orbitals.

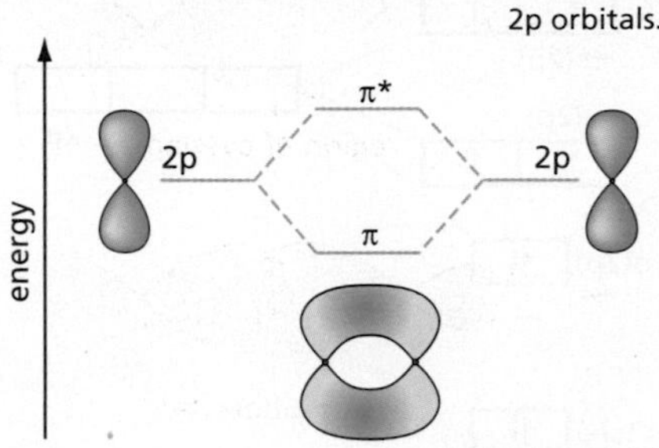

■ **Figure 3.22**
Energy diagram for the formation of π and π* orbitals from the sideways overlap of two 2p orbitals.

Two p orbitals can also undergo sideways overlap. This results in a π bonding orbital that has its main electron density in two regions on opposite sides of the axis between the two bonded atoms (just as a p orbital has its electron density in two regions on opposite sides of the atom). The energy diagram for this orbital is shown in Figure 3.22. The attraction to the two nuclei experienced by the bonding electrons is not as great as in a σ orbital, because the average electron–nucleus distance is larger. The drop in energy on forming the π orbital is therefore less, so the π bond is weaker in energy than the σ bond (see page 70).

This type of bonding occurs in ethene, C_2H_4, carbon dioxide, CO_2, and all other molecules containing double or triple bonds.

3.9 Multiple bonding

Double and triple bonds

Atoms can share more than one electron pair with their neighbours. Sharing two electron pairs produces a **double bond**, and sharing three produces a **triple bond**. The covalencies still conform to those in Table 3.1 (page 58). Examples of multiply bonded oxygen, nitrogen and carbon are shown in Figure 3.23.

■ **Figure 3.23**
Dot-and-cross diagrams for the multiply bonded molecules oxygen, O_2, nitrogen, N_2 and ethene, C_2H_4.

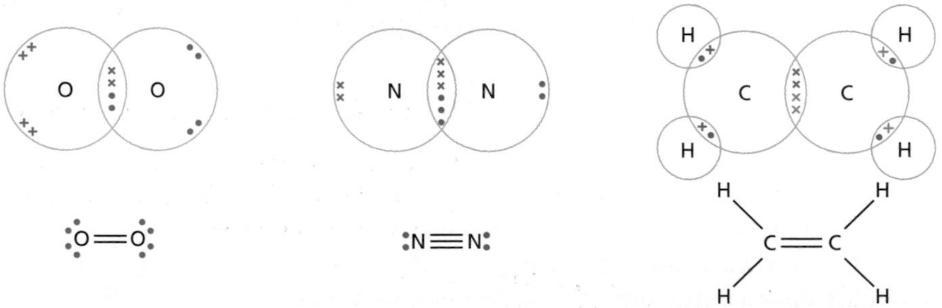

Example Draw a dot-and-cross diagram of the carbon dioxide molecule CO_2.

Answer

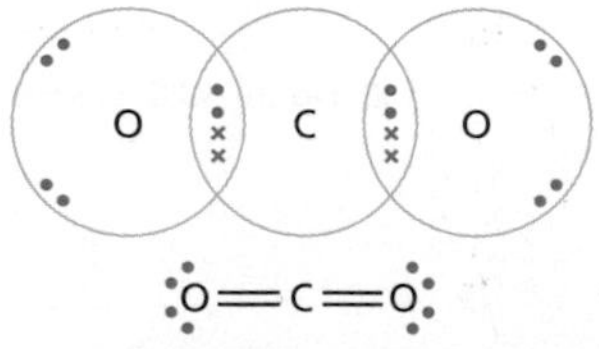

The covalencies of oxygen and carbon are 2 and 4, respectively. The only possible bonding arrangement is therefore O=C=O, and the electrons are shared as shown in Figure 3.24.

■ **Figure 3.24**

Further practice Draw line diagrams and dot-and-cross diagrams to show the bonding in the following molecules: HCN, H_2CO, C_2H_2 (ethyne).

Molecular orbital theory (3): multiple bonds

Earlier panels in this chapter described how s and p orbitals on adjacent atoms can overlap to form σ and π molecular orbitals. These three types of overlap can occur at the same time, to form multiple bonds. This can be illustrated by the diatomic molecules, X_2, of the elements of the second row of the periodic table.

Most of these elements form diatomic molecules, although only N_2, O_2 and F_2 are stable at room temperature. The one 2s orbital and the three 2p orbitals on each atom in the pair can overlap with a corresponding orbital on the other atom. This produces four molecular bonding orbitals, and four antibonding orbitals (see Figure 3.25).

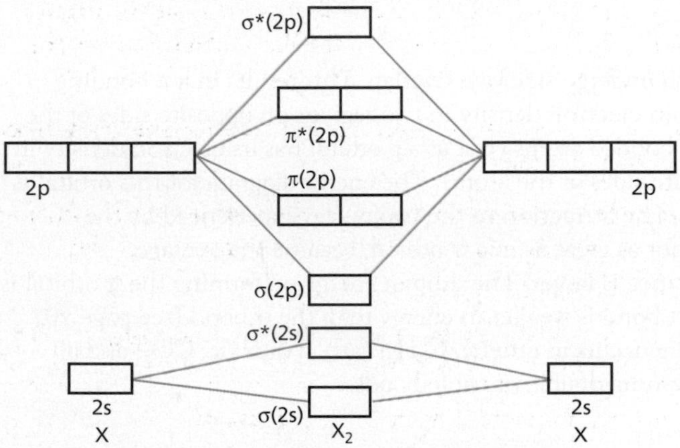

■ **Figure 3.25** Bonding molecular orbitals available in diatomic molecules of the second row of the periodic table.

As we fill these molecular orbitals with electrons, we can now predict the type and approximate strength of the bonding between the atoms in the various X_2 molecules. In Li_2, for example, only the σ (2s) orbital is filled, giving a bond order of one. In Be_2, both the bonding σ (2s) and the antibonding σ* (2s) orbitals would be filled, giving an overall bond order of zero. In fact, Be_2 does not exist. In N_2, four bonding orbitals and one antibonding orbital have been filled, giving an overall bond order of $4 - 1 = 3$. Thereafter, in O_2 and F_2, further antibonding orbitals are being filled, reducing the bond order to 2 and 1, respectively. (See Table 3.2 and Figure 3.26.)

One of the successes of molecular orbital theory was to predict correctly the paramagnetic nature of the O_2 molecule, due to its two unpaired electrons in the π* (2p) orbitals. (**Paramagnetism** is the tendency for molecules to be attracted to a magnetic field, and is usually associated with the presence of one or more unpaired electrons. Most molecules contain fully paired-up electrons, and are not paramagnetic.) The excitation of the electrons in the π* (2p) orbitals of the halogen molecules to the empty σ* (2p) orbital is responsible for their colours (see page 273).

■ **Table 3.2** Bonding and antibonding electrons for some second-row diatomic molecules.

Molecule	Number of bonding electrons	Number of antibonding electrons	Net bond order	Bond enthalpy /kJ mol^{-1}
Li_2	2	0	1	105
Be_2	2	2	0	(0)
N_2	8	2	3	945
O_2	8	4	2	498
F_2	8	6	1	158

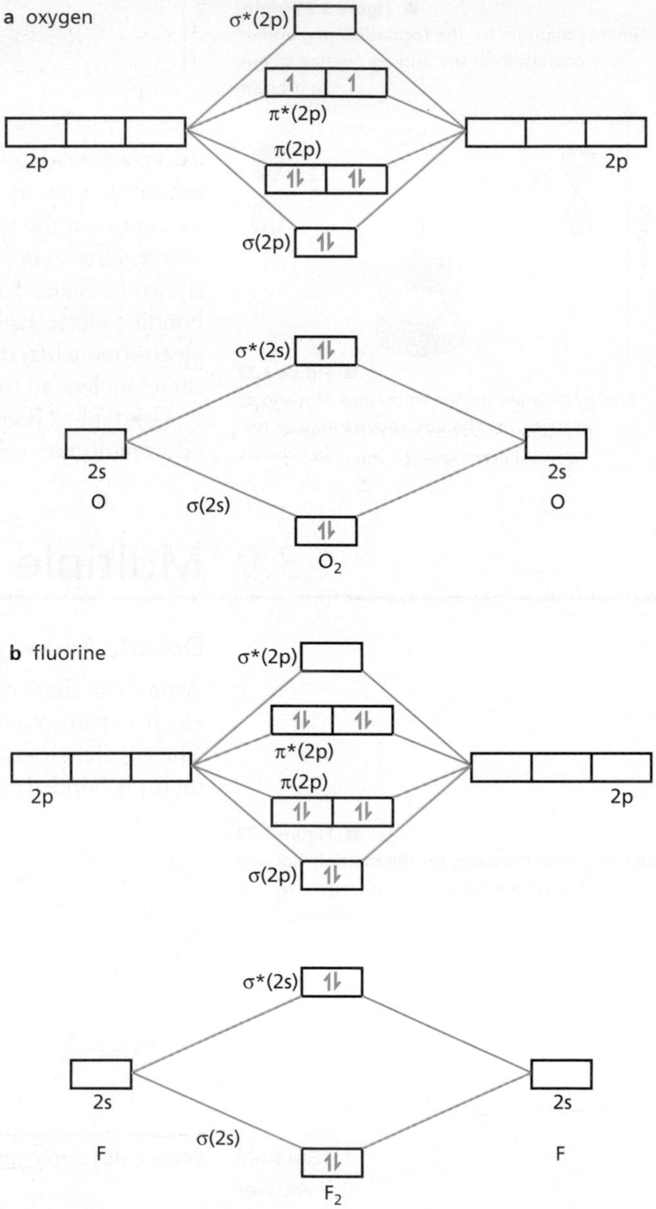

■ **Figure 3.26** Electronic configurations for a oxygen **b** fluorine.

3.10 The shapes of molecules

One of the major advances in chemistry occurred in 1874, when the Dutch chemist Jacobus van't Hoff suggested for the first time that molecules possessed a definite, unique, three-dimensional shape. The shapes of molecules are determined by the angles between the bonds within them. In turn, the bond angles are determined by the arrangement of the electrons around each atom.

VSEPR theory

Because of their similar (negative) charge, electron pairs repel each other. The electrons pairs in the outer (valence) shell of an atom will experience the least repulsion when they are as far apart from each other as possible. This applies both to bonded pairs and to non-bonded (lone) pairs. This principle allows us to predict the shapes of simple molecules. The theory, developed by Nevil Sidgwick and Herbert Powell in 1940, is known as the **valence-shell electron-pair repulsion theory** (or **VSEPR** for short). In order to work out the shape of a molecule, the theory is applied as follows:

1 Draw the dot-and-cross structure of the molecule, and hence count the number of electron pairs around each atom.

2 These pairs will take up positions where they are as far apart from each other as possible. The angles between the pairs will depend on the number of pairs around the atom (see Figure 3.27).

■ **Figure 3.27**
The shapes in which n electron pairs around a central atom arrange themselves, with $n = 2, 3, 4, 5$ and 6.

molecule shape	number of electron pairs	description
	2	linear
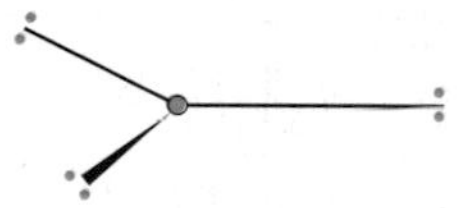	3	triangular planar (trigonal planar)
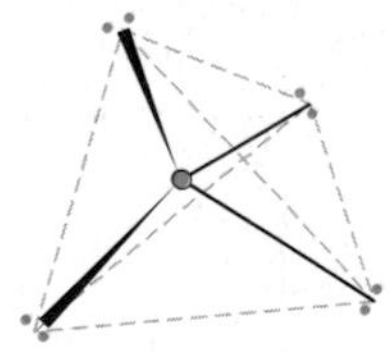	4	tetrahedral
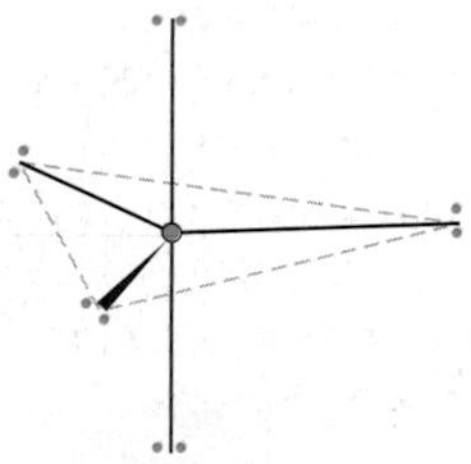	5	triangular bipyramidal (trigonal bipyramidal)
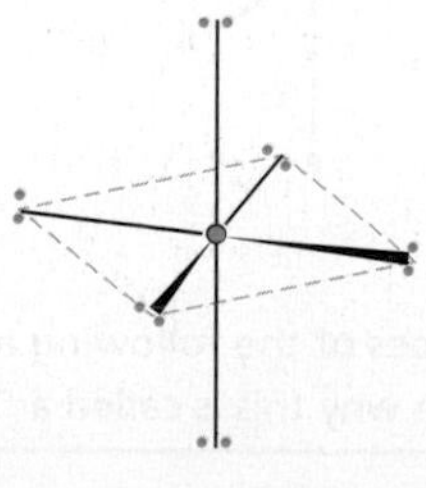	6	octahedral

3 Orbitals containing lone pairs are larger than those containing bonded pairs. They take up more space around the central atom and are, on average, closer to the nucleus. Thus they have a greater repulsive effect on other pairs that surround the atom. This causes the angle between two lone pairs to be larger than the angle between a lone pair and a bonded pair, which in turn is larger than the angle between two bonded pairs:

$$(\text{LP—LP angle}) > (\text{LP—BP angle}) > (\text{BP—BP angle})$$

4 Although they are very important in determining the shape of a molecule, the lone pairs are not included when the molecule's shape is described. For example, the molecules CH_4, NH_3 and H_2O all have four pairs of electrons in the valence shell of the central atom (see Figure 3.28). These four pairs arrange themselves (roughly) in a tetrahedral fashion. But only the methane molecule is described as having a tetrahedral shape. Ammonia is pyramidal (it has only three bonds), and water is described as a bent molecule (it has only two bonds).

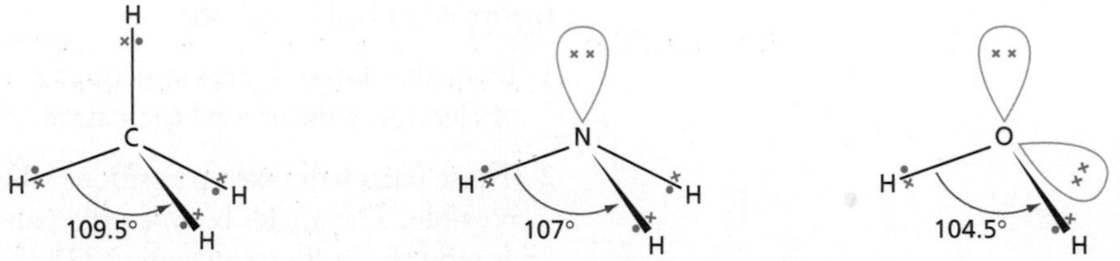

■ Figure 3.28 The shapes of methane, ammonia and water are as predicted from their four pairs of electrons around the central atom. Lone pairs have a greater repulsive effect than bonding pairs, so the angles in ammonia and water are slightly smaller than the tetrahedral 109.5°.

Example Work out the shapes of the following molecules:

a CCl_4 **b** BF_3 **c** PF_5

Answer **a**

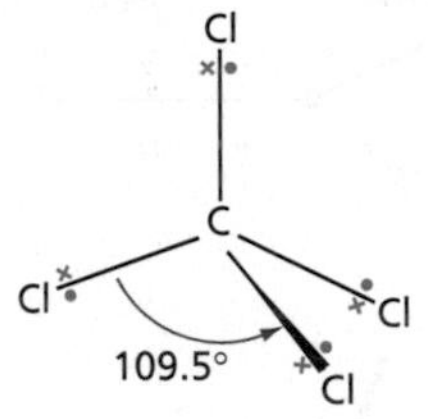

■ Figure 3.29

The valence shell of the carbon atom in CCl_4 contains eight electrons, arranged in four bonded pairs. It therefore takes up a regular tetrahedral shape, just like methane (see Figure 3.29).

b

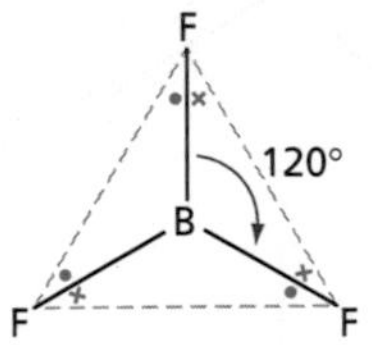

■ Figure 3.30

The valence shell of the boron atom in BF_3 contains six electrons, arranged in three bonded pairs (no lone pairs – see page 55). It therefore takes up a regular triangular planar shape, with all bond angles being equal at 120° (see Figure 3.30).

c

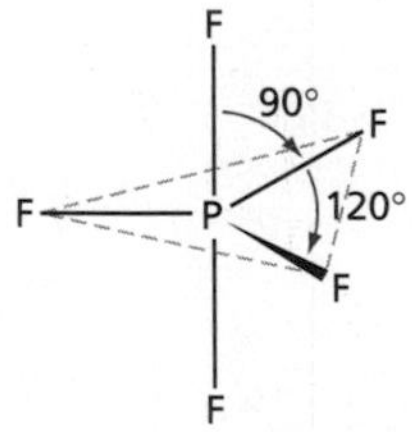

■ Figure 3.31

The valence shell of the phosphorus atom in PF_5 contains ten electrons: five from phosphorus, and one each from the five fluorine atoms (see page 57). The shape will be a regular triangular bipyramid, with bond angles of 120° and 90° (see Figure 3.31).

Further practice Work out the shapes of the following molecules:

BeH_2, ClF_3 (explain why this is called a 'T-shaped' molecule), SF_6.

Double bonds and VSEPR

When a molecule contains a double bond the four electrons in this bond count as one group of electrons, as far as the VSEPR theory is concerned, even though, of course, four electrons exert a greater repulsive force than two. For example, the carbon in carbon dioxide, which is doubly bonded to each of the oxygen atoms, is considered to be surrounded by just two electron groups. Carbon dioxide is therefore predicted to be a linear molecule (see Figure 3.32).

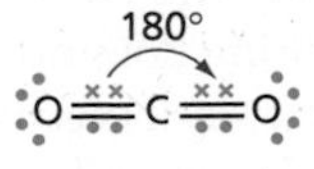

■ **Figure 3.32** Carbon dioxide is a linear molecule.

Example Predict the shape of the sulphur dioxide molecule.

Answer Sulphur shares two of its valence-shell electrons with each of the two oxygen atoms, forming two double bonds. It has two electrons left over. These form a lone pair. There are therefore three groups of electrons around the sulphur atom, and these will arrange themselves approximately into a triangular planar shape. The SO_2 molecule is therefore bent (remember that lone pairs are ignored when the shape is described), with a O—S—O bond angle of about 120°. (In fact, it is slightly less, due to the extra repulsion by the lone pair, Figure 3.33.)

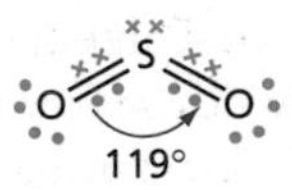

■ **Figure 3.33** Sulphur dioxide is a bent molecule.

Further practice Predict the shapes of the three molecules in the Further practice question on page 61. (Note that a triple bond, like a double bond, can be considered as a single group of electrons.)

The molecular orbital treatment of π bonds in the panel on page 62, coupled with the concept of hybridisation in the panel on pages 71–73, offers an alternative explanation of the shapes of doubly bonded molecules such as carbon dioxide and ethene.

Summary: the shapes of molecules

Table 3.3 summarises how the numbers of bonded and non-bonded electron pairs determine the shapes of molecules.

■ **Table 3.3** The shapes of molecules as determined by the bonding and non-bonding electron pairs.

Total number of electron pairs	Number of bonded pairs	Number of lone pairs	Shape	Example
2	2	0	Linear	BeF_2
3	3	0	Triangular planar	BF_3 (see Figure 3.30)
4	1	3	Linear	HF
4	2	2	Bent	H_2O (see Figure 3.28)
4	3	1	Triangular pyramidal	NH_3 (see Figure 3.28)
4	4	0	Tetrahedral	CH_4 (see Figure 3.28)
5	2	3	Linear	XeF_2 (see Figure 3.34)
5	3	2	T-shaped	ClF_3 (see Figure 3.34)
5	4	1	See-saw	SF_4 (see Figure 3.34)
5	5	0	Triangular bipyramidal	PF_5 (see Figure 3.31)

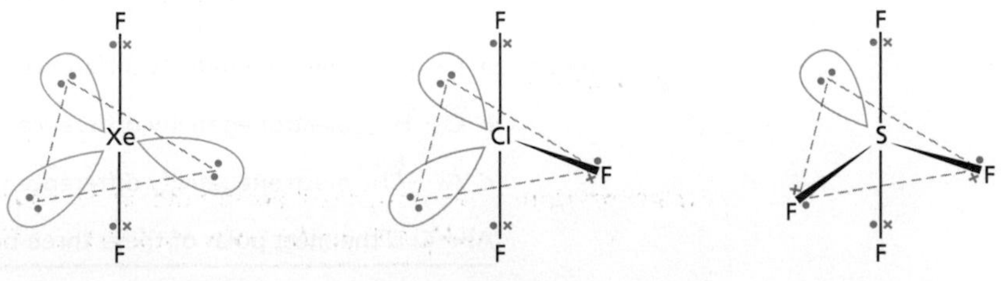

■ **Figure 3.34** The shapes of XeF_2, ClF_3 and SF_4 are all determined by the five pairs of electrons around the central atom.

3.11 Electronegativity and bond polarity

The essential feature of covalent bonds is that the electrons forming the bond are shared between the two atoms. But that sharing does not have to be an equal one. Unless both atoms are the same (for example, in molecules of hydrogen or oxygen), it is more than likely that they will have different **electronegativities**, and this leads to unequal sharing.

> The **electronegativity** of an atom is a measure of its ability to attract the electrons in a covalent bond to itself.

Several chemists have developed quantitative scales of electronegativity values. One of the most commonly used is the scale devised by the American chemist Linus Pauling in 1960. Numerical values of electronegativity on this scale range from fluorine, with a value of 3.98, to an alkali metal such as sodium, with a value of 0.93.

Figure 3.35 shows the electronegativity values of the elements in the first three rows of the periodic table. Notice that the larger the number, the more electronegative the atom is.

■ Figure 3.35 Electronegativities of elements in the first three rows of the periodic table.

H 2.20						
Li 0.98	Be 1.57	B 2.04	C 2.55	N 3.04	O 3.44	F 3.98
Na 0.93	Mg 1.31	Al 1.61	Si 1.90	P 2.19	S 2.58	Cl 3.16

The electronegativity value depends on the effective nuclear charge. This is similar to the way ionisation energies depend on the effective nuclear charge, and the reason is also similar. The nucleus of a fluorine atom (electronegativity value = 3.98), for example, contains more protons and a higher effective nuclear charge, than that of a carbon atom (electronegativity value = 2.55), so is able to attract electrons more strongly than carbon. In the C—F bond, therefore, the electrons will be found nearer on average to fluorine than to carbon. This will cause a partial movement of charge towards the fluorine atom, resulting in a **polar bond** (see Figure 3.36).

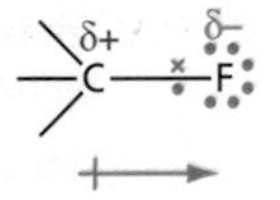

■ Figure 3.36 The carbon–fluorine bond is polar.

This **partial charge separation** is represented by the Greek letter δ (delta), followed by + or − as appropriate. The 'cross-and-arrow' symbol in Figure 3.36 is also sometimes used to show the direction in which the electrons are attracted more strongly.

Most covalent bonds between different atoms are polar. If the electronegativity difference is large enough, one atom can attract the bonded electron pair so much that one electron is transferred and an ionic bond is formed (see the panel opposite and Chapter 4, page 89).

Example

Use the electronegativities in Figure 3.35 to predict the direction of polarisation in the following bonds:

a C—O **b** O—H **c** Al—Cl. Which bond is the most polar?

Answer

a $\overset{\delta+}{C}$—$\overset{\delta-}{O}$ electronegativity difference = 3.44 − 2.55 = 0.89

b $\overset{\delta-}{O}$—$\overset{\delta+}{H}$ electronegativity difference = 3.44 − 2.20 = 1.24

c $\overset{\delta+}{Al}$—$\overset{\delta-}{Cl}$ electronegativity difference = 3.16 − 1.61 = 1.55

Al—Cl is the most polar of these three bonds.

Molecular orbital theory (4): atoms with different electronegativities

As mentioned above, the ionisation energy of a fluorine atom is larger than that of a carbon atom because in fluorine the electrons are more strongly attracted to the nucleus. This means that the 2p orbital of fluorine is lower in potential energy (that is, it is more stable) than is the 2p orbital of carbon. Calculations show that when these two orbitals overlap, the new bonding σ molecular orbital will be closer in energy to fluorine's 2p orbital, and nearer to the fluorine atom than to the carbon atom (see Figure 3.37).

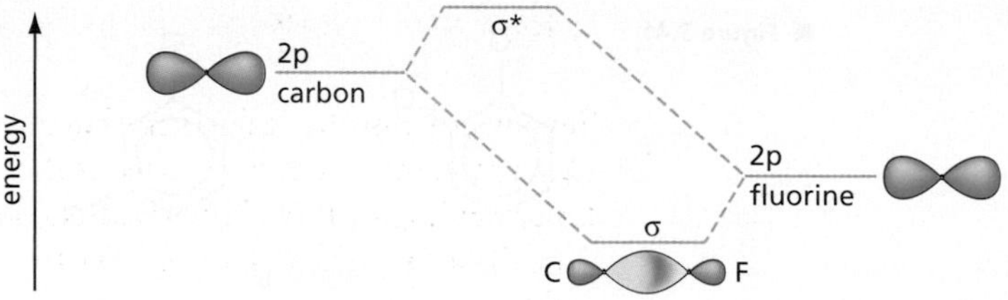

■ **Figure 3.37** The σ bonding orbital formed between carbon and fluorine is closer in position and in energy to fluorine than to carbon.

The drift of bonding electrons towards fluorine (as a result of its higher electronegativity) is therefore predicted from MO theory. If the difference in electronegativity is extremely high, MO theory predicts that the bonding electrons will be very close in space and in energy to a lone pair on the electronegative atom (see Figure 3.38). This represents the complete transfer of the shared bonding pair to the electronegative atom which we see in ionic bonding (see Chapter 4).

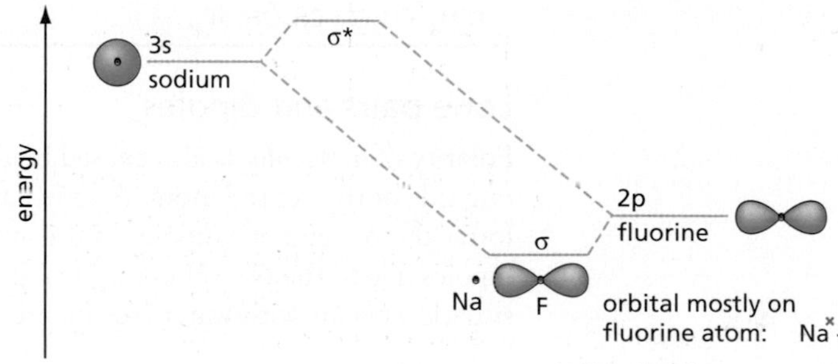

$Na^{\times} + F \rightarrow Na^{+} + F^{-}$

■ **Figure 3.38** When the difference in energy and electronegativity is very great, ionic bonding results.

3.12 The polarity of molecules

Dipoles of molecules

The drift of bonded electrons to the more electronegative element results in a separation of charge, termed a **dipole**. Each of the polar bonds in a molecule has its own dipole associated with it. The overall dipole of the molecule depends on its shape. Depending on the relative angles between the bonds, the individual bond dipoles can either reinforce or cancel each other. If cancellation is complete, the resulting molecule will have no dipole, so will be non-polar. If the bond dipoles reinforce each other, molecules with very large dipoles can be formed.

For example, both hydrogen and fluorine are non-polar molecules, but hydrogen fluoride has a large dipole due to the much larger electronegativity of fluorine (see Figure 3.39).

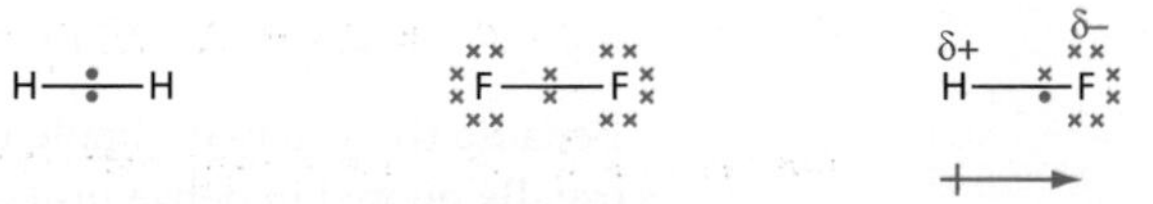

■ **Figure 3.39** The electronegativity difference gives hydrogen fluoride a large dipole.

The two C=O bonds in carbon dioxide are polar, but because the angle between them is 180°, exact cancellation occurs, and the molecule of carbon dioxide is non-polar. A similar situation occurs in tetrachloromethane, but not in chloromethane (see Figure 3.40, overleaf).

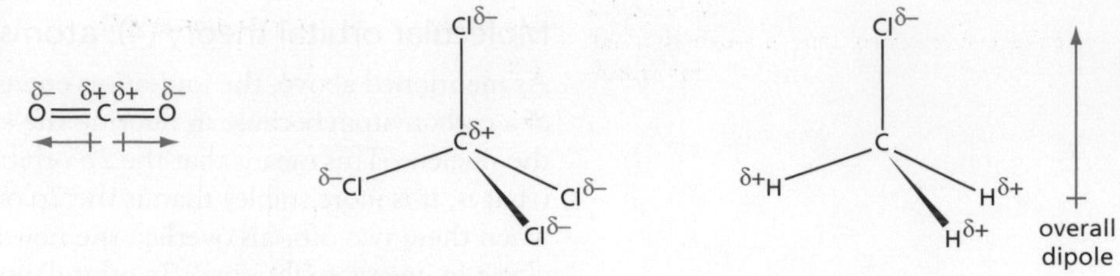

■ **Figure 3.40** The dipoles cancel in carbon dioxide and tetrachloromethane, but not in chloromethane.

Example Which of the molecules in Figure 3.41 has no dipole moment?

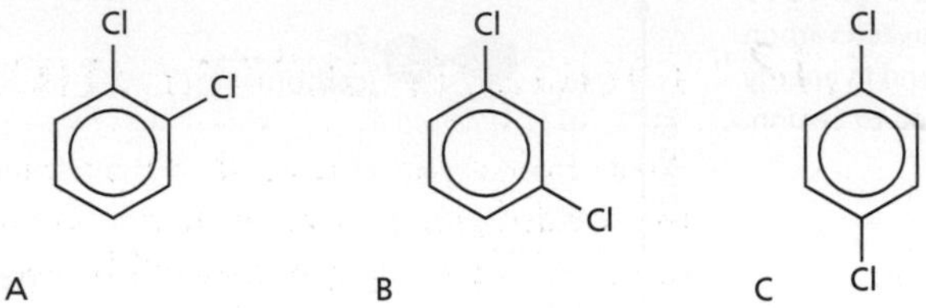

■ **Figure 3.41**

Answer Compound C, 1,4-dichlorobenzene, has no dipole moment. In this compound the C—Cl dipoles are on opposite sides of the planar benzene ring, so their dipoles oppose each other equally. In compounds A and B the C—Cl dipoles are on the same side of the ring.

Further practice

1 Which of the two compounds A and B in Figure 3.41 will have the larger dipole moment, and why?

2 Work out the shapes of the following molecules and decide which of them are polar: C_2H_6, CH_2Cl_2, CH_3OH, SF_4.

Lone pairs and dipoles

Polarity of molecules is also caused by the presence of a lone pair of electrons on one side of the central atom. A lone pair is an area of negative charge and thus forms the δ− end of a dipole. In the ammonia molecule, this dipole reinforces the dipoles due to the N—H bonds, resulting in a highly polar molecule. A similar situation occurs with water (see Figure 3.42).

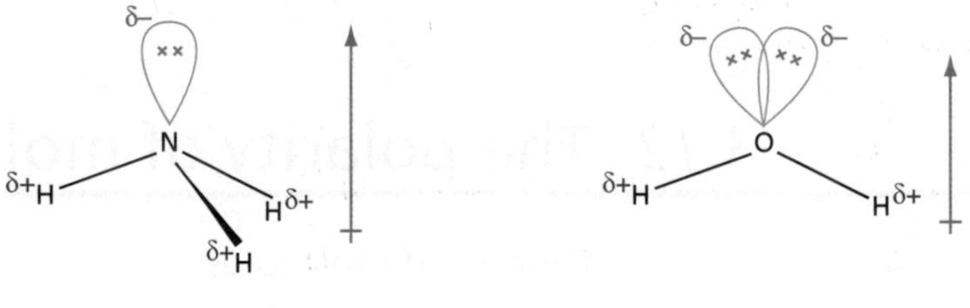

■ **Figure 3.42** Ammonia and water are highly polar molecules.

Measuring dipoles

The strength of a dipole is measured by its **dipole moment**, which is the product of the separated electric charge, δ+ or δ−, and the distance between them. The dipole moment μ associated with the separation of a full electron charge by a distance of 0.15 nm (a typical bond length) can be calculated as follows:

$$\mu = \text{charge} \times \text{distance}$$

charge on $e^- = 1.6 \times 10^{-19}\,C$

$1\,nm = 1 \times 10^{-9}\,m$

$$= (1.6 \times 10^{-19}\,C) \times (0.15 \times 10^{-9}\,m)$$
$$= 2.4 \times 10^{-29}\,C\,m$$

Because the values of dipole moments at the atomic scale are so small, they are usually quoted in debye units, D (named after Peter Debye, a chemist who studied closely the electrical characteristics of molecules). The debye unit has the value $3.34 \times 10^{-30}\,C\,m$, so the dipole moment in the above example has a value of

$$\frac{2.4310^{-29}}{3.34310^{-30}} = 7\,D.$$

■ Table 3.4 The dipole moments of some common covalent molecules.

Molecule	Dipole moment/D
HF	1.91
HCl	1.05
HBr	0.80
HI	0.42
H_2O	1.84
NH_3	1.48

This would be the dipole moment associated with a typical ionic bond, where an electron is completely transferred from one atom to another. Most dipole moments of covalent molecules are considerably less than this (see Table 3.4), showing that the drift in charge is much less than a full electron.

3.13 Induced polarisation of bonds

Apart from a bond's natural polarity caused by the different electronegativities of the bonded atoms, covalent bonds can also have a polarity that is induced (or have their existing polarity enhanced) by the proximity of a strong electric field. This is often followed by the breaking of the bond. The electric field can be due to an area of concentrated electron density, or an electron-deficient atom, or a small highly charged cation.

The following examples of bond polarisation are discussed in more detail in the chapters mentioned:

- polarisation of the bromine molecule by an ethene molecule, as shown in Figure 3.43 (see Chapter 24)

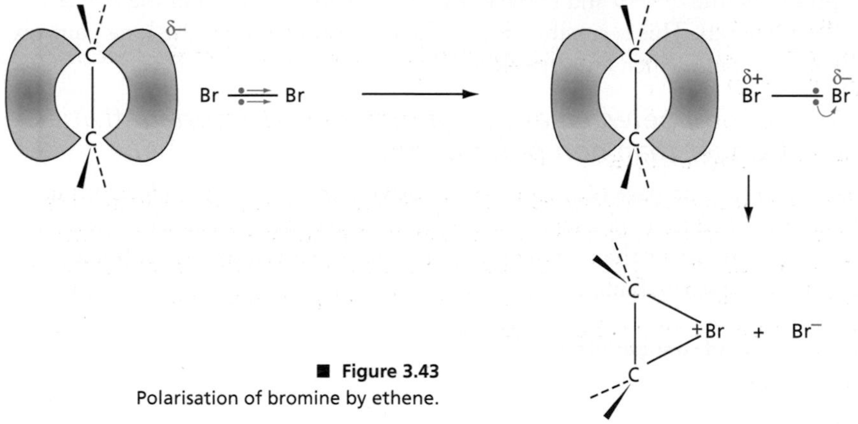

■ Figure 3.43 Polarisation of bromine by ethene.

- polarisation of the chlorine molecule by aluminium chloride, as shown in Figure 3.44 (see Chapter 27)

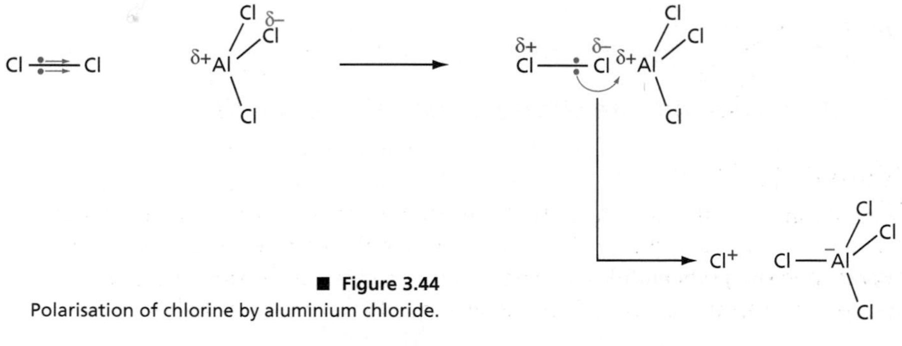

■ Figure 3.44 Polarisation of chlorine by aluminium chloride.

- polarisation of the negative carbonate ion by Group 2 metal ions, as shown in Figure 3.45 (see Chapter 16).

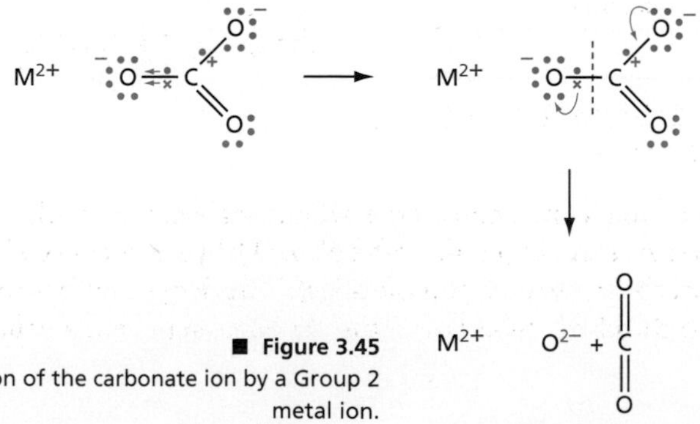

■ Figure 3.45 Polarisation of the carbonate ion by a Group 2 metal ion.

The induced polarisation of **molecules**, and the effect this has on intermolecular attractions, is discussed on page 79.

What determines the strength of a covalent bond?

Bond strengths depend on the amount of orbital overlap there is between the two bonded atoms (the larger the degree of overlap, the stronger the bond) and also on how strongly the electrons in the overlap region are attracted to the nuclei of the atoms. There are three generalisations that can be made.

(1) *Atoms in the first and second periods form stronger bonds than those in the third and later periods*

This is because the 1s and 2p orbitals are small, and close to the nucleus. Electrons in these orbitals do not experience much shielding by inner orbital electrons. Electrons in overlapping orbitals therefore experience a strong pull to both bonding nuclei. Some examples are shown in Table 3.5.

Table 3.5 Atoms in the second period form stronger bonds than those in the third period.

Bond	Bond strength/kJ mol^{-1}
C—C	350
Si—Si	225
H—N	390
H—P	320

(Exceptions are the O—O and F—F bonds, which are weaker than the S—S and Cl—Cl bonds. This is possibly due to the repulsion between the lone pairs on the adjacent atoms, which are very close.)

(2) *Polar bonds, between atoms of different electronegativities, are stronger than non-polar bonds*

As was mentioned on page 66, an electronegative atom attracts the electrons in the bonds attached to it. The more an electron is attracted to a nucleus, the more stable is the system, and the more energy is required to break the bond. Some examples are shown in Table 3.6.

Table 3.6 The bond strength increases as the electronegativity difference increases.

Bond	Bond strength/kJ mol^{-1}
H—N	390
H—O	460
H—F	570

(3) *Multiple bonds are stronger than single bonds*

This is a fairly obvious statement! The more orbitals that overlap, the more electrons can experience the attraction of both nuclei, and hence the stronger the bond. Some examples are shown in Table 3.7.

Table 3.7 Bond strength increases as bond order increases.

Bond	Bond strength/kJ mol^{-1}
C—C	350
C=C	610
C≡C	840
C—O	360
C=O	700
C≡O	1080

3.14 Isoelectronic molecules and ions

Isoelectronic means 'having the same number of electrons'. The term is also used to refer to molecules or ions that contain the same number of bonded and non-bonded electron pairs in their valence shells. Since it is the distribution of electrons that determines the shape of a molecule, isoelectronic molecules or ions all have the same basic shape, often with identical bond angles too. For example:

- the tetrahydridoborate(III) ion, BH_4^-, methane, CH_4, and the ammonium ion, NH_4^+, are all regular tetrahedrons, with H—X—H angles of 109.5°
- the borate ion, BO_3^{3-}, the carbonate ion, CO_3^{2-}, and the nitrate ion, NO_3^-, are all triangular planar, with O—X—O angles of 120°
- the oxonium ion, H_3O^+, and ammonia, NH_3, are both triangular pyramids with H—X—H angles of 107°.

Counting the electrons in the valence shell around an atom is therefore a useful way of predicting the shape of a molecule.

Example

Predict the shapes of

a the methyl cation, CH_3^+

b the methyl anion, CH_3^-

c the nitryl cation, NO_2^+.

Answer

a CH_3^+ is isoelectronic with BH_3 (six electrons in the outer shell), and so it is a triangular planar ion.

b CH_3^- is isoelectronic with NH_3, so it is a triangular pyramidal ion.

c NO_2^+ is isoelectronic with CO_2, so it is a linear ion.

Further practice

Sulphur trioxide is triangular planar, the sulphate(IV) ion (sulphite ion), SO_3^{2-}, is a triangular pyramid and the sulphate(VI) ion, SO_4^{2-}, is tetrahedral. Predict the shapes of the following ions: ClO_3^-, PO_3^-, PO_4^{3-}.

Molecular orbital theory (5): hybridisation of orbitals and the shapes of molecules

Atomic orbitals do not overlap only on adjacent atoms to form molecular orbitals. The orbitals of the same atom can overlap, mix and **hybridise**, to form a new set of orbitals. The **hybrid** orbitals so formed take up a distinctive shape and relative orientation, depending on the number and type of atomic orbitals that have been mixed.

sp hybrid orbitals

The simplest situation is the mixing of an s and a p orbital. Calculations predict that this will result in the formation of two identical hybrid orbitals, directed at 180° to each other, called **sp hybrid orbitals**. Each orbital will have the same energy, which will be intermediate between the energies of the s and p orbitals (see Figure 3.46). The situation can be illustrated by beryllium.

■ Figure 3.46
Forming sp hybrid orbitals from an s orbital and a p orbital.

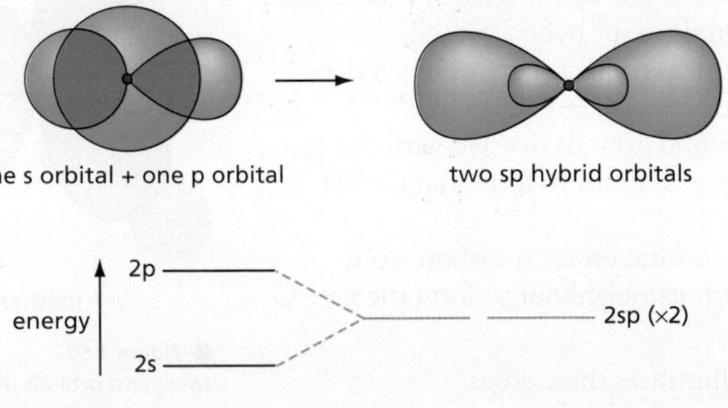

(continued)

The most stable electronic configuration of the beryllium atom is $1s^2 2s^2 2p^0$, that is, with the two second-shell electrons paired in the 2s orbital. But because the two sp hybrid orbitals are now of equal energy, these two electrons will become unpaired, each occupying one of the sp hybrid orbitals (giving the configuration $1s^2 2sp_a{}^1 2sp_b{}^1$). This allows us easily to visualise the formation of the two covalent bonds in the BeF_2 molecule: each results from the sharing of the unpaired electron in the hybrid sp orbital with an unpaired electron in a 2p orbital of fluorine (see Figure 3.47). What is more, the theory predicts a linear molecule, as does the VSEPR theory, and as is measured in practice.

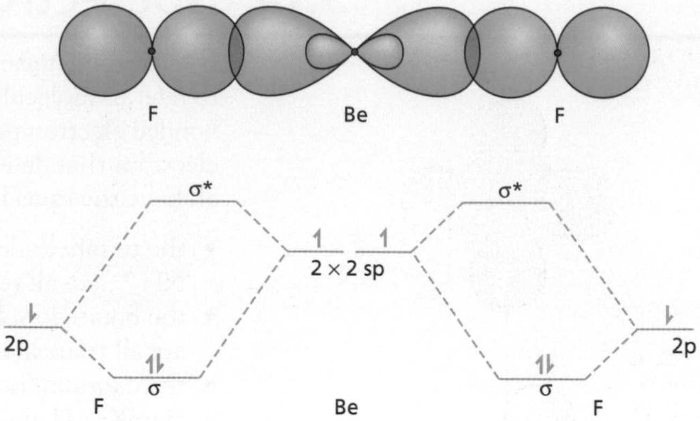

■ **Figure 3.47**
sp hybrid orbitals in the BeF_2 molecule.

sp² hybrid orbitals

A second combination of atomic orbitals that produces hybrid orbitals uses one s and two p orbitals. This results in **sp²** ('ess pee two') **hybrids**. These point towards the corners of an equilateral triangle, and are used to form the bonds in compounds such as boron trifluoride and ethene (see Figure 3.48).

The three electrons in the valence shell of the boron atom therefore occupy the three sp² orbitals singly, allowing bond formation with fluorine (see Figure 3.49).

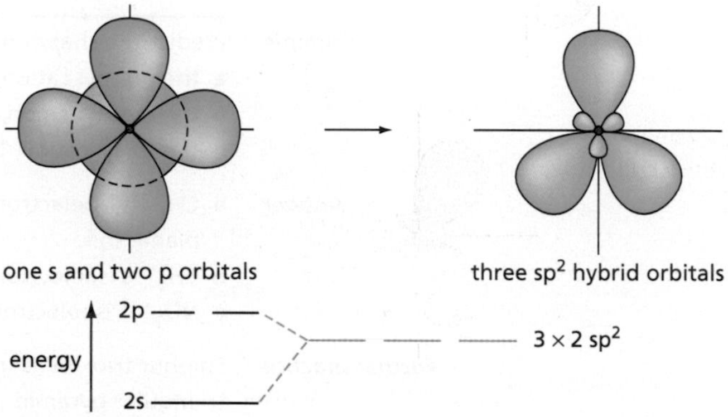

■ **Figure 3.48**
Forming sp² hybrid orbitals from an s orbital and two p orbitals.

sp³ hybrid orbitals

A third possible combination uses all the orbitals in the valence shell – one s and three p orbitals. The resulting **sp³ hybrids** point towards the corners of a regular tetrahedron, and are used to form the bonds in methane, ammonia and water, for example. In methane, the four singly occupied sp³ orbitals overlap with the singly occupied 1s orbitals of four hydrogen atoms. In water, two of the sp³ hybrids contain the two lone pairs, and the other two bond with hydrogen (see Figure 3.50).

In nitrogen, the five valence-shell electrons can occupy sp³ hybridised orbitals, resulting in the electronic configuration $(2sp^3{}_a)^2(2sp^3{}_b)^1(2sp^3{}_c)^1(2sp^3{}_d)^1$. The three singly occupied hybrid orbitals overlap with the 1s orbitals of three hydrogen atoms, whilst the doubly occupied $2sp^3{}_a$ orbital forms the lone pair on the nitrogen atom.

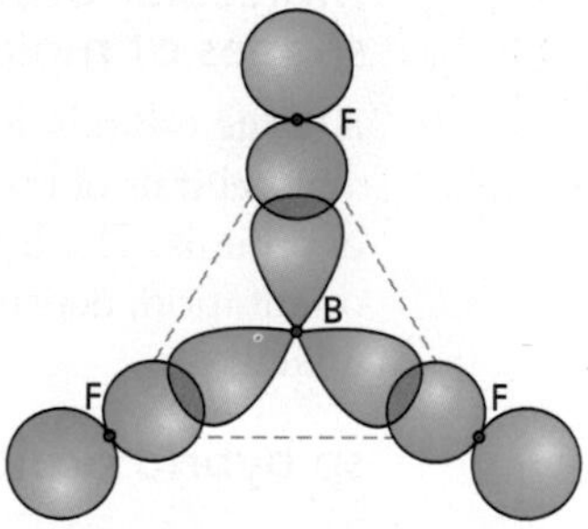

■ **Figure 3.49**
sp² hybrid orbitals in the BF_3 molecule.

Note that in the sp case, two p orbitals remain unhybridised, and in the sp² case, one p orbital remains. We can make use of these p orbitals when we construct double and triple bonds.

For example, the full bonding picture of the ethene molecule (page 61) can be constructed as follows:

1 The 2s orbital and two of the 2p orbitals on each carbon atom are mixed to produce sp² hybrid orbitals.
2 One sp² hybrid orbital on each carbon atom overlaps with its neighbour to form a C—C σ bond.
3 The other four sp² hybrid orbitals overlap with the 1s orbitals on four hydrogen atoms to form four C—H σ bonds.
4 The unhybridised 2p orbital on each carbon atom overlaps sideways with its neighbour to form the C—C π bond.

Figure 3.51 (opposite) illustrates these steps.

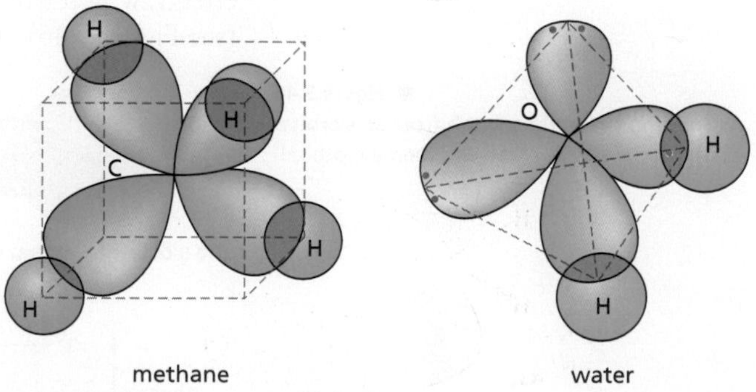

■ **Figure 3.50**
sp³ hybrid orbitals in methane and water.

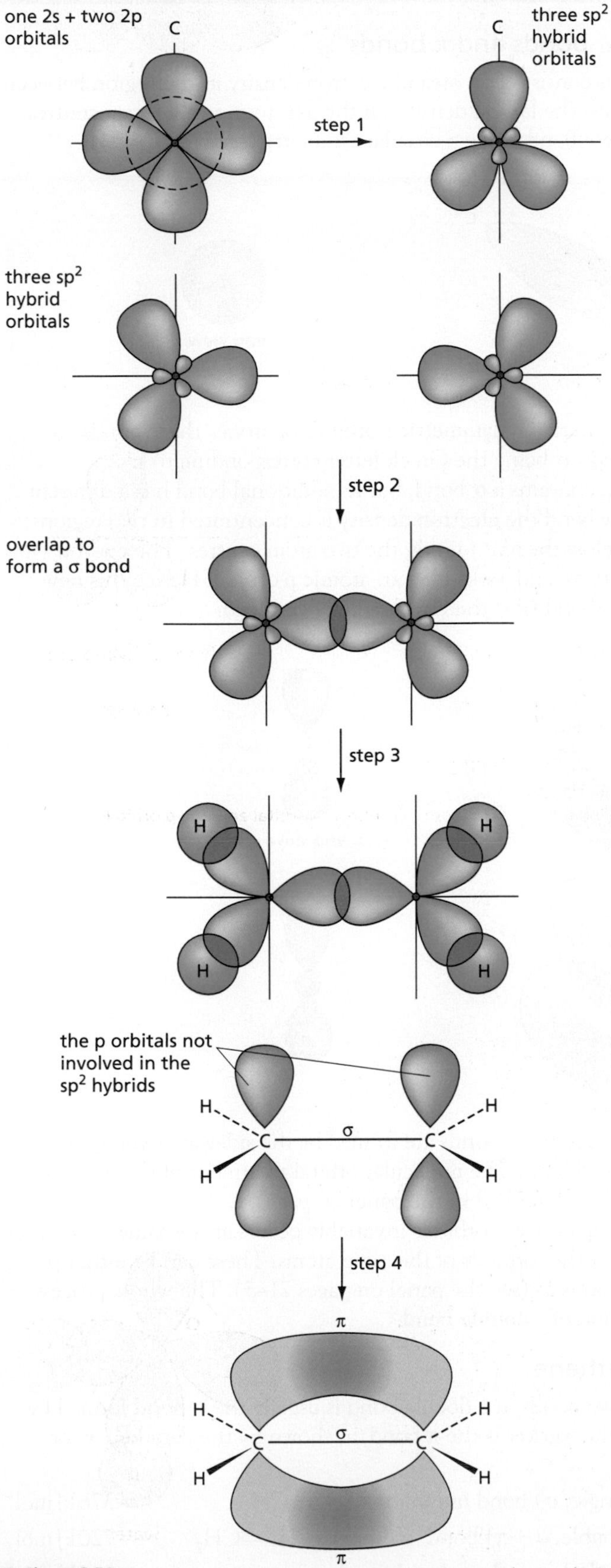

■ **Figure 3.51**
Hybridised orbitals in ethene.

Hybridisation also occurs in the nitrogen molecule, N_2, the ethyne molecule, C_2H_2, and the carbon dioxide molecule, CO_2. In all three cases the nitrogen or carbon atoms are sp hybridised, and σ bonds are formed by the end-on overlap of two sp hybrid orbitals on adjacent atoms. Figure 3.59, page 76, illustrates the result.

Example Describe the bonding in and the shape of the molecules of:

a methanal, CH_2O

b hydrogen cyanide, HCN.

Answer **a** Both the carbon and oxygen atoms are sp^2 hybridised. Of the three sp^2 orbitals around carbon, two are used to form σ bonds by overlapping with the 1s orbitals of the two hydrogen atoms, and one forms the σ bond between carbon and oxygen by overlapping with one of the three sp^2 orbitals on oxygen. The other two sp^2 orbitals house the two lone pairs on oxygen. The unhybridised p orbital on each atom forms the π bond. All four atoms in the molecule lie in a plane, with the H—C—H and H—C—O angles approximately 120° (see Figure 3.52).

■ **Figure 3.52**

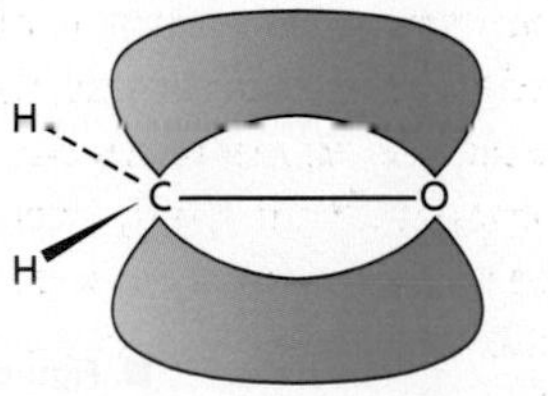

b The left-hand side of the HCN molecule is the same as one end of the ethyne molecule, and the right-hand side is the same as one end of the nitrogen molecule. HCN is linear, with one single bond and one triple bond (see Figure 3.53).

■ **Figure 3.53**

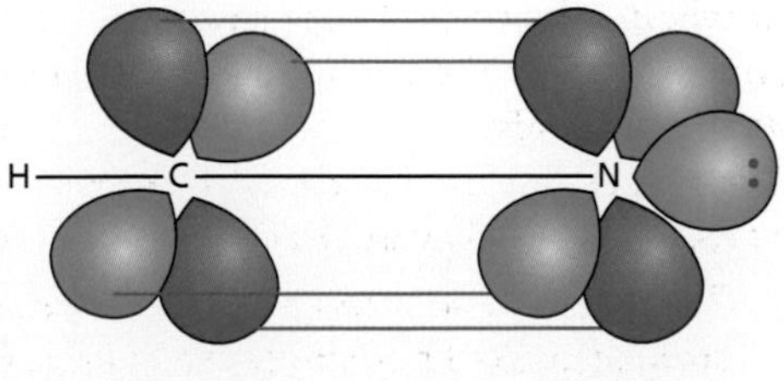

Further practice Describe the bonding in and the shape of the following molecules: hydrogen peroxide, H_2O_2, and propanone, $(CH_3)_2C{=}O$.

Hybridisation is important in delocalised systems too (see pages 76–8). Benzene, the carboxylate ion and the carbonate ion all contain sp^2 hybridised carbon (and oxygen) atoms.

3.15 Localised π bonds

Double bonds – σ bonds and π bonds

Single covalent bonds consist of an area of electron density in the region between the bonded atoms, with the largest density on the axis joining the atom centres. Looked at end-on these bonds have a circular symmetry (see Figure 3.54).

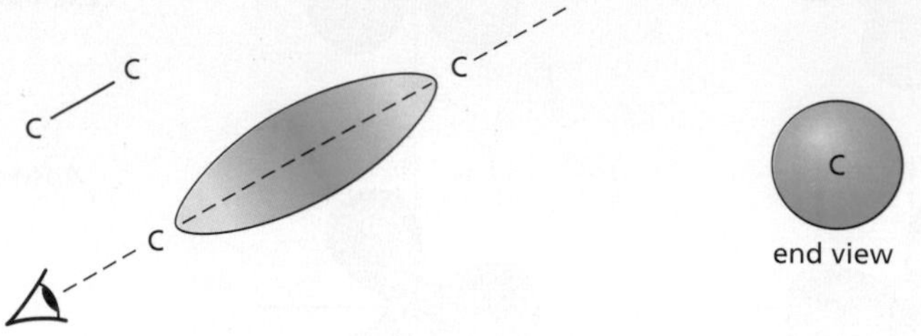

■ **Figure 3.54**
The C—C single bond (a σ bond).

By analogy with the spherically symmetric s orbitals of atoms, these bonds are termed σ (sigma) bonds, σ being the Greek letter corresponding to s.

A double bond also contains a σ bond, but its additional bond has a different symmetry. In this new bond the electron density is concentrated in two regions, one above and one below the axis joining the two atom centres. The end-on view has the same symmetry as, and looks like, an atomic p orbital. Hence this new bond is called a π (pi) bond (π is the Greek equivalent of p).

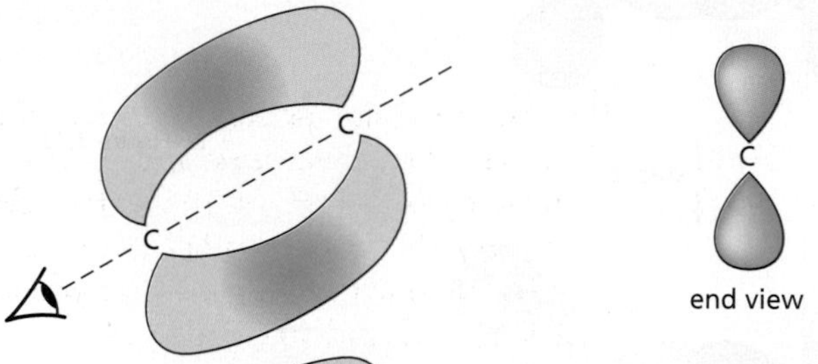

■ **Figure 3.55**
The π bond between two carbon atoms.

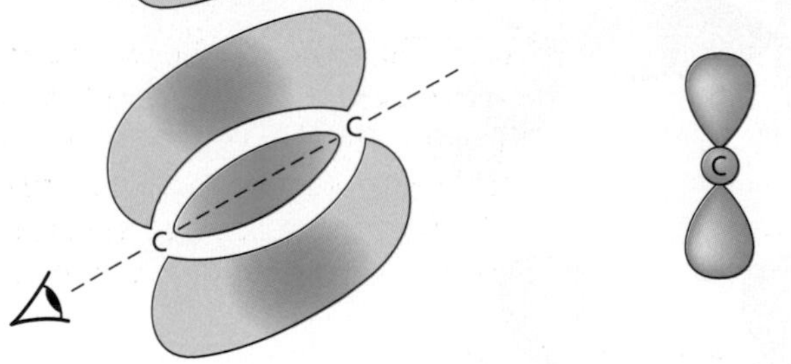

■ **Figure 3.56**
A carbon–carbon double bond.

As we mentioned on page 60, π bonds are formed by the sideways overlap of p orbitals on the bonding atoms. The molecular orbital treatment of the π bonds used in multiple bonds is described in the panel on page 61.

This sideways overlap of two p orbitals invariably occurs at the same time as the end-on overlap of two other orbitals of the same atoms. These can be either p orbitals or sp^n hybrid orbitals (see the panel on pages 71–3). The whole process results in the production of a **double bond**.

The bonding in ethene

The stronger of the two bonds in a double bond is usually the σ bond formed by end-on overlap, and the weaker is the π bond, as shown by the data below for ethane and ethene:

strength of C—C (single, σ) bond in ethane, CH_3—CH_3	$= 376\,kJ\,mol^{-1}$
strength of C=C (double, σ + π) bond in ethene, CH_2=CH_2	$= 720\,kJ\,mol^{-1}$
therefore, extra strength due to the π bond in ethene	$= 344\,kJ\,mol^{-1}$

(assuming the σ bond has the same strength in both compounds).

The bonding picture of the ethene molecule can therefore be constructed as follows. Each carbon atom shares three electrons (one with each of two hydrogen atoms, and one with the other carbon atom). These form σ bonds. The fourth electron on each carbon atom occupies an atomic p orbital. These overlap with each other to form the π bond (see Figure 3.57).

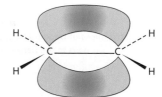

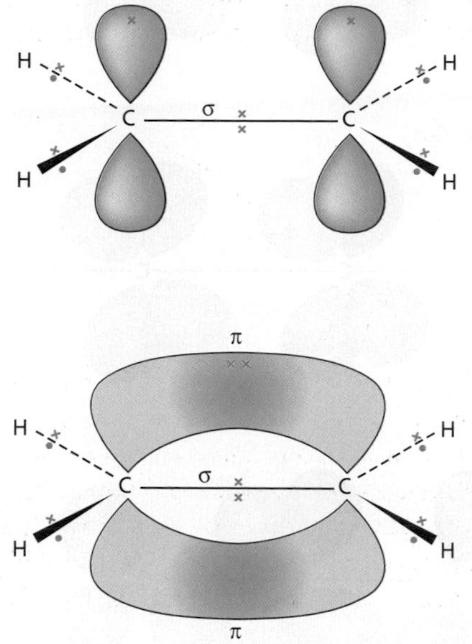

■ **Figure 3.57** The bonding in ethene.

Mutual repulsion of the electrons in the three σ bonds around each carbon atom tends to place them as far apart from each other as possible (see page 63). The predicted angles (H—C—H = H—C—C = 120°) are very close to those observed (H—C—H = 117°; H—C—C = 121.5°).

The π bond confers various properties on the ethene molecule, which are developed further in Chapter 24:

- Being the weaker of the two, the π bond is more reactive than the σ bond. Many reactions of ethene only involve the π bond, leaving the σ bond intact.
- The two areas of overlap in the π bond (both above and below the plane of the molecule) cause ethene to be a rigid molecule, with no easy rotation of one end with respect to the other (see Figure 3.58). This is in contrast to ethane, and leads to the existence of *cis–trans* isomerism in alkenes (see Chapter 22).

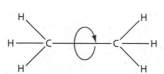

H
H
H
C
C
H
H
H
rotation possible around the single bond
H
H
C
C
H
H
no rotation possible

■ **Figure 3.58** The π bond in ethene prevents rotation around the carbon–carbon double bond.

Triple bonds

Triple bonds are formed when two p orbitals from each of the bonding atoms overlap sideways, forming two π orbitals (in additon to the σ orbital already formed by the end-on overlap of the third p orbital on each atom). This occurs in the ethyne molecule, C_2H_2, and in the nitrogen molecule, N_2 (see Figure 3.59, overleaf). A single carbon atom can also use two of its p orbitals to form π orbitals with two separate bonding atoms. This occurs in the carbon dioxide molecule (see Chapter 18).

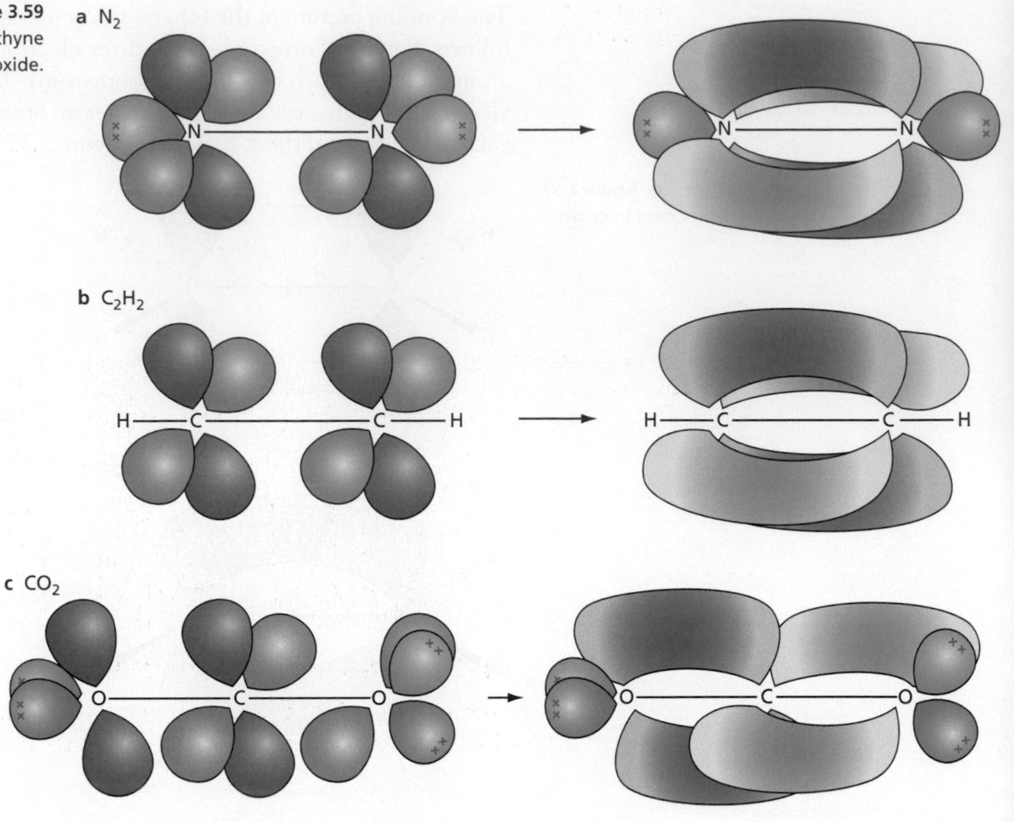

■ **Figure 3.59**
Triple bonds in **a** nitrogen **b** ethyne **c** carbon dioxide.

3.16 Delocalised π bonds

Delocalisation in benzene

Just as two p orbitals can overlap sideways to form a single localised π bond, so two (or more) π bonds can overlap with each other to produce a more extensive π bond. The classic example is benzene, C_6H_6, whose molecule contains a planar, six-membered ring of carbon atoms. A p orbital of each carbon atom can overlap as shown in Figure 3.60**a** to produce three C=C double bonds between atoms 1 and 2, 3 and 4, and 5 and 6 of the ring. Alternatively, the orbitals may overlap as in Figure 3.60**b** to produce double bonds between atoms 2 and 3, 4 and 5, and 6 and 1 of the ring. In fact, both (or neither!) occur.

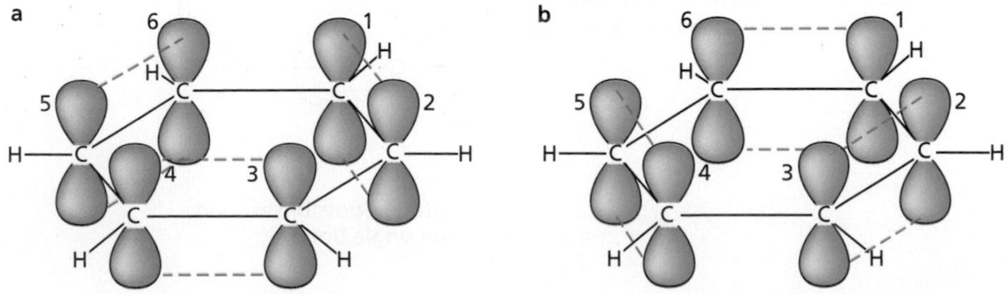

■ **Figure 3.60**
Alternative views of bonding between carbon pairs in the benzene molecule.

All six p orbitals overlap with each neighbour in an equal fashion to produce a set of two doughnut-shaped ring orbitals (Figure 3.61). This set of π orbitals can accommodate six electrons, none of which can be said to be localised between any two particular atoms. These π electrons are described as being **delocalised** around the ring. All six C—C bonds are the same length (shorter than a C—C single bond, but longer than a C=C double bond) and the shape of the ring is a regular hexagon, with every C—C—C angle being 120°. The interesting chemical properties that this ring of delocalised electrons gives benzene are discussed in Chapter 27.

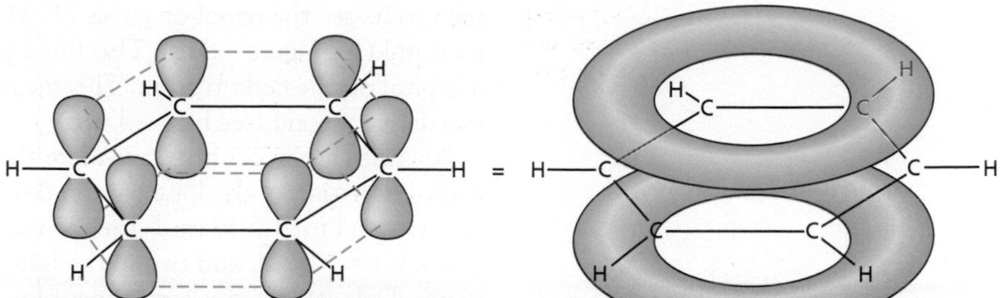

■ **Figure 3.61**
The delocalised bonding in benzene.

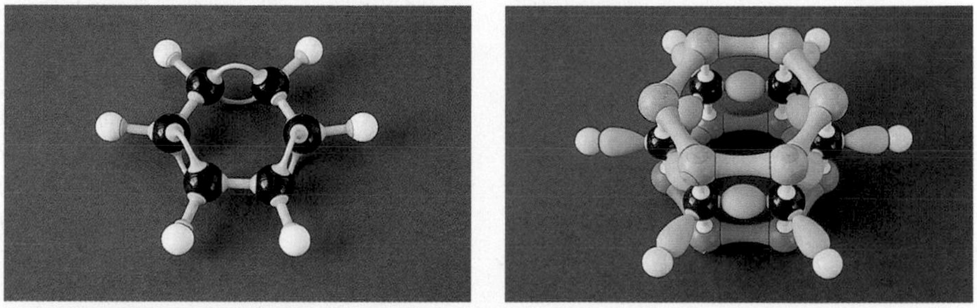

■ **Figure 3.62**
Models of benzene.

Delocalisation in the carboxylate ion

Another example of delocalisation occurs in the carboxylate ion (see Chapter 29), which is sometimes represented as follows:

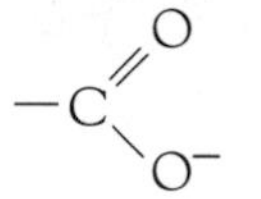

Here a lone pair of electrons in a p orbital on oxygen overlaps with the π orbital of an adjacent carbonyl (C═O) group. Because of the symmetry of the system, the delocalisation is complete, resulting in each oxygen finishing with exactly half an electronic charge, on average. Also each C—O bond has the same length, which is in between the lengths of the C—O single bond and the C═O double bond.

This structure can be derived from the simple localised model as follows. The singly bonded oxygen (which has an extra electron, making it anionic) contains three lone pairs of electrons. If we assume it is sp^2 hybridised, like the carbon atom

■ **Figure 3.63**
Delocalisation in the carboxylate ion.

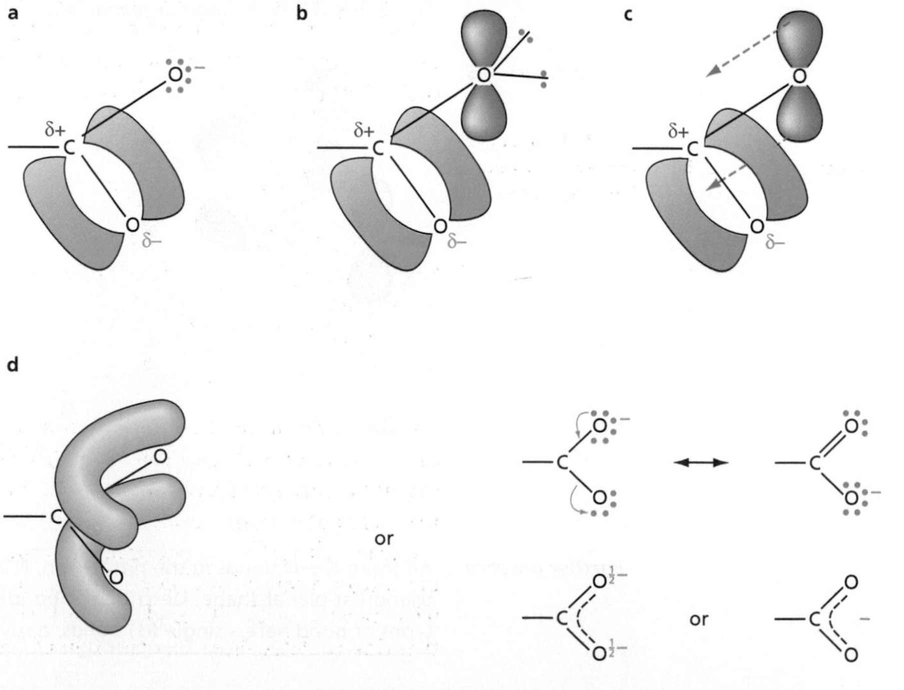

next to it (see the panel on pages 71–3), then one of the lone pairs will be in a p orbital (see Figure 3.63**b**). This filled p orbital can overlap with the end of the π orbital on the carbon atom. This means that the lone pair is partially donated, as a dative π bond (see Figure 3.63**c**).

Because carbon cannot have more than eight electrons in its valence shell, this partial donation of the lone pair on one oxygen will cause the π electrons in the C=O bond to drift towards the second oxygen atom. So the four π electrons (two in the C=O bond, and two in the lone pair) become delocalised over the three atoms. Delocalisation is sometimes represented by two (or more) localised structures joined by double-headed arrows, as in Figure 3.63**d**. The double-headed arrow should be read as 'the actual structure is somewhere between the two extremes drawn here'. The curved arrows represent the movements of electron pairs that convert one extreme into the other.

Example All three C—O bonds in the carbonate ion are the same length. Use the idea of delocalisation to describe the bonding in the CO_3^{2-} ion, and to predict its shape.

Answer If all the bonds in the carbonate ion were localised, it would consist of a carbonyl group to which are attached (by means of single σ covalent bonds) two oxygen atoms, each of which has its valence shell filled by an extra electron, giving it a negative charge (see Figure 3.64).

■ **Figure 3.64**

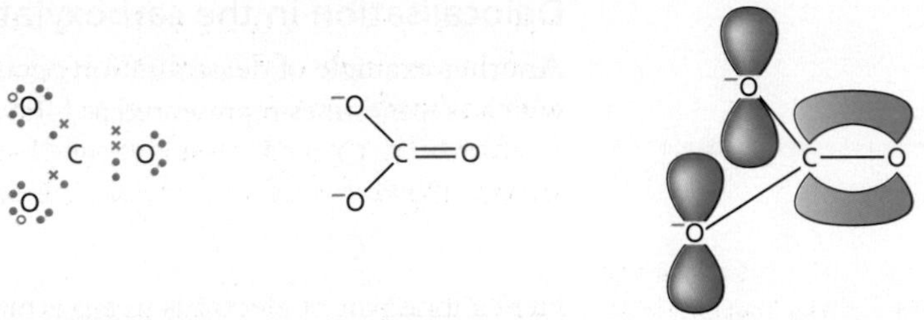

Of the three lone pairs around each of the negatively charged oxygen atoms, two occupy orbitals that are in the same plane as the C—O 'single' bond. They would be expected to be repelled by the bond and also by each other, so as to form angles of 120°. The third lone pair will remain in a p orbital. Each of these p orbitals can overlap with the π orbital of the C=O group, creating a **four-centre delocalised π orbital**. There is nothing to distinguish one C—O bond from the other two, so all three are of equal length, and each oxygen atom carries, on average, $\frac{2}{3}$ of a negative charge (overall charge $= 3 \times -\frac{2}{3} = -2$). The shape will be triangular planar, with all O—C—O bond angles 120° (see Figure 3.65).

■ **Figure 3.65**

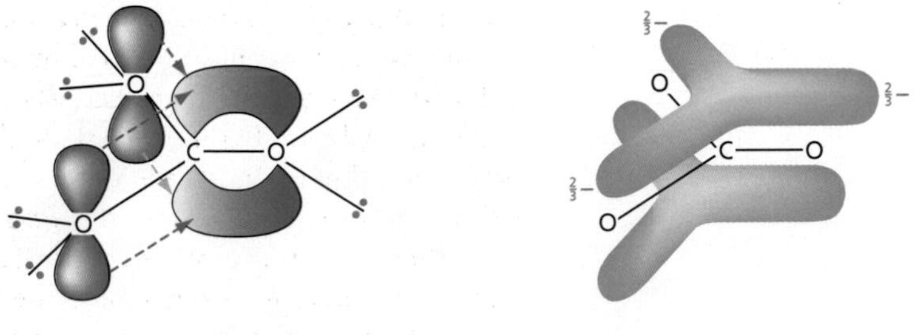

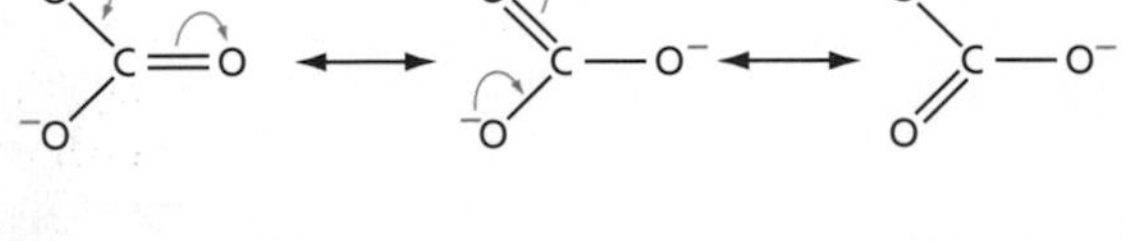

Further practice All three N—O bonds in the nitrate ion, NO_3^-, are the same length, and the ion has a triangular planar shape. Describe the bonding in the nitrate ion. (Hint: there are three types of bond here – single (σ) bonds, dative single (σ) bonds, and delocalised π bonds.)

3.17 Intermolecular forces

Most of the covalently bonded particles we have looked at in this chapter are simple covalent molecules. They differ from giant covalent, ionic and metallic lattices (see Chapter 4) by having very low melting and boiling points. They are often gases or liquids at room temperature. This is because, although the covalent bonding within each molecule is very strong, the **intermolecular attractions** between one molecule and another are comparatively weak. When a substance consisting of simple covalent molecules melts or boils, no covalent bonds within the molecules are broken. The increased thermal energy merely overcomes the weak intermolecular forces. This requires little energy, and hence only a low temperature.

Why are these molecules attracted to each other at all? There are three main categories of intermolecular attraction, which we shall look at in turn. All are electrostatic in origin, as are all forces in chemistry. Strictly speaking, they can all be grouped under the umbrella term **van der Waals forces**, although this term is now often used to describe only the instantaneous dipole force (see below).

Permanent dipole–dipole forces

As we saw on page 67, the uneven distribution of electronic charge within some molecules results in their having a permanent dipole. The positive end of one molecule's dipole can attract the negative end of another's, resulting in an intermolecular attraction in the region of 1–3 $kJ\,mol^{-1}$ (that is, at least 100 times weaker than a typical covalent bond). Such attractions are called **permanent dipole–dipole forces**, and an example is shown in Figure 3.66.

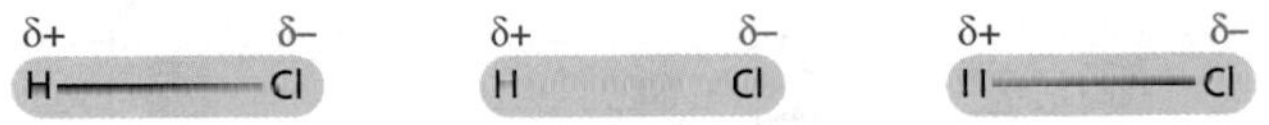

Figure 3.66 Dipole–dipole forces in hydrogen chloride.

Instantaneous dipole forces

Being much lighter, the electrons within a molecule are much more mobile than the nuclei. This constant movement produces a fluctuating dipole. If at any one instant a non-polar molecule possesses such an **instantaneous dipole** (due to there being at that moment more electrons on one side of the molecule than the other), it will induce a corresponding dipole in the molecule next door. This will result in an attraction between the molecules, called an **instantaneous dipole force** or an **induced dipole force** (also termed a **van der Waals force**). The situation is similar to a magnet picking up a steel nail by inducing a magnetic dipole moment in the nail, and hence attracting it. An instant later, the first molecule might change its dipole through another movement of electrons within it. This new dipole will still induce new dipoles in molecules that surround it, and so intermolecular attraction will continue. The strength of the attraction decreases rapidly with distance ($F \propto 1/r^6$) so it is only a very short-range force. Its magnitude depends on the number of electrons within a molecule (the more electrons, the larger the chance of an instantaneous dipole, and the larger that dipole is likely to be) and also on the molecular shape (the greater the 'area of close contact', the larger the force). Instantaneous dipoles occur in all molecules, whether or not they have a permanent dipole, and whether or not they hydrogen bond with each other (see page 80). The magnitude of the instantaneous dipole force, as mentioned above, depends on the size of the molecule, but is approximately 5–15 $kJ\,mol^{-1}$. The effect of this force is seen in the trends in boiling points shown in Table 3.8 (overleaf) – the greater the instantaneous dipole force, the higher the boiling point.

■ **Table 3.8**
Boiling points of some molecular compounds. The greater the instantaneous dipole force, the higher the boiling point.

Molecule	Boiling point/°C	Comments
CH_4	−164	These molecules are the same shape (tetrahedral), but going down the group the number of electrons increases from 10 in CH_4 to 18 in SiH_4.
SiH_4	−112	
F_2	−188	All of these are linear diatomic molecules. Down the group the number of electrons increases from 18 in F_2 to 34 in Cl_2 to 70 in Br_2.
Cl_2	−34	
Br_2	+59	
CH_4	−164	Along the alkane series the molecules are becoming longer (larger surface area of contact) and contain more electrons (an extra 8 for each CH_2 group).
C_2H_6	−89	
C_3H_8	−42	
C_4H_{10}	0	
$CH_3-C(CH_3)_2-CH_3$	+10	These two isomers have the same number of electrons, but the first is tetrahedral – almost spherical – and the second is 'sausage shaped'. The area of close contact is therefore greater in the second one.
$CH_3CH_2CH_2CH_2CH_3$	+36	

Hydrogen bonding

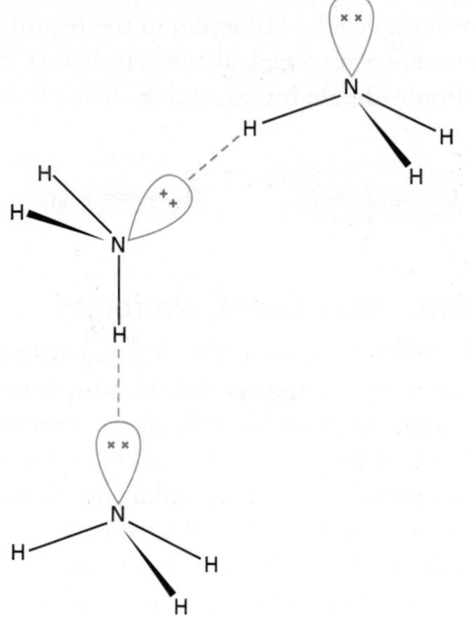

■ **Figure 3.67**
Hydrogen bonding in ammonia.

When a hydrogen atom bonds to another atom, its sole 1s electron is taken up in bonding, and resides in a σ bonding orbital between it and its bonded atom. Its other side, therefore, is devoid of electrons. It is a barren proton, with a significant δ+ charge. The positive charge is even greater if the hydrogen is bonded to an electronegative atom such as nitrogen, oxygen or fluorine. The highly δ+ hydrogen can experience a particularly strong attraction from a lone pair of electrons on an adjacent molecule, creating an intermolecular attractive force of about 20–100 kJ mol^{-1}. Such a force is called a **hydrogen bond**, and it is an order of magnitude stronger than a dipole–dipole force or an instantaneous dipole force – about 10 times weaker than a typical covalent bond. Atoms that contain lone pairs of electrons in orbitals that are small enough to interact with δ+ hydrogen atoms happen also to be nitrogen, oxygen and fluorine. Hydrogen bonding tends to occur in compounds containing these elements.

As a result of hydrogen bonding:

- the boiling points of ammonia, NH_3, water, H_2O, and hydrogen fluoride, HF, are considerably higher than that of methane, CH_4 (see Figure 3.68)
- the boiling points of ammonia, water and hydrogen fluoride are also considerably higher than those of other hydrides in their groups (see Figure 3.69)
- the boiling points of alcohols are considerably higher than those of alkanes with the same number of electrons (see Table 3.9, page 82)
- proteins fold in specific ways, giving them specific catalytic properties
- in DNA, the bases in nucleic acids bond with specific partners, allowing the genetic code to be replicated without errors of transcription (see Figure 3.70).

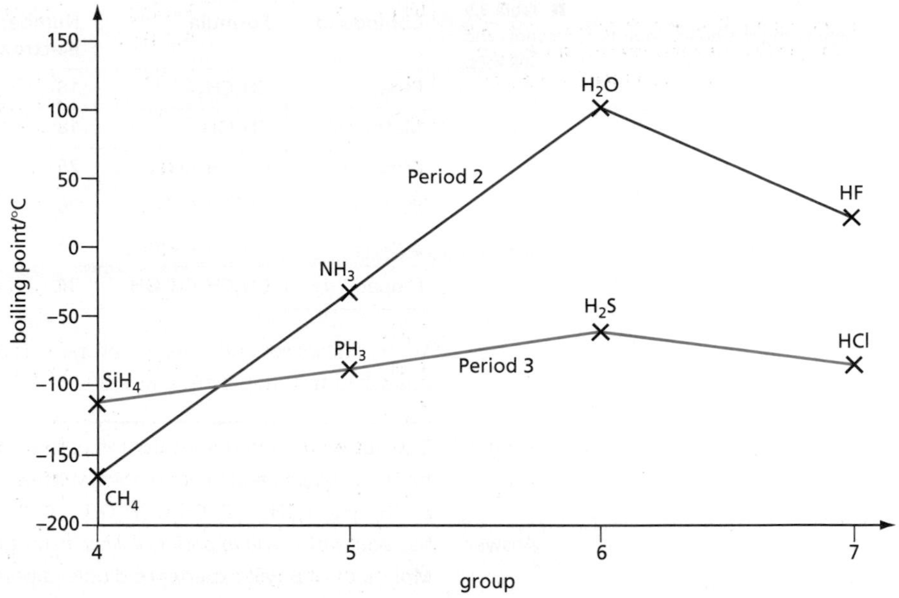

■ Figure 3.68
The enhanced boiling points of NH_3, H_2O and HF.

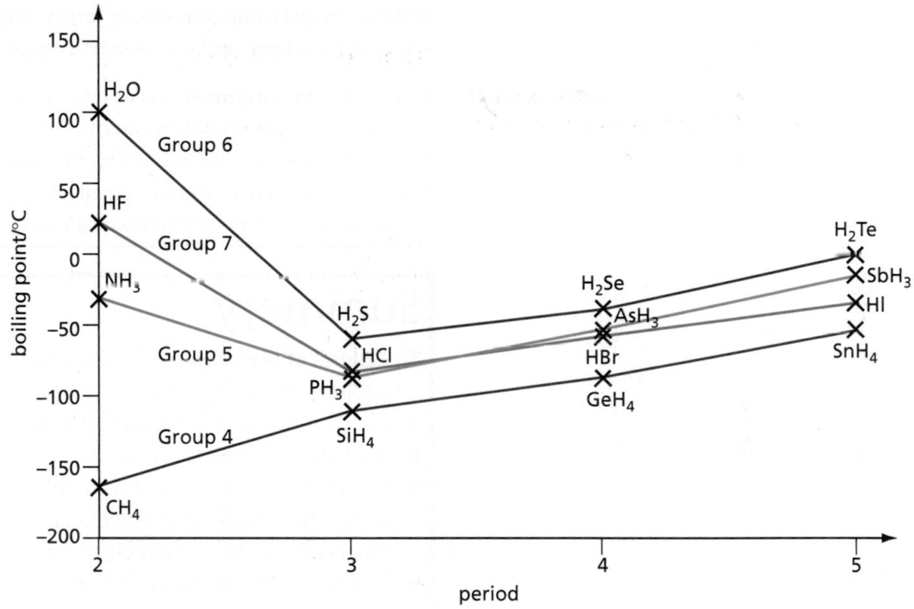

■ Figure 3.69
Boiling points of the hydrides of Groups 4 to 7.

■ Figure 3.70
Thymine, adenine, cytosine and guanine are bases in DNA. The hydrogen bonding between them allows the DNA molecule to replicate (make new copies of itself) and also allows the genetic code to be expressed in the synthesis of proteins.

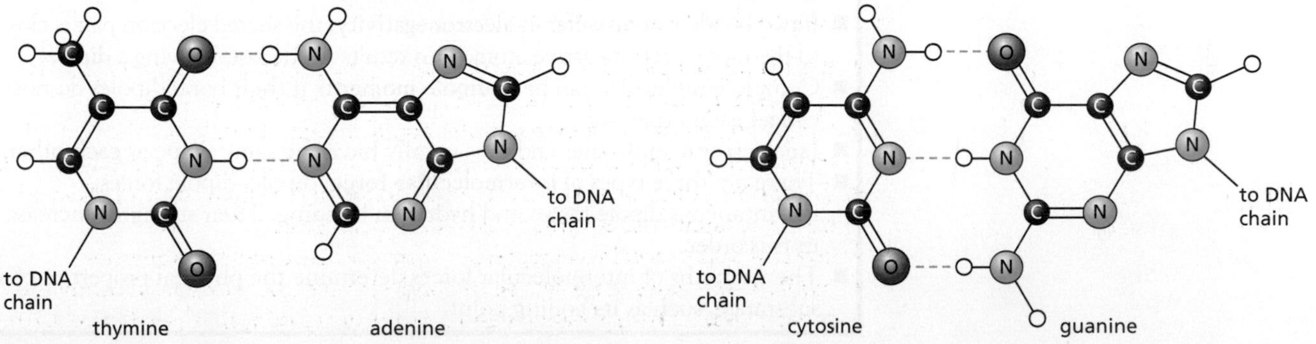

■ **Table 3.9**
Comparing the boiling points of alcohols and alkanes.

Compound	Formula	Number of electrons	Boiling point /°C	Difference /°C
Ethane	CH_3CH_3	18	−89	154
Methanol	CH_3OH	18	+65	
Propane	$CH_3CH_2CH_3$	26	−42	121
Ethanol	CH_3CH_2OH	26	+79	
Butane	$CH_3CH_2CH_2CH_3$	34	0	97
Propan-1-ol	$CH_3CH_2CH_2OH$	34	+97	

Water shows several unusual properties due to its hydrogen bonding. These are described in Chapter 4, pages 86–7.

Example Describe all the intermolecular forces that can occur between the molecules in each of the following liquids, and hence predict which will have the highest, and the lowest, boiling points:
A $CH_3CH_2CH_2CH_3$ B CH_2ClCH_2OH C CH_3CHCl_2.

Answer Molecules of A will experience only instantaneous dipole (van der Waals) forces. Molecules of B will experience dipole–dipole forces, and hydrogen bonding, as well as instantaneous dipole forces. Molecules of C will experience dipole–dipole forces in addition to instantaneous dipole forces. Therefore A will have the least intermolecular attraction, and hence the lowest boiling point, and B will have the highest boiling point.

Further practice

1 Describe the intermolecular forces in the following compounds, and place them in order of increasing boiling points: CH_2OHCH_2OH, CH_2BrCH_2Br, CH_2ClCH_2Cl.
2 Suggest a reason why the differences between the boiling points of the alkanes and alcohols in Table 3.5 become smaller as the chain length increases. (Hint: think of how the intermolecular forces other than hydrogen bonding are changing as chain length increases.)

Summary

- **Chemical bonding** results from the attraction of one atom's electrons by the nucleus of another atom.
- Atoms tend to bond together rather than be free.
- A **covalent bond** is due to two atoms sharing a pair of electrons.
- An atom can often form as many bonds to other atoms as it has electrons in its **valence shell**, with the exception that if the element is in Period 2, it cannot have more than eight electrons in its valence shell.
- Valence-shell electrons not involved in bonding arrange themselves into **lone pairs** around the atom.
- **Dative bonding** occurs when one of the bonded atoms contributes both the bonding electrons.
- Double and triple bonds can occur if two atoms share two or three electron pairs between them.
- The shape of a covalent molecule is determined by the number of electron pairs there are in the valence shell of the central atom, according to **VSEPR theory**.
- If two bonded atoms differ in **electronegativity**, the shared electron pair is closer to the more electronegative atom. This results in the bond having a **dipole**.
- Complete molecules can have **dipole moments** if their bond dipoles do not cancel each other out.
- **Isoelectronic** molecules and ions usually have the same shape as each other.
- There are three types of **intermolecular force**: dipole–dipole forces, instantaneous dipole forces and hydrogen bonding. Their strengths increase in this order.
- The strengths of intermolecular forces determine the physical properties of a substance, such as its boiling point.

4 *Solids, liquids and gases*

In Chapter 3 we saw how atoms bond together covalently to form molecules. We finished with a brief look at how the molecules themselves might attract each other. In this chapter we examine how these attractions between molecules, and between other particles, can help to explain the differences between solids, liquids and gases. We explore the giant structures that atoms and ions can form, in which it is impossible to say where one structural unit finishes and the next one starts. We then take a closer look at gases, and see how the simple assumptions of the kinetic theory of gases lead us to the ideal gas equation, $pV = nRT$. Finally we see how the behaviour of real gases can differ markedly from that of an ideal gas.

The chapter is divided into the following sections.

4.1 The three states of matter

There are three main states or **phases** of matter – solid, liquid and gas. If we consider a substance that we experience every day in all three states, such as water, we can appreciate that the solid phase (ice) occurs at cold temperatures; the liquid phase (water) occurs at intermediate temperatures; and the gaseous phase (steam) occurs at high temperatures. We can see that in order to change a solid into a liquid, or a liquid into a gas, we need to give it energy, in the form of heat (thermal energy). This energy goes into overcoming the attractions between the particles. If we supply heat to a block of ice at a constant rate, and measure its temperature continuously, we obtain a graph like the one shown in Figure 4.1.

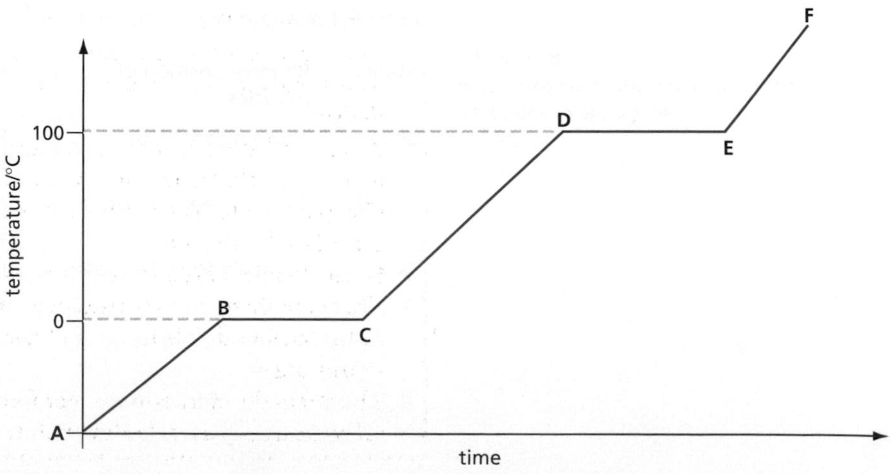

■ **Figure 4.1** The change in temperature over time as a block of ice is heated.

What is happening in the five regions of this graph can be described as follows.

- Between **A** and **B**, the ice is warming up. The molecules of water are fixed in their lattice (see page 85), but are vibrating more and more energetically from **A** to **B**.
- At **B**, the ice begins to melt. Here, the molecules are vibrating so strongly that they can begin to break away from their neighbours and move about independently.
- Between **B** and **C**, ice molecules continue to break away from the lattice. Although heat is still being absorbed by the ice–water mixture, the temperature does not rise. This is because any extra energy that the molecules in the liquid phase may pick up is soon transferred by collision to the ice lattice, where it causes another ice molecule to break away. Not until all the ice has become liquid will the temperature start to rise again.
- From **C** to **D**, the liquid water is warming up. As the temperature increases, the molecules gain more kinetic energy, and move around faster. In water there are small closely knit groups of water molecules containing 5–12 molecules in each group, and the extra energy also causes these to break up.
- At **D**, the water molecules are moving so fast that they can overcome the remaining forces (mostly hydrogen bonds) that hold them together. The water starts to boil. The molecules evaporate to become single H_2O units in the gas phase, separated by distances of many molecular diameters. Just as in the region **B–C**, the temperature remains constant until the last water molecule has boiled away.
- From **E** to **F**, the individual water molecules in the steam gain more and more kinetic energy, so they travel faster and faster. If the steam is in an enclosed volume, this will cause the pressure to increase, as the molecules hit the walls of the container with increasing speed. If, on the other hand, the pressure is allowed to remain constant, the volume of gas will expand.

Example 1 A mole of liquid water has a volume of 18 cm^3, and a mole of steam at 100 °C and room pressure has a volume of 33 dm^3. By how much has the volume increased?

Answer The volume has increased by $\frac{33 \times 10^3}{18} = 1833$ times. (Remember: 1 dm^3 = 10^3 cm^3.)

Example 2 Assuming that all the water molecules in liquid water are touching each other, roughly how far apart are they in steam?

Answer The increase in linear distance between the molecules is the cube root of the increase in volume. So in steam the molecules are $\sqrt[3]{1833} \approx 12$ molecular diameters apart.

Table 4.1 summarises the differences between solids, liquids and gases.

■ **Table 4.1** The positions and movement of the particles in solids, liquids and gases.

State	Relative position of particles	Relative movement of particles
Solid	Touching each other	Fixed in a regular three-dimensional lattice, but may vibrate about fixed positions
Liquid	Touching each other	Moving randomly, often in weakly bonded groups of several molecules at a time
Gas	Far apart from each other	Individual molecules moving randomly at high speeds, colliding with, and bouncing off, each other and the walls of their container

4.2 Lattices

For most substances, the solid and liquid phases usually have about the same density. This is because in both solids and liquids the particles are touching each other, with little space in between them. As we have just mentioned, the major difference between solids and liquids is whether or not the particles have translational motion. In liquids, the particles move fairly randomly and are continually sliding over each other. In solids, on the other hand, the particles do not move around from one place to another. They can only vibrate around fixed positions.

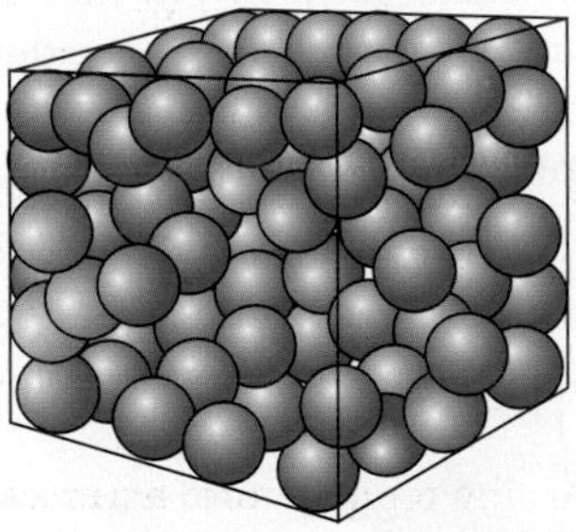

■ **Figure 4.2**
The structure of a non-crystalline (amorphous) solid. The particles are in a random arrangement.

There are two major types of solid – crystalline and non-crystalline solids. In **non-crystalline** (or **amorphous**) solids, the particles are not arranged in any order, or pattern. They are fixed in random positions. If you took a video of a liquid, showing the particles moving around chaotically, and froze a frame from that video, the picture would be indistinguishable from that of a non-crystalline solid. Many non-crystalline solids, for example glass, are often called 'supercooled liquids' (see Figure 4.2).

In **crystalline** solids, the particles are arranged in a regular three-dimensional pattern or array. This is called a **lattice** (see Figure 4.3). Every lattice can be thought of as being made up from lots of subunits or building blocks stacked one on top of the other, in rows and columns. These repeating units are called **unit cells**. Once we have understood the geometry of the unit cell, and how the atoms are joined together in it, we are well on the way to understanding what gives crystalline solids their shapes, and their other physical properties.

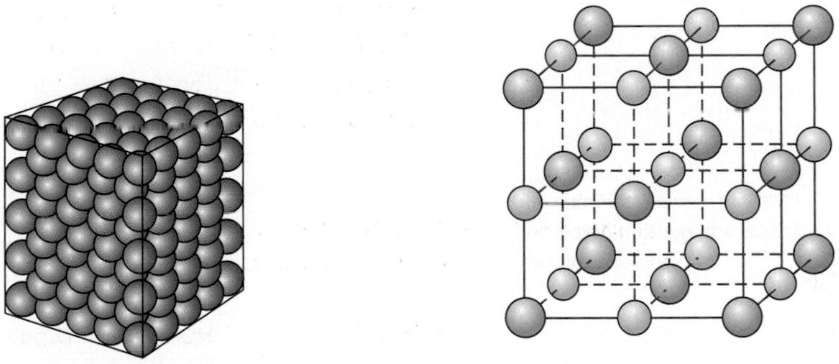

■ **Figure 4.3**
The structure of a crystalline solid, zinc (left). The particles of sodium chloride (right) are in a regular lattice arrangement.

4.3 The building blocks of lattices

The ultimate building blocks of matter are, of course, atoms. The lattices of elements are composed simply of individual atoms. We can distinguish three types of **atomic lattice**:

- **molecular atomic lattices**, such as in solid argon and the other noble gases
- **macromolecular covalent lattices**, such as in diamond and silicon
- **metallic lattices**, such as in iron and all other solid metals.

The lattice building blocks of solid compounds contain more than one element. There are once again three types:

- **simple molecular lattices**, such as in ice or in solid iodine
- **macromolecular covalent lattices**, such as in silicon(IV) oxide
- **ionic lattices**, such as in sodium chloride or calcium carbonate.

Each type of lattice confers different physical properties upon the substance that contains it. We shall explore these properties in the sections that follow.

4.4 Simple molecular lattices

In simple molecular lattices, including molecular atomic lattices, there is no chemical bonding between the particles, just dipole–dipole forces, instantaneous dipole forces or hydrogen bonding attractions, which are comparatively weak. Figure 4.4 shows examples of simple atomic and molecular lattices. The molecules or atoms are packed tightly, but only a small amount of thermal energy is required to overcome the intermolecular forces and break the lattice. Substances containing this type of lattice therefore have low melting points (and low boiling points). As we saw in Chapter 3, page 79, melting points depend on the strength of the intermolecular bonding (see Table 4.2).

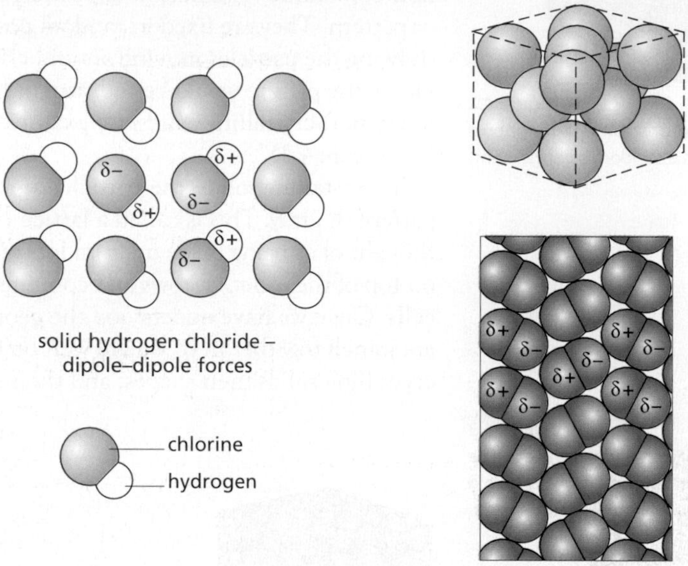

■ Figure 4.4 In solid argon, a molecular atomic lattice, the only intermolecular forces are weak instantaneous dipole forces. In solid hydrogen chloride, a simple molecular lattice, there are dipole–dipole forces. Solid iodine has strong instantaneous dipole forces.

■ Table 4.2 Intermolecular forces and melting points for some substances.

Substance	Formula	Main type of intermolecular attraction	Melting point/°C
Argon	Ar	Weak instantaneous dipole	−189
Hydrogen chloride	HCl	Dipole–dipole	−115
Water	H_2O	Hydrogen bonding	0
Iodine	I_2	Strong instantaneous dipole	114
Sucrose	$C_{12}H_{22}O_{11}$	Strong hydrogen bonding	185

The properties of water are in several ways rather different from those of other simple molecular substances. Unlike the molecules of other compounds capable of forming hydrogen bonds, water has two lone pairs of electrons together with two δ+ hydrogen atoms. This means that it can form two hydrogen bonds per molecule on average (alcohols, ammonia and hydrogen fluoride can only form one hydrogen bond per molecule, on average).

Apart from its higher-than-expected boiling point, water shows the following anomalous properties.

- Liquid water has a high surface tension. The strongly hydrogen-bonded molecules form a lattice across the surface of water, allowing objects which you might expect to sink in water, such as pond skaters and even coins (if you are careful), to 'float' on water (see Figures 4.5 and 4.6).
- Ice is less dense than liquid water. The hydrogen bonds between the water molecules in ice are positioned roughly tetrahedrally around each oxygen atom. This produces an open lattice, with empty spaces between some water molecules. The more random arrangement of hydrogen bonds in liquid water takes up less space (see Figure 4.7).

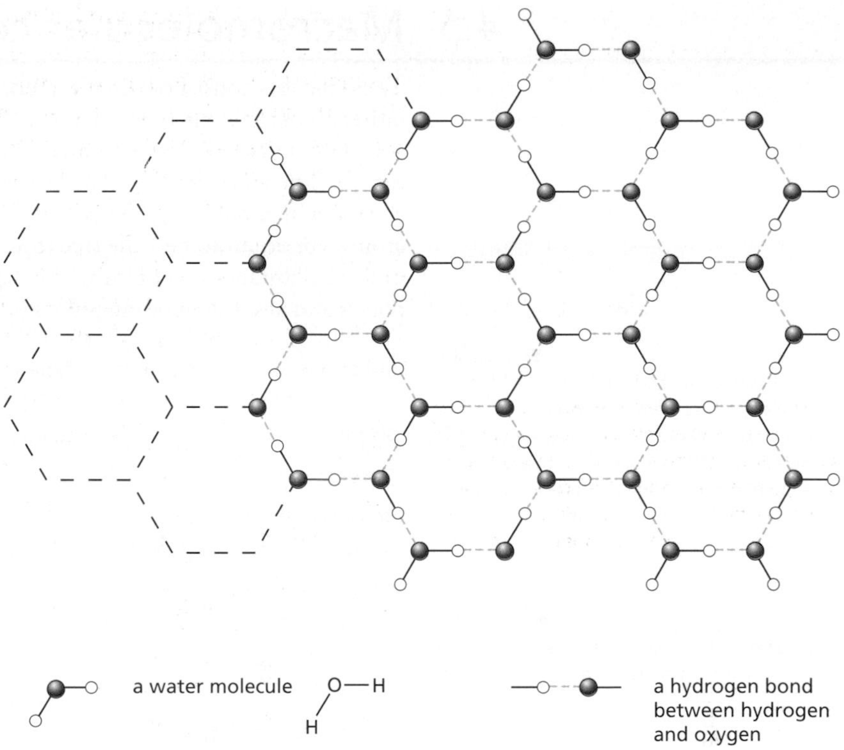

■ **Figure 4.5**
Hydrogen bonding on the surface of water forms a hexagonal array which provides a high surface tension.

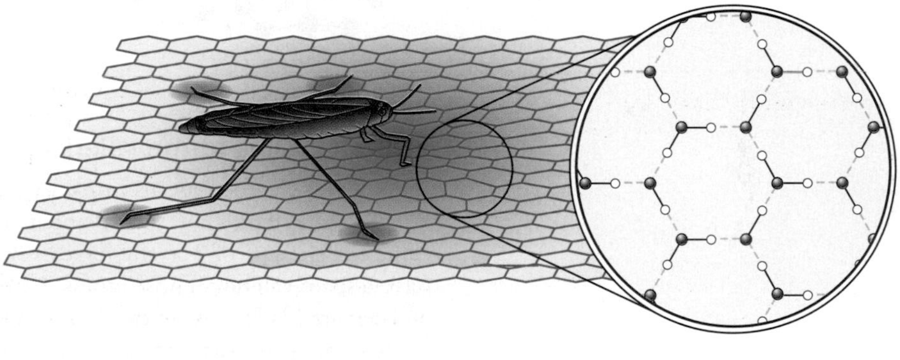

■ **Figure 4.6**
The surface tension of water supports the pond skater/water boatman – the surface is depressed but the hydrogen bonds hold it together.

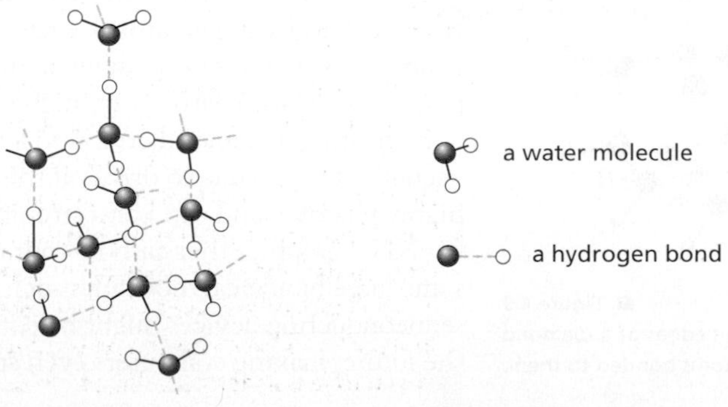

■ **Figure 4.7**
In ice, the molecules are hydrogen bonded in a tetrahedral arrangement, which makes ice less dense than liquid water.

4.5 Macromolecular covalent lattices

These lattices consist of three-dimensional arrays of atoms. These atoms can be either all of the same type, as in the elements carbon (diamond or graphite) and silicon, or of two different elements, such as in silicon(IV) oxide or boron(III) nitride. The atoms are all joined to each other by covalent bonds. So a single crystal of diamond or quartz is in fact a single molecule, containing maybe 1×10^{23} atoms. For atoms to become free from the lattice, these strong bonds have to be broken. Substances containing this type of lattice therefore have very high melting points and also high boiling points (see Table 4.3).

■ Table 4.3 Melting and boiling points are high for substances with a giant covalent lattice. (Note that boron(III) nitride **sublimes** – it changes straight from a solid to a gas when heated, so its melting and boiling points are the same.)

Substance	Formula	Type of interatomic attraction	Melting point/ °C	Boiling point/ °C
Silicon	Si	Giant covalent	1410	2355
Silicon(IV) oxide	SiO_2	Giant covalent	1610	2230
Boron(III) nitride	BN	Giant covalent	3027	3027
Diamond	C	Giant covalent	3550	4827

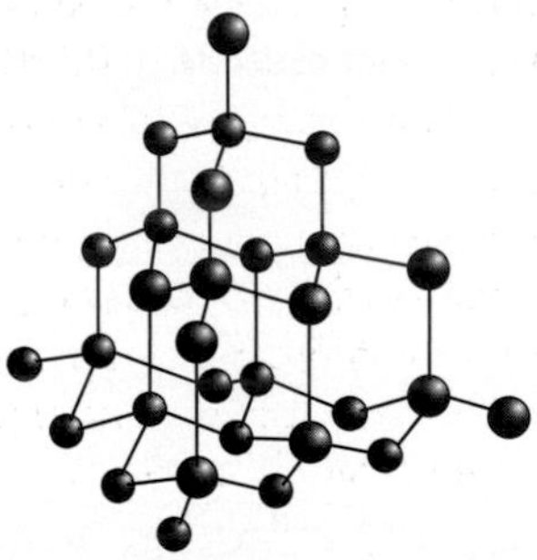

diamond lattice

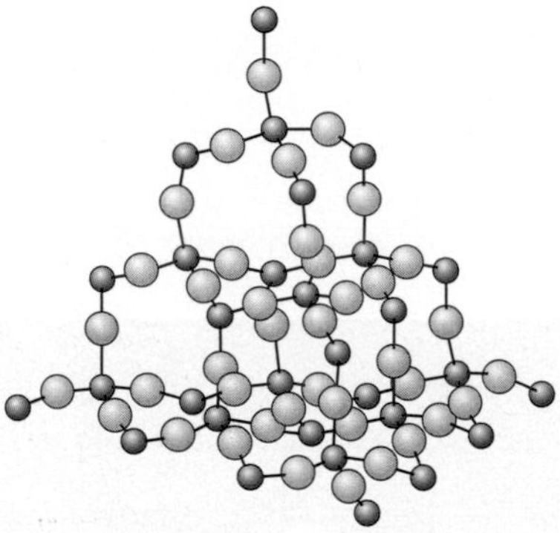

silicon(IV) oxide lattice

■ Figure 4.8 Macromolecular covalent lattices have strong covalent bonds in three dimensions.

What happens to the valencies on the edge of a diamond crystal?

Although the carbon atoms inside a diamond crystal are all joined to four other atoms, and so have their full complement of covalent bonds, those on the flat surface of the side of a crystal have only three carbon atoms joined to them. They have a spare valency. Those atoms occupying positions along a crystal edge or at an apex are likely to have two spare valencies.

How the spare valencies were 'used up' was a mystery for many years. Some thought that they joined up in pairs to form double bonds. However, this would mean that the surface carbon atoms would have to be sp^2 hybridised, rather than sp^3 hybridised as they are in the inside of the crystal. The mystery was solved by studying the surface of very clean diamond crystals using sensitive techniques such as photoelectron spectroscopy and the scanning tunnelling electron microscope.

The spare valencies are used up by bonding with atoms other than carbon. It was discovered that under normal conditions the surface of a diamond crystal is covered with hydrogen atoms, each atom singly bonded to a carbon atom. Diamond is therefore not an element in the true sense, but a hydrocarbon (a polycyclic alkane) with a very high carbon-to-hydrogen ratio!

By heating a diamond crystal to a temperature of 1000 °C in an extremely high vacuum, it is possible to drive off the hydrogen atoms. The resulting surface is highly reactive, and can bond strongly to other atoms, such as oxygen, amine ($—NH_2$) groups, other carbon-containing molecules such as alkenes and even some large biological molecules such as DNA. There is hope that new molecular semiconducting devices might be constructed on wafer-thin diamond surfaces in the future, making computers even smaller and more powerful.

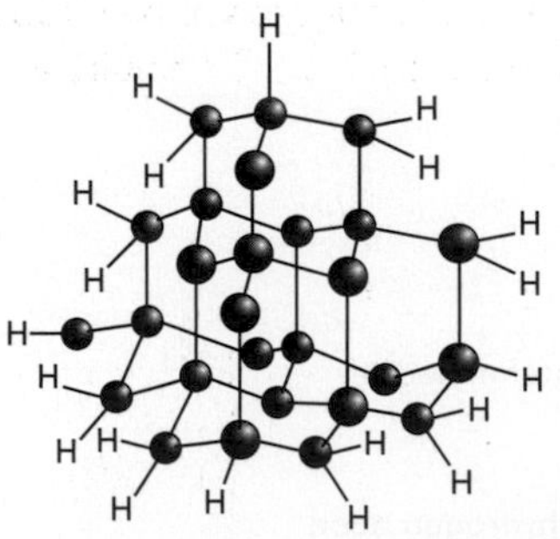

■ Figure 4.9 The carbon atoms on the edges of a diamond crystal have hydrogen atoms bonded to them.

4.6 Ionic bonding – monatomic ions

Ionic bonds are typically formed between a metallic element and a non-metallic element. Unlike covalent bonds, which involve the sharing of one or more pairs of electrons between two atoms, ionic bonding involves one atom (the metal) giving one or more electrons away totally to the non-metallic atom. This results in electrically charged atoms, called **ions**, being formed (see Figure 4.10).

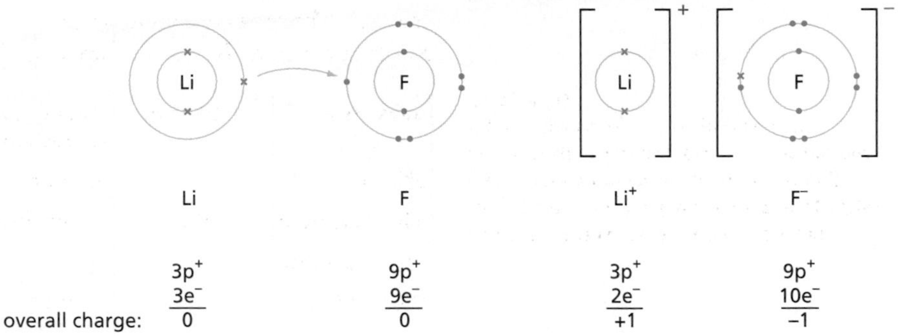

■ **Figure 4.10**
Ionic lithium fluoride is formed by the transfer of an electron.

	Li	F	Li$^+$	F$^-$
	3p$^+$	9p$^+$	3p$^+$	9p$^+$
	3e$^-$	9e$^-$	2e$^-$	10e$^-$
overall charge:	0	0	+1	−1

We shall delay taking a detailed look at the energetics of ionic bond formation until Chapter 10 (though the panel overleaf gives a brief account). It is instructive at this stage to compare the formation of an ionic bond with that of a covalent bond. We saw in Chapter 3, page 66, that bonds formed between atoms of different electronegativities are polar, with an uneven distribution of electrons, and hence of electrical charge. We can consider ionic bonding to be an extreme case – when the electronegativity difference between the two atoms is large enough, the electrons will become totally transferred to the more electronegative atom, and an ionic bond results (see Figure 4.11).

■ **Figure 4.11**
Ionic bonds are formed between atoms of high electronegativity difference.

$$: \overset{\bullet\bullet}{\underset{\bullet\bullet}{F}} : \overset{\bullet\bullet}{\underset{\bullet\bullet}{F}} : \qquad H \overset{\bullet\bullet}{\underset{\bullet\bullet}{\times}} F : \qquad Li^+ \overset{\bullet\bullet}{\underset{\bullet\bullet}{\times}} F :^-$$

electronegativities:	3.98 3.98	2.20 3.98	0.98 3.98
difference:	0	1.78	3.0

When forming ionic bonds, metals usually lose all their outer-shell electrons. Their ionic valency therefore equals their group number. Likewise, non-metals usually accept a sufficient number of electrons to fill their outer shells. Their ionic valency therefore is (8 − group number) (see Table 4.4, page 93, for a comprehensive list of ionic valencies). The resulting ions combine in the correct proportions so as to cancel their charges. The compound is therefore electrically neutral overall.

In lithium oxide, for example, each lithium ion loses the electron in its second shell, and the oxygen atom accepts two electrons (one from each of two lithium atoms) forming a full octet in its second shell (see Figure 4.12). Its empirical formula is therefore Li_2O.

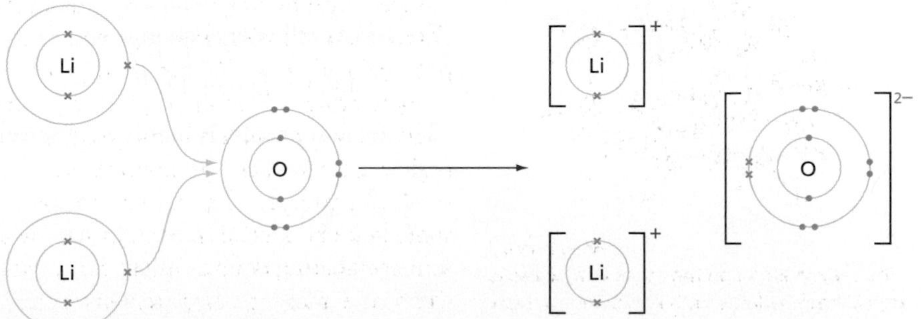

■ **Figure 4.12**
The formation of lithium oxide.

Example Draw dot-and-cross diagrams showing the electronic configuration and charges in magnesium fluoride.

Answer The 12 electrons in magnesium are arranged in the configuration 2.8.2. The Mg^{2+} ion has lost two outer-shell electrons, so has the configuration 2.8.

The nine electrons in fluorine are in the configuration 2.7. The F^- ion has gained an electron, so has the configuration 2.8. We need $2 \times F^-$ (see Figure 4.13).

■ **Figure 4.13**

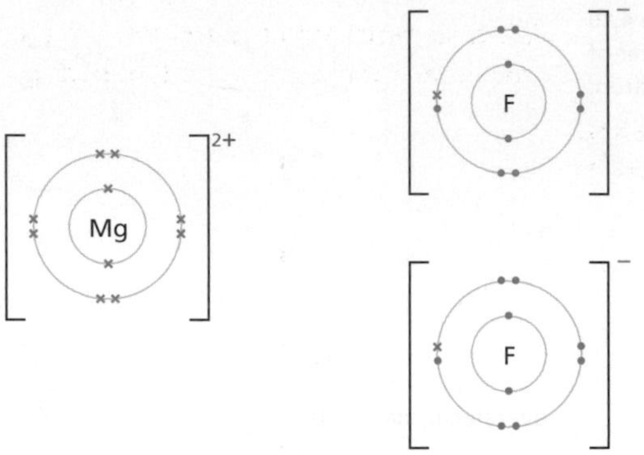

(Note the simplified representation of electronic configuration used here. Magnesium is $1s^2 2s^2 2p^6 3s^2$, shown as 2.8.2 by grouping together the electrons in each shell. Similarly, fluorine is $1s^2 2s^2 2p^5$, which becomes 2.7.)

Further practice Draw dot-and-cross diagrams with charges showing the ionic bonding in magnesium oxide, MgO, lithium nitride, Li_3N, and aluminium oxide, Al_2O_3.

Energetics of ionic bond formation – a brief summary

Ionic bonding involves a separation of charge, which is usually an unfavourable endothermic process. The reason why it becomes more favourable with metals is their comparatively low ionisation energies – the outer electron is fairly easily lost from the lithium atom, and becomes more attracted to the fluorine nucleus than to its original lithium nucleus. The benefits of total electron transfer are completed by the resulting electrostatic attraction between the cation and anion that are formed. Look carefully at the state symbols in the equations below. (For further descriptions of the terms used here, see Chapter 10.)

$$Li(g) \rightarrow Li^+(g) + e^- \qquad \Delta H^\ominus = +513\,kJ\,mol^{-1} \text{ (ionisation energy)}$$
$$F(g) + e^- \rightarrow F^-(g) \qquad \Delta H^\ominus = -328\,kJ\,mol^{-1} \text{ (electron affinity)}$$

Therefore, in the gas phase:

$$Li(g) + F(g) \rightarrow Li^+(g) + F^-(g) \qquad \Delta H^\ominus = 513 - 328 = \mathbf{+185\,kJ\,mol^{-1}}$$

But when the ions form a solid lattice, the ions attract each other, and much energy is released:

$$Li^+(g) + F^-(g) \rightarrow Li^+F^-(s) \qquad \Delta H^\ominus = -1031\,kJ\,mol^{-1} \text{ (lattice energy)}$$

So the overall energy change is:

$$Li(g) + F(g) \rightarrow Li^+F^-(s) \qquad \Delta H^\ominus = 185 - 1031 = \mathbf{-846\,kJ\,mol^{-1}}$$

The overall process is highly exothermic and therefore favourable.

4.7 Ionic bonding – compound ions

The ions that make up ionic compounds are not always monatomic (they do not always contain only one atom). Several common ions contain covalently bound groups of atoms, that have an overall positive or negative charge. Here are a few examples, some of which we came across in Chapter 3.

The ammonium ion, NH_4^+

We first mentioned the ammonium ion in Chapter 3 (pages 59 and 71). It is formed from one nitrogen atom and four hydrogen atoms, minus one electron (see Figure 4.14).

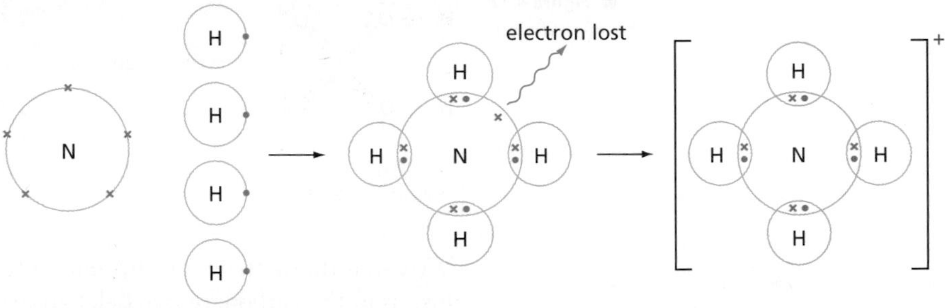

■ Figure 4.14 The formation of the ammonium ion.

An alternative way of looking at its creation is by forming a dative bond from an ammonia molecule to a proton (see page 59). Figure 4.15 shows this approach.

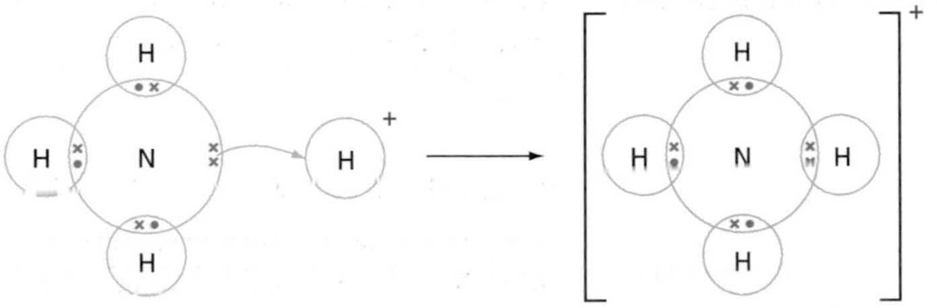

■ Figure 4.15 The formation of the ammonium ion – a different view.

Either description can be used – the first method emphasises that all four bonds are exactly the same, while the second is a better representation of how the ammonium ion is usually formed, by reacting ammonia with an acid:

$$NH_3(g) + HCl(g) \rightarrow NH_4^+Cl^-(s)$$

The carbonate ion, CO_3^{2-}

To understand the bonding in polyatomic anions, it is often easier to look first at their corresponding acids. Chemically, carbonate ions are formed by reacting carbonic acid with a base, that is, by removing hydrogen ions from the acid. If this is done, we can see that the two hydrogen atoms leave their electrons on the carbonate ion, giving it a −2 charge (see Figure 4.16).

The delocalisation that occurs in this ion was discussed in Chapter 3 (page 78).

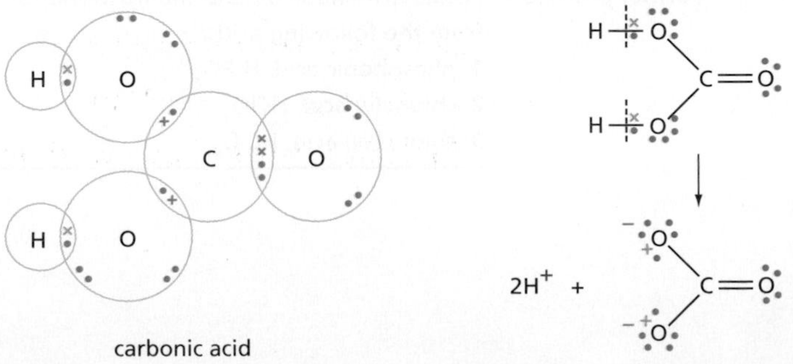

■ Figure 4.16 The carbonate ion.

The sulphate ion, SO_4^{2-}

To study the sulphate ion, let us look first at the corresponding acid. The bonding in sulphuric acid is best described as follows. The sulphur atom has six electrons in its outer shell. It can use four of these to form two double bonds to two oxygen atoms (just as sulphur dioxide, SO_2, see Chapter 3, page 65). The other two electrons in the outer shell of sulphur can form two single bonds to the other two oxygen atoms. These singly-bonded oxygen atoms then in turn form single bonds to two hydrogen atoms (see Figure 4.17). (This bonding arrangement fits in with the respective covalencies given in Table 3.1, page 58.)

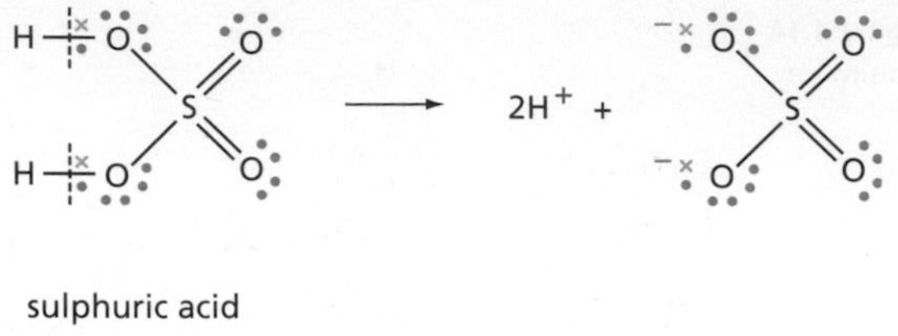

Figure 4.17 The sulphate ion.

We can form sulphate ions by removing two hydrogen ions from sulphuric acid. Just as in the carbonate ion, delocalisation occurs, making all four S—O bonds equivalent. The shape of the SO_4^{2-} ion is a regular tetrahedron (see Figure 4.18).

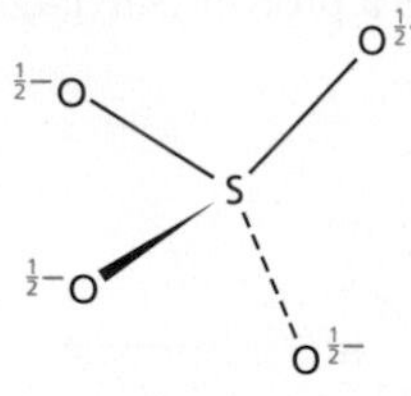

Figure 4.18 In the sulphate ion, the charge is delocalised, and the bonds are all equivalent.

Example

a Deduce the formula of the ion formed from the loss of the H^+ ions from boric acid, H_3BO_3.

b What will be the overall charge on this ion, and hence the charge on each oxygen atom in it?

c Suggest a shape for the ion.

Answer

a If all three hydrogen atoms are lost as H^+ ions, the resulting ion will be BO_3^{3-}.

b The overall charge will therefore be −3, and each oxygen will have a charge of −1.

c Boron uses all its electrons in bonding with the oxygen atoms, so the shape will be triangular planar (see Figure 4.19).

Figure 4.19

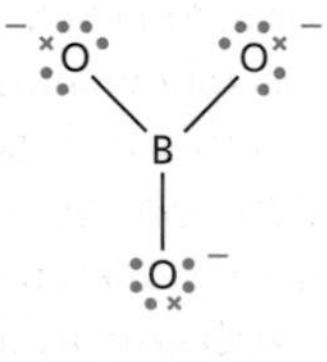

Further practice

Predict the shape of, and the fractional charge on each oxygen atom in, the ions derived from the following acids:

1 phosphonic acid, H_3PO_3
2 chloric(V) acid, $HClO_3$
3 chloric(VII) acid, $HClO_4$.

4.8 The ionic valency table

The ionic valency of a single atom from Groups 1–7 can easily be worked out from its group number. However, it is not so easy to predict the ionic valency of a transition metal, or of a compound ion. Table 4.4 includes all the valencies you will need to know.

Table 4.4
The ionic valencies of some common substances.

Cations		**Anions**	
Name	**Formula**	**Name**	**Formula**
Hydrogen	H^+	Hydride	H^-
Lithium	Li^+	Fluoride	F^-
Sodium	Na^+	Chloride	Cl^-
Potassium	K^+	Bromide	Br^-
Silver	Ag^+	Iodide	I^-
Copper(I)	Cu^+	Oxide	O^{2-}
Magnesium	Mg^{2+}	Sulphide	S^{2-}
Calcium	Ca^{2+}	Nitride	N^{3-}
Barium	Ba^{2+}	Hydroxide	OH^-
Copper(II)	Cu^{2+}	Nitrite [nitrate(III)]	NO_2^-
Zinc	Zn^{2+}	Nitrate [nitrate(V)]	NO_3^-
Iron(II)	Fe^{2+}	Hydrogencarbonate	HCO_3^-
Lead	Pb^{2+}	Hydrogensulphate	HSO_4^-
Aluminium	Al^{3+}	Carbonate	CO_3^{2-}
Iron(III)	Fe^{3+}	Sulphate [sulphate(VI)]	SO_4^{2-}
Chromium	Cr^{3+}	Sulphite [sulphate(IV)]	SO_3^{2-}
Ammonium	NH_4^+	Phosphate	PO_4^{3-}

Ionic inorganic compounds are named according to the following general rules.

1 The name of the compound is usually made up of two words, the first of which is the name of the **cation** (the positive ion), and the second is the name of the **anion** (the negative ion).
2 The **oxidation state (oxidation number)** is the formal charge on a particular element in a compound or ion. Where elements can exist in different oxidation states (see page 146), the particular oxidation state it shows in the compound in question is represented by a Roman numeral. So iron(III) represents iron in its oxidation number of +3 (Fe^{3+}), and so iron(III) chloride is $FeCl_3$. This is particularly useful when referring to compound ions – the nitrogen atom in the nitrate(III) ion, NO_2^-, has an oxidation number of +3, whereas the nitrogen in the nitrate(V) ion, NO_3^-, has an oxidation number of +5.
3 Anions containing just a single atom have names ending in '-ide'. Anions containing oxygen as well as another element have names ending in '-ate'. If there is more than one anion containing a particular element combined with oxygen, the oxidation number of that element is indicated as in rule 2.
4 For some anions, the ending '-ite' is used in the traditional name to indicate that the particular element is combined with some oxygen, but not its maximum amount of oxygen. The recommended name uses the oxidation number instead. For example:

NO_2^- is either nitrite or nitrate(III)
NO_3^- is nitrate or nitrate(V)

SO_3^{2-} is either sulphite or sulphate(IV)
SO_4^{2-} is sulphate or sulphate(VI).

Example Use the ionic valency table (Table 4.4) to predict the formula of:

a zinc nitrate

b aluminium sulphate.

Answer a Zinc ions are Zn^{2+}, nitrate ions are NO_3^-. To obtain a compound that is electrically neutral overall we need two nitrate ions to every one zinc ion:

$$Zn^{2+} + 2NO_3^- \rightarrow \mathbf{Zn(NO_3)_2}$$

b Aluminium ions are Al^{3+}, sulphate ions are SO_4^{2-}, so we need two Al^{3+} (total charge = +6) for every three SO_4^{2-} (total charge = −6).

$$2Al^{3+} + 3SO_4^{2-} \rightarrow \mathbf{Al_2(SO_4)_3}$$

Further practice

1 Use Table 4.4 to write the correct formula for:

a iron(II) fluoride
b magnesium nitride
c copper(I) oxide
d iron(III) hydroxide
e calcium phosphate
f ammonium sulphate
g copper(II) nitrate.

2 Write the names of the following ionic compounds:

a $FeSO_4$
b BaS
c $Mg(HCO_3)_2$
d KNO_2.

3 Which of the following formulae are incorrect? For each incorrect formula, write the correct one.

a AgO
b $Ba(OH)_2$
c $Pb(NO_3)_3$
d AlI_2
e $Fe_2(SO_4)_3$

4.9 Ionic lattices

Coordination number

Now we have seen how individual ions are formed, we shall turn our attention to how they collect together in a lattice. Clearly, oppositely charged ions will attract each other. But we must also remember that ions of the same charge will repel each other. We therefore expect to find that each cation in an ionic lattice will be surrounded by a number of anions as its closest neighbours, and each anion will be surrounded by a number of cations. We do not find two cations, or two anions, adjacent to each other. The number of ions that surround another of the opposite charge in an ionic lattice is called the **coordination number** of that central ion. The coordination number depends on two things – the relative sizes of the ions, and their relative charges.

The relative sizes of the ions

If one of the ions is very small, there will not be room for many oppositely charged ions around it. For Zn^{2+}, for example, the maximum coordination number is usually four. If the cations and anions are nearly equal in size, one ion can be surrounded by eight others. The intermediate case of six neighbours occurs, as might be expected, when one ion is bigger than the other, but not by very much.

The relative charges of the ions

To gain electrical neutrality, a cation with a charge of +2 needs twice as many −1 ions as does a cation of charge +1. We therefore find that the Ca^{2+} ion in calcium chloride, $CaCl_2$, is surrounded by twice as many Cl^- ions as is the Li^+ ion in lithium chloride, LiCl.

These two factors are illustrated in Table 4.5.

■ **Table 4.5**
The coordination number depends on the relative sizes and the charges on the ions involved.

Compound	Cation radius/nm	Anion radius/nm	$\frac{\text{Anion radius}}{\text{Cation radius}}$	Coordination number of cation	Coordination number of anion
ZnS	0.08	0.19	2.4	4	4
NaCl	0.10	0.18	1.8	6	6
MgO	0.07	0.14	2.0	6	6
CsCl	0.17	0.18	1.1	8	8
CaF_2	0.10	0.13	1.3	8	4

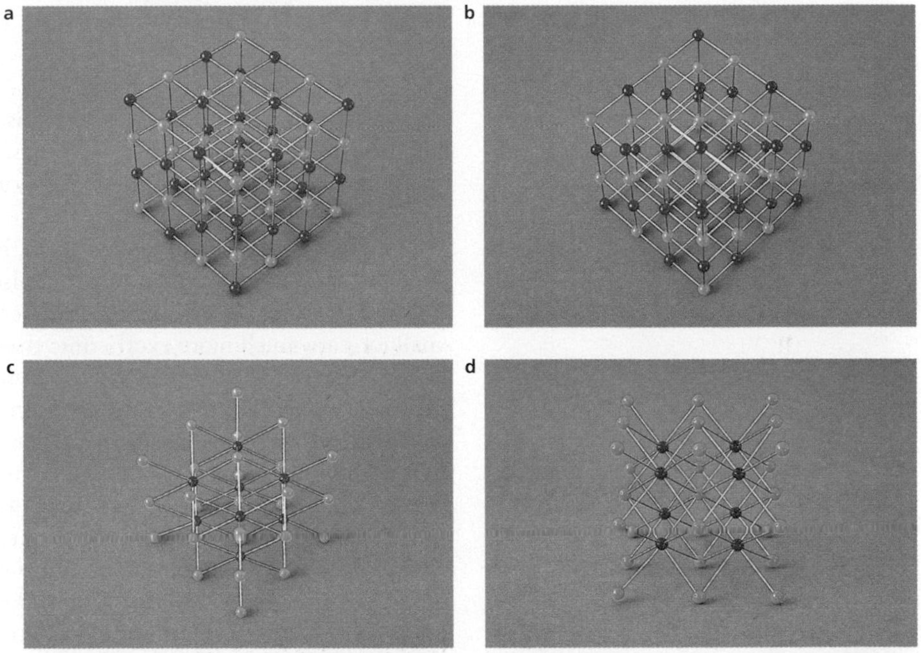

■ **Figure 4.20** Models of the lattices of sodium chloride, NaCl (**a** and **b**) and caesium chloride, CsCl (**c** and **d**); **b** and **d** show more clearly the alternating layers of ions.

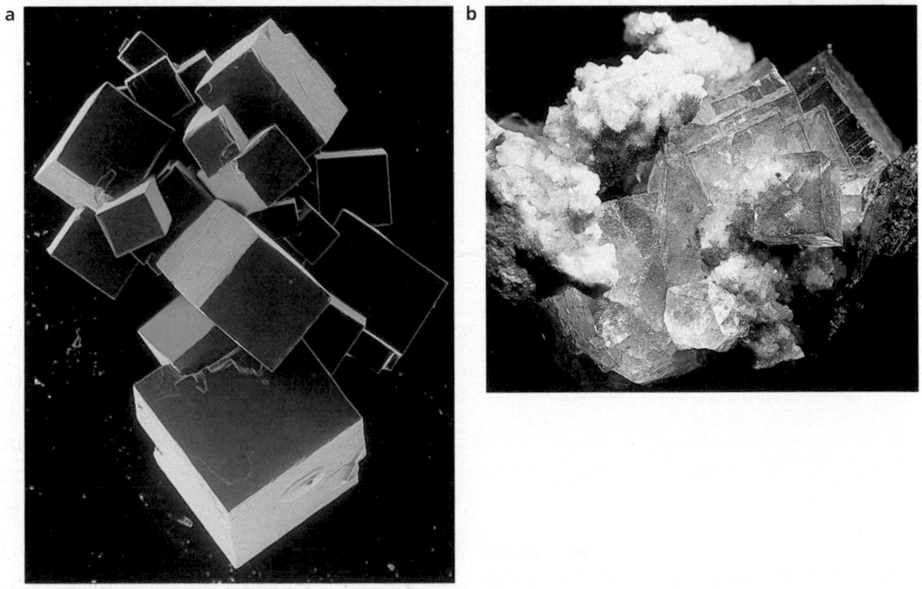

■ **Figure 4.21** Crystals of **a** sodium chloride **b** calcium chloride.

Ion polarisation and charge density

We saw in Chapter 3, page 66, that many covalent bonds involve a slight separation of charge, due to the different electronegativities of the two atoms involved in the bond. This confers a small degree of ionic character on the covalent bond.

Starting from the other extreme, we find that a small degree of covalent character is apparent in some ionic bonds. This is due to polarisation of the (negative) anion by the (positive) cation. It occurs to the greatest extent when a small or highly charged cation is bonded to a large anion. The outer electrons around the large anion are not very firmly held by its nucleus, and can be attracted by the strong electric field around the small highly charged cation. This causes an increase in electron density between the two ions, that is, a degree of localised covalent bonding (see Figure 4.22).

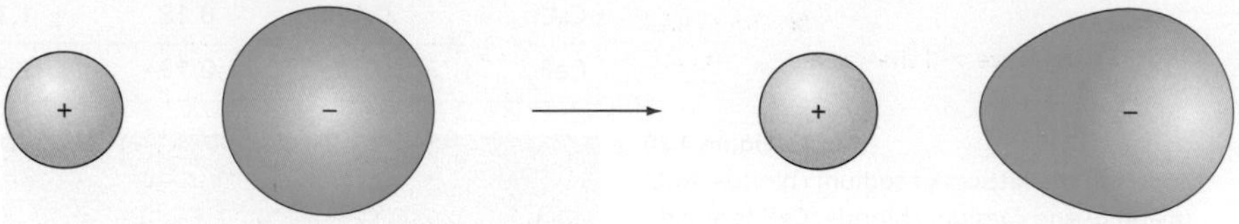

Figure 4.22 Ion polarisation between a small, highly charged cation and a large anion.

Chapter 10, page 191, illustrates how this ion polarisation affects the values of the lattice energies of some ionic compounds. The idea of ion polarisation is also important in explaining the thermal decomposition of metal carbonates and nitrates (Chapter 16, page 312).

The strong electric field around small, highly charged cations attracts not only the outer-shell electrons in anions, but also the lone pairs on polar molecules such as water. We shall see in Chapter 10 that the enthalpy changes of hydration of small ions are much more exothermic than those of larger ions.

The concept of **charge density** is a useful one for explaining the influence an ion (usually a cation) has on the electrons in adjacent ions or molecules.

Imagine two +1 ions, one with a radius of 0.1 nm and one with a radius of 0.2 nm. For each ion, the +1 charge is spread over the surface of a sphere, for which the first ion has a surface area of 0.125 nm^2 ($4\pi r^2$, with $r = 0.1$). The surface area of the second ion is 0.50 nm^2, which is four times as large as the first (area $\propto r^2$).

The **charge density** (sometimes called **surface charge density**) is the amount of charge per unit area of surface. This is $\frac{1}{0.125} = 8$ electron units per nm^2 in the first case, and $\frac{1}{0.5} = 2$ electron units per nm^2 in the second. The extent of polarisation (that is, attraction) of the electrons in adjacent anions, lone pairs and bonds depends on the charge density of the cation. Small, highly charged ions (such as Al^{3+}) have a greater charge density than large, singly charged ions (such as Cs^+). If two ions have similar charge densities, this often results in their having similar properties.

4.10 Properties of ionic compounds

Melting and boiling points

In an ionic lattice, there are many strong electrostatic attractions between oppositely charged ions. We therefore expect that ionic solids will have high melting points. On melting, although the regular lattice is broken down, there will still be significant attractions between the ions in the liquid. This should result in high boiling points also. As we shall see in Chapter 10, the ionic attractions are larger when the ions are smaller, or possess a larger charge. This is apparent from Table 4.6.

Table 4.6 Small, highly charged ions bond strongly, leading to high melting points.

Compound	Cation radius/nm	Anion radius/nm	Cation charge	Anion charge	Melting point/°C
NaCl	0.10	0.18	+1	−1	801
NaF	0.10	0.13	+1	−1	993
MgF_2	0.07	0.13	+2	−1	1261
MgO	0.07	0.14	+2	−2	2852

Strength and brittleness

If an ionic lattice is subjected to a shear or bending force, that is, a force that attempts to break up the regular array of ions, this will inevitably force ions of the same charge closer to each other (see Figure 4.23). The lattice resists this strongly. Ionic lattices are therefore quite hard and strong. If, however, the shearing force is strong enough, the lattice does not 'give', but breaks down catastrophically – the crystal shatters into tiny pieces. The strength of an ionic lattice is an 'all-or-nothing' strength.

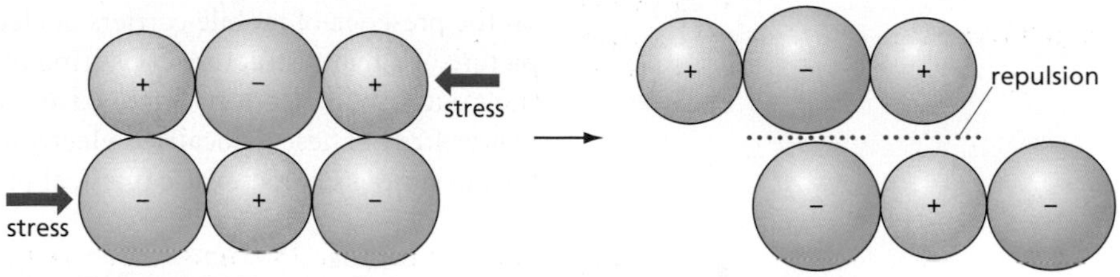

■ **Figure 4.23**
A shearing force will shatter an ionic crystal.

Electrical conductivity

The conduction of electricity is the movement of electrical charge. The individual ions that make up ionic compounds have a net overall electrical charge. If they are able to move, as they can when the compound is either molten or dissolved in water, then an electric current can flow. When in a solid lattice, however, the ions are fixed. Ionic solids do not conduct electricity.

The electrical conductivity of a solution of an ionic compound increases as its concentration increases, for there are more ions to carry the current. Similarly, for the same concentration (say $1\,mol\,dm^{-3}$), a salt solution that contains highly charged ions, such as aluminium chloride, conducts better than a salt solution with ions of a small charge, such as sodium chloride.

The conduction of electricity through metals and other solid conductors is due to the movement of electrons (see below). When moving ions in a solution give up their charge to a solid conductor placed in the solution (an **electrode**), interesting chemical reactions take place. This process, electrolysis, is described in Chapter 13.

4.11 Metallic lattices

All metals conduct electricity. They are also **malleable** (easily beaten or rolled into sheets), **ductile** (easily drawn into rods, wires and tubes) and **shiny**. These properties are due to how the atoms of the metal interact with their neighbours in the lattice.

In a pure metallic element, all atoms are identical. If we look closely at most metallic lattices we find that the atoms are arranged in a very similar way to the argon atoms in solid argon, with each atom being surrounded by 12 others (see Figure 4.4, page 86). But clearly there are greater attractions between them, for the typical melting point of a metal is at least 1000 °C higher than that of argon (see Table 4.7).

■ **Table 4.7**
Metallic bonding is much stronger than the instantaneous dipole forces in argon, but not as strong as bonding in the giant covalent lattice of carbon.

Element	Melting point/°C
Argon	−189
Silver	962
Copper	1083
Iron	1535
Carbon	3550

On the other hand, an element whose atoms are strongly bonded in a macromolecular covalent lattice, such as carbon, has a melting point 2000 °C higher than the typical metal. So metallic bonding is strong, but not as strong as some covalent bonding.

A clue to the nature of metallic bonding comes from looking at the atomic properties of metallic elements – they all have only one, two or three electrons in their outer shells, and have low ionisation energies. The bulk property that is most characteristic of metals, their electrical conductivity, often increases as the number of ionisable outer-shell electrons increases. Electrical conductivity itself depends on the presence of mobile carriers of electric charge. We can therefore build up a picture of a metal structure consisting of an array of atoms, with at least some of their outer-shell electrons removed and free to move throughout the lattice (see Figure 4.24). These **delocalised electrons** (see Chapter 3, page 76) are responsible for the various characteristic physical properties of metals, as detailed below.

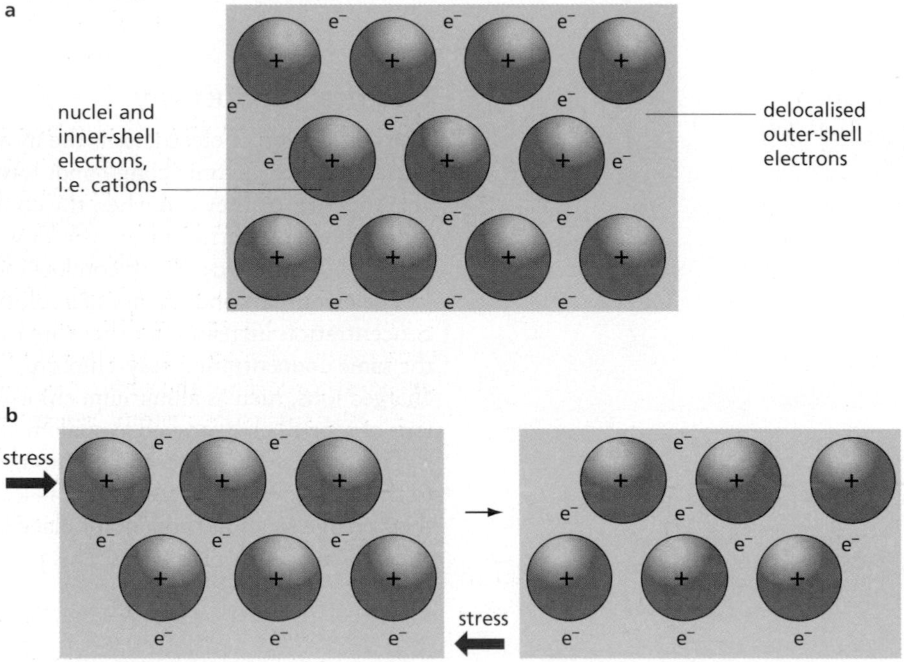

■ **Figure 4.24** The structure of a metallic lattice. When a shearing force is applied, adjacent layers of cations can slide over one another, and the metal lattice can be deformed without shattering.

- After some electrons have been removed from the metal atoms, the atoms are left as positive ions. The attraction between the delocalised electrons and these positive ions is responsible for the strengths of the metallic lattices, and for their fairly high melting points.
- Electrons are very small, and move fast. When a potential difference is applied across the ends of a metallic conductor, the delocalised electrons will be attracted to, and move towards, the positive end. This movement of electrical charge is what we know as an **electric current**.
- The partially ionised atoms in a metal lattice are all positively charged. But they are shielded from each other's repulsions by the 'sea' of delocalised electrons in between them. This shielding is still present no matter how the lattice is distorted. As a result of this, and unlike ionic lattices, metal lattices can be deformed by bending and shearing forces without shattering (see Figure 4.24**b**). Metals are therefore malleable and ductile.
- The shininess and high reflectivity of the surfaces of metals is also a property associated with the delocalised electrons they contain. As a photon of light hits the surface of a metal, its oscillating electric field causes the electrons on the metal's surface to oscillate too. This allows the photon to bounce off the surface without any loss of momentum.

4.12 Graphite and the allotropy of carbon

On page 88 we looked at the macromolecular covalent structure of diamond. Carbon's other common allotrope, graphite, is very different in its properties (see Table 4.8). Carbon also forms fullerenes (see page 101).

Allotropes are two (or more) forms of the same element, in which the atoms or molecules are arranged in different ways.

■ **Table 4.8**
The properties of diamond and graphite.

Property	Diamond	Graphite
Colour	Colourless and transparent	Black and opaque
Hardness	Very hard – the hardest naturally occuring solid	Very soft and slippery – over 500 times softer than diamond
Electrical conductivity	Very poor – a good insulator Resistivity ≈ $1 \times 10^{15}\,\Omega$ cm	Very good along the layers Resistivity ≈ $1 \times 10^{-4}\,\Omega$ cm
Density	3.51 g cm^{-3}	2.27 g cm^{-3}

These differences in properties are due to a major difference in the bonding between the carbon atoms in the two allotropes. In diamond, each carbon atom is tetrahedrally bonded to four others by single, localised covalent bonds. A three-dimensional network results. In graphite, on the other hand, each carbon atom is bonded to only three others (using planar sp^2 orbitals, see the panel on page 72). A two-dimensional sheet of carbon atoms is formed. Each carbon atom has a spare 2p orbital, containing one electron. These can overlap with each other (just as in benzene, see Chapter 3, page 76) to form a two-dimensional delocalised π orbital spreading throughout the whole sheet of atoms. A graphite crystal is composed of many sheets or layers of atoms, stacked one on top of another (see Figure 4.25). Each sheet should be thought of as a single molecule. There is no bonding between the one sheet and the next, but the instantaneous dipole attraction between them is quite substantial, because of the large surface area involved.

■ **Figure 4.25**
The structure of graphite.

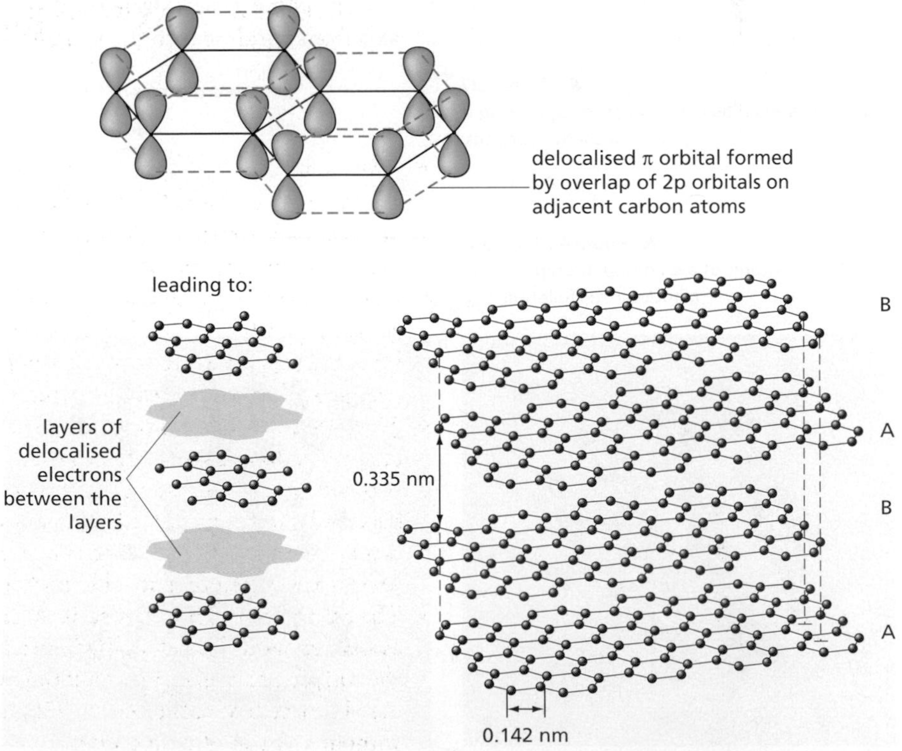

Due to its layered nature, graphite is an **anisotropic** material – its properties are not the same in all directions. These properties of graphite can be clearly related to its structure, and result in a variety of uses, outlined below.

- The use of graphite as an electrical conductor – whether as electrodes for electrolysis or as brushes in electric motors – is well known. Due to its delocalised electrons, graphite is a good conductor along the layers. But because electrons keep to their own layer, and cannot jump from one layer to the next, graphite is a poor conductor in the direction at right angles to the layers. The same is true of the conduction of heat. Graphite therefore finds a use in large crucibles, where the conduction of heat along the walls needs to be encouraged, but the transference of heat directly through the bottom is not wanted (see Figure 4.26).

■ **Figure 4.26**
Graphite conducts heat along its layers.

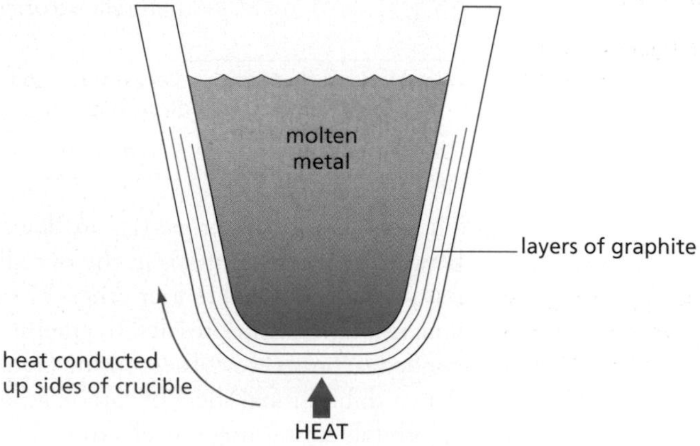

- The weak bonding between the layers allows them to slide over each other easily. One of the first uses of graphite (named from the Greek *grapho*, to draw) made use of this property – layers of carbon atoms could easily be rubbed off a lump of graphite, leaving a black trace on paper or parchment. Apart from its continued use in pencils, graphite is also used today as a component of lubricating greases. The easy sliding of the layers over each other combines with a strong compression strength across the layers. This prevents a weight-bearing axle from grinding into the surface of its bearing (see Figure 4.27).

■ **Figure 4.27**
Graphite layers slide over one another, making it a useful lubricant.

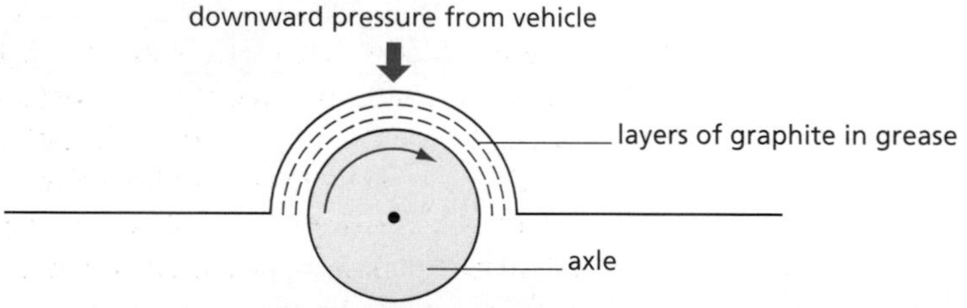

■ **Figure 4.28** (below)
The structures of **a** diamond, **b** graphite and **c** buckminsterfullerine, C_{60}.

a

b

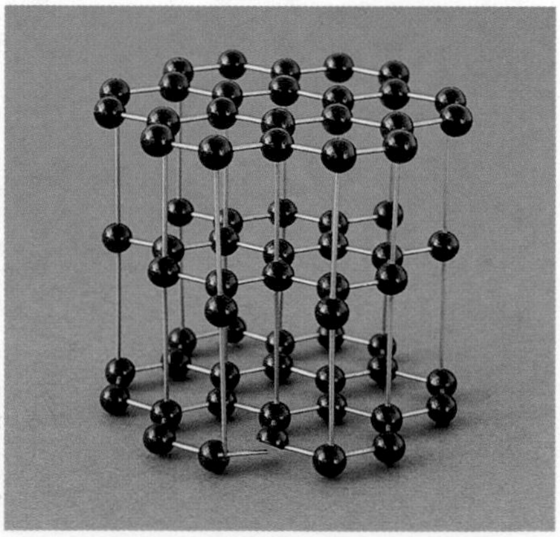

c

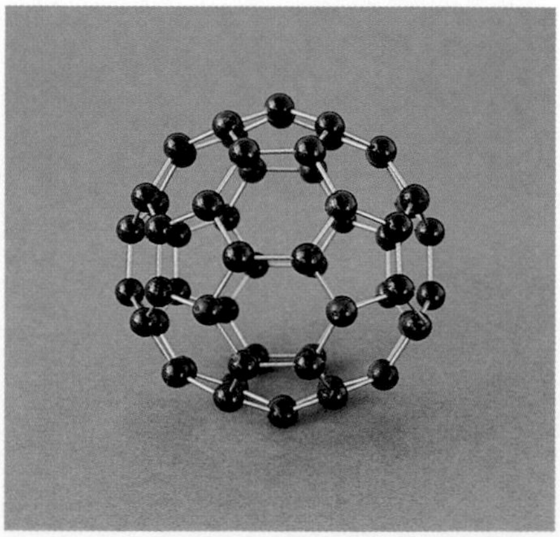

The fullerenes

Apart from the two macromolecular allotropes of carbon, a whole series of simple molecular allotropes are now known. They are the **fullerenes**. The first of these to be discovered was C_{60}. It was obtained by firing a powerful laser at a sample of graphite at a temperature of 10000 °C. It was named **buckminsterfullerene** in honour of the architect R. Buckminster Fuller, who used the principle of the geodesic dome in many of his buildings. (The alternating 5- and 6-membered rings in C_{60} give a bonding pattern similar to the struts in a geodesic dome.) Other fullerenes known have the formulae C_{20}, C_{70}, C_{76}, C_{78}, C_{84}, C_{90} and C_{94}. Many more will be synthesised, to discover whether they will open up exciting new areas of chemistry.

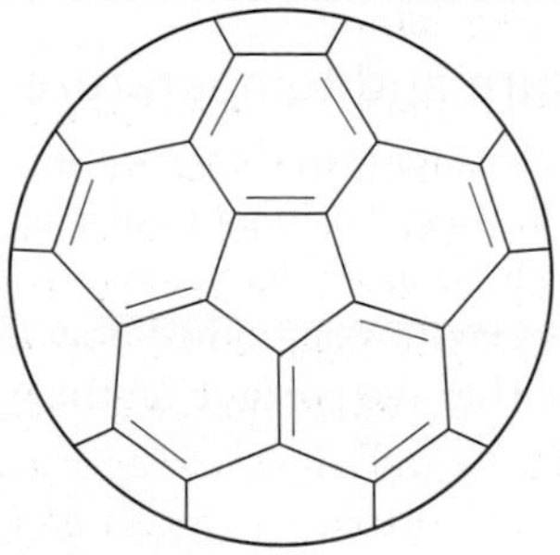

■ **Figure 4.29** (above) The structure of buckminsterfullerene.

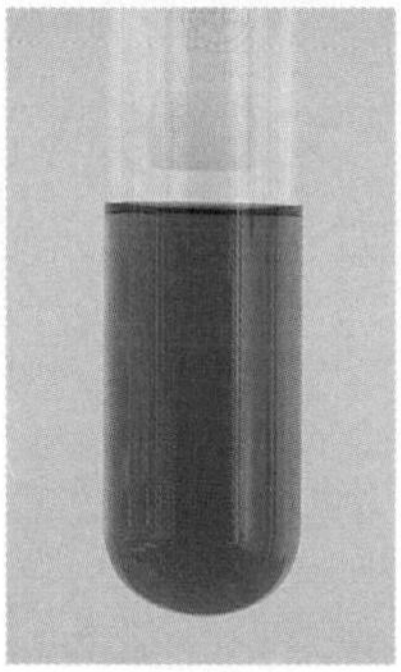

■ **Figure 4.30** (left) Buckminsterfullerene solution.

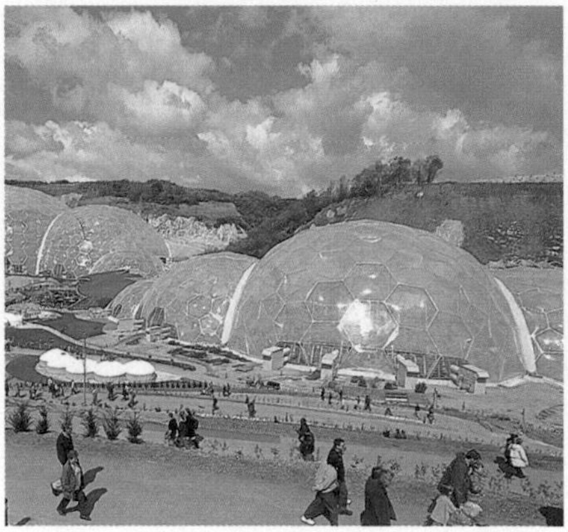

■ **Figure 4.31** (right) Geodesic domes at the Eden Project in Cornwall.

C_{60} itself is a highly symmetrical spherical football-shaped molecule (see Figures 4.28c and 4.29). Most of the higher fullerenes are derived from this basic shape by inserting rings of carbon atoms around the centre. Eventually, nanometre-sized tubes of carbon atoms will be formed. Since the surface of a fullerene is covered by a cloud of delocalised electrons (as in graphite), these nanotubes could have applications as small-scale conductors (see the panel below).

One-dimensional nanoconductors

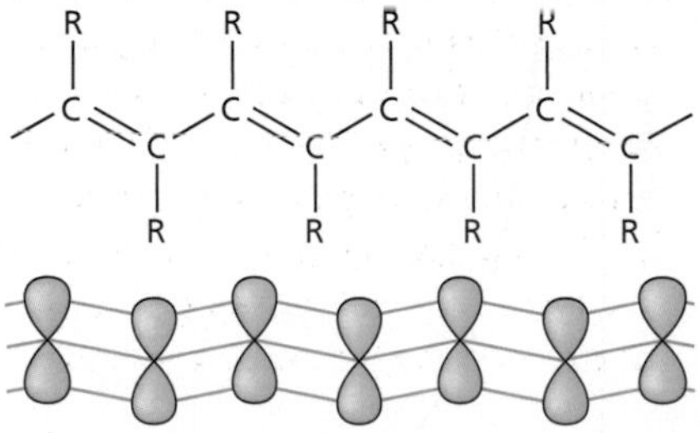

■ **Figure 4.32** The structure of a 'molecular wire'.

There is much research interest today in **nanotechnology**. Molecules are between 0.1 nm (nanometre) and 10 nm in size, and many chemists are interested in constructing molecules that mimic more conventional structures, but on a scale that is a million times smaller. Included in this group are 'molecular wires'. These are long-chain polymers which contain **conjugated double bonds**. This means that the π bonds on adjacent alkene units overlap, and the π electrons become delocalised throughout the whole length of the polymer chain (see Figure 4.32). Just as with the three-dimensional delocalised electrons in metals, or the two-dimensional delocalised electrons in the sheets of graphite, these one-dimensional delocalised electrons can conduct a current when a potential difference is applied to the ends of the polymer. An eventual aim is to construct micro-miniature circuits that can lead electrical signals to 'molecular microchips' in the heart of computing devices that will be thousands of times smaller than today's silicon-based units.

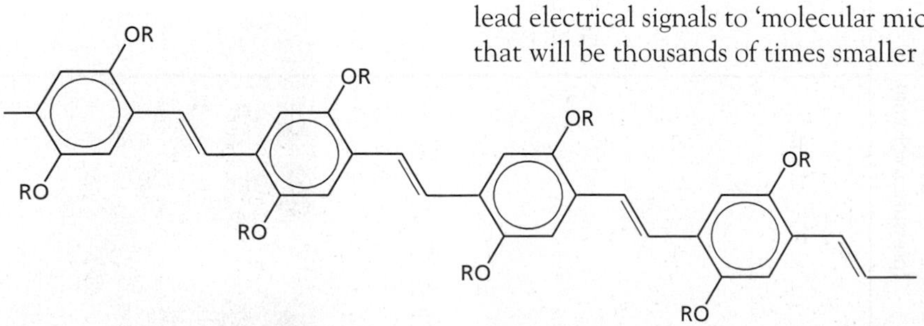

■ **Figure 4.33** The structure of an OP-LED.

An application that is nearly 'in the shops' is the use of similar one-dimensional molecular conductors to create organic polymer light-emitting diodes (OP-LEDs). Here the basic building blocks are ethene and benzene units that alternate along the chain (see Figure 4.33). By altering the nature of the side groups R, the molecules can be made to electroluminesce (that is, emit light when an electrode potential is applied) in any part of the spectrum, from red, through yellow and green, to blue.

4.13 The ideal gas equation

Experiments on gases – volume, pressure and temperature

Some of the first quantitative investigations in chemistry, in the seventeenth and eighteenth centuries, were into the behaviour of gases. Air in particular was readily available, and its volume was easily and accurately measurable. Scientists including Robert Boyle, Jacques Charles and Joseph-Louis Gay-Lussac studied how the volume of a fixed amount of air changed when either the pressure on the gas, or the temperature of the gas, was altered.

Their results are summarised below.

Boyle's Law

The volume V of a fixed mass of gas at a constant temperature is inversely proportional to the pressure p on the gas:

$$V \propto \frac{1}{p} \quad \text{or} \quad pV = \text{constant}$$

The graphs in Figure 4.34 show three ways of plotting the Boyle's Law relationship between the volume and the pressure of a gas.

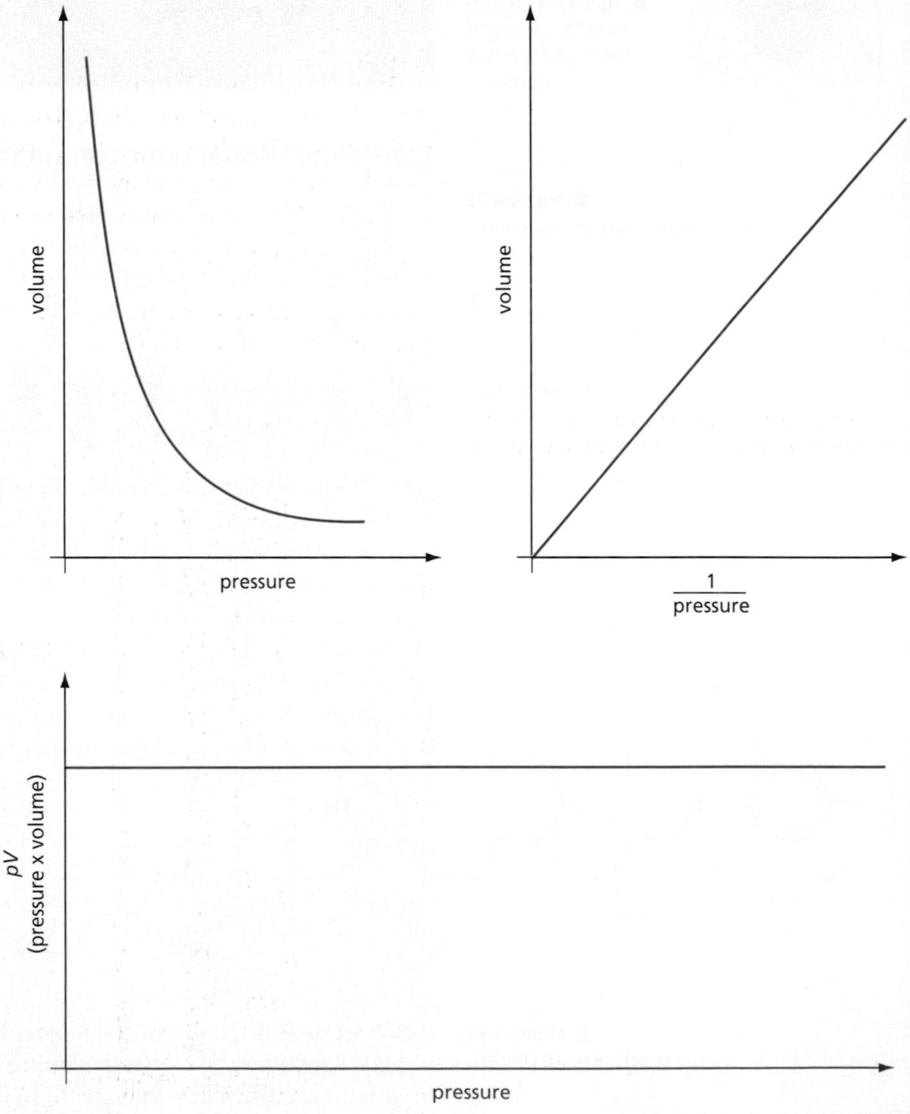

■ **Figure 4.34** These three graphs all show the Boyle's Law relationship.

Charles's Law (also called **Gay-Lussac's Law**)

The volume V of a fixed mass of gas at a constant pressure is directly proportional to its temperature T:

$$V \propto T \quad \text{or} \quad \frac{V}{T} = \text{constant}$$

The units of pressure

The **pressure** of a gas is the force it exerts on a given area of its container. The SI unit of pressure is the **pascal**, Pa, which results from a force of one newton acting on an area of one square metre.

$$1\,\text{Pa} = 1\,\text{N}\,\text{m}^{-2}$$

On this scale, the pressure the atmosphere exerts on the surface of the Earth at sea level is about 1×10^5 Pa. The atmospheric pressure at a particular point on the Earth's surface, however, changes with the weather. It also decreases with height above sea level. The variation with weather can be 10% or so, ranging from 0.95×10^5 Pa during a depression to 1.05×10^5 Pa on a fine, 'high pressure' day.

Scientists have defined **standard atmosphere** as a pressure of 1.032×10^5 Pa. For most purposes, however, the approximation:

$$1\,\text{atm} = 1 \times 10^5\,\text{Pa}$$

is adequate.

Boyle would have measured his pressures with a mercury barometer (see Figure 4.35). The pressure of the atmosphere can support a column of mercury about 0.75 m high. 'Standard atmosphere' was originally defined as the pressure that would support a column of mercury exactly 760 mm high. This is why the conversion factor from atmospheres to pascals is not an exact power of ten.

Both atmospheres and pascals are used today as pressure units. The former is more often used by chemical engineers, whereas laboratory scientists usually use pascals. Two other scales are in common use: car tyre pressures are often measured in pounds per square inch (lb in^{-2}), or in kilograms per square metre (kg m^{-2}).

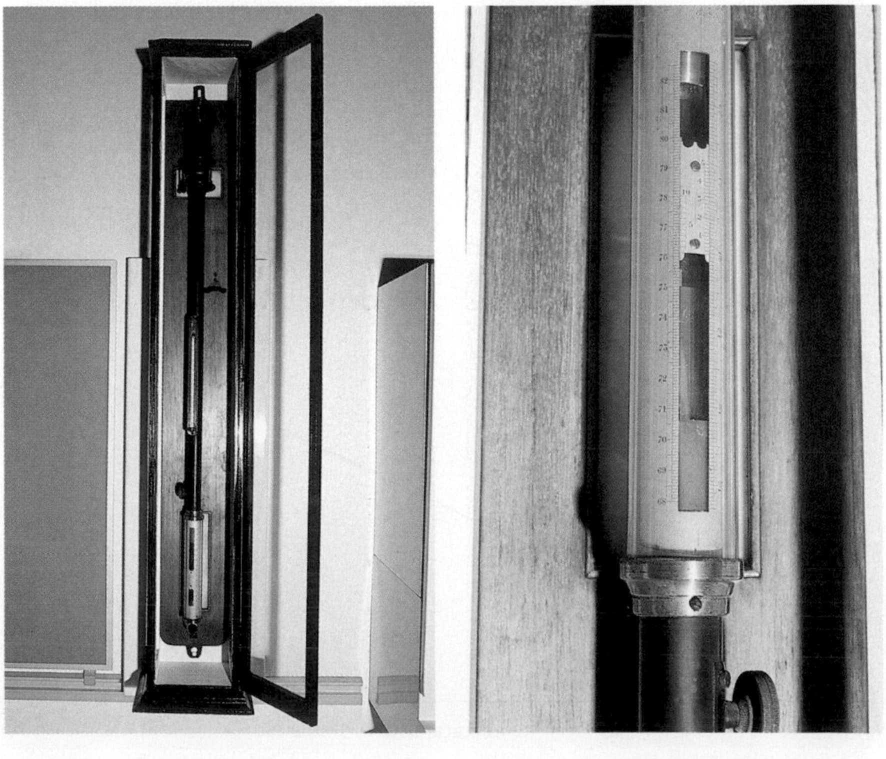

Figure 4.35 A simple barometer (left) and the vernier scale used to read the height to the nearest 0.1 mm.

Absolute zero and the kelvin temperature scale

When the volume of a gas is plotted against its temperature using the Celsius temperature scale, a straight line of positive slope is obtained. If this line is extrapolated back to the point where it crosses the temperature axis, we find the temperature at which the volume of gas would be zero. Accurate measurements show that this point occurs at −273.15 °C (see Figure 4.36). The same temperature is found no matter what volume of gas is used, or at what pressure the experiment is carried out. What is more, the same extrapolated temperature is found no matter what gas we use.

■ Figure 4.36
This graph illustrates Charles's Law and shows the theoretical derivation of absolute zero.

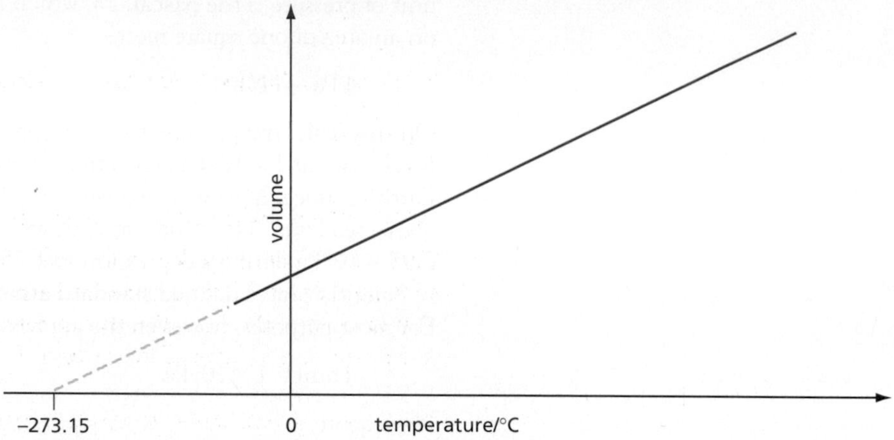

This is a universal and fundamental point on the temperature scale. Below this temperature, Charles's Law predicts that gases would have negative volumes. That is clearly impossible. So presumably it is impossible to attain temperatures lower than −273.15°C. This point is known as the **absolute zero** of temperature, and experiments in many other branches of chemistry arrive at the same conclusion, and the same value for the absolute zero of temperature. Not only is it impossible to attain temperatures lower than absolute zero, it is even impossible to equal it.

Chemists use a temperature scale which starts at absolute zero. As mentioned in Chapter 2 (page 31), it is called the **absolute temperature scale**, and its unit is the kelvin (K). On this scale, each degree is the same size as a degree on the Celsius scale, so the conversion between the two is easy:

$$\text{temperature/K} = \text{temperature/°C} + 273.15$$

or, more usually:

$$\text{temperature/K} = \text{temperature/°C} + 273$$

Figure 4.37 shows the Charles's Law relationship of volume plotted against temperature in both °C and K.

■ Figure 4.37
The relationship between the absolute temperature scale and the Celsius scale.

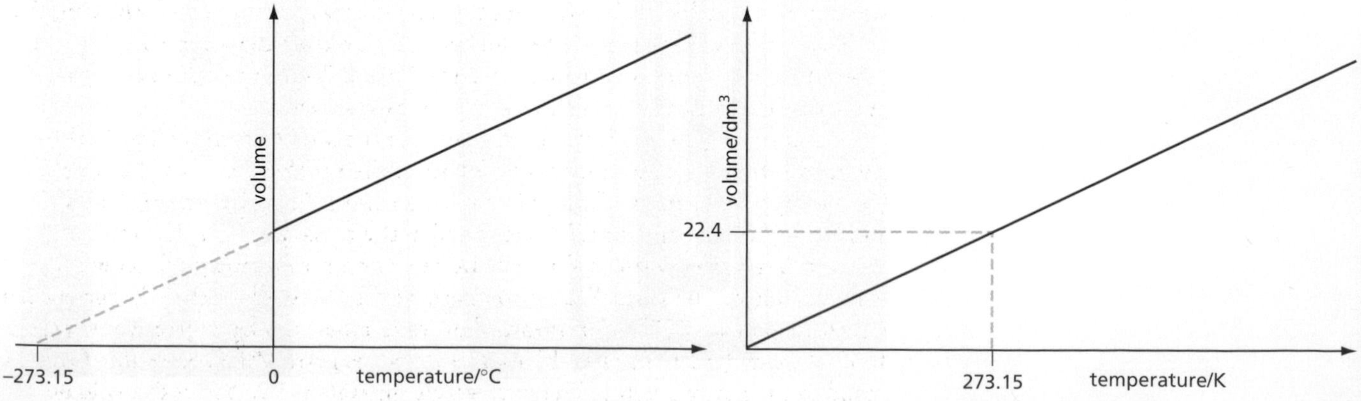

Example What are the following temperatures on the absolute temperature scale?

a 25 °C

b −0.5 °C

c −15 °C

Answer In each case, we add 273 to the temperature in °C.

a Absolute temperature = 273 + 25 = 298 K

b Absolute temperature = 273 − 0.5 = 272.5 K

c Absolute temperature = 273 − 15 = 258 K

Further practice

1 Convert the following temperatures from degrees Celsius to kelvin.

a −197 °C

b +273 °C

2 Convert the following from kelvin to degrees Celsius.

a 198 K

b 500 K

Amount of substance and volume

One other clear influence on the volume of a gas is the amount of gas we are studying. If we perform experiments on 2.0 g of gas rather than 1.0 g, we expect that under the same conditions of temperature and pressure the first sample will have twice the volume of the second. This is found to be the case:

The volume of gas, under identical conditions of temperature and pressure, is proportional to the amount (in moles) of gas present:

$$V \propto n$$

Arriving at the ideal gas equation

We can combine all three influences on the volume of a gas into one relationship:

$$V \propto \frac{1}{p} \qquad V \propto T \qquad V \propto n$$

$$\mathbf{V \propto \frac{nT}{p}}$$

We can convert this relationship into an equation by introducing a constant of proportionality. This is the **gas constant**. Chemists have given it the symbol *R*.

$$V \propto \frac{nT}{p} \qquad V = R\left(\frac{nT}{p}\right) \quad \text{or} \quad \mathbf{pV = nRT}$$

This equation is the **ideal gas equation**. The definition of an **ideal gas** is one which follows this equation exactly, under all conditions of pressure and temperature. The behaviours of many real gases, such as air, and its main components nitrogen and oxygen, together with hydrogen, helium and neon, do fit in with the equation fairly accurately. Some other gases, however, follow the equation only approximately when experiments are carried out at low temperatures or high pressures. The next section will look at why this is the case.

One surprising discovery was that all gases, no matter how they differ in their chemical reactions, or in the sizes or shapes of their molecules, obey the equation (at least approximately). Their points of zero volume ($T_{V=0}$) are all −273.15 °C, and *R* has the same numerical value for all of them, namely $8.31\,J\,K^{-1}\,mol^{-1}$.

To understand why this is the case, we must return to the major difference between liquids and gases. We saw on page 84 that water expands 1833 times when it forms steam at 100 °C. This means that water molecules take up only one part in 1833 of steam (the molecules themselves do not expand when converting from the liquid to the gaseous state). The rest, which amounts to 99.95% of the volume

($= 100 \times \frac{1832}{1833}$) is empty space. This empty space is common to all gases, no matter what sort of molecules they contain. The properties of gases are not the properties of empty space, of course. They arise from the molecules moving about and colliding with each other within that empty space. But as long as the different molecules of various gases move and bump into each other in a similar way, the values of R and $T_{V=0}$ will be the same.

Example Calculate the volume taken up by 1 mol of an ideal gas at room temperature and pressure (25 °C and 1.01×10^5 Pa).

Answer We rearrange the ideal gas equation to:

$$V = \frac{nRT}{p}$$

$n = 1.0$ mol

$R = 8.31\,\mathrm{J\,K^{-1}\,mol^{-1}}$ $[= 8.31\,\mathrm{N\,m\,K^{-1}\,mol^{-1}}]$

$T = 273 + 25 = 298$ K

$p = 1.01 \times 10^5$ Pa $[= 1.01 \times 10^5\,\mathrm{N\,m^{-2}}]$

$$V = \frac{1.0 \times 8.31 \times 298}{1.01 \times 10^5} \quad \left[\text{units: } \frac{(\mathrm{mol}) \times (\mathrm{N\,m\,K^{-1}\,mol^{-1}}) \times (\mathrm{K})}{(\mathrm{N\,m^{-2}})} = \mathrm{m^3}\right]$$

$$= 2.45 \times 10^{-2}\,\mathrm{m^3} = \mathbf{24.5\,dm^3}$$

Note that when carrying out calculations using the ideal gas equation, volumes must be in cubic metres. In calculations we often use the approximate value of 24 $\mathrm{dm^3}$ ($2.4 \times 10^{-2}\,\mathrm{m^3}$) for the volume of 1 mol of an ideal gas at room temperature and pressure.

Further practice How many moles of ideal gas are there in the following volumes?

1 2.8 $\mathrm{dm^3}$ of gas at a pressure of 1.01×10^5 Pa and a temperature of 10 °C.
2 54 $\mathrm{dm^3}$ of gas at a pressure of 5.0×10^5 Pa and a temperature of 600 °C.
3 92 $\mathrm{cm^3}$ of gas at a pressure of 9.5×10^4 Pa and a temperature of 100 °C.

4.14 The behaviour of real gases

Assumptions about ideal gases

It is possible to derive the ideal gas equation from the basic principles of mechanics. The **kinetic theory of gases** starts by making the following assumptions about an ideal gas:

- the molecules of an ideal gas behave as rigid spheres
- there are no intermolecular forces between the molecules of an ideal gas
- collisions between molecules of an ideal gas are perfectly elastic – there is no loss of kinetic energy during collision
- the molecules of an ideal gas have no volume.

The theory then considers that the pressure exerted by the gas is due to the bouncing of the gas molecules off the sides of the container. It calculates the magnitude of this pressure by assuming that the molecules are in constant random motion, and that no kinetic energy is lost during collisions with each other or with the walls of the container.

None of the above four assumptions is 100% true for real gases, however. We saw in Chapter 2 that atoms, and hence molecules, are not rigid, but are rather fuzzy around the edges. And we saw in Chapter 3 that there are various ways in which intermolecular forces can arise. Both of these factors will cause inelastic collisions. Finally, it is clearly the case that molecules do have a volume greater than zero.

How some real gases behave

The behaviour of some real gases is compared with that of an ideal gas in Figure 4.38. As we might expect, the ways in which real gases depart from ideal gas behaviour are different for each gas, because their molecules are different shapes and sizes. In general, though, we can see that the deviations become greater at high pressures.

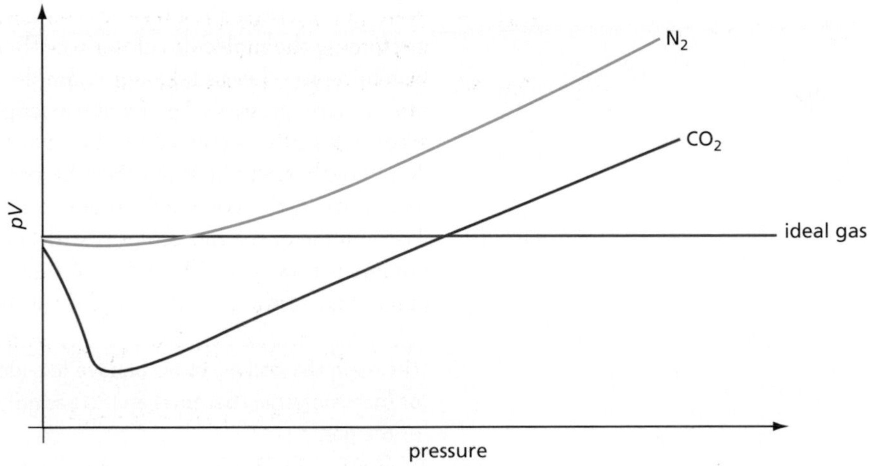

■ **Figure 4.38** Some real gases depart from ideal behaviour, particularly when the pressure is high.

If we now look at how one particular gas behaves as a result of changing the temperature (see Figure 4.39) we see that the deviation is greatest at low temperatures.

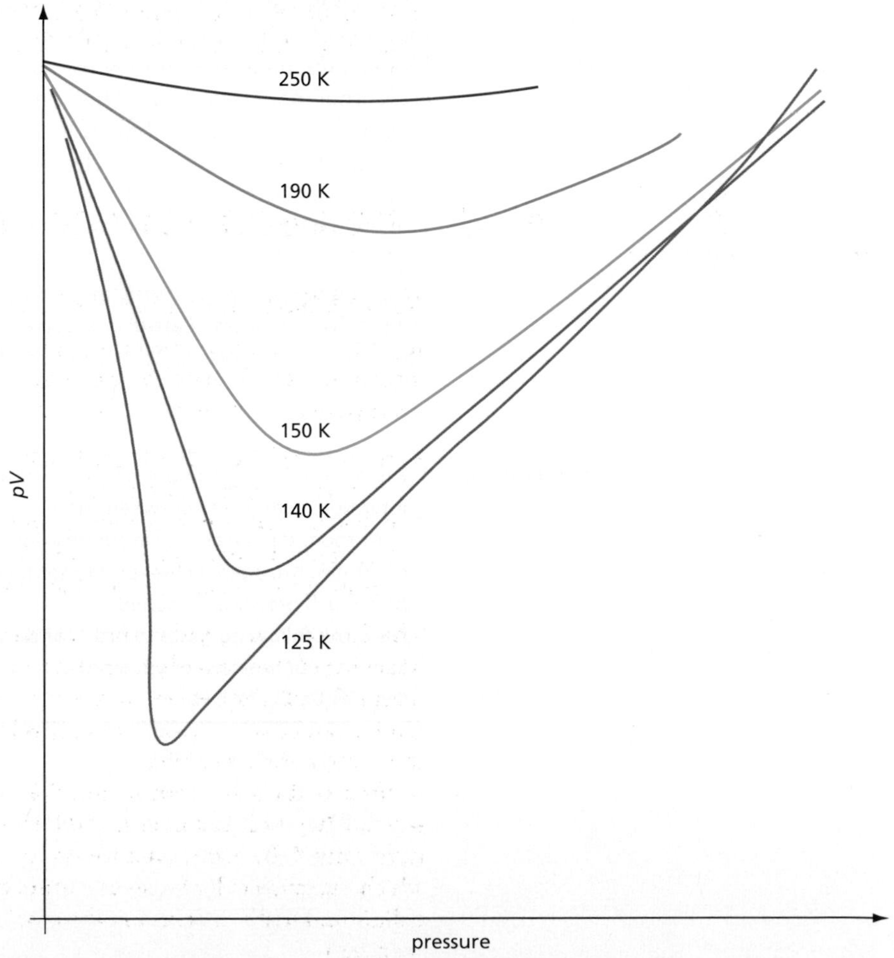

■ **Figure 4.39** The behaviour of nitrogen departs from ideal behaviour to a greater extent as the temperature is lowered.

We can summarise these findings as follows:

The behaviour of real gases is **least** like that of an ideal gas at low temperatures and high pressures.

Explaining deviation from ideal behaviour

When we increase the pressure on a gas, we decrease its volume (Boyle's law). We are forcing the molecules closer together, and the empty space between them is becoming less. Let us take our example of steam, which is 99.95% empty space at atmospheric pressure. 1 g of water occupies 1 cm^3, and this expands to 1833 cm^3 when it forms steam at 100 °C (see page 84). If we compress 1.0 g of steam to 20 atmospheres (which is about the pressure inside the boiler of a steam locomotive), the volume deceases to $\frac{1}{20}$ of its original 1833 cm^3, which is 92 cm^3. The volume of the molecules is still 1 cm^3, which as a percentage of the total volume is now $\frac{1}{92} \times 100 = 1.1\%$. When compressed, the molecules occupy a higher percentage of the gas than the 0.05% they occupy at atmospheric pressure.

The main reason why gases behave less ideally at high pressures is because the volume of their molecules becomes an increasingly significant proportion of the overall volume of the gas.

Looking now at the effect of temperature, we know that decreasing the temperature of a gas also decreases its volume (Charles's Law) and hence reduces the empty space between the molecules. But in addition, decreasing the temperature of a gas also decreases the kinetic energy the molecules possess. They travel slower, and bounce into each other with less force. The less the bouncing force of collision, the more significant will be the forces of attraction that exist between the molecules. If they are moving more slowly, they are likely to stick to each other more effectively. This causes the collisions to become less elastic. Eventually, on further cooling, the intermolecular forces become larger than the bouncing force, and the molecules spend more time sticking together than moving between each other. This is the molecular explanation of why a gas condenses to a liquid when it is cooled to below its boiling point.

The main reason why gases behave less ideally at low temperatures is because the intermolecular forces of attraction become comparable in size to the bouncing forces the molecules experience. This causes the collisions between the molecules to be inelastic.

Further practice

1 Explain the following observation in terms of the sizes of the molecules and the intermolecular forces between them.
At room temperature, carbon dioxide can be liquefied by subjecting it to a pressure of 10 atm. Nitrogen, however, cannot be liquefied at room temperature, no matter how much pressure is applied.

2 Place the following gases in order of decreasing ideality, with the most ideal first. Explain your reasons for your order.
CH_4, CH_3Br, Cl_2, HCl, H_2

Summary

- The major differences between the structures of solids, liquids and gases are whether or not their particles are touching each other, and whether they are fixed in position or moving.
- **Crystalline** solids are composed of **lattices**, which can be **simple molecular**, **macromolecular**, **metallic** or **ionic**. The major physical properties of a solid depend on the type of lattice it contains.
- Simple molecular lattices have low melting points, and are electrical insulators under all conditions.
- Macromolecular lattices have high melting points, and are electrical insulators under all conditions.
- Ionic compounds are formed between (usually) metallic cations and non-metallic anions.
- Ionic lattices have high melting points, are brittle, and only conduct electricity when molten or in solution.
- Metals and graphite conduct electricity because of the delocalised electrons they contain.
- Metallic lattices are **malleable**, **ductile**, and have reasonably high melting points.
- Graphite is an **anisotropic** material, due to the layer nature of its lattice.
- The **ideal gas equation**, $pV = nRT$, can be derived by applying the simple principles of mechanics to a collection of gas particles.
- Although the behaviour of many real gases approximates roughly to that of an ideal gas, real gases are least likely to behave like an ideal gas at low temperatures and at high pressures.

5 Energy changes in chemistry

In this chapter we shall look at the importance of energy changes in chemistry. Chemistry is the study of how atoms combine and recombine to form different compounds, and significant energy changes take place during these reactions. By measuring these energy changes we can find out why chemical reactions take place in the way that they do, and we can also discover the strengths of the bonds that hold atoms together. The concept of energy changes is fundamental to our understanding of the chemical reactions that we observe at the laboratory bench.

The chapter is divided into the following sections.

5.1 Chemical energy revisited – introducing enthalpy

We saw in Chapter 2 (page 31) that chemical energy is made up of kinetic energy and potential energy. The potential energy, due to electrostatic attractions between the particles, is usually a larger component of the total chemical energy than is the kinetic energy of those particles.

Enthalpy and enthalpy changes

The total chemical energy of a substance is called its **enthalpy** (or **heat content**). Values of enthalpy are usually large because the particles attract each other strongly, and we are dealing with the attractions associated with a very large number of particles. It is convenient to consider the energy changes associated with one mole of a substance, so that we always consider the same number of particles (6.0×10^{23}, see Chapter 1, page 8) whatever the substance we are dealing with. In this way a comparison of the enthalpies of different substances gives us a valid comparison of the forces of attraction that exist between the particles in these substances.

In a chemical reaction, we are interested in changes in chemical energy. An **enthalpy change** is given the symbol ΔH. The Greek letter delta, Δ, means 'change', and H is the symbol for enthalpy.

If as a result of the chemical reaction we increase the energy of the system, ΔH is positive. At the start '**the system**' is the reactants, and at the end it is the products. For example, we can represent the pulling apart of one mole of sodium ions and one mole of chloride ions (see page 32) by the following equation:

$$Na^+Cl^-(g) \rightarrow Na^+(g) + Cl^-(g)$$

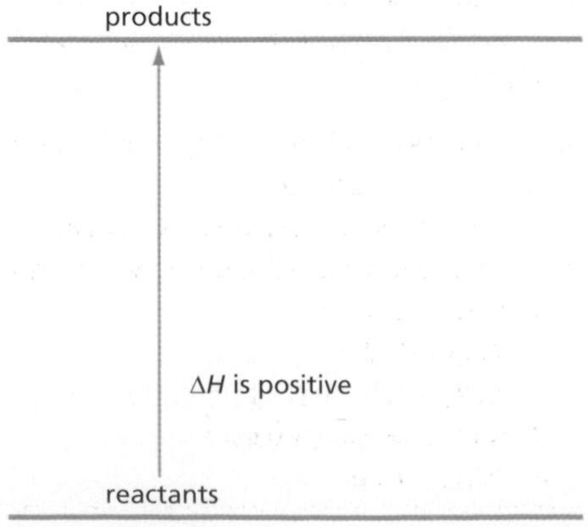

■ **Figure 5.1**
If energy is added to the system, ΔH is positive. The products have more energy than the reactants.

Because we are increasing the potential energy of the system, ΔH is large and positive (see Figure 5.1). It has a value of 500000 joules per mole. For convenience we write this as $\Delta H = +500\,\text{kJ mol}^{-1}$.

If we decrease the energy of the system, ΔH is negative (see Figure 5.2). For example, we can consider the reverse of the process above, that is, bringing a mole of sodium ions and a mole of chloride ions together.

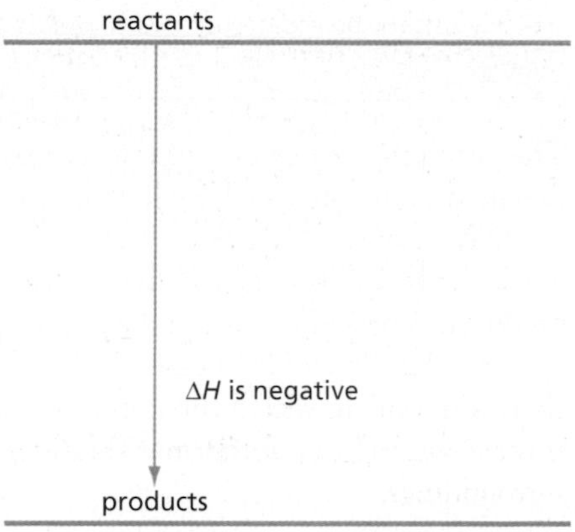

■ **Figure 5.2** If energy is removed from the system, ΔH is negative. The products have less energy than the reactants.

This is conveniently shown by the following equation:

$$Na^+(g) + Cl^-(g) \rightarrow Na^+Cl^-(g) \qquad \Delta H = -500\,\text{kJ mol}^{-1}$$

Gaseous sodium and chloride ions exist only at extremely high temperatures, so they cannot be studied under normal laboratory conditions. We can, however, study them in solution. Dissolving sodium chloride in water can be represented by the following equation:

$$Na^+Cl^-(s) \rightarrow Na^+(aq) + Cl^-(aq) \qquad \Delta H = +4\,\text{kJ mol}^{\;1}$$

Two questions can be asked at this stage.

- How has this value of $+4\,\text{kJ mol}^{-1}$ been measured?
- Why is this enthalpy change so much smaller than the enthalpy change for the gas phase reaction?

We shall answer the first question in the following section. The second question requires us to know how sodium and chloride ions pack together in the solid and how they associate with water molecules when in solution. We shall return to these points in Chapter 16.

Example 1 What type of kinetic energy is present in ice? Explain your reasoning.

Answer Ice is a solid, and the atoms are fixed in position (see Chapter 2, page 31). The only motion is vibration, so the atoms have vibrational kinetic energy.

Example 2 When ice changes to water, state the sign of ΔH. Explain your answer.

Answer When ice melts, some intermolecular bonds are broken, energy is taken in and so ΔH is positive.

Further practice

1 What types of kinetic energy are present in the following substances? In each case explain your reasoning.

a Ar(l) **b** Ar(g) c water **d** steam

2 For the following changes, state the sign of ΔH. Explain your answer.

a steam to water

b $KCl(s) \rightarrow KCl(g)$

c $KCl(g) \rightarrow K^+(g) + Cl^-(g)$

d $Na^+(g) + F^-(g) \rightarrow NaF(s)$

e $2Cl(g) \rightarrow Cl_2(g)$

5.2 Measuring enthalpy changes directly

Exothermic and endothermic reactions

The **Law of Conservation of Energy** states that:

Energy cannot be created or destroyed; it can only be converted into another form of energy.

The enthalpy change of a reaction usually appears as heat. An apparatus that is designed to measure such heat changes is called a **calorimeter**. If there is an enthalpy decrease (ΔH negative), an equivalent amount of heat energy must be given out by the system and the reaction is said to be **exothermic**. The heat produced is passed to the **surroundings**, that is, the environment around the system, where it can be measured. Most chemical reactions are exothermic, but there are some in which the enthalpy increases (ΔH is positive), and the reaction is then said to be **endothermic**. These reactions take in heat from the surroundings.

For **exothermic** reactions, ΔH is negative. The temperature of the surroundings increases and the potential energy of the system decreases.

For **endothermic** reactions, ΔH is positive. The temperature of the surroundings decreases and the potential energy of the system increases.

It is not surprising that most chemical reactions are exothermic. In everyday life we observe changes in which potential energy decreases and is converted into kinetic energy. We expect to see a book falling from the desk to the floor (its potential energy decreases) and would be very surprised if it suddenly rose back up again (its potential energy would increase). Both processes are possible in theory, as they do not violate the Law of Conservation of Energy. We need to introduce an additional idea called **entropy** to explain the one-way process of the conversion of potential energy to other forms of energy. This is discussed in Chapter 10, page 198.

The direction of a chemical change is determined by the relative energy levels of the reactants and products. If that of the reactants is higher than that of the products (exothermic, ΔH negative), the reaction is **thermodynamically** possible. It might not, however, take place because the rate is too slow; it is then said to be **kinetically** controlled. These kinetic factors are considered in Chapter 8.

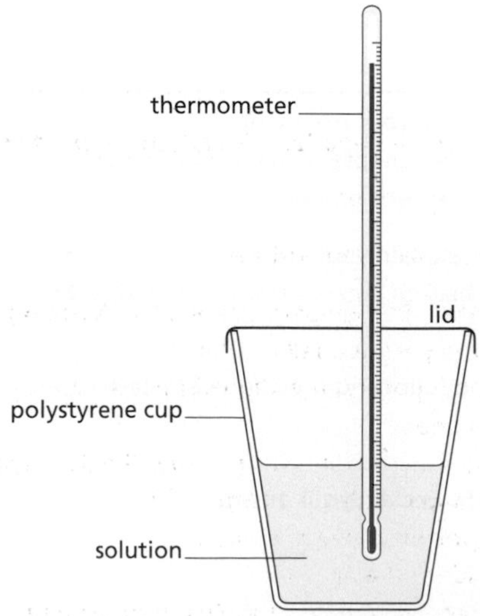

■ **Figure 5.3** A basic calorimeter, used for simple heat experiments.

Measuring temperature changes and calculating ΔH

If we measure the heat given out or taken in during a reaction, we can find the enthalpy change, ΔH. The simplest way of measuring this is to use the energy to heat (or cool) some water. We need to make the following measurements:

- the mass of the reactants
- the mass of water, m
- the rise (or fall) in temperature, ΔT.

We also need to know the amount of energy needed to raise the temperature of water by one degree. This is known as the **specific heat capacity** of water and is given the symbol c. It has the value $4.18\,J\,g^{-1}\,K^{-1}$. (Note that we can measure the temperature change using a thermometer in degrees Celsius, because a change in temperature is the same on either the Celsius or the kelvin scale.)

If we let q be the heat change, we have the following equation:

$$q = mc\Delta T$$

A simple way of measuring the heat change is to use an expanded polystyrene plastic cup, with a lid to make sure that heat losses to the atmosphere are as small as possible (see Figure 5.3). Under these conditions all the heat produced is used to raise the temperature of the contents of the plastic cup which, therefore, behaves as both the system and the surroundings. An apparatus used for measuring transfers of heat in this way is called a calorimeter.

■ *Experiment* To find the enthalpy change of solution of ammonium nitrate

Ammonium nitrate is used in this experiment because the temperature change when it dissolves is quite large. The finely powdered solid is weighed in a beaker and then tipped into water in a polystyrene beaker. The water is stirred until all the solid has dissolved, and the lowest temperature is recorded. The example below shows how the enthalpy change of solution may be calculated from measuring a temperature change in this way.

Example Some powdered ammonium nitrate was added to water in a plastic beaker and the following results obtained. Calculate the **enthalpy change of solution, ΔH_{sol},** of ammonium nitrate.

mass of water = 100 g

specific heat capacity of water = 4.18 J g^{-1} K^{-1}

mass of ammonium nitrate = 7.10 g

initial temperature = 18.2 °C

final temperature = 12.8 °C

Answer

$$q = mc\Delta T$$
$$= 100 \times 4.18 \times (12.8 - 18.2) \quad \text{(remember that '}\Delta\text{' means final – initial)}$$
$$= -2260\,\text{J}$$

$$n(NH_4NO_3) = \frac{m}{M} = \frac{7.10}{18 + 62} = 0.089\,\text{mol}$$

0.089 mol takes in 2260 J

$$\text{so } 1.0\,\text{mol takes in } \frac{2260}{0.089} = 26 \times 10^3\,\text{J}$$

Because heat is taken in, we know that the reaction is endothermic and ΔH_{sol} is positive.

$$\Delta H_{sol} = \mathbf{+26\,kJ\,mol^{-1}}$$

Note that it is usual to show the sign of ΔH, even when it is positive.

Further practice

1 100 cm^3 of water were placed in a plastic beaker. The temperature was 18.7 °C. 16.6 g of solid potassium iodide were added and the temperature fell to 13.4 °C.
   a In this experiment it is necessary to make the potassium iodide dissolve as quickly as possible. Explain why. How is this brought about?
   b Calculate the amount of heat absorbed by the solution. (Assume that the density of the solution is the same as the density of water, which is 1.00 g cm^{-3}.)
   c Calculate the number of moles of potassium iodide weighed out.
   d Calculate ΔH for dissolving potassium iodide in water.

2 75 cm^3 of water were placed in a plastic beaker. When 4.0 g of sodium nitrate were dissolved, the temperature fell by 2.4 K. Calculate ΔH for dissolving sodium nitrate in water.

■ Experiment Measuring other enthalpy changes

To find the **enthalpy change of neutralisation**, ΔH_{neut}, of an acid by a base, a known amount of acid is placed in a polystyrene cup. An equivalent amount of base is added and the rise in temperature recorded.

In a similar way, the temperature change can be measured for a variety of reactions which take place on mixing, and ΔH calculated. If one of the reactants is in excess, it is then only necessary to know the exact amount of the reactant that is completely used up, as it is this reactant that determines the energy change. For example, if magnesium ribbon is dissolved in excess hydrochloric acid, it is the quantity of magnesium that determines the energy change. The actual amount of hydrochloric acid has no effect on the amount of heat given out. The mass of the solution must, however, be known in order to measure the heat evolved, using the formula $q = mc\Delta T$.

Example 50 cm^3 of 2.0 $mol\,dm^{-3}$ sodium hydroxide solution were added to 50 cm^3 of 2.0 $mol\,dm^{-3}$ hydrochloric acid in a polystyrene cup. The following results were obtained. Calculate the enthalpy change of neutralisation, ΔH_{neut}, for this reaction.

initial temperature of HCl = 17.5 °C
initial temperature of NaOH = 17.9 °C
final temperature = 31.0 °C

Answer average temperature of the HCl and NaOH $= \frac{(17.5 + 17.9)}{2} = 17.7\,°C$

(Because the solution is very dilute, its specific heat capacity is taken to be the same as that of water, namely 4.18 $J\,g^{-1}\,K^{-1}$. This approximation is always used when carrying out reactions in dilute aqueous solution.)

$$q = mc\Delta T$$
$$= (50 + 50) \times 4.18 \times (31.0 - 17.7)$$
$$= 5560\,J$$

$n(HCl) = n(NaOH) = c \times V$ (in dm^3) (see page 17)

$$= 2.0 \times \frac{50}{1000} = 0.10\,mol$$

0.10 mol gives out 5600 J

so 1.0 mol gives out $\frac{5600}{0.1} = 56 \times 10^3\,J$

Because heat is evolved, we know that the reaction is exothermic and ΔH_{neut} is negative.

$\Delta H_{neut} = \mathbf{-56\ kJ\,mol^{-1}}$ (i.e. per mole of water formed)

Further practice

1 25 cm^3 of 1.0 $mol\,dm^{-3}$ nitric acid, HNO_3, were placed in a plastic cup. To this 25 cm^3 of 1.0 $mol\,dm^{-3}$ potassium hydroxide, KOH, were added. Both solutions started off at 17.5 °C. The maximum temperature reached after mixing was 24.1 °C.
   - **a** Calculate the heat given out in the reaction.
   - **b** Calculate the number of moles of nitric acid.
   - **c** Hence calculate ΔH for the neutralisation.

2 75 cm^3 of 2.0 $mol\,dm^{-3}$ ethanoic acid, CH_3CO_2H, were placed in a plastic cup. The temperature was 18.2 °C. To this were added 75 cm^3 of 2.0 $mol\,dm^{-3}$ ammonium hydroxide, NH_4OH, whose temperature was 18.6 °C. After mixing, the highest temperature was 31.0 °C. Calculate ΔH for the neutralisation.

3 0.48 g of magnesium ribbon was added to 200 cm^3 of hydrochloric acid in a plastic beaker. The temperature at the start was 20.0 °C, and after the magnesium ribbon had dissolved it rose to 35.2 °C.
   - **a** Write an equation for the reaction of magnesium ribbon with dilute hydrochloric acid.
   - **b** Calculate ΔH for this reaction.

5.3 Enthalpy changes of combustion

A very important measurement is the **enthalpy change of combustion** of a substance. Most organic compounds burn readily and give off a lot of heat. This provides our main source of energy for homes and for industry.

Energy sources

Fossil fuels

Our main source of energy is the combustion of **fossil fuels**, namely coal, oil and natural gas. They give out large amounts of energy when burned, but there are two main disadvantages to their use. The first is that supplies are finite and may run out in the foreseeable future. The second is that this combustion produces carbon dioxide, which contributes to the greenhouse effect (see Chapter 23, page 430), and also other pollutants.

Coal, oil and natural gas are respectively solid, liquid and gaseous fuels. The following considerations are made when choosing which to use.

- Ease of combustion – solid fuels are difficult to ignite and do not burn at a constant rate. For this reason, coal is often powdered before being injected into a furnace.
- Storage and ease of transport – for local use, gas does not need to be stored as it is carried by pipeline from the source. Oil is easily stored and can be distributed by pipeline. Coal has the disadvantage that it needs to be carried by road or rail to where it is to be used. To transport methane without a pipeline, it must be liquefied which requires expensive refrigeration. This makes natural gas of limited use as a vehicle fuel, though propane and butane are more useful as they can be liquefied by pressure alone (and are then called liquefied petroleum gas, LPG). The use of coal for driving steam engines and trains is now of historical interest only.
- Pollution – coal contains many impurities and, when burnt, gives off much sulphur dioxide (see Chapter 15, page 296). The same is true of oil, unless it is carefully refined before use. Crude oil is notorious for the environmental damage caused by spillage from tankers.
- Economic and political considerations – the relative costs of coal, oil and gas must take into account the costs of transport, storage and pollution-reducing measures, and are subject to wide local variations. Fuel taxes raise revenue for governments and the resulting increased cost of fossil fuel may encourage people to reduce their energy consumption.

■ **Figure 5.4** Coal has the highest energy density of the fossil fuels, giving out 76 kJ of energy for each cm^3 of coal burnt.

Because there are serious problems associated with burning fossil fuels, much research has been carried out to find alternative energy sources. These are increasingly important, but it is unlikely that they will provide more than a quarter of our energy demands during the next few years. Many alternative energy sources are **renewable** – they can be replaced as fast as they are used.

Alternative energy sources

Alternative energy sources do not rely on the combustion of fuels. The following are some of those in current use.

- **Hydroelectric power** – there are many places in the world where rivers have been dammed and the resulting difference in water level used to provide energy to generate electricity. The resultant damage to the environment is now often considered unacceptable.

■ **Figure 5.5** Wind farms provide pollution-free energy, but have a visual effect on the landscape.

- **Tidal and wave power** – these are little used, even though the large-scale tidal power generator at La Rance in northern France first produced electricity 50 years ago.
- **Wind power** – the use of wind turbines is increasing, although some people object to the construction of large windmills in isolated regions as they have a dramatic effect on the landscape. The wind speed is not always high enough to turn the blades in summer, but it usually is in winter, when the demand for electricity is highest.
- **Geothermal power** – potentially this could be an easy way to provide energy for home heating, though it can only be carried out at sites with particular geological features. Water is pumped from deep under the ground and comes to the surface at a temperature near to boiling point. It is not usually possible to use this hot water to produce electricity, because superheated steam is needed in order to obtain a reasonable energy conversion.
- **Solar panels** – the direct conversion of sunlight into electricity uses expensive solar cells and is of limited use in the winter in the UK, when energy demand is highest.
- **Nuclear power** – the use of nuclear reactors based on **fission reactions**, in which large nuclei split into two smaller nuclei and give out nuclear energy, has sharply declined as a result of accidents such as that at Chernobyl in 1986. There is much debate over the long-term implications of the disposal of nuclear waste, and most countries have cut back on their nuclear programmes. There is the possibility that reactors using **fusion reactions**, in which lighter nuclei join to form a heavier one, may be developed in the next few decades and these provide the best hope for a long-term solution to the energy problem, because they are 'cleaner' and do not produce radioactive waste.

Renewable fuels

Most of the alternative energy sources listed above are renewable in that they are powered by the Sun, so are always available. Some fuels are made synthetically out of resources that can be replaced quickly. For example, tropical countries can produce ethanol cheaply (see Chapter 26, page 488) and this can be used mixed with petrol in motor vehicles.

Hydrogen is another renewable energy resource. It has the advantage of being pollution free, since its combustion produces water, and can be used in fuel cells (see the panel on page 267). Its great disadvantage is the difficulty of its transport, though there is hope that the use of metal hydrides may overcome this problem (see Chapter 19, page 366).

For special applications, for example in rockets (see the panel on pages 119–20), special synthetic fuels are being developed.

Standard enthalpy changes

We can measure enthalpies of combustion very accurately, and they give us information about the forces of attraction that exist between the atoms in molecules. Because enthalpy measurements can be made to a high degree of accuracy, we must be careful to state the conditions of the reaction very precisely. For example, the following equation shows the combustion of methane:

$$CH_4(g) + 2O_2(g) \rightarrow CO_2(g) + 2H_2O(l)$$

The enthalpy change for this reaction at room temperature is about 10% greater than it is at 200 °C. By convention it has been agreed that **standard conditions** of both reactants and products are at a pressure of 1 atmosphere and, unless otherwise specified, at a temperature of 25 °C. Under these conditions the enthalpy change is known as the **standard enthalpy change of combustion** and is given the symbol $\Delta H^{\ominus}_c$, the 'c' indicating combustion and the '$\ominus$' indicating that it was measured and calculated under standard conditions.

It is also important to specify exactly the amount of substance involved in this change. Two common ways of writing the combustion of hydrogen are as follows:

$$H_2(g) + \tfrac{1}{2}O_2(g) \rightarrow H_2O(l)$$
$$2H_2(g) + O_2(g) \rightarrow 2H_2O(l)$$

The first equation, in which one mole of hydrogen gas undergoes combustion, represents $\Delta H^{\ominus}_c$. The enthalpy change for the second reaction is $2 \times \Delta H^{\ominus}_c$.

The **standard enthalpy change of combustion**, $\Delta H^{\ominus}_c$, is the enthalpy change when 1 mole of the substance is completely burnt in excess oxygen under standard conditions (1 atm pressure and 25 °C).

The bomb calorimeter

Highly accurate values of $\Delta H^{\ominus}_c$ can be found only by using a specially constructed apparatus called a **bomb calorimeter**, shown in Figures 5.6 and 5.7. The 'bomb' is a sealed pressure vessel with steel walls. The fuel is placed in the crucible and the 'bomb' filled with oxygen at a pressure of 15 atm. The 'bomb' is then placed in an insulated calorimeter containing a known mass of water. The fuel is ignited by an electric current and the temperature change measured to within 0.01 K. To eliminate heat losses, the calorimeter is placed in another water bath whose temperature is raised with an electric heater so that it just matches the average temperature in the calorimeter. The apparatus is first calibrated using benzoic acid, the enthalpy of combustion of which is recognised as a standard and which can be readily obtained in a high state of purity. There are a number of small corrections which must be applied to the results, but values accurate to 0.1% can be obtained.

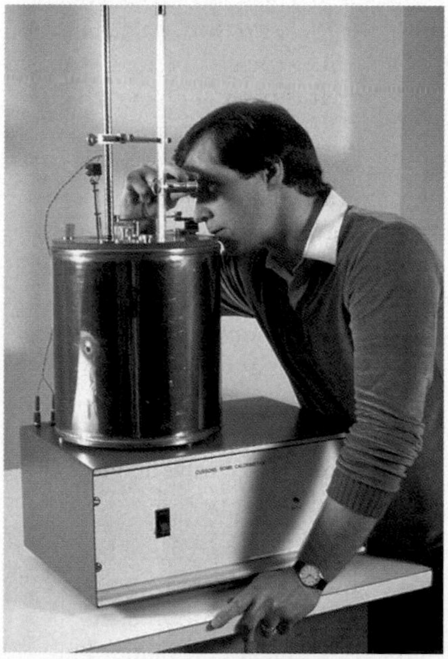

■ **Figure 5.6** (above) Bomb calorimeters are used to measure the energy content not only of fuels but also of foods.

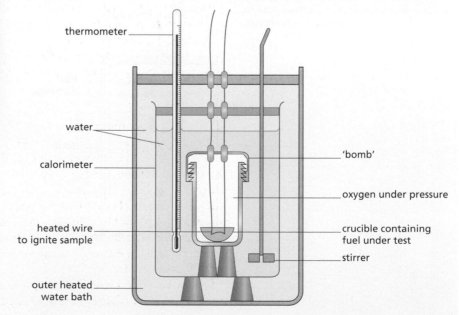

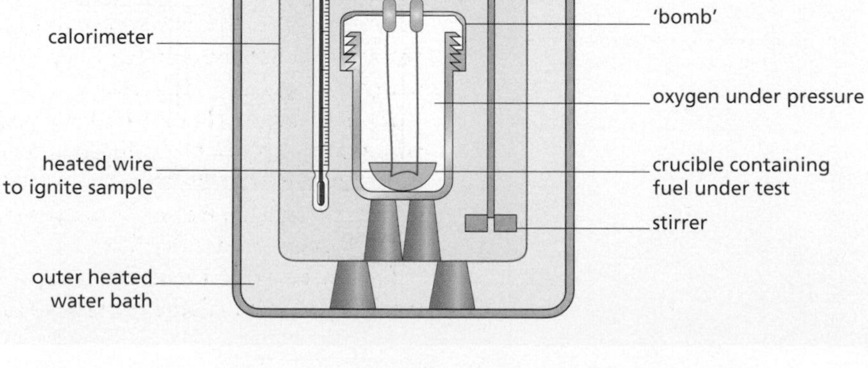

■ **Figure 5.7** (right) In a bomb calorimeter, accurate values of the enthalpy change of combustion can be measured because heat losses to the air are minimised.

■ *Experiment* To measure $\Delta H_c^{\ominus}$

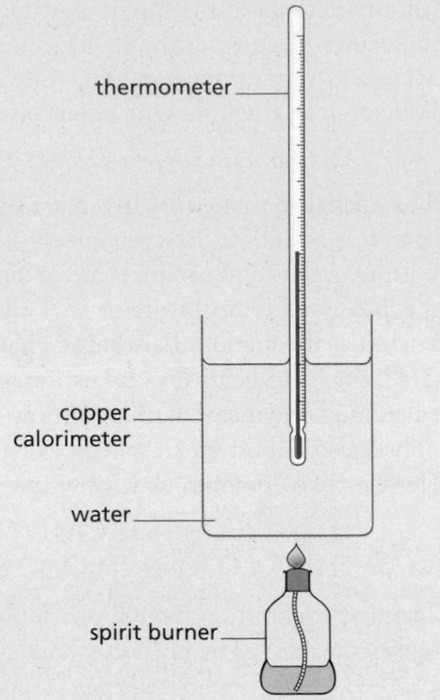

■ **Figure 5.8**
A simple apparatus used to measure enthalpy changes of combustion.

A simple apparatus

Figure 5.8 shows a simple apparatus to measure the enthalpy change of combustion for a fuel such as methanol. A known volume of water is placed in a copper calorimeter and its temperature taken. The calorimeter is clamped so that its base is just a few centimetres above a spirit burner, which contains the fuel. The spirit burner is weighed, placed under the calorimeter and the wick lit. The water in the calorimeter is stirred with the thermometer and when the temperature has risen about 10 °C, the flame is put out and the spirit burner re-weighed.

Thiemann's fuel calorimeter

When determining $\Delta H_c^{\ominus}$ using a copper calorimeter, there are two major sources of error:

- the methanol is not all completely burnt to carbon dioxide and water (some incomplete combustion takes place)
- not all the heat given off is passed to the water.

These errors can be reduced by means of **Thiemann's fuel calorimeter** (see Figure 5.9). The fuel is burnt in a stream of oxygen to ensure complete combustion, and the gases are sucked through a copper spiral placed in water (a heat exchanger) so that very little heat is lost to the air.

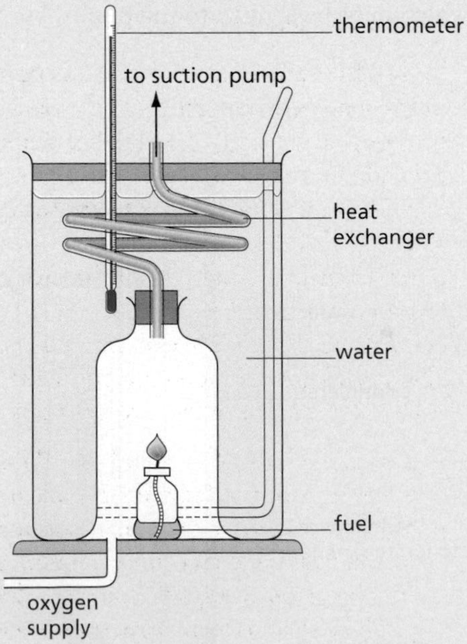

■ **Figure 5.9**
In Thiemann's fuel calorimeter, a supply of oxygen ensures complete combustion, and heat loss to the air is reduced.

Measurements similar to those with the simple copper calorimeter experiment are taken. The cap is replaced on the spirit burner after putting out the flame to reduce losses of fuel by evaporation before re-weighing. The oxygen should be supplied fast enough so that the fuel burns with a clear blue flame. The suction pump is usually fully on, but it may need to be turned down if the suction is so vigorous that the flame is pulled off the spirit burner. Thiemann's apparatus can give results to within 80% of quoted values.

Example In an experiment to determine $\Delta H_c^\ominus$ for methanol, CH_3OH, the following readings were obtained. Calculate $\Delta H_c^\ominus$ for methanol.

mass of water in calorimeter	= 200 g
mass of methanol and burner at start	= 532.68 g
mass of methanol and burner at end	= 531.72 g
temperature of water at start	= 18.3 °C
temperature of water at end	= 29.6 °C

Answer We shall ignore the heat taken in by the calorimeter.

Temperature rise of water = 11.3 K

$q = mc\Delta T$

$= 200 \times 4.18 \times 11.3 = 9400\,\text{J}$

mass of methanol burnt = 0.96 g and M_r (CH_3OH) = 32

so amount of methanol burnt $= \frac{0.96}{32} = 0.03\,\text{mol}$.

Because heat is evolved, we know that the reaction is exothermic and $\Delta H_c^\ominus$ is negative.

$$\Delta H_c^\ominus = \frac{-9400}{0.03} = -310 \times 10^3\,\text{J mol}^{-1} \text{ or } \mathbf{-310\,kJ\,mol^{-1}}$$

Further practice

1 a What is meant by 'standard conditions'?
  b Why is it necessary to specify the conditions of a reaction?
2 A burner containing hexanol, $C_6H_{13}OH$, had a mass of 325.68 g. It was lit and placed under a copper calorimeter containing 250 cm^3 of water. The temperature of the water rose from 19.2 °C to 31.6 °C. Afterwards the burner's mass was 324.37 g. Calculate:
  a the heat evolved
  b ΔH_c for hexanol.
3 a State the two main sources of error in the experiment described in question **2**.
  b Explain how these two errors are made as small as possible in Thiemann's apparatus.

Rocket fuels

To launch a rocket into space, an explosive fuel called a **propellant** is needed. All propellants have two main components:

- a combustible fuel
- an oxidising agent (oxidant).

In the atmosphere, the oxidant is oxygen from the air, but since a rocket operates in space it needs to carry its own oxidising agent. When the fuel combines with the oxidising agent, it gives out a large amount of heat which makes the gases produced expand. This expansion provides thrust for the rocket. The efficiency of the rocket engine is determined by the temperature and volume of the gaseous products.

The fuel and oxidant of a rocket motor make up most of the mass of the rocket. The main function of the propellant is to produce as much energy for a given mass as possible. Cost may be of secondary importance to the heat evolved per kilogram. So, while cars that operate in the atmosphere use a cheap fuel (see Chapter 23), rocket fuels are often expensive chemicals.

Rocket fuels are of two types – liquid and solid.

- Liquid fuels are often relatively cheap, but may need refrigeration so that the rocket can be fuelled only immediately before lift-off.
- Solid fuels have the advantage that they can be stored in the rocket. They are used in small rockets (for example, fireworks) and in large launcher rockets (for example, the space shuttle). They have the disadvantage that they produce solid as well as gaseous products and this limits their power output.

We shall now look at liquid and solid propellants in more detail.

Liquid propellants

Liquid propellants may contain relatively common chemicals, such as liquid oxygen with a hydrocarbon or liquid hydrogen fuel. They must be refrigerated and have a relatively poor power:weight ratio.

More exotic liquid fuels are also used. The lunar module used to transport astronauts from the Apollo command module to the Moon's surface was powered with liquid dinitrogen tetraoxide, N_2O_4, and dimethylhydrazine, $(CH_3)_2NNH_2$. These react as follows:

$$2N_2O_4(l) + (CH_3)_2NNH_2(l) \rightarrow 2CO_2(g) + 3N_2(g) + 4H_2O(g)$$
$$\Delta H^{\ominus} = -1800\,\text{kJ mol}^{-1} = -6980\,\text{kJ kg}^{-1}$$

Fuels with an even higher power:weight ratio have been investigated (for example, fluorine and boron hydride), but they are too toxic for use on the ground.

$$6F_2(g) + B_2H_6(g) \rightarrow 6HF(g) + 2BF_3(g)$$
$$\Delta H^{\ominus} = -2800\,\text{kJ mol}^{-1} = -10\,900\,\text{kJ kg}^{-1}$$

Solid propellants

The most familiar of solid propellants is the gunpowder or 'black powder' used in fireworks. The simplified chemical equation for the reaction is as follows:

$$3C(s) + S(s) + 2KNO_3(s) \rightarrow K_2S(s) + 3CO_2(g) + N_2(g)$$
$$\Delta H^{\ominus} = -280\,\text{kJ mol}^{-1} = -1040\,\text{kJ kg}^{-1}$$

The booster stage of the space shuttle uses a mixture of ammonium chlorate(VII), NH_4ClO_4, and aluminium powder:

$$10Al(s) + 6NH_4ClO_4(s) \rightarrow 5Al_2O_3(s) + 3N_2(g) + 6HCl(g) + 9H_2O(g)$$
$$\Delta H^{\ominus} = -3250\,\text{kJ mol}^{-1} = -9850\,\text{kJ kg}^{-1}$$

Figure 5.10 The solid fuel boosters of the space shuttle burn ammonium chlorate(VII) and aluminium. Clouds of aluminium oxide are produced as well as the gases that propel the shuttle into space.

5.4 Hess's Law and enthalpy change of formation

Introducing Hess's Law

There are very few reactions whose enthalpy change can be measured directly by measuring the change in temperature in a calorimeter. Fortunately we can find enthalpy changes for other reactions indirectly. To do this we make use of **Hess's Law**, which states that the value of ΔH for a reaction is the same whether we carry out the reaction in one step or in many steps, provided initial and final states or conditions are the same.

Hess's Law states that the enthalpy change, ΔH, for a reaction is independent of the path taken.

This law is illustrated by Figure 5.11. If $(\Delta H_1 + \Delta H_2)$ was greater than ΔH, we could get energy for nothing by going round the cycle: reactants, intermediate, products, reactants. This contradicts the Law of Conservation of Energy (page 112).

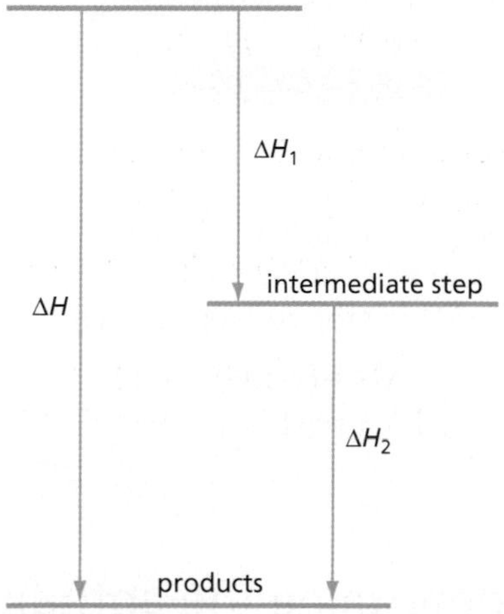

■ **Figure 5.11** Hess's Law: ΔH is independent of the path taken, and $\Delta H = \Delta H_1 + \Delta H_2$.

An example of the use of Hess's Law allows us to find ΔH for the decomposition of calcium carbonate:

$$CaCO_3(s) \rightarrow CaO(s) + CO_2(g)$$

This reaction is slow and requires a high temperature to bring it about. Direct measurement of the temperature change is therefore impracticable. We can, however, carry out two reactions that take place readily at room temperature and use their enthalpy changes to find the value of ΔH that we want. These are the reactions of calcium carbonate and calcium oxide with dilute hydrochloric acid:

$$CaCO_3(s) + 2HCl(aq) \rightarrow CaCl_2(aq) + H_2O(l) + CO_2(g) \qquad \Delta H_1^\ominus = -17\,kJ\,mol^{-1} \quad (1)$$

$$CaO(s) + 2HCl(aq) \rightarrow CaCl_2(aq) + H_2O(l) \qquad \Delta H_2^\ominus = -195\,kJ\,mol^{-1} \quad (2)$$

There are two ways in which we can use these values to find the enthalpy change for the decomposition of calcium carbonate.

Method (1): Subtracting equations

If we subtract equation (2) from equation (1), we have:

$$CaCO_3(s) - CaO(s) \rightarrow CO_2(g) \qquad \Delta H^\ominus = -17 - (-195) = +178\,kJ\,mol^{-1}$$

If we add 'CaO(s)' to both sides, we have the equation we want.

In this method we have taken away '2HCl(aq)' from the left-hand sides of the equations and '$CaCl_2$(aq) + H_2O(l)' from the right-hand sides. These terms are associated with a fixed amount of energy, and we are taking away this same amount of energy from both equations, so this has no effect on the final answer.

Method (2): Constructing a Hess's Law diagram

This method is preferred with more complicated examples. We draw a diagram like that in Figure 5.12 to indicate the two routes by which the reaction can be carried out. The diagram is not intended to give any indication of the actual energy levels (that is why they are not drawn vertically).

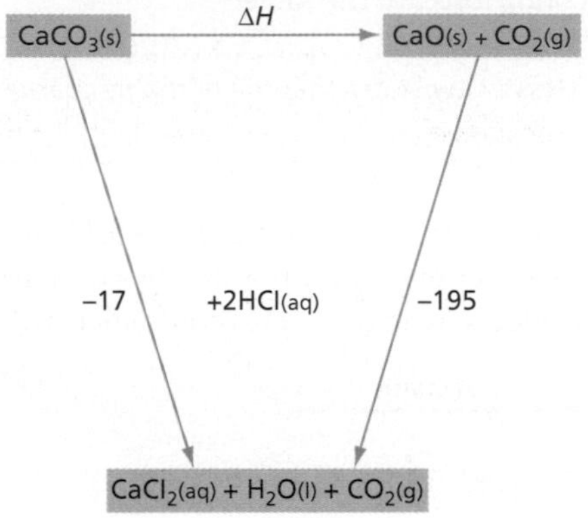

Figure 5.12 Hess's Law cycle for the decomposition of calcium carbonate. All figures are in $kJ\,mol^{-1}$.

In order to go from '$CaCO_3$(s)' to '$CaCl_2$(aq) + H_2O(l) + CO_2(g)', we can either go directly or via 'CaO(s) + CO_2(g)'.

Considering the enthalpy change involved for both routes gives us ΔH:

$$\Delta H + (-195) = -17$$
$$\text{and } \Delta H = -17 - (-195) = +178\,kJ\,mol^{-1}$$

Experiment To determine the enthalpy change when copper(II) sulphate is hydrated

For the hydration of copper(II) sulphate:

$$CuSO_4(s) + 5H_2O(l) \rightarrow CuSO_4.5H_2O(s)$$

the enthalpy change cannot be found directly. This is because if we add five moles of water to one mole of anhydrous copper(II) sulphate, we do not produce hydrated copper(II) sulphate crystals. These can only be made by crystallisation from a solution. The enthalpy change can, however, be found indirectly by determining the enthalpy change of solution of both anhydrous copper(II) sulphate and hydrated copper(II) sulphate (see Figure 5.13).

According to Hess's Law, $\Delta H_1 = \Delta H + \Delta H_2$.

0.10 mol of anhydrous copper(II) sulphate is added to $100\,cm^3$ of water in a plastic cup, fitted with a lid and a thermometer. When the solid has dissolved, the change in temperature can be used to calculate ΔH_1.

The experiment is repeated using 0.10 mol of powdered hydrated copper(II) sulphate, but using only $91\,cm^3$ of water because the hydrated salt already contains $9\,cm^3$ (0.5 mol) of water. The change in temperature when the hydrated copper sulphate dissolves can be used to calculate ΔH_2.

The required enthalpy change of reacton ΔH is given by:

$$\Delta H = \Delta H_1 - \Delta H_2$$

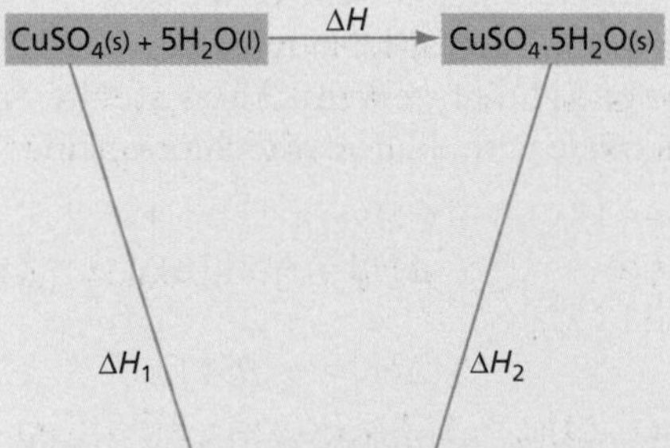

Figure 5.13 Hess's Law cycle to find the enthalpy change when anhydrous copper(II) sulphate crystals are hydrated.

■ Experiment To find the enthalpy change of decomposition of sodium hydrogencarbonate

On heating, sodium hydrogencarbonate decomposes:

$$2NaHCO_3(s) \rightarrow Na_2CO_3(s) + H_2O(l) + CO_2(g)$$

The enthalpy change for this decomposition may be found by measuring the enthalpy change when sodium hydrogencarbonate and sodium carbonate react with hydrochloric acid (see Figure 5.14):

$$NaHCO_3(s) + HCl(aq) \rightarrow NaCl(aq) + H_2O(l) + CO_2(g) \qquad \Delta H_1$$

$$Na_2CO_3(s) + 2HCl(aq) \rightarrow 2NaCl(aq) + H_2O(l) + CO_2(g) \qquad \Delta H_2$$

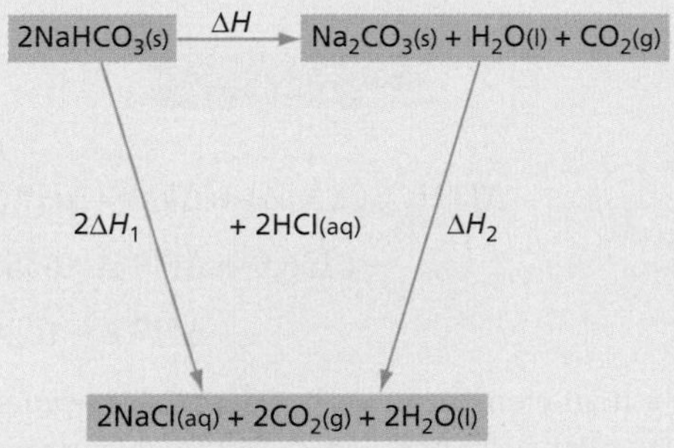

■ **Figure 5.14** Hess's Law cycle for the decomposition of sodium hydrogencarbonate.

Note that ΔH_1 is multiplied by 2 because its equation is multiplied by 2 in order to balance things. 100 cm^3 of dilute hydrochloric acid are placed in a plastic cup, fitted with a lid and thermometer. 0.05 mol of sodium hydrogencarbonate is added. The temperature rise is used to calculate ΔH_1. ΔH_2 is found by repeating the experiment using 0.05 mol of anhydrous sodium carbonate. The required enthalpy change can then be calculated, as $\Delta H = 2\Delta H_1 - \Delta H_2$.

Enthalpy change of formation

Although actual values of the enthalpy contained in individual substances are not known, it is possible to obtain accurate values of a quantity that is related to it, namely the **standard enthalpy change of formation**, $\Delta H^\ominus_f$.

Enthalpy change of formation of a substance
The standard enthalpy change of formation of a substance, $\Delta H^\ominus_f$, is the enthalpy change when one mole of the substance is formed from its elements in their standard states (1 atm pressure, usually 298 K).

The values of $\Delta H^\ominus_f$ for some compounds such as oxides can be determined experimentally, while those for other compounds must be calculated using Hess's Law.

The value of $\Delta H^\ominus_f$ of the oxide often has the same value as $\Delta H^\ominus_c$ of the element. For example, for carbon:

$$C(s) + O_2(g) \rightarrow CO_2(g) \qquad \Delta H^\ominus_f(CO_2(g)) = \Delta H^\ominus_c(C(s)) = -393.5\,kJ\,mol^{-1}$$

and for hydrogen:

$$H_2(g) + \tfrac{1}{2}O_2(g) \rightarrow H_2O(l) \qquad \Delta H^\ominus_f(H_2O(l)) = \Delta H^\ominus_c(H_2(g)) = -285.9\,kJ\,mol^{-1}$$

In other examples, the two are not the same because the two processes are represented by different equations. For example, for aluminium:

$$2Al(s) + 1\tfrac{1}{2}O_2(g) \rightarrow Al_2O_3(s) \qquad \Delta H^\ominus_f = -1675.7\,kJ\,mol^{-1}$$

$$Al(s) + \tfrac{3}{4}O_2(g) \rightarrow \tfrac{1}{2}Al_2O_3(s) \qquad \Delta H^\ominus_c = -837.8\,kJ\,mol^{-1}$$

The value of $\Delta H^\ominus_f(Al_2O_3(s))$ is twice that of $\Delta H^\ominus_c(Al(s))$.

We can use values of $\Delta H_f^\ominus$ to calculate $\Delta H^\ominus$ for a reaction. On page 121 we showed how to measure, indirectly, the value of $\Delta H^\ominus$ for the reaction:

$$CaCO_3(s) \rightarrow CaO(s) + CO_2(g)$$

We can construct a different cycle using standard enthalpy changes of formation to find this enthalpy change, $\Delta H^\ominus$, as shown in Figure 5.15.

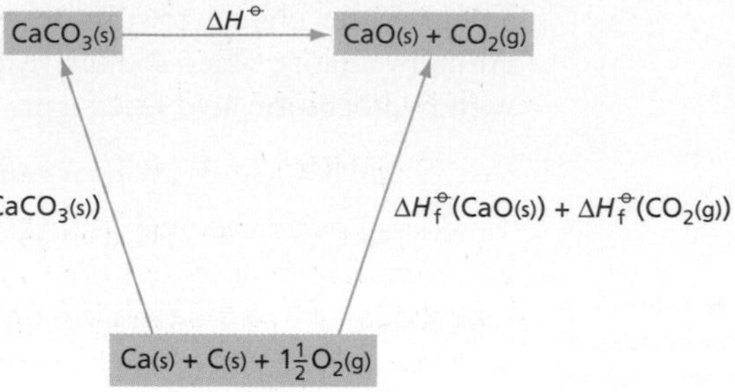

■ **Figure 5.15** Hess's Law cycle to find the enthalpy change for the decomposition of calcium carbonate using standard enthalpy changes of formation.

$$\Delta H_f^\ominus(CaCO_3(s)) + \Delta H^\ominus = \Delta H_f^\ominus(CaO(s)) + \Delta H_f^\ominus(CO_2(g))$$

$$-1206.9 + \Delta H^\ominus = -635.5 + (-393.5)$$

$$\Delta H^\ominus = -1029.0 - (-1206.9) = +177.9\,kJ\,mol^{-1}$$

If an element in its standard state occurs in the equation, it can be ignored in the calculation because its $\Delta H_f^\ominus$ value is zero. For example, to calculate $\Delta H^\ominus$ for the reaction:

$$NO(g) + \tfrac{1}{2}O_2(g) \rightarrow NO_2(g)$$

we have the cycle shown in Figure 5.16.

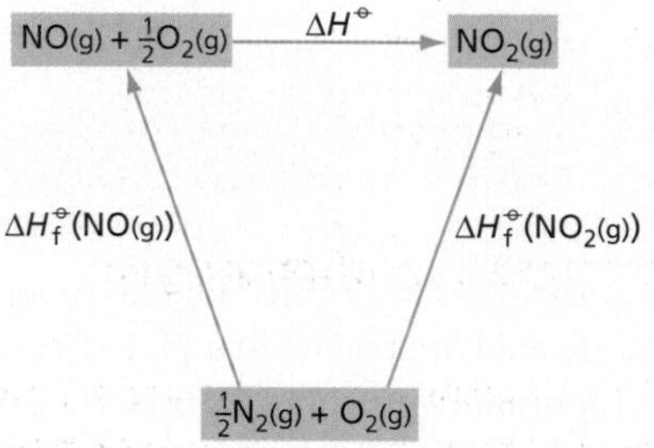

■ **Figure 5.16** Hess's Law cycle for the conversion of nitrogen monoxide to nitrogen dioxide. The enthalpy of formation of the element oxygen is zero, so this term can be ignored.

$$\Delta H_f^\ominus(NO(g)) + \Delta H^\ominus = \Delta H_f^\ominus(NO_2(g))$$

$$+90.4 + \Delta H^\ominus = +33.2$$

$$\Delta H^\ominus = +33.2 - (+90.4) = -57.2\,kJ\,mol^{-1}$$

Further practice

1 State Hess's Law.

2 The following enthalpy changes were measured:

$FeSO_4(s) \rightarrow Fe^{2+}(aq) + SO_4^{2-}(aq)$ $\quad \Delta H_1^\ominus = -69\,kJ\,mol^{-1}$

$FeSO_4.7H_2O(s) \rightarrow Fe^{2+}(aq) + SO_4^{2-}(aq)$ $\quad \Delta H_2^\ominus = +216\,kJ\,mol^{-1}$

Draw a suitable Hess's Law diagram and use it to calculate the enthalpy change when anhydrous iron(II) sulphate is hydrated.

3 Suggest how you could determine the enthalpy changes for the following reactions, which cannot be found directly. In each case draw a suitable Hess's Law diagram and outline what heat measurements you would make.

a $BaO(s) + H_2O(l) \rightarrow Ba(OH)_2(s)$

b $PbCO_3(s) \rightarrow PbO(s) + CO_2(g)$

c $Zn(s) + CuSO_4.5H_2O(s) + 2H_2O(l) \rightarrow ZnSO_4.7H_2O(s) + Cu(s)$
(Harder: this requires three experiments to be carried out.)

4 Using the values of $\Delta H_f^\ominus$ given below, calculate the values of $\Delta H^\ominus$ for the following reactions:

a the thermal decomposition of $BaCO_3$

b $C_2H_6(g) + Br_2(l) \rightarrow C_2H_5Br(l) + HBr(g)$

c $C_2H_4(g) + H_2O(l) \rightarrow C_2H_5OH(l)$

d $TiCl_4(l) + 2Mg(s) \rightarrow Ti(s) + 2MgCl_2(s)$

e $SO_3(l) + H_2O(l) \rightarrow H_2SO_4(l)$

[$\Delta H_f^\ominus$/kJ mol^{-1}: BaO(s) −997.5; $BaCO_3$(s) −1218.8; CO_2(g) −393.5; C_2H_4(g) +52.3; C_2H_6(g) −84.6; C_2H_5Br(l) −92.0; C_2H_5OH(l) −277.7; HBr(g) −36.2; H_2O(l) −285.9; H_2SO_4(l) −814.0; $MgCl_2$(s) −641.8; $TiCl_4$(l) −804.2; SO_3(l) −395.4]

5.5 Bond enthalpies

Average bond enthalpies

When two atoms come together, they may form a bond (see Chapter 3, page 53). To break this bond and separate the two atoms requires energy (ΔH positive). This is because a force has to be moved to pull the atoms apart which increases the potential energy of the system. For example, to break the bond in a chlorine molecule requires 242 kJ mol^{-1}:

$$Cl_2(g) \rightarrow 2Cl(g) \qquad \Delta H = +242\,kJ\,mol^{-1}$$

The situation is more complicated if the molecule contains more than two atoms, but fortunately we find that the energy needed to break a particular type of bond is about the same even though it may be in different molecules. This value is called the **average bond enthalpy** (or **average bond energy**). If a bond joins an atom of X to an atom of Y, the bond enthalpy is represented by the symbol E(X—Y).

The **average bond enthalpy,** E(X—Y), is the enthalpy change when one mole of bonds between atoms of X and atoms of Y are broken in the gas phase:

$$XY(g) \rightarrow X(g) + Y(g) \qquad \Delta H = E(X{-}Y)$$

Explaining reaction enthalpies – the formation of hydrogen chloride

We can use tabulated values of bond enthalpies to explain why some reactions give out so much energy. For example, consider the reaction of hydrogen and chlorine to give hydrogen chloride:

$$H_2(g) + Cl_2(g) \rightarrow 2HCl(g)$$
$$H{-}H + Cl{-}Cl \rightarrow H{-}Cl + H{-}Cl$$

We can break the reaction down into three steps, as shown in Figure 5.17.

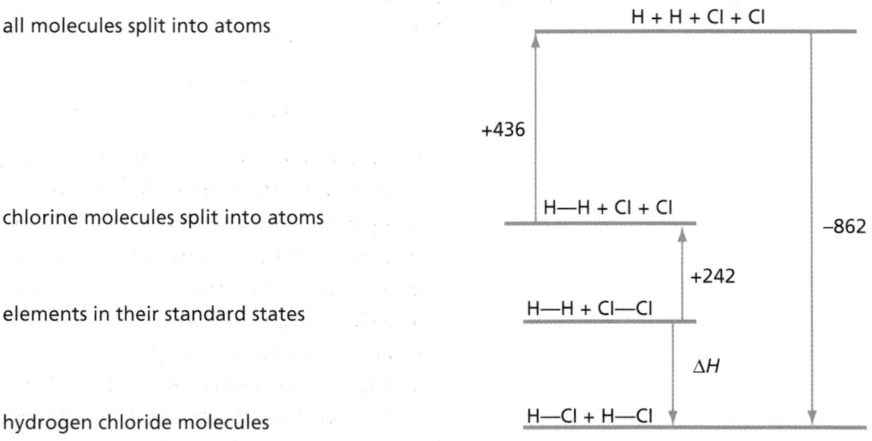

■ **Figure 5.17** The bond-breaking and bond-making steps in the synthesis of hydrogen chloride. The figures are published bond enthalpies, in kJ mol^{-1}.

The changes can be represented by equations:

$$H_2(g) \rightarrow 2H(g) \qquad \Delta H = E(H{-}H) = +436\,kJ\,mol^{-1}$$

$$Cl_2(g) \rightarrow 2Cl(g) \qquad \Delta H = E(Cl{-}Cl) = +242\,kJ\,mol^{-1}$$

$$2H(g) + 2Cl(g) \rightarrow 2HCl(g) \qquad \Delta H = -2E(H{-}Cl) = -2 \times 431 = -862\,kJ\,mol^{-1}$$

We can use Hess's Law to add these three equations, to obtain:

$$H_2(g) + Cl_2(g) \rightarrow 2HCl(g) \qquad \Delta H = -184\,kJ\,mol^{-1}$$

If we study the values of the bond enthalpies above, we see that while the H—H and H—Cl bonds are strong and have similar values, the Cl—Cl is much smaller. It is the weakness of the Cl—Cl bond that is the principal reason why the reaction is so exothermic. The reason why the bonds differ so much in energy is considered in more detail in Chapter 3, page 70.

Notice that in this reaction we broke two bonds (the H—H and Cl—Cl bonds) and formed two bonds (two H—Cl bonds), and that ΔH equalled the bond enthalpies of the bonds broken minus the bond enthalpies of the bonds formed. For most reactions the number of bonds broken equals the number of bonds formed.

> ΔH for a reaction = the bond enthalpies of the bonds broken minus the bond enthalpies of the bonds formed
>
> Number of bonds broken = number of bonds formed

The combustion of hydrogen

Consider a more complicated example, the combustion of hydrogen, whose measured enthalpy change is as follows:

$$2H_2(g) + O_2(g) \rightarrow 2H_2O(g) \qquad \Delta H = -483\,kJ\,mol^{-1}$$

$$H{-}H + H{-}H + O{=}O \rightarrow H{-}O{-}H + H{-}O{-}H$$

Notice that we specify $H_2O(g)$ and not the usual state of water under standard conditions, which is $H_2O(l)$. Liquid water has intermolecular forces between the molecules and these have to be formed as well as the O—H bonds, making the energy of $H_2O(l)$ lower than that of $H_2O(g)$. This is why bond enthalpies are always quoted for the gas phase.

If we consider forming $H_2O(g)$, we can look up the following bond enthalpies:

bonds broken	**bonds formed**
$2E(H{-}H) = 2 \times 436\,kJ\,mol^{-1}$	$4E(O{-}H) = 4 \times 464\,kJ\,mol^{-1}$
$E(O{=}O) = 497\,kJ\,mol^{-1}$	
total 4 bonds	total 4 bonds

ΔH = enthalpy of bonds broken – enthalpy of bonds formed

= 1369 – 1856

$= -487\,kJ\,mol^{-1}$

This calculated value for the enthalpy change is slightly different from the accurate experimental value ($-483\,kJ\,mol^{-1}$), showing that we must not always expect exact agreement between measured energy changes and those calculated from bond enthalpy values. However, this exercise enables us to suggest why the reaction is so exothermic. At first sight all the bonds appear to be of similar strength, but that for oxygen is for two bonds, so that the average for one bond is only $\frac{497}{2}\,kJ\,mol^{-1}$. It is the weakness of the O=O bond that is the principal reason why combustion reactions are so exothermic and therefore so important as sources of energy in everyday life.

The decomposition of hydrogen peroxide

While the average strength of each bond in O=O is $\frac{497}{2}\,kJ\,mol^{-1}$, in actual fact the two bonds do not have the same energy. The O—O single bond is much weaker than this, having a value of only $146\,kJ\,mol^{-1}$. This is the reason why hydrogen peroxide readily decomposes, giving the experimental enthalpy change shown below:

$$2H_2O_2(l) \rightarrow 2H_2O(l) + O_2(g) \qquad \Delta H = -196\,kJ\,mol^{-1}$$

In order to find out why this reaction is so exothermic, we need to consider the bonds made and broken during this reaction:

$$2H_2O_2(g) \rightarrow 2H_2O(g) + O_2(g)$$

that is, ΔH for the production of gaseous water from gaseous hydrogen peroxide rather than the production of liquid water from liquid hydrogen peroxide. If we assume that the energy required to convert one mole of liquid hydrogen peroxide into gas is approximately the same as the energy required to convert one mole of liquid water into gas, ΔH for the gaseous reaction will not be much different from $-196\,\text{kJ mol}^{-1}$.

bonds broken	bonds formed
$4E(\text{O—H})$	$4E(\text{O—H})$
$2E(\text{O—O}) = 2 \times 146 = 292\,\text{kJ mol}^{-1}$	$E(\text{O=O}) = 497\,\text{kJ mol}^{-1}$

We can simplify this calculation by considering only the bonds that are broken and formed – we can ignore the four O—H bonds as they are present in both the reactants and products. So:

$$\begin{aligned}\Delta H &= \text{bonds broken} - \text{bonds formed}\\ &= 292 - 497\\ &= -205\,\text{kJ mol}^{-1}\end{aligned}$$

This value, $-205\,\text{kJ mol}^{-1}$, is close enough to our measured value of $-196\,\text{kJ mol}^{-1}$ to justify using bond enthalpy arguments to discover why the reaction is so exothermic. Bond enthalpies show us that the reason why hydrogen peroxide decomposes is because the O—O single bond is very much weaker than half the strength of the O=O bond.

The simplification of considering only the bonds broken and formed during a reaction is very useful when more complicated examples are being considered.

Example 1

a Explain what is meant by the H—I bond enthalpy.
b Write the symbol for the H—I bond enthalpy.
c Write an equation that shows the change brought about in determining the H—I bond enthalpy.

Answer

a This is the energy required to break one mole of H—I bonds in the gas phase.
b $E(\text{H—I})$
c $HI(g) \rightarrow H(g) + I(g)$

Example 2

a Write an equation to show the breakdown of methane, $CH_4(g)$, into atoms.
b How is ΔH for this reaction calculated using the C—H bond enthalpy?

Answer

a $CH_4(g) \rightarrow C(g) + 4H(g)$
b $\Delta H = 4 \times E(\text{C—H})$

Further practice

1 Consider the reaction:

$$CH_4(g) + Cl_2(g) \rightarrow CH_3Cl(g) + HCl(g)$$

a What bonds are broken and what bonds are formed?
b Calculate:
  i the energy required to break these bonds
  ii the energy released on forming the bonds.

$E(\text{C—H}) = 415\,\text{kJ mol}-1$; $E(\text{Cl—Cl}) = 242\,\text{kJ mol}^{-1}$; $E(\text{C—Cl}) = 338\,\text{kJ mol}^{-1}$; $E(\text{H—Cl}) = 431\,\text{kJ mol}^{-1}$

c Hence calculate ΔH for the reaction.
d Explain why the following reaction would have a similar value of ΔH:

$$C_2H_6(g) + Cl_2(g) \rightarrow C_2H_5Cl(g) + HCl(g)$$

e Explain why the experimental values for ΔH for the chlorination of methane and ethane are not identical in practice.

2 Consider the reaction:

$$ICl(s) \rightarrow I(g) + Cl(g) \qquad \Delta H = +210\,\text{kJ mol}^{-1}$$

Explain why this value of ΔH is not the same as that of the I—Cl bond enthalpy.

Summary

- The **enthalpy change** of a reaction is the change in heat energy brought about by the reaction due to a change in energy of the reacting substances.
- Enthalpy changes can be measured directly by measuring the heat given out during a reaction, using a **calorimeter**.
- The heat q evolved is calculated using the equation $q = mc\Delta T$, where m is the mass of water or solution, c the specific heat capacity of water and ΔT the change in temperature.
- For an exothermic reaction ΔH is negative; for an endothermic reaction ΔH is positive.
- Most reactions are exothermic, with negative ΔH.
- **Hess's Law** states that ΔH for a reaction is independent of the path taken. If ΔH for a reaction cannot be directly measured, it may often be found using Hess's Law.
- The enthalpy change for a reaction is equal to the bond enthalpies of the bonds broken minus the bond enthalpies of the bonds formed.
- A study of bond enthalpies indicates why ΔH for some reactions is so large.

Some enthalpy definitions

- **Standard conditions** are one atmosphere pressure and a specified temperature, usually 298K.
- The **standard enthalpy change** of a reaction, $\Delta H^{\ominus}$, is the enthalpy change when moles of the reactants indicated by the equation are completely converted into products under standard conditions.
- The **enthalpy change of neutralisation**, ΔH_{neut}, is the enthalpy change when one mole of acid is neutralised by one mole of base.
- The **standard enthalpy change of combustion**, $\Delta H^{\ominus}_{c}$, is the enthalpy change when one mole of the substance is completely burnt in oxygen under standard conditions.
- The **standard enthalpy change of formation**, $\Delta H^{\ominus}_{f}$, is the enthalpy change when one mole of the substance is formed from its elements under standard conditions.
- The **average bond enthalpy**, $E(X—Y)$, is the enthalpy change when one mole of bonds between atoms of X and Y are broken in the gas phase:

$$X—Y(g) \rightarrow X(g) + Y(g) \qquad \Delta H = E(X—Y)$$

6 *Acids and bases*

In this chapter we introduce the Brønsted–Lowry theory of acids and bases. This theory describes substances as acids or bases when dissolved not only in water but also in a variety of other solvents. The important quantitative technique of titration is introduced and a detailed explanation given of how to use the concept of the mole to calculate the required result. The importance of the mole is further emphasised with titration calculations from back titrations.

The chapter is divided into the following sections.

6.1 The Arrhenius theory

What is an acid?

Two hundred years ago, substances were called **acids** if they tasted sour (the Latin for sour is *acidus*) or if they changed the colour of some plant extracts. The essential feature of their chemistry was unknown and it was many years before the present-day definition of an acid was recognised. The following is a brief history of how the modern idea of an acid developed.

- 1778 – because most non-metallic oxides dissolved in water to give acidic solutions, Antoine Lavoisier proposed that all acids contain oxygen. This is why he chose the name for the gas oxygen – the word means 'acid producer'.
- 1816 – Humphry Davy showed that Lavoisier's view was incorrect when he proved that hydrochloric acid contained hydrogen and chlorine only.
- 1844 – Justus von Liebig suggested that acids react with metals to give hydrogen.
- 1844 – as a result of his work on the conductivity of electrolytes, Svante Arrhenius defined an acid as a substance that gives hydrogen ions in water.

Arrhenius supposed that the hydrogen ion was simply H^+, but we now know that this ion, a 'bare' proton, exists only in a high vacuum. In water, a water molecule forms a dative bond with a hydrogen ion to give the **oxonium ion**, H_3O^+ (also called the **hydroxonium ion**). This is further hydrated by hydrogen bonding (see Figure 6.1). For this reason it is best to avoid writing just H^+ in chemical equations, though H_3O^+, $H_3O^+(aq)$ and $H^+(aq)$ are perfectly acceptable.

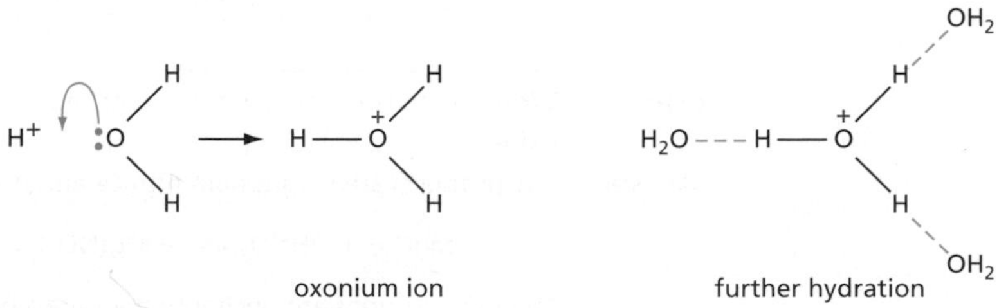

■ **Figure 6.1** The simple hydrogen ion forms a bond with a water molecule to give the oxonium ion, which is then further hydrated by hydrogen bonding.

Bases and alkalis

The question of what constituted a **base** remained more doubtful. Originally anything that reduced the sourness of an acid was considered to be a base. According to the Arrhenius theory, bases reacted with the hydrogen ion to give the non-acidic product water. This restricted bases to including just metal oxides and hydroxides. Other substances were excluded because they produced slightly acidic products, for example, carbonates produced carbon dioxide when reacted with an acid.

The Arrhenius theory of acids and bases

- An **acid** produces hydrogen ions in water.
- A **base** reacts with an acid to give a salt and water only. A base is a metal oxide or hydroxide.

Some Group 1 and Group 2 metal hydroxides dissolve in water to give alkaline solutions. This is also true of their oxides, because they react with water to form hydroxides. These bases that are soluble in water are known as **alkalis**.

$$M_2O(s) + H_2O(l) \rightarrow 2M^+(aq) + 2OH^-(aq) \qquad \text{Group 1}$$
$$MO(s) + H_2O(l) \rightarrow M^{2+}(aq) + 2OH^-(aq) \qquad \text{Group 2}$$

Representing acid–base reactions

We now recognise that in an acid–base reaction, the ions making up the salt are **spectator ions** – they do not take part in a chemical reaction. We therefore represent these reactions by an ionic equation which omits such spectator ions. For example, in the equation for the neutralisation of hydrochloric acid with sodium hydroxide:

$$H^+(aq) + Cl^-(aq) + Na^+(aq) + OH^-(aq) \rightarrow H_2O(l) + Na^+(aq) + Cl^-(aq)$$

we can omit the sodium and chloride ions, leaving:

$$H^+(aq) + OH^-(aq) \rightarrow H_2O(l)$$

Using simplified ionic equations in this way has the advantages of showing:

- which bonds are being broken and which are formed
- that many apparently different acid–base reactions are, in fact, identical.

According to the Arrhenius theory, acid–base reactions can be divided into three main types:

- an acid and an alkali:

$$H^+(aq) + OH^-(aq) \rightarrow H_2O(l)$$

- an acid and an insoluble metal oxide, for example, copper(II) oxide or aluminium oxide:

$$2H^+(aq) + CuO(s) \rightarrow H_2O(l) + Cu^{2+}(aq)$$
$$6H^+(aq) + Al_2O_3(s) \rightarrow 3H_2O(l) + 2Al^{3+}(aq)$$

- an acid and an insoluble metal hydroxide, for example, magnesium hydroxide or iron(III) hydroxide:

$$2H^+(aq) + Mg(OH)_2(s) \rightarrow 2H_2O(l) + Mg^{2+}(aq)$$
$$3H^+(aq) + Fe(OH)_3(s) \rightarrow 3H_2O(l) + Fe^{3+}(aq)$$

Example Write the full equation and the ionic equation for the action of nitric acid on magnesium oxide.

Answer The products are magnesium nitrate and water. The full equation is:

$$MgO(s) + 2HNO_3(aq) \rightarrow Mg(NO_3)_2(aq) + H_2O(l)$$

The $NO_3^-(aq)$ ions are unchanged and can be omitted in the ionic equation, but the Mg^{2+} ion changes from the solid to the aqueous state and must be included:

$$MgO(s) + 2H^+(aq) \rightarrow Mg^{2+}(aq) + H_2O(l)$$

Further practice Write full equations and ionic equations for the following reactions:

1 sulphuric acid with sodium hydroxide
2 hydrochloric acid with iron(III) hydroxide
3 nitric acid with zinc oxide.

6.2 The Brønsted–Lowry theory

Donating and accepting protons

The limitation of the Arrhenius theory is that it can be applied only to reactions in aqueous solution. Many reactions of the acid–base type take place in solvents other than water or under anhydrous conditions. For example, the reaction:

$$CuO(s) + 2HCl(g) \rightarrow CuCl_2(s) + H_2O(l)$$

is very similar to:

$$CuO(s) + 2HCl(aq) \rightarrow CuCl_2(aq) + H_2O(l)$$

and should certainly be classified as an acid–base reaction.

Johannes Brønsted and Thomas Lowry realised that the important feature of an acid–base reaction was the transfer of a proton from an acid to a base. They therefore defined an **acid** as a substance that could donate a proton and a **base** as a substance that could accept a proton.

The Brønsted–Lowry theory of acids and bases

- An **acid** is a proton donor.
- A **base** is a proton acceptor.

According to this theory, we might expect that any substance containing a hydrogen atom could act as an acid. In practice, a substance behaves as an acid only if the hydrogen atom already carries a slight positive charge, $\overset{\delta-}{—X}—\overset{\delta+}{H}$. This is the case if it is bonded to a highly electronegative atom to the right of the periodic table, often oxygen or a halogen.

In order to accept a proton, a base must contain a lone pair of electrons that it can use to form a dative bond with the proton (see Chapter 3, page 59). So, like an acid, a base must contain an atom from the right-hand side of the periodic table, and again this atom is often oxygen.

Conjugate acid–base pairs

When an acid such as hydrochloric acid loses a proton, a base such as Cl^- is formed. A proton could return to this base and re-form HCl. This HCl/Cl^- system is known as a **conjugate acid–base pair**. The changes can be represented by the equations:

$$\underset{\text{acid}}{H{:}Cl} \rightarrow \underset{\text{proton}}{H^+} + \underset{\text{conjugate base}}{{:}Cl^-} \quad \text{and} \quad \underset{\text{base}}{{:}Cl^-} + \underset{\text{proton}}{H^+} \rightarrow \underset{\text{conjugate acid}}{H{:}Cl}$$

Hydrochloric acid can donate one proton – it is called a **monoprotic** acid.

Some acids can donate more than one proton. For example, sulphuric acid is **diprotic** – it can lose two protons in two different steps:

$$H_2SO_4 \rightarrow HSO_4^- + H^+ \quad \text{step 1}$$
$$HSO_4^- \rightarrow SO_4^{2-} + H^+ \quad \text{step 2}$$

Similarly, some bases can accept more than one proton. The carbonate ion is diprotic – it can gain two protons in two different steps:

$$CO_3^{2-} + H^+ \rightarrow HCO_3^-$$
$$HCO_3^- + H^+ \rightarrow H_2CO_3$$

- Acids that can donate a maximum of one, two or three protons are called monoprotic, diprotic or triprotic, respectively. They are also called monobasic, dibasic or tribasic.
- Bases that can receive a maximum of one, two or three protons are called monoprotic, diprotic or triprotic, respectively. They are also called monoacidic, diacidic or triacidic.

In this book we shall use the terms monoprotic, diprotic, etc. to avoid the confusion caused by referring to an acid as a base and vice versa.

Water is an exceptional substance in that it can behave as both an acid and as a base. Such a substance is called an **ampholyte** as it shows **amphoteric** behaviour.

$$H_2O + H^+ \rightarrow H_3O^+ \quad \text{water behaving as a base}$$
$$H_2O \rightarrow H^+ + OH^- \quad \text{water behaving as an acid}$$

Example 1   Draw the structure of the conjugate base of hydrogen sulphide, H_2S.

Answer   Hydrogen sulphide behaves as an acid:

$$H \div S \div H \rightarrow H \div S\!:^- + H^+$$

and HS^- is the conjugate base.

Example 2   Draw the structure of the conjugate acid of the ethanoate ion, $CH_3CO_2^-$.

Answer   The ethanoate ion behaves as a base:

$$CH_3CO \div O\!:^- + H^+ \rightarrow CH_3CO \div O \div H$$

and ethanoic acid, CH_3CO_2H, is the conjugate acid.

Further practice   1   Draw the structures of the conjugate bases of the following acids.
     a HBr   **b** HNO_3 **c** H_2SO_4 **d** HSO_4^- **e** HS^-

2   Draw the structures of the conjugate acids of the following bases.
     a F^- **b** HS^- **c** CO_3^{2-} **d** HSO_4^- **e** H_2O **f** NH_3

6.3   Brønsted–Lowry acid–base reactions

When an acid gives up its proton to a base, a **proton transfer reaction** has taken place. This is a Brønsted–Lowry acid–base reaction.

An example is dissolving hydrogen chloride in water. The hydrogen chloride is the acid, and water acts as a base:

$$\underset{\text{acid}}{HCl} \rightarrow \underset{\text{proton}}{H^+} + \underset{\text{conjugate base}}{Cl^-}$$

$$\underset{\text{base}}{H_2O\!:} + \underset{\text{proton}}{H^+} \rightarrow \underset{\text{conjugate acid}}{H_3O^+}$$

A convenient way of representing this change is by use of the 'curly arrow'. This shows the movement of a pair of electrons. The 'tail' shows where the pair of electrons are at the start of the reaction, and the 'head' shows where they finish. These two changes can be represented as follows:

$$H \div Cl \rightarrow H^+ + :Cl^- \quad \text{and} \quad H_2O\!: \, ^{\frown}H^+ \rightarrow H_3O^+$$

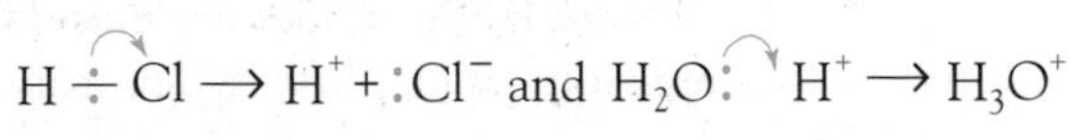

The structure of the H_3O^+ ion can be represented either as:

$$\overset{H}{\underset{H}{>}}O\!: \rightarrow H^+ \text{ (showing a dative bond) or, better, as: } \overset{H}{\underset{H}{>}}O^+\!\!-H$$

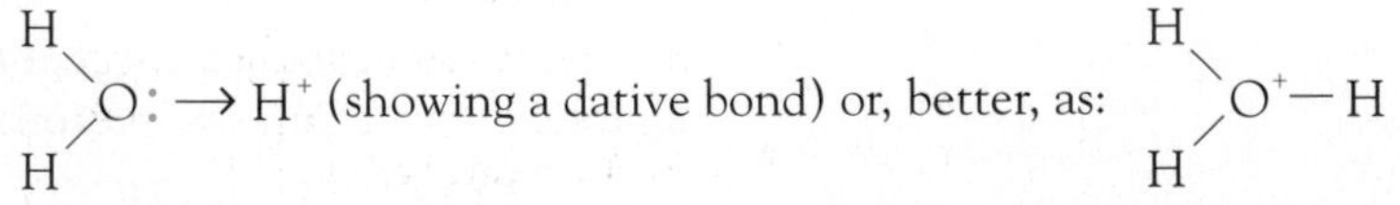

In this book, we largely avoid the use of the arrow showing a dative bond as it suggests that the bond is not a covalent bond (which it is), and also because this arrow may be confused with the curly arrow which indicates more precisely the movement of the electron pair. Also, such an arrow is sometimes used to show the polarity of a bond.

We have now two **half-equations** which represent the proton movements that happen during a reaction. These can be combined together to give the whole equation. We do not specify the states in a half-equation as it does not represent an actual chemical reaction, but we do so in the complete equation:

half-equations:

$$\underset{\text{acid}_1}{HCl} \rightarrow H^+ + \underset{\text{base}_1}{Cl^-} \quad \text{and} \quad \underset{\text{base}_2}{H_2O} + H^+ \rightarrow \underset{\text{acid}_2}{H_3O^+}$$

complete equation:

$$\underset{\text{acid}_1}{HCl(g)} + \underset{\text{base}_2}{H_2O(l)} \rightarrow \underset{\text{base}_1}{Cl^-(aq)} + \underset{\text{acid}_2}{H_3O^+(aq)}$$

Notice that there must be four species in the equation, two conjugate acid–base pairs. If the acid or base is diprotic, there are two possible neutralisation reactions. For example, in the neutralisation of sulphuric acid with sodium hydroxide, the following reactions may happen:

$H_2SO_4 \rightarrow H^+ + HSO_4^-$ reaction 1 with acid:base ratio $\leqslant 1$

$H_2SO_4 \rightarrow 2H^+ + SO_4^{2-}$ reaction 2 with acid:base ratio $\geqslant 2$

This gives two possible neutralisation reactions depending on the proportions of acid and base present in the reaction.

With excess acid:

$$\underset{\text{acid}_1}{H_2SO_4(aq)} + \underset{\text{base}_2}{OH^-(aq)} \rightarrow \underset{\text{base}_1}{HSO_4^-(aq)} + \underset{\text{acid}_2}{H_2O(l)}$$

With excess base:

$$\underset{\text{acid}_1}{H_2SO_4(aq)} + \underset{\text{base}_2}{2OH^-(aq)} \rightarrow \underset{\text{base}_1}{SO_4^{2-}(aq)} + \underset{\text{acid}_2}{2H_2O(l)}$$

Example 1

a The carbonate ion is diprotic. Write two overall equations for the reaction of dilute nitric acid with aqueous sodium carbonate.

b Hence write the relevant half-equations.

Answer

a With excess acid:

$$2H_3O^+(aq) + CO_3^{2-}(aq) \rightarrow H_2CO_3(aq) + H_2O(l)$$

With excess base:

$$\underset{\text{acid}_1}{H_3O^+(aq)} + \underset{\text{base}_2}{CO_3^{2-}(aq)} \rightarrow \underset{\text{acid}_2}{HCO_3^-(aq)} + \underset{\text{base}_1}{H_2O(l)}$$

b With excess acid:

$$H_3O^+ \rightarrow H^+ + H_2O \text{ and } CO_3^{2-} + 2H^+ \rightarrow H_2CO_3$$

With excess base:

$$H_3O^+ \rightarrow H^+ + H_2O \text{ and } CO_3^{2-} + H^+ \rightarrow HCO_3^-$$

Example 2

Write the relevant half-equations and the overall equations for the possible reactions when solid sodium chloride is treated with concentrated sulphuric acid.

Answer

With excess acid:

$$H_2SO_4 \rightarrow HSO_4^- + H^+ \text{ and } Cl^- + H^+ \rightarrow HCl$$

With excess base:

$$H_2SO_4 \rightarrow SO_4^{2-} + 2H^+ \text{ and } Cl^- + H^+ \rightarrow HCl$$

With excess acid:

$$\underset{\text{base}_1}{Cl^-(s)} + \underset{\text{acid}_2}{H_2SO_4(l)} \rightarrow \underset{\text{base}_2}{HSO_4^-(s)} + \underset{\text{acid}_1}{HCl(g)}$$

With excess base:

$$\underset{\text{base}_1}{2Cl^-(s)} + \underset{\text{acid}_2}{H_2SO_4(l)} \rightarrow \underset{\text{base}_2}{SO_4^{2-}(s)} + \underset{\text{acid}_1}{2HCl(g)}$$

Further practice

Write the relevant half-equations and the overall equation for the following reactions:

1 hydrogen bromide gas and solid iron(II) oxide

2 aqueous sulphuric acid and aqueous potassium hydroxide

3 concentrated sulphuric acid and water

4 dilute hydrochloric acid and sodium hydrogencarbonate

5 aqueous sodium hydrogensulphate and aqueous sodium carbonate

6 ammonia gas and hydrogen chloride gas.

6.4 Strong and weak acids

According to the Arrhenius theory, a **strong acid** is one that is virtually completely ionised in aqueous solution, while a **weak acid** is one that is ionised only to a small extent (possibly less than 10%). According to the Brønsted–Lowry theory, a **strong acid** is one that gives up a proton more easily than the H_3O^+ ion does. This means that when a strong acid such as hydrogen chloride is dissolved in water, the following reaction takes place virtually completely:

$$HCl(g) + H_2O(l) \rightarrow H_3O^+(aq) + Cl^-(aq)$$

A similar reaction takes place with other strong acids such as sulphuric acid and nitric acid. If the acid is strong, that is, if it has a great tendency to give up a proton, then its conjugate base must be very weak, as it has very little tendency to accept a proton.

A **weak acid** is therefore one that gives up a proton less easily than the H_3O^+ ion does. So if a weak acid such as ethanoic acid, CH_3CO_2H, is dissolved in water, it barely ionises at all:

$$CH_3CO_2H(l) + H_2O(l) \rightarrow H_3O^+(aq) + CH_3CO_2^-(aq)$$

Since ethanoic acid has little tendency to give up a proton, its conjugate base, the ethanoate ion, must be a strong base, because it has a tendency to accept a proton. By observing that strong acids displace weaker acids, an approximate order of acid strength can be found, as shown in Table 6.1. A quantitative basis for this order is discussed in Chapter 12.

■ **Table 6.1** Some common acids and bases in order of their strengths.

Acid	Strength	Base	Strength
H_2SO_4	Very strong ↑	HSO_4^-	Very weak
HCl		Cl^-	
HNO_3		NO_3^-	
H_3O^+	Fairly strong	H_2O	Weak
HSO_4^-		SO_4^{2-}	
CH_3CO_2H		$CH_3CO_2^-$	
H_2CO_3	Weak	HCO_3^-	Less weak
NH_4^+		NH_3	
NCO_3^-		CO_3^{2-}	↓
H_2O	Very weak	OH^-	Fairly strong

The direction in which an acid–base reaction takes place is governed by energy changes (Chapter 5, page 110), just as with any other chemical reaction. The energy change is greatest when a strong acid reacts with a strong base, and these are the ones that react together most readily. If a weak acid reacts with a weak base, the reaction will not go to completion (see Chapter 9).

The strength of an acid depends on its structure. The importance of the structures of some acids is discussed further in Chapter 29, pages 542–3.

6.5 Solvents other than water

The Brønsted–Lowry theory can be applied to acids in solvents other than water. One of the first solvents to be used instead of water was liquid ammonia, with a boiling point of −33 °C. Ammonia, like water, is an ampholyte and can behave as an acid or a base.

As an acid:

$$NH_3 \rightarrow NH_2^- + H^+$$

As a base:

$$NH_3 + H^+ \rightarrow NH_4^+$$

So in liquid ammonia, ammonium chloride, $NH_4^+Cl^-$, behaves as a strong acid, and sodamide, $Na^+NH_2^-$, behaves as a strong base.

Another solvent that may be used for acid–base reactions is concentrated sulphuric acid. This is used both as an acid and as a solvent in the nitration of benzene (see Chapter 27, pages 503–5).

Sometimes an **aprotic solvent** is used for acid–base reactions, that is, a solvent that does not donate a proton. Under these circumstances, the Brønsted–Lowry theory of acids and bases can no longer apply and the more general **Lewis theory** is used instead (see the panel below).

The Lewis theory

There are many reactions that have much in common with acid–base reactions. For example, the formation of complex ions (Chapter 19, pages 350–1) and the reactions of electrophiles with nucleophiles in organic chemistry (Chapter 25, page 465). All involve the movement of a pair of electrons from a species that contains a lone pair of electrons to a centre that is electron deficient. In 1916 Gilbert Newton Lewis, using the Bohr theory of the atom that had been put forward five years earlier, suggested that the formation of a covalent bond involved the sharing of a pair of electrons. In 1923, he suggested that an acid–base reaction should be viewed as the transfer of a pair of electrons from the base to the acid.

The Lewis theory of acids and bases

- An **acid** receives a pair of electrons.
- A **base** donates a pair of electrons.

The advantage of the Lewis theory is that it emphasises the similarity between acid–base reactions that involve proton transfer and some other reactions that do not. It is useful in explaining why some reactions in organic chemistry that are catalysed by acids can also be catalysed by other substances that can attract a pair of electrons. For example, the nitration of benzene is carried out using sulphuric acid as a strong acid, but the reaction also takes place readily if nitric acid and boron trifluoride, BF_3, are used. Nitric acid acts as a Lewis base and boron trifluoride as a Lewis acid.

Aluminium(III) chloride acts as a Lewis acid in the Friedel–Crafts reaction (see Chapter 27, page 505).

The limitation of the Lewis theory is that it is too general. Oxidation–reduction reactions also involve electron transfer, and there is no clear distinction between a Lewis acid–base reaction and an oxidation–reduction reaction such as lead(II) to lead(IV), both of which involve the transfer of two electrons. It is therefore better to use the Brønsted–Lowry theory for most acid –base reactions, and apply the Lewis concept only if the reaction does not involve proton transfer.

6.6 Acid–base titrations

Calculations involving acid–base reactions

Quantitative information about acid–base reactions is most conveniently found by **titration**. In titration, a known volume of a solution of one reactant is measured using a pipette. Into this is run a solution of the other reactant from a burette, until the reaction between the two substances is just complete. For an acid–base reaction, the completion of the reaction is conveniently found by noting a change in a colour in an indicator that has been added to the reaction mixture.

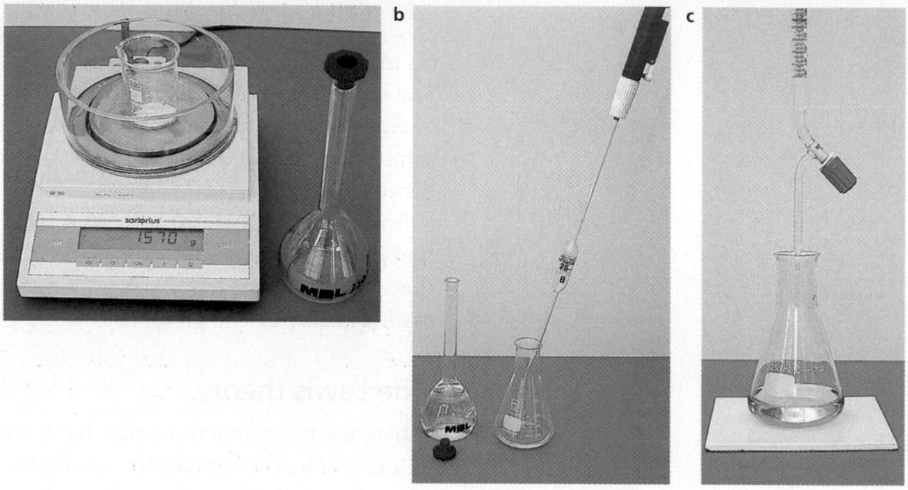

■ **Figure 6.2** The stages in carrying out a titration.
a A standard solution is made up of one reactant.
b A measured volume of this standard solution is put into a flask and a solution of the other reactant is run in from a burette.
c The end-point is detected using an indicator, and the volume read off the burette.

The results of a titration are used in one of two ways:

- for quantitative analysis, for example, to find the concentration of the solution in the burette
- to establish the **stoichiometry** of a chemical reaction, that is, to find how many moles of one substance react with one mole of another substance.

Both of these methods require calculations involving the following two equations, which we met in Chapter 1.

The basic equations used in titrations

$n = \frac{m}{M}$ (see page 7)	n = amount (in mol), m = mass (in g), M = molar mass (in $g\,mol^{-1}$)
$n = cV$ (see page 17)	c = concentration (in $mol\,dm^{-3}$), V = volume (in dm^3)

Note that volumes may sometimes be measured in cm^3 rather than dm^3. Since $1\,cm^3 = \frac{1}{1000}\,dm^3$, the following equation is used:

$$n = c \times \frac{v}{1000}$$

where c is the concentration in $mol\,dm^{-3}$ and v the volume in cm^3. This equation is often used in the form:

$$c = \frac{1000}{v} \times n$$

The first step in a titration is usually to make a **standard solution** (that is, one of a known concentration measured in $mol\,dm^{-3}$). This is done by dissolving a weighed amount of the pure substance in water, and making the solution up to a known volume (often $250\,cm^3$) in a calibrated flask.

Example 1 0.360 g of sodium hydroxide was made up to $250\,cm^3$ in a calibrated flask. What is the concentration of the solution?

Answer The molar mass of sodium hydroxide, NaOH, is $(23+16+1)=40\,g\,mol^{-1}$.

$$n=\frac{m}{M}=\frac{0.360}{40}=9.00\times10^{-3}\,mol$$

$$c=\frac{1000}{V}\times n=\frac{1000}{250}\times9.00\times10^{-3}=\mathbf{3.60\times10^{-2}\,mol\,dm^{-3}}\text{ or }\mathbf{0.0360\,mol\,dm^{-3}}$$

In titration calculations, it is usual to work to 3 significant figures.

Example 2 1.778 g of hydrated ethanedioic acid, $H_2C_2O_4.2H_2O$, were made up to $100\,cm^3$ in a calibrated flask. What is the concentration of the solution?

Answer The molar mass of $H_2C_2O_4.2H_2O$ is $(2+24+64+2\times18)=126\,g\,mol^{-1}$

$$n=\frac{m}{M}=\frac{1.778}{126}=1.41\times10^{-2}\,mol$$

$$c=\frac{1000}{V}\times n=\frac{1000}{100}\times1.41\times10^{-2}$$

$$=\mathbf{1.41\times10^{-1}\,mol\,dm^{-3}}\text{ or }\mathbf{0.141\,mol\,dm^{-3}}$$

Further practice What is the concentration of each of the following solutions?

1 0.730 g of potassium hydroxide made up to $100\,cm^3$.
2 12.230 g of $Na_2CO_3.10H_2O$ made up to $250\,cm^3$.
3 1.367 g of barium hydroxide made up to $250\,cm^3$.
4 0.825 g of ethanoic acid made up to $100\,cm^3$.

Carrying out a titration

Having made a standard solution, the next step is to carry out the titration using this and the solution whose concentration we want to find. One of the solutions (usually the acid) is placed in the burette, and a known volume of the other solution from a pipette (often $25\,cm^3$) is placed in a conical flask. A few drops of the indicator are added to the flask, and the solution from the burette is slowly added, with shaking, until the first permanent colour change is noted (this is the **end-point**). There are several points we should note about the procedure.

- The first and final readings of the burette should be recorded.
- Readings should be recorded to the nearest $0.05\,cm^3$, which is the approximate size of one drop. So readings are recorded as, for example, 25.0, 25.05 and $26.0\,cm^3$.
- During the first titration, the end-point is often missed. However, the reading should still be recorded as a 'rough' titration.
- Further titrations should then be done so that the results of two consecutive titrations differ by no more than $0.1\,cm^3$.

■ Figure 6.3
When reading a burette, the volume noted is the level of the bottom of the meniscus. What volume has been run out of the burette shown in the diagram?

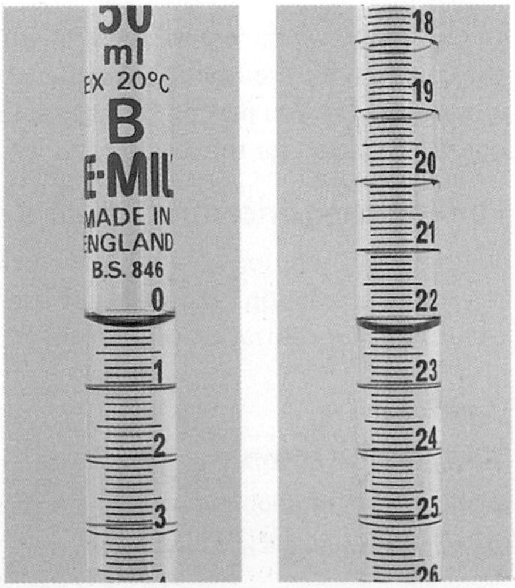

Indicators

The two most common indicators that are used in titrations are methyl orange and phenolphthalein. Their colour changes are shown in Table 6.2.

■ **Table 6.2** The colour changes of two commonly used indicators.

Indicator	Colour change when running acid from the burette	Colour change when running alkali from the burette
Methyl orange	Yellow to first tinge of orange	Red to first tinge of orange
Phenolphthalein	Pink to just colourless	Colourless to just pink

■ **Figure 6.4** The colour changes of methyl orange (**a**→**b** and **c**→**d**) and phenolphthalein (**e**→**f** and **g**→**h**) at the end-point.

In elementary work, you will be told which indicator to use. You will also be given the equation for the reaction if either the acid or base is diprotic. In more advanced work you may be asked to select the appropriate indicator and to work out the equation for the reaction, as we shall see in Chapter 12.

Finding the concentration of a solution

We have two solutions A and B. We know the concentration of solution A (which is a standard solution) and we know the two volumes used in the titration. We want to find the concentration of solution B. The situation is as shown in Table 6.3.

■ **Table 6.3** The basis of a titration calculation.

Solution A	Solution B
Concentration known = c_A	Concentration to be found = c_B = ?
Volume from titration = V_A	Volume from titration = V_B
Amount (in moles) = n_A ⟶	Amount (in moles) = n_B

The balanced chemical equation tells us how many moles of B react with one mole of A.

A table like Table 6.3 will help you to carry out a titration calculation.

- The first step is to work out the amount (in moles) of reactant A:

$$n_A = c_A \times \frac{v_A}{1000}$$

- Then use the chemical equation to find the number of moles of B that react with one mole of A.
- In this way you can calculate the amount (in moles) of solution B used in the titration.
- Finally, work out c_B using the equation:

$$c_B = \frac{1000}{v_B} \times n_B$$

Example 1 25.0 cm^3 of 0.0500 $mol\,dm^{-3}$ sodium hydroxide neutralised 20.0 cm^3 of hydrochloric acid. What is the concentration of the hydrochloric acid?

Answer First we work out $n(NaOH)$:

$$n = c \times \frac{v}{1000}$$

$$= 0.0500 \times \frac{25.0}{1000} = 1.25 \times 10^{-3}\,mol$$

Next we write a balanced equation:

$$NaOH(aq) + HCl(aq) \longrightarrow NaCl(aq) + H_2O(l)$$

This tells us that 1 mol of NaOH reacts with 1 mol of HCl. To find $c(HCl)$, we use:

$$c = \frac{1000}{v} \times n$$

$$= \frac{1000}{20.0} \times 1.25 \times 10^{-3}$$

$$= \mathbf{0.0625\,mol\,dm^{-3}}$$

Example 2 23.65 cm^3 of 0.0800 $mol\,dm^{-3}$ hydrochloric acid were neutralised by 25.0 cm^3 of a solution of sodium carbonate according to the equation:

$$2HCl(aq) + Na_2CO_3(aq) \rightarrow 2NaCl(aq) + H_2O(l) + CO_2(g)$$

What is the concentration of the solution of sodium carbonate?

Answer

$$n(HCl) = c \times \frac{v}{1000} = 0.0800 \times \frac{23.65}{1000} = 1.892 \times 10^{-3}\,mol$$

Since 2 mol of HCl react with 1 mol of Na_2CO_3,

$$n(Na_2CO_3) = \tfrac{1}{2} \times 1.892 \times 10^{-3} = 9.46 \times 10^{-4}\,mol$$

$$\text{so } c(Na_2CO_3) = \frac{1000}{v} \times n$$

$$= \frac{1000}{25.0} \times 9.46 \times 10^{-4}$$

$$= \mathbf{0.0378\,mol\,dm^{-3}}$$

Finding the purity of a substance, or the amount of water of crystallisation

One use of titrations is to find the purity of a chemical. A variation is to find the number of molecules of water of crystallisation in a substance. The following examples illustrate these uses.

Example 1 0.982 g of an impure sample of sodium hydroxide was made up to 250 cm^3. 25.0 cm^3 of this solution were neutralised by 23.5 cm^3 of 0.100 mol dm^{-3} hydrochloric acid. What is the percentage purity of the sodium hydroxide?

Answer

$$n(\text{HCl}) = c \times \frac{v}{1000} = 0.100 \times \frac{23.5}{1000} = 2.35 \times 10^{-3}\,\text{mol}$$

Since 1 mol of NaOH reacts with 1 mol of HCl,

$$n(\text{NaOH}) \text{ in } 25\,\text{cm}^3 = 2.35 \times 10^{-3}\,\text{mol}$$

$$n(\text{NaOH}) \text{ in } 250\,\text{cm}^3 = 2.35 \times 10^{-3} \times \frac{250}{25.0}$$

$$= 2.35 \times 10^{-2}\,\text{mol}$$

Since $M(\text{NaOH}) = 40\,\text{g mol}^{-1}$,

mass of pure NaOH $= 2.35 \times 10^{-2} \times 40 = 0.940\,\text{g}$

percentage purity of NaOH $= \frac{0.940}{0.982} \times 100 =$ **95.7%**

Example 2 Washing soda has the formula $Na_2CO_3.xH_2O$. A mass of 1.4280 g of washing soda was made up to 250 cm^3. 25.0 cm^3 of this solution were neutralised by 20.0 cm^3 of 0.0500 mol dm^{-3} hydrochloric acid. The equation for the reaction is:

$$Na_2CO_3(aq) + 2HCl(aq) \rightarrow 2NaCl(aq) + CO_2(g) + H_2O(l)$$

Find the value of *x*, and the formula of washing soda.

Answer

$$n(\text{HCl}) = c \times \frac{v}{1000} = 0.0500 \times \frac{20.0}{1000} = 1.00 \times 10^{-3}\,\text{mol}$$

Since 1 mol of HCl reacts with $\frac{1}{2}$ mol of Na_2CO_3,

$$n(Na_2CO_3) \text{ in the pipette} = \tfrac{1}{2} \times 1.00 \times 10^{-3} = 5.00 \times 10^{-4}\,\text{mol}$$

$$n(Na_2CO_3) \text{ in } 250\,\text{cm}^3 = 5.00 \times 10^{-4} \times \frac{250}{25.0} = 5.00 \times 10^{-3}\,\text{mol}$$

Since $M(Na_2CO_3) = 106\,\text{g mol}^{-1}$,

mass of anhydrous Na_2CO_3 in 250 cm^3 $= 5.00 \times 10^{-3} \times 106 = 0.530\,\text{g}$
mass of water in the washing soda $= (1.4280 - 0.530) = 0.898\,\text{g}$

Since $M(H_2O) = 18\,\text{g mol}^{-1}$,

$$n(H_2O) = \frac{0.898}{18} = 4.99 \times 10^{-2}\,\text{mol}$$

So 5.00×10^{-3} mol of Na_2CO_3 combine with 4.99×10^{-2} mol of H_2O.

Therefore 1 mol of Na_2CO_3 combines with $\frac{4.99 \times 10^{-2}}{5.00 \times 10^{-3}} = 9.98$ mol of H_2O.

So 1 mol of Na_2CO_3 combines with 10 mol of H_2O, to the nearest whole number. Therefore $\boldsymbol{x = 10}$ and the formula of washing soda is $\mathbf{Na_2CO_3.10H_2O}$.

Further practice

1 1.250 g of concentrated sulphuric acid were made up to 250 cm^3. 25.0 cm^3 of this solution were neutralised by 24.85 cm^3 of 0.102 mol dm^{-3} sodium hydroxide. The equation for the reaction is:

$$H_2SO_4(aq) + 2NaOH(aq) \rightarrow Na_2SO_4(aq) + 2H_2O(l)$$

What is the purity of the concentrated sulphuric acid?

2 Ethanedioic acid has the formula $H_2C_2O_4.xH_2O$. 0.900 g of the acid were made up to 250 cm^3. 25.0 cm^3 of this solution were neutralised by 26.75 cm^3 of 0.0532 mol dm^{-3} sodium hydroxide. The equation for the reaction is:

$$H_2C_2O_4(aq) + 2NaOH(aq) \rightarrow Na_2C_2O_4(aq) + 2H_2O(l)$$

What is the value of *x*, and hence what is the formula of ethanedioic acid?

Finding the stoichiometry of a reaction

Titration can also be used to work out the stoichiometry of an acid–base reaction. If both the acid and the base are monoprotic, then one mole of the acid must react with one mole of the base. But if one of them is diprotic, there are two possible neutralisation reactions. For example, sulphuric acid and sodium hydroxide could react in two ways:

$$H_2SO_4(aq) + NaOH(aq) \rightarrow NaHSO_4(aq) + H_2O(l)$$
$$H_2SO_4(aq) + 2NaOH(aq) \rightarrow Na_2SO_4(aq) + 2H_2O(l)$$

By making standard solutions of both the acid and the base, the stoichiometry can be found by titration.

Example

25.0 cm³ of 0.105 mol dm⁻³ sodium carbonate neutralised 24.50 cm³ of 0.213 mol dm⁻³ hydrochloric acid, using methyl orange as an indicator. Construct an equation for the reaction of sodium carbonate with hydrochloric acid under these conditions.

Answer

$$n(Na_2CO_3) = c \times \frac{v}{1000} = 0.105 \times \frac{25}{1000} = 2.625 \times 10^{-3}\,mol$$

$$n(HCl) = c \times \frac{v}{1000} = 0.213 \times \frac{24.50}{1000} = 5.2185 \times 10^{-3}\,mol$$

So 1 mol of Na_2CO_3 reacts with $\frac{5.2185}{2.625} = 1.988$ mol of HCl

Therefore 1 mol of Na_2CO_3 reacts with 2 mol of HCl (to the nearest whole number). The equation is:

$$Na_2CO_3(aq) + 2HCl(aq) \rightarrow 2NaCl(aq) + CO_2(g) + H_2O(l)$$

Further practice

25.0 cm³ of 0.0765 mol dm⁻³ potassium carbonate, K_2CO_3, neutralised 22.35 dm³ of 0.0850 mol dm⁻³ hydrochloric acid, using phenolphthalein as an indicator. Construct an equation for the reaction between potassium carbonate and hydrochloric acid under these conditions.

6.7 Back titrations

Sometimes there is no definite end-point to a reaction that can be detected using an indicator. An example is in the use of calcium carbonate as the base. Because it is insoluble in water, it reacts slowly with acid and so there is no sharp end-point.

We get round this difficulty by dissolving the calcium carbonate in a known excess of acid, and then finding the acid left over by titration against a soluble base such as sodium hydroxide. The acid used up by the calcium carbonate is the acid added minus the acid left at the end. This method is known as **back titration**.

Back titration
(acid used) = (acid at start) − (acid left at end)

The amount of an acid can also be estimated using back titration. With this method excess base, rather than excess acid, is added and the remainder found by titration with acid.

Example 1

0.765 g of an impure sample of calcium carbonate was dissolved in 25.0 cm³ of 1.00 mol dm⁻³ hydrochloric acid. The resulting solution was made up to 250 cm³ in a calibrated flask. 25.0 cm³ of this solution were neutralised by 24.35 cm³ of 0.050 mol dm⁻³ sodium hydroxide. What is the purity of the calcium carbonate?

Answer

First find the amount of acid at the start:

$$n(HCl)\ added = c \times \frac{v}{1000} = 1.00 \times \frac{25.00}{1000} = 2.50 \times 10^{-2}\,mol$$

Then find the amount of acid left at the end:

$$\text{amount of NaOH used} = c \times \frac{v}{1000} = 0.050 \times \frac{24.35}{1000} = 1.2175 \times 10^{-3}\,\text{mol}$$

Since 1 mol of NaOH reacts with 1 mol of HCl,

$$n(\text{HCl}) \text{ left over in 250 cm}^3 \text{ flask} = \frac{250}{25} \times 1.2175 \times 10^{-3}$$
$$= 1.2175 \times 10^{-2}\,\text{mol}$$

Now find the amount of acid used up in reaction with calcium carbonate:

$$n(\text{HCl}) \text{ used by CaCO}_3 = 2.50 \times 10^{-2} - 1.2175 \times 10^{-2}$$
$$= 1.2825 \times 10^{-2}\,\text{mol}$$

The equation for the reaction is:

$$\text{CaCO}_{3(s)} + 2\text{HCl(aq)} \rightarrow \text{CaCl}_2\text{(aq)} + \text{H}_2\text{O(l)} + \text{CO}_{2(g)}$$

Since 1 mol of HCl reacts with $\frac{1}{2}$ mol of CaCO$_3$,

$$n(\text{CaCO}_3) = \frac{1}{2} \times 1.2825 \times 10^{-2} = 6.413 \times 10^{-3}\,\text{mol}$$
$$M(\text{CaCO}_3) = 40 + 12 + 48 = 100\,\text{g mol}^{-1}$$

So mass of pure CaCO$_3$ in sample $= 6.413 \times 10^{-3} \times 100 = 0.6413\,\text{g}$

and purity of CaCO$_3 = \dfrac{0.6413}{0.765} \times 100 = \textbf{83.8\%}$

Example 2 3.920 g of an oxide of formula MO were completely dissolved in 30.0 cm³ of 2.00 mol dm^{-3} sulphuric acid. The resulting solution was made up to 100 cm³. 25.0 cm³ of this solution were neutralised by 27.50 cm³ of 0.100 mol dm^{-3} sodium hydroxide. What is the relative atomic mass of M? Identify the metal.

Answer First find the amount of acid at the start:

$$n(\text{H}_2\text{SO}_4) \text{ added} = c \times \frac{v}{1000} = 2.00 \times \frac{30.0}{1000} = 6.00 \times 10^{-2}\,\text{mol}$$

Then find the amount of acid left at the end:

$$n(\text{NaOH}) = c \times \frac{v}{1000} = 0.100 \times \frac{27.50}{1000} = 2.75 \times 10^{-3}\,\text{mol}$$

As the H$_2$SO$_4$ is in excess, the equation for the reaction is:

$$\text{H}_2\text{SO}_4\text{(aq)} + 2\text{NaOH(aq)} \rightarrow \text{Na}_2\text{SO}_4\text{(aq)} + 2\text{H}_2\text{O(l)}$$

Since 1 mol of NaOH neutralises $\frac{1}{2}$ mol of H$_2$SO$_4$,

$$n(\text{H}_2\text{SO}_4) \text{ left over in 100 cm}^3 \text{ flask} = \frac{1}{2} \times 2.75 \times 10^{-3} \times \frac{100}{25}$$
$$= 5.50 \times 10^{-3}\,\text{mol}$$

Now find the amount of acid used up in reaction with MO:

$$n(\text{H}_2\text{SO}_4) \text{ used by oxide} = 6.00 \times 10^{-2} - 5.50 \times 10^{-3} = 5.45 \times 10^{-2}\,\text{mol}$$

The oxide MO reacts with H$_2$SO$_4$ as follows:

$$\text{MO(s)} + \text{H}_2\text{SO}_4\text{(aq)} \rightarrow \text{MSO}_4\text{(aq)} + \text{H}_2\text{O(l)}$$

We know that 5.45×10^{-2} mol of H$_2$SO$_4$ reacted with the oxide. Since one mole of acid reacts with one mole of oxide, the amount of oxide used is 5.45×10^{-2} mol, which has a mass of 3.920 g.

$$M = \frac{m}{n} = \frac{3.920}{5.45 \times 10^{-2}} = 71.9\,\text{g mol}^{-1}$$

$$A_r(M) + A_r(O) = 72 \text{ (to nearest whole number)}$$
$$A_r(M) = 72 - 16 = \textbf{56}$$

The metal is **iron**.

Further practice

1 A 0.789 g sample of impure magnesium oxide, MgO, was dissolved in $25.0\,cm^3$ of $1.00\,mol\,dm^{-3}$ sulphuric acid. The resulting solution was made up to $100\,cm^3$. $25.0\,cm^3$ of this solution were neutralised by $27.80\,cm^3$ of $0.100\,mol\,dm^{-3}$ sodium hydroxide. Calculate the percentage purity of the magnesium oxide.

2 A 0.421 g sample of hydrated lithium hydroxide, $LiOH.xH_2O$, was made up to $250\,cm^3$. $25.0\,cm^3$ of this solution were neutralised by $20.05\,cm^3$ of $0.0500\,mol\,dm^{-3}$ hydrochloric acid. What is the value of *x*?

3 (Harder) A 10.00 g sample of a fertilizer which contained ammonium sulphate, $(NH_4)_2SO_4$, was boiled with $25.0\,cm^3$ of $2.00\,mol\,dm^{-3}$ sodium hydroxide. The ammonium sulphate released ammonia as shown by the following equation:

$$(NH_4)_2SO_4(s) + 2NaOH(aq) \rightarrow Na_2SO_4(aq) + 2NH_3(g) + 2H_2O(l)$$

The resulting solution was made up to $250\,cm^3$. $25.0\,cm^3$ of this solution were neutralised by $24.55\,cm^3$ of $0.0500\,mol\,dm^{-3}$ hydrochloric acid. What is the percentage by mass of ammonium sulphate in the fertilizer?

Summary

- According to the **Brønsted–Lowry theory** of acids and bases, an **acid** is a proton donor and a **base** is a proton acceptor.
- A strong acid has a weak **conjugate base**, and vice versa.
- Some acids are **diprotic**, and can donate up to two protons (for example, sulphuric acid). Some bases are diprotic, and can accept up to two protons (for example, the carbonate ion).
- The **end-point** at which an acid is exactly neutralised by a base can be determined using an **indicator** such as methyl orange or phenolphthalein.
- A **standard solution** is a solution made up such that its concentration in $mol\,dm^{-3}$ is known.
- In a **titration**, a standard solution takes part in a neutralisation reaction. The volumes of both solutions are measured, so that, for example, the unknown concentration of the other solution may be calculated.
- In a **back titration**, the titration is used to calculate the amount of acid or base left in a solution after reaction with, for example, a solid reactant.

Key reactions and skills

- You need to know, or be able to work out, equations for the reactions of hydrochloric, nitric and sulphuric acids with:
  - soluble hydroxides, such as sodium hydroxide, NaOH, or ammonium hydroxide, NH_4OH
  - insoluble hydroxides, such as magnesium hydroxide, $Mg(OH)_2$, or iron(III) hydroxide, $Fe(OH)_3$
  - insoluble oxides, such as copper(II) oxide, CuO, or iron(III) oxide, Fe_2O_3
  - soluble carbonates, such as sodium carbonate, Na_2CO_3
  - insoluble carbonates, such as calcium carbonate, $CaCO_3$.
- You need to be able to calculate from experimental results:
  - the concentration of a standard solution
  - the concentration of another solution following titration with a standard solution
  - the number of moles of base that react with one mole of acid
  - an analysis of a back titration.

7 Oxidation and reduction

In Chapter 6, we showed that acid–base reactions involve proton transfer. In this chapter we show that oxidation–reduction (redox) reactions involve electron transfer. In order to find out the oxidation state of an element, it is helpful to use the concept of oxidation number. This enables us to write half-equations, and hence build up whole equations. We can use quantitative analysis to find the number of electrons transferred in a reaction – one convenient method is volumetric analysis using either potassium manganate(VII) or iodine.

The chapter is divided into the following sections.

7.1 What are oxidation and reduction?

Losing and gaining electrons

As long ago as 1741, the word 'reduce' was used to describe the process of extracting metals from their ores. 'Reduction' referred to the fact that the mass decreased when the ore was converted into its metal. Some years later, Lavoisier recognised that these ores were oxides, and by the beginning of the nineteenth century the idea that reactive metals were easily oxidised had become established.

These observations formed the basis of the definitions of oxidation as the addition of oxygen and reduction as the removal of oxygen. But with the use of electrolysis, for example, for the extraction of reactive metals such as sodium and aluminium, this definition became too limited. The reduction of an oxide by heating it with hydrogen or carbon and its reduction in an electrolytic cell are similar chemical processes – both involve the addition of electrons to a positively charged metal ion. The modern definition of **oxidation** is, therefore, the removal of electrons from a substance; conversely, **reduction** is the addition of electrons to a substance. Reduction takes place at the negative electrode (**cathode**) of an electrolytic cell; oxidation takes place at the positive electrode (**anode**).

- **Oxidation** is the removal of electrons.
- **Reduction** is the addition of electrons.

During electrolysis:

- Oxidation takes place at the **anode** – the anode accepts the electrons.
- Reduction takes place at the **cathode** – the cathode provides the electrons.

Half-equations and redox reactions

The processes of oxidation and reduction happen simultaneously in oxidation–reduction reactions (also called **redox reactions**) – one substance is oxidised while the other is reduced. Redox reactions can therefore be broken down into two half-reactions represented by half-equations that show the electron movements. For example, for the combustion of magnesium to form magnesium oxide:

Half-equation 1: oxidation   either $Mg \rightarrow Mg^{2+} + 2e^-$ or $Mg - 2e^- \rightarrow Mg^{2+}$

Half-equation 2: reduction $O_2 + 4e^- \rightarrow 2O^{2-}$

The oxidation half-equation above is written in two ways. The first method shows more clearly that charge is being conserved, while the second method shows more clearly that electron loss is taking place. We will generally use the first method.

To balance the two half-equations together, the number of electrons lost from the magnesium must be the same as the number gained by the oxygen. This can be achieved in two ways:

- add 2 × (half-equation 1) to (half-equation 2) so that 2 mol of magnesium react with 1 mol of oxygen:

$$2Mg(s) + O_2(g) \rightarrow 2MgO(s)$$

- add half-equation 1 to $\frac{1}{2}$ × (half-equation 2) so that 1 mol of magnesium reacts with $\frac{1}{2}$ mol of oxygen:

$$Mg(s) + \tfrac{1}{2}O_2(g) \rightarrow MgO(s)$$

In a redox reaction, the number of electrons lost = the number of electrons gained.

The complete equation for the combustion of magnesium could have been written easily without using half-equations. With more complicated redox examples, however, this would be much harder. Consider, for example, the reduction of aqueous iron(III) sulphate by metallic zinc. The first stage is to identify the oxidised and reduced species. The Fe^{3+}(aq) ions are reduced to Fe(s) while the Zn(s) is oxidised to Zn^{2+}(aq) ions. We can now write the relevant half-equations:

oxidation: $Zn \rightarrow Zn^{2+} + 2e^-$
reduction: $Fe^{3+} + 3e^- \rightarrow Fe$

We must multiply the first equation by 3 and the second equation by 2 in order to balance the electron transfer:

$$3Zn \rightarrow 3Zn^{2+} + 6e^-$$
$$2Fe^{3+} + 6e^- \rightarrow 2Fe$$

Adding these together gives:

$$3Zn(s) + 2Fe^{3+}(aq) \rightarrow 3Zn^{2+}(aq) + 2Fe(s)$$

This reaction proceeds because Zn(s) is a more powerful reducing agent than Fe(s). Alternatively, we can say that Fe^{3+}(aq) is a more powerful oxidising agent than Zn^{2+}(aq). This can be found out qualitatively by carrying out the reaction in a test tube. To compare strengths of oxidising and reducing agents quantitatively, we need to combine the two half-reactions into an electrochemical cell and measure the voltage produced, as we shall see in Chapter 13.

Example 1 In the reaction:

$$Zn(s) + CuSO_4(aq) \rightarrow ZnSO_4(aq) + Cu(s)$$

a What species has been oxidised and what species has been reduced?
b Write the two relevant half-equations.

Answer **a** Zn(s) is oxidised: Cu^{2+}(aq) is reduced (note that it is not $CuSO_4$(aq) that is reduced).
b Oxidation: $Zn \rightarrow Zn^{2+} + 2e^-$; reduction: $Cu^{2+} + 2e^- \rightarrow Cu$

Example 2 In the following reactions identify whether the underlined species have been oxidised, reduced or neither.

a $\underline{Al(s)} + 1\tfrac{1}{2}Br_2(l) \rightarrow AlBr_3(s)$
b $2Mg(s) + \underline{TiO_2(s)} \rightarrow 2MgO(s) + Ti(s)$
c $\underline{Ag^+(aq)} + Cl^-(aq) \rightarrow AgCl(s)$
d $\underline{Cl_2(aq)} + SnCl_2(aq) \rightarrow SnCl_4(aq)$

Answer **a** Oxidised **b** Reduced **c** Neither **d** Reduced

Further practice **1** Write half-equations and then full equations for the reactions that take place between:
a aqueous sulphuric acid and magnesium metal
b liquid bromine and lithium metal
c oxygen gas and aluminium metal
d zinc metal and aqueous silver nitrate.

2 In the following reactions, identify which species have been oxidised and which have been reduced.
   a $Fe(s) + S(s) \rightarrow FeS(s)$
   b $MnO_2(s) + 4HCl(aq) \rightarrow MnCl_2(aq) + Cl_2(g) + 2H_2O(l)$
   c $2FeCl(aq) + Cl_2(g) \rightarrow 2FeCl_3(aq)$
   d $CuO(s) + Cu(s) \rightarrow Cu_2O(s)$
3 In the following reactions, identify whether the underlined species have been oxidised, reduced or neither.
   a $\underline{Cu}O(s) + H_2SO_4(aq) \rightarrow CuSO_4(aq) + H_2O(l)$
   b $\underline{Pb}O_2(s) + 4HCl(aq) \rightarrow PbCl_2(aq) + Cl_2(g) + 2H_2O(l)$
   c $4\underline{Fe}(OH)_2(s) + O_2(g) + 2H_2O(l) \rightarrow 4Fe(OH)_3(s)$
   d $\underline{Zn}(s) + 2V^{3+}(aq) \rightarrow Zn^{2+}(aq) + 2V^{2+}(aq)$

7.2 Oxidation numbers

So far we have restricted oxidation and reduction to elements and their ionic compounds. For redox reactions of this type, it is quite clear which species are losing or gaining electrons. However, many compounds contain covalent bonds and for these a simple ionic treatment is inappropriate.

Oxidation number and simple ionic compounds

To get round this difficulty the concept of **oxidation number** has been introduced. This idea can be applied to both ionic and covalent compounds. For simple ionic compounds, the oxidation number of the element is the same as the charge on the species containing the element. This means that all elements have oxidation number of 0, cations have positive oxidation numbers and anions have negative oxidation numbers. The oxidation number is written in arabic numbers in brackets, immediately after the species (without a space). So $Cl_2(g)$ contains Cl(0), $FeCl_2$ contains Fe(+2) and MgO contains O(−2).

The convention is useful when a metal can exist in more than one **oxidation state**. Iron, for example, has oxidation numbers of +2 and +3 which can be distinguished in their formulae by using roman numerals; for example, $FeSO_4$ is iron(II) sulphate and $Fe_2(SO_4)_3$ is iron(III) sulphate. In the names or formulae, roman numerals are used.

Oxidation number and covalent molecules

The concept of oxidation number can be extended to covalent compounds in the following way. The only covalent compounds that contain bonds with no ionic character at all are those in elements such as hydrogen, H_2, or sulphur, S_8. These atoms are given an oxidation number of 0 by convention. All other covalent bonds have some ionic character, whose magnitude depends on the elctronegativity difference between the two bonded atoms (see Chapter 3, page 66). The convention is to assign oxidation numbers as though the bond were completely ionic, rather than partially ionic. For example, in carbon dioxide, CO_2, oxygen is more electronegative than carbon. Each oxygen is given the oxidation number −2, just as in an ionic metal oxide. In order to maintain electrical neutrality, the carbon must be assigned the oxidation number +4. In this case it is helpful to include the '+' sign, as in other compounds carbon can have other oxidation numbers, including −4. Of course, we do not suggest that carbon dioxide is actually composed of one C^{4+} and two O^{2-} ions, but the convention indicates that the oxidation of carbon to carbon dioxide is equivalent to a four-electron transfer.

With a few exceptions that will be discussed later, oxygen in its compounds always has an oxidation number of −2. Similarly, in most of its compounds, hydrogen has an oxidation number of +1. This often helps in working out the oxidation numbers of other elements in covalent compounds.

- The **oxidation number** of atoms in an element = 0.
- In a compound, the more electronegative element is given a negative oxidation number (fluorine always has oxidation number −1).
- The sum of all the oxidation numbers in a molecule = 0.
- In most compounds, the oxidation number of oxygen is −2 and that of hydrogen is +1.

Example 1 What is the oxidation number of carbon in each of the following compounds?

a CH_4 **b** CCl_4 **c** CH_2Cl_2

Answer

a Carbon is more electronegative than hydrogen. Each hydrogen has oxidation number +1, so that of carbon is −4.

b Chlorine is more electronegative than carbon. Each chlorine has oxidation number −1. Let the oxidation number of carbon be x. Then $x + 4(-1) = 0$, so the oxidation number of carbon is +4.

c $x + 2(+1) + 2(-1) = 0$, so the oxidation number of carbon is 0.

Example 2 What is the oxidation number of the underlined species in the following compounds?

a $\underline{S}O_3$ **b** $\underline{I}Cl$ **c** $\underline{P}_2O_3$

Answer In each case let the unknown oxidation number be x.

a $x + 3(-2) = 0$, giving $x = +6$, so S = +6.

b $x + (-1) = 0$, giving $x = +1$, so I = +1.

c $2x + 3(-2) = 0$, giving $x = +3$, so P = +3.

Further practice What is the oxidation number of the underlined element in each of the following compounds?

1 $\underline{Cl}_2O_7$ **2** $\underline{P}F_3$ **3** $\underline{C}HCl_3$ **4** $\underline{N}_2H_4$ **5** $\underline{N}H_2OH$

Oxidation number and ions containing covalent bonds

Many ions are not simple, but are made up of several covalently bonded atoms. A familiar example is the sulphate ion, SO_4^{2-}. The oxidation number of each oxygen is −2, but it is not immediately clear what the oxidation number of the sulphur atom is. We can, however, work out the oxidation number of sulphur in potassium sulphate, K_2SO_4, because the sum of all the oxidation numbers is 0. If the oxidation number of sulphur is x, then $2(+1) + x + 4(-2) = 0$, giving $x = +6$. This must also be the oxidation number of sulphur in the SO_4^{2-} ion. It can be found directly by letting the sum of the oxidation numbers equal the charge on the ion: $x + 4(-2) = -2$, giving $x = +6$ as before.

In an ion, the sum of the oxidation numbers equals the charge on the ion.

Example What is the oxidation number of: **a** Cr in $Cr_2O_7^{2-}$ **b** V in VO_2^+?

Answer Let the unknown oxidation number be x.

a $2x + 7(-2) = -2$, giving $x = +6$

b $x + 2(-2) = 1$, giving $x = +5$

Further practice What is the oxidation number of:

1 Al in AlO_2^- **2** P in HPO_3^{2-} **3** V in VO^{2+} **4** C in $C_2O_4^{2-}$ **5** Pb in $PbCl_4^{2-}$ **6** Sn in $Sn(OH)_6^{2-}$?

The oxidation numbers of oxygen and hydrogen

Oxygen is the second most electronegative element, and in most of its compounds, it has an oxidation number −2. There are two exceptions.

- As fluorine is more electronegative than oxygen, compounds of oxygen and fluorine are oxygen fluorides rather than fluorine oxides. For example, in the fluoride OF_2, each fluorine has an oxidation number −1 and oxygen has an oxidation number +2.
- Peroxides contain the O—O bond. In both the peroxide ion, O_2^{2-}, and in covalent peroxides, for example hydrogen peroxide, H_2O_2, each oxygen atom has an oxidation number −1.

In most of its compounds, hydrogen is joined to a more electronegative element such as carbon or oxygen. In these compounds the oxidation number of hydrogen is +1. But there are also compounds in which hydrogen is combined with a metal that is less electronegative than hydrogen (for example, lithium hydride, LiH). In these compounds the oxidation number of hydrogen is −1.

When oxidation numbers cease to be useful

There are two main uses for oxidation numbers.

- In quantitative work, particularly volumetric analysis as we shall see later in the chapter, it is essential to know the stoichiometry of the reaction being studied and this is most easily found using oxidation numbers.
- When converting one compound into another, it is essential to know if the reaction involves oxidation and reduction. This is most easily found by using oxidation numbers.

Oxidation numbers can readily be calculated if the compound is not too complicated. For example, the oxidation number of each carbon in ethane, C_2H_6, is −3. However, a problem arises when we calculate the oxidation of carbon in propane, C_3H_8, as $-2\frac{2}{3}$, as the carbon atoms in this compound are in the same chemical environment as those in ethane.

In the conversion of ethane into bromoethane:

$$C_2H_6 + Br_2 \rightarrow C_2H_5Br + HBr$$

very little information is gained by stating that the oxidation number of the carbon atoms has increased from −3 to −2. It does show that a two-electron transfer reaction has taken place, but this is more easily established by noting that the two Br(0) atoms have changed their oxidation number to −1.

In more complicated reactions, which in any case are unlikely to be quantitative, the use of [O] to indicate the addition of oxygen and [H] the addition of hydrogen is an established convention. In this way, the oxidation of ethanol to ethanoic acid can be represented by:

$$C_2H_5OH + 2[O] \rightarrow CH_3CO_2H + H_2O$$

which shows that a four-electron transfer reaction has taken place ($2[O] + 4e^- \rightarrow 2[O(-2)]$). Similarly, the reduction of propene to propane can be represented by:

$$C_3H_6 + 2[H] \rightarrow C_3H_8$$

which shows that a two-electron transfer reaction has taken place ($2[H] \rightarrow 2H(+1) + 2e^-$).

- Use oxidation numbers as an aid in balancing redox reactions.
- Avoid fractional oxidation numbers.

If a simple calculation gives a fractional oxidation number this usually means that some atoms have one integral oxidation number and other atoms have another. For example, in Fe_3O_4 the average oxidation number of iron is $+2\frac{2}{3}$, but a better description is to assign one iron atom as Fe(+2) and two as Fe(+3).

7.3 Balancing redox equations

An important use of oxidation numbers is to help establish the stoichiometry of an unknown redox reaction. The overall reaction can be found by combining the half-equations for the oxidation and reduction reactions.

Writing the half-equations

Some half-equations are easy to work out, for example, the oxidation of iron(II) ions to iron(III) ions:

$$Fe^{2+} - e^- \rightarrow Fe^{3+} \quad \text{or} \quad Fe^{2+} \rightarrow Fe^{3+} + e^-$$

Others are more difficult, for example, the oxidation of nitrite ions to nitrate ions in acid solution. The following stages lead to the relevant half-equation.

- Work out oxidation numbers – nitrite, NO_2^-, is N(+3) and nitrate, NO_3^-, is N(+5).
- Balance for redox – we know that this is a two-electron transfer reaction from the oxidation numbers given: $N(3) \rightarrow N(5) + 2e^-$

  $$NO_2^- \rightarrow NO_3^- + 2e^-$$

- Balance for charge – because the reaction is being carried out in acid solution, we add H^+ ions to one side of the equation. We need $2H^+$ on the right-hand side of the equation so that the overall charge on each side is the same, in this case −1.

  $$NO_2^- \rightarrow NO_3^- + 2e^- + 2H^+$$

  (If the reaction had been carried out in alkaline solution, we would balance by adding OH^- ions to one side of the equation.)
- Balance for hydrogen – this is done by adding water to one side of the equation. We need H_2O on the left-hand side of the equation to balance the $2H^+$ on the right-hand side.

  $$NO_2^- + H_2O \rightarrow NO_3^- + 2e^- + 2H^+$$

- Check for oxygen – the half-equation is now balanced, but it is a good idea to check that it is correct by confirming that the oxygen atoms balance.

To produce a balanced half-equation:

- Work out oxidation numbers.
- Balance for redox – add the correct number of e^- to account for the difference between the two oxidation states.
- Balance for charge – add the correct number of H^+ (or OH^-) to one side of the equation so that the total charge on each side is the same.
- Balance for H – add the correct number of H_2O to one side of the equation so that both sides have the same number of H atoms.
- Check for O – there should now be an equal number of O atoms on each side of the equation.

Example Produce a balanced half-equation for the reduction of manganate(VII) ions to manganese(II) ions in acid solution.

Answer

Work out oxidation numbers:	MnO_4^- is Mn(+7) and Mn^{2+} is Mn(+2).
Balance for redox:	$MnO_4^- + 5e^- \rightarrow Mn^{2+}$
Balance for charge:	$MnO_4^- + 5e^- + 8H^+ \rightarrow Mn^{2+}$
	(In acid solution, xH^+ ions are added to one side of the equation to balance for charge: $-1 + (-5) + x(+1) = +2$, giving $x = 8$.)

Balance for H: $MnO_4^- + 5e^- + 8H^+ \rightarrow Mn^{2+} + 4H_2O$

(This is done by adding yH_2O to one side of the equation. With $8H^+$ on the left-hand side, we need $4H_2O$ on the right-hand side.)

Check for O: 4O 4O

Further practice

Produce half-equations for the following changes:

1 the oxidation of $Cr(OH)_3$ to CrO_4^{2-} in alkaline solution
2 the reduction of VO_2^+ to V^{2+} in acid solution
3 the oxidation of $H_2C_2O_4$ to CO_2 in acid solution
4 the reduction of $Cr_2O_7^{2-}$ to Cr^{3+} in acid solution
5 the reduction of IO_3^- to I_2 in acid solution
6 the reduction of oxygen in alkaline solution
7 the reduction of H_2O_2 to water in acid solution
8 the oxidation of H_2O_2 to oxygen in alkaline solution.

Producing a full equation from two half-equations

A full equation can be written by adding the two half-equations for the oxidised and reduced species together. It is essential that the electrons balance between the two half-equations, and to this end one (or both) of them may have to be multiplied throughout by an appropriate factor.

After adding the equations together, the moles of any substance that appears on both sides of the equation (for example, water) can be deleted.

To produce a full equation:

- Write down the two half-equations.
- Multiply the equations so that the electrons cancel.
- Add the equations together.
- Simplify if possible.
- Check the equation balances for charge.
- Add state symbols.

Example

Produce a balanced equation for the oxidation of iron(II) ions with acidified manganate(VII) ions.

Answer

The two relevant half-equations are:

oxidation: $Fe^{2+} \rightarrow Fe^{3+} + e^-$

reduction: $MnO_4^- + 5e^- + 8H^+ \rightarrow Mn^{2+} + 4H_2O$

We must multiply the first equation by 5 to eliminate the electrons.

$$5Fe^{2+} \rightarrow 5Fe^{3+} + 5e^-$$

On addition, we then have:

$$MnO_4^-(aq) + 5Fe^{2+}(aq) + 8H^+(aq) \rightarrow Mn^{2+}(aq) + 5Fe^{3+}(aq) + 4H_2O(l)$$

There are 17 positive charges on both sides of the equation.

Further practice

1 Produce balanced equations for the following reactions:
   a the oxidation of I^- ions to iodine with iodate(V) ions, IO_3^-, in acid solution
   b the oxidation of NO_2^- to NO_3^- with manganate(VII) ions, MnO_4^-, in acid solution
   c the reduction of VO^{2+} to V^{2+} with metallic zinc in acid solution
   d the oxidation of $Fe(OH)_2$ to $Fe(OH)_3$ by oxygen.

2 (Harder) Write balanced equations for the following reactions:
   a the reduction of $Cr_2O_7^{2-}$ to Cr^{3+} with metallic magnesium in acid solution
   b the reduction of nitrate(V) to ammonia with metallic aluminium in alkaline solution
   c the oxidation of $Cr(OH)_3$ to CrO_4^{2-} by H_2O_2 in alkaline solution.

Disproportionation

If an element has three or more oxidation states, then it can act as its own oxidant (oxidising agent) and reductant (reducing agent). Usually, the higher the oxidation state of the element, the more powerful an oxidant it is. If a compound that contains the element in a high oxidation state is mixed with a compound that contains the element in a low oxidation state, the result is that the intermediate oxidation state is formed. For example, if vanadium(+5) is mixed with vanadium(+3), the result is a compound containing vanadium(+4).

$$VO_2^{+}(aq) + V^{3+}(aq) \rightarrow 2VO^{2+}(aq)$$

If iodine(+5) is mixed with iodine(−1), the product is iodine(0) (see Figure 7.1).

■ Figure 7.1
Oxidation of iodide, I^-, by iodate(V), IO_3^-. The diagram shows that one iodate ion oxidises five iodide ions.

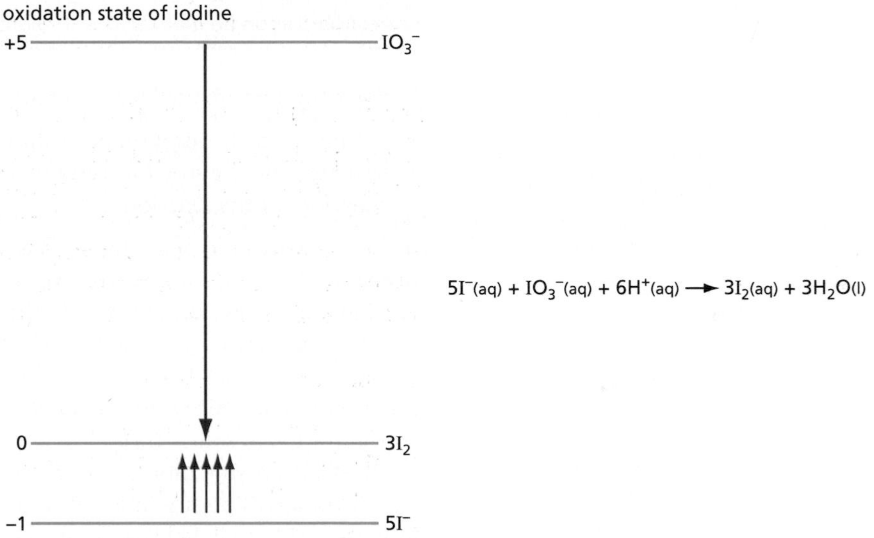

The strength of a substance as an oxidising agent can be measured using electrochemical cells, as we shall see in Chapter 13. Such measurements explain why the high oxidation state oxidises the low oxidation state.

Occasionally, the intermediate oxidation state is a more powerful oxidant than the higher oxidation state. This is because the intermediate state has a structure that makes it unstable. The intermediate state then **disproportionates** and breaks down to a mixture of the substances in the higher and lower oxidation states. A familiar example is hydrogen peroxide, H_2O_2, containing O(−1). This spontaneously breaks down to oxygen, O(0), and water, which contains O(−2) (see Figure 7.2).

Some more examples of disproportionation are described in Chapter 17, pages 330–1.

■ Figure 7.2
The disproportionation of hydrogen peroxide, H_2O_2. The diagram shows that each molecule of H_2O_2 is converted into one H_2O and $\frac{1}{2}O_2$.

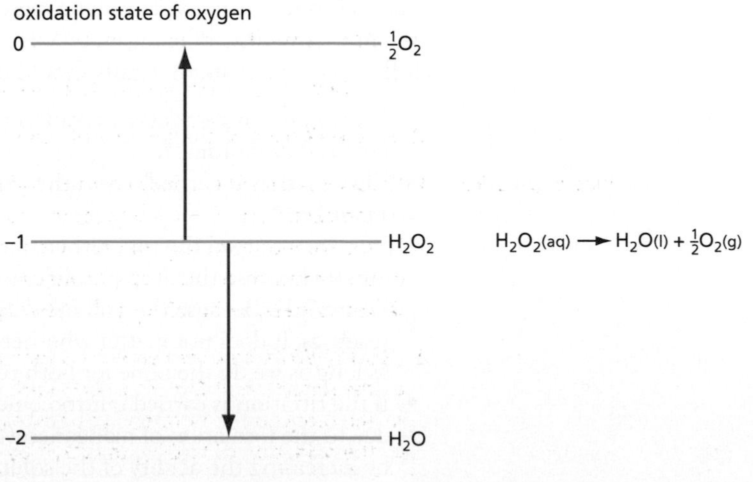

7.4 Redox titrations

Redox titrations involve calculations similar to those described for acid–base titrations (Chapter 6, pages 136–7).

The basic equations used in titrations – a reminder

$$n = \frac{m}{M}$$

$$n = c \times \frac{v}{1000}$$

or $$c = \frac{1000}{v} \times n$$

n = amount (in mol), m = mass (in g), M = molar mass (in g mol^{-1}),
c = concentration (in mol dm^{-3}), v = volume (in cm^3)

Redox titrations are used for two main reasons:

- to find the concentration of a solution
- to determine the stoichiometry of a redox reaction and hence to suggest a likely equation for the reaction.

If the concentration of a solution B is unknown and is to be found by titration, we need a standard solution A (that is, one whose concentration is known) and we need to know how many moles of solution B react with one mole of solution A. After having carried out the titration, the stages in the calculation are as shown in Table 6.3, page 138. The balanced chemical equation allows us to calculate the number of moles of B from the number of moles of A. This then leads us to the concentration of solution B.

If instead the stoichiometry of the reaction is to be found, then the concentration of both solution A and solution B must be known. The volumes from the titration enable us to calculate the number of moles of solution B that react with one mole of solution A.

Although a wide range of oxidising agents can be used in redox titrations, we shall concentrate on two, namely potassium manganate(VII) and iodine.

Titrations using potassium manganate(VII) as the oxidant

Usually potassium manganate(VII) titrations are carried out in acid solution. Under these conditions, the following half-equation is relevant:

$$MnO_4^- + 5e^- + 8H^+ \rightarrow Mn^{2+} + 4H_2O$$

You will meet this equation often in redox reactions, and it is a good idea to learn it.

The deep purple manganate(VII) ion is used as its own indicator. This solution is run into the flask from the burette, and in the course of the reaction the colour changes as nearly colourless manganese(II) ions are produced. At the end-point, the first extra drop of manganate(VII) ions makes the solution turn pink. The following experimental details should be noted.

- Potassium manganate(VII) is not very soluble, and the highest concentration used is 0.02 mol dm^{-3}.
- The reaction is carried out in the presence of sulphuric acid at approximately 1 mol dm^{-3}.
- It is usual to read the top, rather than the bottom, of the meniscus in the burette because the deep purple colour obscures the reading at the bottom (see Figure 7.3). Because the volume delivered is the difference between the readings, it does not matter whether we read the top or bottom of the meniscus, as long as we do the same for both readings.
- If the titration is carried out too quickly, the solution may turn brown. This is due to the formation of manganese(IV) oxide, MnO_2. This can be avoided either by increasing the acidity of the solution or by warming the solution.

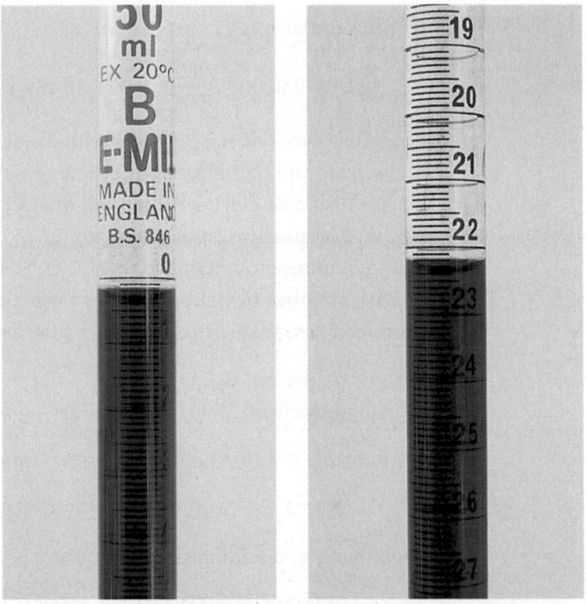

■ **Figure 7.3**
The top of the meniscus is read when using potassium manganate(VII) in a titration. Compare the readings with those in Figure 6.3, page 137.

■ **Figure 7.4**
To avoid the solution turning brown, warm the solution or use plenty of acid.

Example 1 In a titration, 25.0 cm^3 of a solution of iron(II) sulphate required 22.4 cm^3 of 0.0200 $mol\,dm^{-3}$ potassium manganate(VII) solution at the end-point. What is the concentration of the solution of iron(II) sulphate?

Answer

$$n(KMnO_4) = c \times \frac{V}{1000} = 0.0200 \times \frac{22.4}{1000}$$

$$= 4.48 \times 10^{-4}\,mol$$

oxidation: $Fe^{2+} \rightarrow Fe^{3+} + e^-$

reduction: $MnO_4^- + 5e^- + 8H^+ \rightarrow Mn^{2+} + 4H_2O$

To balance the electrons, 5 mol of Fe^{2+} react with 1 mol of $KMnO_4$. Therefore:

$n(Fe^{2+})$ used is $5 \times 4.48 \times 10^{-4} = 2.24 \times 10^{-3}$ mol

$$c(Fe^{2+}) = \frac{1000}{V} \times n$$

$$= \frac{1000}{25.0} \times 2.24 \times 10^{-3}$$

$$= \mathbf{0.0896\,mol\,dm^{-3}}$$

Example 2 25.0 cm^3 of a solution of 0.0200 $mol\,dm^{-3}$ of ethanedioic acid, $H_2C_2O_4$, reacted with 20.0 cm^3 of 0.0100 $mol\,dm^{-3}$ potassium manganate(VII) solution.
How many moles of ethanedioic acid react with 1 mol of potassium manganate(VII)? Suggest a likely equation for the reaction.

Answer $n(H_2C_2O_4) = c \times \frac{V}{1000}$ $n(KMnO_4) = c \times \frac{V}{1000}$

$= 0.0200 \times \frac{25.0}{1000} = 5.00 \times 10^{-4}$ mol $= 0.0100 \times \frac{20.0}{1000} = 2.00 \times 10^{-4}$ mol

Therefore 5 mol of $H_2C_2O_4$ react with 2 mol of $KMnO_4$ (or $2\frac{1}{2}$ mol react with 1 mol). We know from the half-equation for manganate(VII) that 2 mol of $KMnO_4$ receive $2 \times 5 = 10$ mol of electrons, so 5 mol of $H_2C_2O_4$ donate 10 mol of electrons, and 1 mol of $H_2C_2O_4$ donates $\frac{10}{5} = 2$ mol of electrons.

The oxidation number of C in $H_2C_2O_4$ is +3. During the oxidation, 2C lose 2 electrons, so each changes oxidation number from +3 to +4. This suggests that CO_2 may be produced. We shall write the half-equation assuming this:

oxidation: $H_2C_2O_4 \rightarrow 2CO_2 + 2e^- + 2H^+$
reduction: $MnO_4^- + 5e^- + 8H^+ \rightarrow Mn^{2+} + 4H_2O$

We multiply the first half-equation by 5 and the second by 2, and add them together:

$$2MnO_4^- + 5H_2C_2O_4 + 16H^+ \rightarrow 2Mn^{2+} + 8H_2O + 10CO_2 + 10H^+$$

Simplifying and adding state symbols:

$$\mathbf{2MnO_4^-(aq) + 5H_2C_2O_4(aq) + 6H^+(aq) \rightarrow 2Mn^{2+}(aq) + 8H_2O(l) + 10CO_2(g)}$$

(Check the charges balance: four positive charges on both sides.)

Further practice

1 A steel nail with a mass of 2.47 g was dissolved in aqueous sulphuric acid and the solution made up to 250 cm^3 in a standard flask. 25.0 cm^3 of this solution reacted with 18.7 cm^3 of 0.0105 mol dm^{-3} potassium manganate(VII) solution. Calculate:
- a the concentration of the iron solution
- b the mass of iron present in the 250 cm^3 flask
- c the percentage by mass of iron in the steel nail.

2 25.0 cm^3 of acidified 0.0370 mol dm^{-3} sodium nitrate(III), $NaNO_2$, solution reacted with 23.9 cm^3 of 0.0155 mol dm^{-3} potassium manganate(VII) solution.
- a Calculate the number of moles of sodium nitrate(III) that react with 1 mol of manganate(VII).
- b Suggest a likely equation for the reaction.

3 A 1.31 g sample of hydrated potassium ethanedioate, $K_2C_2O_4.xH_2O$, was dissolved in acid and made up to 250 cm^3 in a standard flask. 25.0 cm^3 of this solution reacted with 28.5 cm^3 of 0.0100 mol dm^{-3} potassium manganate(VII) solution. Calculate:
- a the amount of potassium manganate(VII) in 25.0 cm^3 of solution
- b the amount of potassium ethanedioate in 250 cm^3 of solution
- c $M_r(K_2C_2O_4.xH_2O)$
- d the value of *x*.

Titrations using iodine as the oxidant

Iodine is a weak oxidising agent, but it is often used to oxidise the thiosulphate ion, $S_2O_3^{2-}$, to the tetrathionate ion, $S_4O_6^{2-}$:

$$I_2(aq) + 2S_2O_3^{2-}(aq) \rightarrow 2I^-(aq) + S_4O_6^{2-}(aq)$$

This is another reaction that you will meet often, so it is a good idea to learn this equation.

The thiosulphate solution is placed in the burette. The iodine solution is initially brown in colour, and as the thiosulphate is added it fades to pale yellow. Finally at the end-point the solution becomes colourless. The end-point is rather indistinct and it is usual to add a few drops of starch solution. This forms an intense blue colour with iodine. The end-point is then a sharp change from blue to colourless, when all the iodine has been used up. The starch solution is not added until the iodine solution is pale yellow – if it is added at the start of the titration, clumps of blue solid may be formed which are difficult to break up.

■ **Figure 7.5**
The end-point of an iodine titration is made much sharper using starch. A few drops of starch solution are added when the solution is pale yellow; the end-point occurs when the intense blue colour just disappears.

This type of titration is rarely used simply to find the concentration of an iodine solution. There is a large number of oxidising agents that oxidise iodide ions to iodine, and the iodine titration is used mainly to analyse such reactions. If an excess of aqueous potassium iodide is added to the oxidising agent, the amount of iodine liberated is directly related to the amount of oxidising agent used. Table 7.1 lists some of the oxidising agents that may be estimated in this way.

■ **Table 7.1**
Some oxidising agents that can be estimated using iodine/thiosulphate. Each mole of iodine reacts with two moles of thiosulphate.

Oxidising agent	Equation	$\frac{n\text{ (thiosulphate)}}{n\text{ (oxidising agent)}}$
Cl_2	$Cl_2 + 2I^- \rightarrow 2Cl^- + I_2$	2
Br_2	$Br_2 + 2I^- \rightarrow 2Br^- + I_2$	2
IO_3^-	$IO_3^- + 6H^+ + 5I^- \rightarrow 3I_2 + 3H_2O$	6
MnO_4^-	$MnO_4^- + 5I^- + 8H^+ \rightarrow 2\frac{1}{2}I_2 + 4H_2O$	5
Cu^{2+}	$Cu^{2+} + 2I^- \rightarrow CuI + \frac{1}{2}I_2$	1

Thiosulphate oxidation numbers

The thiosulphate ion, $S_2O_3^{2-}$, is a good example of the inappropriate use of oxidation numbers.

The equation:

$$I_2(aq) + 2S_2O_3^{3-}(aq) \rightarrow 2I^-(aq) + S_4O_6^{2-}(aq)$$

shows that the half-equations are as follows:

$$\text{oxidation:} \quad 2S_2O_3^{2-} \rightarrow S_4O_6^{2-} + 2e^-$$
$$\text{reduction:} \quad I_2 + 2e^- \rightarrow 2I^-$$

So each thiosulphate ion receives one electron. Some books suggest that the oxidation number of each sulphur atom has changed from +2 to $+2\frac{1}{2}$. Another suggestion compares the $S_2O_3^{2-}$ ion with the SO_4^{2-} ion, and assigns S(+6) to the central sulphur atom and S(−2) to the other one. On oxidation, the S(−2) atom then becomes S(−1). This may show what is taking place during the reaction more accurately.

In short, oxidation numbers cease to be useful when applied to the thiosulphate ion. All that needs to be known is that each $S_2O_3^{2-}$ ion loses one electron when it reacts with iodine – it is not necessary to know the oxidation number of each sulphur atom.

Example Although the active ingredient in a commercial bleach is chlorate(I) ions, ClO^-, its concentration is often quoted in terms of free chlorine. This is because the concentration is estimated by liberating iodine from potassium iodide. 10.0 cm³ of the bleach were made up to 250.0 cm³ in a standard flask. 25.0 cm³ of this solution were added to an excess of aqueous potassium iodide.

The iodine liberated reacted with 22.0 cm³ of 0.105 mol dm⁻³ thiosulphate solution. Calculate:

a the concentration of 'chlorine' in the diluted bleach solution

b the concentration of 'chlorine' in g dm⁻³ in the commercial bleach. $A_r(\text{Cl}) = 35.5$.

Answer **a** First we calculate the amount of thiosulphate used:

$$n(\text{S}_2\text{O}_3^{2-}) = c \times \frac{V}{1000} = 0.105 \times \frac{22.0}{1000} = 2.31 \times 10^{-3}\,\text{mol}$$

The ratio in Table 7.1 (page 155) tells us that $n(\text{Cl}_2)$ in 25.0 cm³ of diluted bleach solution is $\frac{1}{2} \times 2.31 \times 10^{-3} = 1.155 \times 10^{-3}$ mol. Therefore:

$$c(\text{Cl}_2) = \frac{1000}{V} \times n = \frac{1000}{25.0} \times 1.155 \times 10^{-3}$$

$$= \mathbf{0.0462\,mol\,dm^{-3}}$$

b $M(\text{Cl}_2) = 2 \times 35.5\,\text{g mol}^{-1}$. The original bleach was diluted 25 times. Therefore the concentration of chlorine in the bleach is:

$$25 \times (2 \times 35.5) \times 0.0462 = \mathbf{82.0\,g\,dm^{-3}}$$

Further practice

1 A 25.0 cm³ sample of 0.0210 mol dm⁻³ potassium peroxodisulphate(VI), $K_2S_2O_8$, was treated with an excess of potassium iodide. The iodine liberated reacted with 21.0 cm³ of 0.0500 mol dm⁻³ thiosulphate. Calculate:

a the amount of $S_2O_8^{2-}$ and I^- used

b the amount of I^- that react with 1 mol of $S_2O_8^{2-}$.

c Suggest a likely equation for the reaction between $K_2S_2O_8$ and KI.

2 A 1.00 g sample of brass (an alloy of copper and zinc) was dissolved in nitric acid and made up to 250 cm³ in a standard flask. To a 25.0 cm³ sample of this solution was added an excess of aqueous potassium iodide. The iodine liberated reacted with 27.8 cm³ of 0.0425 mol dm⁻³ thiosulphate. Calculate:

a the concentration of Cu^{2+} in the solution

b the total mass of copper dissolved

c the percentage of copper in the sample of brass.

Summary

- **Oxidation** is the removal of electrons; **reduction** is the addition of electrons.
- In a redox reaction, the number of electrons lost = the number of electrons gained.
- To work out the number of electrons lost or gained, **oxidation numbers** may be used.
- Oxidation numbers can also be used to work out the oxidation and reduction half-equations for a redox reaction. The two half-equations, suitably combined together, give the whole oxidation–reduction equation.
- Titrations can be carried out based on redox reactions. The most important redox titrations use potassium manganate(VII) or iodine.

8 Rates of reaction

In Chapter 5, we explored enthalpy changes for reactions. While the value of ΔH may indicate the direction of chemical change, it does not tell us how fast a reaction takes place. This can be found out only by experiment. The rate of reaction depends on a variety of factors, each of which can be studied separately in order to find out how it affects the rate. We find that for a reaction to take place, not only must the molecules of the reactants collide, but they must collide with sufficient energy to get over an 'energy barrier'. In a typical chemical reaction, only a tiny fraction of collisions are effective in bringing about chemical change.

The chapter is divided into the following sections.

8.1 Why do reactions take place at different rates?
8.2 Measuring rates of reaction
8.3 Making a reaction go faster – increasing the collision rate
8.4 Making a reaction go faster – overcoming the activation energy barrier
8.5 The collision theory
8.6 Catalysts
8.7 The effect of light on chemical reactions

8.1 Why do reactions take place at different rates?

How quickly a reaction goes, or the **rate** of reaction, is defined and measured in terms of how quickly the reactant is used up, or how quickly a product forms. The rate of reaction varies greatly for different reactions, and under different conditions.

Enthalpy of reaction revisited

In Chapter 5, we showed that the sign of ΔH indicates the likely direction in which a reaction will take place. Most reactions are exothermic, that is, ΔH negative, and very few are endothermic, that is, ΔH positive. We might be tempted to think that if a reaction is highly exothermic (there is a large drop in energy associated with it), then it would take place very rapidly. This, however, is not necessarily the case. For example, a mixture of methane and oxygen shows no sign of reaction at room temperature and pressure even though $\Delta H^\ominus$ for the reaction is $-890\,kJ\,mol^{-1}$. On the other hand, the reaction between hydrochloric acid and sodium hydroxide takes place very rapidly, even though $\Delta H^\ominus$ for this reaction is only $-56\,kJ\,mol^{-1}$. The study of the rate of a reaction is called **kinetics**, while the study of charges in energy is called **thermodynamics** (as we saw in Chapter 5).

- The value of ΔH cannot be used to tell us about the **rate** of reaction, or how fast a reaction takes place.
- **Kinetics** is the study of the rates of reactions.

The activation energy barrier

The reason why ΔH cannot be used to tell us about the rate of reaction is because there is an energy barrier between the reactants and products that has to be overcome. This can be thought of as similar to a ridge at the front of a desk, over which a book must be pushed before it will fall to the floor. Before the book will fall, we have to first push it harder for it to rise over the ridge. In chemical reactions, this ridge is called the **activation energy**, and it is given the symbol E_a. The size of the activation energy (the height of the ridge) does not depend on the size of ΔH (the height of the desk off the floor).

We can draw an **energy diagram** for a reaction, as shown in Figure 8.1. Such a diagram is sometimes called a **reaction profile**. The reactants are higher in energy than the products, but higher still is the top of the activation energy peak. There are therefore two factors that determine whether a reaction takes place or not. These are the sign of ΔH and the value of E_a.

■ **Figure 8.1**
Reaction profiles for the combustion of methane and the neutralisation reaction between sodium hydroxide and hydrochloric acid. The *x*-axis is labelled 'progress of reaction', indicating the change from reactants to products over time.

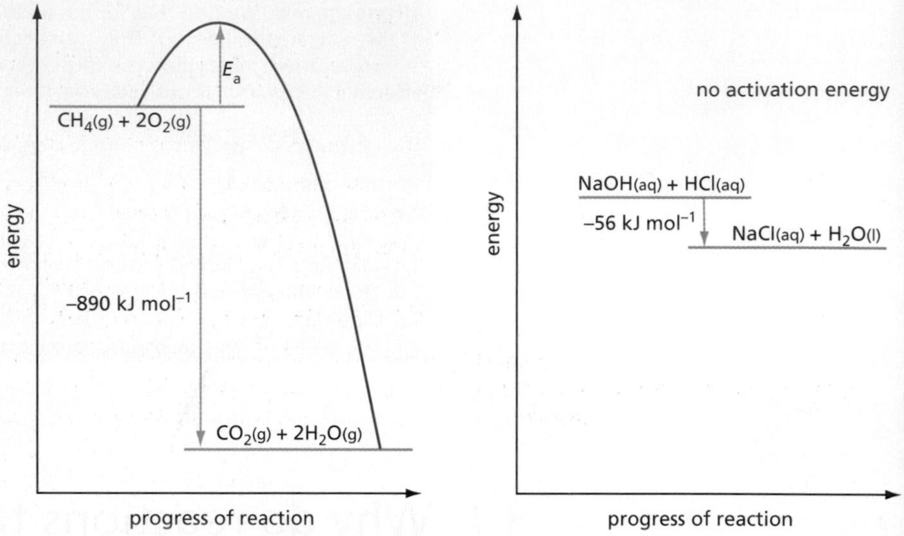

- If ΔH is positive, the reactants are lower in energy than the products. The reactants are said to be **thermodynamically stable** with respect to the products. Such reactions are unlikely to take place.
- Even if ΔH is negative, the reaction may still not be observed to take place because it has a high activation energy. Because of this, the rate of the reaction is negligible at room temperature. The reactants are said to be **kinetically inert** (or **kinetically stable**).

Table 8.1 summarises the effect of the four combinations of ΔH and E_a on the rate of a reaction.

■ **Table 8.1**
The rate of a reaction depends on both the enthalpy change and the activation energy.

ΔH	E_a	Result
+ve	Low	No reaction, reactants thermodynamically stable
+ve	High	No reaction, reactants thermodynamically stable
−ve	Low	Reaction
−ve	High	No obvious reaction, reactants kinetically stable

8.2 Measuring rates of reaction

We study the rate of a reaction for two reasons:

- to tell us the best conditions to make the reaction go as quickly as possible
- to tell us about the mechanism of a reaction, that is, how the reaction actually takes place.

We shall look here at how we measure rates of reaction, in order to find out what factors control those rates. Reaction mechanisms are considered in more detail in Chapter 11, and are particularly important when studying organic reactions.

What to measure?

Simple observation tells us that some reactions go faster than others, but to actually measure the rate we must find out how fast one of the products is being produced or how fast one of the reactants is being used up. This can be done using either a physical method or a chemical method of analysis. A physical method is more convenient than a chemical method, because it does not disturb the reaction we are studying. It depends on there being a change in some physical property such as volume, mass or colour during the course of the reaction.

Physical methods of analysis – measuring the volume of gas given off

The reaction between hydrochloric acid and calcium carbonate may be studied using a physical method of analysis:

$$CaCO_3(s) + 2HCl(aq) \rightarrow CaCl_2(aq) + H_2O(l) + CO_2(g)$$

The reaction gives off a gas, whose volume can be measured. A suitable apparatus is shown in Figure 8.2. If we plot a graph of the volume of carbon dioxide collected against time (Figure 8.3), the rate of the reaction is given by the slope (or gradient) of the graph. The steeper the gradient, the higher the rate. The rate is greatest at the start, then decreases and finally becomes zero at the end of the reaction.

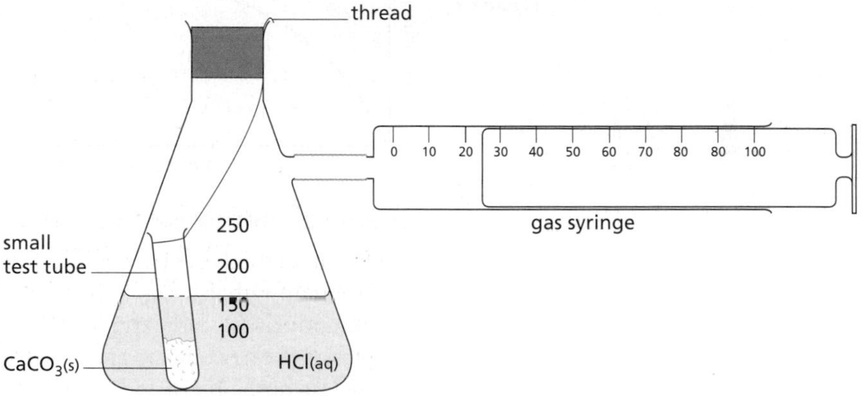

■ **Figure 8.2** Apparatus used to study the rate of a reaction that gives off a gas.

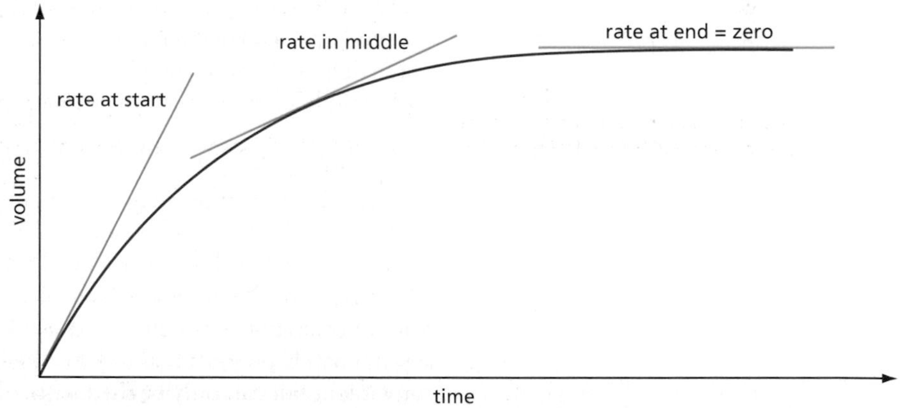

■ **Figure 8.3** A typical graph of volume against time for a reaction that gives off a gas. The gradient gives the rate of reaction at that point. The rate is greatest at the start, then decreases and finally reaches zero at the end of the reaction.

Physical methods of analysis – measuring the decrease in mass

An alternative method of monitoring the same reaction involves carrying out the reaction in an open flask placed on top of a top-pan balance. As the gas is given off, the mass of the flask decreases. If we plot a graph of mass against time, the rate of the reaction is the negative gradient of the graph (see Figure 8.4). Note that rates of reaction must always be positive, as a negative rate would mean that the reaction was going backwards. The gradient in Figure 8.4 is negative, so if we take its negative value, we get a positive rate.

We can measure the rate of this reaction under different conditions, for example, changing the temperature, or the concentration of acid, or the surface area of calcium carbonate, to investigate the conditions under which it will proceed the fastest.

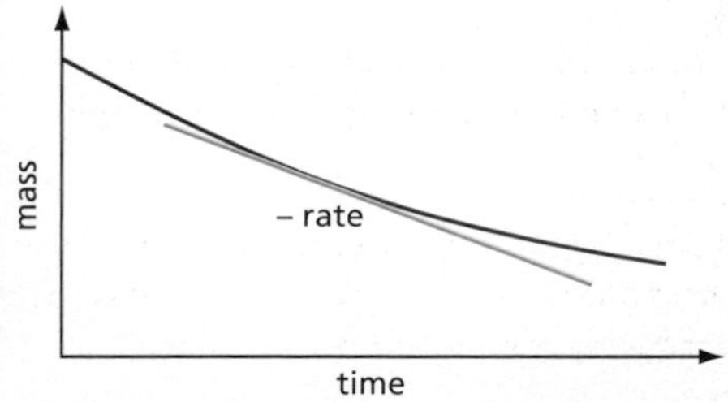

■ **Figure 8.4** A typical graph of mass against time for a reaction that loses a gas to the atmosphere. The rate of the reaction is the negative gradient of the graph.

Physical methods of analysis – measuring a change in colour

We may sometimes be able to use the change in colour of a solution to measure the rate of a reaction. An example of this is the oxidation of aqueous potassium iodide by aqueous potassium peroxodisulphate(VI), $K_2S_2O_8$, during which iodine is released.

$$S_2O_8^{2-}(aq) + 2I^-(aq) \rightarrow 2SO_4^{2-}(aq) + I_2(aq)$$

Iodine dissolves in potassium iodide to give a brown solution. The intensity of the brown colour may be measured using an instrument called a **colorimeter**. In this device, light from a light source passes through a filter that selects just one colour of light. This then shines onto a cell containing the reaction mixture, and a detector monitors the colour of light passing through the solution.

Iodine solution is brown because it absorbs in the blue part of the spectrum and reflects other wavelengths. We therefore select a blue filter for this experiment (see page 358). The colorimeter is first adjusted to zero using water in the cell. The colorimeter provides readings of **absorbance**, which is proportional to the concentration of the iodine. A graph of absorbance against time (see Figure 8.5) is obtained, which is similar in shape to the volume-against-time graph in Figure 8.3.

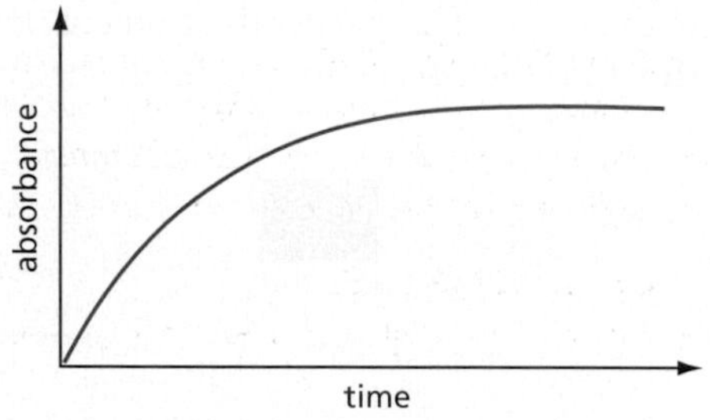

■ **Figure 8.5**
A typical graph of absorbance against time. This is equivalent to the volume-against-time graph in Figure 8.3.

Chemical methods of analysis – sampling and 'clock' reactions

Chemical methods of analysis interfere with the reaction being studied. Because of this, we must either set up a new reaction mixture for each measurement we take, or alternatively we can extract small amounts of the reaction mixture, called **samples**, and carry out the analysis on them.

The rate of the reaction between potassium peroxodisulphate(VI) and potassium iodide described above may be followed using a 'clock' method. A separate experiment is set up for each determination of rate. The two reactants are mixed in the presence of a known amount of sodium thiosulphate, $Na_2S_2O_3$, and a little starch. The sodium thiosulphate reacts with the iodine produced, converting it back to iodide, in the reaction we met in Chapter 7:

$$I_2(aq) + 2S_2O_3^{2-}(aq) \rightarrow 2I^-(aq) + S_4O_6^{2-}(aq)$$

The amount of sodium thiosulphate present is much smaller than the amounts of all the other reagents. When all the sodium thiosulphate has been used up, the iodine (which is still being produced by the reaction) reacts with the starch to give an intense blue colour. For a fixed amount of sodium thiosulphate, the faster the reaction, the shorter the time, t, this takes. The rate of the reaction is then approximately proportional to $\frac{1}{t}$ (see Figure 8.6). A **'clock' reaction** is very convenient but can only be used with a limited number of reactions.

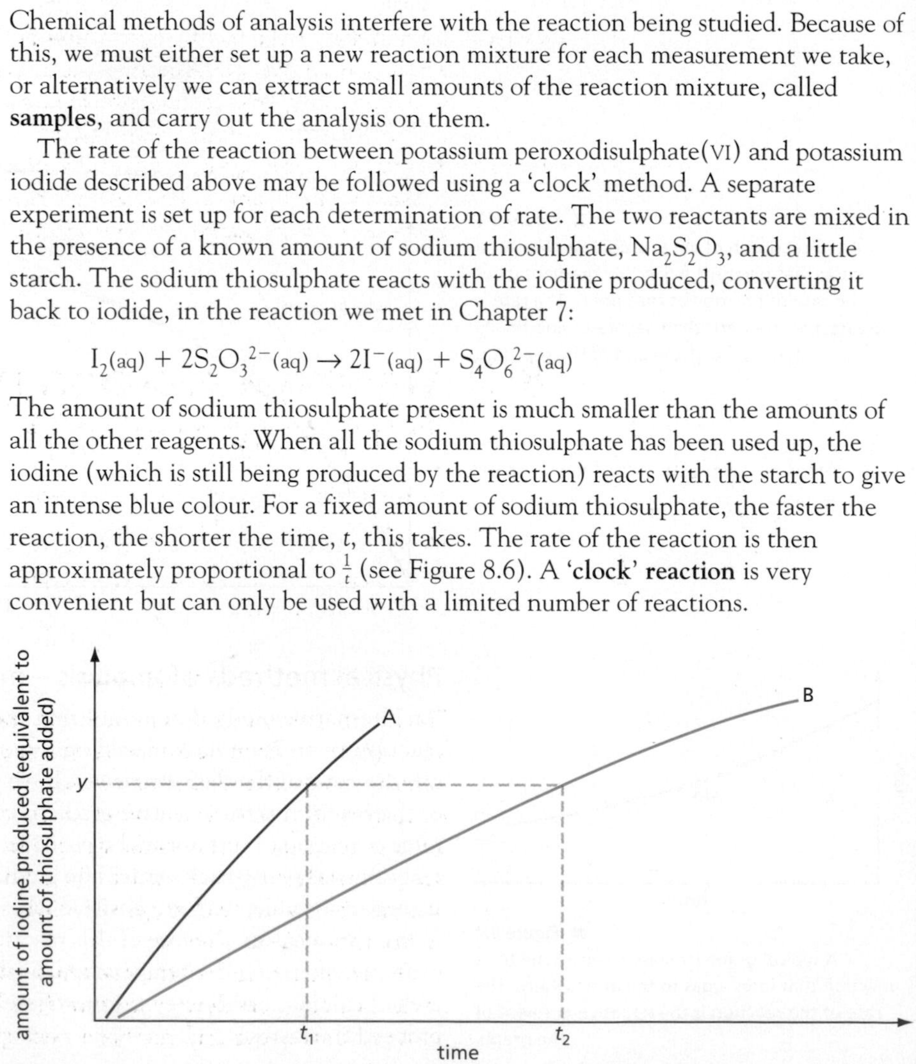

■ **Figure 8.6**
The graphs show two reactions A and B. They may vary in, for example, concentration or temperature. At y, all the sodium thiosulphate has been used up and the iodine reacts with the starch. The rate of reaction A is approximately equal to y/t_1, and the rate of reaction B is approximately equal to y/t_2. This means that the ratio of the times taken for the 'clock' to stop gives the ratio of the rates:

$$\frac{\text{rate of A}}{\text{rate of B}} = \frac{t_2}{t_1}$$

■ Figure 8.7
Carrying out a 'clock' reaction.

Finding the rate of reaction

Physical methods include:

- measuring the volume of gas given off over time
- measuring a change in mass over time
- following a change in colour over time using a colorimeter.

Chemical methods include:

- removing a sample at timed intervals and using chemical analysis, for example titration
- using a 'clock' method.

Example Suggest methods of measuring the rates of the following reactions:

a zinc powder with aqueous copper sulphate
b hydrogen peroxide decomposing to give water and oxygen
c $(CH_3)_3CBr + H_2O \rightarrow (CH_3)_3COH + H^+ + Br^-$

Answer
a Measure the change in colour of the solution using a colorimeter.
b Measure the volume of gas given off, or measure the decrease in mass.
c Measure the conductivity of the solution.

8.3 Making a reaction go faster – increasing the collision rate

At the molecular level, the rate of reaction depends on two factors:

- how often the reactant molecules hit each other
- what proportion of the collisions have sufficient energy to overcome the activation energy.

The effect of concentration

We can increase the rate at which the reactant particles collide in several ways. The simplest is to increase the **concentration** of the reactants. At a higher concentration, the reactant molecules are closer together and so will collide more frequently. The same is true if we increase the **pressure** for gaseous reactions – the volume gets smaller and so the molecules are closer together. However, changing the pressure has virtually no effect on reactions in the solid or liquid phase, because their volume changes very little when put under pressure, so their particles do not move closer together.

The concentrations of the reactants decrease during the course of a reaction and so the rate also goes down with time (see Figure 8.3). At the end of the reaction, at least one of the reactants has been used up completely and the rate is then zero.

If we repeat the reaction using more concentrated reactants, the initial rate of the reaction will be higher. However, it must not be assumed that if, for example, we double the concentration, the rate will be twice as high. The actual size of the increase in rate is impossible to predict and can be found only by experiment. This study of quantitative kinetics is explored in Chapter 11.

Example When 2 g (an excess) of powdered calcium carbonate were added to 50 cm^3 of 0.10 mol dm^{-3} hydrochloric acid, the results shown in Table 8.2 were obtained.

■ **Table 8.2**

Time/s	10	20	30	40	50	60	70	80
Volume of CO_2 given off/cm^3	25	45	60	70	75	78	80	80

a Plot a graph of these results.
b Use the graph to find the rate at the start of the reaction.
c Explain the shape of the graph.

Answer **a**

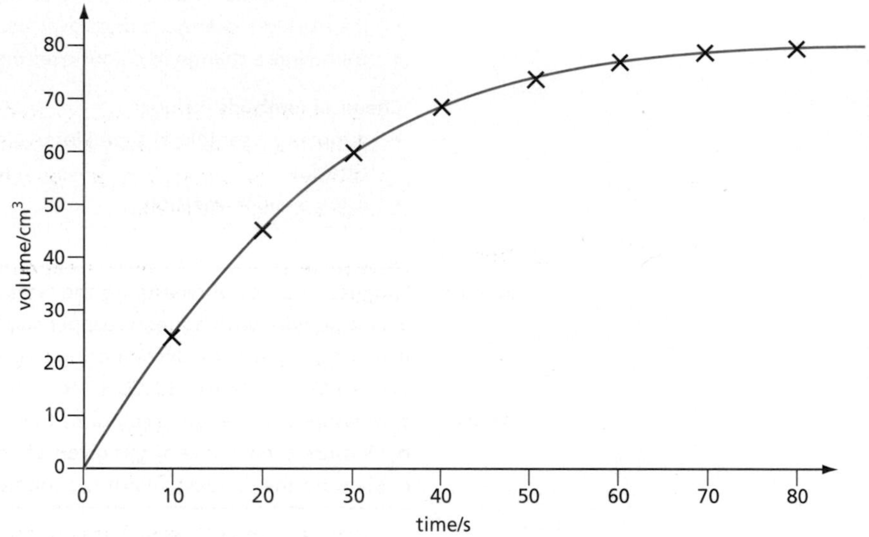

■ **Figure 8.8**

b The gradient of the graph at the beginning shows that 25 cm^3 would be given off in 10 s. This is a rate of **2.5 $cm^3 s^{-1}$**.
c The rate reduces (the reaction slows down) as the reaction proceeds, because the acid becomes more dilute. The reaction finally stops when all the acid has been used up.

Further practice

1 The experiment was repeated using **a** 50 cm^3 of 0.20 mol dm^{-3} hydrochloric acid and then again with **b** 50 cm^3 of 0.050 mol dm^{-3} hydrochloric acid. Draw a sketch of the likely graphs produced and comment on the differences between each graph and that in Figure 8.8.

2 Solutions of potassium iodide and potassium peroxodisulphate(VI) were mixed together and a clock started. A sample of the mixture was placed in the cell of a colorimeter and the absorbance measured at suitable time intervals. The results shown in Table 8.3 were obtained.

■ **Table 8.3**

Time/s	30	60	90	120	150	180	210	240	270
Absorbance	0.072	0.13	0.19	0.25	0.29	0.32	0.34	0.35	0.35

a Plot a graph of absorbance against time.
b Find the rate of the reaction after: **i** 60 s **ii** 120 s.
c Explain why these two reaction rates differ.

3 When hydrochloric acid is added to sodium thiosulphate, $Na_2S_2O_3(aq)$, a fine precipitate of sulphur appears (Figure 8.9). This gradually makes the solution opaque.

$$2HCl(aq) + Na_2S_2O_3(aq) \rightarrow 2NaCl(aq) + H_2O(l) + SO_2(aq) + S(s)$$

■ **Figure 8.9**
The reaction between hydrochloric acid and sodium thiosulphate produces a precipitate of sulphur. This allows us to study the rate of the reaction – when the cross is no longer visible, the clock is stopped.

The rate of the reaction may be determined by measuring how long it takes for the solution to become so opaque that a cross marked on a piece of paper placed under the beaker containing the reaction mixture just becomes invisible. The results shown in Table 8.4 were obtained.

■ **Table 8.4**

Experiment number	Concentration/mol dm^{-3}		Time/s
	HCl	$Na_2S_2O_3$	
1	0.20	0.40	39
2	0.20	0.30	50
3	0.20	0.20	83
4	0.20	0.10	170

a Plot a graph of (1/time) against the concentration of sodium thiosulphate.
b Explain why the rate changes with the concentration of sodium thiosulphate.
c Estimate the concentration of sodium thiosulphate that would give a time of 60 s.

The effect of changing the amount of reactants

If we keep the concentrations of all the reactants the same but, for example, double their amounts, the actual rate of the reaction will be unaffected although the amount of product will be doubled. This means that if we measure the rate from the volume of gas evolved or the mass lost, we will obtain different results for the rate. The actual rate of reaction between calcium carbonate and hydrochloric acid is unaffected if we double the mass of calcium carbonate and double the volume of the acid even though twice the volume of gas is evolved. However, if we are using the apparatus in Figure 8.3 (page 159) the measured rate of reaction will appear to be twice as great because twice as much gas is evolved in the same time. For quantitative work, it is important to define what we mean by the rate of reaction more carefully, and this is considered further in Chapter 11.

The effect of temperature

We can increase the rate of collision by raising the temperature. This causes the particles to move faster (increases their kinetic energy) and so they collide more frequently. It is important to realise that this increased rate of collision has only a small effect, and turns out to be less important than the main effect of increasing temperature, namely increasing the energy rather than the rate of collisions, which we shall consider later in the chapter.

State of division

Many reactions are **homogeneous**, that is, the reactants are uniformly mixed, either as gases or in solution. Others are **heterogeneous** – there is a boundary between the reactants, for example, between a solid and a solution. In such reactions, the rate of collision will increase if we make the **surface area** of this boundary as large as possible. For example, powdered sugar dissolves faster than lump sugar, water drops evaporate faster than a large puddle, and powdered zinc dissolves in acid faster than lumps of zinc do. The size of the boundary between the reactants in a heterogeneous reaction is described by the **state of division** – for example, a powdered solid is described as 'finely divided'. The efficiency of solid catalysts is also increased if they are finely divided (see page 168).

Example When 2 g (an excess) of lump calcium carbonate were added to 50 cm^3 of 0.10 mol dm^{-3} hydrochloric acid, the results shown in Table 8.5 were obtained.

■ Table 8.5

Time/s	10	20	30	40	50	60	70	80
Volume of CO_2 given off/cm^3	5	10	14	18	22	25	28	31

a How does the rate compare with that found in the Example on page 162?
b Explain why the rate is different.
c What would be the final volume of carbon dioxide? Explain your answer.

Answer a The rate is lower.
b The surface area of lump calcium carbonate is smaller than that of powdered calcium carbonate, so there are fewer collisions taking place every second.
c 80 cm^3. The final amount of carbon dioxide is determined by the amount of acid, because the calcium carbonate is present in excess.

8.4 Making a reaction go faster – overcoming the activation energy barrier

We mentioned on page 163 that the energy of the collisions between reacting particles is more important than the rate of collision. Even if a reaction has a small activation energy, only a tiny fraction of the collisions that take place have enough energy to overcome it. These collisions are the **effective collisions**. A typical fraction might be one in ten thousand million (1 in 10^{10}). This is the reason why many reactions are slow at room temperature. There are two ways in which we can increase this fraction:

- give the collisions more energy
- find another route with a lower activation energy.

Giving the collisions more energy

The usual way of increasing the energy of the collisions is to raise the temperature. As mentioned earlier, a minor effect of increasing the temperature is that the rate of collision increases. However, the principal effect is that a higher proportion of the collisions now have enough energy to get over the activation energy hump – there are more effective collisions. If we raise the temperature from 25 °C to 125 °C, for example, the increase in the rate of collision goes up by a factor of only 1.2, but the fraction of reactant molecules able to get over the energy barrier could increase by 1000 times. A useful working rule is to say that 'a 10 °C rise in temperature doubles the rate of reaction'. So, for a rise in temperature from 25 °C to 125 °C, the rate might increase 2^{10} or 1024 times.

- A 10 °C rise in temperature often doubles the rate of reaction.
- Increasing the temperature increases the number of **effective collisions**.

Finding another route with a lower activation energy

The second way of increasing the number of collisions that have sufficient energy to overcome the activation energy barrier is to lower to the height of the barrier. This is most easily achieved by using a catalyst.

A **catalyst** is a substance that speeds up a chemical reaction without being chemically used up itself.

It does this by providing an alternative pathway for the reaction, and this pathway has a lower activation energy than that of the uncatalysed reaction (see Figure 8.10).

■ Figure 8.10 The energy profile for a reaction, with and without a catalyst.

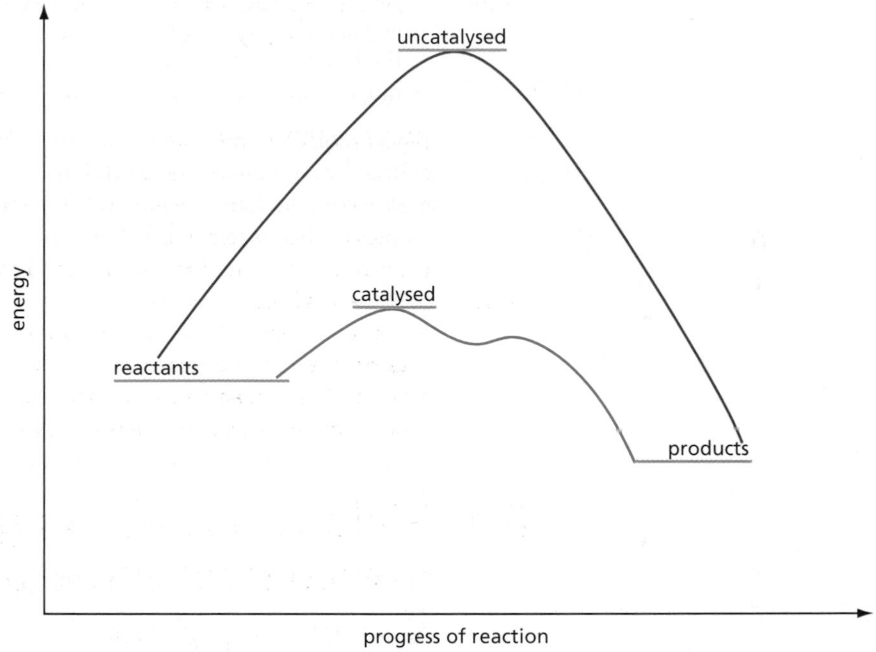

The height of the activation energy barrier may also be changed if a different solvent is used for the reaction. Many inorganic reactions are carried out using water as a solvent, but most organic substances are insoluble in water and so must be dissolved in a different solvent. This has a marked effect on the rate of reaction.

Example 1 Return to your graph plotted from Table 8.2 (page 162) for the reaction between hydrochloric acid and calcium carbonate, which was carried out at 25 °C. On the same axes, sketch the graph you might expect if the experiment were repeated at 35 °C rather than at 25 °C.

Answer The graph should show an initial gradient twice as steep, since the rate will be about twice as fast. The final volume will be the same because there is the same amount of hydrochloric acid.

Example 2 Sketch an energy profile for a slow reaction with a large negative ΔH.

Answer Since the reaction is slow, there will be a high activation energy. Since ΔH is large and negative, the products are much lower in energy than the reactants.

Further practice

1 a Add to your graph from Example 1 above a sketch of the graph you might expect if the experiment were conducted at 15 °C rather than at 25 °C.

b Explain the differences and similarities in the graphs.

2 Sketch energy profiles for the following reactions:

a a fast reaction with ΔH small and negative

b a fairly fast reaction with ΔH small and positive.

8.5 The collision theory

Rates of collision

As we have seen, in order for a reaction to take place, the reactant molecules must first collide and secondly have sufficient combined energy to get over the activation energy barrier.

For gaseous reactions, it is possible to calculate how often the molecules collide. This depends on three main factors:

- the size of the molecules – the bigger the molecules, the greater the chance of collision
- the mass of the molecules – the larger the mass, the slower the molecules move at a given temperature and the fewer the collisions
- the temperature – as the temperature is raised, molecules move faster and the rate of collisions increases.

Bigger molecules have larger masses so the first two factors tend to cancel out. The collision rate therefore depends largely on temperature. So, at a given temperature, most molecules have similar collision rates. At room temperature and pressure, in each cubic decimetre of gas there are about 1×10^{32} collisions every second. If every one of these collisions produced a reaction, all gaseous reactions would be over in a fraction of a second.

For molecules in liquids, there is no simple way to estimate their rate of collision. We would expect the rate to be lower than in gases, because the molecules are moving more slowly and because the solvent molecules get in the way. On the other hand, there are situations when reactant molecules can become trapped in a 'cage' of solvent molecules and this increases their chance of collision. The two effects tend to cancel each other out. Studies on reactions both as gases and in solution suggest that the rates are often not too different from each other.

Energy of collision – the energy distribution curve

The molecules in a sample of gas have a range of energies. In order to estimate what fraction of collisions will have sufficient energy to get over the activation energy barrier, we need to know how many molecules in a sample have various different energies – in other words what their **energy distribution** is. This was first calculated by James Clerk Maxwell and Ludwig Boltzmann. The **energy distribution curve** that results is shown in Figure 8.11. We can see that there are very few molecules with very low energy values. Most molecules have a moderate amount of energy, as shown by the highest point on the graph – this shows the most probable energy. There are a few molecules with very high energy values.

■ **Figure 8.11** The Maxwell–Boltzmann energy distribution curve.

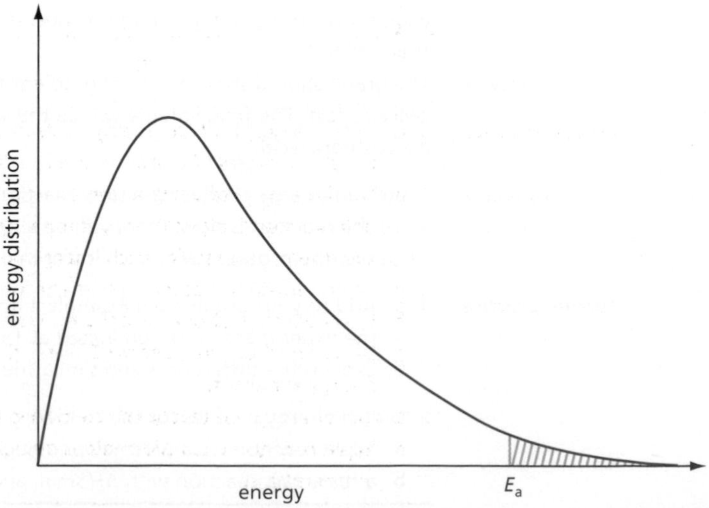

■ **Figure 8.12**
James Clerk Maxwell 1831–79 (left) and Ludwig Boltzmann 1844–1906 (right).

At a given temperature, the shape of the curve is the same for all gases, as it depends only on the kinetic energy of the molecules (and as we have mentioned this depends only on temperature). The area under the graph gives the total number of molecules. The line E_a is marked, showing a high activation energy for a particular reaction. The shaded area shown is the fraction of molecules that have energy greater than this E_a.

If we increase the temperature, the curve has a similar shape but the maximum moves to the right, showing that the most probable energy increases. The energies become more spread out, so there will be fewer molecules with the most probable energy and the maximum is slightly lower. Figure 8.13 shows the effect of raising the temperature by 100 °C. We can see that the shaded area showing the fraction of molecules with energy greater than E_a is much larger at the higher temperature.

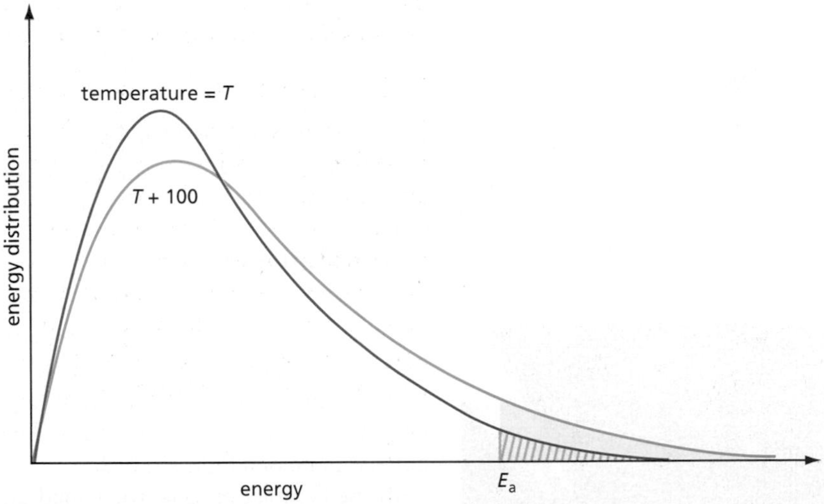

■ **Figure 8.13**
The effect on the energy distribution curve of increasing the temperature by 100 °C. The curve moves to the right as the temperature rises. The curve is slightly flatter, so that the area under it remains constant.

Further practice

1 a Sketch a graph of the energy distribution of gas molecules at room temperature. Label the graph '*T*'. Draw a line to show the activation energy, E_a, of a reaction.
  b On the same axes, sketch the energy distribution at a temperature of about 100 °C below room temperature. Label the graph '*T* – 100'. Shade the areas that show the fractions of molecules with energy greater than E_a at the two temperatures.

2 a Sketch another graph of the energy distribution of gas molecules at room temperature.
  b Draw a line to show the activation energy, E_a, of a reaction.
  c Draw another line to show E_a for a catalysed reaction, in which the activation energy is halved.
  d Explain why the catalysed reaction is so much faster than the uncatalysed reaction.
  e Explain why the collision rate is approximately the same for all gases at room temperature.

8.6 Catalysts

A **catalyst** speeds up a reaction by providing a pathway of lower activation energy. The effect can be readily shown on the energy distribution graph. On Figure 8.11, if E_a is made smaller, the shaded area is much larger and so the rate is increased.

Catalysts may be **homogeneous**, that is, uniformly mixed with the reactants, or **heterogeneous**, in which case the catalyst and the reactants are in different physical states. There is also an important class of biochemical catalysts called **enzymes**. These three classes will be considered separately.

Homogeneous catalysis

Many reactions are catalysed by the $H^+(aq)$ ion. An example of this is the iodination of propanone:

$$CH_3COCH_3(aq) + I_2(aq) \xrightarrow{H^+(aq)} CH_2ICOCH_3(aq) + H^+(aq) + I^-(aq)$$

The mechanism of this reaction is studied in detail in Chapter 11. Another reaction catalysed by the $H^+(aq)$ ion is the formation of an ester, for example methyl methanoate. The overall mechanism is explained in detail in Chapter 29.

Homogeneous catalysis can also take place in the gas phase. The production of the harmful gas ozone from oxygen near the ground is catalysed by nitrogen dioxide, which is produced in car engines:

$$O_2(g) + NO_2(g) \rightarrow NO(g) + O_3(g)$$
$$NO(g) + \tfrac{1}{2}O_2(g) \rightarrow NO_2(g)$$

Nitrogen dioxide may also catalyse the oxidation of sulphur dioxide to sulphur trioxide in the atmosphere, and hence speed up the production of acid rain (see Chapter 15, page 296):

$$NO_2(g) + SO_2(g) \rightarrow SO_3(g) + NO(g)$$
$$NO(g) + \tfrac{1}{2}O_2(g) \rightarrow NO_2(g)$$

The presence of chlorine atoms from CFCs in the upper atmosphere is responsible for the catalytic destruction of ozone (see page 473).

The rates of many redox reactions are increased by the use of catalysts, usually transition metal compounds. Some of these are considered in Chapter 19.

Heterogeneous catalysis

■ **Figure 8.14**
A catalyst in the form of a gauze. This provides a large surface area, but prevents the catalyst being swept away in the stream of gases.

Two important examples of the use of solid heterogeneous catalysis are in the Contact process and the Haber process, described in Chapter 20. The reaction takes place on the surface of the catalyst, so the catalyst should ideally be in the form of a fine powder. This is often impracticable because the catalyst would then be lost in the flowing stream of gas. The catalyst is therefore usually made into pellets, gauze or a fluidised bed.

In the Contact process, the vanadium in the catalyst changes its oxidation state. The mechanism is considered in more detail in Chapter 19, page 366. The Haber process involves a hydrogenation reaction. Other hydrogenation reactions include those of the alkenes (Chapter 24) and carbonyl compounds (Chapter 28). Many transition metals catalyse hydrogenation reactions, for example, platinum and nickel (see Chapter 24, page 444). This is because they are able to absorb large quantities of hydrogen and form **interstitial hydrides**. In these hydrides, hydrogen atoms are held in the spaces between the metal atoms in the lattice. These hydrogen atoms are able to add on to the multiple bond of a molecule that becomes **adsorbed** on to the metal surface nearby. There are three stages in catalysis involving surface adsorption, as follows.

- **Adsorption** – the reactants are first adsorbed onto the surface of the catalyst. The metal chosen must adsorb the reactants easily but not so strongly that the products do not come off again.

- **Reaction** – the reactants are held on the surface in such a position that they can readily react together.
- **Desorption** – the products leave the surface.

The rate of the reaction is controlled by how fast the reactants are adsorbed and how fast the products are desorbed. When the catalyst surface is covered with molecules, there is no increase in reaction rate even if the pressure of the reactants is increased.

The efficiency of the catalyst depends on the nature of the surface of the catalyst. This is shown by the following effects.

- **Poisoning** – many catalysts are rendered ineffective by trace impurities. For example, hydrogenation catalysts are poisoned by traces of sulphur impurities. This is one reason why nickel is preferred to platinum as a catalyst. Nickel is relatively inexpensive, and if a large amount of nickel is used, although some of it may be deactivated by poisoning, enough will remain for it still to be effective. However, platinum is expensive, and if a small quantity of platinum is used, all of it might be poisoned.
- **Promotion** – the spacing on the surface of a catalyst is important. For example, it is known that only some surfaces of the iron crystals act as effective catalysts in the Haber process. The addition of traces of other substances may make the catalyst more efficient by producing **active sites** where the reaction takes place most readily (see Chapter 20, page 371).

The three-way catalytic converter

Ideally, the hydrocarbons in unleaded petrol are converted completely into carbon dioxide and water when burned in a vehicle engine. These are non-polluting products, though the additional carbon dioxide probably has an effect on the Earth's climate (see Chapter 23, page 430). In practice, however, three other products are formed which are immediately harmful:

- unburnt hydrocarbon
- carbon and carbon monoxide, from incomplete combustion of the fuel
- oxides of nitrogen formed from the nitrogen in the air.

The amount of each impurity produced depends on the type and efficiency of the engine. When starting up and when idling, the proportions of impurities are higher than when driving on the open road. Diesel engines produce higher quantities of carbon particles than do petrol engines. To reduce the amounts of these impurities, engines are fitted with **catalytic converters**. These catalyse the following reactions:

$$CO(g) + NO(g) \rightarrow CO_2(g) + \tfrac{1}{2}N_2(g)$$
$$CO(g) + \tfrac{1}{2}O_2(g) \rightarrow CO_2(g)$$
$$\text{hydrocarbons} + O_2(g) \rightarrow CO_2(g) + H_2O(g)$$

The exhaust gases are passed through the converter, which contains metals such as platinum and rhodium supported on a honeycomb support (see Figure 8.15). Because lead poisons this catalyst, it is essential to use unleaded petrol in a car fitted with a catalytic converter.

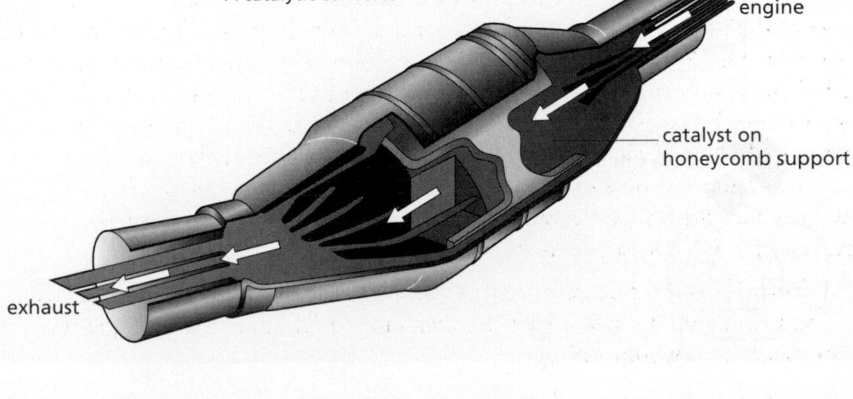

■ **Figure 8.15** A catalytic converter.

Enzymes

Chemical reactions that take place in living cells are remarkably fast, bearing in mind that they usually take place at just above room temperature and at very low concentrations. This is because they are catalysed by very efficient catalysts called enzymes. Enzymes are proteins, which are polymers made up of units called amino acids (see Chapter 30, pages 574–6). These units are joined in a chain, like beads in a necklace. A typical enzyme may contain 500 amino acids and have a relative molecular mass of 50000.

Enzymes are **globular proteins**. In this type of protein, the amino-acid chain coils into an approximately spherical shape (see Figure 8.16). Globular proteins are soluble in water, but if their globular structure is changed, they become insoluble. An example of this process, which is called **denaturing**, is the setting of the white of an egg when boiled. Because globular proteins are water soluble, they are homogeneous catalysts, though their mechanism of action resembles heterogeneous catalysis in that the reaction takes place at a specific point on their surface, called the **active site**.

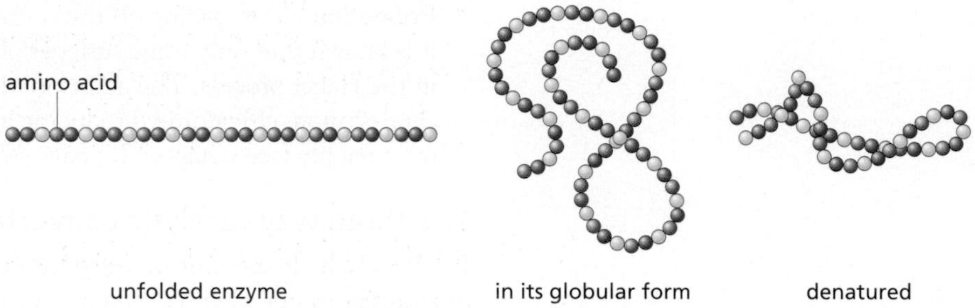

■ **Figure 8.16** Proteins are like a necklace of beads – an enzyme may contain 500 amino acids. The enzyme folds into a globular shape, but it is easily denatured, for example by heat, when it becomes ineffective as a catalyst.

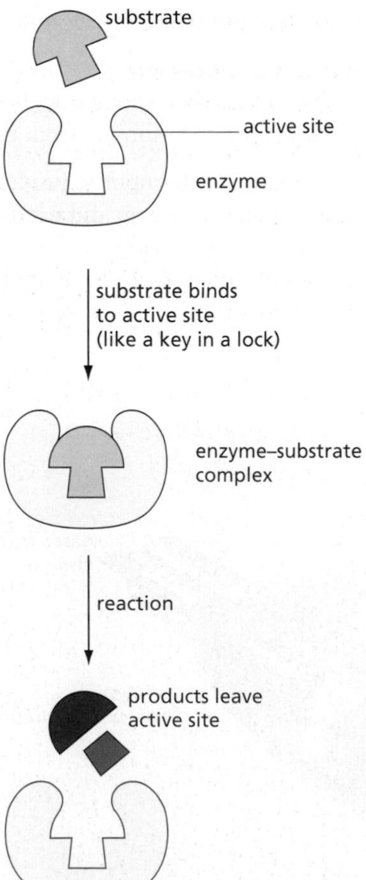

■ **Figure 8.17** Only a molecule of the correct size and shape can attach to the active site and be converted into products.

Most enzymes catalyse only one reaction, or one group of similar reactions. A reactant molecule will only attach to the active site if it has the right size and shape. Once in the active site, the reactant molecule is held in the correct orientation to react and form the products of the reaction. The products leave the active site and the enzyme is then ready to act as a catalyst again. This mechanism of action of an enzyme is described as the **lock and key hypothesis**, illustrated in Figure 8.17. The reactant that fits the enzyme's active site is called the **substrate**.

Some examples of enzyme-catalysed reactions are as follows.

- Yeast contains a mixture of enzymes called **zymase** that converts sucrose into ethanol and carbon dioxide:

$$\underset{\text{sucrose}}{C_{12}H_{22}O_{11}(aq)} + H_2O(l) \rightarrow \underset{\text{ethanol}}{4C_2H_5OH(aq)} + 4CO_2(g)$$

- Hydrogen peroxide is harmful to cells. It is produced in many metabolic reactions but is rapidly decomposed to water and oxygen in the presence of the catalyst **catalase**:

$$H_2O_2(aq) \rightarrow H_2O(l) + \tfrac{1}{2}O_2(g)$$

- When ethanol is metabolised, it is first converted into ethanal by removing two hydrogen atoms in the presence of an enzyme called a **dehydrogenase**:

$$\underset{\text{ethanol}}{C_2H_5OH(aq)} \rightarrow \underset{\text{ethanal}}{CH_3CHO(aq)} + [2H]$$

Every living cell contains at least 1000 different enzymes, each catalysing one of the many different chemical reactions that take place inside the cell. Some enzymes are so efficient that one molecule can catalyse the reaction of 10 000 reactant molecules every second. This means that virtually every collision of the substrate with the active site leads to reaction, showing that the enzyme-catalysed reaction has a very low activation energy.

8.7 The effect of light on chemical reactions

A few reactions are affected by light. These are called **photochemical reactions**, and they fall into two main types:

- endothermic reactions in which light provides the energy input
- exothermic reactions that are started off by light.

Most photochemical reactions are of the first type. They use the energy of the light to bring about the reaction. Examples include the breakdown of silver chloride to silver and chlorine and the bromination of alkanes.

■ **Figure 8.18** Photochromic sunglasses which darken on exposure to light.

Black-and-white photography

The effect of light on silver chloride was observed by Boyle in the seventeenth century. All the silver halides undergo a photochemical reaction in which the halide decomposes into silver and halogen:

$$AgX(s) \rightarrow Ag(s) + \tfrac{1}{2}X_2(g)$$

This reaction forms the basis of black-and-white photography. In a modern black-and-white film, the usual halide is silver bromide, which is suspended in an emulsion of gelatin. On exposure to light, the silver bromide undergoes a photochemical change. It absorbs a photon of light, and an electron is transferred from the bromide ion to the silver ion (see Figure 8.20), leaving the silver bromide in an excited state, represented by AgBr*.

A 'fast' film uses large particles of silver bromide (a coarse emulsion) as they have a greater chance of absorbing a photon of light than small particles. The disadvantage of using a coarse emulsion is that the resolution of the film, and thus the sharpness of the image, suffers as a result.

■ **Figure 8.19** (right)
Grains of silver on a black-and-white film emulsion.

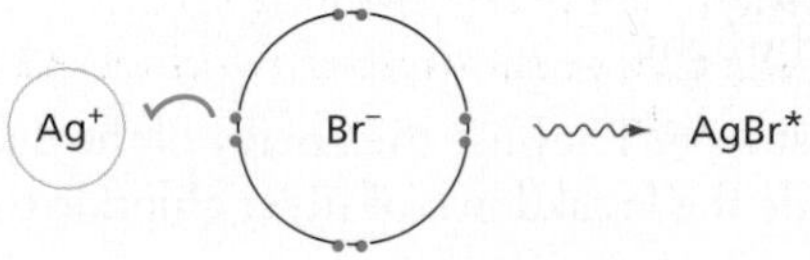

Figure 8.20 (above)
Electron transfer in silver bromide.

When a photograph is taken, the areas of film exposed to light therefore have excited silver bromide on them. The film is then developed. The first stage in the development process is reduction of the excited silver bromide to silver. This is carried out using a reducing agent, and alkaline benzene-1,4-diol (old name hydroquinone) is the most common:

$$2AgBr + HO-\langle\!\bigcirc\!\rangle-OH \rightarrow O=\langle\!\bigcirc\!\rangle=O + 2Ag + 2HBr$$

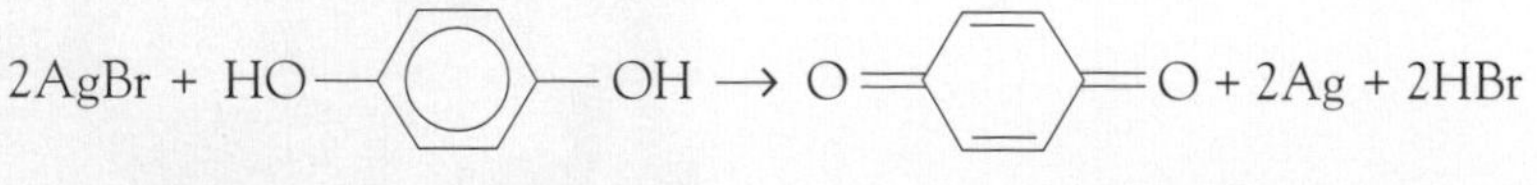

Care must be taken over the timing of this stage of the development process. If the reducing agent is left in contact with the film for too long, some of the unexposed silver bromide may be reduced and the film becomes fogged.

The film is then **fixed** by being treated with a solution of 'hypo', sodium thiosulphate, $Na_2S_2O_3(aq)$. This dissolves any unchanged silver bromide, preventing it from reacting further when the film is exposed to light:

$$AgBr(s) + 2S_2O_3^{2-}(aq) \rightarrow Ag(S_2O_3)_2^{3-}(aq) + Br^-(aq)$$

The film is washed thoroughly at this stage to remove all the silver–thiosulphate complex. What remains is a 'negative' of the image, with light areas of the image showing as dark silver areas.

Chain reactions

A few photochemical reactions are **chain reactions**. These are exothermic reactions that have a high activation energy. The light breaks a bond in one of the reactants, removing an atom. This free atom is energetic enough to react with another reactant molecule and produce another free atom.

The process then repeats, building up into a **chain reaction**. The light has provided the energy to overcome the activation energy barrier. Examples of chain reactions include the reaction of hydrogen with chlorine, and the chlorination of alkanes (see Chapter 23, page 433).

Summary

- The **rate** of a reaction is defined as how fast a product appears or how fast a reactant disappears.
- The rate of a reaction is determined by the height of the **activation energy barrier**, and not by the value of ΔH.
- The rate of reaction at a given time is shown by the gradient of the graph of amount of product against time (or the negative gradient of the graph of amount of reactant against time).
- The rate increases if the reactants are made more concentrated, because the molecules collide more frequently. For the same reason, increasing the pressure of a gaseous reaction increases the rate.
- The rate increases if the temperature is raised, mainly because there are more collisions with sufficient energy to react.
- The area under the **energy distribution curve** shows the fraction of molecules within a certain energy range.
- Only the fraction of molecules with energy greater than the activation energy, E_a, can react. These are shown by the area to the right of E_a on the energy distribution curve.
- A **catalyst** provides an alternative pathway of lower activation energy.
- The rate of a reaction in solution is usually changed if another solvent is used.
- Some reactions go faster in the presence of light.

9 *Equilibria*

Many chemical reactions do not go to completion. They stop at a point called the equilibrium position, where both reactants and products are present. In this chapter, we study how a change in reaction conditions brings about a change in the position of equilibrium of a reaction.

The chapter is divided into the following sections.

9.1 Reversible reactions

Many chemical reactions, such as combustion, appear to go completely, converting all the reactants to products. For example, if we burnt methane in an excess of oxygen, we would not expect to find any methane left over. The reason why reactions such as these go to completion is because they are highly exothermic, that is, ΔH is large and negative. But there are many reactions, particularly in organic chemistry, for which ΔH is small. These reactions may not go to completion – at the end of the reaction a detectable amount of reactant remains, mixed with the product. Such reactions are called **reversible reactions**.

How can we decide whether a reaction is reversible or not? It depends on how carefully we measure the concentrations of reactant and product. A common laboratory method of analysis may detect reactants in the final mixture if they are present in as high a proportion as 1%, for example. More refined techniques may be many times more sensitive. For example, the reaction of a strong acid and a strong base:

$$H^+(aq) + OH^-(aq) \rightarrow H_2O(l)$$

appears to go to completion. However, this is not actually the case, as shown by the fact that pure distilled water has a slight electrical conductivity. This conductivity is due to the ionisation of water:

$$H_2O(l) \rightarrow H^+(aq) + OH^-(aq)$$

indicating that the reverse reaction is taking place to a small extent. Because only two molecules in every 1×10^8 molecules ionise, we can usually ignore it, but it becomes important when we study the pH scale of acidity (Chapter 12).

We consider a reaction as having gone to completion if reactants are present in the final mixture at such low concentrations that they do not affect any tests or reactions that we may want to carry out on the products. In practical terms this means that if the reaction is more than 99% complete, we can usually regard it as having gone to completion.

9.2 The position of equilibrium

Dynamic equilibrium

When a reversible reaction appears to stop, that is, when no overall change is taking place, we say that the system is in **equilibrium**. For example, a reaction may come to equilibrium when 75% of the reactants have changed into products. Under these conditions, the composition of the equilibrium mixture is 75% products and 25% reactants, and this composition is known as the **position of equilibrium**.

At equilibrium, because there is no overall change from reactants to products, it appears as though the reaction has stopped. This is not actually the case. There is no reason why the **forward reaction**, that is, the reaction between the reactants, should not continue, since there are still reactants present in the mixture. The reason why there is no overall change is that the **back reaction**, which turns the products back into reactants, is now going at exactly the same rate. An equilibrium in which the forward and back reactions are both taking place is termed a **dynamic equilibrium.**

For a reversible reaction:

$$\text{reactants} \underset{\text{back reaction}}{\overset{\text{forward reaction}}{\rightleftharpoons}} \text{products}$$

The double-headed arrow $\rightleftharpoons$ indicates a reaction that goes to equilibrium.

At equilibrium:

the rate of the forward reaction = the rate of the back reaction

Demonstrating dynamic equilibrium

A particularly effective way of showing that an equilibrium is dynamic is to use **isotopic labelling**. For example, in the Haber process, nitrogen and hydrogen come to equilibrium with ammonia:

$$N_2(g) + 3H_2(g) \rightleftharpoons 2NH_3(g)$$

Deuterium (a heavy isotope of hydrogen) is substituted for some of the hydrogen gas in an equilibrium mixture of the gases. Some deuterium becomes incorporated into ammonia, as can be shown by analysis of the mixture by mass spectroscopy. This shows that the forward reaction is continuing to take place, and this must be balanced by an equal rate of back reaction in order that the same position of equilibrium is maintained (see Figure 9.1).

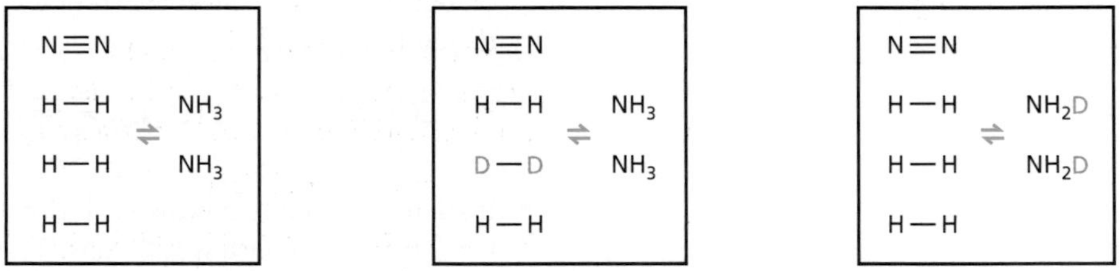

■ Figure 9.1
If deuterium is substituted for some of the hydrogen gas in the equilibrium mixture, some deuterium becomes bonded to nitrogen in ammonia molecules.

Similar experiments may be carried out with radioactive isotopes. For example, solid lead chloride, $PbCl_2$, is only slightly soluble in cold water. Some solid lead chloride is placed in a saturated solution of lead chloride, which is labelled with a radioactive lead isotope, $Pb^{2+}(aq)$. Although the solution is saturated and no more lead chloride can dissolve overall, the solid takes up some of the radioactivity (see Figure 9.2). This shows that some of the radioactive lead ions in the solution have been precipitated into the solid, and an equal number of lead ions from the solid must have dissolved to keep the solution saturated.

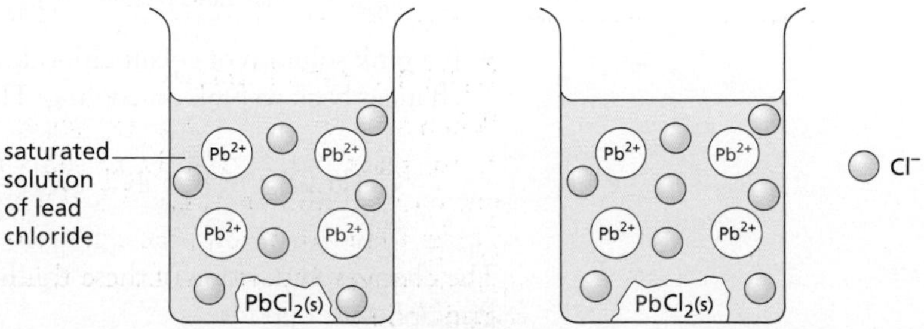

■ Figure 9.2
If solid lead chloride is placed in a saturated solution of lead chloride labelled with radioactive lead ions, Pb^{2+} (shown in red), the solid becomes radioactive.

Example

a What is meant by a reversible reaction? Give an example of a reversible reaction.

b How could you show that it was reversible?

Answer

a A reversible reaction is one in which it is possible to detect unchanged reactants in the equilibrium mixture, and in which both the forward and back reactions take place. One example is the neutralisation of a strong acid with a strong base:

$$H^+(aq) + OH^-(aq) \rightleftharpoons H_2O(l)$$

This must be reversible because pure water conducts electricity, to a small extent.

Further practice

1 State, giving a reason, whether ΔH for a reversible reaction is large or small.

2 a What is meant by dynamic equilibrium?

b Explain why it is reasonable to think that chemical equilibrium is dynamic rather than static.

c Describe an experiment that shows a chemical equilibrium is dynamic rather than static.

Moving the position of equilibrium

In Chapter 8 we studied the effect of changing the following factors on the rate of reaction:

- concentration
- pressure (for gaseous reactions)
- temperature
- state of division (for heterogeneous reactions)
- catalysts.

We shall now consider what effect (if any) these factors may have on the position of equilibrium of a reaction.

Some examples of equilibria

- If acid is added to a yellow solution containing CrO_4^{2-} ions, the solution turns orange owing to the formation of $Cr_2O_7^{2-}$ ions. The reaction can be reversed by the addition of alkali.

$$\underset{\text{yellow}}{2CrO_4^{2-}(aq)} + 2H^+(aq) \underset{OH^-(aq)}{\overset{H^+(aq)}{\rightleftharpoons}} \underset{\text{orange}}{Cr_2O_7^{2-}(aq)} + H_2O(l)$$

- If alkali is added to a brown solution of iodine, the colour fades owing to the formation of colourless IO^- ions. The colour returns on the addition of acid.

$$\underset{\text{brown}}{I_2(aq)} + 2OH^-(aq) \underset{H^+(aq)}{\overset{OH^-(aq)}{\rightleftharpoons}} \underset{\text{colourless}}{2IO^-(aq)} + H_2O(l)$$

- If a mixture of $NO_2(g)$ (brown) and $N_2O_4(g)$ (almost colourless) is suddenly compressed, the mixture darkens and then fades. The initial increase in colour is due to the increase in concentration of $NO_2(g)$, but this decreases as the equilibrium moves in the direction of the $N_2O_4(g)$.

$$\underset{\text{brown}}{2NO_2(g)} \underset{\text{decreased pressure}}{\overset{\text{increased pressure}}{\rightleftharpoons}} \underset{\text{colourless}}{N_2O_4(g)}$$

- If a pink solution of cobalt chloride is warmed, the solution goes blue. It changes back to pink on cooling. The forward reaction is endothermic.

$$\underset{\text{pink}}{Co(H_2O)_6^{2+}(aq)} + 4Cl^-(aq) \underset{\text{cooling}}{\overset{\text{warming}}{\rightleftharpoons}} \underset{\text{blue}}{CoCl_4^{2-}(aq)} + 6H_2O(l)$$

The changes in position of these equilibria can be predicted from Le Chatelier's Principle (see opposite).

9.3 Changing the concentration

If we increase the concentration of one of the reactants in a system in equilibrium, the rate of the forward reaction increases and more products form. The resulting increase in the concentration of the products will then lead to an increased rate of the back reaction until a new position of equilibrium is set up.

The position of equilibrium has moved to the right (see Figure 9.3), with slightly more product being present than at the original position of equilibrium. This is in agreement with **Le Chatelier's Principle**.

> **Le Chatelier's Principle** states that if the conditions of a system in equilibrium are changed (for example, by changing the concentration or pressure or temperature), the position of equilibrium moves so as to reduce that change.

$\text{reactant}_1 + \text{reactant}_2 \rightleftharpoons \text{product}$ system in equilibrium

$\text{reactant}_1 + \text{reactant}_2 \rightleftharpoons \text{product}$ increase in concentration of reactant

$\text{reactant}_1 + \text{reactant}_2 \rightleftharpoons \text{product}$ equilibrium moves to the right

■ **Figure 9.3**
An increase in the concentration of one reactant converts more of the other reactant to product and moves the position of equilibrium to the right.

In a similar way, if the concentration of one or more of the reactants is reduced (or the concentration of the products increased), the position of equilibrium moves to the left. The situation may be summarised as shown in Table 9.1.

■ **Table 9.1**
The effect on the position of equilibrium of changing the concentration of the reactants or the products.

Concentration of reactants	Concentration of products	Equilibrium changes position to the:
Increased		Right
Decreased		Left
	Increased	Left
	Decreased	Right

Example 1 What is the effect of increasing the concentration of hydrogen ions on the following reaction?

$$CH_3CO_2^- + H^+ \rightleftharpoons CH_3CO_2H$$

Answer The equilibrium is pushed over to the right, that is, more ethanoic acid, CH_3CO_2H, is formed.

Example 2 Consider the following equilibrium:

$$\underset{\text{acid}}{CH_3CO_2H} + \underset{\text{alcohol}}{C_2H_5OH} \rightleftharpoons \underset{\text{ester}}{CH_3CO_2H_5} + H_2O$$

Explain why removing water leads to more ester being formed in the equilibrium mixture.

Answer Initially, removing water means that the rate of backward reaction is decreased while that of the forward reaction stays the same. Le Chatelier's Principle predicts that the position of equilibrium will move to oppose the change in the equilibrium mixture. Hence the position of equilibrium moves to the right, producing more ester and water.

Further practice

1 Consider the following equilibrium:

$$H_2(g) + I_2(g) \rightleftharpoons 2HI(g)$$

How will the position of equilibrium alter when each of the following changes is brought about?

a adding more iodine
b adding more hydrogen iodide
c removing hydrogen

9.4 Changing the pressure

In a gaseous reaction, a change in pressure may affect the position of equilibrium in a way similar to changing the concentration in an aqueous reaction. However, for a mixture of gases, an increase in pressure may increase the concentration of reactants and products to the same extent and may therefore have no effect on the position of equilibrium. This occurs if there are the same number of gas molecules on the left-hand and right-hand sides of the equation. An example is the decomposition of hydrogen iodide:

$$2HI(g) \rightleftharpoons H_2(g) + I_2(g)$$

Here there are two molecules of gas on both sides of the equation, so changing the pressure will not change the position of equilibrium.

At a lower temperature, hydrogen iodide decomposes as follows, producing solid iodine:

$$2HI(g) \rightleftharpoons H_2(g) + I_2(s)$$

Now two gas molecules on the left form one gas molecule on the right. Because pressure affects gaseous components of the mixture only, an increase in pressure now will speed up the forward reaction more than the back reaction, so that the equilibrium moves to the right. This again agrees with the prediction from Le Chatelier's Principle, because the increase in pressure will be reduced if the equilibrium moves in the direction in which fewer gas molecules are formed. We can summarise the situation as shown in Table 9.2.

Table 9.2 The effect on the position of equilibrium of increasing the pressure.

Number of gaseous reactant molecules	Number of gaseous product molecules	Direction of movement of the position of equilibrium when the pressure is increased
More	Fewer	Right
Fewer	More	Left
Same	Same	No change

Example What is the effect of an increase in pressure on the synthesis of ammonia?

$$N_2(g) + 3H_2(g) \rightleftharpoons 2NH_3(g)$$

Answer In this reaction, four molecules of gas become two molecules of gas. An increase in pressure favours the products, as these contain fewer gas molecules. So increasing the pressure results in an equilibrium mixture containing more ammonia.

Further practice What effect (if any) will there be on the following equilibria when the pressure is increased?

1 $2HBr(g) \rightleftharpoons H_2(g) + Br_2(g)$
2 $PCl_3(g) + Cl_2(g) \rightleftharpoons PCl_5(g)$
3 $H_2O(l) \rightleftharpoons H_2O(g)$
4 $CaCO_3(s) \rightleftharpoons CaO(s) + CO_2(g)$

9.5 Changing the temperature

Le Chatelier's Principle can also be used to predict the effect of a temperature change on the position of equilibrium. The most important factor is whether the forward reaction is exothermic (ΔH negative) or endothermic (ΔH positive). If the temperature is raised, the equilibrium will move in the direction which will tend to lower the temperature, that is, the endothermic direction. The effect is summarised in Table 9.3.

Table 9.3 The effect on the position of equilibrium of changing the temperature.

Temperature change	Sign of ΔH for forward reaction	Direction of movement of the position of equilibrium
Increase	+ve	Right
Decrease	+ve	Left
Increase	−ve	Left
Decrease	−ve	Right

Careful control of temperature is important in the Contact process and the Haber process, as we shall see in Chapter 20.

Example What is the effect of raising the temperature on the following equilibrium?

$$H_2(g) + I_2(g) \rightleftharpoons 2HI(g) \qquad \Delta H^\ominus = -12\,kJ\,mol^{-1}$$

Answer The reaction is exothermic (gives out heat). On raising the temperature, the equilibrium will move in the direction that reduces the temperature, that is, in the endothermic direction. The equilibrium will therefore move to the left.

Further practice What will be the effect of lowering the temperature on the following equilibria?

1 $N_2(g) + 3H_2(g) \rightleftharpoons 2NH_3(g) \qquad \Delta H^\ominus = -92\,kJ\,mol^{-1}$

2 $H_2O(l) \rightleftharpoons H^+(aq) + OH^-(aq) \qquad \Delta H^\ominus = +58\,kJ\,mol^{-1}$

9.6 State of division and the addition of catalysts

State of division

In a heterogeneous reaction, a small particle size speeds up the rate of a reaction by increasing the surface area. If the reaction is in equilibrium, both the forward and back reactions will be speeded up when the surface area is increased. This increase in speed is the same for both reactions. If, for example, the surface area of a liquid was doubled, the rate at which molecules leave this surface would be doubled, but so too would the rate at which they joined surface again. The position of equilibrium is therefore unchanged (see Figure 9.4).

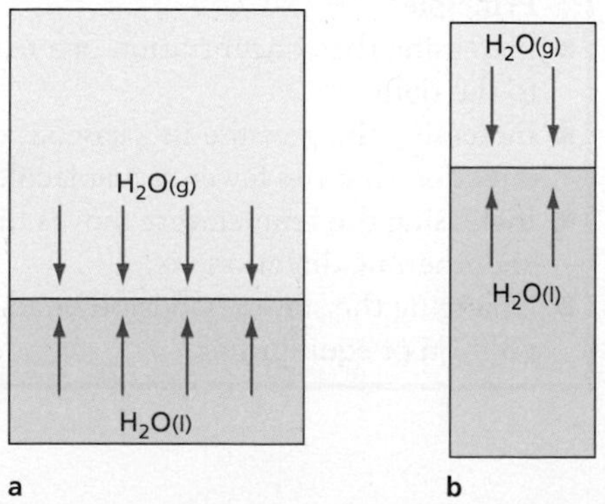

Figure 9.4 In **a** and **b** there is the same amount of water, but the surface area in **a** is twice that in **b**. The rates of evaporation and condensation are both twice as fast in **a** as in **b**, but the position of equilibrium is unchanged.

Catalysts

The effect of adding a catalyst to a mixture in equilibrium is similar to that of increasing the surface area. Both the forward and back reactions are speeded up by the same amount, and the position of equilibrium is unaltered. It may not be immediately obvious why this should be the case, but it can be demonstrated by the following thermodynamic argument. If an exothermic reaction is in equilibrium and then a catalyst is added which makes the reaction faster in the exothermic direction only, energy is obtained. This cannot come from the catalyst which, by definition, is left chemically unchanged at the end. We would have obtained energy from nowhere, which is contrary to the Law of Conservation of Energy (see Figure 9.5).

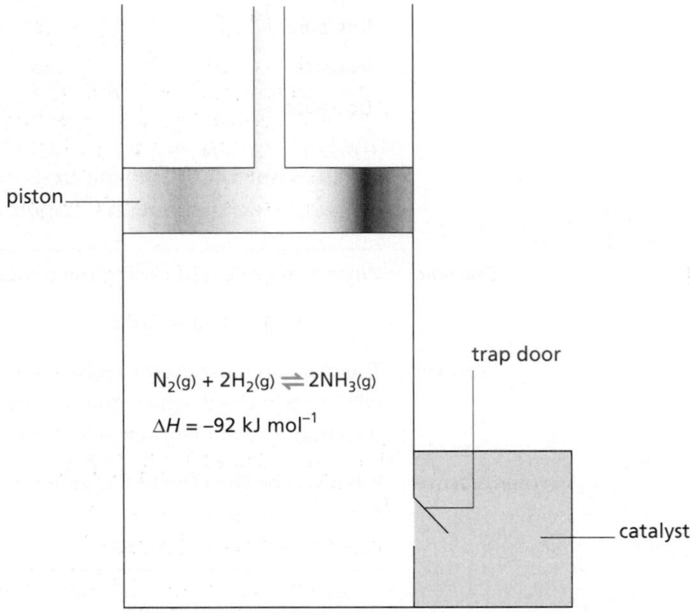

Figure 9.5 A reaction in equilibrium is placed in contact with a catalyst by opening the trap door. If this changed the position of equilibrium, the piston would move up or down, depending on whether the catalyst shifted the equilibrium in the endothermic or exothermic direction. This movement of the piston could be used to close the trap door, and the process would then be repeated. Perpetual motion would have been created, which is impossible.

- Reducing the particle size or increasing the surface area speeds up a reaction, but does not change the position of equilibrium.
- The addition of a catalyst does not affect the position of equilibrium, only how quickly equilibrium is reached.

Summary

- Many chemical reactions do not go to completion – they take up a position of **dynamic equilibrium**.
- This **position of equilibrium** may change if the reaction conditions are changed. The direction of change can be predicted using **Le Chatelier's Principle**.
- Increasing the concentration of a reactant moves the position of equilibrium to the right.
- Increasing the pressure in a gaseous reaction moves the equilibrium in the direction that has fewer gas molecules.
- Increasing the temperature moves the position of equilibrium in the endothermic direction.
- Changing the state of division or adding a catalyst has no effect on the position of equilibrium.

10 Further energy changes

In Chapter 5 we showed how we could measure the standard enthalpy change, $\Delta H^\ominus$, of a reaction. This enthalpy change is determined by the relative strengths of the bonds in the reactants and in the products. In this chapter, we look at how we can measure the strengths of the different types of bond that are met in chemical substances.

In Chapter 5, we also suggested that the sign of $\Delta H^\ominus$ gives an indication of the direction of chemical change. Although most reactions are exothermic ($\Delta H^\ominus$ negative), the existence of endothermic reactions shows that this is not the whole story. The quantity that does show the direction of chemical change is the Gibbs function (Gibbs free energy change), $\Delta G^\ominus$, and this is discussed at the end of the chapter.

The chapter is divided into the following sections.

10.1 Heat and work

Reactions can do work

In Chapter 5, we assumed that all the energy of a chemical reaction was transferred from the system to the surroundings in the form of heat, q. In an exothermic reaction, the enthalpy change, ΔH, is negative and heat is removed from the system and added to the surroundings. But many chemical reactions are used not to give out heat but to drive a turbine in a power station or a piston in a motor car, or to produce electricity in a battery. In this case, some of the energy is transferred as work, w, and the remainder as heat. The connection between ΔH, q and w is given by the equation:

$$\Delta H = q + w$$

Heat and work

- **Heat** is energy transferred between a system and its surroundings solely as a result of a temperature difference between them.
- **Work** is energy transferred between a system and its surroundings when both are at the same temperature. This energy is usually electrical energy, but it can be mechanical energy or electromagnetic radiation.

In a calorimeter, no work is done (provided no gas is evolved at constant pressure), so $\Delta H = q$. If the reaction is endothermic, an amount of heat, $+q$, at constant temperature, is added to the system from the surroundings and the enthalpy change, ΔH, is positive. If the reaction is exothermic, an amount of heat, $-q$, is removed from the system and added to the surroundings and ΔH is negative.

If work is done by a reaction then $\Delta H \neq q$. The ability of a reaction to do work is extremely important and we shall show later in the chapter that it is this ability of a reaction to produce work, rather than its ability to produce heat, that shows the direction of chemical change.

Introducing internal energy

In the panel on page 118, we met the bomb calorimeter. Here all the energy of the reaction appears as heat, but this is not necessarily the case in calorimeters that are open to the atmosphere. If the calorimeter operates at atmospheric pressure, some of the energy may be used in expansion. Any expansion of solids or liquids is insignificant and can be ignored. But gases take up a considerable volume, and if there is a change in volume of the system as a result of a change in the number of gas molecules in the reaction, Δn, some work is done. In this case the energy change with no expansion (at constant volume) is not quite the same as the energy change with expansion (at constant pressure). The energy change at constant pressure is $\Delta H^\ominus$. The energy change at constant volume is called the change in **internal energy**, and is given the symbol $\Delta U^\ominus$. If we assume the gases to be ideal, the connection between $\Delta H^\ominus$ and $\Delta U^\ominus$ is given by:

$$\Delta H^\ominus = \Delta U^\ominus + \Delta nRT$$

where Δn = total number of moles of gas in products − total number of moles of gas in reactants, R is the gas constant (see Chapter 4, page 105) and T is absolute temperature.

Even for reactions in which Δn is significant, the difference between $\Delta H^\ominus$ and $\Delta U^\ominus$ is very small. For example, for the combustion of ethane at room temperature:

$$C_2H_6(g) + 3\tfrac{1}{2}O_2(g) \rightarrow 2CO_2(g) + 3H_2O(l)$$

$$\Delta n = -2.5 \text{ (we ignore the liquid water)}$$

$$\Delta nRT = 2.5 \times 8.3 \times 298 \approx 6000\,J\,mol^{-1}$$

For this reaction, $\Delta H^\ominus = -1561\,kJ\,mol^{-1}$ and $\Delta U^\ominus = -1555\,kJ\,mol^{-1}$, an extremely small difference. For most purposes, the distinction between $\Delta H^\ominus$ and $\Delta U^\ominus$ can be ignored.

The distinction between $\Delta H^\ominus$ and $\Delta U^\ominus$ can cause confusion, for example, when using the concept of the lattice energy of an ionic solid such as sodium chloride. **Lattice energy** is defined as the internal energy change, ΔU, when solid is formed from gaseous ions; for sodium chloride this is ΔU for the reaction:

$$Na^+(g) + Cl^-(g) \rightarrow NaCl(s)$$

The value found using a Born–Haber cycle (see page 190) gives the **lattice enthalpy**, ΔH, for the same reaction. For sodium chloride, ΔU and ΔH differ by $5\,kJ\,mol^{-1}$ which, for accurate work, must be taken into account.

Functions of state

Enthalpy, H, and internal energy, U, are called **functions of state**, because their values are uniquely determined by the external conditions. Standard conditions under which they are measured are 1 mole, a fixed temperature (often 298 K), for H, a fixed pressure of 1 atm, and for U a fixed volume. This means that the change for a particular reaction, $\Delta H^\ominus$ or $\Delta U^\ominus$, also has a unique value independent of how the change is brought about (Hess's Law, see page 121). However, heat and work are not functions of state, because the proportion of each varies depending on how the reaction is carried out. The *maximum* amount of work that can be obtained from a reaction is, however, a function of state – it is called the **free energy** and is discussed further later in the chapter.

10.2 Enthalpy change of atomisation

In considering the enthalpy change of a reaction, we often use published values of standard enthalpy changes for particular changes. One fundamental measurement in chemistry is the **standard enthalpy change of atomisation, $\Delta H^{\ominus}_{at}$**. This is the energy required to produce one mole of gaseous atoms under standard conditions. It is a quantitative indication of the strength of the bonding in the substance. If a solid breaks up into atoms, it may first melt (accompanied by the **standard enthalpy change of fusion, $\Delta H^{\ominus}_{m}$**), then evaporate (accompanied by the **standard enthalpy change of vaporisation, $\Delta H^{\ominus}_{b}$**) and finally, in the gas phase, any remaining bonds break (the sum of all the bond energies). So:

$$\Delta H^{\ominus}_{at} = \Delta H^{\ominus}_{m} + \Delta H^{\ominus}_{b} + \text{bond energy terms}$$

For elements, $\Delta H^{\ominus}_{at}$ is the enthalpy change when one mole of atoms is produced. For compounds, it is the enthalpy change when one mole of the substance is broken down completely into atoms.

$\Delta H^{\ominus}_{at}$ for

- elements is the enthalpy change to produce 1 mole of atoms under standard conditions
- compounds is the enthalpy change to convert 1 mole of the substance into atoms under standard conditions.

The magnitude of the enthalpy changes associated with the different processes depends on the type of substance. For a metal, the enthalpy changes associated with melting and evaporation are the most important, whilst for a covalent substance, the bond energy terms predominate. Ionic substances are a special case because they break up into ions in the gas phase. These will be considered later in the chapter. Table 10.1 gives some examples of $\Delta H^{\ominus}_{at}$, along with $\Delta H^{\ominus}_{m}$ and $\Delta H^{\ominus}_{b}$, for some metallic and covalent substances.

■ **Table 10.1** Enthalpy changes of fusion, vaporisation and atomisation for some substances. For elements, $\Delta H^{\ominus}_{at}$ is the enthalpy change when one mole of atoms is produced. In the gas phase, metals and carbon have broken down almost completely into atoms, so $\Delta H^{\ominus}_{at} \approx \Delta H^{\ominus}_{m} + \Delta H^{\ominus}_{b}$. For compounds, $\Delta H^{\ominus}_{at}$ is the enthalpy change when one mole of the substance is broken down completely into atoms. Simple molecular substances retain their structure in the gas phase, so $\Delta H^{\ominus}_{at} \gg \Delta H^{\ominus}_{m} + \Delta H^{\ominus}_{b}$. The structure of silicon(IV) oxide changes from giant covalent when solid to molecular when gaseous, and this is supported by the figures in the table.

Substance	Structure	$\Delta H^{\ominus}_{m}$/kJ mol^{-1}	$\Delta H^{\ominus}_{b}$/kJ mol^{-1}	$\Delta H^{\ominus}_{at}$/kJ mol^{-1}
Na(s)	Metallic	2.6	89.0	107.3
Mg(s)	Metallic	9.0	128.7	147.7
W(s)	Metallic	35.2	799.1	849.4
Cl_2(g)	Molecular	3.2	10.2	121.3
Br_2(l)	Molecular	5.4	15.0	111.8
I_2(s)	Molecular	7.6	20.8	106.8
C(s)	Giant covalent	(sublimes)	710.9	716.7
CH_4(g)	Molecular	0.9	8.2	1663.6
C_6H_{14}(l)	Molecular	13.1	28.9	7519
$AlBr_3$(s)	Molecular	11.3	23.5	1071
SiO_2(s)	Giant covalent	8.2	605.5	1244

10.3 Bond enthalpy

Finding the bond enthalpy in simple molecules

The enthalpy change of atomisation of a compound includes the **bond enthalpies** (see page 125) for all bonds in the substance. In the gas phase, most covalent substances retain their molecular structure. To convert them into atoms, all the bonds in the molecule must be broken. The amount of energy required to break all the bonds is the sum of all the individual bond enthalpies. Remember that the bond enthalpy is the energy required to break one mole of a particular type of bond, and this means that all bond enthalpies are positive. (The term 'bond enthalpy' is synonymous with the older term 'bond energy'.)

If all the bonds in the molecule are the same, we can calculate the average bond enthalpy from the enthalpy of atomisation of the molecule. This done using the enthalpy change of formation, ΔH_f, of the substance and the enthalpies of atomisation of the atoms formed. For example, for methane, we can construct the Hess cycle shown in Figure 10.1.

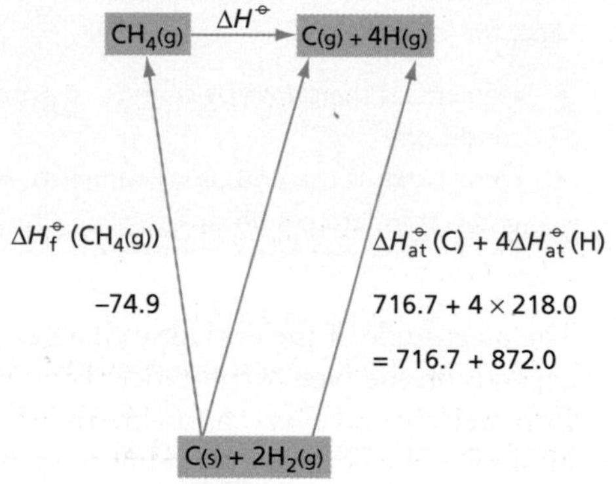

■ Figure 10.1 Hess's Law cycle to find the average bond enthalpy in methane. All figures are in kJ mol^{-1}.

From Figure 10.1 we can see that:

$$-74.9 + \Delta H^\ominus = 716.7 + 872.0$$

$$\text{so } \Delta H^\ominus = 1588.7 - (-74.9)$$

$$= +1663.6\,\text{kJ mol}^{-1}$$

This enthalpy is the energy needed to break four moles of C—H bonds. Because the four bonds in methane are identical, it is reasonable to allocate $\frac{1663.6}{4}$ kJ mol^{-1} for one mole of bonds. This gives a value for the average bond enthalpy of the C—H bond in methane, E(C—H), of 416 kJ mol^{-1} (but see page 187).

We can carry out the same calculation for tetrachloromethane, CCl_4, to find E(C—Cl). We must be careful to break the bonds in gaseous tetrachloromethane, rather than the liquid which is its normal state under standard conditions. Figure 10.2 shows the route.

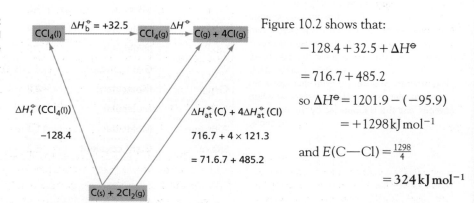

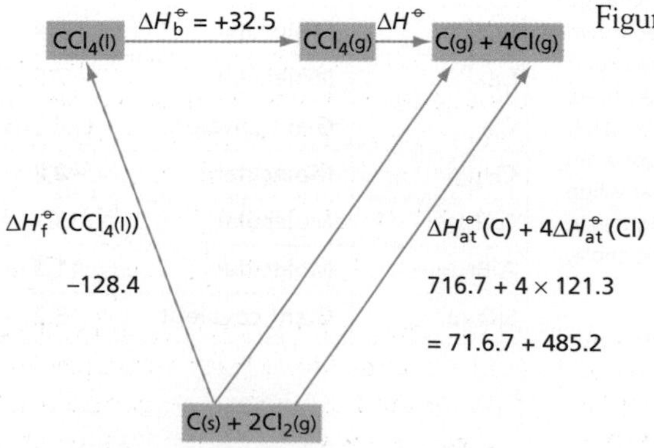

■ Figure 10.2 Hess's Law cycle to find the average bond enthalpy in tetrachloromethane. All figures are in kJ mol^{-1}.

Figure 10.2 shows that:

$$-128.4 + 32.5 + \Delta H^\ominus$$

$$= 716.7 + 485.2$$

$$\text{so } \Delta H^\ominus = 1201.9 - (-95.9)$$

$$= +1298\,\text{kJ mol}^{-1}$$

$$\text{and } E(\text{C—Cl}) = \tfrac{1298}{4}$$

$$= \mathbf{324\,kJ\,mol^{-1}}$$

Many other covalent substances are liquids under standard conditions, and it is essential to remember to include an enthalpy change of vaporisation term in the calculation of bond enthalpy.

Example

a What data are required to calculate the average bond enthalpy of the N—H bond in ammonia, NH_3?

b Calculate the average bond enthalpy, $E(\text{N—H})$.

Answer

a The data required are:

$\Delta H_f^{\ominus}(NH_3(g)) = -46.0\,kJ\,mol^{-1}$; $\Delta H_{at}^{\ominus}(H) = 218.0\,kJ\,mol^{-1}$; $\Delta H_{at}^{\ominus}(N) = 472.7\,kJ\,mol^{-1}$.

b

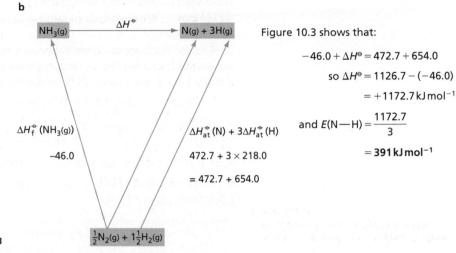

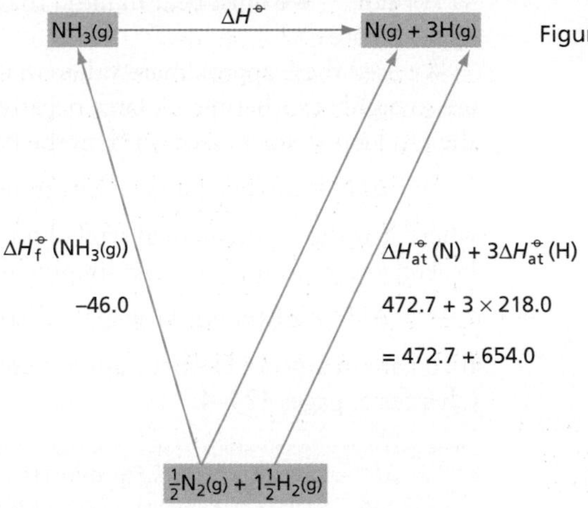

■ **Figure 10.3**

Figure 10.3 shows that:

$$-46.0 + \Delta H^{\ominus} = 472.7 + 654.0$$

$$\text{so } \Delta H^{\ominus} = 1126.7 - (-46.0)$$

$$= +1172.7\,kJ\,mol^{-1}$$

$$\text{and } E(\text{N—H}) = \frac{1172.7}{3}$$

$$= \mathbf{391\,kJ\,mol^{-1}}$$

Further practice

Calculate the average bond enthalpies of the following bonds, using the data given below (all in $kJ\,mol^{-1}$).

1 H—Br **2** O—H **3** P—Cl

[$\Delta H_f^{\ominus}(HBr(g)) = -36.2$; $\Delta H_f^{\ominus}(H_2O(l)) = -285.9$; $\Delta H_f^{\ominus}(PCl_3(l)) = -272.4$; $\Delta H_b^{\ominus}(H_2O) = +44.1$; $\Delta H_b^{\ominus}(PCl_3) = +30.7$; $\Delta H_{at}^{\ominus}(H) = +218.0$; $\Delta H_{at}^{\ominus}(Br) = +111.9$; $\Delta H_{at}^{\ominus}(O) = +249.2$; $\Delta H_{at}^{\ominus}(P) = +333.9$; $\Delta H_{at}^{\ominus}(Cl) = +121.3$]

Alkane	ΔH_c/kJ mol^{-1}	Difference in ΔH_c/kJ mol^{-1}
Methane	−891	
Ethane	−1561	−670
Propane	−2219	−658
Butane	−2878	−659
Pentane	−3536	−658
Hexane	−4195	−659
Heptane	−4856	−661
Octane	−5512	−656
Nonane	−6172	−660
Decane	−6830	−658

■ **Table 10.2** Enthalpy changes of combustion for the first ten straight-chain alkanes in the gaseous state. The constancy in the difference between successive values of enthalpy of combustion suggest that the C—C bond enthalpy has approximately the same value for each hydrocarbon. The same is also true of the C—H bond enthalpy. The value for methane is slightly out of line with the others; the reason for this is discussed on page 189.

Finding a bond enthalpy using other known average bond enthalpies

So far we have considered how to find the bond enthalpy in molecules that contain one type of bond only. Many molecules contain several types of bond, and to calculate the bond enthalpies of all of them we must make an important assumption. This is that the average bond enthalpy of a bond is the same wherever it is found. This is a big assumption and, as we shall see, it is only approximately true.

The best evidence that this assumption is a reasonable one is to look at bond enthalpies in a homologous series (see Chapter 22, page 417), such as the straight-chain alkanes. The difference in the enthalpy change of combustion, ΔH_c, is approximately the same from one hydrocarbon to the next (see Table 10.2). This enthalpy change is associated with the process:

$$\text{—}CH_2\text{—}(g) + 1\tfrac{1}{2}O_2(g) \rightarrow CO_2(g) + H_2O(l)$$

and is a result of the bonding in the products being stronger than the bonding in the reactants. As the bond enthalpies in $1\tfrac{1}{2}O_2(g)$, $CO_2(g)$ and $H_2O(l)$ are fixed terms, it must mean that the bond enthalpies in —CH_2—(g) are also constant.

If we assume that a particular type of bond has the same enthalpy wherever it is found, we can calculate bond enthalpies in more complex systems. For example, consider the processes:

$$C_2H_6(g) \rightarrow 2C(g) + 6H(g) \quad \Delta H_{at} = 2825\,kJ\,mol^{-1} = 1 \times E(\text{C—C}) + 6 \times E(\text{C—H})$$

$$C_3H_8(g) \rightarrow 3C(g) + 8H(g) \quad \Delta H_{at} = 4000\,kJ\,mol^{-1} = 2 \times E(\text{C—C}) + 8 \times E(\text{C—H})$$

If we multiply the first equation by 2 and take away the second equation, we have $4 \times E(C—H) = 1650\,kJ\,mol^{-1}$. So $E(C—H) = \frac{1650}{4} = 412.5\,kJ\,mol^{-1}$.

Substituting this value of $E(C—H)$ into the first equation allows us to calculate $E(C—C) = 350\,kJ\,mol^{-1}$. By carrying out similar calculations with a large number of compounds, general **average bond enthalpy** values are obtained. For the two we have been considering, they are $E(C—C) = 346\,kJ\,mol^{-1}$ and $E(C—H) = 413\,kJ\,mol^{-1}$. We must bear in mind that these are average values and are only approximate.

We used these approximate values in Chapter 5 to find out why some reactions are so highly exothermic. A large negative value of $\Delta H^{\ominus}$ implies that the bonding in the products is much stronger than the bonding in the reactants. In general we have:

$$\Delta H^{\ominus} = \Sigma E(\text{reactants}) - \Sigma E(\text{products})$$

where Σ is the mathematical symbol for 'the sum of'. If only a few bonds are broken and re-formed, we can simplify the calculation by using:

$$\Delta H^{\ominus} = \Sigma E(\text{bonds broken}) - \Sigma E(\text{bonds formed})$$

If you are in any doubt about using these formulae, you should look again at Chapter 5, pages 123–4.

$\Delta H^{\ominus} = \Sigma E(\text{reactants}) - \Sigma E(\text{products})$
$\Delta H^{\ominus} = \Sigma E(\text{bonds broken}) - \Sigma E(\text{bonds formed})$

Example 1 Given that $\Delta H^{\ominus}_f(H_2O_2(l)) = -187.6\,kJ\,mol^{-1}$, $\Delta^{\ominus}_b(H_2O_2(l)) = +51.6\,kJ\,mol^{-1}$, $\Delta H^{\ominus}_{at}(O) = +249.2\,kJ\,mol^{-1}$ and $\Delta H^{\ominus}_{at}(H) = +218.0\,kJ\,mol^{-1}$, calculate $\Delta H^{\ominus}$ for the following atomisation:

$$H_2O_2(g) \rightarrow 2H(g) + 2O(g)$$

Answer

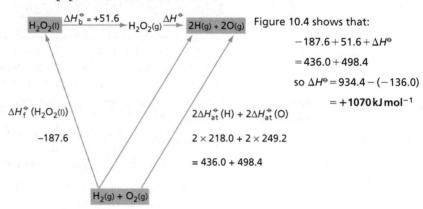

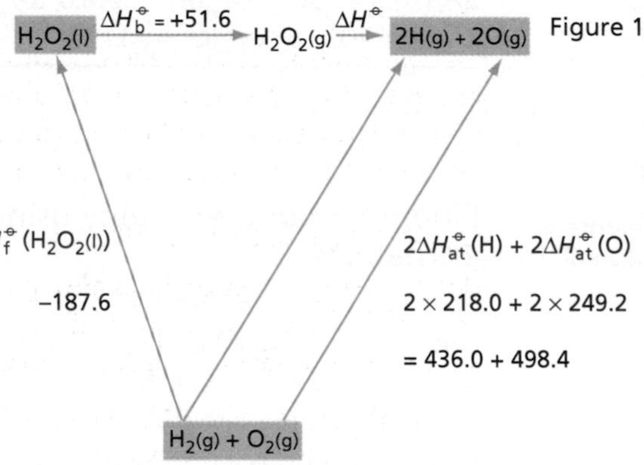

Figure 10.4 shows that:

$-187.6 + 51.6 + \Delta H^{\ominus}$

$= 436.0 + 498.4$

so $\Delta H^{\ominus} = 934.4 - (-136.0)$

$= \mathbf{+1070\,kJ\,mol^{-1}}$

■ **Figure 10.4**

Example 2 Given $E(O—H) = +464\,kJ\,mol^{-1}$, calculate $E(O—O)$.

Answer $2 \times 464 + E(O—O) = 1070$

So $E(O—O) = 1070 - 928 = \mathbf{142\,kJ\,mol^{-1}}$

Further practice

1 a Given that $\Delta H^{\ominus}_f(N_2H_4(g)) = 95.4\,kJ\,mol^{-1}$, $\Delta H^{\ominus}_{at}(N) = +472.7\,kJ\,mol^{-1}$ and $\Delta H^{\ominus}_{at}(H) = +218.0\,kJ\,mol^{-1}$, calculate $\Delta H^{\ominus}$ for the following atomisation:

$$N_2H_4(g) \rightarrow 2N(g) + 4H(g)$$

b Draw the structure of hydrazine, N_2H_4.

c Given that $E(N—H) = 391\,kJ\,mol^{-1}$, calculate $E(N—N)$.

d Hence estimate $\Delta H^{\ominus}$ for the following reaction:

$$N_2H_4(g) + H_2(g) \rightarrow 2NH_3(g)$$

2 (Harder) The following equations show the values of ΔH_{at} of butane and pentane.

$C_4H_{10}(g) \rightarrow 4C(g) + 10H(g)$ $\quad \Delta H = +5174\,kJ\,mol^{-1}$

$C_5H_{12}(g) \rightarrow 5C(g) + 12H(g)$ $\quad \Delta H = +6347\,kJ\,mol^{-1}$

Calculate $E(C—C)$ and $E(C—H)$.

10.4 Bond dissociation enthalpy and average bond enthalpies

Bond dissociation enthalpy – breaking one bond at a time

For methane, the average bond enthalpy, $E(\text{C—H})$, is calculated from the energy needed to break all the bonds homolytically (see Chapter 22), represented by the equation:

$$CH_4(g) \rightarrow C(g) + 4H(g) \qquad \Delta H^{\ominus} = +1663.6\,\text{kJ mol}^{-1}$$

and $E(\text{C—H}) = \frac{1663.6}{4} = 416\,\text{kJ mol}^{-1}$. This does not have the same value as the energy required to break just one bond in $CH_4(g)$, represented by the equation:

$$CH_4(g) \rightarrow CH_3(g) + H(g) \qquad \Delta H = +438.6\,\text{kJ mol}^{-1}$$

This quantity is known as the **bond dissociation enthalpy** and is given the symbol $D(\text{CH}_3\text{—H})$. There are four separate bond dissociation enthalpy terms for methane, represented by the following equations:

$$CH_4(g) \rightarrow CH_3(g) + H(g) \qquad D(\text{CH}_3\text{—H}) = +438.6\,\text{kJ mol}^{-1}$$
$$CH_3(g) \rightarrow CH_2(g) + H(g) \qquad D(\text{CH}_2\text{—H}) = +458.7\,\text{kJ mol}^{-1}$$
$$CH_2(g) \rightarrow CH(g) + H(g) \qquad D(\text{CH—H}) = +425.7\,\text{kJ mol}^{-1}$$
$$CH(g) \rightarrow C(g) + H(g) \qquad D(\text{C—H}) = +340.6\,\text{kJ mol}^{-1}$$

All the bond dissociation enthalpies have different values. The sum of them, however, represents the equation:

$$CH_4(g) \rightarrow C(g) + 4H(g)$$

and must have the same value as the enthalpy change of atomisation of methane, $1663.6\,\text{kJ mol}^{-1}$.

The variation in the values of the different bond dissociation enthalpies can be explained in terms of electron repulsions. As successive hydrogen atoms are removed, the repulsions between the bonding pairs of electrons and the non-bonding electrons left on the carbon atom become less.

Bond dissociation enthalpy – different types of bond

Bond dissociation enthalpies also differ if different types of C—H bond are broken. It makes a significant difference whether a hydrogen atom is removed from a primary, secondary or tertiary carbon atom (see Chapter 23, page 436).

$RCH_3(g)$ primary alkane	$\rightarrow RCH_2(g)$ primary radical	$+ H(g)$	$\Delta H^{\ominus} = +418\,\text{kJ mol}^{-1}$
$RCH_2R(g)$ secondary alkane	$\rightarrow RCHR(g)$ secondary radical	$+ H(g)$	$\Delta H^{\ominus} = +401\,\text{kJ mol}^{-1}$
$R_3CH(g)$ tertiary alkane	$\rightarrow R_3C(g)$ tertiary radical	$+ H(g)$	$\Delta H^{\ominus} = +400\,\text{kJ mol}^{-1}$

This variation is due to the different stabilities of primary, secondary and tertiary radicals. The order of stability is primary $<$ secondary $\approx$ tertiary, similar to the stability of carbocations (see page 447). The more stable the radical, the easier it is to form and the lower the value of the bond dissociation enthalpy.

There are, therefore, three different types of bond enthalpy:

- bond dissociation enthalpy, for example, $D(\text{CH}_3\text{—H})$ for methane $= 438.6\,\text{kJ mol}^{-1}$
- specific average bond enthalpy, for example, $E(\text{C—H})$ for methane $= 416\,\text{kJ mol}^{-1}$
- general average bond enthalpy, for example, $E(\text{C—H}) = 413\,\text{kJ mol}^{-1}$

For diatomic molecules, the three values are identical.

There can be confusion between specific and general average bond enthalpies. The specific bond enthalpy is sometimes called the **mean bond enthalpy** or **bond enthalpy term**. We shall use the term 'average bond enthalpy' and qualify it with 'specific' or 'general' if there is a possibility of confusion.

The value of a bond enthalpy in a particular compound may differ widely from the general average value. There are many reasons for this but the following are the most important.

- The bonds may be of intermediate character.
- The bonds may be strained.

We shall now look at each of these situations.

Bonds with intermediate character

Many molecules contain single bonds with some double-bond character. They contain a σ bond and also have additional p overlap giving a degree of π bonding (see Chapter 3). The length and strength of these bonds is between that of a single and that of a double bond.

The classic example is that of benzene (see pages 76–7). The six C—C bonds in the ring all have the same length, 0.139 nm, nearer to the length of a C=C double bond, 0.134 nm, than to the length of a C—C single bond, 0.154 nm. The observed total enthalpy change of atomisation of benzene, represented by the equation:

$$C_6H_6(g) \rightarrow 6C(g) + 6H(g)$$

is 5527 kJ mol^{-1}. The sum of all the bond enthalpies assuming alternate single and double bonds is 5349 kJ mol^{-1}. The difference between these two values, 178 kJ mol^{-1}, is a measure of the extra stability due to delocalisation.

Double bonds may also have some triple bond character. In methanal, H_2CO, the bond enthalpy of the C=O bond is 695 kJ mol^{-1}. The same bond in carbon dioxide has a bond enthalpy of 804 kJ mol^{-1}. This extra strength shows that the bonds in carbon dioxide have considerable triple bond character (see Chapter 18, page 341).

Bond strain

As we saw in Chapter 3, bonds have natural angles determined by the repulsion of the electron pairs in the valence shell. If the bonds are bent away from this natural angle, the resulting **strain** weakens the bond. The result is that the energy required to break all the bonds in the molecule is less than would be expected if there were no strain.

A classic example is cyclopropane. The C—C—C bond angles in the molecule are 60°, much less than the natural tetrahedral angle of 109.5°. The enthalpy change of atomisation, represented by the equation:

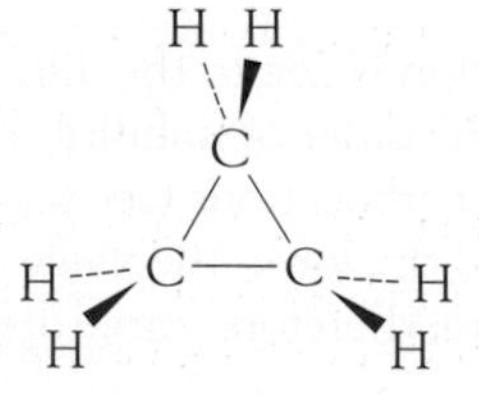

$$\rightarrow \quad 3C(g) + 6H(g)$$

is 3403 kJ mol^{-1}. The sum of the general average bond enthalpies is 3516 kJ mol^{-1}. The difference between these values, 113 kJ mol^{-1}, shows the weakening of the bonds due to strain. Similar strain is found in epoxyethane (see Chapter 24, page 449).

This is the reason why $E(C—H)$ for methane, $416\,kJ\,mol^{-1}$, is slightly larger than the average value for a C—H bond, $413\,kJ\,mol^{-1}$. In methane, the hydrogen atoms are as far away as possible from each other, but in the other alkanes, they sometimes come close together as a result of bond rotation. This close approach leads to repulsion and a slight weakening of the C—H bond. This is why the difference between the enthalpies of combustion of methane and ethane is larger than for other pairs of alkanes (see Table 10.2, page 185).

Further practice

1 a Calculate the enthalpy change of atomisation of gaseous cyclohexane, $C_6H_{12}(g)$, using the following data.
[$\Delta H^{\ominus}_f(C_6H_{12}(l)) = -157.7\,kJ\,mol^{-1}$; $\Delta H^{\ominus}_b(C_6H_{12}) = +34.6\,kJ\,mol^{-1}$;
$\Delta H^{\ominus}_{at}(C) = +716.7\,kJ\,mol^{-1}$; $\Delta H^{\ominus}_{at}(H) = +218.0\,kJ\,mol^{-1}$]

b If $E(C—H) = 413\,kJ\,mol^{-1}$, calculate $E(C—C)$ for cyclohexane.

c This value of $E(C—C)$ indicates that there is no bond strain in cyclohexane, even though the bond angles in a regular hexagon, 120°, are much larger than the natural tetrahedral angle, 109.5°. Suggest why this is so.

2 The bond dissociaton enthalpies (in $kJ\,mol^{-1}$) for the N—H bonds in ammonia are as follows.

$D(NH_2—H) = 448$; $D(NH—H) = 368$; $D(N—H) = 356$

a Write equations which represent these three bond dissociation enthalpies.

b Calculate the value of $E(N—H)$.

10.5 The Born–Haber cycle – finding the strengths of ionic bonds

Lattice energy

We have seen that the following enthalpy changes are used as a measure of the strength of different types of bonding:

- intermolecular forces:

$M(s) \rightarrow M(l)$ enthalpy change of fusion, $\Delta H^{\ominus}_m$

$M(l) \rightarrow M(g)$ enthalpy change of vaporisation, $\Delta H^{\ominus}_b$

- metallic:

$M(s) \rightarrow M(g)$ enthalpy change of atomisation, $\Delta H^{\ominus}_{at}$

- covalent:

$X—Y(g) \rightarrow X(g) + Y(g)$ bond enthalpies, $E(X—Y)$ or $D(X—Y)$

In the same way we could use the enthalpy change of the following process:

$$M^{n+}X^{n-}(s) \rightarrow M^{n+}(g) + X^{n-}(g)$$

as a measure of the strength of ionic bonding. It does not, of course, measure the strength of a 'single' ionic bond between two ions, because there are many ionic bonds of different strengths in all directions in an ionic lattice. Nevertheless, it is a useful comparison. Conventionally it is usual to quote the enthalpy change for the reverse process, that is, the enthalpy change when a mole of ionic solid is formed from its isolated gaseous ions. This quantity is known as the **lattice energy (LE)** (or lattice enthalpy), and it has a high negative value because forming an ionic compound is a strongly exothermic process. For example, for sodium chloride:

$$Na^+(g) + Cl^-(g) \rightarrow NaCl(s) \qquad LE = -787\,kJ\,mol^{-1}$$

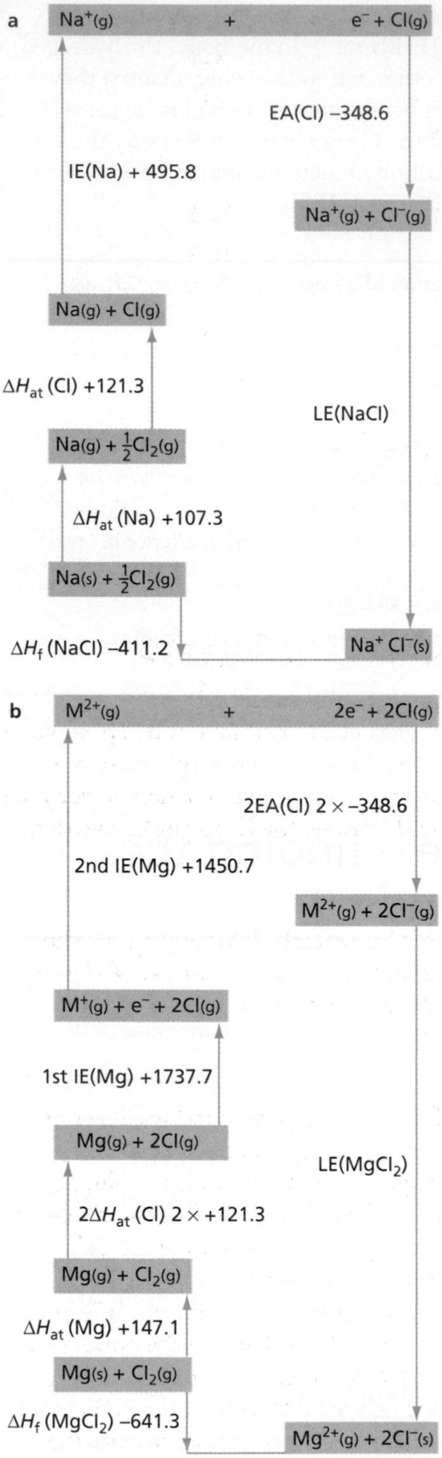

■ **Figure 10.5**
The Born–Haber cycle for **a** sodium chloride and **b** magnesium chloride. The values are in kJ mol^{-1}. The diagrams are not to scale. The values of some terms are highly uncertain and lattice energies are usually given to the nearest whole number.
The enthalpy changes associated with the removal and addition of an electron in the gas phase differ slightly from the IE and EA values. However, for the overall processes $Na(g) + Cl(g) \rightarrow Na^+(g) + Cl^-(g)$ and $Mg(g) + 2Cl(g) \rightarrow Mg^{2+}(g) + 2Cl^-(g)$, $\Delta H = IE + EA$.

The lattice energy is the energy change when 1 mole of the solid is formed from its isolated ions in the gas phase. Lattice energies are found in a similar way to bond enthalpies. In the determination of lattice energies, however, there is an additional factor, namely the transfer of an electron, in our example from the sodium atom to the chlorine atom. The energy associated with this process is the sum of two energies:

- the ionisation energy (IE) of sodium
- the energy of the change: $Cl(g) + e^- \rightarrow Cl^-(g)$.

We met ionisation energies in Chapter 2. The energy for the addition of an electron to a gaseous chlorine atom is called the **electron affinity (EA)** of chlorine. We met electron affinities in Chapter 4. If there is a space in an atom's outer shell for an extra electron, the effective nuclear charge usually attracts an electron sufficiently strongly to make its electron affinity exothermic. However, a second electron affinity is always negative because the charge on the ion repels the second electron. For example, for oxygen:

$$O(g) + e^- \rightarrow O^-(g) \qquad \Delta H = -141\,\text{kJ mol}^{-1}$$
$$O^-(g) + e^- \rightarrow O^{2-}(g) \qquad \Delta H = +790\,\text{kJ mol}^{-1}$$
$$O(g) + 2e^- \rightarrow O^{2-}(g) \qquad \Delta H = +649\,\text{kJ mol}^{-1}$$

Figure 10.5a shows how the lattice energy of sodium chloride may be worked out. Such a cycle constructed to find a lattice energy is called a **Born–Haber cycle**.

The values of all the terms in this Born–Haber cycle are known, except that for the lattice energy. This can be calculated as follows:

$$\Delta H_{at}(Na) + \Delta H_{at}(Cl) + IE(Na) + EA(Cl) + LE(NaCl) = \Delta H_f(NaCl)$$
$$+107.3 + 121.3 + 495.8 + (-348.6) + LE(NaCl) = -411.2$$
$$LE(NaCl) = -411.2 - (+375.8)$$
$$= -787\,\text{kJ mol}^{-1}$$

The situation is slightly more complicated if the ions carry more than one charge. For example, for magnesium chloride, the energy change associated with the process:

$$Mg(g) \rightarrow Mg^{2+}(g) + 2e^-$$

is the sum of the first and second ionisation energies of magnesium (it is not just the second ionisation energy of magnesium). In addition, as the formula is $MgCl_2$, we need to produce two moles of chloride ions. We have:

$$\Delta H_{at}(Mg) + 2\times\Delta H_{at}(Cl) + \text{1st IE}(Mg) + \text{2nd IE}(Mg) + 2\times EA(Cl) + LE(MgCl_2) = \Delta H_f(MgCl_2)$$
$$+147.1 + 2\times\ 121.3 + 737.7 + 1450.7 + (2\times -348.6) + LE(MgCl_2) = -641.3$$
$$LE(MgCl_2) = -641.3 - (+1880.9)$$
$$= -2522\,\text{kJ mol}^{-1}$$

The magnitude of the lattice energy

The value of the lattice energy of magnesium chloride is much larger than that for sodium chloride. There are two reasons for this. The first is that there are more cation-to-anion attractions because there are twice as many chloride ions. The second reason is that each of these attractions is much stronger, because the magnesium ion carries twice the charge of the sodium ion.

There are other factors that determine the size of the lattice energy. The most important of these are the way the ions pack together in the lattice (the lattice type) which in turn determines how closely the ions approach each other (the sum of their ionic radii). The effects on the lattice energy of the distance between the ions and the charges on the ions are illustrated in Table 10.3.

■ **Table 10.3**
For ionic compounds of the same type, the shorter the inter-ionic distance, the greater the value of the lattice energy. Compounds containing doubly or triply charged ions have much larger values of lattice energy.

Compound	Inter-ionic distance/nm	Charges on ions	Lattice energy/kJ mol^{-1}
LiF	0.211	1, 1	−1036
NaCl	0.279	1, 1	−787
CsI	0.385	1, 1	−604
BeF_2	0.167	2, 1	−3505
$MgCl_2$	0.259	2, 1	−2375
BaI_2	0.363	2, 1	−1877
Li_2O	0.210	1, 2	−2797
MgO	0.210	2, 2	−3791
Al_2O_3	0.189	3, 2	−16470

Lattice energies are determined by the magnitudes of:

- the charges on the ions
- the inter-ionic distance
- the type of lattice.

Calculating the lattice energy from the crystal structure

The type of lattice and the inter-ionic distance can be found by X-ray crystallography (see Chapter 14, page 270). Using this information, it is possible to calculate the lattice energy, assuming that the solid is built up of spherical ions (the ionic model). A fairly complicated calculation is needed to obtain an accurate result, but a good approximation can be found by using the Kapustinski equation:

$$\text{LE} = -\frac{107\, n\, z^+ z^-}{(r^+ + r^-)}$$

where n is the number of ions in the formula (for example, 5 for Al_2O_3), z^+ and z^- are the charges on the cation and anion, respectively, and r^+ and r^- are the ionic radii in nanometres. For magnesium chloride, this calculation gives:

$$\text{LE} = -\frac{107 \times 3 \times 2 \times 1}{0.259} = -2480\,\text{kJ mol}^{-1}$$

This is about 5% different from the value found from experimental results using the Born–Haber cycle. More accurate calculation, using a different formula, gives a value of $-2326\,\text{kJ mol}^{-1}$, very close to the experimentally determined value.

An agreement between the experimental and calculated values of lattice energy for a compound shows that the ionic model is a good one for this compound. Sometimes, however, there is a considerable difference, even though the calculation has been carried out as accurately as possible. A significant difference (say 10%) suggests that the ionic model needs modifying, because the bonding in the lattice contains considerable covalent character. This tends to make the experimental value for the lattice energy more negative than the calculated value, and this is most apparent when a cation with a high charge density distorts the anion (see Chapter 3, page 70). Table 10.4 compares the calculated and experimental values for lattice energy for some compounds.

■ **Table 10.4**
Comparison of some calculated values of lattice energy with those obtained experimentally using the Born–Haber cycle. The agreement is good, unless there is enough ionic distortion to give the bonds considerable covalent character.

Compound	Calculated value/ kJ mol^{-1}	Experimental value/ kJ mol^{-1}	Percentage difference
CsCl	−657	−659	0
LiI	−730	−757	4
$MgCl_2$	−2326	−2375	2
AgF	−953	−967	1
AgI	−808	−889	9

Ceramics

Ceramics are substances with giant structures that contain covalent bonds with considerable ionic character. They are hard but brittle. They are hard because of the strong ionic forces that hold the crystal together, and brittle because deformation brings similarly charged ions closer together leading to increased repulsion. If the ions have multiple charges, the lattice energy, melting point and Young modulus (a measure of stiffness) all increase, as Table 10.5 shows.

■ **Table 10.5**
The properties of some ionic substances.

Substance	Charge on ions	Lattice energy/ $kJ\,mol^{-1}$	Melting point/ °C	Young modulus/ $GN\,m^{-2}$
NaCl	1:1	−787	801	44
MgO	2:2	−3791	2852	245
Al_2O_3	3:2	−16470	2072	525

The high melting points mean that ceramics are used for furnace linings. The linings for blast furnaces are made of bricks with a high content of aluminium oxide, and those for steel converters are mainly magnesium oxide, made by heating magnesium carbonate (see Chapter 20, pages 374–6).

As the charge on the ionic substance increases, so does the covalent character of the bonds. Ceramics such as silicon(IV) oxide, SiO_2, silicon(IV) nitride, Si_3N_4, silicon(IV) carbide, SiC, and tungsten carbide, WC, are useful in adding stiffness to plastics or metals, for example:

- glass fibres embedded in epoxy resins (GRP) are used in boat construction
- the tiles covering the space shuttle contain silicon(IV) oxide fibres
- the blades in some jet engines are stiffened with silicon(IV) carbide.

Some ceramics can lose their electrical resistance completely when cooled and act as **superconductors** while still at relatively high temperatures. These ceramics are mixtures of metal oxides including copper oxide. One, with the approximate formula $YBa_2Cu_3O_7$, is a superconductor at 90K. In the future, these superconducting ceramics could be used to transmit electricity over large distances without any loss in energy, or to make highly efficient electrical devices.

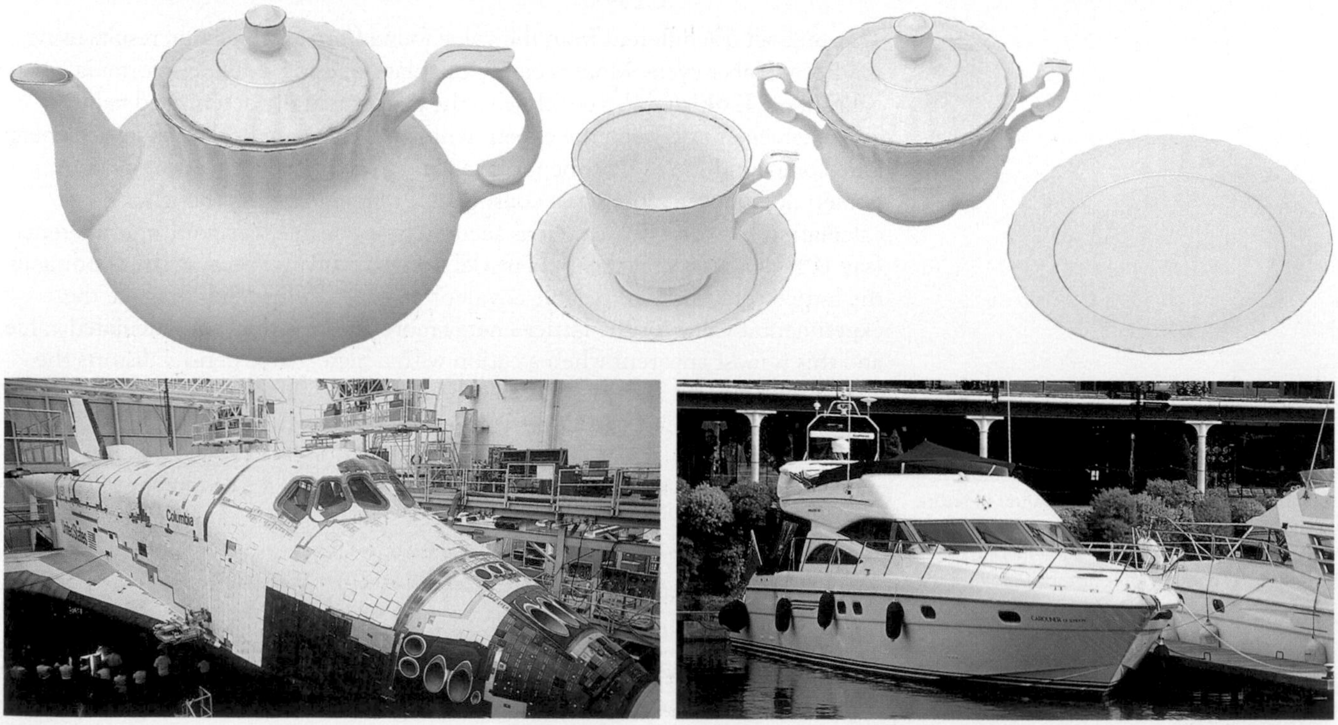

■ **Figure 10.6**
Ceramics provide many useful properties. Fine china is hard but brittle. The tiles covering the space shuttle have an extremely low thermal conductivity, and so insulate the inside of the shuttle from the heat at re-entry. GRP is tough and does not corrode, and so is used extensively in boat-building.

Example 1

a Write an equation that represents the chemical change associated with the lattice energy of calcium fluoride.

b Which two elements, one from Group 1, the other from Group 7, form a compound with the most negative value of lattice energy?

Answer

a $Ca^{2+}(g) + 2F^{-}(g) \rightarrow CaF_2(s)$

b Lithium and fluorine, because they have the smallest ionic radii.

Example 2

a Draw a Born–Haber cycle for rubidium oxide, Rb_2O.

b Use the following values (all in $kJ\,mol^{-1}$) to calculate the lattice energy of rubidium oxide.

$\Delta H_{at}(Rb) = +80.9$; $\Delta H_{at}(O) = +249.2$; $IE(Rb) = +403.0$; $1st\ EA(O) = -146.1$; $2nd\ EA(O) = +795.5$; $\Delta H_f(Rb_2O) = -339.0$

Answer

a

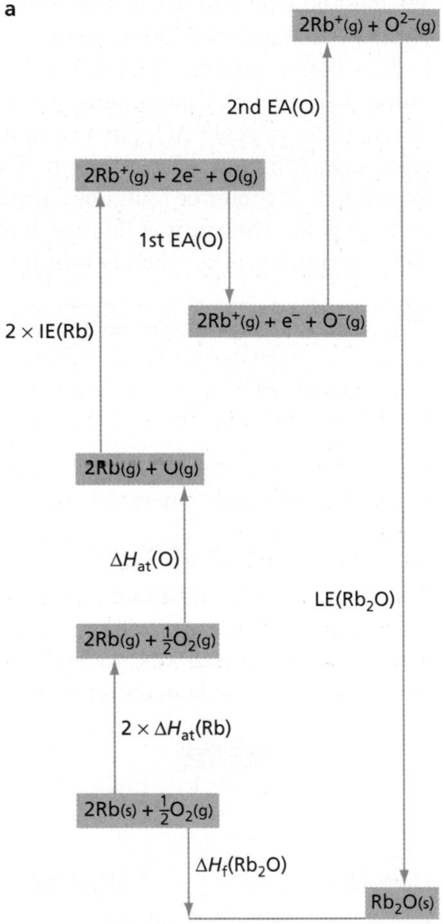

■ **Figure 10.7**

b From Figure 10.7 we can calculate that:

$$(2 \times 80.9) + 249.2 + (2 \times 403.0) - 146.1 + 795.5 + LE(Rb_2O) = -339.0$$
$$LE(Rb_2O) = -339.0 - (+1866.4)$$
$$= \mathbf{-2205.0\,kJ\,mol^{-1}}$$

Further practice

1 a Draw a Born–Haber cycle for barium sulphide, BaS.

b Use the following values (all in $kJ\,mol^{-1}$) to calculate the lattice energy of barium sulphide.

$\Delta H_{at}(Ba) = +180$; $\Delta H_{at}(S) = +249.2$; $1st\ IE(Ba) = +503$; $2nd\ IE(Ba) = 965$; $1st\ EA(S) = -141.4$; $2nd\ EA(S) = +790.8$; $\Delta H_f(BaS) = -595.8$

2 Explain why the following substances have different values of lattice energy from that of sodium chloride.

a Potassium chloride **b** Magnesium sulphide **c** Calcium chloride

3 Suggest why the experimental value of the lattice energy of silver bromide differs from the calculated value.

10.6 Enthalpy changes of solution and hydration

Enthalpy change of solution

The solubilities of salts in water show wide variations, with no obvious pattern. One of the factors determining solubility is whether the **enthalpy change of solution**, ΔH_{sol}, of the salt is positive or negative. This enthalpy change is the energy associated with process:

$$M^{z+}X^{z-}(s) \rightarrow M^{z+}(aq) + X^{z-}(aq)$$

We must be careful to specify the dilution of the final solution when quoting this enthalpy change. On dilution, the ions in a solution become more extensively hydrated (an exothermic process) and also move further apart (an endothermic process). The relative importance of these two effects changes with dilution, affecting the values of ΔH_{sol} in a complicated way. A quoted single value of ΔH_{sol} refers to an infinitely dilute solution. This value cannot be determined directly by experiment and must be found by a process of extrapolation. In practice, there comes a point when further dilution has no measurable effect on the value of ΔH_{sol}, and this may be taken as infinite dilution.

The **enthalpy change of solution**, ΔH_{sol}, is the enthalpy change when one mole of solute dissolves in an infinite volume of water (or enough water so that further dilution has no additional effect).

If ΔH_{sol} is positive, the salt is likely to be insoluble or of very low solubility. If ΔH_{sol} is approximately zero or negative, then the salt is likely to be soluble or very soluble.

Enthalpy change of hydration

The value of ΔH_{sol} can be found by comparing two energy changes. The first is the energy that holds the lattice together, the lattice energy. The second is the energy given out when gaseous ions are hydrated, called the **enthalpy change of hydration**, symbol ΔH_{hyd}. For sodium chloride, we have the cycle shown in Figure 10.8.

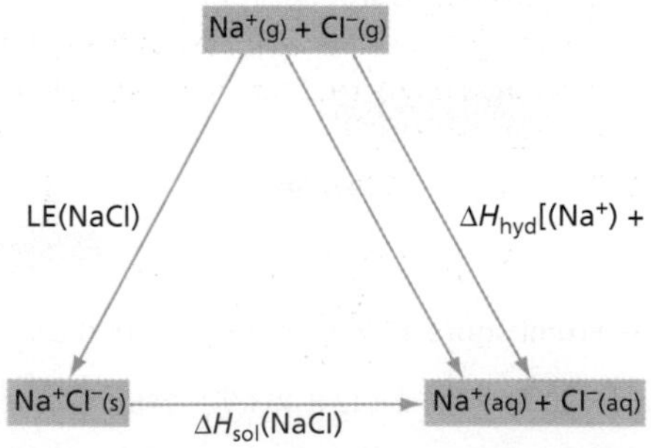

■ **Figure 10.8** Born–Haber cycle for the dissolving in water of sodium chloride.

The value of LE(NaCl)($-787\,kJ\,mol^{-1}$) is found using the Born–Haber cycle and ΔH_{sol} can be found experimentally ($+3.9\,kJ\,mol^{-1}$). We have:

$$\begin{aligned}\Delta H_{hyd} &= LE(NaCl) + \Delta H_{sol}\\ &= -787 + 3.9\\ &= -783\,kJ\,mol^{-1}\end{aligned}$$

We can now see why similar substances show wide variations in solubility. A small change in either the lattice energy or the enthalpy change of hydration has a very large effect on the difference between the two terms. These variations are used to explain the differences in solubility of some compounds of Group 2 in Chapter 16, page 314.

Ion	ΔH_{hyd}/kJ mol^{-1}
H^+	−1120
Li^+	−544
Na^+	−435
Mg^{2+}	−1980
Ca^{2+}	−1650
Al^{3+}	−4750
F^-	−473
Cl^-	−339

■ **Table 10.6**
Absolute enthalpy changes of hydration for some ions.

The Born–Haber cycle does not enable us to find ΔH_{hyd} for the individual ions, because it gives us the sum of the enthalpy changes of hydration for the cation and anion. If we wish to find their individual values, we must assign a value to one of the ions. The value for the H^+ ion is generally agreed to be $-1120\,kJ\,mol^{-1}$ and, by using this value, we can calculate the **absolute enthalpy change of hydration** of ions, as shown in Table 10.6.

The values in Table 10.6 show that the absolute enthalpy change of hydration depends on the ability of the ion to attract water molecules. Small, highly charged ions have the most negative values (see Chapter 4, page 96).

Enthalpy changes of hydration are more negative:

- if the ion is small
- if the ion has two or three units of charge on it.

As with lattice energies, it is possible to calculate theoretical values of absolute enthalpies of hydration. Approximate values can be obtained using the following formulae:

anions:

$$\Delta H_{hyd} = \frac{-70z^- \times z^+}{r^-}$$

cations:

$$\Delta H_{hyd} = \frac{-70z^- \times z^+}{(r^+ + 0.085)}$$

As on page 191, z^- and z^+ are the charges on the anion and cation, respectively, and r^- and r^+ are the ionic radii. Cations hydrate more extensively than anions and have, therefore, much larger hydrated ionic radii. This increase in ionic radius is reflected in the 0.085 term in the expression for cations.

Example

a Draw an energy cycle to show the connection between solid magnesium chloride, its gaseous ions and its aqueous ions.

b If $LE(MgCl_2) = -2375\,kJ\,mol^{-1}$ and $\Delta H_{sol}(MgCl_2) = -155.1\,kJ\,mol^{-1}$, calculate the value of $\Delta H_{hyd}(MgCl_2)$.

c Compare this value with the value obtained from the separate absolute enthalpy changes of hydration for Mg^{2+} and Cl^- in Table 10.6.

Answer

a See Figure 10.9.

b From Figure 10.9:

$$\Delta H_{hyd} = LE + \Delta H_{sol}$$
$$= -2375 - 155.1$$
$$= -2530\,kJ\,mol^{-1}$$

c From Table 10.6:

$$-1980 - 2 \times 339 = -2658\,kJ\,mol^{-1}$$

Fair agreement (within 5%).

■ **Figure 10.9**

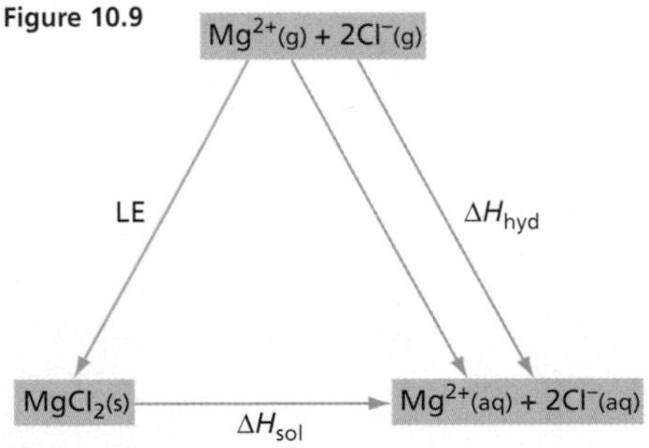

Further practice

1 Draw a Born–Haber cycle for aluminium fluoride, AlF_3(s).

2 Use the following values (all in $kJ\,mol^{-1}$) to calculate the lattice energy of aluminium fluoride.

$\Delta H^\ominus_f(AlF_3) = -1504.1$; $\Delta H^\ominus_{at}(Al) = 324.3$; $\Delta H^\ominus_{at}(Cl) = 121.3$; 1st IE(Al) = 577;
2nd IE(Al) = 1817; 3rd IE(Al) = 2745; EA(Cl) = −348.6

3 Draw an energy diagram to show the connection between AlF_3(s), its gaseous ions and its aqueous ions.

4 Use the figures in Table 10.6 and the lattice energy value obtained in **2** to calculate $\Delta H_{sol}(AlF_3)$.

Enthalpy change of neutralisation

An approximate value for the enthalpy change of neutralisation for the reaction between an acid and a base can be found by simple calorimetry, as we saw in Chapter 5, page 114. Some values for this enthalpy change are given in Table 10.7.

■ **Table 10.7** Values of some enthalpy changes of neutralisation.

Reaction	Acid	Base	Type of reaction	ΔH_{neut}/kJ mol^{-1}
1	HCl	NaOH	Strong acid–strong base	−57.9
2	HNO_3	KOH	Strong acid–strong base	−57.6
3	CH_3CO_2H	NaOH	Weak oxoacid–strong base	−56.1
4	H_3PO_4	NaOH	Weak oxoacid–strong base	−58.0
5	H_2S	NaOH	Weak acid–strong base	−32.2
6	HCN	NaOH	Weak acid–strong base	−15.9

These neutralisation reactions can be divided into three categories, which we shall consider in turn.

Strong acid and strong base

The reaction, in all cases, is:

$$H_3O^+(aq) + OH^-(aq) \rightarrow 2H_2O(l)$$

so the value of ΔH should be constant ($-57.6\,kJ\,mol^{-1}$). This is approximately true (Table 10.7, reactions 1 and 2). The small variation is due to differences in the enthalpy change of hydration of the spectator ions. When the two solutions are mixed, dilution takes place and this is accompanied by slightly different enthalpy changes of dilution.

Weak oxoacid and strong base

Acids such as ethanoic acid, CH_3CO_2H, and phosphoric(V) acid, H_3PO_4, (Table 10.7, reactions 3 and 4) are largely un-ionised. The neutralisation reaction is principally that of the un-ionised acid with OH^- ions, for example:

$$CH_3CO_2H + OH^-(aq) \rightarrow CH_3CO_2^-(aq) + H_2O \qquad \Delta H^\ominus = -56.1\,kJ\,mol^{-1}$$

This reaction can be broken down into two steps:

$$CH_3CO_2H + H_2O \xrightarrow{\Delta H} CH_3CO_2^-(aq) + H_3O^+(aq)$$

$$H_3O^+(aq) + OH^-(aq) \rightarrow 2H_2O(l) \qquad \Delta H^\ominus = -57.6\,kJ\,mol^{-1}$$

Combining these equations together, we have $\Delta H^\ominus = -56.1 - (-57.6) = +1.5\,kJ\,mol^{-1}$. This is the enthalpy change associated with ionising the aqueous acid. Its value is very small, because the reaction involves both the breaking and the making of an O—H bond. It does not explain why ethanoic acid is such a weak acid. This weakness is not primarily associated with the O—H bond strength, but is mainly due to adverse entropy changes which take place when the acid is ionised (see page 201).

Weak acid and strong base

The third class of neutralisation reactions (Table 10.7, reactions 5 and 6) have significantly smaller enthalpy changes of neutralisation. During neutralisation, a different type of bond is being broken from that which is being formed. For these reactions, the large positive value of enthalpy change of ionisation is a major factor in determining why the acid is so weak. For example,

$$H_2S(aq) + H_2O(l) \rightarrow HS^-(aq) + H_3O^+(aq)$$

10.7 The concept of free energy

Maximum work, minimum heat

A **spontaneous change** is one that, in the absence of any barrier such as activation energy, takes place naturally in the direction stated. Not all chemical changes that occur spontaneously are exothermic, and so the principle that ΔH is negative cannot always be used as the criterion of feasibility. In order to find a universal principle, we need to consider the work that can be obtained from a reaction as well as the heat that it produces. Spontaneous processes often take place because the potential energy of the system decreases – for example, a weight falls to the ground or an electric current flows from a point of high potential to one of lower potential. In both these examples, the reduction in potential energy could be used to produce useful work – the weight could be used to turn the hands of a clock, and the electric current could be used to drive a motor. It is this ability to produce work, not the ability to produce heat, that determines whether a reaction is feasible or not.

It is not easy to see how much work a reaction produces. The reaction may be carried out in a bomb calorimeter, when no work is produced (only heat, remember $\Delta H = q + w$), or it may be used to produce electricity in a cell, when a lot of work is produced. The proportion of heat to work depends on how we carry out the reaction. Heat and work are therefore *not* functions of state of the system (see page 182).

The following simple example illustrates the division of energy between heat and work. Two weights w_1 and w_2 are attached to a piece of string over a pulley (see Figure 10.10). If the system has no friction, $w_1 = w_2$. If we allow w_1 to move down a distance d, all its loss in potential energy is transferred into useful work, increasing the potential energy of w_2. If the system has friction, w_1 must be greater than w_2 before w_1 can move downwards. Now, not all the potential energy lost by w_1 is converted into useful work and the remainder appears as heat generated by friction. In the extreme case, if we drop w_1 on to the floor and $w_2 = 0$, no work is done at all and all the energy appears as heat.

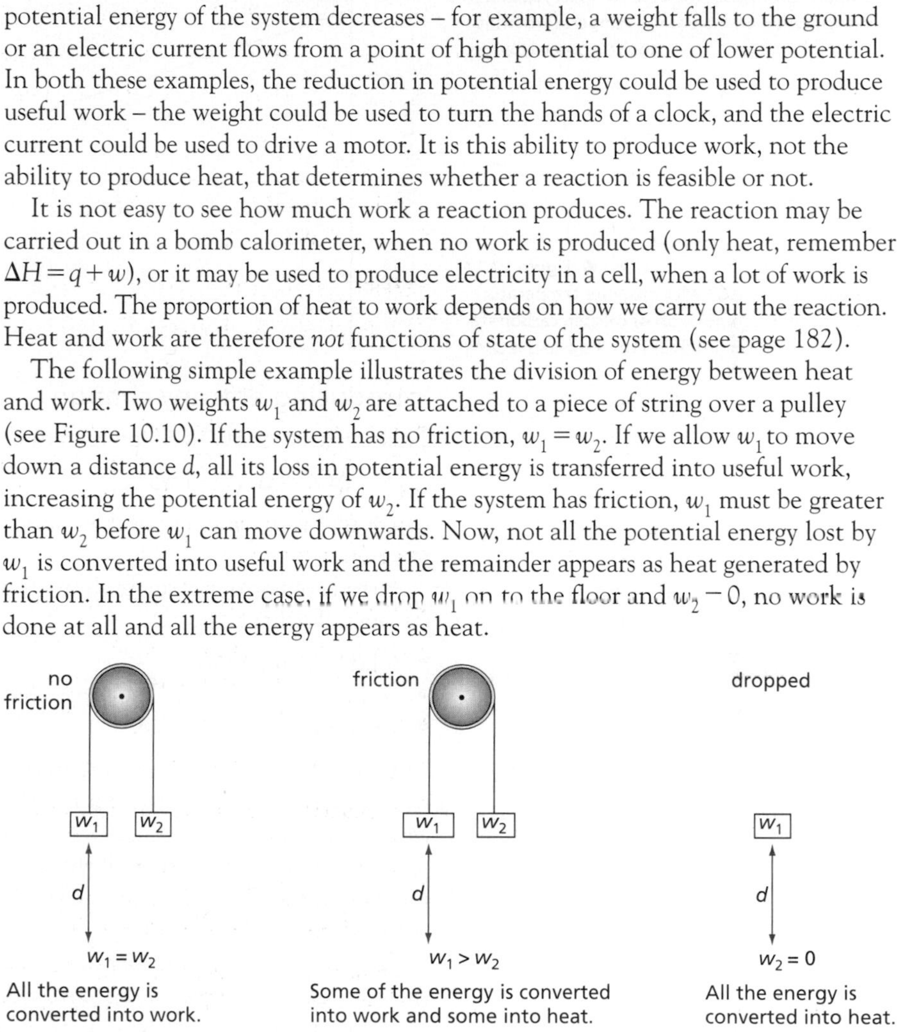
no friction
friction
dropped
w1
w2
d
w1 = w2
w1 > w2
w2 = 0
All the energy is converted into work.
Some of the energy is converted into work and some into heat.
All the energy is converted into heat.

■ **Figure 10.10**
If the pulley has no friction, all the potential energy is converted to work. If the pulley has friction, some of the potential energy is converted into heat. If the weight is dropped, all the potential energy is converted into heat.

Although the proportion of the heat to work produced varies, we can carry out a chemical reaction under special conditions that produce the maximum amount of work and the minimum amount of heat. This corresponds to the first example in Figure 10.10 when there is no friction and the system operates under reversible conditions. A specially designed electrical cell is an important example (see Chapter 13). This maximum amount of work *is* a function of state, and it is called the **free energy** or **Gibbs function**. Under standard conditions it is given the symbol ΔG. We use $\Delta G^\ominus$ rather than $\Delta H^\ominus$ as our criterion for the direction of chemical change.

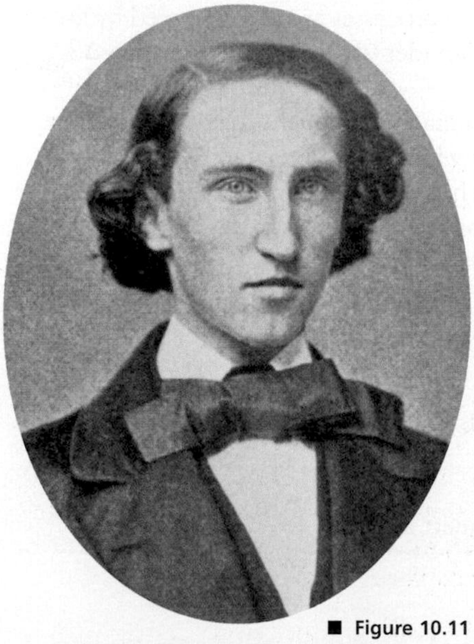

■ **Figure 10.11**
Josiah Willard Gibbs 1839–1903, the father of chemical thermodynamics.

The **standard free energy change, $\Delta G^\ominus$**, is the maximum amount of work that can be obtained from a reaction under standard conditions.

- If $\Delta G^\ominus$ is negative, a reaction can take place spontaneously.
- If $\Delta G^\ominus$ is positive, a reaction cannot take place spontaneously.

Temperature/ K	$\Delta H^\ominus$/ kJ mol^{-1}	$\Delta G^\ominus$/ kJ mol^{-1}
298	+179	+131
1000	+178	+16
1500	+176	−64

■ **Table 10.8**
The effect of temperature on $\Delta H^\ominus$ and $\Delta G^\ominus$ for the thermal decomposition of calcium carbonate. The value of $\Delta H^\ominus$ varies only slightly with temperature. This is the case for most reactions, except when an increase in temperature brings about a phase change.

ΔG and temperature

The value of $\Delta H^\ominus$ does not usually vary much with temperature, but the value of $\Delta G^\ominus$ changes significantly. Data for the decomposition of calcium carbonate show this (Table 10.8).

$$CaCO_3(s) \rightarrow CaO(s) + CO_2(g)$$

This variation of $\Delta G^\ominus$ with temperature is shown in Figure 10.12.

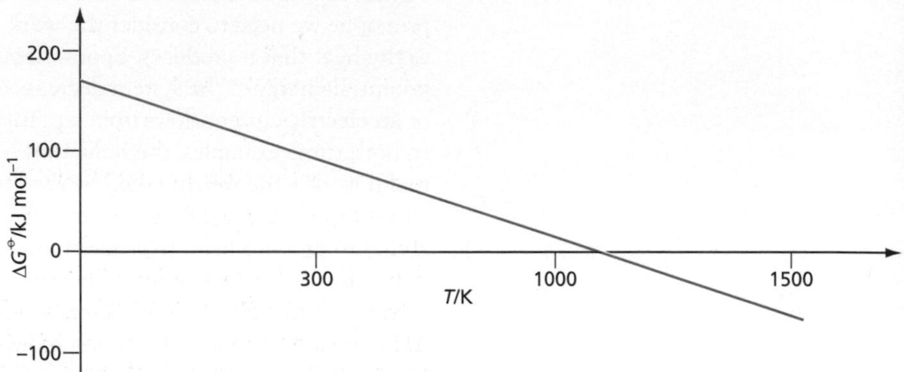

■ **Figure 10.12**
The variation of $\Delta G^\ominus$ with absolute temperature, *T*, for the decomposition of calcium carbonate. The intercept on the *y*-axis is $\Delta H^\ominus$.

Introducing entropy change

This variation of $\Delta G^\ominus$ with temperature may be expressed by the following equation:

$$\Delta G^\ominus = \Delta H^\ominus - T\Delta S^\ominus \qquad (1)$$

where $\Delta S^\ominus$ is the gradient of the graph, and is called the **entropy change** of the reaction.

A study of entropy changes shows that its value is significantly large and positive for the following changes:

- solid to liquid to gas
- an increase in the number of gas molecules
- when mixing takes place.

The particles in a liquid are more **disordered** (that is, less regularly arranged) than those in a solid. The arrangement of the particles in a gas is even more disordered. Mixing two solutions produces a more disordered state than keeping them separate. So we can see that all three of these processes are accompanied by an increase in the disorder of the system. We can identify the entropy change ΔS_{system} with this increase in disorder.

This increase in the *disorder* of the *system* may be sufficient to counterbalance the increase in the *order* of the *surroundings* which is the inevitable result of an endothermic reaction (that is, one for which ΔH is positive). If widely dispersed heat energy is taken in from the surroundings and localised in the system, the entropy of the surroundings, $\Delta S_{surroundings}$, has decreased. The two entropy terms ΔS_{system} and $\Delta S_{surroundings}$ are combined in the term ΔS_{total}, and this must be positive for the change to be spontaneous.

$$\Delta S_{total} = \Delta S_{system} + \Delta S_{surroundings}$$

For a spontaneous change, $\Delta S_{total} > 0$. This is the **Second Law of Thermodynamics.**

Second Law of Thermodynamics
For a spontaneous change, the total entropy must increase.

Thermodynamics shows that:

$$\Delta S_{surroundings} = -\frac{\Delta H}{T}$$

From:

$$\Delta S_{total} = \Delta S_{system} + \Delta S_{surroundings}$$

we have:

$$\Delta S_{total} = \Delta S_{system} - \frac{\Delta H}{T}$$

Multiplying throughout by T:

$$T\Delta S_{total} = T\Delta S_{system} - \Delta H$$

Changing signs throughout:

$$-T\Delta S_{total} = \Delta H - T\Delta S_{system}$$

Comparing this equation with equation (1), we can see that:

$$\Delta G = -T\Delta S_{total}$$

There is therefore a clear relationship between the free energy change of a system and the change in the total entropy of the system plus its surroundings.

Looking again at equation (1), we see that $\Delta G^{\ominus}$ tends to be negative (spontaneous process) if either $\Delta H^{\ominus}$ is negative or $\Delta S^{\ominus}$ is positive. It is the signs of $\Delta H^{\ominus}$ and $\Delta S^{\ominus}$ together that determine whether a reaction is feasible or not.

As we have seen, the value of $\Delta H^{\ominus}$ changes only slightly with temperature and may, as an approximation, be taken as constant. The same is true for $\Delta S^{\ominus}$. But this approximation is not valid if there is a change in phase of one of the reactants or products as the temperature is increased. Under these conditions both $\Delta H^{\ominus}$ and $\Delta S^{\ominus}$ change significantly.

Some examples of entropy changes

Table 10.9 lists values for some entropy changes. When gases mix, there is no enthalpy change and the positive value of $\Delta S^{\ominus}$ (entropy increase) indicates that they mix completely. The same is true of similar liquids that obey Raoult's Law (see Chapter 12, page 238). These two examples in which $\Delta H = 0$, show that it is the overall entropy change, rather than the enthalpy change, that is the important factor. $\Delta S^{\ominus}$ is also positive for melting and evaporation, because both processes lead to a more disordered system. Combination reactions have a negative value of $\Delta S^{\ominus}$ because there is often a reduction in the number of gas molecules. Conversely, decomposition reactions have an entropy increase.

■ **Table 10.9** Some values of entropy changes.

Example	$\Delta S^{\ominus}$/J K^{-1} mol^{-1}	Type of change
Mixing two gases (one mole of each)	+12	Mixing
$H_2O(s) \rightarrow H_2O(l)$	+22	Melting
$C_6H_{14}(l) \rightarrow C_6H_{14}(g)$	+88	Evaporation
$K(s) + \frac{1}{2}I_2(s) \rightarrow KI(s)$	−17	Combination
$N_2(g) + 3H_2(g) \rightarrow 2NH_3(g)$	−198	Combination
$CaCO_3(s) \rightarrow CaO(s) + CO_2(g)$	+160	Decomposition
$NaCl(s) \rightarrow Na^+(aq) + Cl^-(aq)$	+43	Solution
$Ca^{2+}(aq) + CO_3^{2-}(aq) \rightarrow CaCO_3(s)$	−204	Precipitation

ΔG and pressure

$\Delta G^{\ominus}$ is the standard free energy change needed to convert reactants into products at 1 atm pressure and at constant temperature (usually 298 K). We have considered how $\Delta G^{\ominus}$ changes as the temperature is changed, and a reaction that is not feasible at one temperature may become feasible at another. The same is true for changes in pressure. Although $\Delta G^{\ominus}$ cannot change (remember the pressure is specified as

1 atm under standard conditions), ΔG can vary. If the reactants and products are gaseous, an increase in pressure increases their free energy because work is done on them as a result of compression. If the reactants and products have the same volume ($\Delta n = 0$), the free energy of each increase by the same amount and there is no effect on ΔG. However, if the volumes are different, more free energy is added to the reactants or products (whichever have the larger volume). So ΔG becomes more negative in the direction of the change that has the smaller number of molecules. This is what we predict from Le Chatelier's Principle.

Consider the reaction:

$$2NO_2(g) \rightarrow N_2O_4(g)$$

At 1 atm and 60 °C, $\Delta G^\ominus = +1.4\,kJ\,mol^{-1}$ and nitrogen dioxide gas will not spontaneously change completely into dinitrogen tetraoxide. At 10 atm, and 60 °C, $\Delta G = -5.0\,kJ\,mol^{-1}$ and the reaction is then spontaneously feasible. An increase in pressure favours the side of the reaction that contains fewer molecules.

In practice, the reaction does not go to completion, even at 10 atm. This point is considered further on page 204.

Example For the synthesis of ammonia at 300 K,

$$N_2(g) + 3H_2(g) \rightarrow 2NH_3(g) \qquad \Delta H^\ominus = -92\,kJ\,mol^{-1} \text{ and } \Delta G^\ominus = -33\,kJ\,mol^{-1}$$

Assuming $\Delta H^\ominus$ and $\Delta S^\ominus$ do not change with temperature:

a calculate $\Delta G^\ominus$ at 750 K

b explain why it is essential to carry out the Haber process at very high pressure.

Answer **a** $\Delta G^\ominus = \Delta H^\ominus - T\Delta S^\ominus$ so $-T\Delta S^\ominus = \Delta G^\ominus - \Delta H^\ominus$

At 300 K, $-T\Delta S^\ominus = -33 - (-92) = +59\,kJ\,mol^{-1}$ and $\Delta S^\ominus = \frac{59}{300}\,kJ\,K^{-1}\,mol^{-1}$

At 750 K, $-T\Delta S^\ominus = 750 \times \frac{59}{300} = +148\,kJ\,mol^{-1}$

$\Delta G^\ominus = -92 + 148 = \mathbf{+56\,kJ\,mol^{-1}}$

b At 750 K, the reaction is thermodynamically unfavourable, but as Δn for the reaction is -2, high pressure makes it more feasible.

Further practice Heptane is converted into methylbenzene industrially:

$$C_7H_{16}(l) \rightarrow C_7H_8(l) + 4H_2(g)$$

At 300 K, $\Delta H^\ominus = +211\,kJ\,mol^{-1}$ and $\Delta G^\ominus = +110\,kJ\,mol^{-1}$.

Estimate $\Delta G^\ominus$ at:

1 600 K

2 900 K.

3 Explain why your estimated values are unreliable.

10.8 Applying free energy changes

The free energy of a reaction tells us whether the reaction proceeds spontaneously and this depends on the signs and magnitudes of both $\Delta H^\ominus$ and $\Delta S^\ominus$ for the reaction. As $\Delta H^\ominus$ and $\Delta S^\ominus$ can each be positive or negative, there are four possibilities.

$\Delta H^\ominus$ and $\Delta S^\ominus$ both positive

These endothermic reactions, which may not be feasible at room temperature, become feasible if the temperature is raised (Le Chatelier's Principle). The following are some examples:

- melting and boiling
- decomposition reactions
- electrolysis
- dissolving (in some cases).

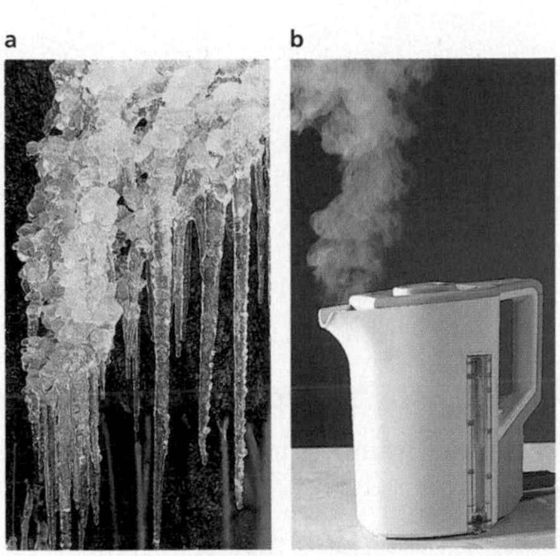

■ **Figure 10.13** ΔH and ΔS both positive. **a** Melting of ice and **b** boiling water.

It is easy to see why both $\Delta H^{\ominus}$ and $\Delta S^{\ominus}$ are positive for melting and boiling. The change is endothermic because intermolecular bonds are being broken. There is an increase in the entropy because disorder increases from solid to liquid to gas.

Most decomposition reactions (for example, the cracking of alkanes and the thermal decomposition of calcium carbonate) are endothermic because the total bond enthalpy in the products is less than that in the reactants. The energy required to break relatively strong bonds is not recovered by the formation of fewer or weaker bonds. Decomposition reactions are accompanied by an increase in entropy because the change in number of molecules, Δn, is positive.

Decomposition may be brought about by electrolysis rather than by heating. The reaction can then take place at a temperature at which $\Delta G^{\ominus}$ is positive, because it is being driven by the passage of an electric current through a potential difference. For the electrolysis of water, the minimum voltage needed to bring about decomposition is 1.23 V. This point is discussed in more detail in Chapter 13.

Liquids of similar polarity (for example, hexane and heptane) mix together in all proportions. If the polarity of the liquids is different, however, $\Delta H^{\ominus}$ becomes so large and positive that they may be only partially miscible (for example, butan-1-ol and water). The solubility then increases if the temperature is raised. This is usually the situation when a covalent solid dissolves in a liquid (for example, benzoic acid in water).

The dissolving of ionic solids in water is an extremely complex process. Because ions hydrate to a greater extent in dilute solution, dissolving an ionic solid in water may be exothermic when the solution is dilute, but endothermic when it becomes saturated. Two entropy terms operate:

- an entropy increase because the ions in the solid are free to move in solution
- an entropy decrease because water molecules that were originally free to move become restricted by hydration of the ions.

This complex interplay between enthalpy and entropy factors makes it very difficult to explain why some ionic compounds are very soluble in water while others are highly insoluble. For ions with a single charge, the overall entropy change is usually positive and this means that compounds of Group 1 are water soluble, even when the enthalpy of solution is positive. For ions with a multiple charge, the overall entropy change is usually negative. Many compounds in Group 2 have a positive enthalpy of solution and combining this with an adverse entropy term makes them insoluble (see Chapter 16, page 314).

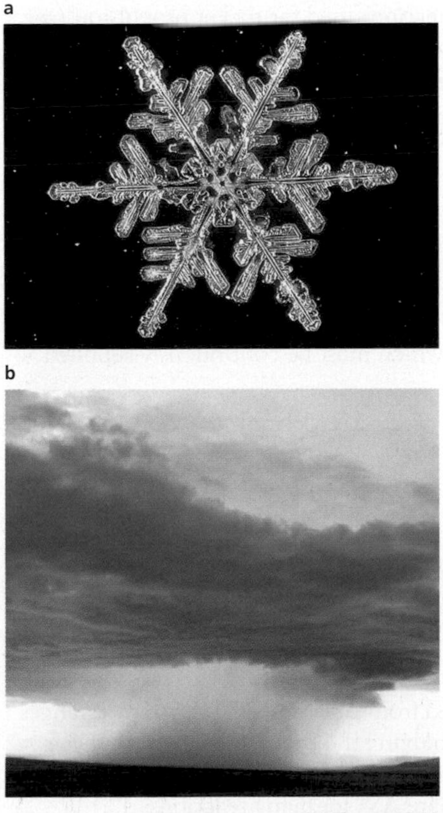

■ **Figure 10.14** ΔH and ΔS both negative. **a** Formation of a snow crystal and **b** rain from clouds.

$\Delta H^{\ominus}$ and $\Delta S^{\ominus}$ both negative

These exothermic reactions are feasible at low temperatures. Here are some common examples:

- condensation and freezing
- addition and combination reactions
- electrochemical cells
- precipitation.

These reactions are the reverse of the reactions in which both $\Delta H^{\ominus}$ and $\Delta S^{\ominus}$ are positive. Although addition and combination reactions are feasible at room temperature, it may be that the rate of reaction is then so slow that a catalyst has to be used. An example of this is the catalytic hydrogenation of an alkene. Sometimes, even with a catalyst, the rate may still be too low for a reasonable yield to be obtained in an acceptable length of time. To increase the rate still further, the pressure may be increased (for example, in the Haber process) or the temperature increased by a carefully controlled amount so that the increase in rate more than compensates for the adverse change in feasibility (see the Contact process and the Haber process in Chapter 20).

■ **Figure 10.15**
ΔH negative and ΔS positive.
a Fireworks and **b** explosive decomposition of ammonium nitrate.

$\Delta H^\ominus$ negative and $\Delta S^\ominus$ positive

These reactions are feasible at all temperatures. The reactants are said to be **metastable** under all conditions because they exist only because the activation energy of the reaction is so high. They are thermodynamically unstable, but kinetically inert. The following are some examples:

- a few decomposition reactions
- organic combustion reactions
- combustion of explosives (see the panel on page 204).

A few substances decompose exothermically because the total bond enthalpy in the products is higher than that in the reactants. Examples include the decomposition of ozone and dinitrogen oxide:

	$\Delta H^\ominus$/kJ mol^{-1}	$\Delta S^\ominus$/J K^{-1} mol^{-1}
$O_3(g) \rightarrow 1\frac{1}{2}O_2(g)$	−142.7	+68.7
$N_2O(g) \rightarrow N_2(g) + \frac{1}{2}O_2(g)$	−82.0	+74.4

This means that ozone and dinitrogen oxide cannot be synthesised directly by the reverse of the reactions above. Ozone is made by the combination of oxygen atoms with oxygen molecules, and dinitrogen oxide is made by the thermal decomposition of ammonium nitrate:

$$NH_4NO_3(s) \rightarrow N_2O(g) + 2H_2O(g)$$

When burnt, organic fuels yield gaseous carbon dioxide and water and the evolution of these gases makes $\Delta S^\ominus$ for the reaction positive. An example is the combustion of octane:

	$\Delta H^\ominus$/kJ mol^{-1}	$\Delta S^\ominus$/J K^{-1} mol^{-1}
$C_8H_{18}(g) + 12\frac{1}{2}O_2(g) \rightarrow 8CO_2(g) + 9H_2O(g)$	−5109	+238

For this reaction, $\Delta S^\ominus$ is positive, because $13\frac{1}{2}$ gas molecules become 17 gas molecules and $\Delta n = 3\frac{1}{2}$. The combustion reactions used for rocket propulsion (see the panels on pages 119–20) also have large entropy increases.

If the reaction rapidly gives off large quantities of gas at high temperature, an explosion may result. An example is the decomposition of propane-1,2,3-triyl trinitrate (commonly called nitroglycerine):

	$\Delta H^\ominus$/kJ mol^{-1}	$\Delta S^\ominus$/J K^{-1} mol^{-1}
$2C_3H_5N_3O_9 \rightarrow 3N_2 + 5H_2O + 6CO_2 + \frac{1}{2}O_2$	−3617	+1840

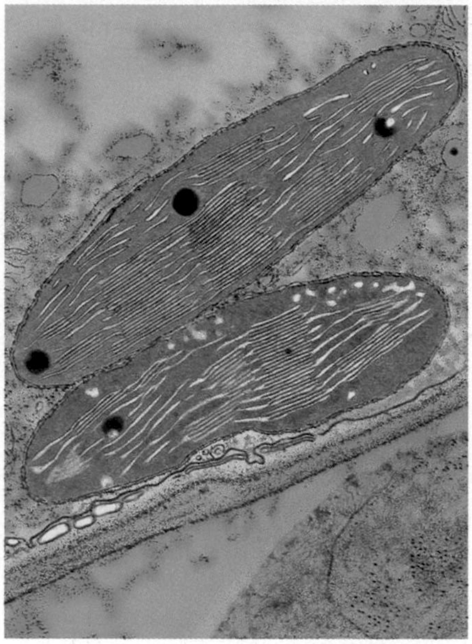

■ **Figure 10.16**
ΔH positive and ΔS negative.
Photosynthesis in chloroplasts.

$\Delta H^\ominus$ positive and $\Delta S^\ominus$ negative

Reactions of this type are not spontaneously feasible and have to be driven. An example is photosynthesis (Figure 10.16), which must be continuously supplied with energy from sunlight.

	$\Delta H^\ominus$/kJ mol^{-1}	$\Delta S^\ominus$/J K^{-1} mol^{-1}
$6CO_2(g) + 6H_2O(l) \rightarrow C_6H_{12}O_6(s) + 6O_2(g)$	+2803	−225

How can we measure $\Delta G^\ominus$?

Sometimes $\Delta G^\ominus$ can be found by directly measuring the maximum work that can be obtained from a reaction. This is achieved with an electrochemical cell operating under special conditions (see Chapter 13, page 256), using the equation:

$$\Delta G^\ominus = -nFE^\ominus$$

if we know $\Delta S^\ominus$, $\Delta G^\ominus$ can also be calculated from $\Delta H^\ominus$ and $\Delta S^\ominus$. Entropy changes may be found either by calculation or by studying the thermal behaviour of the substance at low temperatures. Standard molar enthalpies are recorded for many substances, and these can be used to calculate $\Delta S^\ominus$ for many reactions.

A third method depends on measuring the equilibrium constant K for the reaction (see Chapter 12 – a large equilibrium constant means the position of equilibrium lies to the right-hand side of the equation).

The equilibrium constant can then be used to find $\Delta G^{\ominus}$ using the following equation:

$$\Delta G^{\ominus} = -2.30\,RT\log_{10}K \text{ or } RT\log_{e}K$$

This equation shows us that if $\Delta G^{\ominus}$ is only slightly positive, the reaction proceeds to an equilibrium position well over towards the reactant side of the equation. For example, if $\Delta G^{\ominus} = +10\,\text{kJ mol}^{-1}$, $K = 0.018$. This means that the reaction has proceeded to a small but significant extent. If $\Delta G^{\ominus} = +20\,\text{kJ mol}^{-1}$, $K = 0.0003$, which is insignificant. For most practical purposes, if $\Delta G^{\ominus} > +20\,\text{kJ mol}^{-1}$, the extent to which the reaction has taken place can be ignored. Similarly, if $\Delta G^{\ominus} < -20\,\text{kJ mol}^{-1}$, the reaction can be considered as having gone to completion.

- If $\Delta G^{\ominus} > +20\,\text{kJ mol}^{-1}$, the reaction is insignificant.
- If $-20\,\text{kJ mol}^{-1} < \Delta G^{\ominus} < +20\,\text{kJ mol}^{-1}$, the reaction will reach an equilibrium.
- If $\Delta G^{\ominus} < -20\,\text{kJ mol}^{-1}$, the reaction goes virtually to completion.

We can now see why $\Delta H^{\ominus}$ can often be used as a guide to the feasibility of a reaction. For most reactions, the value of $\Delta S^{\ominus}$ is less than $100\,\text{J K}^{-1}\text{mol}^{-1}$. At room temperature, $T\Delta S = 300 \times 100 = 30\,000\,\text{J mol}^{-1}$, so the effect that $\Delta S^{\ominus}$ has on $\Delta G^{\ominus}$ is less than $30\,\text{kJ mol}^{-1}$. So if $\Delta H^{\ominus}$ is more negative than $-50\,\text{kJ mol}^{-1}$, we can confidently predict that the reaction will go to completion. This is often the case, particularly with combination reactions in inorganic chemistry. On the other hand, if $\Delta H^{\ominus}$ is more positive than $+50\,\text{kJ mol}^{-1}$, we can predict that there will be no significant reaction at room temperature. The situation changes as the temperature is raised because at high temperatures the entropy term becomes increasingly more important.

At room temperature:

- if $\Delta H^{\ominus} > +50\,\text{kJ mol}^{-1}$, reaction is unlikely
- if $-50\,\text{kJ mol}^{-1} < \Delta H^{\ominus} < +50\,\text{kJ mol}^{-1}$, the entropy term is important
- if $\Delta H^{\ominus} < -50\,\text{kJ mol}^{-1}$, reaction is feasible.

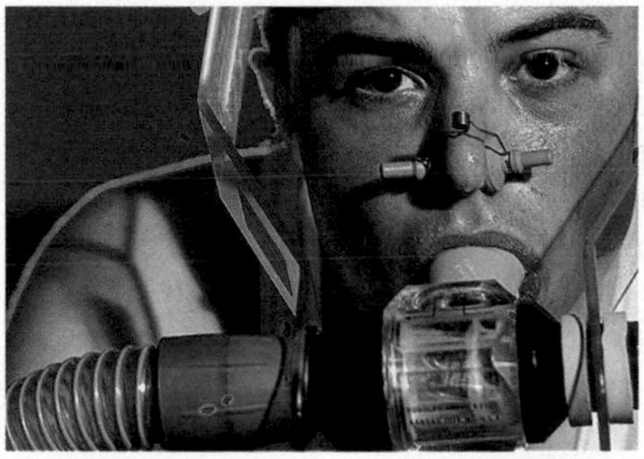

■ **Figure 10.17** Measuring the efficiency of an athlete's lungs, which depends on free energy changes in biochemical reactions.

Below are listed some reactions in which the entropy term cannot be ignored, even at room temperature:

- phase changes
- dissolving and precipitation
- dissociation of weak acids
- many biochemical reactions.

Many organic reactions are accompanied by only a small enthalpy change, and so the entropy term is significant. However, for these reactions it is the rate at which they react, rather than the position of equilibrium, that determines the yield. Kinetics, not thermodynamics, is then the controlling factor (see Chapter 11).

Example For the following processes, explain why $\Delta H^{\ominus}$ and $\Delta S^{\ominus}$ have the signs (positive or negative) shown.

a $Mg(s) + \frac{1}{2}O_2(g) \rightarrow MgO(s)$ $\Delta H^{\ominus}$ −ve, $\Delta S^{\ominus}$ −ve

b $C_2H_5OH(l) \rightarrow C_2H_4(g) + H_2O(l)$ $\Delta H^{\ominus}$ +ve, $\Delta S^{\ominus}$ +ve

Answer **a** $\Delta H^{\ominus}$ is negative because relatively weak bonds in magnesium and oxygen are converted into strong ionic bonds in magnesium oxide. $\Delta S^{\ominus}$ is negative because $\Delta n = -1$.

b $\Delta H^{\ominus}$ is positive because a relatively weak π bond is being formed. $\Delta S^{\ominus}$ is positive because $\Delta n = +1$.

Further practice For the following processes, explain why $\Delta H^{\ominus}$ and $\Delta S^{\ominus}$ have the signs (positive or negative) shown.

1 $C_2H_4(g) + H_2(g) \rightarrow C_2H_6(g)$ $\Delta H^{\ominus}$ −ve, $\Delta S^{\ominus}$ −ve

2 $H_2O(s) \rightarrow H_2O(g)$ $\Delta H^{\ominus}$ +ve, $\Delta S^{\ominus}$ +ve

3 $H_2O_2(l) \rightarrow H_2O(l) + \frac{1}{2}O_2(g)$ $\Delta H^{\ominus}$ −ve, $\Delta S^{\ominus}$ +ve

4 $NH_4NO_3(s) + (aq) \rightarrow NH_4^+(aq) + NO_3^-(aq)$ $\Delta H^{\ominus}$ +ve, $\Delta S^{\ominus}$ −ve

a

b

c

Figure 10.18
a Ariane rocket. Liquid hydrogen is used as fuel in the rocket engine. The two rocket boosters either side of the main body add to the thrust of the engine. b High explosive is used to demolish this warehouse. c Airbags in cars require rapid release of nitrogen to inflate.

Explosives

An explosion is a chemical reaction that produces sound, as well as a great deal of heat and light. Sound is produced if the speed of the ejected gases exceeds the speed of sound, $330\,m\,s^{-1}$ in air, resulting in the propagation of a shock wave.

Some explosions are produced by **propellants**, fuels used to drive rockets or to set in motion a shell or bullet in the barrel of a gun. These reactions must take place quickly and smoothly. This is achieved by using a source of heat near one small part of the propellant to initiate the reaction and then the heat produced by the reaction here sets off the propellant in contact with it. The speed of propagation is comparatively slow, probably only a few metres per second, and lasts all the time the bullet or shell is in the barrel of the gun. Propagation of the explosion by heat is a characteristic of **low explosives**. Common examples of low explosives include gunpowder (used in firework rockets) and cellulose trinitrate (used as cordite in the cartridges of shells or bullets). Low explosives may be set off using a match and fuse, or by being hit with a percussion cap. The simplified equations for the decompositions are as follows:

$$\underbrace{3C(s) + S(s) + 2KNO_3(s)}_{\text{gunpowder}} \rightarrow K_2S(s) + 3CO_2(g) + N_2(g)$$

$$\underset{\text{cellulose trinitrate}}{C_6H_7N_3O_{11}(s)} \rightarrow 4\tfrac{1}{2}CO(g) + 1\tfrac{1}{2}CO_2(g) + 3\tfrac{1}{2}H_2O(g) + 1\tfrac{1}{2}N_2(g)$$

Cellulose trinitrate contains enough oxygen for all the products to be gaseous. The carbon monoxide produced burns to form carbon dioxide in the air around the explosion.

Other types of explosive are set off by a shock wave. This process is called **detonation** and is characteristic of **high explosives**. The whole explosion is nearly instantaneous as the shock wave travels at the speed of sound, which in a solid can be as high as $1000\,m\,s^{-1}$. The explosion produced is much more intense than that from a low explosive. High explosives are used in mining, for the demolition of buildings and in the warheads of shells. A high explosive would be disastrous if used in the cartridge of a rifle – the explosion would be so violent that the gun would explode instead of the bullet being sent out of the barrel.

Common examples of high explosives include TNT (see Chapter 27, page 504) and propane-1,2,3-triyl trinitrate (see page 202). The latter, by itself, is highly dangerous and sensitive to shock. Alfred Nobel made his fame and fortune by showing that when it was absorbed in clay, a much more stable explosive is produced, called **dynamite**. Propane-1,2,3-triyl trinitrate contains enough oxygen for complete combustion, but TNT needs additional oxygen, often supplied by ammonium nitrate. A simplified equation representing the decompositions of TNT and ammonium nitrate is as follows:

$$C_7H_5N_3O_6(s) + 3\tfrac{1}{2}NH_4NO_3(s) \rightarrow 7CO(g) + 5N_2(g) + 9\tfrac{1}{2}H_2O(g)$$

High explosives must be set off with a detonator. This contains a small quantity of a high explosive, such as mercury fulminate or lead azide, that is very sensitive to shock. Detonators can be used to set off either high explosives or low explosives, because the shock wave produced is sufficiently intense to set off a high explosive and enough heat is given out to set off a low explosive.

$$\underset{\text{mercury fulminate}}{Hg(CNO)_2(s)} \rightarrow Hg(l) + 2CO(g) + N_2(g)$$

$$\underset{\text{lead azide}}{Pb(N_3)_2(s)} \rightarrow Pb(s) + 3N_2(g)$$

This rapid release of nitrogen is used to inflate airbags in cars, using sodium azide, NaN_3.

The manufacture of explosives is very dangerous and these chemicals have caused many fatal accidents. On no account should their preparation be attempted in the laboratory.

Summary

Some key definitions

- The **standard enthalpy change of atomisation**, $\Delta H^\ominus_{at}$, of an element is the enthalpy change when 1 mole of atoms is formed from the element in its standard state.
- The **standard enthalpy change of formation**, $\Delta H^\ominus_f$, is the enthalpy change when 1 mole of the substance is formed from its elements in their standard state.
- The **average bond enthalpy**, E(X—Y), is the enthalpy change when 1 mole of bonds is broken in the gas phase.
- The **lattice energy**, LE, is the enthalpy change when 1 mole of the solid is formed from its isolated ions in the gas phase.
- The **ionisation energy**, IE, is the minimum energy change required to remove an electron from a mole of atoms in the gas phase.
- The **electron affinity**, EA, is the energy change when an electron is added to a mole of atoms in the gas phase.
- The **enthalpy change of solution**, $\Delta H^\ominus_{sol}$, is the enthalpy change when 1 mole of solute is dissolved in an infinite volume of water.
- The **enthalpy change of hydration**, $\Delta H^\ominus_{hyd}$, is the enthalpy change when 1 mole of isolated ions in the gas phase is dissolved in an infinite volume of water.
- The enthalpy change of a reaction can produce heat and/or work. In a calorimeter, only heat is produced.
- For a metal, $\Delta H^\ominus_{at}$ is the enthalpy change when all the metallic bonds are broken. For a covalent compound, $\Delta H^\ominus_{at}$ is the energy required to break any intermolecular forces plus the sum of all the bond enthalpies.
- It is found that average bond enthalpies for any one specific bond are approximately the same from one substance to another, unless the molecule contains delocalised or strained bonds.
- The **bond dissociation enthalpy**, D(X—Y), is the energy required to break a specific bond in a molecule.
- Experimental lattice energies can be found using a **Born–Haber cycle**. Lattice energies can also be calculated from the crystal structure, assuming the substance is completely ionic. A marked difference between the experimental and calculated values of lattice energy indicates that the lattice contains significant covalent character.
- The enthalpy change of solution is determined by the lattice energy and the enthalpy change of hydration.
- **Entropy** is a measure of the disorder in a system or its surroundings. In a spontaneous process, the sum of the entropy of the system and the entropy of the surroundings increases.
- This is shown by the equation $\Delta G^\ominus = \Delta H^\ominus - T\Delta S^\ominus$, where $\Delta G^\ominus$ is the **free energy change** and $\Delta S^\ominus$ is the entropy change of the system. $\Delta G^\ominus$ is negative for a spontaneous process.
- The entropy of the surroundings increases in an exothermic reaction, when $\Delta H^\ominus$ is negative. This means that, at low temperatures, the reaction is usually feasible.
- At high temperatures, $\Delta S^\ominus$ becomes increasingly important and reactions in which $\Delta H^\ominus$ and $\Delta S^\ominus$ are positive become feasible.
- Although a reaction may be thermodynamically feasible, it is often not seen to take place because it is very slow.

11 Quantitative kinetics

Chapter 8 gave us an overview of how to measure rates of reaction, and what factors increase the number of collisions that have enough energy to overcome the activation energy barrier. In this chapter, we look at rates of reaction quantitatively. Analysing the effects of changes in concentration on the reaction rate enables us to write a rate equation, and from this we can obtain information about the mechanism of the reaction. A study of the effect of temperature often enables us to calculate the activation energy, E_a. We develop our analysis of the collision theory, the simplest model to explain kinetic data.

The chapter is divided into the following sections.

11.1 The rate of a reaction

An expression for the rate of reaction

In Chapter 8, we used several different measures of the rate of a reaction. Examples included the gradient of a graph of volume of gas produced against time (page 159), and the gradient of a graph of absorbance against time (page 160). In order to study rates of reaction quantitatively, we need a precise definition of rate. This may be expressed either as how the concentration of a product P increases with time, or how the concentration of a reactant R decreases with time:

$$\text{rate} = \frac{\text{change in product concentration}}{\text{time interval}}$$

$$\text{or rate} = \frac{-\text{ change in reactant concentration}}{\text{time interval}}$$

The negative sign in the second expression, reflects the fact that the concentration of the reactant is decreasing and gives us a positive value for the rate. The expressions above may be written:

$$\text{rate} = \frac{\Delta[\text{P}]}{\Delta t} \quad \text{or} \quad \text{rate} = \frac{-\Delta[\text{R}]}{\Delta t}$$

Using calculus notation, this becomes:

$$\text{rate} = \frac{\text{d}[\text{P}]}{\text{d}t} \quad \text{or} \quad \text{rate} = \frac{-\text{d}[\text{R}]}{\text{d}t}$$

Rate is measured in units of concentration per unit time, $\text{mol}\,\text{dm}^{-3}\,\text{s}^{-1}$.

Measuring rates by physical and chemical analysis

Before investigating the rate of a particular reaction, we need to know the overall (stoichiometric) equation so that we can decide what method we will use to follow it. For example, if we were studying the following esterification reaction:

$$\underset{\text{ethanoic acid}}{CH_3CO_2H} + \underset{\text{ethanol}}{C_2H_5OH} \rightarrow \underset{\text{ethyl ethanoate}}{CH_3CO_2C_2H_5} + H_2O$$

we might decide to follow the decrease in the amount of ethanoic acid, measuring this by titration.

The rate at a given time t is then:

$$\frac{-\Delta[CH_3CO_2H]}{\Delta t} \quad \text{or at time } t\text{:} \quad \frac{-d[CH_3CO_2H]}{dt}$$

We could also express the rate as any of the following:

$$\frac{-d[C_2H_5OH]}{dt} \quad \text{or} \quad \frac{+d[CH_3CO_2C_2H_5]}{dt} \quad \text{or} \quad \frac{+d[H_2O]}{dt}$$

which would all give the same value.

However, for the reaction:

$$N_2O_4(g) \rightarrow 2NO_2(g)$$

dinitrogen tetraoxide — nitrogen dioxide

$$\frac{-d[N_2O_4]}{dt} \neq \frac{d[NO_2]}{dt}$$

because two moles of nitrogen dioxide are produced for each mole of dinitrogen tetraoxide used up. When there are different coefficients in the equation like this, it is usual to define the rate in terms of the substance with a coefficient of one. So in this case:

$$\text{rate} = \frac{-d[N_2O_4]}{dt} = \frac{1}{2}\frac{d[NO_2]}{dt}$$

Figure 11.1 illustrates this point.

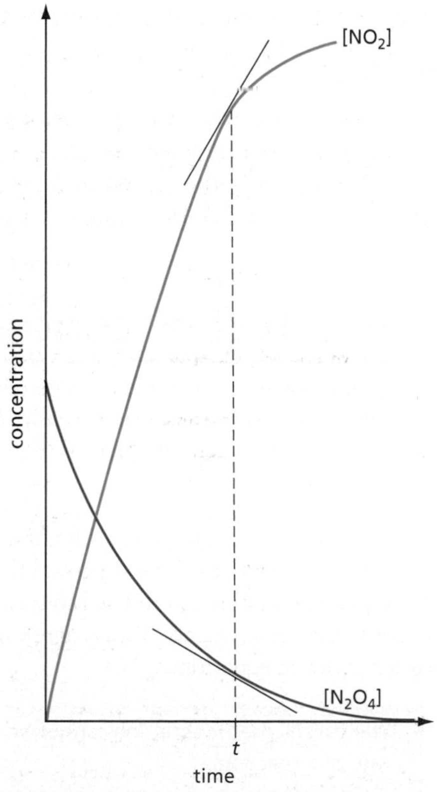

■ **Figure 11.1**
The rate of the reaction at time t is the gradient of the concentration–time graph at that point. The rate is either $-d[N_2O_4]/dt$ or $+\frac{1}{2}d[NO_2]/dt$.

As we mentioned in Chapter 8, although chemical analysis can be used to follow how the amount of one substance changes in a reaction, physical methods are usually preferred. For the reaction above, because the number of moles of gas is changing, we could follow the change in volume if we were carrying out the reaction in the gas phase. Alternatively, if we were carrying out the reaction in solution we could follow the change in colour (dinitrogen tetraoxide is pale yellow, while nitrogen dioxide is dark brown).

The esterification reaction would be difficult to follow using a physical method because there is no obvious change in a physical property as the reaction proceeds. Here we would have to use a chemical method of analysis. Samples of the reaction mixture are extracted at measured time intervals and the amount of ethanoic acid remaining is found by titration with alkali. One problem with this method is that the reaction continues to take place in the sample until the titration is complete, so that the concentration of ethanoic acid at time *t* is difficult to measure accurately. This reaction is carried out using an acid catalyst, and the reaction in the samples can be effectively stopped by immediately adding the exact amount of alkali needed to neutralise the acid catalyst added to the reaction mixture at the start of the reaction. Other reactions can be stopped by appropriate methods, such as rapid cooling of the samples.

Factors affecting the rate of a reaction – a quantitative approach

In Chapter 8, we looked at the various factors that determine the rate of a reaction. In this chapter, we examine quantitatively how some of these factors affect the rate. This quantitative data often gives us information about how the reaction takes place, that is, about the mechanism of the reaction. This allows us to study chemical reactions at the most fundamental level.

We have already discussed qualitatively how the following factors affect the rate of a reaction:

- concentration (or pressure for gas reactions)
- temperature
- catalysts
- state of division
- nature of the solvent
- light.

For convenience, when we carry out kinetics experiments in the laboratory, we usually study homogeneous reactions in aqueous solution. The principal factors that we can study under these conditions are the effect of concentration, temperature and homogeneous catalysts. These will now be considered in some detail.

11.2 The effect of concentration

In this section we shall study a single reaction, the iodination of propanone, but the principles established can be applied to many other reactions.

In acid solution, iodine reacts with propanone as follows:

$$CH_3COCH_3(aq) + I_2(aq) \rightarrow CH_2ICOCH_3(aq) + H^+(aq) + I^-(aq)$$

It would be tempting to predict that the rate of the reaction depends on the concentration of both the iodine and the propanone. As we shall see, this is not the case. Information about the kinetics of a reaction can be found only by experiment, and does not always agree with what might be expected from the stoichiometric equation.

- The stoichiometric equation cannot be used to predict how concentration affects the rate of a reaction.
- Kinetic data must be found by experiment.

We use the stoichiometric equation to decide which substance we are going to follow during the reaction in order to measure the rate. In this case, the concentration of iodine is most easily measured, either using a colorimeter (see Chapter 8, page 160) or by titration with sodium thiosulphate (see Chapter 7, page 154). It is much more difficult to follow the change in concentration of either propanone or the products of the reaction.

Because we are investigating the effect of concentration on the rate of the reaction, all other factors that might affect the rate should be kept constant. Accurate control of temperature is essential. The reaction should be carried out in a water bath whose temperature is controlled to within 1 K. For very accurate work, thermostatic water baths whose temperature varies by less than 0.1 K are available. Once the reaction has started, the reactants are shaken well in order to mix them thoroughly.

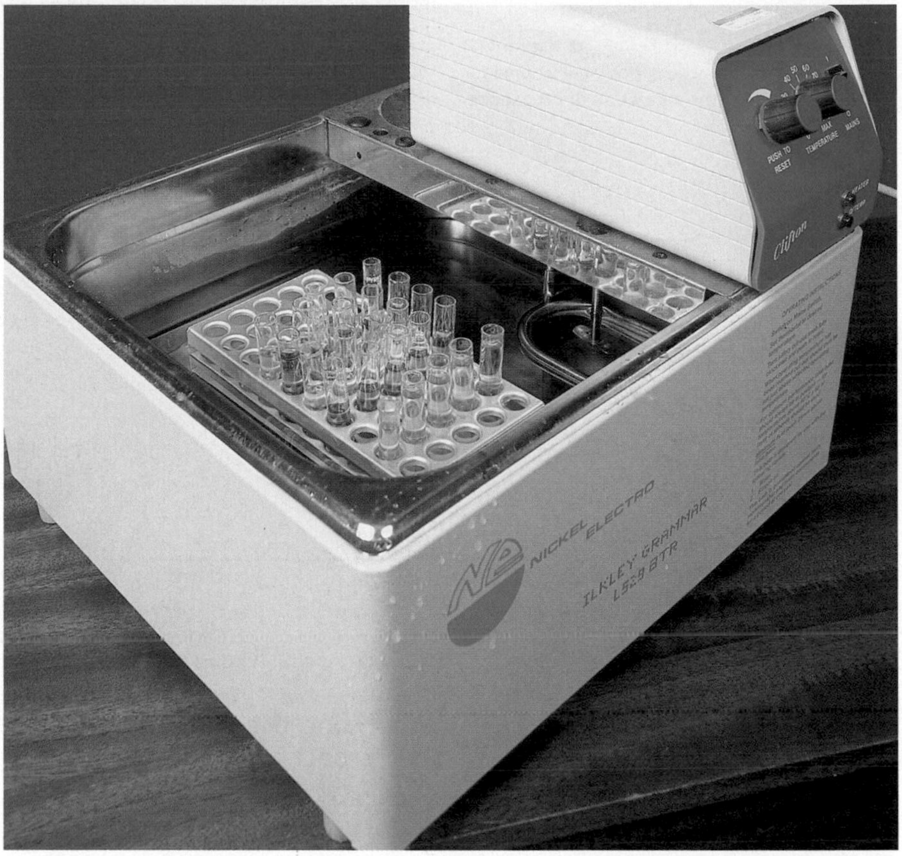

■ Figure 11.2
A thermostatic water bath keeps the temperature of the reaction constant, removing one factor that might affect the rate.

Initial investigations show us that the reaction is not affected by light, but that it is affected by $H^+(aq)$ ions which act as a catalyst. As this is a homogeneous catalyst, we shall treat $H^+(aq)$ ions in the same way that we treat the other two reactants. We now have three substances whose concentration may affect the rate, $I_2(aq)$, $CH_3COCH_3(aq)$ and $H^+(aq)$. In order to find out how each of them affects the rate of the reaction, we must vary each in turn while keeping the other two constant. There are two main ways of designing experiments to achieve this:

- keep all reactants in excess except the one being studied
- the initial rate method.

We shall look at each in turn.

Keeping all reactants in excess except one

If we make the concentrations of propanone and hydrogen ions much higher than that of iodine, then during the course of the reaction, their concentrations vary so little that they may be taken as constant. Table 11.1 illustrates why this is the case.

■ Table 11.1
A typical set of concentrations used to study the iodination of propanone. The concentration of hydrogen ions increases because as well as being a catalyst, it is also being produced by the reaction.

Reactant	Concentration at start/mol dm^{-3}	Concentration at end/mol dm^{-3}
I_2	0.01	0.0
CH_3COCH_3	1.00	0.99
H^+	0.50	0.51

A mixture of propanone and sulphuric acid is placed in a thermostatically controlled water bath. The solution of iodine is also placed in the bath in a separate container. At the start of the experiment, the two solutions are mixed, shaken well and a stopclock started. At regular times, the concentration of iodine is found and a graph plotted of iodine concentration against time (see Figure 11.3).

The graph is a straight line. This shows that the rate, given by the negative gradient of the graph, is constant. The rate does not change as the iodine concentration decreases – the rate is independent of the iodine concentration.

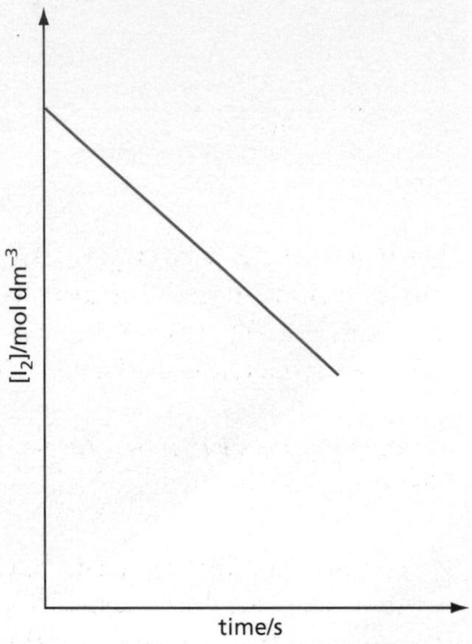

■ **Figure 11.3** Graph of iodine concentration against time for the iodination of propanone.

Analysing the results – the order of the reaction with respect to each reactant

Since the rate of the reaction is independent of the iodine concentration, we can write:

$$\text{rate} = \frac{-\,\mathrm{d}[I_2]}{\mathrm{d}t} = k\text{, a constant}$$

We can indicate that the rate is not affected by the iodine concentration by writing:

$$\text{rate} = k[I_2]^0$$

Any term raised to the power of zero is equal to one, so the term could be omitted altogether. We include it to emphasise that the effect of iodine concentration on the rate has been studied and found to be **zero order**, the **order** being the exponent of the term, in this case zero. This result may seem surprising, and emphasises the point that we cannot predict which reactants determine the reaction rate from the stoichiometric equation.

Of course the rate is not independent of all the reactants, and the concentration of at least one of them must affect the rate. Often we find that the rate varies directly with a particular reactant [R]. We express this as:

$$\text{rate} = k[R]^1$$

and say that the rate is **first order** with respect to the reactant. Occasionally we find that the rate is proportional to the square of [R]. We express this as:

$$\text{rate} = k[R]^2$$

and say that the rate is **second order** with respect to the reactant.

We can determine the order with respect to each reactant by studying concentration-against-time graphs like the one obtained for iodine. The three common types of rate dependence are shown in Figure 11.4.

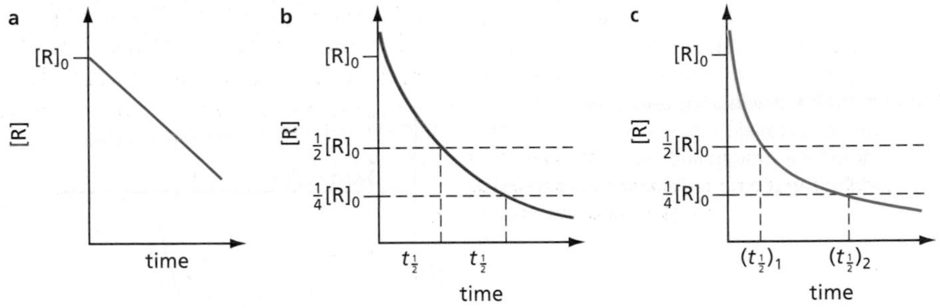

■ **Figure 11.4** The three most common concentration–time graphs. $[R]_0$ is the concentration at time $t = 0$.

The first example is like Figure 11.3 for iodine. The graph is a straight line and the rate is constant. This shows that the rate is zero order with respect to that reactant, so that rate $= k[R]^0$. (Since $[R]^0 = 1$, rate $= k$.)

In the second example, the curve is an exponential, and has a constant half-life, $t_{\frac{1}{2}}$.

The **half-life** of a reaction is the time taken for the concentration of a reactant to decrease to half its initial value.

By 'constant half-life' we mean that the successive half-lives are the same. That is, the time taken for the concentration of the reactant to decrease from $[R]_0$ to $\frac{1}{2}[R]_0$ is the same as the time taken for the concentration to decrease from $\frac{1}{2}[R]_0$ to $\frac{1}{4}[R]_0$, and so on.

In the third example, successive half-lives become longer. This indicates an order greater than 1. Since orders higher than 2 are most unusual, the rate is likely to be second order, that is, rate $= k[R]^2$.

In order to confirm that the reaction is of the order suggested by the concentration–time graph, the results can be used to plot a graph that is a straight line. For a zero-order reaction, the [R]–time graph is a straight line. A first-order reaction gives a straight line for a $\log_{10}$[R]–time graph. A second-order reaction gives a straight-line graph of 1/[R] against time (see Figure 11.5).

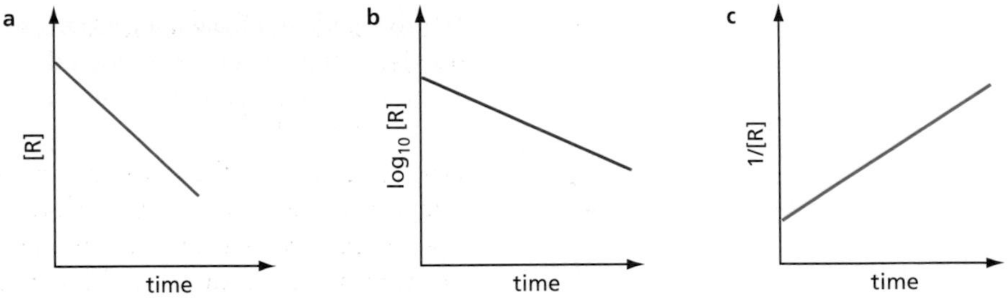

■ **Figure 11.5** Straight-line graphs are obtained for the following plots:
a zero order: [R] against time
b first order: $\log_{10}$ [R] against time
c second order: 1/[R] against time.

Integrated forms (if you know some calculus)

Zero order: $$\frac{-\,d[R]}{dt} = k[R]^0 \quad \text{and} \quad [R] = kt + c$$

First order: $$\frac{-\,d[R]}{dt} = k[R]^1 \quad \text{and} \quad 2.30 \log_{10}[R] = kt + c$$

Second order: $$\frac{-\,d[R]}{dt} = k[R]^2 \quad \text{and} \quad \frac{1}{[R]} = kt + c$$

Another, less reliable, way to check the order is to measure the gradient of the concentration–time graph at various times in order to find the rates. These rates can then be plotted against concentration, and Figure 11.6 shows how this distinguishes between the three possible orders.

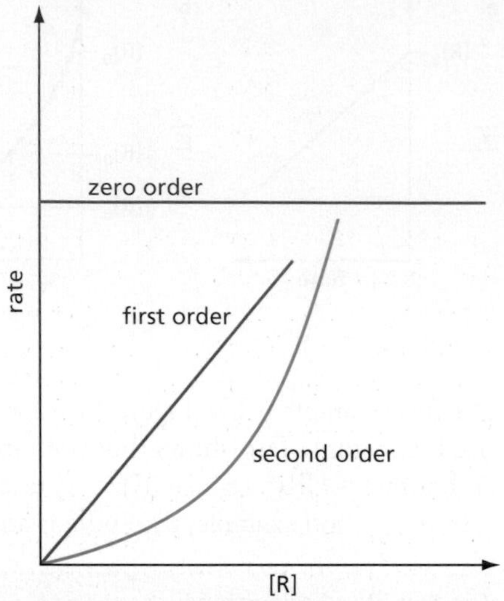

Figure 11.6 Rate–concentration graphs can be used to distinguish between zero-order, first-order and second-order reactions. The zero-order graph is a straight line parallel to the *x*-axis. The first-order graph is a straight line passing through the origin. The second-order graph is a parabola starting at the origin.

The initial rate method

Analysing the effect of concentration of a reactant on the rate of a reaction by the method of keeping all reactants except one in excess has the advantage that a large amount of data can be collected from each experiment. The results may be plotted so that a straight-line graph is obtained, possibly up to the point when the reaction is 90% completed. If this is the case, we can be confident that the order suggested by the results is the one that should be taken as correct.

Sometimes, however, it is not possible to make all the reactants in excess. For example, in the iodination of propanone experiment, the iodine concentration must always be small as this is the reactant whose concentration we are following during the course of the reaction. We then need another method to find the order of reaction with respect to the propanone and the acid.

This is done by carrying out a series of experiments in which the initial concentration of all the reactants is kept the same while the one under investigation is varied. At the start of each experiment, the **initial rate** is found from the gradient of the concentration–time graph (see Figure 11.7). By studying how the initial rate changes when the concentration of this one reactant is varied, we can find the order with respect to that reactant.

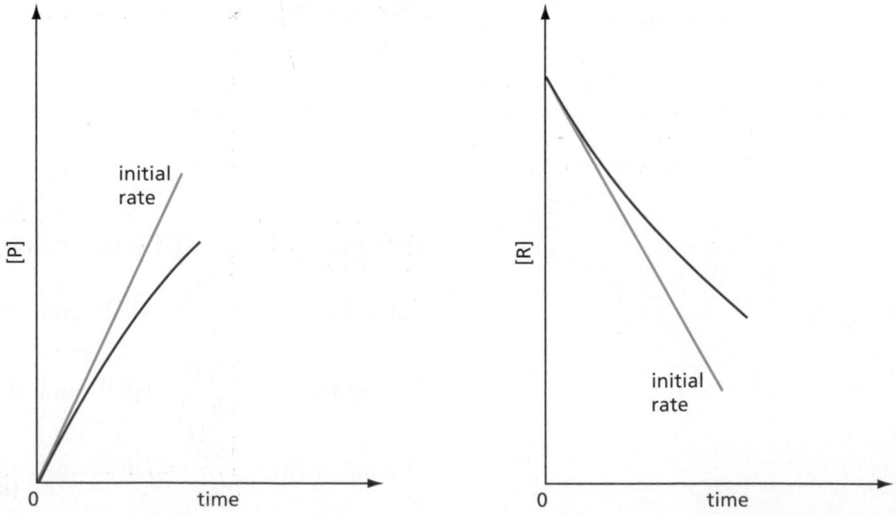

Figure 11.7 The initial rate is found from the gradient at the start of the reaction, that is, when $t = 0$.

■ Table 11.2
A typical set of results for the iodination of propanone. Initial rates were calculated from the iodine concentration–time graph. Because the order of the reaction with respect to iodine has already been found, it is not necessary to vary $[I_2]$. $[I_2]$ is kept smaller than the concentrations of the other reactants because change in $[I_2]$ is being used to monitor the reaction rate. Note that at the start of the reaction, when the initial rate is measured, $[I_2]$ is constant at $0.002\,mol\,dm^{-3}$.

Experiment	$[H^+]$/ $mol\,dm^{-3}$	$[I_2]$/ $mol\,dm^{-3}$	$[CH_3COCH_3]$/ $mol\,dm^{-3}$	Initial rate/ $mol\,dm^{-3}\,s^{-1}$
1	0.50	0.002	0.50	2.0×10^{-5}
2	0.30	0.002	0.50	1.2×10^{-5}
3	0.10	0.002	0.50	4.0×10^{-6}
4	0.50	0.002	0.30	1.2×10^{-5}
5	0.50	0.002	0.10	4.0×10^{-6}

Table 11.2 shows a typical set of results for the iodination of propanone, in which the concentrations of first hydrogen ions and then propanone were varied. From experiments 1, 2 and 3, as $[H^+]$ is reduced in the ratio 5:3:1, so the rate decreases in the ratio 5:3:1. Therefore, the reaction is first order with respect to $[H^+]$.

From experiments 1, 4 and 5, as $[CH_3COCH_3]$ is reduced in the ratio 5:3:1, so the rate again decreases in the ratio 5:3:1. Therefore, the reaction is also first order with respect to $[CH_3COCH_3]$.

In the case of the iodination of propanone experiment, it is easy to obtain an accurate value for the initial rate because the graph of $[I_2]$ against t is a straight line. This is not usually the case, and then the determination of the initial rate, found from the gradient at the origin, is much more difficult to obtain accurately.

Sometimes the initial rate may be found by a 'clock' method. The initial rate of the reactions between peroxodisulphate(VI) ions and iodide ions:

$$S_2O_8^{2-}(aq) + 2I^-(aq) \rightarrow 2SO_4^{2-}(aq) + I_2(aq)$$

can be found by adding a known amount of thiosulphate ions and a little starch. Initially the iodine produced reacts with the thiosulphate ions:

$$I_2(aq) + 2S_2O_3^{2-}(aq) \rightarrow 2I^-(aq) + S_4O_6^{2-}(aq)$$

but, as we saw on page 160, after the thiosulphate ions have been used up, the iodine reacts with the starch to give a blue colour. The time taken, t, for this colour to appear may be used as an approximate measure of the initial rate. If, for example, $[S_2O_3^{2-}] = 0.0030\,mol\,dm^{-3}$, the equation shows that this will be used up when $[I_2] = 0.0015\,mol\,dm^{-3}$. Say the time taken, t, is 60 s, then:

$$\text{initial rate of reaction } \frac{-\Delta[I_2]}{\Delta t} = \frac{0.0015}{60} = 2.5 \times 10^{-5}\,mol\,dm^{-3}\,s^{-1}$$

This is only an approximate value because it assumes that the concentration–time graph is a straight line not only near the origin, but up to the point when the blue colour appeared (see Figure 11.8). The approximation is reasonable provided the reaction is only a small part towards completion.

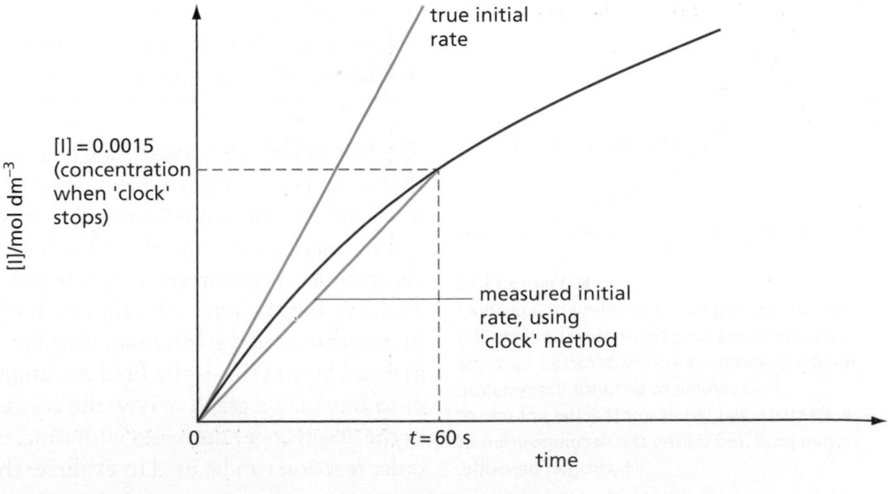

■ Figure 11.8
A 'clock' method gives a value for the initial rate that is lower than the true value, because it assumes that the concentration–time graph is a straight line until the 'clock' stops.

11.3 The rate equation

Arriving at the rate equation

When the order of a reaction with respect to each reactant has been found, the results are combined together in the form of a **rate equation**. For the iodination of propanone, we have the following results:

$$\text{rate} \propto [I_2]^0$$
$$\text{rate} \propto [H^+]^1$$
$$\text{rate} \propto [CH_3COCH_3]^1$$

These may be combined to give the following equation:

$$\text{rate} = k[I_2]^0[H^+]^1[CH_3COCH_3]^1$$

The constant k is known as the **rate constant**. Strictly speaking, it is not necessary to include the $[I_2]^0$ term (which is equal to one), and the '^{1}' after $[H^+]$ and $[CH_3COCH_3]$ could be omitted. It is, however, sometimes useful to include these terms in the rate equation as they emphasise the effect of varying the concentration of each reactant.

The reaction is now described as follows. It is:

- zero order with respect to $[I_2]$
- first order with respect to $[H^+]$
- first order with respect to $[CH_3COCH_3]$
- of total order 2, or **second order overall.**

The total order is the sum of the exponents from the rate equation, in this case, $0+1+1=2$.

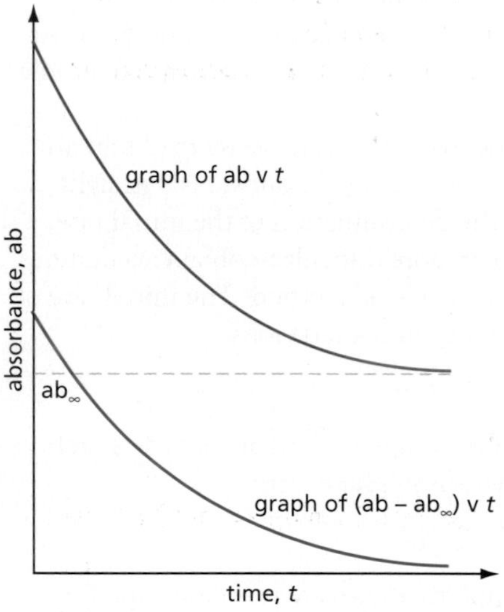

■ **Figure 11.9**
Graph showing how a property of a reactant changes with time for a first-order reaction. The final reading ab_∞ may not be zero, so this value has to be subtracted from all the readings. An example is using a colorimeter to measure the decrease in absorbance of bromine as it reacts with methanoic acid.

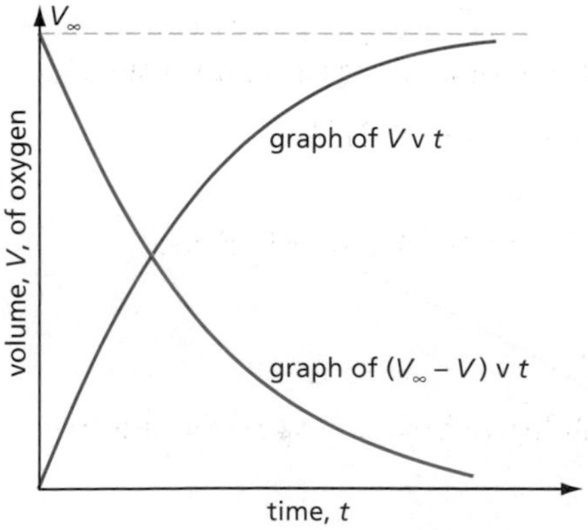

■ **Figure 11.10**
Graph showing how a property of a product changes with time for a first-order reaction. The volume readings are each subtracted from the final volume to produce the required exponential curve. An example is the volume of oxygen produced during the decomposition of hydrogen peroxide.

Finding the rate constant

In order to work out a value for the rate constant, k, we can use a set of readings such as those in Table 11.2. For example, if we use the figures in experiment 1, including the units:

$$\text{rate} = k[I_2]^0[H^+]^1[CH_3COCH_3]^1$$

$$2.0\times10^{-5}\,\text{mol dm}^{-3}\,\text{s}^{-1} = k\times1\,(\text{no units})\times0.50\,\text{mol dm}^{-3}\times0.50\,\text{mol dm}^{-3}$$

$$k = \frac{2.0\times10^{-5}\,\text{mol dm}^{-3}\,\text{s}^{-1}}{1\times0.50\times0.50\,\text{mol}^2\,\text{dm}^{-6}}$$
$$= 8.0\times10^{-5}\,\text{mol}^{-1}\,\text{dm}^3\,\text{s}^{-1}$$

The units for the rate constant depend on the total order of the reaction.

Units of k

- overall first-order reaction: s^{-1}
- overall second-order reaction: $\text{mol}^{-1}\,\text{dm}^3\,\text{s}^{-1}$
- overall third-order reaction: $\text{mol}^{-2}\,\text{dm}^6\,\text{s}^{-1}$

The first-order case is interesting because k has no units of concentration. This means that it can be determined directly from a change in a physical property such as volume or colour. As long as it is known that this property is proportional to the concentration, it is not necessary to use the actual concentration. For example, the rate of decomposition of hydrogen peroxide may be studied by measuring the volume of oxygen evolved, and the rate of bromination of methanoic acid by measuring the decrease in absorbance of the bromine. One key reading must be determined as accurately as possible, and this is the final reading when the reaction is complete. This enables us to construct a graph of how the concentration of a reactant (or the volume of gas, or the absorbance) decreases with time, which finishes at zero. Such a graph for a first-order reaction can be used to evaluate the rate constant (see Figures 11.9 and 11.10).

There are two ways in which the rate constant can be found from these exponential 'concentration'–time graphs. The first is to use the graph to find the half-life, $t_{\frac{1}{2}}$ (see Figure 11.4**b**, page 211). The rate constant is related to the half-life by the following equation:

$$k = \frac{2.30 \log_{10} 2}{t_{\frac{1}{2}}} = \frac{0.693}{t_{\frac{1}{2}}}$$

The second method is to use the points to plot a logarithmic, straight-line graph (see Figure 11.5**b**, page 211). The rate constant is given by 2.30 times the gradient of the graph.

Use of $\log_e$ (if you know some calculus)

A more detailed study of kinetics makes use of **natural logarithms**, or log to the base e, $\log_e$. These logs are also used in Chapters 12 and 13.

The function $y = e^x$ has the property that its differential is the same, so that:

$$x = \log_e y \quad \text{and} \quad \frac{dy}{dx} = e^x$$

Substituting and integrating, we have:

$$\frac{dy}{y} = \int dx = x = \log_e y$$

For a first-order reaction:

$$\frac{-\,d[R]}{dt} = k[R] \quad \text{and} \quad \int \frac{-\,d[R]}{[R]} = \int dt$$

If we integrate from the initial concentration $[R_0]$ to $[R]$ and from $t = 0$ to t, we have:

$$\log_e [R_0] - \log_e [R] = kt \quad \text{or} \quad \log_e \frac{[R_0]}{[R]} = kt$$

If $[R]$ is the amount left after one half-life, $t_{\frac{1}{2}}$, $\frac{[R_0]}{[R]} = 2$ and

$$k = \frac{\log_e 2}{t_{\frac{1}{2}}} = \frac{0.693}{t_{\frac{1}{2}}}$$

which is the equation used above.

Another example of the use of $\log_e$ is in the Arrhenius equation (see page 219).

$$k = Ae^{-E_a/RT} \text{ (often written as } k = A\exp(-E_a/RT))$$
$$\log k = \log_e A - E_a/RT$$

and a plot of $\log_e k$ against $1/T$ is a straight line with gradient $-E_a/R$ and intercept $\log_e k$ A.

$\log_e$ and $\log_{10}$

You may be familiar with $\log_{10}$, but not with $\log_e$. These two types of logarithms are connected by the equation:

$$\log_e x = \log_{10} x \times \log_e 10$$

Because $\log_e 10 = 2.30$,

$$\log_e x = 2.30 \log_{10} x \quad \text{or} \quad \log_{10} x = \frac{\log_e x}{2.30}$$

In the text we have used $2.30 \log_{10} x$ instead of $\log_e x$.

The rate equation shows how the rate changes with the concentration of the reactants (and the concentration of a homogeneous catalyst, if one is present). The rate constant is unaffected by changes in these concentrations. That is why it is called a rate **constant**. If we exclude heterogeneous catalysts and light, the only factor that changes the value of a rate constant is temperature. This effect of temperature is studied in Section 11.5.

- The **rate constant** is the constant of proportionality in a rate equation.
- It is unaffected by changes in concentration.
- It changes with temperature.

Example

In tetrachloromethane at 45 °C, dinitrogen pentaoxide, N_2O_5, decomposes as follows:

$$N_2O_5 \rightarrow 2NO_2 + \tfrac{1}{2}O_2$$

The rate of the reaction was measured at different times. The results are shown in Table 11.3.

Table 11.3 Rates and reactant concentrations at different times. 1 μmol = 10^{-6} mol.

$[N_2O_5]$/mol dm^{-3}	2.21	2.00	1.79	1.51	1.23	0.82
rate/μmol dm^{-3} s^{-1}	22.7	21.0	19.3	15.7	13.0	8.3

a Suggest a method of following the reaction.
b Plot a graph of rate against $[N_2O_5]$.
c Use your graph to find the order of the reaction with respect to N_2O_5.
d Calculate the rate constant, giving its units.
e What will be the shape of the graph of $[N_2O_5]$ against time?

Answer

a Either measure the volume of oxygen given off, or measure the absorbance, since nitrogen dioxide is coloured.

b

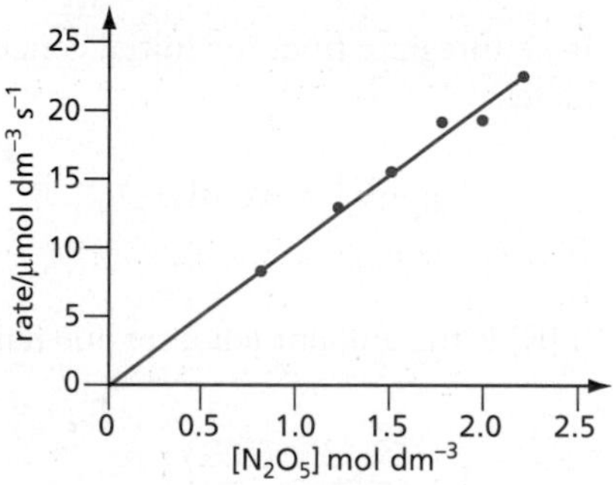

Figure 11.11

c The graph is a straight line passing through the origin, so the reaction is first order.

d Rate = $k[N_2O_5]$ so $k = \dfrac{\text{rate}}{[N_2O_5]}$ = gradient of graph = 1.0×10^{-5} s^{-1}

e The graph will show an exponential decay with a constant half-life.

Further practice

In alkaline solution, iodide ions react with chlorate(I) ions as follows:

$$I^-(aq) + ClO^-(aq) \rightarrow Cl^-(aq) + IO^-(aq)$$

The reaction can be followed by measuring the absorbance of IO^-(aq) ions at 400 nm in a colorimeter. Table 11.4 shows a series of measured initial rates.

Table 11.4

Experiment number	$[I^-(aq)]$/mol dm^{-3}	$[ClO^-(aq)]$/mol dm^{-3}	Initial rate/ mol dm^{-3} s^{-1}
1	0.0010	0.00073	4.5×10^{-5}
2	0.0010	0.0010	6.2×10^{-5}
3	0.0010	0.0014	8.7×10^{-5}
4	0.00073	0.0010	4.6×10^{-5}
5	0.0014	0.0010	8.6×10^{-5}

1 Calculate the order of the reaction with respect to I^-(aq) ions and ClO^-(aq) ions. Explain your answer.
2 Write a rate equation for the reaction.
3 Calculate the rate constant, stating the units.

11.4 Reaction mechanisms

Postulating a mechanism

The rate equation is often used as a basis for postulating a likely mechanism for the reaction. Most reactions can be broken down into a number of steps, one of which has a high activation energy that determines the overall rate of the reaction. This step is called the **rate-determining step**. An analogy is a group of people buying a paper at the local newsagent. If one person arrives, it takes 1 second to pick up the paper, 10 seconds to pay for it and 1 second to leave the shop. If ten people arrive at the same time, it takes each of them 10 seconds to pick up the paper, 100 seconds for them all to pay, but they still take only 1 second each to leave the shop. There will be a queue at the checkout, but not on the way out of the shop. Any step that takes place after the rate-determining step has no effect on the overall rate.

In the rate equation for the iodination of propanone,

$$\text{rate} = k[I_2]^0[H^+]^1[CH_3COCH_3]^1$$

iodine does not appear in the rate equation, so any step involving iodine must come after the rate-determining step. It is also reasonable to postulate that the first step is the reaction of a proton with propanone, as both of these species appear as first-order terms in the rate equation. This reaction is an acid–base reaction. These are usually fast and reversible, so we write:

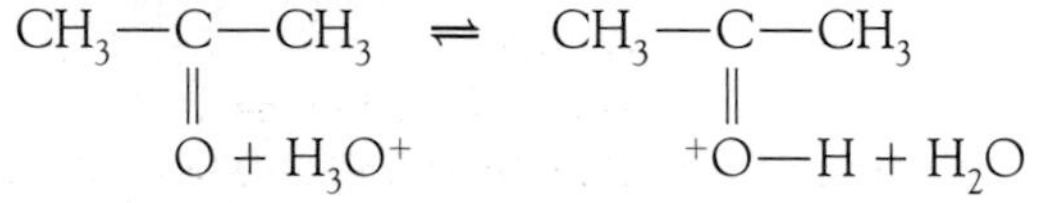

$$CH_3{-}\underset{\underset{O + H_3O^+}{\|}}{C}{-}CH_3 \rightleftharpoons CH_3{-}\underset{\underset{^+O{-}H + H_2O}{\|}}{C}{-}CH_3 \qquad \text{fast equilibrium}$$

The second step probably controls the rate of the reaction and produces H_3O^+ which, being a catalyst, is not used up in the reaction. A possible second step is:

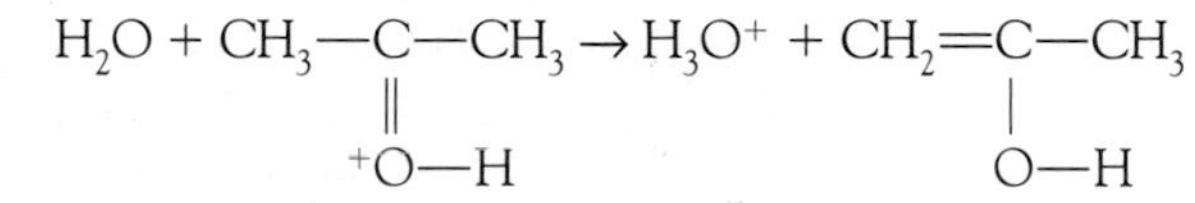

$$H_2O + CH_3{-}\underset{\underset{^+O{-}H}{\|}}{C}{-}CH_3 \rightarrow H_3O^+ + CH_2{=}\underset{\underset{O{-}H}{|}}{C}{-}CH_3 \qquad \text{slow, rate-determining step}$$

Iodine reacts rapidly with compounds containing the $C{=}C$ group, so a fast step follows:

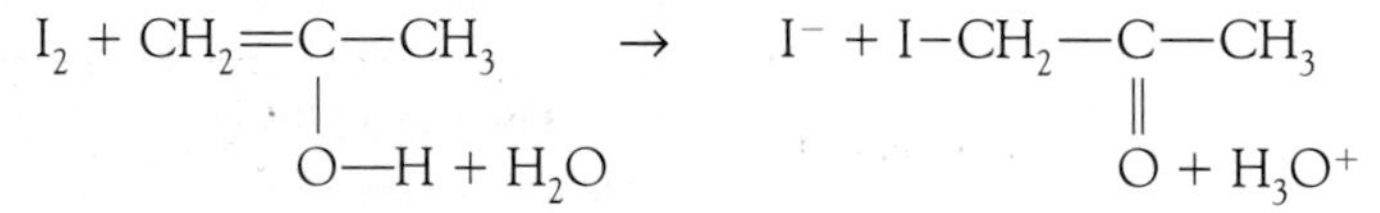

$$I_2 + CH_2{=}\underset{\underset{O{-}H + H_2O}{|}}{C}{-}CH_3 \rightarrow I^- + I{-}CH_2{-}\underset{\underset{O + H_3O^+}{\|}}{C}{-}CH_3 \qquad \text{fast step}$$

If we add up all the mechanistic steps, we arrive at the overall stoichiometric equation:

$$I_2\text{(aq)} + CH_3COCH_3\text{(aq)} \rightarrow CH_2ICOCH_3\text{(aq)} + H^+\text{(aq)} + I^-\text{(aq)}$$

Mechanistically we can define a homogeneous catalyst as a substance that appears in the rate equation but not in the stoichiometric equation. (Actually for the reaction we have been considering this is not strictly true – because the catalyst, H_3O^+, appears in the product, we call this an **autocatalytic reaction**.)

Substances such as:

$$CH_3{-}\underset{\underset{^+O{-}H}{\|}}{C}{-}CH_3 \quad \text{and} \quad CH_2{=}\underset{\underset{O{-}H}{|}}{C}{-}CH_3$$

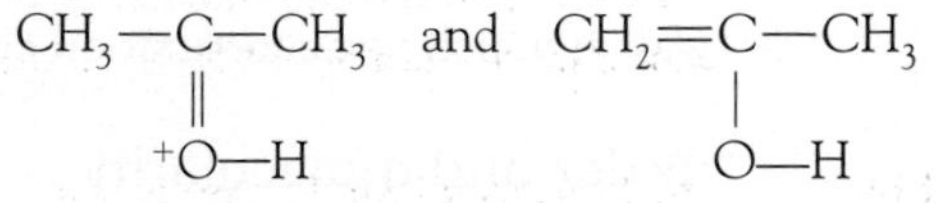

are known as **intermediates**, because they are produced during the reaction but not part of the final products.

Testing the mechanism

We have now postulated a possible mechanism for the reaction that is consistent with the kinetic data. This is very different from saying that this mechanism is the most likely one. We need to carry out further experiments to confirm the mechanism. Some of the techniques used are listed overleaf.

- **Use a wider range of concentrations** – the experimentally determined rate equation probably holds over only a limited range of concentrations. For example, the proposed mechanism for the iodination of propanone predicts that at very low concentrations of iodine, the zero-order dependence of the iodine changes to first-order dependence. This is because the rate of the last reaction will be given by the expression: rate $= k[I_2][CH_2{=}C(OH)CH_3]$, and so this rate will decrease as $[I_2]$ decreases. This can be shown to be the case.
- **Use sophisticated analytic techniques** – these may be able to detect the presence of the postulated intermediates. The use of nuclear magnetic resonance (see Chapter 14) shows that acidified propanone contains about 1 molecule in 10^6 of the enol form, $CH_2{=}C(OH)CH_3$, an intermediate postulated by the mechanism.
- **Do experiments on the intermediates** – some intermediates are stable enough for experiments to be carried out on them. For example, some organic halides form tertiary carbocations (see Chapter 25, page 467) that can be isolated.
- **Use isotopic labelling** – if an atom is labelled with an isotope (not necessarily a radioactive one), the label may indicate which bond has been broken in a reaction. For example, when some esters labelled with ^{18}O are hydrolysed, the ^{18}O appears in the alcohol and not in the acid group. This shows that the acyl oxygen bond, and not the alkyl oxygen bond, is the one that is broken:

$$\underset{\text{ester}}{CH_3C({=}O){-}^{18}O{-}C_2H_5} + H_2O \rightarrow \underset{\text{acid}}{CH_3CO_2H} + \underset{\text{alcohol}}{H^{18}OC_2H_5}$$

- **Kinetic isotope effect** – deuterium behaves slightly differently from hydrogen, for example, the C—D bond is slightly harder to break than the C—H bond. This means that if this bond is broken during the rate-determining step, a compound containing C—D bonds reacts between 5 and 10 times as slowly as one with C—H bonds. This is the case with the iodination of propanone. CD_3COCD_3 reacts considerably more slowly than CH_3COCH_3.

  On the other hand, the fact that C_6H_6 and C_6D_6 nitrate at the same rate shows that the rate-determining step is the initial attack by the NO_2^+ ion (see Chapter 27, page 504), and not the elimination of the proton.
- **Change the solvent** – the rate of ionic reactions changes with the polarity of the solvent. For example, the rate of hydrolysis of 2-bromo-2-methylpropane is raised by the addition of sodium chloride. The sodium chloride increases the polarity of the solvent and increases the ionisation of the bromide:

  $$(CH_3)_3CBr \rightleftharpoons (CH_3)_3C^+ + Br^-$$

  On the other hand, the addition of sodium bromide reduces the overall rate. This is because the ionisation of $(CH_3)_3CBr$ is suppressed by the high concentration of Br^- ions (Le Chatelier's Principle) and this is larger than the positive effect caused by the increased polarity of the solvent.

But no amount of experimental work can ever prove that the mechanism is the correct one. In particular, the role of the solvent is always uncertain as there is no way in which its concentration can be varied without changing the overall polarity.

Order and molecularity

Some complex reactions have orders that are negative or fractional. Other reactions have an order that changes with concentration and only a detailed mathematical treatment shows why this is so (see, for example, the Michaelis equation, page 227).

But while the order of reaction may be non-integral, the molecularity of a reaction is integral. The **molecularity** is the number of species in the rate-determining step. This must be integral, probably one or two. For many reactions, the order and molecularity are the same, and this can cause confusion between the two.

Example Hydrogen cyanide, HCN, adds on to ethanal, CH_3CHO, to give $CH_3CHOHCN$. Two mechanisms have been proposed for this reaction:

1 $CH_3CHO + H^+ \rightarrow [CH_3CHOH]^+$ and $[CH_3CHOH]^+ + CN^- \rightarrow CH_3CHOHCN$
2 $CH_3CHO + CN^- \rightarrow [CH_3CHOCN]^-$ and $[CH_3CHOCN]^- + H^+ \rightarrow CH_3CHOHCN$

The rate equation is rate $= k[CN^-][CH_3CHO][H^+]^0$.

a Which mechanism is consistent with the rate equation? Explain your answer.
b Which step in this mechanism is the rate-determining step?

Answer **a** Mechanism **1** is excluded because $[H^+]$ is in the first equation so would appear in the rate equation. Mechanism **2** is consistent with the rate equation because the step involving H^+ appears after the rate-determining step.
b The first step is the rate-determining step.

Further practice

1 For the following reactions, suggest mechanisms that are compatible with the rate equations:

a $H_2O_2(aq) + 3I^-(aq) + 2H^+(aq) \rightarrow 2H_2O(l) + I_3^-(aq)$
rate $= k[H_2O_2(aq)][I^-(aq)]$

b $ClO^-(aq) + I^-(aq) \rightarrow IO^-(aq) + Cl^-(aq)$
rate $= k[ClO^-(aq)][I^-(aq)][OH^-(aq)]^{-1}$
(Hint: $[OH^-][H^+]$ is a constant.)

c $BrO_3^-(aq) + 5Br^-(aq) + 6H^+(aq) \rightarrow 3H_2O(l) + 3Br_2(aq)$
rate $= k[H^+(aq)][Br^-(aq)][BrO_3^-(aq)]$

2 Suggest why the following reactions have two terms in their rate equations.

a $C_2H_5CO_2CH_3 + H_2O \rightarrow C_2H_5CO_2H + CH_3OH$
rate $= k_1[C_2H_5CO_2CH_3][H^+] + k_2[C_2H_5CO_2CH_3][OH^-]$

b $H_2(g) + I_2(g) \rightleftharpoons 2HI(g)$
rate $= k_1[H_2][I_2] - k_2[HI]^2$
(Hint: what **two** reactions are taking place in each case?)

11.5 The effect of temperature on the rate constant

In Chapter 8, we discussed qualitatively why an increase in temperature speeds up the rate of a reaction. There are two reasons for this:

- an increase in the rate at which the molecules collide (the **collision frequency**)
- an increase in the fraction of molecules with sufficient energy to overcome the activation energy barrier.

The second effect is so much more important than the first that we can usually ignore the change in collision frequency.

In 1875, Svante Arrhenius suggested that the effect of temperature on the rate constant could be expressed quantitatively by the following equation:

$$2.30\log_{10}k = 2.30\log_{10}A - \frac{E}{RT}$$

where A and E are both constants, R is the gas constant and T the absolute temperature. This means that if a graph of $\log_{10}k$ is plotted against $1/T$, a straight line is obtained whose gradient is

$$\frac{-E}{2.30\,R}$$

This equation is usually written in the following form, known as the **Arrhenius equation**:

$$k = Ae^{(-E/RT)}$$

(See page 215 for an explanation of e.)

Experimentally, Arrhenius showed that, for a given reaction, A and E were constant over an appreciable range of temperature. The significance of these two constants became apparent when a model for reaction kinetics was created – in particular, E was equated with the activation energy, E_a.

Example The initial rate constant, *k*, for the following reaction:

$$H_2(g) + I_2(g) \rightarrow 2HI(g)$$

changes with temperature *T* as shown in Table 11.5.

Table 11.5

T/K	600	650	700	750
k/mol^{-1} dm^3 s^{-1}	4.0×10^{-6}	5.0×10^{-5}	5.0×10^{-4}	3.2×10^{-3}

a Plot a graph of $\log_{10} k$ against $1/T$.

b Use the graph to evaluate the constant *E*.

c The bond enthalpy of the I—I bond is 151 kJ mol^{-1}. Suggest the likely first step of the reaction.

d (Harder) Calculate the constant *A*.

Answer Table 11.6 gives the data for the graph in Figure 11.12.

Table 11.6

(1/*T*)/K^{-1}	0.00167	0.00154	0.00143	0.00133
$\log_{10} k$	−5.40	−4.30	−3.30	−2.49

Figure 11.12 a

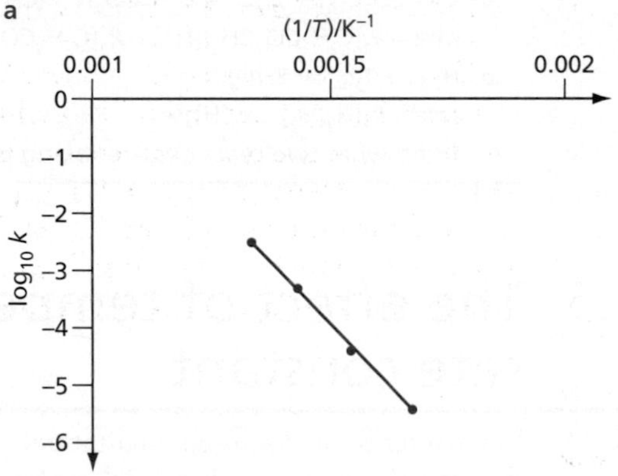

b $\text{Gradient} = \dfrac{-2.49 - (-5.40)}{0.00133 - 0.00167} = \dfrac{2.91}{-0.00034} = -8600$

Since $\text{gradient} = \dfrac{-E}{2.30\,R}$

$$-8600 = \dfrac{-E}{2.30 \times 8.31}$$

and $E = 2.30 \times 8.31 \times (-8.600) =$ **160 kJ mol^{-1}**

c The constant *E* has a value very similar to that of *E*(I—I). This suggests that the first step is: $I_2(g) \rightarrow 2I\bullet(g)$

d $2.30\log_{10}k = 2.30\log_{10}A - \dfrac{E}{RT}$

so using $y = mx + c$, $2.30\log_{10}A$ is *c*, the intercept on the graph (Figure 11.12).

We know that $\dfrac{-E}{R}$ is −8600.

Using data for the first point:
$-5.40 = -8600 \times (0.00167) + \log_{10}A$

$\log_{10}A = 8.96$
$A =$ **9×10^8 mol^{-1} dm^3 s^{-1}**

Further practice

1 If the rate of a reaction doubles for a 10 K rise in temperature, calculate the activation energy, E_a:
   a at 300 K (Hint: combine two equations relating temperature and rate, one at 300 K, and one at 310 K)
   b at 1000 K.
   c Use your result to comment on the statement that 'a 10-degree rise in temperature doubles the rate of a reaction'.

2 The rate of the reaction:

$$C_2H_5Br + OH^- \rightarrow C_2H_5OH + Br^-$$

was measured in a propanone/water mixture. The concentration of both reactants was initially 0.010 mol dm^{-3}. The rate of the reaction varied with temperature as shown in Table 11.7.

■ **Table 11.7**

T/°C	10	20	30	40
Rate/mol dm^{-3} s^{-1}	5.0×10^{-6}	1.7×10^{-5}	5.5×10^{-5}	1.8×10^{-4}

Use these figures to calculate:
   a the activation energy, E_a
   b the pre-exponential factor, A.

11.6 The collision theory revisited

In Chapter 8, we saw that in the collision theory, it is assumed that molecules must collide before they can react. The collision must have a certain minimum energy, the activation energy E_a, in order for the molecules to overcome an energy barrier that exists between the reactants and products. This produces the familiar reaction profile for the reaction (Figure 11.13). If the reaction is exothermic, the activation energy for the back reaction is so high that virtually no product molecules react to re-form reactants, and the reaction goes to completion.

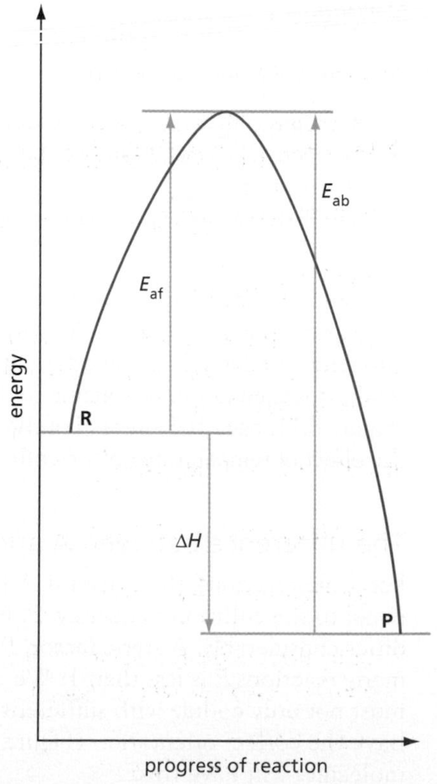

■ **Figure 11.13** Reaction profile for a typical reaction. The activation energy for the forward reaction is E_{af}. The reaction is exothermic, and the activation energy for the back reaction, E_{ab}, is so high that virtually no product molecules re-form reactants. The difference between E_{af} and E_{ab} is ΔH, the enthalpy change for the reaction.

The constant *A* – the collision frequency

In the collision theory, the constant A in the Arrhenius equation is identified with the **collision frequency**, Z. This is the number of collisions that molecules have every second when all the reactants are at unit concentration.

The **collision frequency**, *Z*, is proportional to:

- $\sqrt{T}$ (*T* is the absolute temperature)
- d^2 (*d* is the diameter of the molecule, assumed to be spherical)
- $\frac{1}{M_r}$ (M_r is the relative molecular mass).

Large molecules have a greater chance of colliding than small ones, but they also have a larger mass and move more slowly. So, in the gas phase at a particular temperature, Z is approximately the same for a wide variety of molecules. The value of Z at room temperature and pressure for a mole of gas is about 10^{11} collisions per second. Notice that the assumption we made on page 220, that A is independent of temperature is not quite true, because Z varies with $\sqrt{T}$. Because of this, there is a slight deviation from a straight line when $\log k$ is plotted against $1/T$. The effect is a very small one and is insignificant compared with the main effect of increasing temperature, namely the effect on the rate constant.

The factor $e^{(-E_a/RT)}$ – the fraction of collisions with sufficient energy to react

The factor $e^{(-E_a/RT)}$ represents the fraction of collisions that have energy greater than the activation energy, E_a. A typical value for E_a at room temperature is $60\,kJ\,mol^{-1}$. This gives a fraction of $10^{(-60000/2.3\times8.3\times300)} = 10^{-10.5} = 3\times10^{-11}$. So only one collision in 3×10^{10} has sufficient energy to overcome the activation energy barrier. If we multiply this by $10^{11}\,mol^{-1}\,dm^3\,s^{-1}$, the collision frequency, we arrive at a rate constant of $3\,mol^{-1}\,dm^3\,s^{-1}$, a typical value for a moderately fast reaction.

The effect of E_a on reaction rate

- Reactions with large E_a are slow, but the rate increases rapidly with temperature.
- Reactions with small E_a are fast, but the rate does not increase as rapidly with temperature.
- Catalysed reactions have small values of E_a.

It is often assumed that the fraction of collisions with energy greater than E_a is the same as the fraction of molecules with this energy. This is not strictly true, but the difference is relatively unimportant at high energies when the fraction is very small. It is then not unreasonable to draw the molecular energy distribution curve (Figure 8.11, page 166) instead of the collision distribution curve when showing the effect of temperature on the collision frequency.

The difference between *A* and *Z*

For some reactions, the constant A in the Arrhenius equation is approximately equal to the collision frequency, Z. For most reactions, however, the values differ considerably. A steric factor, *P*, is then introduced so that $PZ = A$. For many reactions *P* is less than 1. We can visualise that the reacting molecules must not only collide with sufficient energy to react, but that they must also have the correct orientation (Figure 11.14). Only a small proportion of reactant molecules will have this.

■ **Figure 11.14**
Two NO_2 molecules approaching each other with sufficient energy to overcome the energy barrier must collide in the correct orientation in order to combine to form N_2O_4.

Unfortunately, some reactions have P greater than 1. It is difficult to find a reasonable interpretation for this result using the collision theory, and we need to interpret the mechanism from a slightly different standpoint.

The collision theory can be extended to reactions in solution. In solution, solvated molecules and ions rather than simple molecules interact; these interactions are known as **encounters** rather than collisions. If the solvent does not play an important part in the reaction, the encounter rate is similar to the collision rate in the gas phase. For example, the encounter rate for the reaction in Figure 11.14 in tetrachloromethane is nearly the same as the collision frequency in the gas phase, at the same temperature.

It would appear, at first sight, that encounter rates should be smaller than collision frequencies because solvent molecules impede collisions. On the other hand, there are situations where encounters are more likely to happen, because the molecules are trapped in a 'cage' of solvent molecules (see Figure 11.15).

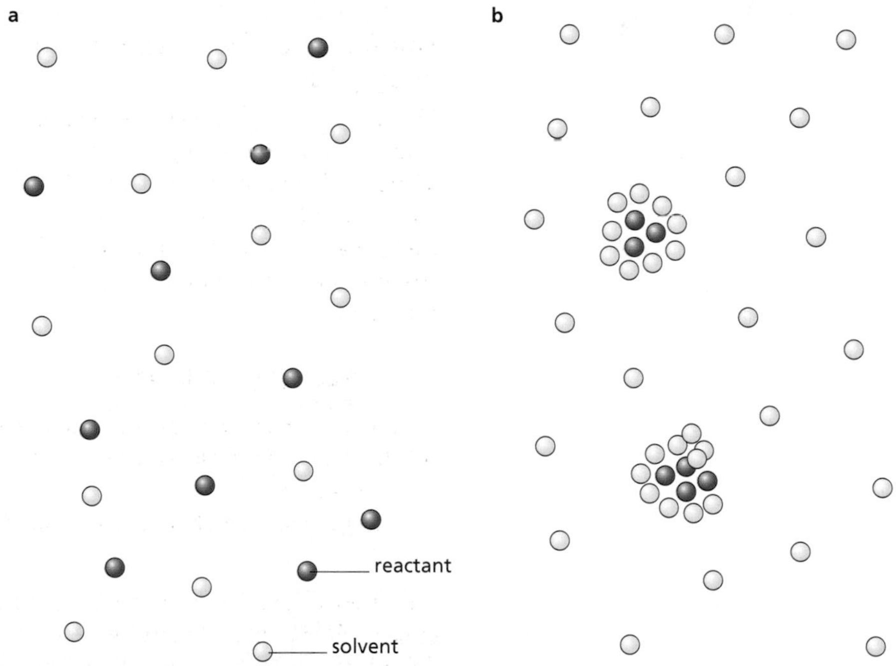

■ **Figure 11.15**
a Sometimes, solvent molecules impede collisions; **b** at other times, the 'cage' effect increases the likelihood of collision.

Example Calculate the fraction of collisions that have energy greater than $50\,kJ\,mol^{-1}$ at:
a 300 K **b** 1000 K.

Answer Fraction $= 10^{-E/(2.3\,RT)}$

a Fraction $= 10^{-50\,000/(2.3\times8.3\times300)} = 10^{-8.7} = 2.0\times10^{-9}$

b Fraction $= 10^{-50\,000/(2.3\times8.3\times1000)} = 10^{-2.6} = 2.4\times10^{-3}$

Further practice

1 Calculate the fraction of collisions that have energy greater than $100\,kJ\,mol^{-1}$ at:
a 300 K **b** 1000 K.

2 Calculate how much faster the reaction is at 1000 K than at 300 K. State any assumptions you made in carrying out this calculation.

3 For a reaction with an activation energy of $50\,kJ\,mol^{-1}$, the increase in rate from 300 K to 1000 K is a factor of 1.3×10^6. Compare this increase with the one calculated in **2**.

The transition state theory

In the collision theory, attention is focused on the colliding molecules. In the **transition state theory**, attention is directed instead to the top of the reaction profile curve. This energy maximum is known as the **transition state** and is given the symbol $TS^{\ddagger}$.

A simple treatment of this theory assumes that there is an equilibrium between the reactant molecules and the transition state. The reaction can then be written as:

$$\text{reactants} \overset{K}{\rightleftharpoons} TS^{\ddagger} \rightarrow \text{products} \quad \text{where } K = \frac{[TS^{\ddagger}]}{[\text{reactants}]}$$

We can apply the equation $\Delta G = -RT \log_e K$ (see Chapter 10, page 202) to the equilibrium.

$$\log_e K = \frac{-\Delta G}{RT}$$

$$K = \frac{[TS^{\ddagger}]}{[\text{reactants}]} = e^{-\Delta G/RT}$$

$$\text{Therefore, } [TS^{\ddagger}] = e^{-\Delta G/RT}[\text{reactants}]$$

where ΔG is the difference between the free energy of the transition state and that of the reactants.

The rate of the reaction depends on the probability that the transition state turns into products rather than reverts to reactants again. Theory shows that this term is RT/Lk, which has a value of $\approx 10^{13}$ at room temperature. So:

$$\text{rate of reaction} = 10^{13} \times e^{\Delta G/RT}[\text{reactants}]$$

Hence the rate constant $k = 10^{13}\, e^{\Delta G/RT}$. If we compare this with the collision theory, in which $k = Ae^{-E_a/RT}$, there are several differences.

- The activation energy, E_a, is now the free energy of activation, ΔG_a.
- The constant of proportionality is larger and is proportional to T rather than to $\sqrt{T}$.

We can use the equation $\Delta G_a = \Delta H_a - T\Delta S_a$ (Chapter 10, page 198, equation (1)) to split ΔG_a into ΔH_a, which is the activation energy E_a, and ΔS_a, the entropy of activation. So from $k = 10^{13}\, e^{\Delta G/RT}$, we have:

$$k = 10^{13} e^{\Delta S/R} e^{-\Delta H/RT}$$

We can equate the steric factor, P, of the collision theory with the entropy of activation, ΔS_a, of the transition state theory. Usually, the entropy of activation is negative, because two reactant molecules come together to form a single species in the transition state. A negative entropy of activation reduces the 10^{13} proportionality term of the transition state theory to a value comparable to A of the collision theory. Sometimes the entropy of activation is fairly small; this is often the case for rearrangement reactions which involve a single species and for some reactions in solution when solvation effects are important. Under these circumstances, the proportionality term remains at 10^{13} and then P appears to be greater than 1.

11.7 Reaction profiles

Intermediates and transition states

If we draw the reaction profile for a multi-step reaction such as that outlined on page 217, it is important to distinguish an **intermediate**, which has an energy minimum, from a **transition state**, at the top of the energy curve, which has an energy maximum. An intermediate is a definite chemical species that exists for a finite length of time. A transition state has no permanent lifetime of its own – it exists for just a few femtoseconds (10^{-15} s) when the molecules are in contact with each other. Even a reactive intermediate, with a lifetime of only a microsecond, has a long lifetime in comparison with the time that colliding molecules are in contact.

A simple one-step reaction, for example, the S_N2 hydrolysis of a primary halogenoalkane (see Chapter 25, page 466), has a single energy maximum (see Figure 11.16).

■ **Figure 11.16**
The reaction profile for a one-step reaction. There is a single transition state, TS‡, at the energy maximum.

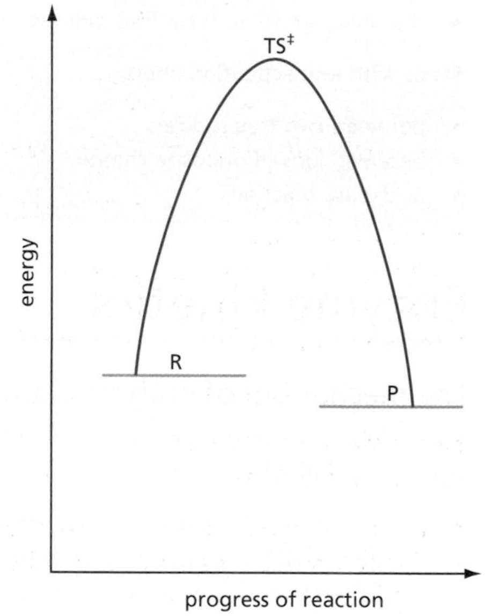

If the reaction has two steps, there is an intermediate, I, and two transition states. An example is the hydrolysis of 2-bromo-2-methylpropane (an S_N1 reaction, see page 466). The first step is rate determining and so has the higher activation energy (see Figure 11.17**a**).

■ **Figure 11.17**
The reaction profile for a two-step reaction. There is an intermediate I and also two transition states. In **a**, the first step is rate determining; in **b**, the second step is rate determining.

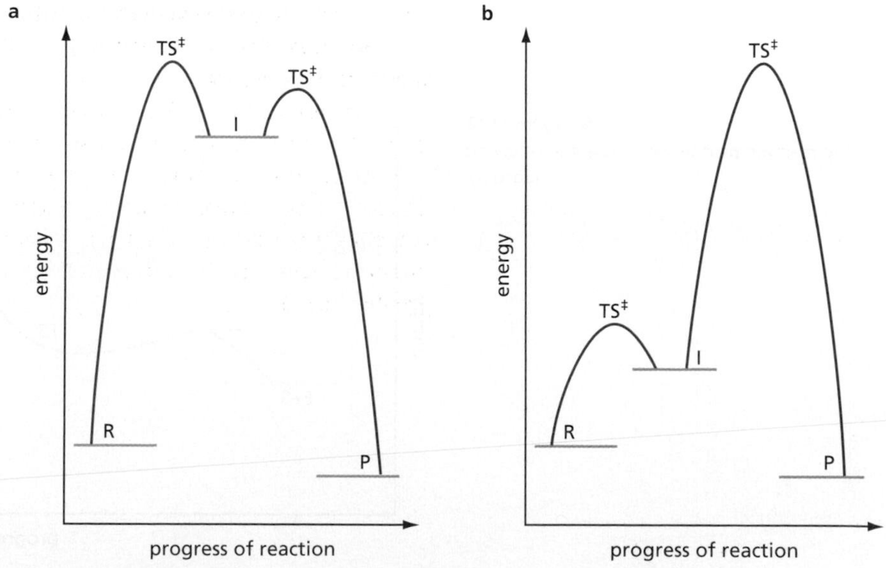

The structure of the transition state

It is possible to guess the structure of the transition state using **Hammond's postulate**. This states that the structure of the transition state will resemble that of the intermediate nearest to it in energy. In Figure 11.17**a**, the structures of the two transition states resemble that of the intermediate, I. In Figure 11.17**b**, the structure of the first transition state resembles that of the intermediate, while that of the second will be somewhere between that of the intermediate and the product.

While it is not possible to find the exact energy of the transition state except by experiment, some common-sense rules can be followed. For example, if the stage involves bond breaking, the activation energy is high, but if the stage is a reaction between ions of opposite charge, the activation energy is low.

Steps with high activation energy:

- between two neutral molecules
- between ions of similar charge
- if a bond breaks to form free radicals

Steps with low activation energy:

- between two free radicals
- between ions of opposite charge
- acid–base reactions

11.8 Enzyme kinetics

The mechanism of enzyme action

The mechanism of enzyme catalysis has been extensively researched. The principal steps are as follows.

$E + S \rightarrow ES$	enzyme + substrate → enzyme–substrate complex
$ES \rightarrow EP$	enzyme–substrate complex → enzyme–product complex
$EP \rightarrow E + P$	enzyme–product complex → enzyme + products

The substrate is first bonded to the active site of the enzyme to form the **enzyme–substrate complex**. Within this complex, the substrate molecule then undergoes chemical change that may involve bond reorganisation and/or attack by other molecules. The complex now comprises the reaction products still joined to the enzyme. This complex then breaks down to release the products and the free enzyme. The conversion of the enzyme–substrate complex to the enzyme–product complex is usually the slowest step, with the highest activation energy (see Figure 11.18).

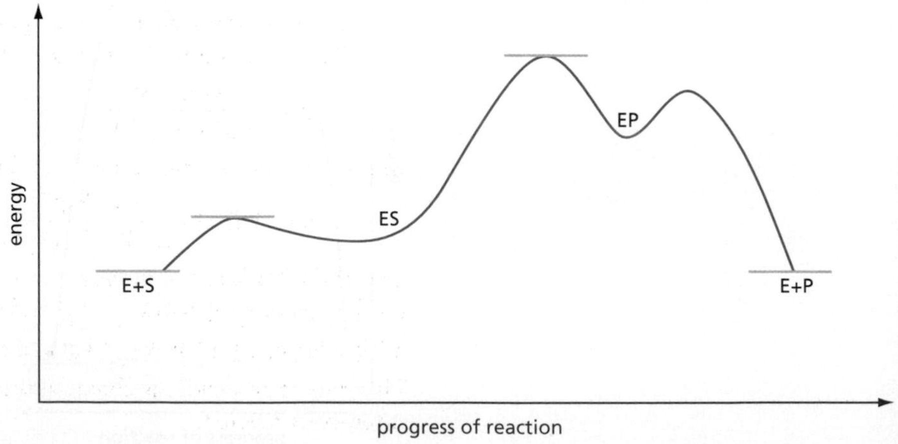

■ **Figure 11.18**
The reaction profile for an enzyme-catalysed reaction.

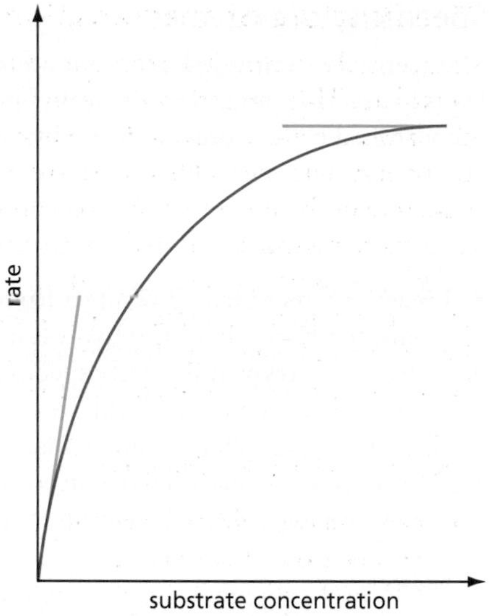

■ **Figure 11.19**
Graph of the rate of an enzyme-catalysed reaction against substrate concentration [S]. At low [S], the rate of reaction increases as [S] increases, but at high [S] the rate is unaffected by changes in [S].

The relative molecular mass of an enzyme is very large (typically 10^4 to 10^5), so the concentration in the cell in terms of molecules (and therefore active sites) is usually very low. When the substrate concentration is low, the rate of formation of enzyme–substrate complex (and therefore the rate of reaction) increases as the substrate concentration increases. However, at high substrate concentrations, all the enzyme has been converted into enzyme–substrate complex and so further increases in substrate concentration do not affect the overall rate (see Figure 11.19). This behaviour is characteristic of enzyme-catalysed reactions.

The Michaelis equation

The rate of an enzyme-catalysed reaction changes with the substrate concentration in a way that is quantitatively expressed by the **Michaelis equation**. The reaction mechanism may be simplified to the following two steps:

$$\mathrm{E + S} \underset{k_{-1}}{\overset{k_1}{\rightleftharpoons}} \mathrm{ES} \qquad \text{fast equilibrium}$$

$$\mathrm{ES} \xrightarrow{k_2} \mathrm{P + E} \qquad \text{slow, rate-determining step}$$

If the initial enzyme concentration (which is always small) is $[\mathrm{E_0}]$, we have:

$$[\mathrm{E_0}] = [\mathrm{E}] + [\mathrm{ES}] \qquad (1)$$

We cannot determine [ES] directly, but it may be calculated as follows.
The rate at which [ES] changes is equal to the rate of formation minus the rate of breakdown:

$$\frac{-\,\mathrm{d[ES]}}{\mathrm{d}t} = k_1[\mathrm{E}][\mathrm{S}] - (k_{-1}[\mathrm{ES}] + k_2[\mathrm{ES}]) \qquad (2)$$

Substituting $[\mathrm{E}] = [\mathrm{E_0}] - [\mathrm{ES}]$ from equation (1), we have:

$$\frac{-\,\mathrm{d[ES]}}{\mathrm{d}t} = k_1([\mathrm{E_0}] - [\mathrm{ES}])[\mathrm{S}] - (k_{-1}[\mathrm{ES}] + k_2[\mathrm{ES}])$$

For most of the reaction [ES] is very small and approximately constant. So:

$$\frac{-\,\mathrm{d[ES]}}{\mathrm{d}t} = 0$$

This is known as the **steady state approximation** and it is widely used to solve complicated equations. This leads us to:

$$[\mathrm{ES}] = \frac{k_1[\mathrm{E_0}][\mathrm{S}]}{k[\mathrm{S}] + k_{-1} + k_2}$$

and the rate of the reaction is given by:

$$\text{rate} = k_2[\mathrm{ES}] = \frac{k_1 k_2[\mathrm{E_0}][\mathrm{S}]}{k_1[\mathrm{S}] + k_{-1} + k_2}$$

This is the **Michaelis equation**.
If [S] is small, $k_1[\mathrm{S}] \ll (k_{-1} + k_2)$ and rate is first order with respect to [S].
If [S] is large, $k_1[\mathrm{S}] \gg (k_{-1} + k_2)$ and rate is zero order with respect to [S].
This type of behaviour is characteristic of enzyme reactions and has been extensively studied.

■ **Table 11.8**
Activation energies for the decomposition of hydrogen peroxide.

Catalyst	E_a/kJ mol^{-1}
Uncatalysed	76
Platinum	50
Catalase	8

Specificity of enzymes

Many enzyme-controlled reactions are so efficient that virtually every collision leads to reaction, and the activation energy is very small (see Table 11.8). The rate is controlled by how quickly the substrate molecules diffuse to the active site. The ability of an enzyme to hold substrate molecules in the correct orientation on the active site is also important. This means that the steric factor *P* for enzyme-catalysed reactions is important, and that enzymes are very specific in their action (Figure 11.20).

- **Lipase** enzymes break down fats (lipids). They are specific for certain groups of atoms, but bring about the hydrolysis of many different ester linkages.
- The enzyme **trypsin** is highly specific. It breaks down proteins by hydrolysing the links between amino acids, but will hydrolyse a peptide link only if it is adjacent to particular amino acids, namely arginine or lysine.
- The enzyme **α-glucosidase** is **stereospecific** – it attacks only one type of optical isomer (see page 407). It hydrolyses starch but does not hydrolyse cellulose.

■ **Figure 11.20**
The specificity of some enzymes.

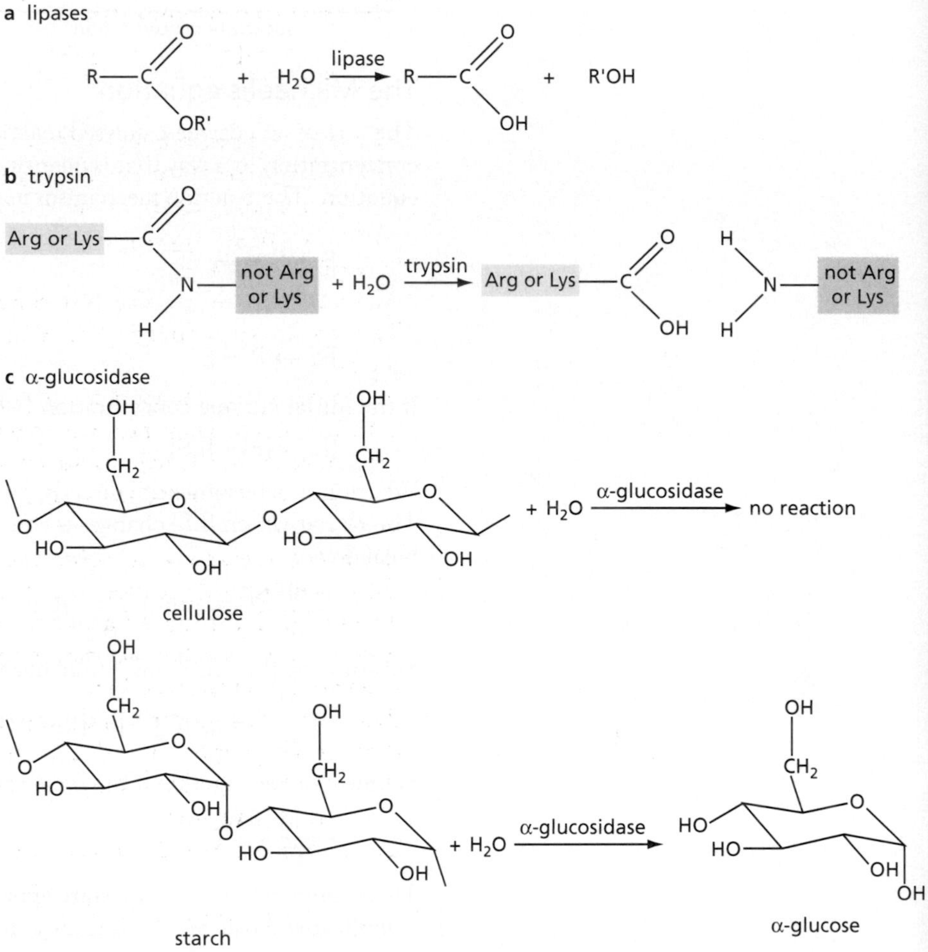

Enzymes are easily poisoned by molecules similar to the substrate. For example, fluoroethanoic acid is very poisonous because it binds firmly to the active sites of enzymes involved in the breakdown of ethanoic acid. Enzymes may also be poisoned by heavy metal ions such as Hg^{2+} which join on to sulphur atoms in the protein chain.

Many enzymes require another molecule or ion in order to become active. Some oxidation–reduction processes need molecules called **coenzymes** which undergo the actual electron transfer. Other oxidation–reduction processes have an iron atom attached to the enzyme and the electron transfer takes place when iron(II) is converted to iron(III). Other enzymes need a metal ion as part of the substrate – for example, the conversion of glucose into glucose-6-phosphate needs Mg^{2+} ions to be present.

Temperature and enzymes

Enzyme-catalysed reactions have low activation energies, so temperature does not have a large effect on their rate. For example, in the presence of catalase, a 10 °C rise in temperature increases the rate of decomposition of hydrogen peroxide by a factor of only 1.1. However, above about 40 °C most enzymes become deactivated because they become denatured. Denaturing has a high activation energy, typically 200–400 kJ mol^{-1}, because the process involves the breaking of a large number of intermolecular forces. The denatured enzyme has a random structure and the active site has been irreversibly destroyed.

Summary

- The **rate** of a reaction is given by the following expressions:

$$\text{rate} = \frac{\Delta[\text{P}]}{\Delta t} \text{ or rate} = -\frac{\Delta[\text{R}]}{\Delta t}$$

  The units of rate are mol dm^{-3} s^{-1}.
- The **order** of the reaction with respect to a reactant shows how the concentration of that reactant affects the rate of the reaction. The order for each reactant is found by experiment and these orders are combined together in a **rate equation**.
- The proportionality constant in the rate equation is called the **rate constant**. The rate constant does not vary with concentration but it does vary with temperature.
- A homogeneous catalyst does not appear in the overall stoichiometric equation, but its concentration does appear in the rate equation.
- If the reactants and their coefficients in the rate equation are the same as those in the stoichiometric equation, the reaction may take place in a single step.
- If the reactants and their coefficients in the rate equation differ from those in the stoichiometric equation, the reaction takes place in more than one step.
- The step with the highest activation energy is the **rate-determining step**. The number of species that take part in the rate-determining step is known as the **molecularity** of the reaction.
- Reactants whose concentrations appear in the rate equation react before or at the rate-determining step. Reactants whose concentrations do not appear in the rate equation but do appear in the stoichiometric equation react after the rate-determining step.
- The **Arrhenius equation** links the temperature with the rate constant:

$$2.30 \log_{10} k = 2.30 \log_{10} A - \frac{E}{RT}$$

  A graph of $\log_{10} k$ against $1/T$ gives a straight line with gradient $-E/2.30R T$ and intercept $\log_{10} A$.
- An increase in temperature increases the fraction of collisions with enough energy to overcome the activation energy. This fraction is $e^{-E_a/RT}$. The collisions must be correctly orientated as well as having enough energy for the reaction to occur.
- **Transition states** are at the maxima in the energy profile of a reaction.
- **Intermediates** are at the minima in the energy profile of a reaction.
- **Enzymes** act as highly specific catalysts.
- An enzyme-controlled reaction is first order at low substrate concentrations, but zero order at high substrate concentrations.

12 Quantitative equilibria

In Chapter 9 we introduced reversible reactions. We saw how the position of equilibrium may change in response to a change in concentration, pressure or temperature, according to Le Chatelier's Principle. In this chapter we study equilibria quantitatively. We shall distinguish between the position of equilibrium, which is affected by the changes mentioned above, and the equilibrium constant, which stays the same (at a given temperature) when concentrations are changed. We shall describe two equilibrium constants for gaseous reactions, K_c and K_p. We also study equilibria in dilute aqueous solution. Under these conditions the concentration of water is fixed, which enables us to simplify calculations of equilibria such as the solubility of electrolytes and the dissociation of weak acids.

The chapter is divided into the following sections.

12.1 The equilibrium constant K_c

In Chapter 9, we showed how the position of equilibrium changes when the external conditions are altered. Changing the state of division or adding a catalyst affects the rate of the reaction, but does not affect the position of equilibrium. The factors that do affect the position of equilibrium are:

- the concentration of the reactants and products
- the temperature.

In this section we shall study quantitatively the effect of changes in concentration on the position of equilibrium.

For a reaction in equilibrium, for example,

$$N_2O_4(g) \rightleftharpoons 2NO_2(g)$$

it can be shown that, at a given temperature,

$$K_c = \frac{[NO_{2(g)}]^2_{eq}}{[N_2O_{4(g)}]_{eq}} \text{ which is usually simplified to } \frac{[NO_2]^2}{[N_2O_4]}$$

where K_c is a constant, called the **equilibrium constant**. The subscript 'c' indicates that this equilibrium constant is calculated using concentrations and can, therefore, be distinguished from another equilibrium constant, K_p, which uses pressures, as we shall see later in the chapter.

The value of the equilibrium constant does not change with concentration, unlike the position of equilibrium. The only factor that affects its value is temperature.

There are several ways in which this equation can be derived. The most fundamental is by means of thermodynamics, but it can be approached through experiment or by way of kinetics.

Deriving equilibrium constants

By experiment

A reaction, for example:

$$CH_3CO_2H + C_2H_5OH \rightleftharpoons CH_3CO_2C_2H_5 + H_2O$$

is studied by making up a range of mixtures containing different amounts of reactants and products. After the mixtures have come to equilibrium, they are analysed, for example, by determining the amount of ethanoic acid present by titration. For a wide range of mixtures, it is found that the ratio:

$$\frac{[CH_3CO_2C_2H_5][H_2O]}{[CH_3CO_2H][C_2H_5OH]}$$

is constant.

Thermodynamic approach

We saw in Chapter 10 (page 202) that $\Delta G = -RT\log_e K$. At a particular temperature, ΔG is constant, which means that K is also a constant.

Kinetic approach

If the kinetics of the forward and back reactions are known, the equilibrium constant can be related to the rate constants. For example, for the reaction:

$$N_2O_4(g) \rightleftharpoons 2NO_2(g)$$

it is found that:

$$\text{rate of forward reaction at equilibrium} = k_f[N_2O_4]$$
$$\text{rate of backward reaction at equilibrium} = k_b[NO_2]^2$$

At equilibrium, the two rates are equal, therefore:

$$k_f[N_2O_4] = k_b[NO_2]^2$$

At a given temperature, the two rate constants are fixed, and:

$$\frac{[NO_2]^2}{[N_2O_4]} = \frac{k_f}{k_b} = K_c$$

For a general reaction:

$$wA + xB + \ldots \rightleftharpoons yC + zD + \ldots$$

$$K_c = \frac{[C]^y[D]^z}{[A]^w[B]^x}\ldots$$

Note that the terms in the square brackets are equilibrium concentrations and strictly should be written as, for example, $[C]^y_{eq}$.

The units of K_c are $(\text{mol dm}^{-3})^{\Delta n}$, where $\Delta n = (y + z + \ldots) - (w + x + \ldots)$. So for the reaction:

$$2Fe^{2+}(aq) + 2I^-(aq) \rightarrow 2Fe^{3+}(aq) + I_2(aq)$$

$$K_c = \frac{[Fe^{3+}]^2[I_2]}{[Fe^{2+}]^2[I^-]^2}$$

and the units are $(\text{mol dm}^{-3})^{3-4} = \text{mol}^{-1}\text{dm}^3$.

Evaluating K_c for a reaction

In order to determine the value of an equilibrium constant, known amounts of reactants are allowed to come to equilibrium. The amount of one of the substances in the equilibrium mixture is determined. The others can then be found from the stoichiometric equation, as described in the following example.

Example 1 2.0 mol of ethanoic acid and 2.0 mol of ethanol are mixed and allowed to come to equilibrium with the ethyl ethanoate and water they have produced. At equilibrium, the amount of ethanoic acid present is 0.67 mol. Calculate K_c.

Answer We start by working out the amounts, and hence the concentrations, of the four substances present at equilibrium from the stoichiometric equation.

	CH_3CO_2H +	C_2H_5OH ⇌	$CH_3CO_2C_2H_5$ +	H_2O
moles at start (given/mol):	2.0	2.0	0	0
moles at equilibrium (given/mol):	0.67			
moles at equilibrium/mol:	0.67	0.67	(2.0 − 0.67)	(2.0 − 0.67)
	0.67 mol	0.67 mol	1.33 mol	1.33 mol
concentrations at equilibrium:	$\frac{0.67}{V}$	$\frac{0.67}{V}$	$\frac{1.33}{V}$	$\frac{1.33}{V}$

where V is the total volume of the mixture.

$$K_c = \frac{[CH_3CO_2C_2H_5][H_2O]}{[CH_3CO_2H][C_2H_5OH]} = \frac{1.33/V \times 1.33/V}{0.67/V \times 0.67/V}$$

$= \mathbf{4.0}$ (no units)

K_c has no units because there are the same numbers of molecules on both sides of the equation. This also means that the volume, V, cancels. The value of K_c, and the position of equilibrium, do not depend on the volume of the reaction mixture.

Example 2 0.30 mol of dinitrogen tetraoxide, N_2O_4, is allowed to come to equilibrium with nitrogen dioxide, NO_2.

The amount of nitrogen dioxide at equilibrium is 0.28 mol.

Calculate:

a the amount of dinitrogen tetraoxide at equilibrium

b the value of K_c at the temperature of the experiment, given that the volume of the containing vessel is 10.0 dm³.

Answer **a**

	$N_2O_4(g)$ ⇌	$2NO_2(g)$
moles at start/mol:	0.30	0
moles at equilibrium/mol:	$(0.30 - \frac{1}{2} \times 0.28)$	0.28

moles of N_2O_4 at equilibrium = **0.16 mol**

b $[N_2O_4] = \frac{0.16}{10}\ \text{mol dm}^{-3}$ $\quad [NO_2] = \frac{0.28}{10}\ \text{mol dm}^{-3}$

$$K_c = \frac{[NO_2]^2}{[N_2O_4]} = \frac{0.028^2}{0.016} = \mathbf{0.049\ mol\ dm^{-3}}$$

Example 3 At 650 °C, K_c for the reaction:

$$N_2(g) + 3H_2(g) \rightleftharpoons 2NH_3(g)$$

is $4 \times 10^{-2}\ \text{mol}^{-2}\,\text{dm}^6$. For the following reactions, write down expressions for the equilibrium constants and hence, at 650 °C, calculate their values.

a $2NH_3(g) \rightleftharpoons N_2(g) + 3H_2(g)$

b $\frac{1}{2}N_2(g) + 1\frac{1}{2}H_2(g) \rightleftharpoons NH_3(g)$

Answer **a** For the reaction $N_2(g) + 3H_2(g) \rightleftharpoons 2NH_3(g)$,

$$K_c = \frac{[NH_3]^2}{[N_2][H_2]^3} = 4 \times 10^{-2}\ \text{mol}^{-2}\,\text{dm}^6$$

For the reaction $NH_3(g) \rightleftharpoons N_2(g) + 3H_2(g)$,

$$\text{equilibrium constant} = \frac{[N_2][H_2]^3}{[NH_3]^2} = \frac{1}{K_c} = \mathbf{25\,mol^2\,dm^{-6}}$$

b For the reaction $\frac{1}{2}N_2(g) + 1\frac{1}{2}H_2(g) \rightleftharpoons NH_3(g)$,

$$\text{equilibrium constant} = \frac{[NH_3]}{[N_2]^{\frac{1}{2}}[H_2]^{1\frac{1}{2}}} = \sqrt{K_c} = \mathbf{0.2\,mol\,dm^{-3}}$$

Further practice

1 2.00 mol of propanoic acid and 3.00 mol of propan-1-ol are mixed and allowed to come to equilibrium. At equilibrium, 0.50 mol of propanoic acid remain.
  a Write an equation for the reaction.
  b Calculate K_c.
2 2.00 mol of hydrogen gas are mixed with 2.00 mol of nitrogen gas and allowed to come to equilibrium with ammonia in a 20.0 dm^3 container. At equilibrium there are 3.00 mol of gas in total. Calculate:
  a the number of moles of each of the three gases at equilibrium
  b K_c.

12.2 Using equilibrium constants

Finding the composition of the equilibrium mixture – a simple example

An equilibrium constant can be used to calculate the composition of an equilibrium mixture. For example, the equilibrium constant for the reaction:

$$CH_3CO_2H + C_2H_5OH \rightleftharpoons CH_3CO_2C_2H_5 + H_2O$$

is 4.0. If we start with 2.0 mol of ethanoic acid and 2.0 mol of ethanol, the composition of the equilibrium mixture can be found by means of some simple algebra as folllows.

Let there be x mol of ethyl ethanoate at equilibrium:

	CH_3CO_2H	C_2H_5OH	$CH_3CO_2C_2H_5$	H_2O
moles at start/mol:	2.0	2.0	0	0
moles at equilibrium/mol:	$(2.0 - x)$	$(2.0 - x)$	x	x
concentrations at equilibrium/mol:	$\frac{(2.0 - x)}{V}$	$\frac{(2.0 - x)}{V}$	$\frac{x}{V}$	$\frac{x}{V}$

$$K_c = \frac{[CH_3CO_2C_2H_5][H_2O]}{[CH_3CO_2H][C_2H_5OH]}$$

The volume terms cancel out. This gives us:

$$\frac{x^2}{(2.0 - x)^2} = 4.0$$

Taking square roots of both sides,

$$x = \pm 2.0(2.0 - x)$$
$$x = 1.3\,\text{mol} \quad \text{or} \quad 4.0\,\text{mol}$$

The second value must be rejected as it is impossible to produce more than 2.0 mol of product. Therefore, at equilibrium there are 1.3 mol of ethyl ethanoate, 1.3 mol of water, 0.7 mol of ethanoic acid and 0.7 mol of ethanol.

Using the quadratic formula

In the reaction between ethanoic acid and ethanol, if the initial amount were 3.0 mol of ethanoic acid and 2.0 mol of ethanol, the expression would be:

$$\frac{x^2}{(3.0-x)(2.0-x)} = 4.0$$

so

$$x^2 = 4.0(6.0 - 5.0x + x^2)$$

or

$$3x^2 - 20.0x + 24.0 = 0$$

This can be solved using the formula for quadratic equations as follows. For the general equation $ax^2 + bx + c = 0$,

$$x = \frac{-b \pm \sqrt{b^2 - 4ac}}{2a}$$

$$x = \frac{20.0 \pm \sqrt{(400 - 288)}}{6}$$

$$x = 5.1\,\text{mol or } 1.6\,\text{mol}$$

The first solution is rejected because the maximum possible yield is 2.0 mol.

Simplifying more complex cases

If the reactants or products are made up of three or more molecules, the equation cannot be solved by simple methods and often an approximation is made to simplify the calculation. For example, for the reaction:

$$N_2(g) + 3H_2(g) \rightleftharpoons 2NH_3(g)$$

at 900 K, $K_c = 7.0 \times 10^{-2}\,\text{mol}^{-2}\,\text{dm}^6$. We can calculate the equilibrium composition when 1 mol of nitrogen is mixed with 3 mol of hydrogen at 900 K in a 300 dm^3 container, as follows.

	N_2	H_2	NH_3
moles at start:	1	3	0
moles at equilibrium:	$(1-x)$	$(3-3x)$	$2x$
concentrations at equilibrium:	$\frac{(1-x)}{300}$	$\frac{(3-3x)}{300}$	$\frac{2x}{300}$

$$K_c = \frac{[NH_3]^2}{[N_2][H_2]^3} \quad \text{and} \quad 7.0 \times 10^{-2} = \frac{4x^2(300)^2}{(1-x)(3-3x)^3}$$

This equation can be solved, but it is much easier if the following simplification is made. Because K_c is small, x is small and $(1-x) \approx 1$. So:

$$(1-x)(3-3x)^3 \approx 1 \times 3^3 = 27.$$

Therefore

$$4x^2(300)^2 = 27 \times 7.0 \times 10^{-2}$$

so

$$x = \pm\frac{\sqrt{1.89}}{600} = 2.3 \times 10^{-3} \text{ (rejecting the negative value)}$$

The equilibrium mixture therefore contains $2x =$ **4.6×10^{-3} mol** of ammonia.

(If the equation is solved accurately the answer is 4.5×10^{-3} mol, within the limits of accuracy quoted for the data.)

Further practice

K_c is 1.50 mol dm^{-3} for the reaction:

$$N_2O_4(g) \rightleftharpoons 2NO_2(g)$$

0.5 mol of dinitrogen tetraoxide are placed in a 5.0 dm^3 flask and allowed to come to equilibrium. What is the composition of the equilibrium mixture?

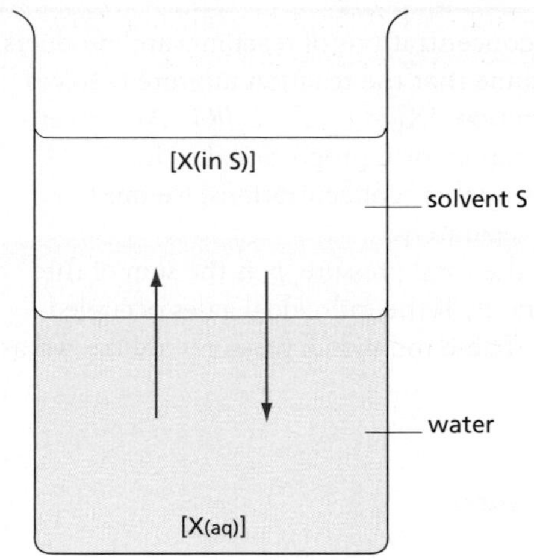

■ **Figure 12.1**
At equilibrium, the rates at which X leaves each layer are equal.

Distribution coefficients

Solvent extraction is a technique used to purify substances dissolved in aqueous solutions. For example, an aqueous solution of a substance X is shaken with another solvent, S, which is **immiscible** with water (that is, they are insoluble in each other) but X is very soluble in solvent S. When equilibrium has been established, the rate at which X leaves the aqueous layer equals the rate at which it leaves solvent S (see Figure 12.1).

We have:

rate of leaving aqueous layer $= k_1$[X(aq)] and
rate of leaving solvent S $= k_2$[X(in S)]

At equilibrium:

$$\frac{[\text{X(in S)}]}{[\text{X(aq)}]} = \frac{k_1}{k_2} = \text{a constant } K_D$$

This constant K_D is known as the **distribution coefficient** (or sometimes the **partition coefficient**).

The aqueous layer is placed in a separating funnel and a small quantity of solvent added. The solvent ethoxyethane (ether) is often used, and the method is then called **ether extraction**. The mixture is shaken to establish equilibrium, the layers allowed to separate and the aqueous layer run off (it is at the bottom, being more dense than ethoxyethane). The ethoxyethane layer is put into a separate container. The process is repeated by shaking the depleted aqueous layer with further small samples of ethoxyethane. The ethoyxyethane layers are combined and dried (the water removed). When the ethoxyethane layer is dry, as shown by it becoming clear, it is distilled off, taking suitable precautions as it is very flammable. Finally pure X is obtained by fractional distillation (see the panel on pages 238–9).

■ **Figure 12.2**
Iodine partitioned between water and hexane.

Example 0.10 mol of ammonia in 100 cm³ of water are shaken with 50 cm³ of ethoxyethane (K_D = 8.0). Calculate:

a the amount of ammonia extracted into the ethoxyethane
b the percentage of ammonia remaining in the water.

Answer **a** Let amount of ammonia extracted into the ethoxyethane be *x* mol. Therefore, amount of ammonia remaining in the water = (0.10 − *x*) mol.

$$\frac{[\text{NH}_3 \text{ in ethoxyethane}]}{[\text{NH}_{3(aq)}]} = 8.0$$

$$\frac{\frac{x}{50} \times 1000}{\frac{0.10 - x}{100} \times 1000} = 8.0$$

Therefore: $20x = 8.0\ (1.0 - 10x)$

$100x = 8.0$

$x =$ **0.080 mol**

b Percentage remaining $= \frac{0.10 - 0.080}{0.10} \times 100 =$ **20%**

12.3 The equilibrium constant K_p

For gaseous equilibria, we can express the concentrations of reactants and products in terms of their partial pressures. If we assume that the reaction mixture behaves as an ideal gas, $pV = nRT$ and the concentration, $[X], = n_X/V = p_X/RT$. At a given temperature, therefore, the pressure of a component is proportional to its concentration. In order to use pressures rather than concentrations, we must consider the effect of pressure on each gas separately.

If we have a mixture of gases, A, B, . . ., the total pressure, p, is the sum of the pressure that each gas would produce separately if the individual gases occupied the same volume at the same temperature. These individual pressures are known as the **partial pressures,** p_A, p_B, ...

$$p = p_A + p_B + \dots$$

This is called **Dalton's Law of Partial Pressures.**

$$[A] = \frac{p_A}{RT}, \quad [B] = \frac{p_B}{RT}, \quad \dots$$

For the general gaseous reaction:

$$wA(g) + xB(g) + \dots \rightleftharpoons yC(g) + zD(g) + \dots$$

because RT is constant, we can define another equilibrium constant:

$$K_p = \frac{(p_C)^y (p_D)^z}{(p_A)^w (p_B)^x}$$

where K_p is the equilibrium constant using partial pressures. The units of K_p are $(\text{pressure})^{(y+z)-(w+x)}$, where pressure may be measured in atmospheres or pascals.

An advantage of using K_p is that the position of equilibrium is often found by pressure measurements. In order to carry out such calculations, a relationship between partial pressures, numbers of moles and total pressure must be established. Assume that we have n mol of gas at a total pressure of p atm. Of these n mol, n_A mol are of gas A, whose partial pressure is p_A atm. The mole fraction, x_A, of gas A is n_A/n.

The ideal gas equation gives $p_A V = n_A RT$ and $pV = nRT$. Because V, p and T are fixed,

$$x_A = \frac{n_A}{n} = \frac{p_A}{p}$$

so

$$p_A = x_A p$$

To calculate partial pressures

The **partial pressure** of a component in a mixture of gases equals its mole fraction multiplied by the total pressure.

$$p_A = x_A p, \quad p_B = x_B p, \quad \text{etc.}$$

Example 2.0 mol of phosphorus pentachloride, PCl_5, is placed in an 82 dm^3 flask at 500 K. If the phosphorus pentachloride did not decompose, the ideal gas equation shows that the pressure would be 1.0 atm. However, at 500 K an equilibrium is established:

$$PCl_5(g) \rightleftharpoons PCl_3(g) + Cl_2(g)$$

and the measured pressure at equilibrium is 1.2 atm.

Calculate:

a K_p for the reaction

b the composition of the equilibrium mixture if the pressure is increased to 5.0 atm by decreasing the volume of the container.

Answer **a** The pressure increases by 20% because the total number of moles has increased by 20%, from 2.0 to 2.4. The equilibrium composition can be calculated as follows.

	PCl_5	PCl_3	Cl_2	total
moles at start/mol:	2.0	0	0	2.0
moles at equilibrium/mol:	$(2.0-x)$	x	x	$(2.0-x)+x+x=2.4$

so $x=0.4$. This value can now be used to calculate K_p for the reaction:

	PCl_5	PCl_3	Cl_2	total
moles at equilibrium/mol:	1.6	0.4	0.4	2.4
mole fraction:	$\frac{1.6}{2.4}$	$\frac{0.4}{2.4}$	$\frac{0.4}{2.4}$	
partial pressure/atm:	0.67×1.2	0.167×1.2	0.167×1.2	

$$K_p=\frac{(p_{PCl_3})(p_{Cl_2})}{(p_{PCl_5})}=\frac{(0.167\times1.2)\times(0.167\times1.2)}{0.67\times1.2}=\frac{0.2\times0.2}{0.8}=\mathbf{0.05\,atm}$$

b

	PCl_5	PCl_3	Cl_2	total
moles at start/mol:	2.0	0	0	2.0
moles at equilibrium/mol:	$(2.0-x)$	x	x	$(2.0+x)$
mole fraction:	$\frac{(2.0-x)}{(2.0+x)}$	$\frac{x}{(2.0+x)}$	$\frac{x}{(2.0+x)}$	
partial pressures/atm:	$\frac{5(2.0-x)}{(2.0+x)}$	$\frac{5x}{(2.0+x)}$	$\frac{5x}{(2.0+x)}$	

$$K_p=\frac{(p_{PCl_3})(p_{Cl_2})}{(p_{PCl_5})}=\frac{25x^2\times(2.0+x)}{5(2.0-x)(2.0+x)^2}=\frac{5x^2}{(4.0-x^2)}=0.05\,\text{atm}$$

so $5.05x^2=0.2$ and $x=\mathbf{0.2}$ (rejecting the negative value).

Notice that Le Chatelier's Principle predicts that increasing the pressure reduces the fraction dissociated. The calculation shows that this reduction is from 0.40 to 0.2.

Further practice 5.0 mol of sulphur dioxide, SO_2, were mixed with 2.5 mol of oxygen, O_2, and brought to equilibrium with sulphur trioxide, SO_3, at 600 °C. Initially the pressure was 3.0 atm, but at equilibrium the pressure dropped to 2.4 atm.

1 Calculate the partial pressures of the three gases at equilibrium.

2 Calculate K_p for the reaction: $2SO_{2}(g) + O_{2}(g) \rightleftharpoons 2SO_{3}(g)$.

K_c and K_p

If a reaction takes place with no change in the number of molecules ($\Delta n=0$), K_c and K_p have the same value (no units). If $\Delta n\neq0$, the connection between them is:

$$\frac{K_p}{K_c}=RT^{\Delta n}$$

If pressures are measured in atmospheres, note that R has the value $0.0820\,dm^3\,atm\,mol^{-1}K^{-1}$ rather than $8.3\,J\,mol^{-1}K^{-1}$, and V is in dm^3 rather than m^3.

Further practice For the reaction

$$2NO_{2}(g) \rightleftharpoons N_2O_{4}(g)$$

show that $\frac{K_p}{K_c}=\frac{1}{RT}$

Fractional distillation

Saturated vapour pressure and boiling point

A liquid in a sealed container evaporates and the vapour creates a vapour pressure. The vapour pressure rises and finally reaches a maximum value. While some liquid remains, this value is constant at a given temperature and is known as the **saturated vapour pressure**. The rate of evaporation equals the rate of condensation, and liquid and gas are in equilibrium.

When a liquid is heated, its saturated vapour pressure increases. It boils if the container is not sealed when its saturated vapour pressure is equal to that of the atmosphere. If liquid A has a higher saturated vapour pressure than liquid B, liquid A boils at a lower temperature than liquid B (see Figure 12.3).

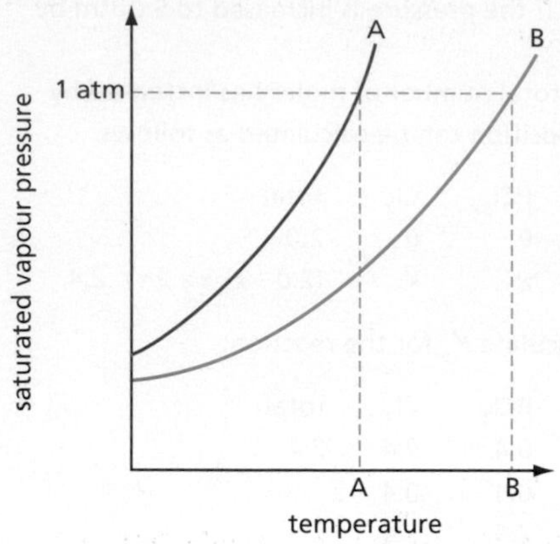

■ **Figure 12.3** Liquid A, with a higher saturated vapour pressure than liquid B, boils at a lower temperature.

Raoult's Law

If the two liquids A and B are mixed, the saturated vapour pressure of A, p_A, is decreased from the value of the pure liquid, p_A°, because there are fewer molecules of A on the liquid surface. The same is true of liquid B. In an ideal case, the reduction in saturated vapour pressure is proportional to the mole fractions. So we have:

$$p_A = p_A^\circ x_A \quad \text{and} \quad p_B = p_B^\circ x_B$$

This is known as **Raoult's Law**.

For two liquids, the total pressure p is equal to $p_A + p_B$, and $x_A + x_B = 1$. The situation is shown in Figure 12.4.

■ **Figure 12.4** (left) Graph of saturated vapour pressure against composition for a mixture of two liquids, A and B.

■ **Figure 12.5** (right) Graph of boiling point against composition for the mixture of A and B.

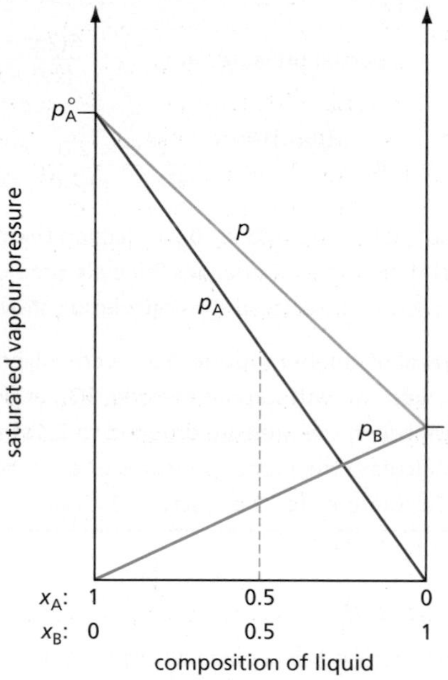

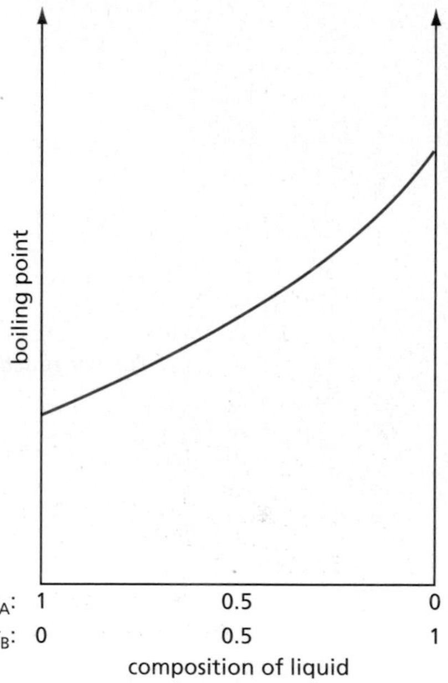

If a mixture of A and B is heated, the boiling point is between those of the two pure components. Liquid A, having the higher saturated vapour pressure, has the lower boiling point (see Figure 12.5).

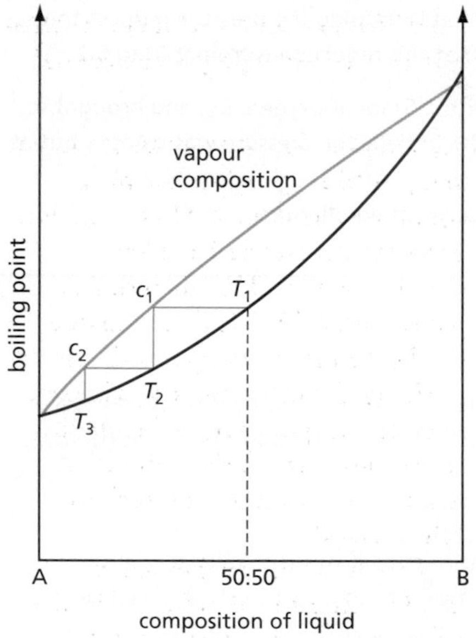

■ **Figure 12.6** Fractional distillation of a mixture of liquids A and B. The 50:50 mixture boils at temperature T_1. The vapour has composition c_1 and condenses at temperature T_2. The process is continued until nearly pure A is obtained as the distillate.

Separating mixtures of liquids by fractional distillation

Two liquids such as A and B can be separated by fractional distillation. If we start with a 50:50 mixture of the two liquids, the vapour contains a greater proportion of A than B. The composition of this vapour is indicated by point c_1 in Figure 12.6. The vapour of composition c_1 given off at temperature T_1 condenses at a lower temperature, T_2, than the 50:50 mixture we started with. Further distillation results in a mixture of composition c_2 and boiling point T_3. The process may be continued until the required degree of purity is obtained.

Example A 50:50 mixture of two components A and B is distilled at a temperature at which the partial vapour pressure of pure A is 120 kPa and that of pure B is 80 kPa. What will be the mole fraction of each component in the vapour that distils off?

Answer

$$p_A = p_A^{\circ}x_A = 120 \times 0.5 = 60\,\text{kPa}$$
$$p_B = p_B^{\circ}x_B = 80 \times 0.5 = 40\,\text{kPa}$$
$$p = p_A + p_B = 60 + 40 = 100\,\text{kPa}$$

As we saw on page 236, for a mixture of gases:

$$\frac{n_A}{n_B} = \frac{p_A}{p_B}$$

Therefore:

$$\text{mole fraction of A in the vapour} = \frac{60}{100} = 0.6$$

$$\text{mole fraction of B in the vapour} = \frac{40}{100} = 0.4$$

By one simple distillation we have increased the mole fraction of A by 20%, from 0.5 to 0.6.

Further practice If the distillate from the first distillation in the example above is condensed and re-distilled, what will be the mole fraction of A in the new distillate?

■ **Figure 12.7** (above) Industrial fractional distillation.

As A is removed by distillation, the mixture left behind becomes richer in B. After many distillations, almost pure A will have been distilled off, leaving almost pure B behind in the distilling flask.

The process can be made continuous by the use of a **fractionating column**. This keeps the vapour in the column for longer, allowing it to condense and boil again repeatedly. The process is used extensively in industrial chemistry, for example, the fractional distillation of crude oil (see Chapter 23, page 425).

A common misconception is that when a mixture of two liquids is fractionally distilled, pure A comes off at its boiling point, leaving pure B behind. This is incorrect, because at the boiling point of A, B has an appreciable vapour pressure. In practice, fractional distillation is used to obtain the components at a particular stated purity. For ordinary laboratory use, this is often 99% pure. Other techniques such as gas chromatography may be used to obtain small samples of highly pure material.

Azeotropic mixtures

Some pairs of liquids cannot be completely separated by fractional distillation. For example, if dilute aqueous ethanol is fractionally distilled, the distillate never contains more than 95.6% of ethanol, because this mixture has a lower boiling point, 78.2 °C, than that of pure ethanol, 78.5 °C. This mixture can be distilled further but will remain unchanged. Such a mixture is known as an **azeotropic mixture**. To obtain anhydrous ethanol, the water is removed using a chemical such as calcium oxide.

If dilute aqueous nitric acid is fractionally distilled, water is distilled off leaving behind an azeotropic mixture that contains 68% nitric acid. This is 'bench concentrated' nitric acid. Pure nitric acid is known as 'fuming' nitric acid.

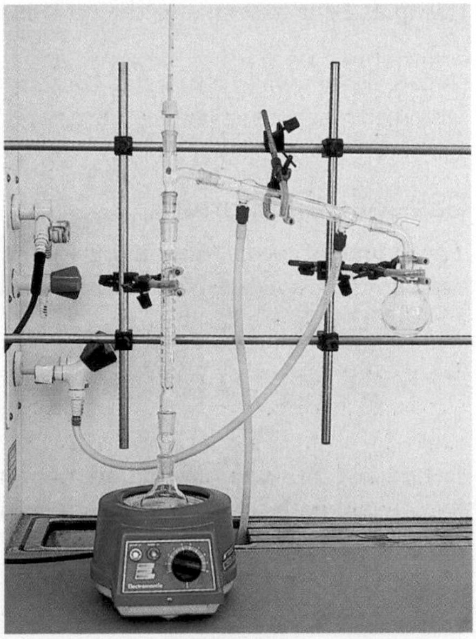

■ **Figure 12.8** Fractional distillation in the laboratory.

12.4 Heterogeneous equilibria

So far in this chapter we have looked at equilibria involving gases, for which K_c or K_p may be calculated, and at equilibria in solution, for which we use K_c. We shall now look at equilibria involving more than one phase.

Gas–liquid equilibrium

At a given temperature, a liquid in a closed container may be in equilibrium with its vapour. For example, if water is in contact with its vapour:

$$H_2O(l) \rightleftharpoons H_2O(g)$$

we may write:

$$K_c = \frac{[H_2O(g)]}{[H_2O(l)]}$$

At a given temperature, the $[H_2O(l)]$ term is a constant, irrespective of the amount of water present. This is because:

$$[H_2O(l)] = \frac{m}{MV} = \frac{\text{density}}{M}$$

and both density of water and its molar mass are constant at a given temperature. For water, the density is $1000\,g\,dm^{-3}$ and $M_p = 18\,g\,mol^{-1}$. So:

$$[H_2O(l)] = \frac{1000}{18 \times 1} = 55.5\,mol\,dm^{-3}$$

This value will be used again later in the chapter.

We can absorb this constant $55.5\,mol\,dm^{-3}$ into K_c, giving us a new constant, $K' = [H_2O(g)]$. Using the ideal gas equation, we have:

$$pV = nRT$$

$$p = \frac{nRT}{V} = [H_2O(g)]RT$$

where p is the **saturated vapour pressure**. Hence $p = K'RT$ and this shows us that, at a given temperature, any liquid has a definite saturated vapour pressure.

The amount of liquid present at equilibrium does not affect the position of equilibrium. What is affected is the rate at which the equilibrium is reached. A large surface area increases the rate – for example, spilt water dries more quickly if it is spread out over an area than if it is left in a deep puddle.

A pure liquid term in K_c or K_p is a constant at a given temperature, and may be absorbed into the equilibrium constant.

Gas–solid equilibrium

The argument used above that any pure liquid can be ignored in K_c or K_p also holds for pure solids. Consider the equilibrium:

$$CaCO_3(s) \rightleftharpoons CaO(s) + CO_2(g)$$

$$K_p = \frac{p_{CaO(s)} \times p_{CO_2(g)}}{p_{CaCO_3(s)}} \quad \text{or} \quad p_{CO_2(g)} = \frac{K_p \times p_{CaCO_3(s)}}{p_{CaO(s)}}$$

The vapour pressures of solids are very small, and constant. Since $p_{CaO(s)}$ and $p_{CaCO_3(s)}$ are both constant, the expression on the right of this equation has a fixed value K'_p, where:

$$K'_p = \frac{K_p \times p_{CaCO_3(s)}}{p_{CaO(s)}}$$

$p_{CO_2(g)}$ therefore has a fixed value:

$$p_{CO_2(g)} = K'_p$$

K_p, and hence K'_p, increase with temperature. At a particular temperature, $p_{CO_2(g)}$ ($=K'_p$) will equal 1 atmosphere (see Table 16.4, page 312).

A pure solid term in K_c or K_p is a constant at a given temperature, and may be absorbed into the equilibrium constant.

Phase diagrams

Figure 12.3 (page 238) shows how the saturated vapour pressure of a liquid varies with temperature. If we cool the liquid, it freezes to a solid. The solid also produces vapour. The solid/vapour curve is similar in shape to that of the liquid/vapour curve.

Because pressure has virtually no effect on the volumes of solids and liquids, the solid/liquid curve is a nearly vertical straight line. These three curves are combined in Figure 12.9, known as a **phase diagram**. The lines show how the equilibria between the three phases vary with temperature and pressure. In the areas to each side of the lines only one phase exists.

■ Figure 12.9
A typical phase diagram for a pure substance.

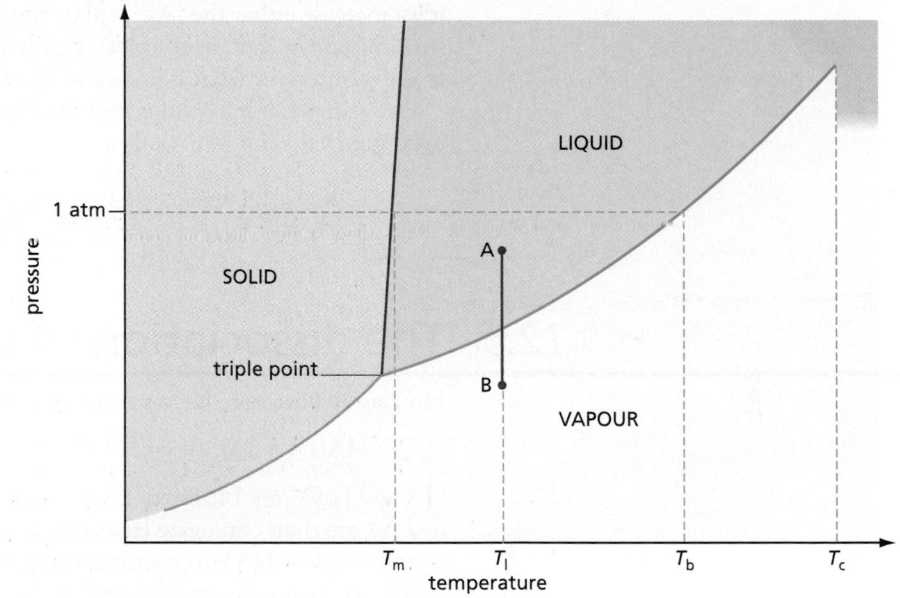

Usually, a gas at temperature T_1 can be liquefield by pressurising it. This can be represented by the vertical line A–B on Figure 12.9.

Above a certain temperature T_c, called the **critical temperature**, a gas will not turn into a liquid whatever the external pressure applied. Above the critical temperature, the gas has the same density as that of the liquid and the two states become indistinguishable. Critical temperatures of some gases are shown in Table 12.1.

Gas	Critical temperature/°C
N_2	−147
O_2	−119
CO_2	31
NH_3	132
SO_2	157
H_2O	374

■ Table 12.1
The critical temperatures of some common gases. The values show why cylinders of compressed nitrogen and oxygen contain gas, while a cylinder of compressed carbon dioxide contains liquid.

If a horizontal line at a pressure of 1 atm is drawn on Figure 12.9, the temperature at which it crosses the solid–liquid boundary line is the melting point of the substance, T_m. The temperature at which the 1 atm line crosses the liquid–gas boundary is the boiling point of the substance, T_b.

The point where the three lines intersect in Figure 12.9 is known as the **triple point**. For water this is at 273.16 K and 611.2 Pa. The value for carbon dioxide is more than five times atmospheric pressure. This means that, at atmospheric pressure, carbon dioxide changes straight from solid to gas, that is, it **sublimes**. In a cylinder, carbon dioxide is liquid because the pressure is much higher, about 70 times atmospheric pressure.

Solubility of salts

Another example of a heterogeneous equilibrium is dissolving an ionic substance in water. Take, for example, solid silver chloride in contact with a saturated solution:

$$Ag^+Cl^-(s) \rightleftharpoons Ag^+(aq) + Cl^-(aq)$$

We can write:

$$K_c = \frac{[Ag^+(aq)][Cl^-(aq)]}{[Ag^+Cl^-(s)]}$$

At a given temperature, the $[Ag^+Cl^-(s)]$ term is a constant, irrespective of the amount of silver chloride present. We can absorb this $[Ag^+Cl^-(s)]$ term in the equilibrium constant and define a new equilibrium constant:

$$K_{sp} = [Ag^+(aq)][Cl^-(aq)]$$

where K_{sp} is called the **solubility product**. For AgCl(s), the value of K_{sp} is 1×10^{-10} mol^2 dm^{-6}. In a saturated solution of silver chloride in water, $[Ag^+(aq)] = [Cl^-(aq)]$, and since their product $= 1 \times 10^{-10}$ mol^2 dm^{-6}, each of them must equal $\sqrt{(1 \times 10^{-10})} = 1 \times 10^{-5}$ mol dm^{-3}.

If we have a saturated solution of silver chloride, its solubility will be decreased if we increase either the $[Ag^+(aq)]$ or the $[Cl^-(aq)]$ term, for example, by adding silver nitrate or sodium chloride. This will cause silver chloride to precipitate out of the solution, by what is known as the **common ion effect**.

For example, if $[Cl^-(aq)]$ is kept at 0.1 mol dm^{-3} by adding sodium chloride, the maximum value for $[Ag^+(aq)]$ is:

$$\frac{K_{sp}}{[Cl^-(aq)]} = \frac{1 \times 10^{-10}}{0.1} = 1 \times 10^{-9}\,\text{mol dm}^{-3}$$

12.5 The dissociation of weak acids and K_a

In Chapter 6 we saw that an acid HA dissociates in water as follows:

$$HA(aq) + H_2O(l) \rightleftharpoons H_3O^+(aq) + A^-(aq)$$

HA and H_3O^+ are Brønsted–Lowry acids (they can donate a proton), and H_2O and A^- are their conjugate bases (they can accept a proton). If the equilibrium lies over to the left, H_3O^+ is a stronger acid than HA and HA is said to be a weak acid.

We may write the equilibrium expression for this reaction as follows.

$$K_c = \frac{[H_3O^+(aq)][A^-(aq)]}{[H_2O(l)][HA(aq)]}$$

We saw on page 240 that pure water has a concentration of 55.5 mol dm^{-3}. In dilute aqueous soution, $[H_2O(l)]$ differs little from this value and so may be taken as a constant. Because $[H_3O^+(aq)] = [H^+(aq)]$, we can now define a new constant, the **acid dissociation constant, K_a**, as follows:

The acid dissociation constant K_a for an acid HA:

$$K_a = \frac{[H^+(aq)][A^-(aq)]}{[HA(aq)]}$$

For ethanoic acid, a typical weak acid, $K_a = 1.76 \times 10^{-5}$ mol dm^{-3}.

We can use K_a values to calculate the concentration of hydrogen ions in a solution of a weak acids. For example, in 0.02 mol dm^{-3} ethanoic acid, we have the following concentrations.

	[HA(aq)]	[H⁺(aq)]	[A⁻(aq)]
initial/$mol\,dm^{-3}$	0.02	0	0
equilibrium/$mol\,dm^{-3}$	$(0.020 - x)$	x	x

$$K_a = \frac{x^2}{(0.020 - x)} = 1.7 \times 10^{-5}\,mol\,dm^{-3}$$

The solution to this equation can be simplified by letting $(0.020 - x) \approx 0.020$. This is usually justified, because the degree of ionisation of the weak acid is very small. So:

$$x^2 = 0.020 \times 1.7 \times 10^{-5}$$
$$x = \sqrt{(0.020 \times 1.7 \times 10^{-5})} = 5.8 \times 10^{-4}\,mol\,dm^{-3}$$

(Without the approximation, the answer would have been $5.75 \times 10^{-4}\,mol\,dm^{-3}$. The difference is insignificant because it is within the limits of the accuracy to which the data is given.)

So for an acid whose dissociation constant is K_a and whose initial concentration is $c\,mol\,dm^{-3}$, we may use the formula $[H^+] = \sqrt{(K_a \times c)}$.

Note that this equation can be used only if there are no added $H^+(aq)$ ions or $A^-(aq)$ ions. It cannot be used for calculations involving buffer solutions (see page 247).

For a weak acid:

$[H^+] = \sqrt{(K_a \times c)}$

Example Calculate the hydrogen ion concentration in aqueous $0.0050\,mol\,dm^{-3}$ hydrofluoric acid, for which $K_a = 5.6 \times 10^{-4}\,mol\,dm^{-3}$.

Answer $[H^+] = \sqrt{(K_a \times c)} = \sqrt{(5.6 \times 10^{-4} \times 5.0 \times 10^{-3})} = \mathbf{1.7 \times 10^{-3}\,mol\,dm^{-3}}$

Further practice

1 Calculate the hydrogen ion concentration in the following aqueous solutions:
  a $0.0036\,mol\,dm^{-3}$ methanoic acid, for which $K_a = 1.6 \times 10^{-4}\,mol\,dm^{-3}$
  b $1.3\,mol\,dm^{-3}$ hydrocyanic acid, for which $K_a = 4.9 \times 10^{-10}\,mol\,dm^{-3}$.
2 (Harder) Calculate the hydrogen ion concentration in $2.0 \times 10^{-3}\,mol\,dm^{-3}$ chloroethanoic acid, for which $K_a = 1.3 \times 10^{-3}\,mol\,dm^{-3}$, using
  a the approximate formula $[H^+] = \sqrt{(K_a \times c)}$
  b the accurate formula $[H^+]^2 = K_a(c - [H^+].)$
  (You will need to use the quadratic formula.)
  c Comment on why the two values differ significantly.

12.6 The ionic product of water, K_w, and the pH scale

The ionic product of water

Pure water conducts electricity slightly, so it must be ionised to a small extent:

$$2H_2O(l) \rightleftharpoons H_3O^+(aq) + OH^-(aq)$$

For this reaction:

$$K_c = \frac{[H_3O^+(aq)][OH^-(aq)]}{[H_2O(l)]^2}$$

Because $H_2O(l)$ is a constant, we can define a new constant called the **ionic product**, K_w, such that $K_w = [H^+][OH^-]$. Experimentally, K_w is found to have a value of $1.0 \times 10^{-14}\,mol^2\,dm^{-6}$ at 25 °C.

The ionic product of water, K_w:

$$K_w = [H^+][OH^-] = 1.0 \times 10^{-14}\,mol^2\,dm^{-6} \text{ at } 25\,°C$$

This shows that an acidic solution contains a few hydroxide ions, and that an alkaline solution contains a few hydrogen ions. Even in a concentrated solution of sodium hydroxide there are still some hydrogen ions – the equilibrium of the reaction:

$$H_3O^+(aq) + OH^-(aq) \rightleftharpoons 2H_2O(l)$$

is displaced to the right, but a few $H_3O^+(aq)$ ions remain.

In a neutral solution, $[H^+] = [OH^-]$. In such a solution,

$$[H^+]^2 = 1.0 \times 10^{-14}\,mol^2\,dm^{-6}$$

so

$$[H^+] = \sqrt{(1.0 \times 10^{-14})} = 1.0 \times 10^{-7}\,mol\,dm^{-3}$$

The pH scale

The hydrogen ion concentration is a measure of the acidity of a solution. In different solutions, the hydrogen ion concentration can have an inconveniently wide range of values (from about $1\,mol\,dm^{-3}$ to $1.0 \times 10^{-14}\,mol\,dm^{-3}$). It was for this reason that the **pH scale** was introduced. This scale is defined by the expression:

$$pH = -\log_{10}[H^+]$$

(or more precisely as:

$$pH = -\log_{10}([H^+]/mol\,dm^{-3})$$

because logarithms can be taken only of a quantity without units).

$$pH = -\log_{10}[H^+]$$

So in a neutral solution, in which $[H^+] = 1.0 \times 10^{-7}\,mol\,dm^{-3}$,

$$\log_{10}[H^+] = -7.0$$
$$pH = 7.0$$

In an acidic solution, for example, one in which $[H^+] = 3.0 \times 10^{-2}\,mol\,dm^{-3}$,

$$\log_{10}[H^+] = -1.52$$
$$pH = 1.52$$

To find the pH of an alkaline solution, for example, one in which $[OH^-] = 2.5 \times 10^{-3}\,mol\,dm^{-3}$, we must first work out $[H^+]$ using K_w:

$$[H^+][OH^-] = 1.0 \times 10^{-14}\,mol^2\,dm^{-6}$$

$$[H^+] = \frac{1.0 \times 10^{-14}}{2.5 \times 10^{-3}} = 4.0 \times 10^{-12}\,mol\,dm^{-3}$$

$$\log_{10}[H^+] = -11.40$$
$$pH = 11.40$$

The pH of strong acids and bases

We can calculate the pH of a solution of a strong acid or a strong base if we assume it to be completely ionised. For a monoprotic acid, that is, one that releases only one H^+ ion per molecule of acid, $[H^+] = [acid]$. For a monoprotic base, that is, one that reacts with only one H^+ ion, $[OH^-] = [base]$.

For a strong diprotic acid such as sulphuric acid, the situation is complicated by the fact that although the first ionisation is that of a strong acid, the second is not:

$$H_2SO_4 \rightleftharpoons H^+ + HSO_4^- \quad K_a \text{ very large}$$
$$HSO_4^- \rightleftharpoons H^+ + SO_4^{2-} \quad K_a = 1.0 \times 10^{-2}\,\text{mol dm}^{-3}$$

As an approximation, we can ignore the second ionisation and treat sulphuric acid as a strong monoprotic acid.

For strong diprotic bases such as barium hydroxide, $Ba(OH)_2$, we may assume complete ionisation and so $[OH^-] = 2 \times [\text{base}]$.

pH and activity

The measured pH of a concentrated solution of a strong acid is never as low as the calculated value, even though the acid may be completely ionised. This is because the ions are close together and interact with each other. This makes their effective concentration, called the **activity**, less than the actual concentration. For example, a solution of $1.0\,\text{mol dm}^{-3}$ hydrochloric acid, which theoretically has pH = 0, has an actual pH of 0.09. Only at concentrations less than $0.01\,\text{mol dm}^{-3}$ do activities and concentrations have similar values.

This activity effect means that concentrated solutions of strong acids never have a pH more negative than −0.3, and concentrated solutions of strong bases never have a pH greater than 14.3. In this chapter, however, we shall ignore this activity correction, even for high concentrations.

pH and pK_a

The symbol 'p' means '$-\log_{10}$'. It can also be used with K_a and other similar equilibrium constants:

$$pK_a = -\log_{10} K_a \text{ (or, more strictly, } -\log_{10} (K_a/\text{mol dm}^{-3}))$$

For ethanoic acid, $K_a = 1.76 \times 10^{-5}\,\text{mol dm}^{-3}$ and $pK_a = 4.75$. Sometimes 'pOH' is used to represent '$-\log_{10} [OH^-]$'. Notice that pH + pOH = 14.

Example Calculate the pH of the following solutions:

a a $0.050\,\text{mol dm}^{-3}$ solution of hydriodic acid, HI (a strong acid)

b a $0.30\,\text{mol dm}^{-3}$ solution of hydrofluoric acid, HF, $K_a = 5.6 \times 10^{-4}\,\text{mol dm}^{-3}$

c a $0.40\,\text{mol dm}^{-3}$ solution of sodium hydroxide.

Answer **a** Because HI is a strong acid, $[HI] = [H^+] = 0.050\,\text{mol dm}^{-3}$.

$\log_{10} [H^+] = -1.3$ and pH = **1.3**

b $[H^+] = \sqrt{(K_a \times c)} = \sqrt{(5.6 \times 10^{-4} \times 0.30)}$
$= 1.30 \times 10^{-2}\,\text{mol dm}^{-3}$

$\log_{10} [H^+] = -1.9$ and pH = **1.9**

c Because $[H^+][OH^-] = 1.0 \times 10^{-14}\,\text{mol}^2\,\text{dm}^{-6}$,

$$[H^+] = \frac{1.0 \times 10^{-14}}{0.40}$$
$$= 2.5 \times 10^{-14}\,\text{mol dm}^{-3}$$

$\log_{10} [H^+] = -13.6$ and pH = **13.6**

Further practice Calculate the pH of the following solutions:

1 $0.0025\,\text{mol dm}^{-3}$ butanoic acid, $K_a = 1.5 \times 10^{-5}\,\text{mol dm}^{-3}$

2 a saturated solution of calcium hydroxide, $Ca(OH)_2$, with a concentration of $0.015\,\text{mol dm}^{-3}$

3 (Harder) a $1.0 \times 10^{-7}\,\text{mol dm}^{-3}$ solution of hydrochloric acid. (Don't ignore the H^+ from the water, and use the charge balance relationship $[H^+] = [Cl^-] + [OH^-]$.)

12.7 The dissociation of weak bases and K_b

A weak base may be either a neutral molecule, such as ammonia, or a negatively charged ion, such as the ethanoate ion, $CH_3CO_2^-$. A weak base forms a weakly alkaline solution:

$$B(aq) + H_2O(l) \rightleftharpoons BH^+(aq) + OH^-(aq)$$
$$B^-(aq) + H_2O(l) \rightleftharpoons BH(aq) + OH^-(aq)$$

By analogy with K_a

$$K_b = \frac{[BH^+][OH^-]}{[B]} \quad \text{or} \quad \frac{[BH][OH^-]}{[B^-]}$$

To calculate the pH of the solution containing a weak base, we use a similar approximation to that used for a weak acid, giving us:

$$[OH^-] = \sqrt{(K_b \times c)}$$

There is a simple relationship between K_b for a base and K_a, the acid dissociation constant of the conjugate acid.

BH^+ is the conjugate acid of B and BH is the conjugate acid of B^-. For the equilibria:

$$BH^+ \rightleftharpoons B + H^+ \quad \text{and} \quad BH \rightleftharpoons B^- + H^+$$

$$K_a = \frac{[H^+][B]}{[BH^+]} \quad \text{or} \quad \frac{[H^+][B^-]}{[BH]}$$

If we multipy K_b by K_a, we have:

$$K_b \times K_a = \frac{[BH^+][OH^-]}{[B]} \times \frac{[H^+][B]}{[BH^+]} \quad \text{or} \quad \frac{[BH][OH^-]}{[B^-]} \times \frac{[H^+][B^-]}{[BH]}$$

$$= [OH^-][H^+] = K_w$$

So if K_b is not given, it is possible to calculate it from K_a for the conjugate acid. For example, K_a for the NH_4^+ ion is 5.6×10^{-10} mol dm^{-3}.

$$K_a \times K_b = K_w, \text{ so } K_b = \frac{K_w}{K_a}$$

$$K_b \text{ for } NH_3 = \frac{10^{-14}}{5.6 \times 10^{-10}} = 1.8 \times 10^{-5} \text{ mol dm}^{-3}$$

The dissociation constant K_b for a weak base B or B^-:

$$K_b = \frac{[BH^+][OH^-]}{[B]} \quad \text{or} \quad \frac{[BH][OH^-]}{[B^-]}$$

$$K_a \times K_b = K_w$$

$$[OH^-] = \sqrt{(K_b \times c)}$$

Example 1 Calculate the pH of a 0.50 mol dm^{-3} solution of ammonia, NH_3, for which $K_b = 1.8 \times 10^{-5}$ mol dm^{-3}.

Answer $[OH^-] = \sqrt{(K_b \times c)} = \sqrt{(1.8 \times 10^{-5} \times 0.50)} = 3.0 \times 10^{-3}$ mol dm^{-3}

$$[H^+] = \frac{K_w}{[OH^-]} = \frac{1.0 \times 10^{-14}}{3.0 \times 10^{-3}} = 3.3 \times 10^{-12} \text{ mol dm}^{-3}$$

$\log_{10}[H^+] = -11.5$ and pH = **11.5**

Example 2 K_a for ethanoic acid is 1.76×10^{-5} mol dm^{-3}.

a Calculate K_b for the ethanoate ion.

b Calculate the pH of a 0.0010 mol dm^{-3} solution of sodium ethanoate.

Answer **a** $K_b = \frac{K_w}{K_a} = \frac{1.0 \times 10^{-14}}{1.76 \times 10^{-5}} = 5.7 \times 10^{-10}$ mol dm^{-3}

b $[OH^-] = \sqrt{(K_b \times c)} = \sqrt{(5.7 \times 10^{-10} \times 0.0010)} = 7.5 \times 10^{-7}$ mol dm^{-3}

$$[H^+] = \frac{K_w}{[OH^-]} = \frac{1.0 \times 10^{-14}}{7.5 \times 10^{-7}} = 1.3 \times 10^{-8} \text{ mol dm}^{-3}$$

$\log_{10}[H^+] = -7.9$ and pH = **7.9**

12.8 Buffer solutions

What is a buffer?

Many experiments, particularly in biochemistry, need to be carried out in solutions of constant pH. Although it is impossible to make a solution whose pH is totally unaffected by the addition of small quantities of acid or alkali, it is possible to make a solution, called a **buffer solution**, whose pH remains *almost* unchanged. A moderately concentrated solution of a strong acid or alkali behaves in this way, and can be used to provide solutions of nearly constant pH in the 0–2 or 12–14 ranges. Dilute solutions of strong acids and bases are useless as buffers between pH 2 and pH 12 because the addition of small quantities of acid or alkali changes their pH considerably. Because this intermediate range of pH is very useful, particularly for biochemical experiments, other means of making buffer solutions have been devised.

A weak acid, by itself, acts as a poor buffer solution because its pH drops sharply when a small quantity of acid is added. A mixture of a weak acid and its salt, however, behaves as a good buffer solution. Because there is a high concentration of $A^-(aq)$, the equilibrium:

$$\underset{\text{reservoir of acid}}{HA(aq)} \rightleftharpoons H^+(aq) + \underset{\text{reservoir of base}}{A^-(aq)}$$

is well over to the left, which increases the pH of the solution. When a small amount of strong acid is added, most of the extra $H^+(aq)$ ions react with the reservoir of $A^-(aq)$, and this tends to minimise the decrease in pH. When a small amount of strong base is added, most of the extra $OH^-(aq)$ ions react with the reservoir of $HA(aq)$, and this tends to minimise the increase in pH.

Example Write equations to show how the buffer solution described above removes:

a added H^+ ions

b added OH^- ions.

Answer **a** $A^-(aq) + H^+(aq) \rightarrow HA(aq)$

b $HA(aq) + OH^-(aq) \rightarrow A^-(aq) + H_2O(l)$

Finding the pH of a buffer solution

To calculate the pH of a buffer solution of known composition, we start with the equation:

$$K_a = \frac{[H^+(aq)][A^-(aq)]}{[HA(aq)]}$$

Because the presence of excess $A^-(aq)$ suppresses the ionisation of HA, the equilibrium concentrations of $A^-(aq)$ and $HA(aq)$ are almost identical to their initial concentrations. We have, therefore,

$$K_a = \frac{[H^+(aq)][A^-(aq)]_{initial}}{[HA(aq)]_{initial}}$$

This is the equation used for buffer solution calculations. Note that the equation $[H^+] = \sqrt{(K_a \times c)}$ cannot be used in this case, because $[H^+] \neq [A^-]$.

Other weak acid/weak base pairs act as buffer solutions. In general, for a buffer solution:

$$[H^+](aq)] = \frac{K_a[acid]_{initial}}{[base]_{initial}}$$

A **buffer solution** is one whose pH remains nearly constant on the addition of small quantities of acid or base. For a buffer solution:

$$[H^+(aq)] = \frac{K_a[acid]_{initial}}{[base]_{initial}}$$

For a buffer solution note that:

$$[H^+(aq)] \neq \sqrt{(K_a \times c)}$$

Example Calculate the pH of a solution made by mixing 100 cm³ of 0.10 mol dm⁻³ ethanoic acid, $K_a = 1.76 \times 10^{-5}$ mol dm^{-3}, with 100 cm^3 of 0.20 mol dm^{-3} sodium ethanoate.

Answer After mixing, the volume is 200 cm^3 and so the solution contains 0.050 mol dm^{-3} ethanoic acid and 0.10 mol dm^{-3} sodium ethanoate.

$$[H^+(aq)] = \frac{K_a[HA(aq)]_{initial}}{[A^-(aq)]_{initial}} = \frac{1.7 \times 10^{-5} \times 0.050}{0.10} = 8.5 \times 10^{-6} \text{ mol dm}^{-3}$$

pH = **5.1**

Further practice

1 A buffer solution is made by mixing 100 cm^3 of 0.20 mol dm^{-3} Na_2HPO_4 with 500 cm^3 of 0.30 mol dm^{-3} NaH_2PO_4. For the equilibrium:

$$H_2PO_4^-(aq) \rightleftharpoons H^+(aq) + HPO_4^{2-}(aq),\ K_a = 6.3 \times 10^{-7} \text{ mol dm}^{-3}$$

Calculate the pH of the solution.

2 A buffer solution is made by mixing 50 cm^3 of 0.10 mol dm^{-3} ammonia with 50 cm^3 of 0.20 mol dm^{-3} ammonium chloride. K_b for ammonia is 1.8×10^{-5} mol dm^{-3}. Calculate:

a K_a for the ammonium ion

b the pH of the buffer solution.

3 (Harder) Calculate the change in pH when 0.050 cm^3 of 1.00 mol dm^{-3} hydrochloric acid is added to:

a 100 cm^3 of a solution of 0.000 10 mol dm^{-3} hydrochloric acid

b 100 cm^3 of a solution that contains 0.10 mol dm^{-3} ethanoic acid and 0.10 mol dm^{-3} sodium ethanoate.

(Hint: assume all the hydrochloric acid reacts with ethanoate ions.)

K_a for ethanoic acid is 1.7×10^{-5} mol dm^{-3}.

Using buffer solutions

There are many situations where it is essential to control the pH. In the following examples, just the addition of an acid or an alkali is sufficient to keep the pH in the required range without buffering.

- In swimming baths, sterilisation is usually carried out using chlorine. This makes the water acidic (see Chapter 17, page 324). Solid calcium hydroxide is added to bring the pH to just above 7.
- Acidic soils are treated with calcium carbonate or calcium hydroxide.
- For acid-loving plants, ammonium sulphate may be added to the soil. This acts as a weak acid and can bring the pH of the soil to pH 4, which is suitable for many azaleas and rhododendrons.

In other cases, more accurate control of pH is essential. In the human body, the production of carbon dioxide by respiration lowers the pH of the blood from about pH 7.5 to pH 7.3. The pH of blood does not fall below this value because blood acts as an efficient buffer. There are several ways in which this buffering is accomplished, of which the following are the most important.

- Proteins are made up of amino acids (see Chapter 30). These contain both acidic and basic groups and can therefore act as buffers:

$$R{-}NH_2 + H^+ \rightarrow R{-}NH_3^+$$
$$R{-}CO_2H + OH^- \rightarrow R{-}CO_2^- + H_2O$$

- The acid H_2CO_3, derived from dissolved carbon dioxide, is buffered by the presence of the hydrogencarbonate ion, HCO_3^-.

Further practice Write equations to show how the H_2CO_3/HCO_3^- buffer system reacts with:
1 added H^+ ions 2 added OH^- ions.

Many other processes require strict pH control, including the use of shampoos and other hair treatments, developing photographs, medical injections and fermentation. This pH control is achieved by the use of an appropriate buffer solution.

12.9 Titration curves, indicators and K_{in}

Titration curves for different acids and alkalis

During a titration, acid is usually run into alkali in the presence of an indicator. This indicator suddenly changes colour at the end-point. This is the point at which the number of moles of acid is exactly balanced by the same amount of alkali. But as we shall see, this does not always mean that the solution at this point has a pH of 7.

Either the acid or the alkali, or both, may be strong or weak, so that there are four possible combinations. If an acid is run into an alkali, the pH changes as shown in Figure 12.11.

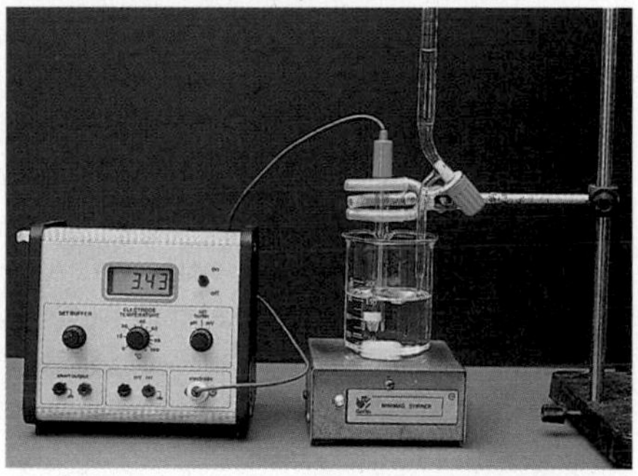

■ **Figure 12.10** (right) In this set-up to record titration curves, the reaction mixture is stirred magnetically and the change in pH recorded on the meter.

■ **Figure 12.11** (below) The graphs show the pH changes as 1.0 mol dm^{-3} acid is run into 100 cm^3 of 0.10 mol dm^{-3} alkali. The end-point comes after 10 cm^3 of acid have been added. The purpose of making the acid ten times as concentrated as the alkali is that there is then very little pH change due to dilution.

a

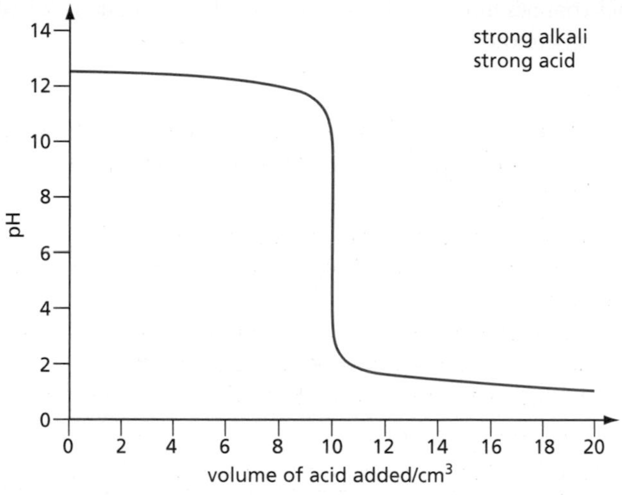

b

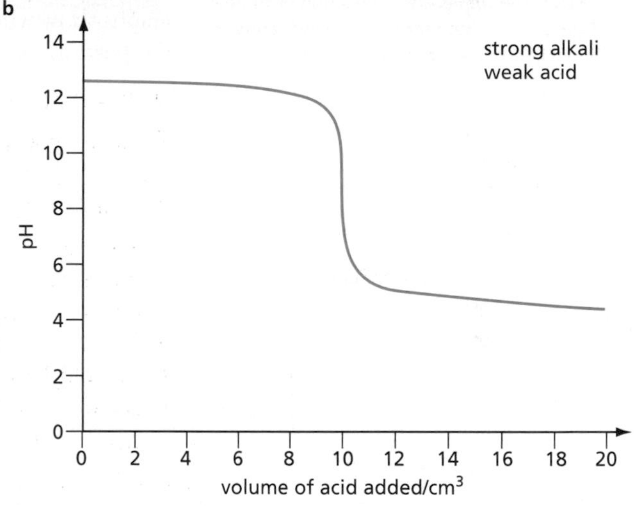

c

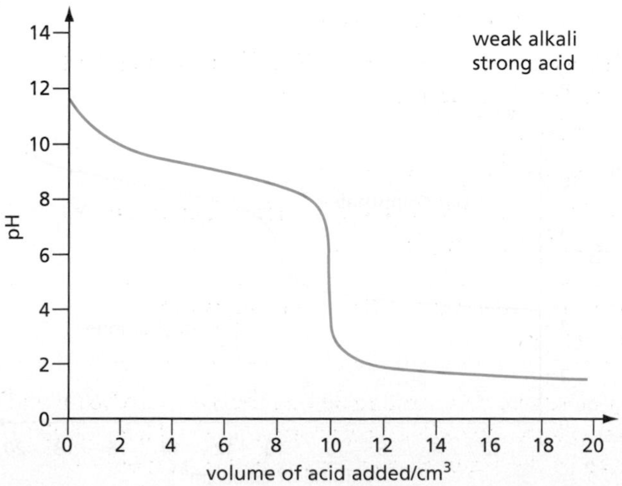

d

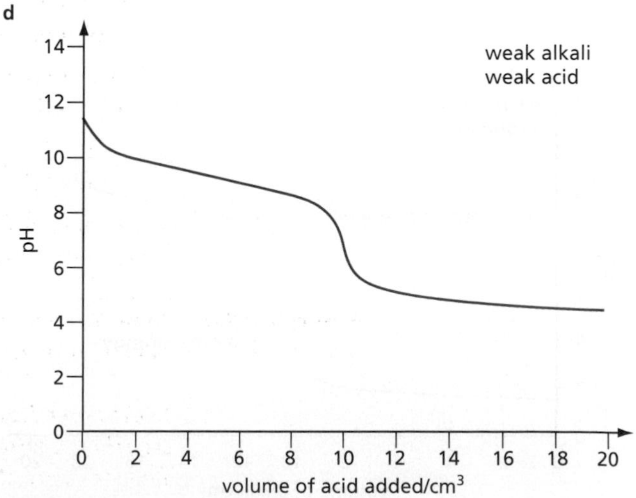

Alkali	Acid	Indicator
Strong	Strong	Methyl orange or phenolphthalein
Strong	Weak	Phenolphthalein
Weak	Strong	Methyl orange
Weak	Weak	Neither

■ **Table 12.2**
The suitability of methyl orange and phenolphthalein as indicators for titrations between strong and weak acids and alkalis.

With a strong alkali, for example sodium hydroxide, the pH starts at 13 and then decreases slowly until $10\,cm^3$ of acid have been added. At this point, the **equivalence point**, there is a sharp drop in pH, to about pH 3 with a strong acid or to about pH 5 with a weak acid. With a weak alkali, for example ammonia, the pH starts at about 11 and falls with an S-shaped curve to about pH 7 until $10\,cm^3$ of acid have been added. At this point there is a drop, sharply with a strong acid to about pH 2 and slowly with a weak acid to about pH 5.

To find the equivalence point accurately, an indicator must be chosen that changes colour when the curve is steepest. The colour will then change when only one further drop of acid is added. Two common indicators are methyl orange and phenolphthalein. Methyl orange changes colour over the pH range 3.2 to 4.4 and phenolphthalein over the pH range 8.2 to 10.0. The titration curves show us that methyl orange is suitable for titrations between a strong or weak alkali and a strong acid, and phenolphthalein is suitable for titrations between a strong alkali and a weak or strong acid. Neither indicator gives a clear end-point, that is when there is a distinct colour change, in a titration between a weak alkali and a weak acid. This is summarised in Table 12.2.

Figure 12.12 shows the titration curves when alkali is added to acid, rather than the other way around. The pH ranges over which methyl orange and phenolphthalein change colour are also shown.

The steep portions of Figures 12.11 and Figure 12.12 are the parts where the pH changes most rapidly with added alkali or acid. It is at these points that an indicator will change colour most rapidly. Conversely, the flat portions of the graphs show where the pH changes most slowly on addition of acid or alkali. These flat portions are, therefore, where the best buffering action occurs.

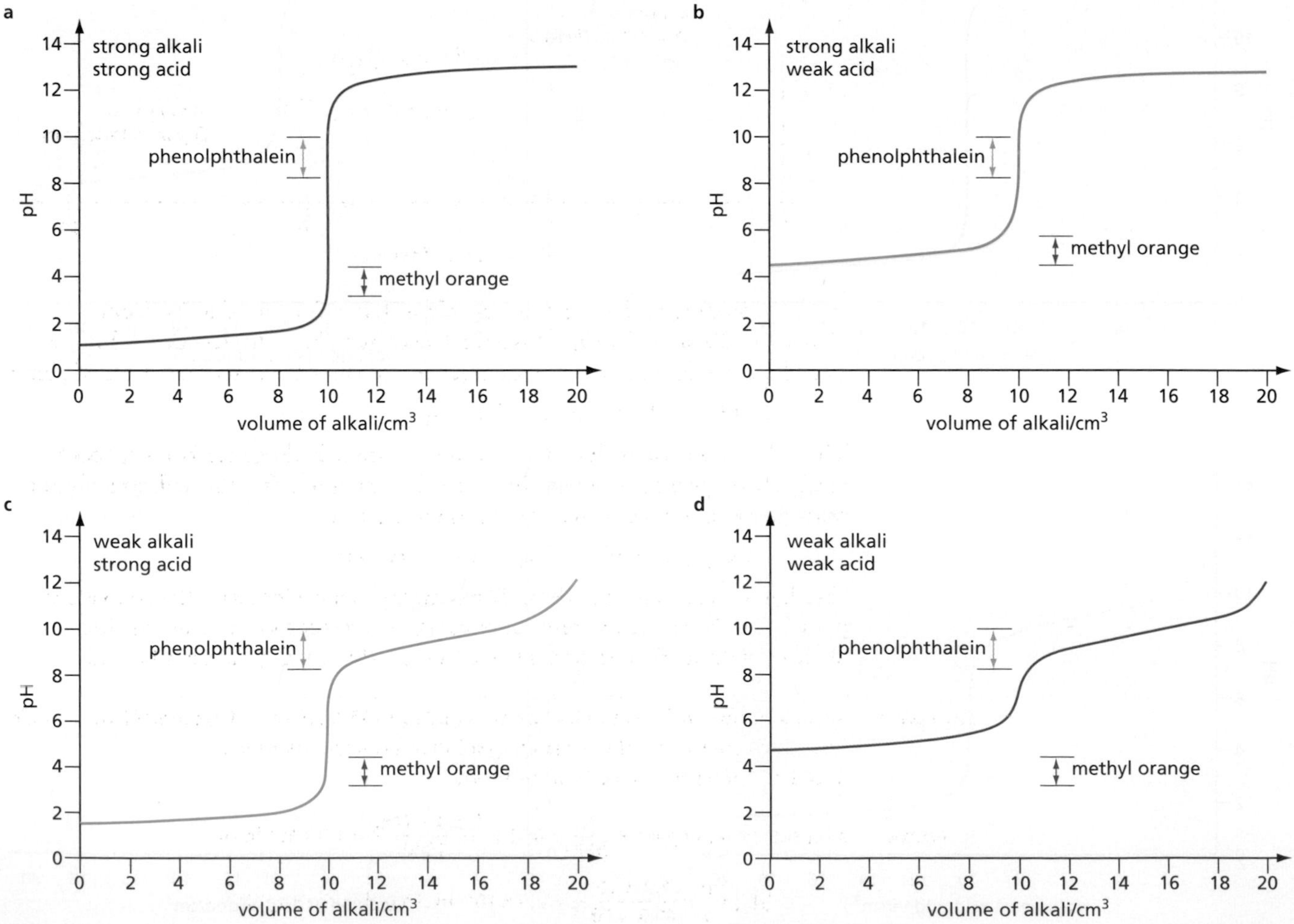

■ **Figure 12.12**
Titration curves starting with $100\,cm^3$ of $0.10\,mol\,dm^{-3}$ acid and adding $1.0\,mol\,dm^{-3}$ alkali. The curves are a reflection of those in Figure 12.11. The ranges of the indicators are shown beside the steep portions of the curves.

The acid dissociation constant for an indicator

Why do methyl orange and phenolphthalein change colour over different pH ranges? An indicator is a weak acid, whose acid form, HIn, is a different colour from its ionised form, In^-. HIn for methyl orange and phenolphthalein have different acid dissociation constants K_a. For an indicator, K_a is termed K_{in}, and pK_a becomes pK_{in}.

	$HIn(aq) \rightleftharpoons H^+(aq) + In^-(aq)$		pK_{in}
methyl orange:	red	yellow	3.7
phenolphthalein:	colourless	pink	9.3

At a pH equal to pK_{in}, the two forms HIn and In^- are present in equal concentrations. The eye does not easily detect this mid-point, but sees instead the first permanent change in colour. This may occur when one form has ten times the concentration of the other form. This ten-fold excess means that the eye detects the colour change over a range of about one pH unit either side of the mid-point.

Titrating a diprotic base

If a diprotic base is titrated, there are two steep portions on the graph. For example, Figure 12.13 shows how the pH changes when sodium carbonate solution is titrated with hydrochloric acid.

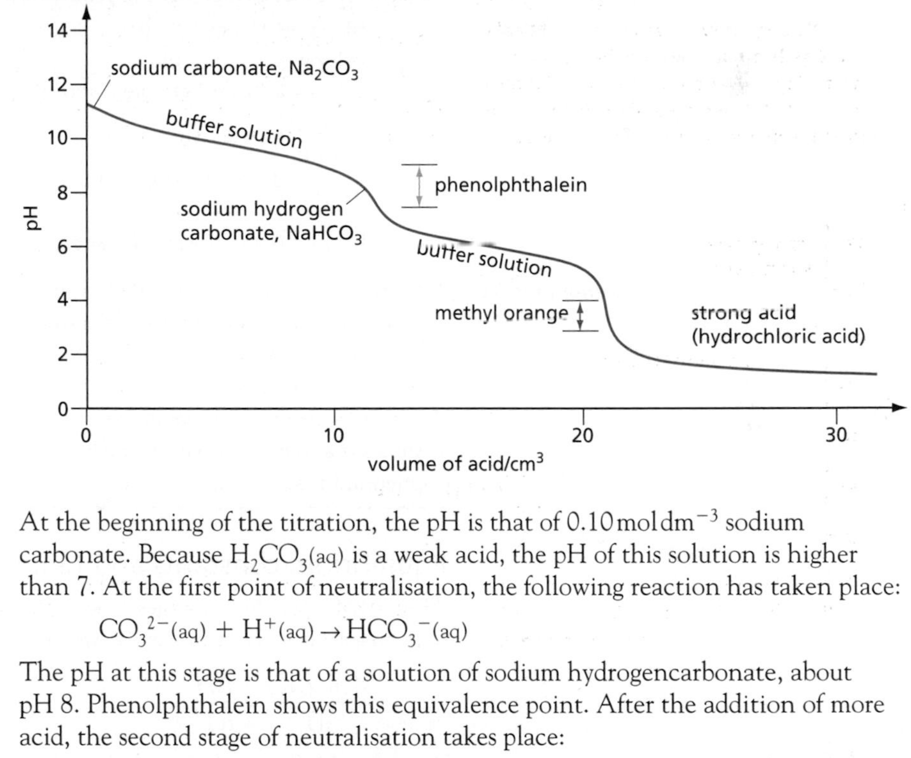

Figure 12.13 Titration curve for the addition of 1.0 mol dm^{-3} hydrochloric acid to 0.10 mol dm^{-3} sodium carbonate. There are two steep portions to the graph.

At the beginning of the titration, the pH is that of 0.10 mol dm^{-3} sodium carbonate. Because $H_2CO_3(aq)$ is a weak acid, the pH of this solution is higher than 7. At the first point of neutralisation, the following reaction has taken place:

$$CO_3^{2-}(aq) + H^+(aq) \rightarrow HCO_3^-(aq)$$

The pH at this stage is that of a solution of sodium hydrogencarbonate, about pH 8. Phenolphthalein shows this equivalence point. After the addition of more acid, the second stage of neutralisation takes place:

$$HCO_3^-(aq) + H^+(aq) \rightarrow CO_2(aq) + H_2O(l)$$

This happens at a much lower pH, and methyl orange indicates the equivalence point. This shows the importance of using the correct indicator, as the titre obtained with methyl orange is twice that obtained with phenolphthalein.

Example 1

Some 0.10 mol dm^{-3} sodium hydroxide is added to 25.0 cm^3 of 0.10 mol dm^{-3} hydrochloric acid. Calculate the pH of the resulting solution after the addition of:

a 24.9 cm^3 of alkali **b** 50.0 cm^3 of alkali.

Answer

a $n(\text{HCl})\text{ remaining} = c \times \frac{V}{1000} = 0.10 \times \frac{(25.0 - 24.9)}{1000} = 1.0 \times 10^{-5}\text{ mol}$

$[H^+] = \frac{n}{V} = \frac{1.0 \times 10^{-5}}{49.9 \times 10^{-3}} = 2.0 \times 10^{-4}\text{ mol dm}^{-3}$ and pH = **3.7**

b The acid has now all reacted, so we need to calculate the pH from the concentration of hydroxide ions in the solution:

$$n(\text{NaOH})\text{ remaining} = c \times \frac{v}{1000} = 0.10 \times \frac{(50.0 - 25.0)}{100} = 2.5 \times 10^{-3}\,\text{mol}$$

volume of solution = 25.0 + 50.0 = 75.0 cm^3

$$[\text{OH}^-] = \frac{n}{v} = \frac{2.5 \times 10^{-3}}{75.0 \times 10^{-3}} = 0.033\,\text{mol dm}^{-3}$$

$$[\text{H}^+] = \frac{K_w}{[\text{OH}^-]} = \frac{10^{-14}}{0.033}$$

$[H^+] = 3.0 \times 10^{-13}$ mol dm^{-3} and pH = **12.5**

Example 2 The indicator bromophenol blue has $pK_{in} = 4.0$ and the indicator thymol blue has $pK_{in} = 8.9$. State which indicator(s) could be used for titrating:

a potassium hydroxide with ethanoic acid

b calcium hydroxide with nitric acid.

Answer **a** For a strong alkali–weak acid titration, the steep part of the titration curve is between pH 12 and pH 6, so choose thymol blue.

b For a strong alkali–strong acid titration, the steep part of the titration curve is between pH 12 and pH 2 so either indicator could be used.

Example 3 Bromophenol blue is yellow in acid and blue in alkali. Calculate the pH of a solution that contains 10 times as much of the yellow form as the blue form.

Answer

$$\underset{\text{yellow}}{\text{HIn(aq)}} \rightleftharpoons \text{H}^+\text{(aq)} + \underset{\text{blue}}{\text{In}^-\text{(aq)}}$$

$pK_{in} = 4.0$ (from *Example 2* above), so $K_{in} = 1.0 \times 10^{-4}$

$$\frac{[\text{H}^+][\text{In}^-]}{[\text{HIn}]} = 1.0 \times 10^{-4}$$

$$[\text{H}^+] = \frac{1.0 \times 10^{-4} \times [\text{HIn}]}{[\text{In}^-]} = \frac{1.0 \times 10^{-4} \times 10}{1}\,\text{mol dm}^{-3}$$

$[H^+] = 1.0 \times 10^{-3}$ mol dm^{-3} and pH = **3.0**

Further practice

1 Some 0.50 mol dm^{-3} hydrochloric acid is added to 25.0 cm^3 of 0.50 mol dm^{-3} potassium hydroxide. Calculate the pH of the solution after the addition of:

a 12.5 cm^3 of acid

b 25.1 cm^3 of acid.

2 When 10 drops of phenolphthalein are added to a solution, the intensity of the colour shows that 6 drops are in the pink form. Calculate the pH of the solution. pK_{in} for phenolphthalein is 9.3.

Equilibrium constants and temperature

The only factor that changes an equilibrium constant is a change in temperature. The quantitative relationship between K and T is given by the **van't Hoff equation**:

$$\frac{d(\log_e K)}{dT} = \frac{\Delta H}{RT^2}$$

Because ΔH does not vary much with temperature, this equation is often used in the integrated form:

$$\log_{10} K_{T_2} - \log_{10} K_{T_1} = \frac{\Delta H}{2.30R}\left(\frac{1}{T_1} - \frac{1}{T_2}\right)$$

where a change in temperature from T_1 to T_2 changes the equilibrium constant from K_{T_1} to K_{T_2}.

When

ΔH is positive, $K_{T_2} > K_{T_1}$

ΔH is negative, $K_{T_2} < K_{T_1}$

as predicted by Le Chatelier's Principle.

Example K_c for the reaction:

$$N_2O_4 \rightleftharpoons 2NO_2$$

is 0.050 mol dm^{-3} at 300 K and $\Delta H = +57$ kJ mol^{-1}. Calculate K_c at 400 K.

Answer

$$\log_{10} K_{T_2} - \log_{10} K_{T_1} = \frac{\Delta H}{2.30R}\left(\frac{1}{T_1} - \frac{1}{T_2}\right)$$

$$\log_{10} K_{T_2} = \frac{57\,000}{2.30 \times 8.31}(0.00333 - 0.00250) + \log_{10} 0.050$$

$$= 2.48 - 1.30$$

$$K_{T_2} = \mathbf{15\,mol\,dm^{-3}}$$

Further practice K_w is 1.008×10^{-14} at 300 K and 7.296×10^{-14} mol^2 dm^{-6} at 330 K. Calculate ΔH for the ionisation of water.

Summary

- The position of equilibrium is affected by changes in concentration/pressure and temperature.
- The **equilibrium constant** changes only with a change in temperature.
- For the general reaction:

$$wA + xB + \ldots \rightleftharpoons yC + zD + \ldots$$

$$K_c = \frac{[C]^y[D]^z\ldots}{[A]^w[B]^x\ldots} \quad \text{and} \quad K_p = \frac{[p_C]^y[p_D]^z\ldots}{[p_A]^w[p_B]^x\ldots}$$

- If the number of moles of the reactants at the start is known, analysis of one of the substances at equilibrium is sufficient to calculate the amounts of all the others using the stoichiometric equation.
- In aqueous solution, the product $[H^+][OH^-]$ is a constant, the **ionic product of water**, K_w, whose value is 1.0×10^{-14} mol^2 dm^{-6} at room temperature.
- **pH** $= -\log_{10} [H^+]$
- For strong acids, $[H^+] = c$, the concentration of the acid. For strong bases, $[OH^-] = c$, the concentration of the base.
- A weak acid HA is incompletely dissociated. Its strength is measured by the **acid dissociation constant**, K_a:

$$K_a = \frac{[H^+][A^-]}{[HA]}\,\text{mol dm}^{-3}$$

  For a weak acid, $[H^+] = \sqrt{(K_a \times c)}$.
- For a weak base BOH, the dissociation constant K_b is given by:

$$K_b = \frac{[B^+][OH^-]}{[BOH]}\,\text{mol dm}^{-3}$$

  For a weak base, $[OH^-] = \sqrt{(K_b \times c)}$.
- $K_w = K_a \times K_b$
- A **buffer solution** resists changes of pH on the addition of small quantities of acid or base.
  A buffer solution is often a mixture of a weak acid and its salt. For such a buffer solution:

$$[H^+] = \frac{K_a[\text{acid}]_{\text{initial}}}{[\text{salt}]_{\text{initial}}}$$

- During a titration there is a sudden change in pH; this causes an indicator to change colour rapidly. The choice of indicator depends on the strengths of the acid and alkali being titrated.

13 *Electrochemistry*

In this chapter, we study how a chemical reaction can produce electricity. The standard electrode potential, $E^{\ominus}$, of an element is a measure of its tendency to form ions, or to change from one oxidation state to another. The size and sign of $E^{\ominus}$ indicates the feasibility of a reaction. In contrast, many reactions that are not otherwise feasible can be carried out by electrolysis – by passing an electric current through the reaction mixture. By choosing appropriate conditions, a wide variety of useful chemicals can be manufactured.

The chapter is divided into the following sections.

13.1 The electrochemical cell

Measuring electrode potentials

If a metal, for example zinc, is placed in a solution containing its ions, an equilibrium is set up:

$$Zn(s) \rightleftharpoons Zn^{2+}(aq) + 2e^-$$

Initially the metal begins to dissolve and becomes negatively charged, but soon an equilibrium is established in which the rate of dissolving is balanced by ions recovering electrons and reforming the metal (see Figure 13.1). The more reactive the metal, the further to the right is the position of equilibrium and the larger the negative charge on the metal. Therefore measuring this charge gives us a measure of the reactivity of the metal.

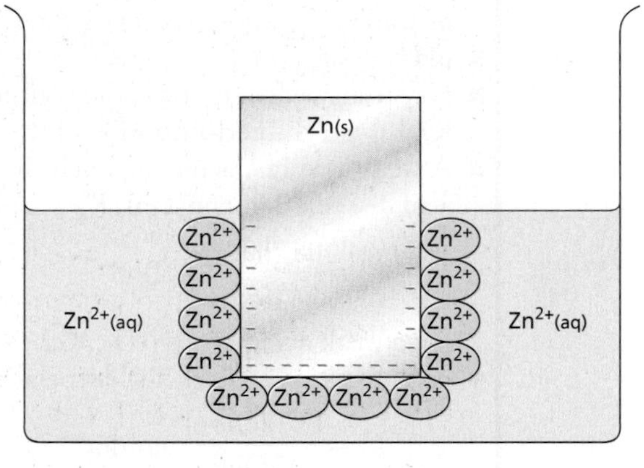

■ Figure 13.1 The equilibrium established by zinc metal in contact with its ions. A double layer of charges is formed at the surface of the metal.

However, we cannot measure this charge simply by connecting one terminal of a voltmeter to the metal, because a voltmeter measures **potential difference** (p.d.) and needs its other terminal to be connected to a second electrode at a different electrical potential. It would be convenient if this second electrode could be made by putting an inert metal, for example platinum, into the solution. Unfortunately this does not help, even if no current flows, for platinum sets up its own potential. This platinum potential varies from experiment to experiment, so inconsistent results are obtained. Another approach must be used.

The solution to the problem is to regard the zinc/zinc ion system as a **half-cell** and connect it to another half-cell. This will allow us to measure the potential difference between the two cells accurately and consistently. This means that we can only compare reactivities, rather than measuring them absolutely. The other half-cell could be, for example, a copper/copper ion system (see Figure 13.2).

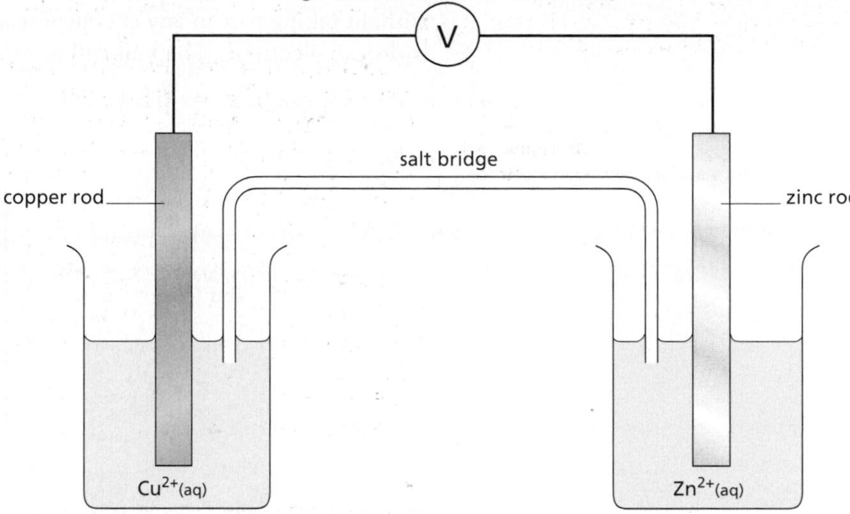

■ **Figure 13.2** Two half-cells are combined to make an electrochemical cell.

The voltage produced by the two half-cells depends on the conditions. If the conditions are standard (298 K, 1.0 atm, solutions of 1.0 mol dm^{-3}), the voltage is the **standard cell e.m.f.**, $E^{\ominus}_{cell}$, for this cell.

The need for a salt bridge

The voltage is measured using a high-resistance electronic voltmeter. This takes very little current so that the voltage measured is the highest possible value for these two half-cells. Even so, some electrons must flow and, because zinc is more reactive than copper, the following half-reactions tend to take place:

at the zinc electrode: $Zn(s) \rightarrow Zn^{2+}(aq) + 2e^-$
at the copper electrode: $Cu^{2+}(aq) + 2e^- \rightarrow Cu(s)$

This means that the zinc *solution* becomes positively charged, with an excess of $Zn^{2+}(aq)$ ions, and the copper *solution* becomes negatively charged with, for example, an excess of $SO_4^{2-}(aq)$ ions. Because of this, the current would cease unless the circuit were completed by electrically connecting the two solutions. This cannot be done with a piece of wire, which passes only electrons, because we need to move positive ions one way and negative ions the other. (It does not matter which ions actually move, because very few of them are transferred compared with those already in the solutions.) The circuit is completed by a **salt bridge** (see Figure 13.3) dipping into the two solutions. This may contain porous plugs that allow ions to flow while minimising the mixing of the electrolytes by diffusion. The electrolyte in the salt bridge is usually potassium nitrate. This is used because all potassium compounds and all nitrates are soluble, and so no precipitate will form with any ions in contact with it.

■ **Figure 13.3** Copper and zinc half-cells connected together.

The standard hydrogen electrode

For the cell described, $E^{\ominus}_{cell} = -1.075\,V$, the negative sign showing that the zinc electrode is more negative than the copper electrode. This fits in with what we know of the relative reactivities of copper and zinc. This value tells us the voltage of a standard zinc half-cell compared with a standard copper half-cell. Conventionally, it has been agreed to compare half-cells against an electrode called a **standard hydrogen electrode**. By international agreement, this has been assigned $E^{\ominus} = 0.00\,V$. The reaction is:

$$H^+(aq) + e^- \rightleftharpoons \tfrac{1}{2}H_2(g)$$

This is a very slow reaction, and to speed it up the hydrogen is bubbled over a platinum catalyst. The platinum surface is **platinised** – it has a spongy layer of platinum electrolysed onto it, which increases its effective surface area. The platinum is also an inert electrode, and transfers electrons to and from the circuit without taking part in any chemical reaction. Figure 13.4 shows the standard hydrogen electrode. The half-cell reaction is represented as:

$$H^+(aq) + e^- \rightleftharpoons \tfrac{1}{2}H_2(g), Pt$$

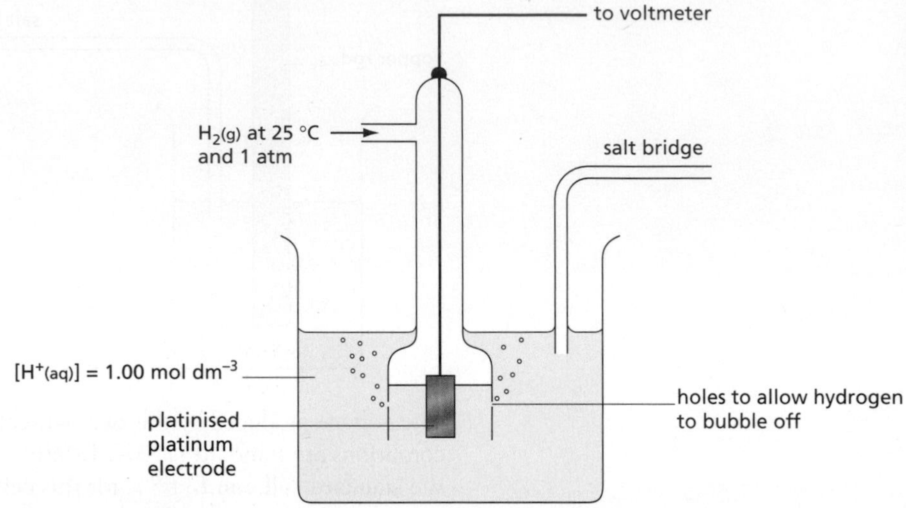

■ **Figure 13.4** The standard hydrogen electrode.

Standard electrode potential, $E^\ominus$

All half-cell reactions are written in the following format, with the electrons on the left:

oxidised form + ne^- ⇌ reduced form

If the voltage produced by a half-cell is measured against a standard hydrogen electrode, the voltage is the **standard electrode potential**, $E^\ominus$, of the half-cell.

A **standard hydrogen electrode** is the half-cell represented by the equation $H^+(aq) + e^- \rightarrow \tfrac{1}{2}H_2(g)$, Pt under standard conditions.

Standard conditions are: $p = 1$ atm, $T = 298$ K and all concentrations = 1 mol dm^{-3}.

The reduced form of the substance is not necessarily the free element – it may be the element in a lower oxidation state. For example, in a solution containing 1.0 mol dm^{-3} of Fe^{2+}(aq) ions and 1.0 mol dm^{-3} of Fe^{3+}(aq) ions, electrons are transferred from the Fe^{2+}(aq) ions to the Fe^{3+}(aq) ions. In order to measure the p.d. for this half-cell, these electrons can be carried away on a platinum wire placed in the solution. A shiny platinum wire is used for this electrode, because it is not acting as a catalyst. The half-cell is written:

$$Fe^{3+}(aq) + e^- \rightleftharpoons Fe^{2+}(aq), Pt$$

and a diagram of it is shown in Figure 13.5.

A common error is to draw this cell with two electrodes, one dipping into Fe^{2+}(aq) ions and the other dipping into Fe^{3+}(aq) ions. However, this setup would not give the required value of $E^\ominus$ because a standard electrode potential must be measured relative to a standard hydrogen electrode.

Many standard electrode potentials have been measured, and a few of the more important ones are listed in Table 13.1.

These standard electrode potentials are sometimes called redox potentials. A large negative value of $E^\ominus$ indicates a highly reactive metal that is easily oxidised. A large positive value of $E^\ominus$ indicates a highly reactive non-metal that is easily reduced. The order of these $E^\ominus$ values matches the observed chemical reactivity of the substances concerned, though the large negative value for lithium is surprising (see Chapter 16, page 318).

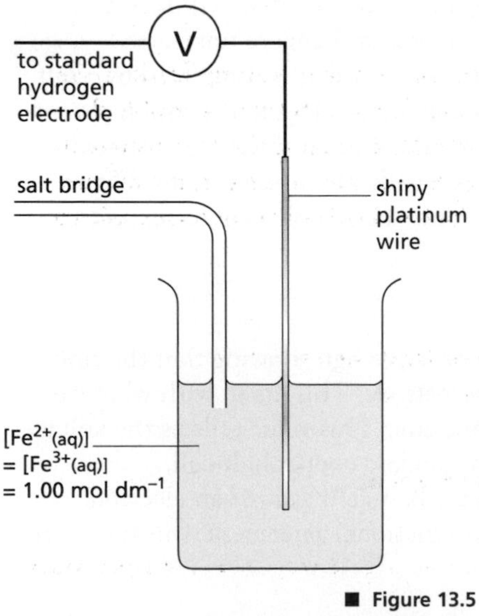

■ **Figure 13.5** The half-cell used to measure $E^\ominus$ for the Fe^{3+}/ Fe^{2+} system. [Fe^{3+}] and [Fe^{2+}] are shown as 1 mol dm^{-3}. This is not essential – the important fact is that they are equal, as we shall see later in the chapter.

Table 13.1
Values of $E^\ominus$ for some common half-reactions. In water, in which $[H^+(aq)] = 10^{-7}\,mol\,dm^{-3}$ (that is, the conditions are non-standard), $E(H^+/\frac{1}{2}H_2) = -0.41\,V$ and $E(\frac{1}{2}O_2/H_2O) = +0.82\,V$. These non-standard values are used later in the chapter.

	Oxidised form		Reduced form	$E^\ominus$/V
weakest oxidising agents ↓		strongest reducing agents ↑		
	$Li^+(aq)$		$Li(s)$	−3.03
	$K^+(aq)$		$K(s)$	−2.92
	$Na^+(aq)$		$Na(s)$	−2.71
	$Mg^{2+}(aq)$		$Mg(s)$	−2.37
	$Al^{3+}(aq)$		$Al(s)$	−1.66
	$Zn^{2+}(aq)$		$Zn(s)$	−0.76
	$Fe^{2+}(aq)$		$Fe(s)$	−0.44
	$H^+(aq)$		$\frac{1}{2}H_2(g)$, Pt	0.00
	$Cu^{2+}(aq)$		$Cu(s)$	+0.34
	$\frac{1}{2}I_2(aq)$		$I^-(aq)$, Pt	+0.54
	$Fe^{3+}(aq)$		$Fe^{2+}(aq)$, Pt	+0.77
	$Ag^+(aq)$		$Ag(s)$	+0.80
	$\frac{1}{2}Br_2(aq)$		$Br^-(aq)$, Pt	+1.09
	$\frac{1}{2}O_2(g)$, Pt + $2H^+(aq)$		$H_2O(l)$	+1.23
	$\frac{1}{2}Cl_2(aq)$		$Cl^-(aq)$, Pt	+1.36
	$MnO_4^-(aq) + 8H^+(aq)$		$Mn^{2+}(aq)$, Pt	+1.51
	$\frac{1}{2}F_2(aq)$		$F^-(aq)$, Pt	+2.87
strongest oxidising agents		weakest reducing agents		

Cell diagrams

Instead of sketching the two half-cells, a **cell diagram** can be drawn. Figure 13.2 (page 255) is a sketch of a cell in which a zinc half-cell is compared with a copper half-cell. The cell diagram for these two half-cells is:

$$Cu(s) \mid Cu^{2+}(aq) \;\vdots\; Zn^{2+}(aq) \mid Zn(s)$$

The solid vertical lines indicate a phase boundary, with the reduced forms on the outside, and the dotted line represents the salt bridge. (Sometimes a double dotted line is used for the salt bridge.) The concentrations of the solutions are assumed to be $1.00\,mol\,dm^{-3}$ unless a different concentration is shown underneath the ions. The e.m.f. of this cell is that of a zinc rod (right-hand half-cell) compared with a copper rod (left-hand half-cell).

Example Write a cell diagram for:

a a magnesium electrode compared with a standard hydrogen electrode

b a $Fe^{3+}(aq)/Fe^{2+}(aq)$ half-cell compared with a zinc half-cell.

Answer **a** Pt, $\frac{1}{2}H_2(g) \mid H^+(aq) \;\vdots\; Mg^{2+}(aq) \mid Mg(s)$

Note that some textbooks write the Pt, $\frac{1}{2}H_2(g)$ as Pt $[\frac{1}{2}H_2(g)]$.

b $Zn(s) \mid Zn^{2+}(aq) \;\vdots\; Fe^{3+}(aq), Fe^{2+}(aq) \mid Pt$

Here $Fe^{3+}(aq)$ and $Fe^{2+}(aq)$ are separated by a comma because they are in the same phase.

Further practice Write cell diagrams for:

1 a $Cr^{3+}(aq)/Cr^{2+}(aq)$ half-cell compared with a copper half-cell

2 a $MnO_4^-(aq)/Mn^{2+}(aq)$ half-cell (include $H^+(aq)$ ions and $H_2O(l)$) compared with a bromine half-cell.

Calculating the standard e.m.f. of a cell, $E^\ominus_{cell}$

The values of $E^\ominus$ shown in Table 13.1 (page 257) can be used to work out the standard e.m.f. of a cell, $E^\ominus_{cell}$. Conventionally this is the voltage of the right-hand half-cell measured against the left-hand half-cell. So:

The standard e.m.f. of a cell

$$E^\ominus_{cell} = E^\ominus(\text{right-hand half-cell}) - E^\ominus(\text{left-hand half-cell})$$

Example 1 Calculate $E^\ominus_{cell}$ for a copper electrode compared with a zinc electrode.

Answer $E^\ominus_{cell} = +0.34 - (-0.76) = \mathbf{+1.10\,V}$

(It is a good idea always to include the sign.)

Example 2 Calculate $E^\ominus_{cell}$ for the following cell:

$$Pt\,|\,Fe^{2+}(aq),\ Fe^{3+}(aq)\ \vdots\ Ag^{+}(aq)\,|\,Ag(s)$$

Answer $E^\ominus_{cell} = +0.80 - (+0.77) = \mathbf{+0.03\,V}$

Further practice Write the cell diagrams and calculate $E^\ominus_{cell}$ for the following cells:

1 a sodium electrode compared with a zinc electrode

2 a chlorine electrode compared with a bromine electrode

3 an oxygen electrode compared with a copper electrode.

13.2 Using $E^\ominus_{cell}$ values to measure feasibility

When the two electrodes of a cell are connected to each other, electrons flow from the negative electrode to the positive electrode. Chemical changes take place at each electrode that reduce the voltage produced by the cell (the terminal p.d.) and finally the reaction stops because the two electrodes are at the same electrical potential. If $E^\ominus_{cell}$ is positive, the reaction corresponding to the cell diagram takes place from left to right. For example, the cell diagram:

$$Zn(s)\,|\,Zn^{2+}(aq)\ \vdots\ Cu^{2+}(aq)\,|\,Cu(s) \quad E^\ominus_{cell} = +1.10\,V$$

corresponds to the reaction:

$$Zn(s) + Cu^{2+}(aq) \rightarrow Zn^{2+}(aq) + Cu(s)$$

and, as $E^\ominus_{cell}$ is positive, the reaction is feasible and should take place if zinc is added to copper sulphate solution. In this case, experiment confirms that the prediction is correct. In other cases, this does not happen. There are two reasons why this may be so, and we shall look at each in turn.

Non-standard conditions

When determining $E^\ominus_{cell}$, all solutions have a concentration of $1.0\,mol\,dm^{-3}$. This is unlikely to be the case if the reaction is carried out in a test tube. For the zinc/copper sulphate reaction, the voltage is so large and positive that changing the concentrations has no effect. However, the situation is different if $E^\ominus_{cell}$ is less than $+0.2\,V$ – under these conditions a fairly small change in concentration could make the voltage become negative and the reaction is then not feasible. This effect is treated quantitatively in Section 13.3.

High activation energy

Many reactions, particularly those involving gases and solutions, have a high activation energy. Although the $E^\ominus_{cell}$ value may indicate that the reaction is feasible, it is often so slow that no change is observed if it is carried out in the test tube. For example, $E^\ominus$ for Cu^{2+}/Cu is $+0.34\,V$. This suggests that hydrogen gas should react with copper sulphate solution, but no reaction is observed. The slowness of the conversion of gases into ions is the reason why platinised platinum is used in the standard hydrogen electrode.

Just as the conversion of H_2(g) to H^+(aq) ions has a high activation energy, so too does the reverse process. This means that the production of hydrogen (and oxygen as well) by electrolysis often requires a greater voltage than is predicted by the $E^\ominus$ values. This **overvoltage** is mentioned again later in this chapter.

Example Use $E^\ominus$ values from Table 13.1 to calculate $E^\ominus_{cell}$ for the following reactions, and hence predict whether they are feasible:

a $Zn(s) + Mg^{2+}(aq) \rightarrow Zn^{2+}(aq) + Mg(s)$

b $\frac{1}{2}Cl_2(aq) + H_2O(l) \rightarrow \frac{1}{2}O_2(g) + Cl^-(aq) + 2H^+(aq)$

Answer a $E^\ominus_{cell} = E^\ominus(Mg^{2+}/Mg) - E^\ominus(Zn^{2+}/Zn) = -2.37 - (-0.76)$

$= \mathbf{-1.61\,V}$: not feasible

b $E^\ominus_{cell} = E^\ominus(\frac{1}{2}Cl_2/Cl^-) - E^\ominus(\frac{1}{2}O_2/H_2O) = +1.36 - (+1.23)$

$= \mathbf{+0.13\,V}$: feasible

Cl^- and H_2O are the reduced forms.

Further practice Use $E^\ominus$ values from Table 13.1 to calculate $E^\ominus_{cell}$ and hence predict whether the following reactions are feasible.

(You may find Chapter 7, pages 144–5, helpful in writing the relevant half-equations.)

1 $\frac{1}{2}H_2(g) + Ag^+(aq) \rightarrow H^+(aq) + Ag(s)$

2 $MnO_4^-(aq) + 5Cl^-(aq) + 8H^+(aq) \rightarrow Mn^{2+}(aq) + 4H_2O(l) + 2\frac{1}{2}Cl_2(aq)$

3 $Fe(s) + 2Fe^{3+}(aq) \rightarrow 3Fe^{2+}(aq)$

4 $Ag^+(aq) + Fe^{2+}(aq) \rightarrow Fe^{3+}(aq) + Ag(s)$

Corrosion

■ **Figure 13.6** An untreated iron gate rusts when exposed to air and water.

All reactive metals **corrode** – they react with oxygen and water in the air. Some form a thin layer of oxide (for example, aluminium and chromium) that protects them from further attack. Iron readily **rusts** to form hydrated iron(III) oxide, $Fe_2O_3.H_2O$ or FeO.OH. Unlike the oxides of aluminium and chromium, this does not stick well to the metal surface. It easily flakes off and exposes more of the surface to further corrosion.

Rusting is a complex electrochemical process that takes place most readily under the following conditions:

- some of the iron is in contact with air while other regions are not
- the iron is in contact with water containing salt or other ionic substances.

In contact with air, the following reaction takes place:

$$1\tfrac{1}{2}O_2 + 3H_2O + 6e^- \rightarrow 6OH^- \qquad E^\ominus = +0.81\,V \text{ at pH 7}$$

In the air-free region, iron dissolves:

$$2Fe \rightarrow 2Fe^{3+} + 6e^- \qquad E^\ominus = -0.04\,V$$

The overall equation is:

$$2Fe(s) + 1\tfrac{1}{2}O_2(g) + H_2O(l) \rightarrow 2FeO.OH(s)$$

The electrons flow in the metal from the air-rich to the air-free region. This takes place more quickly if the resistance of the solution that completes the circuit is lowered by the addition of salt.

Corrosion can be minimised in several ways. These include the following:

- painting or covering in plastic to exclude air
- coating iron with another metal, for example zinc, tin or chromium
- alloying, for example with chromium
- fixing to the iron a more reactive metal, such as magnesium. The reactive metal preferentially dissolves and the process is therefore called **sacrificial protection** or **anodic protection**. Magnesium dissolves, having a more negative $E^\ominus$ value (−2.37 V) than iron, keeping the iron negative and discouraging the reaction:

$$Fe \rightarrow Fe^{3+} + 3e^-$$

13.3 Non-standard conditions

In Figure 13.2 (page 255), we showed a zinc half-cell connected to a copper half-cell. Under standard conditions, $E^{\ominus}_{cell} = +1.10\,V$ and, when the cell passes current, electrons flow from the zinc to the copper rod. If we now reduce the concentration of the zinc ions in the zinc half-cell, the equilibrium:

$$Zn^{2+}(aq) + 2e^- \rightleftharpoons Zn(s)$$

is displaced to the left and the negative charge on the zinc rod becomes bigger. This can be predicted from Le Chatelier's Principle (see Chapter 9, page 177), and it always happens when the concentration of the oxidised form is reduced below the standard value of $1.0\,mol\,dm^{-3}$.

Reducing the concentration of metal ions (the oxidised form) in a half-cell makes its potential more negative.

Example For the following cells, state qualitatively the effect on $E^{\ominus}_{cell}$ of making the change described.

a $Mg(s)\,|\,Mg^{2+}(aq) \,\vdots\, Zn^{2+}(aq)\,|\,Zn(s)$; reducing $[Zn^{2+}(aq)]$

b $Ag(s)\,|\,Ag^{+}(aq) \,\vdots\, \frac{1}{2}Cl_2(aq), Cl^{-}(aq)\,|\,Pt$; adding Cl^- ions to the silver half-cell

Answer **a** When $[Zn^{2+}]$ falls, the zinc electrode becomes more negative. It is the nearer to the potential of the magnesium electrode, so the overall e.m.f. is reduced.

b The addition of Cl^- ions gives a precipitate of silver chloride. This means that $[Ag^+]$ falls, making the silver electrode less positive. Its potential is then further away from the more positive chlorine electrode, so the e.m.f. is increased.

The Nernst equation

The quantitative relationship showing how E_{cell} changes under non-standard conditions is given by the **Nernst equation**:

$$E = E^{\ominus} + \frac{RT}{zF}\log_e [M^{n+}]$$

where M^{n+} is the metal ion and F is the Faraday constant (see page 264). At 25 °C, and using $\log_{10}$, this equation can be written in the form:

$$E = E^{\ominus} + \frac{0.0590}{z}\log_{10} [M^{n+}]$$

For example, a zinc rod dipping into a solution containing $0.0100\,mol\,dm^{-3}$ of $Zn^{2+}(aq)$ ions has an electrode potential given by:

$$E = 0.76 + \frac{0.0590}{2}\log_{10} [0.0100] = -0.76 + \frac{0.0590}{2} \times (-2.0) = -0.82\,V$$

The difference between $E^{\ominus}$ and E is not large, and the fact that conditions are not standard is not usually important when making predictions from electrode potentials. There are, however, two situations when concentration effects should be taken into account:

- if $E^{\ominus}_{cell} < 0.2\,V$
- if the concentration of the ions is very small.

The second case can arise if most of the aqueous ions are precipitated out or made into a complex.

The equations given above apply to metals dipping into solutions of their ions. In general:

$$E = E^{\ominus} + \frac{0.0590}{z}\log_{10} \frac{[oxidised]}{[reduced]}$$

When measuring $E^\ominus$ for a system such as:

$$Fe^{3+}(aq) + e \rightleftharpoons Fe^{2+}(aq), Pt$$

it is not necessary that $[Fe^{3+}(aq)]$ and $[Fe^{2+}(aq)]$ are both $1.0\,mol\,dm^{-3}$. It is sufficient that they are equal. This is sometimes relevant because many substances, for example potassium manganate(VII), have a maximum solubility much less than $1.0\,mol\,dm^{-3}$.

Example Calculate E_{cell} for the following cell measured against a standard hydrogen electrode:

$$Cu^{2+}(aq),\ 0.0050\,mol\,dm^{-3}/Cu(s).$$

Answer

$$E = E^\ominus + \frac{0.0590}{z}\log_{10}[M^{n+}]$$

$$= +0.34 + \frac{0.0590}{2} \times (-2.30)$$

$$= \mathbf{+0.27\,V}$$

Further practice For the following cell:

$$Zn(s)\,|\,Zn^{2+}(aq) \,\vdots\, Cu^{2+}(aq)\,|\,Cu(s)$$

calculate the change in $E^\ominus_{cell}$ when:

1 $[Zn^{2+}(aq)]$ is reduced from $1.0\,mol\,dm^{-3}$ to $0.0010\,mol\,dm^{-3}$
2 both $[Zn^{2+}(aq)]$ and $[Cu^{2+}(aq)]$ are reduced to $0.032\,mol\,dm^{-3}$
3 sodium hydroxide solution is added to the copper half-cell until $[OH^-(aq)] = 0.0010\,mol\,dm^{-3}$. $[K_{sp}(Cu(OH)_2) = 2.0 \times 10^{-19}\,mol^3\,dm^{-9}]$

The pH meter

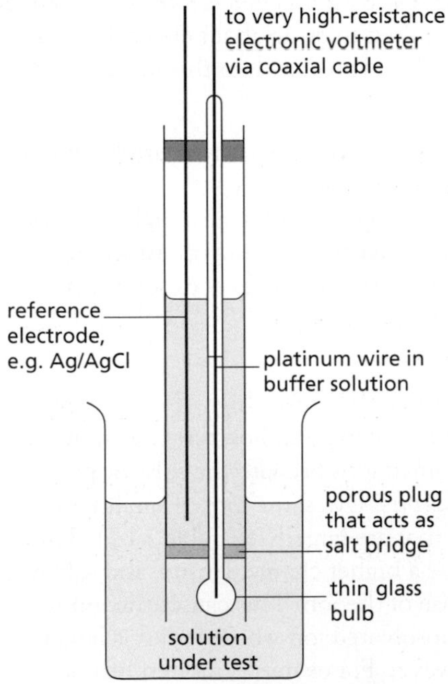

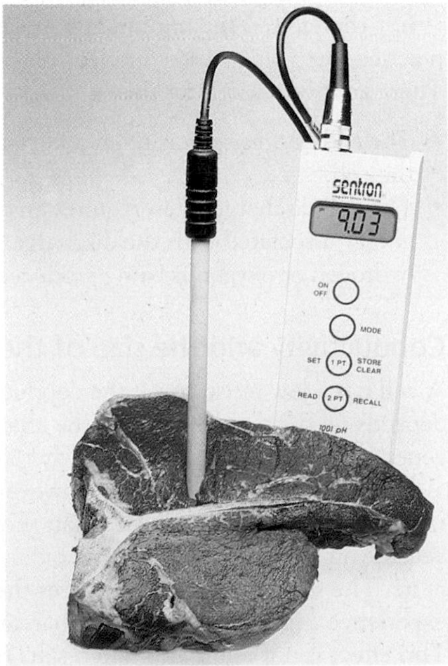

The Nernst equation forms the theoretical basis of the **pH meter**. The probe in a pH meter contains two electrodes. One is a reference electrode, whose potential remains constant – it is often a silver wire dipping into a saturated solution of silver chloride. The other electrode could be a hydrogen electrode, but as this is extremely difficult to set up, a more convenient electrode is often used. A specially constructed thin glass bulb selectively allows the passage of hydrogen ions, and this is the main feature of the **glass electrode** (see Figure 13.7). The two electrodes are connected by a porous plug that acts as a salt bridge.

■ **Figure 13.7** (above left) A pH meter electrode assembly. Because the glass electrode has a very high resistance, the electronic voltmeter has to measure currents less than 1×10^{-9} A.

■ **Figure 13.8** (above right) A pH meter in use.

For the pH meter, the Nernst equation becomes:

$$E = E^\ominus - 0.0590\ pH$$

The 0.0590 term becomes larger if the temperature is increased from 25 °C, and so the pH meter has to be adjusted for temperature changes. The potential of the reference electrode, $E^\ominus$, is eliminated by calibrating the meter with a buffer solution of known pH.

13.4 Electrolysis

Pushing the cell reaction in the opposite direction

If the electrodes of a cell are connected together, electrons flow from the negative to the positive terminal. If the terminals are instead connected to an external voltage supply and the applied voltage is gradually increased, electrons still flow as before until the external voltage equals E_{cell}. At this voltage no current flows. If the external voltage is increased still further, current flows in the opposite direction and **electrolysis** takes place (see Figure 13.9).

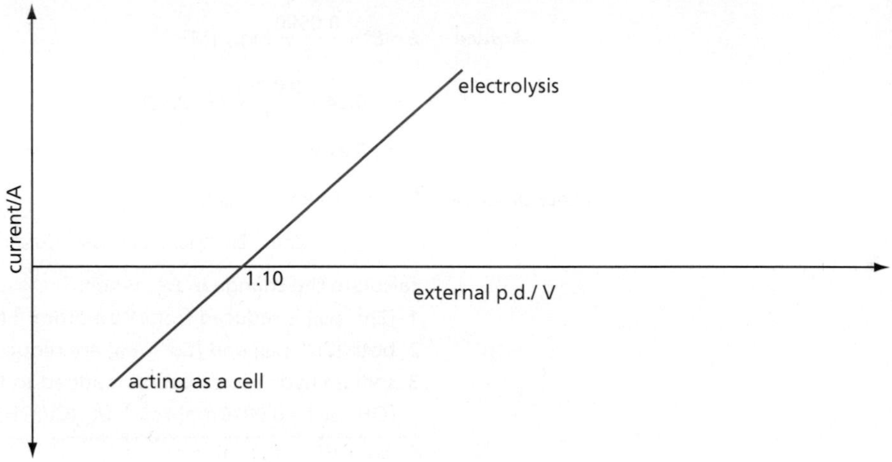

■ **Figure 13.9**
Voltage–current graph for a copper/zinc cell. At 1.10 V, no current flows. Above 1.10 V, electrolysis takes place. The reaction runs in reverse and it is Zn^{2+}(aq) that is reduced.

During electrolysis, positively charged ions, called **cations**, gain electrons at the negatively charged electrode, the **cathode**. At the same time negatively charged ions, called **anions**, give up electrons at the positively charged electrode, the **anode**.

E_{cell} therefore is the minimum voltage required to bring about electrolysis. In practice, the voltage used for electrolysis is always greater than this minimum. There are two reasons for this.

- The cell has resistance, so an additional p.d. is needed to drive current through the cell.
- The cell discharge often requires an overvoltage to overcome a high activation energy associated with the discharge. This is particularly important when hydrogen or oxygen is being produced by electrolysis of aqueous solutions.

Conductivity and the size of the ions

A cell has a low resistance if the conductivity of the ions is high. This conductivity depends on the number of ions, the charge they carry and how fast they move. In general, large ions move more slowly than small ions because the solvent produces a greater viscous drag on them. However, surprisingly, some ions of small ionic radii have lower conductivities than those that are larger (see Table 13.2). This is because an ion of very small ionic radius has a higher charge density, and solvates more. The solvation sphere increases the size of the ion. This causes the ion to experience a greater viscous drag than an unsolvated ion whose radius is larger. This effect is not found in molten salts however. For example, molten lithium chloride has a greater conductivity than molten sodium chloride, the reverse of the situation in aqueous solution.

■ **Table 13.2**
The conductivities of some common ions in aqueous solution. (The unit S is the siemens, the SI unit of conductance.)

Ion	increasing ionic radius →							increasing ionic radius →			
	Li^+	Na^+	K^+	Rb^+	Cs^+	H^+	OH^-	F^-	Cl^-	Br^-	I^-
Molar conductivity/ S cm^2 mol^{-1}	39	50	74	78	77	350	197	55	76	78	77

The very high conductivities of the H^+(aq) ion and the OH^-(aq) ion are surprising. These are the ions derived from the strongly hydrogen-bonded solvent water. It may be that they can transport charge by means of electrons jumping from one solvent molecule to another, rather than by the ions bodily moving through the viscous solvent (see Figure 13.10). In this way their conductivity would be much greater than that expected from their size.

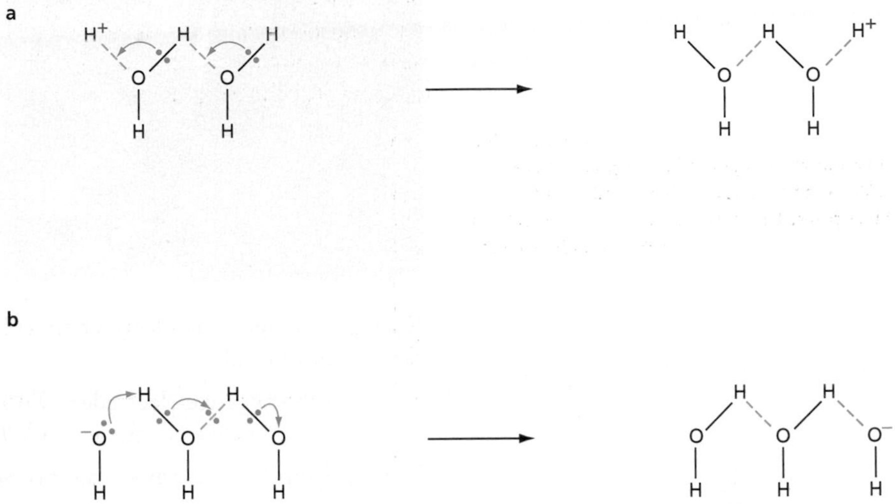

■ **Figure 13.10** The movement of H^+(aq) and OH^-(aq) ions through water by electron shifts.

Selective discharge

In an electrolysis cell, the electrolyte may contain several cations or anions. Under these circumstances, the one that is discharged is the one that requires the least energy. This is called **selective discharge**. For example, if a mixture of copper chloride and zinc chloride is electrolysed, copper rather than zinc is deposited at the cathode. This is because $E^{\ominus}$ for the Cu^{2+}/Cu half-cell is closer in value to $E^{\ominus}$ for the $\frac{1}{2}Cl_2/Cl^-$ half-cell than $E^{\ominus}$ for the Zn^{2+}/Zn half-cell is. This shows that copper requires a smaller potential difference to be discharged than zinc does (Table 13.3).

■ **Table 13.3** In a solution containing copper chloride and zinc chloride, copper rather than zinc is deposited at the cathode because a smaller voltage is needed to bring about the electrolysis.

Cation	Anion	$E^{\ominus}_{cell}$
$E^{\ominus}(Zn^{2+}/Zn) = -0.76\,V$	$E^{\ominus}(\frac{1}{2}Cl_2/Cl^-) = +1.36\,V$	2.12 V
$E^{\ominus}(Cu^{2+}/Cu) = +0.34\,V$	$E^{\ominus}(\frac{1}{2}Cl_2/Cl^-) = +1.36\,V$	1.02 V

In aqueous solutions H_3O^+(aq) and OH^-(aq) ions are present and so selective discharge takes place. If dilute sulphuric acid is electrolysed, for example, hydrogen is liberated at the cathode because the only cation present is the H_3O^+(aq) ion. There are, however, both OH^-(aq) ions and SO_4^{2-}(aq) ions in solution. A study of $E^{\ominus}$ values predicts that oxygen should be liberated at the anode (see Table 13.4):

$$2OH^-(aq) \rightarrow H_2O(l) + \frac{1}{2}O_2(g) + 2e^-$$

■ **Table 13.4** The $E^{\ominus}_{cell}$ values predict that hydrogen and oxygen are evolved in the electrolysis of dilute sulphuric acid.

Cathode	Anode	$E^{\ominus}_{cell}$
$E^{\ominus}(H^+/H) = 0\,V$	$\frac{1}{2}O_2/OH^- = +1.23\,V$ $[OH^-] = 10^{-14}\,mol\,dm^{-3}$	1.23 V
	$\frac{1}{2}S_2O_8^{2-}/SO_4^{2-} = +2.01\,V$	2.01 V

Because both hydrogen and oxygen have a high overvoltage, they are often not discharged even though $E^{\ominus}$ values suggest that they should be. For example, the electrolysis of brine usually yields chlorine rather than oxygen (see Chapter 20, page 380) and it is possible in aqueous solution to electroplate objects with metals (for example, nickel) that have negative values of $E^{\ominus}$ for their half-cells.

The Faraday constant

■ **Figure 13.11**
Michael Faraday 1791–1867. One of his many achievements was to establish a quantitative basis for electrochemistry.

During discharge in an electrochemical cell, the following changes take place in the copper/zinc cell:

at the negative electrode: $Zn(s) \rightarrow Zn^{2+}(aq) + 2e^-$
at the positive electrode: $Cu^{2+}(aq) + 2e^- \rightarrow Cu(s)$

During electrolysis, however, the reverse processes take place:

at the cathode: $Zn^{2+}(aq) + 2e^- \rightarrow Zn(s)$
at the anode: $Cu(s) \rightarrow Cu^{2+}(aq) + 2e^-$

In order to deposit one mole of zinc and dissolve one mole of copper, 2 moles of electrons need to be passed through the cell. One mole of electrons carries a charge of −96 500 C. This is the same charge as the charge on the electron, 1.603×10^{-19} C, multiplied by the Avogadro constant, $6.022 \times 10^{23}\,mol^{-1}$. Numerically, the same amount of charge is carried by one mole of protons; this quantity is called the **Faraday constant** and has value of $+96\,500\,C\,mol^{-1}$.

The **coulomb** (symbol C) is the unit of electrical charge. One coulomb is the amount of charge that is passed when one ampere flows for one second. The amount of a substance dissolved or deposited in an electrolysis cell is given by:

$$\text{amount} = I \times \frac{t}{zF}\ \text{mol}$$

where z is the charge on the ion, I the current in amperes and t the time in seconds.

Example Calculate the mass of copper that dissolves when a current of 0.75 A is passed through a zinc/copper cell for 45 minutes. $A_r(Cu) = 63.5$.

Answer

$$\text{amount} = I \times \frac{t}{zF} = 0.75 \times \frac{45 \times 60}{2 \times 96\,500} = 0.0105\,\text{mol}$$

mass of copper = 0.0105 × 63.5 = **0.64 g**

Further practice

1 Calculate the mass of aluminium produced when a current of 1000 A is passed through a cell containing Al^{3+} ions for 1 hour.

2 a Show that 0.0016 mol of oxygen is produced when a current of 1.0 A is passed through sulphuric acid solution for 10 minutes.
b Calculate the volume of oxygen that should be liberated.
c Suggest why the actual volume collected may be less than this.

3 Explain why:
a hydrogen and oxygen are evolved when aqueous sodium hydroxide is electrolysed
b when aqueous copper sulphate is electrolysed with platinum electrodes, the cathode becomes plated with copper and oxygen is evolved at the anode
c when aqueous copper sulphate is electrolysed with copper electrodes, the anode dissolves.

13.5 Applications of electrolysis

Reactions that are not energetically feasible may be driven by electrolysis. The following are some examples of industrial importance.

Extraction of metals

Reactive metals such as sodium, aluminium, magnesium and lithium are extracted by electrolysis (see Chapter 20, pages 378–9). The use of lightweight alloys containing lithium and magnesium makes the extraction of these metals increasingly important.

Purification of copper

The conductivity of copper increases tenfold when it is more than 99.9% pure. The impure copper is made the anode of an electrolysis cell, and a small strip of pure copper the cathode. A p.d. of about 0.2–0.4 V is applied, and virtually all the energy is used up in passing current through the cell. Any metal ions with $E^{\ominus}$ lower than that for Cu^{2+} dissolve from the anode, but are not discharged and stay in solution. Any metals with $E^{\ominus}$ higher than that for Cu^{2+} do not dissolve from the anode in the first place. They drop off the anode as the copper around them dissolves, and fall to the bottom as 'anode sludge'. This contains rare and useful elements, such as silver, gold and selenium. Subsequently this sludge is removed and the rare elements extracted.

Electroplating

Electroplating has become less important recently, as stainless steel has largely taken over from chromium-plated steel. However, chromium plating is still used for taps and other bathroom fittings, and silver plating is used for decorative items. The silver layer sticks on firmly only if the steel is first plated with nickel (forming electroplated nickel silver, EPNS).

■ **Figure 13.12** Chromium-plated objects. In the past, radiator grills, bumpers and other parts of cars were chromium plated.

In order to produce a thin, even layer of metal during electrolysis, the metal needs to be deposited slowly. This means that the concentration of free metal ions must be low. In chromium plating, the electrolyte is a solution containing $CrO_4^{2-}(aq)$ ions mixed with a low concentration of $Cr^{3+}(aq)$ ions. As the $Cr^{3+}(aq)$ ions are deposited, they are replenished by the reduction of $CrO_4^{2-}(aq)$ ions at the cathode.

at the cathode: $Cr^{3+}(aq) + 3e^- \rightarrow Cr(s)$

$$CrO_4^{2-}(aq) + 8H^+(aq) + 3e^- \rightarrow Cr^{3+}(aq) + 4H_2O(l)$$

In silver plating, a solution containing $Ag(CN)^{2-}(aq)$ ions is used. This is a stable complex and produces a very low concentration of $Ag^+(aq)$ ions. This makes the potential of the silver electrode so much more negative that the solution no longer reacts directly with metals such as copper or nickel.

at the cathode: $Ag(CN)^{2-}(aq) \rightleftharpoons Ag^+(aq) + 2CN^-(aq)$

$$Ag^+(aq) + e^- \rightarrow Ag(s)$$

Manufacture of chemicals

Important chemicals such as sodium hydroxide and sodium chlorate(I) are made by the electrolysis of brine (see Chapter 20, pages 380–2).

Driving reactions by electrolysis

Reactions that are not thermodynamically feasible (that is, those for which either ΔH or ΔG is positive) may be driven by electrolysis. An example is the electrolysis of sodium chloride:

$$NaCl(s) \rightarrow Na(s) + \tfrac{1}{2}Cl_2(g) \quad \Delta H^{\ominus} = +411\,kJ\,mol^{-1}, \Delta G^{\ominus} = +384\,kJ\,mol^{-1}$$

The actual conditions of electrolysis are not standard – the sodium chloride is molten to make it conducting and the temperature is 600 °C. Even so, the free energy change is unfavourable. The minimum external voltage needed to bring about electrolysis can be calculated using the equation:

$$\Delta G^{\ominus} = -zFE^{\ominus}_{cell}$$

If $\Delta G^{\ominus} = +384\,kJ\,mol^{-1}$, $E^{\ominus}_{cell} = 4.0\,V$. In practice a voltage of 6.7 V is used in order to overcome the resistance of the cell.

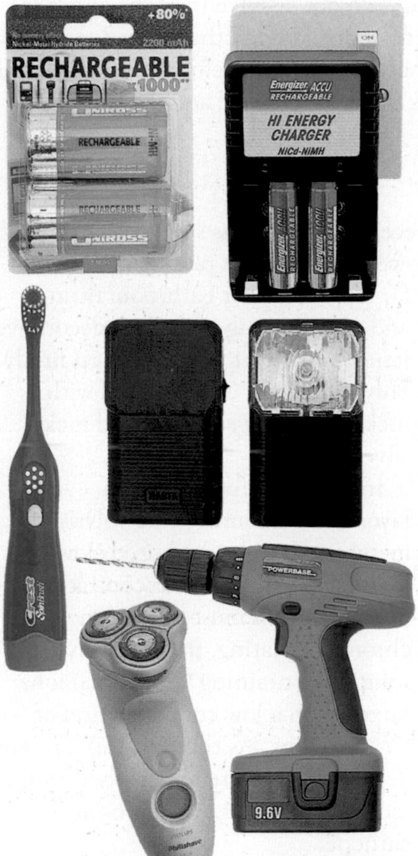

■ **Figure 13.13** Rechargeable batteries. The chemical reactions that take place at each electrode must be reversible. Side reactions limit the number of recharging cycles that can be achieved – for most cells this is about 1000.

Rechargeable batteries

Theoretically, most cells can be recharged when they run down, but in practice there are difficulties if the recharging is carried out quickly. The problem may be that metals are not re-deposited in an even layer, or that gaseous hydrogen and oxygen are given off.

Much money has been spent in producing a high-density, high-energy rechargeable battery to power an electric car, but as yet no completely satisfactory solution has been found.

The familiar lead–acid accumulator (the traditional 'car battery') has lead plates dipping into moderately concentrated sulphuric acid. After the first charging, the positive plate becomes covered with a layer of lead(IV) oxide. During discharge the following reactions take place:

negative plate: $Pb(s) + H_2SO_4(aq) \rightarrow PbSO_4(s) + 2H^+(aq) + 2e^-$

positive plate: $PbO_2(s) + H_2SO_4(aq) + 2H^+(aq) + 2e^- \rightarrow PbSO_4(s) + 2H_2O(l)$

During recharging, the reactions are reversed. Because sulphuric acid is a good conductor of electricity, the internal resistance of the lead–acid cell is very low. This means that it can produce very large currents, making it suitable for powering the starter motor in a car. Its disadvantage is its low power-to-weight ratio.

Another cell being investigated uses molten sulphur and molten sodium. With any cell such as this that operates above room temperature, there are problems of initial heating, and these problems are added to in the sodium–sulphur cell which also uses hazardous chemicals.

There are fewer difficulties when designing small, rechargeable cells. The most familiar are the alkaline cadmium cell and the lithium cell. Their principal defects are their relatively high cost.

For the cadmium/nickel cell:

negative electrode: $Cd + 2OH^- \rightarrow Cd(OH)_2 + 2e^-$ $\quad E^{\ominus} = -0.81\,V$

positive electrode: $NiO_2 + 2H_2O + 2e^- \rightarrow Ni(OH)_2 + 2OH^-$ $\quad E^{\ominus} = +0.49\,V$

Fuel cells

When a fuel is burnt, and the heat is used to drive a piston or to turn the blades of a turbine, the overall efficiency is always low, no matter how well the plant is designed. This is a thermodynamic limitation. The typical efficiency of a motor car is 20%, and that of a power station is 40%. The only way to raise the efficiency is to use a higher operating temperature. This in turn produces fresh problems.

In order to achieve higher efficiencies, efforts have been made to convert the fuel directly into electrical energy by means of a **fuel cell**. The obvious pollution-free fuel is hydrogen, which can be burnt in air to produce water. The reactions are the reverse of the electrolysis of water. In acidic solution, we have:

negative plate: $H_2(g) \rightarrow 2H^+(aq) + 2e^-$

positive plate: $\frac{1}{2}O_2 + 2H^+ + 2e^- \rightarrow H_2O(l)$

and in alkaline conditions:

negative plate: $H_2(g) + 2OH^-(aq) \rightarrow 2H_2O(l) + 2e^-$

positive plate: $\frac{1}{2}O_2 + H_2O(l) + 2e^- \rightarrow 2OH^-(aq)$

The problem is that the reactions at the electrodes are slow, even when the electrodes are made of platinum which acts as a catalyst. There is a further complication for cars in that the hydrogen must be transported, either in heavy cylinders or at very low temperatures as a liquid. Efforts are being made to carry the hydrogen in the form of transition metal hydrides (see Chapter 19, page 366) but these are costly and can absorb only a limited amount of hydrogen.

As an alternative to transporting hydrogen to fuel cars, methanol is a fuel that is easy to carry, and can be broken down to produce carbon dioxide and hydrogen. This introduces an additional stage in the process, and there is also the danger that some poisonous carbon monoxide may be produced in addition to the carbon dioxide. Attempts are therefore being made to produce fuel cells that use methanol directly, but these cells are only partially successful as they have to be operated above room temperature.

In the final analysis, fuel cells are really only another form of storage battery. In the first place, the hydrogen must have been produced chemically, possibly by electrolysis, and this process is not pollution-free. At present, a promising line of development is to use a hybrid system, running the car on electricity in towns where pollution is the main problem, and then, on the open road, use a traditional fuel to recharge the batteries.

Summary

- A metal dipping into a solution of its ions forms a **half-cell**. When this is connected to another half-cell by means of salt bridge, an **electrochemical cell** is set up. Under standard conditions, the voltage set up by this cell is the **standard cell e.m.f.**, $E^\ominus_{cell}$, for this cell.

  $$E^\ominus_{cell} = E^\ominus(\text{right-hand cell}) - E^\ominus(\text{left-hand cell})$$

- If the left-hand half-cell is a **standard hydrogen electrode**, the standard cell e.m.f. is the **standard electrode potential, $E^\ominus$**, for the right-hand half-cell.
- The value of $E^\ominus$ measures the oxidising/reducing power of the half-cell system, and the feasibility of redox reactions can be predicted from $E^\ominus_{cell}$ values.
- A redox reaction that is predicted to be feasible from $E^\ominus_{cell}$ values might not take place because the conditions are non-standard, or because the activation energy is very high.
- **Electrolysis** is the process of driving a reaction that is not thermodynamically feasible, by passing an electric current through it.
- The ions that are discharged in electrolysis are the ones that require the least energy.
- The amount of substance in moles dissolved or deposited during electrolysis is given by:

  $$\text{amount} = I \times \frac{t}{zF}$$

  where I is the current, t the time in seconds, z the charge on the ion and F the Faraday constant.
- Electrolysis has important uses in the chemical industry.

14 Radiation and matter

In this chapter we examine the changes brought about when radiation falls on matter. High-frequency photons have the highest energy and bring about the most energetic changes, often disrupting molecules. Less energetic photons in the infrared part of the spectrum, or photons of radio frequency, are more useful in determining the fine structure of a molecule.

The chapter is divided into the following sections.

14.1 Photons and energy levels

Energy, frequency and wavelength

In Chapter 2 (page 32–3) we introduced the quantum theory. This shows that electromagnetic radiation can be viewed as a wave as well as behaving like a stream of particles. In this chapter, the particle nature of radiation is the more important aspect.

Each photon has associated with it an amount of energy known as a quantum, and the size of this packet of energy is proportional to the frequency of the radiation. This is expressed as:

$$E = hf$$

where E is the energy of the radiation, f (often written as v) is its frequency in hertz, Hz (s^{-1}), and h is the Planck constant (6.62×10^{-34} J s). The frequency, f, of a photon is related to its wavelength, λ, as follows:

$$f = \frac{c}{\lambda}$$

where c is the speed of light ($3 \times 10^{8}\,\mathrm{m\,s^{-1}}$). For example, radiation of 300 nm, in the ultraviolet part of the spectrum, has a frequency of 1×10^{15} Hz:

$$f = \frac{3 \times 10^{8}}{300 \times 10^{-9}} = 1 \times 10^{15}\,\mathrm{Hz}$$

So the energy of a photon of this radiation is given by:

$$E = 6.62 \times 10^{-34} \times 1 \times 10^{15} = 6.62 \times 10^{-19}\,\mathrm{J}$$

This is the energy of just one photon, but we would need a mole of photons to interact with a mole of substance. To find the energy of a mole of photons, we multiply by the Avogadro constant:

$$E = 6.62 \times 10^{-19} \times 6.0 \times 10^{23} = 397\,\mathrm{kJ\,mol^{-1}}$$

This is about the energy of an average chemical bond. Ultraviolet radiation therefore interacts with the outer electrons of atoms, and those in chemical bonds.

The types of interaction brought about by different frequencies of radiation are shown in Figure 14.1.

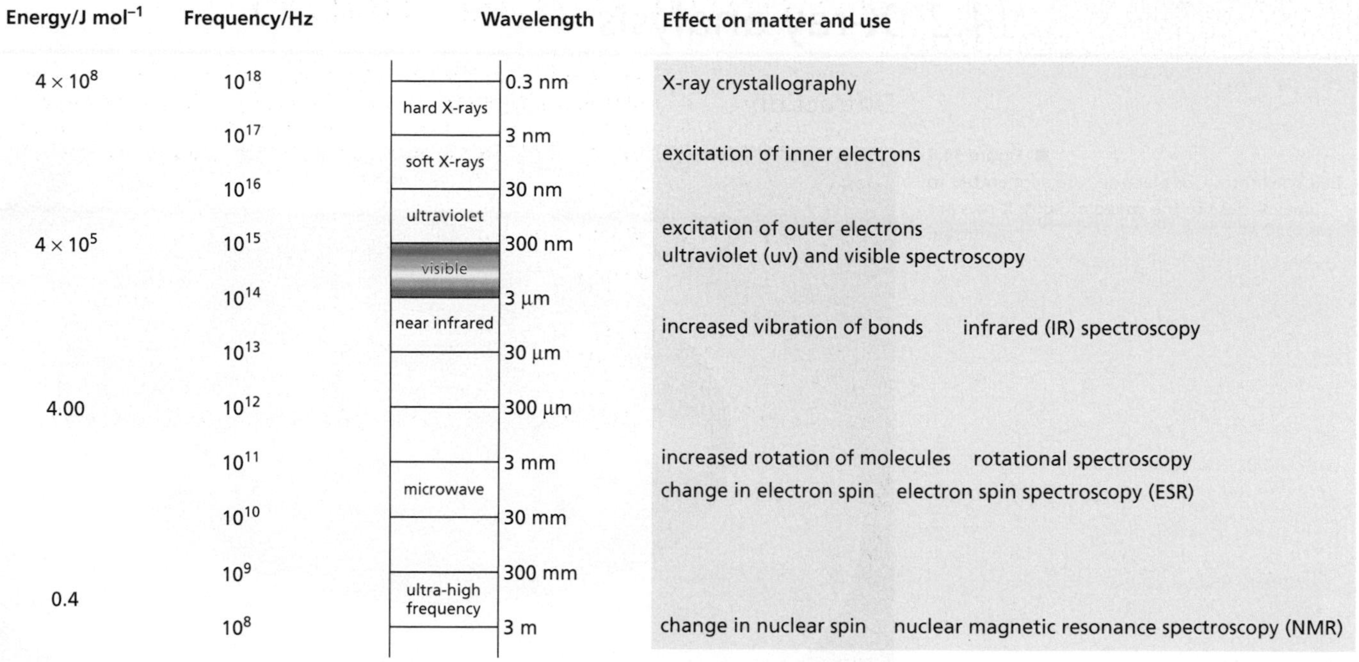
Energy/J mol⁻¹
Frequency/Hz
Wavelength
Effect on matter and use
4 × 10⁸
10¹⁸
0.3 nm
X-ray crystallography
hard X-rays
10¹⁷
3 nm
soft X-rays
excitation of inner electrons
10¹⁶
30 nm
ultraviolet
excitation of outer electrons
ultraviolet (uv) and visible spectroscopy
4 × 10⁵
10¹⁵
300 nm
visible
10¹⁴
3 μm
near infrared
increased vibration of bonds
infrared (IR) spectroscopy
10¹³
30 μm
4.00
10¹²
300 μm
10¹¹
3 mm
increased rotation of molecules
rotational spectroscopy
microwave
change in electron spin
electron spin spectroscopy (ESR)
10¹⁰
30 mm
10⁹
300 mm
ultra-high frequency
0.4
10⁸
3 m
change in nuclear spin
nuclear magnetic resonance spectroscopy (NMR)

■ Figure 14.1 The parts of the electromagnetic spectrum used in chemical analysis. The visible spectrum extends from 400–700 nm.

Irradiating matter

When radiation falls on matter it may be:

- absorbed or
- transmitted or
- reflected.

The relative proportions of the three processes vary depending on the type of radiation and the substance being irradiated. Reflection arises as a result of absorption followed by re-emission. Whether radiation of a certain wavelength will be absorbed by a substance or not depends on the energy levels that exist in the substance. In the simplest case of two energy levels, E_1 and E_2, the frequency of radiation absorbed is given by:

$$f = \frac{E_2 - E_1}{h}$$

and this will also be the frequency of radiation re-emitted (see Figure 14.2).

In practice, the energy levels available in a substance are usually affected by its chemical environment. This means that instead of a single frequency being absorbed and re-emitted in the form of a line, a range of frequencies is absorbed and re-emitted, forming a **band**.

If this band is analysed using high-definition equipment, it may often be **resolved** into a series of lines very close together, showing its **fine structure**.

The frequencies of radiation absorbed by a substance are measured using a spectrometer. Figure 14.3 shows how such a device works. A sample of the substance being investigated is put into the machine and irradiated. The frequencies of radiation that are absorbed by the substance are detected and the resulting spectrum may be printed out or displayed on a screen. There are several different types of spectrometer that use different frequencies of the electromagnetic spectrum, and we shall go on to look at these in this chapter.

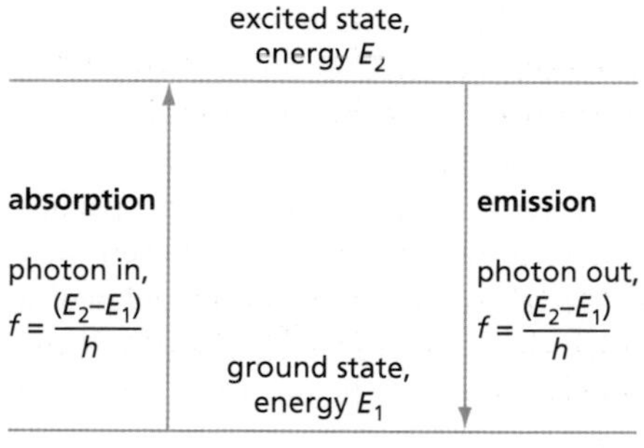

■ Figure 14.2 If a substance has two energy levels, E_1 and E_2, photons of a single frequency are absorbed and the same frequency is re-emitted.

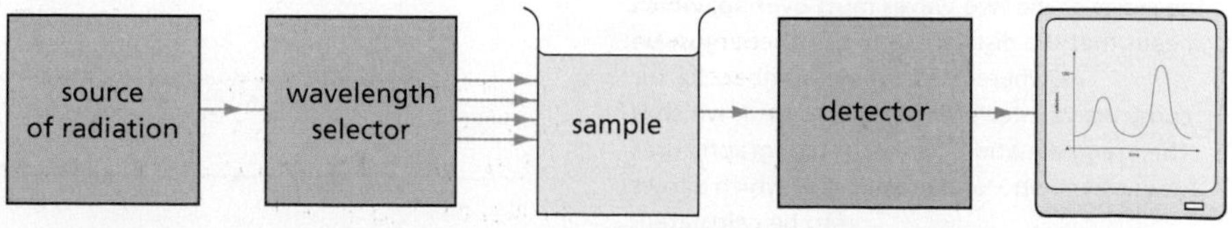

■ Figure 14.3 Block diagram of a colorimeter or spectrometer. The wavelength selector is either a filter or a special selector called a monochromator. The sample is in a suitable container such as a cuvette or test tube. The display may be on a meter, chart recorder or computer.

14.2 X-ray analysis

Diffraction

■ **Figure 14.4**
In a synchrotron, as electrons are accelerated to speeds close to the speed of light, X-rays are emitted.

X-rays have a wavelength of about 0.1 nm, comparable with interatomic distances. If a beam of **monochromatic** X-rays (that is, X-rays of a single wavelength) is shone onto an ionic crystal, the ions in the crystal behave like a diffraction grating. In some directions the waves reinforce each other, and in other directions the waves cancel out, forming an interference pattern. In 1912, by studying the diffraction pattern obtained from simple crystals, William Henry Bragg and his son William Lawrence Bragg worked out the structures of sodium chloride, potassium chloride and zinc sulphide crystals, and also determined the spacing between the ions. They showed that the diffraction could be regarded as reflection by planes of ions, a distance *d* apart, which reflected the X-rays. If the wavelength of the X-rays is λ, they showed that constructive interference occurs (see Figure 14.5) for X-rays striking the crystal planes at an angle θ, when

$$n\lambda = 2d\sin\theta$$

X-rays are shone onto the crystal at a measured angle θ. The wavelength λ of the X-rays is known and so the distance *d* apart of the crystal planes can be calculated. The intensity of the reflections is determined by the electron density in the crystal, which depends on the number of ions in the plane. Combining all the information obtained enables the crystal structure to be determined. The technique is called **X-ray crystallography**.

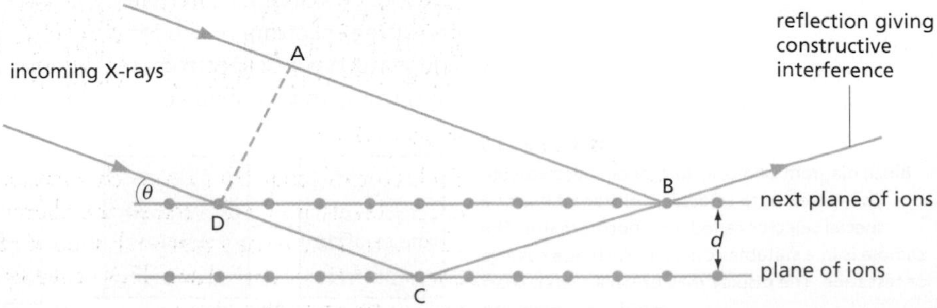

■ **Figure 14.5**
The path difference between the two waves that interfere constructively is BC + CB − AB. Using trigonometry, this distance can be shown to be equal to $2d\sin\theta$. For constructive interference, the crests of the two waves must overlap, which means that the distance between them must be $n\lambda$ where *n* is a whole number. So, for constructive interference, $n\lambda = 2d\sin\theta$, which is the Bragg equation. X-ray crystallography uses X-rays of known λ and measures θ, which allows *d* to be calculated.

Two techniques are used in modern X-ray crystallography.

- **Single crystal** – a small single crystal is rotated, under computer control, in the beam of X-rays. The positions and intensities of the maxima are recorded by detectors placed around the crystal.
- **Powder** – the crystal is powdered and placed on a slide which is tilted, under computer control, in the X-ray beam. Within the powder are crystals in every orientation, and some of these will be at the correct angle to give constructive interference.

Analysis of the results takes a long time, even with a fast computer. However, it is now possible to find the structure of complex molecules such as haemoglobin, as well as that of simple ionic compounds. The main problem with the technique is often in the preparation of a suitably pure crystal of the substance being analysed.

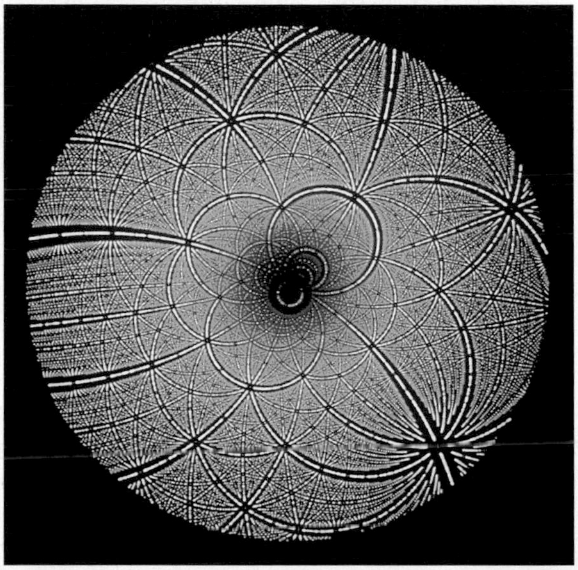

■ **Figure 14.6**
An X-ray diffraction pattern from a crystal of the enzyme ribulose bisphosphate carboxylase/oxygenase which is thought to be the most abundant protein in nature.

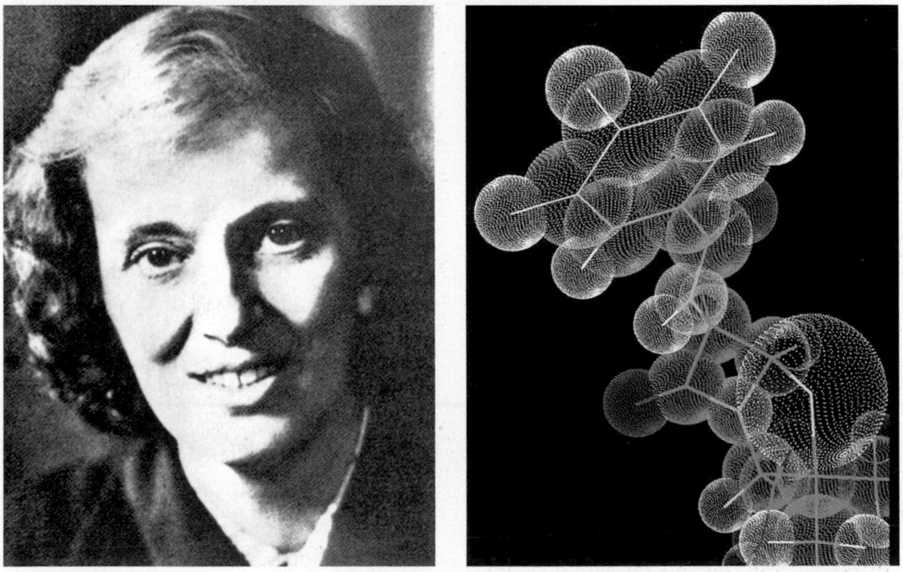

■ **Figure 14.7**
Dorothy Hodgkin (1910–94) was awarded a Nobel Prize for Chemistry for discovering the structures of many important molecules, including penicillin (right), using X-ray crystallography.

Absorption

Each type of atom has its own characteristic inner electronic energy levels. These energy levels are unaffected by the chemical environment, which affects only outer electrons. This means that each type of atom absorbs X-rays at frequencies characteristic of that atom. This provides a sensitive method of analysis, widely used in forensic chemistry.

Identifying atoms by spectroscopy

X-ray spectroscopy

When X-rays interact with a substance, inner-shell electrons in the atom are excited to outer orbitals, or they may be removed completely, ionising the atom. Following this, electrons move down to fill the empty inner orbitals, emitting radiation. The absorption or emission frequencies are characteristic for a particular atom, wherever it is found, because the energy levels of inner electrons are unaffected by bonding. With a suitable apparatus, these characteristic frequencies of X-rays can be measured. The technique is called **X-ray spectroscopy**.

The identification of elements by this method is of wide application. The following are some examples.

- In the manufacture of steel or cement, it is essential to analyse the elements present in the mixture quickly and repeatedly. A complete analysis can be done by X-ray spectroscopy in 15 minutes, which gives an opportunity to adjust the composition of the mixture if it is not of the desired specification.
- Trace elements can be identified in geological specimens.
- The composition of ancient glasses and other artefacts gives information about the method used in their manufacture and their likely place of origin.
- In forensic work, it is possible to match samples of paint or metal accurately from, for example, scratches on the body of a car.
- Small samples of heavy metal pollutants can be identified, for example, lead and barium in the bones of ancient skeletons.

Visible spectroscopy

Many atoms emit or absorb in the visible part of the spectrum. In order to obtain a fine line spectrum, the atoms must be isolated in the gas phase. This means the substance must first be vaporised at high temperature, either by spraying a solution of the substance into a flame or by heating a sample in a furnace.

The spectrum may be studied in absorption or emission. **Emission spectroscopy** requires the substance to be vaporised at a very high temperature, possibly by placing it in a spark gap. More commonly, **absorption spectroscopy** is used. The sample is vaporised and light of a specific frequency passed through the vapour. By suitable calibration, the amount of absorption can be related to the concentrations of various metal ions. The method can be automated and is widely used to measure, for example, the concentrations of ions such as sodium and calcium in the blood. The technique is known as **atomic absorption spectroscopy**.

14.3 Ultraviolet and visible spectra

Absorbing in the ultraviolet and visible regions

All atoms and molecules absorb in the ultraviolet visible region because the photons are energetic enough to excite outer electrons. If the frequency is high enough, **photoionisation** takes place by the loss of an electron.

Most compounds absorb in the ultraviolet rather than the visible region, and as a result are colourless. When white light is shone onto a colourless substance, either as a large crystal or in solution, most of the light is transmitted and the substance appears to be colourless and transparent. When white light is shone onto a colourless substance in the form of a powder, most of the light is randomly reflected by the individual particles and powder appears opaque and white. So a colourless substance appears white when powdered.

When a substance absorbs ultraviolet radiation and becomes electronically excited, this energy is rapidly lost. It may be simply re-radiated, in which case no overall absorption takes place. However, in practice there are a number of possibilities and most of them result in overall absorption (see Figure 14.8).

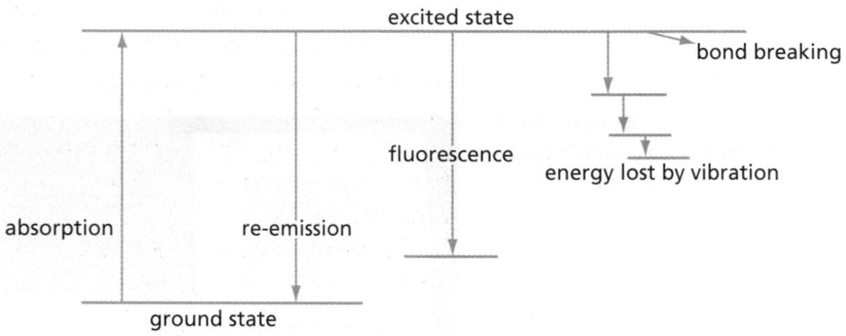

Figure 14.8 When an electron is excited in a molecule, there are several ways in which this energy can be lost. This is because there are vibrational energy levels in the molecule as well as electronic energy levels. The overall result is that radiation is absorbed and the temperature of the substance rises.

The absorption of radiation by chlorine

When a molecule absorbs a photon of radiation, an electron is excited into an orbital of higher energy. For the chlorine molecule, the lowest frequency needed to bring about this excitation is just in the blue end of the spectrum, and as a result chlorine appears yellow-green in colour.

The electronic excitation produces an excited chlorine molecule. This contains a weak bond that is easily broken (see Figure 14.9).

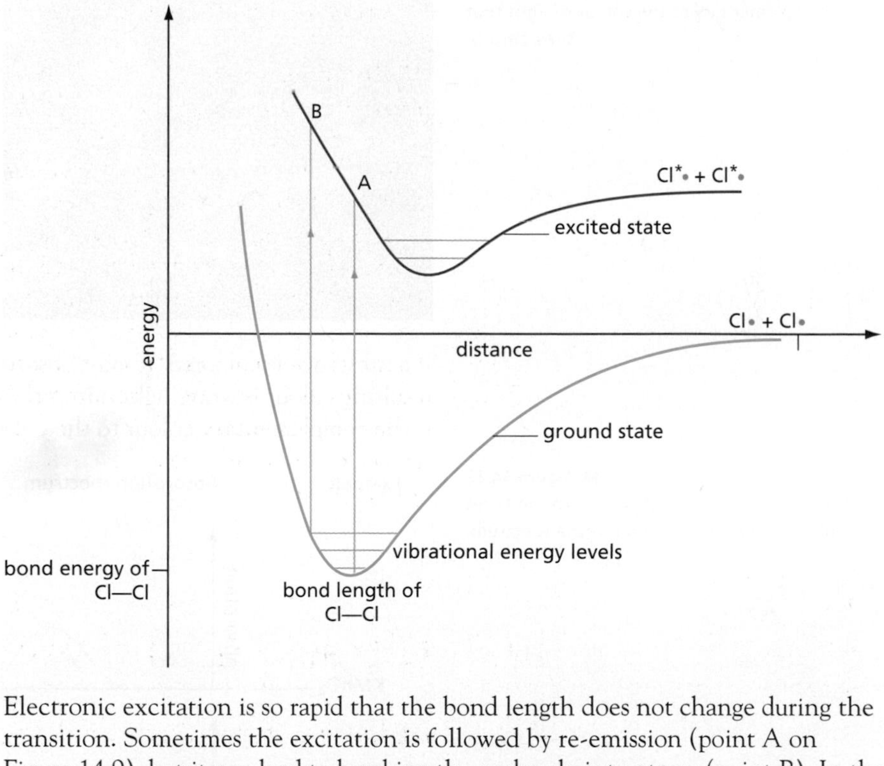

Figure 14.9 Potential energy–distance curves for the ground state and excited state of a chlorine molecule. The excited state is higher in energy and the bond in the molecule much weaker. An electron at point A will emit radiation and return to the ground state (though probably at a higher vibrational energy level). An electron at point B will cause the bond in the excited state to break, producing two 'hot' atoms.

Electronic excitation is so rapid that the bond length does not change during the transition. Sometimes the excitation is followed by re-emission (point A on Figure 14.9), but it can lead to breaking the molecule into atoms (point B). In the latter case, 'hot' chlorine atoms are produced that have large kinetic energies. These are the ones that are responsible for radical substitution reactions (see Chapter 23, pages 433–7).

The production of chlorine atoms is really a two-step process:

$$Cl_2 \xrightarrow{hf} Cl_2^* \quad \text{and} \quad Cl_2^* \rightarrow 2Cl\bullet^*$$

where Cl_2^* is an excited chlorine molecule and $Cl\bullet^*$ a 'hot' chlorine atom.

Ultraviolet and visible spectroscopy

Some substances can be detected by ultraviolet spectroscopy. They absorb strongly in the ultraviolet region because they contain specific groups (see Chapter 31, pages 592–3). The most important of these are listed in Table 14.1.

■ **Table 14.1**
Some groups that absorb strongly in the near ultraviolet, with the wavelengths at which they absorb.

Group	Wavelength/nm
C=C	190
C=O	190, 300
(benzene ring)	190, 200, 250

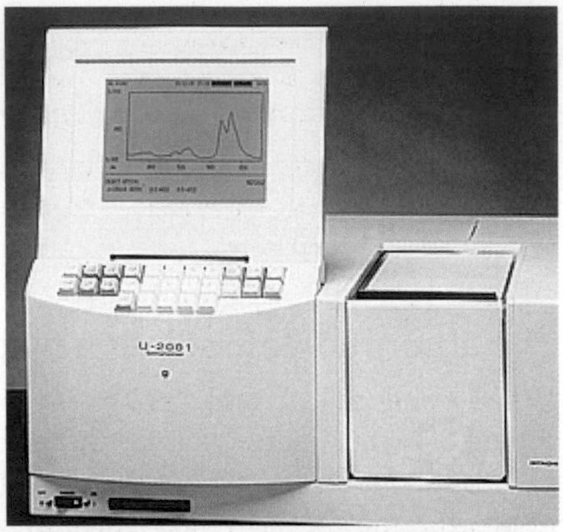

■ **Figure 14.10**
UV spectrometer. The sample chamber on the right has a lid to cut out strong light.

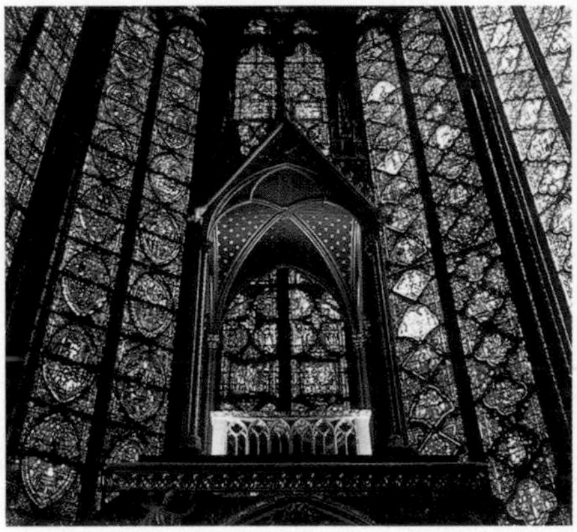

■ **Figure 14.11**
Rose petals and stained glass windows absorb in the visible region. We see the colour complementary to the colour of light that they absorb.

If a substance is coloured, it must absorb in the visible region of the spectrum. The resulting colour is white light minus the colour being absorbed. The colour we see is the **complementary** colour to the colour being absorbed (see Figure 14.12).

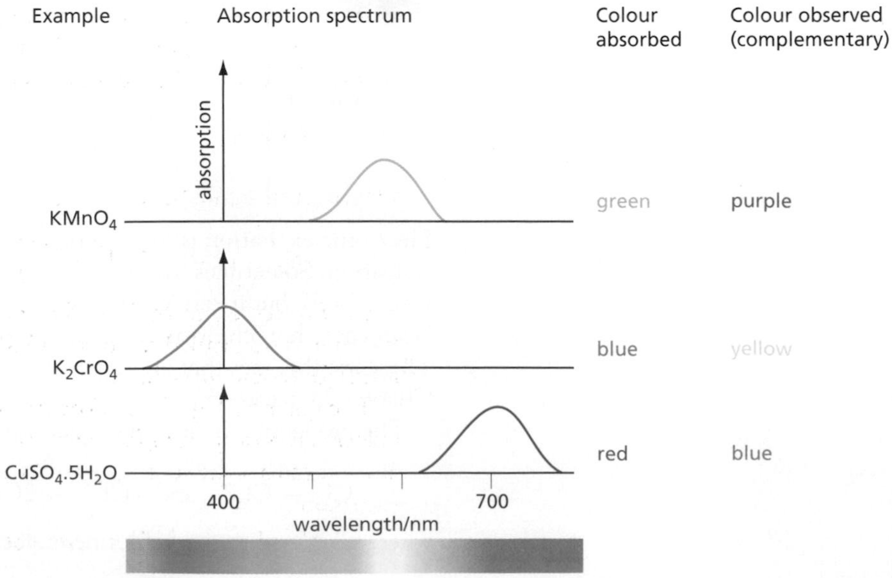

■ **Figure 14.12**
The colours observed when absorption takes place in the visible region of the spectrum.

To absorb in the visible region, the substance must have two energy levels that are very close together. Two main mechanisms bring this about, namely **charge transfer** and **d-to-d transitions**.

Charge transfer

Some substances change their bonding slightly when they absorb visible light. The following are a few examples.

Many solid metal oxides are coloured even though their ions, in solution, are colourless. An example is lead(II) oxide. Other metal oxides have a different colour from their ions in solution, for example, copper(II) oxide. These oxides are essentially ionic, but the metals are of moderate electronegativity so the bonding contains considerable covalent character. The absorption of a photon of visible radiation promotes one of the electrons of the oxide ion into the partially covalent bond by a process called **charge transfer** (see Figure 14.13).

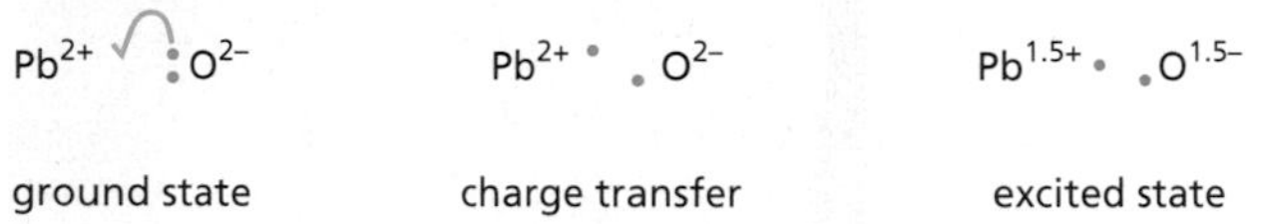

■ **Figure 14.13** Charge transfer – an electron of the oxide ion of lead(II) oxide is promoted by a photon of visible radiation to a bonding orbital. The transfer of one electron from the O^{2-} ion to the bond is equivalent to the transfer of half an electron from the O^{2-} ion to the Pb^{2+} ion.

Another example of charge transfer happens in the chromate(VI) ion, CrO_4^{2-}, which is yellow. Its electronic structure is similar to that of the sulphate(VI) ion, SO_4^{2-}, which is colourless. Both ions absorb radiation and promote an electron from a Cr—O (or S—O) bonding orbital to a higher energy orbital located on the central Cr (or S) atom. The weakly oxidising sulphur(VI) atom accepts an electron with difficulty; it needs a photon of ultraviolet radiation to bring about the electron promotion and so the ion is colourless. The more strongly oxidising chromium(VI) atom requires slightly less energy and the absorption, while still predominately in the ultraviolet region, has a 'tail' in the visible region. The result is that a slight excess of red and green radiation is given out, which the eye detects as pale yellow. The absorption band for the manganate(VII) ion, MnO_4^-, which is a very strong oxidising agent, shifts to the green and so the ion appears an intense purple colour (see Chapter 19, page 356).

The Beer–Lambert Law (if you know some calculus)

We first zero a colorimeter or spectrometer using a cell containing solvent alone. This compensates for any absorption due to the solvent and gives a reading corresponding to 100% **transmission** or zero **absorbance**. The intensity of the radiation passing through the cell under these conditions is I_0.

Inside the cell, radiation of intensity I is passed through a length of δl of solution, and a fraction $\delta I/I$ is absorbed. The size of this fraction depends on the nature of the solute in the solution and its concentration, c. For a given solute:

$$\frac{\delta I}{I} \propto -c\,\delta l$$

On integration, we have:

$$\log_e\left(\frac{I_0}{I}\right) \propto cl \quad \text{or} \quad \log_{10}\left(\frac{I_0}{I}\right) = \varepsilon cl$$

where l is the length of the cell (often 1.0 cm) and ε is a constant, characteristic for each solute, called the molar **absorption coefficient**. The equation is known as the **Beer–Lambert Law**.

The quantity $\log_e(I_0/I)$ is known as the **absorbance** or **optical density**. Most colorimeters or spectrophotometers have an absorbance scale – this has the advantage that the readings are directly proportional to the concentration of the solute:

$$\text{absorbance} = \varepsilon cl \qquad \text{or} \qquad \text{absorbance} \propto c \text{ (for the same solute)}$$

Fluorescent agents in washing powder

Many substances are yellow or brown because the 'tail' of the ultraviolet absorption exends into the violet end of the spectrum. This is certainly true of many stains on clothing, which may be incompletely removed during washing. The manufacturers of washing powders compensate for this by adding a whitener, which is a fluorescent dye, to their products. A fluorescent dye converts the near ultraviolet region of sunlight into blue light. The blue light that is given off by the clothing then masks the absorption from stains that are incompletely removed. Many years ago, the same principle was used by adding a blue dye to the weekly wash.

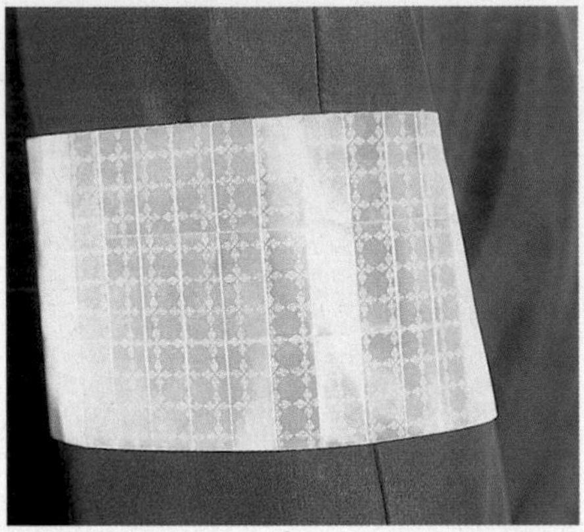

■ **Figure 14.14** A fluorescent armband.

Many dyes contain a conjugated double bond–single bond system. These include the azo dyes (Chapter 30, page 571), indicators such as methyl orange and phenolphthalein, and natural products such as carotene, haemoglobin and chlorophyll (Figure 14.15).

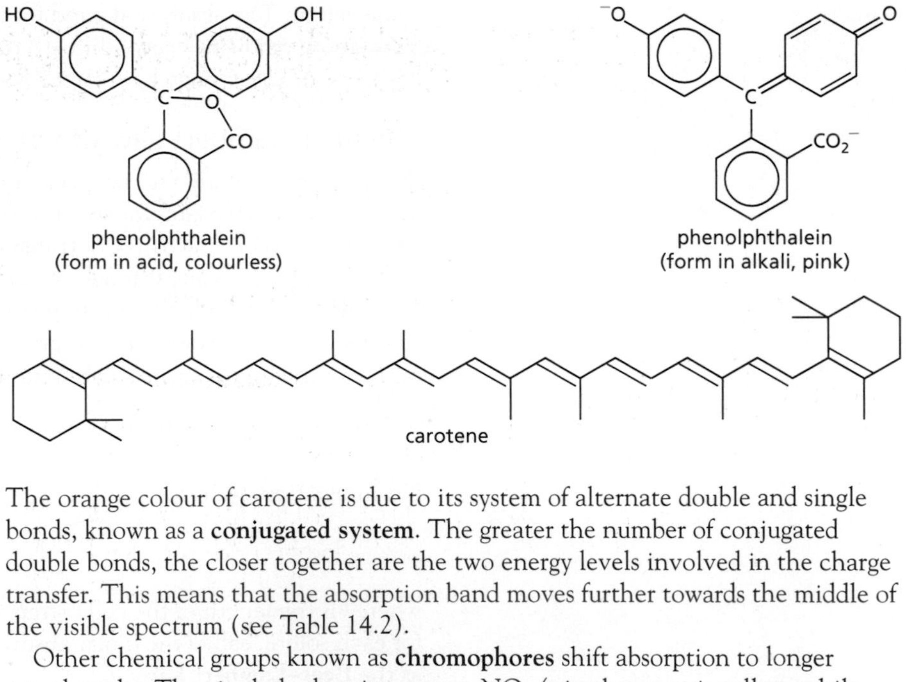

■ **Figure 14.15** The structures of phenolphthalein and carotene.

Number of conjugated double bonds	Wavelength of maximum of absorption/nm
5	342
6	380
7	401
8	411

■ **Table 14.2** The wavelength of the maximum of the absorption band gets larger as the degree of conjugation increases.

The orange colour of carotene is due to its system of alternate double and single bonds, known as a **conjugated system**. The greater the number of conjugated double bonds, the closer together are the two energy levels involved in the charge transfer. This means that the absorption band moves further towards the middle of the visible spectrum (see Table 14.2).

Other chemical groups known as **chromophores** shift absorption to longer wavelengths. They include the nitro group, NO_2 (nitrobenzene is yellow, while benzene is colourless) and the —N=N— group in the azo dyes (see Chapter 30, page 571).

d-to-d transitions

The colour of many transition metal ions is due to transitions from one d orbital to another. This is discussed further in Chapter 19, page 357.

14.4 The hydrogen spectrum

The emission spectrum of hydrogen

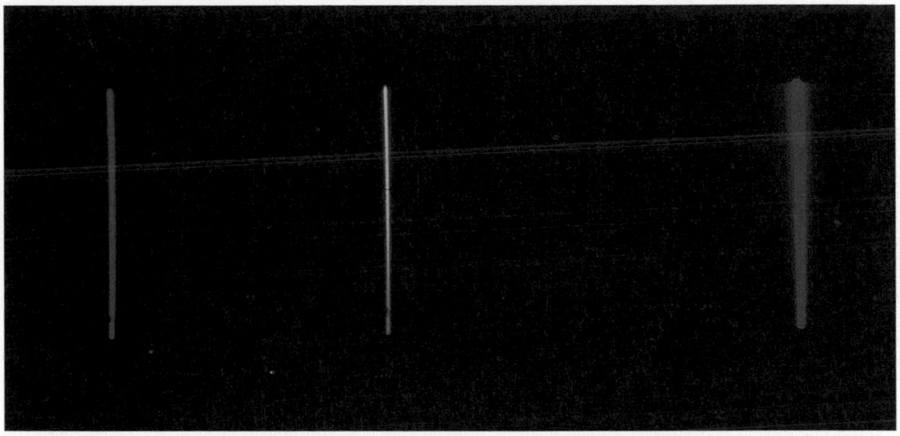

■ **Figure 14.16**
Part of the visible emission spectrum of hydrogen.

A detailed study of the hydrogen spectrum formed the basis for Bohr's theory of the atom, and subsequently the wave mechanics description of its structure.

At low pressures, gases conduct an electric current when a high voltage is applied, and give out characteristic colours. The electrons in the atoms are excited to higher energy levels by the electric current, and as they fall back to lower energy levels they emit a series of lines of characteristic frequencies. This is called an **emission spectrum**. Under these conditions, hydrogen molecules break down to atoms, which glow red. If this spectrum is examined with a spectroscope, lines appear in the visible part of the spectrum and in the near ultraviolet region (see Figure 14.17).

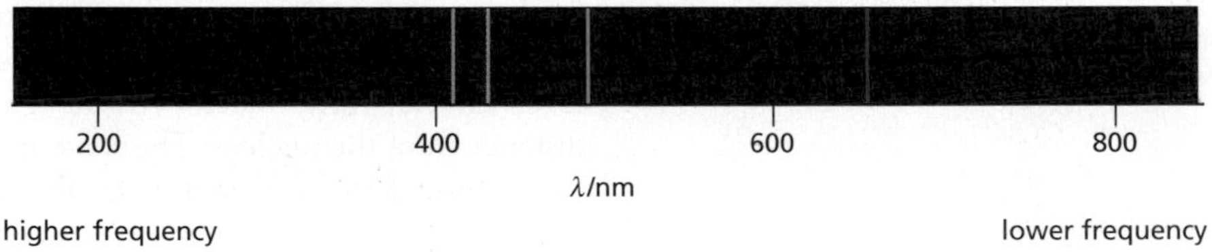

■ **Figure 14.17**
The hydrogen spectrum in the visible and near ultraviolet regions.

As the frequency increases, the lines become closer together. This group of lines is known as the **Balmer series**.

Analysing the spectrum – the model of the atom

Johann Balmer showed that the frequencies, f, of the lines of the hydrogen spectrum fitted the formula:

$$f = R_{\infty}\left(\frac{1}{m^2} - \frac{1}{n^2}\right)$$

where $m = 2$, $n = 3, 4, 5, \ldots$ and R_{∞} is a constant known as the **Rydberg constant**. Subsequently, another series was found in the ultraviolet region at about 100 nm, called the **Lyman series**, with a value of $m = 1$, and three other series followed in the infrared region, called the **Paschen**, **Brackett** and **Pfund series** with $m = 3$, 4 and 5, respectively.

The great achievement of Bohr in 1913 was to devise a model of the hydrogen atom that explained the origin of these lines. He made two assumptions:

- electrons orbited at fixed distances from the nucleus
- their angular momentum had definite values.

His model interpreted the hydrogen spectrum in the following way. The values of n are the energy levels that the electrons start from, and the values of m are the energy levels they finish at (see Figure 14.18).

Figure 14.18 Bohr's interpretation of the lines in the hydrogen spectrum. There are also further series in the far infrared region.

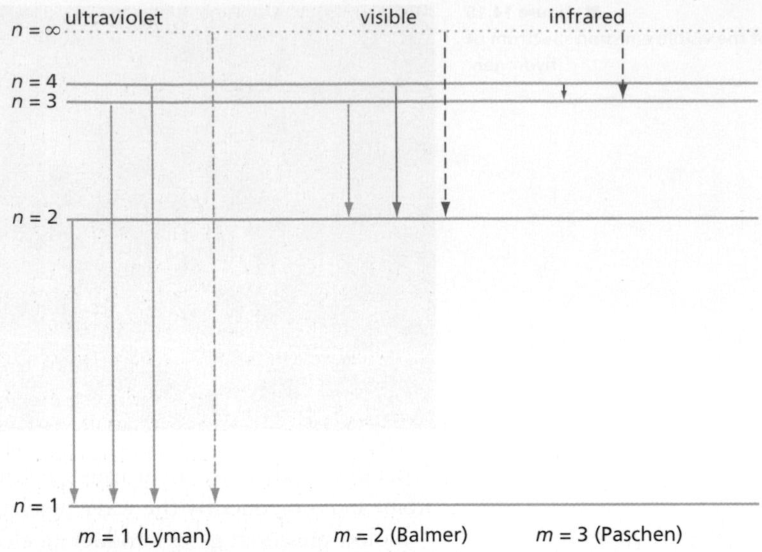

When $n = \infty$, the hydrogen atom is ionised. The lines in each series become closer together at high frequencies and, by extrapolation, the frequency at which they converge can be found. For the Lyman series, this corresponds to the ionisation energy of hydrogen. This use of convergence limits is the most accurate method for the determination of ionisation energies, and is the method generally used.

The Bohr model, in which electrons orbit at fixed distances from the nucleus, has been modified by the modern picture of the atom (see Chapter 4). The important point, which has not been modified, is the idea that electron energy levels have definite values, even though an electron in an orbital may vary in its distance from the nucleus. The wave-mechanical model also has the advantage that it provides an explanation for the assumptions made by Bohr.

14.5 Infrared spectra

Increasing the vibrational energy of bonds

Photons in the infrared region of the spectrum have insufficient energy to promote electrons, but they can bring about an increase in the vibrational energy of the bonds. The frequency of vibration is determined by the stiffness of the bond and the masses of the atoms. The stiffness of a bond is closely related to its bond energy. A light atom attached by a strong bond (for example, O—H) vibrates at a higher frequency, while a heavy atom attached by a weak bond (for example, C—I) vibrates at a lower frequency.

An IR spectrometer is designed along the lines of the general spectrometer shown in Figure 14.3 (page 269). A variation of infrared spectroscopy is **Raman spectroscopy**. An intense laser beam is shone onto the sample and a spectroscope placed at right angles to the beam is used to analyse the frequencies emitted. Apart from the original frequency of the laser, there are additional lines whose frequency differences correspond to infrared vibrations. The existence of the NO_2^+ ion in nitrating mixture was detected in this way (see Chapter 27, page 504).

There are two main applications of infrared spectroscopy. With small molecules, accurate information can be obtained about bond lengths, bond angles and the stiffness of the bonds. This is not possible with a larger molecule, but infrared spectroscopy is used to identify the groups present in the molecule, and to identify the substance by its 'fingerprint'.

IR spectra of small molecules

A molecule absorbs in the infrared region only if the vibration causes a change in its dipole moment (see Chapter 3, page 67). The most common gases in the atmosphere, nitrogen, oxygen and argon, therefore do not absorb. The two most common gases that do absorb are water vapour and carbon dioxide.

Even in the gas phase, the absorption spectra are in the form of bands. The line spectrum due to the vibration frequency has subsidiary lines superimposed on it that arise from changes in rotation of the molecules. A diatomic molecule, for example hydrogen chloride, shows a single vibrational absorption band, while a triatomic molecule such as water may show up to three bands. By analysis of the vibration frequency and the spacing of the fine structure, very accurate information about the bond lengths and the bond angles can be found. The vibration frequency does not give the actual bond strength; instead it gives the **force constant** that measures the stiffness of the bond, which, though related to the bond strength, is not quite the same thing.

IR spectra of larger molecules

If the molecule is fairly large, a particular bond may vibrate fairly independently of the rest of the molecule. This means that certain bonds show characteristic vibration frequencies, irrespective of where they are found. The actual frequency does vary slightly from one molecule to another, so the vibration bands are in a **region** of the infrared spectrum, rather than at a specific frequency.

Some characteristic vibration frquencies are given in Table 14.3. These are **stretching** vibration frequencies produced as the bond becomes longer and shorter. Bonds also vibrate in other, more complex, ways and produce other bands in different parts of the infrared spectrum.

■ **Table 14.3**
Some characteristic stretching vibrations in the infrared region of the spectrum. Because infrared frequencies are very large (1 GHz = 10^9 Hz) and inconvenient to use, they are often quoted as **wavenumbers**. The wavenumber is the number of waves in 1 cm.

Bond	λ/nm	*f*/GHx	Frequency of absorption (wavenumber)/cm^{-1}
C—H	3400	88	2900
C—F	8300	36	1200
C—Cl	14000	21	700
C—Br	16000	18	600
C—I	20000	15	500
C—O	9000	33	1100
C=O	5900	52	1700
O—H	2800	110	3600

There are two main uses for infrared spectra:

- to identify the groups present in a molecule (see Chapter 31, page 594)
- to identify the molecule. Every molecule has its own distinct **fingerprint** that makes its infrared spectrum unique. Comparison of an infrared spectrum with known spectra therefore enables chemists to identify a substance or to assess its purity (see Chapter 31, pages 594–5).

A typical infrared spectrum is that of ethanol (see Figure 14.19). It is possible to identify the vibration bands of the O—H and C—O bonds.

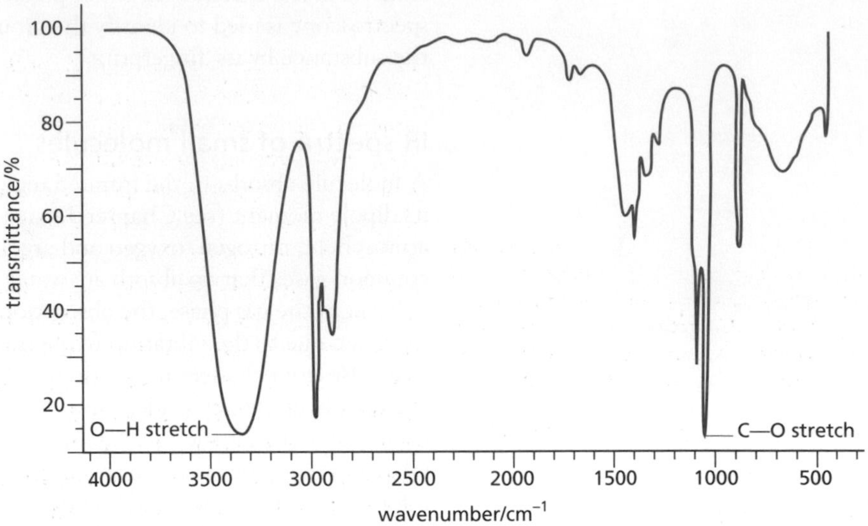

■ **Figure 14.19**
Infrared spectrum of ethanol, CH_3CH_2OH. The stretching vibrations of the O—H and C—O bonds are shown on the diagram. The O—H vibration at 3350 cm^{-1} is broader and of lower wavenumber than that of an isolated O—H bond because of hydrogen bonding in the alcohol. The C—O bond vibration occurs at 1050 cm^{-1}.

Microwave spectra

Some molecules absorb in the microwave region of the spectrum. These absorptions are due to changes in the rotation of the molecule. The most common molecule that absorbs in this way is water. This has a series of strong absorptions at a frequency near 3 GHz (wavelength 10 cm). This absorption is made use of in a microwave oven. The microwaves pass through most substances, but are absorbed by the water in the food. The increased rotation of the water molecules is converted into translational kinetic energy, and the temperature rises (Figure 14.20). The containers used for the food must not absorb microwaves – they can be made of plastic, glass or ceramic (which do not absorb) but must not be made of metals (which absorb strongly).

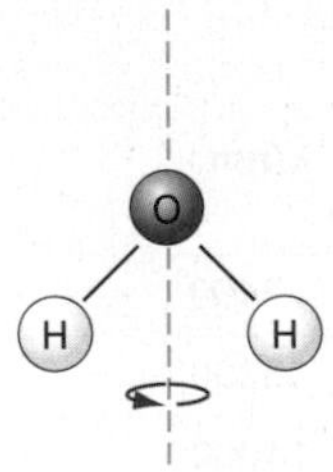

■ **Figure 14.20** (left)
When water molecules absorb microwave radiation, their increased rotation is converted into translational kinetic energy by collision with other molecules.

■ **Figure 14.21** (right)
The water in food is heated by microwaves in an oven. As the microwaves penetrate inert substances, the food is cooked on the inside at the same time as the outside is being heated.

Microwaves are also absorbed by substances that contain unpaired electrons. Such substances are either transition metals (see Chapter 19) or are radicals (see Chapter 23, page 433). An unpaired electron can have two different orientations and in the presence of a magnetic field, these orientations have slightly different energies. In a field of 0.3 T (about 3000 times that of the Earth's magnetic field) the energy difference is 4 J, corresponding to 10 GHz or 3 cm microwaves.

The absorption of microwaves by unpaired electrons is known as **electron spin resonance (ESR)**. The technique is used to detect radicals and to locate the position of the unpaired electron in the molecule.

14.6 Nuclear magnetic resonance spectra

The basis of NMR

Like electrons, nucleons (protons and neutrons) have spin. If an atom has an even number of nucleons, the spins cancel out and there is no overall magnetic moment. If, however, an atom has an odd number of nucleons, there is an overall magnetic moment. As a result, the nucleus can take up one of two orientations, whose energy splits in the presence of an external magnetic field. This splitting of the energy forms the basis of **nuclear magnetic resonance (NMR) spectroscopy**.

The extent of the splitting is proportional to the strength of the external field, and to increase the splitting very large external fields are used. The strength of the applied magnetic field is measured in **teslas, T**; one tesla is about 10 000 times as strong as the Earth's magnetic field. Many machines use a field strength of 9.4 T. For a hydrogen atom, this creates an energy difference of 0.16 J, corresponding to a frequency of 400 MHz, in the UHF region of electromagnetic spectrum. An NMR spectrometer therefore detects the absorption of UHF radiation, in a similar fashion to any other spectrometer (Figure 14.3, page 269). The principal difference is that the sample is also subjected to a strong magnetic field. Compounds containing hydrogen atoms therefore show a nuclear magnetic resonance absorption band at 400 MHz. The frequencies absorbed by other common atoms that have an odd number of nucleons are shown in Table 14.4.

■ **Table 14.4**
The absorption frequencies, in an external field of 9.4 T, for some common atoms studied with NMR. In this section we shall study proton magnetic resonance only, as this is by far the most common type of NMR spectroscopy.

Nucleus	Absorption frequency/MHz
^{1}H	400
^{13}C	101
^{19}F	377
^{31}P	162

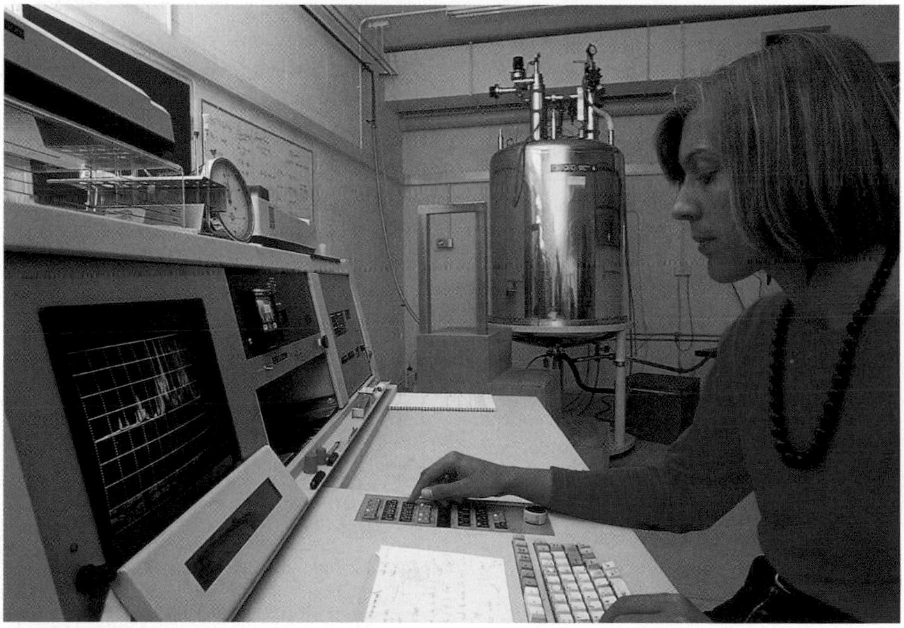

■ **Figure 14.22**
This NMR spectrometer measures the absorbance of UHF radio frequency radiation by hydrogen atoms (which have an odd number of nucleons – just one proton) in an external magnetic field. The absorption frequency varies slightly depending on the chemical environment of the hydrogen atom, allowing identification of hydrogen-containing compounds.

The absorption is very weak, because the populations of atoms at each of the two energy levels are almost the same. At room temperature for a hydrogen atom, the population of the upper level differs from the lower end by only 1 part in 30 000, so that absorption is nearly always cancelled out by re-emission. The situation is even worse for carbon-13, because carbon contains only 1% of this isotope. In order to make the absorbance as intense as possible, the following strategies are adopted:

- cool the sample, which increases the population difference
- use as large a magnetic field as possible to increase the splitting
- measure the absorption many times and average the results using a computer.

NMR is often used to detect the hydrogen atoms in water, and the analysis of water in the human body forms the basis of magnetic resonance imaging (see the panel on page 284).

Chemical shift

Magnetic resonance spectroscopy would be of little value in chemical analysis if all hydrogen atoms absorbed the same frequency of radiation. In practice, their exact absorption frequency depends on the chemical environment in which they are found. The electron cloud around the hydrogen atom partially screens the nucleus from the external magnetic field. The effect is a very small one, but the atom attached to the hydrogen affects the degree of screening, and hence the frequency of the absorption. This effect is described more fully in Chapter 31, pages 600–2.

For convenience, the hydrogen atoms in the compound tetramethylsilane, $(CH_3)_4Si$ (known as TMS) are used as a reference. The sample being investigated is mixed with TMS and the frequency of the absorption is measured relative to that of TMS. There are a number of reasons why this compound is used:

- all the hydrogen atoms in TMS are equivalent, so it gives a single absorption peak
- most other groups absorb at higher frequencies than TMS.

Group	Electronegativity of atom attached to carbon	δ/p.p.m.
$F—CH_3$	3.98	4.5
$Cl—CH_3$	3.16	2.9
$Br—CH_3$	2.96	2.5
$I—CH_3$	2.66	2.0
$H—CH_3$	2.20	0.8

■ **Table 14.5** Chemical shifts vary with the partial charge on the carbon atom attached to the hydrogen. A carbon atom attached to a highly electronegative atom such as fluorine has a larger partial charge than a carbon atom attached only to hydrogen atoms.

The extent of the change in frequency of absorption is called the **chemical shift**, symbol δ. It is defined as:

$$\Delta\delta = \frac{10^6(f - f_{TMS})}{f_{TMS}}$$

Values of δ are quoted in parts per million (p.p.m.). Because chemical shifts are very small, the magnetic field must be identical throughout the sample. To achieve this, the super-conducting magnets used in modern machines are very carefully constructed.

The chemical shift increases as the electronegativity of the atom attached to the hydrogen increases. For example, in a halogenoalkane, a carbon atom attached to a fluorine atom is more electronegative than a carbon atom attached to an iodine atom. This is due to the inductive effect of the halogen atom (see Chapter 22, page 415). The more electronegative carbon atom draws the electron cloud away from an attached hydrogen atom more, reducing the screening and increasing the absorption frequency (see Table 14.5).

The values of some chemical shifts for hydrogens within different organic groups are given in Table 14.6.

■ **Table 14.6** Some chemical shifts. Just as for infrared spectroscopy, protons associated with particular groups absorb in a region of the spectrum rather than at a definite frequency.

Proton type	Groups	δ/p.p.m.	Range
$Si(CH_3)_4$	Tetramethylsilane (TMS)	0.0	0
$C—CH_3$	End of alkyl chain	0.9	±0.5
$C—CH_2—C$	Middle of alkyl chain	1.4	±0.5
=C—CH—	Adjacent to C=C	1.9	±0.5
—C(O)—CH—	Adjacent to C=O (ketones, esters, acids)	2.3	±0.5
$C_6H_5—CH—$	Adjacent to arene ring	2.5	±0.5
O—CH—	Adjacent to oxygen (alcohols, ethers, esters)	3.3	±0.5
=C—H	Alkenyl	5.5	±1
$C_6H_5—H$ (benzene)	Aryl	7.5	±1.5
—C(O)—H	Aldehydic	10.0	±0.5
—O—H	Alcohols	3	±2*
Ar—O—H	Phenols	5.5	±1*
—CO—O—H	Carboxylic acids	11.0	±2*

* These δ values are very dependent on the nature and acidity of the solvent and the extent of hydrogen bonding between molecules and the solvent.

Note: all values are approximate – relevant protons should be found within the stated range of the above values. The total chemical shift experienced by a proton between two functional groups is often approximately the sum of the effect of each group separately.

The NMR spectrum for ethanol

Ethanol has three different types of hydrogens, each absorbing at a different chemical shift (see Figure 14.23). The areas under the three peaks are in the ratio 1:2:3.

Under high resolution, the three peaks split into a fine structure. This can be used to give further information about the structure (see Chapter 31).

■ **Figure 14.23**
The NMR spectrum for ethanol. In **a** the hydrogen of the OH group appears at 5.1 p.p.m. as it is hydrogen bonded. In **b** this hydrogen appears at 2.6 p.p.m. – the ethanol in solution in carbon tetrachloride is less hydrogen bonded. The areas of the three peaks are in the ratio 1:2:3, which shows the proportions of the three types of hydrogen atom present.

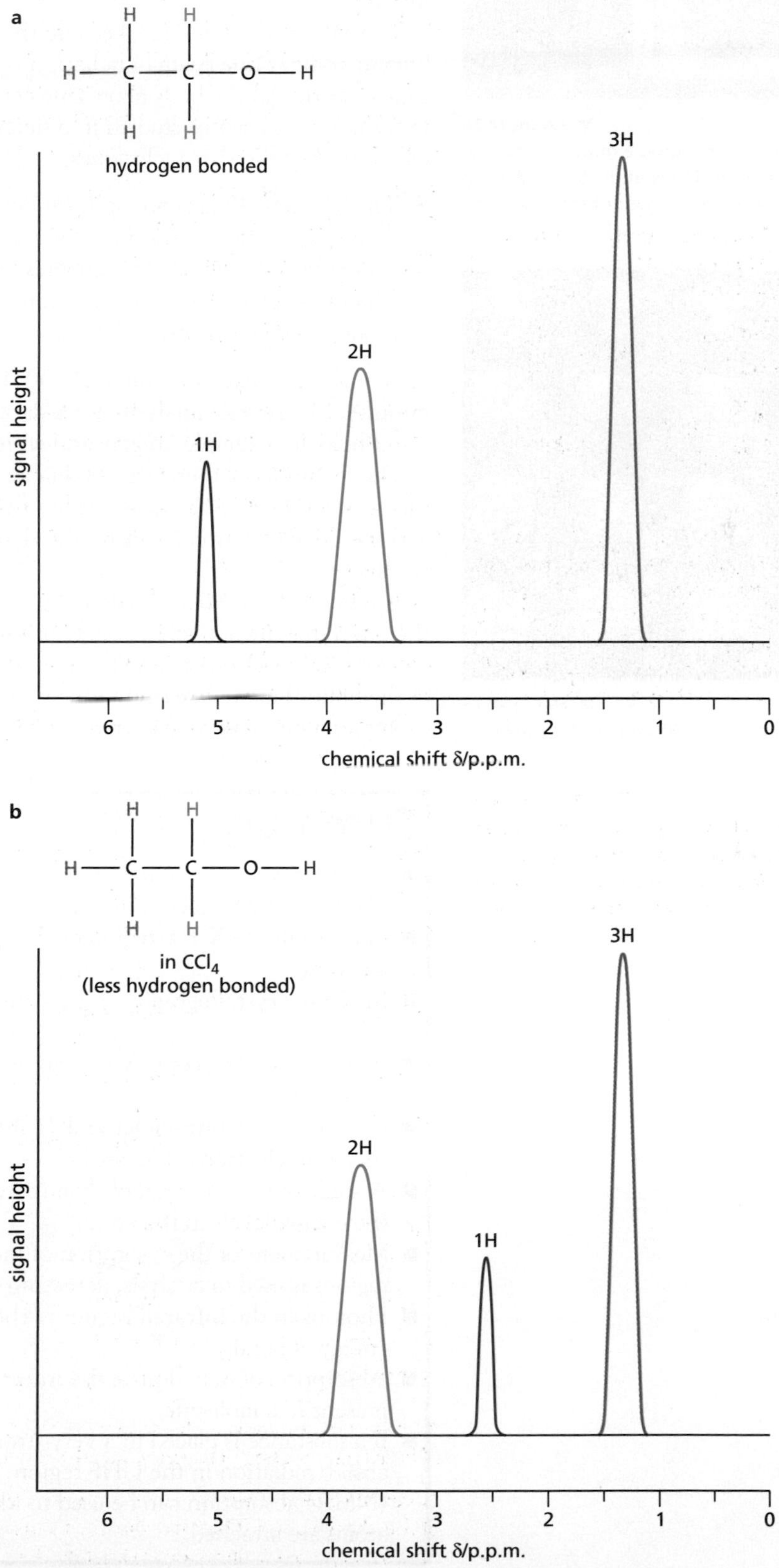

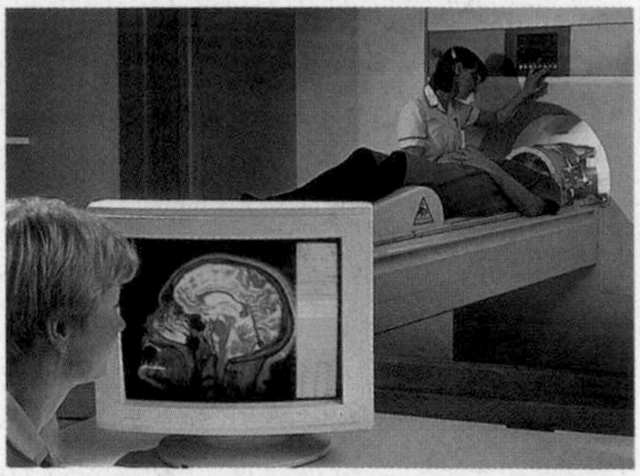

■ **Figure 14.24**
An MRI scanner gives a three-dimensional picture of the inside of the body, allowing non-invasive diagnosis of many diseases.

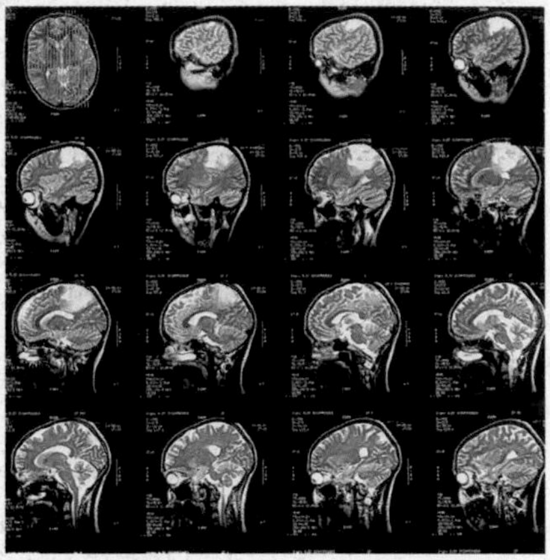

■ **Figure 14.25**
Images from an MRI scan of a human head.

Magnetic resonance imaging (MRI)

Because the human body is made up mostly of water, it responds to nuclear magnetic resonance. By suitable scanning, an image of the water distribution in the body can be built up, which is invaluable in the diagnosis of various illnesses, in particular brain disorders. The word 'nuclear' has been dropped from the name of the technique, to avoid suggesting to patients that nuclear radiation is involved.

The MRI scanner is large, because the magnetic field must pass through the human body. A fine beam of radiation is applied, giving the absorption pattern of a cross-section of the body about one centimetre thick. Within this cross-section, water molecules can be studied in different parts of the body. This is done by means of the following techniques.

- The magnetic field is not uniform but varies from one side to another. As the frequency of the radiation is changed, water at different depths inside the body responds to the signal. This enables a one-dimensional picture to be built up.
- The radiation beam is rotated through 360°. This enables a computer to produce a two-dimensional image – a 'slice'.

A typical brain scan containing 20–30 slices can be obtained in less than ten minutes. More subtle analysis of the data makes it possible to distinguish between water held, for example, in grey and white tissue or in cancerous and normal cells.

The technique is therefore invaluable in, for example, the diagnosis of brain tumours or Alzheimer's disease. It has the great advantage of being non-invasive, and the UHF radiation is much safer than X-rays, which are used in alternative techniques.

A refinement of MRI is to detect phosphorus-31 rather than protons. Areas of the brain that are actively in use require ATP (adenosine triphosphate) for the biochemical reactions that take place. It is therefore possible to locate the regions of the brain that are most actively involved when different mental processes (for example, sight, reasoning or spatial work) are carried out.

Summary

- The energy E of a photon of electromagnetic radiation is given by $E = hf$, where f is the frequency and h is the Planck constant.
- Photons in the **X-ray region** of the spectrum interact with inner electrons in an atom.
- In **X-ray crystallography**, X-rays are used to find the structures of crystalline solids.
- In **X-ray spectroscopy**, X-rays are used to identify which elements are present in a substance.
- Photons in the **ultraviolet and visible regions** of the spectrum interact with the outer electrons of an atom.
- A study of the spectrum of atomic hydrogen gives detailed information about the energy levels in the atom.
- Measurement of the absorption of radiation in the ultraviolet and visible regions is used in analysis, detecting specific groups in a molecule.
- Photons in the **infrared region** of the spectrum increase the vibrational energy of bonds.
- Absorption of radiation in the infrared region is used to identify the groups present in a molecule.
- If a substance is placed in a very strong magnetic field, its hydrogen atoms absorb radiation in the UHF region. This **nuclear magnetic resonance (NMR)** absorption can be used to identify the groups to which the hydrogen atoms are attached.

INORGANIC CHEMISTRY

15 *Periodicity*

One of the triumphs of nineteenth-century chemistry was the periodic classification of the elements. In the periodic table, elements are arranged according to their atomic number. Elements with similar properties are placed in columns called groups. With increasing atomic number, there are systematic changes in the properties of the elements. In this chapter, we explore these changes in properties associated with crossing a row or period for the lighter elements and their simple compounds.

The chapter is divided into the following sections.

15.1 Periodic classification

Any curious person tries to find patterns in an apparently random collection of information. In the nineteenth century, when about 50 elements had been discovered and characterised, attempts were made to find patterns in their properties. Three such attempts are particularly significant.

- In 1829, Johann Döbereiner pointed out that some similar elements formed 'triads', with the middle one showing properties intermediate between the other two. Examples of these triads included lithium, sodium and potassium; calcium, strontium and barium; and chlorine, bromine and iodine.
- In 1864, John Newlands published his 'Law of Octaves'. In this he showed that if the elements are arranged in order of their relative atomic masses, many show properties similar to the element that is eight places further on. Examples include sodium and potassium; magnesium and calcium; boron and aluminium; oxygen and sulphur; nitrogen and phosphorus; and fluorine and chlorine.
- The first periodic table was independently produced by Lothar Meyer and by Dmitri Mendeleev in 1870. Lothar Meyer was the first to put his ideas down on paper, but publication of his manuscript was delayed for two years, and so Mendeleev is often given sole credit.

■ Figure 15.1 Dmitri Mendeleev (1834–1907) collected information about the elements and wrote them down on cards, which he arranged according to relative atomic mass and to the properties of the elements. Along with Lothar Meyer, he formed the periodic table we use today.

Two features of Mendeleev's classification were particularly significant. The first was that he left gaps in his table for elements that were then unknown. He successfully predicted the existence of the elements scandium, gallium and germanium, which were discovered a few years later in 1879, 1875 and 1886, respectively. The second important feature was that for the elements cobalt and nickel, and for the elements tellurium and iodine, he reversed their orders from the order expected based solely on their relative atomic masses. Subsequent work showed that this reversal was completely justified; he had correctly assigned them according to their atomic numbers and the discrepancy in the order of their relative atomic masses was due to the relative proportions of isotopes in the elements (see Chapter 2, page 24).

15.2 A modern form of the periodic table

Groups and periods

A modern form of the periodic table is shown in Figure 15.2.

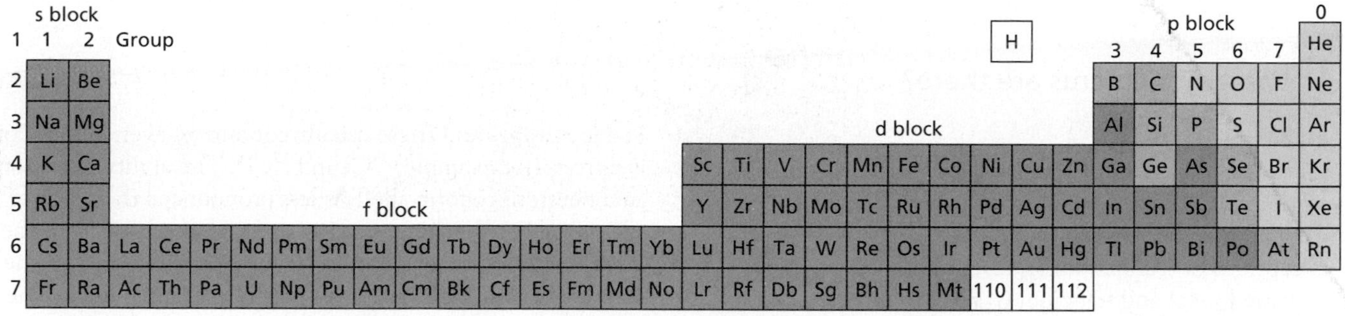

■ **Figure 15.2**
A modern form of the periodic table. The elements shown in orange are good conductors of electricity, and the ones shown in blue are poor conductors. The purple shading indicates that the principal form of the element is a semiconductor. As only a few atoms of the elements at the end of the periodic table have been made, these have been left white.

Elements placed under each other in a column are in the same **group** and show many similarities in their physical and chemical properties. Elements are also arranged in rows or **periods** across the table. The elements in a period have different physical and chemical properties, but trends become apparent in these properties as we move across a period.

s, p, d and f blocks

The first two groups form the **s block** (Groups 1 and 2). These have the outer electronic configuration ns^1 and ns^2, respectively, where n is the number of the shell. The last six groups form the **p block** (Groups 3 to 0) in which the p subshell is being progressively filled. These elements have the outer electronic configuration $ns^2\ np^1$ to $ns^2\ np^6$.

In between the s and p blocks is the **d block**, in which the d subshell is being progressively filled. Our study of the d block is largely restricted to the elements scandium to zinc, and these elements have outer electronic configurations $4s^2\ 3d^1$ to $4s^2\ 3d^{10}$, as we shall see in Chapter 19. When we refer to other d-block elements that are lower in the periodic table, we say that they belong to a **subgroup** – for example, the elements copper, silver (Ag) and gold (Au) belong to the copper subgroup.

About a quarter of the known elements belong to the **f block**. The first row, the 4f, includes all the elements from cerium (Ce) to lutetium (Lu) inclusive. In the past, these elements were called the 'rare earths', but as their abundance in the Earth's crust is comparable to that of lead, this was a misnomer. They are now called the **lanthanides** as their properties are similar to those of the element lanthanum (La) that precedes them in the periodic table. Although they are not particularly rare, the lanthanides are difficult to purify from each other because their chemical properties are nearly identical. The elements of the second row, the 5f, are called the **actinides**. All these elements are radioactive, and most have to be made artificially.

Trends across the periodic table

If the elements are arranged in order of their atomic numbers, several repeating patterns become apparent. For convenience these will be discussed under four headings:

- elements
- oxides
- chlorides
- hydrides.

Example Give the full electronic configurations of:
a beryllium b phosphorus.

Answer a $1s^2\ 2s^2$ b $1s^2\ 2s^2\ 2p^6\ 3s^2\ 3p^3$

Further practice 1 Give the full electronic configurations of:
a sodium b barium.
2 How many elements are there that have at least one stable isotope?

How many elements are there?

Stable nuclei

Until the Manhattan Project was set up in 1940 to make the first atomic bomb, there were 88 known elements. These were all the elements up to uranium ($_{92}$U) with the exceptions of technetium ($_{43}$Tc), promethium ($_{61}$Pm), astatine ($_{85}$At) and francium ($_{87}$Fr), which have no stable isotopes. The elements after bismuth ($_{83}$Bi) have no long-lived isotopes and the ones that are found naturally exist only because they are derived from long-lived isotopes of uranium and thorium that have half-lives of thousands of millions of years.

The question arises as to why there are only 81 elements that have a stable isotope. Nuclear stability is governed by the same general principles as chemical stability, namely that the potential energy of the system tends to a minimum and, at equilibrium, there is a balance between attractive and repulsive forces. The main principles applying to nuclear stability are as follows.

- Protons, being positively charged, repel each other by electrostatic repulsion.
- Nucleons attract each other by the **strong nuclear force** which, at very short distances, is much stronger than the electrostatic force.
- Protons and neutrons, like electrons, tend to form pairs and make up shells.

For the light elements, electrostatic repulsion between protons is minimised by the presence of neutrons in the nucleus. Neutrons increase the number of attractions through the strong nuclear force, but do not add to electrostatic repulsion. Up to calcium, the numbers of protons and neutrons in the nucleus are approximately equal, so the mass number is about twice the atomic number. After calcium, the electrostatic repulsion resulting from the increased number of protons can only be counterbalanced by making the neutron:proton ratio greater than 1. Finally, after bismuth ($_{83}$Bi), the electrostatic repulsion between the protons becomes so great that the nuclei break down, often by α-emission or by nuclear fission (see Chapter 2, page 26).

Paired protons and 'magic numbers'

Because protons form pairs, elements of even atomic number are more stable than those of odd atomic number, and this is shown by the fact that they have many more stable isotopes. Nuclei with an odd atomic number have only one or two stable isotopes and these usually contain an even number of neutrons (for example, ^{35}Cl and ^{37}Cl). The ability of protons and neutrons to form shells is less pronounced than that of electrons, but there are 'magic numbers' which indicate the number of protons or neutrons required to fill a shell. If the number of protons or neutrons equals a 'magic number' the nucleus shows additional stability. The 'magic numbers' are not quite the same as the number of electrons that make up complete shells, being 2, 8, 20, 50, 114, 126 and 184. Some nuclei have extra stability due to double 'magic numbers', such as helium (^{4}He), oxygen (^{16}O) and calcium (^{40}Ca). The extra stability of tin ($_{50}$Sn) is shown by the fact that it has nine stable isotopes. Bismuth, which is just at the edge of stability, has only one stable isotope and that has 126 neutrons.

Unstable nuclei

Over the last 50 years many new elements have been made. These include the technetium, promethium, astatine and francium missing in nature, and also some 20 elements heavier than uranium. Thousands of tonnes of one of these, plutonium ($_{94}$Pu), have been manufactured for atomic warheads. With a half-life of 24 360 years, this highly toxic material represents a severe hazard for generations to come.

In general, the half-life of the element becomes shorter as the atomic number increases, though the elements of even atomic number are slightly less unstable than those with an odd number of protons. Most of the elements have been made in large enough quantities to study their detailed chemistry. The actinides have properties similar to those of the lanthanides. The difference in energy btween the 5f and 6d orbitals is less than the difference between the 4f and 5d orbitals, and this means that the earlier actinides show higher oxidation states than the corresponding lanthanides. At lawrencium ($_{103}$Lr) and rutherfordium ($_{104}$Rf), a new d block is started and these elements show chemical similarities with lutetium ($_{71}$Lu) and hafnium ($_{72}$Hf) above them in the periodic table.

Most interest has been given to a predicted 'island of stability', associated with 114 or 126 protons and 184 neutrons. Recently an isotope of element 114 has been synthesised with a half-life of 30 s, much longer than the half-life of the preceding elements. It is possible that new elements with even longer half-lives will be made in the future.

15.3 Periodic trends in the elements

In this and in subsequent sections, we shall look at the trends in the properties of the elements and their compounds in the second and third periods of the periodic table.

Appearance

The elements on the left of the periodic table have low values of ionisation energy and electronegativity, and so they show the properties associated with metallic bonding (see Chapter 4, page 97) – for example, they are shiny and conduct electricity. In the middle of the periodic table, elements that are semiconductors show a dull shine and are poor conductors of electricity (typically 10^{-12} times that of a metal). The elements at the right of the periodic table are dull and are such poor conductors that they are used as electrical insulators (their conductivities are virtually nil, being typically 10^{-18} times that of a metal).

Table 15.1 Some properties of the elements of the second period. The atomic radii in **bold** are metallic radii, those in *italics* are van der Waals radii and the rest are covalent radii (see Figure 15.4).

	Li	Be	B	C	N	O	F	Ne
Melting point/°C	181	1278	2300	3527	−210	−218	−220	−246
Boiling point/°C	1347	2970	3658	4827	−196	−183	−188	−246
Electrical conductivity	⟵ good ⟶		⟵ fair ⟶		⟵ nil ⟶			
Atomic radius/nm	**0.152**	**0.113**	0.088	0.077	0.070	0.066	0.058	*0.160*
First ionisation energy/kJ mol^{-1}	513	899	801	1086	1402	1314	1681	2081
Electronegativity	0.98	1.57	2.04	2.55	3.04	3.44	3.98	—

Table 15.2 Some properties of the elements in the third period. Once again, the atomic radii in **bold** are metallic radii, those in *italics* are van der Waals radii and the rest are covalent radii.

	Na	Mg	Al	Si	P	S	Cl	Ar
Melting point/°C	98	649	660	1410	44	113	−101	−189
Boiling point/°C	883	1090	2467	2355	280	445	−34	−186
Electrical conductivity	⟵ good ⟶			poor	⟵ nil ⟶			
Atomic radius/nm	**0.154**	**0.160**	**0.143**	0.117	0.110	0.104	0.099	*0.191*
First ionisation energy/kJ mol^{-1}	496	738	577	787	1012	1000	1251	1520
Electronegativity	0.93	1.31	1.61	1.90	2.19	2.58	3.16	—

Melting and boiling points

Metals form giant lattices and their melting and boiling points are largely determined by the strength of the metallic bonding. Metallic bonds are strong if:

- the bonds are short, so that delocalised electrons are strongly attracted to the nuclei of the atoms
- several electrons are available to be added to the 'sea' of electrons.

Group 1 metals therefore have low melting points, because only one electron is available for bonding. The melting points of Group 2 and Group 3 metals are higher because more electrons are available for bonding and their atomic radii are smaller. The highest melting points are found in the d block, whose atoms have small metallic radii and can use both s and d electrons for bonding.

The semiconductors in the middle of the periodic table have high melting and boiling points, because they form giant covalent structures that contain strong bonds with some metallic character.

At the right of the periodic table, the non-metallic elements form small, discrete molecules and have very low melting and boiling points. These properties are determined by the strength of their intermolecular forces rather than the strength of the bonds between their atoms. A common error is to confuse intermolecular forces with interatomic forces. Remember that bond energies are irrelevant when discussing the melting and boiling points of substances with covalent molecules. In general, larger molecules (for example S_8 and I_2) have the highest melting and boiling points because they have the largest intermolecular forces. This is because:

- they have many electrons in the electron cloud
- this electron cloud is easily distorted, because it is far away from the influence of the nucleus.

There is, therefore, a periodic trend in melting and boiling points. They rise from a low value at the left of the periodic table, reach a maximum in the middle, followed by a rapid fall to very low values on the right (see Figure 15.3).

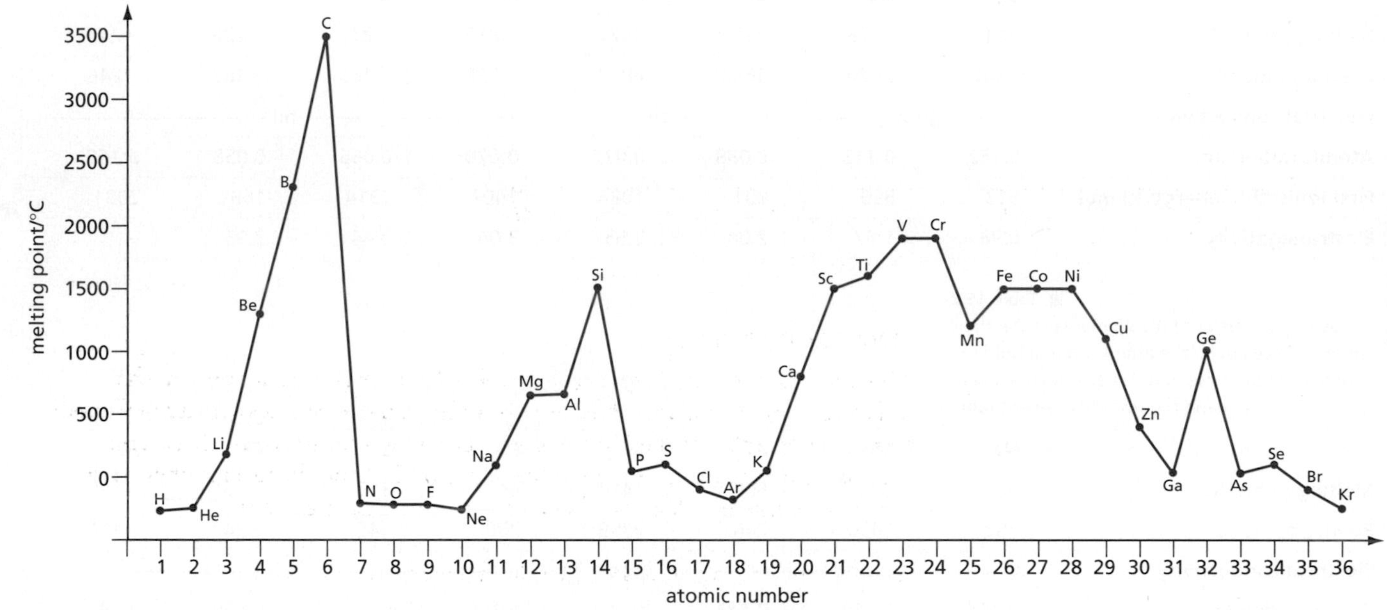

■ **Figure 15.3** Melting points for the first 36 elements of the periodic table.

Atomic radius

There is no single measurement that can be taken as the atomic radius for all atoms (see Chapter 2, page 50). At the left of the periodic table, the atomic radius is that of the atom in the metal (the **metallic radius**), and at the right it is half the distance between the nuclei at the ends of the covalent bond (the **covalent radius**). The lighter Group 0 elements form very few compounds, so their atomic radius is that of an isolated atom (the **van der Waals radius**). Figure 15.4 illustrates these different measures of atomic radius.

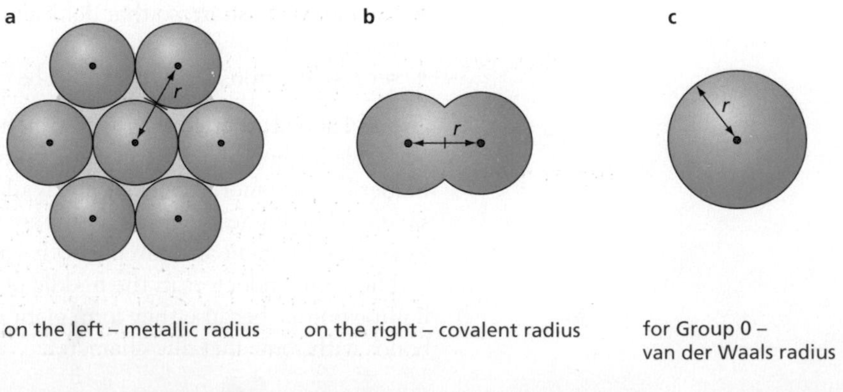

■ **Figure 15.4** The three measurements of atomic radius used across a period. In **a** and **b**, the radius is half the intermolecular distance.

In general, atomic radius decreases on crossing the period. There are two effects that act in opposition:

- the increasing nuclear charge makes the inner electron clouds contract
- the addition of electrons to the outer shell leads to increased repulsion and a slight expansion.

The first effect is larger than the second, so overall there is a contraction in atomic radius on crossing a period (see Figure 15.5).

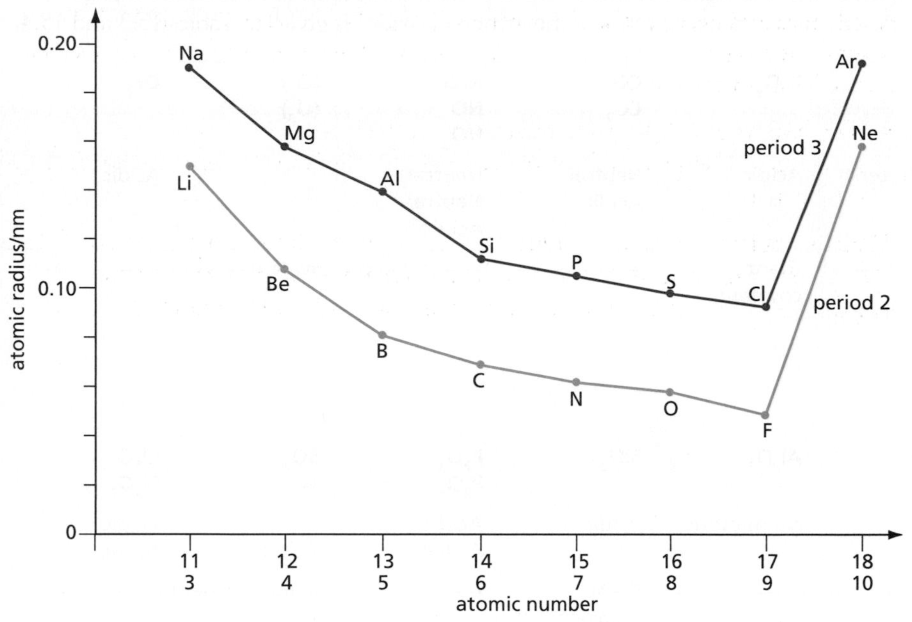

■ **Figure 15.5** The atomic radius decreases on crossing the second and third periods. The large jump to Ar and Ne is because the points show the van der Waals radii, while the other points show metallic or covalent radii (see Tables 15.1 and 15.2).

Ionisation energy

This was discussed in Chapter 2, pages 47–9. As we go across a period, the ionisation energy increases because each electron experiences a greater effective nuclear charge due to the increased number of protons in the nucleus. The increase is not the same for electrons in s, p and d orbitals, because they each experience different degrees of shielding by the inner electrons.

The main trends on crossing a period for the elements

The elements:

- change from metals to semiconductors to non-metals
- show a maximum in melting and boiling points in the middle of the period
- show a decrease in atomic radius
- show a general increase in first ionisation energy.

Example 1 State and explain which element in the second period has:

a the highest melting point **b** the lowest melting point.

Answer **a** Carbon, because it has a giant covalent lattice containing strong bonds.

b Neon – being monatomic, its van der Waals forces are very weak.

Example 2 Explain why the atomic radius of oxygen is smaller than that of carbon.

Answer The nuclear charge has increased from 6 in carbon to 8 in oxygen and this makes the inner shell of electrons contract.

Further practice

1 Which elements in the third period are metals?

2 What is the most reliable way to identify an element as a metal?

3 Which element(s) in the third period form diatomic molecules?

4 Explain why the metallic radius of aluminium is smaller than that of magnesium.

5 Explain why the ionisation energy of carbon is greater than that of boron.

6 Explain why the ionisation energy of sulphur is smaller than that of phosphorus.

15.4 Periodic trends in the oxides

On crossing the periodic table, the oxides of the elements change from basic to amphoteric to acidic in character. Some non-metals (for example, carbon and sulphur) form more than one oxide. In these cases it is found that the oxide of the element in the higher oxidation state is the more acidic (see Chapter 6, page 131). A summary of the properties of the principal oxides is given in Tables 15.3 and 15.4.

Table 15.3
Properties of the principal oxides of the second period.

	Li_2O	BeO	B_2O_3	CO CO_2	N_2O NO NO_2	(O_2) (O_3)	OF_2
Type of oxide	Alkaline	Amphoteric	Acidic	Neutral Acidic	Neutral Neutral Acidic		Acidic
Type of bonding	← Ionic →		Giant covalent	← Simple covalent →			

Table 15.4
Properties of the principal oxides of the third period.

Formula	Na_2O	MgO	Al_2O_3	SiO_2	P_2O_3 P_2O_5	SO_2 SO_3	Cl_2O Cl_2O_7
Type of oxide	Alkaline	Basic	Amphoteric	Acidic	Acidic Acidic	Acidic Acidic	Acidic Acidic
Type of bonding	← Ionic →			Giant covalent	← Simple covalent →		

Alkaline and basic oxides

The chemistry of each oxide is determined by its type – alkaline, basic, amphoteric or acidic. Lithium and sodium oxides are alkaline, dissolving in water to give solutions with a pH greater than 7:

$$M_2O(s) + H_2O(l) \rightarrow 2MOH(aq) \quad (M = Li \text{ or } Na)$$

while magnesium oxide is basic (it is almost insoluble in water, but reacts with acids):

$$MgO(s) + 2H^+(aq) \rightarrow Mg^{2+}(aq) + H_2O(l)$$

Amphoteric oxides

The amphoteric oxides of beryllium and aluminium react extremely slowly with aqueous acids or alkalis, but their hydroxides react readily:

with acids

$$Be(OH)_2(s) + 2H^+(aq) \rightarrow Be^{2+}(aq) + 2H_2O(l)$$
$$Al(OH)_3(s) + 3H^+(aq) \rightarrow Al^{3+}(aq) + 3H_2O(l)$$

with alkalis

$$Be(OH)_2(s) + 2OH^-(aq) \rightarrow Be(OH)_4^{2-}(aq)$$
$$Al(OH)_3(s) + OH^-(aq) \rightarrow Al(OH)_4^-(aq)$$

This similar behaviour of two elements that are diagonally adjacent to each other in the periodic table is known as the **diagonal relationship**. The reason for this similar behaviour is because they have similar electronegativities. Electronegativities increase on moving across a period, but decrease on moving down a group. Therefore, a move of one across and one down means that there is only a small change in electronegativity.

Acidic oxides

Another pair of elements that show this diagonal relationship are boron and silicon. Both have giant covalent oxides that react very slowly with aqueous alkalis. When mixed with solid sodium hydroxide and melted, the following reactions take place:

$$B_2O_3(l) + 2NaOH(l) \rightarrow 2NaBO_2(l) + H_2O(g)$$

$$SiO_2(l) + 2NaOH(l) \rightarrow Na_2SiO_3(l) + H_2O(g)$$

The acids boron(III) hydroxide, $B(OH)_3$, and silicon(IV) hydroxide, $Si(OH)_4$, are such weak acids that they have no effect on universal indicator paper.

Oxides of carbon

Carbon monoxide is the 'anhydride' of methanoic acid, but because it reacts very slowly even with molten sodium hydroxide, it is usually regarded as a neutral oxide. The molecule is isoelectronic (see Chapter 3, page 71) with nitrogen and its very high bond energy ($1075\,kJ\,mol^{-1}$) shows that it contains a triple bond ($^-C{\equiv}O^+$).

Carbon dioxide dissolves in water to give a weakly acidic solution. This solution contains mainly hydrated carbon dioxide and less than 0.1% is present as carbonic acid, $H_2CO_3(aq)$:

$$CO_2(aq) + H_2O(l) \rightleftharpoons H_2CO_3(aq)$$

The reason why carbon dioxide is a gas and silicon dioxide a solid is discussed in Chapter 18, page 341.

Oxides of nitrogen

The formulae of the oxides of nitrogen are unusual. Dinitrogen oxide, N_2O, is isoelectronic with carbon dioxide and has a similar structure ($N^-{=}N^+{=}O$). Nitrogen monoxide, NO, (see Figure 15.6) has a structure that is in-between that of nitrogen (triple bond) and oxygen (double bond). It can be considered as having $2\frac{1}{2}$ bonds and is an 'odd' molecule, with 11 electrons in its outer shell. Nitrogen dioxide, NO_2, is also an unusual molecule – at room temperature, two NO_2 molecules combine together (**dimerise**) to form dinitrogen tetraoxide, N_2O_4 (see Figure 15.7). Only nitrogen dioxide and its dimer, dinitrogen tetraoxide, are acidic. They dissolve in water to give a mixture of nitrous and nitric acids. This is an example of diproportionation (see Chapter 7, page 150).

$$2NO_2(g) + H_2O(l) \rightarrow \underset{\text{nitrous acid}}{HNO_2(aq)} + \underset{\text{nitric(V) acid}}{HNO_3(aq)}$$

■ **Figure 15.6** (below) **a** Nitrogen monoxide is colourless, while **b** nitrogen dioxide is brown.

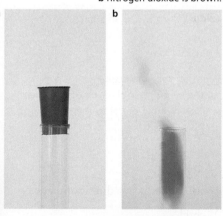

a b

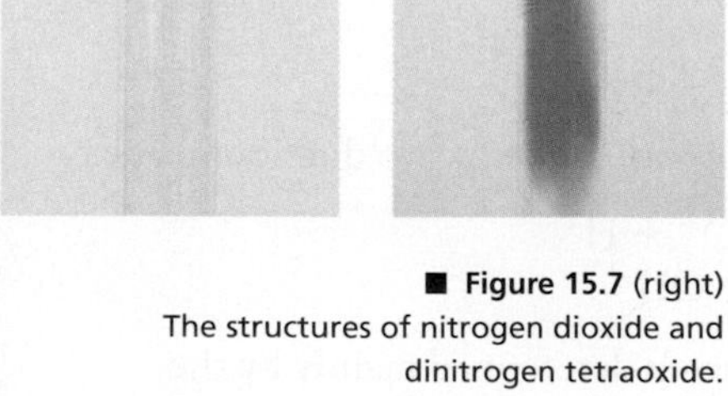

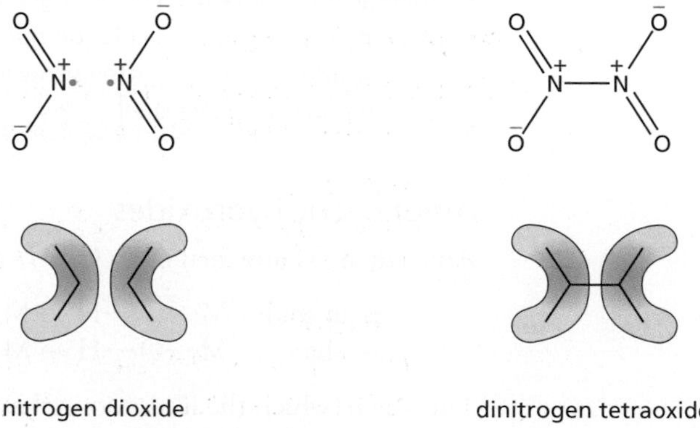

nitrogen dioxide dinitrogen tetraoxide

■ **Figure 15.7** (right) The structures of nitrogen dioxide and dinitrogen tetraoxide.

Oxides of phosphorus and sulphur

Phosphorus(III) oxide, empirical formula P_2O_3, has the molecular formula P_4O_6 and phosphorus(V) oxide, empirical formula P_2O_5, has the molecular formula P_4O_{10}, and either formula may be used in each case. The structures of their molecules are based on the regular tetrahedron of the P_4 molecule (see Figure 15.8). They dissolve in water to give the weak acid phosphonic acid, H_3PO_3, and the slightly stronger acid phosphoric(V) acid, H_3PO_4.

$$P_2O_3(s) + 3H_2O(l) \rightarrow 2H_3PO_3(aq)$$
$$P_2O_5(s) + 3H_2O(l) \rightarrow 2H_3PO_4(aq)$$

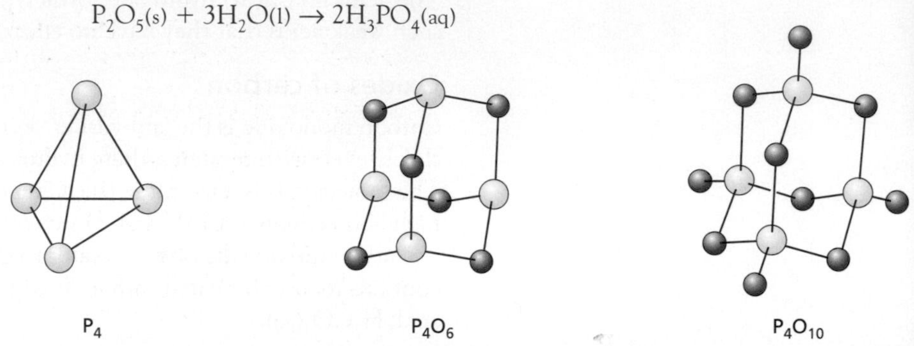

■ **Figure 15.8** The structures of P_4, P_4O_6 and P_4O_{10}.

The molecules of the gases sulphur dioxide, SO_2, and sulphur trioxide, SO_3, are bent and triangular planar, respectively (see Figure 15.9). They dissolve in water to give the weak sulphurous acid, H_2SO_3, and the very strong sulphuric acid, H_2SO_4.

$$SO_2(g) + H_2O(l) \rightarrow H_2SO_3(aq)$$
$$SO_3(g) + H_2O(l) \rightarrow H_2SO_4(aq)$$

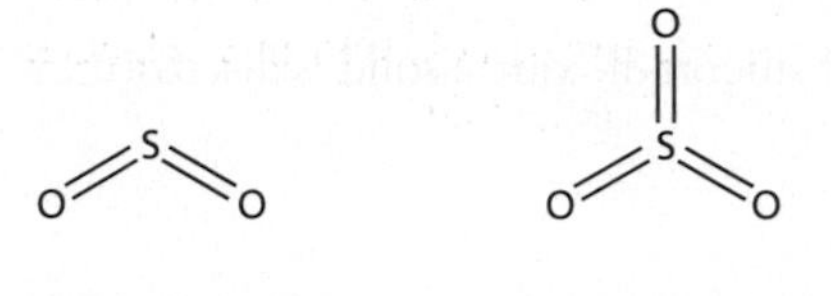

■ **Figure 15.9** The structures of sulphur dioxide and sulphur trioxide.

Oxides of the halogens

Oxygen difluoride, OF_2, reacts with water to give hydrogen fluoride and hydrogen peroxide. Dichlorine oxide, Cl_2O, gives the very weak acid chloric(I) acid, HOCl.

$$OF_2(g) + H_2O(l) \rightarrow 2HF(aq) + H_2O_2(aq)$$
$$Cl_2O(g) + H_2O(l) \rightarrow 2HOCl(aq)$$

The main trends on crossing a period for the oxides

The oxides:

- change from basic to amphoteric to acidic
- change in structure from ionic to giant covalent to small molecular
- show an increase in the oxidation number of the elements in their oxides. In their highest oxides, elements often show an oxidation number equal to the number of electrons in their outer shell.

Amphoteric hydroxides

An element M attached to an O—H group can ionise in two different ways:

as an acid: $M—O—H \rightarrow M—O^- + H^+$

as a base: $M—O—H \rightarrow M^+ + OH^-$

The way in which the ionisation takes place is determined mainly by the electronegativity of M (see page 291). The electronegativity difference between oxygen and hydrogen is 1.24, so that the OH bond has considerable ionic character, $^{\delta-}O—H^{\delta+}$. If M is more electronegative than hydrogen, the M—O bond has less ionic character than the O—H bond does, and the substance behaves principally as an acid. This is generally the case when M is nitrogen, phosphorus, sulphur or chlorine.

If the electronegativity of M is lower than 1.5, the M—O bond has greater ionic character than the O—H bond does, and the substance behaves as a base. This is the case when M is in Group 1 or Group 2 (with the exception of beryllium).

If the electronegativity of M is slightly lower than that of hydrogen, the substance could behave as an acid or a base, that is, be amphoteric. This happens when M is beryllium or aluminium. It is also the case for tin and lead (see Chapter 18) and for many of the elements in the 3d block (see Chapter 19). Some elements that might be expected to be amphoteric are not, for example, iron and nickel. No satisfactory explanation has been given for this behaviour, but it may be because their oxides and hydroxides are so highly insoluble that it is impossible to make them dissolve in excess alkali.

Amphoteric behaviour of M—O—H

Electronegativity of M	Acid–base character
<1.5	basic
1.5–2.5	amphoteric
>2.5	acidic

In the above discussion, it has been assumed that the solvent is water. In other solvents, the results may be different; for example, nitric acid becomes a base if sulphuric acid is used as a solvent (see Chapter 27, page 504).

Example 1

a Explain what is meant by an alkaline oxide.

b Write an equation to show that lithium oxide is an alkaline oxide.

Answer

a An alkaline oxide is an oxide that dissolves in water to give a solution that is strongly alkaline (pH ≈ 13).

b $Li_2O(s) + H_2O(l) \rightarrow 2LiOH(aq)$

Example 2

a Explain what is meant by an amphoteric oxide.

b Write equations to show that aluminium hydroxide is amphoteric.

Answer

a An amphoteric oxide is an oxide that reacts with both acids and bases.

b $Al(OH)_3(s) + 3H^+(aq) \rightarrow Al^{3+}(aq) + 3H_2O(l)$
and $Al(OH)_3(s) + OH^-(aq) \rightarrow Al(OH)_4^-(aq)$

Example 3

a What is meant by a 'weak acid'?

b Name one oxide that dissolves in water to give a very weak acid, and write an equation to show this reaction.

Answer

a A weak acid is an acid that is incompletely ionised in water.

b Carbon dioxide or dichlorine oxide:

$$CO_2(g) + H_2O(l) \rightarrow H_2CO_3(aq) \quad \text{or} \quad Cl_2O(g) + H_2O(l) \rightarrow 2HClO(aq)$$

Example 4

What change takes place in the type of bonding in the oxides in crossing the second and third periods?

Answer

The oxides change from ionic to covalent across a period.

Further practice

1 Write equations to show that beryllium oxide is amphoteric.

2 a What is meant by a neutral oxide?
  b Give the name and formula of one neutral oxide.

3 What changes take place in the type of oxide in crossing the second and third periods?

4 a Predict the formula of the oxide of gallium.
  b Suggest what type of oxide it is, and also what kind of bonding it contains.

5 a Predict the formulae of two oxides of selenium.
  b Write equations for their reactions with water and suggest whether they form weak or strong acids.

Source	Mass of sulphur dioxide produced/ M tonnes per year
Volcanoes	20–40
Dimethyl sulphide	30
Power stations	16
Cars	<1

■ **Table 15.5**
Amounts of sulphur dioxide produced globally from different sources. Dimethyl sulphide is produced from organisms in the sea, and is readily oxidised to sulphur dioxide.

Source	Mass of NO_x produced/ M tonnes per year
Thunderstorms	10
Power stations	8
Cars	6
Industry	6

■ **Table 15.6**
Amounts of NO_x produced globally from different sources.

■ **Figure 15.10**
As well as the physical destruction caused by a volcanic eruption, huge amounts of sulphur dioxide are also given out into the atmosphere.

■ **Figure 15.11** (above)
Lightning provides the energy to turn nitrogen and oxygen in the atmosphere to nitrogen oxides.

■ **Figure 15.12** (right)
The effects of acid rain on a statue.

Acid rain

Sources of sulphur dioxide and nitrogen oxides

Sulphur dioxide is produced whenever sulphur-containing compounds are burnt. Three-quarters of it is produced naturally, as Table 15.5 shows.

Nitrogen combines with oxygen only at high temperatures or in the presence of sparks, such as lightning. Initially nitrogen monoxide is formed, but this reacts slowly with oxygen to give nitrogen dioxide. The mixture of nitrogen oxides is often written as NO_x to indicate its variable composition. Most of this is produced by human activity, as Table 15.6 shows.

Pollution and acid rain

The presence of dry sulphur dioxide and nitrogen dioxide gases in the atmosphere is harmful. The principal adverse effects are as follows:

- respiratory problems, particularly to babies, the elderly and bronchitis sufferers
- damage to trees and other vegetation
- nitrogen dioxide catalyses the formation of ozone, which is a dangerous pollutant at ground level.

Sulphur dioxide is readily oxidised by moist air to give the strong acid sulphuric acid:

$$SO_2(g) + \tfrac{1}{2}O_2(g) + H_2O(l) \rightarrow H_2SO_4(aq)$$

Nitrogen dioxide dissolves in water to give a mixture of nitrous and nitric acids:

$$2NO_2(g) + H_2O(l) \rightarrow HNO_2(aq) + HNO_3(aq)$$

Nitric and sulphuric acids are strongly acidic. When present in rainwater they produce **acid rain**. This is rainfall with a pH lower than pH5. Nitrogen dioxide may exacerbate acid rain formation from sulphur dioxide by catalysing its oxidation to sulphur trioxide:

$$SO_2(g) + NO_2(g) \rightarrow SO_3(g) + NO(g)$$
$$NO(g) + \tfrac{1}{2}O_2(g) \rightarrow NO_2(g)$$

Acid rain causes the following problems:

- It causes aluminium ions in the soil to dissolve. A high concentration of these ions is poisonous to most forms of aquatic life, in particular fish. However, algae flourish under these conditions, particularly if the water also contains high concentrations of phosphate and nitrate ions as a result of excessive use of fertilisers. When the algae die, their decomposition uses up all the oxygen in the water, so that it can no longer support any form of life. The water is said to have undergone **eutrophification**.

- The stonework of buildings is attacked by the acids. Much renovation work has been necessary on old buildings recently because of pollution generated over only the last 100 years.

The harmful effects of acid rain pollution can be minimised by taking the following precautions:

- remove sulphur dioxide from the emissions of power stations as they pass up the chimney stacks
- remove sulphur from diesel fuel and petrol
- use catalytic converters in cars to destroy NO_x (see Chapter 8, page 169).

15.5 Periodic trends in the chlorides

On crossing the second and third periods, the chlorides of the elements change in structure from ionic to covalent. The main properties are summarised in Tables 15.7 and 15.8.

■ Table 15.7 Properties of the principal chlorides of the second period.

	LiCl	$BeCl_2$	BCl_3	CCl_4	NCl_3	Cl_2O	ClF
Type of bonding	Ionic	←			Covalent		→
Action of water	Dissolves	← Hydrolyses →		Nil	←	Hydrolyses	→

■ Table 15.8 Properties of the principal chlorides of the third period.

	NaCl	$MgCl_2$	$AlCl_3$	$SiCl_4$	PCl_3 PCl_5	S_2Cl_2 SCl_2	Cl_2
Type of bonding	← Ionic →		Border-line	←	Covalent		→
Action of water	← Dissolves →		←		Hydrolyses		→

Ionic chlorides

The chlorides of lithium, sodium and magnesium are ionic and dissolve in water, without chemical reaction taking place, and as a result their solutions are neutral.

Aluminium chloride

The structure of beryllium chloride is discussed in Chapter 16, page 309.

At room temperature, aluminium chloride forms a layer lattice. This can be regarded as being made up of layers of close-packed Cl^- ions with alternate layers empty and the other layers two-thirds full of Al^{3+} ions. The Al^{3+} ion has a high charge density (see Chapter 4, page 96); it distorts the Cl^- ions so that the bonding is intermediate between ionic and covalent. Just below its melting point, the solid expands and changes into a structure containing covalent Al_2Cl_6 molecules. These persist in the gas phase, but if the temperature is raised to 800 °C, they break down to monomeric $AlCl_3$ (see Figure 15.13).

■ Figure 15.13 The structure of aluminium chloride.

a solid aluminium chloride, $AlCl_3(s)$

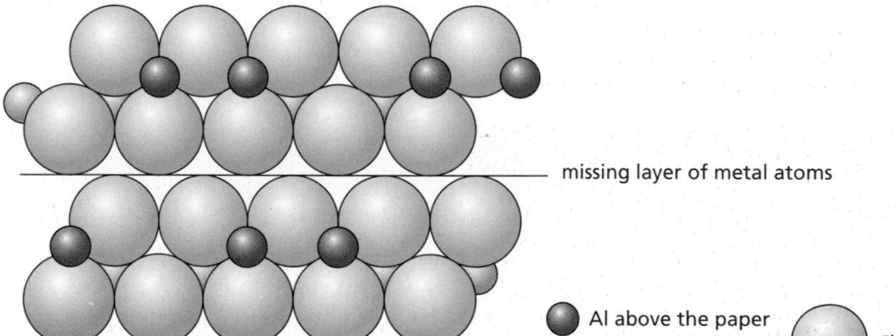

b $Al_2Cl_6(g)$, 200 °C

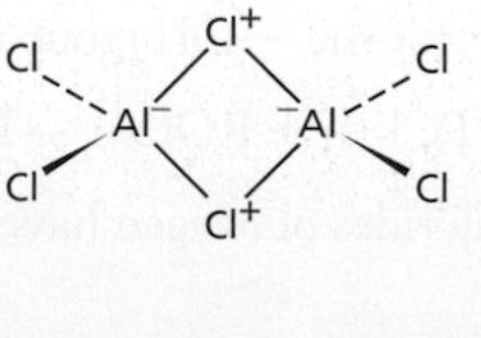

c $AlCl_3(g)$, 800 °C

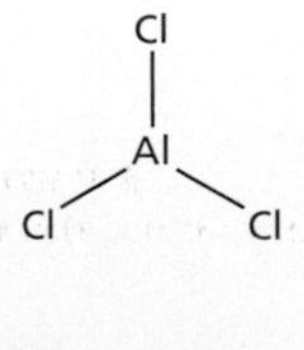

Chlorides and hydrolysis

Both beryllium and aluminium chlorides react rapidly when moistened with water:

$$BeCl_2(s) + H_2O(l) \rightarrow BeO(s) + 2HCl(g)$$
$$2AlCl_3(s) + 3H_2O(l) \rightarrow Al_2O_3(s) + 6HCl(g)$$

This reaction with water is called **hydrolysis**. It contrasts with the behaviour of lithium, sodium and magnesium chlorides, whose ions hydrate but undergo no chemical reaction.

If anhydrous beryllium and aluminium chlorides are dissolved in a large quantity of water, hydrated ions are formed:

$$BeCl_2(s) + 4H_2O(l) \rightarrow Be(H_2O)_4^{2+}(aq) + 2Cl^-(aq)$$
$$AlCl_3(s) + 6H_2O(l) \rightarrow Al(H_2O)_6^{3+}(aq) + 3Cl^-(aq)$$

The resulting solutions are acidic because the hydrated ions have a high charge density and donate a proton to water:

$$Be(H_2O)_4^{2+}(aq) + H_2O(l) \rightarrow Be(H_2O)_3OH^+(aq) + H_3O^+(aq)$$
$$Al(H_2O)_6^{3+}(aq) + H_2O(l) \rightarrow Al(H_2O)_5OH^{2+}(aq) + H_3O^+(aq)$$

Boron and silicon chlorides are rapidly hydrolysed to give the hydroxides, which as we saw on page 293 are extremely weak acids (see Figure 15.14).

$$BCl_3(l) + 3H_2O(l) \rightarrow B(OH)_3(s) + 3HCl(aq)$$
$$SiCl_4(l) + 4H_2O(l) \rightarrow Si(OH)_4(s) + 4HCl(aq)$$

■ **Figure 15.14**
The mechanism of hydrolysis of boron and silicon chlorides.

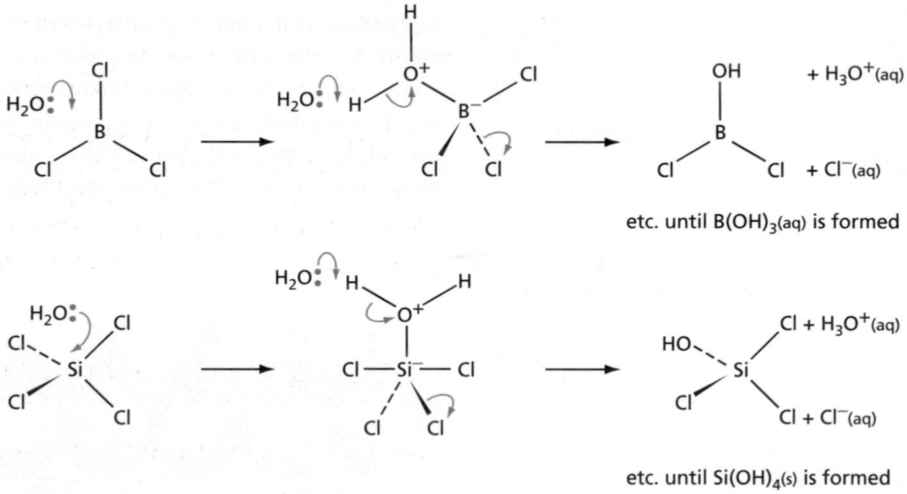

Carbon tetrachloride, CCl_4, is not hydrolysed by cold water. This inertness is discussed in Chapter 18, page 341.

On hydrolysis, nitrogen chloride, NCl_3, gives ammonia and chloric(I) acid, HClO. This unusual reaction probably results from the similar electronegativities of nitrogen and chlorine.

With water, phosphorus(III) chloride gives the weak acid phosphonic acid, H_3PO_3, and phosphorus(V) chloride gives the moderately strong acid phosphoric(V) acid, H_3PO_4.

$$PCl_3(l) + 3H_2O(l) \rightarrow H_3PO_3(aq) + 3HCl(aq)$$
$$PCl_5(s) + 4H_2O(l) \rightarrow H_3PO_4(aq) + 5HCl(aq)$$

With only a limited supply of water, phosphorus(V) chloride produces phosphorus trichloride oxide, PCl_3O. This is the product when phosphorus(V) chloride is used as a test for the —OH group in organic chemistry (see Chapter 26, page 482).

$$PCl_5(s) + ROH(l) \rightarrow RCl(l) + PCl_3O(l) + HCl(g)$$

The chlorides of oxygen have already been considered on page 294.

Disulphur dichloride, S_2Cl_2, and sulphur dichloride, SCl_2, hydrolyse slowly to give a variety of sulphur-containing compounds as well as hydrochloric acid.

The fluorides of chlorine, ClF and ClF_3, are even more reactive than fluorine itself. They react violently with water to give hydrogen fluoride, hydrogen chloride, oxygen and ozone.

The reversible reaction of chlorine with water is considered in Chapter 17, page 330.

The main trends on crossing a period for the chlorides

The chlorides:

- change from ionic to covalent
- become more readily hydrolysed
- show an increase in the oxidation number of the elements in their chlorides.

Example 1

a Distinguish between dissolving and hydrolysis.

b Name one chloride that dissolves in water, and one chloride that undergoes hydrolysis.

Answer

a Dissolving involves hydration but no chemical reaction, while hydrolysis is a chemical reaction that produces new substances.

b Lithium, sodium or magnesium chloride dissolve. Any of the other chlorides undergo hydrolysis except carbon tetrachloride.

Example 2 State how the bonding in the chlorides changes on crossing the second and third periods.

Answer From ionic (through borderline) to covalent.

Example 3 State two ways in which you would know that a reaction had taken place when a few drops of water are added to aluminium chloride.

Answer Heat is given out and hydrogen chloride gas is evolved.

Further practice

1 Write an equation that shows why a solution of beryllium chloride in water is acidic.

2 Predict the shape of the PCl_3 molecule and suggest a likely bond angle.

3 Iron(III) chloride, $FeCl_3$, forms dimers in the gas phase similar to those of aluminium chloride. Draw the likely structure of these dimers.

4 Write a balanced equation for the hydrolysis of $NCl_{3(l)}$.

15.6 Hydrogen and the hydrides

The position of hydrogen in the periodic table

The placing of hydrogen in the periodic table presents a problem. Its electronic configuration is $1s^1$, suggesting that it should be placed in Group 1. However, its ionisation energy is so high (see Table 15.9) that it cannot be considered to be a metal, though there is some evidence of a metallic form at very high pressure. There are no compounds containing the H^+ ion, as this minute ion would have such a high charge density that it would strongly distort any anion near to it. Hydrogen has some similarities to the Group 7 elements. It exists as diatomic molecules and, since it is one electron short of an inert gas configuration, forms compounds containing the H^- ion.

It is probably best to place hydrogen by itself in the periodic table, half-way between Groups 1 and 7. Such a position places hydrogen near boron, and this is appropriate because both elements have similar electronegativities. Helium can then be placed in its natural position above neon in Group 0.

■ Table 15.9 Some properties of hydrogen.

Ionisation energy/kJ mol^{-1}	1312
Electron affinity/kJ mol^{-1}	−73
Covalent radius/nm	0.037
H^+ radius/nm	1×10^{-6}
H^- radius/nm	0.154
Electronegativity	2.20

Trends in the hydrides of the second and third periods

■ Table 15.10 Properties of the principal hydrides of the second period.

	LiH	BeH_2	B_2H_6	CH_4	NH_3 N_2H_4	H_2O H_2O_2	HF
Type of hydride	← Alkaline →		← Neutral →		Basic	Neutral	Acidic
Type of bonding	Ionic	Giant covalent	← Covalent →				
Reaction in dry air	Stable	Stable	Catches fire	← Stable →			

■ Table 15.11 Properties of the principal hydrides of the third period.

	NaH	MgH_2	AlH_3	SiH_4	PH_3	H_2S	HCl
Type of hydride	← Alkaline →			← Neutral →		← Acidic →	
Type of bonding	← Ionic →		Giant covalent	← Covalent →			
Reaction in dry air	← Stable →			← Catches fire →		← Stable →	

s-block hydrides and the H^- ion

Hydrogen reacts directly with the s-block elements to produce hydrides containing the hydride ion, H^-. This is a large ion and is easily polarised by any cation near to it. It behaves as a strong base, abstracting a proton from water and giving hydrogen gas:

$$H^-(s) + H_2O(l) \rightarrow OH^-(aq) + H_2(g)$$

The hydride ion is also a very strong nucleophile and reducing agent, and the hydrides $NaBH_4$ and $LiAlH_4$ act as a source of H^- ions to reduce the C=O group in organic chemistry (see Chapter 28, page 529, and Chapter 29, page 547).

The covalent hydrides

Beryllium and aluminium hydrides have to be made indirectly. They are polymeric and readily react with water, giving the hydroxides and hydrogen. Boron forms a number of hydrides. They are covalent and catch fire in air. The simplest hydride has the formula B_2H_6 (not BH_3) and has a bridged structure similar to that of Al_2Cl_6.

The hydrides of carbon are well known and are discussed in detail in Chapters 23 and 24. In air, they are thermodynamically unstable but fail to catch fire at room temperature because the strength of the C—H bond (413 kJ mol^{-1}) makes the reaction kinetically inert.

Silanes, the silicon analogue of alkanes, are known up to Si_9H_{20}. They catch fire in air because the Si—H bond is comparatively weak (318 kJ mol^{-1}). There are no hydrides containing multiple bonds, because silicon does not form stable π bonds (see Chapter 18, page 336).

As well as ammonia, NH_3, nitrogen forms hydrazine, N_2H_4. Both these hydrides are weak bases and in water exist mainly as hydrated molecules. A derivative of N_2H_4, 2,4-dinitrophenylhydrazine, is used as a test reagent in organic chemistry (see Chapter 28, page 532).

Phosphine, PH_3, is neutral and almost insoluble in water because it cannot form hydrogen bonds with it. Phosphine catches fire in air.

Hydrogen peroxide, H_2O_2, readily decomposes to form water and oxygen because of the weakness of the O—O bond (see Chapter 5, page 126). This decomposition is catalysed by many substances, including manganese(IV) oxide, finely divided platinum and the enzyme catalase.

Hydrogen sulphide, H_2S, smells of rotten eggs and is highly poisonous. Its smell should not be confused with that of sulphur dioxide produced from burning sulphur. Hydrogen sulphide is a weak acid and is appreciably soluble in water:

$$H_2S(g) + H_2O(l) \rightleftharpoons HS^-(aq) + H_3O^+(aq)$$

Hydrogen fluoride and hydrogen chloride are considered in Chapter 17, pages 325–9.

The main trends on crossing a period for the hydrides

The hydrides:

- change from ionic to covalent
- change from basic to neutral to acidic
- in the middle of the period, easily burn in oxygen.

The hydrides of the d block are mentioned in Chapter 19, page 366.

Example 1 Write equations for the following reactions:

a the action of water on hydrogen chloride
b the combustion of silane, SiH_4
c the hydrolysis of magnesium hydride, MgH_2.

Answer

a $HCl(g) + H_2O(l) \rightarrow H_3O^+(aq) + Cl^-(aq)$
b $SiH_4(g) + 2O_2(g) \rightarrow SiO_2(s) + 2H_2O(g)$
c $MgH_2(s) + 2H_2O(l) \rightarrow Mg(OH)_2(s) + 2H_2(g)$

Example 2 Explain why phosphine, PH_3, catches fire in air while ammonia, NH_3, does not.

Answer P—H bonds are long and weak, while N—H bonds are short and strong. The combustion of ammonia has, therefore, a high activation energy (it is *not* energetically unfavourable).

Further practice

1 Write equations for the following reactions:
  a the combustion of B_2H_6
  b the complete combustion of hydrogen sulphide, H_2S
  c the action of hydrochloric acid with hydrazine, N_2H_4.

2 a What is the likely formula of selenium hydride?
  b Suggest what type of hydride it is.
  c Suggest, with a reason, its likely bond angle.

Summary

- In the periodic table, elements are arranged in order of increasing atomic number. Elements with similar properties are placed under each other, in **groups**. Elements with increasing atomic number show periodic changes in their properties and are arranged in **periods**.
- The elements in the second and third periods change across the periods from metallic to non-metallic.
- The oxides of the elements in the second and third periods change across the periods from being ionic and basic to being covalent and acidic.
- The chlorides of the elements in the second and third periods change across the periods from being ionic and unreactive towards water to being covalent and hydrolysed by water.

16 *The s block*

In this chapter we shall see that the elements of the s block, Groups 1 and 2, show typical metallic behaviour. They form compounds containing M^+ and M^{2+} ions, respectively, and the reactivity of the metals increases down each group. The chemical behaviour of the elements in the groups will be seen to be related to fundamental atomic parameters such as the nuclear charge and atomic and ionic size.

The chapter is divided into the following sections.

16.1 Properties of the elements

	Group 1						Group 2					
	Li	**Na**	**K**	**Rb**	**Cs**	**Fr**	**Be**	**Mg**	**Ca**	**Sr**	**Ba**	**Ra**
Melting point/°C	181	98	64	39	28	27	1278	649	839	769	729	700
Boiling point/°C	1347	883	774	688	679	677	2970	1090	1484	1384	1637	1140
Density/g cm^{-3}	0.53	0.97	0.86	1.53	1.87	—	1.85	1.74	1.55	2.54	3.59	5.0
Metallic radius/nm	0.152	0.154	0.227	0.248	0.265	0.27	0.113	0.160	0.197	0.215	0.217	0.223
M^+ ionic radius/nm	0.078	0.098	0.133	0.149	0.165	0.180	0.034	0.079	0.106	0.127	0.143	0.152
ΔH_{at}/kJ mol^{-1}	160	107	89	81	76	73	325	148	178	164	180	159
First IE/kJ mol^{-1}	513	496	419	403	376	400	899	738	590	550	503	509
Second IE/kJ mol^{-1}	—	—	—	—	—	—	1757	1451	1145	1064	965	979
$E^\ominus$/V	−3.04	−2.73	−2.92	−2.92	−2.92	−3.1	−1.97	−2.36	−2.84	−2.89	−2.92	−2.92
Electronegativity	0.98	0.93	0.82	0.82	0.79	0.7	1.57	1.31	1.00	0.95	0.89	0.89

■ **Table 16.1**
The main properties of the elements in Groups 1 and 2. The values for francium and radium are uncertain, and some data is unknown.

a b c

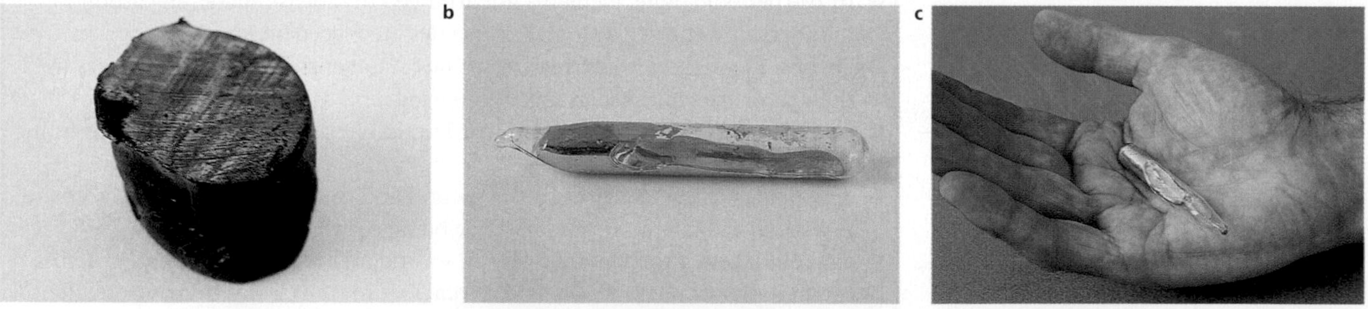

■ **Figure 16.1** (above)
Group 1 – the alkali metals, lithium and caesium. **a** Lithium is covered with a black film of the nitride, Li_3N. **b** Caesium is so reactive it is kept in a sealed ampoule and its melting point is so low that it melts when the ampoule is held in the hand, **c**.

a b

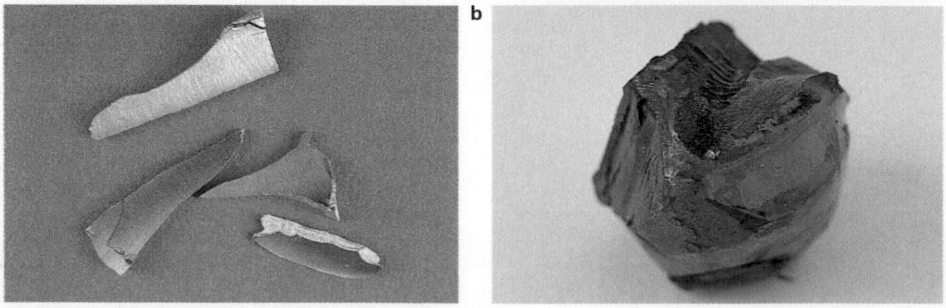

■ **Figure 16.2** (right)
Group 2 – the alkaline earth metals, **a** beryllium and **b** barium.

All the elements in Groups 1 and 2 are metals. They are good conductors of electricity and, when free of an oxide layer, are shiny. On descending Group 1, the melting and boiling points decrease as the metallic bonds become longer and weaker. The same general trend is also present in Group 2 (see Figure 16.3), but there are irregularities in this behaviour, possibly related to the different crystal structures of the elements. This decrease in the strength of metallic bonding explains why the enthalpy changes of atomisation decrease down the groups (see Figure 16.4).

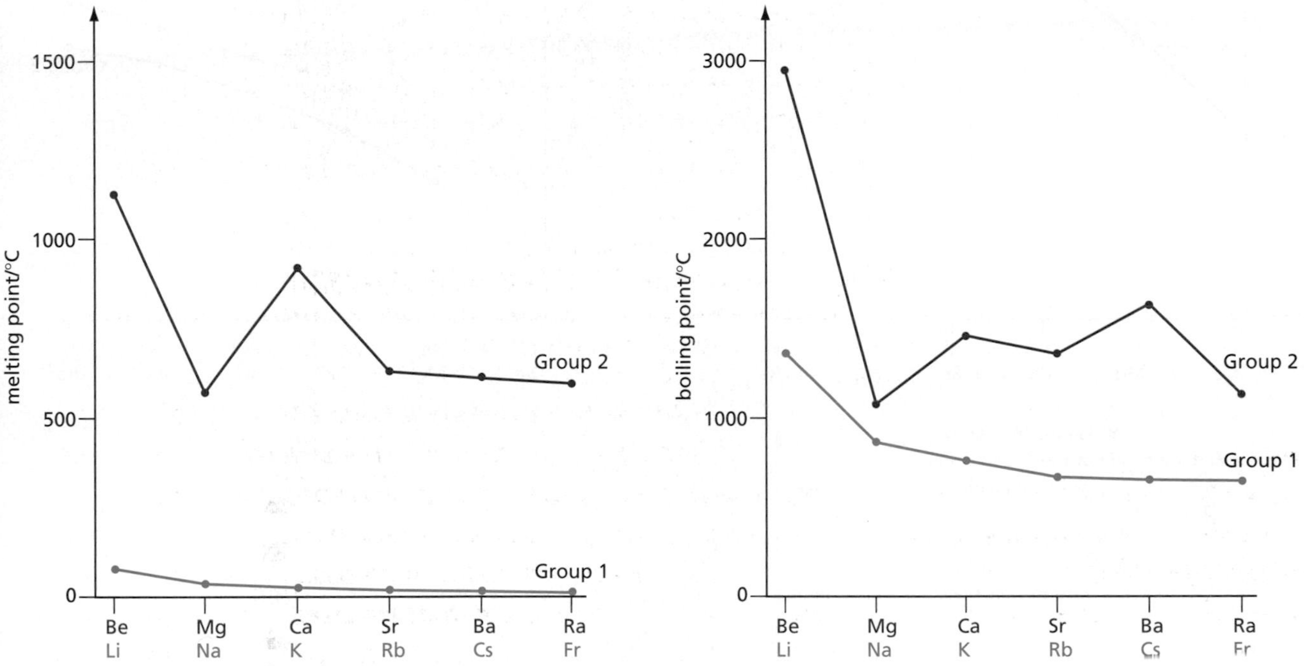

■ **Figure 16.3** (above)
Melting and boiling points of the Group 1 and Group 2 elements.

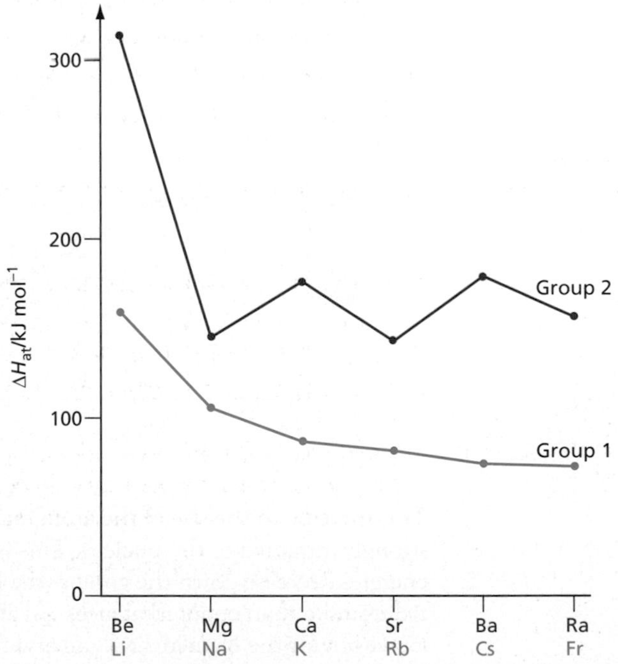

■ **Figure 16.4** (right)
Enthalpy changes of atomisation of the Group 1 and Group 2 elements.

The metallic radii become larger as we go down each group, because an additional shell is being added to the atom, which more than compensates for the decrease in size of the inner shells caused by the increasing nuclear charge. There is a large decrease in size when the ion is formed because the whole of the outer shell (ns^1 or ns^2) is removed, leaving behind the much smaller core (see Figure 16.5, overleaf).

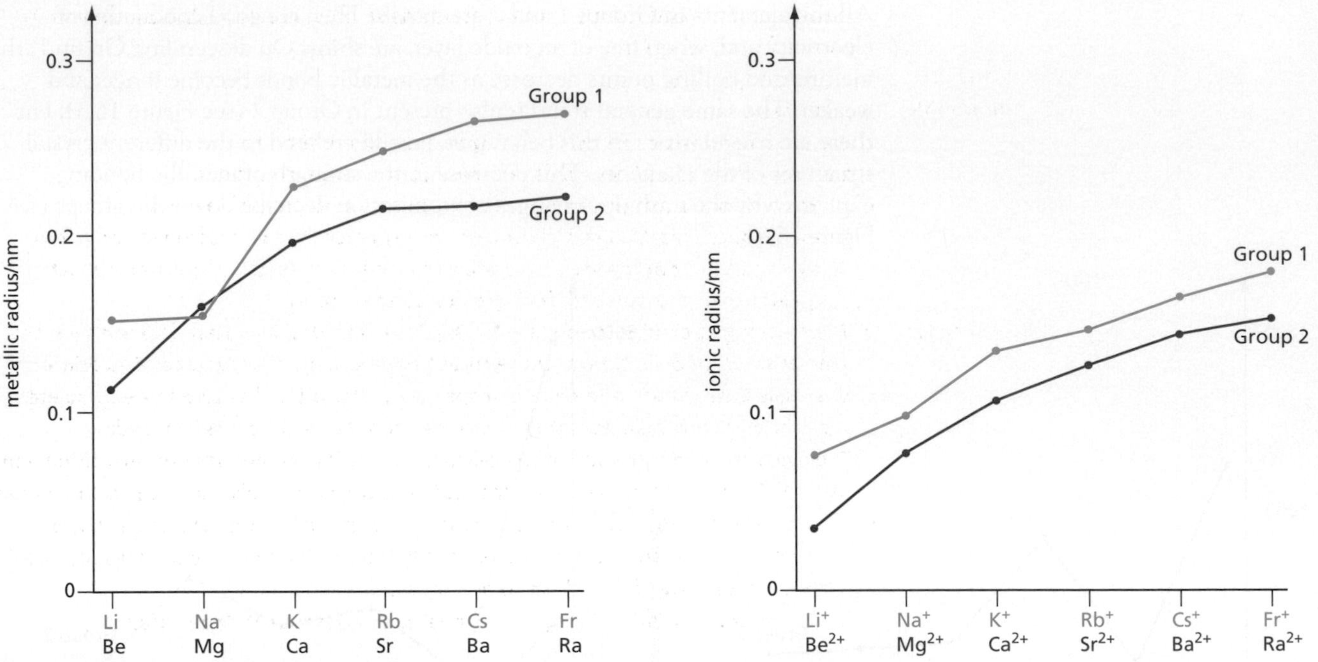

■ **Figure 16.5** (above)
Metallic and ionic radii of the Group 1 and Group 2 elements.

■ **Figure 16.6** (right)
First ionisation energies of the Group 1 and Group 2 elements.

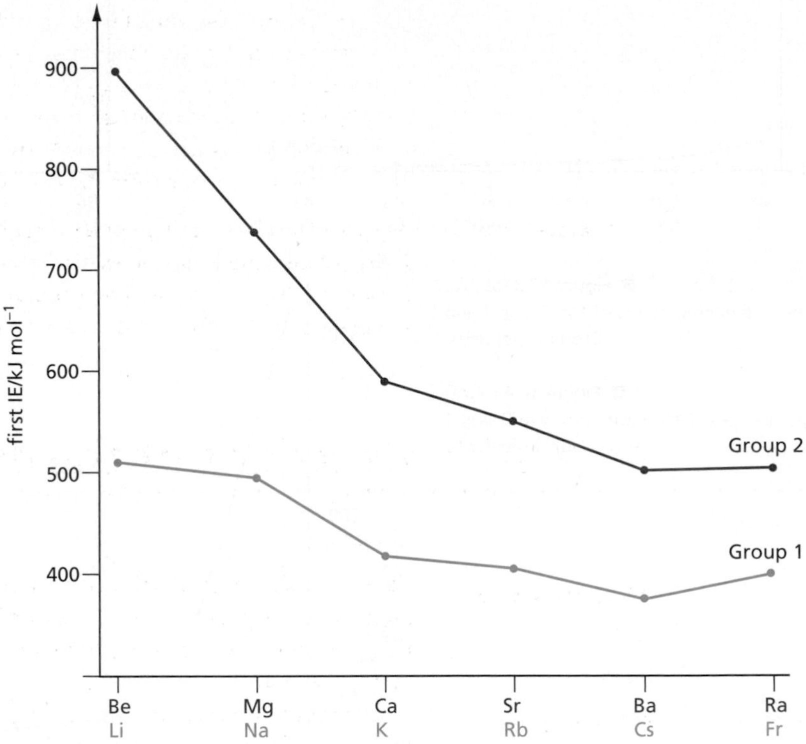

This increase in the size of the atom means that the outer electrons are less strongly attracted to the nucleus. This is the principal reason why ionisation energies decrease down the groups (see Figure 16.6) – the increase in metallic radius more than counterbalances any increase in the effective nuclear charge. However, for the elements radium and francium, the increase in nuclear charge becomes the predominant factor and the first ionisation energies of francium and radium are actually slightly higher than those of caesium and barium, respectively.

Although the ionisation energies decrease in a regular pattern, the standard electrode potentials, $E^{\ominus}$, show less regular behaviour. In particular, the value for lithium is more negative than the values for most of the other elements in Group 1. This point is discussed further on pages 317–18.

Most Group 1 and Group 2 compounds have characteristic flame colours (see Chapter 21, page 387).

Example Explain why:

a the metallic radius of potassium is bigger than that of lithium

b the Ca^{2+} ionic radius is smaller than the metallic radius of calcium

c i the first ionisation energy of calcium is smaller than that of magnesium

ii the second ionisation energy of calcium is greater than its first ionisation energy

iii the increase from the second to the third ionisation energy of calcium is much greater than the increase from the first to the second.

Answer a Their electronic configurations are K: [Ar] $4s^1$ and Li: [He] $2s^1$. (Here [Ar] and [He] are used as shorthand to show that the cores have the same electronic configurations as the noble gases argon and helium, respectively.) Potassium has two more complete shells of electrons than has lithium and therefore has a larger metallic radius.

b The electronic configuration of the calcium atom is [Ar] $4s^2$ and that of the calcium ion is [Ar] $4s^0$. The loss of the 4s subshell makes the calcium ion smaller than the calcium atom.

c i The metallic radius of magnesium is smaller than that of calcium, so the outer electrons are nearer to the nucleus in magnesium and are more strongly attracted. This makes it harder for them to be removed.

ii The second ionisation energy is the energy associated with the process:

$$Ca^{+}(g) \rightarrow Ca^{2+}(g) + e^{-}$$

The second electron is harder to remove than the first because it is being removed from an ion that is already positively charged, rather than from a neutral atom.

iii The third ionisation energy of calcium involves the removal of an electron from the 3p subshell, which is lower in energy than the 4s subshell. The energy required to remove an electron from the 3p subshell is therefore greater than that required to remove an electron from the 4s subshell.

Further practice

1 Explain why the melting point of magnesium is higher than that of sodium.

2 Explain why the second ionisation energy of caesium is much greater than the first.

3 Explain why the first ionisation energy of potassium is less than that of sodium.

4 Explain why the ionic radius of rubidium is smaller than its metallic radius.

16.2 Reactions of the elements

With air

Metal	Principal products
Li	Li_2O, Li_3N
Na	Na_2O_2
K, Rb, Cs	M_2O_2, MO_2
Be, Mg, Ca, Sr	MO, M_3N_2
Ba	BaO, BaO_2, Ba_3N_2

■ **Table 16.2**
The principal products when the Group 1 and 2 metals react with air. Products such as Na_2O_2 contain the peroxide ion, O_2^{2-}.

Except for the most reactive metals in Group 1, all the s-block elements require heating before they will combine directly with the two major elements in air. Reaction with oxygen is slightly more vigorous. Table 16.2 gives a summary of the main products.

Reactivity increases on descending the groups. This is not because the products are thermodynamically more stable, but because the activation energy of the reaction becomes lower.

The simple oxides react with water to give a solution containing hydroxide ions:

$$O^{2-}(s) + H_2O(l) \rightarrow 2OH^{-}(aq)$$

though there is little obvious reaction with magnesium and beryllium oxides because their hydroxides are only slightly soluble (see page 314). The peroxides react with water to produce hydrogen peroxide and hydroxide ions:

$$O_2^{2-}(s) + H_2O(l) \rightarrow H_2O_2(aq) + 2OH^{-}(aq)$$

The nitrides react with water too, giving ammonia:

$$N^{3-}(s) + 3H_2O(l) \rightarrow NH_3(g) + 3OH^{-}(aq)$$

With water

The Group 1 elements all react with cold water to give solutions that are strongly alkaline. Sodium, for example, reacts as follows:

$$Na(s) + H_2O(l) \rightarrow NaOH(aq) + \frac{1}{2}H_2(g)$$

The heat generated during the reaction often melts the metal and ignites the hydrogen (see Table 16.3).

Metal	State	Speed of reaction
Li	Solid	Slow
Na	Liquid	Vigorous, may ignite
K	Liquid	Very vigorous, ignites
Rb	Liquid	Ignites violently
Cs	Liquid	Explosive

■ **Table 16.3**
Summary of the reactivity of the Group 1 metals with water.

■ **Figure 16.7**
Potassium burns with a lilac flame as it reacts with water.

The Group 2 metals are much less reactive. Beryllium and magnesium react with steam at red heat.

$$Be(s) + H_2O(g) \rightarrow BeO(s) + H_2(g)$$
$$Mg(s) + H_2O(g) \rightarrow MgO(s) + H_2(g)$$

The reaction of magnesium with cold water is very slow, producing a solution that is only just alkaline (because magnesium hydroxide is only slightly soluble in water, see page 314). The other Group 2 metals react increasingly more quickly down the group with cold water, to give solutions that are strongly alkaline. Barium, for example, reacts as follows:

$$Ba(s) + 2H_2O(l) \rightarrow Ba(OH)_2(aq) + H_2(g)$$

Because calcium hydroxide is only slightly soluble, a white precipitate is often formed when calcium reacts with water.

With acids

As might be expected, the reactions of the s-block metals with acids are much more vigorous than their reactions with water. Hydrogen gas is once more evolved, and the salt of the metal that corresponds to the anion of the acid is produced. For example, beryllium slowly dissolves in dilute acids with the evolution of hydrogen:

$$Be(s) + H_2SO_4(aq) \rightarrow BeSO_4(aq) + H_2(g)$$

The other Group 2 (and Group 1) metals react increasingly more quickly down the groups, though strontium and barium stop reacting with sulphuric acid because of the insolubility of their sulphates.

The evolution of hydrogen with sodium is sometimes used as a test for the hydroxyl group in organic chemistry (see Chapter 26, page 481).

With chlorine and hydrogen

All the metals react directly with chlorine (and other halogens) to give solid anhydrous chlorides. Potassium and magnesium, for example, react as follows:

$$K(s) + \tfrac{1}{2}Cl_2(g) \rightarrow KCl(s)$$
$$Mg(s) + Cl_2(g) \rightarrow MgCl_2(s)$$

All the chlorides except for beryllium chloride, $BeCl_2$, have ionic structures.

Except for beryllium, all the metals react directly with hydrogen to give solid ionic hydrides that contain the H^- ion. Lithium, for example, reacts as follows:

$$Li(s) + \tfrac{1}{2}H_2(g) \rightarrow LiH(s)$$

Ionic hydrides react rapidly with water, liberating hydrogen:

$$H^-(s) + H_2O(l) \rightarrow OH^-(aq) + H_2(g)$$

Example Write equations for the following reactions:

a water with calcium
b oxygen with lithium
c hydrochloric acid with magnesium.

Answer

a $Ca(s) + 2H_2O(l) \rightarrow Ca(OH)_2(aq) + H_2(g)$
b $2Li(s) + \tfrac{1}{2}O_2(g) \rightarrow Li_2O(s)$
c $Mg(s) + 2HCl(aq) \rightarrow MgCl_2(aq) + H_2(g)$

Further practice Write equations for the following reactions:

1 water with rubidium
2 iodine with calcium
3 nitrogen with lithium
4 oxygen with sodium

16.3 The halides

Group 1 metal halides

The Group 1 metals all form anhydrous halides. Lithium, whose ion has a high charge density, also forms hydrated solid halides such as $LiCl.H_2O(s)$. All the s-block halides dissolve in water to give neutral solutions containing hydrated ions. Lithium forms $Li(H_2O)_4^+$ ions, and the other s-block halides give mainly $M(H_2O)_6^+$ ions.

The crystal structures are of two types. The structures are largely determined by two factors:

- having as many oppositely charged ions in contact with each other as possible
- preventing like charged ions from touching each other.

Metal elements, which have atoms of the same size, have a maximum coordination number of 12 (see Chapter 4, pages 94 and 97), but this is impossible in an ionic solid because it would bring like charges into contact with each other. The maximum coordination number of ionic compounds is usually 8, and this coordination number is found if the cations and anions are of comparable size (for example, in caesium chloride, see Figure 16.8, overleaf). For most of the Group 1 halides, the cation is much smaller than the anion. If there were 8 anions around each cation, the anions would be in close contact with each other. The coordination number is therefore found to be 6 – this is the sodium chloride structure, shown in Figures 16.9 and 16.10 (overleaf).

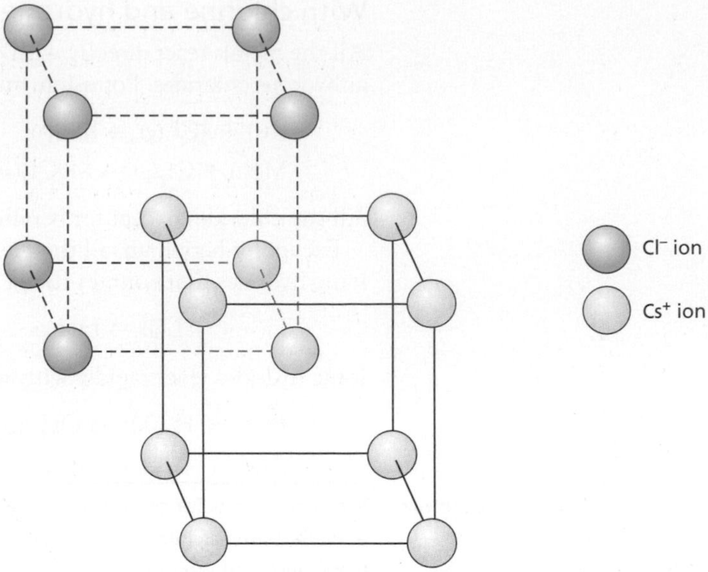

■ **Figure 16.8**
The caesium chloride structure. Each Cs^+ ion is surrounded by eight Cl^- ions at the corners of a cube, and similarly each Cl^- ion is surrounded by eight Cs^+ ions. The structure is described as 8:8 and is formed by two cubic lattices penetrating each other.

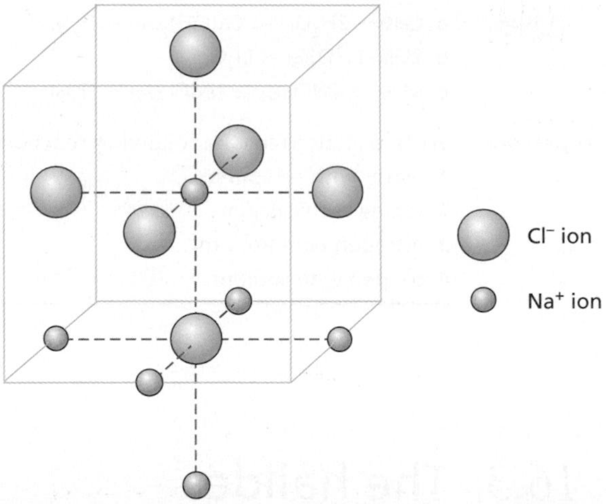

■ **Figure 16.9**
The sodium chloride structure. Each Na^+ ion is surrounded by six Cl^- ions on the faces of a cube, and similarly each Cl^- ion is surrounded by six Na^+ ions. The structure is described as 6:6 and is formed by two face-centred cubic lattices penetrating each other.

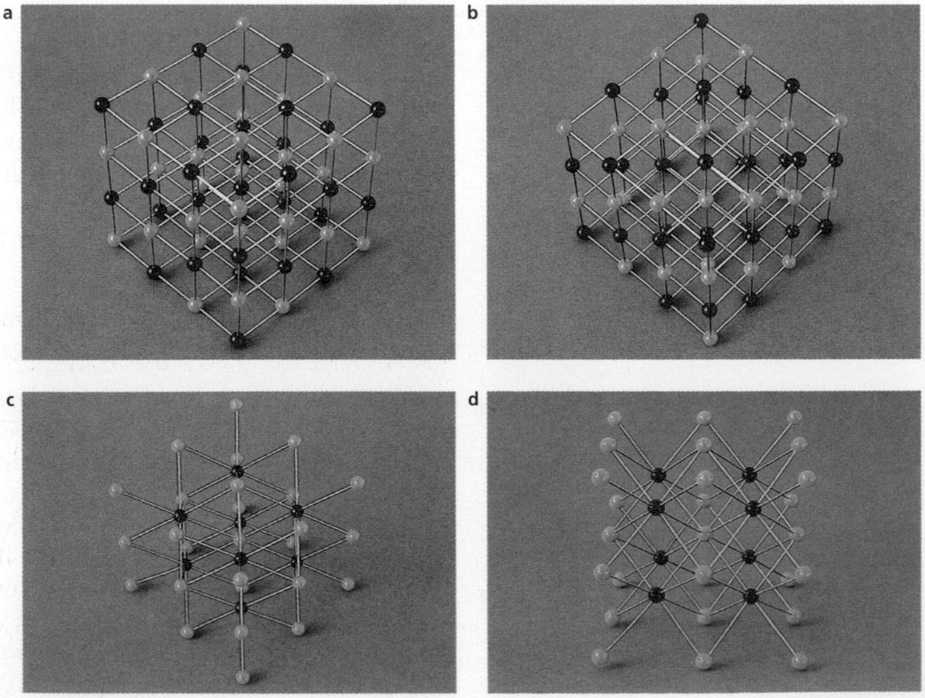

■ **Figure 16.10**
The sodium chloride (**a** and **b**) and caesium chloride (**c** and **d**) structures. **b** and **d** show the layers of Na^+ or Cs^+ ions and the layers of Cl^- ions.

a

b

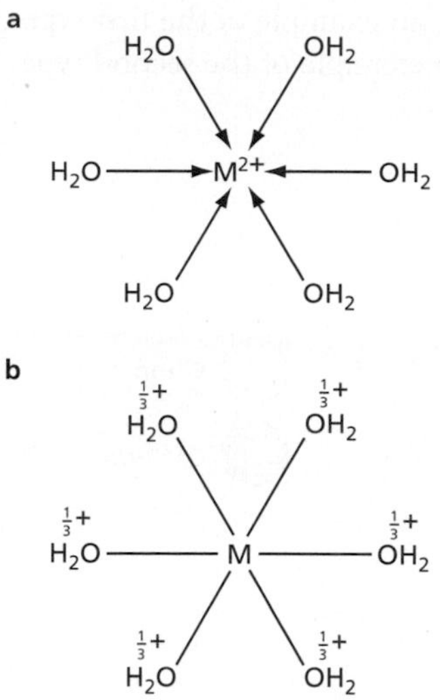

■ **Figure 16.11** (above)
The formation of an $M(H_2O)_6^{2+}$ ion. The overall
charge is spread over the outside of the ion, so
that structure **b** is more representative than
structure **a**.

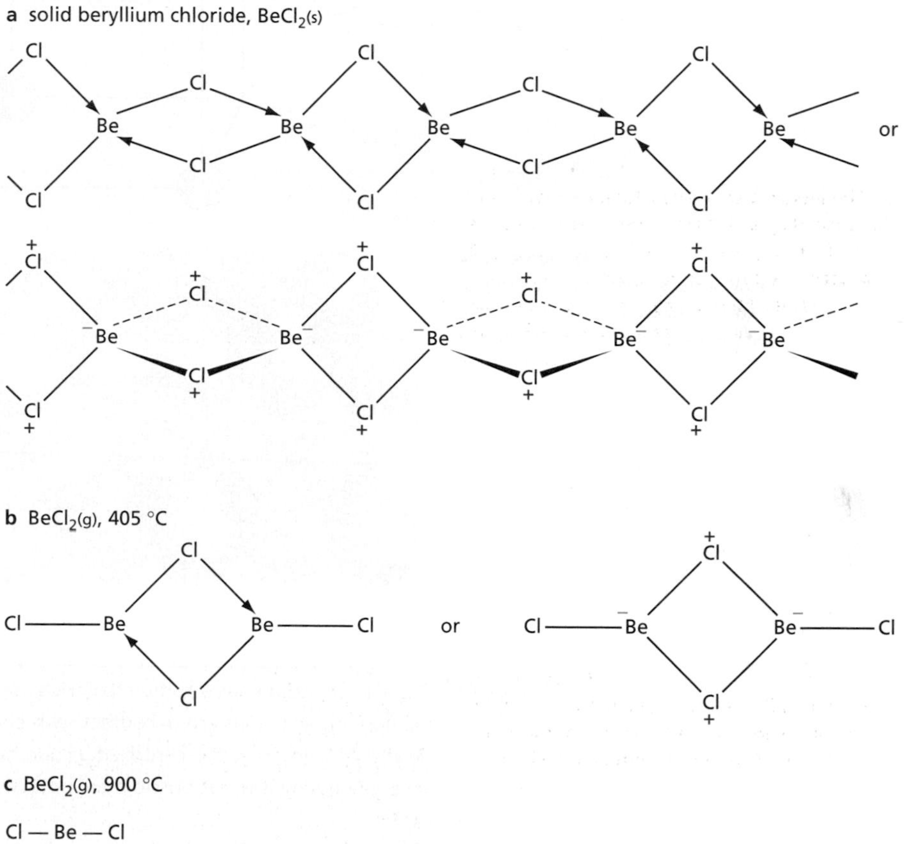

■ **Figure 16.12** (right)
The structures of beryllium chloride.

Group 2 metal halides

As we have mentioned, the Group 2 metals form anhydrous halides by direct
combination. Because their ions have a high charge density, they also form
hydrated solid chlorides, for example, $MgCl_2.6H_2O(s)$. This contains the
$Mg(H_2O)_6^{2+}$ ion, which is also present when the substance dissolves in water. The
bonding in this ion, and in other similar cations, is dative covalent, formed when
pairs of electrons are donated from six water molecules to the metal ion. Because
oxygen is much more electronegative than the metal, the electron donation is
incomplete and the overall charge is spread over the outside of the ion (see Figure
16.11).

Anhydrous beryllium chloride has an unusual bridged structure in the solid
state which, in the vapour, breaks down to Be_2Cl_4 and finally to $BeCl_2$ (see
Figure 16.12).

a solid beryllium chloride, $BeCl_2(s)$

b $BeCl_2(g)$, 405 °C

c $BeCl_2(g)$, 900 °C

Cl — Be — Cl

Hydrated beryllium chloride decomposes on heating, being hydrolysed by its water
of crystallisation:

$$BeCl_2.4H_2O(s) \rightarrow BeO(s) + 2HCl(g) + 3H_2O(g)$$

A similar reaction takes place when hydrated magnesium chloride is heated:

$$2MgCl_2.6H_2O(s) \rightarrow Mg_2OCl_2(s) + 2HCl(g) + 11H_2O(g)$$

so that the anhydrous salt must be made by direct combination or by heating the
hydrate in a stream of dry HCl.

The structures of Group 2 halides are complicated because there must be an
equal number of 'holes' as cations in the lattice (see page 85) in order to achieve
electrical neutrality. These 'holes' are formed in two ways:

- by omitting every other cation in the lattice (most of the fluorides and
  chlorides)
- by leaving out a whole layer of metal ions (most of the bromides and iodides).

Figure 16.13 shows the structure of calcium fluoride, an example of the first type of structure, and Figure 16.14 shows calcium iodide, an example of the second type.

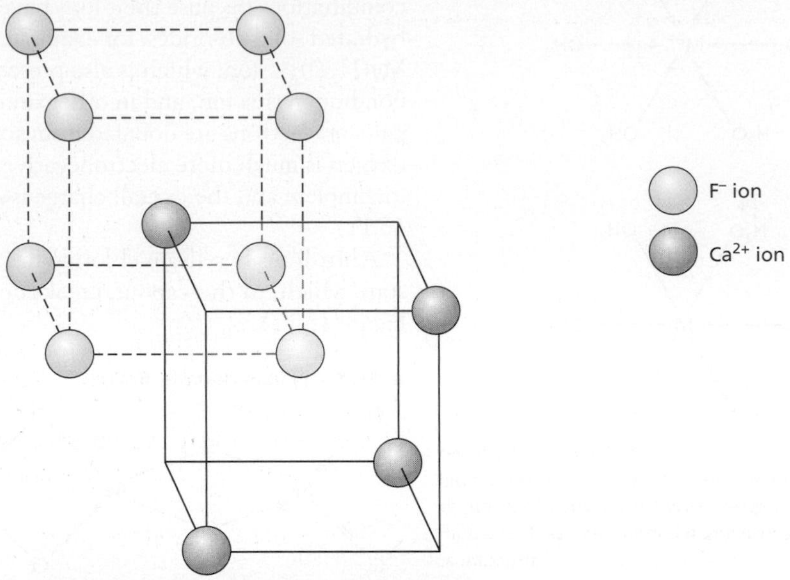

■ **Figure 16.13**
The structure of calcium fluoride, which is similar to that of caesium chloride but with missing alternate metal ions. It has 8:4 coordination.

F⁻ ion

Ca²⁺ ion

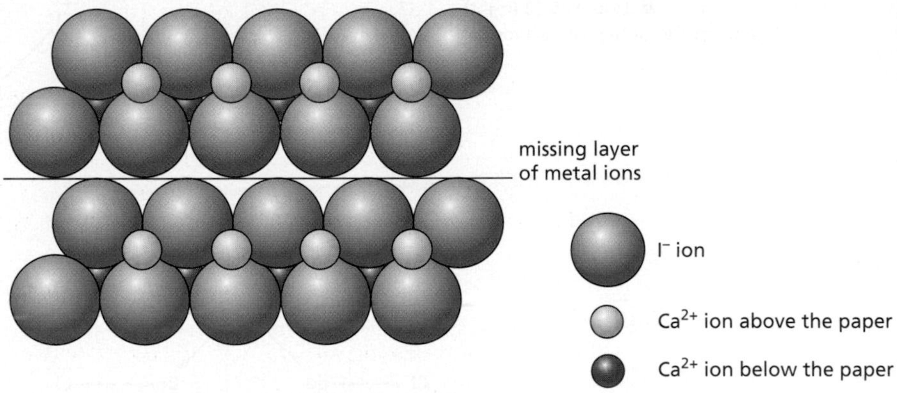

■ **Figure 16.14**
The structure of calcium iodide, which is similar to that of sodium chloride with each alternate layer of metal ions missing. It has 6:3 coordination.

missing layer of metal ions

I⁻ ion

Ca²⁺ ion above the paper

Ca²⁺ ion below the paper

Example 1   Potassium iodide has a sodium chloride structure.
  a How many I⁻ ions are in contact with each K⁺ ion?
  b If the K⁺ ion is at the centre of a cube, where are the I⁻ ions located?
  c Explain why it is not possible for potassium iodide to form a caesium chloride structure.

Answer   **a** 6
  b At the centre of each face.
  c The potassium ion is small, so the iodide ions would be in close contact with each other.

Example 2   **a** How many electrons are there around each beryllium atom in $BeCl_{2(g)}$?
  b Draw a dot-and-cross diagram to show the electronic configuration of $BeCl_{2(g)}$.

Answer   **a** 4

  b $\overset{\times\times}{\underset{\times\times}{\times}}Cl\overset{\times}{\underset{\bullet}{—}}Be\overset{\bullet}{\underset{\times}{—}}\overset{\times\times}{\underset{\times\times}{Cl}}\times$

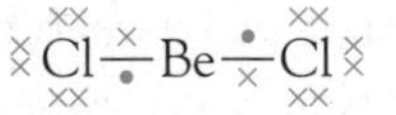

Further practice   **1** Caesium iodide has a caesium chloride structure.
  a How many I⁻ ions are in contact with Cs⁺ ion?
  b If the Cs⁺ ion is at the centre of a cube, where are the I⁻ ions located?
  c Explain why, for caesium iodide, the caesium chloride structure is more stable than the sodium chloride structure.
  d Suggest why caesium fluoride forms a sodium chloride structure rather than a caesium chloride structure.
  2 a How many electrons are there around each beryllium atom in $Be_2Cl_{4(g)}$?
  b Draw a dot-and-cross diagram to show the electronic configuration of $Be_2Cl_{4(g)}$.

16.4 Thermal stability of the carbonates and nitrates

How we use the elements and compounds of the s block

Because Group 1 and Group 2 metals are very reactive, they are all extracted by the electrolysis of a molten salt, usually the chloride. The extraction and uses of these metals is discussed in Chapter 20, page 378.

Sodium hydroxide (from the electrolysis of brine) and sodium carbonate have many important uses, as we shall see in Chapter 20.

Calcium carbonate occurs as **limestone** and finds many important uses. The large deposits of calcium cabonate found in the Earth's crust were formed largely from the shells of marine organisms that lived millions of years ago. This calcium cabonate, originally in the form of **aragonite**, may dissolve as calcium hydrogencarbonate, $Ca(HCO_3)_2$ (see page 315) and be precipitated as **chalk**. Alternatively it may be converted by heat and pressure into metamorphic rock such as **marble**. Calcium carbonate is also present in bones, along with calcium phosphate, $Ca_3(PO_4)_2$.

The main uses of limestone are as follows:

- treating acidic soils
- in the manufacture of iron and steel (see Chapter 20, pages 375–6).
- making cement.

The decomposition of limestone is also industrially important:

$$CaCO_3(s) \rightarrow CaO(s) + CO_2(g) \qquad \Delta H = +180\,kJ\,mol^{-1}$$

This highly endothermic reaction is carried out in kilns heated to 900 °C.

Calcium oxide, called **quicklime**, readily takes up water to form calcium hydroxide or **slaked lime**, $Ca(OH)_2$:

$$CaO(s) + H_2O(l) \rightarrow Ca(OH)_2(s) \qquad \Delta H = -65\,kJ\,mol^{-1}$$

The main uses of slaked lime include:

- treating acidic soils and lakes
- as a constituent of mortar in brickwork.

A saturated solution of calcium hydroxide is **limewater**, used in testing for carbon dioxide, as we shall see later in the chapter (page 315).

Figure 16.15 Powdered limestone and slaked lime are both used to neutralise acidity in lakes caused by acid rain.

Carbonates

Of the Group 1 carbonates, only that of lithium breaks down readily at the temperature of a Bunsen burner:

$$Li_2CO_3(s) \rightarrow Li_2O(s) + CO_2(g)$$

The Group 2 carbonates become more stable to heat as the group is descended, and only beryllium, magnesium and calcium carbonates break down at the temperature of a Bunsen burner (see Table 16.4):

$$MCO_3(s) \rightarrow MO(s) + CO_2(g)$$

Table 16.4 Temperature of decomposition of the Group 2 carbonates at 1 atm pressure.

M	Temperature of decomposition/°C
Be	250
Mg	540
Ca	900
Sr	1289
Ba	1360

This trend in the ease of thermal decomposition can be explained in several ways. The simplest explanation considers the sizes of the cations. Cations are small at the top of the group, and so have a high charge density. This high charge density distorts the carbonate ion, so that it more readily breaks down to the oxide ion and carbon dioxide (see Figure 16.16).

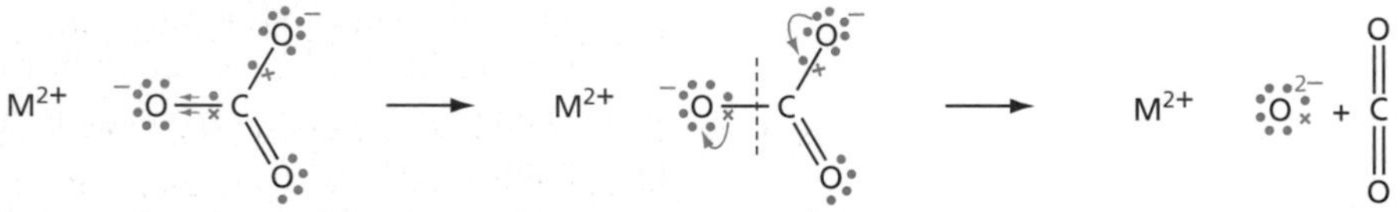

Figure 16.16 The high charge density of the beryllium and magnesium cations distorts the carbonate ion, so it breaks down at a lower temperature.

Energetics of the thermal decomposition of the Group 2 carbonates

The trend in the ease of thermal decomposition of the Group 2 carbonates can also be explained using arguments based on lattice energies (see Chapter 10, page 189). The lattice energy of a Group 2 carbonate is mainly determined by the large size of the carbonate ion, and varies only slightly as the cation becomes larger. On the other hand, the lattice energies of the Group 2 oxides are mainly determined by the size of the cation, so these become much less negative on descending the group. The enthalpy change for the decomposition of the carbonate depends on the difference between the lattice energy for the carbonate and that for the oxide, so the oxide is favoured at the top of the group, and the carbonate at the bottom (see Figure 16.17 and Table 16.5). The enthalpy change of decomposition of the carbonate, ΔH_2, is given by:

$$\Delta H_2 = -\mathrm{LE}(MCO_3) + \Delta H_1 + \mathrm{LE}(MO)$$

ΔH_1 (for the breakdown of the carbonate ion) has a constant value ($+778\,\mathrm{kJ\,mol^{-1}}$) for all the carbonates, so that the stability of a particular carbonate depends on the difference in lattice energy between the carbonate and the oxide.

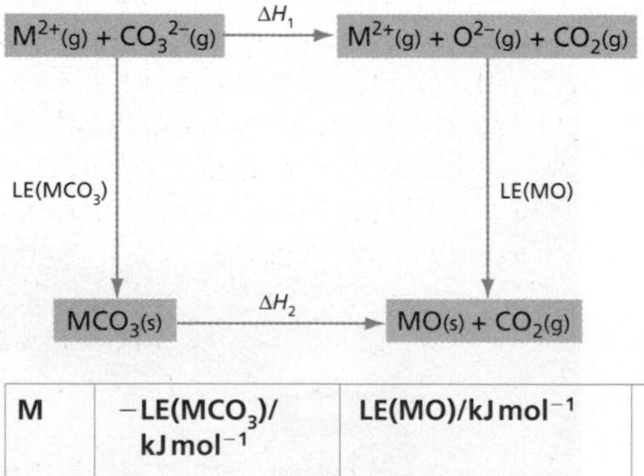

Figure 16.17 Thermodynamic cycle for the decomposition of Group 2 carbonates.

Table 16.5 Thermochemical data for the thermal decomposition of the Group 2 carbonates. The lattice energies for the carbonates change by 1142 kJ mol^{-1} from beryllium to barium. The change in lattice energies for the oxides is greater (1389 kJ mol^{-1}), so that beryllium oxide is more stable compared with beryllium carbonate than is barium oxide compared with barium carbonate. So beryllium carbonate decomposes at a lower temperature than does barium carbonate. (The actual value of this decomposition temperature depends on entropy changes as well as enthalpy changes, but these entropy changes are likely to be very similar for each of the decomposition reactions, so they can be ignored in this discussion of trends rather than actual values.)

M	$-\mathrm{LE}(MCO_3)$/ kJ mol^{-1}	LE(MO)/kJ mol^{-1}	ΔH_1	ΔH_2	Temperature of decomposition of $MCO_3(s)$/°C
Be	3686	−4443	+778	+21	250
Mg	3113	−3791	+778	+100	540
Ca	2810	−3410	+778	+178	900
Sr	2679	−3223	+778	+234	1289
Ba	2544	−3054	+778	+268	1360

Nitrates

With the exception of lithium nitrate, the Group 1 nitrates melt on heating and evolve oxygen to leave the nitrites (nitrate(III)):

$$MNO_3(l) \rightarrow MNO_2(l) + \tfrac{1}{2}O_2(g) \qquad (M = Na, K, Rb, Cs)$$

Lithium nitrate resembles the Group 2 nitrates in decomposing to form the oxide, giving off nitrogen dioxide as well as oxygen:

$$2LiNO_3(l) \rightarrow Li_2O(s) + 2NO_2(g) + \tfrac{1}{2}O_2(g)$$
$$M(NO_3)_2(l) \rightarrow MO(s) + 2NO_2(g) + \tfrac{1}{2}O_2(g) \qquad (M = Be, Mg, Ca, Sr, Ba)$$

If strontium and barium nitrates are heated carefully, it is possible to isolate the nitrites before they decompose further to the oxides. In this way their behaviour resembles that of the Group 1 nitrates.

The ease of decomposition of Group 2 nitrates follows the same trend as that shown by the carbonates and for the same reason – the NO_3^- ion is distorted by the high charge density of the cations at the top of the group.

Example 1 Which of the following carbonates decompose at the temperature of a Bunsen burner?
lithium carbonate, potassium carbonate, calcium carbonate, strontium carbonate

Answer Lithium carbonate and calcium carbonate (just).

Example 2 Write equations for the thermal decomposition of:
a caesium nitrate
b lithium carbonate
c magnesium nitrate.

Answer
a $CsNO_3(l) \rightarrow CsNO_2(l) + \tfrac{1}{2}O_2(g)$
b $Li_2CO_3(s) \rightarrow Li_2O(s) + CO_2(g)$
c $Mg(NO_3)_2(l) \rightarrow MgO(s) + 2NO_2(g) + \tfrac{1}{2}O_2(g)$

Further practice Explain qualitatively why magnesium carbonate is less thermally stable than calcium carbonate.

16.5 Solubilities of the compounds

Trends in solubility – sulphates and hydroxides

It is often stated that all Group 1 compounds are soluble in water. However, some Group 1 compounds are only slightly soluble, for example, lithium fluoride, LiF, and potassium chlorate(VII) (potassium perchlorate), $KClO_4$. These illustrate a general trend found in solubilities. The solubilities of the fluorides (small anion) increase from lithium to caesium, while the solubilities of the perchlorates (large anion) decrease from lithium to caesium. In general, small anions tend to form insoluble compounds with small cations, and large anions tend to form insoluble compounds with large cations.

The same effect is found in Group 2. The solubilities of the hydroxides (small anion) increase from beryllium to barium, while the solubilities of the sulphates (large anion) decrease from beryllium to barium (see Table 16.6).

Table 16.6 Solubilities of the Group 2 hydroxides and sulphates.

M	Solubility of $M(OH)_2$/moles per 100 g of water	Solubility of MSO_4/moles per 100 g of water
Be	8.0×10^{-7}	2.4×10^{-1}
Mg	1.6×10^{-5}	2.2×10^{-1}
Ca	2.5×10^{-3}	1.5×10^{-3}
Sr	3.4×10^{-3}	7.1×10^{-4}
Ba	4.1×10^{-2}	1.1×10^{-6}

Energetics of the solubilities of the Group 2 sulphates and hydroxides

Sulphates become less soluble down the group

The enthalpy change of solution for the Group 2 sulphates can be calculated using the thermochemical cycle shown in Figure 16.18 (see Chapter 10, page 194).

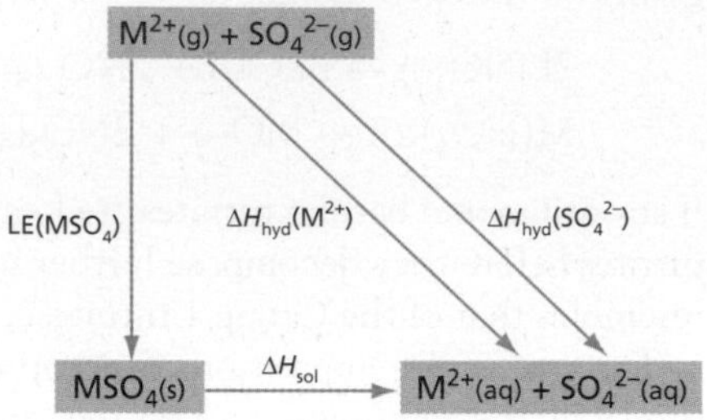

■ Figure 16.18 Thermochemical cycle for the solubility of Group 2 sulphates.

This shows us that:

$$\Delta H_{sol} = +\Delta H_{hyd}(M^{2+} + SO_4^{2-}) - LE(MSO_4)$$

Similar arguments can be used to explain the solubility trends as were used to explain the decomposition of the carbonates. For the sulphates, the values of lattice energies are dominated by the large sulphate ion, so there is a comparatively small change from beryllium to barium. The enthalpy changes of hydration change to a greater extent from the small beryllium ion to the much larger barium ion (see Table 16.7). ΔH_{hyd} (SO_4^{2-}) has the constant value $-1160\,kJ\,mol^{-1}$.

■ Table 16.7 Thermochemical data for the solubilities of the Group 2 sulphates. The lattice energies vary by only $574\,kJ\,mol^{-1}$ from beryllium sulphate to barium sulphate. The variation in the values of $\Delta H^{\ominus}_{hyd}$ for the Group 2 metal ions is greater ($687\,kJ\,mol^{-1}$), so that the general trend is for $\Delta H^{\ominus}_{sol}$ to become more positive (by $113\,kJ\,mol^{-1}$) going down the group. This is the principal reason for the decreasing solubility. (The actual value of each solubility also depends on entropy changes, but once again the argument based on enthalpy changes is adequate as an explanation of the general trend.)

M	$-LE(MSO_4)$/ $kJ\,mol^{-1}$	$\Delta H_{hyd}(M^{2+})$/ $kJ\,mol^{-1}$	$\Delta H_{hyd}(SO_4^{2-})$/ $kJ\,mol^{-1}$	ΔH_{sol}	Solubility in 100 g of MSO_4/mol
Be	3033	−1960	−1160	−87	2.4×10^{-1}
Mg	2959	−1890	−1160	−91	2.2×10^{-1}
Ca	2704	−1562	−1160	−18	1.5×10^{-3}
Sr	2572	−1414	−1160	−2	7.1×10^{-4}
Ba	2459	−1273	−1160	+26	1.1×10^{-6}

Hydroxides become more soluble down the group

For the hydroxides:

$$\Delta H_{sol} = \Delta H_{hyd}(M^{2+}) + 2\Delta H_{hyd}(OH^-) - LE(MOH)$$

$2 \times \Delta H_{hyd}(OH^-)$ has the constant value $-1100\,kJ\,mol^{-1}$. The lattice energies for the hydroxides change more down the group than do the lattice energies for the sulphates, and this is the reason for the reversed trend in solubility (see Table 16.8).

■ Table 16.8 Thermochemical data for the solubilities of the Group 2 hydroxides. Because the hydroxide ion is small, the lattice energies of the Group 2 hydroxides are sensitive to changes in size of the cation and vary in value by $800\,kJ\,mol^{-1}$ from beryllium hydroxide to barium hydroxide. The changes in $\Delta H^{\ominus}_{hyd}(M^{2+})$ are less than this ($687\,kJ\,mol^{-1}$). So in contrast to the sulphates, $\Delta H^{\ominus}_{sol}$ becomes more negative, by $113\,kJ\,mol^{-1}$, on descending the group and the solubility increases.

M	$-LE(M(OH)_2)$/ $kJ\,mol^{-1}$	$\Delta H_{hyd}(M^{2+})$/ $kJ\,mol^{-1}$	$2 \times \Delta H_{hyd}(OH^-)$/ $kJ\,mol^{-1}$	ΔH_{sol}	Solubility in 100 g of $M(OH)_2$/mol
Be	3120	−1960	−1100	+60	8.0×10^{-7}
Mg	2993	−1890	−1100	+3	1.6×10^{-5}
Ca	2644	−1562	−1100	−18	2.5×10^{-3}
Sr	2467	−1414	−1100	−47	3.4×10^{-3}
Ba	2320	−1273	−1100	−53	4.1×10^{-2}

Further practice

1 What two factors determine the enthalpy of solution of an ionic solid?
2 State qualitatively how these two factors change for the hydroxides of Group 2.
3 Explain qualitatively why, on descending Group 2, the values of lattice energies of the chromates change less than do their values of enthalpies of hydration.

■ **Figure 16.19**
Stalactites and stalagmites form as dissolved calcium hydrogencarbonate turns back into calcium carbonate as the water drips from the roof of the cave.

Carbonates

The carbonates of Group 1 are all soluble, while the carbonates of Group 2 are all insoluble. The insolubility of calcium carbonate forms the basis of the limewater test for carbon dioxide:

$$Ca(OH)_2(aq) + CO_2(g) \rightarrow CaCO_3(s) + H_2O(l)$$

As more carbon dioxide bubbles through the solution, the precipitate dissolves to form calcium hydrogencarbonate, $Ca(HCO_3)_2$, which is soluble.

$$CaCO_3(s) + H_2O(l) + CO_2(g) \rightarrow Ca(HCO_3)_2(aq)$$

Calcium hydrogencarbonate is one of the substances responsible for **hard water**.

Hard water

The dirt on everyday objects, such as crockery and clothes, is often embedded in grease that makes it insoluble in water. In order to wash off the dirt, it must first be made soluble. This is done by adding a **detergent**. Detergent molecules have a hydrocarbon 'tail', that attracts the grease, and an ionic 'head', that makes it soluble in water. An insoluble globule of grease is surrounded by detergent molecules and becomes soluble. The types of detergent and how they work are discussed in Chapter 27, page 508.

Soaps (see Chapter 29, page 553) are anionic detergents. A common soap is sodium octadecanoate, $C_{17}H_{35}CO_2^-Na^+$. A synthetically made anionic detergent is sodium dodecylbenzenesulphonate:

$$C_{12}H_{23}-\langle\bigcirc\rangle-SO_3^-Na^+$$

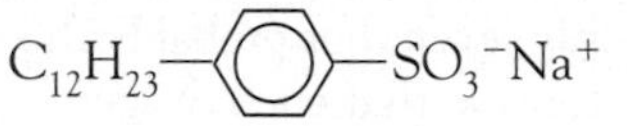

This is more effective than soap, but has the disadvantage of being more acidic, and it is not biodegradable.

Anionic detergents form a precipitate with hard water. Hard water contains ions such as $Ca^{2+}(aq)$, $Mg^{2+}(aq)$ and $Fe^{2+}(aq)$, either as sulphates or hydrogencarbonates. With sodium octadecanoate and calcium ions, for example, the following reaction occurs:

$$Ca^{2+}(aq) + 2C_{17}H_{35}CO_2^-(aq) \rightarrow Ca(C_{17}H_{35}CO_2)_2(s)$$

The reaction produces an unpleasant **scum**, and also means that more soap must be used. There are several ways to overcome the problem.

- Boil the water. This converts soluble hydrogencarbonate ions into insoluble carbonates, for example:

$$Ca^{2+}(aq) + 2HCO_3^-(aq) \rightarrow CaCO_3(s) + H_2O(l) + CO_2(g)$$

This removes calcium ions from solution – it destroys **temporary hardness**. Soluble sulphates remain after boiling, and constitute **permanent hardness**.
- Add sodium carbonate. This precipitates metal ions as insoluble carbonates, for example, with magnesium ions:

$$Mg^{2+}(aq) + CO_3^{2-}(aq) \rightarrow MgCO_3(s)$$

This destroys both temporary and permanent hardness.
- Add a softening agent. The softening agents added to detergents are usually polyphosphates, for example, $Na_5P_3O_{10}$. The polyphosphate ion combines with metal ions, making them negatively charged. This means that they are no longer attracted by anionic surfactants.

$$Ca^{2+}(aq) + P_3O_{10}^{5-}(aq) \rightarrow CaP_3O_{10}^{3-}(aq)$$

Another softening agent is edta, which is discussed in Chapter 19, page 354.

The problem of hard water is minimised if a neutral or cationic surfactant is used, as these do not form a precipitate with metal ions. They are more expensive than anionic detergents, but are often used for washing more delicate materials or hair.

Further practice

1 Copy and complete:

An ionic solid with a large anion is likely to be less soluble if the cation is (large/small).

2 Would you expect barium chromate, $BaCrO_4$, to be more or less soluble than calcium chromate, $CaCrO_4$?

16.6 Oxidation states

In virtually all the compounds of the Group 1 and Group 2 elements, the metals are in oxidation numbers +1 and +2, respectively. At first sight, the metals in peroxides and superoxides appear to have higher oxidation numbers, but more detailed analysis shows that this is not the case.

- Peroxides, for example, Na_2O_2 and BaO_2, contain the $^{-}O—O^{-}$ ion. In peroxides, oxygen is in the −1 oxidation state and both of the oxygen atoms have eight electrons in their outer shells. As a result, both sodium and barium retain their usual oxidation numbers of +1 and +2, respectively.
- Superoxides, for example, KO_2, contain the $O—O^{-}$ ion. These compounds contain an odd number of electrons. Only the larger cations in Groups 1 and 2 create sufficient space in the lattice to accommodate the O_2^{-} ion. It appears that one oxygen atom has eight electrons in its outer shell and the other has only seven, but probably the single charge is delocalised over both oxygen atoms so that each oxygen atom has effectively $7\frac{1}{2}$ electrons with an oxidation number of $-\frac{1}{2}$.

No compounds of Group 1 have been made that contain higher oxidation states than +1. This is not surprising, because the second ionisation energies are very high – they involve removing an electron from the inner p subshell. This extra energy might be compensated for by a larger lattice energy term, and calculations suggest that this is indeed the case for the theoretical salt caesium(II) fluoride, CsF_2, whose enthalpy change of formation is estimated to be about $-125\,kJ\,mol^{-1}$. However, the enthalpy change of formation of caesium(I) fluoride, CsF, is more negative ($-554\,kJ\,mol^{-1}$), with the result that any CsF_2 formed would immediately break down:

$$CsF_2(s) \rightarrow CsF(s) + \tfrac{1}{2}F_2(g) \qquad \Delta H = -554 - (-125) = -429\,kJ\,mol^{-1}$$

Similar arguments apply to the non-existence of compounds such as MF_3 in Group 2. The non-existence of compounds such as magnesium(I) chloride, MgCl, is harder to explain. The loss of one electron from the magnesium atom requires much less energy than the loss of two electrons, but there is also a corresponding reduction in lattice energy, resulting from the reduction in charge from +2 to +1. The enthalpy change of formation of MgCl is estimated to be $-125\,kJ\,mol^{-1}$. However, the enthalpy change of formation of $MgCl_2$ is much more negative ($-641\,kJ\,mol^{-1}$) and any MgCl that might be formed would immediately undergo the following disproportionation reaction:

$$2MgCl(s) \rightarrow Mg(s) + MgCl_2(s) \qquad \Delta H = -641 - (2 \times -125) = -391\,kJ\,mol^{-1}$$

This disproportionation reaction takes place only because the enthalpy change of formation of $MgCl_2(s)$ is more than twice as negative as that for MgCl(s).

Further practice

1 Explain qualitatively why the following compounds do not exist.

a $NaCl_2$

b CaCl

2 Some Group 1 compounds contain ions that do not have eight electrons in their outer shell. Give the formulae of two such ions.

Solutions of metals in liquid ammonia

The Group 1 metals dissolve in liquid ammonia to give blue solutions. These solutions contain solvated free electrons, and are useful reducing agents. Substances called **cryptands** may be added to these solutions. Cryptands complex strongly with M^+ ions, which has the effect of increasing the number of free electrons in the solution:

$$M \rightarrow M^+ + e^-$$

Some of these free electrons combine with metal atoms to produce the M^- ion:

$$M + e^- \rightarrow M^-$$

and compounds can be isolated from such solutions that contain this ion. The M^- ion has the outer electronic configuration ns^2 and the oxidation state -1.

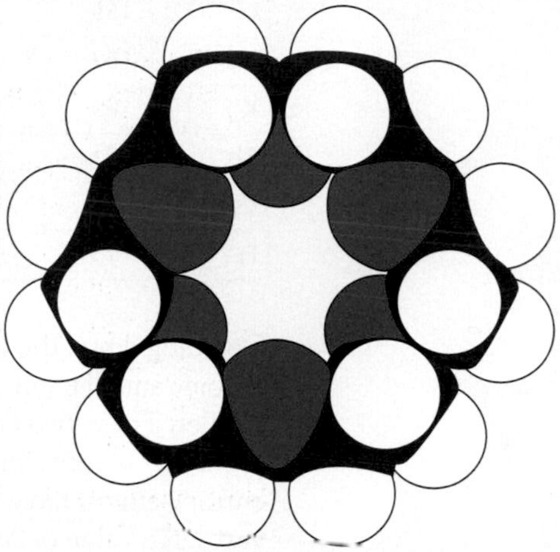

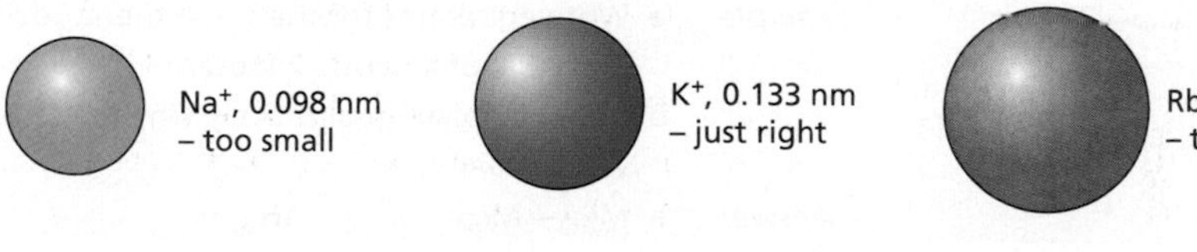

■ **Figure 16.20** The structure of a cryptand that complexes strongly with K^+ ions.

16.7 Standard electrode potentials

In spite of differences in their reactivities, the standard electrode potentials (see page 256) of the Group 1 metals are very similar. A surprise is that lithium has the most negative potential, and yet it is the least reactive metal. There are two points to be considered here.

- Reactivity is a kinetic property, while $E^\ominus$ is a thermodynamic property. The two properties are not necessarily connected.
- $E^\ominus$ values are dependent on several thermodynamic quantities of the metal that may operate in opposite directions.

$E^\ominus$ for a Group 1 metal is related to the enthalpy change associated with the following reaction:

$$M^+(aq) + \tfrac{1}{2}H_2(g) \rightarrow M(s) + H^+(aq)$$

This can be written as two half-reactions:

$$M^+(aq) + e^- \rightarrow M(s)$$
$$H^+(aq) + e^- \rightarrow \tfrac{1}{2}H_2(g)$$

By definition, the second half-reaction is assigned $E^{\ominus} = 0\,V$, but the value of the enthalpy change for this half-reaction is $-498\,kJ\,mol^{-1}$. The reverse of the first half-reaction may be broken down into three steps, each of which is a recognisable thermodynamic quantity:

1	$M(s) \rightarrow M(g)$	ΔH_{at}
2	$M(g) \rightarrow M^{+}(g) + e^{-}$	first IE
3	$M^{+}(g) \rightarrow M^{+}(aq)$	$\Delta H_{hyd}(M^{+})$
	$M(s) \rightarrow M^{+}(aq) + e^{-}$	

The values of these three steps for the Group 1 metals are given in Table 16.9.

■ **Table 16.9** Some thermodynamic data for the Group 1 metals.

Metal	$\Delta H^{\ominus}_{at}$/ kJ mol^{-1}	First IE/ kJ mol^{-1}	$\Delta H^{\ominus}_{hyd}(M^{+})$/ kJ mol^{-1}	$\Delta H^{\ominus}(H^{+}/\frac{1}{2}H_2)$/ kJ mol^{-1}	Σ	$E^{\ominus}$/V
Li	159	513	−499	−498	−325	−3.04
Na	107	496	−390	−498	−285	−2.73
K	89	419	−305	−498	−295	−2.92
Rb	81	403	−281	−498	−295	−2.92
Cs	76	376	−248	−498	−294	−2.92
Fr	73	400	−230	−498	−255	−3.1

On going down the group, the positive values of $\Delta H^{\ominus}_{at}$ and first ionisation energy become smaller, but at the same time, the negative values of $\Delta H^{\ominus}_{hyd}(M^{+})$ become smaller. These two effects work in opposition to each other so that the sum of the terms (Σ) is very similar. The correlation of Σ with $E^{\ominus}$ values is not perfect because entropy effects have been ignored. The anomalous value for lithium is due to its very large value of $\Delta H^{\ominus}_{hyd}(M^{+})$, a result of its small ionic size.

Example

a Write equations for the three thermodynamic terms that specify the standard electrode potential of a Group 2 metal.

b Hence, explain qualitatively why the $E^{\ominus}$ values of calcium, strontium and barium are very similar.

Answer

a $M(s) \rightarrow M(g)$ ΔH_{at}

$M(g) \rightarrow M^{2+}(g) + 2e^{-}$ Σ (1st IE + 2nd IE)

$M^{2+}(g) \rightarrow M^{2+}(aq)$ $\Delta H_{hyd}(M^{2+})$

b The first two terms become more positive from calcium to strontium to barium, but the last term becomes more negative.

The two effects tend to cancel out so that the sum of the three terms is almost constant.

Further practice

1 Explain why the values of $\Delta H^{\ominus}_{at}$, first IE and $\Delta H^{\ominus}_{hyd}(M^{+})$ in Table 16.9 (above) change as they do.

2 Explain why caesium reacts much more violently with water than potassium even though their $E^{\ominus}$ values are the same.

Summary

- All the s-block elements are metals.
- Their atomic (metallic) and ionic radii increase down each group.
- Their ionisation energies and electronegativities decrease down each group.
- Group 1 metals form M^+ ions and Group 2 metals form M^{2+} ions.
- The s-block nitrates and carbonates become thermally more stable at the bottom of each group.
- The solubility of Group 2 hydroxides increases down the group, while that of the sulphates decreases.
- The chemistry of the s-block elements can be explained in terms of the energy changes involved.

Key reactions you should know

- Oxygen with metal:

$2Li(s) + \frac{1}{2}O_2(g) \rightarrow Li_2O(s)$	
$2Na(s) + O_2(g) \rightarrow Na_2O_2(s)$	
$M(s) + \frac{1}{2}O_2(g) \rightarrow MO(s)$	M = Be, Mg, Ca, Sr, Ba

- Chlorine with metal:

$M(s) + \frac{1}{2}Cl_2(g) \rightarrow MCl(s)$	M = Li, Na, K, Rb, Cs
$M(s) + Cl_2(g) \rightarrow MCl_2(s)$	M = Be, Mg, Ca, Sr, Ba
	Cl_2 could be F_2, Br_2 or I_2

- Water with metal:

$M(s) + H_2O(l) \rightarrow MOH(aq) + \frac{1}{2}H_2(g)$	M = Li, Na, K, Rb, Cs
$M(s) + 2H_2O(l) \rightarrow M(OH)_2(s) + H_2(g)$	M = Be, Mg, Ca, Sr,
	(Ba gives $Ba(OH)_2(aq)$).

- Oxide and hydroxide with acid:

$M_2O(s) + 2H^+(aq) \rightarrow 2M^+(aq) + H_2O(l)$	M = Li, Na, K
$Na_2O_2(s) + 2H^+(aq) \rightarrow 2Na^+(aq) + H_2O_2(aq)$	
$MO(g) + 2H^+(aq) \rightarrow M^{2+}(aq) + H_2O(l)$	M = Be, Mg, Ca, Sr, Ba
$MOH(s) + H^+(aq) \rightarrow M^+(aq) + H_2O(l)$	M = Li, Na, K, Rb, Cs
$M(OH)_2(s) + 2H^+(aq) \rightarrow M^{2+}(aq) + H_2O(l)$	M = Be, Mg, Ca, Sr, Ba

- Carbon dioxide with hydroxides:

$M(OH)_2(aq) + CO_2(g) \rightarrow MCO_3(s)$	M= Be, Mg, Ca, Sr, Ba
$MCO_3(s) + CO_2(g) + H_2O(l) \rightarrow M(HCO_3)_2(aq)$	M= Mg, Ca (in hard water)

- Heat on carbonates and nitrates:

$Li_2CO_3(s) \rightarrow Li_2O(s) + CO_2(g)$	
$M_2CO_3(s)$ no reaction	M = Na, K, Rb, Cs
$MCO_3(s) \rightarrow MO(s) + CO_2(g)$	M = Be, Mg, Ca, Sr, Ba; easier down the group
$2LiNO_3(s) \rightarrow Li_2O(s) + 2NO_2(g) + \frac{1}{2}O_2(g)$	
$MNO_3(s) \rightarrow MNO_2(s) + \frac{1}{2}O_2(g)$	M = Na, K, Rb, Cs
$M(NO_3)_2(s) \rightarrow MO(s) + 2NO_2(g) + \frac{1}{2}O_2(g)$	M = Be, Mg, Ca, Sr, Ba

- Solubility of hydroxides and sulphates:

  The solubilities of the hydroxides increase from Be to Ba.

  The solubilities of the sulphates decrease from Be to Ba.

17 Group 7 and Group 0

In this chapter, we study the trends in the properties of the Group 7 elements, the halogens. Being non-metals, they show decreasing reactivity with increasing atomic number. The noble gases, Group 0, are also considered.

The chapter is divided into the following sections.

17.1 Properties of the halogens

■ **Table 17.1**
The main properties of the elements in Group 7.

	F	Cl	Br	I	At
Melting point/°C	−220	−101	−7	114	302
Boiling point/°C	−188	−34	59	184	337
Covalent radius/nm	0.058	0.099	0.114	0.133	—
van der Waals radius/nm	0.135	0.181	0.195	0.215	—
X^- ionic radius/nm	0.133	0.181	0.196	0.222	0.227
ΔH_{at}/kJ mol^{-1}	79	122	112	107	—
E(X—X)/kJ mol^{-1}	159	242	193	151	110
First IE/kJ mol^{-1}	1681	1251	1140	1008	930
First EA/kJ mol^{-1}	−328	−349	−325	−295	−270
$E^{\ominus}$/V	−2.87	−1.36	−1.09	−0.54	+0.2
Electronegativity	3.98	3.16	2.96	2.66	2.2

Melting points and boiling points

On descending Group 7, the melting points and boiling points increase (see Figure 17.1). This is because the intermolecular forces become stronger as the number of easily polarisable electrons rises (see Chapter 3, page 79). (It is not due to any changes in the strength of the bonding between the atoms in the molecules, because these are not broken during melting or boiling.)

■ **Figure 17.1**
Melting and boiling points of the Group 7 elements.

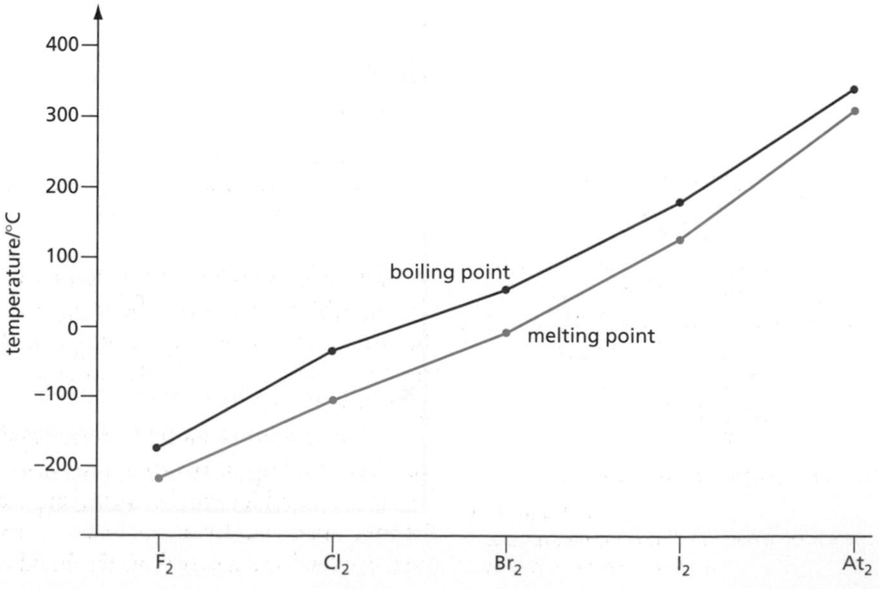

Radii and bond enthalpies

The extra shells cause the covalent, van der Waals and ionic radii to increase down the group (see Figure 17.2). Adding each extra shell more than compensates for the decrease in size of the inner shells caused by the increasing nuclear charge. The addition of an extra electron to form the X^- ion is to an outer shell already containing seven electrons, so this has very little effect on the size of the atom as measured by the van der Waals radius.

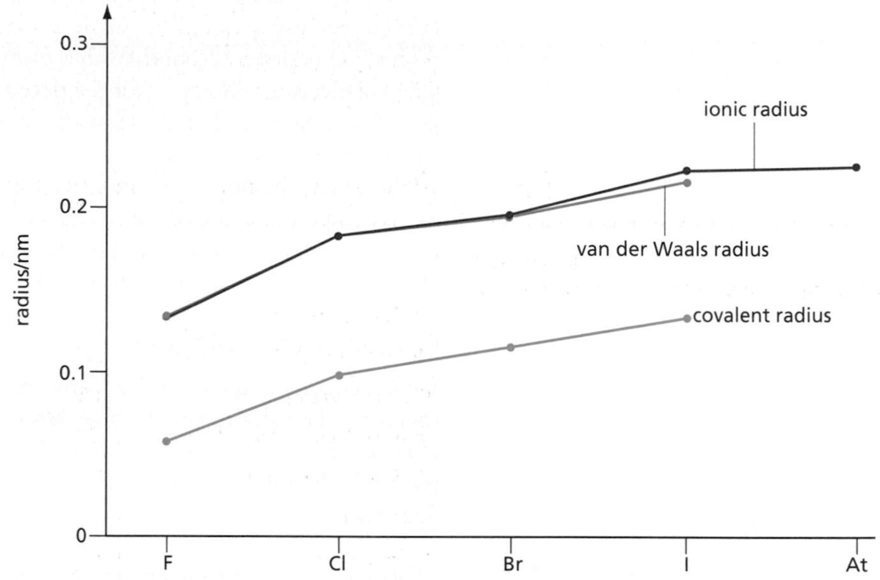

Figure 17.2 Covalent, van der Waals and ionic radii of the Group 7 elements.

Increasing atomic radius normally leads to weaker covalent bonding. An exception to this rule is found with fluorine, whose bond strength is about 150 kJ mol^{-1} weaker than would be predicted from the trend established with the other halogens (see Figure 17.3). The weakness of the F—F bond is one of the main reasons why the element is so reactive.

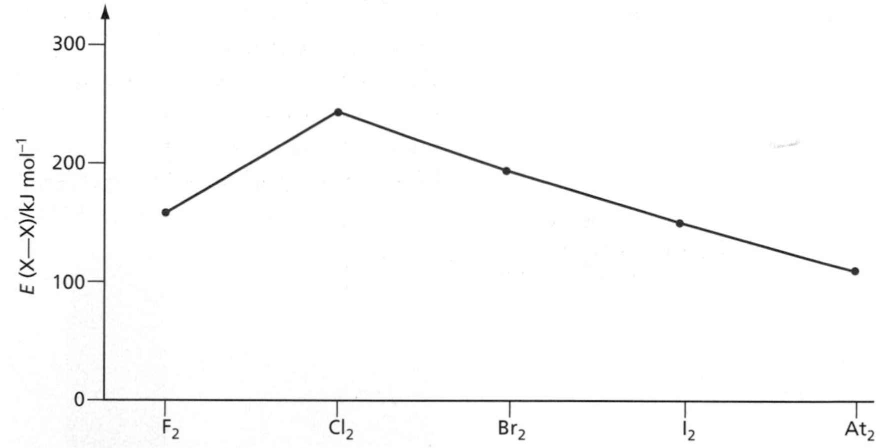

Figure 17.3 Bond enthalpies of the Group 7 elements.

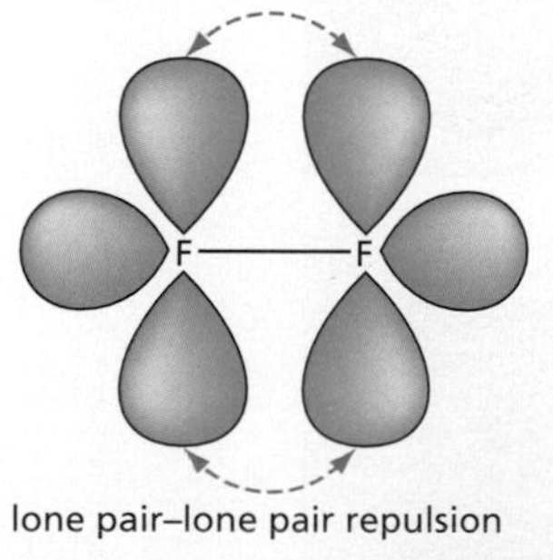

Figure 17.4 Lone pair–lone pair repulsion may be the reason why the F—F bond is so weak.

Several suggestions have been made to explain the weakness of the F—F bond. One possibility is that the lone pairs on each fluorine atom are so close together that they strongly repel each other (see Figure 17.4). This theory also explains why the N—N and O—O single bonds are also much weaker than would be suggested by their bond lengths.

For the gaseous elements chlorine and fluorine, the enthalpy change of atomisation, ΔH_{at}, is one-half the bond enthalpy, E(X—X). This is not the case for bromine and iodine, because ΔH_{at} also includes the energy required to break the intermolecular forces that exist in the liquid and solid states. These energies are considerably smaller than the bond enthalpies (see Chapter 10).

Ionisation energies and electron affinities

On descending the group, the increasing atomic radius takes the outer shell of electrons further away from the nucleus and so the ionisation energies decrease. The increased nuclear charge does not have an appreciable effect on the ionisation energy, because the inner electrons effectively screen the electrons in the outer shell from the increased attraction of the nucleus.

The low electron affinity of fluorine is surprising, because we would expect a small atom to attract an electron strongly. It may be that the strong attraction between an electron and the nucleus is counterbalanced by the repulsion the electron experiences from the other electrons in the small outer shell. This low value of electron affinity is not restricted to fluorine; other elements in the second period also have smaller electron affinities than the elements immediately below them in the periodic table.

Fluorine is the most electronegative element and its oxidation number is -1 in all its compounds. The other elements in the group commonly have oxidation number -1 but can also have positive oxidation numbers in compounds with elements of higher electronegativity, for example, oxygen.

Colour of the halogens

Each halogen has an absoption band in the near ultraviolet or visble region of the spectrum. The absorption involves excitation of an antibonding π electron to an antibonding σ orbital which is higher in energy (see Chapter 3, page 62). The position of the maximum of this band determines the colour of the halogen (see Table 17.2).

Table 17.2 Colours and wavelengths of the absorption bands in the near ultraviolet and visible regions for the halogens.

Halogen	Wavelength of absorption maximum/nm	Colour
Fluorine	300	Pale yellow
Chlorine	350	Yellow-green
Bromine	400	Red-brown
Iodine	500	Purple

The colour of iodine in solution varies with the solvent. In non-polar solvents (for example, tetrachloromethane) it is purple, the same colour as the gas. In more polar solvents (for example, ethanol) it is brown. This is because the solvent affects the energy levels in the iodine molecule and shifts the position of the absorption band towards the violet end of the spectrum. In solvents of moderate polarity (for example, ethyl ethanoate), iodine has a red colour. In water, in the presence of $I^-(aq)$ ions (see page 324), the brown $I_3^-(aq)$ ion is formed.

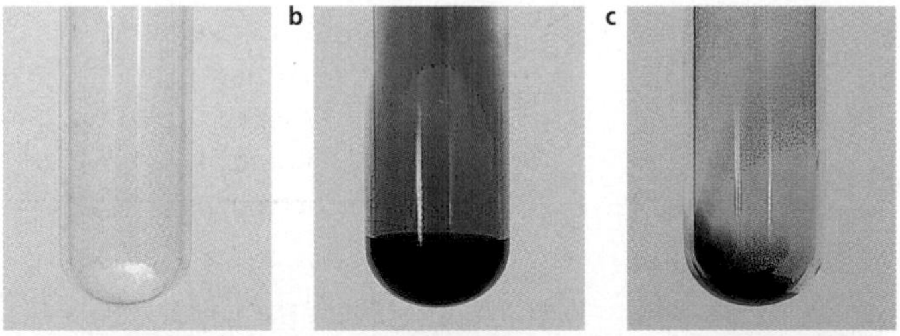

Figure 17.5 The colours of the halogens: **a** chlorine **b** bromine **c** iodine.

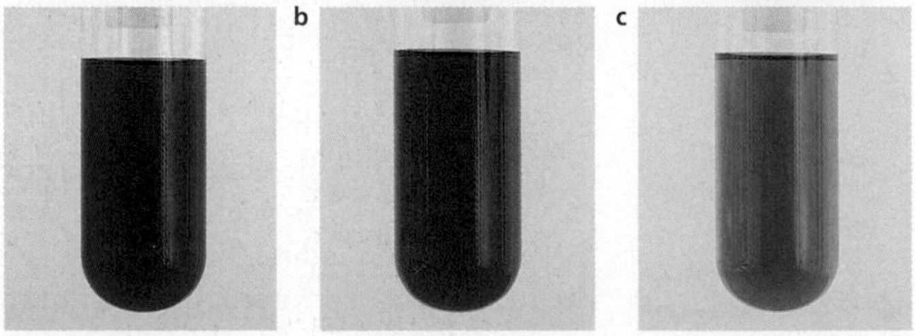

Figure 17.6 Iodine is **a** purple in tetrachloromethane, **b** red in ethylethanoate, and **c** brown in ethanol.

Uses of the elements and their compounds

Sodium fluoride is added to toothpaste and, controversially, to drinking water to reduce dental decay. It acts on a constituent of tooth enamel, converting hydroxyapatite, $Ca_5(PO_4)_3(OH)$, into fluoroapatite, $Ca_5(PO_4)_3F$. Fluoroapatite is less prone to attack by the acids produced by plaque bacteria. In higher concentrations, fluorides are toxic. Fluorine is used to make sulphur hexafluoride, SF_6, a gas which is compressed to form an insulating liquid used in high voltage electrical equipment. It is also a constituent of PTFE (page 457) and chloroflurocarbons (page 473). Hydrogen fluoride is used to etch glass.

Chlorine and sodium chlorate(I), NaClO, are made by the electrolysis of brine and are used as disinfectants and bleaches (see Chapter 20, page 381–2). The use of chlorinated hydrocarbons and chlorofluorocarbons (CFCs) is now restricted because of their harmful environmental effects (see Chapter 25, page 473). Another large-scale use of chlorine is in the manufacture of polychloroethene (PVC, see Chapter 25, page 457).

Bromine is made by displacement from sea water by chlorine (see page 324). In the past, the principal use of bromine was in the manufacture of 1,2-dibromoethane, an additive to leaded petrol, but this is now much less important. The largest use of bromine compounds at present is in the manufacture of flame retardants.

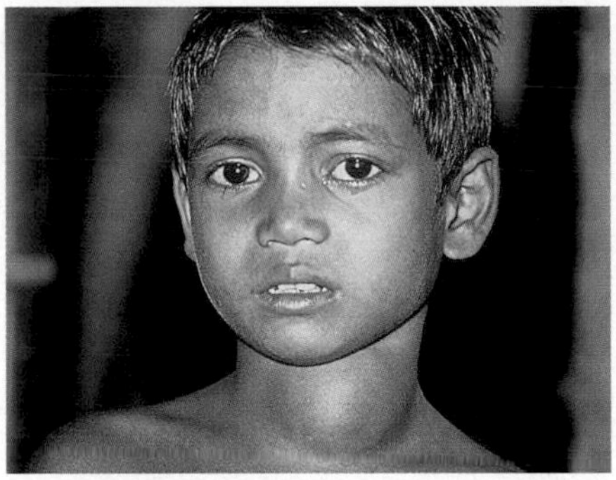

Figure 17.7 People with goitre have an enlarged thyroid gland, as it struggles to overcome iodine deficiency. Goitre is uncommon in the UK but still occurs in other countries where nutrition is poor.

All the halogens are toxic – they damage living tissue. A dilute solution of iodine is not so harmful to humans, however, and is sometimes used as a relatively safe method of sterilising water or for disinfecting wounds. A deficiency of the hormone thyroxin, which contains iodine, causes goitre. To guard against this, in some areas iodides are added to drinking water. They are also used in animal feedstuffs.

Example

Briefly explain the following.

a The bond strength of iodine is less than that of bromine.
b The boiling point of chlorine is higher than that of fluorine.
c The ionisation energy of fluorine is higher than that of chlorine.
d The electron affinity of bromine is lower than that of chlorine.

Answer

a The I—I bond is longer than the Br—Br bond, so the bonding pair of electrons is further away from the nuclei and less strongly attracted to them.
b The intermolecular forces in chlorine are greater than those in fluorine, because chlorine has a greater number of easily polarisible electrons in its electron cloud.
c The 2p electrons of fluorine are closer to the nucleus than the 3p electrons of chlorine. The outer electrons in fluorine are therefore more strongly attracted and require more energy to remove them.
d The 4p electrons of bromine are less firmly bound than the 3p electrons of chlorine. The addition of a further 4p electron, therefore, gives out less energy than the addition of a 3p electron.

Further practice

1 Briefly explain each of the following.
   a The melting point of chlorine is lower than that of bromine.
   b The covalent radius of bromine is larger than that of chlorine.
   c The Cl^- ionic radius is the same as the van der Waals radius of chlorine.
   d ΔH_{at} for chlorine is $\frac{1}{2}E(Cl—Cl)$.
2 (Harder) Suggest a reason for each of the following.
   a ΔH_{at} for fluorine is less than ΔH_{at} for chlorine.
   b $E^\ominus$ for fluorine is very much more negative than $E^\ominus$ for chlorine.
3 Estimate, giving your reasons, the value of:
   a ΔH_{at} for astatine
   b the covalent radius of astatine.

17.2 Reactions of the halogens

Displacement reactions

The oxidising strength of the halogens is in the order $F_2 > Cl_2 > Br_2 > I_2$.
A halogen that is a more powerful oxiding agent displaces one that is less powerful. For example, chlorine displaces bromine:

$$Cl_2(aq) + 2Br^-(aq) \rightarrow 2Cl^-(aq) + Br_2(aq)$$

This reaction forms the basis of the extraction of bromine from sea water.

With hydrogen

The halogens react with hydrogen to give hydrogen halides, for example:

$$H_2(g) + Cl_2(g) \rightarrow 2HCl(g)$$

The reactivity with hydrogen decreases down the group as the strength of the H—X bond decreases (see Table 17.3). The reactivity of fluorine is particularly high as the weak F—F bond breaks and two very strong H—F bonds form.

■ **Table 17.3** Reaction conditions for a mixture of a halogen with hydrogen.

Halogen	Reaction conditions
Fluorine	Explodes at the boiling point of hydrogen (−253 °C)
Chlorine	Explodes when exposed to ultraviolet light
Bromine	Slow reaction when heated
Iodine	Reversible reaction when heated

With water

The relative ability of the halogens to combine with hydrogen is also shown by their reactions with water (see Table 17.4).

■ **Table 17.4** Reactions of the halogens with water.

Halogen	Equation	Comment
Fluorine	$F_2(g) + H_2O(l) \rightarrow 2HF(aq) + O_2(g)$	Violent reaction, O_3 produced as well
Chlorine	$Cl_2(g) + H_2O(l) \rightleftharpoons HCl(aq) + HClO(aq)$	Reversible reaction
Bromine	$Br(l) + H_2O(l) \rightleftharpoons HBr(aq) + HBrO(aq)$	Equilibrium lies to the left
Iodine	$I_2(s) + H_2O(l) \rightleftharpoons HI(aq) + HIO(aq)$	Iodine is almost insoluble

The halogens also react directly with hydrocarbons (see Chapter 23, page 433, and Chapter 24, page 444).

Iodine is almost insoluble in water but readily dissolves in a solution of potassium iodide, forming the deep brown $I_3^-(aq)$ ion. It is this ion that is liberated when solutions of potassium iodide are oxidised (see Chapter 7, page 155).

$$2I^-(aq) - 2e^- \rightarrow I_2(aq)$$

$$I_2(aq) + I^-(aq) \rightleftharpoons I_3^-(aq)$$

This latter reaction is readily reversed, so that the $I_3^-(aq)$ ion behaves as $I_2(aq)$ in its rections. When $I_2(aq)$ appears in an equation, the predominant species is usually $I_3^-(aq)$.

With metals

The decreasing reactivity trend from fluorine to iodine is also shown in the reactions with metals. Many metals catch fire in fluorine, but may have to be heated in order to initiate reaction with the other halogens. If the metal has more than one oxidation state, reaction with fluorine brings out the higher state while reaction with iodine brings out the lower. The product with chlorine or bromine depends on the relative stability of the two oxidation states. The following equations illustrate this behaviour.

$$Mn(s) + 1\tfrac{1}{2}F_2(g) \rightarrow MnF_3(s) \quad \text{but} \quad Mn(s) + Cl_2(g) \rightarrow MnCl_2(s)$$
$$Fe(s) + 1\tfrac{1}{2}Cl_2(g) \rightarrow FeCl_3(s) \quad \text{but} \quad Fe(s) + I_2(s) \rightarrow FeI_2(s)$$

Example Write balanced equations with state symbols for the following reactions:

a hydrogen with iodine
b sodium with bromine
c aluminium with iodine
d bromine and tin to give tin(IV) bromide
e bromine with aqueous sodium iodide.

Answer

a $H_2(g) + I_2(g) \rightleftharpoons 2HI(g)$
b $Na(s) + \tfrac{1}{2}Br_2(l) \rightarrow NaBr(s)$ or $2Na(s) + Br_2(l) \rightarrow 2NaBr(s)$
c $Al(s) + 1\tfrac{1}{2}I_2(s) \rightarrow AlI_3(s)$ or $2Al(s) + 3I_2(s) \rightarrow 2AlI_3(s)$
d $Sn(s) + 2Br_2(l) \rightarrow SnBr_4(s)$
e $Br_2(aq) + 2I^-(aq) \rightarrow 2Br^-(aq) + I_2(aq)$

Further practice

1 Write balanced equations with state symbols for the following reactions:
  a rubidium with iodine
  b tin and iodine to give tin(II) iodide
  c boron and chlorine
  d nitrogen and fluorine
  e chlorine and aqueous potassium iodide.

2 Suggest:
  a why bromine is more soluble in aqueous sodium bromide than in water
  b an equation for the action of astatine with water.

17.3 The hydrogen halides

Boiling points

Figure 17.8 shows the boiling points of the hydrogen halides, and clearly brings out the importance of hydrogen bonding in hydrogen fluoride (see Chapter 4, page 80).

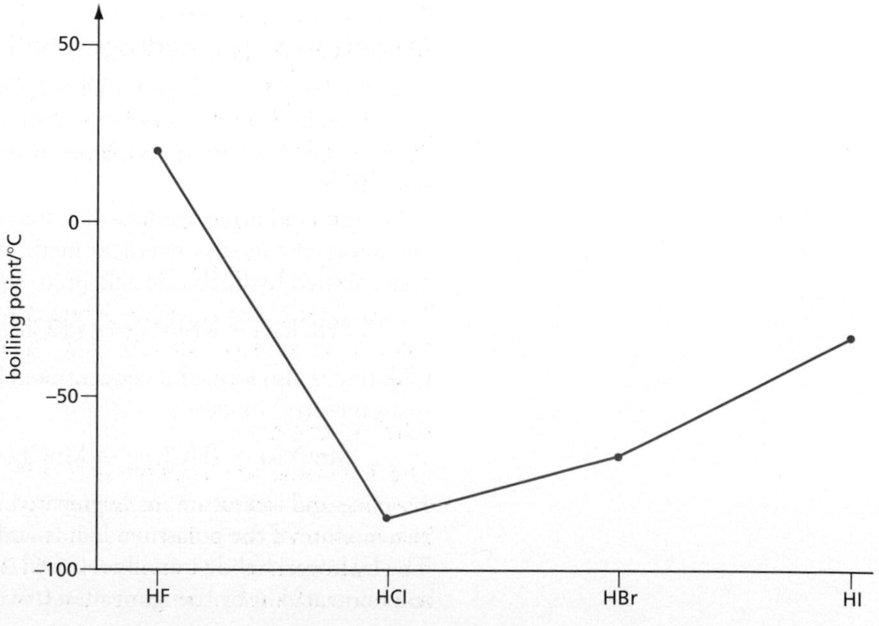

■ **Figure 17.8** Boiling points of the hydrogen halides.

Preparation of hydrogen halides

Industrially, large amounts of hydrogen chloride are produced as a by-product in the manufacture of chlorinated hydrocarbons, but in the laboratory, use of the free halogens is inconvenient. Hydrogen chloride and hydrogen fluoride are therefore prepared by warming one of their salts with concentrated sulpuric acid:

$$NaF(s) + H_2SO_4(l) \rightarrow NaHSO_4(s) + HF(g)$$
$$CaCl_2(s) + 2H_2SO_4(l) \rightarrow Ca(HSO_4)_2(s) + 2HCl(g)$$

In the case of hydrogen bromide and hydrogen iodide, the hot concentrated sulphuric acid is a strong enough oxidising agent to convert them into bromine and iodine respectively:

$$NaBr + H_2SO_4(l) \rightarrow NaHSO_4 + HBr(g)$$
$$2HBr(g) + H_2SO_4(l) \rightarrow Br_2(l) + 2H_2O(l) + SO_2(g)$$

The yield of hydrogen bromide is improved if the acid is first diluted with an equal volume of water. This forms the basis of one method for the preparation of bromoalkanes (see Chapter 26, page 483).

Hydrogen iodide is so easily oxidised that very little of it is produced. The sulphuric acid is reduced not only to sulphur dioxide, but further to sulphur and hydrogen sulphide.

In order to obtain a better yield of hydrogen bromide or hydrogen iodide, anhydrous phosphoric(V) acid, H_3PO_4, can be used. This is a poor oxidising agent. The mixture must be heated to liberate the hydrogen halide gas, and as a result the yield of hydrogen iodide is reduced by reversible decomposition.

$$KI(s) + H_3PO_4(l) \rightarrow KH_2PO_4(s) + HI(g)$$
$$2HI(g) \rightleftharpoons H_2(g) + I_2(g)$$

The characteristic behaviour of solid chlorides, bromides and iodides towards hot concentrated sulphuric acid can be used in analysis (see Chapter 21, page 390).

Another way to prepare hydrogen bromide or hydrogen iodide is to add water to the phosphorus trihalide, prepared by direct combination of the halogen with red phosphorus:

$$1\tfrac{1}{2}I_2(s) + P(s) \rightarrow PI_3(s)$$
$$PI_3(s) + 3H_2O(l) \rightarrow H_3PO_3(aq) + 3HI(g)$$

This reaction is also used to make bromo- or iodoalkanes (see Chapter 26, page 483).

Oxidation of the hydrogen halides to the free halogens

The standard electrode potential of fluorine is so positive that it is impossible to make the element by chemical oxidation of hydrogen fluoride. Fluorine is prepared by electrolysis of potassium fluoride dissolved in anhydrous liquid hydrogen fluoride.

Powerful oxidising agents will convert hydrogen chloride to chlorine. In the laboratory, the most convenient method for the preparation of chlorine is to drop concentrated hydrochloric acid on to solid potassium manganate(VII).

$$8HCl(aq) + KMnO_4(s) \rightarrow KCl(aq) + MnCl_2(aq) + 4H_2O(l) + 2\tfrac{1}{2}Cl_2(g)$$

Chlorine is also formed if concentrated hydrochloric acid is warmed with manganese(IV) oxide:

$$MnO_2(s) + 4HCl(aq) \rightarrow MnCl_2(aq) + 2H_2O(l) + Cl_2(g)$$

Bromine and iodine are easily prepared by dropping concentrated sulphuric acid on to a mixture of the potassium halide and manganese(IV) oxide (see Figure 17.9). The hydrogen halide initially released is then oxidised, either by the concentrated sulphuric acid or by the manganese(IV) oxide.

■ **Figure 17.9** a
a Bromine and **b** iodine can both be prepared by dropping concentrated sulphuric acid onto the solid halide.

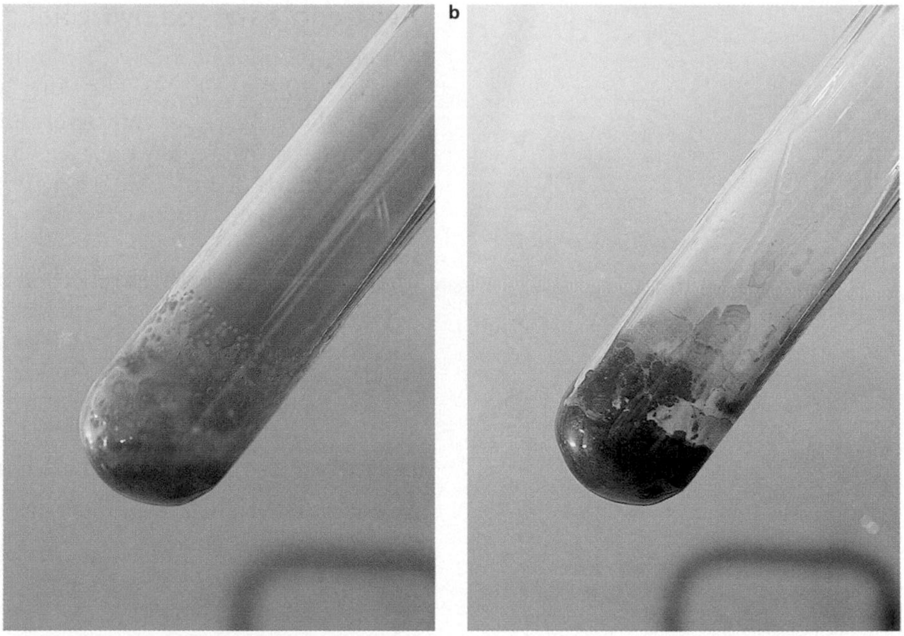

Reactions of the hydrohalic acids

Hydrogen halides are readily soluble in water, dissolving to give hydrohalic acids. These acids show typical behaviour with metals and their oxides, hydroxides and carbonates (see Chapter 6, page 132). For example, hydrobromic acid reacts with magnesium metal as follows:

$$2HBr(aq) + Mg(s) \rightarrow MgBr_2(aq) + H_2(g)$$

It is sometimes more convenient to use ionic equations to represent the reactions of acids:

$$2H^+(aq) + Mg(s) \rightarrow Mg^{2+}(aq) + H_2(g)$$
$$2H^+(aq) + O^{2-}(s) \rightarrow H_2O(l)$$
$$H^+(aq) + OH^-(aq) \rightarrow H_2O(l)$$
$$2H^+(aq) + CO_3^{2-}(s) \rightarrow H_2O(l) + CO_2(g)$$

Hydrogen fluoride is unique in attacking glass, and is used in etching apparatus and for making decorative glass:

$$SiO_2(s) + 4HF(l) \rightarrow SiF_4(g) + 2H_2O(l)$$

The reason why this reaction takes place is because the Si—F bond is very strong. $E(\text{Si—F}) = 582\,\text{kJ}\,\text{mol}^{-1}$. It is the strongest single covalent bond known. Its strength is due to its short length and its high polarity.

■ **Figure 17.10**
Hydrogen fluoride reacts with glass, and is used for etching.

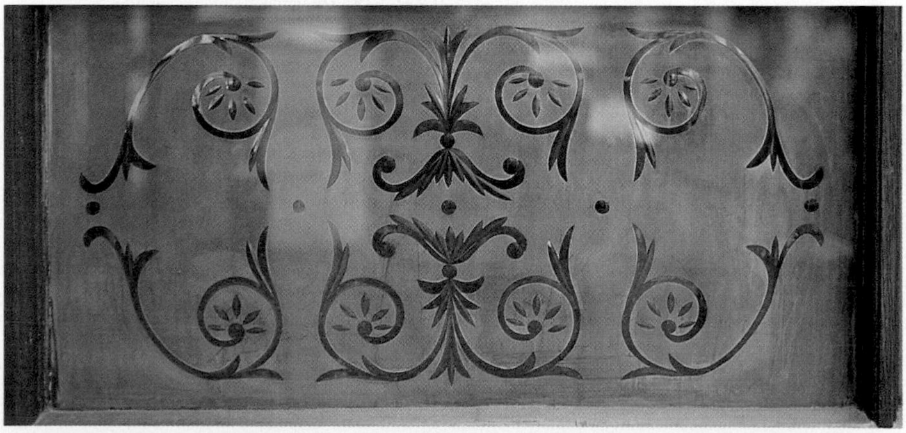

The strengths of the hydrohalic acids

Although it is highly corrosive, hydrofluoric acid is a weak acid ($K_a = 5.6 \times 10^{-4}$ mol dm^{-3}). The other hydrohalic acids are all considered to be strong acids, being almost completely ionised in dilute aqueous solution. The concentrated acids are, however, largely un-ionised. Measurement of the degree of ionisation shows that the acid strength is in the order HCl(aq) < HBr(aq) < HI(aq). This trend is usually attributed to the decreasing strength of the H—Hal bond as it becomes longer.

The very high strength of the H—F bond is often quoted as the reason for the weakness of hydrogen fluoride as an acid. However, this is only one of the factors involved in ionisation. Careful analysis shows that hydrogen bonding in hydrogen fluoride and the low electron affinity of fluorine are more significant factors (see the panel below).

Energetics of the strength of the hydrohalic acids

The ionisation of a hydrohalic acid may be broken down into several steps, and the enthalpy of the change ΔH_{ion} calculated using a Hess cycle (see Figure 17.11).

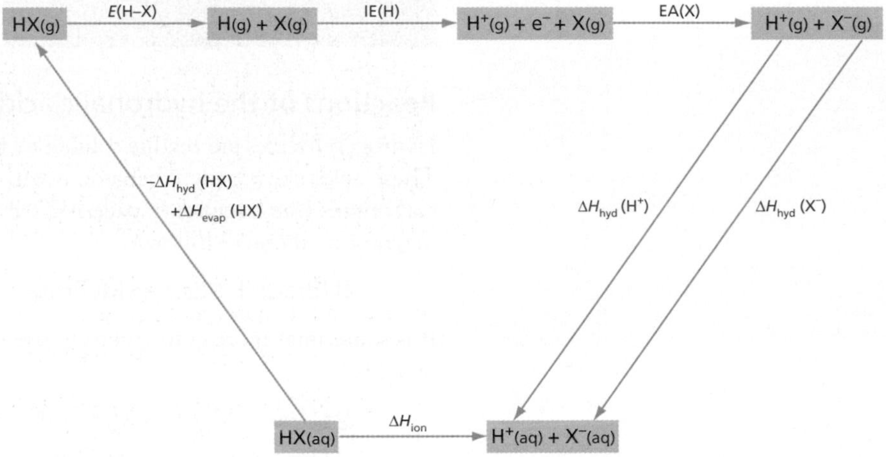

■ **Figure 17.11** Thermodynamic cycle for the ionisation of a hydrohalic acid.

Values for these various energy changes are given in Table 17.5.

■ **Table 17.5** Energy changes in kJ mol^{-1} for the ionisation of hydrohalic acids.

	HF	HCl	HBr	HI
ΔH_{evap} (HX)/kJ mol	50	21	8	8
E(H—X)/kJ mol	568	432	366	298
IE(H)/kJ mol	1310	1310	1310	1310
EA(X)/kJ mol	−328	−349	−325	−295
$\Delta H_{hyd}(H^+)$/kJ mol	−1120	−1120	−1120	−1120
$\Delta H_{hyd}(X^-)$/kJ mol	−473	−339	−306	−262
ΔH_{ion}/kJ mol	+7	−45	−67	−61

The high value of E(H—F) is counterbalanced by the high enthalpy of hydration of the F^- ion. The principal factor that makes ΔH_{ion} positive for hydrofluoric acid is the lower value of electron affinity for fluorine. This is also enhanced by the high ΔH_{evap} of hydrofluoric acid as a result of extensive hydrogen bonding in solution.

As we have seen, the strengths of the hydrohalic acids is in the order HF ≪ HCl < HBr < HI. These strengths as measured by K_a are related to ΔG_{ion} rather than ΔH_{ion} (see Chapter 10), and this is why there is not complete agreement between the relative values of ΔH_{ion} and the actual order of acid strength.

A test for hydrohalic acids

Like other halides, hydrohalic acids can be identified using aqueous silver nitrate (see Table 17.6).

■ Table 17.6 Identification of the hydrohalic acids using aqueous silver nitrate.

Acid	Appearance of precipitate	Solubility of precipitate in $NH_3(aq)$
HF	None, AgF is soluble in water	—
HCl	White, AgCl	Soluble in dilute $NH_3(aq)$
HBr	Cream, AgBr	Soluble in concentrated $NH_3(aq)$
HI	Yellow, AgI	Insoluble in concentrated $NH_3(aq)$

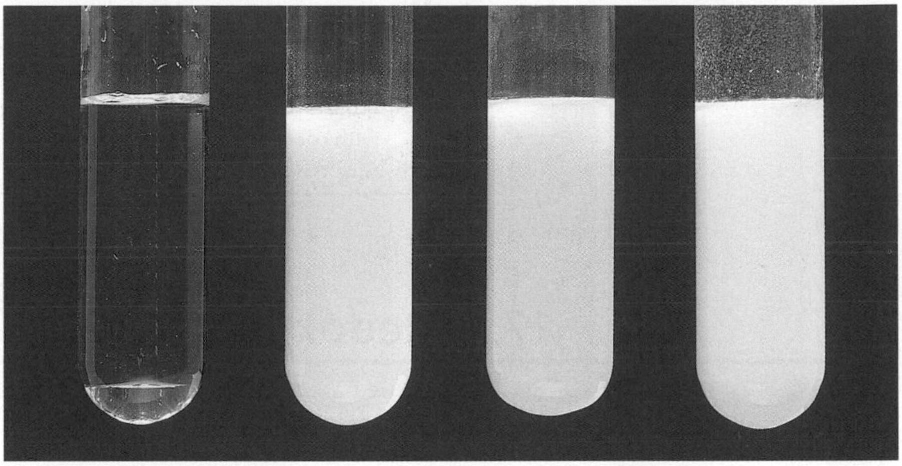

■ Figure 17.12 The colours of the silver halides are used in tests for the hydrohalic acids and also for halide salts. Silver fluoride is soluble, silver chloride is white, silver bromide cream and silver iodide yellow.

The precipitates formed are silver halides. For example, with hydrobromic acid the following reaction takes place:

$$Ag^+(aq) + Br^-(aq) \rightleftharpoons AgBr(s)$$

This precipitate dissolves in concentrated ammonia solution, as the Ag^+ ion forms a complex ion, displacing the above equilibrium to the left.

$$Ag^+(aq) + 2NH_3(aq) \rightleftharpoons Ag(NH_3)_2{}^+(aq)$$

The overall equation can be represented by:

$$AgBr(s) + 2NH_3(aq) \rightleftharpoons Ag(NH_3)_2{}^+(aq) + Br^-(aq)$$

The solubilities of the silver halides are in the order AgF > AgCl > AgBr > AgI. Silver fluoride is so soluble that fluorides do not give a precipitate with aqueous silver nitrate. The precipitates of the other silver halides dissolve in ammonia, to form the complex ion shown above, if the halide concentration can be reduced to below that of a saturated solution of the silver halide in question. This is achieved by dilute ammonia solution with silver chloride, concentrated ammonia solution with silver bromide but not at all with silver iodide.

The solubilities of the silver halides can be attributed to their increasing degree of covalent character as we go down the group and the halogen atom becomes less electronegative. The increase in covalent character is shown by an increasing discrepancy between the experimental values of lattice energies and the values calculated on the assumption that the bonding is 100% ionic (see page 191 and Table 17.7).

■ Table 17.7 Differences between experimental and calculated values of lattice energies for the silver halides.

Halide	Experimental LE/kJ mol^{-1}	Calculated LE/kJ mol^{-1}	% difference
AgF	−967	−953	1
AgCl	−915	−864	6
AgBr	−904	−830	8
AgI	−889	−808	9

The increasing covalent character from silver fluoride to silver iodide is also the reason why their colour changes from white to yellow. Charge transfer requires less energy if the bond has considerable covalent character (see Chapter 14, page 275).

Further practice

1 Write ionic equations for the following reactions:
   a hydriodic acid with silver nitrate
   b hydrochloric acid with strontium hydroxide
   c hydrobromic acid with barium carbonate
   d hydrobromic acid with aluminium metal.

2 Write equations, with state symbols, for the following reactions:
   a phosphoric(V) acid heated sodium bromide
   b phosphorus tribromide and water
   c concentrated sulphuric acid with sodium iodide to give iodine, sulphur and water
   d manganese(IV) oxide heated with hydrobromic acid.

3 a List the formulae of all the sulphur compounds formed when sodium iodide reacts with concentrated sulphuric acid.
   b Give the oxidation number of sulphur in each of these compounds.

17.4 Redox reactions

As we have seen, fluorine is the most electronegative element, and its oxidation number in all its compounds is −1. The other halogens may show oxidation numbers as high as +7 in compounds with the more electronegative elements oxygen and fluorine.

Oxidation number +1: halic(I) acids

When these halogens dissolve in water, halic(I) acids, HXO, are formed (see Table 17.4, page 324); for example, with chlorine:

$$Cl_2(g) + H_2O(l) \rightleftharpoons HCl(aq) + HClO(aq)$$

$$\text{oxidation number:}\quad 2Cl(0) \qquad Cl(-1) \qquad Cl(+1)$$

This change in oxidation numbers shows that this is a disproportionation reaction (see Chapter 7, page 151).

Chloric(I) acid is a powerful oxidising agent, acting as a bleach and killing bacteria. This is the reason why chlorine gas bleaches moist litmus paper, and why chlorine is used to disinfect drinking water and swimming pools. On standing, chloric(I) acid decomposes:

$$HClO(aq) \rightarrow HCl(aq) + \tfrac{1}{2}O_2(g)$$

The reaction is catalysed by light.

All the halic(I) acids are very weak acids (K_a/mol dm^{-3}: $HClO = 3 \times 10^{-8}$, $HBrO = 2 \times 10^{-9}$, $HIO = 2 \times 10^{-11}$).

If the halogens are treated with aqueous alkali, halates(I) are formed; for example, with chlorine:

$$Cl_2(g) + 2OH^-(aq) \rightarrow Cl^-(aq) + ClO^-(aq) + H_2O(l)$$

This is another disproportionation reaction.

Sodium chlorate(I) is the active ingredient in household bleaches and disinfectants. It is very dangerous to add acid (for example, vinegar) to bleach because chlorine, which is very toxic, is given off.

$$Cl^-(aq) + ClO^-(aq) + 2H^+(aq) \rightarrow H_2O(l) + Cl_2(g)$$

Oxidation number +5: halate(V) compounds

If a concentrated solution containing halate(I) ions is warmed, halate(V) ions are formed. For example, with chlorate(I) ions:

$$3ClO^-(aq) \rightarrow ClO_3^-(aq) + 2Cl^-(aq)$$
$$\text{oxidation number:}\quad 3Cl(+1) \qquad Cl(+5) \qquad 2Cl(-1)$$

The ease of this disproportionation is in the order $ClO^- < BrO^- < IO^-$. Iodate(I) ions readily disproportionate on standing at room temperature.

Sodium chlorate(V) has been used as a weedkiller, but its use is now restricted as it also forms an ingredient in homemade explosives. It is a powerful oxidising agent and gives off oxygen on heating:

$$NaClO_3(s) \rightarrow NaCl(s) + 1\tfrac{1}{2}O_2(g)$$

Oxidation number +7

The maximum oxidation number for the halogens is +7. This is found in the oxide Cl_2O_7, and in the very strong acid chloric(VII) acid, $HClO_4$, which is formed when Cl_2O_7 dissolves in water. Iodine forms, with fluorine, a compound IF_7 (see the panel below).

Interhalogen compounds

Different halogens combine together to give **interhalogen compounds** of four main types (see Table 17.8).

■ Table 17.8 The bonding and shapes of some interhalogen compounds.

Formula type	Example	Number of bonding electron pairs	Number of non-bonding electron pairs	Shape
AX	ICl	1	3	Linear
AX_3	ClF_3	3	2	T-shaped
AX_5	BrF_5	5	1	Square pyramidal
AX_7	IF_7	7	0	Distorted octahedral

Interhalogen compounds readily add on to alkenes. The less electronegative halogen acts as the electrophile (compare the addition of hydrogen halides in Chapter 24).

They also form salts with alkali metals, for example $Cs^+IBr_4^-$, and cationic species such as $[ClF_2^+]$, $[SbF_6^-]$. See if you can predict the shapes of these halogen-containing species.

Further practice

1 What are the oxidation numbers of chlorine in:
a Cl_2O b ClF_3 c Cl_2O_7 d BrCl?

2 Write equations, with state symbols, for the following reactions:
a iodine with aqueous sodium hydroxide in the cold
b iodine with aqueous sodium hydroxide on warming
c liquid Cl_2O_7 and water
d the action of heat on sodium iodate(V).

3 By considering the oxidation numbers of chlorine in the following substances, state which of these reactions involve disproportionation.
a $KClO_3(s) \rightarrow KCl(s) + 1\tfrac{1}{2}O_2(g)$
b $4KClO_3(s) \rightarrow 3KClO_4(s) + KCl(s)$
c $Cl_2O(g) + H_2O(l) \rightarrow 2HClO(aq)$
d $ClO^-(aq) + 2H^+(aq) \rightarrow H_2O(l) + Cl_2(aq)$

17.5 Group 0

Properties of the noble gases

The elements in Group 0 were once called the inert gases ('inert' meaning unreactive). However, since a few compounds of these elements have been made, they are now termed the **noble gases**.

The elements are all monatomic. Their ionisation energies are far too high for them to form metallic lattices. The formation of any covalent bond requires the use of d orbitals which are much higher in energy than the p orbitals that hold the electrons, and this is only possible if the bond formed is strong enough to compensate for this. This is rare (except with the comparatively strong bonds to fluorine and oxygen), so the elements exist as single atoms.

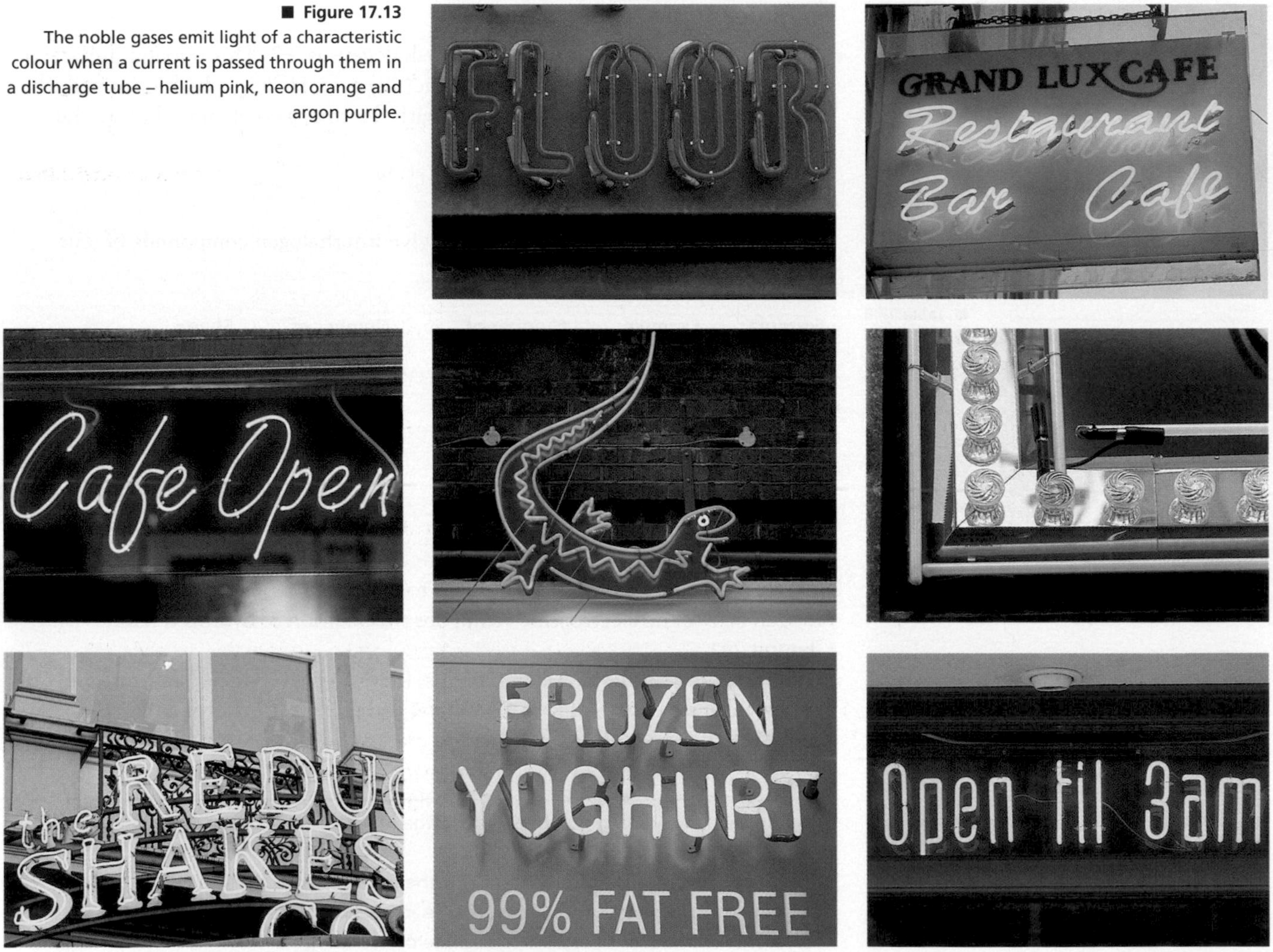

■ **Figure 17.13** The noble gases emit light of a characteristic colour when a current is passed through them in a discharge tube – helium pink, neon orange and argon purple.

The melting and boiling points of the noble gases are very low because the van der Waals forces are extremely small. They increase as the group is descended because there are more electrons that are easily polarised on the outside of the atoms (see Table 17.9).

■ **Table 17.9** Some properties of the noble gases. Helium becomes a solid only under a pressure of 100 atm.

	He	Ne	Ar	Kr	Xe	Rn
Melting point/°C	−269	−249	−189	−157	−112	−71
Boiling point/°C	−269	−246	−186	−152	−107	−62
IE/kJ mol^{-1}	2372	2081	1520	1351	1170	1040

Reactions of the noble gases

The chemistry of the noble gases is largely that of xenon. Krypton forms the fluoride KrF_2, and no doubt there would be a more extensive chemistry of radon if it could be studied more conveniently (the half-life of the longest-lived radon isotope is 3.8 days). Recently a compound of argon, HArF, has been made at low temperatures and it is thought that similar compounds of neon and perhaps helium may also be synthesised.

The simplest compounds of xenon are the fluorides XeF_2, XeF_4 and XeF_6, and the oxides XeO_3 and XeO_4. Only when xenon is bonded to the small atoms of fluorine or oxygen are the bonds strong enough to supply the energy necessary to use 5d orbitals. The shapes of these molecules are consistent with those predicted by the VSEPR theory of Chapter 3, page 63 (see Figure 17.14).

■ Figure 17.14 The molecular shapes of the fluorides and oxides of xenon.

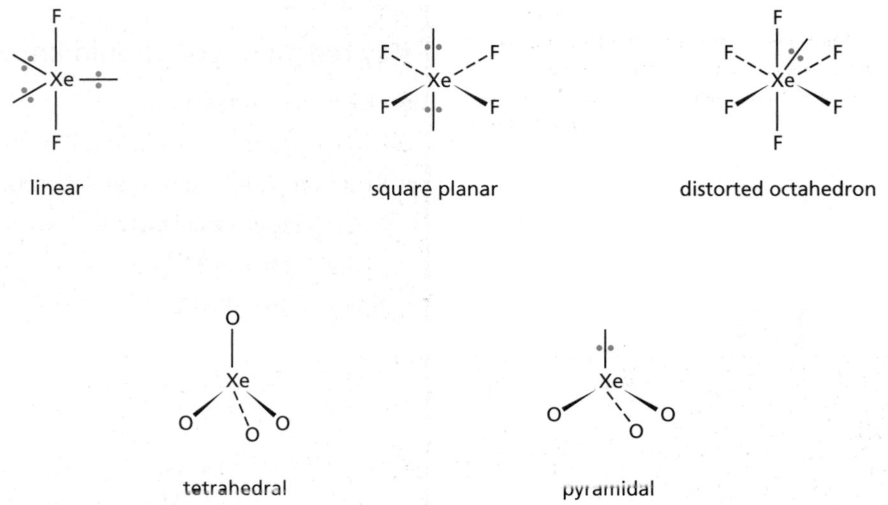

The fluorides all decompose in contact with water. The oxides are dangerously explosive.

Example

a Explain why xenon does not form any compounds with chlorine.

b By considering the number of lone pairs and bonding pairs in the molecule KrF_2, predict its shape.

c Write an equation for the action of water on XeF_6 to give XeO_3.

Answer

a The Xe—Cl bond would be long and therefore not strong enough to provide the energy for an electron in the 5p orbital to use a 5d orbital.

b There are three lone pairs and two bonding pairs. The molecule is linear, with the bonding pairs in the axial position.

c $XeF_6(s) + 3H_2O(l) \rightarrow XeO_3(aq) + 6HF(aq)$

Further practice

1 Explain why the boiling point of krypton is higher than that of argon.

2 Suggest why the melting and boiling points of the noble gases are very similar.

3 If a little water is added to XeF_6, $XeOF_4$ is formed.

a Write an equation for this reaction.

b By considering the number of lone pairs and bonding pairs in the $XeOF_4$ molecule, predict its shape.

Summary

- The halogens are non-metals that become less reactive with increasing atomic number.
- The properties of chlorine, bromine and iodine are very similar. Some of the properties of fluorine are atypical.
- Hydrochloric, hydrobromic and hydriodic acids behave as typical strong acids. Hydrofluoric acid is a weak acid.
- The ease of oxidation of the hydrogen halides is $HF < HCl < HBr < HI$.
- Chlorine, bromine and iodine exhibit a variety of oxidation states with the more electronegative elements fluorine and oxygen.
- Chlorine is widely used as a disinfectant and bleaching agent.
- The noble gases form very few compounds, and only with fluorine and oxygen.

Key reactions you should know

- Displacement reactions:
  $Cl_2(aq) + 2Br^-(aq) \rightarrow 2Cl^-(aq) + Br_2(aq)$ — Cl_2 displaces Br_2 displaces I_2
- Reaction with hydrogen and metals:
  $X_2(g) + H_2(g) \rightarrow 2HX(g)$ — X = F, Cl, Br, I
  $\frac{1}{2}X_2(g) + M(s) \rightarrow MX(s)$ — X = F, Cl, Br, I M = Group 1
  $X_2(g) + M(s) \rightarrow MX_2(s)$ — X = F, Cl, Br, I; M = Group 2, some d block
  $1\frac{1}{2}X_2(g) + Al(s) \rightarrow AlX_3(s)$ — X = F, Cl, Br, I Also B, some d block
- Reaction with water:
  $F_2(g) + H_2O(l) \rightarrow 2HF(aq) + O_2$
  $X_2(g) + H_2O(l) \rightleftharpoons HOX(aq) + H^+(aq) + X^-(aq)$ — $Cl > Br > I$ (disproportionation reaction)
- Concentrated sulphuric acid on solid halide:
  $X^-(s) + H_2SO_4(l) \rightarrow HX(g) + HSO_4^-(s)$ — X = F, Cl, Br, I
  $2HX(g) + H_2SO_4(l) \rightarrow X_2(g) + SO_2(g) + 2H_2O(l)$ — X = Br, I
- Disproportionation reaction of $XO^-(aq)$:
  $3XO^-(aq) \rightarrow XO_3^-(aq) + 2X^-(aq)$ — $X = Cl < Br < I$
- Solubility of silver halides:
  Solubilities of halides decrease down the group.

18 Group 4

In this chapter we study the elements in Group 4 of the periodic table. They show a change in properties down the group from the non-metallic element carbon, through the semiconductors silicon and germanium, to the metals tin and lead. In their simple compounds, the elements usually exhibit the group oxidation state of +IV, but tin and lead also exhibit the +II oxidation state.

The chapter is divided into the following sections.

18.1 Properties of the elements

Table 18.1
Some properties of the elements in Group 4.

	C (graphite)	Si	Ge	Sn	Pb
Melting point/°C	3527	1410	937	232	328
Boiling point/°C	4827	2355	2830	2270	1740
Density/g cm^{-3}	2.26	2.33	5.32	7.31	11.35
ΔH_{at}/kJ mol^{-1}	717	456	377	302	195
E(X—X)/kJ mol^{-1}	348	226	163	195	100
Atomic radius/nm	0.077	0.117	0.123	0.141	0.175
First IE/kJ mol^{-1}	1086	787	762	709	716
Electronegativity	2.55	1.90	2.01	1.96	2.33

Atomic properties

The trends in most of the properties of the Group 4 elements with increasing atomic number can be explained in terms of the increasing atomic radius. The interatomic bonds become weaker as the bonds become longer, and this is reflected by the overall trend of decreasing melting and boiling points and by the decreasing enthalpy changes of atomisation.

The ionisation energies also decrease as the atoms become larger but, for lead, the increase in size is outweighed by the very large increase in nuclear charge. Between tin and lead the nuclear charge rises by 32 units, and as a result the ionisation energy of lead is slightly higher than that of tin.

The electronegativities are similar and show no clear trend. The value for lead is very uncertain because the element forms no covalent substances that are stable enough for their thermochemistry to be reliably determined.

Structure of the elements

Carbon

Carbon has several **allotropes**, that is, different forms of the same element (see Figure 18.1, overleaf). The structures of **diamond** and **graphite** were described in Chapter 4 (pages 88 and 99). Because graphite has a layer structure with a large distance between the layers, its density is lower than that of diamond. This, and the fact that the change:

$$\text{C(graphite)} \rightarrow \text{C(diamond)} \qquad \Delta H = +3\,\text{kJ mol}^{-1}$$

is endothermic, mean that, by Le Chatelier's Principle, high temperature and high pressure favour the conversion from graphite to diamond. However, the conversion has a very high activation energy because strong carbon–carbon bonds must be broken to bring it about. Diamonds are manufactured by heating graphite under high pressure in the presence of a catalyst.

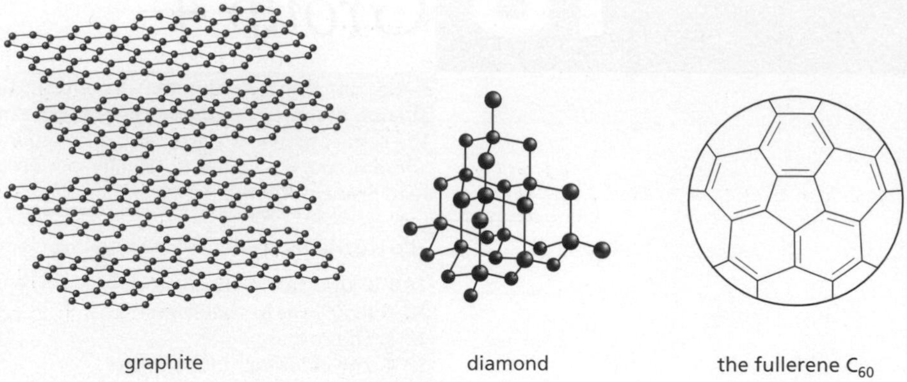

■ **Figure 18.1** The structures of graphite, diamond and C_{60}.

■ **Figure 18.2** Diamonds.

Recently large diamonds have been made by vacuum deposition of carbon. These diamonds will have many industrial applications, for example, as detectors of ultraviolet radiation and as heat sinks in electronic devices.

Another allotrope of carbon is the series of **fullerenes**, described on page 101. Unlike the other allotropes, they are soluble in organic solvents, forming dark red solutions. They will doubtless find many applications in the future, possibly as the basis of high-temperature superconductors.

Silicon and germanium

Silicon and germanium do not form π bonds, because their atoms are too large for effective p-to-p orbital sideways overlap. They therefore exist only as the diamond structure. They are both semiconductors, used extensively in the electronics industry (see the panel below).

Band theory and semiconductors

Bands of energy levels

We saw on page 54 that when the electron clouds of two atoms of hydrogen overlap, the two 1s orbitals form two molecular orbitals, the σ bonding orbital and the σ* antibonding orbital. When four atoms combine together, four molecular orbitals are formed. The molecular orbitals of highest and lowest energy result from the overlap of the orbitals of adjacent atoms, while the intermediate levels arise from overlap of the orbitals of atoms further away.

In a solid containing *L* atoms, *L* molecular orbitals are formed with a range of energies. These form a **band** with effectively a continuous range of energy levels (see Figure 18.3).

■ **Figure 18.3** The origin of a band of orbitals from *L* atoms in a lattice.

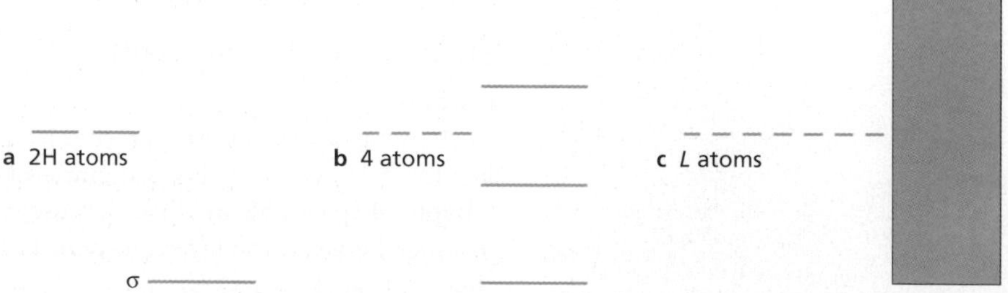

Each molecular orbital can contain two electrons of opposite spin, so a band formed from *L* atoms is able to contain 2*L* electrons. For a solid to be able to conduct electricity and behave like a metal, a band must be partially filled with electrons. If the band is empty, there are no electrons to carry to the current; if the band is full, there are no empty energy levels for the electrons to move into.

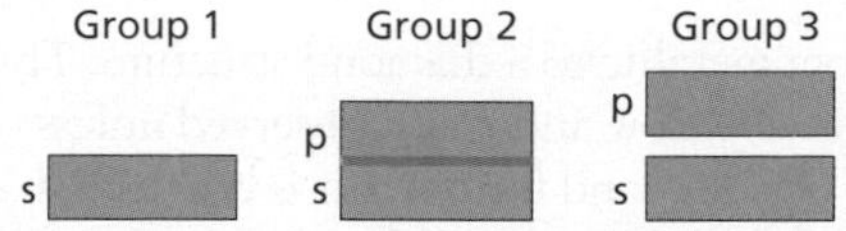

■ **Figure 18.4**
Bands in solids of Group 1, Group 2 and Group 3 elements.

Good conductors

Solids containing bands that are half full are good conductors of electricity. This is the case with Group 1 and Group 3 metals, and sodium and aluminium, for example, are good conductors of electricity. It might be expected that Group 2 elements would be insulators, because the band derived from the s orbitals is completely full. However, this band overlaps with the band derived from the p orbitals and, as this is empty, good conduction is also possible for Group 2 metals (see Figure 18.4).

A similar situation exists in the 3d block, in which the bands derived from the 3d orbitals overlap with the band derived from the 4s orbitals.

Semiconductors

The elements in Group 4 show an interesting gradation in properties from insulator to semiconductors to metals. The bands derived from the s orbital and one of the p orbitals are full. The bands from the other two p orbitals are slightly higher in energy and are empty. This is the situation with diamond, which is normally considered to be an insulator. Further down the group, the gap between the full p band (the **valency band**) and the empty p band (the **conduction band**) becomes smaller, because the energy levels of the atomic orbitals become closer together (see Figure 2.33, page 42). With tin and lead, the orbitals overlap and metallic lattices are formed (see Figure 18.5).

■ **Figure 18.5**
Valency and conduction bands in the Group 4 elements.

With silicon and germanium, the gaps are sufficiently close that electrons from the valency band are easily excited to the conduction band. This may be brought about in several ways.

- Raising the temperature – as the temperature is raised, a larger fraction of electrons is excited to the valency band. (This fraction is given by $e^{-E/RT}$, where E is the size of the energy gap.) This means that, unlike metals, semiconductors have an increased conductivity when the temperature rises.
- Radiation – a photon of radiation with sufficient energy promotes an electron and increases the conductivity. Use is made of this in photocells, using selenium, and in ultraviolet detectors, using diamond.

Pure silicon and germanium have a low conductivity at room temperature, because their energy gaps are close together. This is called their **intrinsic conductivity**. For use in electronics, the pure material is **doped** – tiny amounts of impurity are added. The addition of a trace of a Group 5 element, for example arsenic or antimony, makes the material a better conductor because it now has an excess of electrons in the lattice. This kind of semiconductor is called an *n*-type material (*n* for negative). The addition of a trace of a Group 3 element, for example gallium or indium, leads to a deficit of electrons in the lattice and produces a *p*-type material (*p* for positive). The increase in conductivity due to doping is called the **extrinsic conductivity**.

Electronics components are increasingly made of compounds formed from an element in Group 3 with an element in Group 5, for example, gallium arsenide, GaAs. These show similar semiconducting properties to silicon or germanium. In theory, a compound such as sodium chloride would also behave similarly, but the energy gap is too large for this behaviour to be of any practical use. Solid sodium chloride does conduct electricity very slightly, but this is not due to the movement of electrons. Ionic solids contain **holes** or **defects** in the lattice that allow limited movement of ions, but the resulting conductivity is minute compared with their conductivity when molten or in solution.

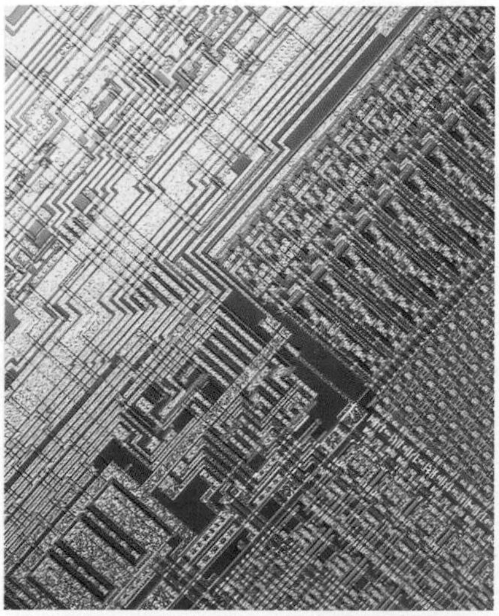

■ **Figure 18.6**
Many modern 'silicon chips' are made of etched gallium arsenide.

Tin

Below 19 °C, the structure of tin changes from metallic to a diamond structure. This change from one solid structure to another is very slow and is not observed unless the temperature is much lower than 19 °C. The diamond form of tin is brittle and loses its metallic character. It is said that the bronze buttons of Napoleon's army fell off in the cold during the retreat from Moscow, and that the tin coating on the food cans used in Scott's expedition to the Antarctic suffered a similar fate.

Example Suggest explanations for the following observations.

a The atomic radius of lead is greater than that of silicon.
b The ionisation energy of silicon is less than that of carbon.
c Silicon has no allotrope corresponding to graphite.
d The Si—Si bond strength is less than the C—C bond strength.

Answer

a There are three more shells of electrons in lead than in silicon, and their increase in size more than compensates for the contraction of the inner shells of electrons due to the increasing nuclear charge.
b The outer electrons in silicon are further away from the nucleus than those in carbon, and are less strongly attracted to the nucleus. They are therefore easier to remove.
c A graphite structure requires the formation of multiple bonds. Silicon does not form multiple bonds because the p orbitals are too far away from each other to form stable π bonds.
d The Si—Si bond is longer and therefore weaker than the C—C bond.

Further practice Suggest explanations for the following observations.

1 The first ionisation energy of lead is greater than that of tin.
2 The melting point of lead is lower than that of silicon.
3 The density of lead is much higher than that of graphite.
4 The conversion of graphite to diamond needs a catalyst.

18.2 Reactions of the elements

With acid

Tin and lead are metallic enough to dissolve in dilute hydrochloric acid to give tin(II) or lead(II) chloride and hydrogen:

$$Sn(s) + 2HCl(aq) \rightarrow SnCl_2(aq) + H_2(g)$$

At room temperature, the reaction with lead stops because lead(II) chloride is insoluble and protects the lead from further attack. At higher temperatures, the reaction proceeds more readily as the chloride is much more soluble under these conditions. Lead(II) iodide, PbI_2, is also much more soluble in hot water than in cold; it is a yellow solid, compare silver iodide (Chapter 17, page 329), but the solution is colourless.

These reactions of tin and lead show them exhibiting their lowest oxidation state of +II. Only their p electrons are used in bonding, leaving the s electrons as part of the core. The incorporation of the s electrons into the core is known as the **inert pair effect**. Presumably, as the elements become more metallic and there is a greater tendency to form ionic rather than covalent bonds, the lower oxidation state is favoured because the higher oxidation state would require the formation of highly charged ions that are energetically difficult to form. The effect is not restricted to Group 4; for example thallium, in Group 3, forms Tl^+ ions and bismuth, in Group 5, forms Bi^{3+} ions.

Carbon, silicon and germanium do not react with dilute acids but dissolve in either hot concentrated sulphuric acid or hot concentrated nitric acid to give the dioxides, for example:

$$C(s) + 2H_2SO_4(l) \rightarrow CO_2(g) + 2SO_2(g) + 2H_2O(g)$$

With oxygen

When heated, carbon, silicon and germanium combine with the oxygen in the air to give dioxides, for example:

$$Si(s) + O_2(g) \rightarrow SiO_2(s)$$

In a limited supply of air, carbon also forms the monoxide:

$$2C(s) + O_2(g) \rightarrow 2CO(g)$$

Carbon monoxide is the main reducing agent in the blast furnace during the extraction of iron (see Chapter 20, pages 375–6). It is highly toxic as it forms a stable complex with haemoglobin in the blood, preventing it from acting as an oxygen carrier.

When oxygen is passed through molten lead, 'red lead', Pb_3O_4, is formed.

$$3Pb(l) + 2O_2(g) \rightarrow Pb_3O_4(s)$$

This behaves if it were a mixture of two oxides, $2PbO.PbO_2$. If 'red lead' is treated with aqueous nitric acid, the lead(II) oxide dissolves, leaving brown lead(IV) oxide.

$$Pb_3O_4(s) + 4HNO_3(aq) \rightarrow 2Pb(NO_3)_2(aq) + PbO_2(s)$$

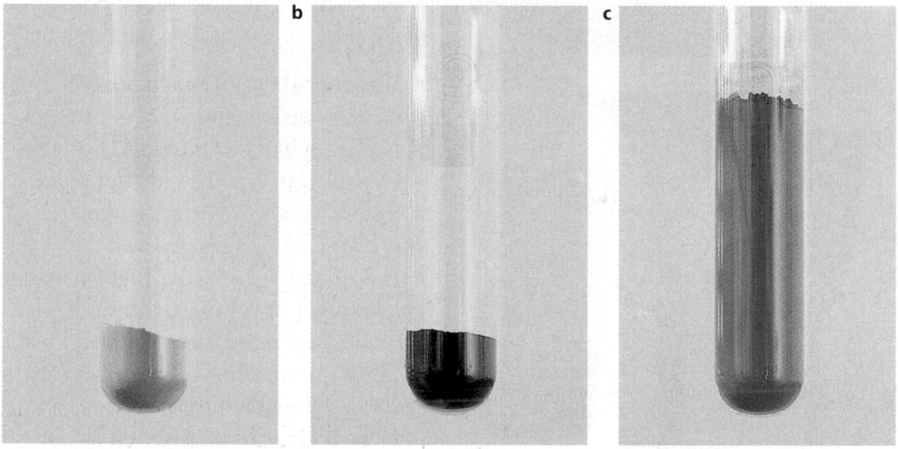

■ **Figure 18.7** **a** Three lead oxides – one-tenth of a mole of **a** lead(II) oxide, **b** lead(IV) oxide and **c** 'red lead'.

18.3 The oxides

Acid–base behaviour

As mentioned in Chapter 15 (page 293), carbon monoxide is a neutral oxide while carbon dioxide and silicon(IV) oxide are weakly acidic. When silicon(IV) oxide is heated with oxides or carbonates, **silicates** are formed which are the basis of glass (see the panel overleaf). It is also the basis of slag formation in the blast furnace (see page 376). For example, with calcium carbonate:

$$CaCO_3(s) + SiO_2(s) \rightarrow CaSiO_3(l) + CO_2(g)$$

The very slow reverse process takes place in the weathering of rocks.

Lead(II) oxide dissolves in most dilute acids to give Pb^{2+} ions, though the reaction with sulphuric acid stops because of the insolubility of lead(II) sulphate. Both lead(II) oxide and lead(IV) oxide are amphoteric. When aqueous sodium hydroxide is added to a solution containing $Pb^{2+}(aq)$ ions, a white precipitate is formed that dissolves in an excess of the reagent (see Chapter 21, page 388).

$$Pb^{2+}(aq) + 2OH^-(aq) \rightarrow Pb(OH)_2(s)$$
$$Pb(OH)_2(s) + 2OH^-(aq) \rightarrow Pb(OH)_4^{2-}(aq)$$

Lead(IV) oxide dissolves in cold concentrated hydrochloric acid to give lead(IV) chloride, and in hot concentrated sodium hydroxide solution to give $Pb(OH)_6^{2-}$ ions.

$$PbO_2(s) + 4HCl(aq) \rightarrow PbCl_4(aq) + 2H_2O(l)$$
$$PbO_2(s) + 2OH^-(aq) + 2H_2O(l) \rightarrow Pb(OH)_6^{2-}(aq)$$

Glasses

In everyday usage, the word 'glass' refers to the transparent silica-based material used, for example, to make windows and bottles. However, strictly speaking, **glass** is a state of matter rather than a particular material, and this state of matter is not confined to material derived from silica.

Most liquids, when cooled, freeze at a definite temperature and turn into a crystalline solid. Some liquids, particularly those that contain large molecules, become so viscous that the molecules are not able to take on a definite crystalline arrangement. These materials form an **amorphous** solid, called a **glass** (see Chapter 4, page 85). There is therefore no definite freezing point and the resulting supercooled liquid is unstable, having higher energy than the true crystalline form.

■ **Figure 18.8** Some examples of crystal glass.

Silicate materials readily form glasses. Pure silicon(IV) oxide forms a glass, but it needs to be heated to at least 1600 °C in order to obtain a transparent material. The addition of other substances lowers this temperature to below 1000 °C, which makes fabrication much easier.

Glasses are made by heating silica, SiO_2, in the form of sand, with metal oxides, often made by thermal decomposition of the carbonates. The materials used to make some of the more common glasses are as follows.

- Window glass – this is the most common glass, made by heating silicon(IV) dioxide with sodium carbonate and calcium carbonate. It contains 30% Na_2SiO_3, 20% $CaSiO_3$ and 50% SiO_2.
- Crystal glass – if the calcium oxide used to make window glass is replaced by lead(II) oxide, a glittering material of high refractive index is produced. This is used for decorative glassware.
- Borosilicate glass – the high-melting-point, low-expansion glass used for laboratory apparatus contains boron(III) oxide, B_2O_3, in addition to silicon(IV) oxide. Metal ions are provided by aluminium(III) oxide, Al_2O_3.
- Silica – pure silica glass does not expand when heated and the red-hot material can be plunged into cold water without cracking. It is also transparent to ultraviolet light and is used in optical components in spectrometers (see Chapter 14).

■ **Figure 18.9** Changing the composition of glass alters its properties considerably. The soft glass tube melts at the temperature of a bunsen flame while a hard glass tube does not.

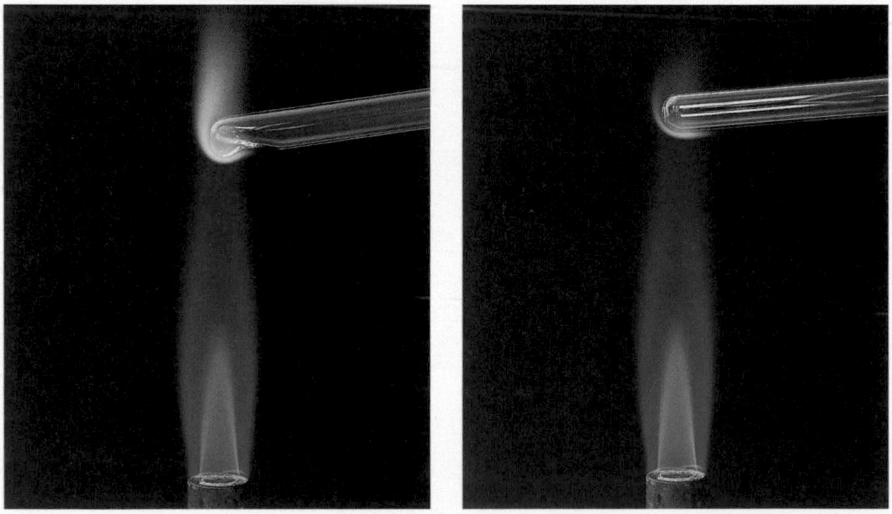

If a glass is heated at its softening temperature for a long time, it may start to crystallise. This change is known as **devitrification**, and it makes the glass opaque and very brittle. The same change may also occur in very old glass at room temperature.

Many non-crystalline plastics form glasses. Some, such as polystyrene (polyphenylethene) and Perspex, are glasses at room temperature, while others, such as rubber, have to be cooled before they lose their elasticity.

Structures

The structures of carbon dioxide, CO_2(g), and silicon(IV) oxide, SiO_2(s), differ because of the relative strengths of the single and double bonds (see Figure 18.10).

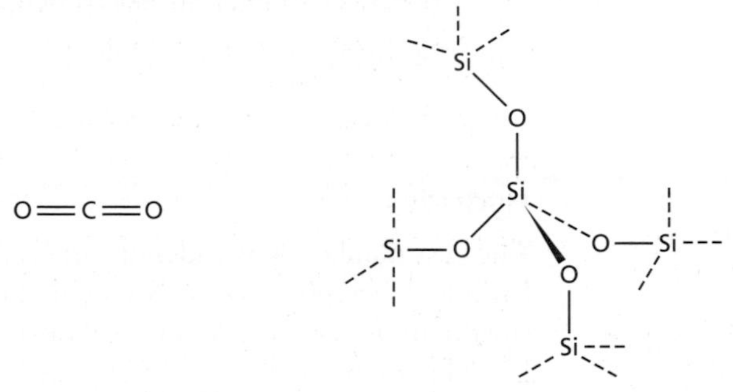

■ Figure 18.10
The structures of carbon dioxide and silicon(IV) oxide. The most stable form of silicon(IV) oxide, α-quartz, forms a helix, making the solid optically active.

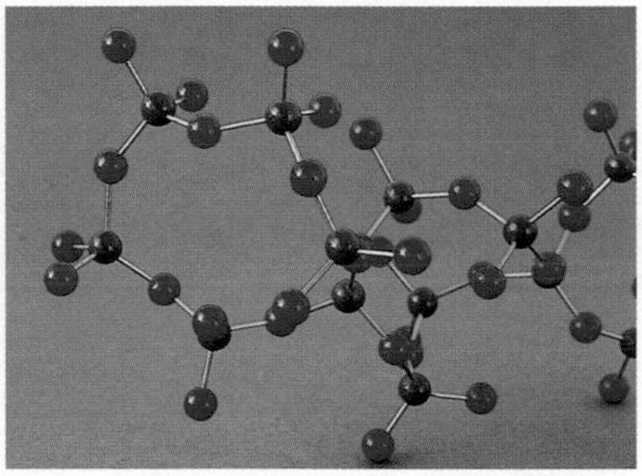

■ Figure 18.11
A model of the silicon(IV) oxide structure.

The double bonds in the carbon dioxide molecule are more than twice as strong (804 kJ mol^{-1}) as those of a C—O bond (358 kJ mol^{-1}). The double bond in methanal, H_2CO, is only 695 kJ mol^{-1} and this suggests that the bonds in carbon dioxide contain some triple character formed as a result of delocalisation of the π bonds over the whole molecule. The result is that, for carbon dioxide, a multiply bonded structure is favoured over a giant structure containing single bonds.

The situation is different for silicon(IV) oxide. As has already been stated, silicon does not form stable π bonds, so that a giant singly bonded structure is the result.

Lead(II) oxide forms a layer lattice, showing that it is a borderline ionic/covalent. Unlike silicon(IV) oxide, which shows 4:2 coordination, lead(IV) oxide shows 6 : 3 coordination, presumably due to the larger size of the lead atom 'Red lead', Pb_3O_4, can be thought of as dilead(II) lead(IV) oxide, $(Pb^{2+})_2(PbO_4^{\,4-})$.

18.4 The chlorides and hydrides

Chlorides

Carbon tetrachloride, CCl_4 (also known as the organic derivative tetrachloromethane, see Chapter 23) cannot be made directly from its elements, even though the reaction is energetically favourable ($\Delta H = -128$ kJ mol^{-1}). This shows the importance of kinetic factors in a reaction that involves breaking strong C—C bonds. Carbon tetrachloride can be made by the chlorination of methane (see Chapter 23, pages 433–5). It is very resistant to hydrolysis, which is again due to kinetic inertness, as $\Delta H = -356$ kJ mol^{-1}, almost the same value as that for the corresponding hydrolysis of silicon(IV) chloride which proceeds rapidly:

$$SiCl_4(l) + 2H_2O(l) \rightarrow SiO_2(s) + 4HCl(aq) \qquad \Delta H = -369\ \text{kJ mol}^{-1}$$

There are two main reasons why carbon tetrachloride is kinetically so inert.

- The four large chlorine atoms shield the small carbon atom from attack by water molecules.
- Carbon, unlike silicon, has no d orbitals to receive electrons donated by water molecules.

The likely mechanism for the hydrolysis of silicon(IV) chloride is shown in Figure 18.12.

■ Figure 18.12
The likely mechanism for the hydrolysis of silicon(IV) chloride.

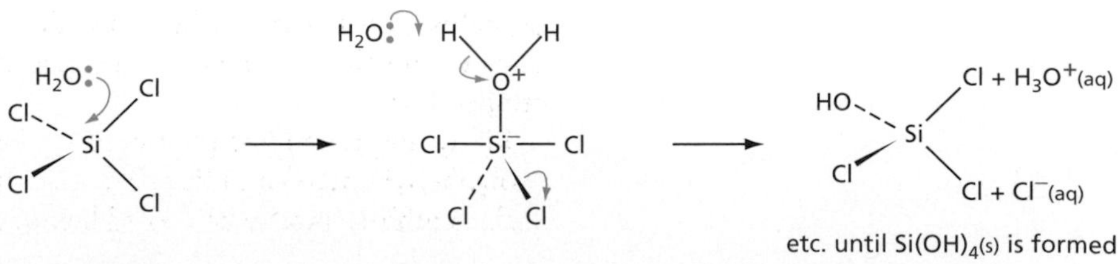

Lead forms two chlorides, the ionic lead(II) chloride, $PbCl_2$, and the covalent lead(IV) chloride, $PbCl_4$. The latter is made (together with H_2PbCl_6) when lead(IV) oxide is dissolved in cold concentrated hydrochloric acid:

$$PbO_2(s) + 6HCl(aq) \rightarrow H_2PbCl_6(aq) + 2H_2O(l)$$

On warming, the solution rapidly decomposes:

$$PbCl_4(aq) \rightarrow PbCl_2(s) + Cl_2(g)$$

Hydrides

The vast number of hydrides of carbon forms the basis of organic chemistry. Silicon hydrides corresponding to alkanes are known, the largest containing at least six silicon atoms. Because of the weakness of the Si—H and Si—Si bonds, these hydrides are spontaneously inflammable in air. For example silane, SiH_4, reacts as follows:

$$SiH_4(g) + 2O_2(g) \rightarrow SiO_2(s) + 2H_2O(l)$$

Because silicon does not form π bonds, there are no silicon hydrides corresponding to ethene or benzene.

There is no firm evidence for a lead hydride.

Example 1 Write equations for the following reactions:

a the action of hot dilute hydrochloric acid on lead

b the action of carbon dioxide on carbon to give carbon monoxide.

Answer a $Pb(s) + 2HCl(aq) \rightarrow PbCl_2(aq) + H_2(g)$

b $CO_2(g) + C(s) \rightarrow 2CO(g)$

Example 2 Suggest explanations for the following observations.

a Silicon does not form a stable oxide SiO.

b Carbon does not react directly with chlorine to give carbon tetrachloride, CCl_4, although $\Delta H_f(CCl_4) = -128\,kJ\,mol^{-1}$.

Answer a SiO would contain π bonds. Silicon does not form π bonds because the atoms are too large to allow effective p-to-p sideways overlap.

b The direct formation of carbon tetrachloride involves breaking strong C—C bonds, so has a high activation energy. The reaction is therefore too slow to be observed.

Example 3 State, with a reason, the shape of the $Pb(OH)_4^{2-}$ ion.

Answer The lead atom in this ion has four pairs of bonded electrons in its outer shell. The shape is therefore a regular tetrahedron.

Further practice

1 Write equations for the following reactions:

a the action of heat on red lead, Pb_3O_4, to give lead(II) oxide and oxygen

b the combustion of the silane Si_2H_6 in air

c the action of carbon dioxide on sodium hydroxide solution (two equations)

d the reaction of silicon(IV) oxide with molten sodium hydroxide.

2 Suggest explanations for the following observations.

a The bond lengths in graphite are in between those for a C—C single bond and those for a C—C bond in benzene.

b At room temperature, lead(II) chloride is a solid while lead(IV) chloride is a liquid.

3 State, with a reason, the shape of the $PbCl_6^{2-}$ ion.

Silicones

Bond	Bond enthalpy/ kJ mol⁻¹	Bond length/ nm	Electronegativity difference
C—C	348	0.154	0
Si—Si	226	0.235	0
C—O	358	0.143	0.89
Si—O	453	0.166	1.54

■ **Table 18.2**
Comparison of the bonding for carbon and silicon.

The longer Si—Si bond is much weaker than the short C—C bond. However, the Si—O bond is actually stronger than the C—O bond, even though it is slightly longer. This is because the Si—O bond has considerable ionic character, owing to the large electronegativity difference between the silicon and oxygen atoms.

This means that compounds containing the Si—O—Si linkage are formed readily and are very stable. The giant structure of silicon(IV) oxide has already been mentioned and a vast range of polysilicates, such as clay and mica, are known. A useful class of materials containing an Si—O—Si backbone are the **silicones** (note the spelling). These have alkyl groups attached to the spare bonds on the silicon atoms.

The simplest silicones are oils that have linear molecules. These are obtained by the hydrolysis of the dichlorosilicon derivative, for example, dichlorodimethylsilane, $(CH_3)_2SiCl_2$. These, like silicon(IV) chloride, are rapidly attacked by water, to give hydroxides initially and then silicones.

$$(CH_3)_2SiCl_2 + 2H_2O \rightarrow (CH_3)_2Si(OH)_2 + 2HCl$$

$$n\text{HO}-\underset{\underset{CH_3}{|}}{\overset{\overset{CH_3}{|}}{Si}}-\text{OH} + n\text{HO}-\underset{\underset{CH_3}{|}}{\overset{\overset{CH_3}{|}}{Si}}-\text{OH} \rightarrow \left[\underset{\underset{CH_3}{|}}{\overset{\overset{CH_3}{|}}{Si}}-\text{O}-\underset{\underset{CH_3}{|}}{\overset{\overset{CH_3}{|}}{Si}}-\text{O} \right]_n + n\text{H}_2\text{O}$$

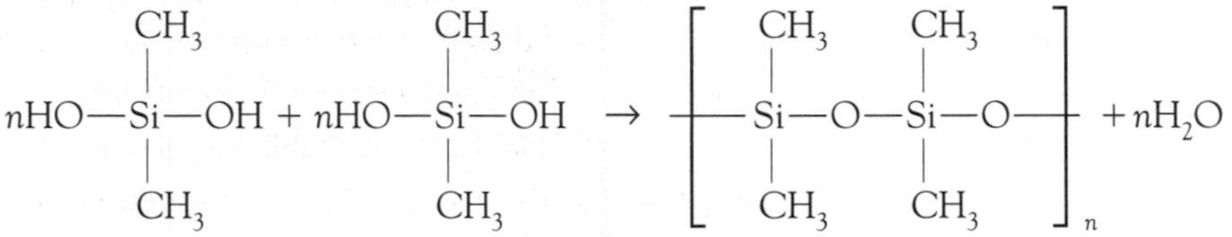

silicone

The reaction is catalysed by a trace of sulphuric acid. The length of the chain can be controlled by adding small quantities of trimethylchlorosilane, $(CH_3)_3SiCl$, which seals the ends of the chains. Because the alkyl groups are hydrophobic, silicone oils are used to waterproof fabrics. Silicone oils can be used at much higher temperatures than hydrocarbon oils, and find a use in high-temperature hydraulic systems.

If the chains are very long, silicone oils become very viscous. If this viscous liquid is heated with a trace of an oxidising agent, cross-linking takes place between the methyl groups and a rubber is formed:

$$\left[\underset{\underset{CH_3}{|}}{\overset{\overset{CH_3}{|}}{Si}}-\text{O} \right] \begin{array}{c} + \\ [O] \\ + \end{array} \left[\underset{\underset{CH_3}{|}}{\overset{\overset{CH_3}{|}}{Si}}-\text{O} \right] \longrightarrow \left[\underset{\underset{CH_2}{|}}{\overset{\overset{CH_3}{|}}{Si}}-\text{O} \right] \begin{array}{c} | \\ CH_2 \\ | \end{array} \left[\underset{\underset{CH_3}{|}}{\overset{\overset{CH_2}{|}}{Si}}-\text{O} \right] + \text{H}_2\text{O}$$

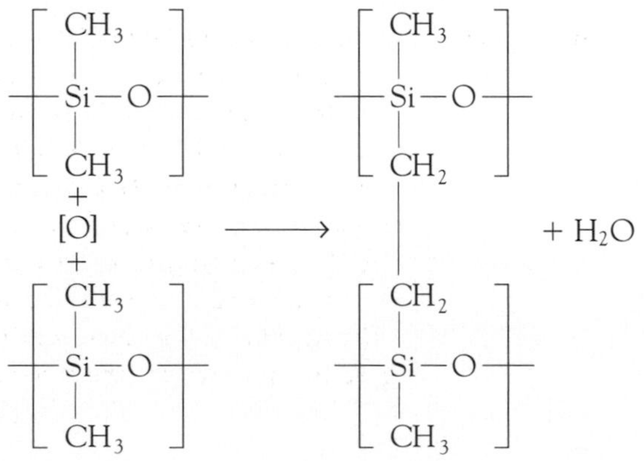

■ **Figure 18.13**
Some uses of silicones – potty putty, a power ball and waterproofing.

These silicone rubbers are the constituents of 'potty putty' and bouncy balls that return nearly to their original height when dropped. More controversially, silicone rubbers have been used for breast implants, as they were thought to be chemically inert. This has not proved to be the case and now many implants are being removed only a few years after being inserted.

Summary

- The Group 4 elements change in character from non-metallic, to semiconductors, to metallic as the group is descended.
- At the top of the group the +4 oxidation number is stable; at the bottom the +2 oxidation number is more stable.
- All Group 4 elements form dioxides, which are acidic at the top of the group and amphoteric at the bottom.
- All Group 4 elements form covalent tetrachlorides, which are stable at the top of the group and unstable at the bottom.
- The inertness of many compounds of carbon is due to the strength of the C—C bond.
- Only carbon forms stable multiple bonds.

Key reactions you should know

- Reaction with oxygen and chlorine:

$C(s) + \frac{1}{2}O_2(g) \rightarrow CO(g)$

$X(s) + O_2(g) \rightarrow XO_2(g)$ X = C, Si gives $SiO_2(s)$

$3Pb(l) + 2O_2(g) \rightarrow Pb_3O_4(s)$

$Si(s) + 2Cl_2(g) \rightarrow SiCl_4(l)$

- Action of water on chlorides:

$CCl_4(l)$ no reaction

$SiCl_4(l) + 4H_2O(l) \rightarrow Si(OH)_4(s) + 4HCl(aq)$

$PbCl_2(s)$ insoluble but dissolves on warming

$PbCl_4(aq) + 2H_2O(l) \rightarrow PbO_2(s) + 4HCl(aq)$

- Acid–base reactions of oxides:

$CO(g)$ no reaction	neutral oxide
$CO_2(g) + Ca(OH)_2(aq) \rightarrow CaCO_3(s) + H_2O(l)$	test for $CO_2(s)$; acidic oxide
$SiO_2(s) + Na_2CO_3(s) \rightarrow Na_2SiO_3(l) + CO_2(g)$	making glass; acidic oxide
$PbO(s) + 2HNO_3(aq) \rightarrow Pb(NO_3)_2(aq) + H_2O(l)$	(as a base) } amphoteric oxide
$Pb(OH)_2(s) + 2OH^-(aq) \rightarrow Pb(OH)_4^{2-}(aq)$	(as an acid) } amphoteric oxide

19 The 3d block

In this chapter we study the 3d block, which includes many familiar metals such as iron and copper. The chemical properties of these metals contrasts with those of the metals in the s and p blocks; for example, their ions are coloured and they exist in several oxidation states. The extremely complicated chemistry of the 3d block elements can be rationalised in terms of their electronic configurations, in particular the closeness in the energy levels of the 3d, 4s and 4p orbitals.

The chapter is divided into the following sections.

19.1 Introduction

The elements from scandium to zinc inclusive comprise the **3d block**. This block contains ten elements, because the 3d subshell contains five orbitals, each able to accommodate two electrons (see Chapter 3.) The electronic configuration of scandium is [Ar] $3d^1$ $4s^2$ and that of zinc is [Ar] $3d^{10}$ $4s^2$. A 3d block element is sometimes defined as one of the elements in which the 3d subshell is being progressively filled. This is not strictly speaking correct, because copper has the electronic configuration [Ar] $3d^{10}$ $4s^1$. It is more accurate to state that the 3d block contains elements with electronic configurations from [Ar] $3d^1$ $4s^2$ to [Ar] $3d^{10}$ $4s^2$ inclusive.

Originally this block was known as the 'transition metals', because some of their properties show a gradual change between those of the reactive metal calcium in Group 2 to the much less reactive metal gallium in Group 3. The term **transition metal** is now reserved for those metals in the block that show properties characteristically different from those in the s and p blocks. Two examples of these different properties are that transition metals can exist in more than one oxidation state, and that their ions are often coloured. We must exclude both scandium and zinc from the class of transition metals, for the following reasons.

1 Scandium forms only the colourless Sc^{3+} ion, isoelectronic with the Ca^{2+} ion, with no electrons in the 3d subshell.
2 Zinc forms only the colourless Zn^{2+} ion, isoelectronic with the Ga^{3+} ion, with 10 electrons in the 3d subshell.

It is the elements that form ions with some electrons in the 3d subshell that exhibit these special properties of transition metals.

The 3d elements and the transition metals

- The **3d block** contains elements in which the 3d subshell is being progressively filled.
- The 3d block includes all the elements with the electronic configurations [Ar] $3d^1$ $4s^2$ to [Ar] $3d^{10}$ $4s^2$ inclusive.
- The **transition metals** form some compounds containing ions with an incomplete d subshell.

The reason why the 4s subshell is filled before the 3d subshell is because it is lower in energy (see Chapter 2, page 42), even though it is part of a shell whose average distance is further away from the nucleus. The difference in energy is very small, which means that both s and d electrons may be involved in bonding.

The order of energy levels 4s < 3d holds only as far as calcium. With the increasing nuclear charge from 21 in scandium to 30 in zinc, the energy levels of both the 4s and 3d orbitals decrease, but the 3d level decreases faster than the 4s (see Figure 19.1 and Figure 2.33, page 42). At scandium their energy levels are almost the same, but subsequently the order changes so that 3d is slightly less than 4s. So when an ion is formed, the 4s electrons are removed before the 3d electrons. The reason for this reversal in energy levels is discussed in the next section.

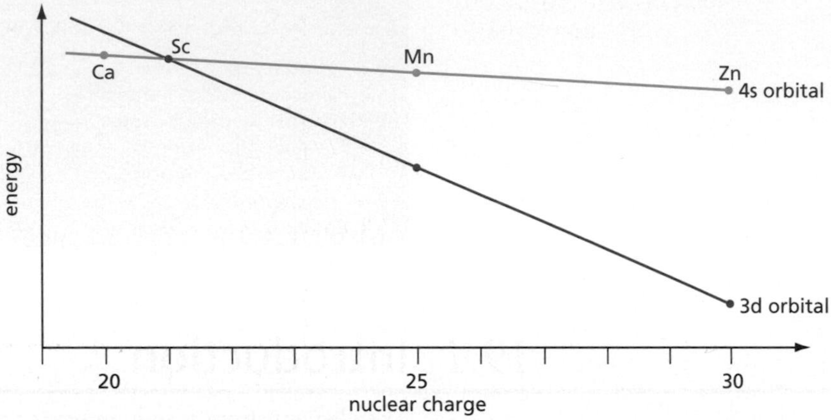

■ **Figure 19.1** The 3d and 4s energy levels on crossing the 3d block. The graph is not to scale.

19.2 Properties of the metals

	Sc	Ti	V	Cr	Mn	Fe	Co	Ni	Cu	Zn
Electronic configuration [Ar]	$3d^1 4s^2$	$3d^2 4s^2$	$3d^3 4s^2$	$3d^5 4s^1$	$3d^5 4s^2$	$3d^6 4s^2$	$3d^7 4s^2$	$3d^8 4s^2$	$3d^{10} 4s^1$	$3d^{10} 4s^2$
Melting point/°C	1541	1660	1887	1857	1244	1535	1495	1453	1083	420
Boiling point/°C	2831	3287	3377	2672	1962	2750	2870	2732	2567	907
Metallic radius/nm	0.161	0.145	0.132	0.125	0.124	0.124	0.125	0.125	0.128	0.133
M^{2+} radius/nm	—	0.080	0.072	0.084	0.091	0.082	0.082	0.078	0.072	0.083
M^{3+} radius/nm	0.083	0.069	0.065	0.064	0.070	0.067	0.064	0.062	—	—
ΔH_{at}/kJ mol^{-1}	378	470	514	397	281	416	425	430	338	131
First IE/kJ mol^{-1}	631	658	650	653	717	759	760	737	745	906
Second IE/kJ mol^{-1}	1235	1310	1414	1592	1509	1561	1646	1753	1958	1733
Third IE/kJ mol^{-1}	2389	2652	2828	2987	3248	2957	3232	3393	3554	3833
$E^\ominus$ (M^{2+}/M)/V	—	−1.63	−1.13	−0.90	−1.18	−0.44	−0.28	−0.26	+0.34	−0.76
$E^\ominus$ (M^{3+}/M^{2+})/V	—	−0.37	−0.26	−0.42	+1.5	+0.77	+1.92	—	—	—
Electronegativity	1.36	1.54	1.63	1.66	1.55	1.83	1.88	1.91	1.90	1.65

■ **Table 19.1** The main properties of the elements in the 3d block.

Electronic configuration

As the atomic number increases by one unit, an extra electron is usually added to the 3d subshell. There are two exceptions to this general trend:

- chromium is [Ar] $3d^5 4s^1$ not [Ar] $3d^4 s^2$
- copper is [Ar] $3d^{10} 4s^1$ not [Ar] $3d^9 4s^2$.

This suggests that $3d^5$ (with a half-filled subshell) and $3d^{10}$ (with a full subshell) are preferred configurations. Both these configurations have a symmetrical 3d cloud of electrons that screens the nucleus more effectively than other configurations.

Melting and boiling points

The melting points and boiling points of the 3d block metals are generally much higher than those of the s and p block metals. This indicates that not only the

4s electrons but also the 3d electrons are involved in the metallic bonding. When melting and boiling points are plotted against atomic number, both graphs have two maxima, with an intermediate minimum at manganese (see Figure 19.2). These minima suggest that the half-filled subshell of 3d electrons is not readily available for metallic bonding. We might expect the same effect in chromium, but it occurs to a lesser degree because the nuclear charge is lower.

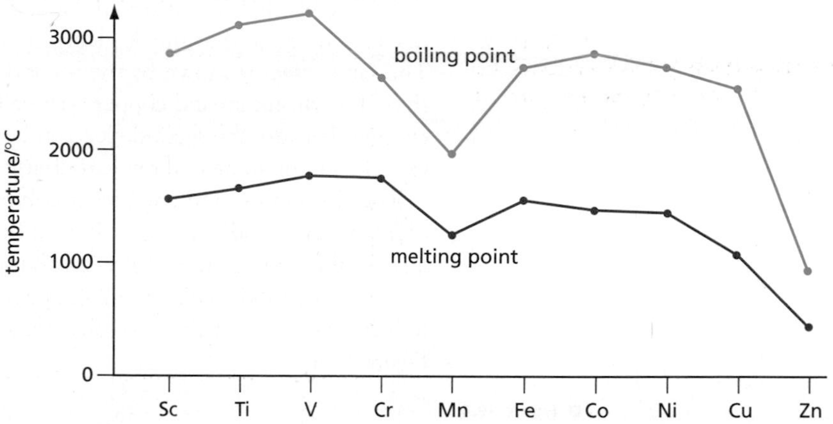

■ Figure 19.2 Melting and boiling points of the 3d-block metals.

The values of ΔH_{at} are determined by the strength of the metallic bonding, and so they also produce a similarly shaped graph with two maxima and a minimum at manganese.

Metallic and ionic radii

The metallic radii tend to decrease across the block as the increasing nuclear charge attracts the outer electrons more strongly. There are minor variations to this trend that can be rationalised in terms of the different strengths of the metallic bonding – for example, the metallic radii of both copper and zinc are larger than those of the preceding metals, and their metallic bonding is weaker as shown by their comparatively low melting points.

There is no obvious trend in the M^{2+} ionic radii, but the M^{3+} ionic radii tend to decrease across the block as the nuclear charge increases and attracts the d electrons more strongly. This decrease in ionic radius is accompanied by an increase in charge density of the ion, leading to an increased stability of the complexes formed towards the right of the block, as we shall see later in the chapter.

Ionisation energies

A graph of ionisation energies against atomic number (see Figure 19.3) shows three main features.

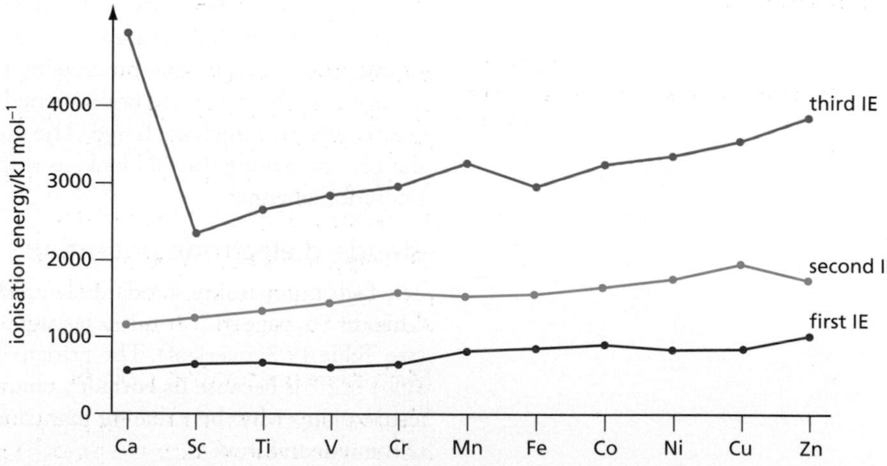

■ Figure 19.3 Ionisation energies of the 3d-block metals.

- The first and second ionisation energies increase only slightly across the block.
- The second ionisation energies are only slightly higher than the first.
- The third ionisation energies are significantly higher, and show a characteristic d subshell pattern.

The first ionisation energies involve the removal of a 4s electron. This is outside the 3d subshell and is partially screened by it. As the nuclear charge increases across the block, the additional d electrons shield the effect of the increasing nuclear charge so that the 4s electron experiences only a small extra attraction. The same effect is shown by the second ionisation energies, with the exception of those for chromium and copper – these have rather higher second ionisation energies because the second electron being removed is a 3d electron, which does experience the increased nuclear charge.

The third ionisation energies involve the removal of a 3d electron. The pattern of five values steadily rising, followed by a drop and then five more steadily rising, mirrors the ionisation energies involving the removal of 2p electrons on crossing the second period (see Figure 2.42, page 49). The drop from manganese to iron is a reflection of the fact that two electrons in the same orbital repel each other (see Figure 19.4).

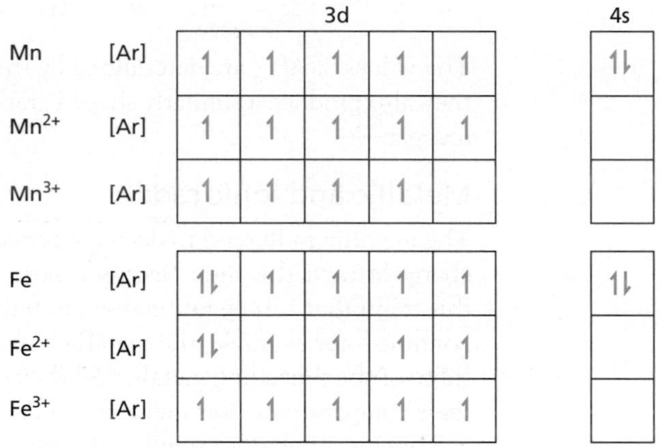

■ **Figure 19.4** The electronic configurations of manganese and iron. The formation of the Fe^{3+} ion involves the removal of an electron from a doubly occupied 3d orbital; this electron is more easily removed because it is repelled by the other electron in the orbital.

3d and 4s energy levels

The relative energies of the 3d and 4s orbitals change on crossing the block (see Figure 19.1, page 346). Up to calcium, the energy of the 4s orbital is lower than that of the 3d, even though, on average, it is further away from the nucleus. This is because the 4s electron spends some time very near to the nucleus, where it experiences the full attraction of the unscreened nuclear charge. A 4s electron is said to be more **penetrating** than a 3d electron.

After calcium, the 4s electrons are screened from the effect of the increasing nuclear charge by the addition of the 3d electrons, which are nearer to the nucleus. The energy of the 4s electrons therefore decreases only slightly on crossing the 3d block. This small decrease in energy explains the small increase in first and second ionisation energies on crossing the 3rd block from scandium to zinc. The 3d electrons, however, are not screened to the same extent and so experience a greater effective nuclear charge. The energy of the 3d electrons therefore decreases sharply on crossing the 3d block, as is shown by the sharp rise in the third ionisation energies.

Standard electrode potentials

The factors governing standard electrode potentials have already been discussed in Chapter 16, page 317. It is interesting to compare the values for zinc and copper (see Table 19.2, overleaf). The principal reason why zinc has such a large negative value of $E^{\ominus}$ is because its enthalpy change of atomisation is so low. The same effect also explains why the value for manganese is out of line with its neighbours, chromium and iron.

■ **Table 19.2**
Energy changes associated with $E^{\ominus}(M^{2+}/M)$ for copper and zinc.

Metal	ΔH_{at}/ kJ mol^{-1}	1st + 2nd IE/ kJ mol^{-1}	$\Delta H_{hyd}(M^{2+})$/ kJ mol^{-1}	$\Delta H(2H^{+}/H_2)$/ kJ mol^{-1}	Σ	$E^{\ominus}(M^{2+}/M)$/V
Cu	338	2703	−2100	−896	+45	+0.34
Zn	131	2640	−2045	−896	−171	−0.76

The ease of oxidation of M^{2+}(aq) to M^{3+}(aq), as shown by $E^{\ominus}(M^{3+}/M^{2+})$, approximately follows the trend of the third ionisation energies (see Figure 19.5).

■ **Figure 19.5**
Third ionisation energies and $E^{\ominus}(M^{3+}/M^{2+})$ for the elements chromium, manganese, iron and cobalt.

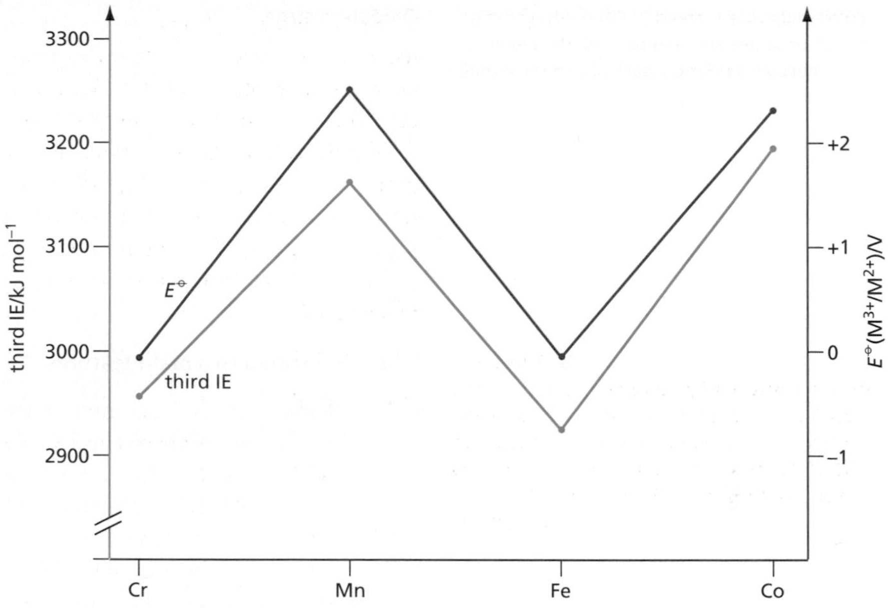

Example 1 Using [Ar] to represent the argon core, give the electronic configurations of the following:

a Cr **b** Cu^{2+} **c** V^{2+}.

Answer **a** [Ar]$3d^5 4s^1$ **b** [Ar]$3d^9$ **c** [Ar]$3d^3$

Example 2 Briefly explain the following observations.

a The first ionisation energy of cobalt is only slightly larger than the first ionisation energy of iron.

b The third ionisation energy of iron is much lower than the third ionisation energy of manganese.

c The metallic radius of vanadium is smaller than that of titanium.

Answer **a** The increase in nuclear charge from 26 to 27 is screened by the addition of an extra d electron. The effective nuclear charge, and hence the attraction of the 4s electrons, therefore increases only slightly.

b The third electron is being removed from a d^6 configuration, that has two electrons in one d orbital. These two electrons repel each other, making each one easier to remove.

c The increase in nuclear charge from 22 to 23 makes the inner clouds of electrons contract, and the radius of the atom decreases.

Further practice

1 Using [Ar] to represent the argon core, give the electronic configurations of the following:

a Cu **b** Co^{2+} **c** Ti^{3+}.

2 Briefly explain the following.

a The second ionisation energy of copper is higher than the second ionisation energy of zinc.

b It is difficult to oxidise Mn^{2+}(aq) to Mn^{3+}(aq).

c On passing from scandium to titanium, the increase in the third ionisation energy is much larger than the increase in the first ionisation energy.

19.3 Variable oxidation states

Table 19.3 shows the more familiar oxidation states of the 3d elements. Many other oxidation states are known which can be stabilised under special conditions. Reference to some of these is made in the notes below the table, and in the description of their detailed chemistry later in this chapter.

■ **Table 19.3**
The more familiar oxidation states of the 3d elements are marked ×. Other less common oxidation states are marked •. All the elements have an oxidation state of 0 in the metal.

Oxidation state	Sc	Ti	V	Cr	Mn	Fe	Co	Ni	Cu	Zn
(I)			•	•	•	•	•	•	×	
(II)		•	•	•	×	×	×	×	×	×
(III)	×	•	•	×	•	×	•	•	•	
(IV)		×	×	•	•	•	•	•		
(V)			×	•	•	•	•			
(VI)				×	•	•				
(VII)					×					

Table 19.3 shows two main features.

- The elements at the beginning of the block show their highest oxidation state equal to the sum of the 3d and 4s electrons. Examples of this are Sc^{3+}, TiO_2, V_2O_5, CrO_4^{2-} and MnO_4^-. Apart from Sc^{3+}, compounds are not formed by the loss of all the 3d and 4s electrons. The high oxidation states come about because all the 3d and 4s electrons are used in covalent bonding.
- The elements at the end of the block form M^{2+} ions by loss of the $4s^2$ electrons. With the exception of Fe^{3+} ion and Cr^{3+}, M^{3+} ions are not easily formed because the M^{3+}/M^{2+} potential is too positive, as we saw in Figure 19.5 (page 349).

Specific points about the oxidation states of each element are discussed under their individual chemistry.

19.4 Complex formation

Ligands

The 3d metal ions are relatively small and have a high charge density. They therefore attract groups containing lone pairs of electrons. These groups are known as **ligands**. Ligands are bases, and also nucleophiles (see Table 25.1, page 465), and their ability to be attracted to metal ions follows a similiar pattern to the nucleophilic strength found in organic chemistry. For common ligands an approximate order of attraction is as follows:

$$H_2O < \text{halide ions} < NH_3 < CN^-$$

The ligand bonds to the metal ion to form a **complex ion**. Usually six ligands combine with one metal ion, so the formation of a complex ion may be represented as the donation of six pairs of electrons to the metal ion. A dative covalent bond (coordinate bond) is formed between each ligand and the metal ion.

Ligand atoms are p-block elements that are more electronegative than the metal atom. This means that the ligand attracts the electrons of the bond away from the metal. The result is that the metal atom is approximately neutral, and the charge is spread to the outside of the ion (see Figure 19.6).

■ **Figure 19.6**
The formation of a complex ion. Water is shown as the ligand in this example. The metal ion could be M^{2+} rather than M^{3+}.

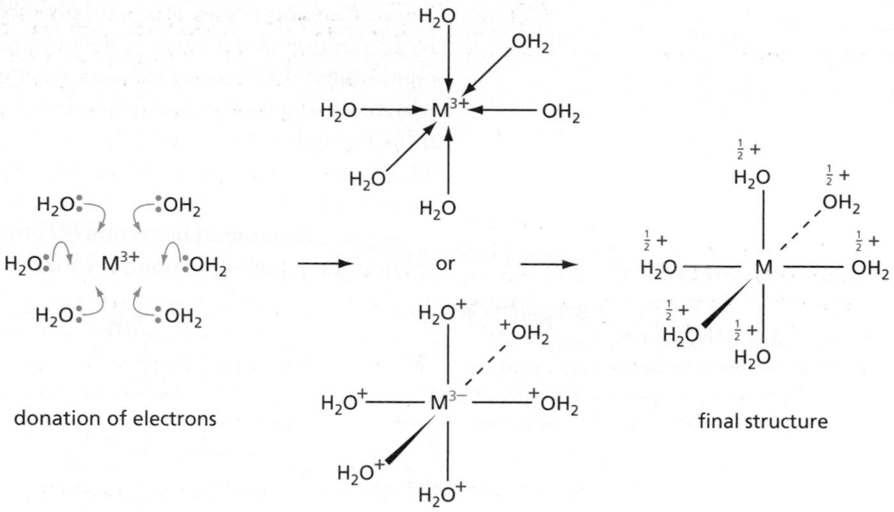

In the example in Figure 19.6, the oxidation number of the metal is simply the same as the charge on the complex ion. If the ligand is charged, for example, a Cl^- ion, the oxidation number of the metal is found in the same way as in any other ion (see Chapter 7, page 146). For example, if the oxidation number of cobalt in the $CoCl_4^{2-}$ ion is x, then:

$$x + 4 \times (-1) = -2 \quad \text{and} \quad x = +2$$

The oxidation number of cobalt in this complex ion is +2.

A **complex ion** is formed when **ligands** donate a pair of electrons to a metal ion.

The 3d block metals form stable complex ions for the following reasons.

- Their ions are small and have a high charge density.
- They have 3d orbitals of low energy level that can accommodate electrons donated by the ligands.

The shapes of complex ions

■ **Figure 19.7**
The shapes of an octahedral complex ion, for example, $Cr(H_2O)_6^{3+}$, and a tetrahedral complex ion, for example, $CoCl_4^{2-}$.

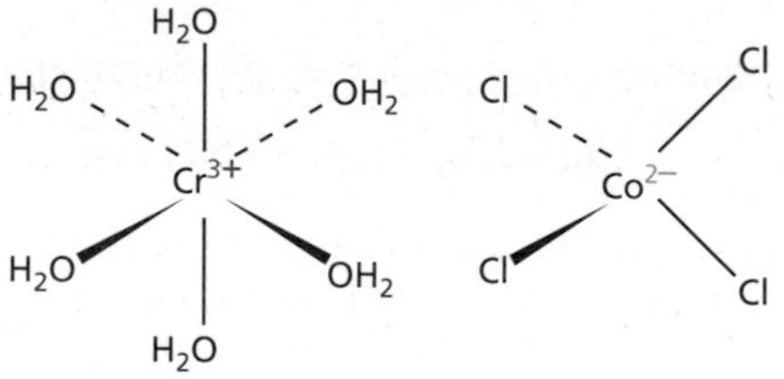

Using the VSEPR arguments (see Chapter 3, page 63), in a complex of formula ML_6^{3+} there are six pairs of electrons used in bonding. The shape of the complex ion is that of a regular octahedron. Some complexes, such as $CoCl_4^{2-}$, have only four ligands. These are usually tetrahedral in shape (see Figure 19.7).

Bonding in complex ions

If ammonia is added to a solution containing Cr^{3+} ions, a purple complex ion is formed. The Cr^{3+} ion is a d^3 ion, that is, it has three d electrons. This leaves two empty d orbitals as well as one 4s and three 4p orbitals that can receive electrons from ammonia to form the $Cr(NH_3)_6^{3+}$ complex ion (see Figure 19.8).

■ **Figure 19.8**
The formation of the $Cr(NH_3)_6^{3+}$ complex ion.

		3d					4s	4p		
Cr	[Ar]	↑	↑	↑	↑	↑	↑			
Cr^{3+}	[Ar]	↑	↑	↑						
$Cr(NH_3)_6^{3+}$	[Ar]	↑	↑	↑	↑↓	↑↓	↑↓	↑↓	↑↓	↑↓

The formation of a Cr(III) complex ion with six ligands therefore requires no reorganisation of the existing d electrons. However, the situation is more complicated if an ion contains more than three d electrons, for example, Fe(III) which is a d^5 ion. This has five half-filled orbitals (see Figure 19.9). In order to form a complex ion with six ligands, two orbitals must be cleared of electrons. This may be done in one of two ways. With ligands that are only weakly attracted, for example, H_2O, two of the d electrons in the Fe^{3+} are promoted to other higher-energy level orbitals. The resulting complex ion still has five half-filled orbitals and is known as a **high-spin complex ion**.

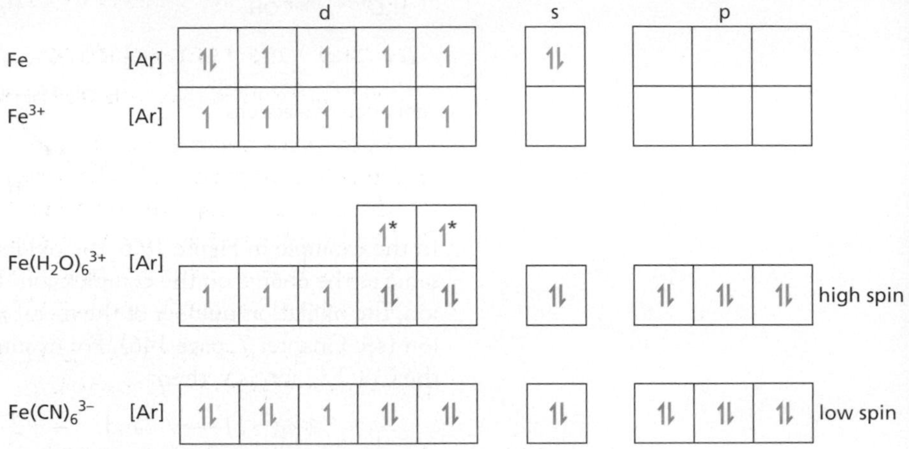

■ **Figure 19.9** The formation of the $Fe(H_2O)_6^{3+}$ and $Fe(CN)_6^{3-}$ ions. The electrons indicated as * are excited to the next highest energy orbital (see the 'Ligand field theory' panel on page 360).

With ligands that are strongly attracted, for example, CN^- ions, two of the d electrons in the Fe^{3+} pair off with electrons in other d orbitals. The resulting complex ion has only one half-filled orbital, and is known as a **low-spin complex ion**.

Ligand exchange reactions

In practice, formation of ions such as $Fe(CN)_6^{3-}$ in aqueous solution involves the displacement of water molecules from $Fe(H_2O)_6^{3+}$ by the stronger CN^- ligands. This is an example of a **ligand exchange reaction**, the substitution of one ligand by another:

$$Fe(H_2O)_6^{3+}(aq) + 6CN^-(aq) \rightarrow Fe(CN)_6^{3-}(aq) + 6H_2O(l)$$

Sometimes not all of the ligands undergo substitution, for example:

$$Cu(H_2O)_6^{3+}(aq) + 4NH_3(aq) \rightarrow [Cu(NH_3)_4(H_2O)_2]^{2+}(aq) + 4H_2O(l)$$

It is common practice to enclose the formula of the complex ion in square brackets if there is any doubt about which groups are attached to the metal ion.

Deprotonation reactions

It is important to take account of the water molecules in hydrated ions. At first sight, the formation of iron(III) hydroxide, $Fe(OH)_3(s)$, when sodium hydroxide solution is added to solution containing $Fe^{3+}(aq)$ ions, looks like a simple precipitation reaction:

$$Fe^{3+}(aq) + 3OH^-(aq) \rightarrow Fe(OH)_3(s)$$

This suggests that the reaction is similar to the precipitation, for example, of barium sulphate when sulphate ions and Ba^{2+} ions are mixed together:

$$Ba^{2+}(aq) + SO_4^{2-}(aq) \rightarrow BaSO_4(s)$$

However, such a treatment ignores the role of the water molecules. The $Fe(H_2O)_6^{3+}$ ion acts as a weak acid, $K_a = 10^{-5}\,mol\,dm^{-3}$, because it undergoes an acid–base reaction with water molecules:

$$Fe(H_2O)_6^{3+}(aq) + H_2O(l) \rightarrow [Fe(H_2O)_5OH]^{2+}(aq) + H_3O^+(aq)$$

If the H_3O^+ ions are removed by the addition of a strong base such as hydroxide ions, further acid–base reactions can take place:

$$[Fe(H_2O)_5OH]^{2+}(aq) + H_2O(l) \rightarrow [Fe(H_2O)_4(OH)_2]^+(aq) + H_3O^+(aq)$$
$$[Fe(H_2O)_4(OH)_2]^+(aq) \rightarrow Fe(OH)_3(s) + H_3O^+(aq) + 2H_2O(l)$$

The overall reaction with hydroxide ions is:

$$Fe(H_2O)_6^{3+}(aq) + 3OH^-(aq) \rightarrow Fe(OH)_3(s) + 7H_2O(l)$$

The net result is that the $Fe(H_2O)_6^{3+}$ ion has been **deprotonated** by a series of acid–base reactions and water ligands have been released.

Complex ions nomenclature

When naming a complex ion, the following rules apply.

- A cation has the usual metal name, for example, copper.
- An anion has the metal name with an 'ate' ending, for example, chromate.
- The oxidation state is indicated in the usual way, for example, iron(III).
- The ligands are given specific names, for example, chloro (Cl^-), aqua (H_2O), hydroxo (OH^-), ammine (NH_3), cyano (CN^-).
- The number of ligands is indicated by the prefixes di, tri, tetra, penta, hexa.

So the complex ion $Fe(H_2O)_6^{3+}$ is the hexaaquairon(III) ion, and $Cu(NH_3)_4^{2+}$ is tetraamminecopper(II), for example.

Example 1

a State the number of d electrons in the Cu^{2+} ion.
b Show, using curly arrows, the formation of the $Cu(H_2O)_6^{2+}$ ion.
c Write an equation for the conversion of $Cu(H_2O)_6^{2+}$ into the $[Cu(NH_3)_4(H_2O)_2]^{2+}$ ion on the addition of aqueous ammonia.
d State the type of reaction represented by this change.

Answer

a 9
b

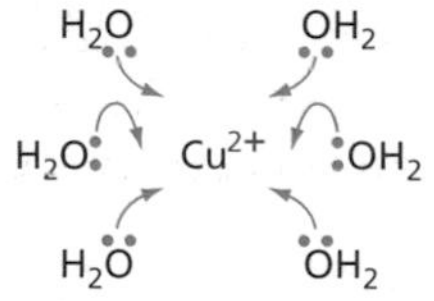

■ **Figure 19.10**

c $Cu(H_2O)_6^{2+}(aq) + 4NH_3(aq) \rightarrow [Cu(NH_3)_4(H_2O)_2]^{2+}(aq) + 4H_2O(l)$
d This is a ligand exchange reaction.

Example 2

a Write an equation for the action of aqueous sodium hydroxide on the $Cu(H_2O)_6^{2+}$ ion.
b State the type of reaction that is represented by this change.

Answer

a $Cu(H_2O)_6^{2+}(aq) + 2OH^-(aq) \rightarrow Cu(OH)_2(s) + 6H_2O(l)$
b This is a deprotonation or acid–base reaction.

Example 3

Name the following complex ions.
a $Fe(CN)_6^{3-}$ b $[Cu(NH_3)_4(H_2O)_2]^{2+}$ c $[Fe(H_2O)_5OH]^{2+}$

Answer

a hexacyanoferrate(III) ion
b tetraamminediaquacopper(II) ion
c pentaaquahydroxoiron(III) ion.

Further practice

1 a State the number of d electrons in the Fe^{2+} ion.
b Show, using curly arrows, the formation of the $Fe(H_2O)_6^{2+}$ ion.
c Write an equation for the conversion of $Fe(H_2O)_6^{2+}$ ion into the $Fe(CN)_6^{4-}$ ion on the addition of aqueous cyanide ions.
d State the type of reaction represented by this change.
e The $Fe(CN)_6^{4-}$ ion is a low-spin complex. How many unpaired electrons does it contain?

2 a Write an equation for the action of aqueous sodium hydroxide on the $Fe(H_2O)_6^{2+}$ ion.
b State the type of reaction that is represented by this change.

3 Name the following complex ions.
a $Cr(NH_3)_6^{3+}$ b $CuCl_4^{2-}$ c $Zn(OH)_4^{2-}$ d $[CrCl_2(H_2O)_4]^+$

Chelates

Bidentate ligands

Ligands such as H_2O and CN^- are attached by one coordinate bond to the metal ion. If the ligand contains two groups that have a lone pair of electrons, it may form two bonds to the metal atom, forming a ring. Such a ligand is called a **chelate**, a name derived from the Greek word for a crab's claw. Stable complexes result if five- or six-membered rings are produced by the chelate and the metal ion.

Two ligands that readily form chelates are 1,2-diaminoethane, $H_2NCH_2CH_2NH_2$, and the ethanedioate ion, $^-O_2CCO_2^-$. These form five-membered rings (see Figure 19.11) and are called **bidentate** ligands because they join by two bonds.

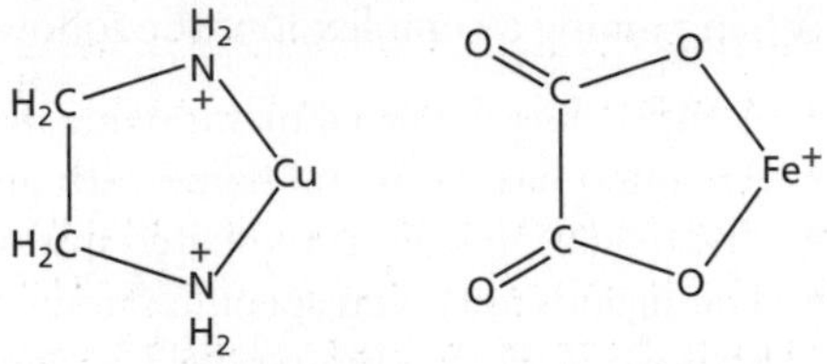

■ **Figure 19.11** Five-membered rings formed between 1,2-diaminoethane and the Cu^{2+} ion, and between the ethanedioate ion and the Fe^{3+} ion.

Chelates form particularly stable complex ions, partly because they form strong bonds to the metal ion, but also because there is an additional entropy effect that adds to their stability. For example, a chelate is formed in which three ethanedioate ions bond to Fe^{3+}. Four species become seven after the reaction so the formation of this chelate is accompanied by an increase in entropy:

$$Fe(H_2O)_6^{3+} + 3C_2O_4^{2-} \rightarrow Fe(C_2O_4)_3^{3-} + 6H_2O$$

Another way of visualising the increase in entropy is to consider the effect after one end of the chelate has become bonded to the metal ion. Once this end is secured, it becomes much more likely that the other end will be in the right position to bond too (see Figure 19.12).

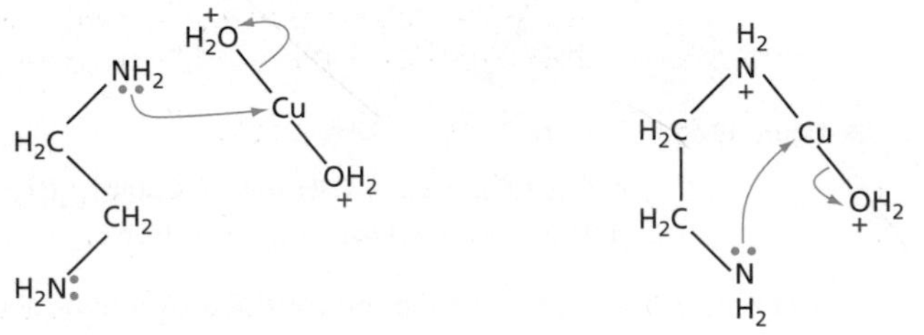

■ **Figure 19.12** Once the first end of a chelate has become bonded to the metal ion, there is an increased probability that the second end will bond.

Multidentate ligands – edta

Some chelates form more than two bonds with the metal ion. A particularly important one is 1,2-diaminoethanetetraethanoic acid, abbreviated as edta, which is used in the form of its disodium salt, containing $(edtaH_2)^{2-}$:

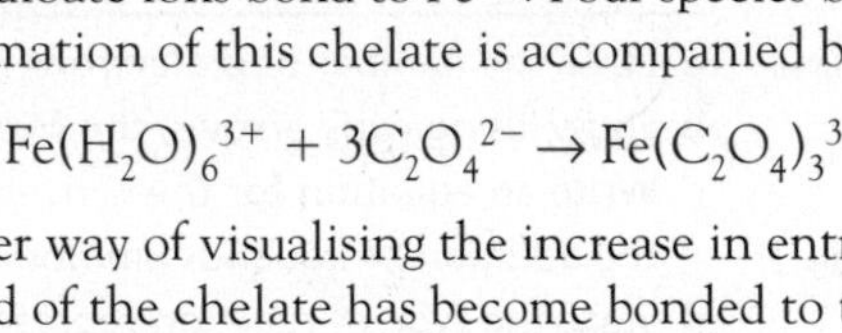

This has six pairs of electrons able to bond to a metal ion, and so forms a hexadentate chelate. edta is used for treating hard water (see Chapter 16, page 315) – it traps Ca^{2+} and Mg^{2+} ions and renders them ineffective as hard water agents.

$$Ca(H_2O)_6^{2+}(aq) + (edtaH_2)^{2-}(aq) \rightarrow Ca(edta)^{2-}(aq) + 2H^+(aq) + 6H_2O(l)$$

This trapping of metal ions is called **sequestering**. It alters the chemical properties of the metal ions, and can be used to counteract the effect of poisoning by heavy metal ions such as lead.

Analysing tap water

The formation of stable complexes between edta and Mg^{2+} and Ca^{2+} ions is used to estimate the hardness of water by titration with edta. A sample of the water being tested is mixed with a buffer solution at pH 10. A few drops of an indicator, solochrome black, are added and a solution of edta is run in from the burette. Solochrome black forms a red complex with magnesium ions. If solochrome black is represented as $HIn^{2-}(aq)$, the reaction is:

$$MgIn^-(aq) + (edtaH_2)^{2-}(aq) \rightarrow Mg(edta)^{2-}(aq) + H^+(aq) + HIn^{2-}(aq)$$

red complex   blue solution

During the titration, the free calcium and magnesium ions first react with the edta:

$$Mg(H_2O)_6^{2+}(aq) + (edtaH_2)^{2-}(aq) \rightarrow Mg(edta)^{2-}(aq) + 2H^+(aq) + 6H_2O(l)$$

When all the free calcium and magnesium ions have reacted with edta, the colour changes from red to blue as the $MgIn^-(aq)$ complex breaks down.

Example

A few drops of solochrome black indicator were added to $50.0\,cm^3$ of tap water. A $0.0100\,mol\,dm^{-3}$ solution of edta in a buffer at pH 10 was added from a burette and the indicator changed from red to blue after the additon of $7.55\,cm^3$.

a Suggest two substances that could be used to make the buffer solution.

b Calculate the total concentration of calcium and magnesium ions in the sample of hard water.

Answer

a Ammonia and an ammonium salt (for example, ammonium chloride) – see Chapter 12, page 247.

b $n(edta) = c \times \dfrac{v}{1000}$

$\qquad = 0.0100 \times \dfrac{7.55}{1000}$

$\qquad = 7.55 \times 10^{-5}\,mol$

This amount of edta reacted with $50.0\,cm^3$ of tap water. Since one mole of calcium or magnesium ions reacts with one mole of edta, the concentration of calcium and magnesium ions in the tap water is given by:

$c\,(Ca^{2+} + Mg^{2+}) = 1000 \times \dfrac{n}{v}$

$\qquad = 1000 \times \dfrac{7.55 \times 10^{-5}}{50.0}$

$\qquad = \mathbf{1.51 \times 10^{-3}\,mol\,dm^{-3}}$

Further practice

$50.0\,cm^3$ of the same tap water was boiled and allowed to cool. The solution was filtered and the precipitate washed. The filtrate and washings were titrated with $0.0100\,mol\,dm^{-3}$ edta and the solochrome black indicator changed from red to blue after the addition of $3.4\,cm^3$.

1 Write an ionic equation to show the reaction that takes place on boiling.

2 Why was the precipitate washed?

3 Calculate the percentage of permanent hardness in the water.

Isomerism in complex ions

Some of the types of isomerism found in organic chemistry (see Chapter 22, pages 404–11) are also found in complex ions. An example of structural isomerism occurs in compounds of formula $CrCl_3.6H_2O$. Three such compounds are known with different properties (see Table 19.4). In particular, they have different numbers of free Cl^- ions (as shown by their reactions with Ag^+ ions) and free water molecules (as shown by the number of water molecules that are easily removed by dehydration).

Table 19.4 The properties of the three isomers of $CrCl_3.6H_2O$.

Colour	Number of free Cl^- ions	Number of free H_2O molecules	Structure
Purple	3	0	$Cr(H_2O)_6{}^{3+}3Cl^-$
Blue-green	2	1	$[Cr(H_2O)_5Cl]^{2+}2Cl^-.H_2O$
Green	1	2	$[Cr(H_2O)_4Cl_2]^+Cl^-.2H_2O$

Complexes can also show stereochemistry. A good example of this is the complex ion $[Co(en)_2Br_2]^+$, where 'en' is used as an abbreviation for 1,2-diaminoethane, $NH_2CH_2CH_2NH_2$. This complex can exist in *cis* and *trans* forms. The *cis* form (but not the *trans*) has a chiral centre (the cobalt atom) and can be resolved into optical isomers (see Figure 19.13).

Figure 19.13 The stereochemistry of $[Co(en)_2Br_2]^+$.

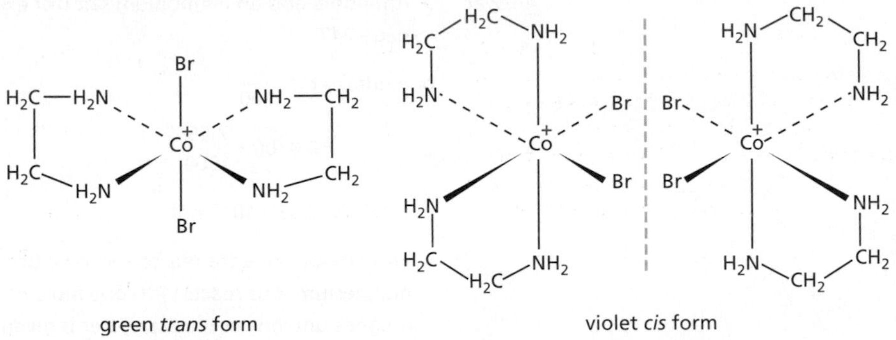

Figure 19.14 The colours of the two forms of $[Co(en)_2Br_2]^+$. **a** Green *trans* form and **b** purple *cis* form.

19.5 Colour in the d block

A compound that contains two electronic energy levels that are close together may absorb radiation in the visible region of the spectrum and therefore appear coloured (see Chapter 14, page 274). There are two ways in which a compound can have two energy levels that are close together:

- **charge transfer** – an electron from one of the atoms of the bond is excited to a higher energy level
- **d-to-d transitions.**

Both these mechanisms operate in the d block.

Charge transfer

Outside the d block, charge transfer is largely restricted to solid compounds such as oxides whose bonding is borderline between ionic and covalent. However, in the d block, the transfer of charge can take place in solution as well. For example, the colours of solutions containing TiO_3^{2-}, VO_2^{+}, CrO_4^{2-} and MnO_4^{-} ions become progressively more pronounced as the ions become more powerful oxidising agents. A powerful oxidising agent attracts electrons strongly and as a result, charge transfer takes place more readily. This means that the movement of electrons requires less energy and the charge transfer absorption band moves to longer wavelengths in the visible region of the spectrum.

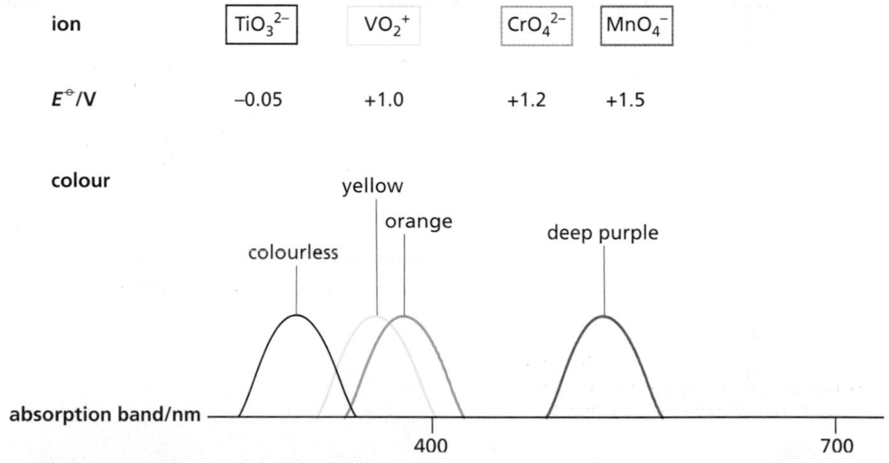

■ **Figure 19.15** As the ion becomes a more powerful oxidising agent, the absorption band moves to longer wavelengths in the visible region of the spectrum. The MnO_4^{-} ion is very deeply coloured because the whole of its absorption band is in the visible region of the spectrum.

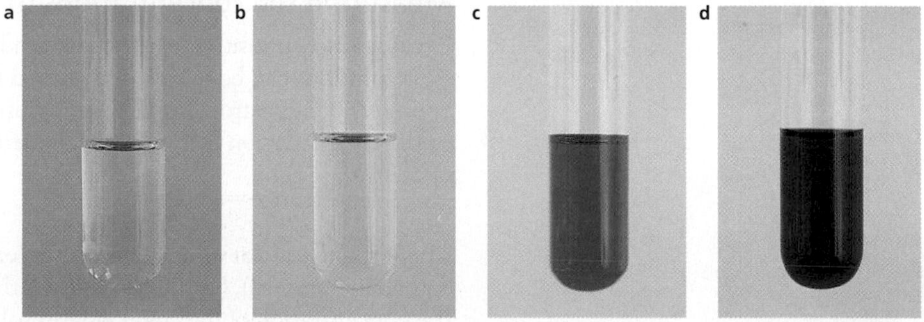

■ **Figure 19.16** Solutions containing compounds in different oxidation states. **a** Ti (IV), **b** V (V), **c** Cr (VI) and **d** Mn (VII).

oxidation state	(IV)	(V)	(VI)	(VII)
absorption maximum/nm	300	360	380	530
colour	colourless	yellow	orange	deep purple

The V (V) and Cr (VI) ions are yellow because they absorb slightly in the blue end of the spectrum. The Mn (VII) ion absorbs strongly in the green part of the spectrum and the resulting mixture of blue and red light appears as deep purple.

Charge transfer complexes are often highly coloured – examples include Prussian blue, usually formulated as $KFe_2(CN)_6$, which is deep blue, and the $[Fe(SCN)(H_2O)_5]^{2+}$ ion, which is deep red (see page 364).

d-to-d transitions

In an isolated atom, the five 3d orbitals have the same energy. In a complex ion, however, the electrostatic field produced by the donated lone pairs on the ligands splits the five orbitals into a group of three and a group of two. These two groups have slightly different energies and the difference depends on the strength of the field produced by the ligands. There is then the possibility that an electron in a lower d orbital energy level can absorb radiation and be promoted into the higher orbital energy level (see Chapter 14, page 269).

Ions that have no d electrons, for example Sc^{3+}, are colourless because there are no d electrons to promote. Similarly d^{10} ions, for example Zn^{2+}, are also colourless because there is no empty space in the upper levels to receive an extra electron (see Figure 19.17).

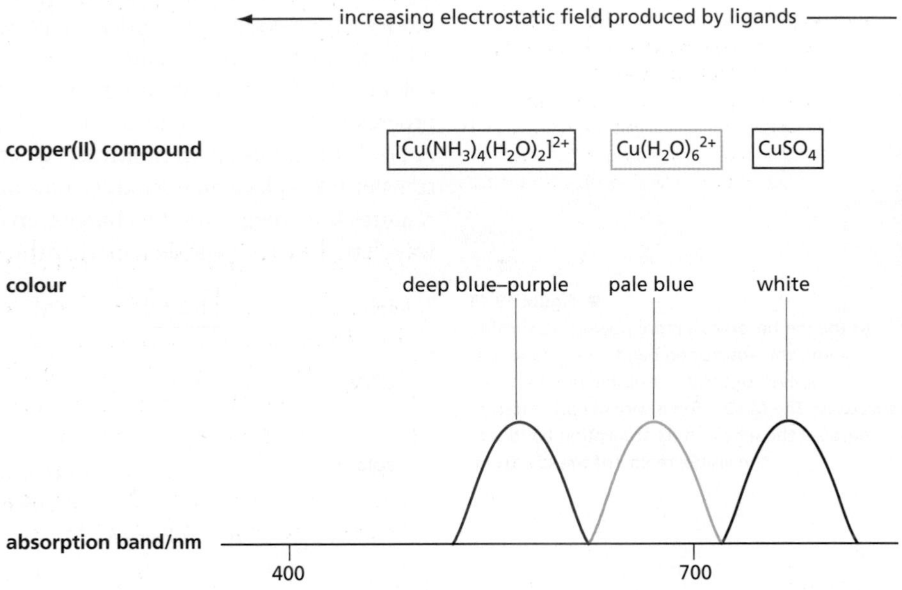

■ **Figure 19.17** Colours of some copper(II) compounds. As the ligand field becomes stronger, the splitting between the d orbitals increases and the absorption moves from the infrared into the visible region of the spectrum.

Measuring the concentrations of coloured complexes

Because many transition metal compounds are coloured, the concentrations of their solutions can be readily determined using a colorimeter (see the panel opposite) or a spectrometer. The absorbance measured by the colorimeter is related to the concentration by the Beer–Lambert Law (see Chapter 14, page 275):

$$\text{absorbance} = \varepsilon c l$$

where ε is the **molar absorption coefficient** (a measure of how strongly a particular compound absorbs), l is the path length through the sample (usually 1 cm) and c is the concentration of the compound in solution.

If ε and l are constant (which they are if we are studying just one compound, and using an identical cell each time), the absorbance is proportional to the concentration. Hence the concentration of the compound can be measured.

If the colour of a complex ion is different from that of the aqueous ion, a colorimeter can be used to find its formula. A filter is selected for the colorimeter that gives maximum absorbance for the complex ion and minimum absorbance for the aqueous ion. By varying the ratio of ligand to metal ion, the point of maximum absorption is found. At this point the ratio of (moles of ligand):(moles of metal ion) gives the number of ligand molecules used in forming the complex.

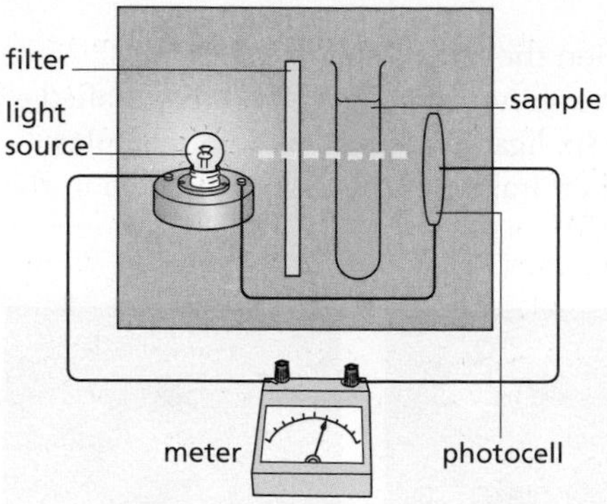

■ **Figure 19.18**
Light from a bulb passes through a filter that selects out a band of colour from the spectrum. This band of colour is strongly absorbed by the coloured solution placed in its path. The amount of light getting through depends on the concentration of the coloured species in the solution. This light is detected by a photocell and displayed on a meter.

Use of the colorimeter

The amount of substance in a coloured solution can be measured using a **colorimeter**. The solution is coloured because it is absorbing light of a complementary colour (see Chapter 14, page 274). If light of this complementary colour is passed through the solution, the concentration of the coloured substance is proportional to the **absorbance**, that is, the fraction of the light that is absorbed by the solution.

An outline of the construction of a colorimeter is shown in Figure 19.18.

It is essential to use the correct filter, that is, one that selects the colour that is absorbed most strongly by the solution. The colorimeter is first adjusted to zero ('zeroed') with a colourless solution, and the solution under test is placed in an identical test tube or cuvette. For some experiments (for example, the one described on page 160) it is not necessary to know the actual concentration of the substance; this is also true when a colorimeter is used in the following experiment to find the formula of a complex ion.

■ *Experiment* To determine the formula of a complex ion

Nickel ions form a blue complex with edta (see page 354). A solution of nickel sulphate, $NiSO_4(aq)$, is mixed with different proportions of a solution of edta and the absorbance measured with a colorimeter, using a red filter. The solution with the highest absorbance has all the nickel ions in the form of the blue complex. A typical set of results is shown in the following example.

Example Solution A is 0.10 mol dm^{-3} $NiSO_4(aq)$ and solution B is 0.20 mol dm^{-3} edta. The following mixtures were made and their absorbance measured, using a 600 nm filter.

Tube number	1	2	3	4	5	6	7	8	9	10	11
Volume of A/cm³	0	1	2	3	4	5	6	7	8	9	10
Volume of B/cm³	10	9	8	7	6	5	4	3	2	1	0
Absorbance	0.00	0.08	0.13	0.19	0.24	0.30	0.35	0.37	0.28	0.19	0.09

■ **Table 19.5**

Use these results to plot absorbance against tube number. Extrapolate the two straight-line portions to find the maximum of the graph, and use this to find the formula of the complex.

Answer

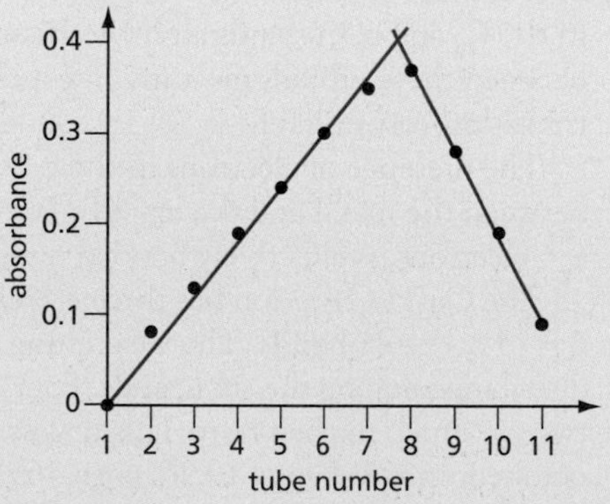

■ **Figure 19.19**

The maximum corresponds to 6.7 cm³ of 0.10 mol dm^{-3} Ni^{2+} ions and 3.3 cm³ of 0.20 mol dm^{-3} edta.

$$n(Ni^{2+}) = c \times \frac{V}{1000}$$

$$n(Ni^{2+}) = 0.10 \times \frac{6.7}{1000} = 6.7 \times 10^{-4}\ \text{mol}$$

$$n(\text{edta}) = 0.20 \times \frac{3.3}{1000} = 6.6 \times 10^{-4}\ \text{mol}$$

Therefore 1 mol of Ni^{2+} ions reacts with 1 mol of edta. If the formula is represented as $(\text{edta}H_2)^{2-}$, the equation for the reaction is:

$$Ni^{2+}(aq) + (\text{edta}H_2)^{2-}(aq) \rightarrow Ni(\text{edta})^{2-}(aq) + 2H^{+}(aq)$$

and the formula of the complex is $\mathbf{Ni(edta)^{2-}}$.

Ligand field theory

As mentioned on page 357, in a complex ion the five d orbitals are usually represented as belonging to two groups. One group contains two orbitals, called e_g, that have their lobes directed towards the six ligands of an octahedral complex. The other group, called t_{2g}, contains three orbitals with their lobes directed in the spaces between the ligands (see Figure 19.20).

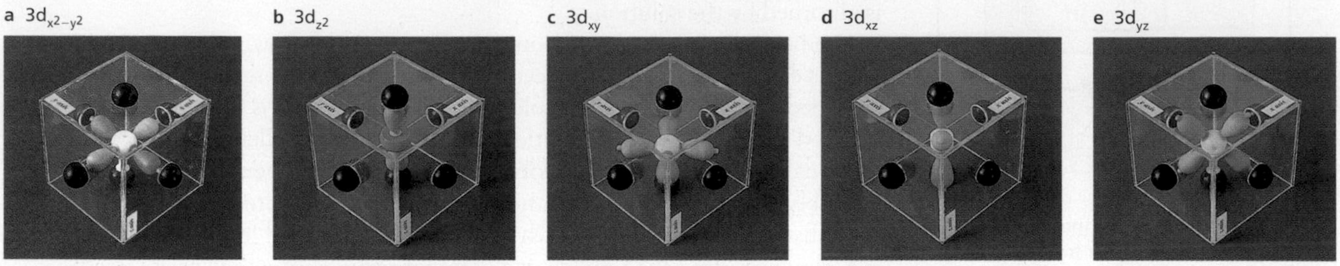

■ **Figure 19.20**
In the electrostatic field created by an octahedral complex, the five 3d orbitals split into two groups; the two e_g orbitals point towards the ligands and the three t_{2g} orbitals point in-between the ligands. Any electrons already in the e_g orbitals are repelled by the lone pairs on the ligands, while those in the t_{2g} orbitals are almost unaffected. The two e_g orbitals are **a** and **b**. The three t_{2g} orbitals are **c**, **d** and **e**.

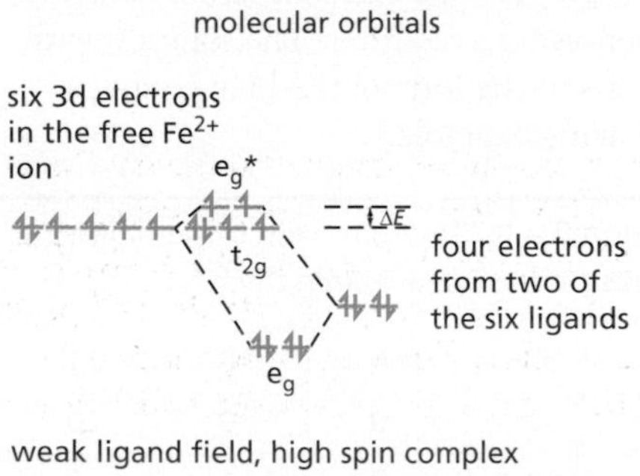

■ **Figure 19.21**
Molecular orbital diagram for the Fe^{2+} ion. For simplicity, only four electrons from the two ligands that interact with six electrons of the five 3d orbitals are shown; the other four ligands donate eight electrons into the one empty 4s orbital, and the three empty 3p orbitals. With a weak ligand field, the ten electrons are accommodated as shown above.

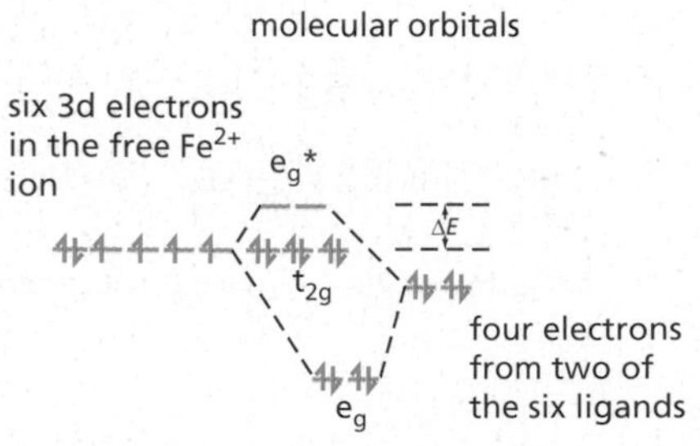

■ **Figure 19.22**
Molecular orbital diagram for the Fe^{2+} ion, showing the accomodation of electrons when the ligand field is strong.

When a complex is formed, the e_g orbitals, being directed towards the ligands, are used to form strong bonds by accepting lone pairs from the ligands. If there are any electrons already in these orbitals, they will be strongly repelled by the lone pairs of the ligands. These electrons must be accommodated in orbitals of higher energy so that the e_g orbitals are available for bonding.

If the atom that donates electrons from the ligand is very electronegative (for example, the oxygen atom in water), the bonding between the ligand and metal has a large degree of ionic character and the ligand field is weak. Under these circumstances the two e_g* orbitals are used to accommodate the highest energy electrons (Figure 19.21) and a high spin complex is formed.

If the atom that donates electrons from the ligand is slightly less electronegative (for example, the carbon atom in the cyanide ion), the bonding between the ligand and the metal is largely covalent and the ligand field is strong. Under these circumstances the energy gap, ΔE, is greater than the electron repulsion when electrons are paired in the t_{2g} orbitals; the result is the formation of a low spin complex (Figure 19.22).

The degree to which the energy levels of the d orbitals are split when a ligand binds can be measured by finding the wavelength of the absorption bands. The d-to-d transitions that are responsible for the colours of most complexes are to the e_g* orbitals. The larger the splitting, ΔE, the further the absorption band moves to the blue end of the spectrum (see Figure 19.17, page 358). If the t_{2g} and e_g* orbitals are all empty, in a d^0 ion, or if they are all full, in a d^{10} ion, there can be no d-to-d transition and the ion is colourless, unless charge transfer is involved. A d^5 ion, for example Mn^{2+} or Fe^{3+}, is very weakly coloured because the electrons in the t_{2g} and e_g* orbitals all have the same spin. This means that a transition between these orbitals must involve a change in spin, and this makes the transition very unlikely.

The presence of electrons in the e_g* antibonding orbitals weakens the binding between the metal and the ligands. For example, the $Fe(H_2O)_6^{2+}$ ion, with two e_g* electrons, is effectively bonded by $6-(2\times\frac{1}{2})=5$ bonds.

The $Cu(H_2O)_6^{2+}$ ion has three e_g* electrons and is effectively bonded by $6-(3\times\frac{1}{2})=4\frac{1}{2}$ bonds. The weakening of the bonds in this complex is not evenly distributed among the six ligands. Four ligands are bonded firmly, but the other two are much further away. This makes the shape of the complex a distorted octahedron (see Figure 19.26, page 365).

19.6 The chemistry of individual elements

General reactions

Most of the 3d metals dissolve in dilute acids, liberating hydrogen and forming M^{2+}(aq) ions. For example, with iron:

$$Fe(s) + 2HCl(aq) \rightarrow FeCl_2(aq) + H_2(g)$$

This reaction illustrates the general principle that, given the choice, it is the lower oxidation states that are formed by reducing agents in acidic conditions.

> The action of an acid on a 3d block metal usually produces a salt with the lower oxidation state.

The +II oxidation states of titanium, vanadium and chromium are so easily oxidised by air to the +III oxidation states that these metals normally form M^{3+}(aq) ions with acids. Scandium forms only the Sc^{3+}(aq) ion. The standard electrode potential of copper is positive, and the metal does not react with non-oxidising acids.

The reactions of the metals with halogens have already been discussed in Chapter 17, page 324.

Scandium

Scandium forms only the Sc^{3+} ion, which is colourless because it contains no d electrons.

Titanium

Titanium metal is light, strong and resistant to chemical attack. It therefore finds use in building aircraft and for implants in surgery. It is made by reduction of titanium(IV) chloride, $TiCl_4$ (see Chapter 20, page 378). Titanium(IV) oxide, TiO_2, is colourless and as it is highly opaque is used for whitening paint, plastics and paper.

The purple $Ti(H_2O)_6{}^{3+}$(aq) ion is relatively stable. However, compounds containing titanium(II) are known only in the solid state, as they react with water liberating hydrogen.

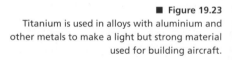

■ **Figure 19.23** Titanium is used in alloys with aluminium and other metals to make a light but strong material used for building aircraft.

Reaction	$E^{\ominus}$/V
Vanadium(5)/vanadium(4)	+1.00
Vanadium(4)/vanadium(3)	+0.34
Vanadium(3)/vanadium(2)	−0.26

Table 19.6
The standard electrode potentials of vanadium in acidic solution.

Oxidation number	Example	Colour
Vanadium(5)	VO_2^+	Yellow
Vanadium(4)	VO^{2+}	Blue
Vanadium(3)	V^{3+}(aq)	Blue-green
Vanadium(2)	V^{2+}(aq)	Lilac

Table 19.7
Colours of the oxidation states of vanadium. The $V(H_2O)_6^{3+}$ ion is purple, but in solution it forms the $[V(H_2O)_5OH]^{2+}$ ion, which is blue-green.

Vanadium

The standard electrode potentials of vanadium become more negative as the oxidation number decreases (see Table 19.6). When vanadium(5) is reduced to vanadium(2) by a powerful reducing agent such as zinc powder or zinc amalgam (zinc dissolved in mercury), it is possible to isolate the intermediate oxidation states.

The usual starting material for the reduction is ammonium vanadate, NH_4VO_3. In acidic solution, this forms the VO_2^+ ion. This reaction is similar to the formation of the NO_2^+ ion from nitric acid in the nitration of benzene (see Chapter 27, page 503).

When ammonium vanadate is reduced, the V^{4+}(aq) ion is not formed because its charge density is too high. It is deprotonated to form the blue VO^{2+}(aq) ion, even in acidic conditions:

$$[V(H_2O)_6]^{4+}(aq) + 2H_2O(l) \rightarrow [V(H_2O)_5O]^{2+}(aq) + 2H_3O^+(aq)$$

The V^{3+} and V^{2+} ions also have characteristic colours (see Table 19.7 and Figure 19.24). These changes illustrate the principle that lower oxidation states are formed by reducing agents in acidic conditions.

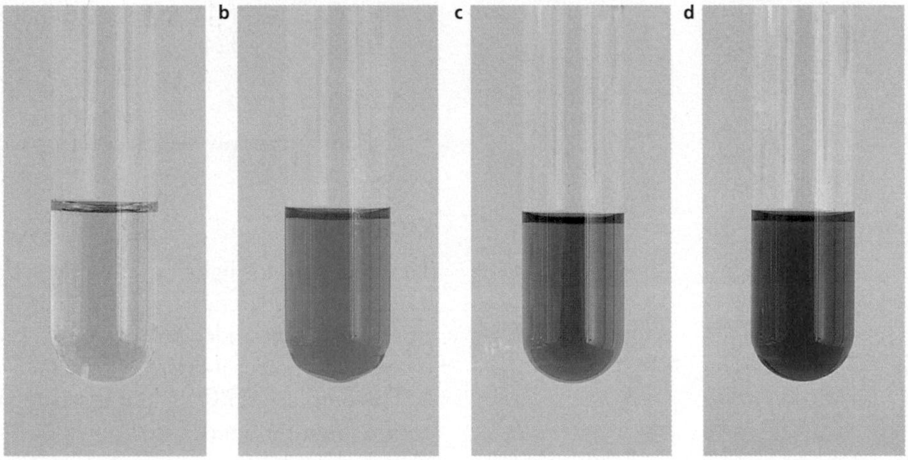

Figure 19.24
The colours of the VO_2^+, VO^{2+}, V^{3+}(aq) and V^{2+}(aq) ions (see Table 19.7).

Chromium

Chromium is resistant to oxidation by the air, and hence is used to electroplate cheaper metals. The major use of chromium is now to make stainless steel, the most common variety of which contains 74% iron, 18% chromium and 8% nickel.

The pale blue Cr^{2+} ion is difficult to prepare because it liberates hydrogen from water. It may be made by reducing the $Cr_2O_7^{2-}$ ion in the presence of zinc and sulphuric acid. Initially the Cr^{3+} ion is formed, and in order to reduce this to the Cr^{2+} ion, air must be excluded. The $Cr(H_2O)_6^{3+}$ ion is deep purple (the colour of 'chrome alum', $KCr(SO_4)_2.12H_2O$), but solutions containing chromium(III) are often green because other ligands displace the water molecules. For example, when aqueous potassium dichromate(VI), acidified with sulphuric acid, is reduced by warming with an alcohol, the product is green chromium(III) (see Chapter 26, page 486). This is because, on warming, SO_4^{2-} ions act as a ligand and displace some of the water molecules from the purple $Cr(H_2O)_6^{3+}$ ion that is initially formed.

On addition of aqueous sodium hydroxide to a solution containing chromium(III) ions, a pale green precipitate is formed. This dissolves in excess sodium hydroxide to give a green solution which may contain the $Cr(OH)_4^-$ ion:

$$Cr(H_2O)_6^{3+}(aq) + 3OH^-(aq) \rightarrow Cr(OH)_3(s) + 6H_2O(l)$$

deprotonation to form pale green precipitate

$$Cr(OH)_3(s) + OH^-(aq) \rightarrow Cr(OH)_4^-(aq)$$

precipitate dissolves in excess

On the addition of an oxidising agent (for example, hydrogen peroxide), the solution turns yellow as it is oxidised to CrO_4^{2-} ions. This illustrates the principle that higher oxidation states are formed by oxidising agents in alkaline conditions.

$$Cr(OH)_4^-(aq) + 1\tfrac{1}{2}H_2O_2(aq) + OH^-(aq) \rightarrow CrO_4^{2-}(aq) + 4H_2O(l)$$

If acid is added to this yellow solution, it turns orange owing to the formation of $Cr_2O_7^{2-}$ ions. The reaction can be reversed by the addition of alkali.

$$\underset{\text{yellow}}{2CrO_4^-(aq)} + 2H^+(aq) \rightleftharpoons \underset{\text{orange}}{Cr_2O_7^{2-}(aq)} + H_2O(l)$$

Chromium also forms compounds in which its oxidation number is +4 or +5. They cannot usually be isolated because they disproportionate to chromium(III) and chromium(VI). They are thought to be formed as intermediates during the oxidation of alcohols with acidified aqueous potassium dichromate(VI).

Manganese

The very pale pink $Mn(H_2O)_6^{2+}$ ion is very resistant to oxidation. This stability is associated with the high third ionisation energy of manganese (see Figure 19.3, page 347). In the presence of aqueous sodium hydroxide, however, an off-white precipitate of manganese(II) hydroxide, $Mn(OH)_2$, is formed that rapidly darkens in air as it is oxidised to brown MnO.OH:

$$Mn(H_2O)_6^{2+}(aq) + 2OH^-(aq) \rightarrow Mn(OH)_2(s) + 6H_2O(l)$$

deprotonation to form off-white precipitate

$$2Mn(OH)_2(s) + \tfrac{1}{2}O_2(g) \rightarrow 2MnO.OH(s) + H_2O(l)$$

air oxidation to give dark brown precipitate

Powerful oxidation, under alkaline conditions, finally produces manganate(VII), MnO_4^-.

$$Mn^{2+}(aq) + 3OH^-(aq) + 2\tfrac{1}{2}[O] \rightarrow MnO_4^-(aq) + 1\tfrac{1}{2}H_2O(l)$$

This is the reverse of the reduction of manganate(VII) in acidic conditions that is used in titration (see Chapter 7):

$$MnO_4^-(aq) + 8H^+(aq) + 5e^- \rightarrow Mn^{2+}(aq) + 4H_2O(l)$$

Standard electrode potentials indicate that MnO_4^- ions should oxidise water to oxygen. In practice, an aqueous solution of potassium manganate(VII) can be kept for several weeks without appreciable decomposition, because the reduction of MnO_4^- in neutral solution is kinetically very slow unless a catalyst is present.

The only common compound containing manganese with an oxidation number intermediate between +2 and +7 is manganese(IV) oxide, MnO_2.

Iron

The third ionisation energy of iron is fairly low because the d^6 configuration of iron(II) has two electrons in the same orbital that repel each other (see Figure 19.4, page 348). This means that compounds containing iron(II) are easily oxidised to iron(III), with a d^5 configuration. Oxidation takes place very readily in alkaline solution to give a hydrated oxide, FeO.OH.

$$Fe(H_2O)_6^{2+}(aq) + 2OH^-(aq) \rightarrow Fe(OH)_2(s) + 6H_2O(l)$$

deprotonation to form pale green precipitate

$$2Fe(OH)_2(s) + \tfrac{1}{2}O_2(g) \rightarrow 2FeO.OH(s) + H_2O(l)$$

air oxidation to give a brown precipitate

The $Fe(H_2O)_6^{3+}$ ion is pale pink-purple (compare the $Mn(H_2O)_6^{2+}$ ion), but when dissolved in water it becomes brown as it is hydrolysed:

$$\underset{\text{pale pink-purple}}{Fe(H_2O)_6^{3+}(aq)} + H_2O(l) \rightarrow \underset{\text{brown}}{[Fe(H_2O)_5OH]^{2+}} + H_3O^+(aq)$$

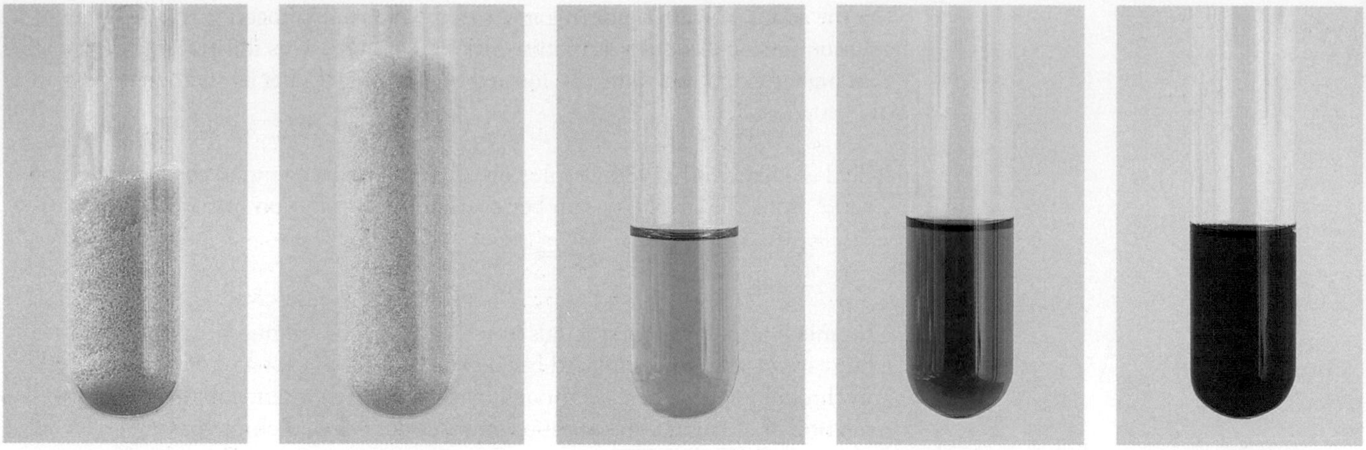

iron(II) sulphate | aluminium iron(III) sulphate | iron(III) chloride (aq) | Prussian blue | iron(III) thiocyanate complex

■ **Figure 19.25** The colours of iron in its various oxidation states.

In alkaline solution, powerful oxidising agents convert iron compounds into iron(VI), containing the FeO_4^{2-} ion. This ion is dark purple in colour and is an extremely powerful oxidising agent.

Although iron has a $d^6 s^2$ configuration and could, theoretically, form compounds with oxidation number +8, none have been prepared as they would be such powerful oxidising agents as to be unstable. The metals below iron in the periodic table do show this high oxidation state, in the oxides RuO_4 and OsO_4.

Two highly coloured complexes of iron are often used as tests.

- If Fe^{2+} ions are added to $K_3[Fe(CN)_6]$ or Fe^{3+} ions added to $K_4[Fe(CN)_6]$, a deep blue precipitate, called Prussian blue, is formed. This has the formula $KFe_2(CN)_6$ and contains iron(II) atoms joined to iron(III) atoms by CN bridges. The deep blue colour is due to charge transfer from the iron(II) atoms to the iron(III) atoms via the CN groups.
- If Fe^{3+} ions are added to thiocyanate ions, CNS^-, a deep blood-red solution is formed, containing the $[Fe(CNS)(H_2O)_5]^{2+}$ complex. This is intensely coloured due to charge transfer from the CNS group to iron(III).

Cobalt

The $Co(H_2O)_6^{2+}$ ion is pale pink. On the addition of Cl^- ions, the solution turns deep blue owing to the formation of the $CoCl_4^{2-}$ ion. The reverse reaction, when water molecules replace Cl^- ions, is the basis of blue cobalt chloride paper turning pink when exposed to water vapour.

The addition of aqueous sodium hydroxide to Co^{2+}(aq) ions gives a precipitate that is initially pink but which turns blue on standing.

Nickel

The $Ni(H_2O)_6^{2+}$ ion is green. The addition of aqueous sodium hydroxide to Ni^{2+}(aq) ions gives a green precipitate that does not dissolve in excess sodium hydroxide.

Copper

As the standard electrode potential of copper is positive, it is not attacked by non-oxidising acids. It does, however, react with hot concentrated sulphuric acid, and also with nitric acid:

$$Cu(s) + \underset{\text{hot, concentrated}}{2H_2SO_4(l)} \rightarrow CuSO_4(s) + 2H_2O(l) + SO_2(g)$$

$$Cu(s) + \underset{\text{cold, concentrated}}{4HNO_3(aq)} \rightarrow Cu(NO_3)_2(aq) + 2NO_2(g) + 2H_2O(l)$$

$$3Cu(s) + \underset{\text{cold, dilute}}{8HNO_3(aq)} \rightarrow 3Cu(NO_3)_2(aq) + 2NO(g) + 4H_2O(l)$$

Ligand	$E^{\ominus}$/V Cu(II)/Cu(I)	$E^{\ominus}$/V Cu(I)/Cu(0)
H_2O	+0.16	+0.52
NH_3	+0.10	−0.10
CN^-	+1.12	−0.44
Cl^-	+0.54	+0.14

■ **Table 19.8**
Standard electrode potentials for some copper complexes.

The familiar blue $Cu(H_2O)_6^{2+}$ ion is suprisingly stable. There are no stable divalent aqueous ions of the elements silver or gold, which lie below copper in the periodic table. In most other complexes, copper(I) is the preferred oxidation state (see Table 19.8).

If copper(II) chloride, $CuCl_2$, is dissolved in concentrated hydrochloric acid, yellow $CuCl_4^{2-}$ ions are formed. These dissolve metallic copper to give the $CuCl_2^-$ ion. On the addition of water, white copper(I) chloride, CuCl, is precipitated. Compounds containing copper(I), with a d^{10} configuration, are colourless, unless charge transfer is involved. An example of colour due to charge transfer is red copper(I) oxide, Cu_2O, formed by reduction of copper(II) complexes in alkaline solution (see Chapter 28, page 534).

The shape of the blue $Cu(H_2O)_6^{2+}$ ion is a distorted octahedron (see Figure 19.26).

When ammonia is added to a solution containing Cu^{2+}(aq), a pale blue precipitate of copper(II) hydroxide, $Cu(OH)_2$, is first formed. This dissolves in excess ammonia solution to give a deep purple-blue solution containing $[Cu(NH_3)_4(H_2O)_2]^{2+}$ ions. This has a similar shape to the $Cu(H_2O)_6^{2+}$ ion, with ammonia occupying the four planar positions with the shorter bond lengths. The ion is sometimes described as square planar, because the four Cu—N bonds are much shorter than the two Cu—O bonds.

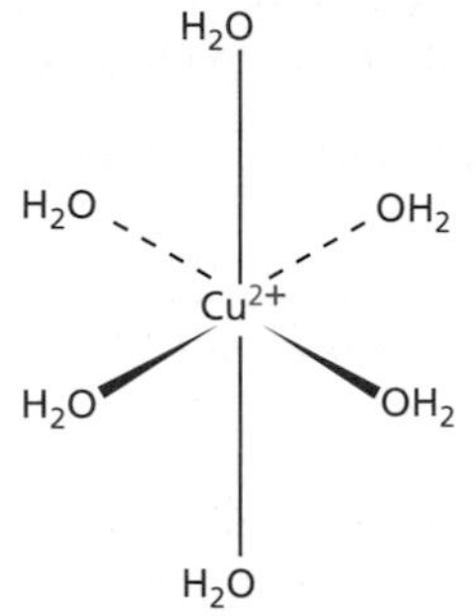

■ **Figure 19.26** (above)
The shape of the $Cu(H_2O)_6^{2+}$ ion.

ligand	Cl^-	en	NH_3	H_2O
absorption maximum/nm	390	530	620	800
colour	yellow	purple	deep blue	pale blue

■ **Figure 19.27** (right)
The colours of some copper(II) complexes. [en = 1,2-diaminoethane] The hydrated ion is pale blue because most of the absorption band is in the infrared part of the spectrum. The anhydrous ion is colourless because it absorbs totally in the infrared.

■ **Figure 19.28**
Brass is an alloy of copper and zinc.

Zinc

Zinc is widely used for galvanising and for making brass. It forms only Zn^{2+} compounds, and they are all colourless because they contain the d^{10} ion. Solutions containing $Zn^{2+}(aq)$ ions give a white precipitate with both sodium hydroxide and ammonia solutions, and this precipitate dissolves in an excess of either reagent.

$$Zn(H_2O)_6^{2+}(aq) + 2OH^-(aq) \rightarrow Zn(OH)_2(s) + 6H_2O(l)$$

deprotonation to form a white precipitate

$$Zn(OH)_2(s) + 2OH^-(aq) \rightarrow Zn(OH)_4^{2-}(aq)$$

precipitate dissolves in excess NaOH(aq)

$$Zn(OH)_2(s) + 4NH_3(aq) \rightarrow Zn(NH_3)_4^{2+}(aq) + 2OH^-(aq)$$

precipitate dissolves in excess $NH_3(aq)$

The $Zn(NH_3)_4^{2+}(aq)$ is tetrahedral as it has only four groups around the central metal atom.

Example Write equations for the following reactions, and state the type of reaction that takes place.

a aqueous sulphuric acid and iron
b chlorine and iron
c hydrochloric acid and zinc oxide
d $CNS^-(aq)$ ions and $Fe^{3+}(aq)$ ions
e NaOH(aq) and $Ni^{2+}(aq)$

Answer

a $Fe(s) + H_2SO_4(aq) \rightarrow FeSO_4(aq) + H_2(g)$; redox
b $Fe(s) + 1\frac{1}{2}Cl_2(g) \rightarrow FeCl_3(s)$; redox
c $2HCl(aq) + ZnO(s) \rightarrow ZnCl_2(aq) + H_2O(l)$; acid–base
d $Fe(H_2O)_6^{3+}(aq) + CNS^-(aq) \rightarrow [Fe(H_2O)_5(CNS)]^{2+}(aq) + H_2O(l)$; ligand exchange
e $Ni(H_2O)_6^{2+}(aq) + 2OH^-(aq) \rightarrow Ni(OH)_2(s) + 6H_2O(l)$; deprotonation

Further practice

1 Write equations for the following reactions, and state the type of reaction that takes place.
  a CN^- ions and $Fe^{2+}(aq)$ ions
  b hydrochloric acid and nickel carbonate
  c aqueous nitric acid and copper(II) oxide
  d chlorine and chromium
  e $NH_3(aq)$ and $Fe^{2+}(aq)$

2 Suggest explanations for the following observations.
  a Copper(I) chloride is white but copper(I) oxide is red.
  b Manganese(III) fluoride reacts with water to give oxygen.

19.7 Catalytic properties

Homogeneous catalysis

The rate of aqueous redox reactions is often increased by the addition of catalysts which are usually compounds from the d block. It is the ease of conversion from one oxidation state to another that is responsible for the catalytic effect of these compounds. A good example of this is the oxidation of iodide ions by peroxodisulphate(VI) ions (see Chapter 8, page 160).

$$2I^-(aq) + S_2O_8^{2-}(aq) \rightarrow 2SO_4^{2-}(aq) + I_2(aq)$$

The reaction is normally quite slow, presumably because the negatively charged ions repel each other. The reaction is catalysed by the addition of a number of d-block metal ions, for example, $Fe^{2+}(aq)$. A possible mechanism is:

$$2Fe^{2+}(aq) + S_2O_8^{2-}(aq) \rightarrow 2Fe^{3+}(aq) + 2SO_4^{2-}(aq)$$

$$2Fe^{3+}(aq) + 2I^-(aq) \rightarrow 2Fe^{2+}(aq) + I_2(aq)$$

Theory and experiment can be used to support this mechanism.

■ **Table 19.9**
Relevant values of $E^\ominus$ for catalysis by Fe^{2+} ions.

System	$E^\ominus$/V
$S_2O_8^{2-}/2SO_4^{2-}$	+2.01
Fe^{3+}/Fe^{2+}	+0.77
$I_2/2I^-$	+0.54

- The relevant values of standard electrode potentials are given in Table 19.9. These show that the postulated mechanism is thermodynamically feasible.
- The two reactions can be tested experimentally. If a solution of $S_2O_8^{2-}$ ions is added to Fe^{2+} ions, Fe^{3+} ions are produced. If Fe^{3+} ions are added to I^- ions, iodine is liberated.

$$2Fe^{2+}(aq) + S_2O_8^{2-}(aq) \rightarrow 2Fe^{3+}(aq) + S_2O_4^{2-}(aq) \qquad E^\ominus_{cell} = +1.34\,V$$
$$2Fe^{3+}(aq) + 2I^-(aq) \rightarrow 2Fe^{2+}(aq) + I_2(aq) \qquad E^\ominus_{cell} = +0.23\,V$$

Another example is the decomposition of hydrogen peroxide, catalysed by the addition of Co^{2+} ions. In acidic solution, no reaction is observed when the reactant and catalyst are mixed, but in the presence of ammonia there is frothing and a rapid evolution of oxygen. A possible mechanism is:

$$H_2O_2(aq) + 2Co(NH_3)_6^{2+}(aq) \rightarrow 2Co(NH_3)_6^{3+}(aq) + 2OH^-(aq)$$
$$2Co(NH_3)_6^{3+}(aq) + 2OH^-(aq) + H_2O_2(aq) \rightarrow 2Co(NH_3)_6^{2+}(aq) + O_2(g) + 2H_2O(l)$$

This mechanism is also supported by values of standard electrode potential (see Table 19.10).

■ **Table 19.10**
Relevant values of $E^\ominus$ for catalysis by cobalt ions in the presence of ammonia.

System	$E^\ominus$/V
$Co(H_2O)_6^{3+}/Co(H_2O)_6^{2+}$	+1.81
H_2O_2/OH^-	+0.87
$Co(NH_3)_6^{3+}/Co(NH_3)_6^{2+}$	+0.1
O_2/H_2O_2	−0.07

Further practice

What are the $E^\ominus_{cell}$ values for the two equations above?

Heterogeneous catalysis

Many d-block elements and compounds act as heterogeneous catalysts. A good example of this is the Contact process for the manufacture of sulphuric acid (see Chapter 20, pages 370–1). The vanadium(V) oxide is thought to work as follows:

$$V_2O_5(s) + SO_2(g) \rightarrow V_2O_4(s) + SO_3(g)$$
$$\tfrac{1}{2}V_2O_4(s) + O_2(g) \rightarrow V_2O_5(s)$$

This mechanism can be supported by experiment. If vanadium(V) oxide is heated with sulphur dioxide, it turns blue and forms V_2O_4. This is converted back to vanadium(V) oxide by heating in air. This catalytic effect is again due to the ability of d-block elements to exist in more than one oxidation state.

Many hydrogenation reactions are catalysed by metals that form **interstitial hydrides**. In these, hydrogen atoms are held in the spaces between metal atoms in the lattice, forming a type of alloy whose properties differ only slightly from that of the parent metal. Interstitial hydrides, like other alloys, have no fixed formula (they are said to be **non-stoichiometric**) – the ratio of hydrogen atoms to metal depends on the conditions, such as the external pressure. Metals of the d block readily form interstitial hydrides as their electronegativities are only slightly lower than that of hydrogen and the spaces in their lattices are large enough to accommodate the small hydrogen atoms. If the spaces are large enough (for example, in palladium), hydrogen can diffuse right through the metal and escape. This is one of the disadvantages of using very high pressures in the Haber process (see Chapter 20, pages 372–3).

The mechanism of hydrogenation of an alkene

The hydrogenation of an alkene using a nickel catalyst (see Chapter 24, page 444) probably involves adsorption of the alkene on the surface of the metal followed by the addition of two hydrogen atoms, not necessarily at the same time (see Figure 19.29).

The addition of the hydrogen atoms occurs on the same side of the alkene – for example, 1,2-dimethylcyclohexene gives *cis*-1,2-dimethylcyclohexane (see Figure 19.30).

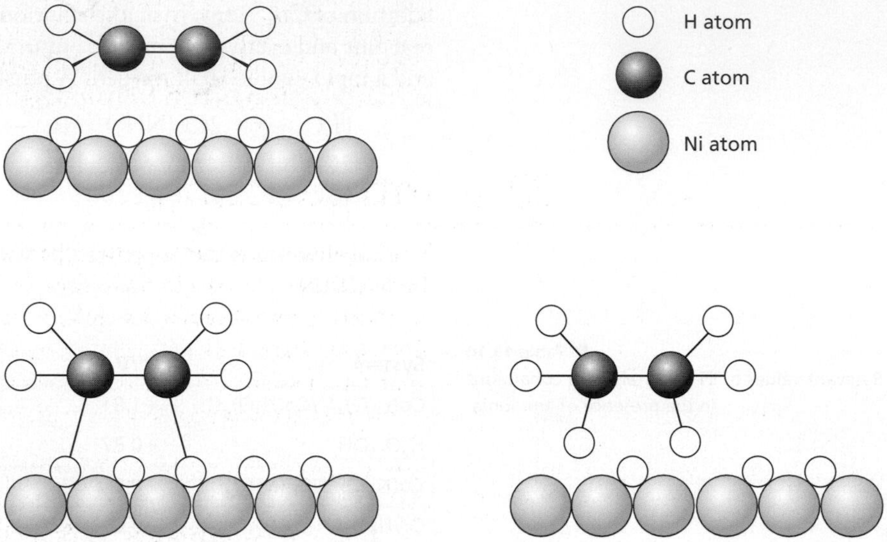

■ **Figure 19.29**
A possible mechanism for the hydrogenation of an alkene using a nickel catalyst.

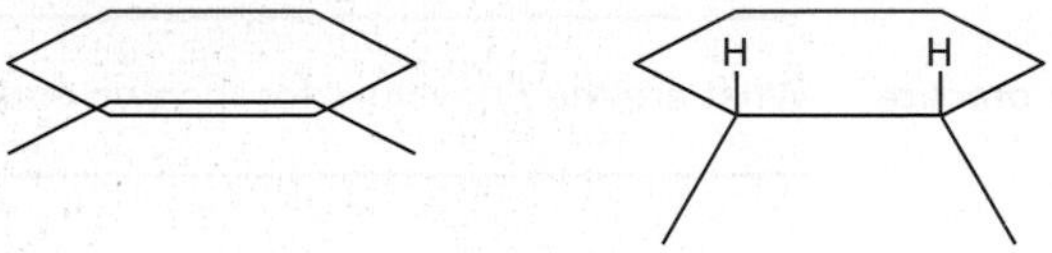

■ **Figure 19.30**
Reduction of 1,2-dimethylcyclohexene to give *cis*-1,2-dimethylcyclohexane. The ring is puckered but is shown flat here for clarity.

Summary

- The **3d block** includes the elements from scandium to zinc inclusive.
- The elements titanium to copper show features associated with **transition metals**, that is, variable oxidation states and coloured ions.
- The elements of the 3d block have high melting and boiling points.
- The first and second ionisation energies increase only slightly across the block from scandium to zinc, as 4s electrons are being removed which are shielded from the nuclear attraction by the inner 3d electrons. The third ionisation energies increase more rapidly as a 3d electron is being removed.
- Most of the elements form M^{2+} ions by loss of the 4s electrons, and some form M^{3+} ions as well.
- These ions of the 3d block elements form **complex ions** by receiving 12 electrons from six **ligands**. These complex ions have an octahedral shape.
- The elements from scandium to manganese have a maximum oxidation number equal to the sum of the 3d and 4s electrons.
- d^0 and d^{10} ions are colourless. Other d-block ions are coloured because of **d-to-d transitions** in the visible region of the spectrum.
- Many d-block ions act as homogeneous catalysts because they readily change from one oxidation state to another.
- Many d-block ions act as heterogeneous catalysts for hydrogenation because they readily form **interstitial hydrides**.

20 Industrial inorganic chemistry

In this chapter, we study some of the manufacturing processes used to make important inorganic chemicals. While laboratory preparation and industrial methods use the same basic chemistry, it is not just a question of an increase in scale when converting one process to the other. Manufacturing methods are governed by considerations of cost that are not so relevant on the laboratory scale. This factor plays a key role in the design of chemical plants.

This chapter is divided into the following sections.

20.1 Introduction

In a laboratory preparation, the apparatus is easily dismantled and the product extracted at each stage in the synthesis. Such **batch** methods are used in industry only for the manufacture of small quantities of specialist chemicals, for example pharmaceuticals. Most manufacturing processes are **continuous**, the raw materials being fed in at one end and the products extracted at the other. Apart from closure for routine maintenance work, the plant runs 24 hours a day, 365 days a year.

■ **Figure 20.1** An industrial batch plant.

In the manufacture of chemicals, economic considerations are of prime importance, the aim being to provide useful chemicals at a competitive price. The plant for batch processes is relatively cheap to build but, because batch methods are very labour intensive, running costs are high. Continuous processes require a huge capital outlay but, being automatic and computer controlled, require the minimum number of workers to operate them. The key economic consideration is how quickly the capital outlay can be paid back, out of the profit made from selling the tonnes of chemicals produced as cheaply as possible.

There are many factors affecting cost that must be considered when designing the plant. The following are some of the more important.

- **Raw materials** – with the exception of air and water, all raw materials cost money and their price may fluctuate with demand.
- **Transport** – in order to reduce the costs of bringing in the raw materials and taking the products to where they are needed, a chemical plant must be appropriately sited. It may be built near the source of chemicals (for example, lime and limestone in Utah) or near a shipping port (for example, Port Said in Egypt).
- **Design and construction of the plant** – the costs will be minimised if standard materials can be used (for example, stainless steel). This is more likely to be achieved by avoiding the use of corrosive chemicals and by keeping the operating temperature and pressure as low as possible.
- **Running costs** – if a plant is operated above room temperature, good insulation is needed to ensure that heat is not wasted. If the reaction is exothermic, a heat exchanger is used to transfer the heat produced to the incoming gases. If possible, high pressures are avoided, as they require large pumps that use a lot of energy. Catalysts may reduce the necessity for high temperatures and pressures and will therefore be used even if their capital outlay is very high. Running costs will also be reduced if the **throughput** of the plant can be increased. This means pumping liquids and gases faster round the plant. However, there is a limit to the rate of flow of liquid and gases through pipes – at high speeds, steady **laminar** flow changes to disorderly **turbulent** flow. Turbulent flow may be deliberately employed when mixing is required, but otherwise it is avoided as it reduces the overall flow rate.

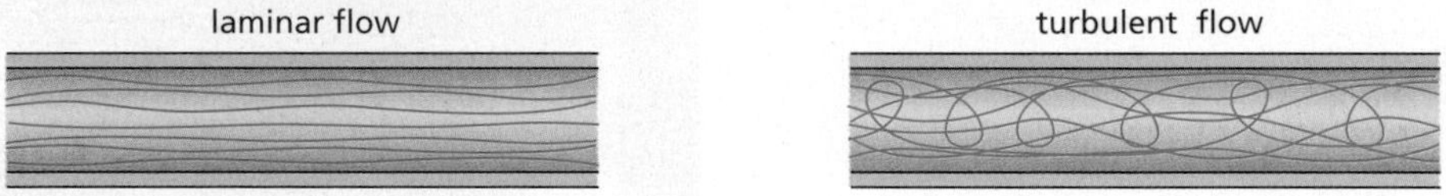

■ **Figure 20.2** If some suspended particles are added to a liquid, the type of flow of the liquid through a tube can be seen. Under ideal conditions the flow is smooth (laminar flow) and the liquid flows quickly. If the flow becomes turbulent, the rate is reduced.

- **Waste products** – most processes produce unwanted by-products. In the past these often polluted the atmosphere or poisoned the local river. Now more stringent controls make sure that pollution is kept to a minimum. From the manufacturer's point of view, the safe disposal of unwanted products increases running costs. Processes are therefore designed to minimise the formation of by-products, or to develop ways in which they may be re-used or recycled.

20.2 Gas phase reactions

In the laboratory, gas phase reactions are difficult to contain and, where possible, are avoided. In industry, however, gas phase reactions are extensively used because they can be readily adapted to continuous production methods, including purification by fractional distillation (see Chapter 12, page 238). Gas phase reactions can be used only with covalent substances that have a relatively low boiling point. They are therefore used extensively in the manufacture of organic chemicals, and their use in inorganic chemistry is restricted mainly to the manufacture of sulphuric acid (the **Contact process**) and the manufacture of ammonia (the **Haber process**).

The Contact process

The essential reactions of the Contact process are as follows.

$$S(g) + O_2(g) \rightarrow SO_2(g) \qquad \Delta H^\ominus = -308\,kJ\,mol^{-1}$$
$$SO_2(g) + \tfrac{1}{2}O_2(g) \rightleftharpoons SO_3(g) \qquad \Delta H^\ominus = -96\,kJ\,mol^{-1}$$
$$SO_3(g) + H_2O(l) \rightarrow H_2SO_4(l) \qquad \Delta H^\ominus = -130\,kJ\,mol^{-1}$$

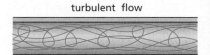

In the first stage, sulphur, obtained from the USA or Poland, is burnt in air at 1000 °C. The heat evolved is used to produce steam, which in turn is used to generate electricity.

In the second stage, which is reversible, a catalyst of vanadium(V) oxide, V_2O_5, is used. In order to improve the yield, the equilibrium is displaced to the right by mixing the sulphur dioxide with about three times as much air as is required by the equation. According to Le Chatelier's Principle, the highest conversion of sulphur dioxide to sulphur trioxide is obtained at high pressures and low temperatures. In practice, the cost of using high pressures is uneconomic and a pressure only slightly above atmospheric pressure is used, just high enough to make the gases circulate freely. A very low temperature cannot be used because the vanadium(V) oxide catalyst becomes effective only above 400 °C.

In order to improve the percentage conversion, the reactant gases are passed through four different catalyst beds. Each bed starts at 400 °C, but the heat evolved in the reaction raises the temperature by several hundred degrees. The gases are therefore cooled back to 400 °C before being passed to the next bed. Using this technique, 99.5% conversion of sulphur dioxide to sulphur trioxide is achieved.

The third stage cannot be carried out by passing the gas directly into water, because the enormous amount of heat produced would vaporise the water. The sulphur trioxide reacts with the water vapour producing a mist of sulphuric acid which is difficult to condense. The sulphur trioxide is therefore passed into 98% sulphuric acid. This has a negligible water vapour pressure. As the sulphur trioxide dissolves, water is added to keep the concentration constant. It is also necessary to cool the acid, because the reaction is exothermic. Unfortunately, this stage is at too low a temperature for the heat produced to be of any use and so it has to be wasted.

The tiny amount of unchanged sulphur dioxide from the second stage is removed from the product and the residual gases are vented to the atmosphere. The high conversion during the second stage keeps the amount of 'acid rain' produced to a minimum (see Chapter 15, page 296).

The main uses of sulphuric acid are in the manufacture of the following:

- paints and pigments
- detergents and soaps
- dyestuffs.

Smaller amounts are widely used in many different processes, including the manufacture of fertilisers, to clean steel before it is galvanised and in car batteries.

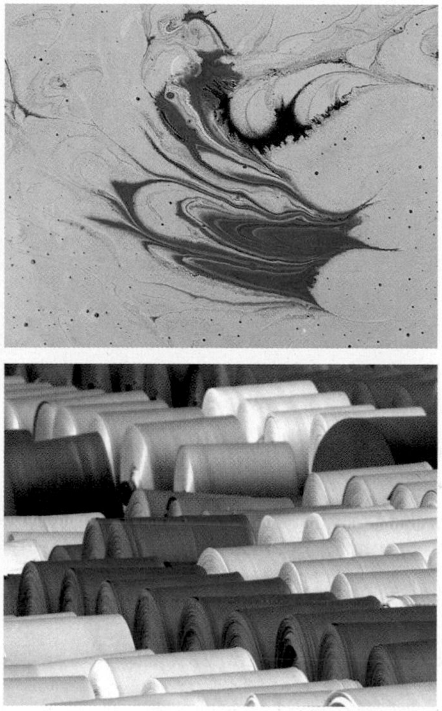

■ Figure 20.3 Sulphuric acid is used in the manufacture of paints, pigments and dyestuffs.

Summary of the Contact process

Sulphur is burnt in air at 1000 °C.
The sulphur dioxide produced is mixed with more air and passed over a vanadium(V) oxide catalyst at about 450 °C.
The sulphur trioxide produced is dissolved in 98% sulphuric acid and water added.

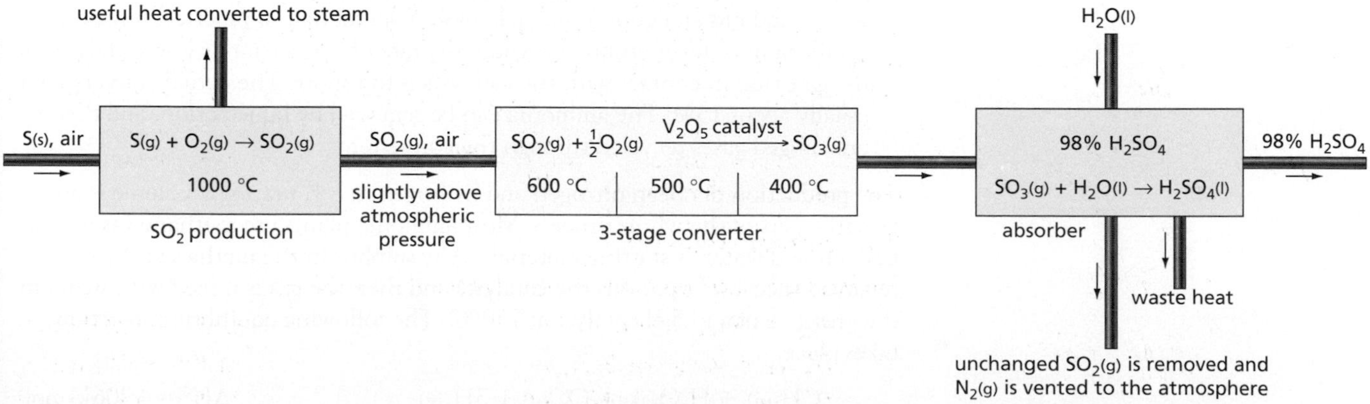

■ Figure 20.4 Flow diagram showing the stages of the Contact process.

The Haber process

The essential reaction for the manufacture of ammonia is as follows.

$$N_2(g) + 3H_2(g) \rightleftharpoons 2NH_3(g) \qquad \Delta H = -92\,kJ\,mol^{-1}$$

Application of Le Chatelier's Principle shows that the highest yield of ammonia is obtained from this equilibrium at low temperatures and high pressures. This is shown in Figure 20.5.

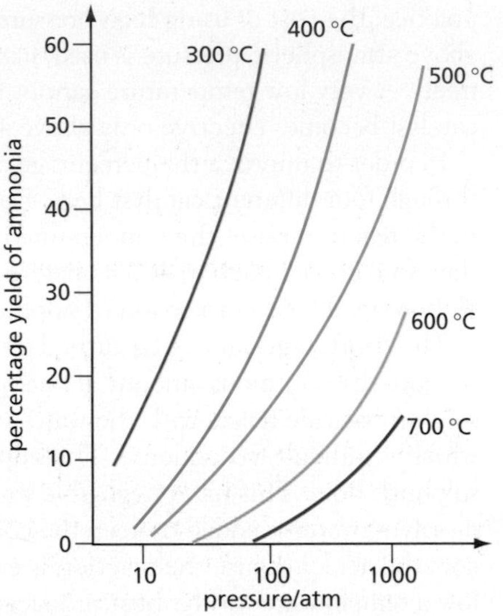

■ **Figure 20.5** Equilibrium percentage yield of ammonia at different temperatures and pressures.

In practice, it is the rate at which equilibrium is reached that is the determining factor, rather than the percentage of ammonia at equilbrium. Equilibrium is reached most quickly at high temperatures, high pressures and in the presence of a catalyst. The reasons for these conditions are explained below.

- The catalyst is made of iron, obtained by reducing iron oxide with hydrogen. There are also traces of potassium oxide and aluminium oxide in the catalyst, which act as **promoters**, that is, substances that improve the efficiency of the catalyst.
- Both thermodynamic and kinetic considerations favour high pressures. While some plants operate at 1000 atm, most plants use a lower pressure of 250 atm. Very high pressures demand a large expenditure of energy for compression and, more importantly, very thick walls for the chemical plant. These are costly, and hydrogen tends to diffuse through many metals at high temperatures and pressures.
- Even the most efficient catalyst, working at high pressures, does not work fast enough to obtain a reasonable conversion at room temperature. A compromise temperature is used – one at which the reaction is fast enough to produce a reasonable yield in a short time, and also gives a reasonable equilibrium percentage conversion. A typical plant, operating at 250 atm pressure, is run at 400 °C and has an equilibrium yield of 40% ammonia. In practice, however, equilibrium is never attained because the rate of conversion is low and the time the gases are in contact with the catalysts is too short. The actual conversion is usually about 15%. The ammonia can be removed by liquefaction, and then the unchanged gases recycled through the converter.

The production of cheap nitrogen and hydrogen in a 1:3 ratio by volume is an essential part of the whole process. Most ammonia plants use methane (as natural gas), air and water as starting materials. Any sulphur in the methane is first removed (because it poisons the catalyst) and then the gas is mixed with steam in the presence of a nickel catalyst at 750 °C. The following equilibrium reaction takes place.

$$CH_4(g) + H_2O(g) \rightleftharpoons CO(g) + 3H_2(g) \qquad \Delta H^\ominus = +206\,kJ\,mol^{-1}$$

The carbon monoxide also poisons the catalyst, so must be removed. This is achieved by the **shift reaction**. The gases are mixed with more steam and passed firstly over an iron(III) oxide catalyst at 400 °C, and then over a copper catalyst at 220 °C.

$$CO(g) + H_2O(g) \rightarrow CO_2(g) + H_2(g) \qquad \Delta H^{\ominus} = -41\,kJ\,mol^{-1}$$

The carbon dioxide is removed by first absorbing it into aqueous potassium carbonate. The hydrogencarbonate which is formed is later decomposed by heat back to the carbonate. The carbon dioxide released in this process may be used later on, in combination with ammonia, for the manufacture of urea, $CO(NH_2)_2$.

$$CO_2(g) + K_2CO_3(aq) + H_2O(l) \rightarrow 2KHCO_3(aq)$$

$$2KHCO_3(s) \xrightarrow{\text{heat}} K_2CO_3(s) + H_2O(g) + CO_2(g)$$

$$CO_2(g) + 2NH_3(g) \xrightarrow{\text{heat}} CO(NH_2)_2(s) + H_2O(l)$$

The oxygen in the air is removed by combination with some of the hydrogen produced.

$$2H_2(g) + O_2(g) \rightarrow 2H_2O(g) \qquad \Delta H^{\ominus} = -484\,kJ\,mol^{-1}$$

The processes are adjusted so that the gases passing into the converter contain 74.3% hydrogen and 24.7% nitrogen (a ratio of 3:1), the remaining 1% being methane and other gases from the air.

The exothermic and endothermic reactions almost balance, and by careful design and the use of heat exchangers, no overall heating is needed. An overall equation for the process may be written as follows:

$$3\tfrac{1}{2}CH_4(g) + 5H_2O(g) + O_2(g) + 4N_2(g) \rightarrow 8NH_3(g) + 3\tfrac{1}{2}CO_2(g)$$

$$\Delta H^{\ominus} = -274\,kJ\,mol^{-1}$$

■ **Figure 20.6**
An ammonia plant.

After the gases have passed through the iron catalyst, they are cooled. The ammonia condenses out as a liquid. The gases are then returned to the main gas stream and recycled through the catalyst. It is necessary to remove argon (from the air) and methane that become concentrated as the gas is recycled. A typical plant produces nearly a million tonnes of ammonia a year.

The main uses of ammonia are as follows.

- In acidic soils, it is injected directly into the ground as a fertiliser.
- It is catalytically oxidised to nitric acid (see below).
- It is converted into ammonium sulphate, ammonium nitrate or urea for use as a fertiliser.
- Ammonium nitrate is also used in explosives.

Summary of the Haber process

Starting materials are methane, steam and air.
Hydrogen and nitrogen are mixed in a 3:1 ratio.
The gases are passed through an iron catalyst.
The pressure is 250 atm, the temperature 450 °C and the percentage conversion to ammonia is 15%.
Ammonia is liquefied and the unchanged reactant gases recycled.

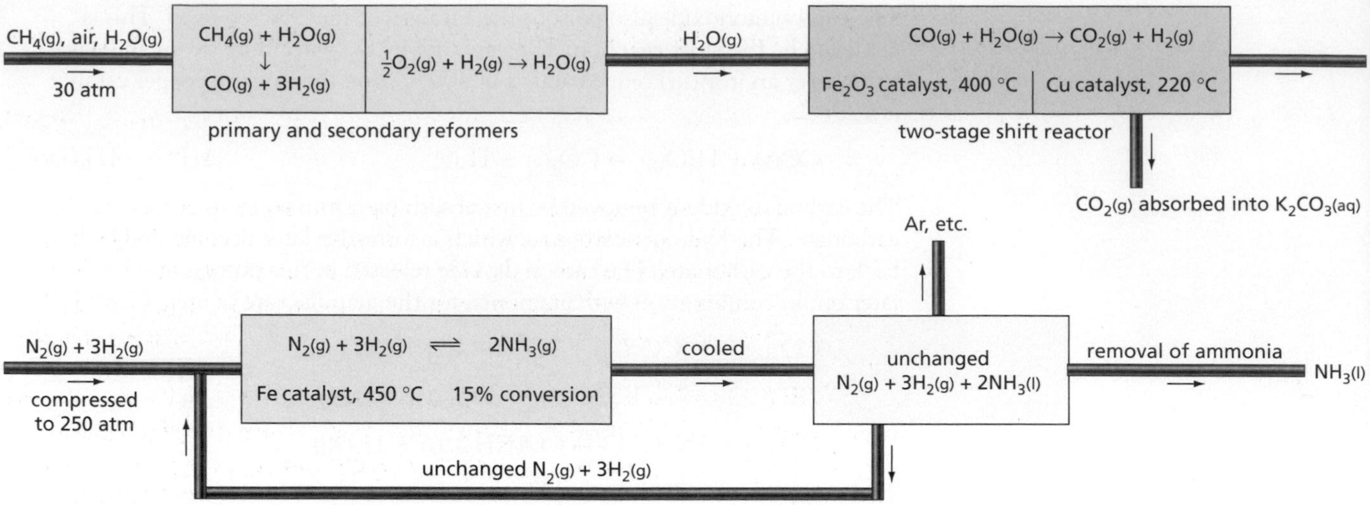

Figure 20.7 Flow diagram showing the stages of the Haber process.

Nitric acid

In the presence of a platinum/rhodium catalyst, ammonia is oxidised by air to nitric acid. There are several stages, but the overall equation is as follows:

$$NH_3(g) + 2O_2(g) \rightarrow HNO_3(l) + H_2O(l) \qquad \Delta H^\ominus = -409\,kJ\,mol^{-1}$$

The principal uses of nitric acid are in the manufacture of the following:

- ammonium nitrate fertiliser
- nylon
- polyurethane plastics and adhesives
- TNT and other explosives.

Details of the production of nitric acid

The primary oxidation of NH_3 produces NO:

$$4NH_3(g) + 5O_2(g) \overset{Pt}{\rightleftharpoons} 4NO(g) + 6H_2O(g) \qquad \Delta H^\ominus = -900\,kJ\,mol^{-1}$$

Theoretically, low temperature favours the production of NO. In practice a temperature of about 900 °C is used, because air oxidation is too slow at lower temperatures, even at elevated pressure (10 atmospheres) and in the presence of a platinum/rhodium catalyst. Care must be taken not to let the pressure or temperature rise too high or the even more exothermic oxidation to nitrogen takes place:

$$4NH_3(g) + 3O_2(g) \rightleftharpoons 2N_2(g) + 6H_2O(g) \qquad \Delta H^\ominus = -1636\,kJ\,mol^{-1}$$

The hot gases are cooled and the heat evolved used to produce steam. The gases, when mixed with air, first yield $NO_2(g)$:

$$2NO(g) + O_2(g) \rightleftharpoons 2NO_2(g) \qquad \Delta H^\ominus = -115\,kJ\,mol^{-1}$$

On further cooling, $N_2O_4(g)$ is formed:

$$2NO_2(g) \rightleftharpoons N_2O_4(g) \qquad \Delta H^\ominus = -58\,kJ\,mol^{-1}$$

The N_2O_4 reacts slowly with water to give a mixture of nitrous and nitric acids (disproportionation):

$$N_2O_4(g) + H_2O(l) \rightleftharpoons HNO_2(aq) + HNO_3(aq)$$

The HNO_2 then rapidly disproportionates into NO and N_2O_4:

$$4HNO_2(aq) \rightleftharpoons N_2O_4(g) + 2NO(g) + 2H_2O(l)$$

After further oxidation the absorption process is repeated.

Because nitric acid forms an azeotrope with water (see page 239) it cannot be concentrated by distillation to above 68% by mass. This is 'concentrated' nitric acid. Pure nitric acid, prepared by chemical dehydration, is known as 'fuming' nitric acid.

20.3 Producing metals by the reduction of ores

Introduction

Many metals are found in nature as oxides or sulphides. To convert these to the metal, carbon is the preferred reducing agent, being cheaper than hydrogen and also more effective at high temperatures. If the oxide is not readily reduced by carbon, another metal can be used as the reducing agent (for example, in the production of titanium) or an electrolytic process may be employed, as we shall see in the next section of this chapter.

Sulphide ores are first converted to oxides by roasting in air. This conversion is carried out because the reduction of oxides by carbon is much more favourable than the reduction of sulphides, as shown by the following reactions.

$$2CuO(s) + C(s) \rightarrow 2Cu(s) + CO_2(g) \qquad \Delta H^\ominus = -239\,kJ\,mol^{-1}$$

$$2CuS(s) + C(s) \rightarrow 2Cu(s) + CS_2(l) \qquad \Delta H^\ominus = +137\,kJ\,mol^{-1}$$

The blast furnace – the production of iron

The reduction of iron oxides by carbon has been carried out for 3000 years. Originally charcoal, made by the partial burning of wood in the minimum of air, was used for the reduction. Coal cannot be used because it contains so many impurities that the iron produced is too brittle to be of any use. The great achievement of Abraham Darby in the eighteenth century was to use coke, made by heating coal in the absence of air, as the reducing agent. Many of the impurities in the coal are removed in this process, leaving a solid that is almost pure carbon.

■ **Figure 20.8** Abraham Darby's process for smelting iron ore using coke instead of the more expensive charcoal formed a cornerstone of the Industrial Revolution. His grandson (also Abraham) built the first iron bridge over the River Severn at Coalbrookdale.

A modern **blast furnace** used to produce iron (see Figure 20.9) may be 70 m high. Although it is operated as only a semi-continuous process, the furnace is never allowed to cool down, except for maintenance work and to reline the inner walls with fireproof bricks. At the top of the furnace a mixture of iron ore (containing at least 60% of iron by mass), limestone and coke are added. At the bottom of the furnace air enriched with oxygen is preheated to 1300 °C, and injected through water-cooled nozzles called **tuyères**. Hydrocarbons such as methane may also be injected at the same time.

At the bottom of the furnace, where the temperature may be 1500 °C, the oxygen in the air combines with the carbon and methane as follows:

$$C(s) + \tfrac{1}{2}O_2(g) \rightarrow CO(g)$$

$$CH_4(g) + \tfrac{1}{2}O_2(g) \rightarrow CO(g) + 2H_2(g)$$

As the ore descends from the top of the furnace, reduction starts when the temperature reaches 500 °C. Typical reactions are as follows:

$$Fe_2O_3(s) + 3CO(g) \rightarrow Fe(s) + 3CO_2(g)$$
$$FeO(s) + H_2(g) \rightarrow Fe(s) + H_2O(g)$$

Further down the furnace, at 900 °C, a series of endothermic reactions takes place.

$$CO_2(g) + C(s) \rightarrow 2CO(g)$$
$$CaCO_3(s) \rightarrow CaO(s) + CO_2(g)$$
$$H_2O(g) + C(s) \rightarrow CO(g) + H_2(g)$$

Near the bottom, where the temperature is 1200 °C, the iron melts. This is below the normal melting point of pure iron (1535 °C) because it contains up to 5% carbon dissolved in it.

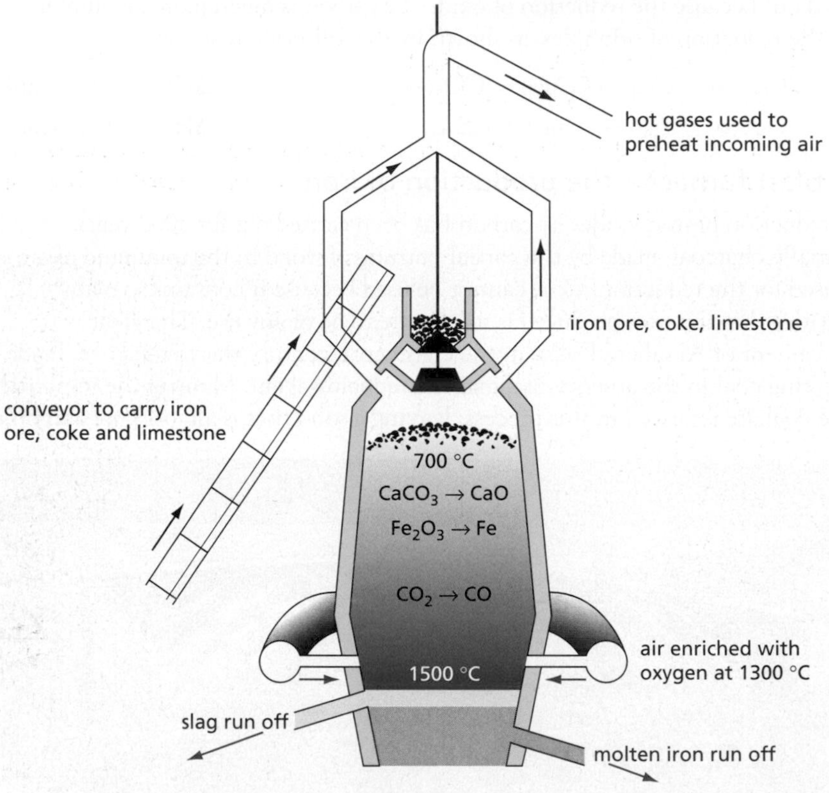

■ **Figure 20.9** (above left) The reactions inside a blast furnace.

■ **Figure 20.10** (above right) A blast furnace.

The other principal impurities in the iron are silicon, sulphur and phosphorus oxides. These combine with the basic calcium oxide and any manganese metal present to form a molten **slag** that floats on top of the molten iron. The slag is run off and when solidified is used to make cement or for road-building material.

$$CaO + SiO_2 \rightarrow CaSiO_3$$
$$Mn + S \rightarrow MnS$$
$$3CaO + P_2O_5 \rightarrow Ca_2(PO_4)_3$$

Periodically, the molten iron is run off into torpedo-shaped containers, each holding 300 tonnes of liquid iron, and transferred directly to the steel conversion plant.

Steel

The iron from the blast furnace may be cooled to form **cast iron**, which is hard but brittle. Most of the iron produced is converted to **steels**, alloys of iron with various metals which improve its properties.

In order to make steel, the carbon content of the iron must be reduced. Other impurities such as sulphur and phosphorus must also be removed, and metals such as vanadium, chromium, manganese and nickel are added to produce alloys.

The molten iron from the blast furnace, together with scrap steel and limestone, is placed in a converter. Pure oxygen is blown into the melt from a water-cooled lance and this converts the carbon into carbon monoxide. Air cannot be used for this because the nitrogen it contains combines with iron to form a nitride that makes the steel brittle. The sulphur and phosphorus oxide impurities combine with the calcium oxide from the limestone and can be removed as a slag which has a composition similar to that produced in the blast furnace.

The percentage of carbon in the steel is important. If it is less than 0.5%, the steel is malleable and easily bent; if it is above 1%, the steel is hard but brittle.

Lead

The principal lead ore is galena, lead sulphide. This is first roasted in air to give lead(II) oxide, which is subsequently reduced by carbon monoxide in a small-scale blast furnace.

$$PbS(s) + 1\tfrac{1}{2}O_2(g) \rightarrow PbO(s) + SO_2(g)$$

$$PbO(s) + CO(g) \rightarrow Pb(s) + CO_2(g)$$

The principal uses of lead are for roofing and in batteries. Lead used to be used for water pipes and in additives to petrol, but these uses have been phased out because of the toxicity of the metal.

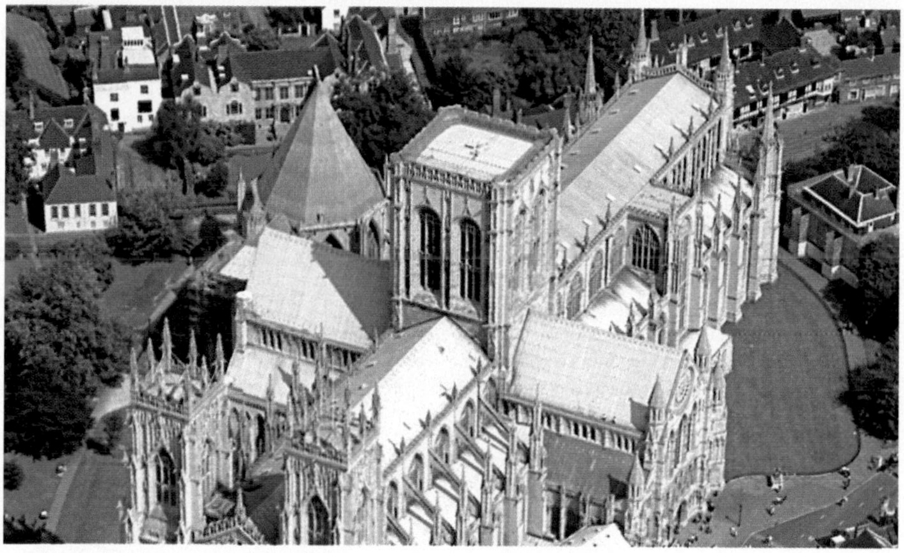

■ **Figure 20.11** Lead has been used for the roofing of York Minster.

Copper

Copper can be obtained directly from its copper sulphide ore by careful roasting in air:

$$CuS(s) + O_2(g) \rightarrow Cu(s) + SO_2(g)$$

It is also extracted by treating the ore with a solution of iron(III) chloride and electrolysing the $CuCl_2^-$ complex ion produced.

The principal uses of copper are as follows:

- electric wiring and machinery
- pipes and roofing.

Nowadays very little copper is used for so-called 'copper' coinage, as this is now made of steel, electroplated with copper. The copper for electrical components must be ultra-pure and is refined by electrolysis (see Chapter 13, page 265).

Zinc

Although some zinc is made by the reduction of zinc oxide using carbon monoxide, the bulk is extracted by electrolysis (see page 381).

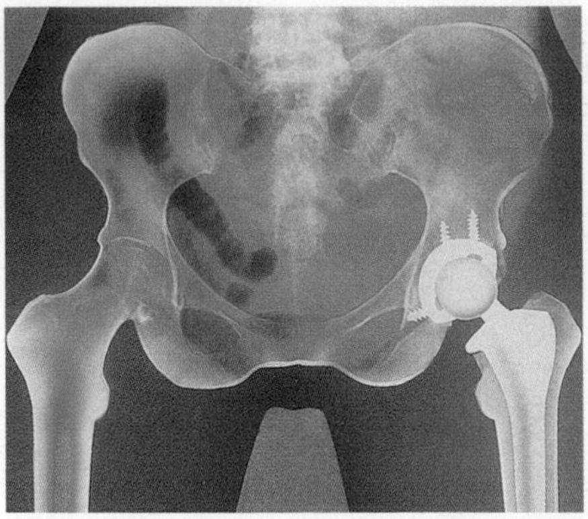

■ **Figure 20.12**
A replacement hip made from titanium, which is unreactive and light.

Titanium

Titanium(IV) oxide cannot economically be reduced by carbon. It is first converted to titanium(IV) chloride by treatment with chlorine and carbon at 1000 °C.

$$TiO_2(s) + C(s) + 2Cl_2(g) \rightarrow TiCl_4(g) + CO_2(g)$$

The chloride is reduced using either magnesium or sodium at 500–1000 °C. The product is treated with hydrochloric acid to remove any magnesium or sodium chloride, and is then purified by vacuum distillation.

$$TiCl_4(s) + 2Mg(s) \rightarrow Ti(s) + 2MgCl_2(s)$$

Titanium is strong, of low density and resistant to chemical attack. Although it is expensive to produce, it is used for the following:

- aerospace industry
- implants in medicine.

Ellingham diagrams

A useful way of showing whether carbon is an effective means of reducing an oxide to the metal is using an **Ellingham diagram**. This shows how the standard free energy change of formation of a metal oxide varies with temperature. If this free energy change is higher than the free energy change of formation of carbon monoxide, then reduction is possible. In order to provide a direct comparison, the graphs are shown per mole of oxygen. Figure 20.13 shows that below about 700 °C, carbon monoxide is a better reductant than carbon, while above that temperature, the reverse is true. This explains the processes that take place at different levels in the blast furnace.

■ **Figure 20.13**
Ellingham diagram for carbon, lead and magnesium. The diagram shows that lead oxide is easily reduced by carbon, but that reduction of magnesium oxide is only feasible above about 1750 °C, an uneconomically high temperature. Magnesium is therefore extracted by electrolysis.

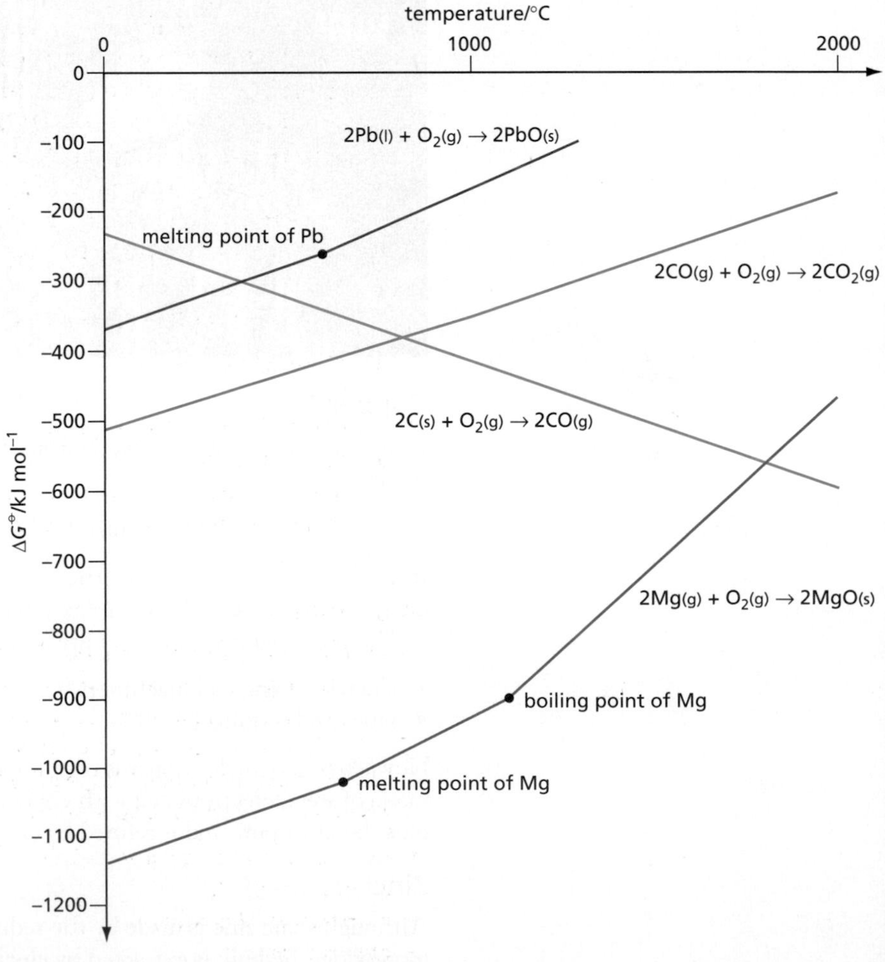

20.4 Electrolytic processes

Very reactive metals such as sodium, lithium, magnesium and aluminium cannot be extracted by reduction of their oxides. They are therefore produced by the electrolysis of their molten salts. The less reactive metal zinc is obtained by electrolysis of aqueous zinc sulphate. The electrolysis of aqueous sodium chloride produces hydrogen, chlorine and sodium hydroxide, all of which are useful chemicals.

Sodium and lithium

■ **Figure 20.14**
Top of reactor core, cooled with liquid sodium.

Sodium metal is produced by the electrolysis of molten sodium chloride, the melting point of which is lowered to 600 °C by the addition of calcium chloride. Molten sodium is produced at the steel cathode and chlorine at the carbon anode:

at the cathode: $2Na^+ + 2e^- \rightarrow 2Na(l)$
at the anode: $2Cl^- \rightarrow 2Cl_2(g) + 2e^-$

A typical cell voltage is 6 V, with a current of 40 000 A.

In the past, most sodium metal was used in the manufacture of tetraethyl lead(IV) (see Chapter 23, page 428), but demand for this has decreased as unleaded petrol takes the place of leaded petrol. At present some sodium is used as a coolant in nuclear reactors, and it is also needed to make chemicals such as sodium tetrahydridoborate(III), $NaBH_4$, and sodamide, $NaNH_2$, which have some industrial importance.

Lithium is produced in a similar way to sodium, and is finding increasing use in lightweight alloys.

Magnesium

Magnesium is obtained from sea water or brine (concentrated salt solution). Magnesium hydroxide is precipitated out under controlled conditions so that the more soluble calcium hydroxide remains in solution (see Chapter 16, page 313). The magnesium hydroxide is converted to anhydrous magnesium chloride, with precautions to prevent hydrolysis to Mg_2OCl_2 (see Chapter 16, page 309). Molten magnesium chloride is electrolysed using a steel cathode and carbon anode. The process requires a lot of electricity and is uneconomic in the UK. Most of the magnesium produced in Europe is made in Norway, where there are abundant supplies of hydroelectric power.

The principal uses of magnesium are as follows:

- in lightweight castings
- in alloys
- as sacrificial anodes to protect iron and steel from rusting
- in the production of titanium (see page 378)
- in flares used to signal distress at sea
- in the manufacture of chemicals such as Grignard reagents (see Chapter 25, page 473).

Aluminium

Most molten aluminium compounds, for example, aluminium chloride, do not conduct electricity. Molten aluminium oxide, Al_2O_3, is a conductor, so its electrolysis yields aluminium. However, it has an extremely high melting point (2060 °C). The oxide is therefore dissolved in molten **cryolite**, sodium hexafluoroaluminate, Na_3AlF_6, which has a melting point lower than 1000 °C.

There are three main stages in the manufacture of aluminium:

- extraction and purification of the ore, **bauxite**, which is hydrated aluminium oxide
- preparation of cryolite
- electrolysis.

Bauxite is a relatively abundant ore. Its principal impurity is hydrated iron(III) oxide, but it may contain silica and titanium oxide as well. The bauxite is dissolved in 10% aqueous sodium hydroxide under 4 atm pressure at 150 °C. The impurities are largely insoluble and can be filtered off as a sludge. Aluminium hydroxide, $Al(OH)_3$, is then precipitated from the clear solution by cooling it for up to three days. This precipitation is accelerated by 'seeding' the solution with a crystal of aluminium hydroxide. The hydroxide precipitate is filtered off and heated to convert it into pure aluminium oxide.

■ **Figure 20.15**
Paul Heroult (1863–1914) and Charles Hall (1863–1914), designers of the electrolytic process to produce aluminium.

$$Al_2O_3\text{(hydrated)} + 2OH^-(aq) + 3H_2O(l) \rightarrow 2Al(OH)_4^-(aq)$$ other impurities are insoluble

$$Al(OH)_4^-(aq) \rightleftharpoons Al(OH)_3(s) + OH^-(aq)$$ precipitation on cooling

$$2Al(OH)_3(s) \rightarrow Al_2O_3(s) + 3H_2O(g)$$ dehydration on heating

The sludge is composed of iron oxide and other impurities. It is washed free of sodium hydroxide before being buried. This treatment is important as untreated sludge is an unattractive brown colour and leaves the soil too alkaline for plants to grow. The sodium hydroxide produced during the precipitation is recycled and used to dissolve more aluminium oxide.

Cryolite is made by dissolving $NaAl(OH)_4$ in hydrofluoric acid and precipitating the product with sodium carbonate:

$$NaAl(OH)_4(aq) + 6HF(aq) + Na_2CO_3(s) \rightarrow Na_3AlF_6(s) + 5H_2O(l) + CO_2(g)$$

The electrolysis is carried out in a steel box whose floor is lined with carbon. This acts as the cathode connection (see Figure 20.16).

■ **Figure 20.16** (right)
Outline diagram of the electrolytic cell used to produce aluminium.

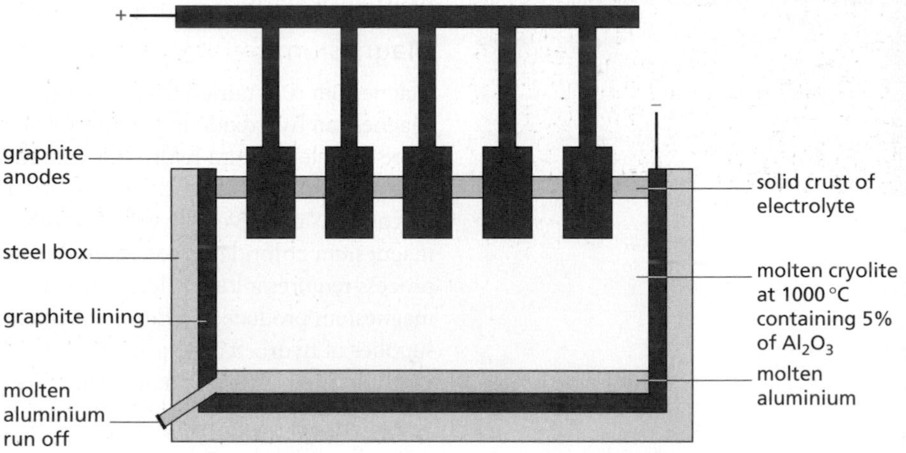

■ **Figure 20.17** (below)
Aluminium is the most abundant metal in the Earth's crust, occurring largely as the ore bauxite. Its high extraction costs mean that recycling is a much more economical way of producing 'new' metal.

The anodes are blocks of graphite suspended in the molten cryolite. Because aluminium is more dense than cryolite, it sinks to the bottom of the cell as it is formed, creating a molten pool which acts as the cathode. Periodically the aluminium is removed and more aluminium oxide added to maintain a concentration of 5%.

The exact nature of the electrolyte is unknown – it has been suggested that the principal ions present are AlO^+ and AlO_2^-. The reactions taking place at the electrodes may be written as follows, although this is certainly a simplification:

at the cathode: $2Al^{3+} + 6e^- \rightarrow 2Al(l)$

at the anode: $3O^{2-} \rightarrow 1\frac{1}{2}O_2(g) + 6e^-$

The oxygen given off at the anode reacts with the carbon, producing carbon dioxide. The graphite blocks therefore have to be renewed regularly.

The process requires enormous amounts of electricity – up to 15 kWh are needed to produce 1 kg of metal. A typical cell operates at 4.5 V and 3×10^5 A, and two-thirds of the energy goes in heating the cell. Fortunately aluminium is easy to recycle and this requires only 5% of the energy required to produce it from bauxite. Up to 60% of aluminium used in Europe is from recycled material.

Aluminium is the second most used metal after iron. It has a good strength:weight ratio and is resistant to corrosion, especially when anodised, that is covered with a thin protective oxide film by anodic oxidation. Its principal uses are as follows:

- in lightweight alloys to build cars, ships, aeroplanes, etc.
- for building work, such as window frames
- in overhead electric cables
- to make packaging, such as cans, foil, etc.

Zinc

Some zinc is made by the reduction of the oxide, but most is now made by electrolysis. The zinc sulphide ore is first roasted in air to make the oxide, which is then dissolved in sulphuric acid:

$$ZnS(s) + 1\tfrac{1}{2}O_2(g) \rightarrow ZnO(s) + SO_2(g)$$
$$ZnO(s) + H_2SO_4(aq) \rightarrow ZnSO_4(aq) + H_2O(l)$$

The solution is electrolysed using lead anodes and aluminium cathodes. Zinc is preferentially discharged because of the high overvoltage for hydrogen (see Chapter 13, page 263).

at the cathode: $Zn^{2+}(aq) + 2e^- \rightarrow Zn(s)$

at the anode: $2OH^-(aq) \rightarrow \tfrac{1}{2}O_2(g) + H_2O(l) + 2e^-$

The principal uses of zinc are as follows:

- for galvanising iron
- in making brass, which is an alloy of copper and zinc (see Chapter 19, page 365).

■ **Figure 20.18** Galvanised iron bucket.

Iron is **galvanised** by covering it with a thin layer of zinc. This is done either by dipping the iron into molten zinc, or by spraying molten zinc onto the iron. Galvanised iron is more resistant to corrosion than iron plated with tin. If a plated surface is scratched, the exposed iron starts to rust. With tin plate, because iron is more reactive than tin, the process continues and finally the tin layer falls off. However, zinc is more reactive than iron, so if galvanised iron is scratched the zinc dissolves and forms an oxide layer along the scratch which protects the iron from further attack.

The electrolysis of brine

If concentrated aqueous sodium chloride is electrolysed, hydrogen is given off at the cathode and chlorine at the anode. Although the standard electrode potential of oxygen is less positive than that of chlorine, OH^- ions are not normally discharged because oxygen has a high overvoltage (see Chapter 13, page 263).

at the cathode: $2H^+(aq) + 2e^- \rightarrow H_2(g)$

at the anode: $2Cl^-(aq) \rightarrow Cl_2(g) + 2e^-$

The chlorine reacts with the sodium hydroxide solution that is produced by the electrolysis.

at the cathode: $Cl_2(aq) + 2OH^-(aq) \rightarrow ClO^-(aq) + Cl^-(aq) + H_2O(l)$

In order to achieve a good yield of sodium chlorate(I), a saturated cold solution of sodium chloride is electrolysed using a platinum anode and an iron cathode. The solution is not stirred, as this would bring additional hydroxide ions near the anode and so increase the likelihood that they would be discharged. A typical household bleach contains 5% sodium chlorate(I), 40% sodium hydroxide and 15% sodium chloride by mass.

If a saturated solution of sodium chloride is electrolysed at 80 °C using platinum electrodes, the chlorate(I) ion disproportionates. On cooling, chlorate(V) crystallises out.

$$3NaClO(aq) \rightarrow NaClO_3(aq) + 2NaCl(aq)$$

Sodium chlorate(V) is used as a weedkiller and in fireworks.

For most purposes, chlorine and sodium hydroxide need to be isolated separately. This is accomplished using a **diaphragm cell**. This contains an asbestos diaphragm that allows the ions to flow from the anode compartment to the cathode compartment. This flow is established by having the level of the electrolyte higher on the anode side of the diaphragm than on the cathode side (see Figure 20.19).

■ **Figure 20.19**
Outline diagram of the diaphragm cell used for the electrolysis of brine.

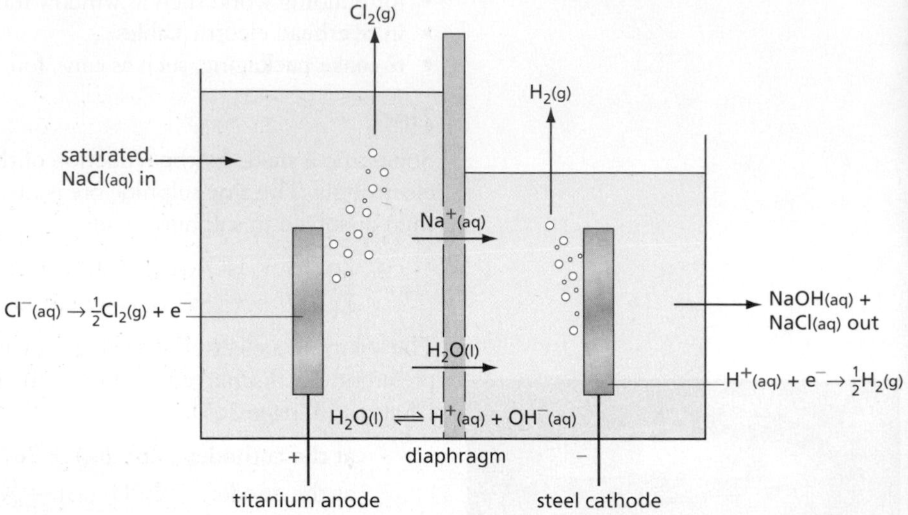

The incoming solution is saturated brine containing about 25% sodium chloride. The outgoing solution contains about 10% sodium hydroxide and 15% sodium chloride. This solution is evaporated to about one-fifth of its volume, when nearly all the sodium chloride crystallises out, leaving a solution containing 50% of almost pure sodium hydroxide. This can be used as a solution or evaporated to dryness to give the solid product.

Modern plants use a membrane rather than a diaphragm in the cell (Figure 20.20). The membrane is made of a polymer, containing PTFE (see page 457), which allows cations but not anions to pass through it. The resulting solution is pure sodium hydroxide, uncontaminated by sodium chloride.

The chlorine and hydrogen gases are collected separately.

The principal uses of the three chemicals are as follows.

Sodium hydroxide:

- for making soap and detergents
- converted into sodium carbonate
- used in refining aluminium oxide (see page 380)
- for making paper.

■ **Figure 20.20**
Chlor-alkali factories produce many secondary products widely used in the chemical industry. All of these products are obtained through the initial process of electrolysis of brine. This modern plant uses membrane cell technology rather than diaphragm cells.

Hydrogen:

- in making ammonia
- for the hydrogenation of oils (see Chapter 24, page 444)
- in welding using an oxy-hydrogen flame.

Chlorine:

- for sterilisation and bleaching
- to make insecticides
- to make solvents such as dichloromethane and tetrachloromethane
- to make CFCs (see Chapter 25, pages 462 and 473)
- to make PVC (see Chapter 24, page 457).

The demand for the three products of the electrolysis varies, and it is difficult to provide the right proportions of each product to meet this changing demand. A surplus of sodium hydroxide can be converted into sodium carbonate. A surplus of chlorine can be converted into hydrogen chloride by combining it with hydrogen. Alternatively, if there is a surplus of hydrogen chloride produced by the chlorination of hydrocarbons, then the hydrogen chloride may be reconverted to chlorine by being oxidised by air in the presence of a catalyst.

The Solvay (ammonia–soda) process

Some sodium carbonate is manufactured by treating sodium hydroxide with carbon dioxide. Most of it is made by the **Solvay process**, using salt and calcium carbonate as the starting materials. Calcium carbonate is insoluble in water, so the reaction must be carried out in several stages. The overall conversion may be represented as follows:

$$2NaCl(s) + CaCO_3(s) \rightarrow CaCl_2(s) + Na_2CO_3(s)$$

but ammonia is also used and recovered, hence the name of the process.

The sodium carbonate may be produced as the anhydrous salt, or as the decahydrate $Na_2CO_3.10H_2O$, commonly known as washing soda.

The principal uses of sodium carbonate are as follows:

- in making glass
- to soften hard water.

Sodium hydrogencarbonate, $NaHCO_3$, is also produced by the process. The principal uses of this are as follows:

- as baking powder
- in anti-acid medicines.

Summary

- Manufacturing processes are governed by economic and environmental considerations. They are usually **continuous** rather than **batch processes**.
- Conditions of temperature, pressure and catalysts are chosen to optimise the throughput.
- Normal pressure and a compromise temperature is used in the Contact process.
- High pressure and a compromise temperature is used in the Haber process.
- Many metals, for example iron, are made by the reduction of oxide ores with carbon.
- More reactive metals, for example aluminium, are made by electrolysis.
- Electrolysis is also used to make important chemicals from brine.

21 Inorganic tests

In this chapter, we study the familiar 'wet' tests for finding out which cations and anions are present in an unknown sample in the test tube. Such methods are very rarely used in an analytical laboratory because modern physical methods are much more sensitive and selective, and can readily be automated, making them less labour intensive. These tests, however, bring together much inorganic chemistry that has already been met, so they provide a useful alternative way at looking at the chemistry of metals and their compounds.

The chapter is divided into the following sections.

21.1 Appearance

Most inorganic compounds are made up of separate ions, and so they show characteristic reactions associated with each type of ion they contain. Analysing a compound therefore involves the separate analysis of cations and of anions.

Being ionic, inorganic compounds are usually solid at room temperature. Much information can be found from their appearance. Usually the substance is in the form of a powder. It may be either amorphous, which suggests that it is insoluble in water and has been made by precipitation, or it may be made up of small crystals, which show that it must be soluble, having been prepared by crystallisation from solution. Crystals may absorb moisture from the air – they may by **hygroscopic**. If they are so hygroscopic that they dissolve in the moisture in the air, they are termed **deliquescent**. Some crystals become covered with powder due to **efflorescence**, that is, the loss of water of crystallisation in dry atmosphere (see Table 21.1).

■ **Table 21.1** Information that can be gained from observing the appearance of a powder.

Type of powder	Inference	Possible type of substance
Amorphous	Made by precipitation	Insoluble, e.g. oxide or carbonate
Crystalline	Made by crystallisation	Soluble, e.g. Group 1 compound, nitrates, most sulphates and chlorides
Hygroscopic	Cation of high charge density	Li^+ and most double and triply charged cations
Efflorescent	Contains much water of crystallisation	$Na_2CO_3.10H_2O$, alums

If the substance is made up of colourless crystals or a white powder, we may infer that we are dealing with a compound from the s block, or possibly from the p or d block. A coloured compound indicates the presence of a transition metal (see Chapter 19, page 357) or possibly an oxide with a low energy charge transfer transition (see Chapter 14, page 275). Table 21.2 gives further details of the information that can be obtained from colour.

■ Table 21.2
Information that can be gained by observing the colour of a compound. The colours of d-block cations are due to d-to-d transitions. The colours of the MnO_4^-, CrO_4^{2-} and $Cr_2O_7^{2-}$ ions, oxides and sulphides are due to charge transfer.

Colour	Possible ions or substance present
Colourless	s-block element; Al^{3+}; Pb^{2+}; d^0 or d^{10} ions, e.g. Cu(I) or Zn^{2+}
Blue	Cu^{2+}, VO^{2+}, Co^{2+}
Pale green	Fe^{2+}, Ni^{2+}
Dark green	Cu(II), Cr^{3+}
Purple	Cr^{3+}, MnO_4^-
Pale violet	Fe^{3+} (goes yellow-brown in solution)
Pink	Mn^{2+}, Co^{2+}
Yellow	CrO_4^{2-}, PbO, S
Orange	$Cr_2O_7^{2-}$, PbO
Red	Pb_3O_4, Cu_2O
Brown	PbO_2, Fe^{3+}, Fe_2O_3
Black	CuO, MnO_2

21.2 Detection of gases

A common analytical test is to observe whether a gas is given off during a reaction, and if so, to identify it. It is important to consider how the gas was produced and, if possible, to write an equation for the reaction. Table 21.3 lists the most likely gases produced. Some of the tests are illustrated in Figure 21.1 (overleaf).

■ Table 21.3
The most common gases produced during analysis. Other less common gases include N_2, CO, SO_3, N_2O, NO, Br_2 and I_2.

Gas	Appearance/Properties	Test
O_2	Colourless, odourless	Glowing splint glows more brightly and may be rekindled
H_2	Colourless, odourless	When lit, burns at mouth of test tube*: $H_2(g) + \frac{1}{2}O_2(g) \rightarrow H_2O(l)$
CO_2	Colourless, almost odourless	Limewater turns milky: $Ca(OH)_2(aq) + CO_2(g) \rightarrow CaCO_3(s) + H_2O(l)$
H_2O	Condenses on cooler part of test tube	Blue cobalt chloride paper turns pink: $CoCl_4^{2-}(aq) + 6H_2O(l) \rightarrow Co(H_2O)_6^{2+}(aq) + 4Cl^-(aq)$
SO_2	Colourless, choking	Acidic gas, acidified $K_2Cr_2O_7$ turns green: $Cr_2O_7^{2-}(aq) + 2H^+(aq) + 3SO_2(g) \rightarrow 2Cr^{3+}(aq) + H_2O(l) + 3SO_4^{2-}(aq)$
HCl†	Colourless, choking, acidic gas; steamy fumes in moist air	Forms white fumes with ammonia gas: $NH_3(g) + HCl(g) \rightarrow NH_4Cl(s)$
Cl_2	Pale green, very choking	Moist litmus paper is bleached (it may turn red first): $Cl_2(g) + H_2O(l) \rightarrow HCl(aq) + HClO(aq)$
NO_2	Brown, choking, acidic	Moist litmus paper turns red
NH_3	Colourless, choking	The only common alkaline gas, forms white fumes with HCl: $NH_3(g) + HCl(g) \rightarrow NH_4Cl(s)$

* When pure, hydrogen burns with a quiet, colourless flame. It is nearly always mixed with air and hence burns with a mild explosion. The flame is often tinged with yellow from the sodium ions in the glass of the test tube.

† HBr and HI give the same tests as HCl, but are nearly always given off in the presence of either bromine or iodine, which are easily identifed by their colours.

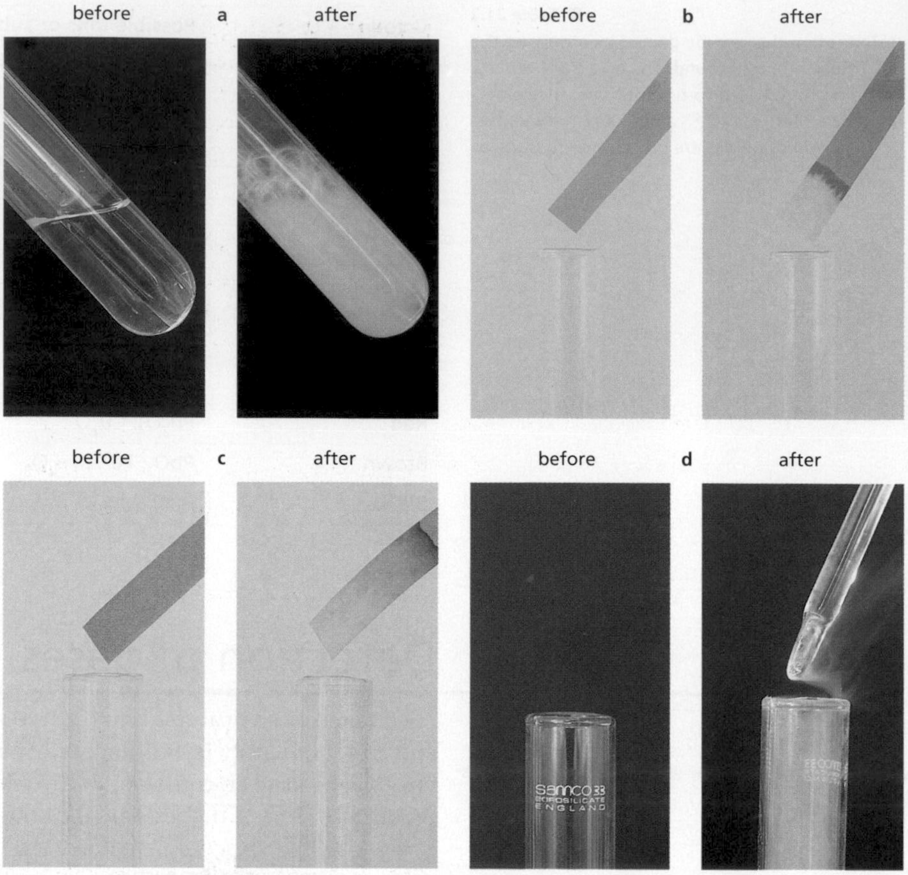

■ **Figure 21.1**
Positive tests for **a** carbon dioxide, **b** water vapour, **c** chlorine and **d** ammonia.

The source of the gases is important because it tells us something about the substance under investigation. Table 21.4 lists the most common sources of the gases in Table 21.3.

■ **Table 21.4**
Possible sources of some common gases.

Gas	Likely source	Typical equation
O_2	Heating Group 1 nitrates	$NaNO_3(s) \rightarrow NaNO_2(s) + \frac{1}{2}O_2(g)$
$O_2 + NO_2$ (H_2O if nitrate is hydrated)	Heating other nitrates	$Ca(NO_3)_2(s) \rightarrow CaO(s) + \frac{1}{2}O_2(g) + 2NO_2(g)$
H_2	HCl(aq) on metal	$Mg(s) + 2HCl(aq) \rightarrow MgCl_2(aq) + H_2(g)$
CO_2	Heating carbonates (not Group 1)	$CuCO_3(s) \rightarrow CuO(s) + CO_2(g)$
	HCl(aq) on carbonates	$CaCO_3(s) + 2HCl(aq) \rightarrow CaCl_2(aq) + H_2O(l) + CO_2(g)$
H_2O	Heating hydrated salts	$CuSO_4.5H_2O(s) \rightarrow CuSO_4(s) + 5H_2O(g)$
SO_2	HCl(aq) on sulphites	$Na_2SO_3(s) + 2HCl(aq) \rightarrow 2NaCl(aq) + H_2O(l) + SO_2(g)$
	Heating sulphates	$2FeSO_4(s) \rightarrow Fe_2O_3(s) + SO_2(g) + SO_3(g)$
HCl	$H_2SO_4(l)$ on chlorides	$KCl(s) + H_2SO_4(l) \rightarrow KHSO_4(s) + HCl(g)$
Cl_2	HCl(aq) with oxidising agents	$4HCl(aq) + MnO_2(s) \rightarrow MnCl_2(aq) + Cl_2(g) + 2H_2O(l)$
NH_3	NH_4^+ and NaOH(aq)	$(NH_4)_2SO_4(aq) + 2NaOH(aq) \rightarrow Na_2SO_4(aq) + 2NH_3(g) + 2H_2O(l)$

21.3 Flame tests

Some cations, particularly those of the s block, give characteristic colours in the flame of a Bunsen burner. A nichrome wire or porcelain rod is first cleaned by dipping it into concentrated hydrochloric acid and heating it in the blue part of Bunsen burner flame until there is no persistent yellow colour (due to sodium impurites). A speck of the solid to be tested is mixed with a drop of concentrated hydrochloric acid on the end of the wire or rod and heated. The function of the hydrochloric acid is to convert non-volatile substances such as oxides and carbonates into the more volatile chlorides. Typical flame colours are listed in Table 21.5 and shown in Figure 21.2.

Table 21.5 Some typical flame test colours.

Colour of flame	Likely ion present
Intense scarlet red	Li^+ or Sr^{2+}
Persistent, intense yellow	Na^+
Lilac	K^+
Intermittent brick-red	Ca^{2+}
Pale green	Ba^{2+}
Blue-green	Cu^{2+}

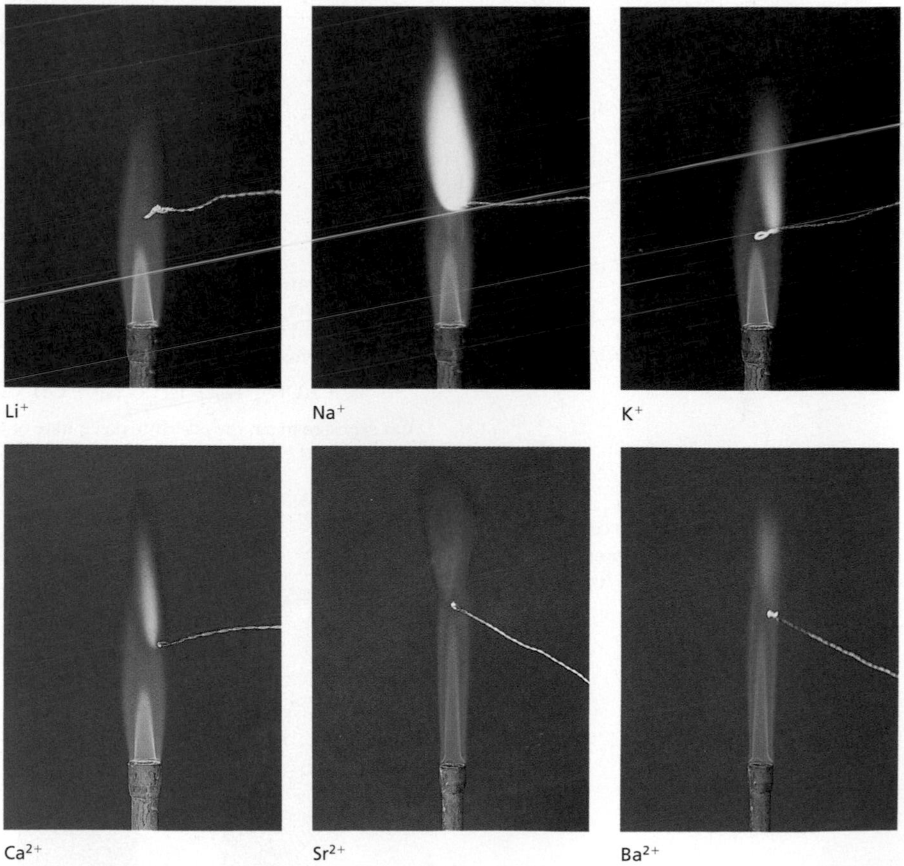

Figure 21.2 Flame tests.

All substances emit light at various visible frequencies, which appear as lines in the spectrum (see Chapter 14, page 277). Often the lines are distributed across the spectrum and there is no overall predominant colour. Sodium has a single line at 590 nm, which can easily be identified with a direct vision spectroscope. Many substances contain sodium as an impurity and it is difficult to eliminate the yellow colour entirely. Its effect can be partially masked by viewing the colours through blue cobalt glass.

21.4 Precipitation reactions for cations

Hydroxides, except those of Group 1 and ammonium, are insoluble in water and often have characteristic appearances. Some are amphoteric and dissolve in excess sodium hydroxide; others dissolve in excess ammonia to form a complex ion. The addition of aqueous sodium hydroxide (Table 21.6) or aqueous ammonia (Table 21.7) to an aqueous solution of the unknown (the tests are useless on the solid) gives valuable information about the cations present.

Reactions with aqueous sodium hydroxide

■ **Table 21.6**
A few drops of aqueous sodium hydroxide are added to a solution of the unknown. If a precipitate forms, further sodium hydroxide is added until it is in excess.

	Cation	On addition of a few drops of NaOH(aq)	On addition of excess NaOH(aq)
	NH_4^+	No ppt, but smell of NH_3 especially on warming	—
Group 1	Li^{+}*, Na^+, K^+,	No ppt	—
Group 2	Mg^{2+}, Ca^{2+} Sr^{2+}*, Ba^{2+}*	White ppt No ppt	Not soluble —
Groups 3 and 4	Al^{3+}, Pb^{2+}	White ppt	Soluble
d block	Cr^{3+}	Green ppt	Soluble†
	Mn^{2+}	Off-white ppt‡	Not soluble
	Fe^{2+}	Pale green ppt‡	Not soluble
	Fe^{3+}	Brown ppt	Not soluble
	Co^{2+}	Pink or blue ppt	Not soluble
	Ni^{2+}	Green ppt	Not soluble
	Cu^{2+}	Blue ppt	Not soluble
	Zn^{2+}	White ppt	Soluble
	Ag^+	Brown ppt	Not soluble

* If the solutions are concentrated, Li^+, Sr^{2+} and Ba^{2+} may give a slight precipitate because their hydroxides are not very soluble (see Chapter 16, page 313).

† If the solution formed by treating the substance with excess sodium hydroxide is treated with hydrogen peroxide, the solution turns yellow:

$$Cr(OH)_4^-(aq) + 1\tfrac{1}{2}H_2O_2(aq) + OH^-(aq) \rightarrow CrO_4^{2-}(aq) + 4H_2O(l)$$

‡ On exposure to air, the off-white precipitate of $Mn(OH)_2$ turns to brown MnO.OH, and the green precipitate of $Fe(OH)_2$ turns to brown FeO.OH.

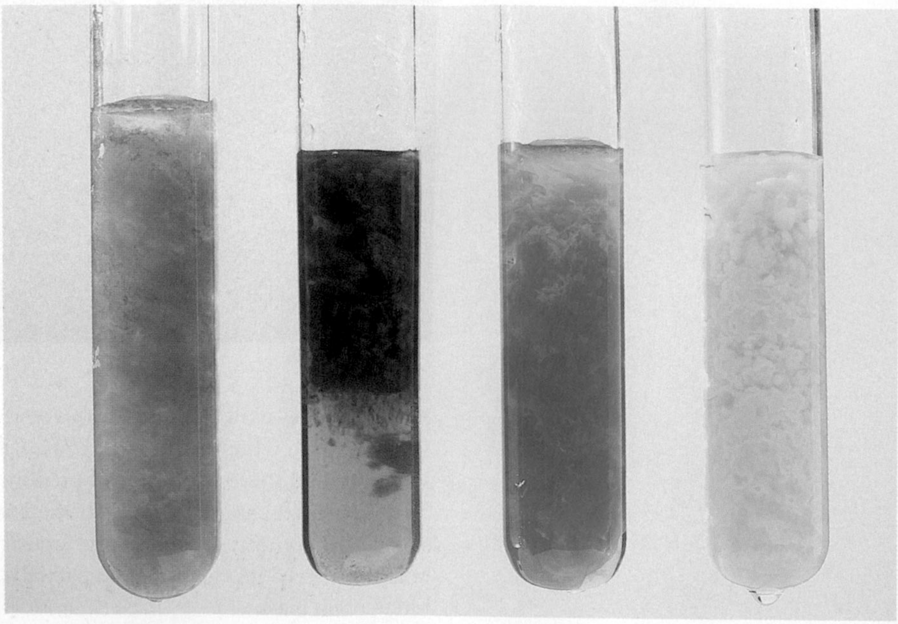

■ **Figure 21.3**
Hydroxide precipitate colours. Can you determine the cation present in each of these solutions?

Summary of equations

For NH_4^+ ions: $NH_4^+(aq) + OH^-(aq) \rightarrow NH_3(g) + H_2O(l)$

For Ag^+ ions: $2Ag^+(aq) + 2OH^-(aq) \rightarrow Ag_2O(s) + H_2O(l)$

For divalent M^{2+} ions: $M^{2+}(aq) + 2OH^-(aq) \rightarrow M(OH)_2(s)$

if the precipitate is soluble in excess: $M(OH)_2(s) + 2OH^-(aq) \rightarrow M(OH)_4^{2-}(aq)$

For trivalent M^{3+} ions: $M^{3+}(aq) + 3OH^-(aq) \rightarrow M(OH)_3(s)$

if the precipitate is soluble in excess: $M(OH)_3(s) + OH^-(aq) \rightarrow M(OH)_4^-(aq)$

Reactions with aqueous ammonia

■ Table 21.7
A few drops of aqueous ammonia are added to a solution of the unknown. If a precipitate forms, further ammonia solution is added until it is in excess.

	Cation	On addition of a few drops of $NH_3(aq)$	On addition of excess $NH_3(aq)$
Group 1	Li^+, Na^+, K^+	No ppt	—
Group 2	Mg^{2+}	White ppt	Not soluble
	Ca^{2+}, Sr^{2+}, Ba^{2+}*	No ppt	—
Groups 3 and 4	Al^{3+}, Pb^{2+}	White ppt	Not soluble
d block	Cr^{3+}	Green ppt	Slightly soluble
	Mn^{2+}	Off-white ppt	Not soluble
	Fe^{2+}	Pale green ppt	Not soluble
	Fe^{3+}	Brown ppt	Not soluble
	Co^{2+}	Pink or blue ppt	Soluble
	Ni^{2+}	Green ppt	Soluble
	Cu^{2+}	Pale blue ppt	Soluble
	Zn^{2+}	White ppt	Soluble
	Ag^+	Brown ppt	Soluble

* Because the OH^- concentration is low, Sr^{2+} and Ba^{2+} ions do not produce a precipitate. A concentrated solution of Ca^{2+} ions may give a faint precipitate.

The ammonia acts in two ways:

- as a weak base, giving a low concentration of OH^- ions, though not sufficient to produce hydroxy complexes
- as a complexing agent, with NH_3 acting as the ligand.

Summary of equations

For Ag^+ ions:

$$2Ag^+(aq) + 2OH^-(aq) \rightarrow Ag_2O(s) + H_2O(l)$$

with excess aqueous ammonia:

$$Ag_2O(s) + 4NH_3(aq) + H_2O(l) \rightarrow 2Ag(NH_3)_2^+(aq) + 2OH^-(aq)$$

For divalent M^{2+} ions:

$$M^{2+}(aq) + 2OH^-(aq) \rightarrow M(OH)_2(s)$$

with excess aqueous ammonia:

$$\underset{\text{air oxidation to Co(III)}}{2Co(OH)_2(s) + 12NH_3(aq) + H_2O(l) + \tfrac{1}{2}O_2(g)} \rightarrow \underset{\text{brown solution}}{2Co(NH_3)_6^{3+}(aq)} + 6OH^-(aq)$$

$$Ni(OH)_2(s) + 6NH_3(aq) \rightarrow \underset{\text{blue-purple solution}}{Ni(NH_3)_6^{2+}(aq)} + 2OH^-(aq)$$

$$Cu(OH)_2(s) + 4NH_3(aq) + 2H_2O(aq) \rightarrow \underset{\text{deep blue solution}}{[Cu(NH_3)_4(H_2O)_2]^{2+}(aq)} + 2OH^-(aq)$$

$$Zn(OH)_2(s) + 4NH_3(aq) \rightarrow \underset{\text{colourless solution}}{Zn(NH_3)_4^{2+}(aq)} + 2OH^-(aq)$$

For trivalent M^{3+} ions:

$$M^{3+}(aq) + 3OH^-(aq) \rightarrow M(OH)_3(s)$$

with excess aqueous ammonia:

$$Cr(OH)_3(s) + 6NH_3(aq) \rightarrow \underset{\text{purple solution}}{Cr(NH_3)_6^{3+}(aq)} + 3OH^-(aq)$$

21.5 Action of acids

Hydrochloric acid

Table 21.8 lists the most common gases given off as a result of treating a solid with aqueous hydrochloric acid.

Gas evolved	Likely anion present	Typical equation
CO_2	Carbonate, CO_3^{2-} or hydrogencarbonate, HCO_3^-	$CuCO_3(s) + 2HCl(aq) \rightarrow CuCl_2(aq) + CO_2(g) + H_2O(l)$ $NaHCO_3(s) + HCl(aq) \rightarrow NaCl(aq) + CO_2(g) + H_2O(l)$
SO_2	Sulphite, SO_3^{2-}	$Na_2SO_3(s) + 2HCl(aq) \rightarrow 2NaCl(aq) + SO_2(g) + H_2O(l)$
$SO_2 + S$	Thiosulphate, $S_2O_3^{2-}$	$Na_2S_2O_3(s) + 2HCl(aq) \rightarrow 2NaCl(aq) + SO_2(g) + S(g) + H_2O(l)$
$NO_2 + NO$	Nitrite, NO_2^-	$2KNO_2(s) + 2HCl(aq) \rightarrow 2KCl(aq) + NO_2(g) + NO(g) + H_2O(l)$

■ **Table 21.8**
Likely gases evolved when a solid is treated with aqueous hydrochloric acid.

Hydrogencarbonates can be distinguished from carbonates as heating solid hydrogencarbonate gives water as well as carbon dioxide:

$$2NaHCO_3(s) \rightarrow Na_2CO_3(s) + H_2O(l) + CO_2(g)$$

The colourless nitrogen monoxide obtained when nitrite is treated with aqueous hydrochloric acid reacts with oxygen in the air to give more brown nitrogen dioxide:

$$NO(g) + \tfrac{1}{2}O_2(g) \rightarrow NO_2(g)$$

If the solid is insoluble in water but dissolves in aqueous hydrochloric acid, it may be a carbonate (if carbon dioxide is given off), a sulphite (if sulphur dioxide is given off) or it may be an insoluble basic oxide. The colour of the resulting solution may help identify the cation.

Sulphuric acid

If there is no reaction with aqueous hydrochloric acid, a fresh sample of the solid may be treated with a few drops of concentrated sulphuric acid. The reaction must be carried out in a fume cupboard. The mixture may be carefully warmed if there is no reaction in the cold. Table 21.9 shows the likely gases that may be detected.

Gas	Likely anion present	Typical equation
HCl	Chloride, Cl^-	$MgCl_2(s) + 2H_2SO_4(l) \rightarrow Mg(HSO_4)_2(s) + 2HCl(g)$
$HBr + Br_2 + SO_2$	Bromide, Br^-	$KBr(s) + H_2SO_4(l) \rightarrow KHSO_4(s) + HBr(g)$ $2HBr(g) + H_2SO_4(l) \rightarrow Br_2(g) + SO_2(g) + 2H_2O(l)$
$HI + I_2 + SO_2 + S + H_2S$	Iodide, I^-	$NaI(s) + H_2SO_4(l) \rightarrow NaHSO_4(s) + HI(g)$ $2HI(g) + H_2SO_4(s) \rightarrow I_2(g) + SO_2(g) + 2H_2O(l)$ and other reactions
HNO_3	Nitrate, NO_3^-	$KNO_3(s) + H_2SO_4(l) \rightarrow KHSO_4(s) + HNO_3(g)$
CO	Methanoate, HCO_2^-	$HCO_2H(l) \rightarrow CO(g) + H_2O(l)$
$CO + CO_2$	Ethanedioate, $C_2O_4^{2-}$	$H_2C_2O_4(s) \rightarrow CO(g) + CO_2(g) + H_2O(l)$
CH_3CO_2H	Ethanoate, $CH_3CO_2^-$	$CH_3CO_2K(s) + H_2SO_4(l) \rightarrow KHSO_4(s) + CH_3CO_2H(g)$

■ **Table 21.9**
Likely gases evolved when a solid is treated with concentrated sulphuric acid.

In these reactions, concentrated sulphuric acid acts as a strong non-volatile acid, displacing more volatile acids as gases. With bromides and iodides, it also acts as an oxidising agent. With methanoates and ethanedioates, it also acts as a dehydrating agent.

Bromine gas is red–brown and iodine vapour is purple. Nitric acid vapour is usually pale brown as it decomposes slightly to nitrogen dioxide. It condenses on the cooler part of the test tube as an oily liquid. Carbon monoxide is colourless and burns quietly with an intense blue flame. Ethanoic acid, CH_3CO_2H, can be identified by its smell of vinegar.

21.6 Precipitation reactions for anions

Aqueous barium chloride

Aqueous barium chloride (or barium nitrate) gives a precipitate with a large number of ions, including carbonate, sulphite and sulphate. In the presence of aqueous hydrochloric acid (or aqueous nitric acid), only sulphate ions give a dense white precipitate. So if the addition of aqueous hydrochloric acid and aqueous barium chloride to a solution of the unknown substance gives a dense white precipitate, sulphate ions are present.

$$Ba^{2+}(aq) + SO_4^{2-}(aq) \rightarrow BaSO_4(s)$$

Aqueous silver nitrate

Aqueous silver nitrate gives precipitates with a large number of ions, including carbonate, chromate(VI) (CrO_4^{2-}), hydroxide, chloride, bromide and iodide. However, in the presence of aqueous nitric acid as well, only chlorides, bromides and iodides give precipitation (see Table 21.10). The colours of these precipitates and their solubilities in aqueous ammonia distinguish them from each other.

Table 21.10 The action of silver ions on halide ions.

Anion	Precipitate	Effect of $NH_3(aq)$ on precipitate
Cl^-	White	Soluble in dilute $NH_3(aq)$
Br^-	Cream or pale yellow	Soluble in concentrated $NH_3(aq)$
I^-	Deep yellow	Insoluble in concentrated $NH_3(aq)$

Typical equations

$$Cl^-(aq) + Ag^+(aq) \rightarrow AgCl(s)$$
$$AgCl(s) + 2NH_3(aq) \rightarrow Ag(NH_3)_2^+(aq) + Cl^-(aq)$$

The different colours and solubilities of the silver halides are discussed in Chapter 17, page 329.

Example Substance **A** is a white amorphous powder. It is insoluble in water. **A** dissolves in aqueous hydrochloric acid, with effervescence, to give a colourless solution **B**. The gas evolved, **C**, turns limewater milky. When aqueous sodium hydroxide was added to solution **B**, a faint white precipitate was formed that did not dissolve in excess. On addition of aqueous ammonia to solution **B**, no precipitate was formed. Substance **A** coloured a Bunsen burner flame green. Explain all these observations, identify **A**, **B** and **C**, and give equations for the reactions that take place.

Answer (It is relatively easy to identify the substance as barium carbonate. However, to answer the question fully, it is important to explain all the observations.)

Table 21.11

Observation	Inference
White, amorphous powder	Not a transition metal
Insoluble in water	Not Group 1, not NO_3^-
Dissolves in HCl with effervescence	Suggests CO_3^{2-} (or HCO_3^-)
Gas turns limewater milky	CO_2 produced from CO_3^{2-} (or HCO_3^-)
Solution **B** + NaOH(aq) gives a faint ppt	Li^+, Sr^{2+} or Ba^{2+} present
Solution **B** + $NH_3(aq)$ gives no ppt	Li^+, Sr^{2+} or Ba^{2+} present
Flame test green	Ba^{2+}

A is barium carbonate, $BaCO_3(s)$
B is barium chloride solution, $BaCl_2(aq)$
C is carbon dioxide, $CO_2(g)$.

Note that the evidence suggests that **A** could also be barium hydrogencarbonate, $Ba(HCO_3)_2$. However, it is more likely to be the carbonate because Group 2 hydrogencarbonates are known only in solution. The equations for the reactions are as follows:

$$BaCO_3(s) + 2H^+(aq) \rightarrow H_2O(l) + CO_2(g) + 2Ba^{2+}(aq)$$
$$CO_2(g) + Ca(OH)_2(aq) \rightarrow CaCO_3(s) + H_2O(l)$$
$$Ba^{2+}(aq) + 2OH^-(aq) \rightarrow Ba(OH)_2(s)$$
(faint ppt)

Further practice

1 **D** is a hygroscopic brown solid. It is readily soluble in water to give a green solution. With aqueous sodium hydroxide, this solution gives a dark green precipitate, **E**, that does not dissolve in an excess of the reagent. **E** turns brown on exposure to air. **D** shows no reaction with aqueous hydrochloric acid, but with concentrated sulphuric acid it frothed and gave off an acidic gas, **F**, as well as a red–brown gas, **G**. When moist air is blown over the mouth of the test tube, **F** gives steamy fumes, and it also gives dense white fumes when a drop of ammonia on a glass rod is brought near. The addition of aqueous nitric acid and silver nitrate to a solution of **D** produced a precipitate that turned green when aqueous ammonia was added.

Explain all these observations, identify **D**, **E**, **F** and **G**, and write equations for the reactions that take place.

2 **P** is a white amorphous powder. It is insoluble in water but dissolves readily in aqueous hydrochloric acid, with no evolution of gas, to give a colourless solution **Q**. To a portion of **Q**, aqueous sodium hydroxide is added. A white precipitate forms that dissolves in an excess of the reagent. To a separate portion of **Q**, aqueous ammonia is added. This also gives a white precipitate that dissolves in an excess of the reagent. A flame test on a sample of **P** is negative.

Explain all these observations, identify **P** and **Q**, and write equations for the reactions that take place.

3 **J** is an orange crystalline solid that is readily soluble in water to give an orange solution. When aqueous sodium hydroxide is added to a sample of the solution, the solution turns yellow and gives off an alkaline gas, **K**. **K** gives dense white fumes with the vapour from concentrated hydrochloric acid. When a solution of **J** is acidified with sulphuric acid and treated with zinc powder, the solution turns green. On addition of sodium hydroxide solution to this green solution, a green precipitate, **L**, forms which dissolves in an excess of the reagent. On heating a sample of **J**, a violent reaction takes place with the evolution of steam and a gas, **M**, that gives no positive reactions to the usual tests. The residue from the reaction is a green powder that does not dissolve in water but dissolves in sulphuric acid to give a green solution. The addition of aqueous sodium hydroxide to this green solution gives the precipitate **L** that had been produced in an earlier test.

Identify **J**, **K** and **L**. Write a balanced ionic equation for the reaction of solution **J** with zinc powder. Suggest the identity of gas **M** and hence write a balanced equation for the action of heat on **J**.

Summary

- Cations can be identified by:
  - colour of the salt
  - flame tests
  - action of aqueous sodium hydroxide and aqueous ammonia on a solution.
- Anions can be identified by:
  - action of heat on the solid
  - action of aqueous hydrochloric and concentrated sulphuric acids on the solid
  - action of $Ag^+(aq)$ ions or $Ba^{2+}(aq)$ ions on solutions.

ORGANIC CHEMISTRY

22 Introduction to organic chemistry

This chapter and those that follow look at the reactions and properties of the vast number of organic compounds – the covalent compounds of carbon. In this chapter we explain how the large variety of organic compounds arises, and how to name them and draw their formulae. Finally, we introduce the way in which organic reactions can be classified.

The chapter is divided into the following sections.

22.1 Introduction

At one time there were thought to be two entirely different classes of chemical substance. Those compounds that had been isolated from the living world, from organisms, were called **organic** compounds. On the other hand, those elements and compounds occurring in rocks and minerals, or made from them by chemical reactions, were called **inorganic** substances.

It was clear that both types of compound were made from the same elements. For example, the carbon dioxide produced by the fermentation of glucose:

$$C_6H_{12}O_6 \rightarrow 2C_2H_5OH + 2CO_2$$

and that obtained from burning the alcohol that is also produced during fermentation:

$$C_2H_5OH + 3O_2 \rightarrow 2CO_2 + 3H_2O$$

were the same compound as that obtained by heating limestone:

$$CaCO_3 \rightarrow CaO + CO_2$$

Carbon dioxide is an inorganic compound. It was accepted that inorganic compounds could be made from organic compounds, but for many years the reverse process was considered to be impossible. It was believed, without much justification, that a vital 'life' force was needed to make organic compounds, so they could only be synthesised within living organisms.

Credit for the shattering of this belief is normally given to Friedrich Wöhler, who in 1828 transformed the essentially inorganic compound ammonium cyanate into the organic compound carbamide (urea) by simply heating the crystals gently:

$$NH_4^+\ OCN^- \xrightarrow{\text{heat}} NH_2{-}CO{-}NH_2$$

The two branches of chemistry still remain distinct however, and are still studied separately at A level and beyond. Organic compounds are essentially those containing more than one carbon atom per molecule. Today, although several important organic chemicals are still isolated from the living world (penicillin is an example), the majority of the organic compounds we use have been synthesised artifically. Sometimes the starting point is an organic chemical, but most often it is a component of crude oil or coal. Both of these are the remains of living organisms, long since dead, decayed and chemically transformed into hydrocarbons.

The ultimate origin of all organic chemicals, in nature and in the laboratory, is the inorganic compound carbon dioxide. This is converted into glucose by the transformations of photosynthesis, the details of which biochemists and organic chemists have only recently unravelled. The overall process is:

$$6CO_2 + 6H_2O \rightarrow \underset{\text{glucose}}{C_6H_{12}O_6} + 6O_2$$

followed by:

$$\text{glucose} \xrightarrow{\text{biochemical pathways}} \text{polysaccharides, proteins, nucleic acids, etc.}$$

22.2 The stability and variety of organic compounds

Today, organic chemicals (artificially made or naturally occurring) outnumber inorganic ones by 80:1, and number well over 10 million different compounds. Hundreds of thousands of new compounds are being made each year. Why are there so many of them? There are two main reasons:

- carbon atoms form strong bonds with other carbon atoms
- generally organic compounds are kinetically stable.

We shall look at each of these reasons in turn.

Table 22.1 The stability of the carbon–carbon bond.

Bond	Bond enthalpy/kJmol^{-1}
C—C	346
N—N	167
O—O	142
Si—Si	222
P—P	201
S—S	226

The ability of carbon to form strong bonds to other carbon atoms

Table 22.1 shows how much more stable bonds between carbon atoms are than bonds between the atoms of other elements.

Carbon–carbon bonding allows chains and rings of carbon atoms to be produced by a process called **catenation** (from the Latin *catena*: chain). What is more, the 4-valent nature of the carbon atom allows branched chains to occur, and the attachment of other atoms, increasing further the number of structures possible.

There are four different ways in which we can arrange four linked carbon atoms (see Figure 22.1).

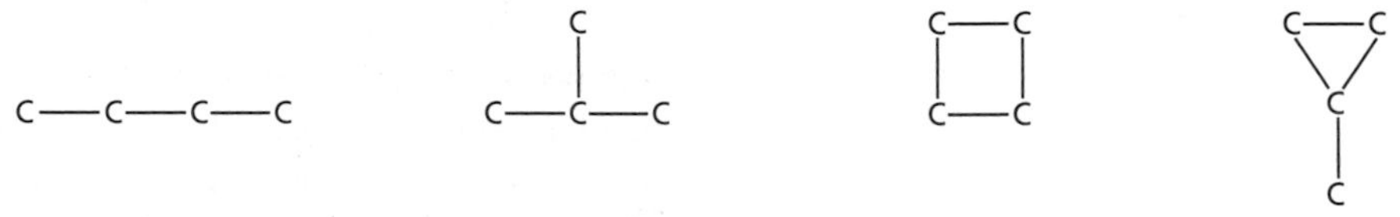

Figure 22.1 Four ways of linking four carbon atoms.

With five carbon atoms there are ten different arrangements, and with six carbon atoms there are no fewer than 17. With a molecule containing 30 carbon atoms, there are more than 4×10^9 different ways of arranging them.

Example Of the ten possible ways of bonding five carbon atoms together, three are open-chain structures (that is, they do not contain a ring of carbon atoms). Draw diagrams of these three.

Answer One structure will contain a chain of five carbon atoms:

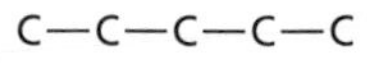

If we take one carbon atom from the end, and join it to the middle of the chain, we produce another structure:

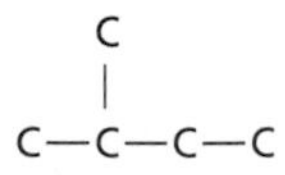

If we do this once more, we arrive at the third of the three open-chain structures:

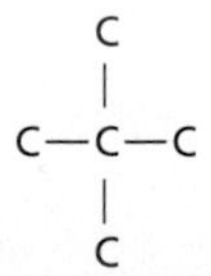

Further practice Draw out the five open-chain structures containing six carbon atoms.

The stability of organic compounds

The second reason why there are so many organic compounds is a more subtle one – once they have been formed, organic compounds are remarkably stable. This stability is more kinetic than thermodynamic. For example, although methane is thermodynamically stable with respect to its elements:

$$CH_4(g) \rightarrow C(s) + 2H_2(g) \qquad \Delta H^{\ominus} = +75\,kJ\,mol^{-1}$$

it is unstable in the presence of oxygen:

$$CH_4(g) + 2O_2(g) \rightarrow CO_2(g) + 2H_2O(l) \qquad \Delta H^{\ominus}_c = -890\,kJ\,mol^{-1}$$

But methane, along with other hydrocarbons (and most organic compounds), does not react with oxygen unless heated to quite a high temperature – the **ignition point**. This low rate of reaction at room temperature is a result of the high activation energy that many organic reactions possess (see Chapter 8, page 164 and Chapter 11, page 221). This in turn is due to the strong bonds that carbon forms to itself and to other elements. These bonds have to be broken before reaction can occur, and breaking strong bonds requires much energy. Thus methane in oxygen is kinetically stable. The first stage in the combustion of methane, for example, involves the breaking of a C—H bond:

$$CH_3{-}H \rightarrow CH_3 + H$$

This involves the input of 439 kJ of energy per mole.

Organic compounds are thermodynamically unstable not only with respect to oxidation, but also with respect to other transformations. For example, the hydration of ethene to ethanol is exothermic, as is the joining of three molecules of ethyne to form benzene:

$$CH_2{=}CH_2 + H_2O \longrightarrow C_2H_5OH \qquad \Delta H^{\ominus} = -43\,kJ\,mol^{-1}$$

$3CH{\equiv}CH \longrightarrow$ 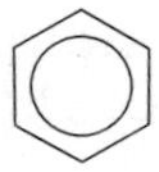 $\Delta H^{\ominus} = -635\,kJ\,mol^{-1}$

And yet ethene gas can be collected over water, and ethyne has no tendency to form benzene under normal conditions. If the right conditions and catalysts can be found, however, it is possible to covert one 'stable' organic compound into another 'stable' compound. It is even possible to use the same reagent (hydroxide ions) on the same compound (a bromoalkane) to produce two different products, depending on the conditions of solvent used (see Chapter 25, page 471, for details):

$$CH_3CHBrCH_3(l) + NaOH(aq) \xrightarrow{\text{warm in water}} NaBr + CH_3CH(OH)CH_3 \quad (1)$$

$$CH_3CHBrCH_3(l) + NaOH(\text{in ethanol}) \xrightarrow{\text{heat in ethanol}} NaBr + H_2O + CH_2{=}CHCH_3 \quad (2)$$

Organic compounds can be considered as occupying stable low-energy 'valleys', with high-energy 'hills' in between them. One catalyst, or set of conditions, can afford a low activation energy 'pass' to a valley next door, whereas another, different catalyst, can open up a new low-energy route into another valley on the other side.

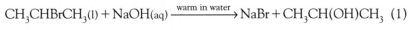

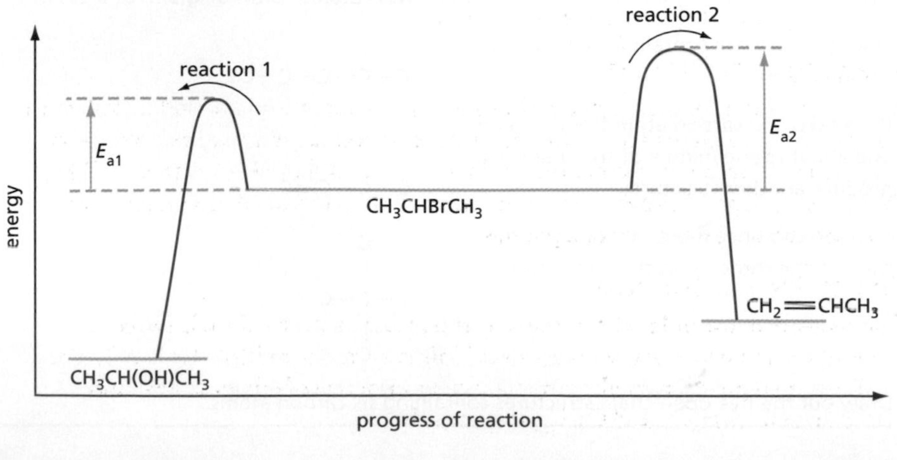

■ **Figure 22.2** Reaction profiles for 'typical' organic reactions. E_{a1} and E_{a2} are the activation energies for reactions 1 and 2, respectively. The activation energy is the energy that the reactants need in order to undergo the reaction.

22.3   The shapes of organic molecules

After Wöhler's conceptual breakthrough in 1828, the next 50 years saw a tremendous increase in activity by organic chemists, making many new compounds and determining their empirical and molecular formulae. However, it was not until after the work of Joseph Le Bel, Jacobus van't Hoff and Friedrich Kekulé in the period 1860–75 that chemists started to think of organic compounds as being made up of separate, three-dimensional molecules. After this period chemists appreciated that the molecular formula was merely a summary of the structure of the molecule, rather than telling the whole story.

We now firmly believe that all organic (and inorganic) molecules possess a three-dimensional structure, and that the various atoms are fixed in their relative positions. The shapes of organic molecules depend ultimately on bond lengths and angles. Bond angles are determined by the mutual repulsion of electron pairs (see Chapter 3, page 63). (The state of hybridisation of the carbon atoms depends on the bond angles – see the panel on page 71.)

■ **Table 22.2**
Bond angles in some organic compounds.

Bond angle	Examples	Hybridisation (see the panel on page 71)
180°	$O{=}C{=}O$ $\qquad$ $H{-}C{\equiv}N$	sp
120°	$\begin{array}{c} H \quad\quad H \\ \diagdown\quad\diagup \\ C = C \\ \diagup\quad\diagdown \\ H \quad\quad H \end{array}$	sp²
109.5°	$\begin{array}{c} H \\ \mid \\ C \\ \diagup \mid \diagdown \\ H \;\; H \\ H \end{array}$	sp³

In each of the molecules carbon dioxide and hydrogen cyanide, the three atoms are collinear (they all lie on the same straight line). In the molecule ethene, all six atoms are coplanar (they all lie in the same plane). Ethene is a 'two-dimensional' molecule. Methane, however, is a three-dimensional molecule, and its shape cannot be easily represented on a two-dimensional page. We use a **stereochemical formula** to show this, as in the last row of Table 22.2. The dotted line represents a bond behind the plane of the paper, whereas the solid wedge represents a bond coming out in front of the plane of the paper. Ordinary lines represent bonds that are in the plane of the paper.

22.4   The different types of formula

There are six types of formula that are used to represent organic compounds. We shall illustrate them by using ethanoic acid, the acid occurring in vinegar, and 'iso-octane', one of the constituents of petrol.

The empirical formula

The information that the **empirical formula** gives about a compound is limited to stating which elements are contained in the compound, and their relative ratios. The empirical formula of ethanoic acid is CH_2O, and that of iso-octane is C_4H_9. We saw in Chapter 1, page 11, how empirical formulae can be calculated from mass or percentage data for compounds.

The molecular formula

The **molecular formula** tells us the actual number of atoms of each element present in a molecule of the compound. This is always a multiple ($\times 1$, $\times 2$, $\times 3$, etc.) of the empirical formula. The molecular formulae of ethanoic acid and iso-octane are $C_2H_4O_2$ and C_8H_{18}, respectively.

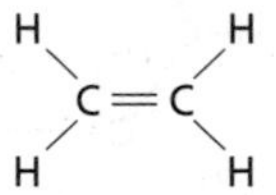

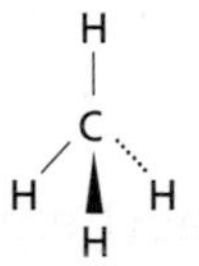

The structural formula

Structural formulae show us which atoms are joined to which in the molecule. The structural formula of ethanoic acid is CH_3CO_2H, and that of iso-octane is $(CH_3)_3CCH_2CH(CH_3)_2$. Structural formulae are used to distinguish between structural isomers (see Section 22.6). C═C double bonds are always shown in structural formulae, and sometimes also C═O and C—C bonds, where this makes the structure clearer.

The displayed formula

Similar to the structural formula, the **displayed formula** is more complete. It shows not only which atoms are bonded to which, but also the type of bonds (single, double, dative). The displayed formulae for ethanoic acid and iso-octane are shown in Figure 22.3.

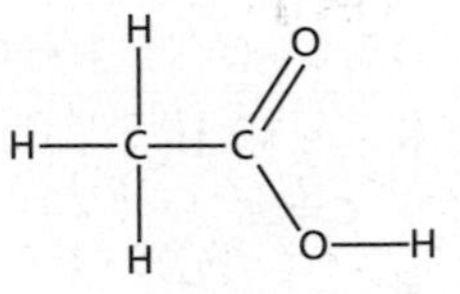

ethanoic acid

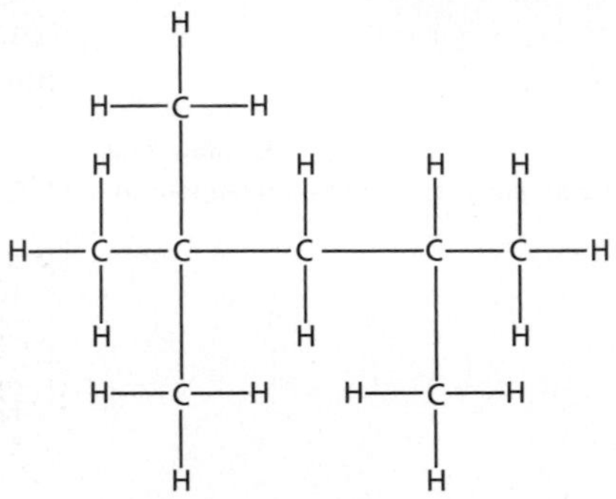

iso-octane

■ **Figure 22.3** Displayed formulae show the bonding in the molecule.

The stereochemical formula

Although the displayed formula tells us much about a molecule, it indicates nothing about its shape. It is difficult to represent the shape of a three-dimensional molecule on a two-dimensional page, so we use the conventions of **stereochemical formulae** illustrated in Figure 22.4.

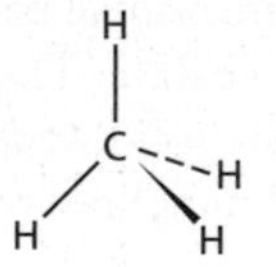

methane, CH_4

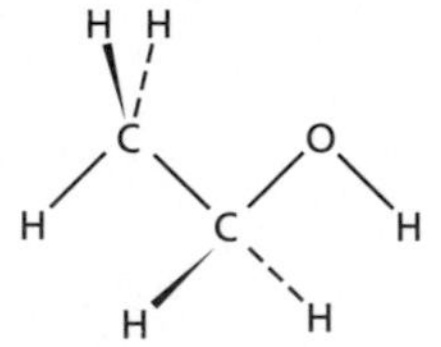

ethanol, C_2H_5OH

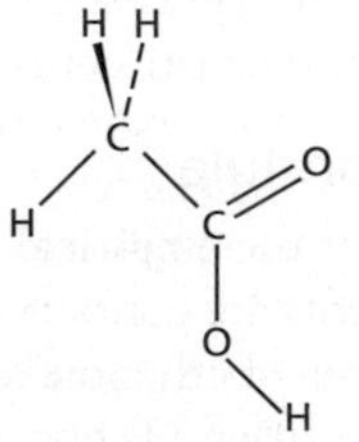

ethanoic acid, CH_3CO_2H

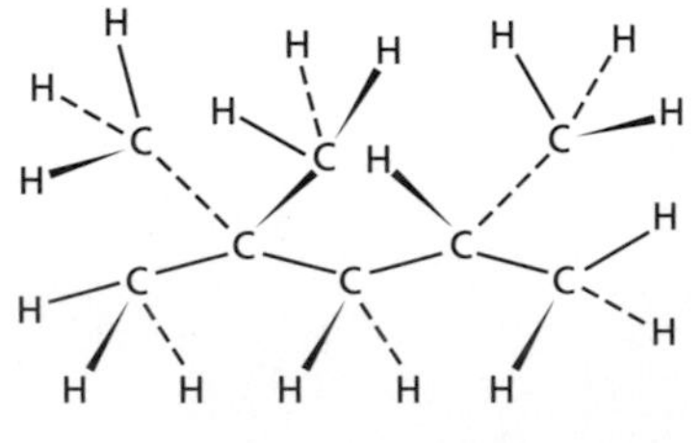

iso-octane, $(CH_3)_3CCH_2CH(CH_3)_2$

■ **Figure 22.4** Stereochemical formulae represent the three-dimensional shape of a molecule. The dashed line represents a bond to an atom *behind* the plane of the paper, and the 'wedge-shaped' line represents a bond to an atom *in front of* the plane of the paper.

The stereochemical formula for iso-octane in Figure 22.4 illustrates how this type of formula can become too complicated and unwieldy for any molecule with more than about five carbon atoms. Quite often, however, you will come across a displayed formula of which a part has been drawn out to show the stereochemistry around a particular atom.

The skeletal formula

The previous five types of formulae have become increasingly more complicated. The **skeletal formula** moves in the other direction. It pares the molecular structure down to its bare minimum, omitting all carbon atoms and all hydrogen atoms joined to carbons. It is an excellent way of describing larger and more complicated molecules. The true skeletal formula shows only the carbon–carbon bonds, but, especially before you become familiar with their use, it is clearer if you include a big dot to represent each carbon atom. Figure 22.5 shows the skeletal formula of three of the four compounds shown in Figure 22.4, together with pictures of the corresponding molecular models. The skeletal formula of iso-octane shows more clearly than any of the other formulae how the carbon atoms are joined together. This type of formula will be used in later chapters to show the shapes of, and bonding in, the larger molecules we shall come across. Its use for smaller molecules is severely limited, however. The skeletal formula for methane is a dot, and that for ethane is a straight line!

■ **Figure 22.5** Skeletal formulae and models of **a** ethanol, **b** ethanoic acid and **c** iso-octane.

a

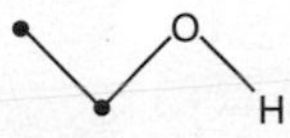

ethanol

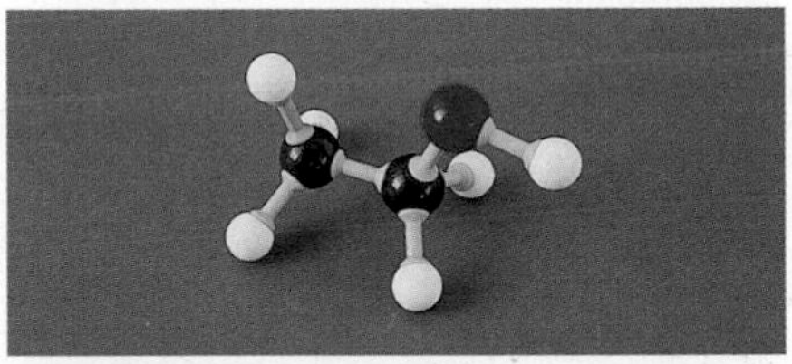

b

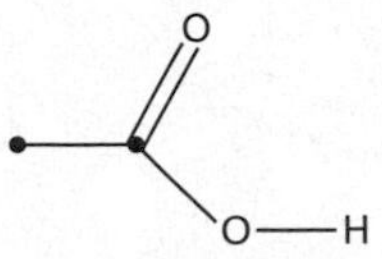

ethanoic acid

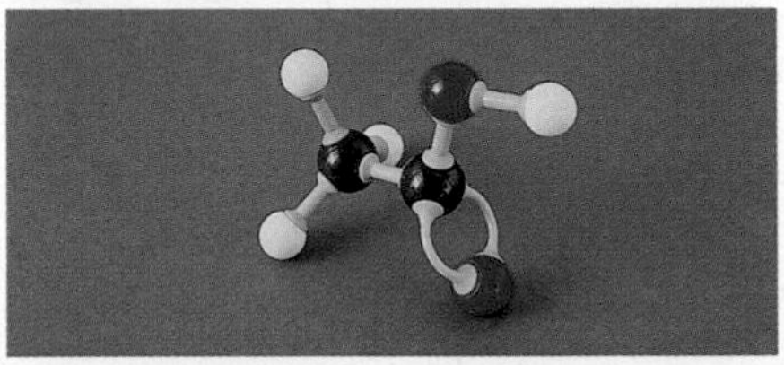

c

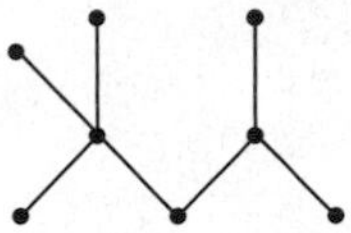

iso-octane

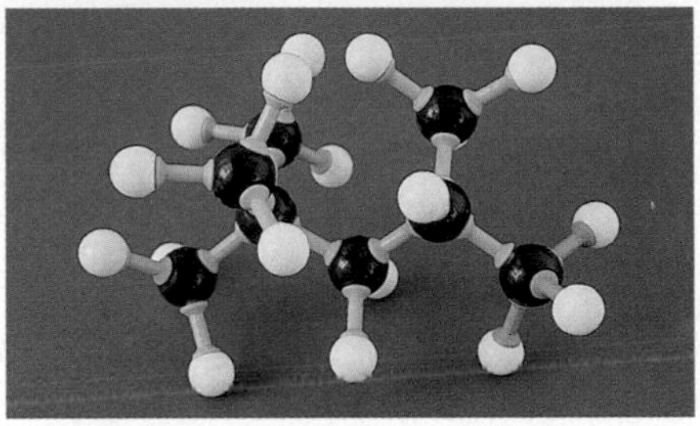

Example 1 What are the molecular and empirical formulae of the following compounds?

a propane, $CH_3CH_2CH_3$
b ethane-1,2-diol (glycol), $HOCH_2CH_2OH$
c dichlorethanoic acid, $CHCl_2CO_2H$

Answer

a For propane, the molecular formula is C_3H_8, which does not reduce further, so the empirical formula is also C_3H_8.
b For ethane-1,2-diol, the molecular formula of $C_2H_6O_2$ reduces to CH_3O as the empirical formula.
c For dichloroethanoic acid, the molecular formula of $C_2H_2O_2Cl_2$ reduces to the empirical formula CHOCl.

Example 2 What are the structural and molecular formulae of the compounds shown in Figure 22.6?

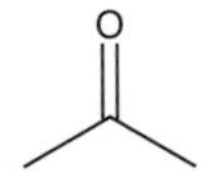

a

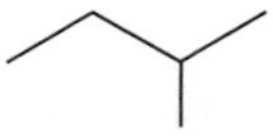

b

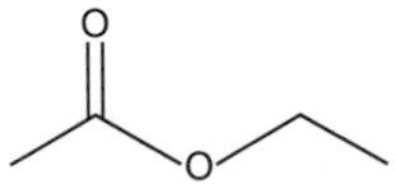

c

■ **Figure 22.6**

Answer

a CH_3COCH_3 C_3H_6O
b $CH_3CH_2CH(CH_3)_2$ C_5H_{12}
c $CH_3CO_2CH_2CH_3$ $C_4H_8O_2$

Further practice Work out the structural, molecular and empirical formulae of the compounds shown in Figure 22.7.

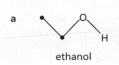

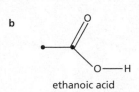

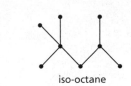

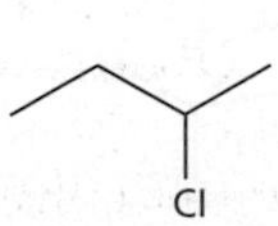

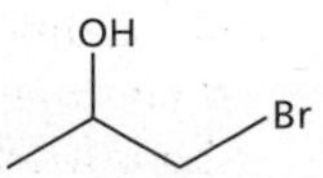

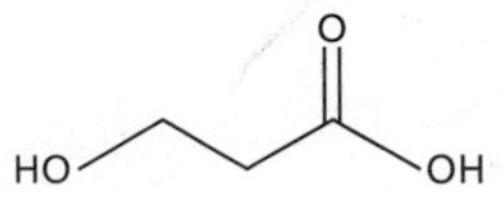

■ **Figure 22.7**

The representation of carbon rings in formulae

Rings of atoms are often represented by their skeletal formulae. Common examples are shown in Figure 22.8.

■ **Figure 22.8**
Skeletal formulae and models of **a** cyclopropane, **b** cyclopentane, **c** cyclohexane, **d** cyclohexene and **e** benzene.

a
cyclopropane

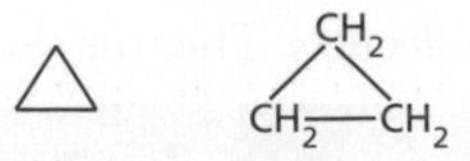

b
cyclopentane

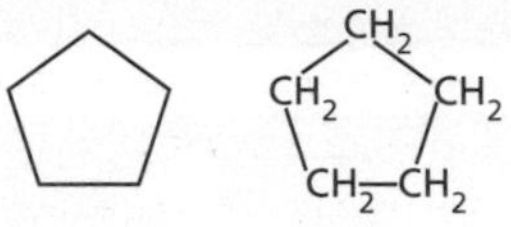

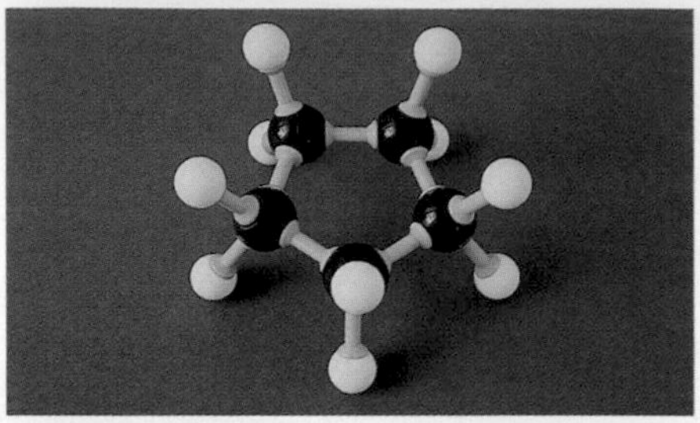

c
cyclohexane

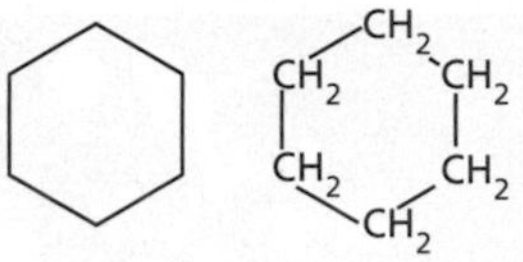

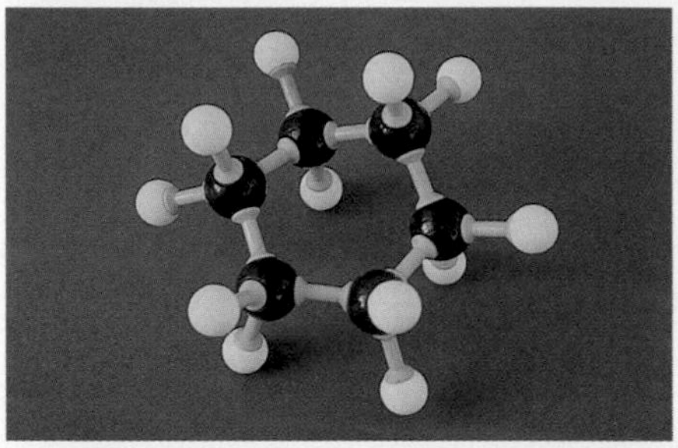

d
cyclohexene

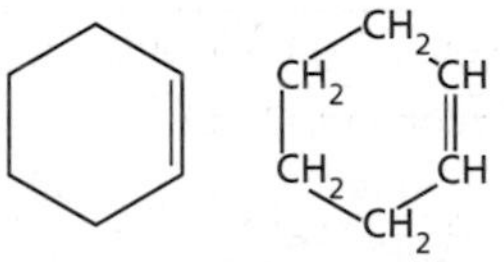

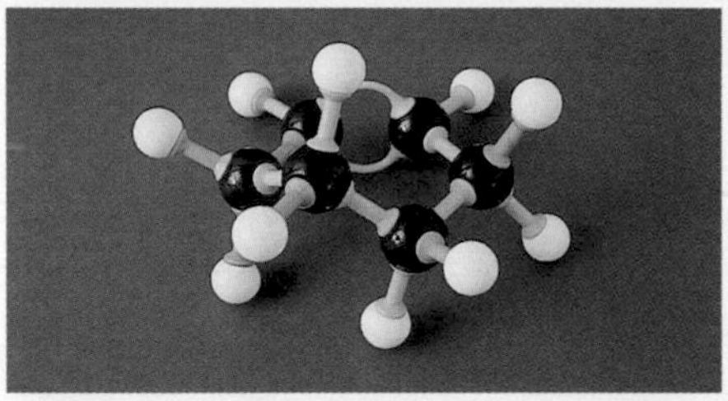

e
benzene

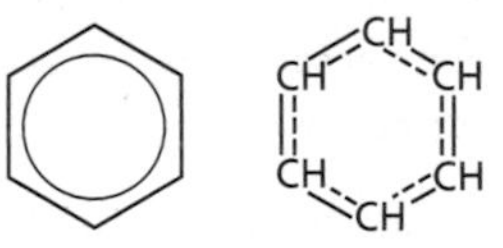

Benzene and cyclopropane are planar rings, but the other three rings in Figure 22.8 are all puckered to some degree. The most common configuration of cyclohexane is called the 'chair' form, which allows the C—C—C bond angles to be the preferred 109.5° (see Figure 22.9).

■ **Figure 22.9**
The 'chair' form of cyclohexane minimises bond strain.

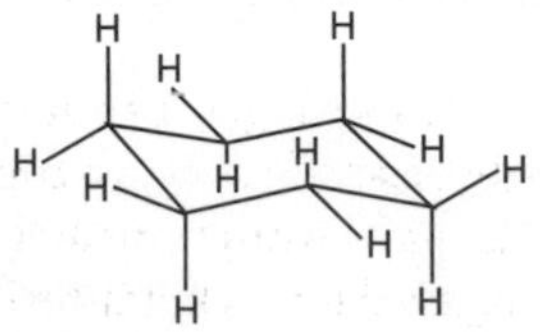

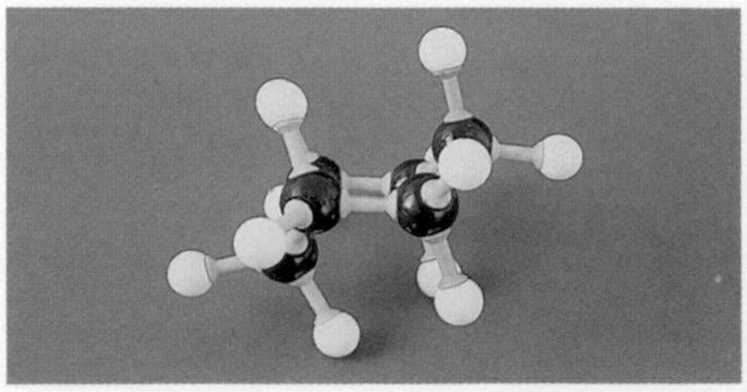

Cyclopropane is a highly strained (and hence reactive) compound (see Chapter 10, page 188). Its bond angles are constrained to be 60° – far removed from the preferred 109.5°. No such problems arise with benzene, however. The internal angles in a (planar) regular hexagon are 120°, which is ideal for carbon atoms surrounded by three electron pairs (sp^2 hybridisation – see the panel on page 71).

Example Give the molecular formulae of the following:

a

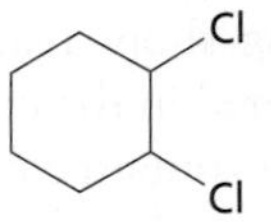

b

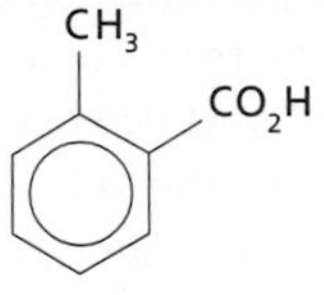

Answer

a Cyclohexane is $(CH_2)_6$, or C_6H_{12}. Two hydrogen atoms have been replaced by chlorines, so the formula is $C_6H_{10}Cl_2$.

b Benzene is C_6H_6. One hydrogen atom has been replaced by CH_3, and another by CO_2H. The molecular formula is $C_8H_8O_2$.

Further practice Give the molecular formulae of the following:

1

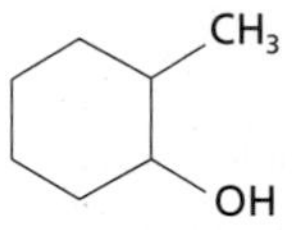

2

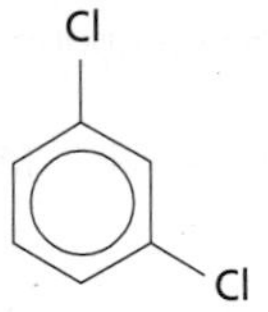

3

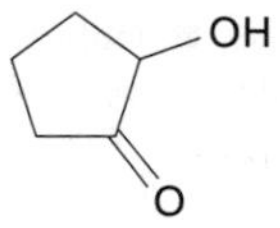

22.5 Naming organic compounds

As was mentioned earlier, organic compounds were originally extracted from living organisms. The sensible way of naming them incorporated the name (often the Latin version) of the plant or animal concerned. Some common examples are given in Table 22.3.

■ **Table 22.3** The origins of the names of some natural organic compounds.

Common name	Structural formula	Isolated from	Latin name
Formic acid	HCO_2H	Ants	*Formica*
Malic acid	CO_2H – $CH(OH)$ – CH_2 – CO_2H	Apples	*Malus*
Menthol	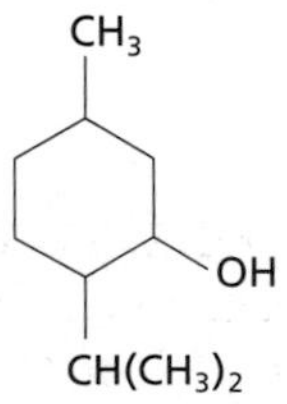	Peppermint	*Mentha piperita*
Oxalic acid	CO_2H – CO_2H	Rhubarb, wood sorrel	*Oxalis*

As more and more organic compounds were discovered, this system became increasingly unsatisfactory for two reasons – firstly, it could not be applied to those compounds made entirely artificially in the laboratory, and secondly, the feat of memory required for a chemist to be able to relate a structure to each name was becoming too great.

Stem name	Number of carbon atoms
Meth-	1
Eth-	2
Prop-	3
But-	4
Pent-	5
Hex-	6
Hept-	7
Oct-	8

■ **Table 22.4**
Stem names and the numbers of carbon atoms they represent. From 5 upwards, the stem is an abbreviation of the corresponding Greek numeral.

So, in the latter part of the nineteenth century, chemists began to devise a systematic, logical form of nomenclature. Using this method, it was possible for a chemist who had not previously come across a particular compound to translate the name into the structural formula and vice versa. The system has been refined over the years, and has been adopted in its latest version by the International Union of Pure and Applied Chemistry (IUPAC). It is (with only a few modifications) completely international. Chemists throughout the world use it, and understand each other, no matter what their first language is. (It has to be said, however, that the **true** language of organic chemistry is the skeletal formula. Sit next to a couple of organic chemists talking 'shop' in a café, and you will see that in no time at all their napkins will be covered in the curious hieroglyphics of skeletal formulae.)

All organic compounds contain carbon, and virtually all of them also contain hydrogen. A systematic name for an organic compound is often based on the 'parent' hydrocarbon. Each name consists of the components listed below (not all may be present, however; it depends on the complexity of the compound).

1 A **stem** – this specifies the number of carbon atoms in the longest carbon chain (see Table 22.4). If the carbon atoms are joined in a ring, the prefix 'cyclo' is added.

2 A **stem-suffix** – this indicates the type of carbon–carbon bonds that occur in the compound (see Table 22.5).

■ **Table 22.5**
Stem-suffixes and their meanings.

Stem-suffix	Meaning
-an	All C—C single bonds
-en	One C=C double bond
-dien	Two C=C double bonds
-trien	Three C=C double bonds
-yn	One C≡C triple bond

So:

$CH_3—CH_3$ is ethane
$CH_3—CH=CH_2$ is propene
$CH_2=CH—CH=CH_2$ is butadiene.

3 A **suffix** – this indicates any oxygen-containing group that might be present (see Table 22.6).

■ **Table 22.6**
Suffixes and their meanings.

Suffix	Meaning	
-al	An aldehyde group:	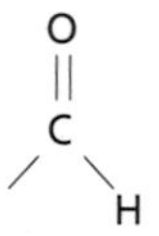
-one	A ketone group:	
-ol	An alcohol group:	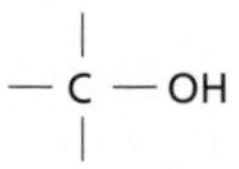
-oic acid	A carboxylic acid group:	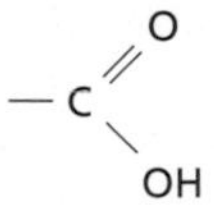

So:

$CH_3—CHO$ is ethanal
$CH_3—CH_2—CO_2H$ is propanoic acid
$CH_3—CO—CH_2—CH_3$ is butanone.

4 One or more prefixes – these indicate non-oxygen-containing groups, which may be other carbon (alkyl) groups, or halogen atoms (see Table 22.7). The prefixes may also include subsidiary oxygen-containing groups.

■ Table 22.7 Prefixes and their meanings.

Prefix	Meaning
Methyl	$CH_3—$
Ethyl	$CH_3CH_2—$
Propyl	$CH_3CH_2CH_2—$
...	...
Chloro	Cl—
Bromo	Br—
Hydroxy	HO—

Prefixes can be combined and they might in turn be prefixed by 'di' or 'tri' to show multiple substitution.

The position of a substituent along the carbon chain is indicated by an Arabic numeral before the substituent. The atoms in the longest carbon chain are numbered successively from the end that results in the smallest overall numbers. So $CH_3—CHCl—CH_2Cl$ is 1,2-dichloropropane, numbering from the right, and not 2,3-dichloropropane (numbering from the left).

These four simple part-names allow us to describe a multitude of compounds. The following examples will show how the parts are pieced together.

a $Cl—CH_2—CH_2—CHO$ is

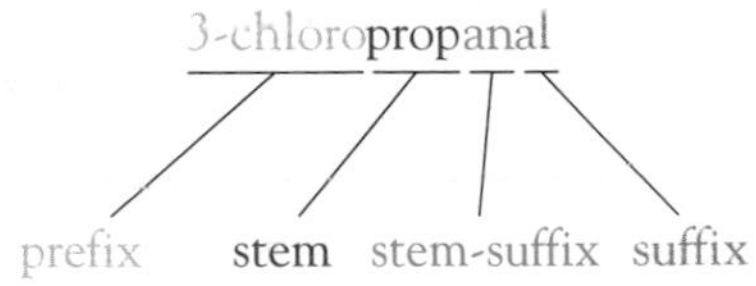

b $CH_3—CH(OH)—CH_3$ is

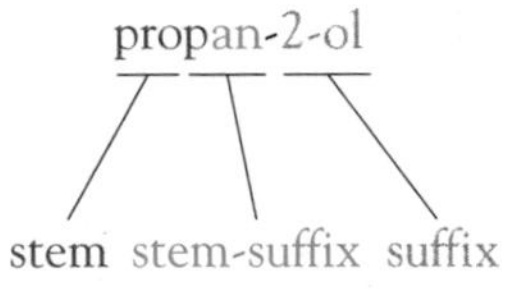

c $CH_3—CHCl—CH(CH_3)_2$ is

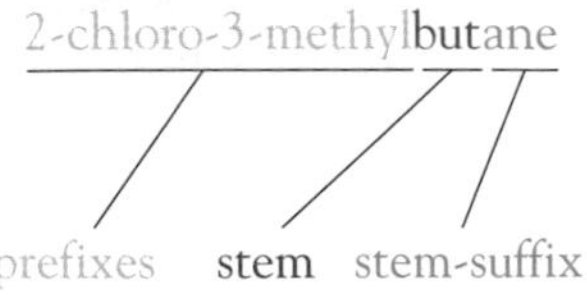

d $(CH_3)_2C=CH—CH_2OH$ is

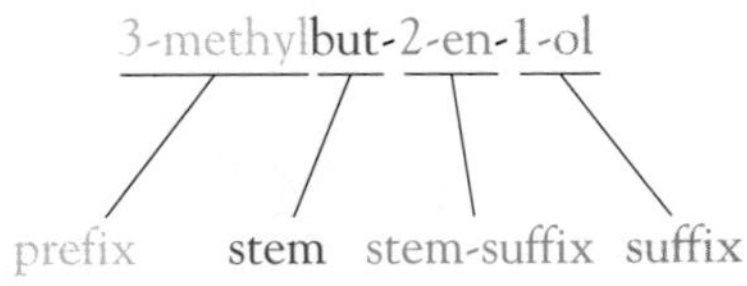

Note that **a** is not called 1-chloropropanal. If a compound is named as an aldehyde, the aldehyde group is always at carbon atom number 1. Note also that **c** could equally well have been called 3-chloro-2-methylbutane (that is, numbering from the right). But the name 3-methyl-2-chlorobutane would have been incorrect, because the substituents should be listed alphabetically.

Example 1 What are the systematic names of the following?

a $(CH_3)_3COH$ b 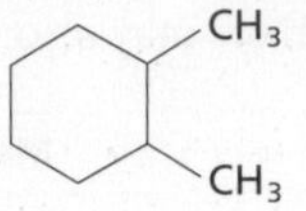 c $CH_3-CH_2-CH(OH)-CH_3$

Answer a 2-methylpropan-2-ol
b 1,2-dimethylcyclohexane
c butan-2-ol

Example 2 Write the structural formulae of the following:
a 3-methylbutan-1-ol
b 1,3-dichloropropane.

Answer a $(CH_3)_2CHCH_2CH_2OH$
b $ClCH_2CH_2CH_2Cl$

Further practice 1 Name the following:
a $(CH_3)_3CCH(OH){-}CH_2CH_3$
b $CH_3{-}CH{=}CH{-}CH_3$
c

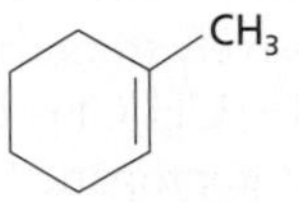

2 Draw structural formulae for the following:
a 2-methyl-2,3-dibromopentane
b 2,2-dimethylpropane.

22.6 Isomers

Isomerism is the property of two or more compounds (called **isomers**) having the same molecular formula, but different arrangements of atoms, and hence different structural formulae.

Isomerism in organic compounds can be classified into five different types, according to the scheme shown in Figure 22.10. Some of these are dealt with in detail when they arise in later chapters, but all are included here in summary form.

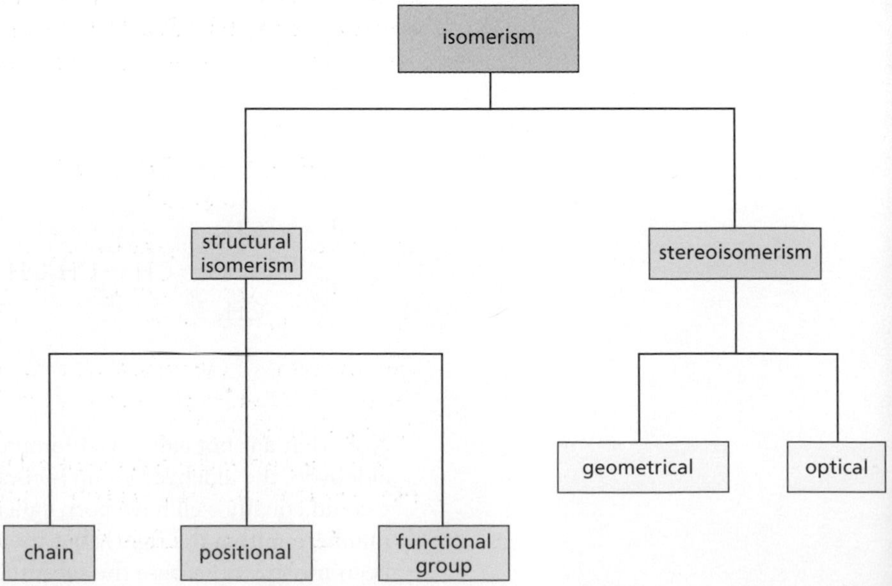

■ **Figure 22.10**
The five types of isomerism.

Structural isomerism

Structural isomers differ as to which atoms in the molecule are bonded to which. There are three types of structural isomerism.

Chain isomerism

Chain isomers have the same number of carbon atoms as each other, but their carbon backbones are different. For example, pentane and 2-methylbutane are isomers with the formula C_5H_{12}:

$$CH_3{-}CH_2{-}CH_2{-}CH_2{-}CH_3 \qquad \begin{array}{l} CH_3{-}\underset{\displaystyle CH_3}{\underset{|}{CH}}{-}CH_2{-}CH_3 \end{array}$$

Note that:

$$CH_3{-}CH_2{-}CH_2{-}\underset{\displaystyle CH_3}{\underset{|}{CH_2}}$$

is not another isomer: this is just another way of writing the formula of pentane. Note also that:

$$CH_3{-}CH_2{-}\overset{\displaystyle CH_3}{\overset{|}{\underset{\displaystyle CH_3}{\underset{|}{CH}}}}$$

is not another isomer. This is another way of writing the formula of 2-methylbutane. Remember that carbon chains are naturally zig-zag arrangements, and that there is completely free rotation around any C—C single bond; however structural formulae are often drawn as flat, with 90° bond angles. The formulae in Figure 22.11 all represent different **conformations** of the molecule of pentane, and at room temperature a particular molecule will be constantly changing its shape from one conformation to another.

Figure 22.11 Conformations of pentane.

Positional isomerism

This is often included under the general umbrella of structural isomerism. **Positional isomers** have the same carbon backbone, but the positions of their functional group(s) differ. Their systematic names reflect this. For example, there are two alcohols containing three carbon atoms:

$$\underset{\text{propan-1-ol}}{CH_3{-}CH_2{-}CH_2{-}OH} \qquad \underset{\text{propan-2-ol}}{CH_3{-}\overset{\displaystyle OH}{\overset{|}{CH}}{-}CH_3}$$

and there are two compounds with the formula $C_2H_4Cl_2$:

$$\underset{\text{1,2-dichloroethane}}{Cl{-}CH_2{-}CH_2{-}Cl} \qquad \underset{\text{1,1-dichloroethane}}{CH_3{-}CHCl_2}$$

(Remember that the carbon chain is numbered from the end that produces the smallest number prefixes in the name, so $CH_3{-}CHCl_2$ is not called 2,2-dichloroethane.)

When a carbon chain contains a double bond, the position of the bond can give rise to isomerism. There are two butenes, for example:

$$\underset{\text{but-1-ene}}{CH_3{-}CH_2{-}CH{=}CH_2} \qquad \underset{\text{but-2-ene}}{CH_3{-}CH{=}CH{-}CH_3}$$

Positional isomerism also occurs in ring systems, as shown in Figure 22.12.

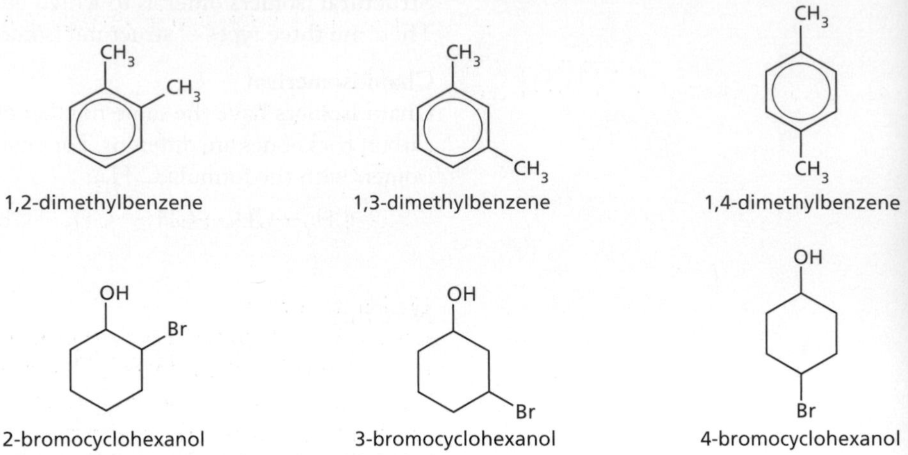

■ **Figure 22.12** Positional isomers in ring molecules. Note that the carbon atoms in the rings are numbered starting with one of the substituted ones, and that in cyclohexanol it is the carbon atom attached to the –OH group that is carbon number 1.

Functional group isomerism

In **functional group isomerism**, the isomers are more radically different than in the types of isomerism considered so far. Compounds that contain different functional groups often have very different reactions, as we shall see in Section 22.8. However, if their molecular formulae are the same, then they are classed as isomers.

For example, the carboxylic acid $CH_3CH_2CO_2H$ (propanoic acid), the ester $CH_3CO_2CH_3$ (methyl ethanoate) and the ketone–alcohol CH_3COCH_2OH (hydroxypropanone) are all isomers, having the molecular formula $C_3H_6O_2$.

Stereoisomerism

Stereoisomers are very closely related to each other. They contain the same atoms bonded to one another, and the bonding and functional groups are identical. They differ only in the way the atoms are arranged in three-dimensional space ('stereo' is derived from the Greek word *stereos*, meaning solid). There are two types of stereoisomerism – **geometrical** and **optical**.

Geometrical isomerism

This occurs with alkenes that do not have two identical groups on either end of the C=C double bond. It arises because, unlike the case of a C—C single bond, it is not possible to rotate one end of a C=C double bond with respect to the other. The following two forms of but-2-ene are therefore distinct and separate isomers:

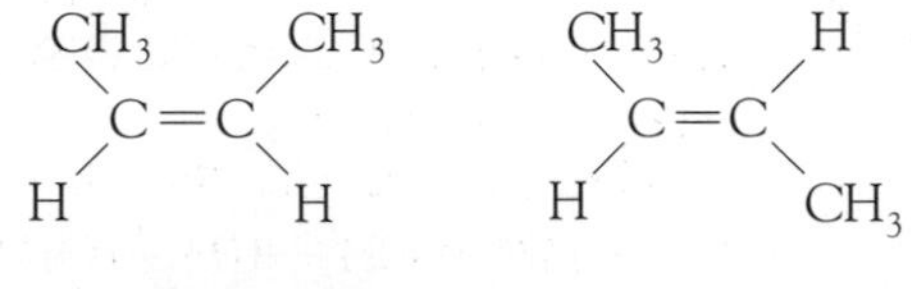

cis-but-2-ene   *trans*-but-2-ene

The isomer with both methyl groups on the same side of the double bond is called the *cis* isomer, and that with the methyl groups on opposite sides is called the *trans* isomer. An alternative name for geometrical isomerism is ***cis–trans* isomerism**. Note that the carbon atoms on both ends of the double bond must each contain two non-identical groups. The following compounds do **not** exhibit geometrical isomerism because one end, or both ends, of the double bond have two identical atoms or groups attached:

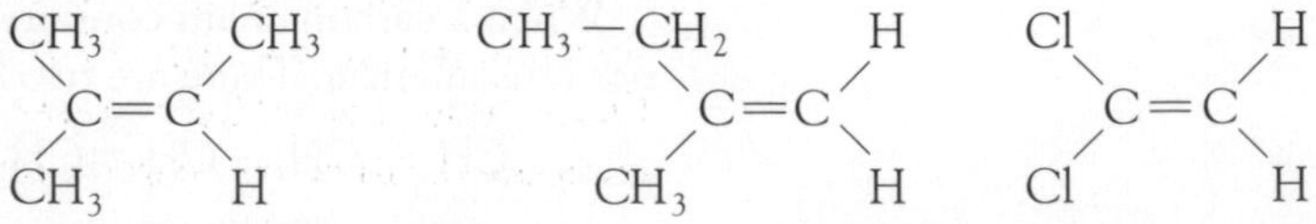

Geometrical isomerism also occurs in ring systems. For example, there are two isomers of 1,3-dimethylcyclobutane, shown in Figure 22.13. (Rotation around a single bond is prevented by linkage in a ring.)

■ **Figure 22.13**
Geometrical isomers of 1,3-dimethylcyclobutane.

cis-1,3-dimethylcyclobutane *trans*-1,3-dimethylcyclobutane

Optical isomerism

Optical isomerism is the most subtle of the five forms of isomerism, and virtually every physical or chemical property of the isomers concerned is identical. The one way in which they can be distinguished is by their effect on the passage of plane-polarised light, hence the name 'optical'. The panel below describes how this effect is measured.

The polarimeter

A **polarimeter** is used to distinguish optical isomers. The device consists of six parts, shown in Figure 22.14. A monochromatic light source (that is, one that produces light at a single wavelength) and a slit produce a thin beam of light, which then passes through a **polariser**. This may be a piece of Polaroid (the material that Polaroid sunglasses are made of) – it contains long straight molecules all lined up in the same direction. It only allows through photons whose electric field is oscillating in the same direction as its molecules.

■ **Figure 22.14**
Outline diagram of a polarimeter.

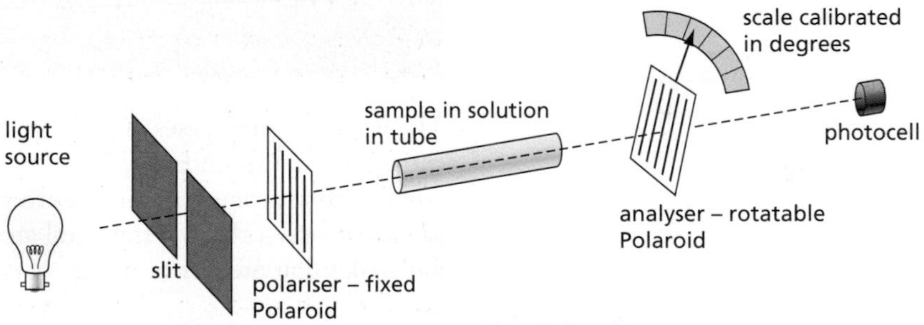

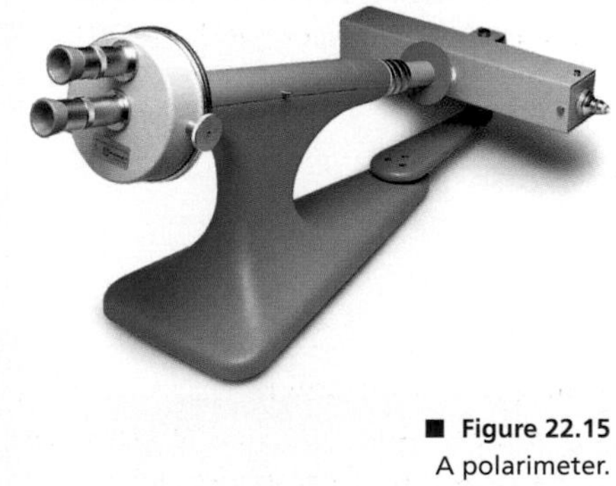

■ **Figure 22.15**
A polarimeter.

The polarised beam now enters the sample. This is a solution of the compound under investigation, held in a tube 10 cm long. As it comes out of the other end, the beam passes through another piece of Polaroid, the **analyser**, and then into the photocell, which produces an electric current proportional to the intensity of light that falls upon it.

The instrument is calibrated by filling the sample tube with pure solvent, and turning the analyser around its axis until the output from the photocell is a minimum. This occurs when the analyser is at right angles to the polariser. (Electronically, it is more accurate to adjust to a minimum output than to a maximum output.) The angle on the scale is recorded.

The sample tube is now filled with a solution of the compound under investigation, and the analyser is turned until the output from the photocell is again a minimum. The new angle on the scale is recorded. The rotation caused by the compound is the difference between the two measured angles.

The actual rotation depends on four factors – the optical path length (the distance through the solution), the concentration of the compound in the solution, the wavelength of light used, and to a small extent, the temperature. Once these have all been allowed for, the intrinsic rotating power of the compound can be compared with that of others.

Optical isomerism arises from the inherent asymmetry of the molecules that make up each isomer. If a molecule has no plane or centre of symmetry, it is called a **chiral** (pronounced 'kai-ral', from the Greek word meaning 'handed') molecule. Chiral molecules are not identical to each other, but are related as a mirror image is to its object.

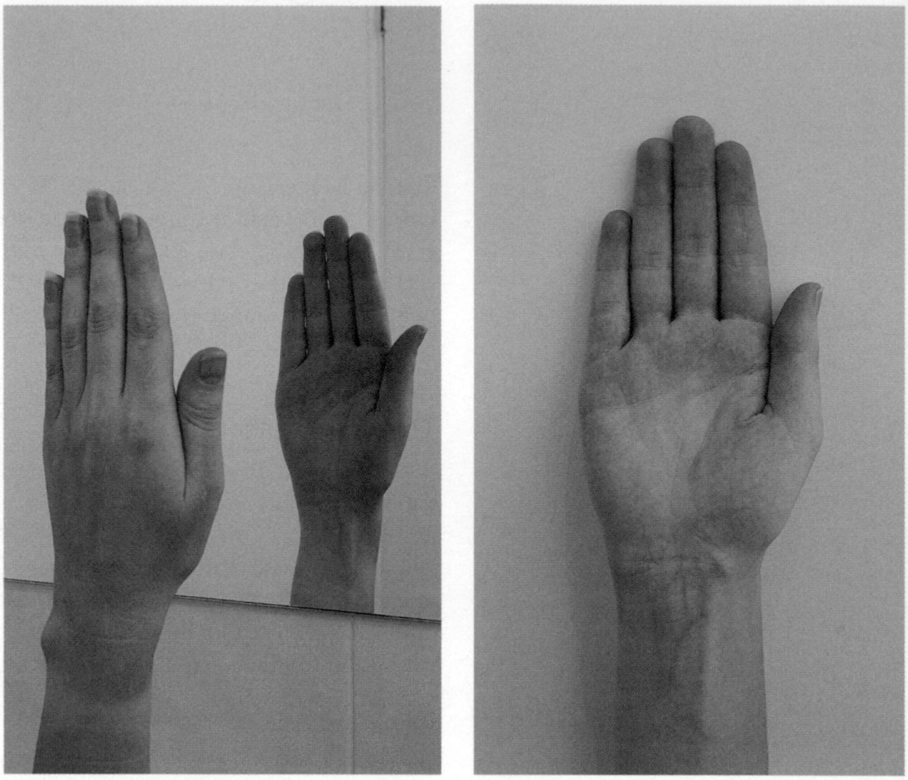

■ **Figure 22.16**
Your hands are chiral. They are not superimposable on each other, or on their mirror images, but one is superimposable on the mirror image of the other. Try this experiment. Hold your left hand, outstretched and palm away from you, in front of a mirror. Now hold your right hand, outstretched and palm towards you, next to the mirror. The image of your left hand now looks identical to (is superimposable on) your real right hand!

The most common cases of chirality occur when a molecule contains a tetrahedral carbon atom surrounded by four different groups of atoms. Such atoms are termed chiral carbon atoms, or chiral centres. There are two ways of arranging the four groups around a chiral tetrahedral atom, and these two arrangements produce molecules that are mirror images of each other (see Figure 12.17).

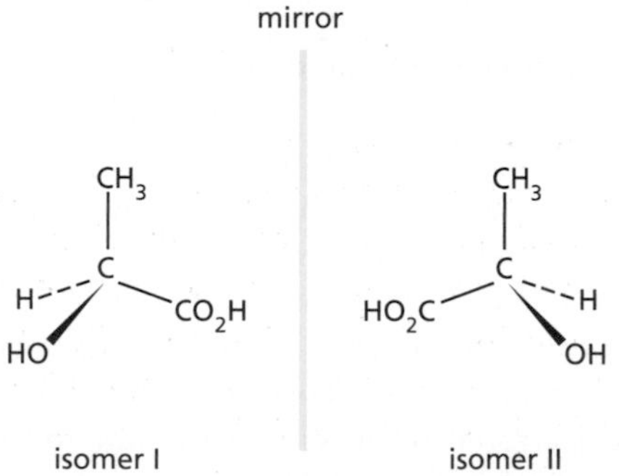

■ **Figure 22.17**
The two molecules are optical isomers – mirror images of each other.

Optical isomerism most commonly occurs when a carbon atom has four different atoms or groups attached to it.

Not all chiral molecules contain an atom attached to four different groups. The panel opposite illustrates some of these. It also includes two examples of compounds we might have expected to be chiral, but are not because their molecules contain a plane of symmetry.

Some unusual molecules that are chiral, and some that are not!

Evidence that it is the molecule as a whole that has to be chiral, rather than any identifiable atom within it, comes from the two contrasting cases of the allenes and the dibromobutanes.

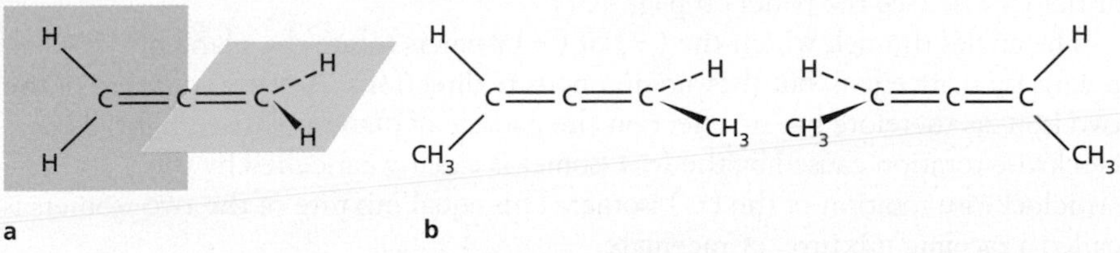

■ **Figure 22.18**
a Allene and b 1,3-dimethylallene.

Allenes

Allene is the common name of prop-1,2-diene. The two double bonds share the central carbon atom, and the planes passing through the two CH_2 groups are at right angles (see Figure 22.18**a**).

Allene itself is not chiral. The plane passing through one CH_2 group exactly bisects the other one, and the two sides are exactly the same as each other. In other words, it has a plane of symmetry (in fact, it has two planes of symmetry, at right angles to each other).

1,3-dimethylallene, however, has no plane of symmetry. It is a chiral molecule in which the groups can be arranged in two ways, one way producing the non-superimposable mirror image of the other (see Figure 22.18**b**). They both rotate the plane of polarised light, in equal but opposite directions.

Dibromobutanes

2,3-Dibromobutane contains two tetrahedral carbon atoms (carbons 2 and 3), each of which is surrounded by four different groups. These groups can be arranged in two different ways around the atom. There are therefore four possible arrangements:

2-*R*, 3-*R* 2-*R*, 3-*S* 2-*S*, 3-*R* 2-*S*, 3-*S*

■ **Figure 22.19**
Optical isomers of 2,3-dibromobutane.

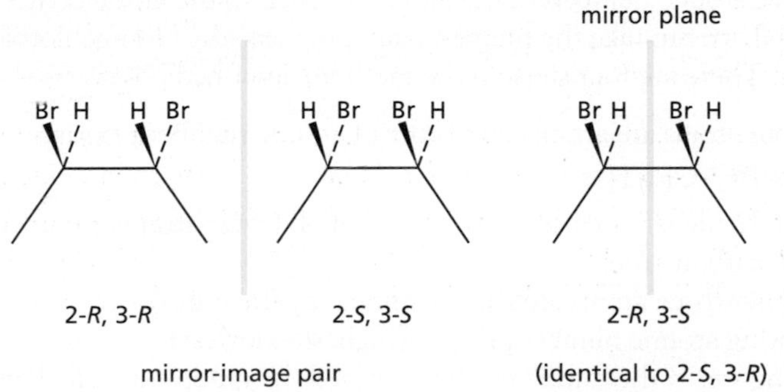

(See the panel overleaf for the *R*/*S* convention). The first and the fourth of these are chiral molecules, non-superimposable mirror images of each other. The second and third, however, are one and the same molecule. There is a mirror plane within the molecule. One end is the mirror image of the other, so overall the molecule is not chiral, and it has no effect on polarised light.

Double ring systems

Some double ring systems show a similar stereochemical relationship to that of the allenes. Compounds **A** and **B** in Figure 22.20 are mirror images. Single ring systems also show some interesting stereochemistry. There are three 1,2-dichloro-cyclopropanes (**C**, **D** and **E**). Compound **C**, *cis*-1,2-dichlorocyclopropane, has a plane of symmetry, as shown. Neither of compounds **D** and **E** (both called *trans*-1,2-dichlorocyclopropane) has a plane of symmetry. They are chiral mirror images of each other.

■ **Figure 22.20**
Stereoisomers of ring compounds.

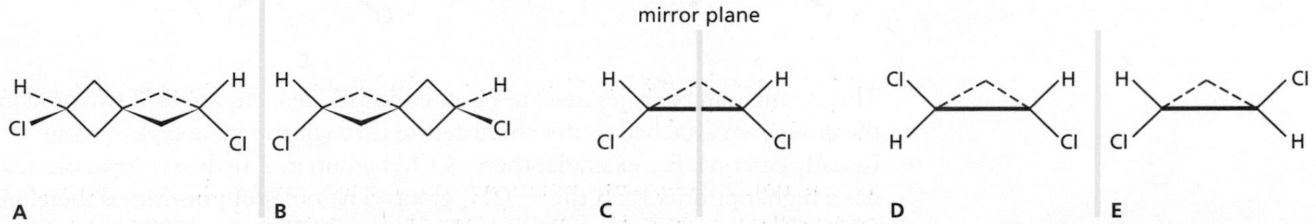

Two molecules that are non-superimposable mirror images of each other are called **enantiomers**. They are distinguished by the prefixes (+)- or (−)-, or the letters *d*- and *l*-, in front of their names. The letters derive from the Latin words *dextro* and *laevo*, meaning right and left, respectively. This relates to the direction in which each isomer rotates the plane of polarised light – either clockwise or anticlockwise (see the panel on page 407).

The angles through which the (+) or (−) isomers rotate the plane of polarisation are equal, but they are in opposite directions. A 50:50 mixture of the two isomers therefore has no effect on the passage of plane-polarised light. The clockwise rotation caused by the (+) isomer is exactly cancelled by the anticlockwise rotation of the (−) isomer. This equal mixture of the two isomers is called a **racemic mixture**, or **racemate**.

Describing optical isomers: the Cahn–Ingold–Prelog notation

For many years chemists had no idea of the relationship between the direction in which a particular isomer rotated polarised light, and the actual configuration around the chiral carbon atom. For example, no one knew whether it was isomer I of 2-hydroxypropanoic acid (Figure 22.17) that had a rotation of +3.3°, or isomer II. Then, in 1949, the Dutch chemist Johannes Bijvoet used a special type of X-ray analysis to determine the absolute configuration of a chiral compound. Since that time, chemists have been able to relate measured rotations to the exact spatial arrangements of the groups around an atom that they can represent on paper. In this way it was deduced that isomer I was the clockwise (+)-rotating isomer.

Once chemists were able to determine the absolute configuration of a chiral compound, a concise method of describing this configuration was needed, without having to draw its stereochemical formula every time. The three chemists R.S. Cahn, Christopher Ingold and Vladimir Prelog suggested a method of priorities based on the relative atomic numbers of the atoms attached to the chiral centre. To illustrate the method, we can take the simplest pair of isomers, the bromochloroiodomethanes, CHBrClI. There are four steps to the method, described below.

1 The four atoms are arranged in order of atomic number priority:

   I > Br > Cl > H

2 The molecule is visualised with the atom of lowest atomic number behind the central carbon atom.

3 The other three atoms are then 'scanned' by the mind's eye in order of decreasing atomic number priority (highest → lowest).

4 If during step 3 the mind's eye has scanned the groups in a clockwise direction, the configuration is specified as *R* (from the Latin *rectus* meaning right). If the groups were scanned in an anticlockwise direction, the configuration is specified as *S* (from the Latin *sinister* meaning left).

Figure 22.21 shows how this applies to the bromochloroiodomethanes.

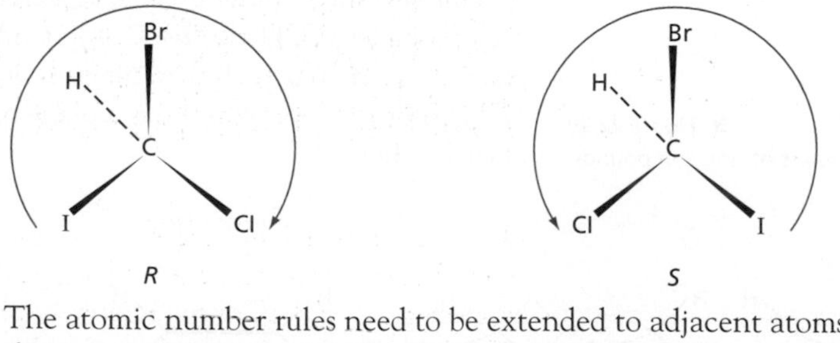

■ **Figure 22.21** Describing bromochloroiodomethanes as *R* and *S* isomers.

The atomic number rules need to be extended to adjacent atoms if two or more of the groups are attached to the chiral centre through the same type of atom (usually carbon). For example, the $—CO_2H$ group in 2-hydroxypropanoic acid has a higher priority than the $—CH_3$ group. The order of priorities is therefore $OH > CO_2H > CH_3 > H$, and isomer I of Figure 22.17 (page 408) can be assigned the configuration *S*-(+)-2-hydroxypropanoic acid.

Sugars, amino acids and most other biologically important molecules are chiral. Usually only one of the two isomers ((+) or (−)) is made in nature (but see the panel below for some exceptions). Many pharmaceutical drugs that are designed to interact with living systems also have chiral centres. Enzymes will usually react with only one of the mirror-image isomers. However, when a compound is made in the laboratory, it is usually produced as the racemic mixture of the two mirror-image forms. Not only is this a potential waste of 50% of an expensive product, it can also cause unforeseen problems if the mixture is administered (see the panel on thalidomide overleaf). Much research effort is therefore put into producing drugs containing molecules of just one optical isomer. Optical purities of >99.5% are the aim.

Some naturally occurring enantiomers

The carvones

You will know the taste of spearmint, whether in the form of chewing gum or toothpaste. You may also know the taste of caraway seeds. The same compound, carvone, is responsible for both tastes (see Figure 22.22).

■ **Figure 22.22** Enantiomers of carvone.

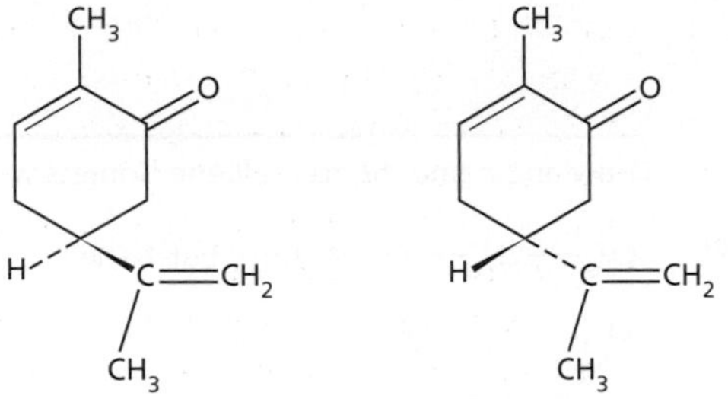

(–)-carvone (spearmint) (+)-carvone (caraway)

The lactic acids

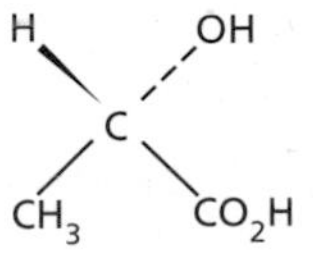

■ **Figure 22.23** (+)-lactic acid is produced in muscle.

The common name for 2-hydroxypropanoic acid is lactic acid. It derives its name from the Latin for milk (*lactis*). It is formed when milk is fermented by certain microorganisms. Unusually, the acid seems to be produced as a racemic mixture of the two enantiomers. This is in contrast to the lactic acid produced in our muscles when we exercise anaerobically. The product here is pure (+)-lactic acid (see Figure 22.23).

The amino acids

Except for glycine, all naturally occurring amino acids are chiral. In almost all cases, the amino acids have the same configuration around their chiral carbon atom (see Figure 22.24).

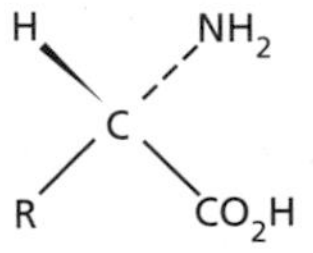

■ **Figure 22.24** Amino acids are chiral molecules (R = side chain, see Chapter 30).

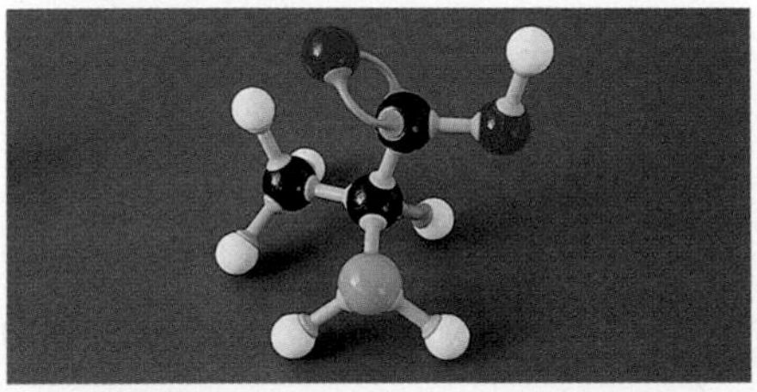

(+)-alanine

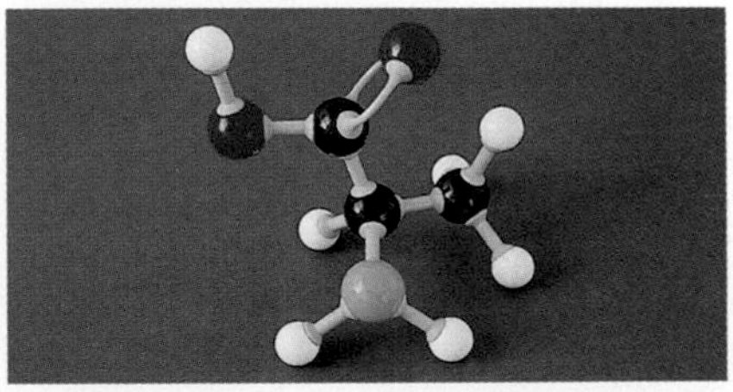

(−)-alanine

The cell walls of some bacteria consist of chains of mucopeptides. These are co-polymers of amino acids and sugars. Some of the amino acids used by bacteria, however, are of the opposite configuration to those that normally occur in nature. This has turned out to be their Achilles heel. The antibiotic penicillin interferes with the synthesis of these mucopeptides, by competing with amino acids of the opposite configuration for the active sites of enzymes. The inhibited enzymes can no longer carry out their function of repairing damaged cell walls or synthesising new ones, so the bacteria die.

The thalidomide tragedy

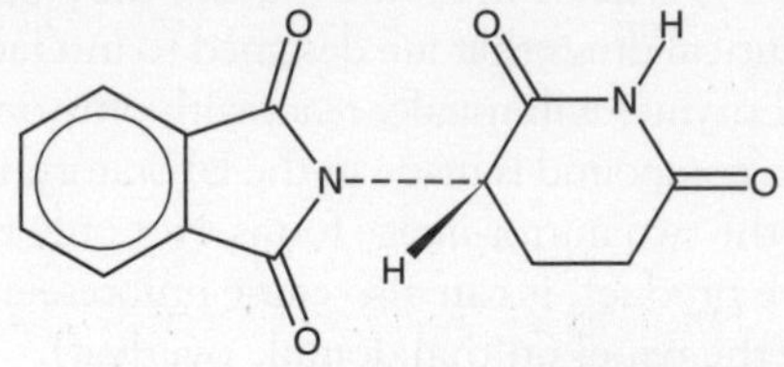

■ **Figure 22.25** Thalidomide.

There have been many tranquilisers developed whose molecules have been based on the barbituric acid ring system. One of these was thalidomide, shown in Figure 22.25. It seemed to be a very successful, non-toxic sedative with no known side-effects. In the early 1960s, however, it was discovered that women who had been prescribed the sedative to treat the nausea ('morning sickness') in early pregnancy had produced badly deformed babies. The drug was immediately withdrawn.

Research showed that whilst the (+) isomer was an effective and safe tranquiliser, the (−) isomer was the culprit in damaging the fetus. Originally it was thought that the problem had been due to contamination by the (−) isomer in some production batches, but subsequent research has shown that racemisation of the pure (+) isomer occurs rapidly as soon as it enters the blood stream. Within 10 minutes racemisation is over 50% complete. Thalidomide will never be marketed again, although analogues that are chirally stable are being investigated as possible anti-inflammatory drugs.

Example 1 Draw and name the four alkene isomers with the molecular formula C_4H_8.

Answer

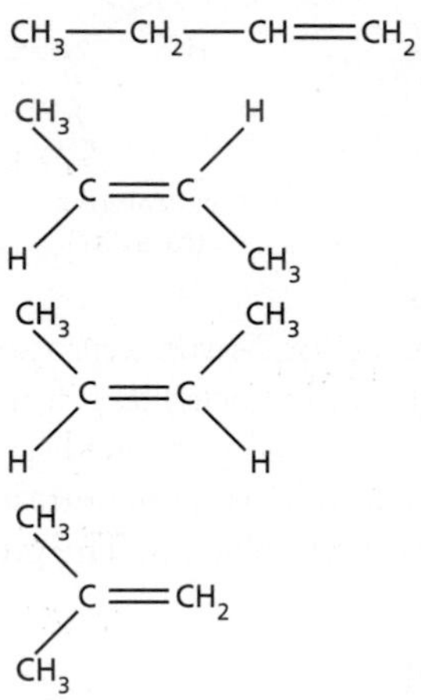

but-1-ene

trans-but-2-ene

cis-but-2-ene

2-methylpropene

■ **Figure 22.26**

Example 2 There are four structural or positional isomers of C_4H_9OH. Draw them, name them, and specify which one of them can exist as a pair of optical isomers (that is, which one is chiral).

Answer

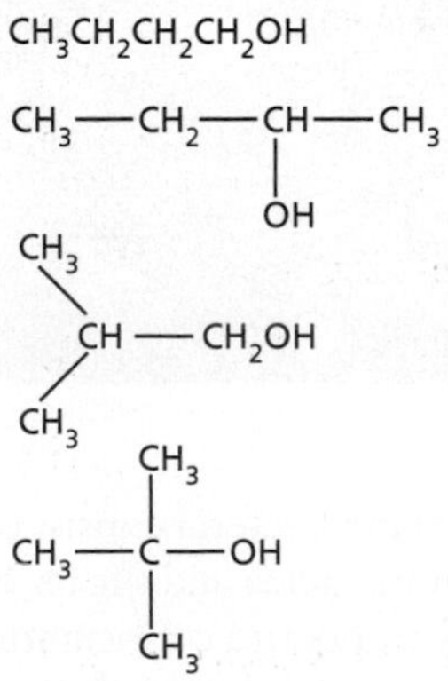

butan-1-ol

butan-2-ol (this is chiral)

2-methylpropan-1-ol

2-methylpropan-2-ol

■ **Figure 22.27**

Further practice

1 There are six alkenes with the molecular formula C_5H_{10}. Draw their skeletal formulae and name them.

2 Draw the skeletal formulae of the four positional isomers of dichloropropane ($C_3H_6Cl_2$). Name them, and specify which contain a chiral centre.

22.7 The different types of organic reaction

Most organic compounds can be represented as:

R—Fg

where Fg represents the functional group, and R represents the rest of the molecule. Most organic reactions are those of functional groups. By definition, the rest of the molecule does not change under the conditions of the reaction that changes the functional group (if it did change, it would have become a functional group itself!). Many functional groups have characteristic reactions, which are covered in detail in the following chapters.

Table 22.8 lists the types of organic reaction we shall be covering.

	Reaction type	Functional group	Example
1	Radical substitution	Alkane	$CH_4 + Cl_2 \longrightarrow CH_3Cl + HCl$
2	Electrophilic substitution	Arene	(benzene) $+ HNO_3 \longrightarrow$ (nitrobenzene, NO_2) $+ H_2O$
3	Nucleophilic substitution	Halogenoalkane	$OH^- + CH_3Br \longrightarrow CH_3OH + Br^-$
		Alcohol	$CH_3OH + HBr \longrightarrow CH_3Br + H_2O$
4	Electrophilic addition	Alkene	$CH_2\!=\!CH_2 + Br_2 \longrightarrow BrCH_2 - CH_2Br$
5	Nucleophilic addition	Aldehyde or ketone (carbonyl)	$(CH_3)_2C\!=\!O + HCN \longrightarrow (CH_3)_2C(OH)(CN)$
6	Hydrolysis	Ester	$CH_3CO_2CH_3 + H_2O \longrightarrow CH_3CO_2H + CH_3OH$
		Amide	$CH_3CONH_2 + H_2O \longrightarrow CH_3CO_2H + NH_3$
7	Condensation	Aldehyde or ketone (carbonyl)	$(CH_3)_2C\!=\!O + H_2N\!-\!R \longrightarrow (CH_3)_2C\!=\!N\!-\!R + H_2O$
8	Elimination	Halogenoalkane	$CH_3CH_2Br + OH^- \longrightarrow CH_2\!=\!CH_2 + Br^- + H_2O$
		Alcohol	$CH_3CH_2OH \longrightarrow CH_2\!=\!CH_2 + H_2O$
9	Reduction	Aldehyde or ketone (carbonyl)	$(CH_3)_2C\!=\!O + 2[H] \longrightarrow (CH_3)_2CH(OH)$
10	Oxidation	Alcohol	$CH_3OH + [O] \longrightarrow CH_2O + H_2O$
		Aldehyde	$CH_2O + [O] \longrightarrow HCO_2H$

■ **Table 22.8** Reaction types in organic chemistry. [O] means an oxygen atom is given by the oxidising agent and [H] means a hydrogen atom is given by the reducing agent.

The meanings of some of the terms used in Table 22.8 are given overleaf.

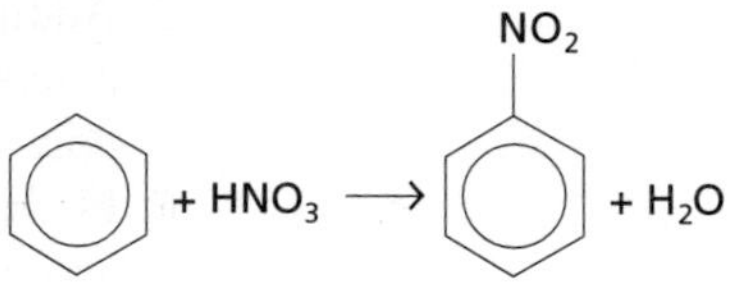

1 **Radical** – this is an atom or group of atoms that has an unpaired electron (that is, it has an odd number of electrons). Radicals (sometimes called free radicals) are highly reactive, and are intermediates in many of the reactions of alkanes. They have the same number of protons as electrons, so they are electrically neutral (unlike cations or anions).
2 **Electrophile** – this is an atom or group of atoms that reacts with electron-rich centres in other molecules. Electrophiles possess either a positive charge or an empty orbital in their valence shell, or they contain a polar bond producing an atom with a partial charge, shown as $\delta+$. Electrophiles are also Lewis acids (see Chapter 6, page 135).
3 **Nucleophile** – this is an atom or group of atoms that reacts with electron-deficient centres in molecules (such as the $C^{\delta+}$ in $C^{\delta+}$—$Cl^{\delta-}$ compounds). All nucleophiles possess a lone pair of electrons, and many are anions. Nucleophiles are called ligands when they react with transition metals (see Chapter 19), and are also Lewis bases (see Chapter 6, page 135).
4 **Substitution** – this is a reaction in which one atom or group replaces another atom or group in a molecule.
5 **Addition** – this is a reaction in which an organic molecule (usually containing a double bond) reacts with another molecule to give only one product: $A + B \rightarrow C$ only.
6 **Hydrolysis** – this is a reaction in which a molecule is split into two by the action of water (often helped by OH^- or H^+ as catalysts).
7 **Condensation** – this is the opposite of hydrolysis. Two molecules come together to form a bigger molecule, with the elimination of a molecule of water (or other small molecule such as HCl).
8 **Elimination** – this is a reaction that forms an alkene by the removal of a molecule of H_2O from an alcohol, or a molecule of HCl from a chloroalkane.
9 **Reduction and oxidation** – these have their usual meanings. Reduction is a reaction in which the sum of the oxidation numbers of the atoms in the functional group decreases. Oxidation is the reverse.

Example Classify the following reactions into one (or more!) of the following six categories: substitution, addition, hydrolysis, reduction, condensation, oxidation.

a $CH_3CONH_2 + H_2O \rightarrow CH_3CO_2H + NH_3$
b $(CH_3)_2C{=}O + H_2N{-}NH_2 \rightarrow (CH_3)_2C{=}N{-}NH_2 + H_2O$
c $CH_3CHO + 2[H] \rightarrow CH_3CH_2OH$

Answer **a** This is the splitting of a molecule into two by the action of water – a hydrolysis reaction. It could also be considered as a substitution reaction, in which —NH_2 is replaced by —OH.
b This reaction is the reverse process – it is a condensation reaction.
c This is both an addition reaction and a reduction.

Further practice The scheme in Figure 22.28 is an outline of a synthesis of the local anaesthetic benzocaine, starting from 4-nitromethylbenzene. Classify each of the reactions I, II and III into one of the above six categories.

■ **Figure 22.28**

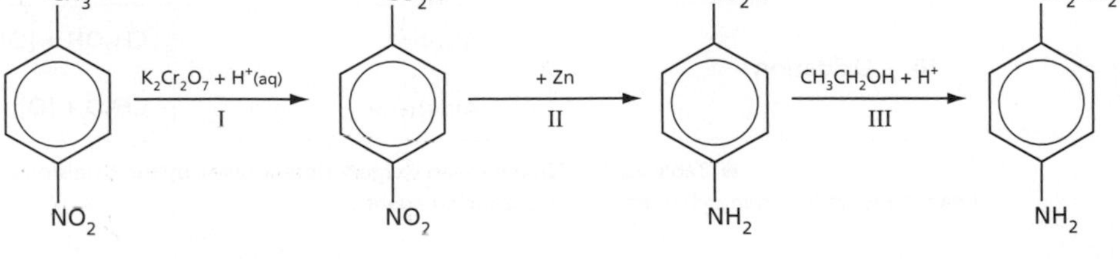

22.8 The different ways of breaking bonds, and how they are represented

Chemical reactions involve the breaking of bonds and the formation of new bonds. In the chapters that follow, we shall be looking at the mechanisms of some organic reactions. By **mechanism** we mean a detailed description of which bonds break, and which form, and in what order these processes occur. A mechanism also includes a description of how any catalyst that might be involved in a reaction works. There are three different ways in which a covalent bond to a carbon atom may break. Each way produces a different carbon species.

Homolysis

Bond homolysis, often called **homolytic fission**, results in carbon radicals (see Figure 22.29). Homolysis (the Greek word *homo* means the same) is the splitting of a bond giving an equal share of bonding electrons to each particle. In Figure 22.29, the carbon and hydrogen atoms take one electron each. The curly arrows with only half a head ('fish-hook arrows') represent how each electron in the bond moves when the bond breaks. The carbon atom in the methyl radical $CH_3\bullet$ has only seven electrons around it. Methyl radicals are highly reactive, and readily form bonds with other atoms or molecules to regain their full octet of electrons.

■ **Figure 22.29** Bond homolysis.

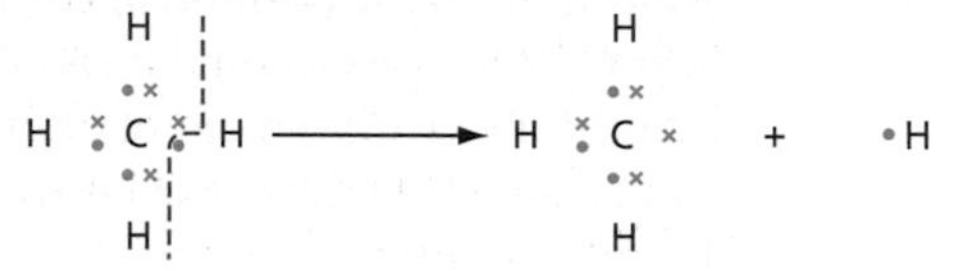

or

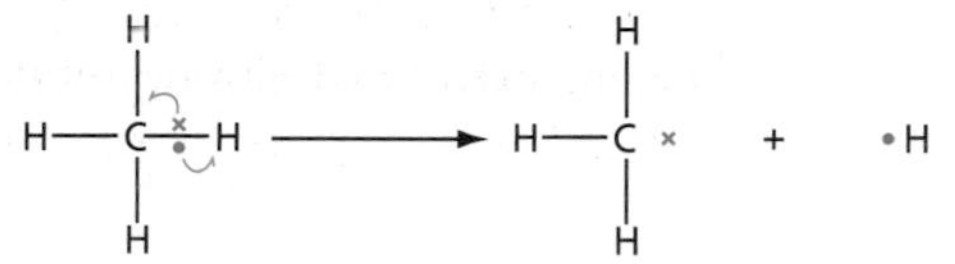

Bond heterolysis forming carbocations

Heterolysis (the Greek *hetero* means other, or different) involves the splitting of a bond giving an unequal share of bonding electrons to each particle. In Figure 22.30, the chlorine atom has taken both the bonding electrons, leaving the carbon atom with none. The curly arrow describes the movement of the bonded pair of electrons to the chlorine atom. Chlorine, being more electronegative than carbon, has already attracted the bonding electrons partially, forming a bond dipole $^{\delta+}C\div Cl^{\delta-}$ (see Chapter 3, page 66). This movement of the electron pair results in the carbon atom in the methyl **carbocation** having only six electrons in its outer shell, and a single positive charge. It is a strong electrophile.

■ **Figure 22.30** Bond heterolysis forming a carbocation.

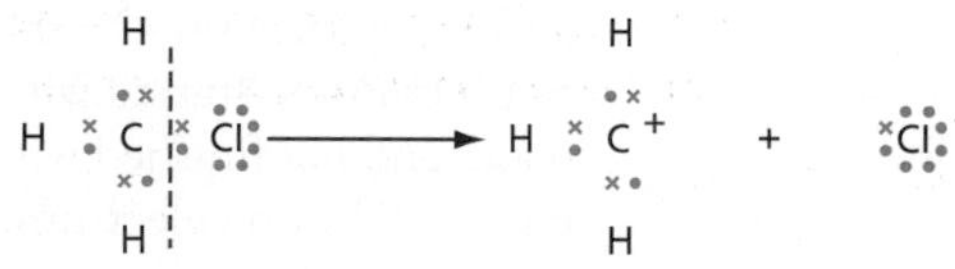

or

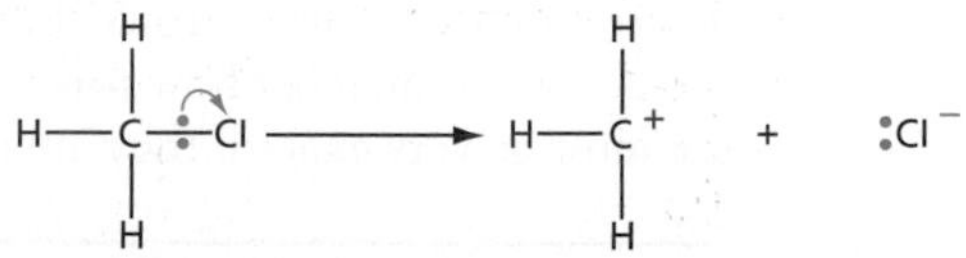

Bond heterolysis forming carbanions

This method of splitting a bond to carbon is much less common than that forming carbocations. Carbon is quite electropositive. Only when it is bonded to an even more electropositive element will its bonds split to form **carbanions**. Metals are highly electropositive, and some of them form covalent bonds with carbon. Methyl lithium is the simplest example (see Figure 22.31).

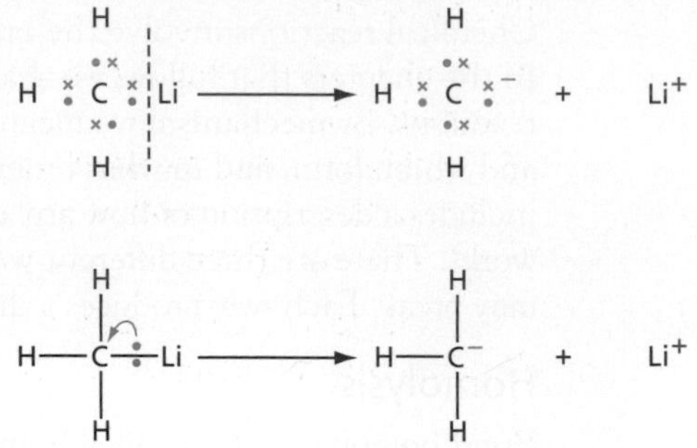

■ **Figure 22.31** Bond heterolysis forming a carbanion.

The methyl carbanion formed contains a full octet of electrons in the outer shell of carbon. Its strong negative charge makes it a highly reactive nucleophile. It will react with virtually everything that contains a $\delta+$ atom. The curly arrow describes the movement of the bonded pair of electrons to the carbon atom.

The use of curly arrows

The symbol ⌒ is a useful shorthand which represents the movement of a pair of electrons from the position at the tail of the arrow to the position at the head. Its meaning is very exact, and it should not be used to imply anything more, or less, than the above movement of electrons.

An example of the use of curly arrows is the reaction that occurs when hydrogen chloride gas dissolves in water:

$$HCl + H_2O \rightarrow H_3O^+ + Cl^-$$

The movement of electrons in this reaction may be represented as follows:

$$H-\ddot{O}(H): \curvearrowright H-Cl \rightarrow H-\overset{+}{O}(H)-H + :Cl^-$$

This equation is a highly condensed way of saying:

> A pair of electrons moves from a lone pair of oxygen in water to the space between that oxygen and the hydrogen of the HCl molecule, to form a new O—H bond. At the same time, the two electrons in the H—Cl bond move towards the chlorine atom and become a lone pair on that atom. Since oxygen started off with 'full use' of a lone pair, and has finished up with only a half-share of the electron pair making up the O—H bond, its charge has decreased by one electronic unit (0 to +1). Since the hydrogen being transferred started with a half-share of the electron pair making up the H—Cl bond, and has finished with a half-share of the electron pair making up the new O—H bond, its charge has remained the same (zero). Since chlorine started with a half-share of the electron pair in the H—Cl bond, and has finished with full use of its own lone pair, its charge has increased by one electronic unit (0 to −1).

From the length of this paragraph you will appreciate how much shorter is the description using curly arrows! But every time you write out mechanisms using them, try to remember how this paragraph was translated into curly arrows. Their use must be very exact if they are to describe electron movements correctly.

Further practice Use curly arrows to describe what happens to the electrons when (ionic) ammonium chloride is made from ammonia and hydrogen chloride:

$$NH_3 + HCl \rightarrow NH_4Cl$$

22.9 Homologous series

Compounds containing the same functional group often form **homologous series.** In such a series:

- the molecular formulae of adjacent members of the series differ by a fixed unit (usually CH_2)
- the physical properties, such as boiling point and density, vary regularly from one compound to another
- the molecular formulae of members of the series fit the same general formula.

Two examples of homologous series are shown in Table 22.9. Their boiling points show a regular trend, as shown in Figure 22.32.

■ **Table 22.9** Some properties of the lower alkanes and alcohols.

Alkanes (general formula C_nH_{2n+2})			Alcohols (general formula $C_nH_{2n+1}OH$)		
Name	**Formula**	**Boiling point/°C**	**Name**	**Formula**	**Boiling point/°C**
Methane	CH_4	−164	Methanol	CH_3OH	65
Ethane	C_2H_6	−89	Ethanol	C_2H_5OH	78
Propane	C_3H_8	−42	Propanol	C_3H_7OH	97
Butane	C_4H_{10}	0	Butanol	C_4H_9OH	118
Pentane	C_5H_{12}	36	Pentanol	$C_5H_{11}OH$	138

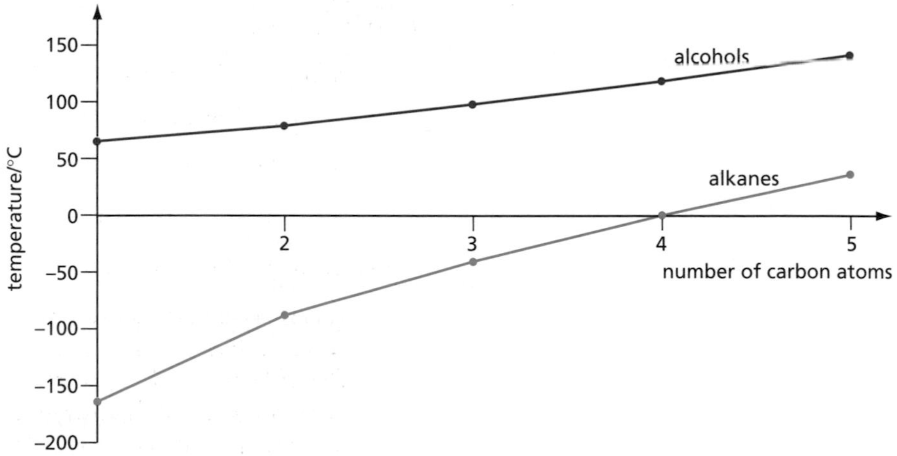

■ **Figure 22.32** Boiling points of alkanes and alcohols.

Example The first and third members of a homologous series have the molecular formulae C_2H_2 and C_4H_6.

a What is the formula of the fourth member of the series?

b What is the general formula of the series?

Answer **a** The difference between the two molecular formulae is C_2H_4, which is $2 \times CH_2$. The fourth member will therefore have the formula $C_4H_6 + CH_2$, which is C_5H_8.

b The general formula is C_nH_{2n-2}.

Further practice The first two members of the following homologous series have the formulae stated:

series A: CH_2O, C_2H_4O

series B: C_2H_3N, C_3H_5N

series C: C_6H_6, C_7H_8

For each series, state:

1 the molecular formula of the fifth member

2 the general formula.

22.10 Calculating yields in organic reactions

When carried out in the laboratory, organic reactions do not always go to completion. This may be due either to the reaction reaching equilibrium, or to it being so slow that it would take an age to complete. Some organic reactions compete with each other (see page 396), and so by-products are formed along with the major product. Even if a reaction goes to completion, that is, every molecule of starting material is converted into a molecule of product, the recovered yield will not be 100%. This is due to material being lost during the purification procedures. For example, if a solid has been purified by recrystallisation, some will inevitably be left in the cold saturated solution after the crystals have been filtered off. If a liquid has been distilled, a small amount of it will remain in the distillation flask, or on the sides of the condenser.

The **percentage yield** is a comparison of the actual yield with the **theoretical yield** – the yield that might have been expected if the reaction took place according to the stoichiometric equation, and there was no loss of product in the purification stages.

$$\text{Percentage yield} = \frac{\text{actual yield} \times 100\%}{\text{theoretically possible yield}}$$

For well designed reactions, in the hands of an able experimental chemist, yields of >95% are possible. More usually, yields of 60–80% are obtained.

Example The reaction of 10.0 g of benzoyl chloride with concentrated ammonia solution produced 5.63 g of pure recrystallised benzamide. Calculate the percentage yield.

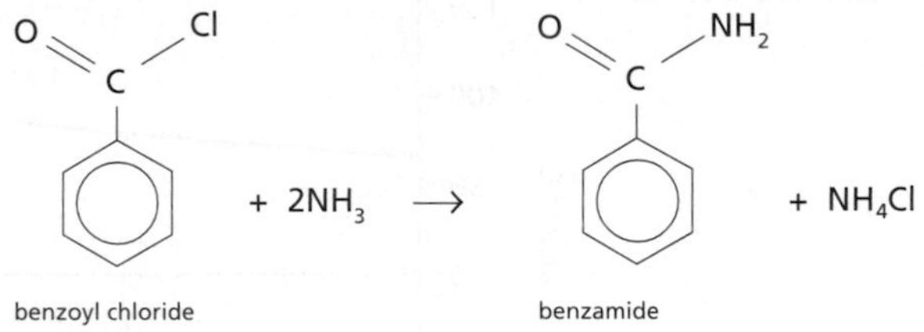

Answer We first calculate the number of moles of benzoyl chloride used (see Chapter 1, page 7).

$$M_r(C_6H_5COCl) = (7 \times 12) + (5 \times 1) + 16 + 35.5 = 140.5$$

$$n = \frac{m}{M}$$

$$n(C_6H_5COCl) = \frac{10.0}{140.5} = 0.071\,17\,\text{mol}$$

From the equation, 1 mol of benzoyl chloride produces 1 mol of benzamide. Therefore:

$$n(C_6H_5CONH_2) = 0.071\,17\,\text{mol}$$

$$m = n \times M$$

$$M_r(C_6H_5CONH_2) = (7 \times 12) + (7 \times 1) + 16 + 14 = 121$$

so

$$m(C_6H_5CONH_2) = 0.071\,17 \times 121 = \mathbf{8.61\,g}$$

and

$$\text{percentage yield} = \frac{\text{actual yield} \times 100\%}{\text{theoretical yield}} = \frac{5.63 \times 100\%}{8.61} = \mathbf{65.4\%}$$

Further practice Calculate the percentage yields in the following cases.

1 On oxidation, 5.0 g of ethanol produced 4.5 g of ethanoic acid:

$$C_2H_5OH + 2[O] \rightarrow CH_3CO_2H + H_2O$$

2 Reacting 10.0 g of propanol with an excess of hydrogen bromide produced 12.5 g of bromopropane:

$$C_3H_7OH + HBr \rightarrow C_3H_7Br + H_2O$$

Summary

- Carbon forms strong bonds to itself and to many other atoms. This allows organic compounds to be kinetically stable, due to the high activation barrier to reaction, even though they may be thermodynamically unstable.
- There are six types of formula used to represent organic compounds – **empirical**, **molecular**, **structural**, **displayed**, **stereochemical** and **skeletal**.
- All organic compounds can be named systematically, using a logical structure based on the number of carbon atoms in their longest chain.
- There are five types of **isomerism** shown by organic compounds – **chain**, **positional**, **functional group**, **geometrical** and **optical**.
- Most organic reactions can be classified as one of ten general types.
- Many organic reactions do not give a 100% yield of product. The **percentage yield** is a measure of how efficient a reaction is.

23 *Alkanes*

Functional group:

C—C—H

This chapter looks at the properties and reactions of the simplest of the homologous series, the alkanes. It describes their extraction from crude oil, how they are transformed into fuels and feedstocks for the chemical industry, and some of their reactions with oxygen, chlorine and bromine. Their characteristic reaction of radical substitution is explained in detail. Finally, their infrared spectra are described.

The chapter is divided into the following sections.

23.1 Introduction

The **alkanes** are the simplest of the homologous series found in organic chemistry. They contain only two types of bond, both strong and fairly non-polar (see Table 23.1).

■ **Table 23.1**
Carbon–carbon and carbon–hydrogen bonds are strong and non-polar.

Bond	Bond enthalpy/kJ mol^{-1}	Polarity (difference in electronegativity)
C—C	346	0
C—H	413	0.35 ($H^{\delta+}$—$C^{\delta-}$)

Consequently, they are not particularly reactive. Their old name (the 'paraffins') reflected this 'little (parum) affinity'. What few reactions they have, however, are of immense importance to us. They constitute the major part of crude oil, and so are the progenitors of virtually every other organic compound made industrially. Apart from their utility as a feedstock for the chemical industry, their major use is as fuels in internal combustion engines, jet engines and power stations. We shall be looking at their role in combustion reactions in detail later in this chapter.

The straight-chain alkenes form a homologous series with the general formula C_nH_{2n+2}. As the number of carbon atoms in the molecule increases, the van der Waals attractions between the molecules become stronger (see Chapter 3, page 79). This has a clear effect on the boiling points and the viscosities of members of the series (see Figure 23.1 and Table 24.1, page 440).

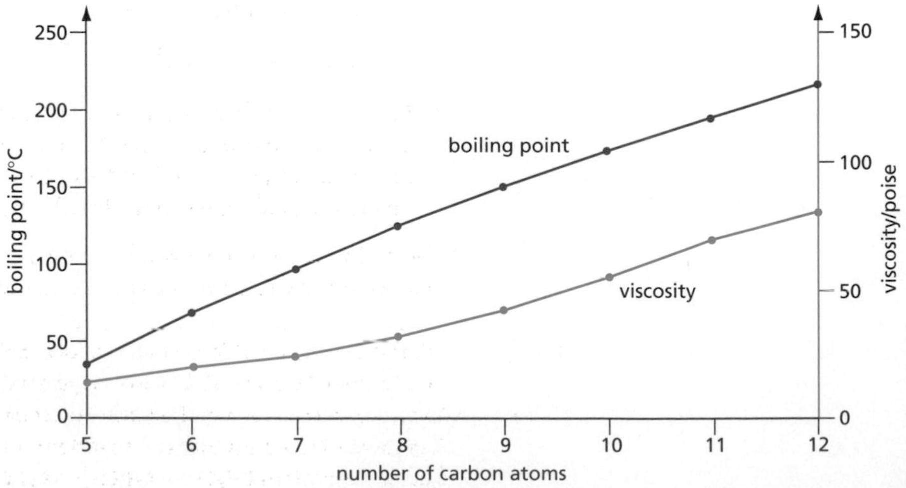

■ **Figure 23.1**
The boiling points and viscosities of the alkanes.

■ Figure 23.2 Space filling and ball-and-stick models of ethane, C_2H_6 (**a** and **b**) and butane, C_4H_{10} (**c**–**f**). The space filling models show the overall shape of the molecule but the ball-and-stick models show the position of the atoms more clearly. Note that **e** and **f** are identical to **c** and **d**, as they are formed by free rotation of the central C—C bond.

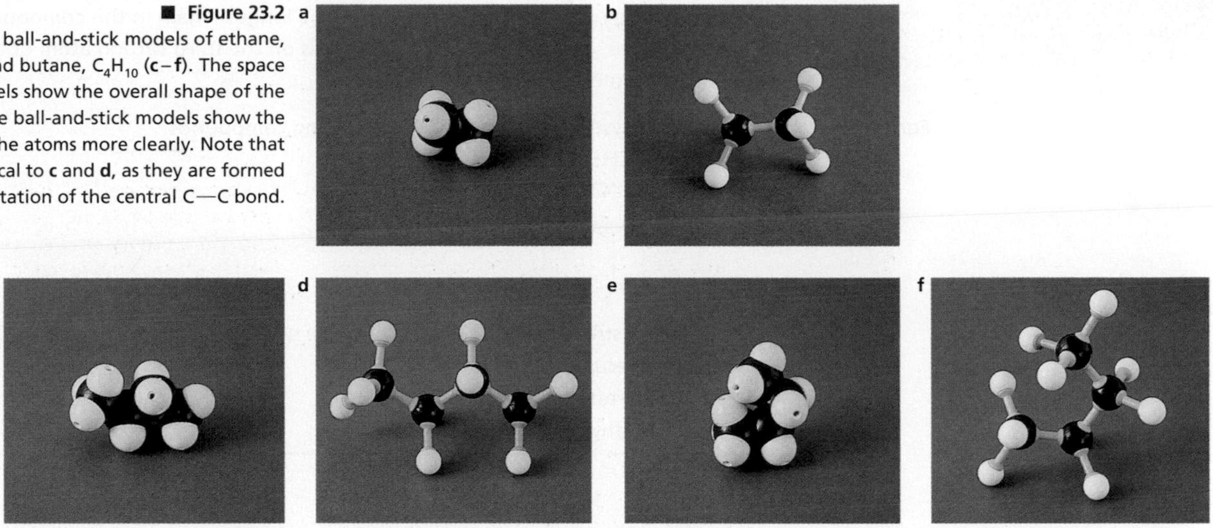

23.2 Isomerism and nomenclature

As was mentioned in Chapter 22, the first step in naming an alkane is to identify the longest carbon chain, which gives the stem. Side branches are then identified by prefixes. Each prefix is derived from the stem name denoting the number of carbon atoms it contains, and is preceded by a number denoting the position it occupies on the longest carbon chain. Two examples will make this clear:

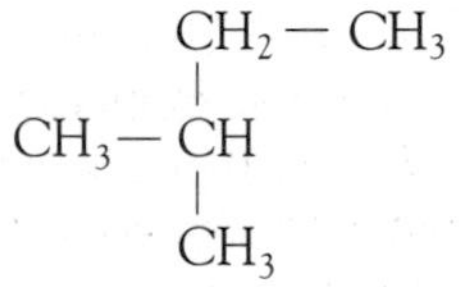

$$\begin{array}{c} CH_2-CH_3 \\ | \\ CH_3-CH \\ | \\ CH_3 \end{array}$$

- The longest carbon chain contains four carbon atoms, so the alkane is a derivative of **butane**.
- There is just one side branch, containing one carbon atom (and hence called **methyl**).
- The branch is joined to the second carbon atom of the butane chain.
- The name of the alkane is **2-methylbutane**.

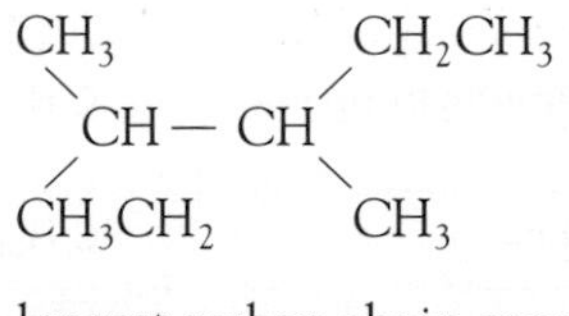

$$\begin{array}{cc} CH_3 & CH_2CH_3 \\ \diagdown CH-CH \diagup \\ CH_3CH_2 \diagup \quad \diagdown CH_3 \end{array}$$

- The longest carbon chain consists of six carbon atoms (**hexane**).
- There are two one-carbon side branches (two methyl groups, **dimethyl**)
- These are on positions 3 and 4 of the hexane chain.
- The name of alkane is **3,4-dimethylhexane**.

The major isomerism shown by the alkanes is chain isomerism, although some more complicated structures are in theory capable of existing as optical isomers.

Example What is the alkane of lowest M_r (relative molecular mass) that contains a chiral centre (see Chapter 22, page 408)? Give its systematic name and molecular formula.

Answer To be a chiral centre, a carbon atom must have four different groups attached to it. The four lowest M_r groups that are all different to each other are H, CH_3, C_2H_5 and C_3H_7. The alkane is therefore $CH_3CH_2-CH(CH_3)-CH_2CH_2CH_3$, whose molecular formula is C_7H_{16}.

Applying the rules listed on pages 402–3, the longest chain in this compound contains six carbon atoms, and the side group (methyl) is on the third carbon atom of the chain. Hence its name is **3-methylhexane.**

Further practice

1 Give the systematic names of the following compounds.

a $(CH_3)_2CH—CH_2—CH(CH_3)_2$

b $C_2H_5CH(CH_3)C_2H_5$

c

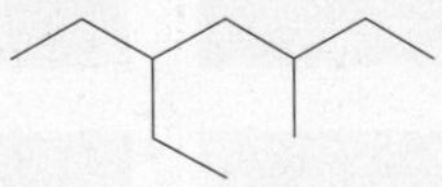

2 Draw the structural and skeletal formulae of the following compounds, and also give their molecular formulae.

a 2,2,3,3-tetramethylpentane

b 3,3-diethylhexane

As the number of carbon atoms in an alkane increases, so does the number of possible structural isomers. There are two isomers of C_4H_{10}, three of C_5H_{12}, and five of C_6H_{14} (see Table 23.2). By the time we reach $C_{10}H_{22}$, there are no fewer than 75 isomers, and for $C_{20}H_{42}$ the number of isomers has been calculated to be in excess of 300 000!

Structural (chain) isomers of alkanes have virtually the same chemical reactions. They differ slightly in two respects, however, as follows.

- The more branches an isomer has, the weaker is the van der Waals bonding between the molecules. This is because the branched molecule is more compact, with a smaller surface area of contact between molecules. The consequence of the weaker intermolecular forces is that the boiling points of branched isomers are *lower* than their straight-chain counterparts (see Table 23.2).
- The more branched isomers do not pre-ignite so easily in internal combustion engines. They are therefore much preferred for use in modern high compression petrol engines (see the panel opposite).

■ Table 23.2 The boiling points of some isomers of C_5H_{12} and C_6H_{14}.

Compound	**Molecular formula**	**Skeletal formula**	**Boiling point /°C**
Pentane	C_5H_{12}		36
2-Methylbutane	C_5H_{12}	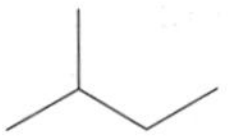	28
2,2-Dimethylpropane	C_5H_{12}		9
Hexane	C_6H_{14}		69
3-Methylpentane	C_6H_{14}	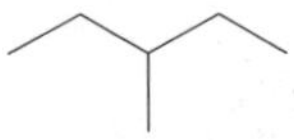	63
2-Methylpentane	C_6H_{14}	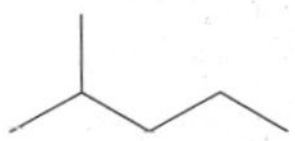	60
2,3-Dimethylbutane	C_6H_{14}	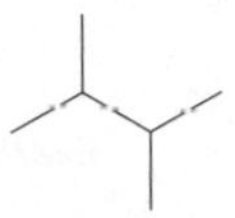	58
2,2-Dimethylbutane	C_6H_{14}	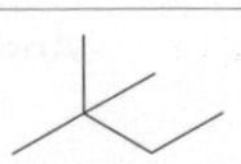	50

The petrol engine

The petrol engine usually used in motor transport today uses the **four-stroke cycle** developed by Nikolaus Otto in the nineteenth century. The key requirements of the fuel at various stages are as follows:

First downstroke of the piston

The fuel must be volatile enough to have formed a homogeneous fuel–air mixture in the carburettor, prior to being sucked into the cylinder through the inlet value. In a fuel-injection car, the first downstroke sucks in only air.

First upstroke

During this stage, the petrol–air mixture is compressed, usually to about one-tenth of its volume. This causes the mixture to heat up by several hundred degrees. It is important, however, that this self-heating (which is due to van der Waals bonds forming between the gas molecules as they are pushed closer together) does not cause pre-ignition of the fuel, before the piston has reached the top of the cylinder. In a fuel-injection car, the fuel is injected near the top of the upstroke.

Second downstroke

When the spark plug sparks, the fuel–air mixture ignites and expands, pushing the piston down the cylinder. The burning of the fuel must be smooth, but all over within about 0.005 s, which is the time it takes for the piston to travel down to the bottom of the cylinder.

Second upstroke

Ideally, the fuel will have completely burnt to form gaseous carbon dioxide and water. However, if there has been insufficient oxygen (air) drawn in with the fuel, or if it has not mixed completely, or if it has not had time to burn completely before being expelled from the cylinder, other substances will be produced.

Carbon (soot) can be a problem, blocking valve seatings and shorting the sparking plug, and causing a smoky exhaust. Unburnt hydrocarbons (C_xH_y) and carbon monoxide, through the incomplete combustion of the fuel, can be components of the exhaust gases. These gaseous pollutants, along with nitrogen oxides, can be removed at a later stage by using a catalytic converter (see Chapter 8, page 169).

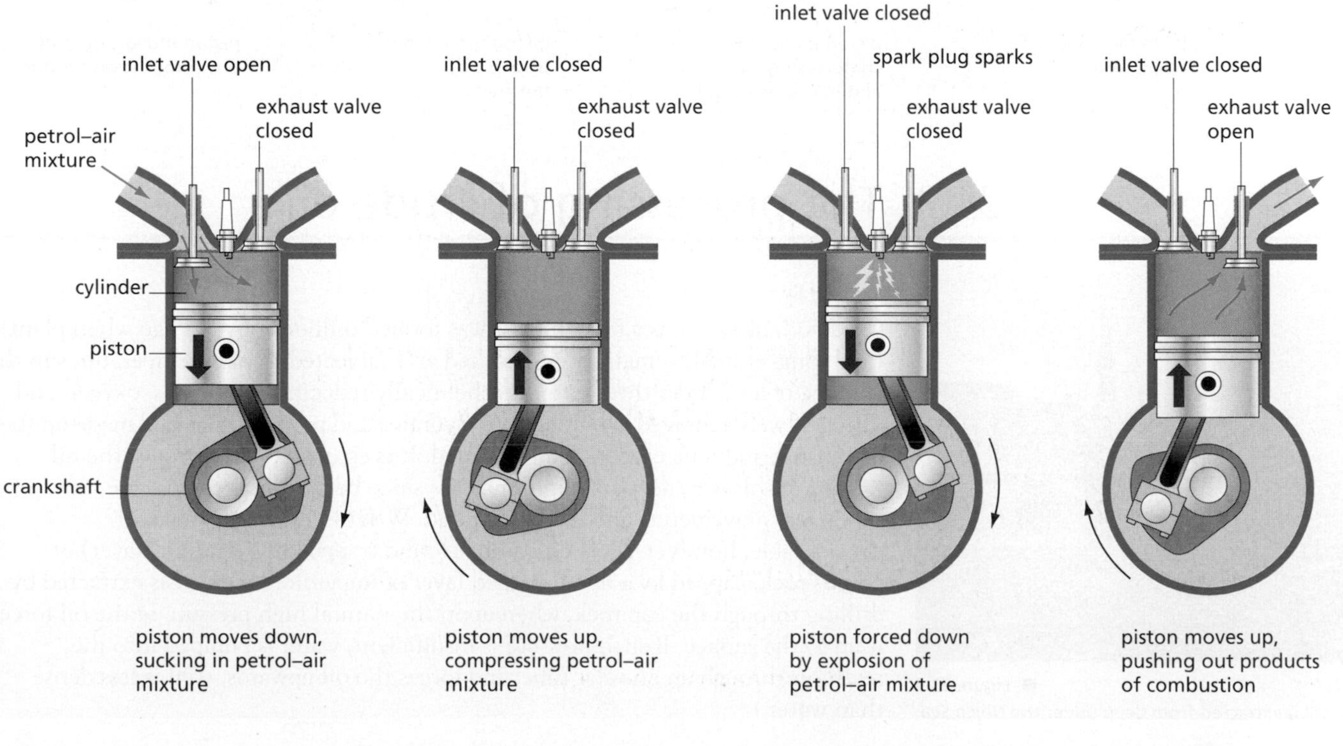

■ **Figure 23.3**
The four-stroke cycle. Movement of the piston up and down causes the crankshaft to rotate. This turns the wheels of the vehicle.

The diesel engine

The main alternative to the four-stroke Otto cycle petrol engine used in motor transport is the four-stroke compression ignition engine developed by Rudolph Diesel in 1890. The principle is similar to the Otto engine, in some respects, but differs significantly in the first two strokes of the cycle:

First downstroke and first upstroke
Only air is admitted (the diesel engine has no carburettor), which is then compressed to only one-twentieth of its volume on the upstroke. This high compression causes a large rise in temperature. Near the top of the upstroke, the fuel is injected into the top of the cylinder under pressure. It immediately ignites in the hot compressed air, pushing the piston down the cylinder on the second downstroke.

The pollutants produced by diesel engines are similar in nature to those produced in a petrol engine. They tend to include less carbon monoxide, but more carbon particles and nitrogen oxides.

■ **Figure 23.4**
The diesel cycle.

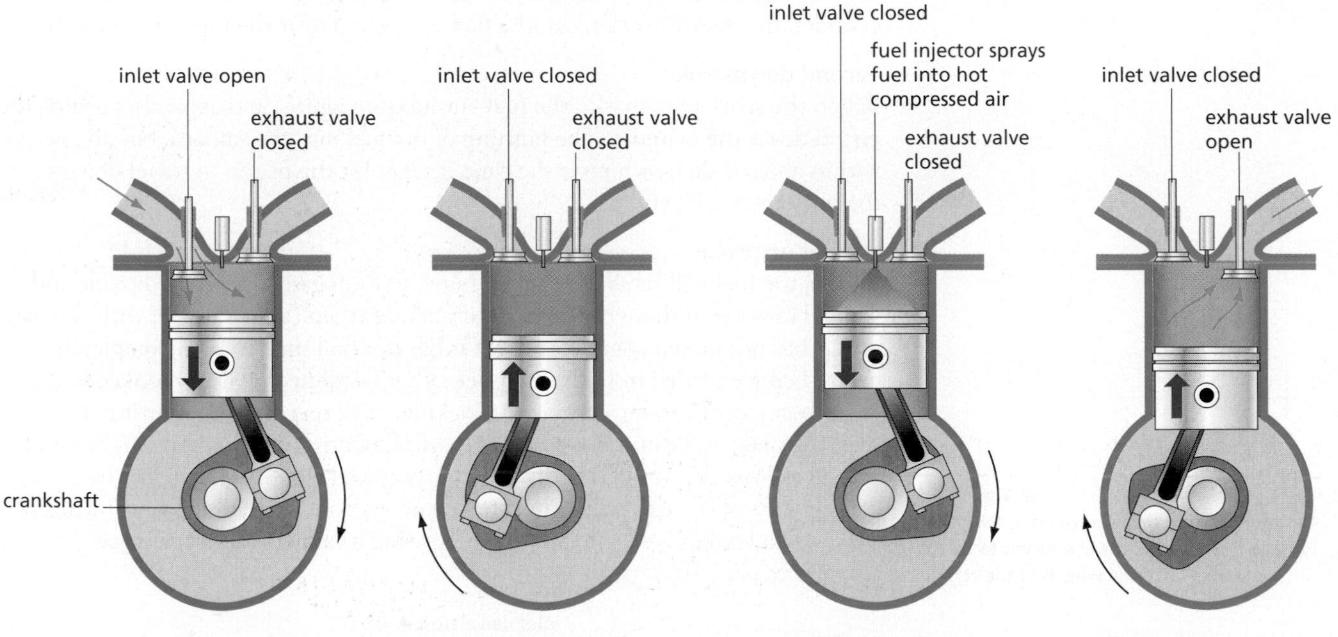

23.3 The processing of crude oil

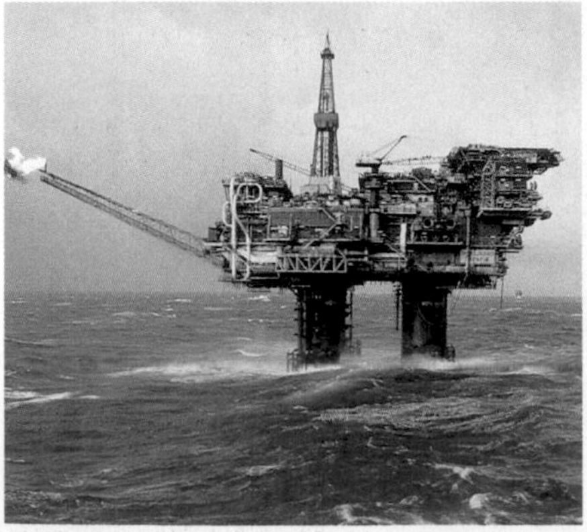

■ **Figure 23.5**
Oil is extracted from deep under the North Sea.

Origins

Crude oil, like coal, is a fossil fuel. It was formed millions of years ago when plant (and some animal) remains were crushed and subjected to high temperatures in the absence of air. Under these extreme chemically reducing conditions, oxygen and nitrogen were removed from the carbohydrates and proteins that had made up the organisms, and long carbon chains formed. It is estimated that most of the oil formed in this way, deep underground, has since been transported to the surface by geological movements, and has evaporated. What is left is (or was) still considerable, however. Most crude oil is found trapped in a stratum (layer) of porous rock, capped by a dome-shaped layer of impervious rock. It is extracted by drilling through the cap rock, whereupon the natural high pressure of the oil forces it up to the surface. If such pressure is insufficient, water is pumped into the reservoir through an annular tube, and forces the oil upwards. (Oil is less dense than water.)

Fractional distillation

Crude oil is a mixture of many hundreds of hydrocarbons, ranging in size from one to 30 or 40 carbon atoms per molecule. The first stage in its processing is a fractional distillation (see Chapter 12, page 238), carried out at atmospheric pressure. The boiling point of crude oil varies continuously from below 20 °C to over 300 °C, and many fractions could be collected, each having a boiling point range of only a few degrees. In practice, the most useful procedure is to collect just four fractions from the primary distillation. Each fraction can then be further purified, or processed, as required. Table 23.3 lists these fractions (and the residue) together with their major uses.

■ **Figure 23.6**
The fractional distillation of crude oil.

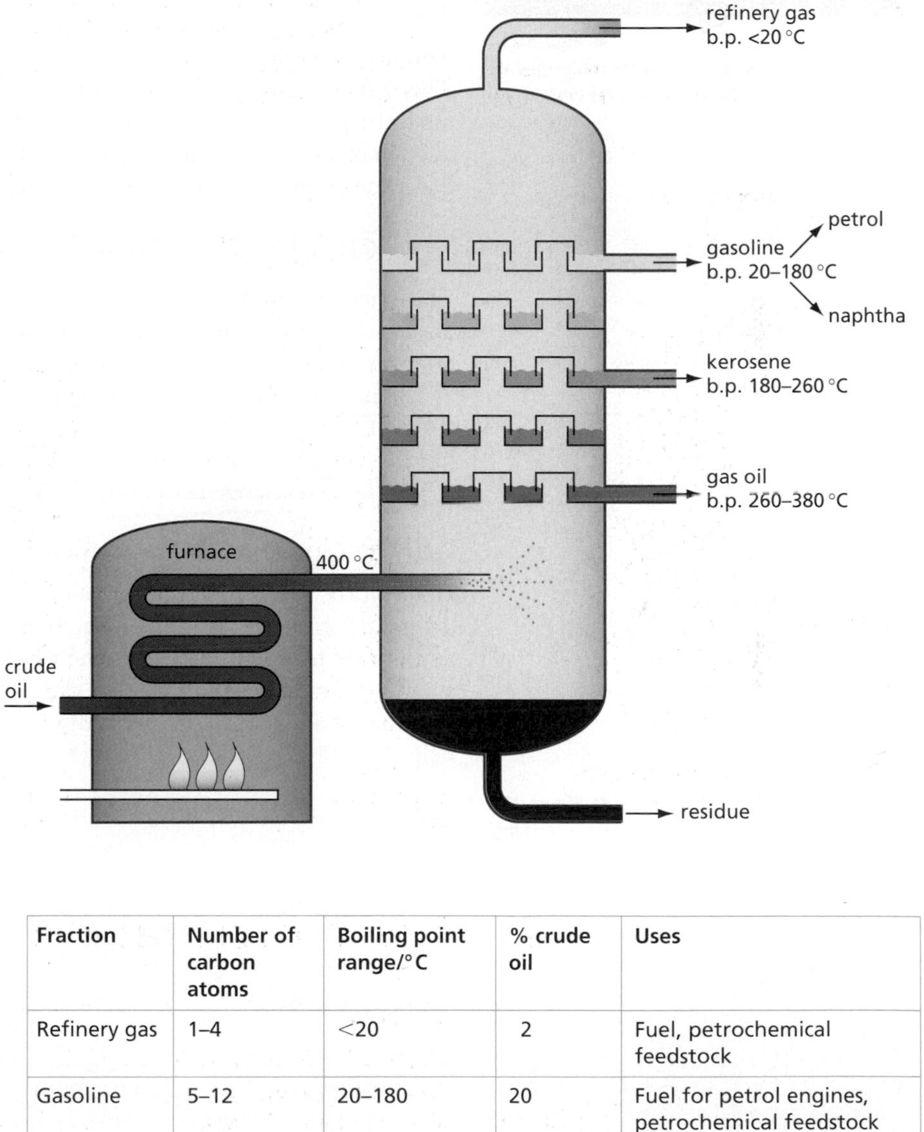

■ **Table 23.3**
Fractions from the distillation of crude oil.

Fraction	Number of carbon atoms	Boiling point range/°C	% crude oil	Uses
Refinery gas	1–4	<20	2	Fuel, petrochemical feedstock
Gasoline	5–12	20–180	20	Fuel for petrol engines, petrochemical feedstock
Kerosene	10–16	180–260	13	Aeroplane fuel, paraffin, cracked to give more gasoline
Gas oil	12–25	260–380	20	Fuel for diesel engines, cracked to give more gasoline
Residue	>25	>380	45	Lubricating oil, power station fuel, bitumen for roads, cracked to give more gasoline

■ **Figure 23.7**
In a cat cracker, long-chain molecules are converted into shorter molecules which have more uses.

Cracking – breaking long-chain molecules into shorter ones

Over 90% of crude oil is burnt as fuels of one sort or another. Only 10% is used as a feedstock to produce the host of organic chemicals (plastics, fibres, dyes, paints, pharmaceuticals) that are derived from crude oil. The relative demand for petrol (and hence for the gasoline fraction) far outstrips the supply from the primary distillation – about twice as much is required as is contained in crude oil. Consequently, methods have been developed to convert the higher boiling, and more abundant, fractions into gasoline. This is done by **cracking** the long-chain molecules into shorter units. Two alternative processes are used – thermal cracking and catalytic cracking. We can illustrate the difference between these using as an example dodecane (boiling point 216 °C), a typical component of the kerosene fraction.

Thermal cracking

Thermal cracking involves heating the alkane mixture to about 800 °C and at moderate pressure, in the absence of air, but in the presence of steam. After only a fraction of a second at this temperature, the mixture is rapidly cooled. By this means dodecane might typically be broken into hexane (boiling point 69 °C) and ethene:

$$CH_3(CH_2)_{10}CH_3 \xrightarrow{\text{heat with steam at 800 °C}} CH_3(CH_2)_4CH_3 + 3CH_2{=}CH_2$$

The ethene by-product is not wasted. Indeed, since it is a key feedstock for the plastics, fibres and solvents industries, the conditions for thermal cracking are often chosen so as to optimise ethene production.

The heat energy at 800 °C is sufficient to break the C—C bonds into two carbon free radicals. Long-chain radicals readily split off ethene units, eventually producing shorter-chain alkanes and alkenes.

Very little rearrangement of the chains occurs during thermal cracking.

How it happens (1): the radical mechanism of thermal cracking

Homolytic C—C bond fission produces alkyl radicals, with only seven electrons around the end carbon atom of the chain. These alkyl radicals tend to split apart at the bond next but one to this end carbon atom, producing ethene and leaving a new alkyl radical with two fewer carbon atoms. This can then undergo further splitting, or a hydrogen atom can be transferred from another radical, if two radicals were to collide.

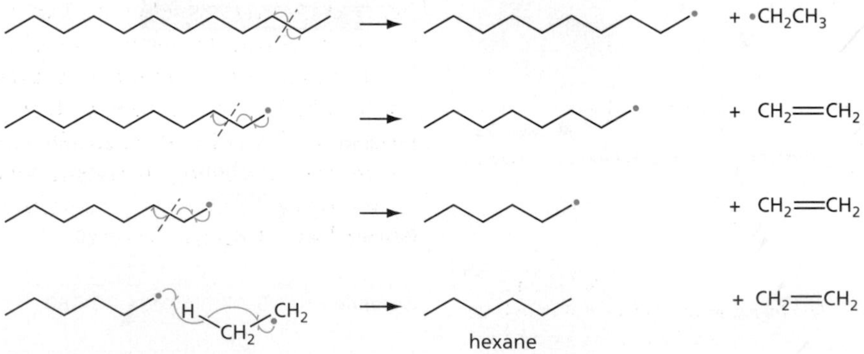

■ **Figure 23.8**
The radical mechanism of thermal cracking.

Catalytic cracking

Catalytic cracking involves heating the alkane mixture to a temperature of about 500 °C and passing it under slight pressure over a catalyst made from a porous mixture of aluminium and silicon oxides (called zeolites). The catalyst aids the production of carbocations from the alkanes, possibly by absorbing hydrogen atoms onto its surface. Carbocations (also called carbonium ions) are not quite as reactive as the free radicals formed during thermal cracking, and they can undergo internal rearrangements before forming the final products.

How it happens (2): the carbocation mechanism of catalytic cracking

The transfer of a hydrogen atom (together with its bonding electrons) to the surface of the catalyst produces an electron-deficient carbocation (its end carbon atom is surrounded by only six electrons). In a similar way to the free radicals formed during thermal cracking, C—C bond fission occurs at the next-but-one bond to this positively charged carbon atom, producing a molecule of ethene and a new alkyl carbocation with two fewer carbon atoms. The shift of a hydrogen atom (with its bonding electrons) from a carbon in the centre of the chain to the carbon of the primary carbocation produces the more stable secondary carbocation (see the panel on page 437). Further rearrangements produce the desired branched-chain alkanes.

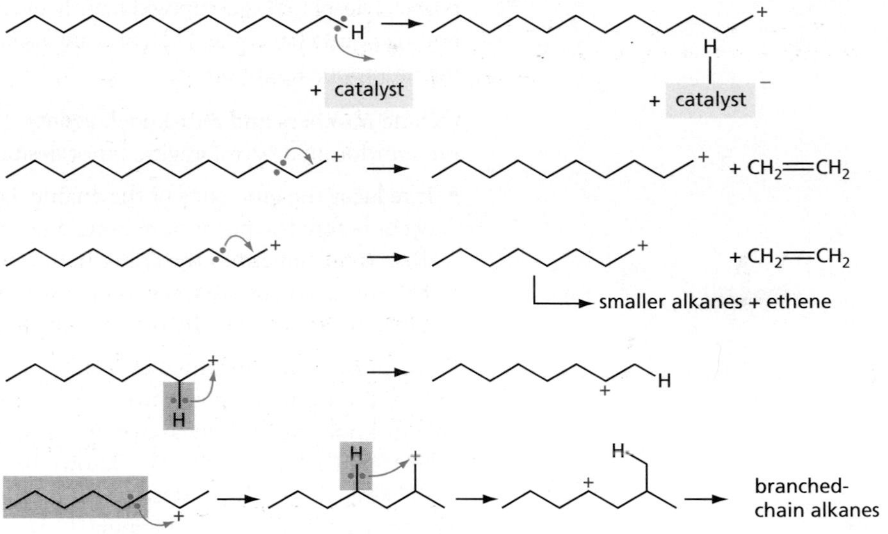

■ **Figure 23.9** The carbocation mechanism of catalytic cracking.

The branched-chain alkanes produced by catalytic cracking are useful components of high octane petrol (see the panel overleaf).

Re-forming

The demand for branched-chain, cyclic and aromatic hydrocarbons as components of high octane petrol is further satisfied by **re-forming** straight-chain alkanes. The vaporised alkane mixture is passed over a platinum-coated aluminium oxide catalyst at 500 °C and moderately high pressure. For example, heptane can be re-formed as shown in Figure 23.10.

In general, re-forming changes straight-chain alkanes into branched-chain alkanes and cyclic hydrocarbons without the loss of any carbon atoms, but often with the loss of hydrogen atoms.

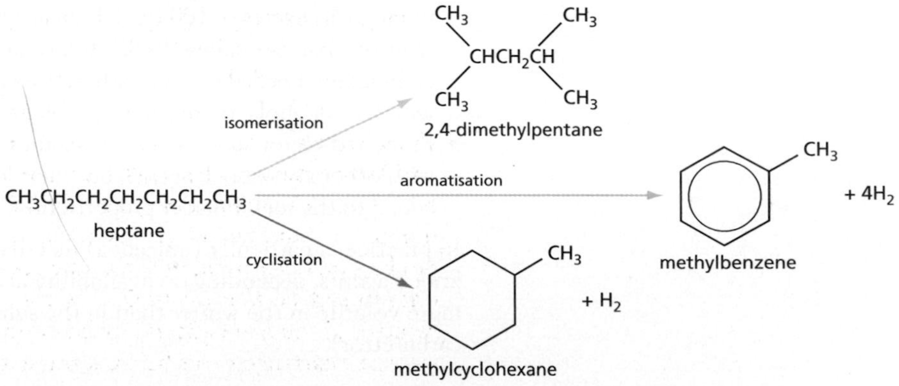

■ **Figure 23.10** Straight-chain alkanes are re-formed into branched-chain, cyclic and aromatic hydrocarbons.

Fuels for petrol and diesel engines

Volatility and pre-ignition

Fuels for petrol and diesel engines must conform to very different requirements.

Because petrol is vaporised before ignition, it needs to have a fairly low boiling point. Diesel fuel, however, is injected directly into the cylinder, so does not need to be as volatile.

Petrol must be blended so that it does not suffer pre-ignition (or auto-ignition) when its mixture with air is compressed. Diesel fuel, on the other hand, must be designed with the exact opposite requirement in mind. It must be able to be auto-ignited on compression, without the aid of a spark.

These two requirements, luckily, go hand in hand with the properties of certain alkanes. The long, straight-chain alkanes auto-ignite well, and have higher boiling points. Diesel fuel is composed mainly of these. Branched-chain alkanes have lower boiling points (see page 422), and are also resistant to auto-ignition. Petrol has a higher proportion of these.

Octane numbers and anti-knock agents

Pre-ignition in a petrol engine is not desirable for two reasons.

- It reduces the efficiency of the engine, because it causes the petrol–air explosion to occur before the piston is moving down the cylinder. In this way it opposes the desired movement of the crankshaft, rather than helps it.
- Because it produces a force in opposition to the movement of the piston, pre-ignition can damage the piston, cranks and associated bearings in the engine.

Pre-ignition is accompanied by a metallic 'knocking' sound, and components added to petrol to minimise pre-ignition are called **anti-knock agents**. The efficiency of an anti-knock agent is measured on a scale called the **octane scale**. The compound 2,2,4-trimethylpentane, C_8H_{18}, known to petroleum chemical engineers as 'iso-octane', is highly resistant to auto-ignition. It is given an **octane rating** of 100. The straight-chain alkane heptane, C_7H_{16}, on the other hand, is very easy to auto-ignite. It is given an octane rating of 0. If a given fuel blend auto-ignites under the same conditions as an 80:20 iso-octane:heptane mixture, it is therefore described as having an octane number of 80. Fuels for most modern petrol engines need to have octane numbers of 95 or more. There are several types of anti-knock agent that can be added to a fuel to increase its octane number.

- Tetraethyl lead(IV) (lead tetraethyl), $(C_2H_5)_4Pb$. This is highly efficient – the addition of only 1 g per litre of petrol will raise the octane number by 10 points. But its use is now severely restricted because of the poisonous nature of airborne lead compounds found in the exhaust from cars running on 'leaded petrol'.
- Aromatic hydrocarbons such as benzene, C_6H_6, and methylbenzene, C_7H_8. These are reasonably efficient, and do not form harmful combustion products. But benzene, which can be present in petrol to the extent of 50 g per litre, is carcinogenic (cancer-producing), and is harmful by absorption through both the skin and the lungs.
- Oxygen-containing compounds such as methanol, CH_3OH, ethanol, C_2H_5OH, and MTBE (methyl tertiary-butyl ether), $(CH_3)_3C—O—CH_3$, have very high octane numbers in excess of 100 (which means that they are even more resistant to auto-ignition than iso-octane itself). If too much of these is added to fuels, however, changes are needed to cars' carburettors (the fuel–air ratio needs increasing) and with the alcohols, there can be problems of miscibility if the petrol becomes damp.
- Branched-chain alkanes or cycloalkanes. These have few of the problems associated with other anti-knock agents, but their lower effectiveness means that they must be added to the fuel in larger proportions. This adds to the expense of the fuel.

In practice, a particular (unleaded) fuel may contain a blend of several different anti-knock agents, depending on availability and time of year (petrols are blended to be more volatile in the winter than in the summer, to allow easier evaporation in a cold carburettor).

23.4 The combustion of fuels

Modern society has an almost inexhaustible thirst for energy, in the form of electricity, space heating for homes and offices, and energy for transport, by road, rail, sea and air.

As we saw in Chapter 5, the energy sources available to us can be divided into two categories – renewable and non-renewable. The former includes the renewable fuel ethanol, obtained from the fermentation of starch-rich plants such as sugar cane; and the alternative energy sources such as wind power, wave power and tidal power. They produce no chemical pollution. Their disadvantages are that they are often seasonal and can be unreliable, and they are not readily transportable, so cannot be used for mobile vehicles. Their energy also cannot be stored for future use. Energy from these alternative sources is usually transformed immediately into electrical energy.

Progress is being made to solve these problems, however. One possibility is to use alternative sources of energy to split water into hydrogen and oxygen (see Chapter 13, page 266). Moreover, hydrogen can be transported and stored for future use as a pollution-free fuel.

Ethanol from fermentation can be used as a renewable fuel for cars and, conceivably, in power stations, just like the hydrocarbon fossil fuels. In such situations, similar types of pollution could result from its burning. Nevertheless, in practice, it tends to be a 'cleaner' fuel than hydrocarbons.

The non-renewable energy sources include the hydrocarbons natural gas and crude oil, and the carbon-rich coal. These are called **fossil fuels**, having been produced many millions of years ago and buried beneath the Earth's surface.

Uranium, the primary fuel for nuclear reactors, was produced an even longer time ago, when the elements of the Solar System were synthesised in the centres of dying stars billions of years ago. In a sense it too is a non-renewable energy source (although the 'fast breeder' reactors of a few decades ago would now be creating more nuclear fuel than they consumed, had there not been insurmountable problems with their development). Nuclear reactors produce little chemical pollution, but the risks of radiation pollution from their use and from the reprocessing of their spent fuel has caused them to fall from favour.

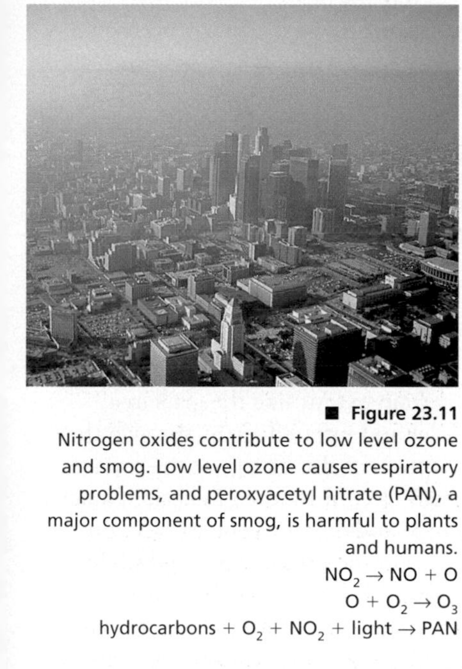

Figure 23.11
Nitrogen oxides contribute to low level ozone and smog. Low level ozone causes respiratory problems, and peroxyacetyl nitrate (PAN), a major component of smog, is harmful to plants and humans.
$NO_2 \rightarrow NO + O$
$O + O_2 \rightarrow O_3$
hydrocarbons + O_2 + NO_2 + light → PAN

There are five main pollutants that are formed when fossil fuels are burned:

- carbon dioxide
- carbon monoxide
- unburnt hydrocarbons
- nitrogen oxides (see Figure 23.11)
- sulphur dioxide.

With the exception of carbon dioxide and sulphur dioxide, these can all be removed from vehicle emissions by a catalytic converter (see Chapter 8, page 169). **Flue gas desulphurisation (FGD)** is used in power stations to reduce the amount of sulphur dioxide reaching the atmosphere. In this process the flue gases are passed through a slurry of calcium carbonate. Calcium carbonate (limestone) is cheap, and the calcium sulphate produced can be sold as gypsum, a component of plaster.

$$CaCO_3(s) + SO_2(g) \rightarrow CaSO_3(s) + CO_2(g)$$
$$CaSO_3(s) + \tfrac{1}{2}O_2(g) \rightarrow CaSO_4(s)$$

The use of the more expensive magnesium carbonate allows the sulphur dioxide to be regenerated and used to make sulphuric acid.

$$MgCO_3(s) \xrightarrow{\text{heat}} MgO(s) + CO_2(g)$$

$$MgO(s) + SO_2(g) \longrightarrow MgSO_3(s)$$

$$MgSO_3(s) \xrightarrow{\text{heat}} MgO(s) + SO_2(g)$$

The greenhouse effect and carbon dioxide

What is the greenhouse effect?

Apart from its obvious role of protecting the contents from the weather, the glass in a greenhouse also warms the inside by means of the **greenhouse effect**. Glass is transparent to visible light, and allows most of the sunlight through. The light is absorbed by objects within the greenhouse (if the objects happen to be plants, useful photosynthesis results – but even a plant-free greenhouse still shows the greenhouse effect). This absorption of light energy causes the objects to warm up. Warm objects emit radiation of a much longer wavelength than the visible spectrum – they give out infrared radiation. Glass is opaque to infrared radiation, so absorbs it, and warms up in the process. So much of the energy from the visible light is trapped inside the greenhouse, which becomes warmer and warmer.

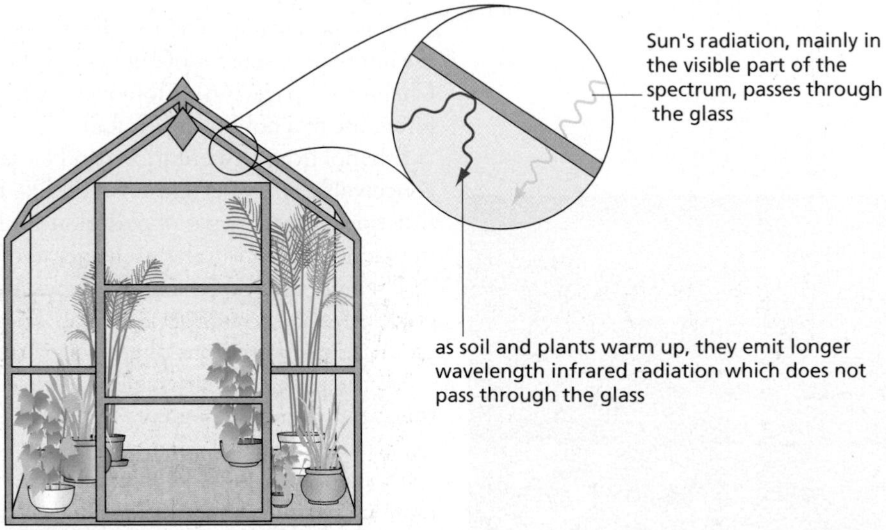

Sun's radiation, mainly in the visible part of the spectrum, passes through the glass
as soil and plants warm up, they emit longer wavelength infrared radiation which does not pass through the glass

■ **Figure 23.12**
It is warmer inside a greenhouse than outside.

The atmosphere around the Earth acts to a certain extent like the glass in a greenhouse. The Sun's rays – visible, ultraviolet and infrared – pass through the atmosphere to the surface of the Earth. Absorption of this energy at the surface causes the land and sea to warm up. They begin to emit infrared radiation, much of which passes straight through the atmosphere and is lost to outer space. But a proportion is absorbed by the gaseous molecules in the atmosphere, and so the energy is not lost, but re-radiated back to the Earth's surface. By this means the surface is kept at a reasonably constant, warm temperature day and night, summer and winter (contrast the situation on Mars, which has little atmosphere – there night-time temperatures can be 100 °C lower than those during the day).

Increasing the greenhouse effect of the atmosphere

The greenhouse effect of the atmosphere is essential to provide the right stable environment for life. But there is a fine balance to be struck between not enough of the effect, in which case more of the incident solar energy will be lost, cooling the Earth to another ice age; and too much greenhouse effect, in which case the Earth will become too warm. This could result in melting icecaps, raised sea levels and changes in climatic air circulation.

Carbon dioxide, produced from the burning of fossil fuels for heating, transport and electricity generation, is the major atmospheric pollutant that increases the greenhouse potential of the atmosphere. Many of the other pollutant gases that we throw into the atmosphere have an even greater potential – for example, CFCs are over 5000 times as effective as greenhouse gases as is carbon dioxide, but they are produced in much smaller quantities. (Their main environmental effect, however, is in destroying the stratospheric ozone layer, as we shall see in Chapter 25.)

More intensive agriculture, especially cattle-rearing, also produces greenhouse gases such as methane, which is ten times as effective as carbon dioxide. But methane is also produced in smaller quantities than carbon dioxide. The case against carbon dioxide is not totally proven, but seems increasingly firm.

The present concentration of carbon dioxide in the atmosphere is only 0.035%. But 300 years ago it was only 0.028%. This 25% increase has contributed to an average increase in global temperature of 0.5 °C. The other major natural greenhouse gas in the atmosphere is water vapour. It has been suggested that a 'runaway' effect might begin to operate – as global warming increases, the warmer atmosphere will be be able to absorb more water vapour, which in turn will lead to a greater greenhouse effect, increasing global warming even more. There are many natural ways of reducing atmospheric carbon dioxide:

- water in the seas dissolves millions of tonnes (but less now than it did, since the average ocean temperature has increased by 0.5 °C in the last 100 years, and gases are less soluble in hot than in cold water)
- plankton can fix the dissolved carbon dioxide into their body mass by photosynthesis
- trees fix more atmospheric carbon dioxide per acre than do grass and other vegetation.

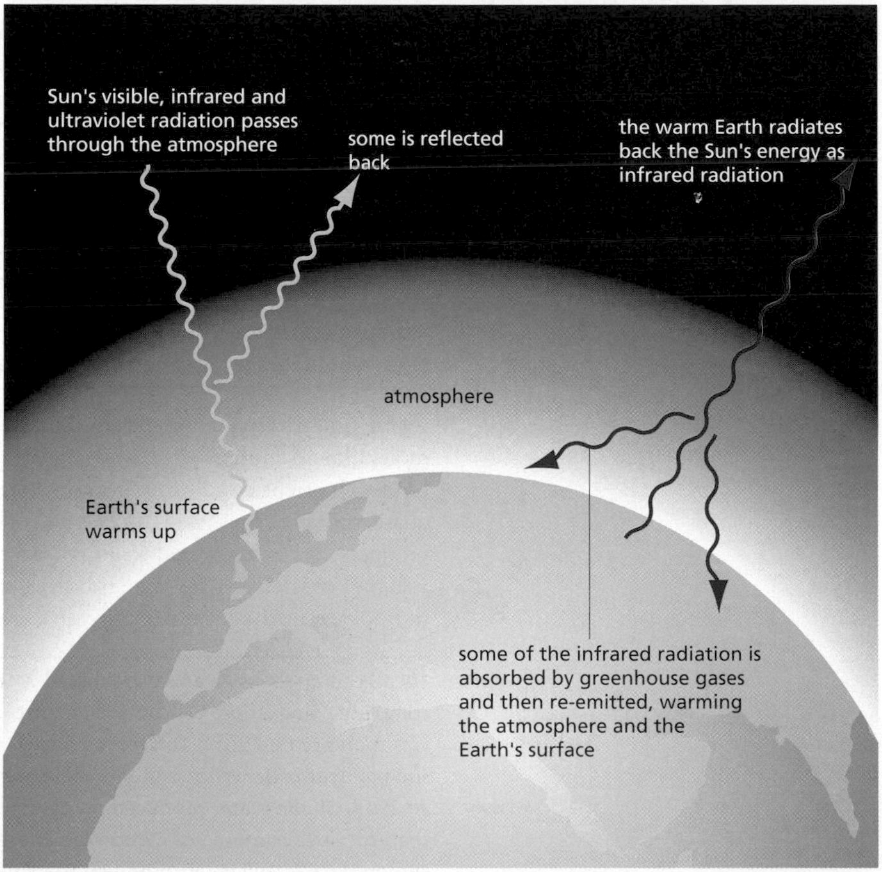

Sun's visible, infrared and ultraviolet radiation passes through the atmosphere
some is reflected back
the warm Earth radiates back the Sun's energy as infrared radiation
atmosphere
Earth's surface warms up
some of the infrared radiation is absorbed by greenhouse gases and then re-emitted, warming the atmosphere and the Earth's surface

■ **Figure 23.13** The greenhouse effect warms the Earth.

■ **Figure 23.14** Carbon dioxide and water molecules absorb infrared radiation by vibrating their bonds. This is how they increase the greenhouse effect.

23.5 Reactions of alkanes

Apart from cracking and re-forming, there are two important reactions of alkanes, namely combustion and halogenation. We shall look at each in turn.

Combustion

Fuels

As mentioned at the start of the chapter, most of the alkanes that are produced from crude oil are burned. Their complete combustion is a highly exothermic process:

$$CH_4(g) + 2O_2(g) \rightarrow CO_2(g) + 2H_2O(l) \qquad \Delta H_c^\ominus = -890\,kJ\,mol^{-1}$$

methane
(natural gas)

$$C_{20}H_{42}(l) + 30.5O_2(g) \rightarrow 20CO_2(g) + 21H_2O(l) \qquad \Delta H_c^\ominus = -13\,368\,kJ\,mol^{-1}$$

eicosane
(fuel for oil-powered power stations)

Example Methane (natural gas) is replacing crude oil and coal as the fuel in some power stations, in an attempt to cut down on greenhouse gas emissions. What is the percentage reduction in carbon dioxide produced per kJ when natural gas replaces fuel oil in a power station?

Answer For methane, the equation tell us that 890 kJ of energy are released when 1 mol of carbon dioxide is formed.

$$\text{carbon dioxide produced per kJ} = \frac{1}{890} = 1.12 \times 10^{-3}\,mol\,kJ^{-1}$$

For fuel oil, 13 368 kJ of energy are released when 20 mol of carbon dioxide are formed.

$$\text{carbon dioxide produced per kJ} = \frac{20}{13\,368} = 1.50 \times 10^{-3}\,mol\,kJ^{-1}$$

The percentage reduction is therefore given by:

$$100 \times \frac{1.50 - 1.12}{1.50} = \mathbf{25\%}$$

Further practice The combustion of coal in a coal-fired power station can be represented as follows:

$$C(s) + O_2(g) \rightarrow CO_2(g) \qquad \Delta H_c^\ominus = -394\,kJ\,mol^{-1}$$

assuming coal is 100% carbon. What would be the percentage reduction in carbon dioxide produced per kJ if natural gas replaces coal in a power station?

Using quantitative combustion data in the determination of empirical and molecular formulae

Under the right conditions, combustion of a hydrocarbon can be carried out quantitatively (that is, in 100% yield). The quantities of the reactants and products can be measured to a high degree of precision, by measuring either their volumes or their masses. The following example shows how this can be put to use in finding out the formulae of hydrocarbons.

Example 10 cm^3 of a gaseous alkane were mixed with 100 cm^3 of oxygen and sparked to cause complete combustion. On cooling to the original temperature (290 K), the volume of gas was found to be 75 cm^3. This reduced to 35 cm^3 on shaking with concentrated aqueous sodium hydroxide. What is the formula of the alkane?

Answer At 290 K, all the water produced on combustion will be a liquid. The 75 cm^3 of gas must therefore be composed of carbon dioxide, along with excess unreacted oxygen. The carbon dioxide will be absorbed by the aqueous sodium hydroxide, so 40 cm^3 (75−35) of carbon dioxide were produced. Using Avogadro's Law, if 10 cm^3 of alkane produce 40 cm^3 of carbon dioxide, then 1 molecule of alkane will produce 4 molecules of carbon dioxide. Therefore the alkane contains 4 carbon atoms, so its formula is C_4H_{10}.

Further practice 10 cm^3 of a gaseous alkane were mixed with 150 cm^3 of oxygen and sparked. On cooling to the original temperature (290 K), the volume of gas was found to be 120 cm^3. This reduced to 70 cm^3 on shaking with concentrated aqueous sodium hydroxide. What is the formula of the alkane?

Halogenation

Alkanes undergo substitution reactions with halogens:

$$CH_4 + X_2 \rightarrow CH_3X + HX$$

Methane reacts vigorously with fluorine, even in the dark and at room temperature. With chlorine and bromine, no reaction occurs unless the reactants are heated or exposed to ultraviolet light. With iodine, no reaction occurs at all. The trend in the calculated ΔH of the reactions matches that of the C—X bond energies (Table 23.4).

■ Table 23.4
Bond enthalpies and enthalpies of reaction for halogen substitution reactions with methane.

Halogen (X)	C—X bond enthalpy/kJ mol^{-1}	ΔH for the reaction/kJ mol^{-1}
F	484	−475
Cl	338	−114
Br	276	−36
I	238	+27

The mechanism of the halogenation reaction

The halogenation reaction involves radicals. When initiated by ultraviolet light, experiments have shown that one photon can cause the production of many thousands of molecules of chloromethane from chlorine and methane. Once the photon has started the reaction off, it can continue on its own for a long time. Studies on the effectiveness of various frequencies of ultraviolet light (see the panel on page 435) have suggested that the photon initiates (starts) the reaction by splitting the chlorine molecule homolytically:

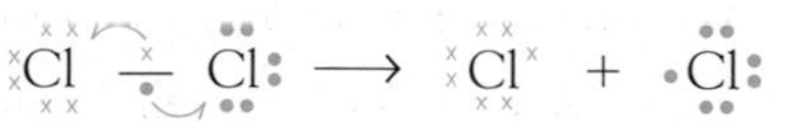

$$\text{Cl—Cl} \longrightarrow \text{Cl} \cdot + \cdot \text{Cl} \qquad (1)$$

We can predict the next step of the reaction by simple bond enthalpy calculations. The chlorine atoms formed by the splitting of the chlorine molecule are highly reactive. They are likely to react with the next molecule they collide with. In the mixture we have just chlorine molecules and methane molecules. Reaction of a chlorine atom with a chlorine molecule:

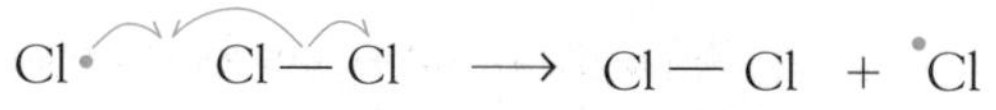

$$\text{Cl}\cdot \quad \text{Cl—Cl} \longrightarrow \text{Cl—Cl} + \cdot\text{Cl} \qquad (2)$$

might well occur, but it does not lead anywhere. It merely uses up one chlorine atom to produce another.

Reaction with methane could conceivably occur in one of two ways:

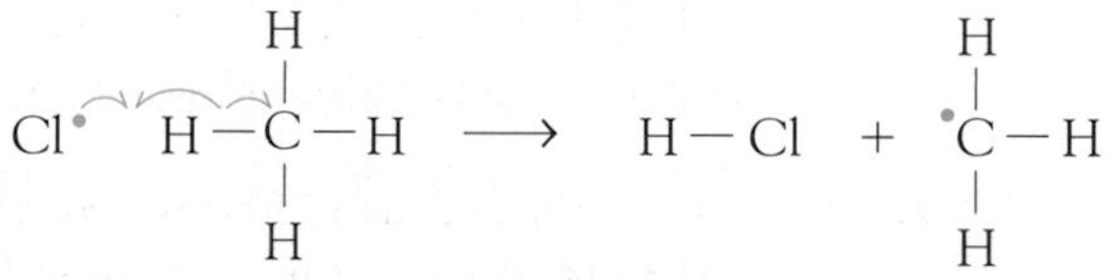

$$\text{Cl}\cdot \quad H{-}\overset{H}{\underset{H}{C}}{-}H \longrightarrow H{-}Cl + \cdot\overset{H}{\underset{H}{C}}{-}H \qquad (3)$$

$$\text{Cl}\cdot \quad \overset{H}{\underset{H}{C}}H \longrightarrow Cl{-}\overset{H}{C}H + H\cdot \qquad (4)$$

In reaction (3), a C—H bond has been broken, and a H—Cl bond formed. $E(\text{C—H}) = 413\,\text{kJ mol}^{-1}$, and $E(\text{H—Cl}) = 431\,\text{kJ mol}^{-1}$. Overall $\Delta H_{(3)} = 413 - 431 = -18\,\text{kJ mol}^{-1}$.

In reaction (4), a C—H bond has been broken, and a C—Cl bond formed. $(E(\text{C—Cl}) = 338\,\text{kJ mol}^{-1})$. Overall $\Delta H_{(4)} = 413 - 338 = +75\,\text{kJ mol}^{-1}$.

Reaction (3) is slightly exothermic, whereas reaction (4) is highly endothermic. Reaction (3) is therefore much more likely to take place.

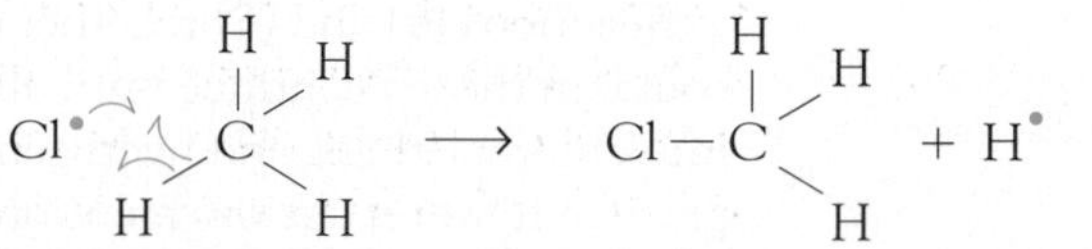

Further practice Use the following bond enthalpies to calculate the ΔH values of reactions (3) and (4) using bromine instead of chlorine. Is reaction (3) still more likely to occur than (4) with bromine?

$E(\text{C—Br}) = 276\,\text{kJ mol}^{-1}$
$E(\text{H—Br}) = 366\,\text{kJ mol}^{-1}$

The methyl radical formed in reaction (3) is equally as reactive as a chlorine atom. It is likely to react with the first molecule it collides with. If it collides with a chlorine molecule, the following reaction is possible:

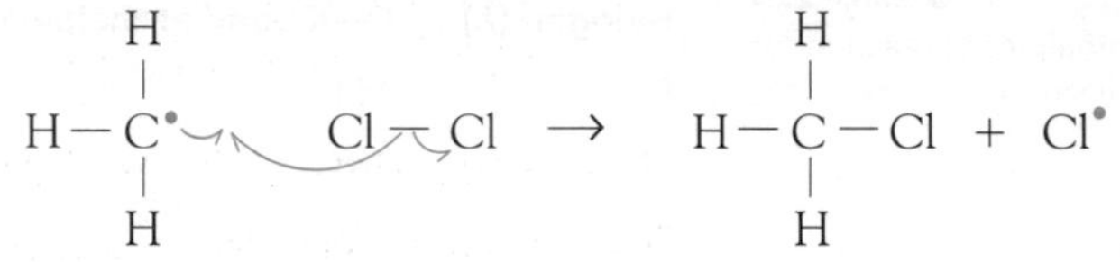
(5)

$$\Delta H = E(\text{Cl—Cl}) - E(\text{C—Cl}) = 242 - 338 = -96\,\text{kJ mol}^{-1}$$

The reaction is highly exothermic, so it is likely to occur. If the methyl radical collides with a methane molecule, the following 'reaction' could occur:

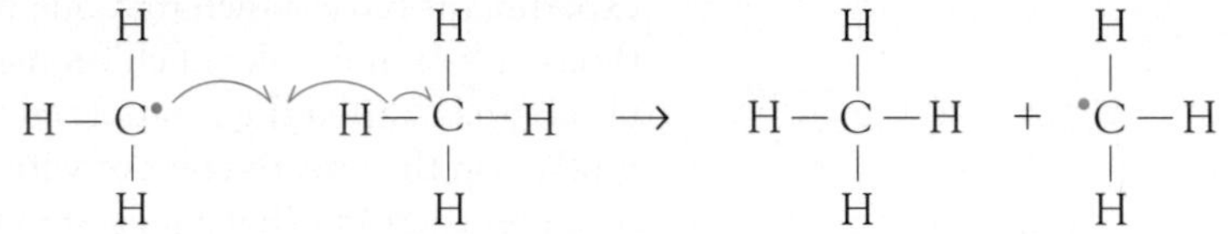
(6)

Reaction (6), like reaction (2), does not lead us anywhere. A hydrogen atom has been transferred from a methane molecule to a methyl radical, forming a methane molecule and another methyl radical!

Let us now take stock of the situation. In the initiation reaction (1), we have produced two reactive chlorine atoms. On energetic grounds, the two most likely reactions after that are (3) and (5):

$$Cl\bullet + CH_4 \rightarrow CH_3^\bullet + HCl \qquad (3)$$
$$CH_3^\bullet + Cl_2 \rightarrow CH_3Cl + Cl\bullet \qquad (5)$$

After these two reactions have taken place we have:

- used up one molecule of methane, and one molecule of chlorine
- produced one molecule of chloromethane, and one molecule of hydrogen chloride
- used up a chlorine atom, but regenerated another one.

In other words, the chlorine atom has been a catalyst for the reaction:

$$CH_4 + Cl_2 \rightarrow CH_3Cl + HCl$$

The chlorine atom has acted as a homogeneous catalyst (see Chapter 8). It has taken part in the reaction, but has not been used up during it.

Reactions (3) and (5) together constitute a never-ending **chain reaction**. They could, in theory, continue until all the methane and chlorine had been converted into chloromethane and hydrogen chloride. This situation does not occur in practice however. As was mentioned above, one photon initiates the production of many thousands of molecules of chloromethane – but only thousands, not millions. Eventually the chain reaction comprising reactions (3) and (5) stops, because there is a (small) chance that two radicals could collide with each other, to form a stable molecule:

$$Cl\bullet + Cl\bullet \rightarrow Cl_2 \qquad (7)$$
$$CH_3^\bullet + Cl\bullet \rightarrow CH_3Cl \qquad (8)$$
$$CH_3^\bullet + CH_3^\bullet \rightarrow CH_3\text{—}CH_3 \qquad (9)$$

Any of these reactions would use up the 'catalysts' Cl• and $CH_3^\bullet$, and so stop the chain reaction. It would require another photon to split another chlorine molecule in order to restart the chain. Overall, the chlorination of methane is an example of a **radical substitution reaction**. The different stages of the chain reaction are named as follows:

$$Cl_2 \rightarrow 2Cl\bullet \qquad \textbf{initiation}$$

$$\left.\begin{array}{l} Cl\bullet + CH_4 \rightarrow CH_3^\bullet + HCl \\ CH_3^\bullet + Cl_2 \rightarrow CH_3Cl + Cl\bullet \end{array}\right\} \textbf{propagation}$$

$$\left.\begin{array}{l} Cl\bullet + Cl\bullet \rightarrow Cl_2 \\ CH_3^\bullet + Cl\bullet \rightarrow CH_3Cl \\ CH_3^\bullet + CH_3^\bullet \rightarrow CH_3{-}CH_3 \end{array}\right\} \textbf{termination}$$

The frequency of light that breaks the Cl—Cl bond

Planck's equation allows us to calculate the minimum frequency (or maximum wavelength) of light that will initiate reaction (1):

$$E(\text{Cl—Cl}) = 242\,\text{kJ mol}^{-1}$$

$$\text{energy required to dissociate one molecule} = \frac{242 \times 10^3}{6 \times 10^{23}}\,\text{J}$$

$$= 4 \times 10^{-19}\,\text{J}$$

$$E = hf \text{ so } f = \frac{E}{h}$$

$$h = 6.62 \times 10^{-34}\,\text{J Hz}^{-1}$$

$$f = \frac{4 \times 10^{-19}}{6.62 \times 10^{-34}} = 6.0 \times 10^{14}\,\text{Hz (yellow light)}$$

A similar calculation for bromine gives $f = 4.8 \times 10^{14}$ Hz (red light).

In practice, a higher frequency than the minimum is necessary (see Chapter 14, page 273).

By-products of the halogenation reaction

As the reaction takes place, the concentration of methane is being reduced, and the concentration of chloromethane is increasing. Chlorine atoms are therefore increasingly likely to collide with molecules of chloromethane, causing the following reactions to take place:

$$Cl\bullet + CH_3Cl \rightarrow HCl + \bullet CH_2Cl$$
$$\bullet CH_2Cl + Cl_2 \rightarrow CH_2Cl_2 + Cl\bullet$$

Further substitution, giving dichloromethane, trichloromethane and eventually tetrachloromethane, is therefore to be expected. A mixture is in fact formed.

Another interesting set of by-products are chlorinated ethanes. These arise from the ethane produced in termination reaction (9) undergoing a similar set of substitution reactions:

$$Cl\bullet + CH_3CH_3 \rightarrow HCl + CH_3CH_2\bullet$$
$$CH_3CH_2\bullet + Cl_2 \rightarrow CH_3CH_2Cl + Cl\bullet$$

These chloroethanes are produced in only very small amounts, as might be expected, since only a small quantity of ethane is produced in reaction (9). But their presence amongst the reaction products adds weight to the mechanism described here: without the formation of ethane from two methyl radicals, chlorinated ethane could not have been formed.

Commercially, the chlorination of methane is important for the production of degreasing and cleansing solvents. The mixture of products is readily separated by fractional distillation (see Table 23.5, overleaf). If only one product is required, it can be made to predominate by mixing Cl_2 and CH_4 in suitable proportions.

■ **Table 23.5** The different boiling points of the chloromethanes allow them to be separated by fractional distillation.

Compound	Formula	Boiling point/°C
Chloromethane	CH_3Cl	−24
Dichloromethane	CH_2Cl_2	40
Trichloromethane	$CHCl_3$	62
Tetrachloromethane	CCl_4	77

Halogenation of higher alkanes

A useful way of classifying organic halides, alcohols, radicals and carbocations is to place them in one of the three categories **primary**, **secondary** and **tertiary**, depending on the number of alkyl groups joined to the central carbon atom. Table 23.6 describes this classification.

■ **Table 23.6** Primary, secondary and tertiary centres. The methyl group CH_3— is also classified as a primary centre.

Number of alkyl groups	Type of centre	General formula	Examples
1	Primary	R— CH_2 —	CH_3CH_2Cl chloroethane $CH_3CH_2CH_2^+$ the prop-1-yl cation
2	Secondary	R \\ CH — / R	CH_3CH_2 \\ CH — OH butan-2-ol / CH_3 CH_3 \\ CH• the prop-2-yl radical / CH_3
3	Tertiary	R | R— C — | R	CH_3 | CH_3— C — Cl 2-chloro-2-methylbutane | CH_2CH_3 CH_3 | CH_3— C• the 2-methylprop-2-yl radical | CH_3

There are two monochloropropane isomers, 1-chloropropane, $CH_3CH_2CH_2Cl$, and 2-chloropropane, $CH_3CHClCH_3$. When propane reacts with chlorine, a mixture of the two monochloro- compounds is produced. This is because the chlorine atom can abstract either one of the primary (end-carbon) hydrogens or one of the secondary (middle-carbon) hydrogens:

$$Cl\bullet + CH_3{-}CH_2{-}CH_3 \rightarrow CH_3{-}CH_2{-}CH_2^{\bullet} + H{-}Cl$$

or

$$Cl\bullet + CH_3{-}CH_2{-}CH_3 \rightarrow CH_3{-}C^{\bullet}H{-}CH_3 + H{-}Cl$$

Because the ratio of (primary hydrogens) : (secondary hydrogens) is 6 : 2, or 3 : 1, it might be expected on probability grounds that the 1-chloropropane : 2-chloropropane ratio in the product should also be 3 : 1. However, this assumes that the abstraction of a primary hydrogen and of a secondary hydrogen are both equally likely.

In fact, it is found that the 1-chloropropane : 2-chloropropane ratio is nearly 1 : 1, which suggests that the secondary hydrogen atoms are three times more likely to be abstracted, and this cancels out the probability effect.

With bromine, the effect is even more pronounced. Here the ratio 1-bromopropane : 2-bromopropane is 1 : 30, showing that the bromine atom is far more selective as to which hydrogen atom it abstracts. Because of the **inductive effect** of the alkyl groups (see the panel below), the secondary prop-2-yl radical is more stable than the primary prop-1-yl radical, and is therefore more likely to form:

$$CH_3 \longrightarrow C{\cdot}H \longleftarrow CH_3 \qquad CH_3 \text{—} CH_2 \longrightarrow CH_2^{\bullet}$$

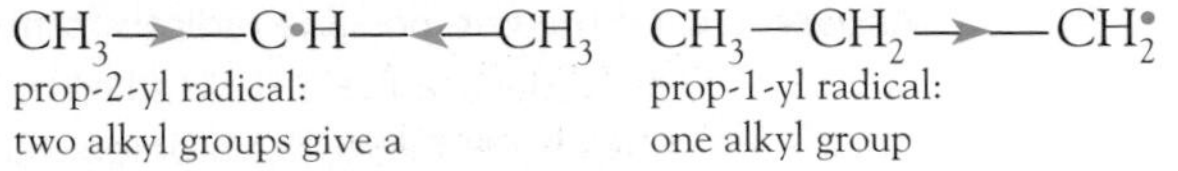

prop-2-yl radical:
two alkyl groups give a
double inductive effect

prop-1-yl radical:
one alkyl group

The inductive effect of alkyl groups

C—C bonds are essentially non-polar, due to the equal electronegativities of the carbon atoms at each end. If one of the carbon atoms is electron deficient in some way (for example, by being surrounded by only seven electrons, as in a radical), the other carbon atom can partially compensate for this deficiency by acting as an electron 'reservoir'. This reservoir effect is not as apparent with C—H bonds, because the bonding electrons are much closer to the hydrogen nucleus.

So the methyl radical:

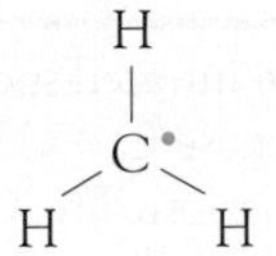

is much less stable than the primary ethyl radical: $CH_3 \longrightarrow \overset{\bullet}{C} \big\langle \begin{smallmatrix} H \\ H \end{smallmatrix}$

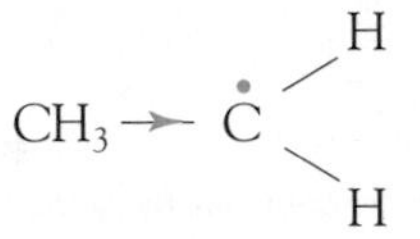

which in turn is less stable than the secondary prop-2-yl radical: $CH_3 \longrightarrow {\cdot}C \big\langle \begin{smallmatrix} H \\ CH_3 \end{smallmatrix}$

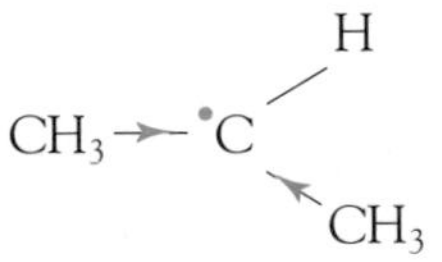

(The arrow on the bond represents the drift of electrons away from the CH_3 group.)

The tertiary 2-methylprop-2-yl radical:

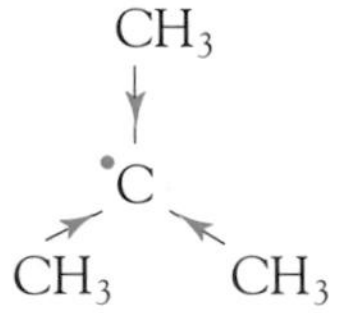

experiences an even greater stabilisation due to the inductive effort of *three* alkyl groups.

The same effect is noticed with carbocations.

For example, it is much easier to form:

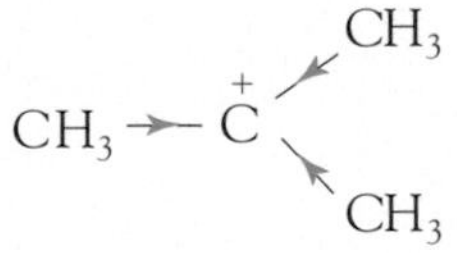

than it is to form:

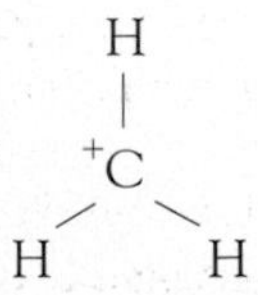

(See Chapter 24, page 447, and Chapter 25, page 467.)

Example Predict the major product formed during the monobromination of:

a $CH_3CH_2CH_2CH_3$

b $(CH_3)_2CH{-}CH_3$.

Answer a Of the two possible radicals formed by hydrogen abstraction from butane, the secondary radical $CH_3CH_2CH{\bullet}CH_3$ is more stable than the primary radical $CH_3CH_2CH_2CH_2{\bullet}$, so 2-bromobutane is the most likely product.

b Although the ratio (primary hydrogens) : (tertiary hydrogen) is 9 : 1 in 2-methylpropane, the tertiary radical $(CH_3)_3C{\bullet}$ is much more stable than the primary radical $(CH_3)_2CH{-}CH_2{\bullet}$, so 2-bromo-2-methylpropane is the most likely product.

Further practice Draw all the possible monobromoalkanes derived from the reaction between bromine and 2-methylbutane, $(CH_3)_2CH{-}CH_2CH_3$, and predict which one will be the major product.

23.6 The infrared spectra of alkanes

The infrared spectrum (see Chapters 14 and 31) of an alkane is extremely simple. Because C—C bonds are essentially non-polar, they do not absorb infrared radiation. The C—H bonds can undergo two basic types of vibration: stretching and bending. These can couple with the vibration of adjacent bonds to give slightly different absorption frequencies (see Table 23.7).

■ Table 23.7
IR absorption frequencies for some C—H bonds (see Chapter 14, page 279).

Vibration	Description	Frequency of absorption (wavenumber)/cm^{-1}
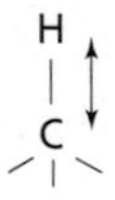	Simple C — H stretch	2880
	Symmetrical C — H stretch	2940
	Asymmetrical C — H stretch	2900
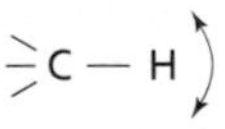	C — H bending	1450
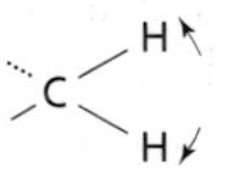	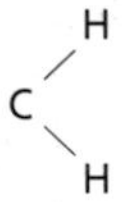 'wagging'	720

The spectra of butane and hexane in Figure 23.15 illustrate these vibrations. Since virtually every functional group has its own characteristic IR absorption, the absence of other peaks in the IR spectrum confirms the presence of an alkane.

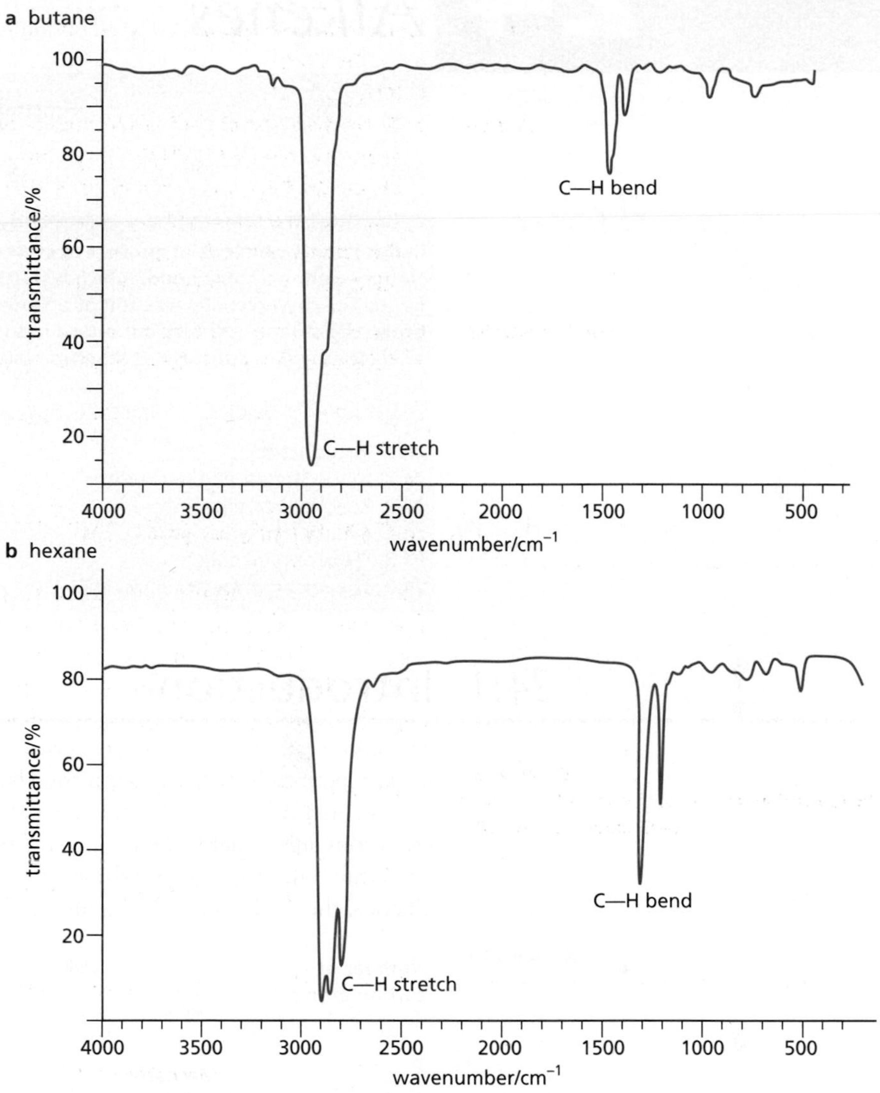

■ **Figure 23.15** IR spectra of **a** butane and **b** hexane.

Summary

- **Alkanes** form the homologous series of general formula C_nH_{2n+2}.
- They show structural (chain), and in some cases optical, isomerism.
- They can be extracted from crude oil by fractional distillation.
- They can be **cracked** and **re-formed** to produce useful fuels and petrochemical feedstocks.
- Their main use is as a source of energy by combustion.
- They react with chlorine and bromine in a **free radical substitution reaction**, giving chloro- or bromoalkanes.
- There are three stages to a **radical chain reaction** – **initiation**, **propagation** and **termination**.

Key reactions you should know

- Combustion:

$$C_nH_y + \left(n + \frac{y}{4}\right)O_2 \rightarrow nCO_2 + \frac{y}{2}H_2O$$

- Radical substitution:

$$C_nH_{2n+2} + X_2 \rightarrow C_nH_{2n+1}X + HX \quad (X = Cl \text{ or } Br)$$

24 *Alkenes*

Functional group:

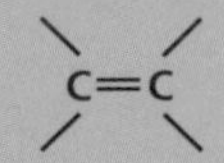

In this chapter we look at another class of hydrocarbon, the alkenes. Alkenes contain a carbon–carbon double bond, which is much more reactive than a carbon–carbon single bond. The characteristic reaction of alkenes is electrophilic addition. Many alkenes, especially ethene, are very important industrial chemicals. The addition polymerisation of alkenes and of substituted ethenes results in many useful plastics.

The chapter is divided into the following sections.

24.1 Introduction
24.2 Isomerism and nomenclature
24.3 Reactions of alkenes
24.4 Addition polymerisation
24.5 Preparing alkenes
24.6 The infrared spectra of alkenes

24.1 Introduction

The **alkenes** form a homologous series with the general formula C_nH_{2n}. They have many reactions in common with the alkanes (alkenes with three or more carbon atoms contain both C—H and C—C bonds, just like alkanes), and their physical properties are virtually identical (see Table 24.1). Reactions of the double bond are very different, however, and these are the dominant feature of the chemistry of alkenes, the double bond being much more reactive than a C—C single bond.

■ Table 24.1
Boiling points of some alkenes and alkanes.

Number of carbon atoms	Alkane		Alkene	
	Name	Boiling point/°C	Name	Boiling point/°C
1	Methane	−162		
2	Ethane	−88	Ethene	−102
3	Propane	−42	Propene	−48
4	Butane	0	But-1-ene	−6
			trans-But-2-ene	1
			cis-But-2-ene	4
	2-Methylpropane	−12	2-Methylpropene	−7
5	Pentane	36	Pent-1-ene	30
6	Hexane	69	Hex-1-ene	64

As we saw in Chapter 3, page 81, the bonding in ethene consists of a σ-bonded framework of two carbon atoms bonded to each other and to the four hydrogen atoms (see Figure 24.1**a**).
This leaves a p orbital on each carbon atom, which can then overlap sideways, giving the π bond (see Figure 24.1**b**).

For a more complete view of the bonding in ethene, see the panel on page 62.

The presence of the π bond confers two special characteristics on the structure and reactivity of alkenes:

- hindered rotation
- reaction with electrophiles.

We shall look at each in turn.

■ **Figure 24.1**
Two sp^2 carbon atoms can form a σ bond and a π bond in ethene.

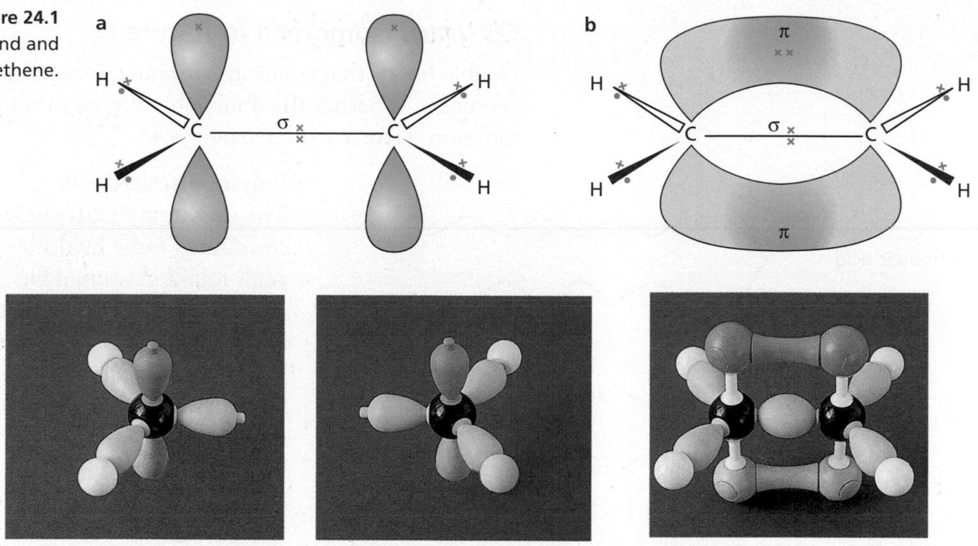

Hindered rotation

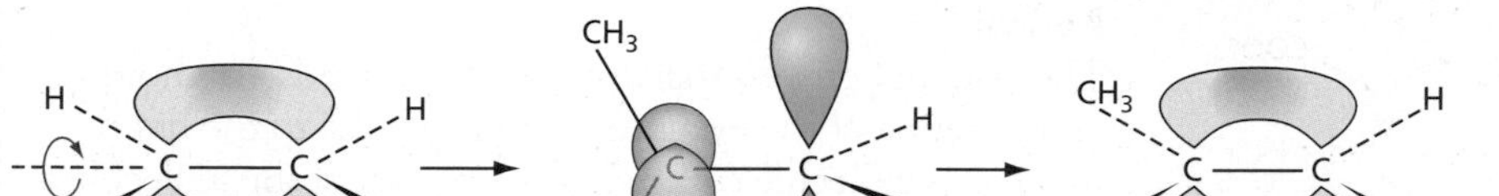

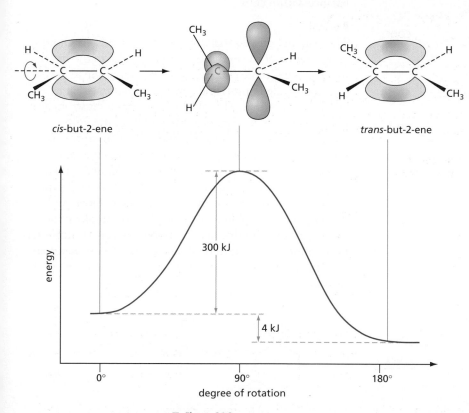

■ **Figure 24.2**
Hindered rotation around the C=C double bond in the but-2-enes.

Unlike the C—C σ bond in ethane, the π bond in ethene fixes the two ends of the molecule in their relative orientations. In order to twist one end with respect to the other, it would be necessary to break the π bond and re-form it again. This would require an input of about 300 kJ mol^{-1}, which is far beyond the energy available at normal temperatures. Therefore, no rotation occurs around the C=C double bond (see Figure 24.2). This large energy input contrasts with a barrier to rotation of only 12 kJ mol^{-1} in ethane. This is readily available to molecules from thermal energy at room temperature, so there is virtually total freedom of rotation about the C—C bonds in alkanes.

Hindered rotation allows the existence of geometrical (*cis–trans*) isomers (see Chapter 22, page 406, and also the next section in this chapter). Notice from Figure 24.2 that the *trans* isomer is more stable than the *cis* isomer by 4 kJ mol^{-1}. This is the case with most isomers of alkenes, and is significant in some natural alkenes.

Reaction with electrophiles

The ethene molecule is flat – both carbon atoms and all four hydrogen atoms lie in the same plane. The electrons in the π bond protrude from this plane, both above and below (see Figure 24.1**b**). This ready availability of electron density attracts electrophiles (see Chapter 22, page 414) to an alkene molecule, allowing a much greater range of reactivity than occurs with alkanes.

Cis–trans isomerism in nature

Double bonds that occur in particular compounds in nature are usually of a fixed geometry. Whether the double bond is *cis* or *trans* is important to its biological function. Here are three examples.

Polyunsaturated oils

The unsaturated fatty acids in certain vegetable oils are considered to be healthier for us than the saturated fatty acids found in animal fats. However, it is not so much the degree of unsaturation that is important, but the type of isomerism that exists around the double bonds. *Cis* double bonds are considered to be more healthy than *trans* ones. Fatty acids containing *trans* double bonds are thought to be no healthier than the fully saturated acids. One of the most healthy is *cis-cis*-linoleic acid, which can be converted in the body into an unsaturated prostaglandin by the route shown in Figure 24.3.

The correct balance of the various prostaglandins found in the body is essential for healthy circulation, effective nerve transmission and for maintaining the correct ionic balance within cells.

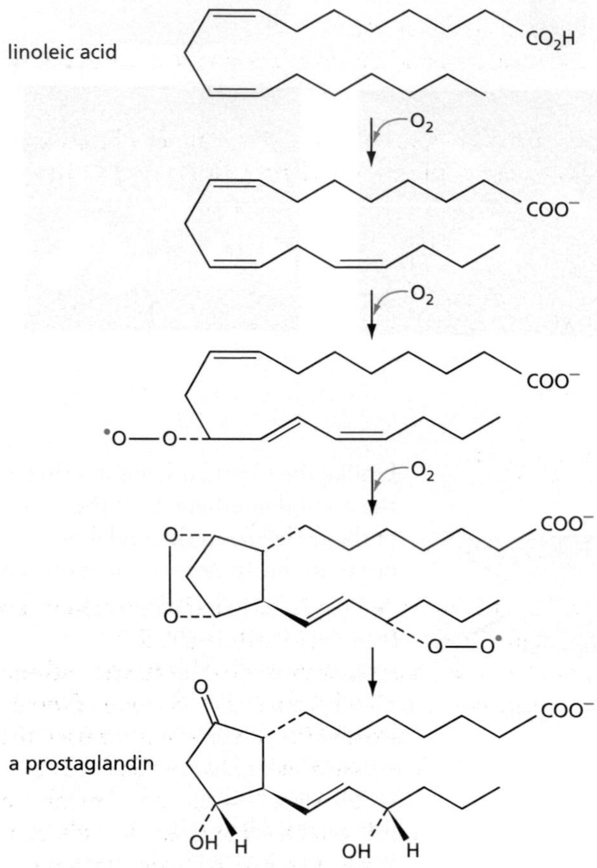

Figure 24.3 Conversion of *cis-cis*-linoleic acid to an unsaturated prostaglandin.

Retinal

The essential process that converts the arrival of a photon on the retina of the eye into a nerve impulse to the brain involves a molecule called rhodopsin. Part of the molecule is a protein (called opsin) and part is an unsaturated aldehyde called 11-*cis*-retinal. This aldehyde absorbs light strongly at a wavelength of 500 nm, near the middle of the visible spectrum. Within a few picoseconds of the absorption of a photon, the *cis* isomer of retinal isomerises to the *trans* isomer (see Figure 24.4). This causes the aldehyde end of the molecule to move about 0.5 nm, which in turn blocks the movement of sodium ions through cell membranes, and triggers a nerve impulse. (A single photon can block the movement of more than a million sodium ions.) Once it has been formed, the *trans*-retinal diffuses away from the opsin protein, and takes several minutes to be re-isomerised to the *cis* isomer.

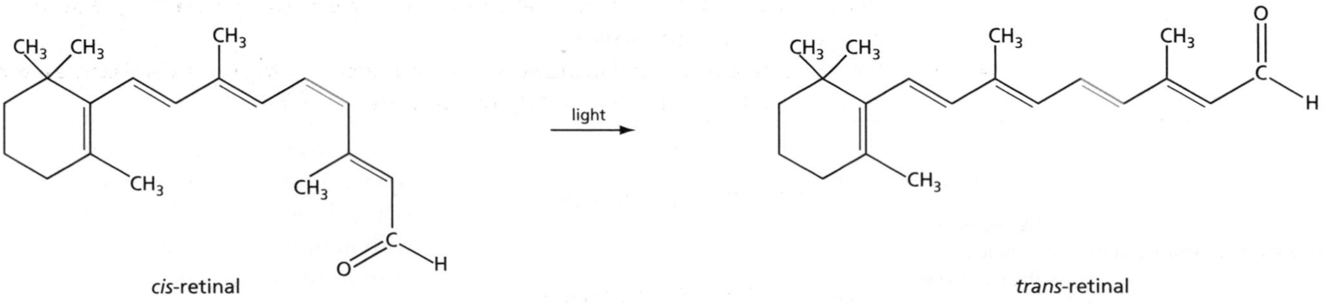

Figure 24.4 Conversion of *cis*-retinal to *trans*-retinal by the absorption of a photon of light.

Rubber and gutta-percha

These naturally occurring unsaturated polymers have quite different physical properties. The double bonds in rubber are *cis*, whereas those in gutta-percha are *trans*. The panel on pages 456–7 describes their structures and properties.

24.2 Isomerism and nomenclature

The principles of nomenclature for alkenes are similar to those used to name alkanes. The longest carbon chain that contains the $C=C$ bond is first identified, and then the side branches are named and positioned along the longest chain. There are two additional pieces of information that the name needs to convey:

- the position of the double bond along the chain
- if appropriate, whether the groups around the double bond are in a *cis* or a *trans* arrangement.

The position of the double bond is designated by including a number before the '*-ene*' at the end of the name, corresponding to the chain position of the first carbon in the $C=C$ bond. So:

$$
\begin{array}{ccc}
H & & H \\
\diagdown & & \diagup \\
& C = C & \\
\diagup & & \diagdown \\
CH_3 & & CH_2CH_3
\end{array}
$$

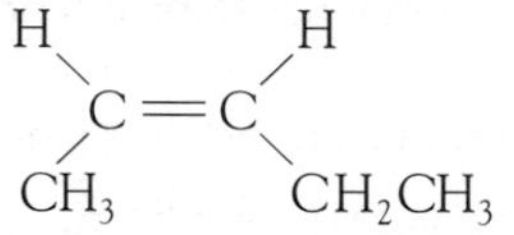

is *cis*-pent-2-ene, and:

$$
\begin{array}{ccc}
H & & H \\
\diagdown & & \diagup \\
& C = C & CH_2CH_2CH_3 \\
\diagup & & \diagdown \diagup \\
H & & CH \\
& & \diagdown \\
& & CH_2CH_2CH_3
\end{array}
$$

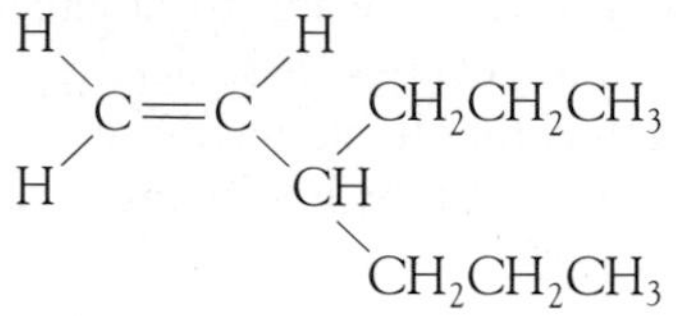

is 3-propylhex-1-ene. (Note that although the longest carbon chain contains seven atoms, this chain does not contain the $C=C$ bond, so is disregarded.)

In addition to the chain (and optical) isomerism shown by the alkanes, alkenes show two other types of isomerism: positional and geometrical. Because of these two additional types of isomerism, there are many more alkene isomers than alkane isomers with a given number of carbon atoms. For example, although there are only two alkanes with the formula C_4H_{10}, there are no fewer than four alkenes with the formula C_4H_8 (see Figure 24.5).

■ **Figure 24.5**
Isomers of four-carbon alkanes and alkenes.

C$_4$ alkanes

butane

2-methylpropane

C$_4$ alkenes

but-1-ene   trans-but-2-ene   cis-but-2-ene   2-methylpropene

---

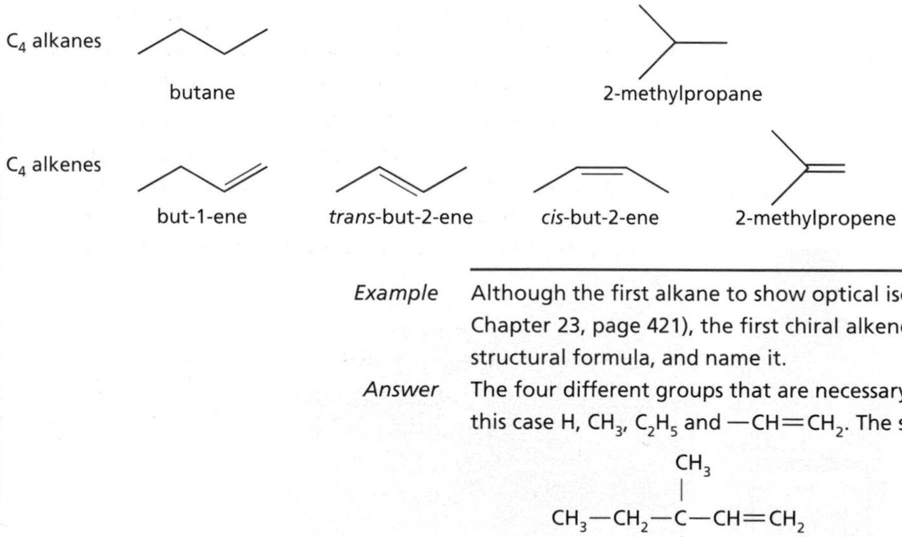

Example   Although the first alkane to show optical isomerism has the molecular formula C_7H_{16} (see Chapter 23, page 421), the first chiral alkene has the molecular formula C_6H_{12}. Draw its structural formula, and name it.

Answer   The four different groups that are necessary for a carbon atom to be a chiral centre are in this case H, CH_3, C_2H_5 and $-CH=CH_2$. The structure is therefore:

$$
\begin{array}{c}
CH_3 \\
| \\
CH_3-CH_2-C-CH=CH_2 \\
| \\
H
\end{array}
$$

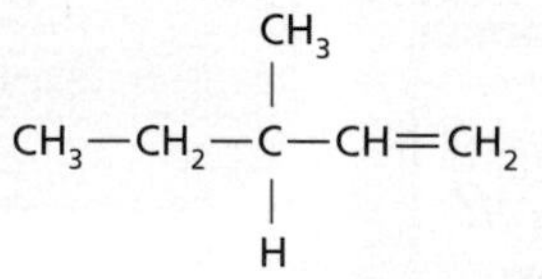

and its name is 3-methylpent-1-ene.

Further practice   Apart from 3-methylpent-1-ene, there are 14 other alkene isomers with the molecular formula C_6H_{12}.

1 Three of these have the same carbon skeleton as 3-methylpent-1-ene. Draw their structural formulae and name them.

2 Of the other 11, five have the same carbon skeleton as hexane, five have the same carbon skeleton as 2-methylpentane, and one has the same carbon skeleton as 2,2-dimethylbutane. Draw out the skeletal formulae of these other 11 isomers.

24.3 Reactions of alkenes

The double bond in alkenes is composed of a strong σ bond and a relatively weak π bond. What is more, as mentioned on page 441, the electrons that make up the π bond are readily available and on the surface of the molecule. Alkenes react by using these π-bond electrons to form bonds to other atoms, leaving the σ bond intact. Normally, this results in only one molecule of product being produced from two molecules of reactant, that is, **addition reactions** (see Chapter 22, page 414).

$$CH_2{=}CH_2 + A{-}B \rightarrow A{-}CH_2{-}CH_2{-}B$$

Addition reactions of alkenes are invariably exothermic. The energy needed to break the π bond and the A—B bond is less than that given out when the C—A and C—B bonds are formed. For example:

$$CH_2{=}CH_2 + H{-}H \rightarrow CH_3{-}CH_3 \qquad \Delta H^{\ominus} = -137\,kJ\,mol^{-1}$$

$$CH_2{=}CH_2 + Br{-}Br \rightarrow BrCH_2{-}CH_2Br \qquad \Delta H^{\ominus} = -90\,kJ\,mol^{-1}$$

Because they are able to 'absorb' (that is, react with) further hydrogen atoms in this way, alkenes are sometimes referred to as **unsaturated hydrocarbons**. Alkanes, on the other hand, having all of their carbon atoms' spare valencies used up by bonding with hydrogen atoms, are termed **saturated hydrocarbons**.

Hydrogenation

Alkenes can be hydrogenated to alkanes by reacting with hydrogen gas. The reaction does not occur readily without a catalyst, but with platinum it proceeds smoothly at room temperature and pressure. Commercially, alkenes themselves are never hydrogenated to alkanes (the alkenes are valuable feedstocks for further reactions, whereas the alkanes produced have value only as fuels). But the reaction is important in the manufacture of solid or semi-solid fats (margarine) from vegetable oils. In the commercial reaction the less effective, but cheaper, metal nickel is used as a catalyst, under a higher pressure.

$$R{-}CH{=}CH{-}R' + H_2 \xrightarrow{\text{Pt or Ni at 10 atm}} R{-}CH_2{-}CH_2{-}R'$$

■ **Figure 24.6**
Partial hydrogenation of sunflower oil produces the semi-solid spread. The mechanism of the catalysis is described in Chapter 19, page 368.

Addition of bromine

Unlike alkanes, alkenes react with bromine at room temperature, and even in the dark. The reaction takes place either with bromine water, or with a solution of bromine in an organic solvent such as hexane or trichloroethane.

$$CH_2{=}CH_2 + Br_2 \xrightarrow[\text{in hexane}]{\text{room temperature}} Br{-}CH_2{-}CH_2{-}Br$$

With bromine water, the major product is 2-bromoethanol, $BrCH_2CH_2OH$. The reaction is very fast – the orange bromine solution is rapidly decolorised. This is an excellent test for the presence of a C=C double bond, since few other compounds decolorise bromine water so rapidly.

The mechanism of the addition of bromine to an alkene

The reaction is an **electrophilic addition**, like most of the reactions of alkenes. As we have seen, the protruding π bond of ethene is an electron-rich area. This δ− area induces a dipole in an approaching bromine molecule, by repelling the electrons of the Br—Br bond (see Figure 24.7).

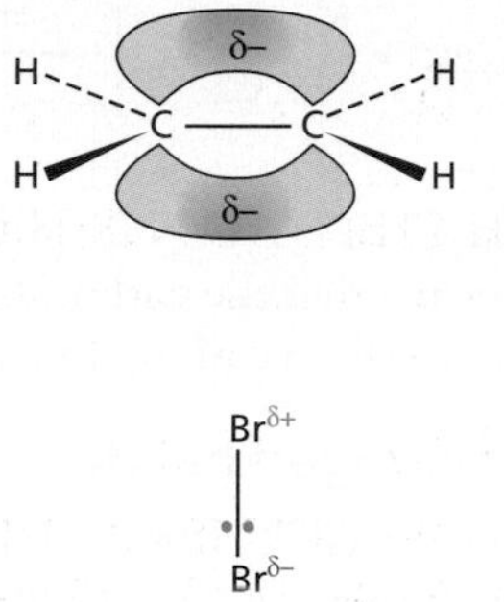

■ Figure 24.7 The electron-rich π bond induces a dipole in the bromine molecule.

Eventually, the Br—Br bond breaks completely, heterolytically, to form a bromide ion. The π electrons rearrange to form a σ bond from one of the carbon atoms to the nearest bromine. The movement of electrons away from the other carbon atom results in the production of a carbocation (see Figure 24.8).

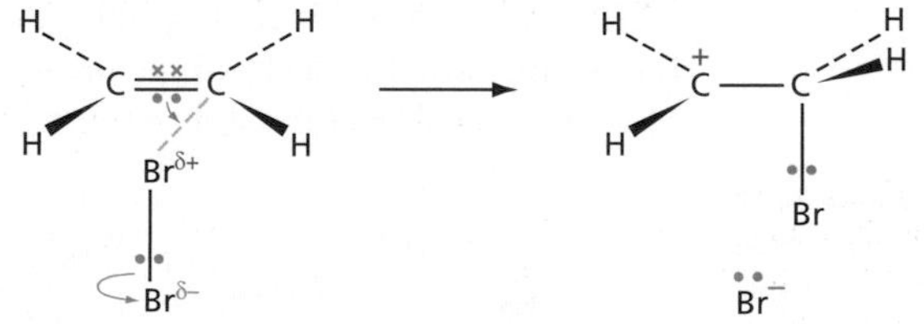

■ Figure 24.8 The Br—Br bond breaks, and a carbocation is formed.

There is a small amount of evidence to suggest that in some cases a lone pair of electrons on the bromine atom can form a dative bond to the cationic carbon atom, spreading out the charge (see Figure 24.9).

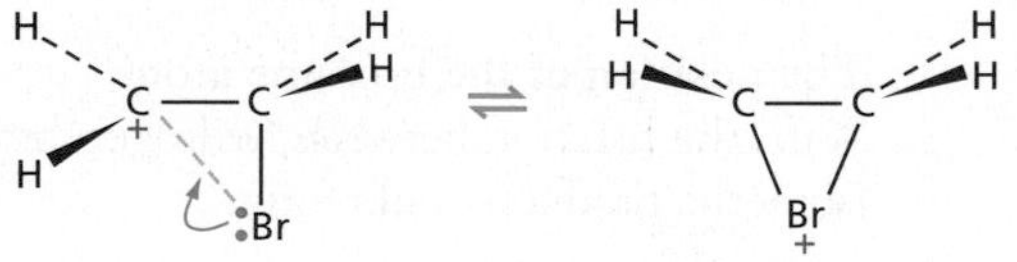

■ Figure 24.9 The lone pair on bromine can form a dative bond.

Finally, the bromide ion acts as a nucleophile and forms a (dative) bond to the carbocation (see Figure 24.10).

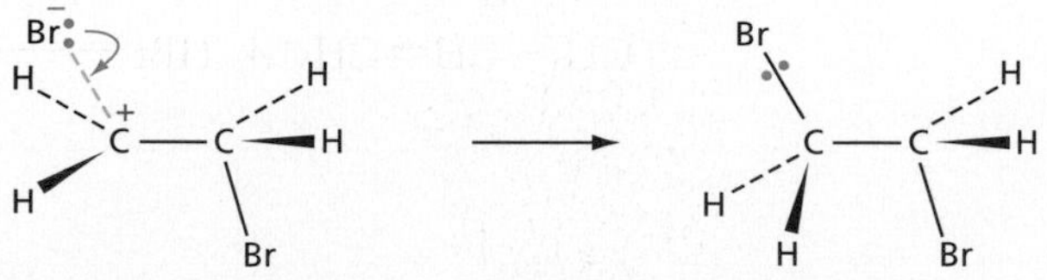

■ Figure 24.10 The free bromide ion bonds to the carbocation.

Other nucleophiles can also attack the carbocation. In bromine water, for example, the water molecule is present in a much higher concentration than the bromide ion, so is more effective as a nucleophile (see Figure 24.11).

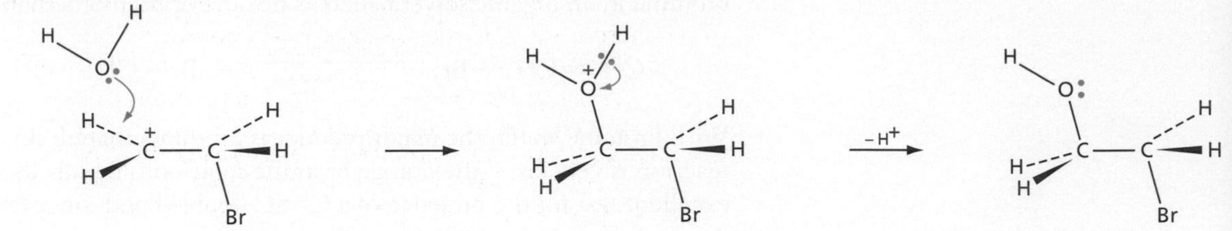

■ **Figure 24.11** In bromine water, water acts as the nucleophile.

The incorporation of a water molecule, and the consequent production of the bromoalcohol, is good evidence for this suggested mechanism via an intermediate carbocation. Another piece of evidence that supports the two-step mechanism is the incorporation of 'foreign' anions when bromination is carried out in an aqueous solution containing a mixture of various salts:

$$CH_2{=}CH_2 + \left.\begin{matrix} Na^+NO_3^- \\ Na^+Cl^- \end{matrix}\right\} \xrightarrow[\text{with } Br_2]{\text{in water}} \left\{\begin{matrix} O_2N{-}O{-}CH_2{-}CH_2{-}Br \\ HO{-}CH_2{-}CH_2{-}Br \\ Br{-}CH_2{-}CH_2{-}Br \\ Cl{-}CH_2{-}CH_2{-}Br \end{matrix}\right.$$

The ratio $(ClCH_2CH_2Br):(O_2NOCH_2CH_2Br)$ reflects the ratio $[Cl^-]:[NO_3^-]$ in the solution, showing that the carbocation picks up the first anion it collides with, as would be expected for a cation of such high reactivity.

Addition of hydrogen halides

The hydrogen halides HCl, HBr and HI all react with alkenes to give halogenoalkanes. Either the (gaseous) hydrogen halide is bubbled directly through the alkene, or ethanoic acid is used as a solvent. (Aqueous solutions of hydrogen halides such as hydrochloric acid cannot be used for this reaction because they cause the production of alcohols, as we shall see shortly.)

$$CH_2{=}CH_2(g) + HBr(g) \rightarrow \underset{\text{bromoethane}}{CH_3{-}CH_2Br(l)}$$

The mechanism of this electrophilic addition reaction is similar to that of bromination, but in this case the electrophile already has a permanent dipole. The $H^{\delta+}$ end of the H—Br molecule is attracted to the π bond in the initial step shown in Figure 24.12.

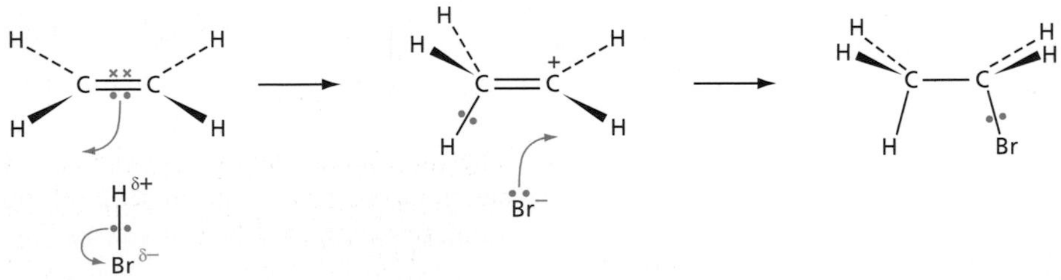

■ **Figure 24.12** The electrophilic addition of hydrogen bromide to ethene.

The position of the bromine atom

With the reaction between hydrogen bromide and propene, two different, isomeric, products could form:

$$CH_3{-}CH{=}CH_2 + HBr \rightarrow \underset{\text{2-bromopropane}}{CH_3{-}CH(Br){-}CH_3}$$

or

$$CH_3{-}CH{=}CH_2 + HBr \rightarrow \underset{\text{1-bromopropane}}{CH_3{-}CH_2{-}CH_2{-}Br}$$

When the reaction is carried out, it is found that 2-bromopropane is by far the major product. Virtually no 1-bromopropane is formed. This can be explained as follows.

The position of the bromine atom along the chain is determined by the position of the + charge in the intermediate carbocation.

$$CH_3-CH^+-CH_3 + Br^- \rightarrow CH_3-\overset{\overset{\displaystyle Br}{|}}{CH}-CH_3$$
secondary prop-2-yl carbocation

$$CH_3-CH_2-CH_2^+ + Br^- \rightarrow CH_3-CH_2-CH_2-Br$$
primary prop-1-yl carbocation

This, in turn, is determined by which end of the double bond the $H^{\delta+}$ of the HBr attaches itself to. We saw in the panel on page 437 that the inductive electron-donating effect of methyl groups causes the prop-2-yl radical to be more stable than the prop-1-yl radical. The effect is even more pronounced with carbocations, which have a carbon atom that is surrounded by only six electrons. The secondary prop-2-yl carbocation is much more stable than the primary prop-1-yl carbocation, and so the $H^{\delta+}$ adds on to position 1 of the double bond, rather than position 2 (see Figure 24.13).

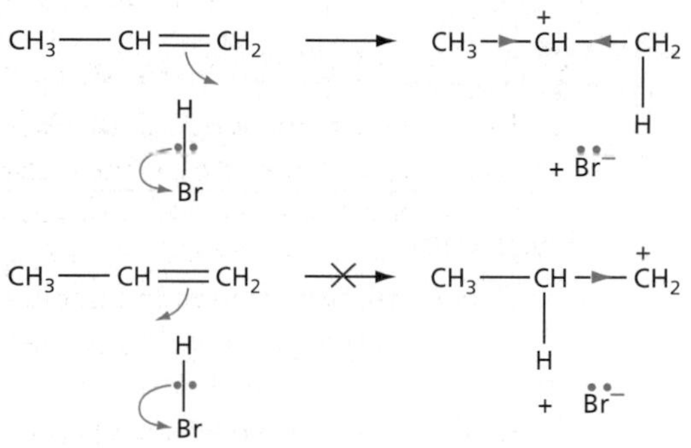

■ **Figure 24.13**
When the hydrogen adds to position 1, the more stable secondary carbocation is formed.

Tertiary carbocations such as

$$CH_3 \rightarrow \overset{+}{\underset{\underset{\displaystyle CH_3}{\uparrow}}{C}} \leftarrow CH_3$$

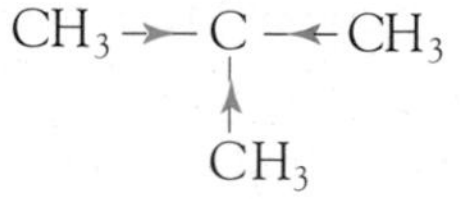

are even more stable than secondary carbocations. So 2-methylpropene reacts with hydrogen bromide to give 2-bromo-2-methypropane:

$$\underset{CH_3}{\overset{CH_3}{>}}C=CH_2 + HBr \rightarrow \underset{CH_3}{\overset{CH_3}{>}}C\underset{Br}{\overset{CH_3}{<}}$$

Markovnikov's rule

In 1869 the Russian chemist Vladimir Markovnikov formulated a rule to predict the orientation of the addition of a hydrogen halide to an unsymmetrical alkene (that is, one in which the two ends of the double bond are not the same). He stated that when HX adds to an unsymmetrical alkene, the hydrogen attaches itself to the least substituted end of the C=C double bond (the end that already has the most hydrogen atoms). A better formulation, based on what we now know of the mechanism, states that an electrophile adds to an unsymmetrical alkene in the orientation that produces the most stable intermediate carbocation.

> **Markovnikov's rule**
>
> When HX adds to a double bond:
>
> - H attaches to the carbon that already has the most Hs, or
> - the electrophile adds in the orientation that produces the most stable intermediate cation.

Example Predict and name the products of the following reactions:

a $CH_2=CH-CH_2-CH_3 + HCl \rightarrow$

b $(CH_3)_2C=CH-CH_3 + HBr \rightarrow$

Answer a The reaction could produce either $CH_3-C^+H-CH_2CH_3$ or $^+CH_2-CH_2-CH_2CH_3$ as the intermediate cation. The first one is more stable, being a secondary carbocation. This will lead to the formation of $CH_3CHClCH_2CH_3$, 2-chlorobutane, as product.

b The tertiary carbocation $(CH_3)_2C^+-CH_2CH_3$ is more stable than the secondary carbocation $(CH_3)_2CH-C^+H-CH_3$, so the product will be $(CH_3)_2CBr-CH_2CH_3$, 2-bromo-2-methylbutane.

Further practice Predict and name the products of the following reactions.

1

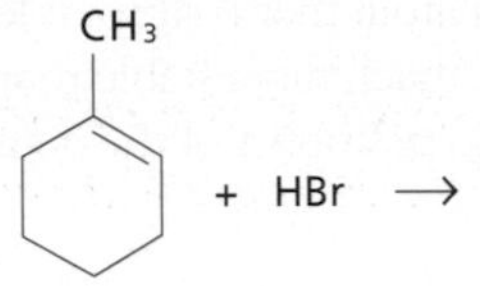

CH₃

+ HBr $\longrightarrow$

2 $CH_3CH_2CH_2CH=C$ 〈 $^{CH_3}_{CH_2CH_3}$ + HI $\longrightarrow$

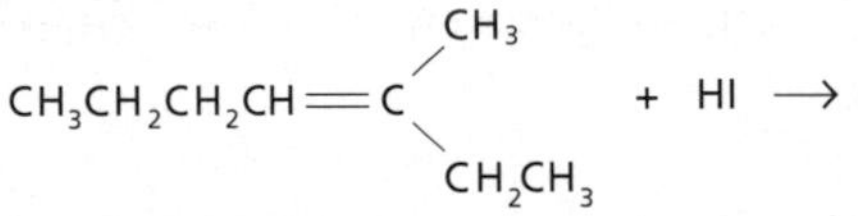

Hydration

In the presence of an acid, water undergoes an electrophilic addition reaction with alkenes to produce alcohols. In the laboratory this is carried out in two stages. Ethene (for example) is absorbed in concentrated sulphuric acid, with cooling, to form ethyl hydrogensulphate:

$$C_2H_4 + H_2SO_4 \rightarrow C_2H_5-O-SO_3H$$

This is then added to cold water:

$$C_2H_5-O-SO_3H + H_2O \rightarrow CH_3CH_2OH + H_2SO_4$$

The sulphuric acid therefore acts as a catalyst. The sodium salts of long-chain alkyl hydrogensulphates find a use as anionic detergents (see Chapter 27, page 508).

Industrially, the hydration of alkenes is the major way of manufacturing ethanol, propan-2-ol and butan-2-ol. The alkene and steam are passed over a phosphoric acid catalyst absorbed onto porous pumice (to give it a large surface area) at a pressure of 70 atm and a temperature of 300 °C.

$$CH_2=CH_2 + H_2O \rightarrow CH_3-CH_2OH$$

Phosphoric acid is used because, like sulphuric acid, it is a non-volatile acid. Although sulphuric acid is the cheaper of the two, it is an oxidising agent, so some by-products are formed with it. The mechanism is an electrophilic addition, with H^+ (from the acid) as the initial electrophile (see Figure 24.14).

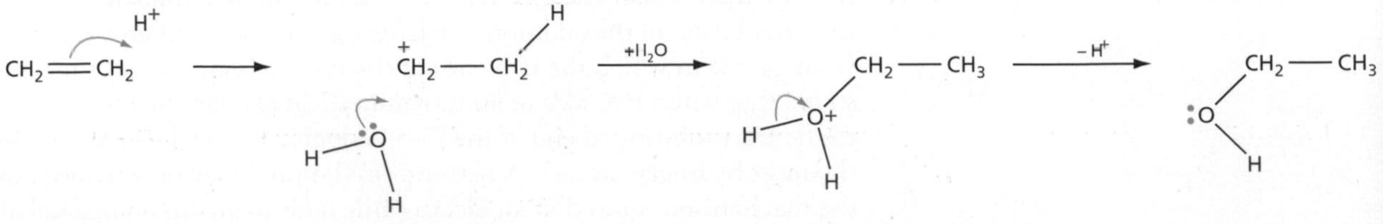

■ **Figure 24.14**
The hydration of ethene.

H^+ is regenerated in the last step, so can be used again.

Example The hydration reaction follows Markovnikov's rule. What alcohol will be formed from the hydration of:

a propene

b 2-methylbut-2-ene?

Answer **a** The most stable carbocation formed from $CH_3-CH=CH_2$ and H^+ is $CH_3-C^+H-CH_3$, so propan-2-ol will be produced.

b Likewise, $CH_3CH_2-C^+(CH_3)_2$ is the most stable carbocation formed from 2-methylbut-2-ene, so 2-methylbutan-2-ol will be the product.

Further practice Predict which alcohols will be formed from the hydration of:

1 but-2-ene

2 2-methylpropene.

As can be seen from example **a** above, the hydration of alk-1-enes always produces alkan-2-ols, rather than alkan-1-ols. The primary alcohols have to be made by a different route. This is reflected in their cost – primary alcohols cost twice as much per litre as do the corresponding secondary alcohols.

Oxidative addition of oxygen

An important industrial reaction of ethene is its catalytic oxidation by air:

$$CH_2=CH_2 + \tfrac{1}{2}O_2 \xrightarrow[250\,°C]{\text{silver catalyst}} \underset{\displaystyle \diagdown_{\textstyle O}\diagup}{CH_2-CH_2}$$

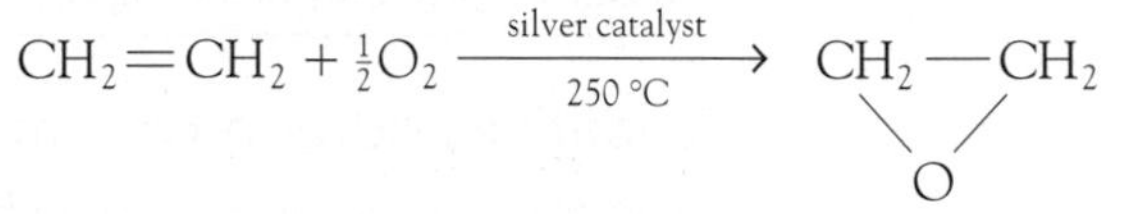

The product epoxyethane (or ethylene oxide) is an important intermediate for making many useful industrial chemicals:

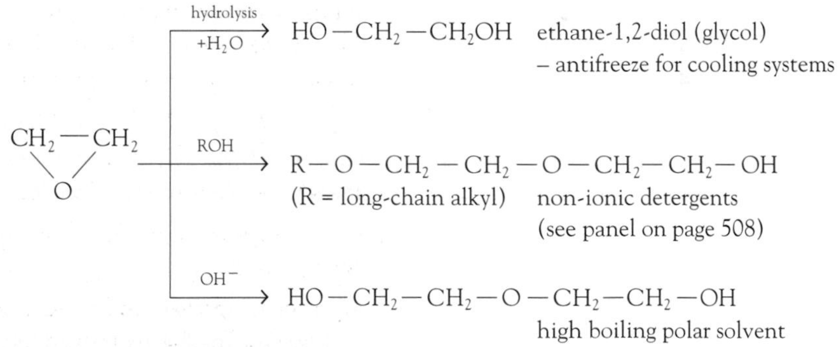

$$\underset{\displaystyle \diagdown_{\textstyle O}\diagup}{CH_2-CH_2}$$

– **hydrolysis** $+H_2O$ → $HO-CH_2-CH_2OH$ ethane-1,2-diol (glycol) – antifreeze for cooling systems

– **ROH** → $R-O-CH_2-CH_2-O-CH_2-CH_2-OH$ (R = long-chain alkyl) non-ionic detergents (see panel on page 508)

– **OH⁻** → $HO-CH_2-CH_2-O-CH_2-CH_2-OH$ high boiling polar solvent

Oxidative addition of ozone (ozonolysis)

When ozone gas is passed into a solution of an alkene, the O_3 molecule adds onto the double bond to give an **ozonide**. Treatment of the ozonide with zinc dust and water produces carbonyl compounds (see Chapter 28). The reaction (see Figure 24.15) has little commercial or preparative value, but is useful for determining the structures of unknown alkenes.

■ **Figure 24.15**
The ozonolysis of an alkene. R, R′ and R″ are alkyl groups.

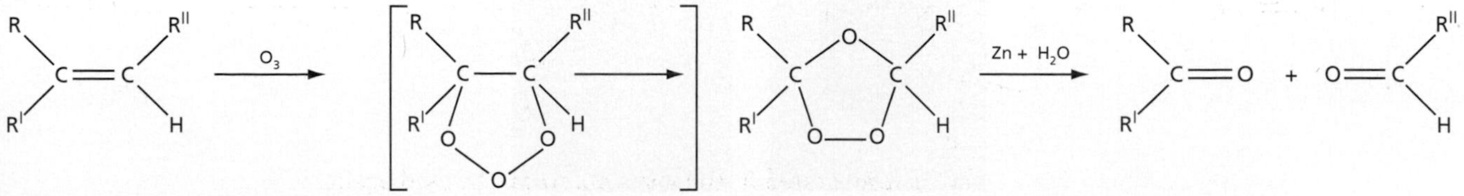

Example **A** and **B** are two isomers with the formula C_5H_{10}. On treatment with ozone, followed by zinc dust and water, compound **A** gave a mixture of ethanal, CH_3CHO, and propanone, CH_3COCH_3. Under the same conditions compound **B** gave a mixture of methanal, CH_2O, and butanone, $CH_3COCH_2CH_3$. What are the structures of **A** and **B**?

Answer Each end of the double bond in an alkene ends up as a C═O group with ozonolysis. We can work backwards from the two products, and put the pieces of the jigsaw together as shown in Figure 24.16.

■ **Figure 24.16**

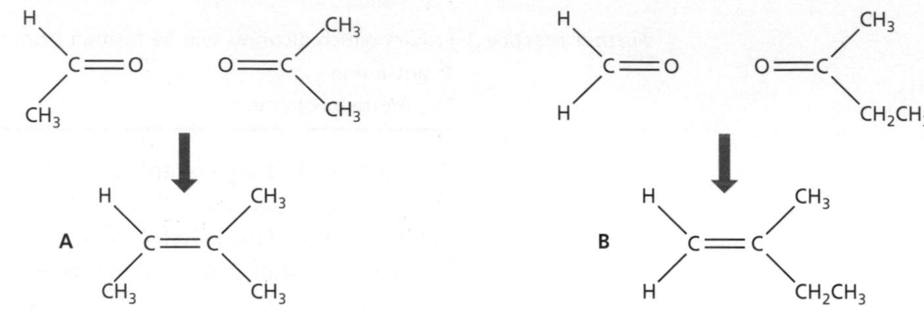

A is therefore 2-methylbut-2-ene and B is 2-methylbut-1-ene.

Further practice Work out the structures of the two compounds **C** and **D** from the following information.

1 C (C_7H_{12}) gave CH_3CHO, $(CH_3)_2CO$ and OHC—CHO in equimolar amounts on ozonolysis.

2 D (C_6H_{12}) gave only one compound $(CH_3)_2CO$ on ozonolysis.

Oxidative addition with potassium manganate(VII)

Apart from the decolorisation of bromine water, another very good test for the presence of a carbon–carbon double bond is the reaction between alkenes and dilute acidified potassium manganate(VII), $KMnO_4(aq)$. When alkenes are shaken with the reagent, its purple colour disappears. Identification of the product shows that a diol has been formed:

$$CH_2{=}CH_2 + H_2O + [O] \rightarrow HO{-}CH_2{-}CH_2{-}OH$$

([O] represents an oxygen atom that has been given by the manganate(VII).)

The reaction occurs rapidly at room temperature, whereas potassium manganate(VII) will only oxidise other compounds (such as alcohols and aldehydes) slowly under these conditions. It is therefore a useful test for the presence of an alkene.

Tests for alkenes

- Alkenes rapidly decolorise bromine water (see page 444).
- Alkenes rapidly decolorise dilute potassium manganate(VII).

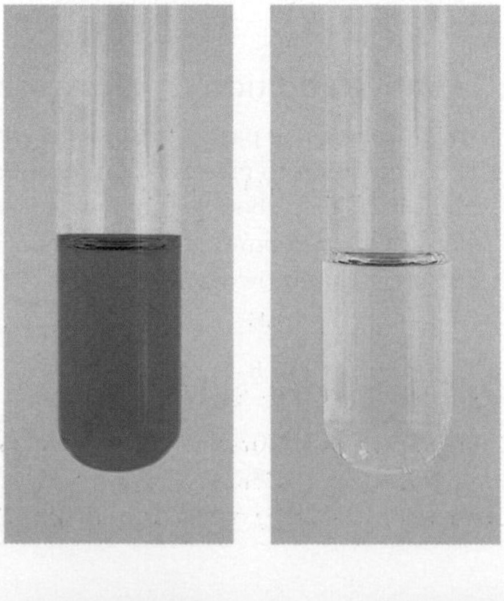

■ **Figure 24.17** Alkenes decolorise bromine water.

Oxidative cleavage with hot potassium manganate(VII)

When heated with acidified potassium manganate(VII), alkenes undergo a similar cleavage reaction to their reaction with ozone. The products from the oxidation depend on the degree of substitution on the C=C bond (see Table 24.2).

■ **Table 24.2** Products from the cleavage of double bonds by potassium manganate(VII).

If the double bond carbon is:	Then the oxidised product will be:
$CH_2=$	CO_2
$R(H)C=$	$R-C(=O)OH$
$R(R)C=$	$R(R)C=O$

An example is shown in Figure 24.18.

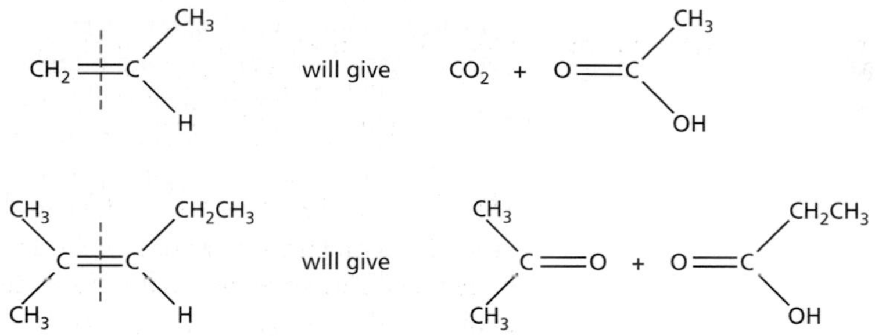

■ **Figure 24.18** Products of oxidative cleavage by potassium manganate(VII).

Example Use Table 24.2 to predict the products of the potassium manganate(VII) oxidative cleavage of:

a

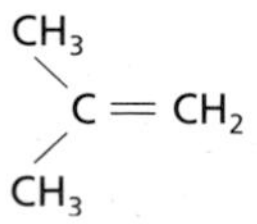

b

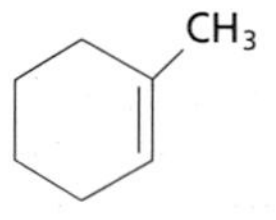

Answer **a** Splitting the double bond in the middle gives:

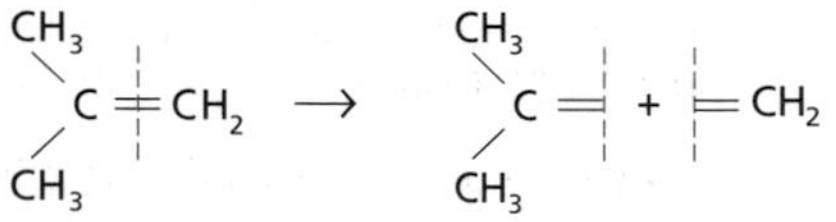

which, according to Table 24.2, will give propanone, $(CH_3)_2CO$, and carbon dioxide, CO_2.

b Similar splitting gives:

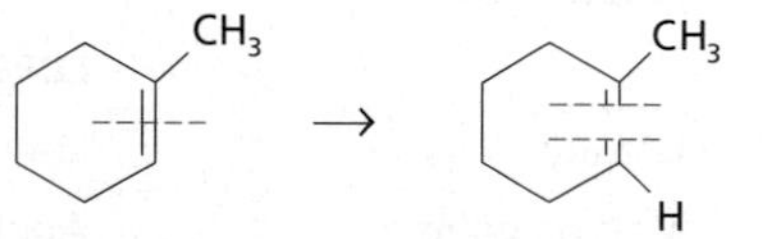

which eventually will produce:

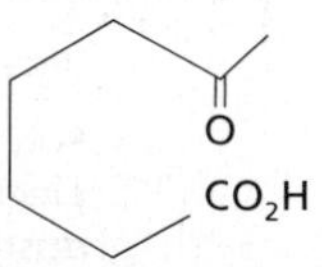

or $CH_3-CO-CH_2CH_2CH_2CH_2CO_2H$

Further practice Geraniol is an alcohol that occurs in oil of rose and other flower essences. Its structural formula is $(CH_3)_2C=CHCH_2CH_2C(CH_3)=CHCH_2OH$. What are the structures of the three compounds formed by oxidation with hot acidified potassium manganate(VII)?

24.4 Addition polymerisation

Formation of poly(ethene)

In 1933, chemists working at ICI in Britain discovered by chance that when ethene was subjected to high pressures in the presence of a trace of oxygen, it produced a soft white solid.

This solid was poly(ethene), and this was the start of the enormous industry based on addition polymers.

$$n(CH_2{=}CH_2) \xrightarrow[\text{1000 atm and 200 °C}]{\text{trace of oxygen}} \left[\begin{array}{cc} \overset{\displaystyle H}{\underset{\displaystyle H}{\vert}} & \overset{\displaystyle H}{\underset{\displaystyle H}{\vert}} \\ -C & -C- \\ \end{array}\right]_{n\ (n\,>\,5000)}$$

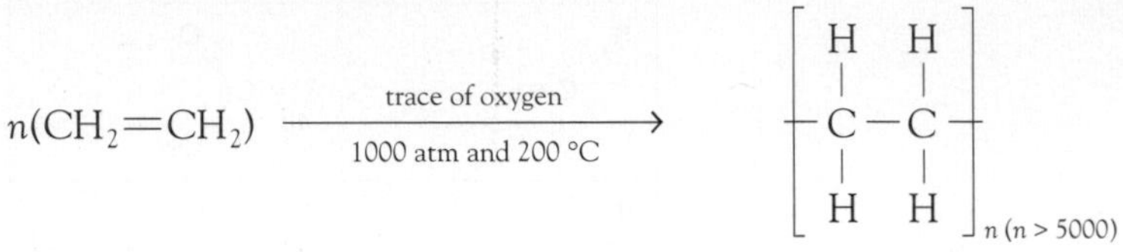

poly(ethene) (LDPE)

Poly(ethene), commonly called polythene, is termed an **addition polymer** because one molecule is made from many ethene units (**monomers**) adding together, without the co-production of any small molecule (contrast the process of condensation polymerisation – see Chapter 29, page 554). Just as in all other addition reactions of ethene, the C=C double bond is replaced by a C—C single bond, and two carbon atoms make two new single bonds to other atoms. The original reaction conditions are still used with little modification to produce **low-density poly(ethene)** or (**LDPE**). Sometimes an organic peroxide is used instead of oxygen as an initiator of the radical chain reaction (see the panel opposite). Using a different catalyst, developed by Karl Ziegler and Giulio Natta in 1953, ethene can be polymerised at much lower pressures to produce **high-density poly(ethene) (HDPE)**:

$$n(CH_2{=}CH_2) \xrightarrow[\text{in heptane at 50 °C and 2 atm}]{Al(C_2H_5)_3\,+\,TiCl_4} \left[\begin{array}{cc} \overset{\displaystyle H}{\underset{\displaystyle H}{\vert}} & \overset{\displaystyle H}{\underset{\displaystyle H}{\vert}} \\ -C & -C- \\ \end{array}\right]_{n\ (n\,\cong\,100\,000)}$$

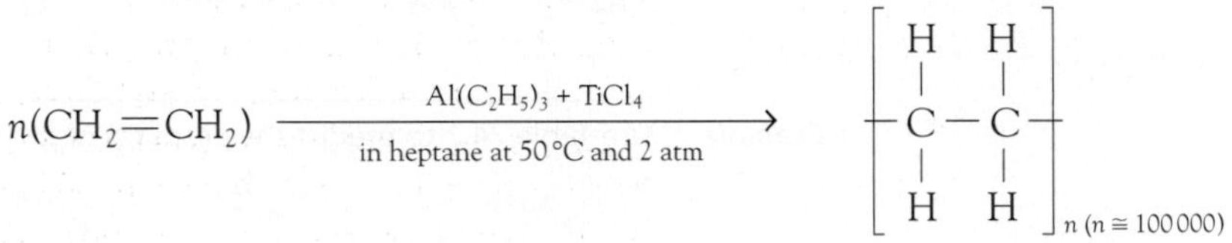

poly(ethene) (HDPE)

The major differences between the structures of LDPE and HDPE are that:

- the average chain length of LDPE is much shorter than that of HDPE
- the chains of LDPE are branched, whereas the chains of HDPE are unbranched.

These differences have an effect on the physical properties, and hence the uses, of the two different types of poly(ethene) (see Table 24.3).

■ Table 24.3
Properties and uses of LDPE and HDPE.

Property/use	Polymer	
	LDPE	**HDPE**
Density	Low	High
Melting point	Approx 130°C	Approx 160°C
Tensile strength	Low	Higher
Flexibility	Very flexible	Much more rigid
Uses	Polythene bags, electrical insulation, dustbin liners	Bottles, buckets, crates

So by altering the chain length and the degree of branching we can alter the physical properties of poly(ethene) to suit the purpose for which it is being used. But chemists and materials scientists have another chemical property they can vary – the structure of the monomer. We shall look at this on page 454.

How it happens: the mechanisms of polymerisation

The radical mechanism, using oxygen or peroxides

Oxygen or alkoxy radicals, produced by gently heating an organic peroxide, can add onto an ethene molecule:

$$R-O-O-R \xrightarrow{\text{heat}} R-O^\bullet + {}^\bullet O-R$$

$$R-O^\bullet + \underset{H}{\overset{H}{\diagdown}}C=C\underset{H}{\overset{H}{\diagup}} \rightarrow R-O-\overset{\displaystyle H}{\underset{\displaystyle H}{C}}-\overset{\displaystyle H}{\underset{\displaystyle H}{C}}{}^\bullet$$

$$\underset{\substack{\text{(oxygen} \\ \text{molecule)}}}{{}^\bullet O-O^\bullet} + \underset{H}{\overset{H}{\diagdown}}C=C\underset{H}{\overset{H}{\diagup}} \rightarrow {}^\bullet O-O-\overset{\displaystyle H}{\underset{\displaystyle H}{C}}-\overset{\displaystyle H}{\underset{\displaystyle H}{C}}{}^\bullet$$

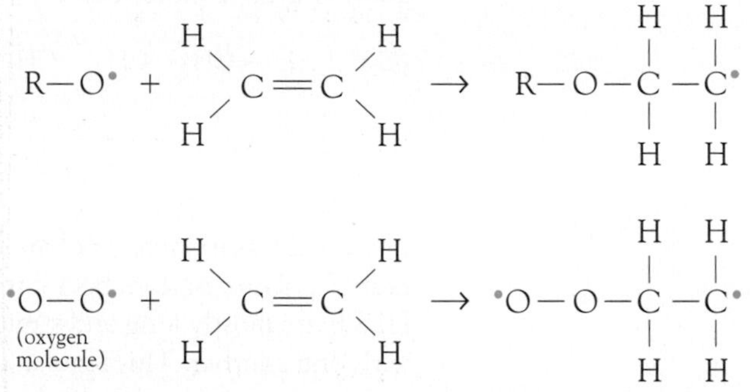

The organic radical formed in this way can add onto another ethene molecule:

$$R-O-\overset{\displaystyle H}{\underset{\displaystyle H}{C}}-\overset{\displaystyle H}{\underset{\displaystyle H}{C}}{}^\bullet + \underset{H}{\overset{H}{\diagdown}}C=C\underset{H}{\overset{H}{\diagup}} \rightarrow R-O-\overset{\displaystyle H}{\underset{\displaystyle H}{C}}-\overset{\displaystyle H}{\underset{\displaystyle H}{C}}-\overset{\displaystyle H}{\underset{\displaystyle H}{C}}-\overset{\displaystyle H}{\underset{\displaystyle H}{C}}{}^\bullet$$

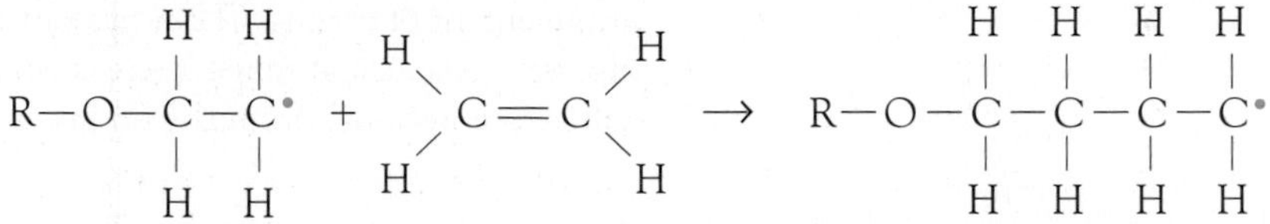

Eventually very long chains of ethene units form.

One particular feature of radical polymerisation is due to the high reactivity of organic free radicals. This can give rise to the phenomenon called **back-biting**. The growing polymer chain folds back on itself, and the primary radical abstracts a hydrogen atom from a carbon atom in the middle of the chain. This results in the formation of a secondary radical (recall from Chapter 23 that secondary radicals are more stable than primary radicals):

$$R-\underset{\substack{\diagdown \\ CH_2-CH_2}}{\overset{H}{CH}}\overset{{}^\bullet CH_2}{\underset{\diagup}{\diagdown}CH_2} \rightarrow R-\underset{\substack{\diagdown \\ CH_2-CH_2}}{\overset{CH_3}{\overset{\bullet}{C}H}}\overset{\diagdown}{\underset{\diagup}{CH_2}}$$

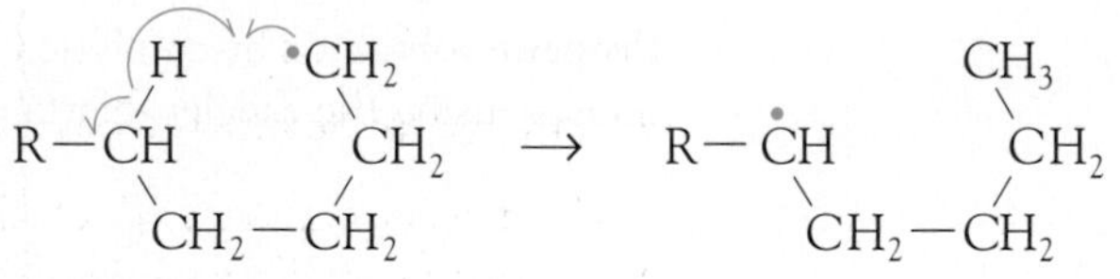

The secondary radical can then continue the chain polymerisation, adding onto further ethene molecules:

$$R-\underset{\displaystyle (CH_2)_3CH_3}{\overset{\displaystyle \bullet}{C}H} + \underset{H}{\overset{H}{\diagdown}}C=C\underset{H}{\overset{H}{\diagup}} \rightarrow R-\underset{\displaystyle (CH_2)_3CH_3}{CH}-CH_2-CH_2{}^\bullet \xrightarrow{+ C_2H_4} \cdots$$

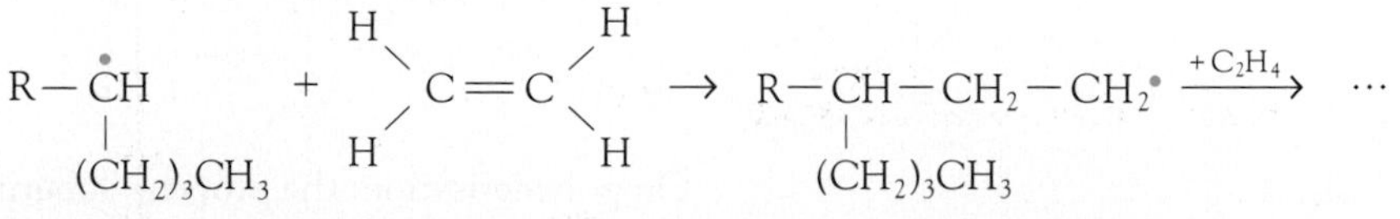

This explains how the chains become increasingly branched. This is the key structural feature of LDPE. Chain termination occurs when two growing chains join together:

$$R-CH_2-CH_2{}^\bullet + {}^\bullet CH_2-CH_2-R \rightarrow R-CH_2-CH_2-CH_2-CH_2-R$$

or when one radical abstracts a hydrogen atom from another radical:

$$R-CH_2-CH_2^\bullet \qquad\qquad R-CH_2-CH_3$$
$$\overset{\displaystyle H}{\underset{\displaystyle R-CH-CH_2}{\big|}} \rightarrow$$
$$R-CH=CH_2$$

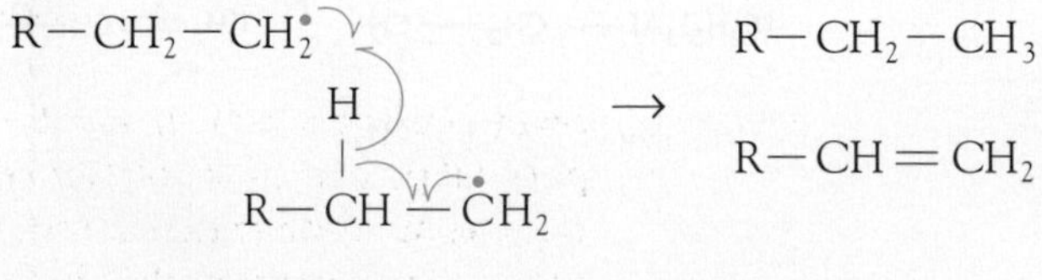

The ionic mechanism, using Ziegler–Natta catalysts

Triethylaluminium is a Lewis acid (see Chapter 6) and an electrophile. It can accept the pair of π electrons of ethene to form a carbocation which is very similar to that found in the usual electrophilic addition reactions:

$$(C_2H_5)_3Al \quad CH_2 = CH_2 \ \rightarrow \ (C_2H_5)_3Al^- - CH_2 - CH_2^+$$

This carbocation can then act as an electrophile in an addition reaction with another ethene molecule:

$$(C_2H_5)_3Al^- - CH_2 - CH_2^+ \quad CH_2 = CH_2 \rightarrow (C_2H_5)_3Al^- - CH_2 - CH_2 - CH_2 - CH_2^+$$

$$\downarrow {\scriptstyle + C_2H_4}$$

etc.

Eventually, as in radical polymerisation, long chains of ethene units form. Unlike radical polymerisation, back-biting does not occur. This means that the chains in HDPE are mostly long and straight. They can align in a parallel arrangement with one another. This regular arrangement allows stronger van der Waals forces to occur between molecules (giving a higher melting point), and it gives a more compact arrangement (and a higher density). The areas in the polymer where this alignment occurs are termed 'crystalline', because of the highly ordered arrangement of chains. (They are not true crystals, in the sense of having a distinct geometrical shape.) Areas where the chains are arranged randomly are called 'amorphous' (from the Greek word meaning 'without shape').

Addition polymerisation of other alkenes

> When monomers join together without the formation of any other products, the resulting polymer is called an **addition polymer**.

Propene (obtained by catalytic cracking) can be polymerised in a similar way to ethene, using the Ziegler–Natta catalyst:

$$n(CH_3 - CH = CH_2) \xrightarrow{\ (C_2H_5)_3Al\,+\,TiCl_4\ } \left[\begin{array}{cc} CH_3 & H \\ | & | \\ C & - C \\ | & | \\ H & H \end{array} \right]_n$$

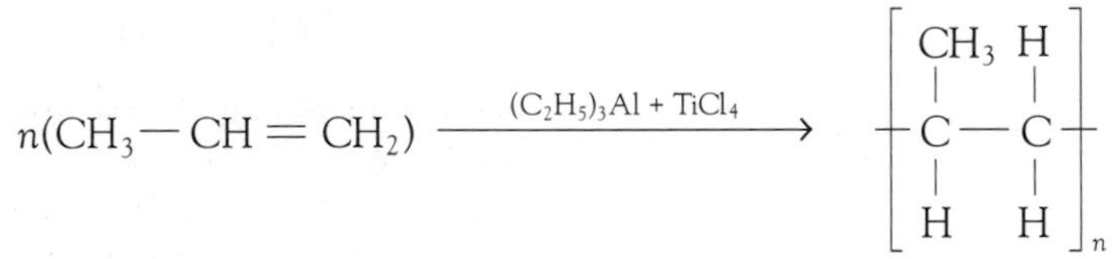

poly(propene)

On polymerisation, the propene subunits line up in a regular head-to-tail arrangement, because as each propene molecule is added to the growing chain, the more stable secondary carbocation is formed as intermediate (see Figure 24.19).

■ **Figure 24.19**
The propene units line up head to tail.

$$(C_2H_5)_3\bar{Al} - CH_2 - \overset{+}{C}H \qquad CH_2 = CH - CH_3 \rightarrow (C_2H_5)_3\bar{Al} - CH_2 - CH - CH_2 - \overset{+}{C}H$$
$$\phantom{(C_2H_5)_3\bar{Al} - CH_2 - }CH_3 \qquad\qquad\qquad\qquad\qquad\qquad\qquad\qquad CH_3 \qquad\quad CH_3$$

head · · · · · head · · · · · tail · · · · · tail

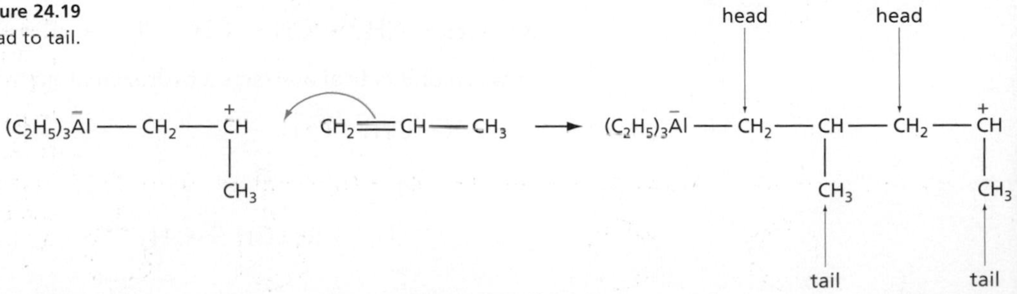

Even with a head-to-tail arrangement, there are many different ways of arranging the methyl side groups along the chain. One of the most useful (and the one that is usually formed under Ziegler–Natta conditions) is the arrangement in which all the methyl groups are on the same side of the chain:

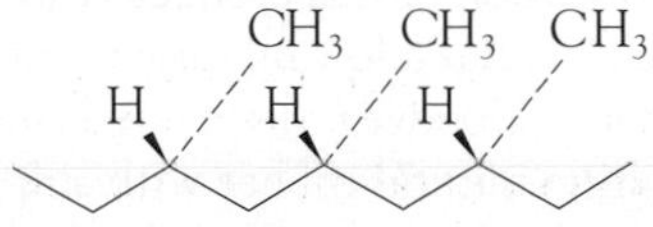

This **isotactic** arrangement (from the Greek word meaning 'equal-ordered') allows the chains to align in crystalline regions, giving the poly(propene) a high melting point and strength. Its main use is in moulded articles such as furniture and car bumpers. When spun into fibres its strength and hard-wearing properties find a use in carpets and ropes.

Two other arrangements of the methyl side chains are known. Which of these occurs in a particular polymer depends on the conditions used (the proportions of the two catalysts, and the temperature).

- A **syndiotactic** polymer has its methyl groups arranged alternately, to one side and to the other:

- An **atactic** polymer has a random arrangement of methyl groups (but note that they are still on alternate carbon atoms of the chain):

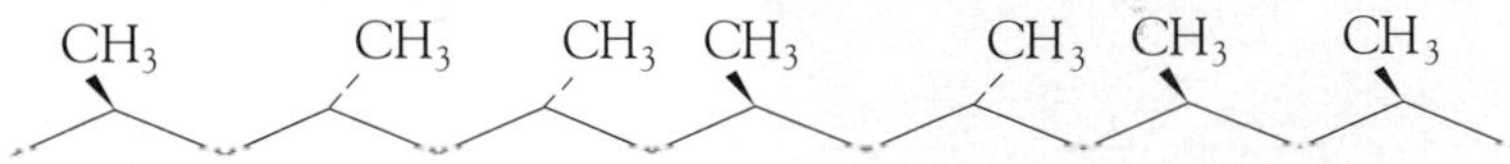

Phenylethene can be polymerised to form poly(phenylethene), otherwise known as polystyrene:

$$nCH_2 = C\diagdown_H \xrightarrow{\text{heat with an organic peroxide}} \left[\begin{array}{c} H \ H \\ | \ | \\ -C-C- \\ | \\ H \end{array}\right]_n$$

poly(phenylethene)
(polystyrene)

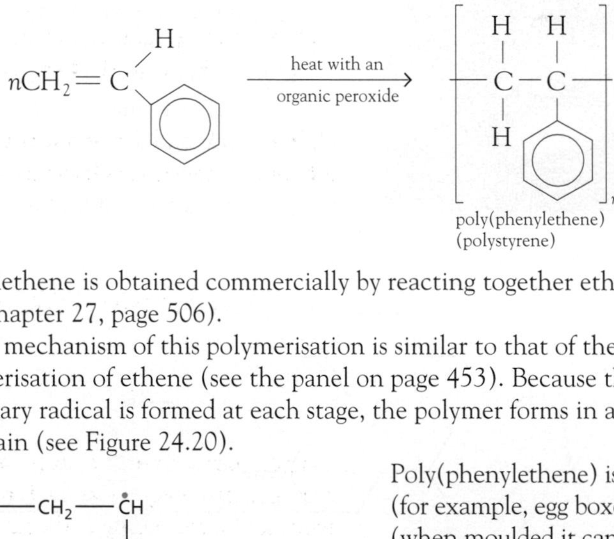

Phenylethene is obtained commercially by reacting together ethene and benzene (see Chapter 27, page 506).

The mechanism of this polymerisation is similar to that of the radical polymerisation of ethene (see the panel on page 453). Because the more stable secondary radical is formed at each stage, the polymer forms in a regular head-to-tail chain (see Figure 24.20).

■ **Figure 24.20**
In phenylethene, the monomer units line up head to tail.

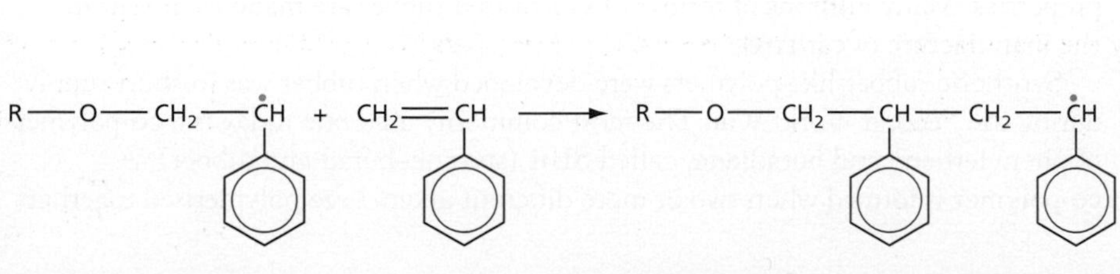

$$R-O^{\bullet} + CH_2=CH \longrightarrow R-O-CH_2-\overset{\bullet}{C}H$$

$$R-O-CH_2-\overset{\bullet}{C}H + CH_2=CH \longrightarrow R-O-CH_2-CH-CH_2-\overset{\bullet}{C}H$$

Poly(phenylethene) is used for packaging (for example, egg boxes), and model kits (when moulded it can depict fine details exactly). When an inert gas is bubbled through the phenylethene as it is being polymerised, the familiar white, very light solid known as **expanded polystyrene** is formed. This is used as a protective packaging, and for sound and heat insulation. Phenylethene is an important component of the co-polymers SBR and ABS (see the panel overleaf).

Rubber – natural and synthetic copolymers

The rubber tree is indigenous to South America. When its bark is stripped, it oozes a white sticky liquid called latex, which is an emulsion of rubber in water. Purification of this produces rubber. In its natural state, rubber is not particularly useful. It has a low melting point, is sticky and has a low tensile strength. In 1839, Charles Goodyear discovered the process of **vulcanisation**, which involves heating natural rubber with sulphur. This produced a substance that had a higher melting point, and was much stronger. Rubber is an addition polymer of 2-methylbutadiene (isoprene) (see Figure 24.21).

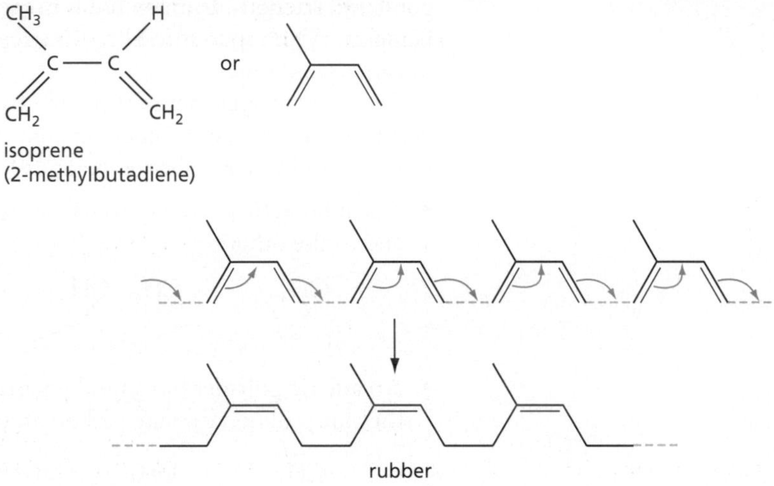

■ **Figure 24.21** (right) Isoprene undergoes addition polymerisation to form rubber.

■ **Figure 24.22** (below) A rubber tree being tapped. The white latex is converted to rubber by the process of vulcanisation.

Note than when a diene undergoes addition polymerisation, a double bond is still present in the product. This double bond may be *cis* or *trans*. In natural rubber, all the double bonds are in the *cis* configuration. The naturally occurring substance called gutta percha is an isomer of rubber, in which all double bonds are *trans*. Gutta percha is harder than rubber, and is used as a coating for golf balls.

The presence of the double bond in rubber allows further addition reactions to take place. This is what happens during the process known as **vulcanisation**. Sulphur atoms add across the double bonds in different chains, cross-linking the rubber molecules (see Figure 24.23). This stops the chains from moving past each other, and gives the material more rigidity.

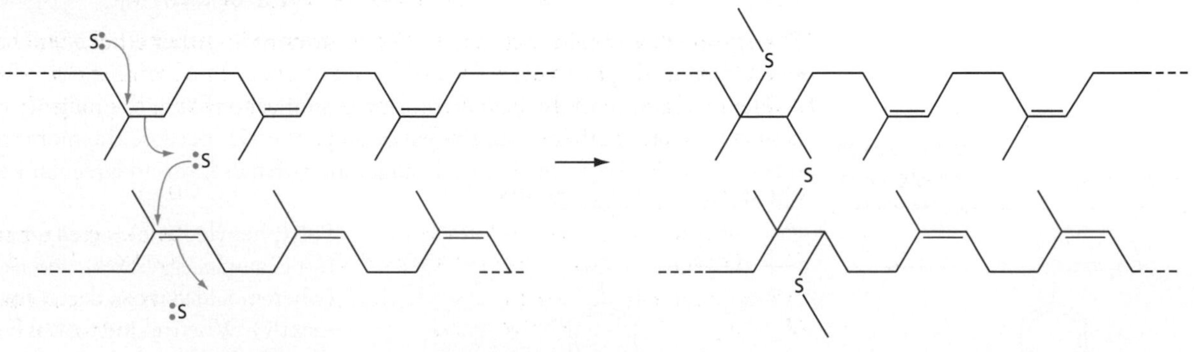

■ **Figure 24.23** Vulcanisation of rubber.

Not all double bonds have sulphur added to them. That would make the substance too hard. About 5% sulphur by mass is adequate to give the desired properties. Many millions of tonnes of vulcanised rubber are made each year for the manufacture of car tyres.

Synthetic rubber-like polymers were developed when rubber was in short supply during the Second World War. The most commonly used one today is a co-polymer of phenylethene and butadiene, called **SBR** (styrene–butadiene rubber). A **co-polymer** is formed when two or more different alkenes are polymerised together.

Even if the ratio of monomers is 50 : 50, there is no guarantee that the monomer fragments will alternate along the chain. The order is fairly random (in contrast to condensation co-polymerisation – see Chapter 29, page 554).

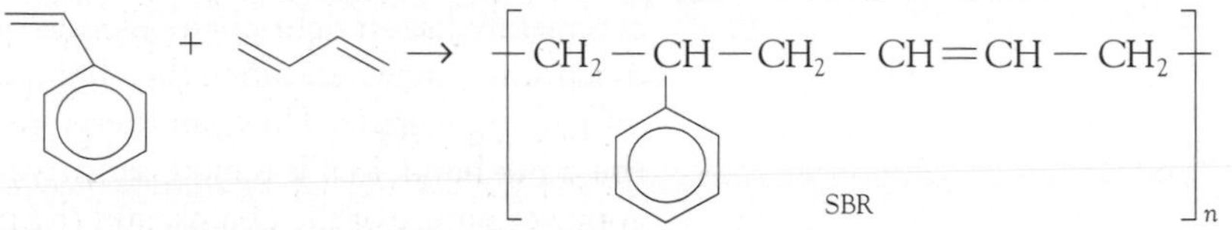

The product SBR still contains a double bond, so can be vulcanised just like natural rubber.

Another co-polymer involving phenylethene and butadiene is the tough, rigid plastic **ABS**, or acrylonitrile–butadiene–styrene:

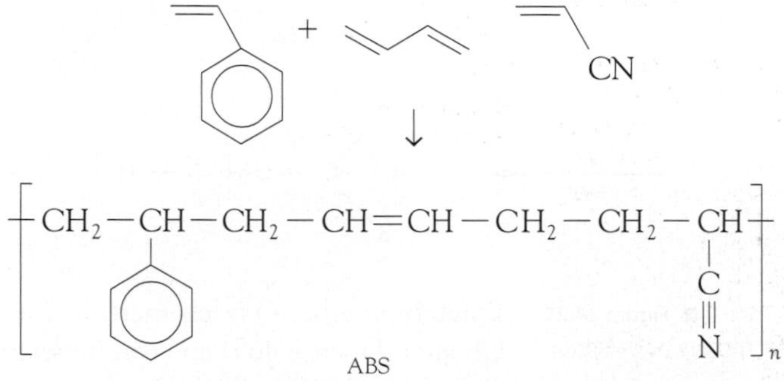

Once again, the monomers join together in a fairly random manner, so the drawing of a particular 'repeat unit' does not imply a regular order. ABS is used for suitcases, telephones and other objects that need to be strong and hard, but not too brittle.

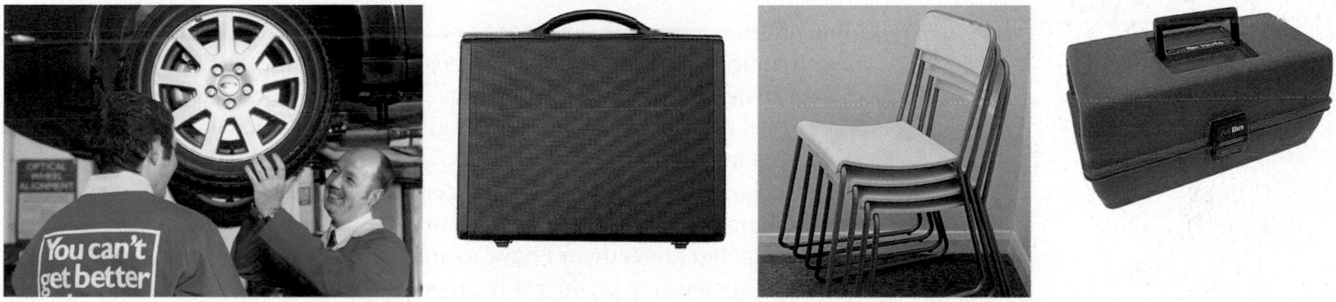

■ **Figure 24.24** Products made from rubber, SBR and ABS.

Addition polymerisation of other ethene derivatives

Most compounds containing the >C=C< group will undergo polymerisation. Several important polymers are made from monomers in which some or all of the hydrogen atoms in ethene have been replaced by other atoms or groups. Some of them, with their uses, are listed in Table 24.4.

■ **Table 24.4** Structures and uses of some addition polymers.

Monomer	Polymer	Uses
Chloroethene $CH_2{=}CHCl$	Poly(chloroethene) (polyvinylchloride, PVC)	Guttering, water pipes, windows, floor coverings
Tetrafluoroethene $CF_2{=}CF_2$	Poly(tetrafluoroethene) (Teflon, PTFE)	Non-stick cookware, bridge bearings
Methyl 2-methylpropenoate $CH_2{=}C(CH_3)CO_2CH_3$	Poly(methyl 2-methylpropenoate) (Perspex)	Protective 'glass', car rear lights, shop signs
Methyl 2-cyanopropenoate $CH_2{=}C(CN)CO_2CH_3$	Poly(methyl 2-cyanopropenoate) (cyanoacrylate)	Instant 'superglue'
Cyanoethene $CH_2{=}CH{-}CN$	Poly(cyanoethene) (acrylic)	Co-polymerised with components of 'acrylic' fibres; also ABS

The disposal of polymers

Polyalkenes, being effectively very long-chain alkanes, are chemically inert. They biodegrade only very slowly in the environment. Although there are bacteria that can metabolise straight-chain alkanes once they have been 'functionalised' by oxidation somewhere along the chain, they find branched-chain alkanes more difficult to degrade. They can therefore attack HDPE more easily than LDPE. On the other hand, LDPE is more sensitive to being broken down by sunlight. The less compact nature of the chains, and the presence of the weaker tertiary C—H bonds (see Table 24.5), allows photo-oxidation to take place more readily (see Figure 24.25).

Table 24.5 Tertiary C—H bonds are weaker than primary C—H bonds.

Bond	**Bond enthalpy/kJ mol^{-1}**
$R_3C—H$	380
$H_3C—H$	435

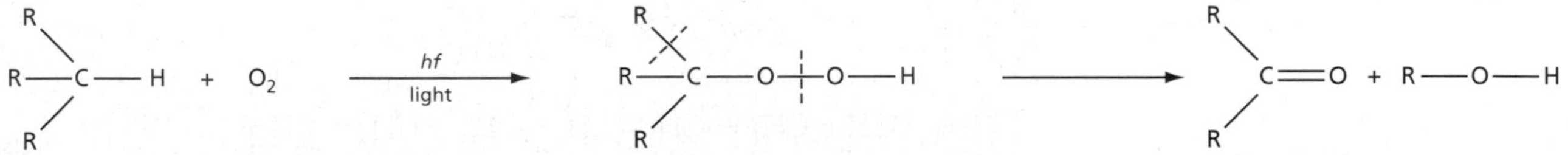

Figure 24.25 Photo-oxidation of tertiary polyalkanes.

Poly(chloroethene) is less inert. Its C—Cl bond can be attacked by alkalis (see Chapter 25, page 465) and can undergo rupture in ultraviolet light (C—Cl bond enthalpy = 338 kJ mol^{-1}). Despite these possible methods of degradation, most addition polymers can potentially cause environmental problems when their useful lives are over. There are four methods that are used, or are being developed, to effect their safe disposal.

1. **Incineration** – the enthalpy change of combustion of all polymers is strongly exothermic. Under the right conditions, incineration can provide useful energy for space heating or power generation. The process requires a high, carefully controlled temperature if the formation of pollutants such as the poisonous gases carbon monoxide and phosgene, $COCl_2$, is to be avoided. Poly(chloroethene) is a particular problem. The highly dangerous 'dioxin' (see Chapter 27, page 515) can be formed unless the conditions are carefully controlled. The chlorine atoms in poly(chloroethene) have to end up somewhere, of course. The preferred combustion product is hydrogen chloride gas, which can subsequently be absorbed in water.
2. **Recycling** – many polymers can be melted and re-moulded. For this process to produce materials that industry can use, at a price that is competitive with newly made material, the recycled polymer has to be clean, and of one type (for example, all polyalkene, or all poly(chloroethene)).
3. **Depolymerisation** – heating polymers to high temperatures in the absence of air or oxygen (pyrolysis) can cause their chains to break up into alkene units. The process is similar to the cracking of long-chain hydrocarbons into alkenes (see Chapter 23). The monomers produced can be separated by fractional distillation and re-used.
4. **Bacterial fermentation** – under the right conditions, and with an optimal strain of bacteria, some polymers (mainly the polyalkenes and the polyesters, see Chapter 29, page 554) can be degraded quite rapidly. The solid polymer is usually first fragmented into very small pieces, and partially photolysed. Useful combustible gases are produced during the fermentation, which can be used as an energy source.

24.5 Preparing alkenes

Cracking of alkanes

The smaller alkenes ethene, propene and butadiene are made commercially by cracking, as we saw in Chapter 23. The method is not of great use for preparing a specific alkene in the laboratory, because it gives a random selection of alkenes, for example:

$$CH_3(CH_2)_{10}CH_3 \xrightarrow{\text{heat with steam at 800 °C}} CH_3(CH_2)_3CH_3 + CH_3CH{=}CH_2 + 2CH_2{=}CH_2$$

■ **Figure 24.26** Alkenes are produced by the cracking of long-chain alkanes (see Chapter 23, page 426). Crude oil is heated and passed through these catalytic cracking towers.

Dehydration of alcohols

This is the reverse of the hydration reaction on page 448. Heating ethanol with concentrated sulphuric acid to 180 °C produces ethene. Concentrated phosphoric acid may also be used. Alcohols also dehydrate when their vapours are passed over strongly heated aluminium oxide:

$$CH_3{-}CH_2OH \xrightarrow[\substack{\text{or conc. } H_3PO_4 \text{ at 200 °C} \\ \text{or pass vapour over strongly heated } Al_2O_3}]{\text{conc. } H_2SO_4 \text{ at 180 °C}} CH_2{=}CH_2 + H_2O$$

Elimination of HCl or HBr from chloro- or bromoalkanes

This reaction is described in Chapter 25, page 469.

$$KOH + CH_3{-}CHCl{-}CH_3 \xrightarrow[\text{dissolved in ethanol}]{\text{heat with KOH}} CH_3{-}CH{=}CH_2 + KCl + H_2O$$

Sometimes an elimination reaction like the dehydration or elimination of a hydrogen halide described above results in more than one alkene product:

$$CH_3CH_2CHBrCH_3 \xrightarrow[\text{dissolved in ethanol}]{\text{heat with KOH}} \underset{80\%}{CH_3CH{=}CHCH_3} + \underset{20\%}{CH_3CH_2CH{=}CH_2}$$

Usually the more substituted alkene, that is, the one with the more alkyl groups attached to the C=C double bond, is more stable than the less substituted alkene. So it is the more substituted alkene that is most likely to be formed, and 80% of the product is but-2-ene. (See also the panel on page 472.)

24.6 The infrared spectra of alkenes

The C═C double bond, like the C—C single bond, is not polar, so the infrared absorption it gives rise to is not strong. Symmetrical, or near-symmetrical, alkenes such as but-2-ene or hex-2-ene show only a weak C═C absorption band. But with alk-1-enes such as hex-1-ene the band is of medium intensity and occurs at 1600–1680 cm^{-1} (see Figure 24.27).

■ Figure 24.27
IR spectra of **a** hexane **b** hex-1-ene.

a hexane

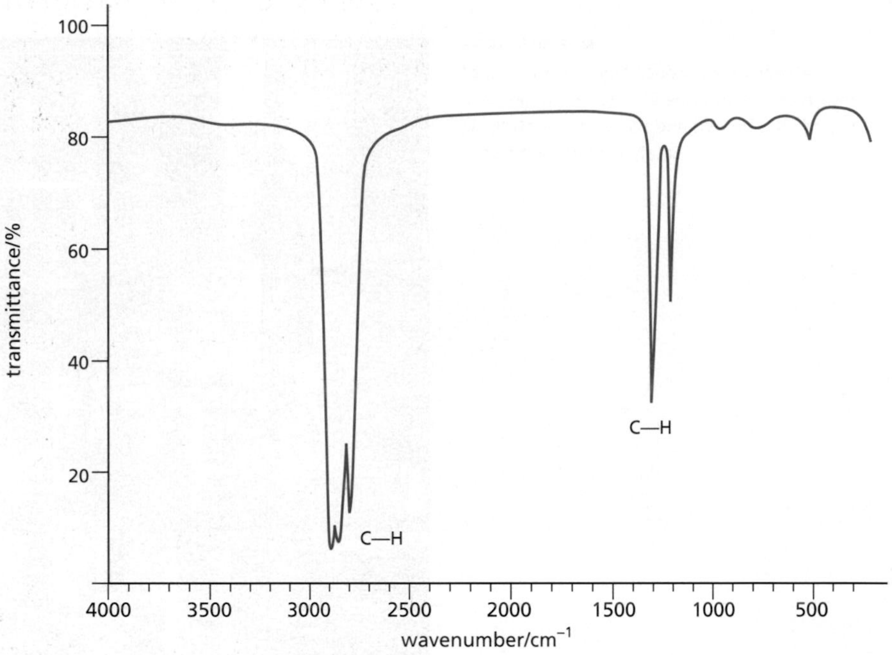

b hex-1-ene

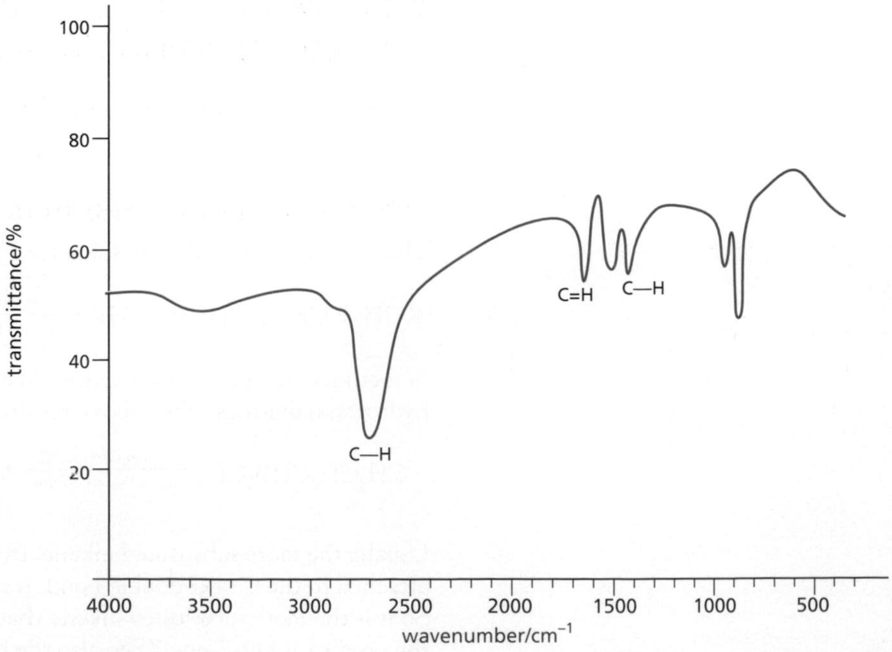

Summary

- Alkenes contain a C=C double bond, and have the general formula C_nH_{2n}.
- In addition to chain isomerism, alkenes show positional isomerism, and also geometrical (*cis–trans*) isomerism due to the restricted rotation around the double bond.
- Their characteristic reaction is **electrophilic addition**. Unsymmetrical electrophiles add in accordance with **Markovnikov's rule**, with the most stable carbocation being formed as intermediate.
- Tests for the presence of a double bond include the decolorisation of bromine water, and the decolorisation of dilute aqueous potassium manganate(VII).
- Ethene and other alkenes form important **addition polymers** such as polythene, polystyrene and Perspex.

Key reactions you should know

- Catalytic hydrogenation (addition):

$$CH_2{=}CH{-}R + H_2 \xrightarrow{\text{Ni}} CH_3{-}CH_2{-}R$$

- Electrophilic substitutions:

$$CH_2{=}CH{-}R + Br_2 \xrightarrow{\text{in the dark}} CH_2Br{-}CHBr{-}R$$

$$CH_2{=}CH{-}R + HBr \longrightarrow CH_3{-}CHBr{-}R$$

$$CH_2{=}CH{-}R + H_2O \xrightarrow{H_3PO_4/70\text{ atm}/300\,°C} CH_3{-}CH(OH){-}R$$

- Oxidations:

$$CH_2{=}CH{-}R + \tfrac{1}{2}O_2 \xrightarrow{\text{Ag}}$$

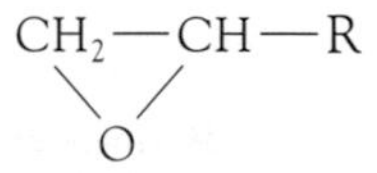

$$CH_2{=}CH{-}R + [O] + H_2O \xrightarrow{\text{cold } KMnO_4} CH_2(OH){-}CH(OH){-}R$$

$$CH_2{=}CH{-}R + 2[O] \xrightarrow{O_3} CH_2O + OHC{-}R$$

$$CH_2{=}CH{-}R + 5[O] \xrightarrow{\text{hot } KMnO_4} CO_2 + HO_2C{-}R\ (+\ H_2O)$$

- Polymerisation:

$$nCH_2{=}CH{-}R \xrightarrow[\text{or } AlR_3]{O_2 \text{ at } 1000\text{ atm}} (CH_2{-}CHR)_n$$

25 *Halogenoalkanes*

Functional group:

$$-\overset{|}{\underset{|}{C}}{}^{\delta+}-X^{\delta-}$$

X = F, Cl, Br, I

This chapter looks at the first of several functional groups that contain atoms other than carbon and hydrogen (termed hetero-atoms). The halogenoalkanes are a reactive and important class of compound much used in organic synthesis. Their characteristic reaction is nucleophilic substitution.

The chapter is divided into the following sections.

25.1 Introduction
25.2 Isomerism and nomenclature
25.3 Reactions of halogenoalkanes
25.4 Chlorofluorocarbons and the ozone layer
25.5 Preparing halogenoalkanes
25.6 The infrared spectra of halogenoalkanes

25.1 Introduction

There are only a few naturally occurring organic halogen compounds. The most important is thyroxine, the hormone secreted by the thyroid gland. Synthetic halogeno compounds have found many uses, however. One of the most important is their general use as intermediates in organic syntheses; they are readily formed from common materials, but can equally easily be transformed into many different functional groups.

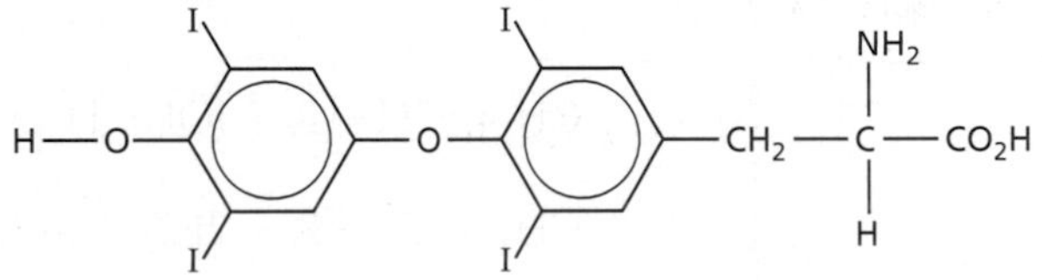

Figure 25.1 Thyroxine.

CFCs (chlorofluorocarbons) are inert, volatile liquids that at one time found favour as the circulating fluids in refrigerators and air-conditioning units. They are now being phased out of use because of their destructive effect on the ozone layer, as we shall see in Section 25.4. Halothane, $CF_3CHBrCl$, is an important anaesthetic, a successor to chloroform, $CHCl_3$, which was one of the first anaesthetics, introduced in the early nineteenth century. The halons, such as $CBrClF_2$, are useful fire extinguishers. Many polychloroalkanes are used as solvents. The use of certain organochlorine compounds as herbicides and insecticides is described in Chapter 27, pages 514–15.

Halogenoalkanes are generally colourless liquids, immiscible with and heavier than water, with sweetish smells. Compared with the alkanes, their boiling points are higher, with the boiling point rising from chloroalkane to the corresponding iodoalkane (see Figure 25.2).

The main intermolecular force in halogenoalkanes is the van der Waals induced dipole force. Their boiling points therefore increase with chain length, just like those of the alkanes.

Furthermore, as we would expect, the trend in boiling point with halogen ($RCl < RBr < RI$) reflects the greater number of electrons available for induced dipole attraction from chlorine to iodine (see Chapter 17). In addition, there is a small amount of permanent dipole–dipole attraction between molecules. This will be most important for the chloroalkanes, becoming less so with bromo- and iodoalkanes (see Figure 25.3).

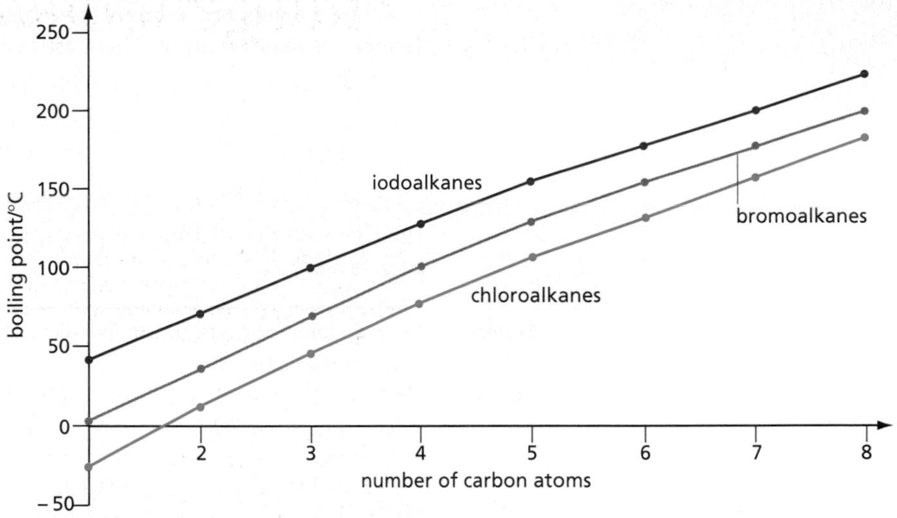

■ **Figure 25.2**
Boiling points of the halogenoalkanes.

■ **Figure 25.3**
Permanent dipole–dipole attractions in halogenoalkanes.

25.2 Isomerism and nomenclature

Halogenoalkanes show chain, positional and optical isomerism. There are, for example, five isomers with the formula C_4H_9Br, shown in Figure 25.4.

■ **Figure 25.4**
Isomers of C_4H_9Br.

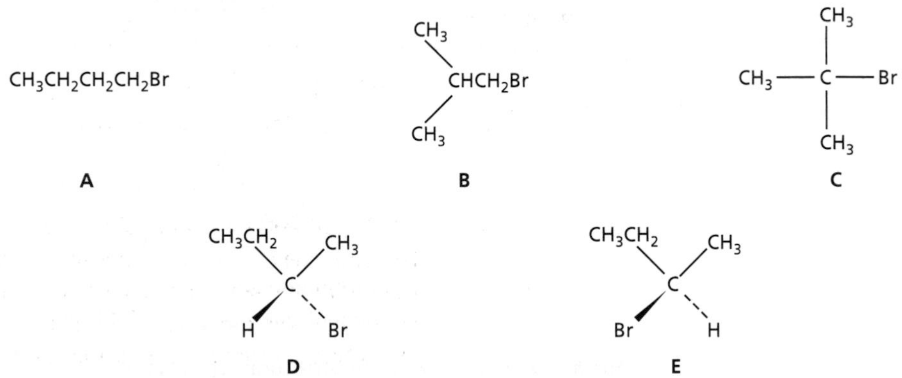

A and the **D/E** pair are positional isomers of each other. **B** and **C** are also positional isomers of each other. **B** and **C** are chain isomers of the other three. **D** and **E** are optical isomers of each other. They form a non-superimposable mirror-image pair (see Chapter 22, page 408).

A halogenoalkane is classed as **primary**, **secondary** or **tertiary**, depending on how many alkyl groups are attached to the carbon atom joined to the halogen atom.

$$R—CH_2—Br \qquad R—\overset{\displaystyle R}{\underset{}{C}}H—I \qquad R—\overset{\displaystyle R}{\underset{\displaystyle R}{C}}—Cl$$

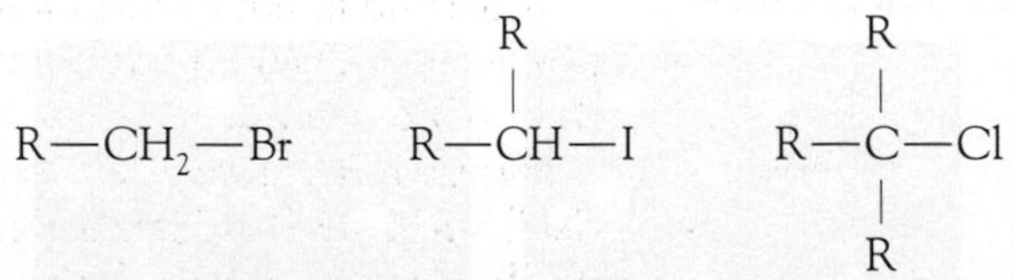

a primary bromoalkane   a secondary iodoalkane   a tertiary chloroalkane

In Figure 25.4, isomers **A** and **B** are primary, **D** and **E** are secondary, and **C** is tertiary.

When a compound contains two halogen atoms, the number of positional isomers increases sharply. There are only two chloropropanes:

$$CH_3—CH_2—CH_2—Cl \qquad CH_3—\underset{}{\overset{Cl}{\overset{|}{C}H}}—CH_3$$

1-chloropropane 2-chloropropane

But there are five dichloropropanes, two of which form mirror-image pairs with each other.

Example Draw skeletal formulae of the five dichloropropanes, and explain which two form a mirror-image pair.

Answer

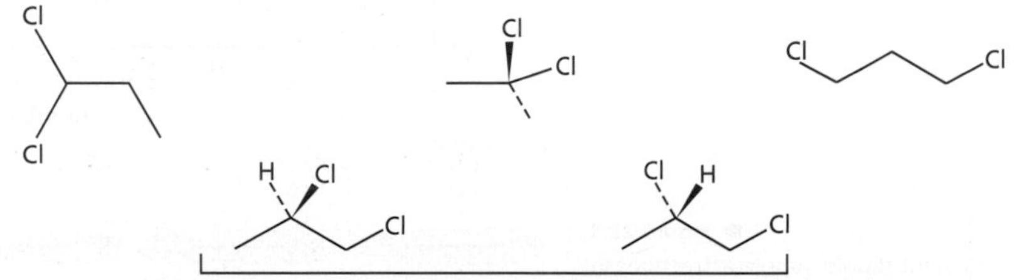

(+) and (–) forms

■ **Figure 25.5**

Further practice How many positional isomers are there of chlorobromopropane, C_3H_6BrCl? How many of these are capable of forming optical isomers? Draw their structures.

The naming of halogenoalkanes follows the rules given in Chapter 22. The longest carbon chain that contains the halogen atom is taken to be the stem. Other side chains are added, and finally the prefix 'fluoro', 'chloro', 'bromo' or 'iodo' is included, together with a numeral showing the position of the halogen atom in the chain:

$CH_3CH_2CHClCH_3$ is 2-chlorobutane
$ClCH_2CH(CH_3)CH_2CH_3$ is 1-chloro-2-methylbutane
$CH_3—CCl_3$ is 1,1,1-trichloroethane.

Example Name the following compounds:

a **b**

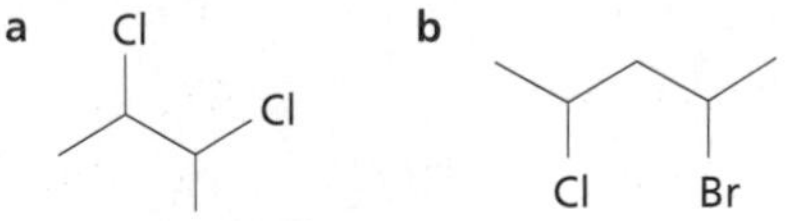

Answer **a** 2,3-dichlorobutane
b 2-bromo-4-chloropentane or 4-bromo-2-chloropentane

Further practice Draw the structural formulae of:
1 1,1,2-trichloropropane
2 2-chloro-2-methylpropane.

■ **Figure 25.6**
Ball-and-stick models of the five isomers of C_4H_9Cl. The more spherical the molecule, the lower the boiling point. What type of isomerism is exhibited by **b** and **c**? And by **d** and **e**?

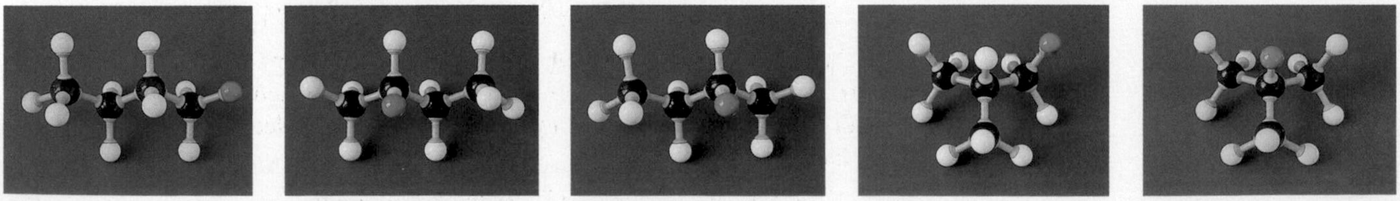

a 1-chlorobutane
b (+)-2-chlorobutane, bp 78° C
c (–)-2-chlorobutane, bp 78° C
d 1-chloro-2-methylpropane, bp 69° C
e 2-chloro-2-methylpropane, bp 50° C

25.3 Reactions of halogenoalkanes

Nucleophilic substitution reactions

Halogens are more electronegative than carbon, and so the C—Hal bond is strongly polarised $C^{\delta+}$—$Hal^{\delta-}$. During the reactions of halogenoalkanes, this bond breaks heterolytically to give the halide ion:

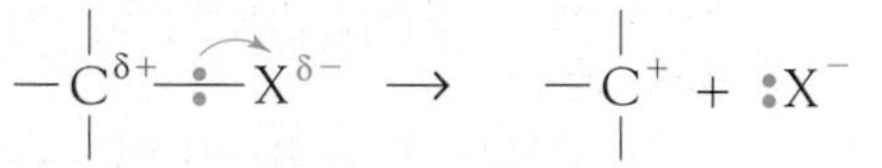

$$-\overset{|}{\underset{|}{C}}^{\delta+}\!\!:\!\!\overset{\frown}{X}^{\delta-} \rightarrow \quad -\overset{|}{\underset{|}{C}}^{+} + :X^- \tag{1a}$$

The C^+ formed is then liable to attack by nucleophiles – anions or neutral molecules that possess a lone pair of electrons.

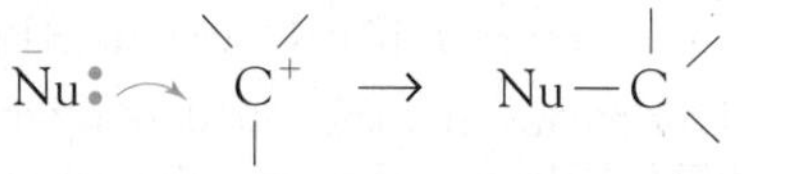

$$\bar{N}u\!:\!\overset{\frown}{} \overset{\diagdown\diagup}{\underset{|}{C}}^+ \rightarrow \quad Nu\!-\!\overset{|}{C}\overset{\diagup}{\diagdown} \tag{1b}$$

Often the nucleophile $\overset{..}{N}u^-$ will attack the $C^{\delta+}$ carbon before the halide ion has left:

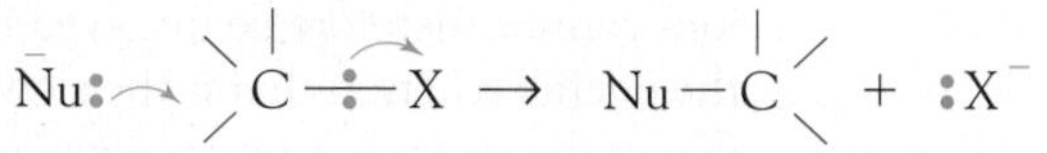

$$\bar{N}u\!:\!\overset{\frown}{} \overset{\diagdown}{\underset{\diagup}{C}}\!\!:\!\!\overset{\frown}{X} \rightarrow \quad Nu\!-\!\overset{|}{C}\overset{\diagup}{\diagdown} + :X^- \tag{2}$$

The dominant reaction is therefore **nucleophilic substitution**. The halide ion is replaced by a nucleophile. Whether mechanism (1a) + (1b) is followed, or mechanism (2), depends on the structure of the halogenoalkane (primary, secondary or tertiary), the nature of the solvent, and the nature of the nucleophile. (The panel overleaf describes the mechanisms of nucleophilic substitution reactions.) Table 25.1 illustrates how some common nucleophiles react with halogenoalkanes.

■ **Table 25.1**
Common nucleophiles that react with halogenoalkanes.

Nucleophile		Products when reacted with bromoethane
Name	**Formula**	
Water	$H_2\overset{..}{O}$	$CH_3CH_2OH + HBr$
Hydroxide ion	$H\overset{..}{O}^-$	$CH_3CH_2OH + Br^-$
Ammonia	$\overset{..}{N}H_3$	$CH_3CH_2NH_2 + HBr$
Cyanide ion	$^-\overset{..}{C}\!\equiv\!N$	$CH_3CH_2CN + Br^-$
Methoxide ion	$CH_3\!-\!\overset{..}{O}^-$	$CH_3CH_2\!-\!O\!-\!CH_3 + Br^-$
Methylamine	$CH_3\!-\!\overset{..}{N}H_2$	$CH_3CH_2\!-\!NH\!-\!CH_3 + HBr$

Hydrolysis

Some halogenoalkanes are reactive enough to be hydrolysed to form alcohols just by heating with water (in practice, a mixed ethanol–water solvent is used because halogenoalkanes are immiscible with water):

$$CH_3\!-\!\overset{\overset{\displaystyle CH_3}{|}}{\underset{\underset{\displaystyle CH_3}{|}}{C}}\!-\!Cl + H_2O \longrightarrow CH_3\!-\!\overset{\overset{\displaystyle CH_3}{|}}{\underset{\underset{\displaystyle CH_3}{|}}{C}}\!-\!OH + HCl$$

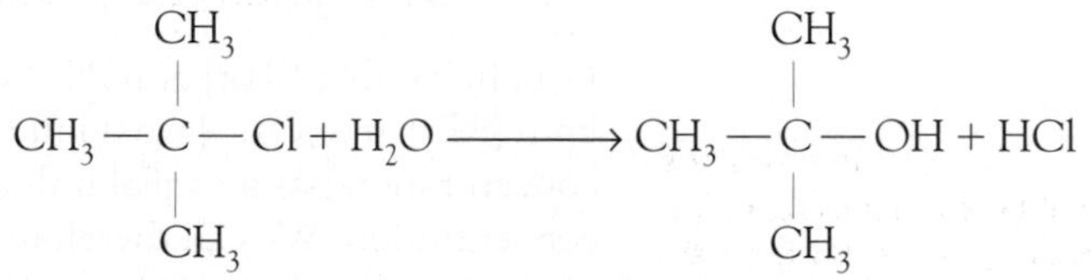

But the reaction is quicker if it is carried out in hot aqueous sodium hydroxide:

$$CH_3CH_2Br + NaOH_{(aq)} \xrightarrow{\text{boil under reflux}} CH_3CH_2OH + NaBr_{(aq)}$$

The relative rates of hydrolysis of chloro-, bromo- and iodoalkanes can be studied by dissolving the halogenoalkane in ethanol and adding aqueous silver nitrate. As the halogenoalkane slowly hydrolyses, the halide ion is released, and forms a precipitate of silver halide:

$$R\!-\!X + H_2O + Ag^+ \rightarrow R\!-\!OH + H^+ + AgX_{(s)}$$

Table 25.4 (page 468) shows the results obtained.

How it happens: the different mechanisms of nucleophilic substitution reactions

If we carry out kinetics experiments of the kind described in Chapter 11 on the hydrolysis of bromoalkanes by sodium hydroxide, we find two different extremes.

The S_N1 reaction

The hydrolysis of 2-bromo-2-methylpropane is a **first-order** reaction:

$$CH_3-\underset{\underset{CH_3}{|}}{\overset{\overset{CH_3}{|}}{C}}-Br + OH^- \rightarrow CH_3-\underset{\underset{CH_3}{|}}{\overset{\overset{CH_3}{|}}{C}}-OH + Br^-$$

$$\text{rate} = k_1[\text{RBr}] \quad (\text{where RBr is } (CH_3)_3CBr)$$

This means that the rate doubles if we double [RBr], but if we double (or halve) [OH$^-$], the rate does not change at all. The rate depends only on the concentration of the bromoalkane. It is independent of the hydroxide concentration. Hydroxide ions cannot therefore be involved in the **rate-determining step** – that is, the step in the overall reaction that is the slowest, that limits the overall rate of reaction. The first of the two mechanisms given on page 465 (equation 1) fits the kinetics equation shown. The mechanism is shown in Figure 25.7.

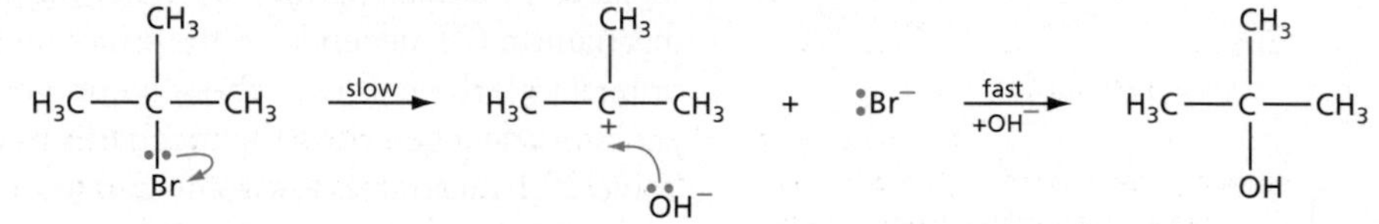

■ **Figure 25.7**
The S_N1 hydrolysis of 2-bromo-2-methylpropane.

$$H_3C-\underset{\underset{Br}{|}}{\overset{\overset{CH_3}{|}}{C}}-CH_3 \xrightarrow{\text{slow}} H_3C-\overset{\overset{CH_3}{|}}{\underset{+}{C}}-CH_3 \;+\; :Br^- \quad \underset{+OH^-}{\xrightarrow{\text{fast}}} \quad H_3C-\underset{\underset{OH}{|}}{\overset{\overset{CH_3}{|}}{C}}-CH_3$$

$$\overset{\cdot\cdot}{O}H^-$$

The first step involves only the heterolysis of the C—Br bond, forming the carbocation and a bromide ion. This is the slow step in the reaction, and hydroxide ions do not take part in it. If [OH$^-$] were doubled, the rate of the second step (carbocation +OH$^-$) might also double. But this second step is already faster than the first one, so the rate of the overall reaction is not affected. This sequence of events is called the S_N1 mechanism (Substitution, Nucleophilic, unimolecular). The reaction profile of the S_N1 reaction is shown in Figure 25.8.

The activation energy of the first step, E_{a1} is high – this is why it is the slow step. That of the second step, E_{a2}, is low – oppositely charged ions attract each other strongly.

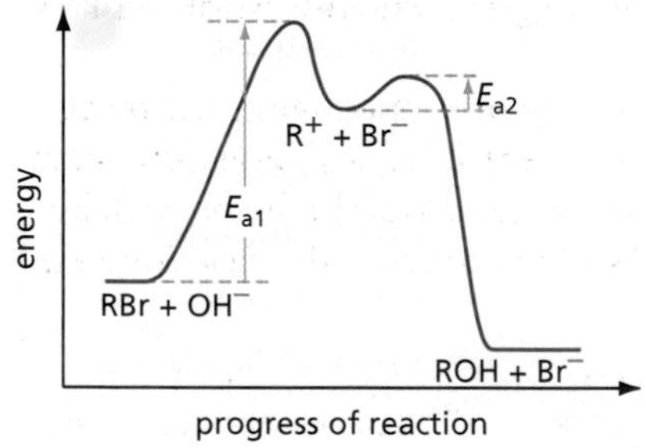

■ **Figure 25.8**
Reaction profile for the S_N1 hydrolysis of 2-bromo-2-methylpropane.

The S_N2 reaction

The hydrolysis of bromomethane is a **second-order** reaction:

$$CH_3-Br + OH^- \rightarrow CH_3-OH + Br^-$$

$$\text{rate} = k_2[\text{RBr}][\text{OH}^-] \quad (\text{where RBr} = CH_3Br)$$

Doubling either [RBr] or [OH$^-$] will double the rate of this reaction. (Doubling both [RBr] and [OH$^-$] would increase the rate four-fold.) Hydroxide ion concentration has an equal influence on the rate as does bromomethane concentration. We can therefore deduce that both bromomethane and hydroxide are involved in the rate-determining step.

The second mechanism given on page 465 (equation 2) fits this kinetic relationship (see Figure 25.9).

■ **Figure 25.9**
The S_N2 hydrolysis of bromomethane.

$$H-\overset{\cdot\cdot}{O}: \quad \overset{H}{\underset{H}{C}}-Br \longrightarrow \left[H-O\cdots\overset{H}{\underset{H}{C}}\cdots Br \right]^{(-)} \longrightarrow H-O-\overset{H}{\underset{H}{C}} \;+\; :Br^-$$

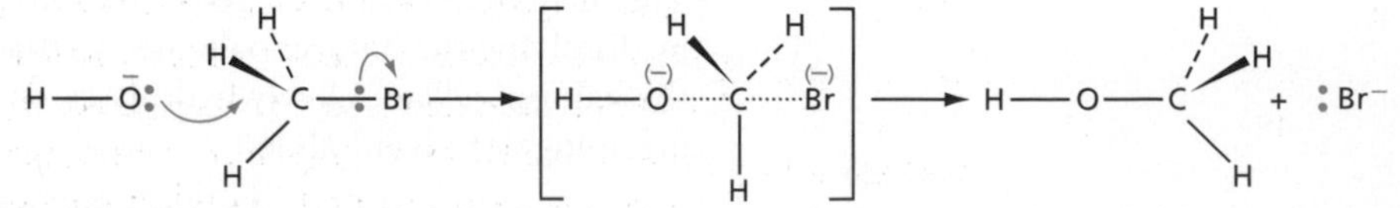

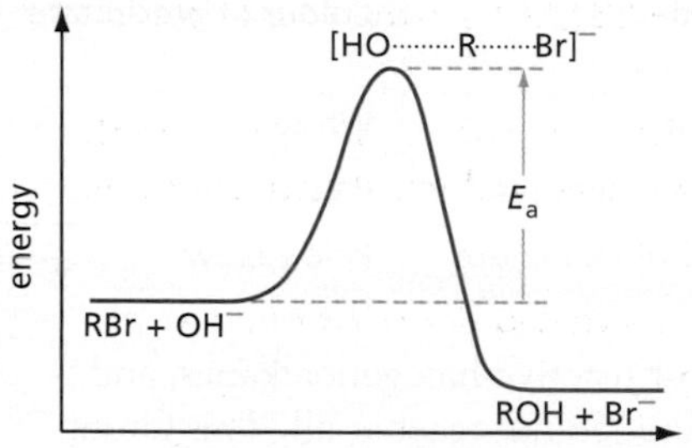

■ **Figure 25.10**
Reaction profile for the S_N2 hydrolysis of bromomethane.

The reaction is a continuous one-step process. The complex shown in square brackets is not an intermediate (like the carbocation in the S_N1 reaction) but a **transition state**. It is a half-way stage in the reaction.

This sequence of events is called the S_N2 mechanism (Substitution, Nucleophilic, bimolecular). The energy profile of the S_N2 reaction is shown in Figure 25.10.

How the relative rates of S_N1 and S_N2 reactions depend on structure

As we saw in Chapter 23, page 437, and Chapter 24, page 447, alkyl groups donate electrons to carbocations and radicals by the inductive effect. The stability of carbocations increases in the order 1° < 2° < 3°. As the carbocation becomes more stable, the activation energy for the reaction leading to it also decreases (see Figure 25.11). We therefore expect that the rate of the S_N1 reaction will increase in the order 1° < 2° < 3°. (See Chapter 8 for a description of the relationship between the magnitude of the activation energy and the rate of a reaction.)

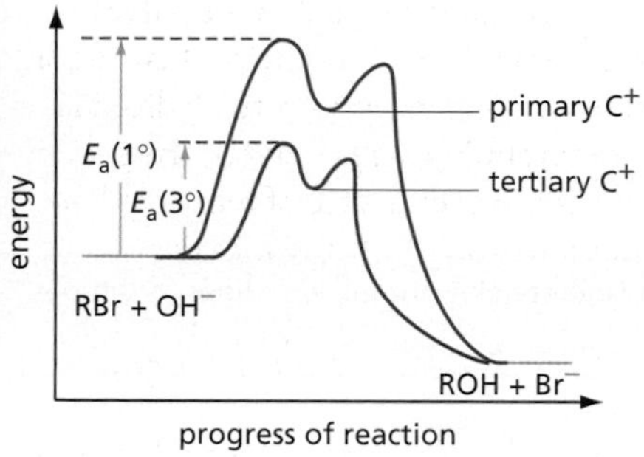

■ **Figure 25.11**
Reaction profiles for the S_N1 hydrolysis of primary and tertiary bromoalkanes.

Table 25.2
The S_N1 reaction is fastest with a tertiary bromoalkane.

Compound	Type	Relative rate of S_N1 reaction
CH_3CH_2Br	Primary	1
$(CH_3)_2CHBr$	Secondary	26
$(CH_3)_3CBr$	Tertiary	60 000 000

No carbocations are formed during the S_N2 reaction. Other factors now come into play. The transition state has five groups arranged around the central carbon atom. It is therefore more crowded than either the starting bromoalkane or the alcohol product, each of which have only four groups around the central carbon atom.

Hydrogen atoms are much smaller than alkyl groups. We therefore expect that the more alkyl groups there are around the central carbon atom, the more crowded will be the transition state (see Figure 25.12), and the higher will be the activation energy E_a (Figure 25.10). This will slow down the reaction (see Table 25.3).

■ **Figure 25.12**
The S_N2 transition state is more crowded with a tertiary bromoalkane.

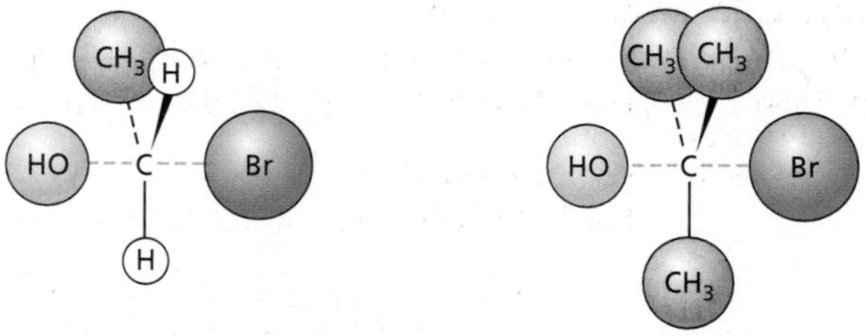

■ **Table 25.3**
The S_N2 reaction is fastest with a primary bromoalkane.

Compound	Type	Relative rate of S_N2 reaction
CH_3CH_2Br	Primary	1000
$(CH_3)_2CHBr$	Secondary	10
$(CH_3)_3CBr$	Tertiary	1

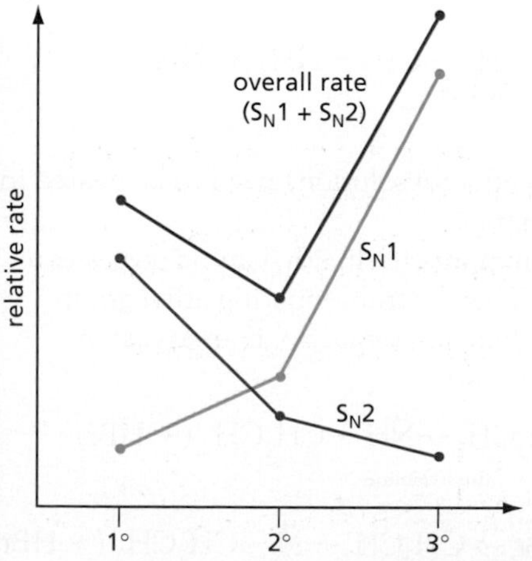

■ **Figure 25.13**
S_N1 and S_N2 hydrolysis reactions for primary, secondary and tertiary halogenoalkanes.

These two effects reinforce each other – the S_N1 reaction is faster with tertiary halides than with primary halides, whereas the S_N2 reaction is faster with primary halides than with tertiary halides. Overall, we expect primary halides to react predominantly by the S_N2 mechanism, tertiary halides to react predominantly by the S_N1 mechanism, and secondary halides to react by a mixture of the two (see Figure 25.13).

■ Table 25.4
Silver halide precipitates formed with halogenoalkanes. The colours of these precipitates are shown in Figure 17.12, page 329.

Compound	Observation on reaction with $AgNO_3 + H_2O$ in ethanol	Colour of precipitate
$(CH_3)_2CH—Cl$	Slight cloudiness after 1 hour	White
$(CH_3)_2CH—Br$	Cloudiness appears after a few minutes	Pale cream
$(CH_3)_2CH—I$	Thick precipitate appears within a minute	Pale yellow

Iodoalkanes are therefore seen to be the most reactive halogenoalkanes, and chloroalkanes the least reactive. Fluoroalkanes do not react at all. This Group 7 trend corresponds to the strengths of the C—X bond. The weaker the bond (that is, the smaller the bond enthalpy), the easier it is to break (see Table 25.5).

■ Table 25.5
Comparing the carbon–halogen bond strengths.

Bond	Bond enthalpy/$kJ\,mol^{-1}$
C—F	485
C—Cl	338
C—Br	285
C—I	213

> Aqueous silver nitrate can be used to distinguish between organic chlorides, bromides and iodides.

The bromoalkanes are the most often used halogenoalkanes in organic reactions. Chloroalkanes react too slowly, and iodoalkanes can be too reactive. They slowly decompose unless stored in the dark. In addition, iodine is a much more expensive element than bromine. We shall use bromoethane to illustrate subsequent reactions of halogenoalkanes.

With ammonia

The ammonia molecule contains a lone pair of electrons. The lower electronegativity of nitrogen compared with oxygen makes ammonia a stronger nucleophile than water. The reaction is as follows:

$$\overset{..}{N}H_3 \quad CH_2 \overset{..}{\longrightarrow} Br \quad \rightarrow \quad CH_3 — CH_2 — \overset{+}{N}H_3 + :Br^-$$
$$\underset{CH_3}{|}$$

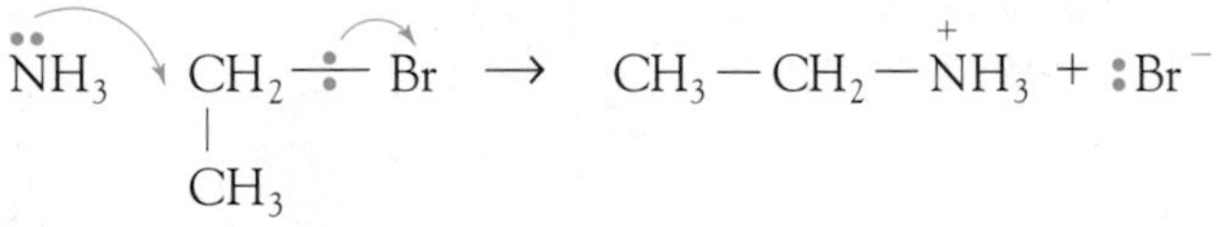

$$CH_3 — CH_2 — \overset{+}{N}H_3 + NH_3 \rightleftharpoons CH_3 — CH_2 — NH_2 + \overset{+}{N}H_4$$
$$\text{ethylamine}$$

Because ammonia is a gas, the reactants (in ethanol solution) need to be heated in a sealed tube, to prevent the ammonia escaping.

The product is called **ethylamine**. Like ammonia itself, ethylamine possesses a lone pair of electrons on its nitrogen atom. The electron-donating ethyl group makes ethylamine even more nucleophilic than ammonia. So, if an excess of bromoethane is used, further reactions can occur:

$$CH_3CH_2\overset{..}{N}H_2 + CH_3CH_2Br \rightarrow CH_3CH_2 — \overset{..}{N}H — CH_2CH_3 \; (+ HBr)$$
$$\text{diethylamine}$$

$$CH_3CH_2 — \overset{..}{N}H — CH_2CH_3 + CH_3CH_2Br \rightarrow CH_3CH_2 — \overset{..}{N} — CH_2CH_3 \; (+ HBr)$$
$$\underset{CH_2CH_3}{|}$$
$$\text{triethylamine}$$

Finally, even triethylamine possesses a lone pair of electrons on its nitrogen atom. It can react as follows:

$$(CH_3CH_2)_3N\!: + CH_3CH_2Br \rightarrow \underset{\text{tetraethylammonium bromide}}{(CH_3CH_2)_4N^+Br^-}$$

To avoid these further reactions, an excess of ammonia is used. The properties and reactions of amines are described in Chapter 30.

With cyanide ions

When a halogenoalkane is heated under reflux with a solution of sodium (or potassium) cyanide in ethanol, a nucleophilic substitution reaction occurs and the halogen is replaced by the cyano group:

$$CH_3CH_2Br + Na^+CN^- \rightarrow CH_3CH_2{-}C{\equiv}N + Na^+Br^-$$

The product is called **propanenitrile**. Its name shows that it contains three carbon atoms, although we started with bromoethane which has just two. The use of cyanide is a good method of increasing the length of a carbon chain by one carbon atom. Nitriles can be hydrolysed to carboxylic acids (see Chapter 29) by heating under reflux with dilute sulphuric acid:

$$CH_3CH_2{-}C{\equiv}N + 2H_2O + H^+ \rightarrow \underset{\text{propanoic acid}}{CH_3CH_2CO_2H} + NH_4^+$$

Heating under reflux means to heat a flask so that the solvent boils continually. A condenser is placed vertically in the neck of the flask to condense the solvent vapour (see Figure 25.14). This method allows the temperature of the reaction mixture to be kept at the boiling point of the solvent without the solvent evaporating away.

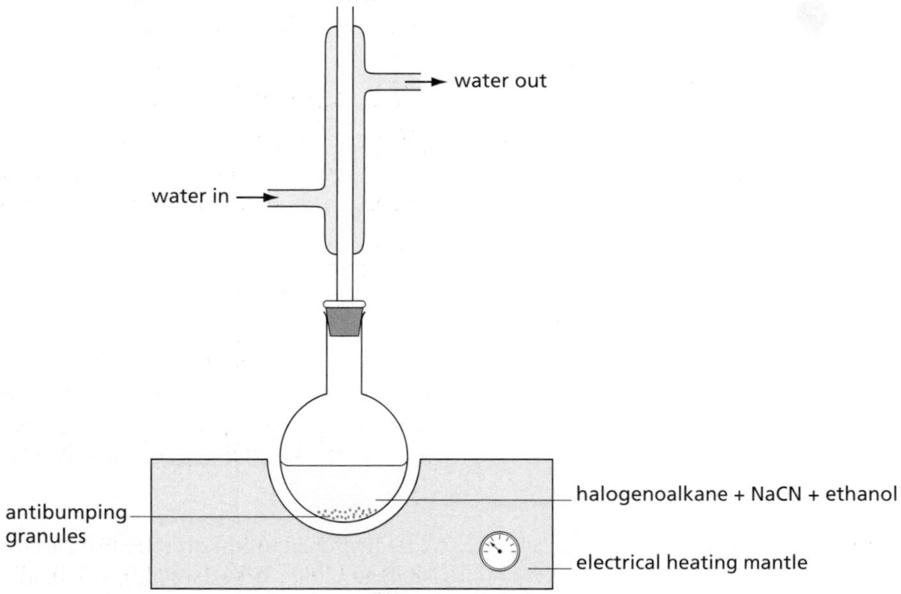

■ **Figure 25.14** A reflux apparatus.

Elimination reactions of halogenoalkanes

Nucleophiles are electron-pair donors. So are bases. They donate a pair of electrons to a hydrogen atom.

The hydroxide ion can act as a nucleophile:

$$HO^-\!: \;\; CH_3 \vdots Br \rightarrow HO{-}CH_3 + :Br^-$$

or it can act as a base:

$$HO^-\!: \;\; H \vdots Cl \rightarrow HO{-}H + :Cl^-$$

Bases and nucleophiles

The strength of an electron pair donor as a nucleophile is not necessarily related to its strength as a base. For example, the iodide ion is an excellent nucleophile towards tetrahedral $C^{\delta+}$ carbon atoms, reacting much faster than the hydroxide ion:

$$I^- + CH_3CH_2Br \rightarrow CH_3CH_2{-}I + Br^-$$
$$OH^- + CH_3CH_2Br \rightarrow CH_3CH_2{-}OH + Br^-$$

But the basicity of the iodide ion is very much lower than that of the hydroxide ion:

$$I^- + H^+ \rightleftharpoons H{-}I \qquad K = 1 \times 10^{-9}\,mol^{-1}\,dm^3$$
$$OH^- + H^+ \rightleftharpoons H{-}OH \qquad K = 6 \times 10^{15}\,mol^{-1}\,dm^3$$

The reason is a subtle one. In the first case, of nucleophilicity, we are measuring the relative rates of the two reactions. For organic reactions with high activation energies, rate is very dependent on the degree of 'advanced bonding' that can occur between the nucleophile and the $C^{\delta+}$ in the transition state. Larger, more diffuse and polarisible anions such as I^-, Br^- and HS^- can achieve this. The activation energies of their transition states are therefore lower than those for the smaller more compact ions such as F^- and OH^-. The overall enthalpy changes of the reactions with OH^- and F^- are, however, more exothermic, because of the strengths of the C—O and C—F bonds (see Figure 25.15).

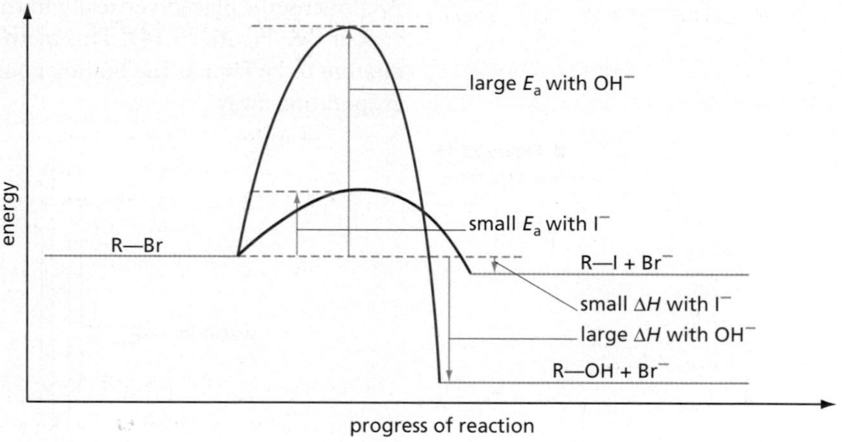

■ **Figure 25.15** Reaction profiles comparing basicity and nucleophilicity.

Basicity, on the other hand, is a thermodynamic property. It is a measure of the position of the acid–base equilibrium. Most bonds to hydrogen form or break very quickly (an exception is the C—H bond), because the activation energy barriers are low. Kinetic differences therefore become less important.

The hydrogen atoms attached to carbon are slightly δ+. They are therefore slightly acidic. When a halogenoalkane is reacted with hydroxide ions under certain conditions, an acid–base reaction can occur, at the same time as the carbon–halogen bond is broken:

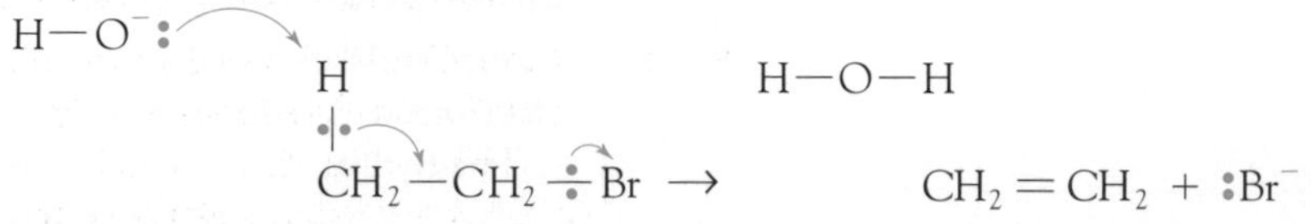

An alkene has been formed by the **elimination** of hydrogen bromide from the bromoalkane. The same reagent (hydroxide ion) can therefore carry out two different types of reaction (substitution or elimination) when reacted with the same bromoalkane. Both reactions do in fact occur at the same time, but the proportion of bromoalkane molecules that undergo elimination can be varied. It depends on three factors:

- the nature of the bromoalkane (primary, secondary or tertiary)
- the strength and physical size of the base
- the solvent used for the reaction.

In general, nucleophilic substitution to form alcohols or ethers is favoured by:

- primary bromoalkanes
- bases of weak or medium strength, and small size
- polar solvents such as water
- low temperature.

On the other hand, elimination to form alkenes is favoured by:

- tertiary bromoalkanes
- strong, bulky bases (such as potassium 2-methylprop-2-oxide, $(CH_3)_3C—O^-K^+$)
- less polar solvents such as ethanol
- high temperature.

The most commonly used reagent is potassium hydroxide dissolved in ethanol and heated under reflux. Potassium hyroxide is used because it is more soluble in ethanol than is sodium hydroxide. Table 25.6 shows some examples.

Table 25.6 Elimination/substitution in bromoalkanes.

	Formula of bromoalkane	Type of bromoalkane	Conditions	$\frac{\text{Elimination}}{\text{substitution}}$ ratio
1	$(CH_3)_2CHBr$	Secondary	2.0 mol dm^{-3} OH^- in 60% ethanol*	1.5
2	$(CH_3)_2CHBr$	Secondary	2.0 mol dm^{-3} OH^- in 80% ethanol*	2.2
3	$(CH_3)_2CHBr$	Secondary	2.0 mol dm^{-3} OH^- in 100% ethanol	3.8
4	$(CH_3)_3CBr$	Tertiary	2.0 mol dm^{-3} OH^- in 100% ethanol	13.0

* the remainder is water

Comparing rows 1–3 of Table 25.6, we can see that the elimination/substitution ratio increases as the solvent becomes richer in ethanol. Comparing rows 3 and 4, we can see that elimination/substitution ratio increases when we replace a secondary bromide by a tertiary one. In a mixed solvent, substitution will give a mixture of the alcohol and an ether (see Chapter 26, page 480).

Example

For the following reagents and conditions, suggest whether:

- substitution will predominate
- substitution and elimination will both occur
- elimination will predominate.

a heating $CH_3CH_2CH_2Br$ with NaOH(aq)
b heating $(CH_3)_3CBr$ with NaOH in ethanol
c heating $(CH_3)_2CHBr$ with $(CH_3)_3C—O^-K^+$

Answer

a Substitution predominates with a primary bromoalkane with OH^- in a polar solvent.
b Elimination predominates with a tertiary bromoalkane with OH^- in a non-polar solvent.
c Elimination predominates with a secondary bromoalkane with a bulky base.

Further practice

Give the structural formula of the main product of each of the following reactions:

1 $(CH_3)_2CH—CH_2Br$ + NaOH in ethanol
2 $CH_3CH_2CHBrCH_2CH_3$ + NaOH(aq)

Which alkene is formed by elimination?

Apart from the competition between elimination and substitution, there is also the question of which alkene is formed by elimination. For example, treatment of 2-bromobutane with potassium hydroxide in ethanol could produce three possible alkenes – but-1-ene and *cis-* or *trans*-but-2-ene (see Figure 25.16).

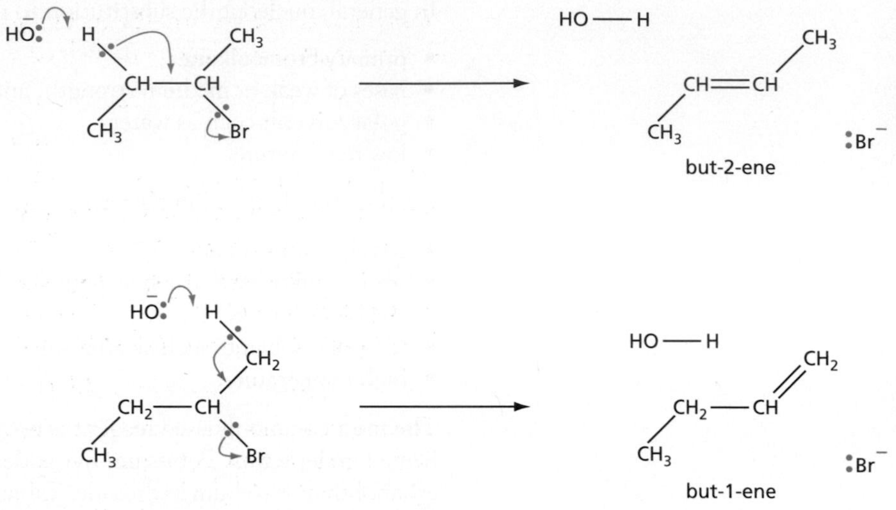

■ Figure 25.16
Possible products of the elimination reaction of 2-bromobutane.

In practice, the most stable alkene, *trans*-but-2-ene, predominates. The least stable is but-1-ene:

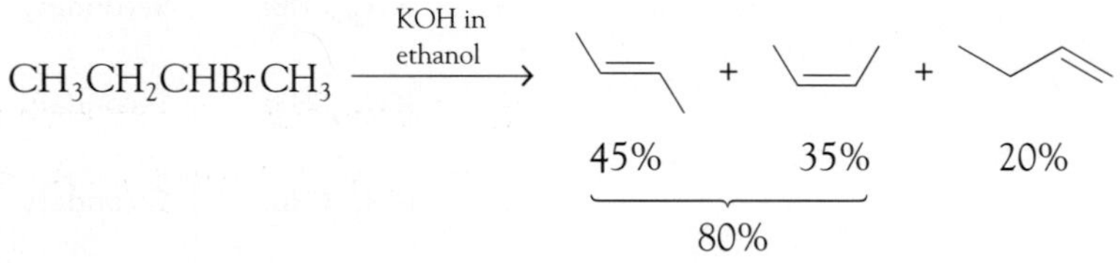

Example Suggest the structural formula of the possible alkenes produced when 2-chloro-2-methylbutane is treated with KOH in ethanol. Which of the possible product alkenes will be found in greatest yield?

Answer

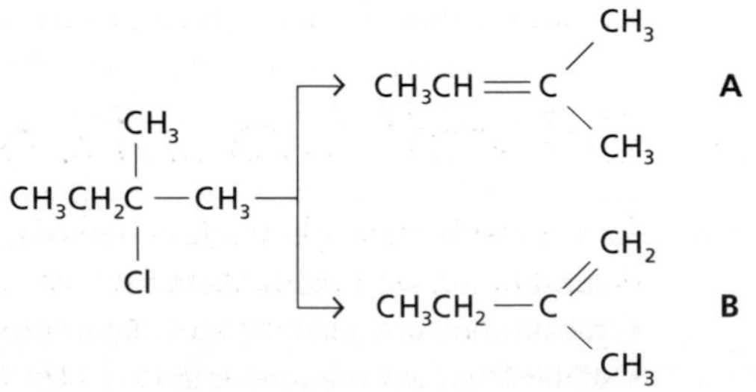

The more substituted alkene (that is, the one with the more alkyl groups on the double bond) is 2-methylbut-2-ene (**A**). This will be found in greater yield than 2-methylbut-1-ene (**B**). Note that neither **A** nor **B** can exist as *cis–trans* isomers, because in each case the right-hand end of the double bond has two identical groups (two methyl groups or two hydrogens).

The reaction of halogenoalkanes with magnesium (Grignard reagents)

In 1901 the French chemist Victor Grignard observed that fine turnings of magnesium metal reacted with a warmed solution of bromoethane in dry ether. A colourless solution was formed. Ether, being a non-polar solvent, does not dissolve ionic compounds. The magnesium had therefore formed a covalent compound. This unusual reaction was the first preparation of what has become an important class of **organometallic compounds**, the **Grignard reagents**.

$$CH_3CH_2Br + Mg \xrightarrow{\text{dry ether}} \underset{\text{ethylmagnesium bromide}}{CH_3CH_2{-}Mg{-}Br}$$

All halogenoalkanes, except fluoroalkanes, form Grignard reagents. Iodides are the most reactive and chlorides the least. Halogenoarenes (see Chapter 27, page 513) also react with magnesium to give Grignard reagents.

Because magnesium is more electropositive than carbon, the carbon–magnesium bond is polarised $C^{\delta-}{-}Mg^{\delta+}$. This causes Grignard reagents to act as nucleophiles in their reactions, acting rather like carbanions (see Chapter 22, page 415–6):

$$CH_3CH_2{-}MgBr \rightleftharpoons CH_3CH_2^{\;-} + MgBr^+$$

Grignard reagents react with water to give alkanes:

$$2CH_3CH_2MgBr + 2H_2O \rightarrow 2C_2H_6 + MgBr_2 + Mg(OH)_2$$

They react with carbonyl compounds to give the magnesium salts of alcohols (see pages 488 and 532). For example:

$$CH_3{-}CHO + C_2H_5MgBr \rightarrow \begin{array}{c} CH_3{-}CH{-}O^-MgBr^+ \\ | \\ C_2H_5 \end{array}$$

The free alcohol is then liberated by the addition of dilute acid:

$$\begin{array}{c} CH_3{-}CH{-}O^-MgBr^+ \\ | \\ C_2H_5 \end{array} + H^+ \rightarrow \begin{array}{c} CH_3{-}CH{-}O{-}H \\ | \\ C_2H_5 \end{array} + Mg^{2+} + Br^-$$

Further practice

Predict the products **X**, **Y** and **Z** of the following reactions.

1 $(CH_3)_2CHMgBr + H_2O \rightarrow$ **X**

2 $(CH_3)_2CHMgBr + (CH_3)_2C{=}O \xrightarrow[\text{(2) add } H^+(aq)]{\text{(1) mix in ether}}$ alcohol **Y** $[C_6H_{14}O]$ $\xrightarrow[\text{conc. } H_2SO_4]{\text{heat with}}$ alkene **Z** $[C_6H_{12}]$

25.4 Chlorofluorocarbons and the ozone layer

The **chlorofluorocarbons** (CFCs) have almost ideal properties for use as aerosol propellants and refrigerant heat-transfer fluids. They are chemically and biologically inert (and hence safe to use and handle), and they can be easily liquefied by pressure a little above atmospheric pressure. Although fairly expensive to produce (compared to other refrigerants like ammonia), they rapidly replaced other fluids that had been used. In the early 1970s, however, concern was expressed that their very inertness was a global disadvantage. Once released into the environment, CFCs remain chemically unchanged for years. Being volatile, they can diffuse throughout the atmosphere, and eventually find their way into the stratosphere (about 20 km above the Earth's surface). Here they are exposed to the stronger ultraviolet rays of the Sun. Although the carbon–fluorine bond is very strong, the carbon–chlorine bond is weak enough to be split by ultraviolet light (see Table 25.5). This forms atomic chlorine radicals, which can upset the delicately balanced equilibrium between ozone formation and ozone breakdown.

$$CF_3Cl \xrightarrow{\text{ultraviolet light}} CF_3{}^\bullet + Cl^\bullet$$

Some stratospheric chemistry

1 Production of oxygen atoms: $O_2 \rightarrow 2O$
(by absorption of UV light at 250 nm wavelength)

Once oxygen atoms have formed, they can react with oxygen molecules to produce ozone, which by absorption of ultraviolet light decomposes to reform oxygen atoms and oxygen molecules. An equilibrium is set up:

2 Natural ozone formation: $O + O_2 \rightarrow O_3$
3 Natural ozone depletion: $O_3 \rightarrow O_2 + O$
(by absorption of UV light at 300 nm wavelength)

Chlorine atoms disrupt the equilibrium by acting as a homogeneous catalyst (see Chapter 8) for the destruction of ozone:

4 $Cl^{\bullet} + O_3 \rightarrow ClO^{\bullet} + O_2$
5 $ClO^{\bullet} + O \rightarrow Cl^{\bullet} + O_2$

Reaction (5), in which chlorine atoms are regenerated, involves the destruction of the oxygen atoms needed to make more ozone by the natural formation reaction (2). In this way, chlorine atoms have a doubly depleting effect on ozone.

It has been estimated that one chlorine atom can destroy over 10^5 ozone molecules before it eventually diffuses back into the lower atmosphere. There it can react with water vapour to produce hydrogen chloride, which can be flushed out by rain as dilute hydrochloric acid.

Once the 'ozone hole' above the Antarctic was discovered in 1985, global agreements were signed in Montreal (1989) and London (1990). As a result, the global production of CFCs has been drastically reduced. In many of their applications they can be replaced by hydrocarbons such as propane. It will still take several decades for natural regeneration reactions to allow the ozone concentration to recover. Other stratospheric pollutants such as nitric oxide from high-flying aircraft also destroy ozone.

■ **Figure 25.17** (below)
The ozone 'hole' is an area over the Antarctic where ozone levels are depleted, particularly in spring. A similar depletion is occurring over the Arctic as well. Ozone in the upper atmosphere absorbs short-wavelength ultraviolet radiation from the Sun's rays, reducing the level of this radiation at the Earth's surface.

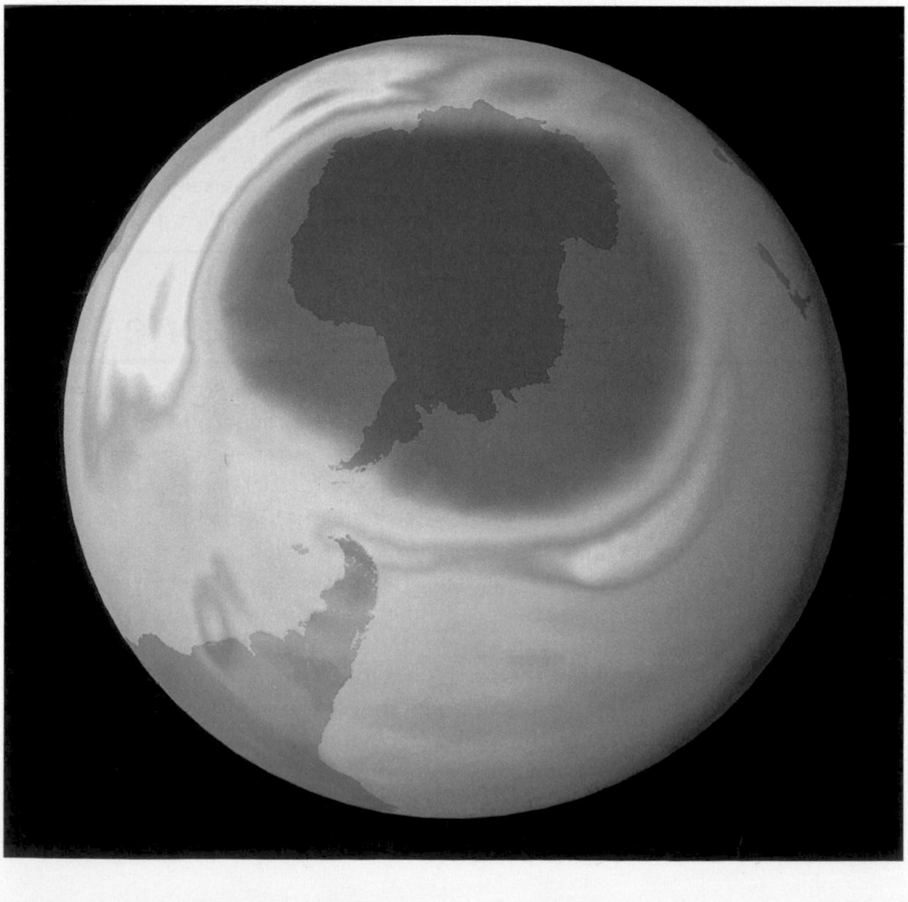

Figure 25.18 (above)
Depletion of the ozone layer puts humans at increased risk of skin cancer and damage to the eyes by short-wavelength UV radiation (UV-B).

25.5 Preparing halogenoalkanes

From alkanes, by radical substitution

As we saw in Chapter 23, pages 433–6, this reaction works well with chlorine and bromine. It has limitations in organic synthesis, however, since it is essentially a random process. It gives a mixture of isomers, and multiple substitution often occurs. Nevertheless it is often used industrially, where fractional distillation can separate the different products. Alkanes that can form only one isomer of a monochloroalkane are the most suitable starting materials.

$$CH_4 + Cl_2 \xrightarrow{\text{light}} CH_3Cl + HCl$$

$$C_2H_6 + Cl_2 \longrightarrow CH_3CH_2Cl + HCl$$

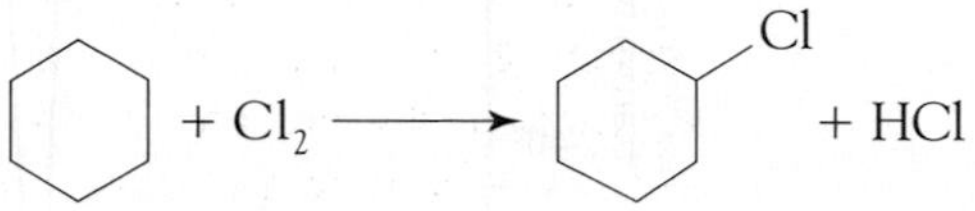

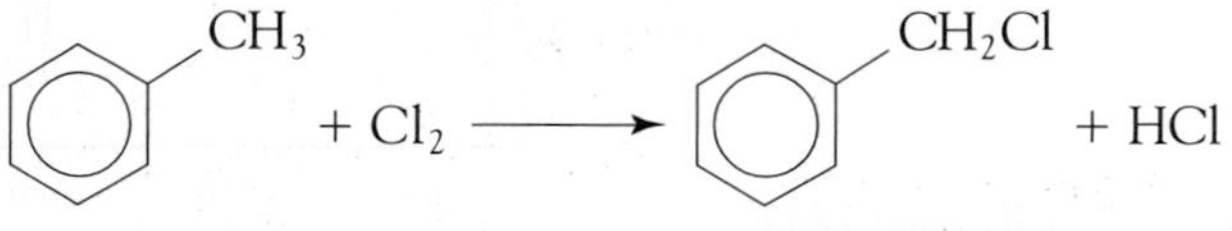

From alkenes, by electrophilic addition

We saw in Chapter 24, pages 444–6, that the addition of chlorine or bromine to alkenes gives 1,2-dihaloalkanes.

$$CH_2{=}CH_2 + Br_2 \rightarrow Br{-}CH_2{-}CH_2{-}Br$$

The addition of hydrogen halides to alkenes gives monohaloalkanes. The orientation of the halogen follows Markovnikov's rule (page 448). The reaction works with hydrogen iodide as well as with hydrogen chloride and hydrogen bromide:

$$CH_3{-}CH{=}CH_2 + HI \rightarrow CH_3{-}CHI{-}CH_3$$

From alcohols, by nucleophilic substitution

These reactions are covered in detail in Chapter 26, pages 482–3. There are several reagents that can be used:

$$CH_3CH_2OH + HCl \xrightarrow{\text{conc. HCl + ZnCl}_2\text{ + heat}} CH_3CH_2Cl + H_2O$$

$$CH_3CH_2OH + PCl_5 \xrightarrow{\text{warm}} CH_3CH_2Cl + HCl + POCl_3$$

$$CH_3CH_2OH + SCl_2O \xrightarrow{\text{warm}} CH_3CH_2Cl + HCl + SO_2$$

$$CH_3CH_2OH + HBr \xrightarrow{\text{conc. H}_2\text{SO}_4\text{ + NaBr + heat}} CH_3CH_2Br + H_2O$$

$$3CH_3CH_2OH + PI_3 \xrightarrow{\text{heat with P + I}_2} 3CH_3CH_2I + H_3PO_3$$

25.6 The infrared spectra of halogenoalkanes

Bond	Frequency of absorption (wavenumber)/cm^{-1}
C—F	1200
C—Cl	700
C—Br	600
C—I	500

Table 25.7
IR absorption frequencies for C—Hal bonds.

Except for the C—F bond, the C—Hal bond tends to be weak, and the halogens (again, excepting fluorine) are increasingly heavy elements. Both effects tend to reduce the frequency of absorption of the C—Hal bond in the infrared (see Chapter 31). Table 25.7 shows typical ranges of absorption. Those for C—Br and C—I bonds are beyond the usual range of an infrared spectrometer, so are not of use in identification.

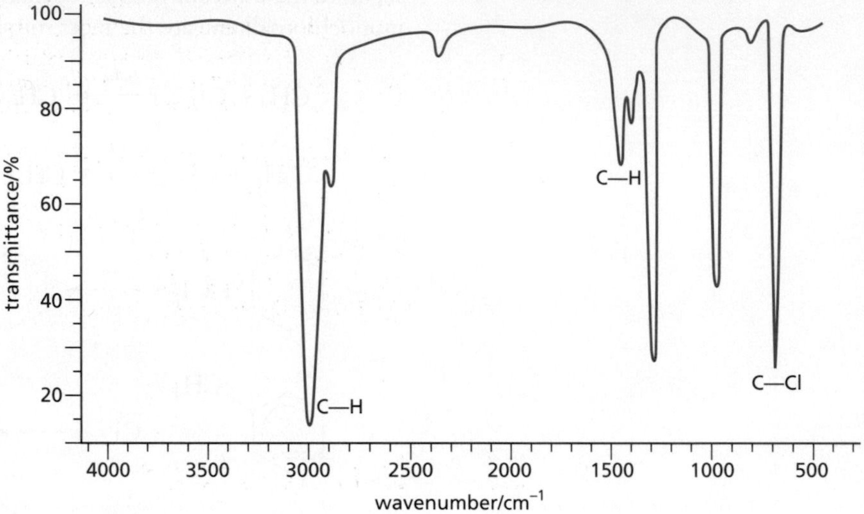

Figure 25.19
IR spectrum of chloroethane.

Summary

- **Halogenoalkanes** contain a δ+ carbon atom.
- Their most common reaction is **nucleophilic substitution**. They also undergo **elimination** reactions to give alkenes.
- They form **Grignard reagents** with magnesium metal.
- Their reactivity increases in order CF < CCl < CBr < CI.
- Fluoro- and chloroalkanes are used as solvents, aerosol propellants and refrigerants. CFCs damage the ozone layer by being photolysed to chlorine atoms, which initiate chain reactions destroying ozone.
- Halogenoalkanes can be made from alkanes, alkenes and alcohols.

Key reactions you should know

(R = primary, secondary or tertiary alkyl unless otherwise stated. All reactions also work with Cl instead of Br, but much more slowly.)

- Elimination:

$$R{-}CH_2{-}CH_2{-}Br + OH^- \xrightarrow[\text{in ethanol}]{\text{heat with NaOH}} R{-}CH{=}CH_2 + H_2O + Br^-$$

- Nucleophilic substitutions:

$$R{-}Br + OH^- \xrightarrow{\text{heat with NaOH in water}} R{-}OH + Br^-$$

$$R{-}Br + 2NH_3 \xrightarrow{NH_3\text{ in ethanol under pressure}} R{-}NH_2 + NH_4Br$$

$$R{-}Br + CN^- \xrightarrow{\text{heat with NaCN in ethanol}} R{-}C{\equiv}N + Br^-$$

$$R{-}C{\equiv}N \rightarrow R{-}CO_2H \qquad R{-}C{\equiv}N \rightarrow R{-}CH_2NH_2$$

- Grignard reagents:

$$R{-}Br + Mg \xrightarrow{\text{Mg in dry ether}} R{-}MgBr \rightarrow \text{various}$$

26 *Alcohols*

Functional group:

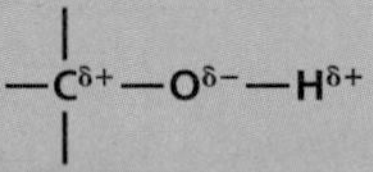

The —OH (hydroxyl) group occurs in many organic compounds, including sugars, carboxylic acids, phenols and alcohols. In this chapter we look at the reactions of the hydroxyl group when it is the only functional group in the molecule, in the alcohols. The hydrogen atom attached to oxygen is slightly acidic, and the δ+ nature of the carbon atom makes it susceptible to nucleophilic substitution reactions. Primary and secondary alcohols also undergo oxidation.

The chapter is divided into the following sections.

26.1 Introduction

The **alcohols** form a homologous series with the general formula $C_nH_{2n+1}OH$. They occupy a central position in organic functional group chemistry. They can be readily interconverted with aldehydes and carboxylic acids by oxidation and reduction, and to halogenoalkanes by nucleophilic substitution. They are useful solvents in their own rights, but are also key intermediates in the production of esters, which are important solvents for the paints and plastics industries.

The polar —OH group readily forms hydrogen bonds to similar groups in other molecules. This accounts for the following major differences between the alcohols and the corresponding alkanes.

- The lower alcohols (C_1, C_2, C_3 and some isomers of C_4) are totally miscible with water, due to hydrogen bonding between the alcohol molecules and water molecules (see Figure 26.1).

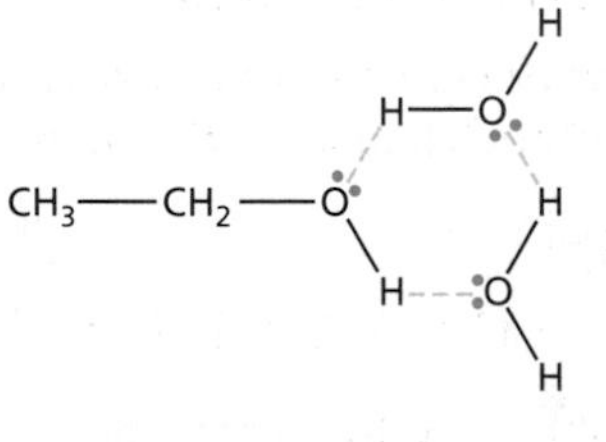

■ Figure 26.1
Hydrogen bonding between ethanol and water.

As the length of the alkyl chain increases, van der Waals attraction predominates between the molecules of the alcohol, and so the miscibility with water decreases.

- The boiling points of the alcohols are all much higher than the corresponding (isoelectronic) alkanes, due to strong intermolecular hydrogen bonding (see Figure 26.2).

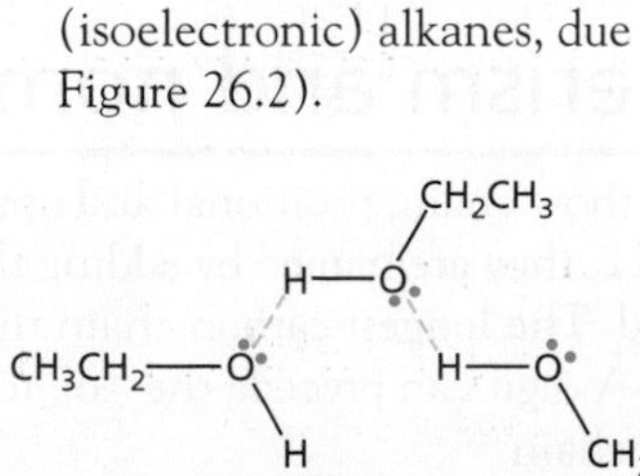

■ Figure 26.2
Intermolecular hydrogen bonding in ethanol.

Table 26.1 and Figure 26.3 show this large difference. As is usual in homologous series, the boiling points increase as the number of carbon atoms increases, due to increased van der Waals attractions between the longer alkyl chains. The enhancement of boiling point due to hydrogen bonding also decreases, as the chains become more alkane-like in character.

■ Table 26.1
Boiling points of some alcohols and alkanes.

Number of electrons in the molecule	Alkane		Alcohol		Enhancement of boiling point/°C
	Formula	Boiling point/°C	Formula	Boiling point/°C	
18	C_2H_6	−88	CH_3OH	65	153
26	C_3H_8	−42	C_2H_5OH	78	120
34	C_4H_{10}	0	C_3H_7OH	97	97
42	C_5H_{12}	36	C_4H_9OH	118	82
50	C_6H_{14}	69	$C_5H_{11}OH$	138	69
58	C_7H_{16}	98	$C_6H_{13}OH$	157	59

■ Figure 26.3
Boiling points of some alcohols and alkanes.

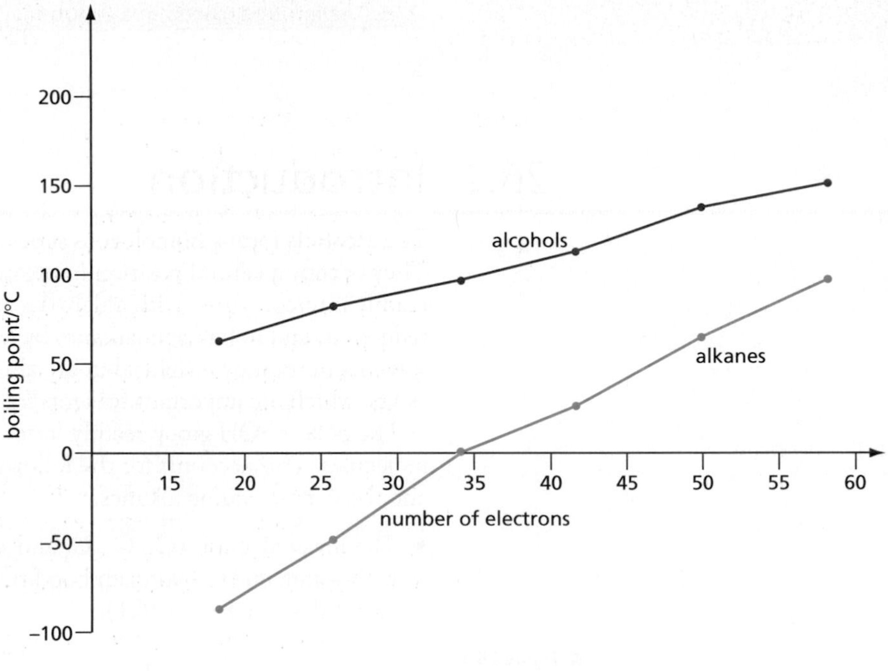

The geometry around the oxygen atom in alcohols is similar to that of the water molecule, with two lone pairs of electrons (see Figure 26.4).

■ Figure 26.4
The oxygen atom in an alcohol has two lone pairs.

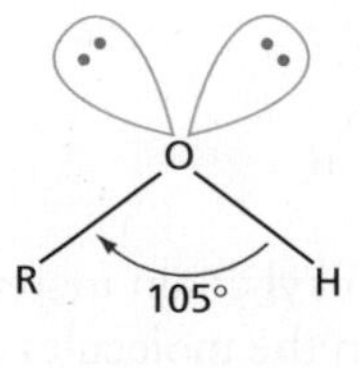

26.2 Isomerism and nomenclature

Alcohols show chain, positional and optical isomerism. As was mentioned in Chapter 22, they are named by adding the suffix '-ol' to the alkane stem of the compound. The longest carbon chain that contains the —OH group is chosen as the stem. A digit can precede the '-ol' to describe the position of the —OH group along the chain.

■ **Figure 26.5** Molecular models of **a** methanol, **b** ethanol, **c** propan-1-ol, **d** propan-2-ol, **e** butan-1-ol.

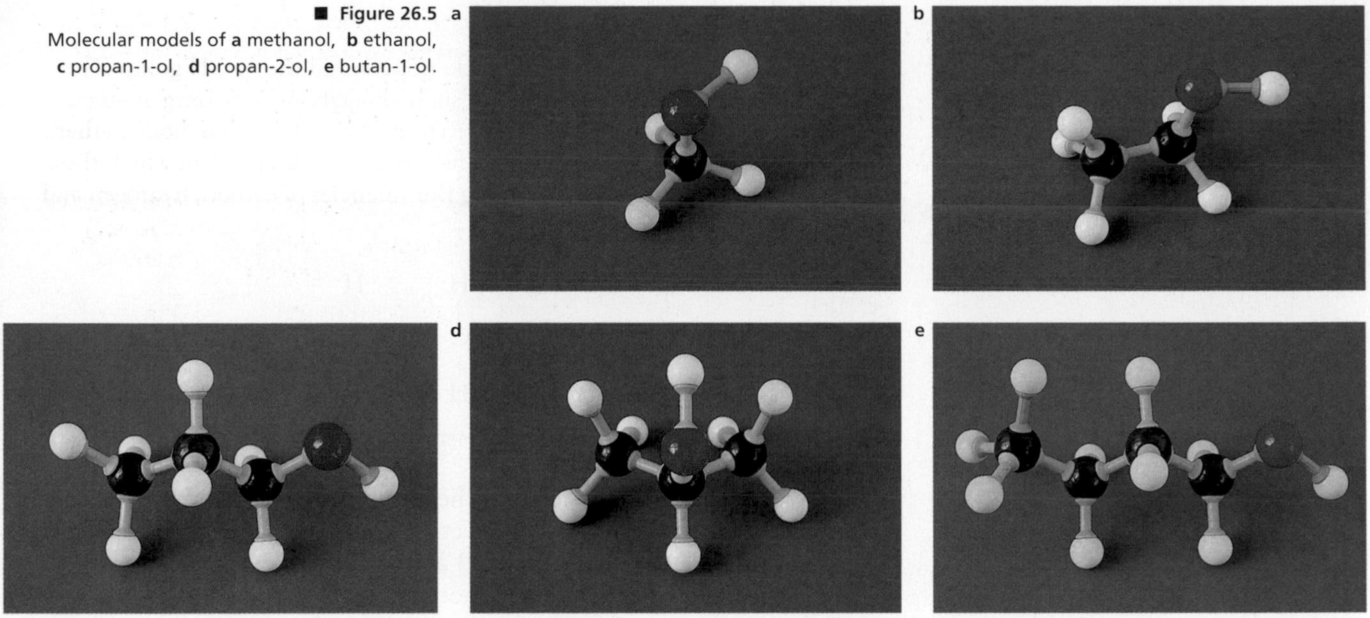

Example Name the following compounds.

a $CH_3—CH(OH)—CH_2—CH_3$

b $CH_2{=}CH—CH_2OH$

c $CH_3—CH_2—CH(CH_3)—CH_2OH$

Answer a The longest (and only) carbon chain contains four carbon atoms. The —OH is on the second atom, so the alcohol is butan-2-ol.

b The three-carbon base is propene. Taking the —OH group to be on position 1, the name is prop-2-en-1-ol.

c The longest chain contains four carbon atoms, with a methyl group on position 2, and the —OH group on position 1 – the compound is 2-methylbutan-1-ol.

Alcohols are classed as primary, secondary or tertiary, depending on the number of alkyl groups there are attached to the carbon atom joined to the —OH group. Of the five isomers with the formula C_4H_9OH, two (**A** and **B** below) are primary alcohols, two (**C** and **D**) are secondary alcohols, and one (**E**) is a tertiary alcohol:

$CH_3CH_2CH_2CH_2OH$ — **A**

$(CH_3)_2CHCH_2OH$ — **B**

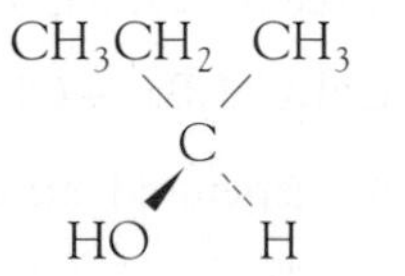

C

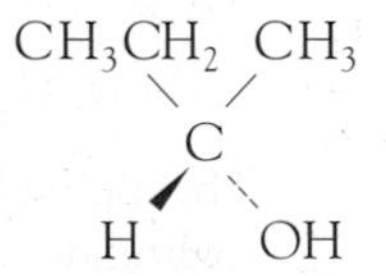

D

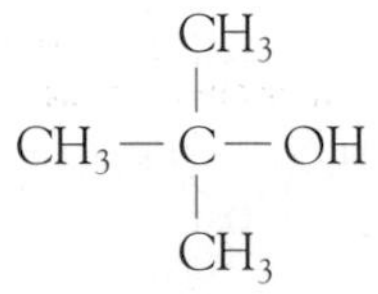

E

Methanol, CH_3OH, is also classified as a primary alcohol.

Further practice 1 Describe the type of isomerism shown by the following alcohols, whose structures are given above.

a A and B

b B and E

c A and C

d C and D

2 Name these five alcohols (note that alcohols C and D will have the same name).

Further isomerism – ethers

Ethers are isomers of alcohols, but they do not contain the —OH group. The ethanol molecule has two carbon atoms, six hydrogens and one oxygen atom. These atoms can be arranged differently to form the molecule **methoxymethane** (also called dimethyl ether). (This is, in fact, the only other way in which these nine atoms can be arranged, still keeping the valencies of carbon, hydrogen and oxygen as 4, 1 and 2, respectively.)

$$H-\overset{\overset{\displaystyle H}{|}}{\underset{\underset{\displaystyle H}{|}}{C}}-\overset{\overset{\displaystyle H}{|}}{\underset{\underset{\displaystyle H}{|}}{C}}-O-H \qquad\qquad H-\overset{\overset{\displaystyle H}{|}}{\underset{\underset{\displaystyle H}{|}}{C}}-O-\overset{\overset{\displaystyle H}{|}}{\underset{\underset{\displaystyle H}{|}}{C}}-H$$

ethanol   methoxymethane (dimethyl ether)

Ethers have very different physical and chemical properties from those of alcohols. Their molecules are polar (due to the electronegativity of the oxygen atom and the non-linear C—O—C angle of 109°) and hence they experience a small dipole–dipole interaction. But they cannot form intermolecular hydrogen bonds with each other, because they do not contain a strongly polar $H^{\delta+}$ atom. Their boiling points, although higher than the corresponding (isoelectronic) alkanes, are therefore much lower than those of the isomeric alcohols (see Table 26.2).

■ Table 26.2
Boiling points of an alkane, an ether and an alcohol.

Compound	Formula	Boiling point/°C
Propane	$CH_3-CH_2-CH_3$	−42
Dimethyl ether	CH_3-O-CH_3	−24
Ethanol	CH_3-CH_2-OH	78

The lower ethers are quite soluble in water due to hydrogen bonding with water molecules. Ethoxyethane (diethyl ether, $C_2H_5-O-C_2H_5$) dissolves to the extent of 8% by mass. The cyclic ethers tetrahydrofuran and dioxan are completely miscible with water in all proportions (see Figure 26.6). All three are useful organic solvents. Being quite polar, they dissolve a wide variety of compounds.

Ethers are very inert, showing almost none of the reactions of alcohols. This inertness is another factor that makes them useful as solvents. Totally dry (anhydrous) ethoxyethane is used as a solvent for reduction reactions with lithium tetrahydridoaluminate(III) (see Chapter 28, page 529), and for the synthesis of Grignard reagents (see Chapter 25, page 473).

Ethers are produced when alkoxide ions undergo substitution reactions with halogenoalkanes (see Chapter 25, page 471).

$$CH_3-O^- + CH_3CH_2Br \rightarrow CH_3-O-CH_2CH_3 + Br^-$$

The alkoxide ion can be generated either by reacting the alcohol with sodium hydroxide:

$$CH_3-OH + OH^- \rightleftharpoons CH_3-O^- + H_2O$$

or by reacting the alcohol with sodium metal:

$$CH_3-OH + Na \rightarrow CH_3-O^-Na^+ + \tfrac{1}{2}H_2$$

The latter is the usual method employed. Reaction with aqueous sodium hydroxide is an equilibrium process ($K_{eq} \approx 1$), so much alcohol is formed as a by-product from the reaction of hydroxide with the halogenoalkane. Elimination reactions also compete (see Chapter 25, page 471).

ethoxyethane
(diethyl ether)

tetrahydrofuran

dioxan

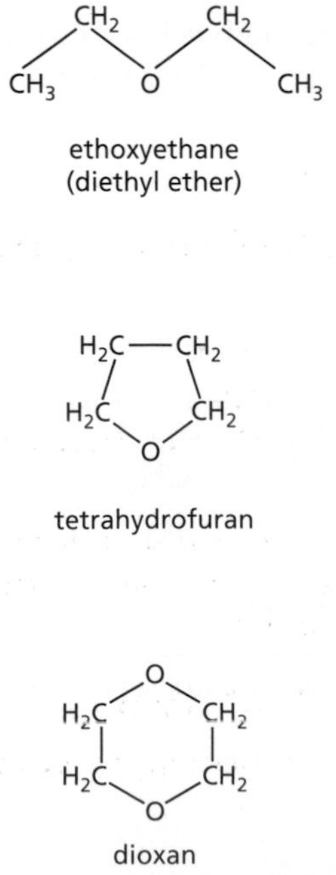

■ Figure 26.6
The structures of some ethers.

26.3 Reactions of alcohols

The use of ethanol as a fuel has been mentioned in Chapter 23. All alcohols burn well in air, but only the combustion of ethanol is of everyday importance.

$$C_2H_5OH + 3O_2 \rightarrow 2CO_2 + 3H_2O$$

For burning in stoves, and for use as a common solvent, ethanol is marketed as 'methylated spirit'. This contains about 90% ethanol, together with 5% water, 5% methanol, and pyridine. The poisonous methanol is added to make it undrinkable, and hence exempt from the large excise duty charged on spirits. The pyridine gives it a bitter taste, which makes it unpalatable.

The other reactions of alcohols can be divided into three groups:

- reactions involving the breaking of the O—H bond
- reactions involving the breaking of the C—O bond
- oxidation of the $>C(H)OH$ group (not given by tertiary alcohols which do not contain an H atom attached to the C atom)

Reactions involving the breaking of the O—H bond

With sodium metal

The hydroxyl hydrogen in alcohols is slightly acidic, just like the hydrogen atoms in water (see Chapter 12, page 243) but to a lesser extent:

$$\text{H—O—H} \rightleftharpoons \text{H—O}^- + \text{H}^+ \qquad K_w = 1.0 \times 10^{-14}\,\text{mol}^2\,\text{dm}^{-6}$$
$$\text{R—O—H} \rightleftharpoons \text{R—O}^- + \text{H}^+ \qquad K = 1.0 \times 10^{-16}\,\text{mol}^2\,\text{dm}^{-6}$$

Alcohols liberate hydrogen gas when treated with sodium metal:

$$CH_3CH_2OH + Na \rightarrow CH_3CH_2O^-Na^+ + \tfrac{1}{2}H_2$$

The reaction is less vigorous than that between sodium and water. The product is called sodium ethoxide (compare sodium hydroxide, from water). This is a white solid, soluble in water (to give a strongly alkaline solution) and in ethanol.

All compounds containing —OH groups liberate hydrogen gas when treated with sodium. The fizzing that ensues is a good test for the presence of an —OH group (as long as water is absent).

Esterification

Esters are compounds that contain the group: $-\overset{\overset{\large O}{\|}}{C}-O-$

They can be obtained by reacting alcohols with carboxylic acids, acyl chlorides or acid anhydrides.

$$CH_3\overset{\overset{O}{\|}}{C}-OH + CH_3CH_2OH \xrightarrow[\text{as catalyst}]{\text{heat with conc. } H_2SO_4} CH_3-\overset{\overset{O}{\|}}{C}-O-CH_2CH_3 + H_2O$$

The methods of ester preparation and their mechanisms will be dealt with in detail in Chapter 29. Esters are useful solvents, and several are also used, in small quantities, as flavouring agents in fruit drinks and sweets.

Reactions involving the breaking of the C—O bond

Although the C—O bond is polarised $C^{\delta+}$—$O^{\delta-}$, it is a strong bond, and does not easily break heterolytically:

$$-\underset{|}{\overset{|}{C}}-OH \rightarrow -\underset{|}{\overset{|}{C^+}} + {}^-OH$$

If, however, the oxygen is protonated, or bonded to a sulphur or phosphorus atom, the C—O bond is much more easily broken. As a result, alcohols undergo several **nucleophilic substitution** reactions.

Reaction with phosphorus(V) chloride or disulphur dichloride oxide
Both of these reagents convert alcohols into chloroalkanes (see Chapter 25, page 475), and in both cases hydrogen chloride is evolved. This fizzing with phosphorus(V) chloride can be used as a test for the presence of an alcohol in the absence of water. Both reactions occur on gently warming the reagents.

$$CH_3CH_2OH + PCl_5 \rightarrow CH_3CH_2Cl + PCl_3O + HCl$$
$$CH_3CH_2OH + SCl_2O \rightarrow CH_3CH_2Cl + SO_2 + HCl$$

Tests for an alcohol

- Alcohols produce hydrogen gas with sodium.
- Alcohols produce hydrogen chloride gas with phosphorus(V) chloride.

How it happens: the reactions of ethanol with PCl_5 and $SOCl_2$

These reactions involve the initial formation of an O—P or O—S bond, which is the result of a nucleophilic substitution on the hetero-atom by the lone pair of electrons on the ethanol's oxygen atom. This is followed by Cl^- acting as a base, abstracting a proton; and lastly a nucleophilic substitution on the carbon atom by another Cl^- ion (see Figures 26.7 and 26.8).

Both reactions are driven to completion by the high strengths of the P=O and S=O bonds that are formed.

Figure 26.7
The reaction of ethanol with PCl_5.

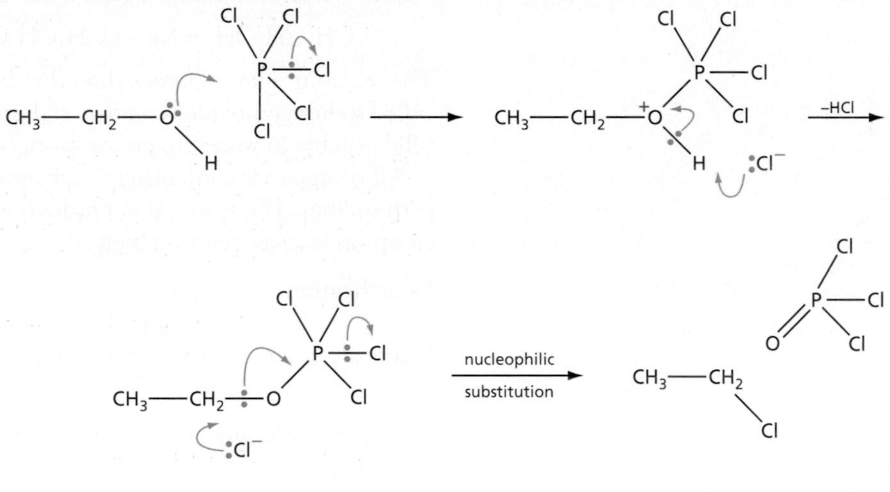

Figure 26.8
The reaction of ethanol with $SOCl_2$.

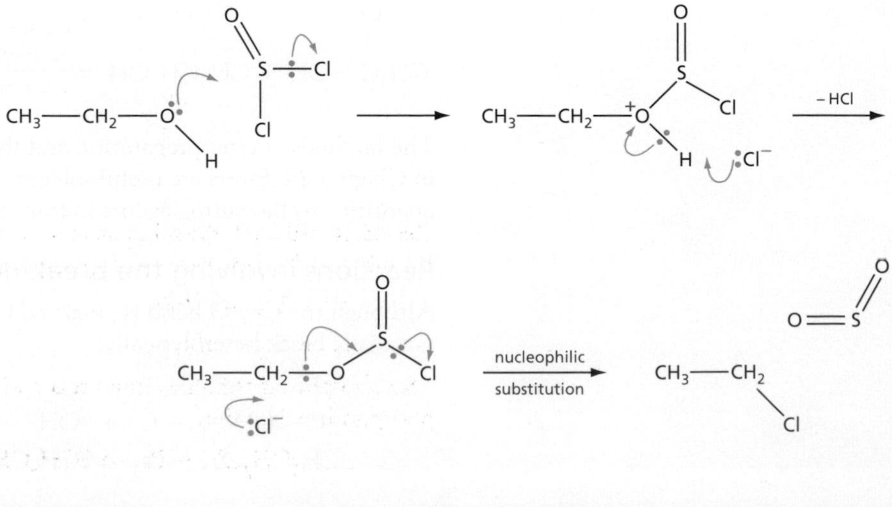

Reaction with hydrochloric acid

Tertiary alcohols react readily with concentrated hydrochloric acid on shaking at room temperature:

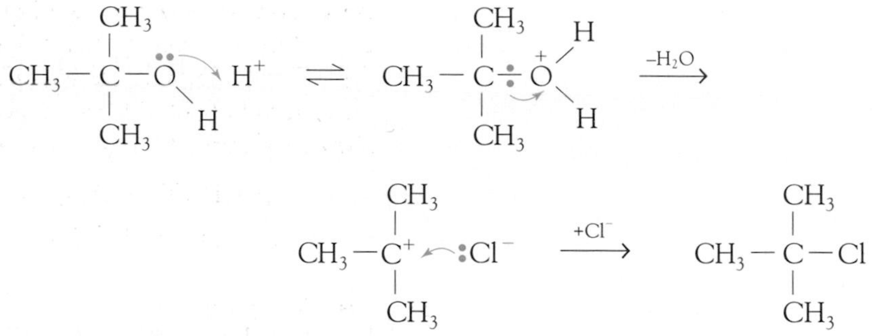

Secondary and primary alcohols also react with concentrated hydrochloric acid, but more slowly. Anhydrous zinc chloride is added as a catalyst, and the mixture requires heating:

$$CH_3CH_2OH + HCl \xrightarrow{ZnCl_2,\ heat} CH_3CH_2Cl + H_2O$$

This reaction is the basis of the **Lucas test** to distinguish between primary, secondary and tertiary alcohols. It relies on the fact that alcohols are soluble in the reagent (concentrated hydrochloric acid and zinc chloride), whereas chloroalkanes are not, and therefore produce a cloudiness in the solution (see Table 26.3).

■ Table 26.3 The Lucas test.

Type of alcohol	Observation on adding conc. HCl + $ZnCl_2$
R_3COH (tertiary)	Immediate cloudiness appears in the solution
R_2CHOH (secondary)	Cloudiness apparent within 5 minutes
RCH_2OH (primary)	No cloudiness apparent unless warmed

Reaction with hydrogen bromide

Alcohols can be converted into bromoalkanes by heating with either concentrated hydrobromic acid or a mixture of sodium bromide, concentrated sulphuric acid and water (50%). These react to give hydrogen bromide:

$$2NaBr + H_2SO_4 \longrightarrow Na_2SO_4 + 2HBr$$

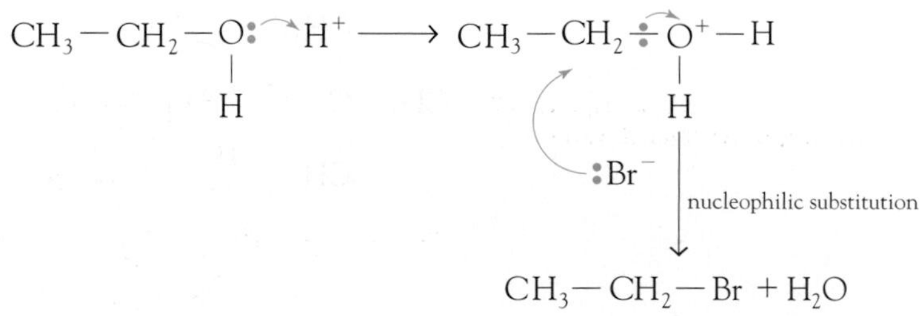

Reaction with red phosphorus and bromine or iodine

Bromine and iodine react with phosphorus to give the phosphorus trihalides:

$$2P + 3Br_2 \rightarrow 2PBr_3$$
$$2P + 3I_2 \rightarrow 2PI_3$$

The phosphorus trihalides then react with alcohols as follows, for example, with phosphorus triiodide:

$$3CH_3CH_2OH + PI_3 \rightarrow 3CH_3CH_2I + P(OH)_3$$

Reaction with concentrated sulphuric acid (1): dehydration

Ethanol and concentrated sulphuric acid react together to give the compound ethyl hydrogensulphate:

$$CH_3-CH_2-OH + H-O-\overset{\overset{O}{\|}}{\underset{\underset{O}{\|}}{S}}-O-H \rightleftharpoons$$

$$CH_3-CH_2-O-\overset{\overset{O}{\|}}{\underset{\underset{O}{\|}}{S}}-O-H + H_2O$$

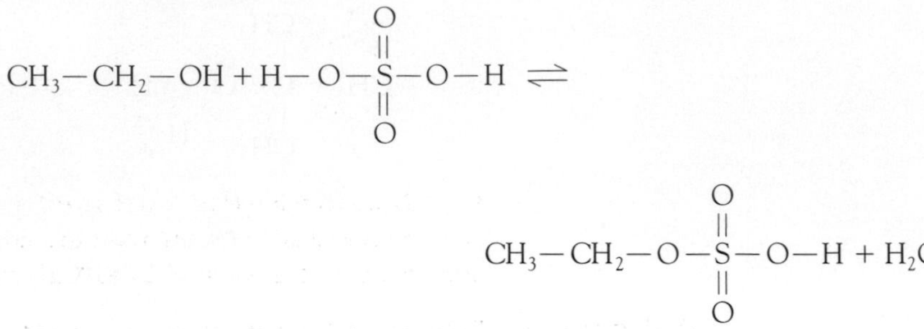

Just like halogenoalkanes (see Chapter 25, pages 469–71), the ethyl hydrogensulphate can undergo either elimination reactions or substitution reactions.

It undergoes elimination to form alkenes:

$$CH_2-CH_2-O-SO_3H \xrightarrow{\text{heat to 170°C}} CH_2{=}CH_2 + H^+ + HSO_4^-$$
$$\downarrow$$
$$H_2SO_4$$

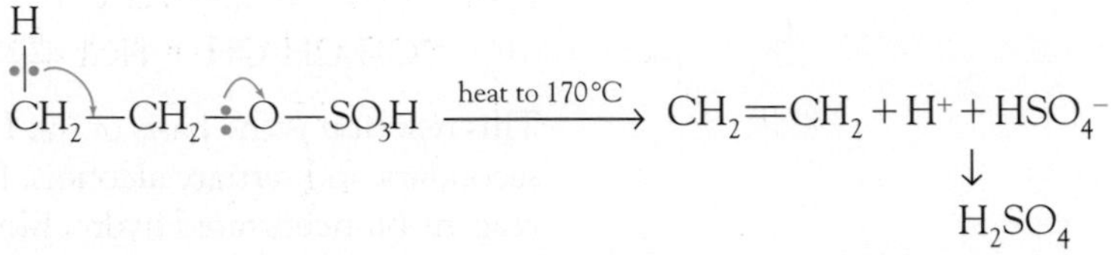

The sulphuric acid is regenerated on elimination, so it is essentially a catalyst for the conversion of the alcohol to the alkene. (Were it not diluted during the reaction, it could be recovered unchanged at the end.)

$$CH_3CH_2OH \rightarrow CH_2{=}CH_2 + H_2O$$

The reaction can also be carried out by heating the alcohol with concentrated phosphoric(v) acid (this is often preferred because, unlike sulphuric acid, it is not also an oxiding agent, and so the formation of by-products is minimised):

$$CH_3CH_2OH \xrightarrow{\text{heat with } H_3PO_4} CH_2{=}CH_2 + H_2O$$

A third method is to pass the vapour of the alcohol over strongly heated aluminium oxide:

$$CH_3CH_2OH \xrightarrow{Al_2O_3 \text{ at } 350°C} CH_2{=}CH_2 + H_2O$$

Tertiary alcohols dehydrate very easily on warming with an acid. The reaction goes via the carbocation which then loses a proton:

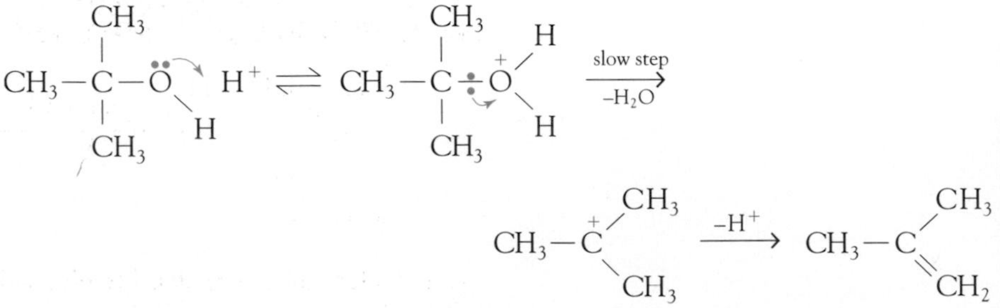

Example Dehydrating longer-chain alcohols can give a mixture of alkenes. Suggest structures for the alkenes possible from the dehydration of the alcohol 1-methylcyclohexanol. Which alkene will be the most stable?

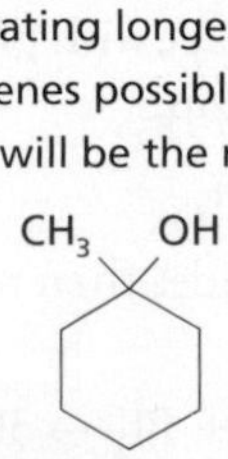

Answer The H and the OH of the water which is eliminated come from adjacent carbon atoms, forming the C=C between them. We can draw three possibilities:

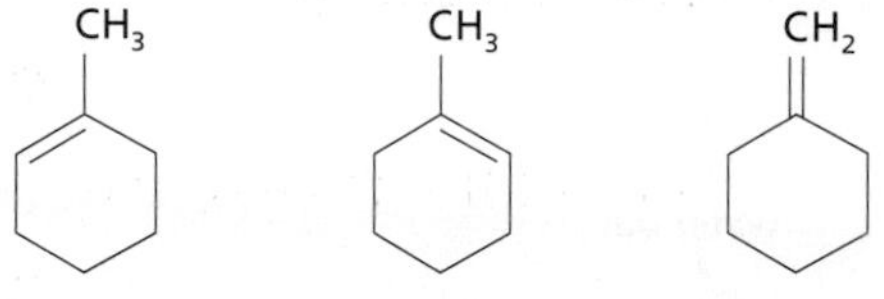

A   B   C

A and **B** are identical (rotating **A** by 180° around a vertical axis produces **B**), so there are two possible alkenes. Alkene **A/B** is the most stable because it has three alkyl groups around the double bond, whereas **C** has only two (see also Chapter 24, page 459).

Further practice Draw all the possible alkenes that could be obtained by the dehydration of 3-methylhexan-3-ol. Which is likely to be the least stable of these alkenes?

Reaction with concentrated sulphuric acid (2): formation of ethers

If ethyl hydrogensulphate is warmed in the presence of an excess of ethanol, a nucleophilic substitution reaction occurs:

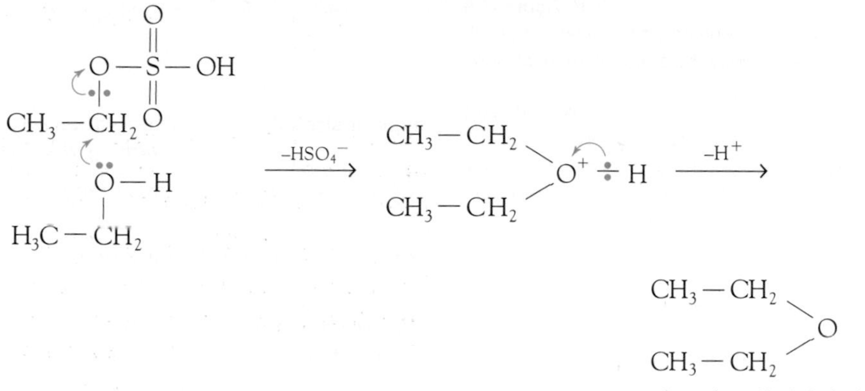

ethoxyethane (diethyl ether)

Symmetrical ethers (those in which the two alkyl groups are identical) are often made this way.

Reactions of the >CH(OH) group

Oxidation

Primary and secondary alcohols are easily oxidised by heating them with an acidified solution of potassium dichromate(VI). The orange dichromate(VI) ions are reduced to green chromium(III) ions. Tertiary alcohols are not easily oxidised. Secondary alcohols are oxidised to ketones (see Chapter 28):

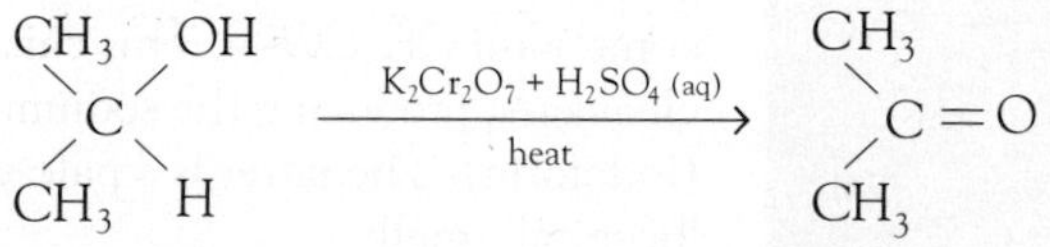

propan-2-ol   propanone

Primary alcohols are oxidised to aldehydes, which in turn are even more easily oxidised to carboxylic acids:

$$CH_3CH_2CH_2OH \xrightarrow[\text{heat}]{K_2Cr_2O_7 + H_2SO_4(aq)} CH_3CH_2CH{=}O \xrightarrow{\text{more oxidant}}$$

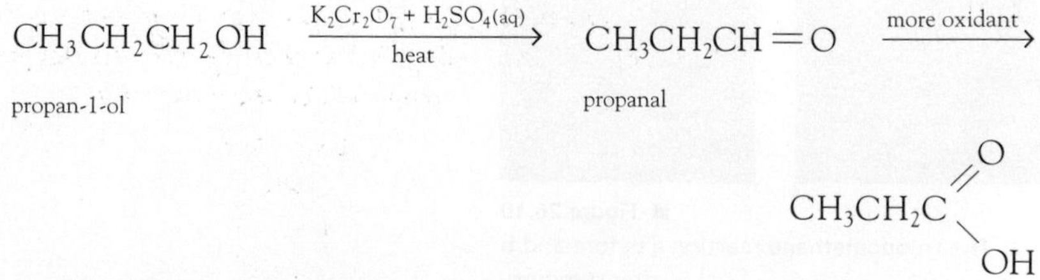

propan-1-ol   propanal

propanoic acid

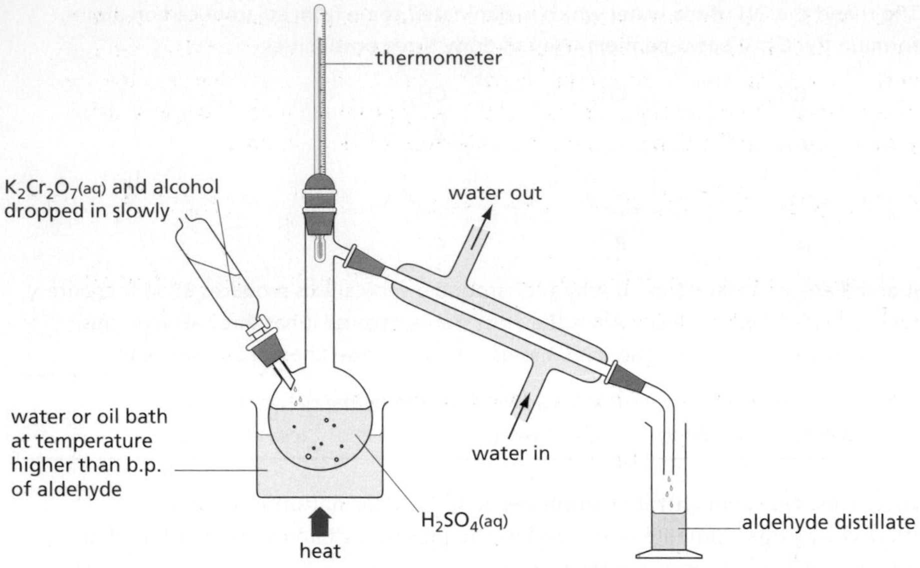

■ **Figure 26.9**
Apparatus for distilling off the aldehyde as it is formed by the oxidation of an alcohol.

Special experimental techniques must be used if the aim is to stop the oxidation at the aldehyde stage. One such method makes use of the higher volatility of the aldehyde (lack of hydrogen bonding). The reaction mixture is warmed to a temperature that is above the boiling point of the aldehyde, but below that of the alcohol. The aldehyde distils out as soon as it is formed, avoiding any further contact with the oxidising agent (see Figure 26.9).

On the other hand, an excess of oxiding agent, and heating the reaction under reflux before distillation (see page 469), allows the alcohol to be oxidised all the way to the carboxylic acid. This oxidation reaction can be used to distinguish between primary, secondary and tertiary alcohols (see Table 26.4).

■ **Table 26.4**
The use of potassium dichromate(VI) and sulphuric acid to distinguish between primary, secondary and tertiary alcohols.

Type of alcohol	Observation on warming with reagent	Effect of distillate on universal indicator
R_3COH (tertiary)	Stays orange	Neutral (stays green) – water only produced
R_2CHOH (secondary)	Turns green	Neutral (stays green) – ketone produced
RCH_2OH (primary)	Turns green	Acidic (goes red) – carboxylic acid produced

Tests for primary, secondary and tertiary alcohols

- The Lucas test (page 483).
- Oxidation with acidified potassium dichromate(VI).

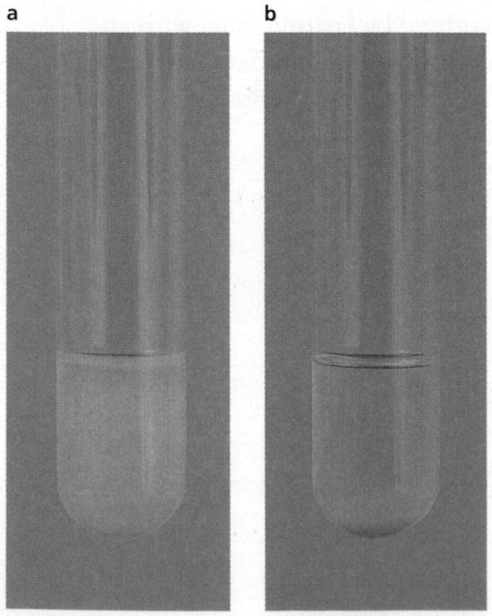

■ **Figure 26.10**
The triiodomethane reaction **a** before and **b** after standing.

The triiodomethane (iodoform) reaction

Alcohols that contain the group $CH_3CH(OH)$—, that is those that have a methyl group and a hydrogen atom on the same carbon atom that bears the OH group, can be oxidised by alkaline aqueous iodine to the corresponding carbonyl compound CH_3CO—. This can then undergo multiple substitution and cleavages, producing the sodium salt of a carboxylic acid and triiodomethane (iodoform). The latter is a pale yellow insoluble solid, with a sweet, antiseptic 'hospital' smell:

$$CH_3CH_2-\underset{}{\overset{OH}{\overset{|}{C}}}H-CH_3 \xrightarrow{I_2\ +\ OH^-(aq)} \left[CH_3CH_2\overset{O}{\overset{\|}{C}}CH_3 \longrightarrow CH_3CH_2\overset{O}{\overset{\|}{C}}CI_3\right]$$

$$\xrightarrow{+\ OH^-(aq)} CH_3CH_2C\begin{matrix}\nearrow O \\ \searrow O^-\end{matrix} \quad + \underset{\text{triiodomethane}}{CHI_3(s)}$$

The mechanism for this reaction is described in the panel on page 536. Except for ethanol, all the alcohols that undergo this reaction are secondary alcohols, with the OH group on the second carbon atom of the chain, that is, they are alkan-2-ols. The exception, ethanol, is the only primary alcohol to give a pale yellow triiodomethane precipitate with alkaline aqueous iodine:

$$CH_3CH_2OH \xrightarrow{I_2+OH^-(aq)} [CH_3{-}CHO \rightarrow CI_3{-}CHO] \rightarrow CHI_3(s) + HCO_2^-(aq)$$

> The **triiodomethane (iodoform) reaction** is a test for the $CH_3CH(OH)$— group or the $CH_3C{=}O$ group (see Chapter 28, page 536).

Example Which of these alcohols will undergo the iodoform reaction?

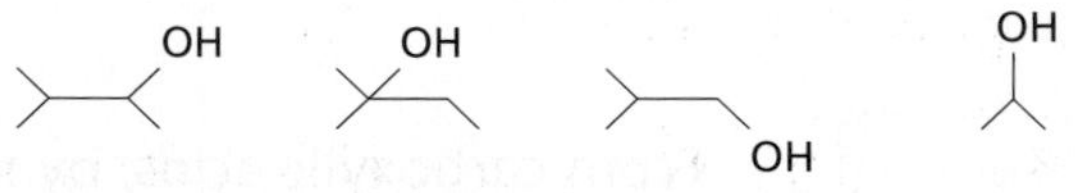

A B C D

Answer Only alcohols **A** and **D** contain the grouping $CH_3CH(OH)$—, so these are the only two to give iodoform. **B** is a tertiary alcohol, whilst **C** is a primary alcohol.

Further practice **P**, **Q** and **R** are three isomeric alcohols with the formula $C_5H_{11}OH$. All are oxidised by potassium dichromate(VI) and aqueous sulphuric acid, but only **P** gives an acidic distillate. When treated with alkaline aqueous iodine, **Q** gives a pale yellow precipitate, but **P** and **R** do not react. What are the structures of **P**, **Q** and **R**?

26.4 Preparing alcohols

As befits their central position in organic synthesis, alcohols can be prepared by a variety of different methods.

From halogenoalkanes, by nucleophilic substitution

See Chapter 25, page 465, for details of this reaction.

$$CH_3CH_2Br + OH^- \rightarrow CH_3CH_2OH + Br^-$$

From alkenes, by hydration

See Chapter 24, page 448, for details. This is the preferred method of making ethanol industrially.

$$CH_2{=}CH_2 + H_2O \xrightarrow[\text{H}_3\text{PO}_4\text{ at }300\,^\circ\text{C and }70\,\text{atm}]{\text{pass vapours over a catalyst of}} CH_3{-}CH_2{-}OH$$

The hydration can also be carried out in the laboratory by absorbing the alkene in concentrated sulphuric acid, and diluting with water:

$$CH_2{=}CH_2 + H_2SO_4 \rightarrow CH_3{-}CH_2{-}OSO_3H$$
$$CH_3{-}CH_2{-}OSO_3H + H_2O \rightarrow CH_3{-}CH_2{-}OH + H_2SO_4$$

From aldehydes or ketones, by reduction

See Chapter 28, page 529, for details. There are three common methods of reducing carbonyl compounds:

- hydrogen on a nickel catalyst:

$$CH_3{-}CH_2{-}CHO \xrightarrow{H_2+Ni} CH_3{-}CH_2{-}CH_2{-}OH$$

- sodium tetrahydridoborate(III) (sodium borohydride) in alkaline methanol is also effective:

$$CH_3-CH_2-CO-CH_3 \xrightarrow[\text{in methanol}]{NaBH_4+OH^-} CH_3-CH_2-CH(OH)-CH_3$$

- lithium tetrahydridoaluminate(III) (lithium aluminium hydride) in dry ether can also be used:

$$CH_3-CO-CH_3 \xrightarrow[\text{dry ether}]{LiAlH_4 \text{ in}} CH_3-CH(OH)-CH_3$$

Lithium tetrahydridoaluminate(III) is a dangerous reagent that can catch fire with water, and ether is also a hazardous solvent, being highly volatile and flammable. Therefore the second method, using sodium tetrahydridoborate(III), is the method of choice.

From carboxylic acids, by reduction

See Chapter 29, page 547, for details.

Lithium tetrahydridoaluminate(III) can reduce carboxylic acids to alcohols. (Carboxylic acids are more difficult to reduce than aldehydes or ketones, and sodium tetrahydridoborate(III) is not a powerful enough reducing agent to reduce acids.)

$$R-CO_2H \xrightarrow[\text{dry ether}]{LiAlH_4 \text{ in}} R-CH_2OH$$

From aldehydes or ketones, by the use of Grignard reagents

See Chapter 25, page 473, and Chapter 28, page 532. Adding a carbonyl compound to a solution of a Grignard reagent in dry ether, followed by dilute acid, produces alcohols. The method is especially useful for making tertiary alcohols.

$$(CH_3)_2C{=}O + C_2H_5MgBr \xrightarrow[\text{then add } H^+_{(aq)}]{\text{mix in dry ether}} (CH_3)_2C(OH)C_2H_5 + Mg^{2+} + Br^-$$

Making ethanol by fermentation

In many parts of the world, the fermentation of sugar or starch is used only to produce aqueous solutions of ethanol for beverages. But in some countries where petroleum is scarce or expensive (Brazil is the classic example), alcohol made by fermentation is used as a fuel for vehicles, either on its own, or as a 25% mixture with petrol. This could become increasingly important for the rest of the world, once oil reserves become depleted.

Yeasts are microorganisms of the genus *Saccharomyces*. They contain enzymes that not only break glucose down into ethanol and carbon dioxide, but also break down starch or sucrose (cane or beet sugar) into glucose. They can therefore convert a variety of raw materials into ethanol:

$$\underset{\text{starch}}{(C_6H_{10}O_5)_n} + nH_2O \xrightarrow{\text{yeast}} \underset{\text{glucose}}{nC_6H_{12}O_6}$$

$$\underset{\text{sucrose}}{C_{12}H_{22}O_{11}} + H_2O \rightarrow \underset{\text{glucose}}{C_6H_{12}O_6} + \underset{\text{fructose}}{C_6H_{12}O_6}$$

$$\underset{\text{fructose}}{C_6H_{12}O_6} \rightarrow \underset{\text{glucose}}{C_6H_{12}O_6}$$

and finally:

$$\underset{\text{glucose}}{C_6H_{12}O_6} \rightarrow \underset{\text{ethanol}}{2C_2H_5OH} + \underset{\text{carbon dioxide}}{2CO_2}$$

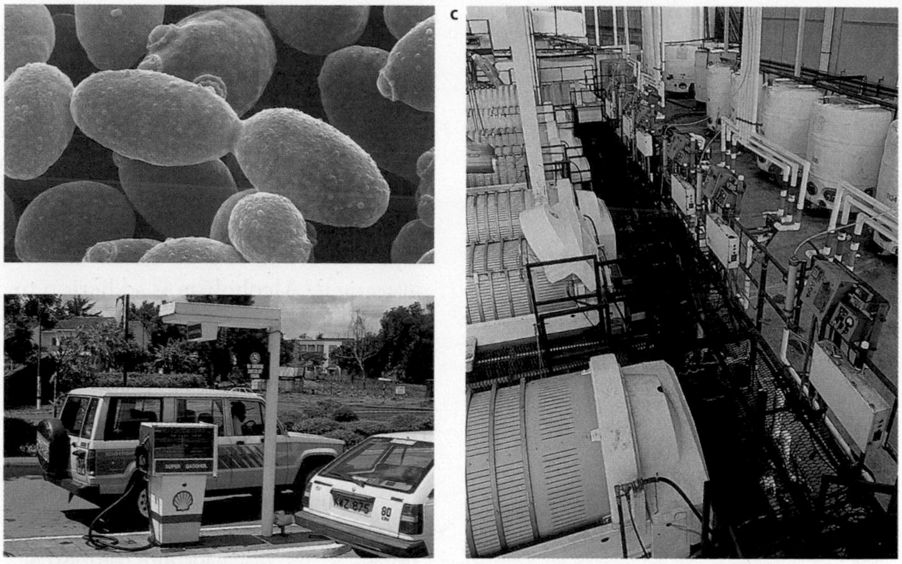

■ **Figure 26.11** **a** An electron micrograph of yeast cells. **b** Ethanol can be used solely or as a petrol additive to power vehicles. **c** Wine production.

The conditions required for successful fermentation are:

- yeast
- water
- yeast nutrients (ammonium phosphate is often used)
- warmth (a temperature of 30 °C is ideal)
- absence of air (with oxygen present, the ethanol can be oxidised partially to ethanoic acid, or completely to carbon dioxide).

The reaction is carried out in aqueous solution, and, assuming enough glucose is present, stops when the ethanol concentration reaches about 15%. Above this concentration, the yeast cells become dehydrated, and the yeast dies. After filtering off the dead yeast cells, the solution is fractionally distilled. The distillate consists of 95% ethanol and 5% water. The remaining water cannot be removed by distillation by even the most efficient fractionating column, because the 95% ethanol is an azeotropic mixture (see Chapter 12, page 239). If required, the remaining 5% water can be removed chemically, by adding quicklime (calcium oxide):

$$CaO(s) + H_2O(l) \rightarrow Ca(OH)_2(s)$$

or metallic magnesium:

$$Mg(s) + 2H_2O(l) \rightarrow Mg(OH)_2(s) + H_2(g)$$

For most chemical and industrial purposes, however, 95% ethanol is perfectly acceptable.

Apart from the economic advantage of producing alcohol (and hence fuel) by this method in areas where crude oil is scarce and expensive, the use of ethanol as a fuel also has an environmental advantage. As was mentioned in Chapter 23, ethanol made by fermentation is a renewable fuel. What is more, its production by photosynthesis and fermentation uses up exactly the same number of molecules of carbon dioxide as its combustion releases:

$$6CO_2 + 6H_2O \xrightarrow{\text{photosynthesis}} C_6H_{12}O_6 + 6O_2 \qquad (1)$$

$$C_6H_{12}O_6 \xrightarrow{\text{fermentation}} 2C_2H_5OH + 2CO_2 \qquad (2)$$

$$2C_2H_5OH + 6O_2 \xrightarrow{\text{combustion}} 6H_2O + 4CO_2 \qquad (3)$$

The carbon dioxide balance is therefore maintained. Indeed, if we add together the three equations (1), (2) and (3) we find no net change in any substance. We have obtained the energy we require entirely from the Sun, albeit indirectly.

Ethanol, the social catalyst

Ever since humans discovered many thousands of years ago that certain plant juices could suffer fermentation when infected with wild yeasts, alcoholic drinks have played their part in both the religious and the secular areas of many people's lives. Whilst the pleasurable aspects of alcohol consumption are well known to many, the unpleasant effects of the gross abuse of the drug are sometimes underestimated. Drunkenness and alcoholism are on the increase, and it is important for all of us to know how to use alcohol safely, and how to avoid misusing it. **Alcoholism** is addiction to alcohol. An alcoholic has a physical dependence on the drug, and needs a large daily alcohol intake to satisfy the craving. Alcoholics eventually develop irreversible problems with their liver, brain, heart, kidneys and circulation.

Beer, wine and spirits

There are three main types of alcoholic drink, depending on the percentage of alcohol they contain (see Table 26.5).

■ **Table 26.5** The percentage of alcohol in some alcoholic drinks.

Drink	Alcohol content/%	Comments
Beer, lager, cider	3–6	Some ciders and lagers can exceed 6% alcohol.
Wine	8–16	Some white wines contain less than 10% alcohol. Most red wines contain 12–13%. Fortified wines such as vermouths, Madeira, sherry and port contain 16–18% alcohol.
Spirits	25–60	Some liqueurs contain only 15% alcohol. Most whiskies, brandies and vodkas contain 40–50%.

Depending on the strain of yeast, the highest concentraton of alcohol that can be achieved through fermentation alone is about 15%. At that concentration, most yeasts die, poisoned by the product of their own metabolism. Spirits are produced by the fractional distillation of the initial liquid produced by fermentation. Fortified wines are produced by adding a spirit (usually brandy) to a wine. The spirit is often added before the natural fermentation has finished (for example, brandy is added to a port wine when it has reached only 10% alcohol) so as to avoid 'stale' flavours that might develop if the wine is left on the yeast for too long.

■ **Figure 26.12** Yeast cannot live in high ethanol concentrations, so whisky and other spirits are distilled to increase their alcohol content.

Beer is made by fermenting malted barley. **Malting** is the process that takes place when barley is moistened and kept in a warm environment for several days – the barley begins to germinate, and enzymes within the grain break down the starch into maltose, a dimer of glucose. Whisky is made by distilling the product of fermenting malted barley. Cider is made by fermenting the sugars in apple juice. Wine is the fermented juice of grapes. The colour (and the tannins – organic acids with a dry bitter taste) in red wine comes from the skin of 'black' grapes. Gin and vodka are derived from fermented potatoes (a vegetable rich in starch). Gin is flavoured with juniper berries.

The effects of alcohol on the body

Whatever the final concentration of alcohol, it is not so much the volume of the beverage that decides its effect on the individual, but the amount of ethanol it contains. This is measured in **units**, where one unit of alcohol is about 12 g. Table 26.6 shows how many units various quantities of common beverages contain.

Beverage	Quantity	Units
Beer ('bitter')	1 pint	2
Lager or strong beer	1 pint	2.5
Wine	1 glass (125 cm³)	1
Whisky	1 measure (25 cm³)	1

■ **Table 26.6** The alcoholic content of drinks is measured in units.

Alcohol is absorbed very quickly through the walls of the stomach and intestines. It circulates through the bloodstream to the liver, where is it eventually oxidised, first to ethanal (the substance responsible for the dry mouth and throbbing headache symptoms of a 'hangover') and eventually to carbon dioxide. The amount of alcohol in the blood, along with the body mass and the physiological tolerance of the drinker, determines the effect of the drug on the behaviour of the individual.

The initial effect is usually one of well-being and increased sociability. Alcohol is a central nervous system depressant, however. Further alcohol intake can result in drowsiness, a lack of coordination, and in some individuals increased aggressiveness. Even greater amounts cause unconsciousness and can lead eventually to death. Alcohol is a very powerful and potentially lethal drug. Had it been discovered within the last 50 years or so (rather than longer than 5000 years ago) it would most likely have been declared as illegal as heroin or cocaine.

A blood alcohol level of 0.05% allows the more pleasant social effects to become apparent. This occurs in an average man after the consumption of about 2 units of alcohol, or after the consumption of 1.5 units by the average woman. The legal limit for driving in the UK is a blood alcohol level of 0.08%. Three units would therefore take most people up to the limit. A concentration of twice this limit often results in the drinker feeling sick, and four times the limit will usually result in unconsciousness.

As was mentioned above, alcohol is oxidised in the liver. The rate at which this occurs depends on the individual, but it normally takes about 7 hours to reduce the blood alcohol concentration by 0.10%, which is about 4 units. The alcohol in the bloodstream that arises from drinking 4 pints of beer (8 units) in an evening will start at about 0.20%, and it will take about 8 hours to reduce its level to below the legal limit of 0.08%.

Three methods have been used to measure blood alcohol levels speedily and reliably. All three depend on a correlation between the alcohol level in the blood and that in either the breath or the urine.

- The **breathalyser** tests the dichromate-reducing capability of a fixed volume of breath (the orange dichromate(VI) turns green when ethanol is present).
- A sample of urine can be taken, and its ethanol content measured by **gas–liquid chromatography** and compared with a standard.
- More recently, a sample of breath is placed in an **infrared spectrometer**, and the intensity of its C—O absorption measured.

Other industrial methods for preparing alcohols

Methanol

Methanol is prepared from methane using a nickel catalyst:

$$CH_4 + H_2O \xrightarrow{\text{nickel catalyst at 800 °C}} CO + 3H_2$$

$$CO + 2H_2 \xrightarrow{\text{nickel catalyst at 400 °C, 20 atm}} CH_3OH$$

Methanol is a useful anti-knock petrol additive, and is an important feedstock for several industrially useful compounds, such as ethanoic acid:

$$CH_3OH + CO \xrightarrow{\text{RhCl}_3\text{+HI at 175 °C, 20 atm}} CH_3CO_2H$$

Propanol

Propanol is prepared industrially from ethene:

$$C_2H_4 + CO + H_2 \xrightarrow{\text{cobalt carbonyl catalyst at 150°C, 200 atm}} CH_3CH_2CHO$$

$$CH_3CH_2CHO \xrightarrow{H_2\text{+nickel}} CH_3CH_2CH_2OH$$

26.5 The infrared spectra of alcohols

Alcohols contain two polar bonds that are strongly infrared active. The $C^{\delta+}—O^{\delta-}$ bond provides a strong absorbance in the 1050 cm^{-1} to 1150 cm^{-1} region of the spectrum (tertiary alcohols absorb at the higher frequencies, primary alcohols at the lower frequencies).

The $O^{\delta-}—H^{\delta+}$ bond absorbs at a higher frequency, consistent with the small mass of the hydrogen atom. In dilute solutions in an organic solvent (where the alcohol molecules are far apart from each other) the bond absorbs sharply at about 3600 cm^{-1}. Under normal conditions, however, extensive intermolecular hydrogen bonding occurs, broadening the absorbance, and lowering the frequency to the 3200–3500 cm^{-1} region.

■ Figure 26.13
IR spectra of **a** hexane and **b** hexan-1-ol.

a hexane

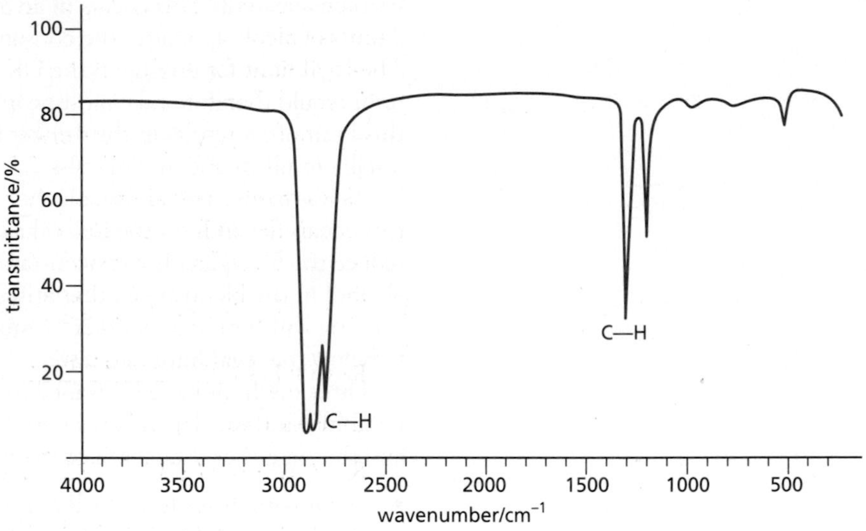

b hexan-1-ol

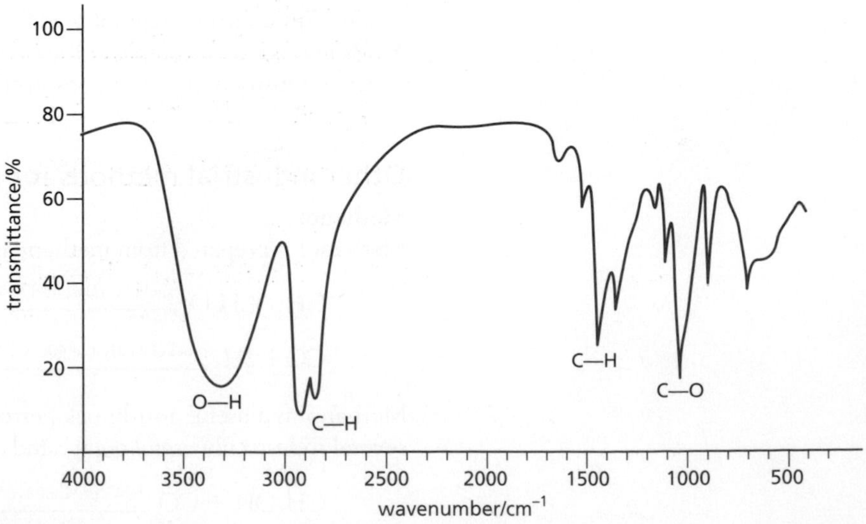

Summary

- Alcohols react in three different ways: by breaking the O—H bond, by breaking the C—O bond, and by breaking the C—H bond next to the —OH group.
- Two tests for the —OH group are the effervescence of hydrogen chloride when reacted with phosphorus(V) chloride, and the effervescence of hydrogen when reacted with sodium metal.
- Primary, secondary and tertiary alcohols can be distinguished either by their different oxidation products with potassium dichromate(VI) and aqueous sulphuric acid, or by using the **Lucas reagent** (concentrated hydrochloric acid and zinc chloride).
- Alcohols are important intermediates in organic synthesis and they are useful solvents. The lower members are used as additives in petrol.
- Alcohols can be prepared from halogenoalkanes, alkenes, carbonyl compounds and carboxylic acids.
- An important source of ethanol is the fermentation of starch or sucrose by yeast.

Key reactions you should know

(R, R′ = primary, secondary or tertiary alkyl unless otherwise stated)

- Combustion:
  $C_2H_5OH + 3O_2 \rightarrow 2CO_2 + 3H_2O$
- Redox:
  $R{-}OH + Na \rightarrow R{-}O^-Na^+ + \frac{1}{2}H_2$
- Esterification:
  $R{-}OH + HO_2CR' \xrightarrow{\text{heat with conc. } H_2SO_4} R{-}OCOR' + H_2O$
- Nucleophilic substitutions:
  $R{-}OH + PCl_5 \rightarrow R{-}Cl + PCl_3O + HCl$
  $R{-}OH + SCl_2O \rightarrow R{-}Cl + SO_2 + HCl$
  $R{-}OH + HCl(\text{conc.}) \xrightarrow{ZnCl_2} R{-}Cl + H_2O$ (best with R = tertiary alkyl)
  $R{-}OH + HBr \xrightarrow{\text{heat with } NaBr+H_2SO_4/H_2O} R{-}Br + H_2O$
- Eliminations:
  $R{-}CH_2{-}CH_2OH \xrightarrow{Al_2O_3 \text{ at } 350\,°C} R{-}CH{=}CH_2 + H_2O$
  $R{-}CH_2{-}CH_2OH \xrightarrow{H_2SO_4 \text{ at } 180\,°C} R{-}CH{=}CH_2 + H_2O$
- Oxidations:
  $RCH_2OH \xrightarrow{\text{heat with } Na_2Cr_2O_7+H_2SO_4(aq)} RCH{=}O \rightarrow RCO_2H$
  $R_2CHOH \xrightarrow{\text{heat with } Na_2Cr_2O_7+H_2SO_4(aq)} R_2C{=}O$
  [$R_3COH \xrightarrow{\text{heat with } Na_2Cr_2O_7+H_2SO_4(aq)}$ no reaction]
- Iodoform reaction:
  $R{-}CH(OH){-}CH_3 \xrightarrow{I_2+OH^-(aq)} RCO_2^- + CHI_3(s)$

27 Arenes and phenols

Functional groups:

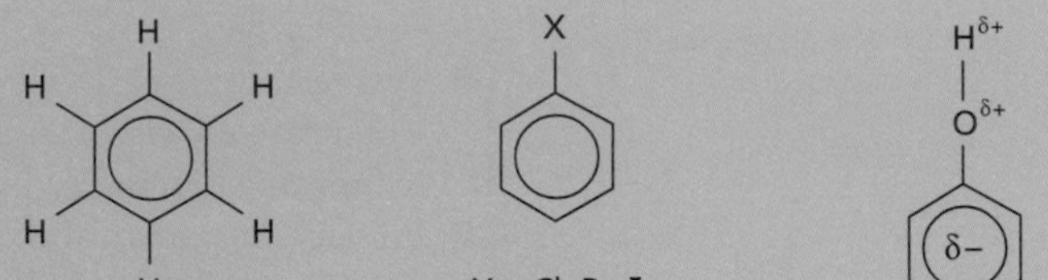

X = Cl, Br, I

arenes **halogenoarenes** **phenols**

The arenes are a group of hydrocarbons whose molecules contain the benzene ring. Although unsaturated, they do not show the typical reactions of alkenes. Their characteristic reaction is electrophilic substitution. Phenols contain an hydroxy group (—OH) joined directly to a benzene ring. The presence of the ring modifies the typical alcohol-like reactions of the —OH group. Likewise, the presence of the —OH group modifies the typical arene-like reactions of the benzene ring.

The chapter is divided into the following sections.

27.1 Introduction

The structure of the benzene ring

Arenes are hydrocarbons that contain one or more benzene rings. The structure of benzene was described in detail in Chapter 3, page 76. It consists of six carbon atoms, arranged in a regular hexagon, each joined to a hydrogen atom and to its neighbours by σ bonds. There are six spare p orbitals, all parallel to each other and perpendicular to the plane of the ring. Each p orbital overlaps equally with both its neighbours, forming a delocalised six-centre molecular π orbital (see Figure 27.1).

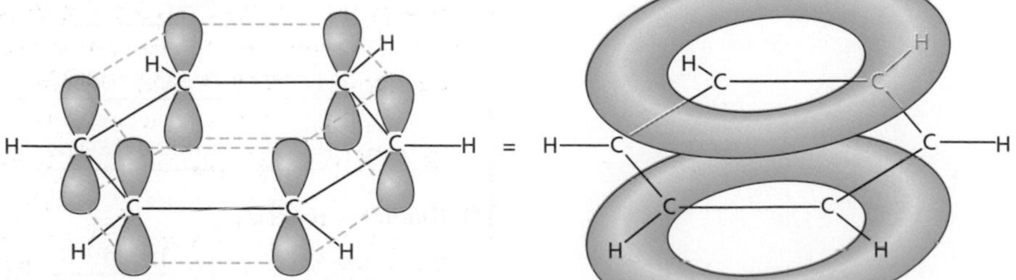

■ **Figure 27.1**
The delocalised π bond in benzene.

All the bond angles in benzene are 120°. All the C—C bonds have the same length, 0.139 nm. This is intermediate between the length of the C—C bond in an alkane (0.154 nm) and the C=C double bond in an alkene (0.134 nm).

The usual representation used for benzene is the skeletal formula. Figure 27.2 shows this and the corresponding structural formula.

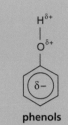

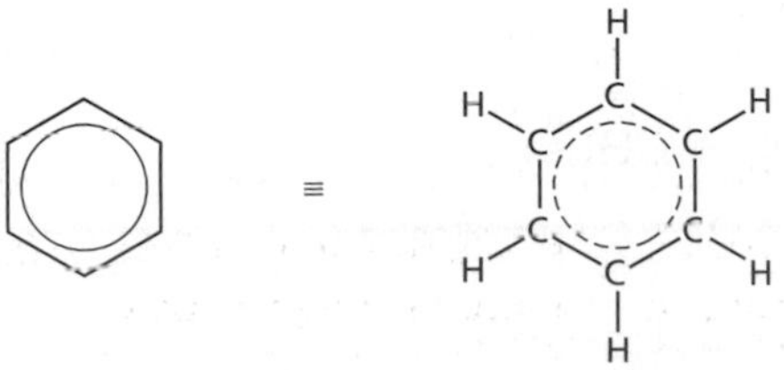

a b

■ **Figure 27.2**
a The skeletal formula and **b** the structural formula for benzene.

In the skeletal formula, each corner of the hexagon is assumed to contain a carbon atom, to which is attached a hydrogen atom, unless otherwise stated. For example, chlorobenzene can be written as in Figure 27.3.

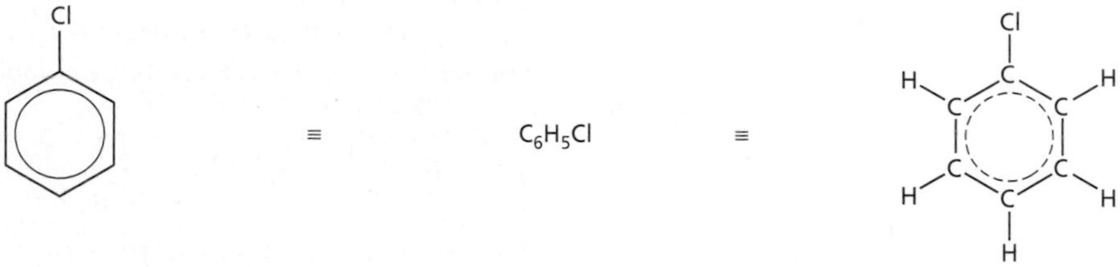

■ **Figure 27.3** Formulae for chlorobenzene.

■ **Figure 27.4** (above) The German chemist Friedrich August Kekulé von Stradonitz (1829–96) was the first to suggest that benzene had a ring structure.

Friedrich Kekulé, in 1865, was the first chemist to suggest the ring structure of benzene. His formula described benzene as a ring of alternating double and single bonds:

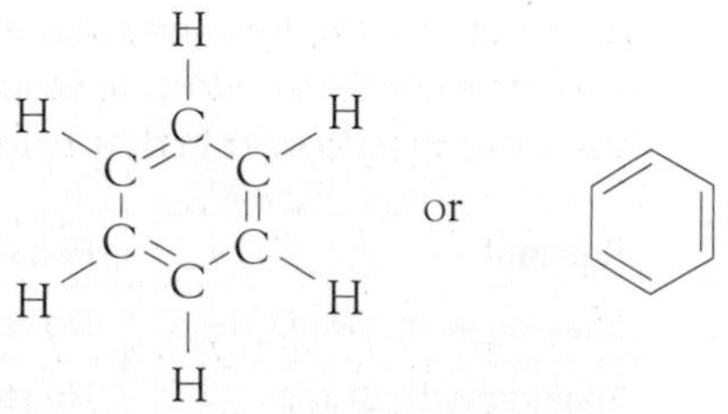

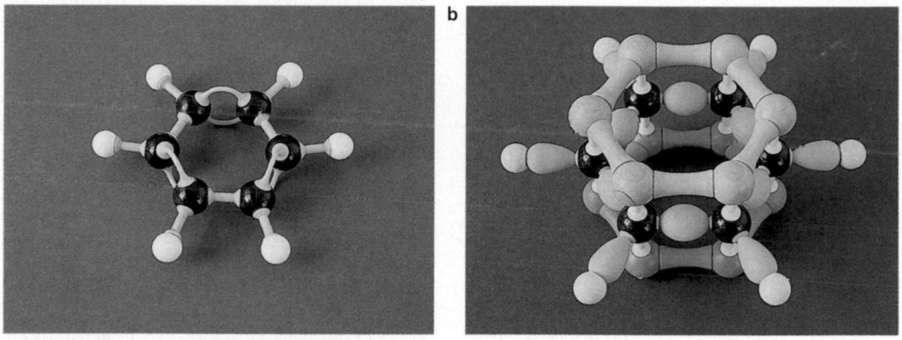

■ **Figure 27.5** (right) **a** Ball-and-stick and **b** electron density molecular models of benzene.

This structure helped to explain the following facts about benzene that were known at that time.

- When a hydrogen atom in benzene is replaced by another atom, only one isomer can ever be made. For example, there is only one compound (chlorobenzene) with the formula C_6H_5Cl. This means that all six hydrogen atoms must be identical in their environment.
- If two hydrogens are replaced by chlorines, to form dichlorobenzene, three (and only three) isomers of $C_6H_4Cl_2$ can be made (see page 499). Kekulé's structure did not quite account for this observation, because we might expect that there would be two isomers of 1,2-dichlorobenzene:

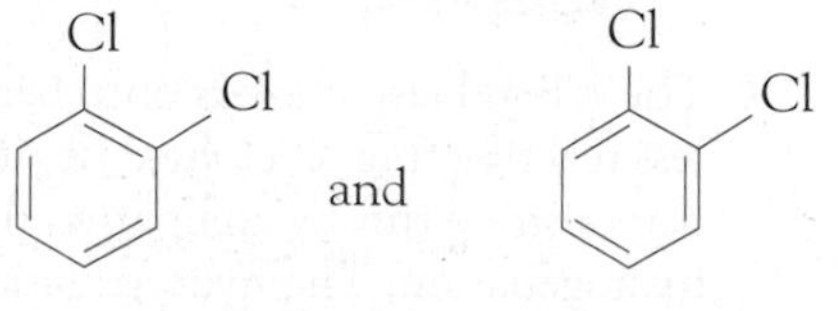

In fact, there is only one 1,2-dichlorobenzene. To overcome this disagreement with his structure, Kekulé proposed that the bonding in the benzene ring alternated between two equivalent structures:

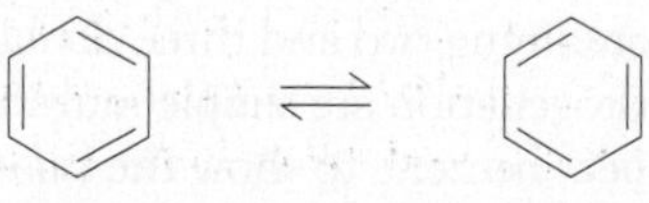

Now that we accept that π electrons can become delocalised in many-atom π orbitals, formed by the overlapping of p orbitals on adjacent atoms, we have no need to postulate this equilibrium between the two Kekulé forms. It is both wrong and misleading. The true bonding in benzene has sometimes been described as being in between the two extremes represented by the two Kekulé forms. We represent this in-between state by a double-headed arrow joining the two formulae:

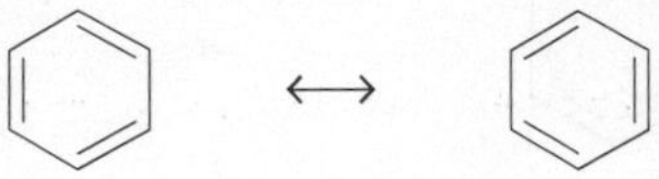

We shall usually represent the structure of benzene by the skeletal formula, as in Figure 27.2a. When we want to show in detail how the π electrons move during reactions, however, we shall use a Kekulé structure.

The Kekulé structure fitted in with the chemical bonding ideas of the 1860s, but was clearly incorrect in one important respect. It predicts that benzene is highly unsaturated – the formula suggests that the molecule contains three double bonds. It ought to undergo addition reactions readily, just like an alkene. In fact, benzene is inert to most reagents that readily add on to alkenes (see Table 27.1).

■ Table 27.1
A comparison of some reactions of benzene and cyclohexene.

Reagent	Benzene	Cyclohexene
Shaking with $KMnO_4(aq)$	No reaction	Immediate decolorisation
Shaking with $Br_2(aq)$	No reaction	Immediate decolorisation
$H_2(g)$ in the presence of nickel	Very slow reaction at 100 °C and 100 atm	Rapid reaction at 20 °C and 1 atm

Under more severe conditions, benzene can be made to react with one of these reagents, bromine. But now the reaction follows a different course – a substitution reaction occurs, rather than addition, and the necessary reaction conditions are much more extreme.

$$+ \ Br_2(l) \xrightarrow[\text{in the absence of water}]{AlCl_3 + heat} \ \overset{\displaystyle Br}{\bigcirc} \ + \ HBr$$

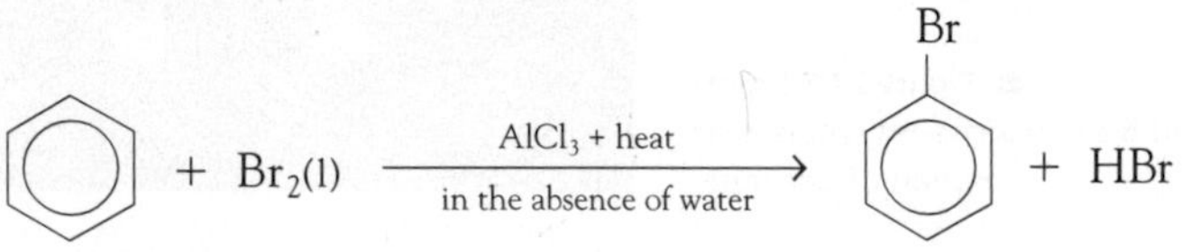

bromobenzene

The stability of benzene

The six-centre delocalised π bond is responsible for the following physical and chemical properties of benzene:

- It causes all C—C bond lengths to be equal, creating a planar, regular hexagonal shape.
- It prevents benzene undergoing any of the normal addition reactions that alkenes show.

The π bond also confers onto benzene an extra stability. Not only is benzene much less reactive than cyclohexene, it is also thermodynamically more stable. We can demonstrate this by comparing the actual and the 'calculated' enthalpy changes of hydrogenation. The hydrogenation of alkenes is an exothermic process:

$$CH_2{=}CH_2 + H_2 \rightarrow CH_3{-}CH_3 \qquad\qquad \Delta H^\ominus = -137\,kJ\,mol^{-1}$$

For many higher alkenes, the enthalpy changes of hydrogenation are very similar to each other, and average about $118\,kJ\,mol^{-1}$. Furthermore, for dienes and trienes (containing two and three double bonds, respectively), the enthalpy changes of hydrogenation are simple multiples of this value (see Table 27.2). So we might expect benzene to show the same trend.

■ **Table 27.2** Enthalpy changes of hydrogenation of some alkenes.

Alkene	Formula	$\Delta H^{\ominus}_{hydrogenation}$/ $kJ\,mol^{-1}$	$\Delta H^{\ominus}$ per C=C/ $kJ\,mol^{-1}$
cis-But-2-ene	$CH_3-CH=CH-CH_3$	–119	–119
Cyclohexene	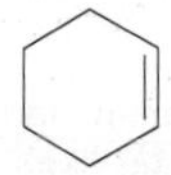	–118	–118
Butadiene	$CH_2=CH-CH=CH_2$	–236	–118
Cyclohexadiene	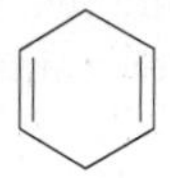	–232	–116
'Cyclohexatriene'	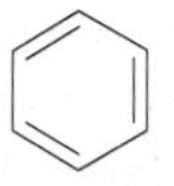	[–354] (predicted)	[–118] (assumed)

If we measure the experimental enthalpy change of hydrogenation of benzene, we find that it is far less exothermic than that predicted for 'cyclohexatriene' in Table 27.2:

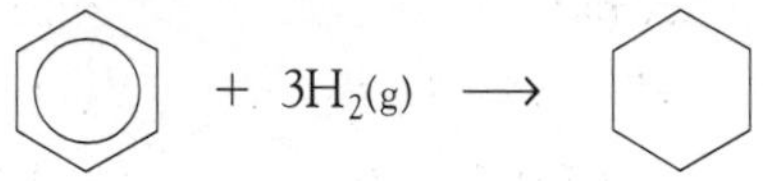

$$+\ 3H_2(g) \longrightarrow \qquad \Delta H^{\ominus}_{hydrogenation} = -205\ kJ\ mol^{-1}$$

So benzene is more stable than 'cyclohexatriene' by 354 – 205 = 149 $kJ\,mol^{-1}$ (see Figure 27.6).

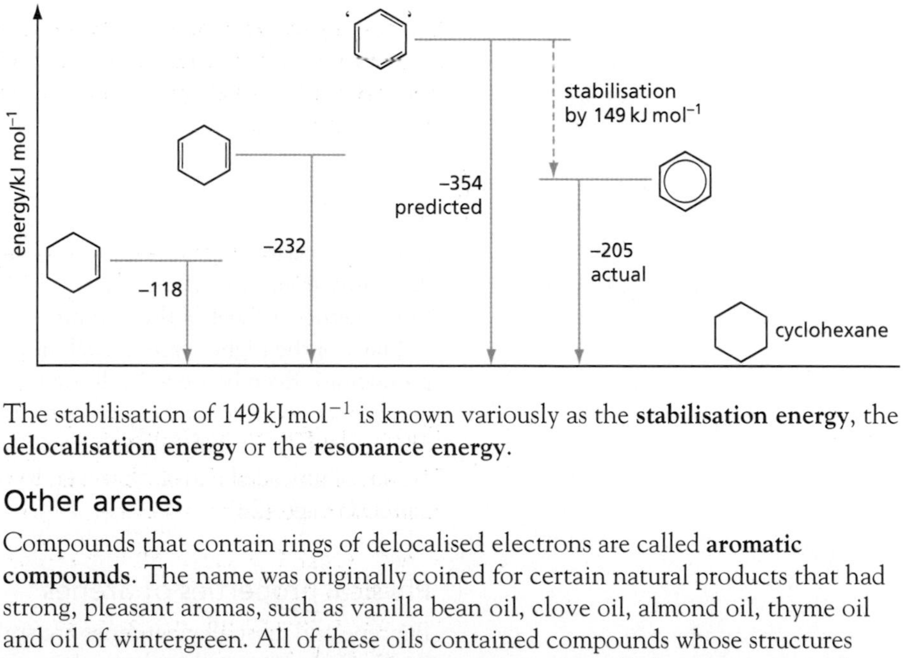

■ **Figure 27.6** Enthalpy changes of hydrogenation of benzene and the cyclohexenes.

The stabilisation of 149 $kJ\,mol^{-1}$ is known variously as the **stabilisation energy**, the **delocalisation energy** or the **resonance energy**.

Other arenes

Compounds that contain rings of delocalised electrons are called **aromatic compounds**. The name was originally coined for certain natural products that had strong, pleasant aromas, such as vanilla bean oil, clove oil, almond oil, thyme oil and oil of wintergreen. All of these oils contained compounds whose structures were found to include a benzene ring. The name eventually became associated with the presence of the ring itself, whether or not the compound had a pleasant aroma (see Figures 27.7 and 27.8, overleaf).

■ **Figure 27.7** Some pleasant-smelling naturally occurring aromatic compounds.

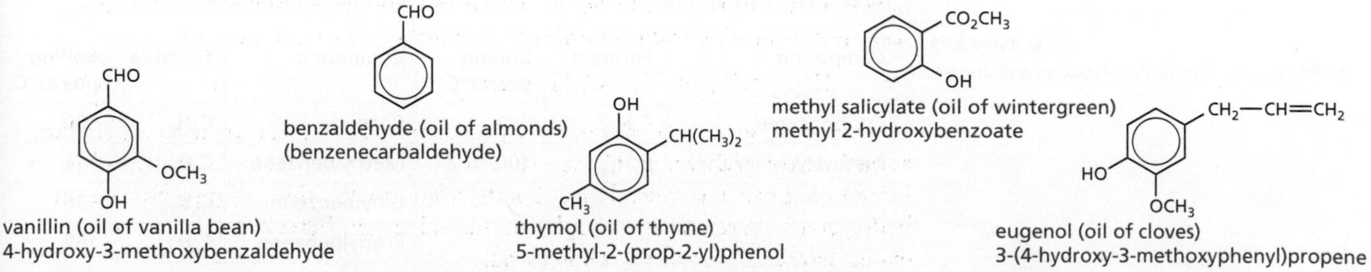

■ **Figure 27.8**
Some not-so-pleasant-smelling aromatic compounds made in the laboratory.

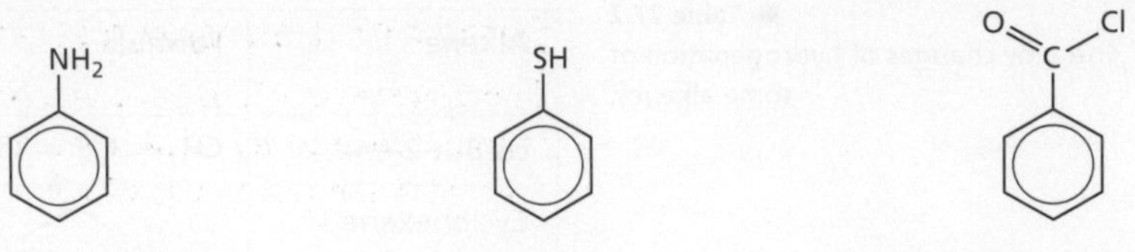

phenylamine (aniline) (musty, tar-like) | thiophenol (burnt rubber) | benzoyl chloride (benzenecarbonyl chloride) (acidic and nauseating)

There are two ways in which benzene rings can join together. Two rings could be joined by a single bond, as in biphenyl:

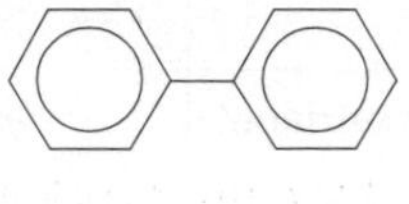

biphenyl, $C_{12}H_{10}$, a 12 π-electron system

Or the two rings could share two carbon atoms in common (with their π electrons), as in naphthalene:

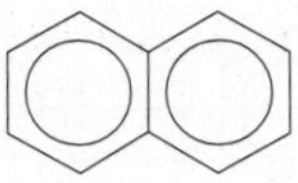

naphthalene, $C_{10}H_8$, a 10 π-electron system

More rings can fuse together, giving such compounds as anthracene and pyrene:

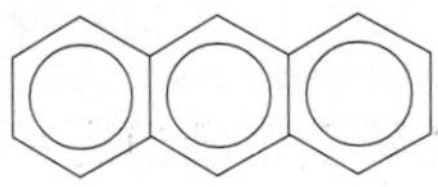

anthracene, $C_{14}H_{10}$

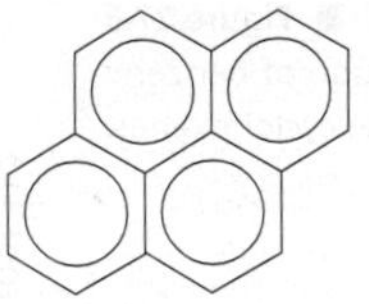

pyrene, $C_{16}H_{10}$

Notice that with each successive ring fused together, the hydrogen : carbon ratio decreases, from 1 : 1 in benzene to 5 : 8 in pyrene. Eventually, as many more rings fuse together, a sheet of the graphite lattice would result (see Chapter 4).

Many of the higher arenes, such as pyrene, are strongly carcinogenic (cancer-producing). Even benzene itself is a highly dangerous substance, causing anaemia and cancer on prolonged exposure to its vapour. At one time it was used as a laboratory solvent, but its use is now severely restricted. It is still added to some brands of unleaded petrol, however, to increase their anti-knock rating (see the panel on page 428).

Physical properties of arenes

Benzene and most alkylbenzenes are strongly oily-smelling colourless liquids, immiscible with, and less dense than, water. They are non-polar, and the only intermolecular bonding is due to the induced dipoles of van der Waals forces. Their boiling points are similar to those of the equivalent cycloalkanes (see Table 27.3), and increase steadily with relative molecular mass as expected.

■ **Table 27.3**
Boiling points of some cyclohexanes and arenes.

Compound	Formula	Boiling point/°C	Compound	Formula	Boiling point/°C
Cyclohexane	C_6H_{12}	81	Benzene	C_6H_6	80
Methylcyclohexane	C_7H_{14}	100	Methylbenzene	C_7H_8	111
			Ethylbenzene	C_8H_{10}	136
			Propylbenzene	C_9H_{12}	159

Some non-benzenoid aromatics

There are several compounds that show 'aromatic character' (that is, they are more thermodynamically stable than expected, and undergo substitution reactions rather than addition reactions) which do not contain benzene rings (see Figure 27.9).

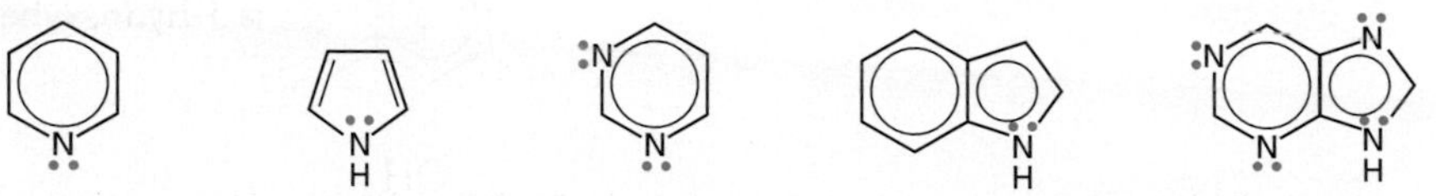

pyridine	pyrrole	pyrimidine	indole	purine
C_5H_5N	C_4H_5N	$C_4H_4N_2$	C_8H_7N	$C_5H_4N_4$

■ **Figure 27.9** Non-benzenoid aromatic molecules of biological importance.

Some of these are of biological importance:
- the pyridine ring occurs in the vitamin niacin
- the pyrrole ring occurs in haemoglobin and chlorophyll
- the indole ring occurs in the amino acid tryptophan and the plant growth hormone indoleacetic acid (IAA)
- the pyrimidine and purine rings occur in the bases of the nucleic acids DNA and RNA.

Many totally new aromatic systems have been synthesised in laboratories. Two are shown in Figure 27.10.

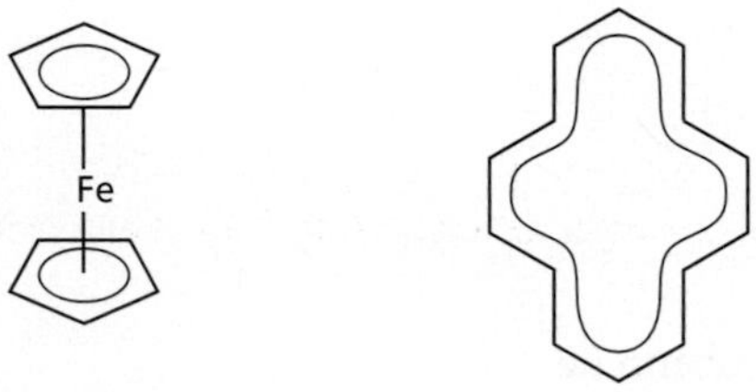

ferrocene $C_{10}H_{10}Fe$

[14]-annulene $C_{14}H_{14}$

■ **Figure 27.10** Synthetic aromatics.

27.2 Isomerism and nomenclature

Aromatic compounds with more than one substituent on the benzene ring can exist as positional isomers. There are three dichlorobenzenes:

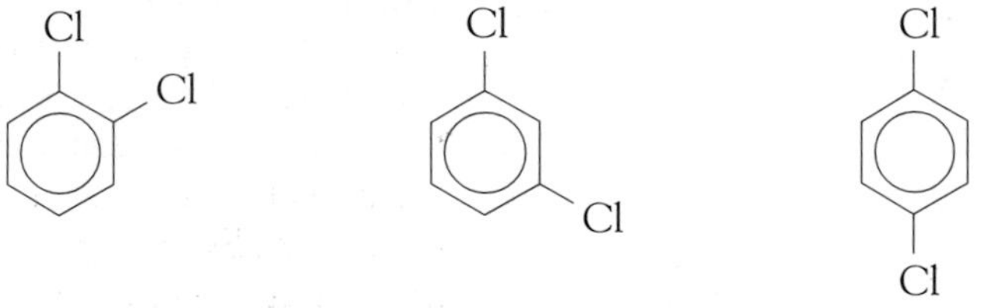

1,2-dichlorobenzene (*ortho*-dichlorobenzene) — 1,3-dichlorobenzene (*meta*-dichlorobenzene) — 1,4-dichlorobenzene (*para*-dichlorobenzene)

The terms *ortho-*, *meta-* and *para-* are sometimes used as prefixes to represent the relative orientations of the groups (see Table 27.4).

■ **Table 27.4** Orientations of substituents on the benzene ring.

Orientation	Prefix	Abbreviation	Example
1,2-	*ortho-*	*o-*	o-nitromethylbenzene 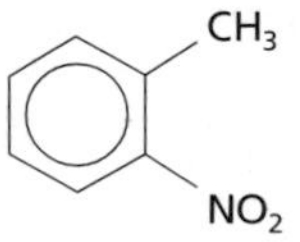
1,3-	*meta-*	*m-*	*m*-chlorobenzoic acid 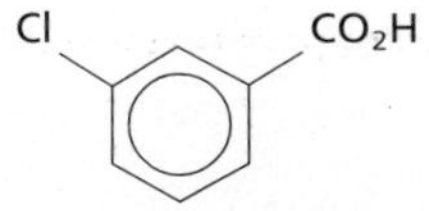
1,4-	*para-*	*p-*	*p*-methylphenylamine 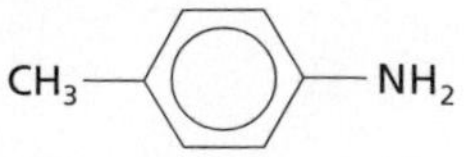

If the two substituents are different, one of them is deemed the 'root' group, in the order of precedence $—CO_2H$, $—OH$, $—CH_3$, —halogen, $—NO_2$. For example:

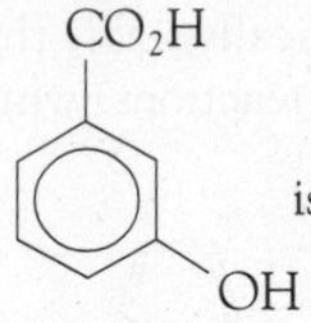

is 3-hydroxybenzoic acid

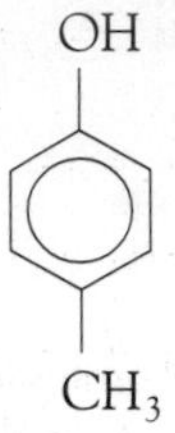

is 4-methylphenol (phenol is 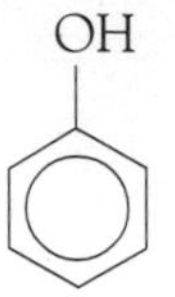)

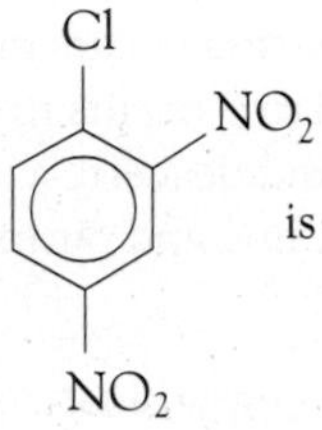

is 2,4-dinitrochlorobenzene

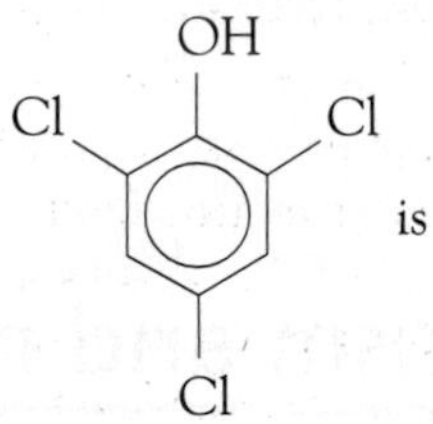

is 2,4,6-trichlorophenol (the antiseptic TCP)

Example Draw out all possible positional isomers of $C_6H_3Br_2OH$ and name them.

Answer There are six isomers. Their names and formulae are as follows:

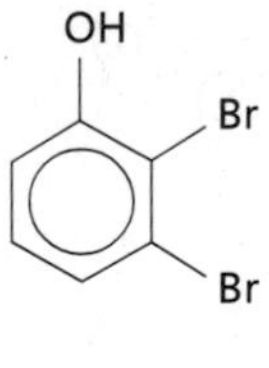

2,3-dibromophenol

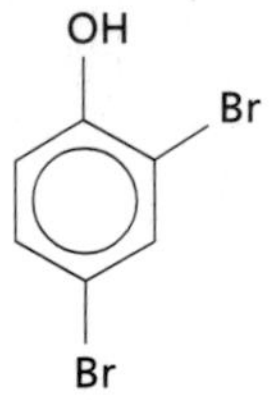

2,4-dibromophenol

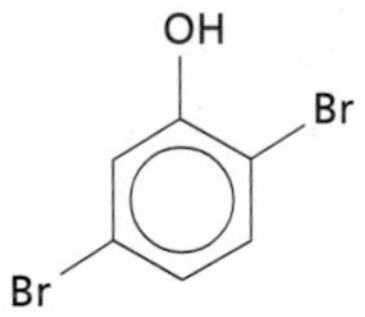

2,5-dibromophenol

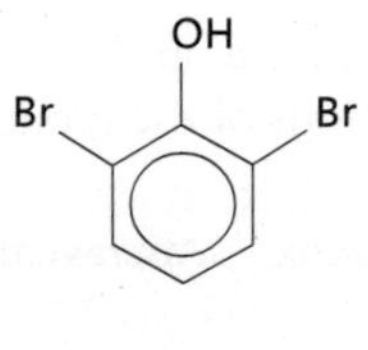

2,6-dibromophenol

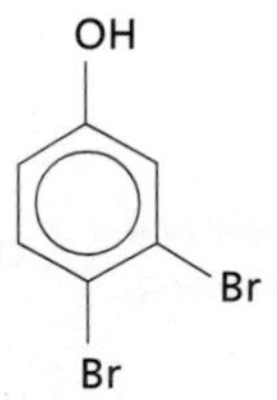

3,4-dibromophenol

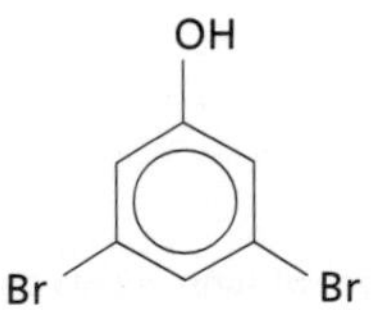

3,5-dibromophenol

Further practice How many isomers are there of trichloromethylbenzene, $C_6H_2CH_3Cl_3$? What are their names?

If the benzene ring is a 'substituent' on an alkyl or alkenyl chain, it is given the name **phenyl**:

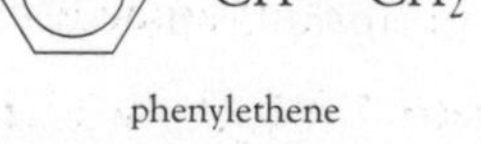

phenylethene

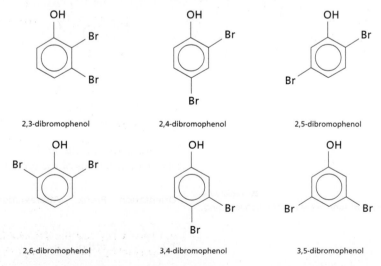

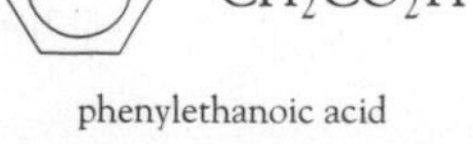

phenylethanoic acid

27.3 Reactions of arenes

Combustion

Benzene and methylbenzene (toluene) are components of many brands of unleaded petrol. In sufficient oxygen, they burn completely to carbon dioxide and steam:

$$C_6H_6 + 7\tfrac{1}{2}O_2 \rightarrow 6CO_2 + 3H_2O$$

If liquid arenes are set alight in the laboratory, they burn with very smoky flames. Much soot is produced because there is insufficient oxygen for complete combustion. A smoky flame is an indication of a compound with a high C : H ratio.

$$C_6H_6 + 1\tfrac{1}{2}O_2 \rightarrow 6C(s) + 3H_2O$$

Just as with hydrogenation, the enthalpy of combustion of benzene is less exothermic than expected, because of the stability due to the six delocalised π electrons.

Electrophilic substitution in benzene

In a similar way to the π bond in alkenes, the delocalised π bond in benzene is an area of high electron density, above and below the six-membered ring. Benzene therefore reacts with electrophiles. Because of the extra stability of the delocalised electrons, however, the species that react with benzene have to be much more powerful electrophiles than those that react with ethene. Bromine water and aqueous acids have no effect on benzene.

The electrophiles that react with benzene are all positively charged, with a strong electron-attracting tendency. The other major differences between benzene and alkenes is what happens after the electrophile has attacked the π bond. In alkenes, an anion 'adds on' to the carbocation intermediate. In benzene, the carbocation intermediate loses a proton, so as to re-form the ring of π electrons. This demonstrates how stable the delocalised system is:

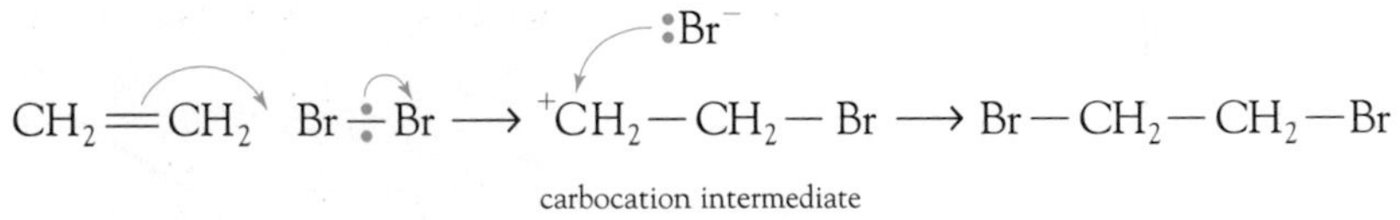

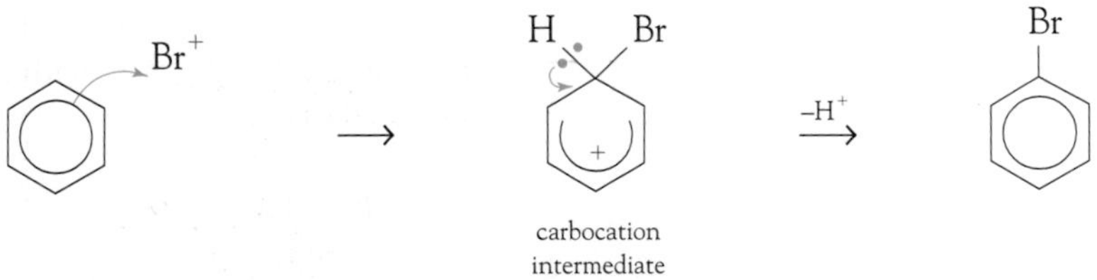

> Alkenes react by **electrophilic addition**.
> Arenes react by **electrophilic substitution**.

Bromination

Benzene will react with non-aqueous bromine on warming in the presence of anhydrous aluminium chloride or aluminium bromide, or iron(III) chloride, or even just iron metal. In the latter case, an initial reaction between iron and bromine provides the iron(III) bromide catalyst:

$$2Fe(s) + 3Br_2(l) \rightarrow 2FeBr_3(s)$$

Anhydrous aluminium or iron(III) halides are Lewis acids (see Chapter 6). They are electron pair acceptors. They can react with the bromine molecule by accepting one of the lone pairs of electrons on bromine:

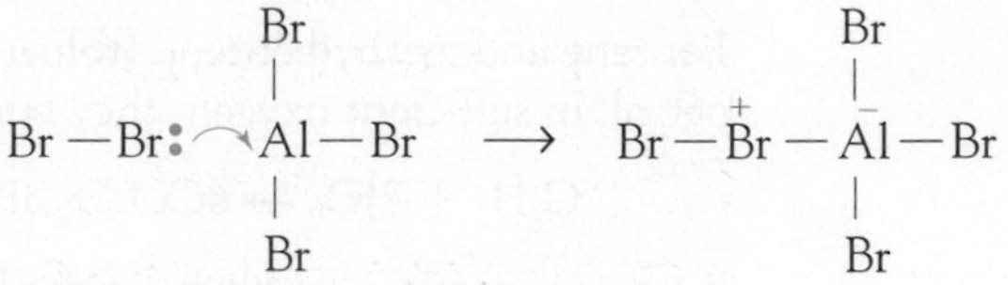

(1a)

This causes strong polarisation of the Br—Br bond, weakening it, and eventually leading to its heterolytic breaking:

(1b)

The bromine cation that is formed is a powerful electrophile. It becomes attracted to the π bond of benzene. It eventually breaks the ring of electrons and forms a σ bond to one of the carbon atoms of the ring:

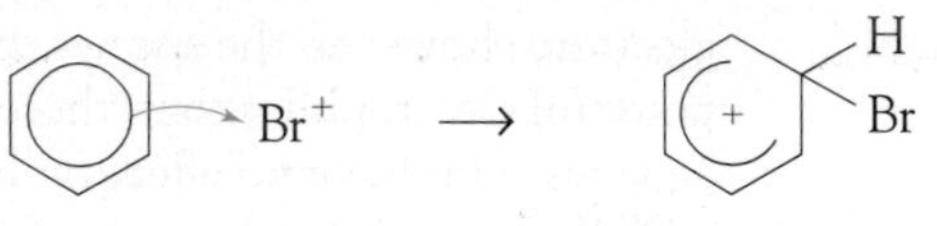

(2)

Two of the six π electrons are used to form the (dative) bond to the bromine atom. The other four π electrons are spread over the remaining five carbon atoms of the ring, in a five-centre delocalised orbital (see Figure 27.11).

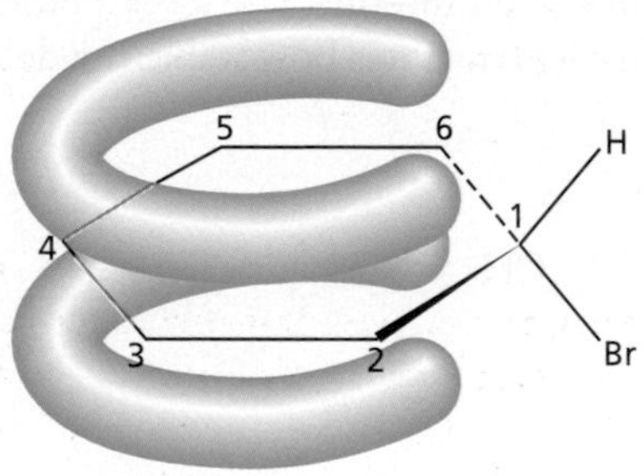

Figure 27.11 The four π electrons are delocalised over five carbons.

The distribution of the four π electrons is not even. They are associated more with carbon atoms 3 and 5 than with atoms 2, 4 and 6. The positive charge is therefore distributed over atoms 2, 4 and 6.

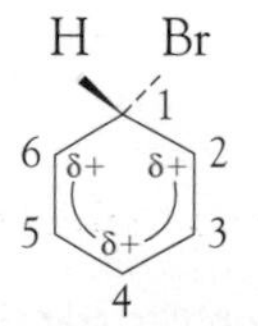

We shall return to this feature of the intermediate carbocation when we look at orientation effects on page 509.

The intermediate carbocation then loses a proton, to re-form the sextet of π electrons:

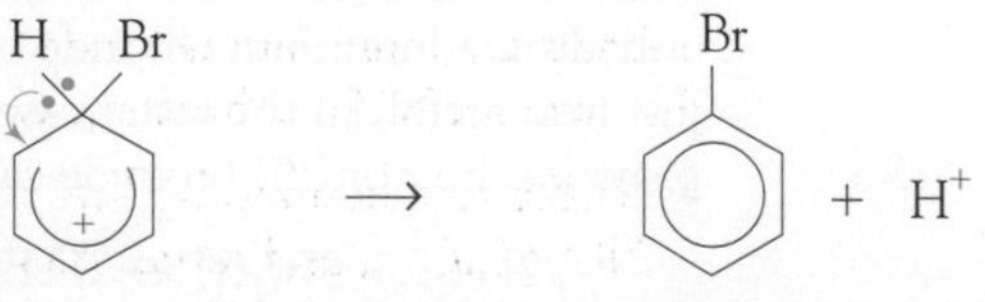

(3)

The final stage regenerates the catalyst, by the reaction between the proton formed in step (3) with the $[AlBr_4]^-$ formed in step (1b):

$$H^+ \quad Br - \overset{\underset{\displaystyle |}{Br}}{\underset{\underset{\displaystyle |}{Br}}{Al}} - Br \quad \longrightarrow \quad H - Br \; + \quad \overset{Br}{\underset{Br}{\diagdown}} Al - Br$$

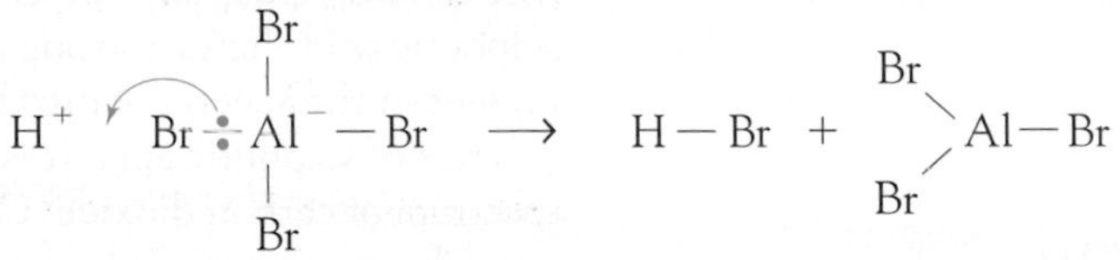

Further substitution can occur with an excess of bromine, to form dibromobenzene, and even tribromobenzene. For the reactions of bromobenzene, see page 513.

Chlorination

Just as with bromine, chlorine in the presence of a Lewis acid such as aluminium chloride will substitute into the benzene ring:

$$\text{benzene} \; + \; Cl_2 \; \xrightarrow[\text{warm}]{AlCl_3} \; \text{chlorobenzene (Cl)} \; + \; HCl$$

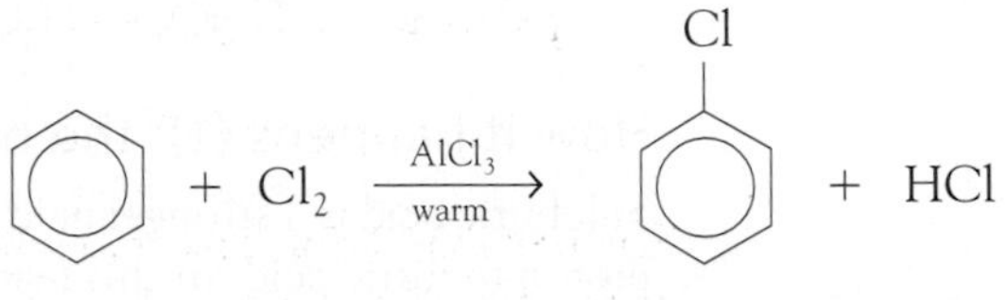

The properties and reactions of chlorobenzene are discussed on page 513.

Further practice   Suggest a mechanism for the chlorination of benzene.

Nitration

Benzene does not react with nitric acid, even when concentrated. But a mixture of concentrated nitric and concentrated sulphuric acids produces nitrobenzene:

$$\text{benzene} \; + \; HNO_3 \; \xrightarrow{\text{conc. } H_2SO_4 \text{ at} <55\,°C} \; \text{nitrobenzene (NO}_2\text{)} \; + \; H_2O$$

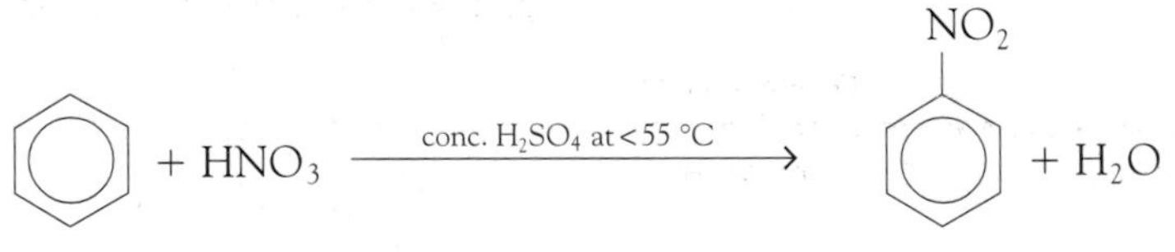

nitrobenzene

Further nitration to dinitrobenzene (and also some oxidation) can occur unless the temperature is controlled at below about 55 °C.

The role of the concentrated sulphuric acid in this reaction is to produce the powerful electrophile needed to nitrate the benzene ring. By analogy with bromination:

$$\text{benzene} \; + \; Br^+ \; \longrightarrow \; \text{bromobenzene (Br)} \; + \; H^+$$

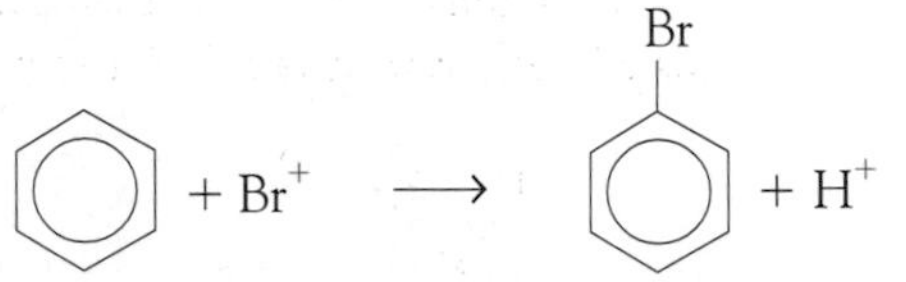

we might expect the electrophile for nitration to be NO_2^+:

$$\text{benzene} \; + \; NO_2^+ \; \longrightarrow \; \text{nitrobenzene (NO}_2\text{)} \; + \; H^+$$

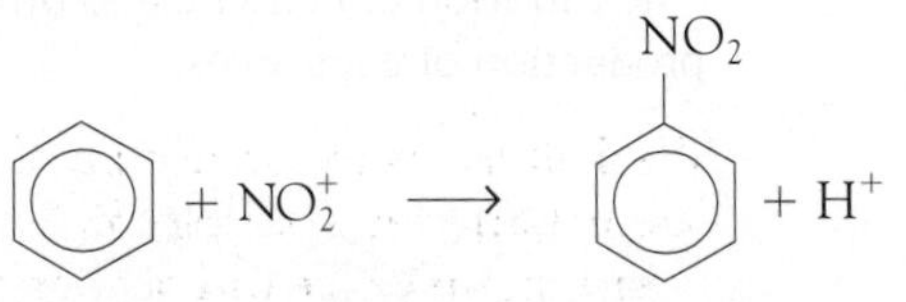

and this indeed seems to be the case.

Over the years various pieces of evidence have been collected to substantiate the claim of NO_2^+ (the nitryl cation, or the nitronium ion) to be the electrophile:

- Stable salts containing the nitryl cation exist, for example nitryl chlorate(VII), $NO_2^+ClO_4^-$, and nitryl tetrafluoroborate(III), $NO_2^+BF_4^-$. Each of these, when dissolved in an inert solvent, nitrates benzene smoothly and in high yield to give nitrobenzene.

- The infrared spectrum (see Chapter 14, page 278) of a solution of nitric acid in sulphuric acid shows a strong absorbance at 1400 cm^{-1}. This absorbance is also present in the spectra of nitryl salts, but is not present in nitric acid in the absence of sulphuric acid. It is almost identical to a similar absorbance in the spectrum of carbon dioxide. CO_2 and NO_2^+ are isoelectronic linear molecules (see Chapter 3, page 71).
- Various physico-chemical data (for example, the depression of freezing point) show that when one molecule of HNO_3 is dissolved in concentrated sulphuric acid, four particles are formed.

These three pieces of evidence suggest that the following reaction occurs when the two acids are mixed:

$$2H_2SO_4 + HNO_3 \rightarrow H_3O^+ + NO_2^+ + 2HSO_4^-$$

How it happens (1): the nitration of benzene

Sulphuric acid is a stronger acid than nitric acid. So strong is it, that it donates a proton to nitric acid. In this reaction nitric acid is acting as a base!

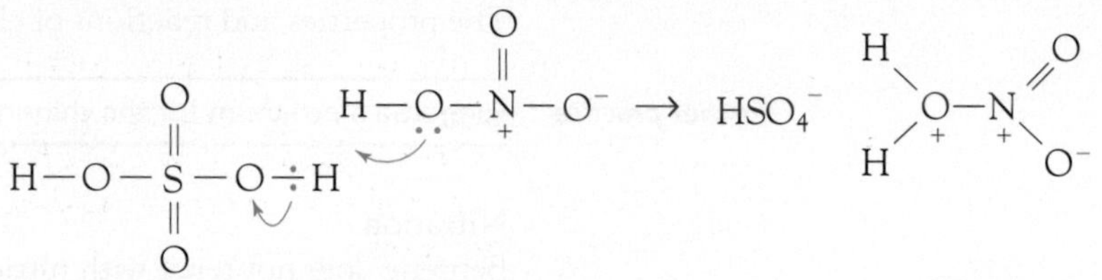

The protonated nitric acid then loses water:

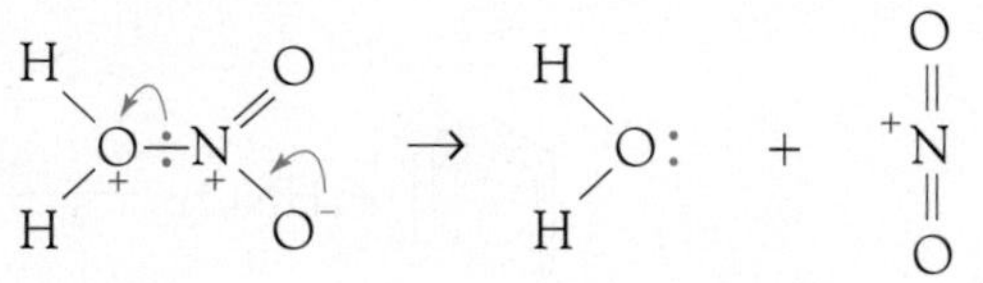

The water is then protonated by another molecule of sulphuric acid:

$$H_2SO_4 + H_2O \rightarrow HSO_4^- + H_3O^+$$

The nitronium ion then attacks the benzene ring in the usual way, forming the carbocation intermediate, which subsequently loses a proton:

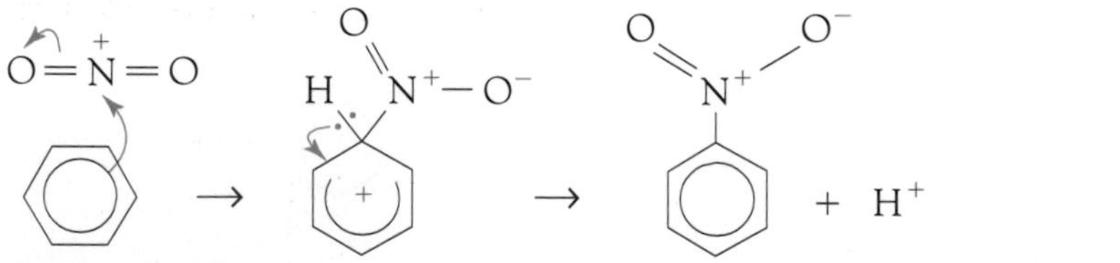

The nitration of benzene and other arenes is an important reaction in the production of explosives:

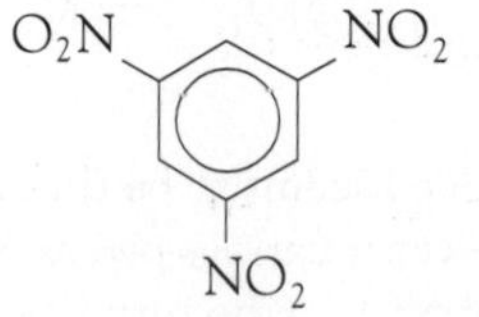

trinitrobenzene (TNB)

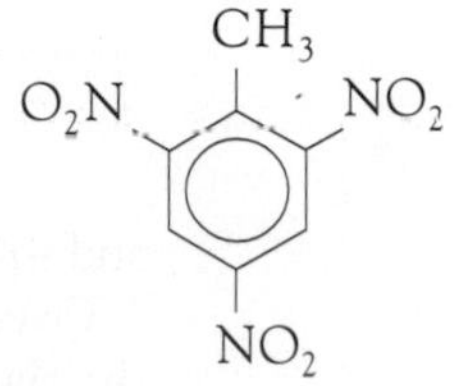

trinitrotoluene (TNT)

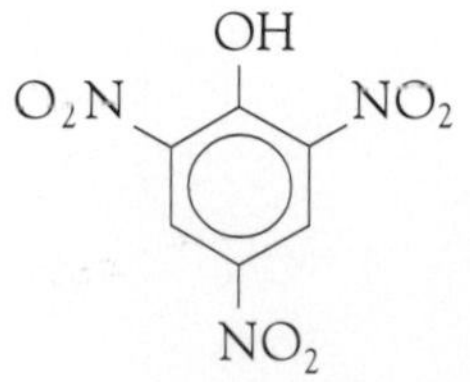

trinitrophenol (picric acid)

Also, via subsequent reduction, nitration is an important route to aromatic amines, which are used to make a variety of dyes (see Chapter 30):

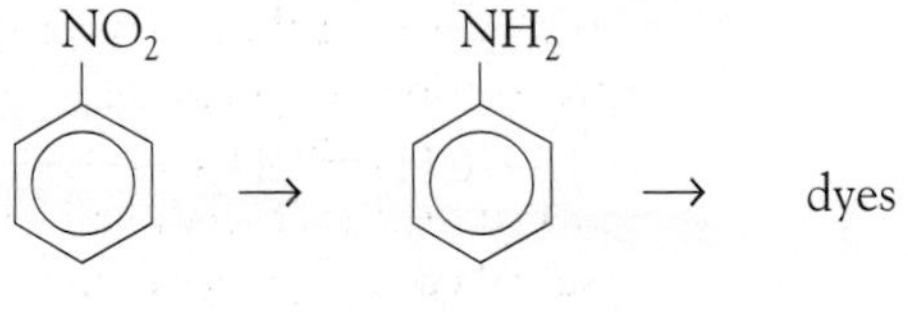

phenylamine

Alkylation (the Friedel–Crafts reaction)
When benzene is heated with a chloroalkane in the presence of aluminium chloride, the alkyl group is attached to the benzene ring:

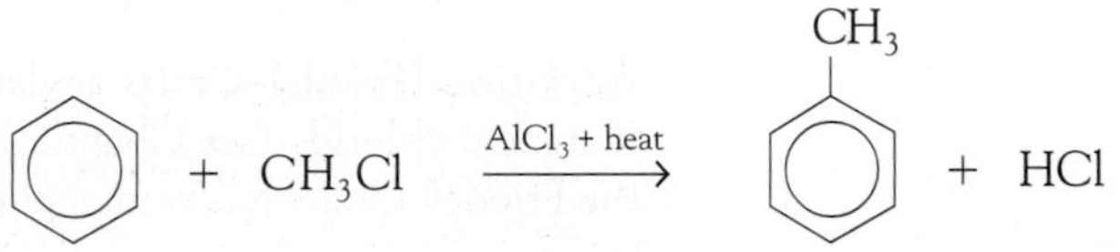

Because of the inductive effect (see Chapter 23, page 437) pushing electrons into the ring from the methyl group, the π electrons in the ring in methylbenzene are more available than those of benzene. This makes the product even more susceptible to electrophilic attack, and di- or tri-substitution is sometimes difficult to avoid:

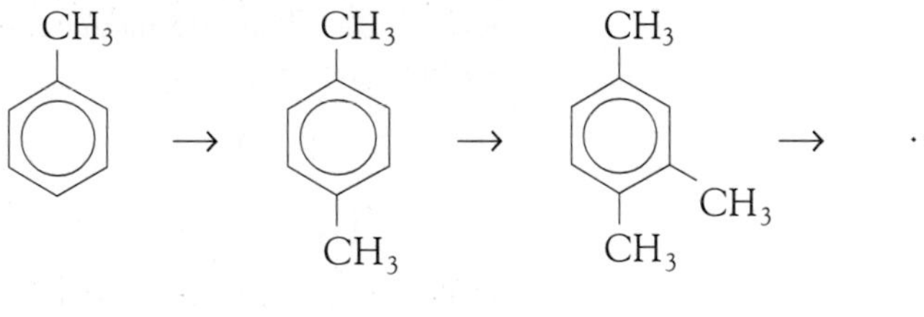

The reaction goes via the formation of an intermediate carbocation:

$$CH_3-Cl\colon \curvearrowright AlCl_3 \rightarrow CH_3 \div \overset{+}{Cl}-\overset{-}{Al}Cl_3 \rightarrow CH_3^+ + [AlCl_4]^-$$

The carbocation is the electrophile:

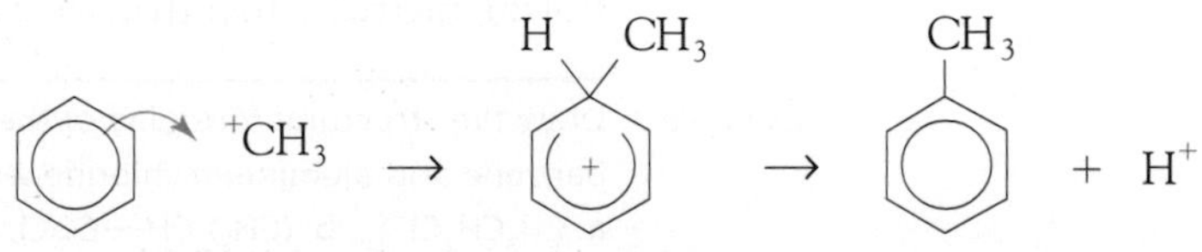

The alkylation reaction can also be carried out using alkenes, in the presence of hydrogen chloride gas and aluminium chloride. The intermediate carbocation is produced as follows:

$$H-Cl\colon \curvearrowright AlCl_3 \rightarrow H-\overset{+}{Cl}-\overset{-}{Al}Cl_3$$

$$CH_2{=}CH_2 \curvearrowright H-\overset{+}{Cl}-\overset{-}{Al}Cl_3 \rightarrow \overset{+}{C}H_2-CH_3 + [AlCl_4]^-$$

In this way, benzene and ethene react to form ethylbenzene, many tonnes of which are used each year in the production of polystyrene (see Chapter 24, page 455):

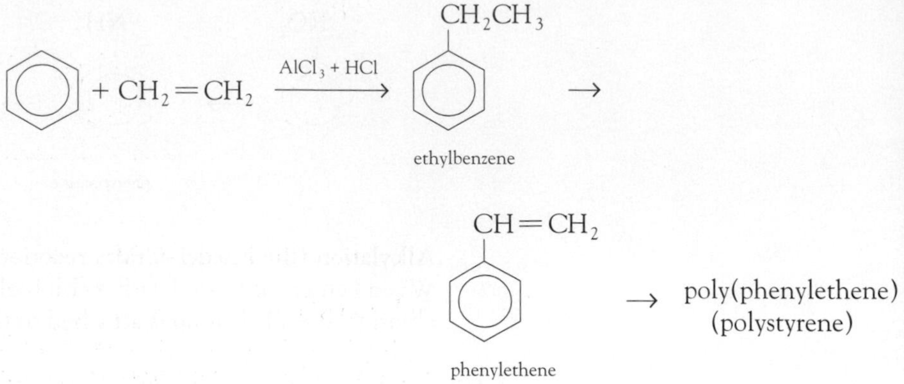

Acylation (Friedel–Crafts acylation)

If an acyl chloride (see Chapter 29, page 549) is used instead of a chloroalkane in the Friedel–Crafts reaction, a phenylketone is produced:

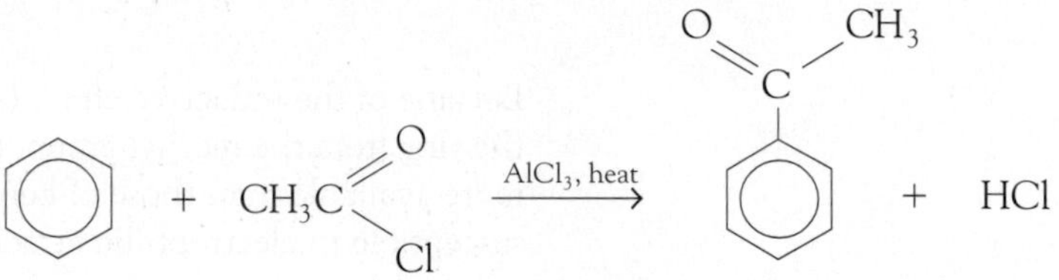

The reaction between an acyl chloride and aluminium chloride produces the acylium ion. The intermediate is formed from the attack of the acylium ion on the benzene ring:

$$CH_3-\overset{\overset{\Large O}{||}}{C}-Cl + AlCl_3 \rightarrow \underset{\text{acylium ion}}{CH_3-\overset{+}{C}=O} + [AlCl_4]^-$$

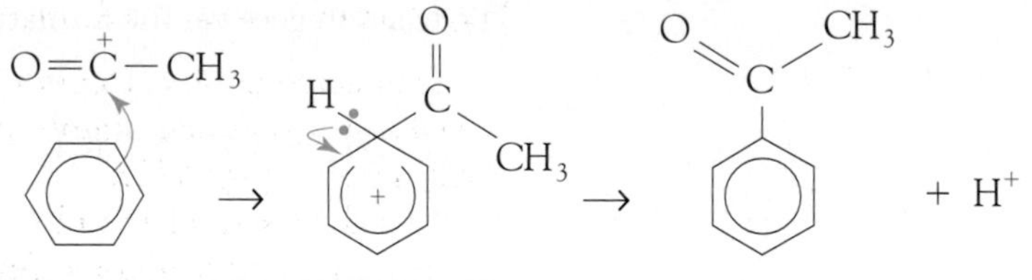

As the product is a Lewis base it will form a 1 : 1 complex with $AlCl_3$ so that a stoichiometric, rather than a catalytic, amount of $AlCl_3$ is needed.

Example Draw the structural formulae of the products you would expect from the reaction of benzene and aluminium chloride with:

a CH_3CH_2Cl **b** $(CH_3)_2CH-COCl$.

Answer **a**

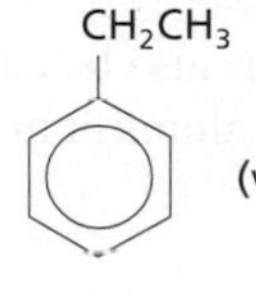

(via the carbocation $CH_3CH_2^+$)

b

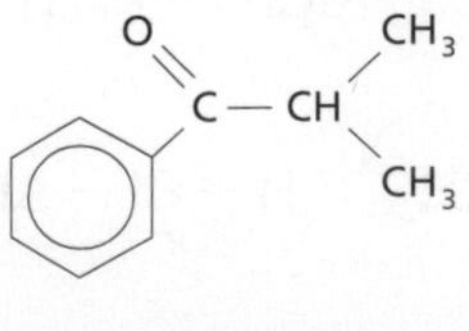

(via the acyl cation $(CH_3)_2CH-\overset{+}{C}=O$)

Further practice What organochlorine compounds are needed to synthesise the following compounds from benzene?

1

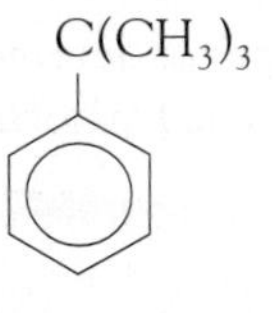

2

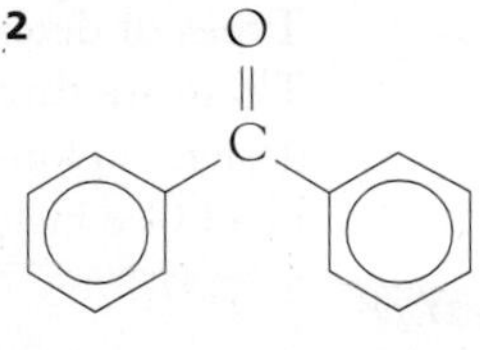

Sulphonation

Concentrated sulphuric acid, often with a few per cent of sulphur trioxide added to it (giving 'oleum'), converts benzene into benzenesulphonic acid:

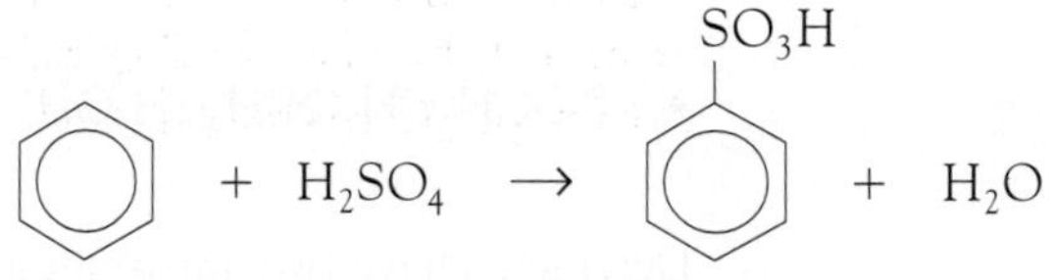

The electrophile is the neutral sulphur trioxide molecule (which is an electrophile because of a partial positive charge on the sulphur atoms due to polar bonding with oxygen):

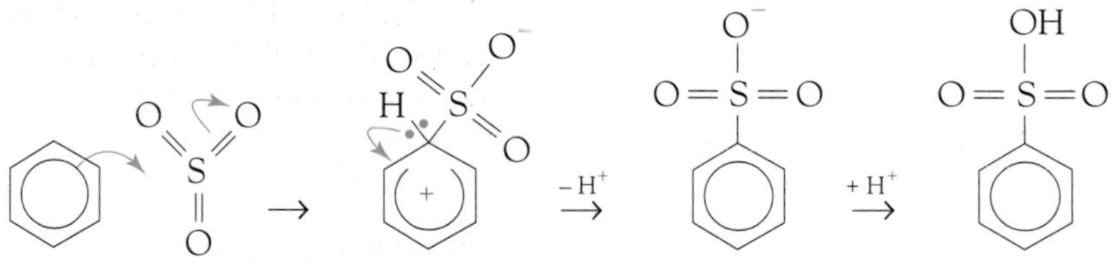

Sulphonation is industrially important in two respects.

- The incorporation of sulphonic acid groups into the molecules of dyes helps to make the dyes water soluble, and also more easily complexed by mordants (see Chapter 30, page 572).
- The sodium salts of arylsulphonic acids incorporating long alkyl chains on the ring are important anionic detergents. They are manufactured by the following pathway:

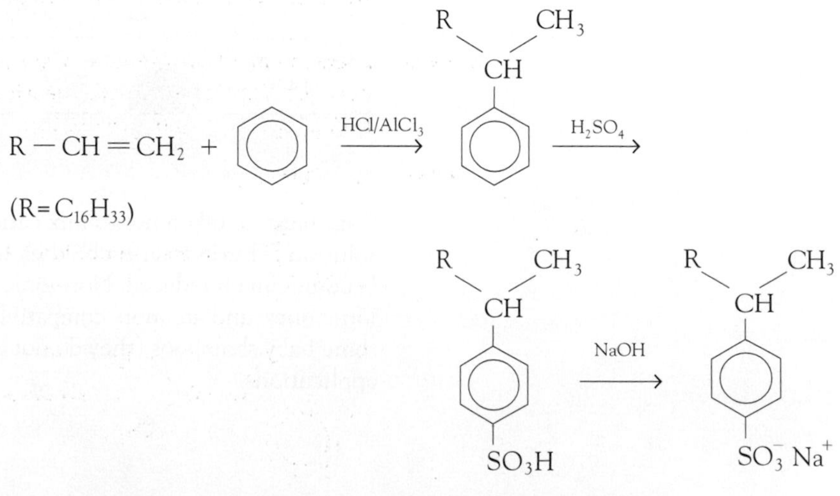

Detergents

Types of detergent

There are three main types of detergent. All three consist of the same molecular design – a long hydrophobic hydrocarbon tail, and a hydrophilic water-soluble head (see Figure 27.12).

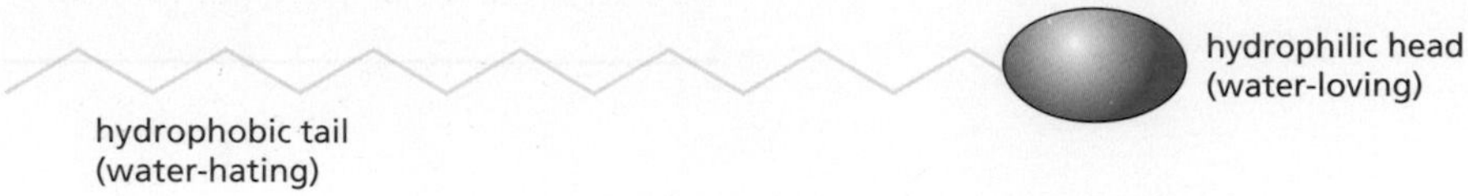

■ **Figure 27.12**
Generalised structure of a detergent molecule.

The three types differ in the nature of the hydrophilic head:

- $RCO_2^-Na^+$, $RSO_3^-Na^+$ — anionic detergents (see above and page 553)
- $RN^+(CH_3)_2CH_2CH_2OHCl^-$ — cationic detergents (see page 570)
- $ROCH_2CH_2OCH_2CH_2OH$ — non-ionic detergents (see page 449)

Detergents have two important applications – as **surfactants** (or wetting agents) and as cleaning agents.

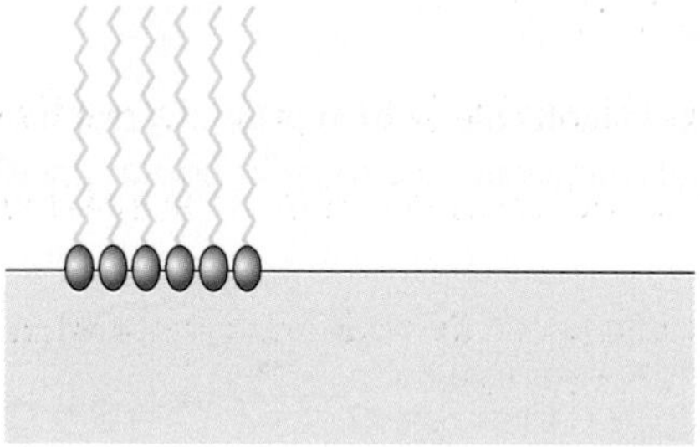

■ **Figure 27.13**
Detergents act as surfactants.

Surfactant activity

When a detergent is dissolved in water, the detergent molecules on the surface of the liquid line up so that their hydrophobic tails are outside the bulk of the liquid (see Figure 27.13).

The liquid surface now takes on a hydrocarbon character, with a much reduced surface tension. The hydrophilic heads break up the surface film of hydrogen-bonded water molecules (see Chapter 4, page 86). This also lowers the surface tension. Water boatmen would not be able to stand on a detergent solution.

Cleaning agents

Water-soluble dirt on the skin or on clothes is not a problem – it can just be rinsed off. Much dirt and grime is water insoluble, however, being hydrophobic and non-polar. When grimy articles are immersed in a detergent solution, the hydrophobic tails of detergent molecules dissolve in the non-polar medium of the grime, leaving their hydrophilic heads in the water. Eventually the grime particle surrounded by detergent molecules, called a **micelle**, floats free of the skin or material.

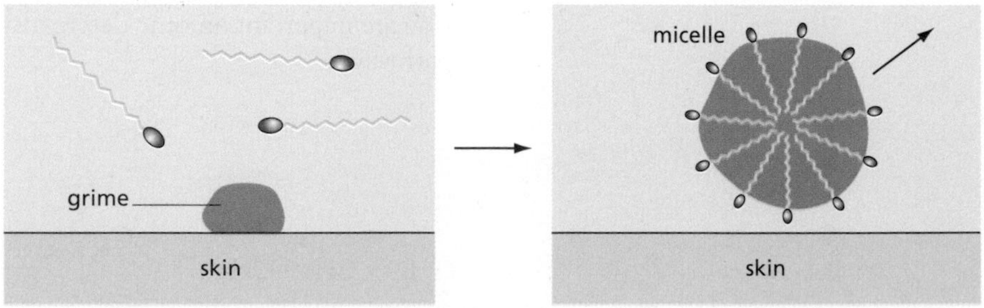

■ **Figure 27.14**
Detergent molecules surround grease particles to form micelles.

Care must be taken not to mix cationic and anionic detergents in the same solution. They attract each other, forming 'salts', and their detergent activity becomes much reduced. Non-ionic detergents are milder in their action than ionic ones, and are more compatible with biological tissue. They find a use in some baby shampoos (they do not cause stinging of the eyes) and in medical applications.

Addition reactions of the benzene ring

When it reacts with electrophiles, benzene always undergoes substitution. There are two ways in which it can be forced to undergo addition reactions.

Addition of hydrogen

Ethene adds on hydrogen over a nickel catalyst readily at room temperature and slight pressure. Benzene, being more stable, requires an elevated temperature and pressure:

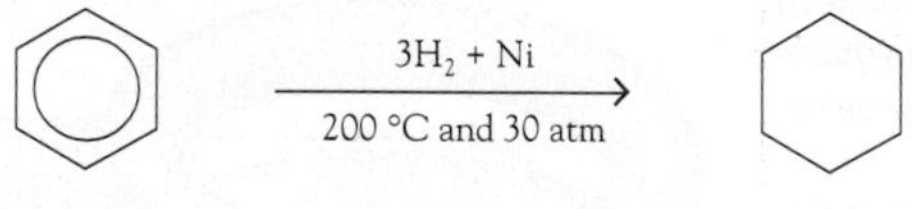

cyclohexane

Addition of chlorine

In the presence of ultraviolet light, chlorine gas reacts with benzene to give 1,2,3,4,5,6-hexachlorocyclohexane ('benzene hexachloride' or BHC):

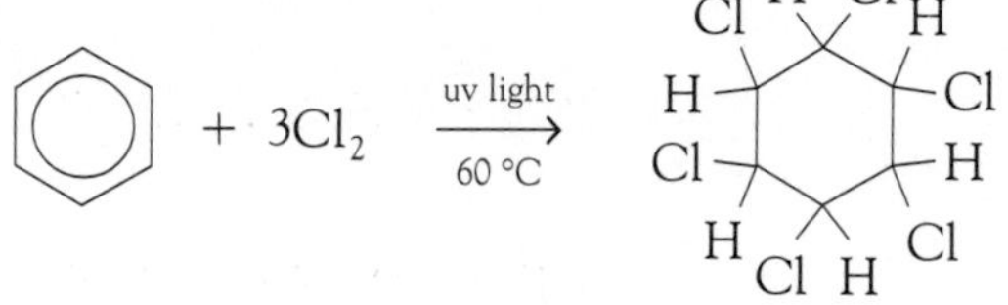

BHC was once much used as an organochlorine insecticide (see the panel on page 514).

Electrophilic substitution in substituted arenes – the orientation of the incoming group

When methylbenzene is treated with nitric and sulphuric acids, the three possible mono-nitro compounds are formed in the following ratios:

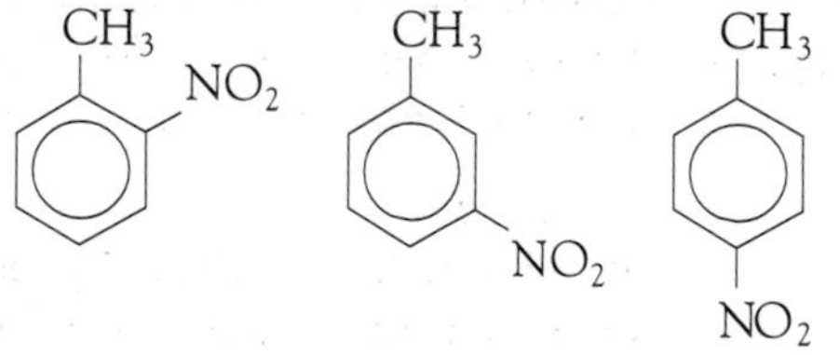

58% 4% 38%

If the NO_2^+ electrophile had attacked the ring in a purely random way, the distribution should have been 40 : 40 : 20 (there being two *ortho* positions and two *meta* positions, but only one *para* position). This non-random attack is seen in other reactions too (see Table 27.5).

The data in Table 27.5 can be interpreted as follows.

- The orientation of the incoming group (NO_2 or Br) depends on the substituent already in the ring, and not on the electrophile.
- Some substituents favour both 2- and 4-substitution, whereas other substituents favour 3-substitution, at the expense of both 2- and 4-substitution.

■ **Table 27.5** Orientation in some electrophilic substitution reactions.

Reaction	**% of product**			**Ratio (*ortho* + *para*)/(*meta*)**
	***ortho* (1,2)**	***meta* (1,3)**	***para* (1,4)**	
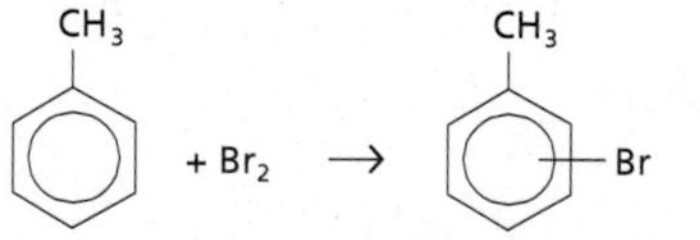	33	1	66	99:1
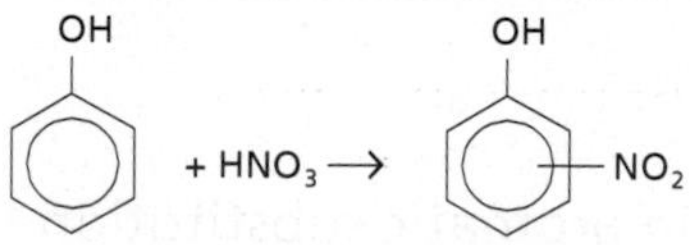	50	1	49	99:1
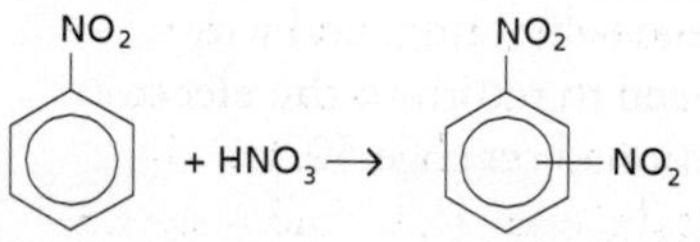	6	93.5	0.5	1:14
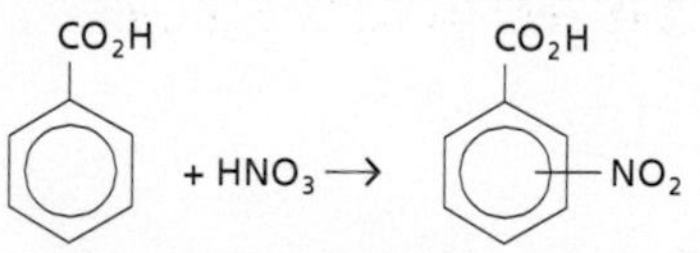	19	80	1	1:4

If we look closely at the type of substituents that are 2,4-directing, we find that either they are capable of donating electrons to the ring inductively, or they have a lone pair of electrons on the atom joined to the ring. This lone pair can be incorporated into the π system by sideways overlap of p orbitals (see Figure 27.15).

On the other hand, all those substituents that favour 3-substitution have a δ+ atom joined directly to the ring (see Table 27.6).

■ **Figure 27.15** (left)
Delocalisation of the lone pair in *ortho–para*-directing substituents.

■ **Table 27.6** (right)
Substituents and their effects on the benzene ring.

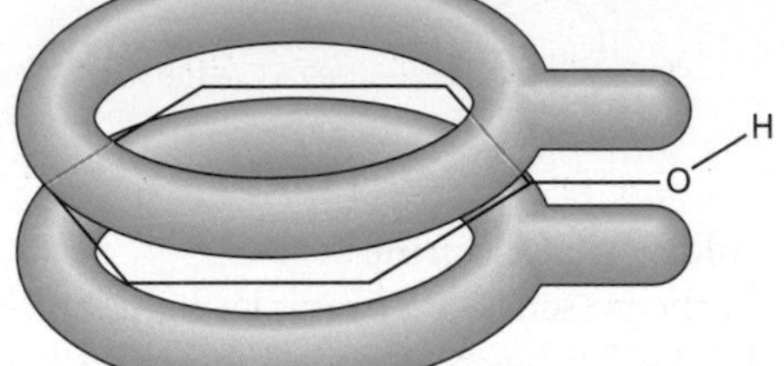

2- and 4-directing substituents	3-directing substituents
$CH_3 \rightarrow Ar$	$\overset{O}{\underset{O}{\overset{\parallel}{N}}}^{+}\!\!-\!Ar$
$H\!-\!\overset{\cdot\cdot}{O}\!-\!Ar$	$\overset{\delta-}{O}\!\!\diagdown\overset{\delta+}{\underset{H}{C}}\!-\!Ar$
$H_2\overset{\cdot\cdot}{N}\!-\!Ar$	$\overset{\delta-}{N}\equiv\overset{\delta+}{C}\!-\!Ar$
	$\overset{\delta-}{O}\!\!\diagdown\overset{\delta+}{\underset{R}{C}}\!-\!Ar$

Substitution in the side chain

Methylbenzene and other alkylbenzenes very readily undergo radical substitution with chlorine or bromine in the presence of ultraviolet light (just like alkanes – see Chapter 23, page 433), or by boiling in the absence of ultraviolet light.

CH₃ ⬡ + Cl₂ →(uv light or boiling)→ CH₂Cl ⬡ + HCl

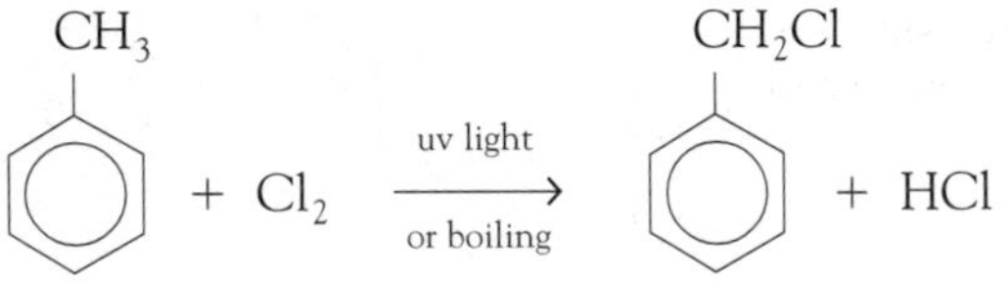

Just as with alkanes, more than one chlorine atom can be incorporated. But substitution occurs only in the side chain during this reaction, not in the ring.

In ethylbenzene, the hydrogen atoms on the carbon atom adjacent to the benzene ring are substituted much more readily than the three on the other carbon, so the following reaction gives a good yield of the product:

CH₂CH₃ ⬡ + Cl₂ →(uv light or boiling)→ CHClCH₃ ⬡ + HCl

>90%

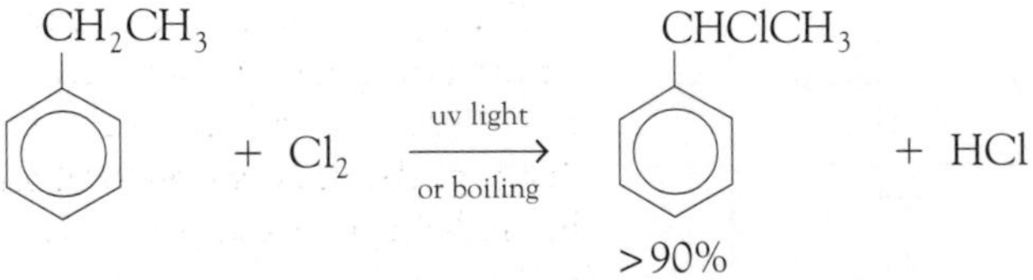

How it happens (2): orientation effects in aromatic substitution

To explain why electron-donating substituents are 2,4-directing, and why electron-withdrawing groups are 3-directing, we need to return to the electron distribution in the carbocation intermediate, mentioned on page 502.

When an electrophile E⁺ attacks a benzene ring, the intermediate can be represented as:

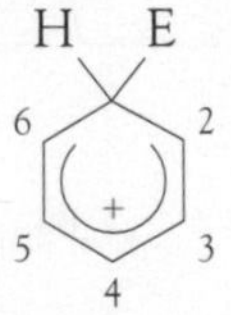

The + charge (that is, the electron deficiency) is due to only four π electrons being spread over five carbon atoms. But the + charge is not evenly spaced. It resides mainly on carbon atoms 2, 4 and 6.

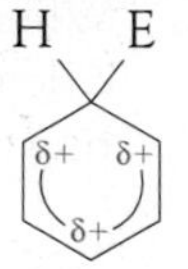

Electron-donating groups at these positions will therefore be more effective at stabilising the intermediate, by spreading out its charge, than if the electron-donating group were in position 3 or 5, one carbon removed from the + charge:

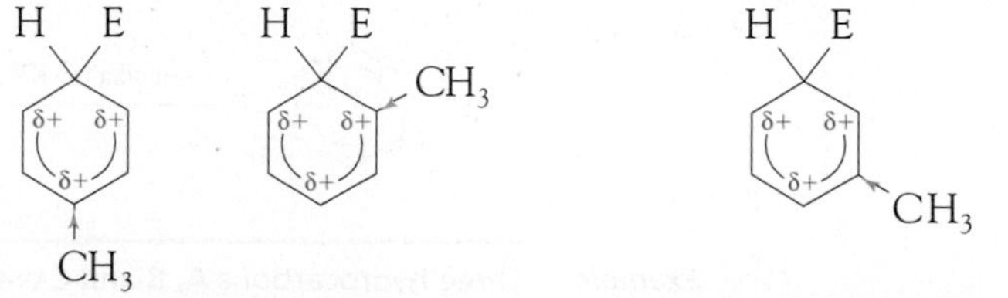

If the methyl group is situated in the 2 or 4 position relative to the electrophile E, the intermediate is <u>much</u> more stable than the intermediate in the substitution of benzene.

If the methyl group is situated in the 3 position relative to the electrophile E, the intermediate is only a <u>little</u> more stable than the intermediate in the substitution of benzene.

This extra stabilisation of the intermediate will have a similar effect on the activation energy. Lower activation energy means a faster rate of reaction (see Figure 27.16). The rate of nitration of methylbenzene is 25 times faster than the rate of nitration of benzene.

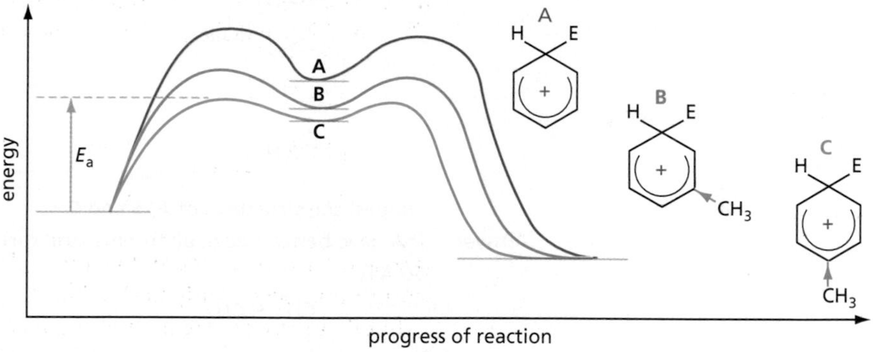

■ **Figure 27.16** Reaction profiles for the electrophilic substitution of benzene and methylbenzene, showing the effect of the substituent on the activation energy.

2- and 4-substitution in methylbenzene therefore goes faster than 3-substitution, and that is why the 2- and 4-products predominate.

The opposite is the case with electron-withdrawing groups. No matter what carbon atom they are attracted to, they will *de*stabilise the carbocation intermediate, and hence slow down the reaction (nitrobenzene reacts with electrophiles at about one-millionth the rate of benzene). But they will have a *less* destabilising effect at positions 3 and 5 than at positions 2, 4 or 6. Hence they are 3-directing.

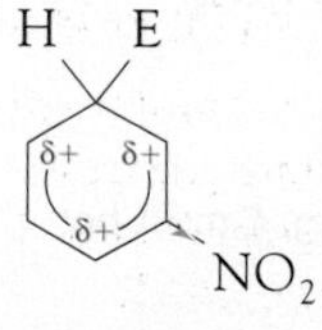

The electron-withdrawing nitro group makes this intermediate <u>less</u> stable than the intermediate in the substitution of benzene.

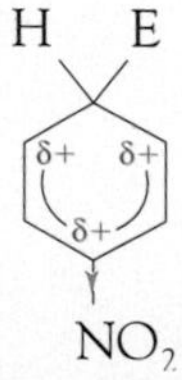

The electron-withdrawing nitro group makes this intermediate <u>much less</u> stable than the intermediate in the substitution of benzene.

Oxidation of the side chain

When alkylbenzenes are treated with hot alkaline potassium manganate(VII), oxidation of the whole side chain occurs, leaving the closest carbon atom to the ring as a carboxylic acid group:

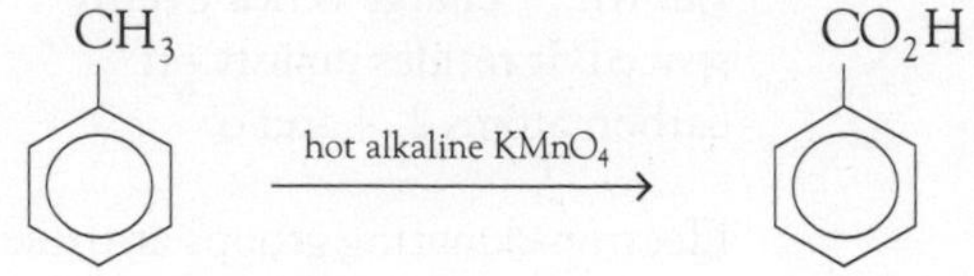

benzoic acid (systematic name: benzenecarboxylic acid)

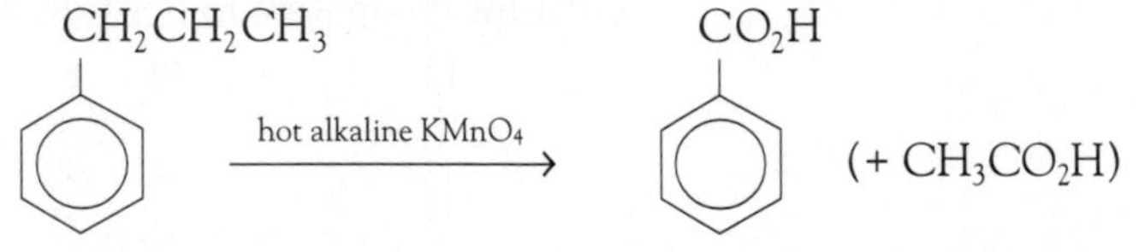

Example Three hydrocarbons **A**, **B** and **C** with the formula C_9H_{12} were oxidised by hot potassium manganate(VII).

Hydrocarbon **A** gave benzoic acid, $C_6H_5CO_2H$.

B gave benzene-1,2-dioic acid:

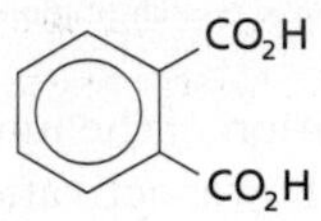

C gave benzene-1,2,4-trioic acid:

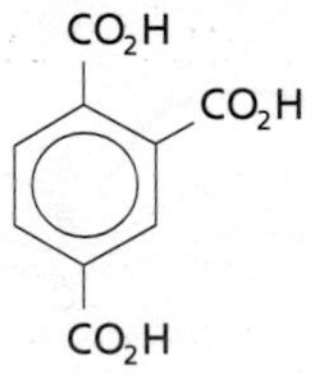

Suggest the structures of **A**, **B** and **C**.

Answer If **A** gave benzoic acid, all three 'extra' carbon atoms must be in the same side chain. So **A** is:

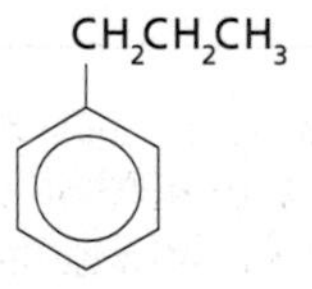

or

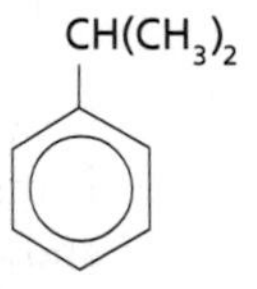

Compound **B** must contain two side chains, since two carboxylic acid groups are left after oxidation. What is more, the chains must be on adjacent carbons in the ring, as a 1,2-dicarboxylic acid is formed. So **B** is:

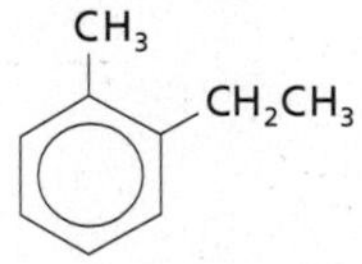

By similar reasoning, **C** must be:

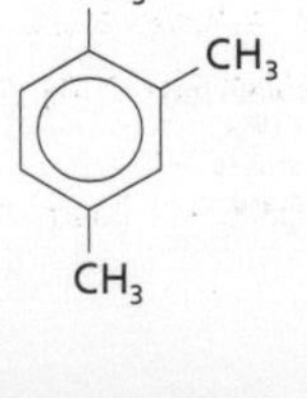

Further practice Suggest what carboxylic acids might be produced when the following compounds (all isomers with the molecular formula $C_{11}H_{14}$) are oxidised by hot potassium manganate(VII).

1 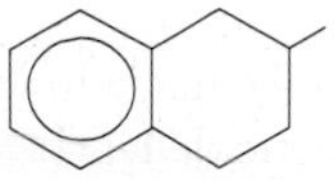

2 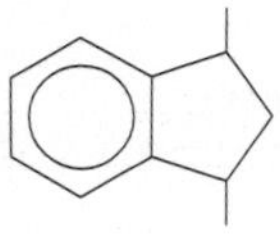

3 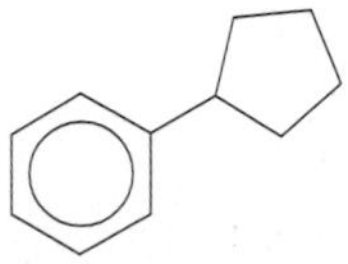

27.4 Halogenoarenes

We saw on pages 502–3 how bromobenzene and chlorobenzene can be made from benzene. The reactions of the ring in halogenobenzenes are similar to those of benzene, while there are some reactions of the halogen atom which show similarities with halogenoalkanes.

Halogenoarenes undergo electrophilic substitution, and can be nitrated or sulphonated:

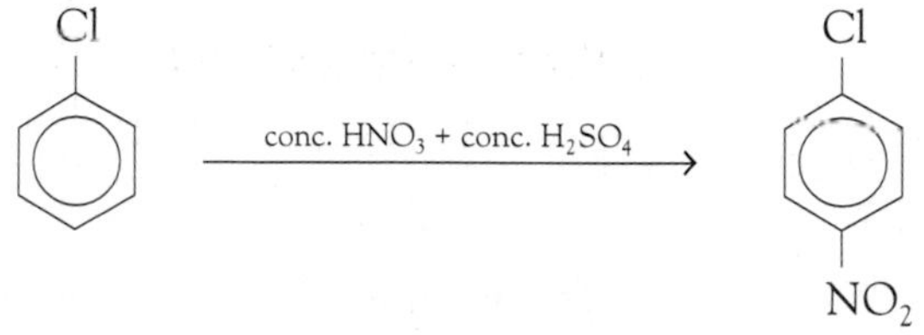

Bromobenzene also forms a Grignard reagent (see Chapter 25, page 473):

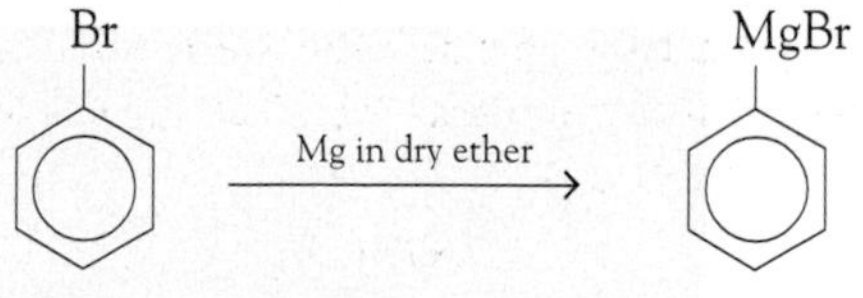

However, unlike halogenoalkanes, halogenoarenes cannot be hydrolysed, even by boiling in aqueous sodium hydroxide. The carbon–halogen bond is stronger in halogenoarenes than it is in halogenoalkanes, possibly due to an overlap of p electrons similar to that in phenol that we saw on page 510 (see Figure 27.17).

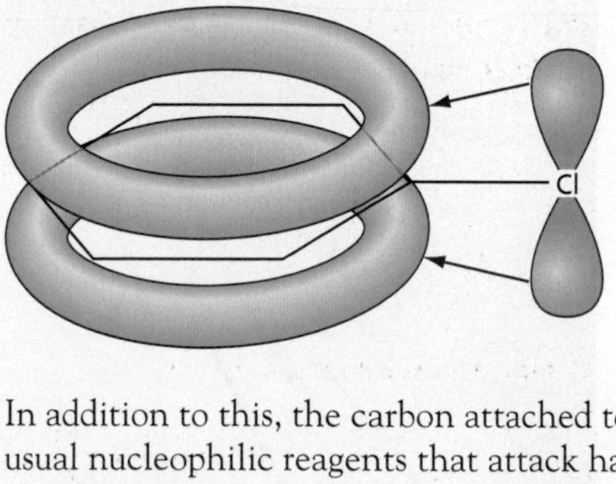

■ **Figure 27.17** Delocalisation of the lone pair in chlorobenzene.

In addition to this, the carbon attached to the halogen is not accessible to the usual nucleophilic reagents that attack halogenoalkanes, since its δ+ is shielded by the negative π cloud of the ring. This means that halogenoarenes are inert to all nucleophiles.

Certain halogenoarenes find important uses as insecticides and herbicides (see the panel overleaf).

Organochlorine insecticides and herbicides

Insecticides

Chlorobenzene used to be made in large quantities as an intermediate in the production of the insecticide DDT:

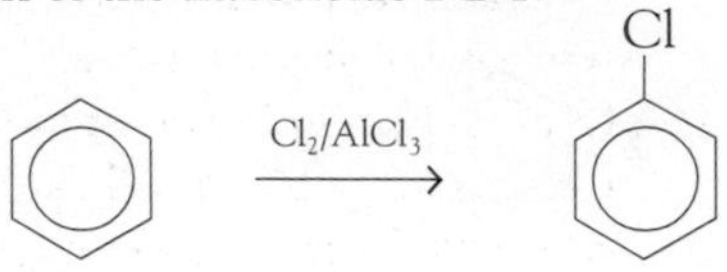

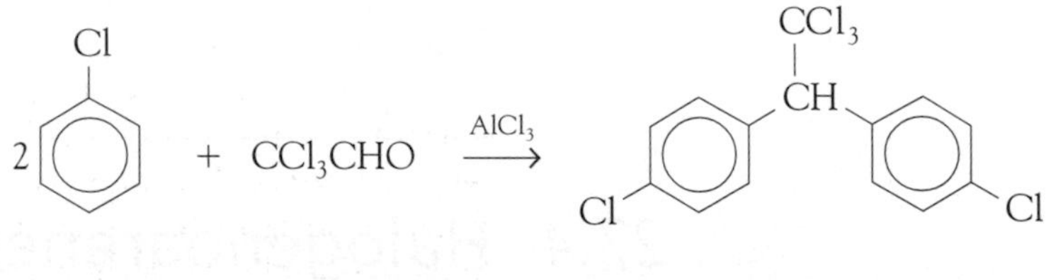

trichloroethanal

4,4-dichlorodiphenyltrichloroethane (DDT)

Another insecticide made from benzene is BHC:

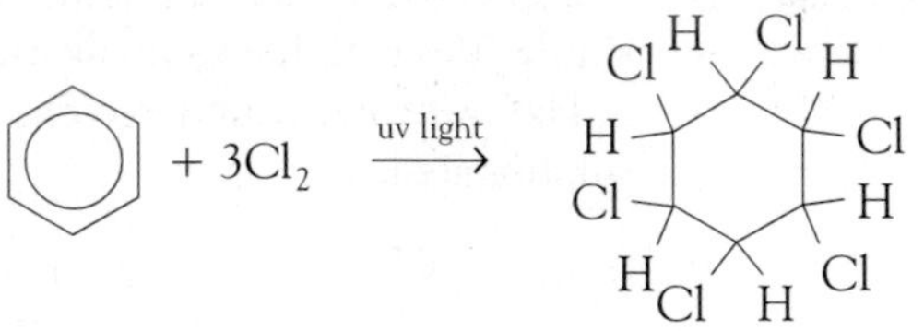

benzene hexachloride (BHC)

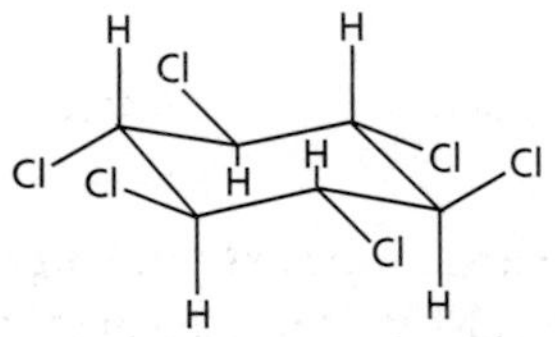

■ **Figure 27.18** Gammexane.

The molecule of BHC contains six chiral centres, and so there are many optical isomers that could be formed in this reaction. The biologically active one is called gamma-BHC, or gammexane (see Figure 27.18).

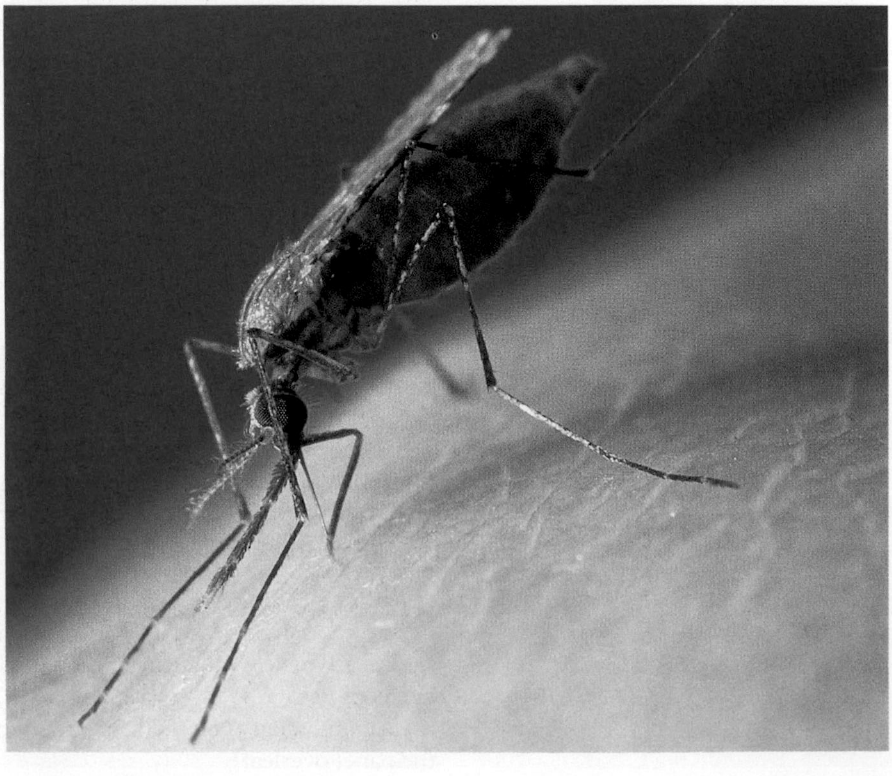

■ **Figure 27.19** Malaria is caused by a microscopic parasite that is spread by the bite of the anopheles mosquito. Malaria affects over 300 million people worldwide and kills around 3 million a year. DDT treatment of the anopheles mosquito saves millions of human lives, so the environmental price is considered worth paying.

The use of chlorinated insecticides is now banned in many countries. These compounds are quite inert to biodegradation, so they stay in the environment for many years. What is more, being non-hydrogen-bonded covalent substances, they are insoluble in water, but soluble in fats and oils. They tend to concentrate in the fatty tissue of animals, especially those higher up in the food chain. Birds of prey were especially vulnerable. At one time the shells of their eggs were so thin that they cracked during attempted incubation, with disastrous consequences for their populations. DDT is still used in parts of Africa and Asia to control the malaria mosquito. Here the advantages (to humans) are considered to outweigh the disadvantages (to the environment).

Herbicides

The compounds 2,4-dichlorophenoxyethanoic acid (2,4-D) and 2,4,5-trichlorophenoxyethanoic acid (2,4,5-T) have been successfully used for many years as selective weedkillers for broad-leaved weeds. They have little effect on narrow-leaved plants such as grass, so can be used as lawn weedkillers. Their action mimics that of the natural plant growth hormone indole ethanoic acid (indoleacetic acid, IAA). They are members of a group of compounds called hormone weedkillers. Their structures are shown in Figure 27.20.

■ Figure 27.20 The structures of some hormone weedkillers (herbicides).

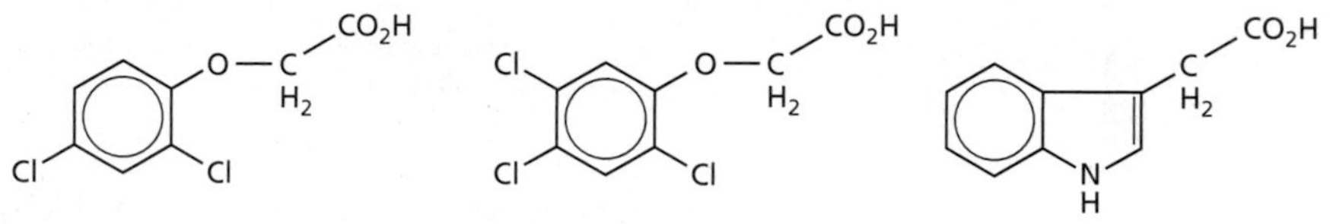

2,4-D 2,4,5-T IAA

2,4-D and 2,4,5-T were components of 'Agent Orange', sprayed as a jungle defoliant by the American army during the Vietnam war. It was during this large-scale use that an unforeseen problem came to light. During the manufacture of the intermediate chlorinated phenol, a small quantity of a highly toxic impurity, tetrachlorodibenzodioxin (or dioxin) was also produced (see Figure 27.21). The temperature of the reaction has to be controlled very carefully if the amount of dioxin is to be kept at a low level.

■ Figure 27.21 Dioxin – a highly toxic molecule.

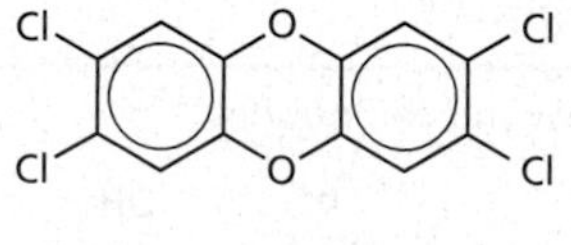

tetrachlorodibenzo-1,4-dioxin
(TCDD, 'dioxin')

Unlike 2,4-D or 2,4,5-T, dioxin is not soluble in water. If ingested, or absorbed through the skin or lungs, it becomes concentrated in the fatty tissues of the body. It has now been discovered that dioxin is one of the most poisonous chemicals known – it is strongly carcinogenic (cancer producing), teratogenic (produces malformation of the foetus) and mutagenic (causes mutations). It causes skin burns and ulcers that heal only very slowly. Improved methods of production of 2,4-D and 2,4,5-T are now in place, but the risk of accidentally producing large quantities of dioxin is still present, should a chemical plant go out of control. Other non-chlorine-containing hormone weedkillers are preferred – they are equally effective, but more expensive.

The use of chlorinated herbicides is now severely restricted, just as is the use of chlorinated insecticides.

27.5 Phenols

Phenols contain two functional groups – the —OH group of the alcohols, and the phenyl ring of the arenes. Their reactions are for the most part the sum of the two sets, but with significant modifications.

Nomenclature

Many of the compounds illustrated in Figures 27.7 and 27.8 (pages 497 and 498) are phenols. If the only other groups on the benzene ring are halogen atoms, nitro, amino or alkyl groups, the compounds are named as derivatives of phenol itself:

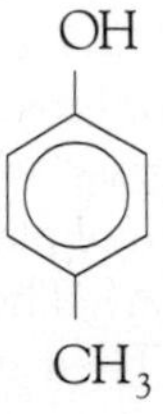

4-methylphenol
(*p*-methylphenol)

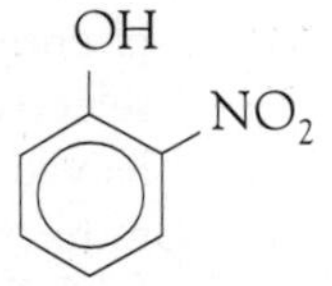

2-nitrophenol
(*o*-nitrophenol)

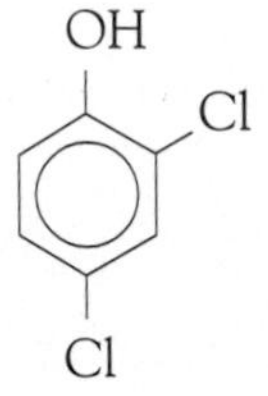

2,4-dichlorophenol

If, however, the other group is an aldehyde, ketone or carboxylic acid group, the phenolic —OH becomes a 'hydroxy' substituent:

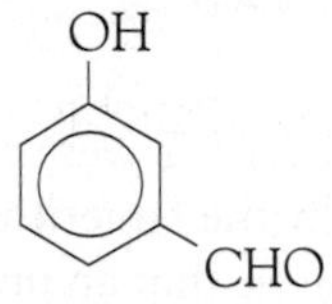

3-hydroxybenzaldehyde
(also known as
3-hydroxybenzenecarbaldehyde)

4-hydroxybenzoic acid
(also known as
4-hydroxybenzenecarboxylic acid)

Example Name the following compounds.

a

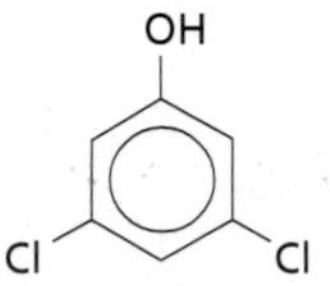

b

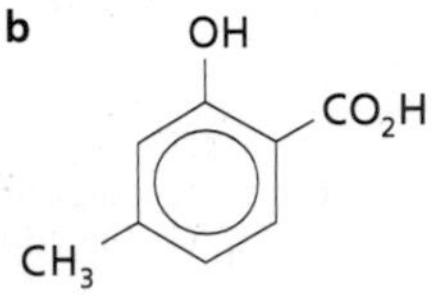

Answer **a** 3,5-dichlorophenol
b 2-hydroxy-4-methylbenzoic acid

Further practice Draw the structural formulae of the following:
1 2,4,6-trimethylphenol
2 3,4,5-trihydroxybenzoic acid.

Reactions of the —OH group

The C—O bond in phenol is very strong, as a result of the delocalisation of the lone pair of electrons on oxygen over the arene ring. There are no reactions in which it breaks, unlike the situation with the alcohols (see Chapter 26, pages 481–5).

Reactions of the O—H bond

Acidity

Phenols are more acidic than alcohols:

$$R{-}O{-}H \rightleftharpoons RO^{-} + H^{+} \qquad K_a = 1.0 \times 10^{-16}\,\text{mol dm}^{-3}$$

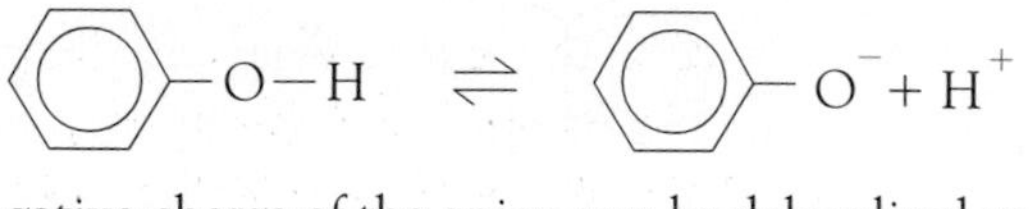

$K_a = 1.3 \times 10^{-10}\,\text{mol dm}^{-3}$

The negative charge of the anion can be delocalised over the benzene ring. Figure 27.22 shows various ways in which this can be represented.

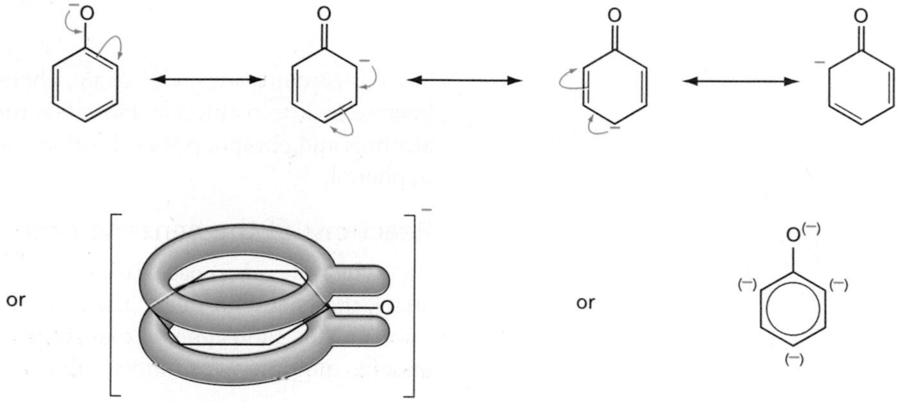

■ **Figure 27.22** Representations of the phenol anion.

Consequently, phenol not only reacts with sodium metal, giving off hydrogen gas:

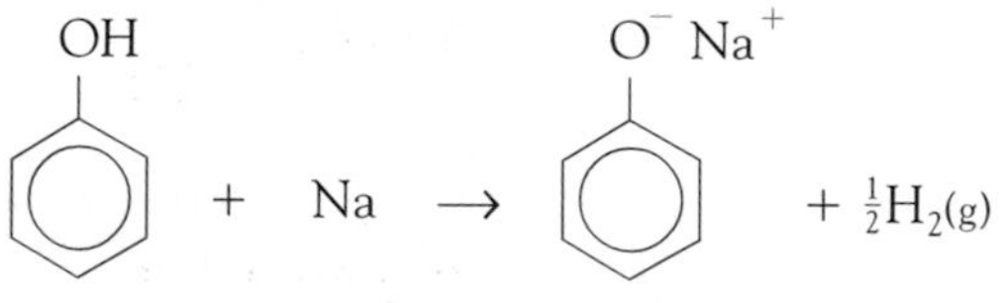

sodium phenoxide
(a white solid)

but, unlike alcohols, it also dissolves in aqueous sodium hydroxide:

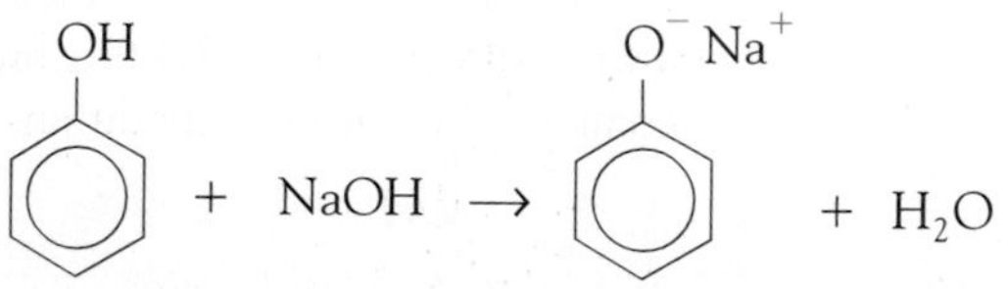

Phenol is visibly acidic. The pH of a 0.1 mol dm^{-3} solution in water is 5.4, so it will turn universal indicator solution yellow. An old name for phenol is carbonic acid.

Esterification

Because of the delocalisation over the ring of the lone pair on the oxygen atom, phenol is not nucleophilic enough to undergo esterification in the usual way, that is, by heating with a carboxylic acid and a trace of concentrated sulphuric acid (for the mechanism of esterification, see Chapter 29).

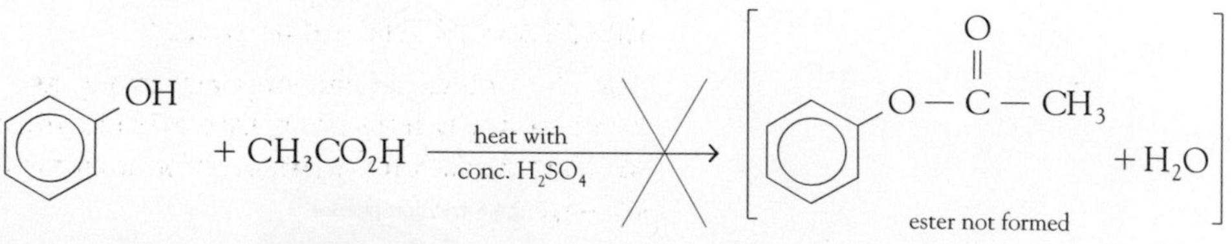

Phenol can, however, be esterified by adding an acyl chloride or an acid anhydride:

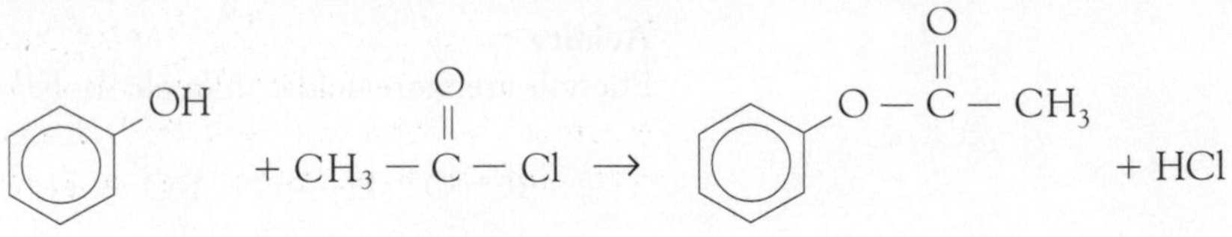

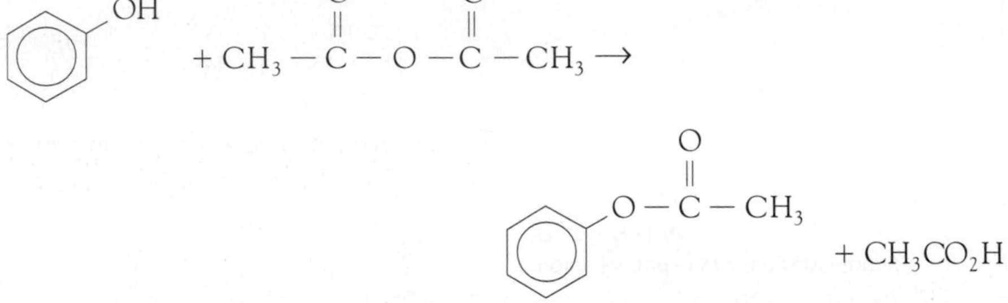

Since it contains an —OH group, phenol reacts with phosphorus(V) chloride to liberate hydrogen chloride. However, the reaction is slower than that between alcohols and phosphorus(v) chloride due to the limited nucleophilicity of the OH in phenol.

Reactions of the benzene ring

As we mentioned on page 510, phenols are much more susceptible to electrophilic attack than benzene, due to the delocalisation of the lone pair of electrons on oxygen. This allows phenol to react with reagents that are more dilute, and also to undergo multiple substitution with ease.

Nitration

When treated with dilute aqueous nitric acid (no sulphuric acid is needed) phenol gives a mixture of 2- and 4-nitrophenols:

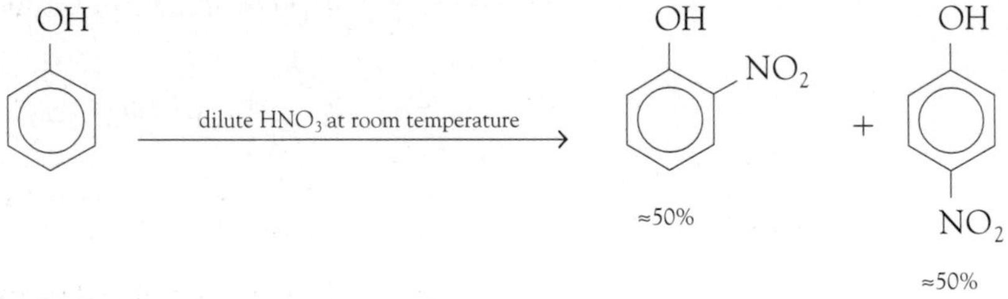

Bromination

Phenol decolorises a dilute solution of bromine in water at room temperature, giving a white precipitate of 2,4,6-tribromophenol. No aluminium bromide is needed – compare this with the conditions needed for the bromination of benzene on page 501.

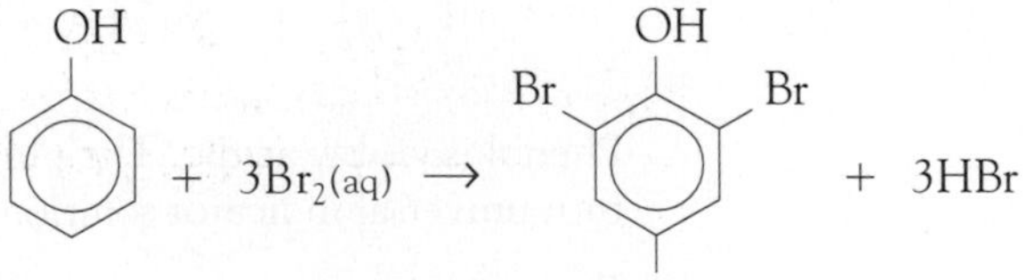

A similar product, formed by the action of chlorine water on phenol, is used (in dilute solution) as the antiseptic TCP (see page 520).

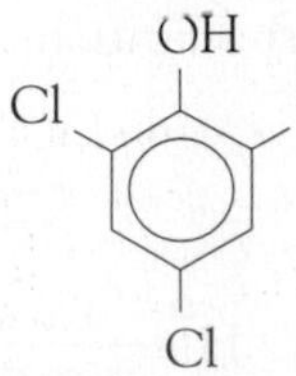

2,4,6-trichlorophenol

The commercial production of phenol

We saw on page 510 that the hydrogen atoms on the CH_2 group next to the ring in ethylbenzene are replaced by chlorine in preference to those on the CH_3 group. This is because the radical on a carbon adjacent to a benzene ring is more stable (due to delocalisation on the ring):

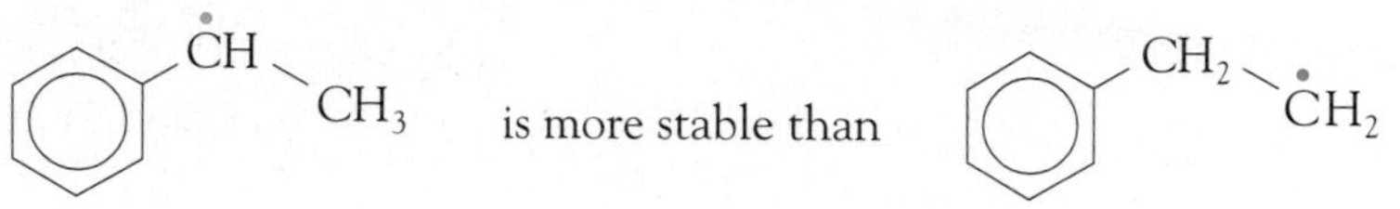

Use is made of this fact in the industrial synthesis of phenol.

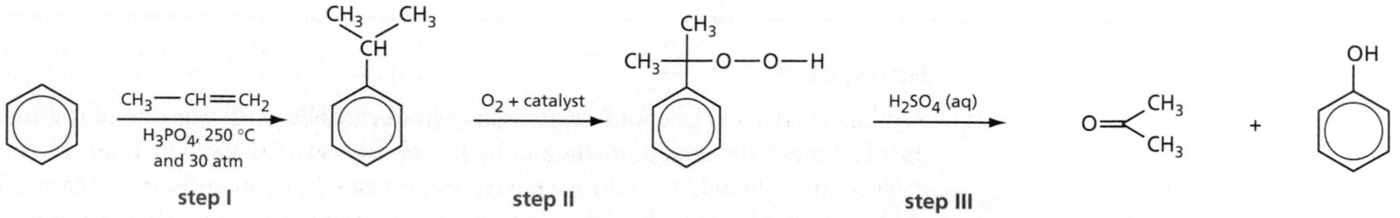

Step I is an electrophilic substitution reaction (note that the more stable secondary carbocation is formed).

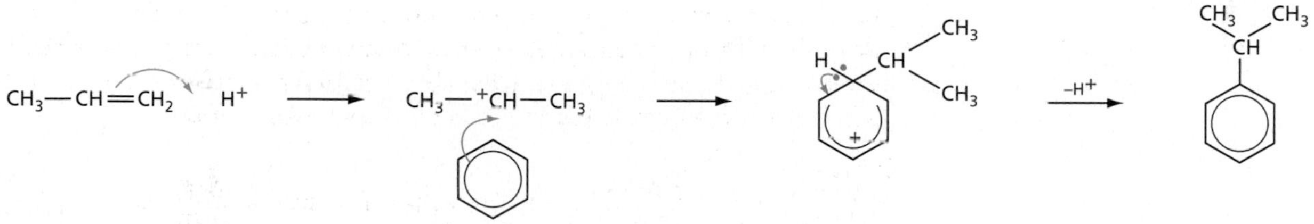

Step II is a radical oxidation (the oxygen molecule is acting as a diradical).

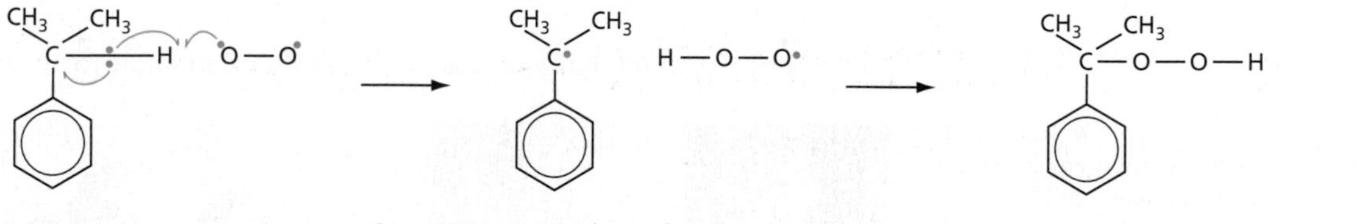

Step III is an acid-catalysed rearrangement reaction.

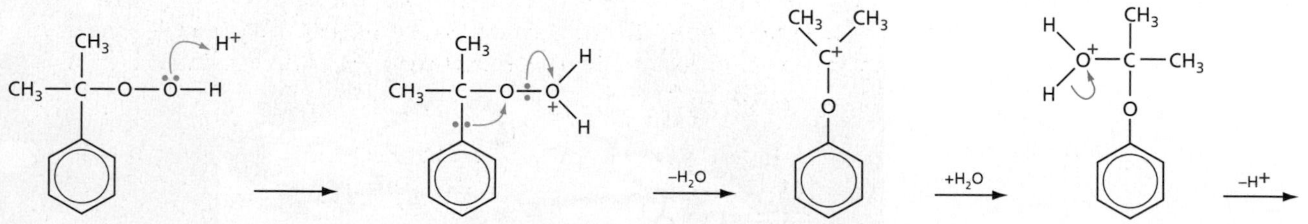

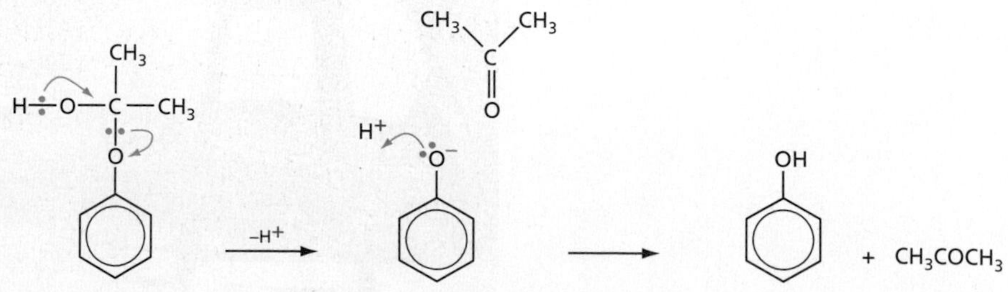

Many tonnes of phenol are made in this way, for use in the manufacture of dyes and resins. The other product, propanone, is a useful solvent.

A specific test for phenols
When a dilute solution of iron(III) chloride is added to a dilute solution of a phenol, a coloured complex is formed. With phenol, a purple coloration is observed:

$$3\ \underset{(aq)}{\text{C}_6\text{H}_5\text{OH}} + Fe^{3+} \longrightarrow \left(\text{C}_6\text{H}_5\text{O}\right)_3 Fe + 3H^+_{(aq)}$$

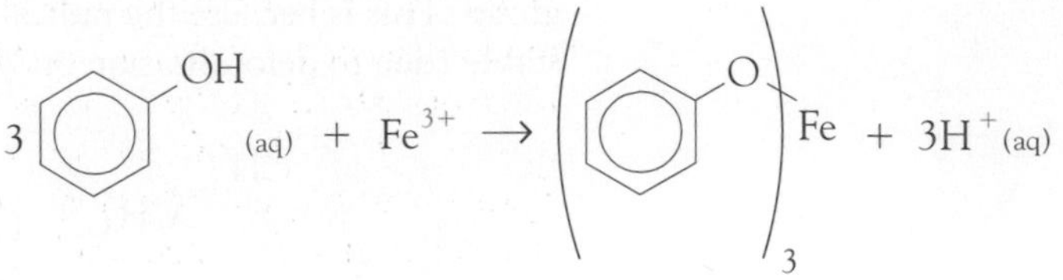

27.6 Some important phenols

Antiseptics

A dilute solution of phenol in water (known as carbolic acid) was one of the first disinfectants to be used in medicine, by Joseph Lister in Glasgow in 1867. Phenol itself is unfortunately too corrosive to be of general use as an antiseptic. Many chloro derivatives have been found to be more potent antiseptics than phenol itself. They can be used in much lower concentrations, which reduces their corrosive effect. Two of the most common are trichlorophenol and chloroxylenol:

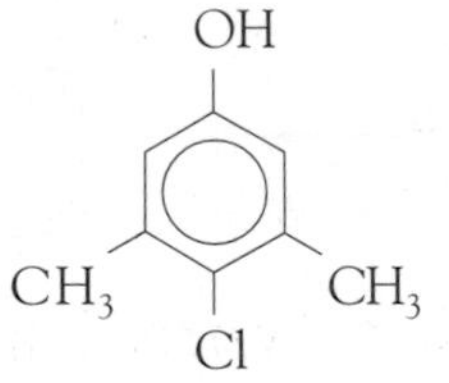

chloroxylenol ('Dettol')
(4-chloro-3,5-dimethylphenol)

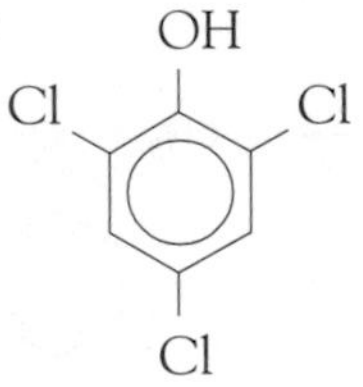

2,4,6-trichlorophenol
(TCP)

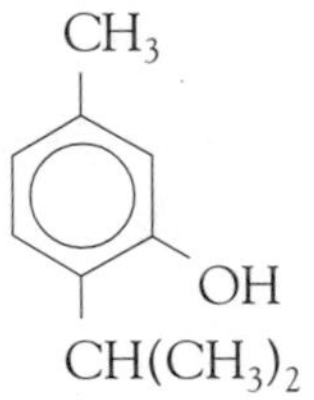

thymol
(5-methyl-2-(prop-2-yl)phenol)

Thymol occurs in oil of thyme, and is an excellent and non-toxic antiseptic.

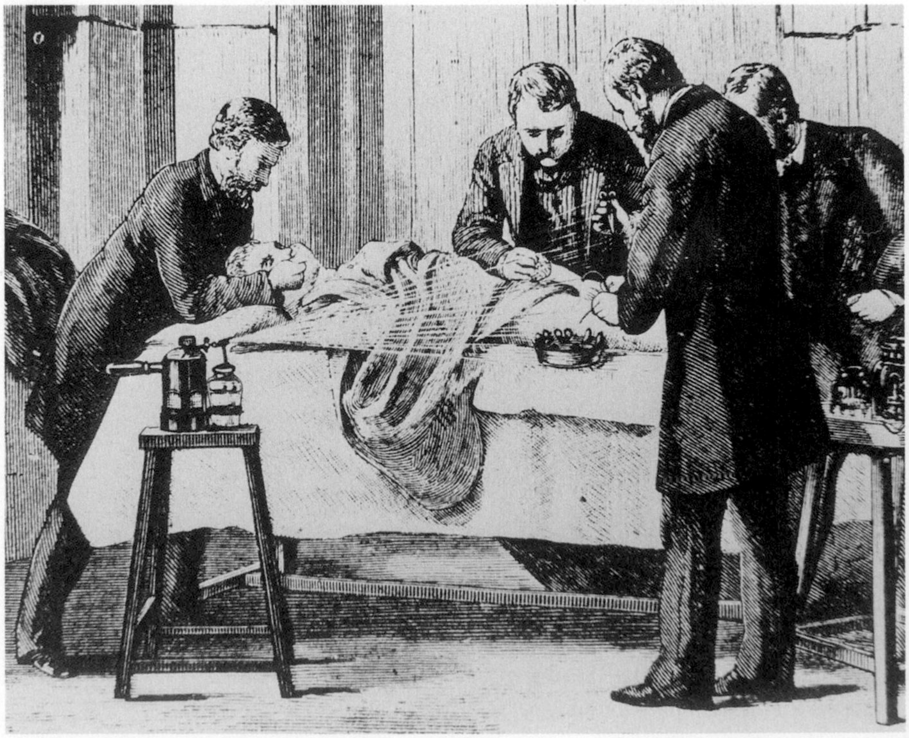

■ **Figure 27.23**
The Scottish doctor Joseph Lister was the first to use phenol as an antiseptic during surgery. Deaths from infections following operations were much reduced as a result, though the corrosive nature of phenol did not help the skin to heal.

Analgesics

A significant number of pharmaceutical drugs contain phenolic groups, or are derived from phenols. The painkilling and fever-reducing properties of an extract of willow bark have been known since at least the sixteenth century. In the nineteenth century, the active ingredient, salicylic acid (2-hydroxybenzoic acid) was isolated and purified (the name 'salicylic' derives from the Latin name for willow: *Salix*). The therapeutic use of salicylic acid was limited, however, because it caused vomiting and bleeding of the stomach. In 1893 an ester derived from salicylic acid and ethanoic acid was found to have much fewer side-effects:

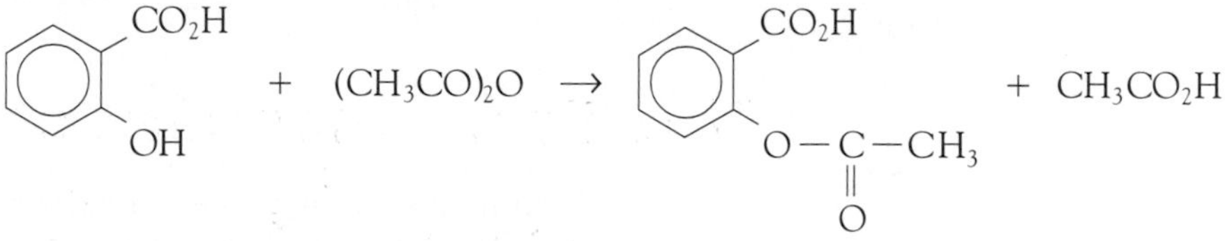

salicylic acid
(2-hydroxybenzoic acid)

aspirin
(ethanoyl 2-hydroxybenzoic acid)

Aspirin is the most widely used of all analgesics. However, it still retains some of the stomach-irritating effects of salicylic acid. A less problematic painkiller is paracetamol, another phenol.

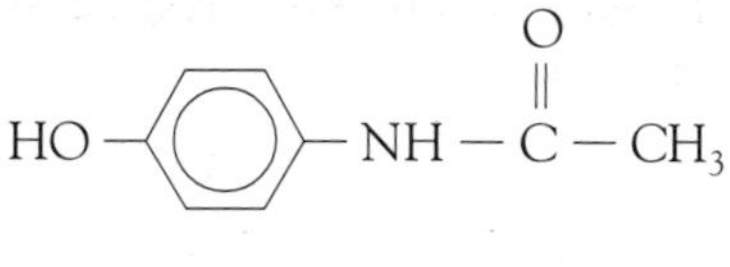

paracetamol

Some opium-based painkillers

More effective than aspirin and paracetamol as painkillers are compounds derived from morphine. Morphine was first isolated from the opium poppy in a pure form in 1805, closely followed by its methyl ether, codeine (see Figure 27.24).

■ Figure 27.24 Two opium-based painkillers.

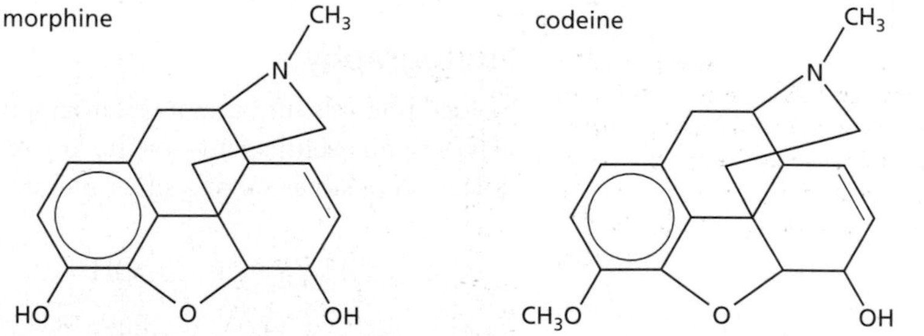

The prolonged use of morphine results in physical addiction. Codeine, although less potent as a painkiller, does not have this disadvantage. Heroin, the most addictive of the opiates, is the di-ethanoyl ester of morphine (see Figure 27.25).

■ Figure 27.25 Heroin is the di-ethanoyl ester of morphine.

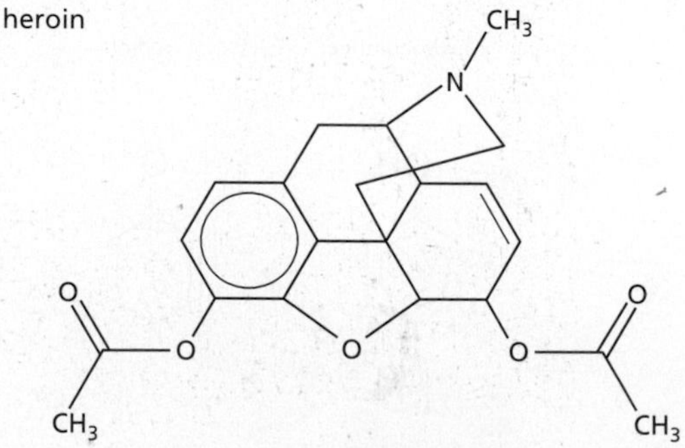

Adrenaline, amphetamines and salbutamol

The hormone adrenaline is a diphenolic amine. It is a central nervous system stimulant produced by the adrenal gland in response to stress. Its physiological reactions are mimicked by amphetamines.

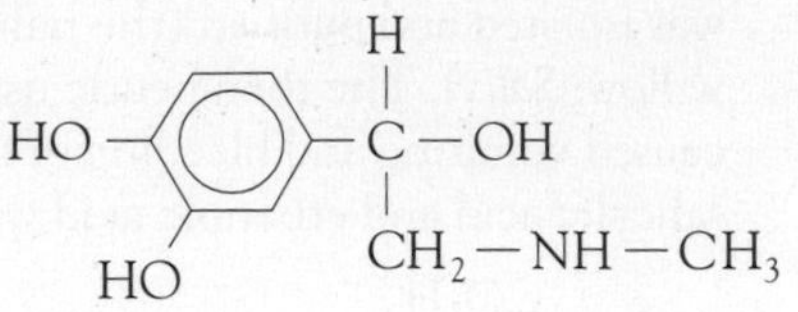

adrenaline

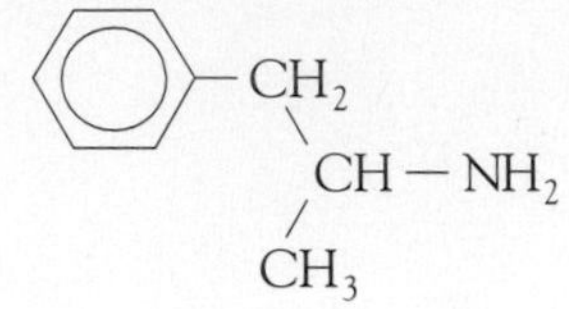

an amphetamine (benzedrine)

Apart from stimulating the central nervous system, another important physiological effect of adrenaline is as a bronchodilator – an agent that allows an easy passage of air through the bronchioles into the lungs. During an asthma attack, the bronchioles become congested and narrow, and breathing becomes difficult. The most effective drug to relieve these symptoms is salbutamol. This molecule mimics the bronchodilator activity of adrenaline, without causing its other physiological effects (which include increasing the heart rate and the blood pressure). Note the similarities in structure between the molecules of salbutamol and adrenaline.

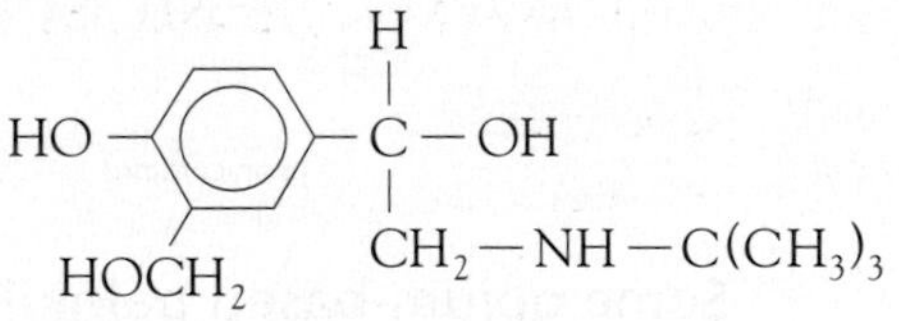

salbutamol

Photography

Several phenols find uses as photographic developers. Their ease of oxidation makes them useful agents for the key reaction in photographic development, the reduction of silver salts to silver metal.

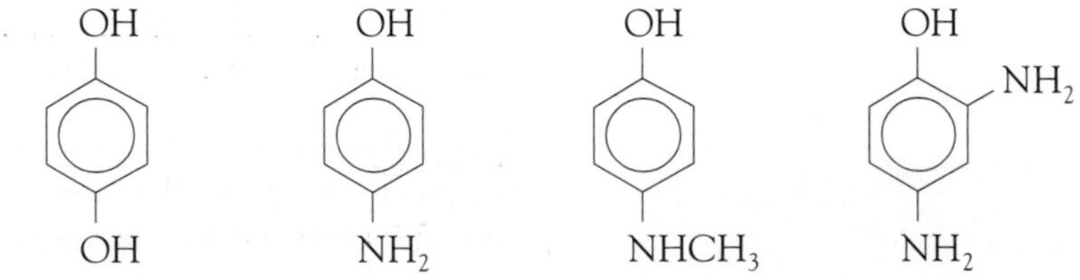

hydroquinone rodinol metol amidol

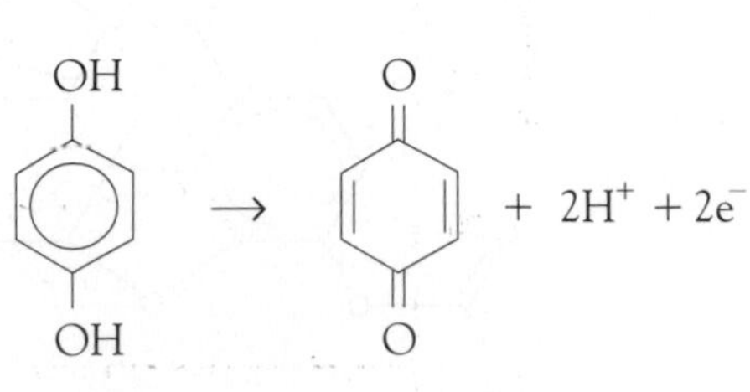

hydroquinone quinone

27.7 Preparing arenes and phenols

- Benzene and methylbenzene are formed during the cracking and re-forming of fractions from the distillation of crude oil (see Chapter 23, page 427).
- Alkylbenzenes can be obtained by the Friedel–Crafts reaction (see page 505).
- The —OH group of phenols can be introduced onto an arene ring by the following sequence (see Chapter 30 for details):

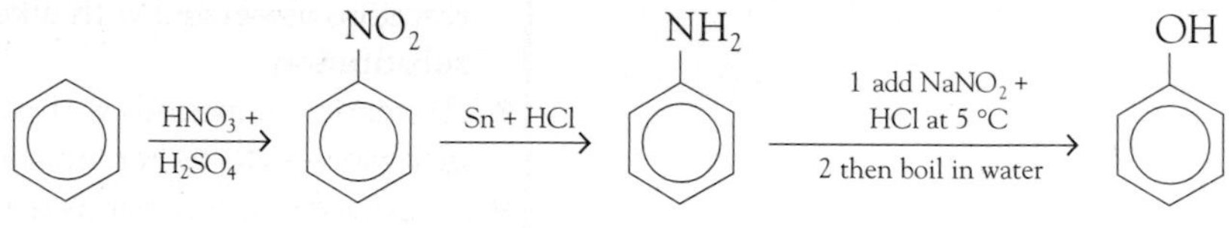

27.8 The infrared spectra of arenes and phenols

Just as with the carbon–carbon double bond in alkenes, the carbon–carbon bond in arenes absorbs in the infrared only if it is unsymmetrical in some way. Bands at 1500 cm^{-1} and 1600 cm^{-1} are associated with the carbon–carbon bonds of the benzene ring. The C—H bonds absorb at 3000–3100 cm^{-1} (in a similar place to the C—H bonds in alkenes).

The O—H bond in phenols absorbs in the same region as the O—H bond in alcohols: 3600 cm^{-1} if monomeric; 3200–3500 cm^{-1} if hydrogen bonded.

Figure 27.26
IR spectra of **a** benzene and **b** phenol.

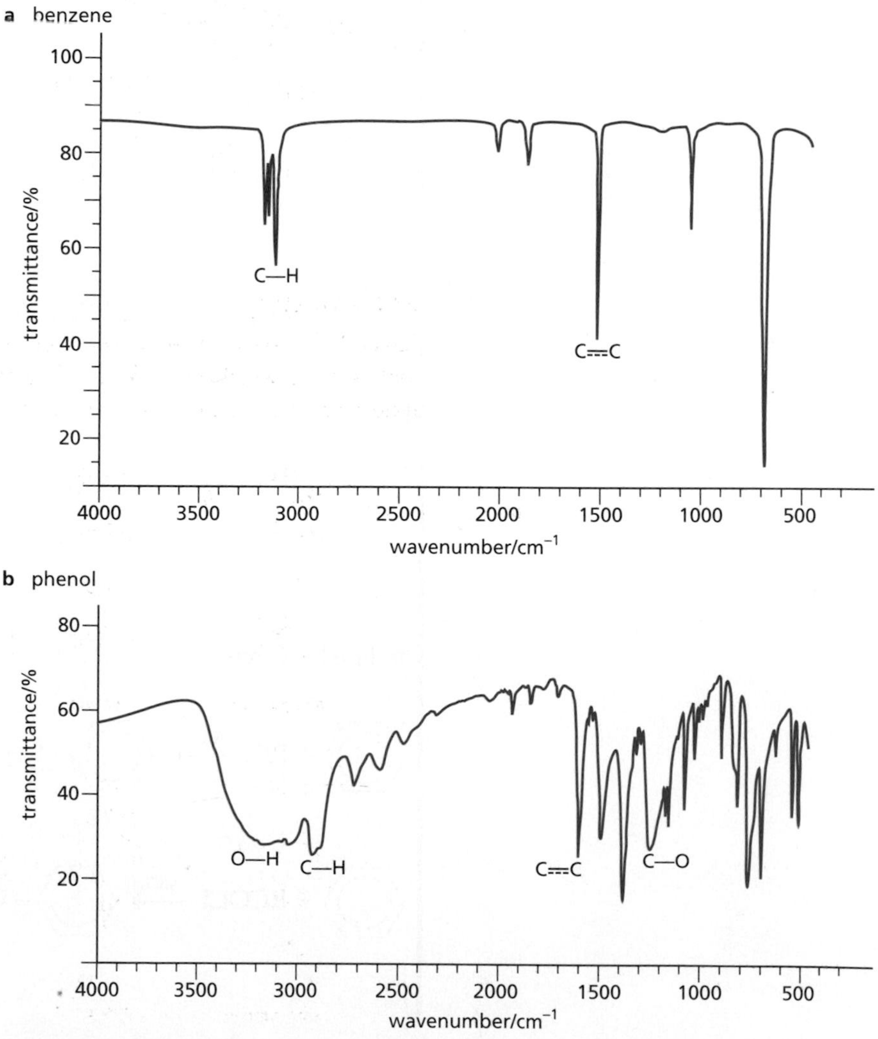

Summary

- **Arenes** are hydrocarbons that contain one or more benzene rings.
- The general formula for compounds containing one ring, C_nH_{2n-6}, shows that arenes are considerably unsaturated.
- The benzene ring contains a delocalised sextet of π electrons. This confers great stability on the system.
- Despite their unsaturation, arenes do not undergo the usual addition reactions associated with alkenes. Their preferred reaction is **electrophilic substitution**.
- Aromatic compounds with alkyl side chains can be oxidised by potassium manganate(VII) to benzenecarboxylic acids.
- Under forcing conditions with hydrogen and with chlorine, the benzene ring can undergo addition rather than substitution.
- Some of the reactions of phenols are similar to those of the —O—H group in alcohols.
- The C—O bond in phenols is very strong, and no reactions occur in which it breaks.
- The benzene ring in phenol is much more susceptible to electrophilic attack than is the ring in benzene itself.

Key reactions you should know

- Electrophilic substitutions:

$$C_6H_6 + Br_2 \xrightarrow{AlCl_3} C_6H_5{-}Br + HBr$$

$$C_6H_6 + Cl_2 \xrightarrow{AlCl_3} C_6H_5{-}Cl + HCl$$

$$C_6H_6 + HNO_3 \xrightarrow{\text{conc } H_2SO_4,\ T<55°C} C_6H_5{-}NO_2 + H_2O$$

$$C_6H_6 + H_2SO_4 \xrightarrow{+\,SO_3} C_6H_5{-}SO_3H + H_2O$$

- Friedel–Crafts:

$$C_6H_6 + RCl \xrightarrow{AlCl_3} C_6H_5{-}R + HCl$$

$$C_6H_6 + RCOCl \xrightarrow{AlCl_3} C_6H_5{-}COR + HCl$$

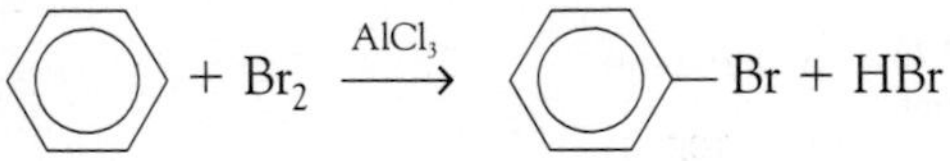

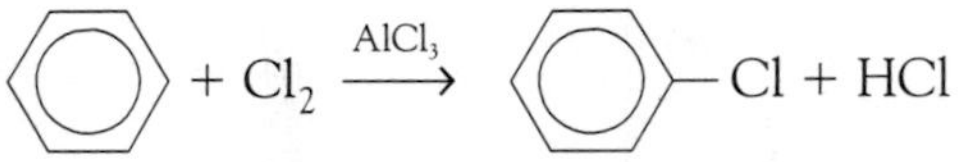

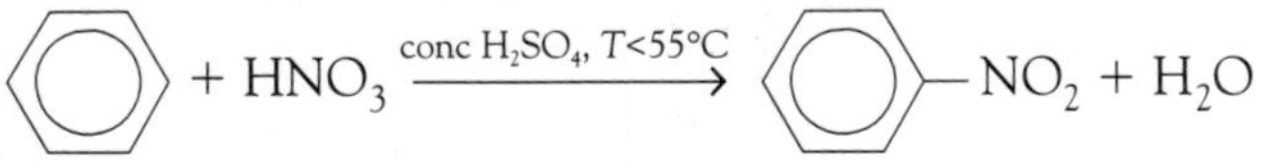

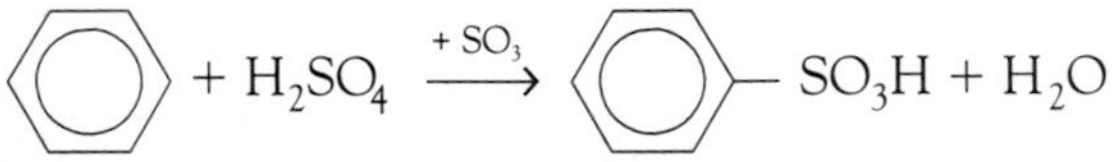

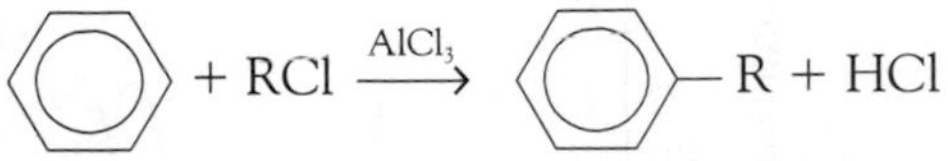

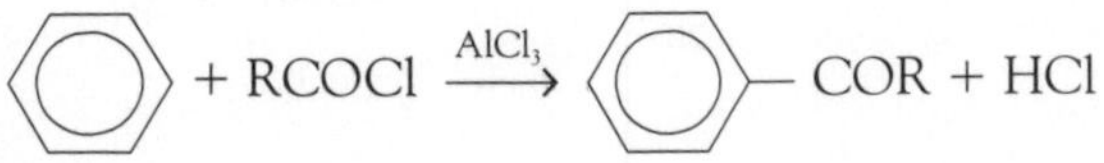

■ Side-chain reactions:

$$\text{C}_6\text{H}_5-\text{CH}_3 + \text{Cl}_2 \xrightarrow{\text{light}} \text{C}_6\text{H}_5-\text{CH}_2\text{Cl} + \text{HCl}$$

$$\text{C}_6\text{H}_5-\text{CH}_3 + 3[\text{O}] \xrightarrow{\text{heat with KMnO}_4 + \text{OH}^-\text{ (aq)}} \text{C}_6\text{H}_5-\text{CO}_2\text{H} + \text{H}_2\text{O}$$

■ Reactions of phenols:

$$\text{C}_6\text{H}_5-\text{OH} + \text{Na} \longrightarrow \text{C}_6\text{H}_5-\text{O}^-\text{Na}^+ + \tfrac{1}{2}\text{H}_2$$

$$\text{C}_6\text{H}_5-\text{OH} + \text{NaOH} \longrightarrow \text{C}_6\text{H}_5-\text{O}^-\text{Na}^+ + \text{H}_2\text{O}$$

$$\text{C}_6\text{H}_5-\text{OH} + \text{RCOCl} \longrightarrow \text{C}_6\text{H}_5-\text{OCOR} + \text{HCl}$$

$$\text{C}_6\text{H}_5-\text{OH} + \text{HNO}_3 \text{ (dil)} \longrightarrow \text{O}_2\text{N}-\text{C}_6\text{H}_4-\text{OH} + \text{H}_2\text{O}$$

$$\text{C}_6\text{H}_5-\text{OH} + 3\text{Br}_2\text{(aq)} \longrightarrow \text{Br}-\text{C}_6\text{H}_2(\text{Br})_2-\text{OH} + 3\text{HBr}$$

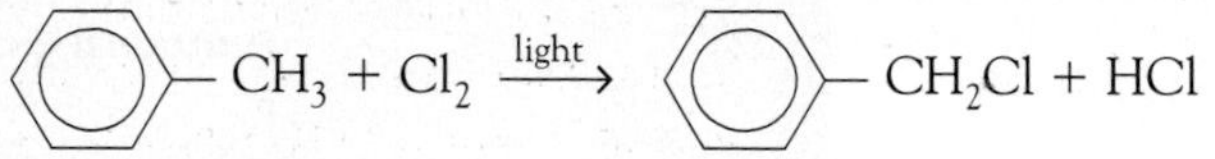

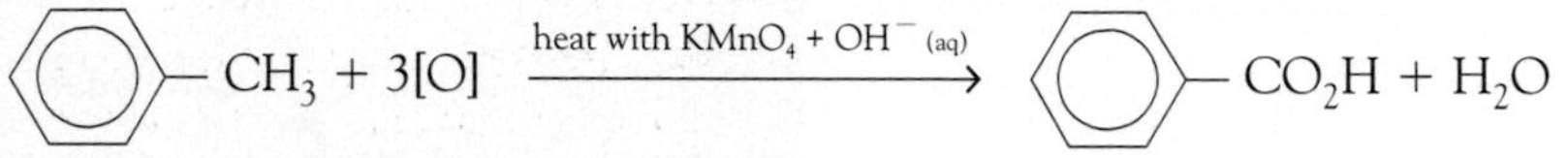

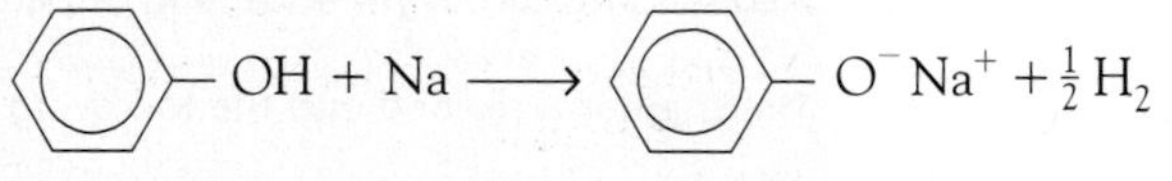

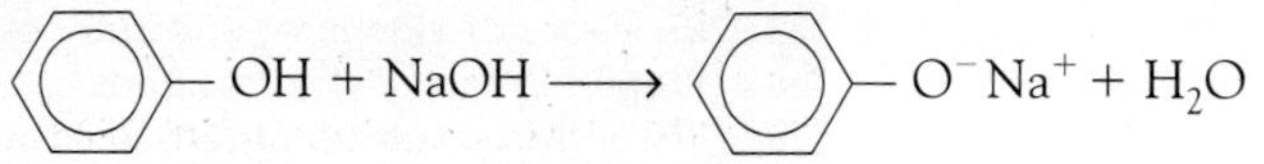

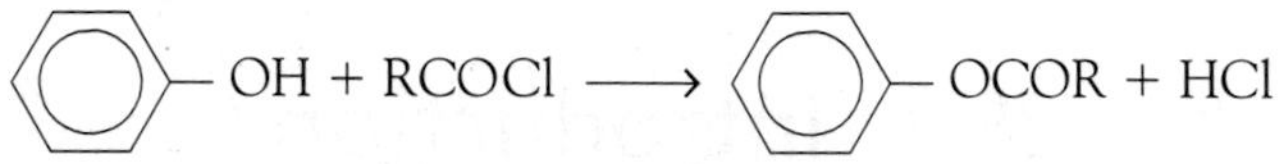

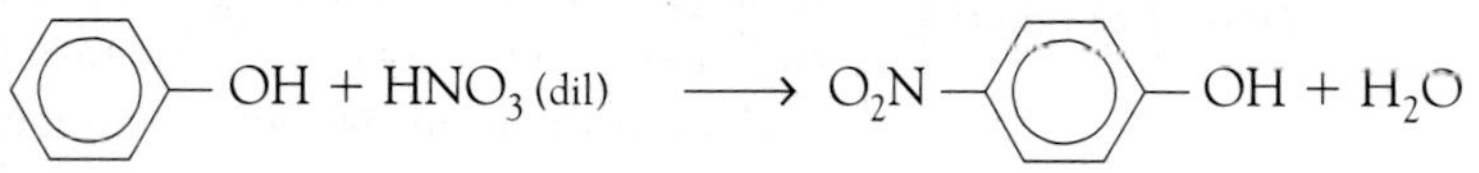

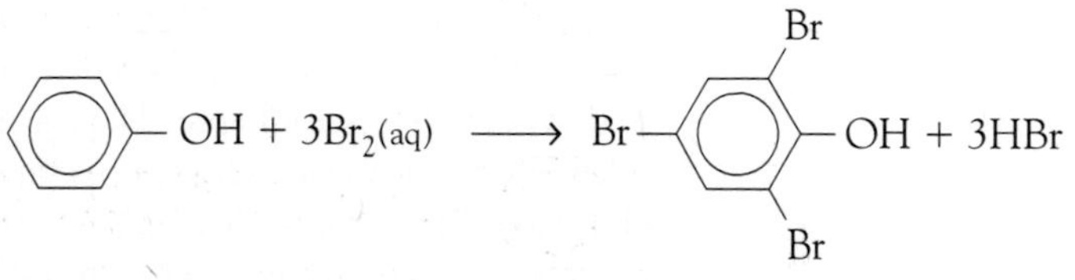

28 *Aldehydes and ketones*

Functional groups:

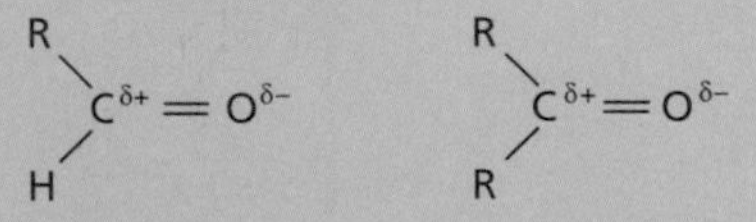

aldehydes **ketones**

The C═O group that is contained in carbonyl compounds is both more polarised, and more polarisable, than the C—O single bond in alcohols and phenols. This chapter introduces the last of our four main mechanisms, nucleophilic addition. This characteristic way in which carbonyl compounds react is due to the highly δ+ carbon atom that they contain. Aldehydes are intermediate in oxidation state between alcohols and carboxylic acids, and can undergo both oxidation and reduction reactions.

The chapter is divided into the following sections.

28.1 Introduction

Group	Class of compounds
R—C(═O)—OH	Carboxylic acid
R—C(═O)—OR'	Ester
R—C(═O)—NH_2	Amide
R—C(═O)—Cl	Acyl chloride
R—C(═O)—O—C(═O)—R	Acid anhydride

■ **Table 28.1**
Some functional groups containing the carbonyl group.

Collectively, aldehydes and ketones are known as **carbonyl compounds**. The carbonyl group, >C═O, is a subunit of many other functional groups (see Table 28.1). But the term carbonyl compounds is reserved for those compounds in which it appears on its own.

The properties of aldehydes and ketones are very similar. Almost all the reactions of ketones are also shown by aldehydes. But aldehydes show additional reactions associated with their lone hydrogen atom. The bonding of the carbonyl group is similar to that of ethene (see Chapter 24, page 440). The double bond is formed by the sideways overlap of two adjacent p orbitals, one on carbon and one on oxygen. Because of its larger electronegativity, oxygen attracts the bonding electrons (in both the σ and the π bonds). This creates an electron-deficient carbon atom (see Figure 28.1).

■ **Figure 28.1**
Oxygen attracts the bonding electrons away from the carbon atom.

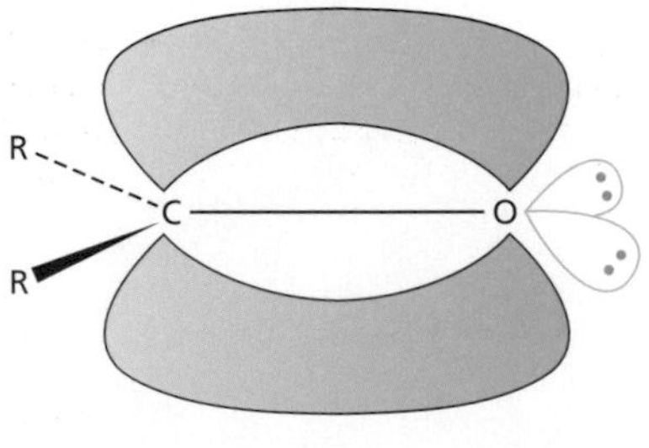

This unequal distribution of electrons is responsible for the two ways in which carbonyl compounds react.

1 The oxygen of the C═O bond can be protonated by strong acids (although carbonyl compounds are not alkaline in aqueous solution, but neutral).
2 The carbon of the C═O bond, on the other hand, is susceptible to attack by nucleophiles, Nu:

$$>C^{\delta+}{=}O^{\delta-} \; H^+ \rightleftharpoons >C^+{-}O{-}H$$

$$Nu^{-}\!: \; >C^{\delta+}{=}O^{\delta-} \rightarrow Nu{-}C{-}O^-$$

The molecules of carbonyl compounds cannot attract each other through intermolecular hydrogen bonds, because they do not contain hydrogen atoms that have a large enough δ+ charge. They can, however, interact with water molecules through hydrogen bonding. The lower members of the series are therefore quite soluble in water (although less so than the corresponding alcohols).

The boiling points of carbonyl compounds are also lower than those of the corresponding alcohols, though significantly higher (by about 40–50 °C) than those of the corresponding alkenes. This is due to the dipole–dipole attractions between the molecules (see Figure 28.2).

Table 28.2 lists some of these properties.

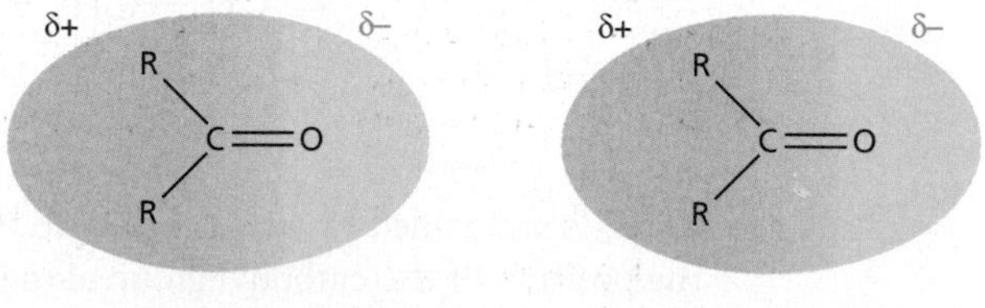

■ **Figure 28.2** Dipole–dipole attractions in carbonyl compounds.

Alkene		Corresponding carbonyl compound			Corresponding alcohol		
Formula	**Boiling point/°C**	**Formula**	**Boiling point/°C**	**Solubility in 100 g of water/g**	**Formula**	**Boiling point/°C**	**Solubility in 100 g of water/g**
$CH_3CH{=}CH_2$	−48	CH_3CHO	20	∞	CH_3CH_2OH	78	∞
$CH_3CH_2CH{=}CH_2$	−6	CH_3CH_2CHO	49	16	$CH_3CH_2CH_2OH$	97	∞
$CH_3CH_2CH_2CH{=}CH_2$	30	$CH_3CH_2CH_2CHO$	76	7	$CH_3CH_2CH_2CH_2OH$	118	8
$(CH_3)_2C{=}CH_2$	−7	$(CH_3)_2CO$	56	∞	$(CH_3)_2CHOH$	82	∞

■ **Table 28.2** Properties of some aldehydes and ketones compared with alkenes and alcohols.

Carbonyl compounds have distinctive smells. Ketones smell sweeter than aldehydes – almost like esters. Aldehydes have an astringent smell, but often with a fruity overtone – for example, ethanal smells of apples. The simplest aldehyde, methanal, has an unpleasant, choking smell. Its concentrated aqueous solution, formalin, is used to preserve biological specimens. Aldehydes and ketones of higher molecular mass have an important use as flavouring and perfuming agents. Figure 27.7 (page 497) includes two carbonyl compounds with benzene rings, vanillin and benzaldehyde, that are used as flavouring agents. Some other carbonyl compounds with pleasant smells are shown in Figure 28.3.

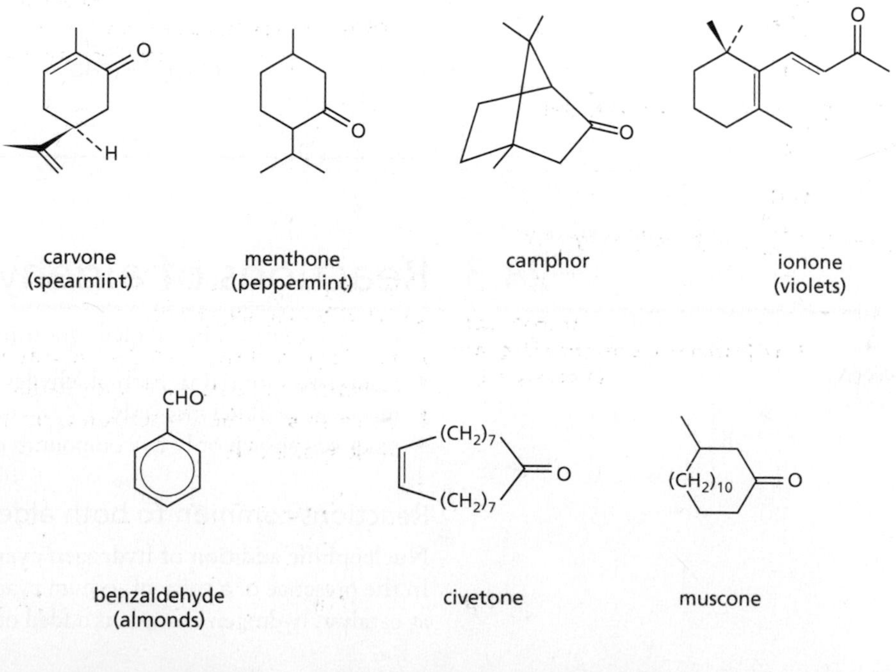

■ **Figure 28.3** Some pleasant-smelling carbonyl compounds, and two naturally occurring musks (bases for perfumes).

28.2 Isomerism and nomenclature

In one respect, aldehydes and ketones can be considered as positional isomers of each other. In propanal, CH_3CH_2CHO, the functional group (carbonyl) is on position 1 of the chain, whereas in propanone, CH_3COCH_3, it is on position 2.

Because of significant differences in their reactions, however, aldehydes and ketones are named differently, and are considered as two separate classes of compound.

Aldehydes are named by adding '-al' to the hydrocarbon stem:

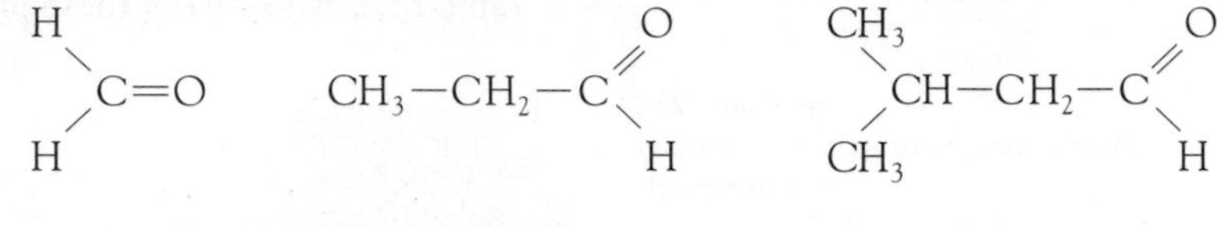

methanal   propanal   3-methylbutanal

Ketones are named by adding '-one' to the hydrocarbon stem. In higher ketones, the position of the carbonyl group along the chain needs to be specified:

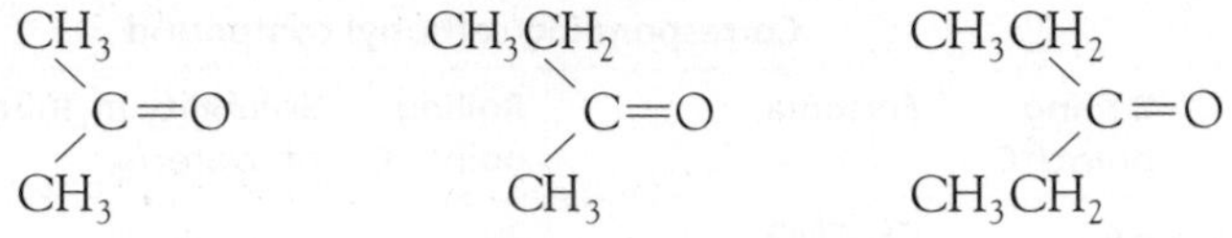

propanone   butanone   pentan-3-one

Example   Draw structures for the following:

a 2-methylbutanal

b pentan-2-one.

Answer   **a** The four-carbon chain is numbered starting at the aldehyde end:

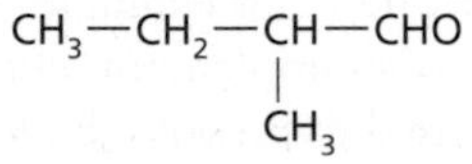

$$CH_3-CH_2-CH-CHO$$
$$|$$
$$CH_3$$

b $$CH_3-CH_2-CH_2-C-CH_3$$
$$\|$$
$$O$$

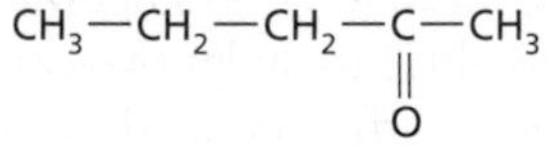

Further practice   Name the following compounds:

1

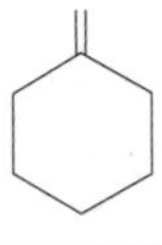

2 $(CH_3)_2C-CHO$

3 $(CH_3CH_2CH_2)_2C=O$

28.3 Reactions of aldehydes and ketones

These reactions can be divided into three groups:

• reactions common to both aldehydes and ketones
• reactions of aldehydes only
• reactions shown only by compounds containing certain extra structural features.

Reactions common to both aldehydes and ketones

Nucleophilic addition of hydrogen cyanide

In the presence of a trace of sodium cyanide (or a base such as sodium hydroxide) as catalyst, hydrogen cyanide is added on to carbonyl compounds:

$$CH_3{-}CHO + HCN \xrightarrow{\text{trace NaCN(aq)}} CH_3{-}\underset{\displaystyle H}{\overset{\displaystyle OH}{C}}{-}C{\equiv}N$$

The product is called a **cyanohydrin**, in this case 2-hydroxypropanenitrile. Like the reaction between sodium cyanide and halogenoalkanes (see Chapter 25, page 469), the reaction is a useful method of adding a carbon atom to a chain. Cyanohydrins can be hydrolysed to 2-hydroxycarboxylic acids (see Chapter 29) by heating with dilute sulphuric acid:

$$CH_3{-}\overset{\displaystyle OH}{CH}{-}CN + 2H_2O + H^+ \rightarrow CH_3{-}\overset{\displaystyle OH}{CH}{-}CO_2H + NH_4^+$$

The cyanohydrin can also be reduced to a hydroxyamine (see Chapter 30).

$$CH_3{-}\overset{\displaystyle OH}{CH}{-}CN + 2H_2 \xrightarrow{\text{nickel catalyst}} CH_3{-}\overset{\displaystyle OH}{CH}{-}CH_2NH_2$$

The mechanism of the addition of hydrogen cyanide to carbonyl compounds

Hydrogen cyanide is neither a nucleophile nor a strong acid. Therefore it cannot react with the carbonyl group by either of the routes shown on page 527. However, as we saw in Chapter 25, the cyanide ion is a good nucleophile. It can attack the $\delta+$ carbon of the C=O bond:

The intermediate alkoxide anion is a strong base. It can abstract a proton from an un-ionised molecule of hydrogen cyanide:

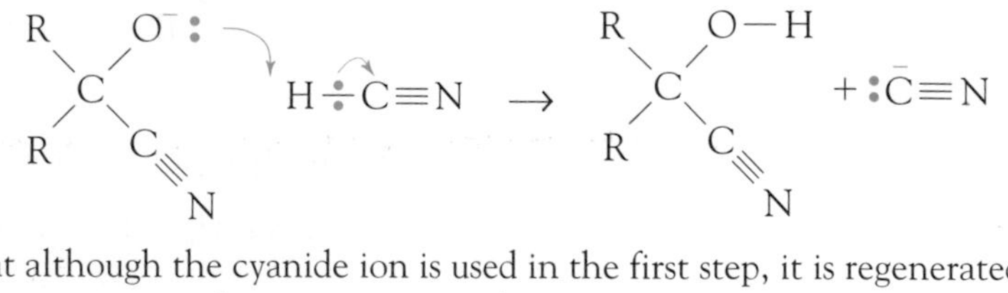

So we see that although the cyanide ion is used in the first step, it is regenerated in the second. It is acting as a homogeneous catalyst.

The reduction of carbonyl compounds by complex metal hydrides

Either sodium tetrahydridoborate(III), $NaBH_4$ (sodium borohydride), or lithium tetrahydridoaluminate(III), $LiAlH_4$ (lithium aluminium hydride), reduces carbonyl compounds to alcohols.

Aldehydes give primary alcohols:

$$CH_3{-}CHO \xrightarrow{\text{NaBH}_4\text{ in alkaline aqueous methanol}} CH_3{-}CH_2OH$$

whilst ketones give secondary alcohols:

$$CH_3{-}CO{-}CH_3 \xrightarrow{\text{LiAlH}_4\text{ in dry ether}} CH_3{-}\overset{\displaystyle OH}{CH}{-}CH_3$$

Because the use of $LiAlH_4$ (which catches fire with water) and ether (which forms highly explosive mixtures in air) is hazardous, the use of $NaBH_4$ is preferred.

How it happens (1): the mechanism of complex metal hydride reduction

The BH_4^- anion approaches the carbonyl group during the reduction stage. It then donates one of its $H^{\delta-}$ atoms, together with the two electrons of the B—H bond, to the $C^{\delta+}$ end of the C=O bond. Nucleophilic addition has taken place:

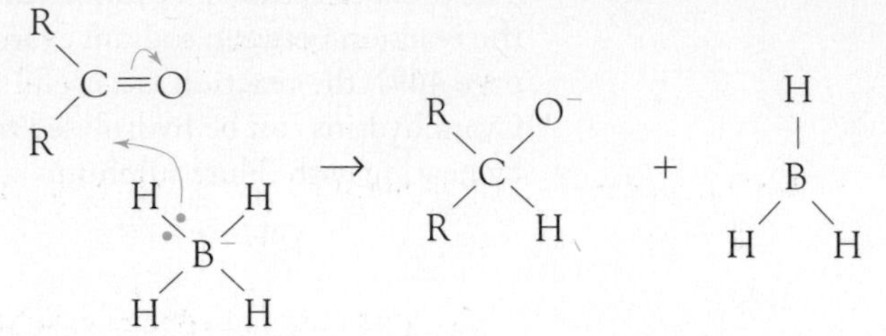

This **hydride transfer mechanism** is similar to that of some other reactions in which a carbonyl group is reduced (see the panel on the Cannizzaro reaction opposite).

The resulting alkoxide ion is protonated by water:

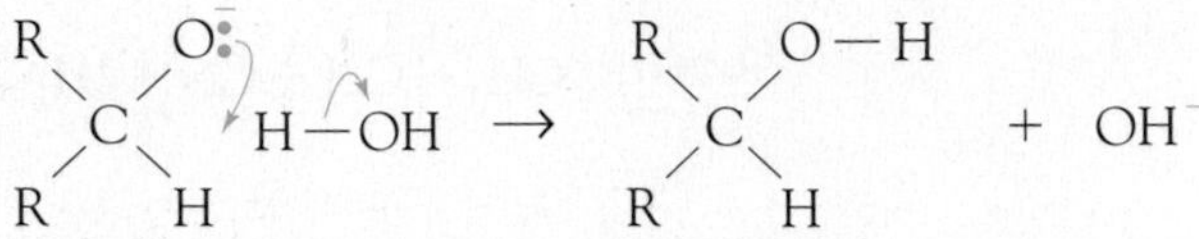

The hydroxide ion formed during this step then reacts with the BH_3 produced in the first step, giving a boron-containing anion. This can then add another of its hydrogen atoms to a second carbonyl group:

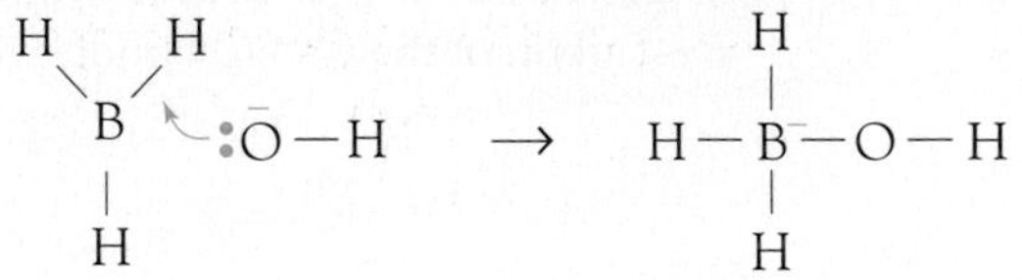

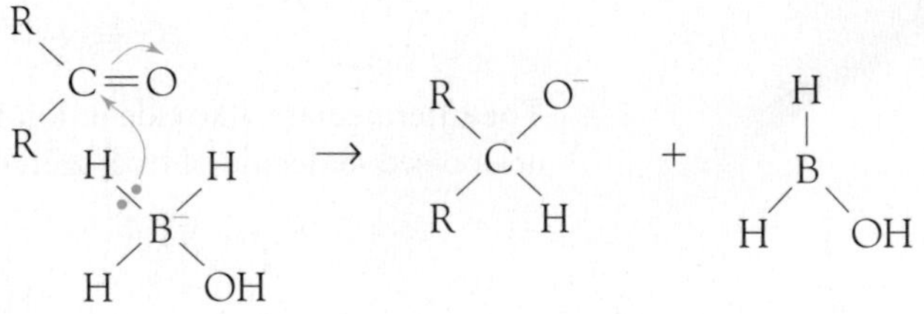

Eventually all four hydrogen atoms in $NaBH_4$ can be transferred to carbonyl groups:

$$NaBH_4 + 4CH_3CHO + H_2O + 2NaOH \rightarrow 4CH_3CH_2OH + Na_3BO_3$$

Catalytic hydrogenation

Like alkenes, carbonyl compounds can be reduced by hydrogen gas over a nickel, palladium or platinum catalyst. The mechanism is similar to heterogeneous catalysis described in Chapter 8, pages 168–9 and Chapter 19, page 367.

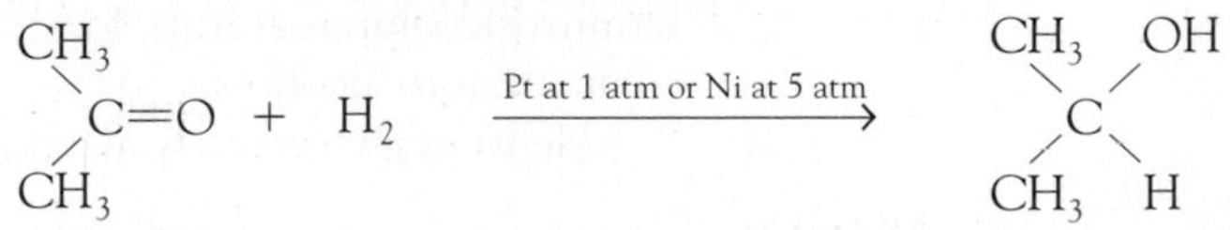

Example Suggest the structural formulae of the intermediates and products of the following reactions:

a

$$CH_3CH_2CHO + HCN \xrightarrow{NaCN} W \xrightarrow{\text{heat with } H_2SO_4(aq)} X$$

b

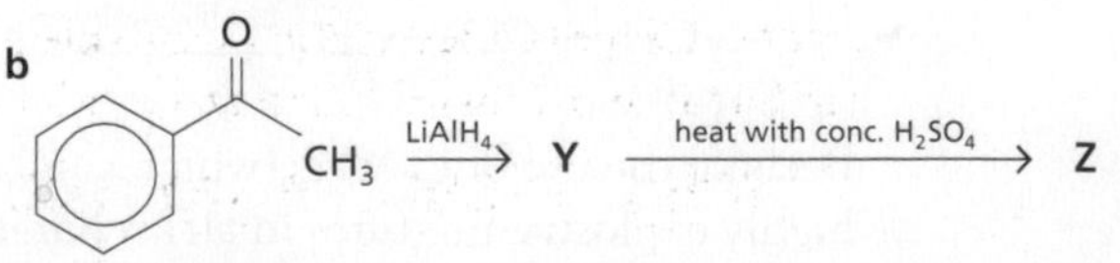

Answer **a**

W is 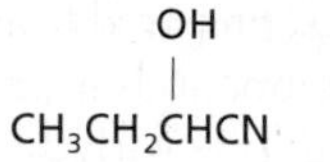 **X** is 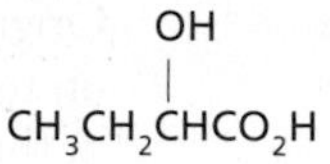

b

Y is 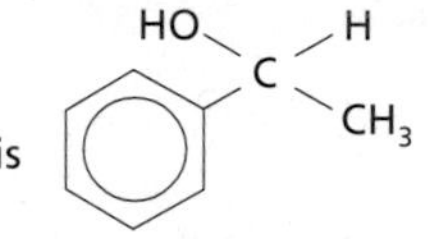 **Z** is 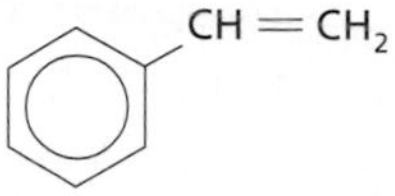

Further practice Suggest reagents for the following transformations:

1

$$(CH_3)_2C=O \xrightarrow{I} \mathbf{A} \xrightarrow{II} CH_3-CH=CH_2$$

2

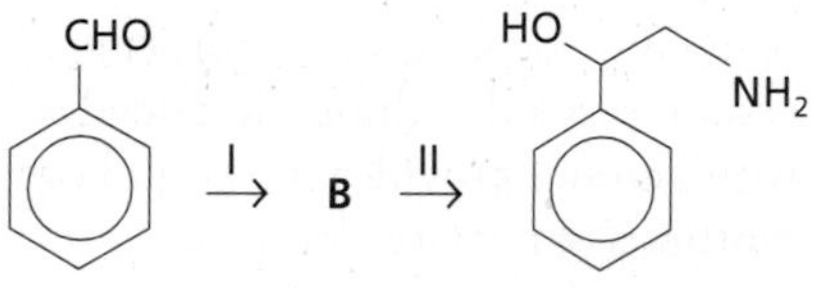

The Cannizzaro reaction

When aryl aldehydes such as benzaldehyde are treated with concentrated potassium hydroxide solution, they undergo a remarkable reaction. One molecule of aldehyde oxidises another aldehyde molecule, and in the process is reduced to the alcohol:

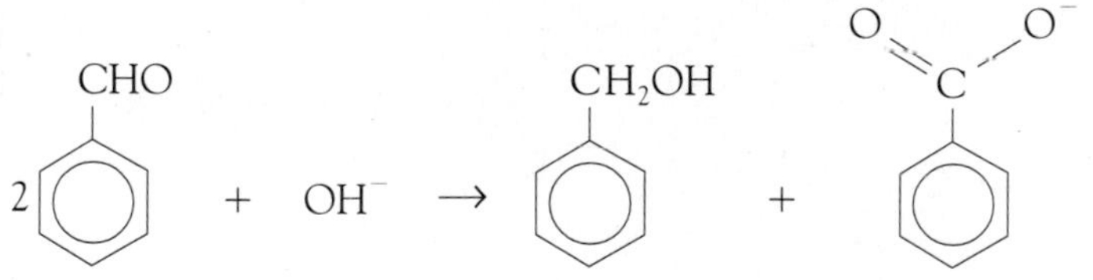

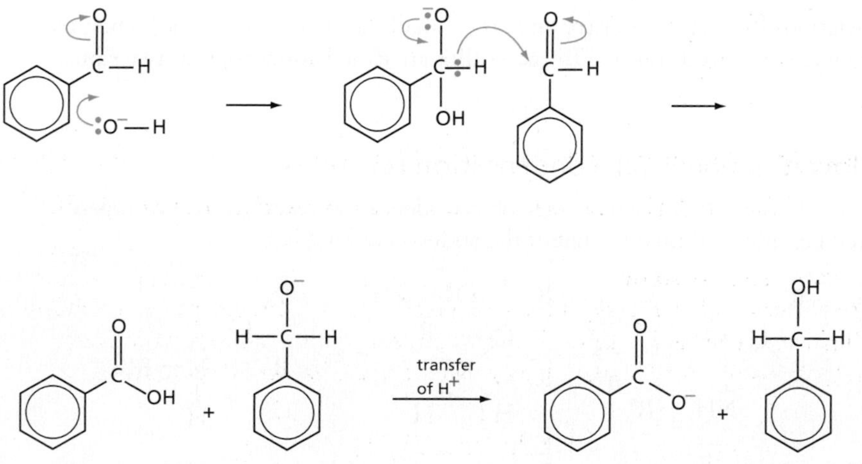

■ **Figure 28.4**
The mechanism of the Cannizzarro reaction.

The mechanism of this diproportionation reaction (see Chapter 7, page 151) is similar to the reduction of carbonyl compounds using complex metal hydrides. A hydride ion (H:⁻) is transferred from one aldehyde molecule to another (see Figure 28.4).

The overall yield is very good: in excess of 95%.

The Cannizzaro reaction occurs with any aldehyde that lacks a hydrogen atom on the carbon next to the C═O group. (If such a hydrogen atom were present, it would ionise in a strong base, and other reactions would compete.) So $(CH_3)_3C{-}CHO$ and H_2CO also undergo the reaction. One interesting use of the reaction is for the reduction of suitable aldehydes to alcohols by the use of methanal in a 'crossed Cannizzaro' reaction:

$$(CH_3)_3C{-}CHO + H_2CO + OH^- \rightarrow (CH_3)_3C{-}CH_2OH + HCO_2^-$$

Nucleophilic addition of Grignard reagents

Grignard reagents, prepared from halogenoalkanes (see Chapter 25, page 473), add on to carbonyl compounds in dry ether solution to give salts of alcohols. Treatment of the reaction mixture with dilute aqueous acid produces the alcohols. Methanal forms primary alcohols, other aldehydes secondary alcohols, whilst ketones form tertiary alcohols.

$$CH_3CHO \xrightarrow[\text{in ether}]{CH_3CH_2MgBr} CH_3 - \underset{\underset{CH_2CH_3}{|}}{\overset{\overset{OMgBr}{|}}{C}} - H \xrightarrow[H_2SO_{4(aq)}]{\text{dilute}} CH_3 - \underset{\underset{CH_2CH_3}{|}}{\overset{\overset{OH}{|}}{C}} - H + MgSO_4$$

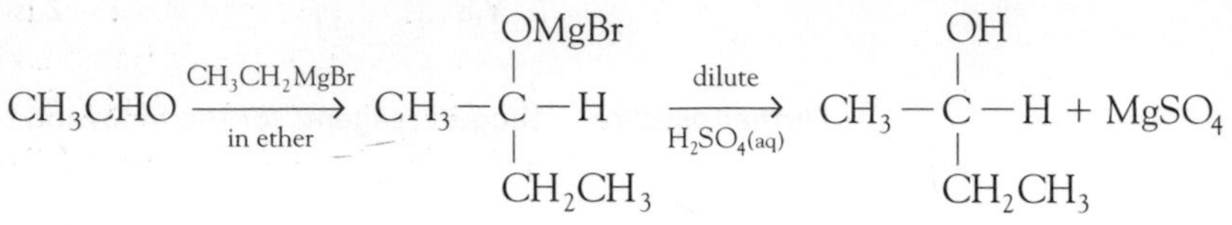

Good yields are obtained. The reaction with aldehydes provides an alternative way of making secondary alcohols (the other being the reduction of ketones by complex metal hydrides described on page 529). The reaction of Grignard reagents with ketones provides the best of only a few methods that are available for the synthesis of tertiary alcohols.

Condensation reactions

Nitrogen nucleophiles readily add on to carbonyl compounds. But usually, the initially formed addition compounds cannot be isolated. They easily lose water to give stable compounds containing a C=N bond:

$$\underset{R}{\overset{R}{\diagdown}}C=O \;+\; H_2N-R' \;\rightarrow\; \left[\underset{R}{\overset{R}{\diagdown}}\underset{\underset{\underset{H}{|}}{N-R'}}{\overset{OH}{C}} \right] \;\rightarrow\; \underset{R}{\overset{R}{\diagdown}}C=N\underset{R'}{\diagup} \;+\; H_2O$$

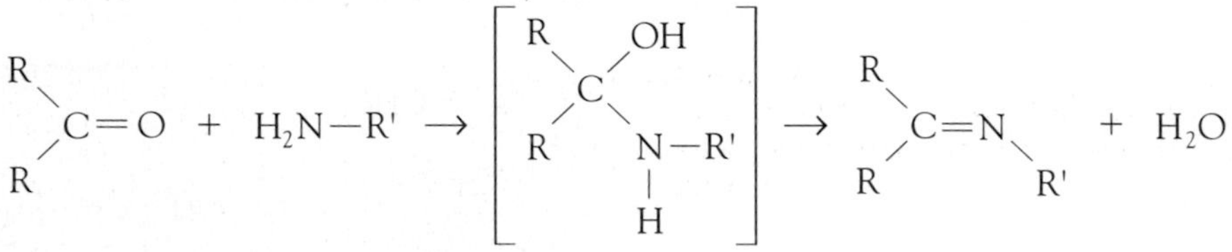

Reactions that form water (or any other small inorganic compound) when two organic compounds react with each other are called **condensation reactions**.

How it happens (2): condensation reactions

The mechanism of the first stage of a condensation reaction with nitrogen nucleophiles is similar to that of the addition of HCN:

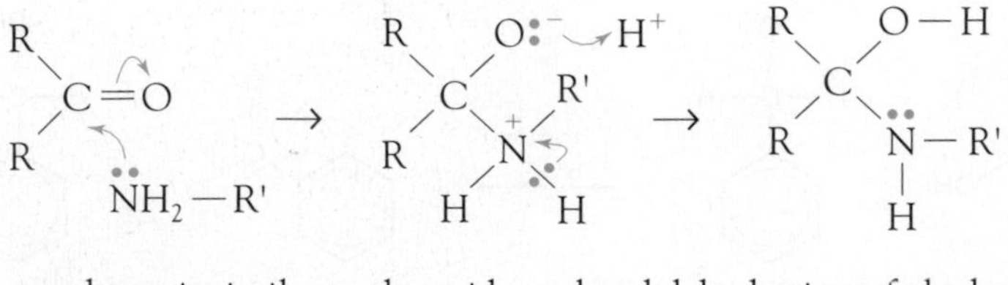

The second step is similar to the acid-catalysed dehydration of alcohols to form alkenes:

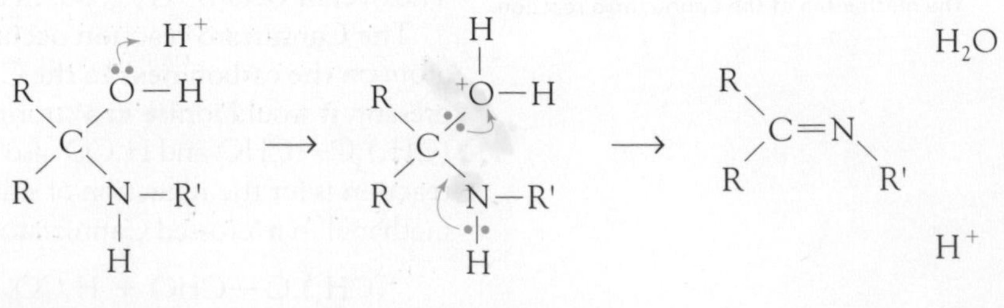

The most important condensation reaction of carbonyl compounds is that with 2,4-dinitrophenylhydrazine (2,4-DNPH). The products are highly crystalline orange solids, which precipitate out of solution rapidly.

> The formation of an orange precipitate when a solution of 2,4-DNPH is added to an unknown compound is a good test for the presence of a carbonyl compound.

If the melting point of a recrystallised sample is taken and compared with tables of known values, the carbonyl compound can be uniquely identified. This is especially useful if the carbonyl compounds have similar boiling points, and so would otherwise be easily confused (see Table 28.3).

$$CH_3\backslash C=O + H_2N-NH-\underset{NO_2}{\underset{|}{\bigcirc}}-NO_2 \rightarrow CH_3\backslash C=N-NH-\underset{NO_2}{\underset{|}{\bigcirc}}-NO_2 + H_2O$$

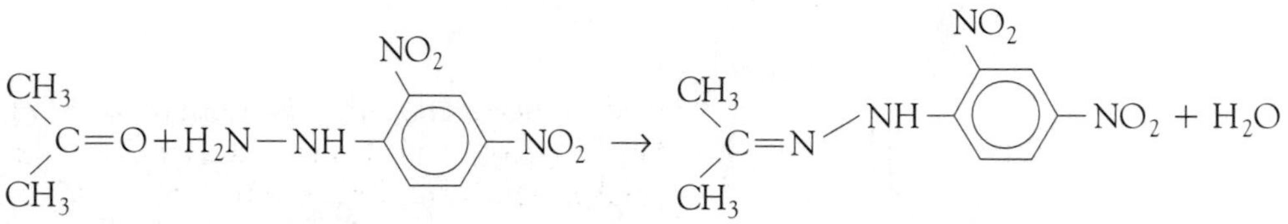

2,4-DNPH

■ **Table 28.3**
The melting points of some 2,4-DNPH derivatives of carbonyl compounds.

Carbonyl compound	Boiling point/°C	Melting point of 2,4-dinitrophenylhydrazone derivative/°C
$CH_3CH_2CH_2CH_2CHO$	103	108
$CH_3CH_2COCH_2CH_3$	102	156
$CH_3CH_2CH_2COCH_3$	102	144

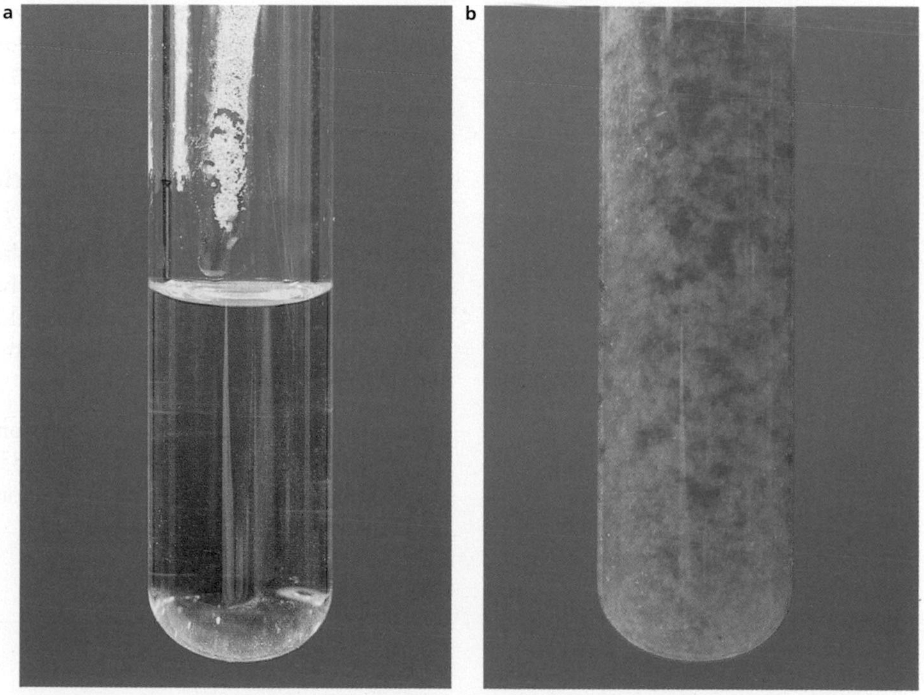

■ **Figure 28.5** **a**
a Ethanal reacting with 2,4-DNPH at the start of the reaction; this forms an orange 2,4-DNPH precipitate, **b**.

 Another condensation reaction of carbonyl compounds is with hydroxylamine. The products are called **oximes**. They can be reduced by hydrogen over a nickel catalyst to primary amines (see Chapter 30).

$$CH_3\backslash C=O + H_2N-OH \xrightarrow{-H_2O} CH_3\backslash C=N\diagup OH \xrightarrow{H_2/Ni} CH_3\diagup C\backslash H \diagup NH_2 + H_2O$$

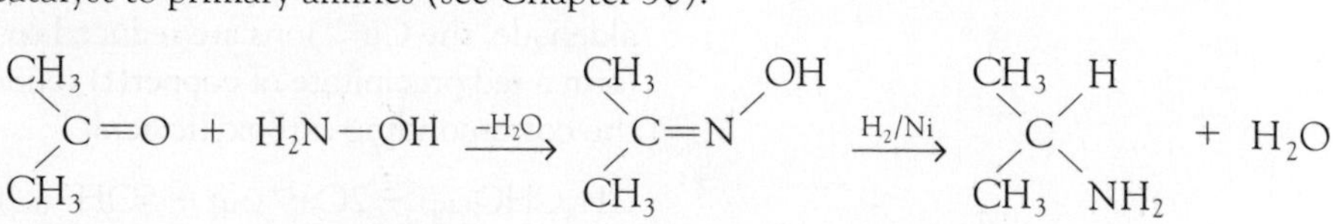

hydroxylamine

Example Predict the products of the following reactions.

a

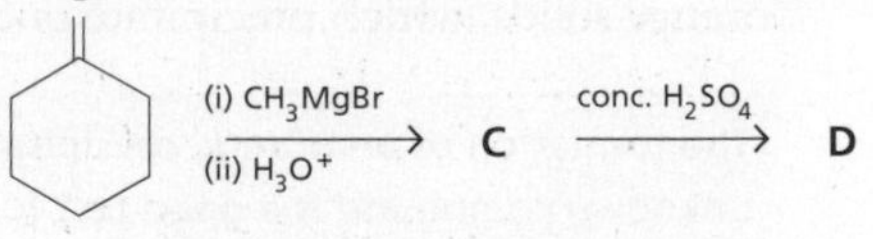

b

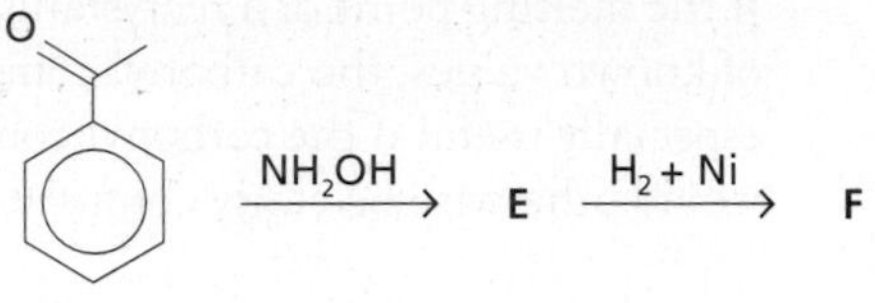

Answer **a**

C is 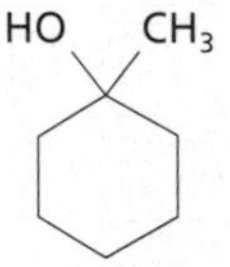

D is 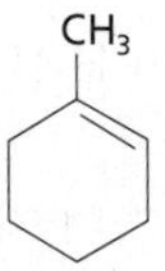 or 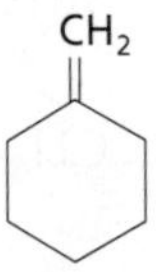

b

E is 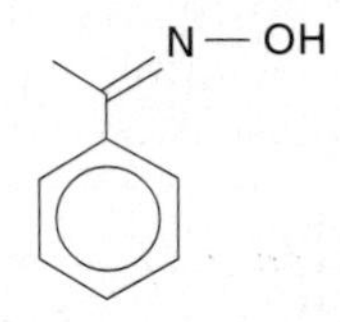

F is 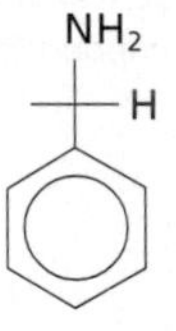

Further practice Two isomers G and H have the molecular formula C_4H_8O. G forms an orange precipitate with 2,4-DNPH whereas H does not. H decolorises bromine water, but G does not. On treatment with hydrogen and nickel, both G and H give the same compound, butan-1-ol. Suggest structures for G and H.

Reactions undergone only by aldehydes

Aldehydes are distinguished from ketones by having a hydrogen atom directly attached to the carbonyl group. This hydrogen atom is not in any way acidic, but the C—H bond is significantly weaker than usual. The —CHO group is readily oxidised to a —COOH group, making aldehydes mild reducing agents. This oxidation, specific to aldehydes, has been used to design the following tests that will distinguish aldehydes from ketones.

Oxidation by acidified potassium dichromate(VI) solution

It was mentioned in Chapter 26, page 485, that aldehydes are more easily oxidised than alcohols. The reaction takes place on gentle warming, and the colour of the reagent changes from orange to green.

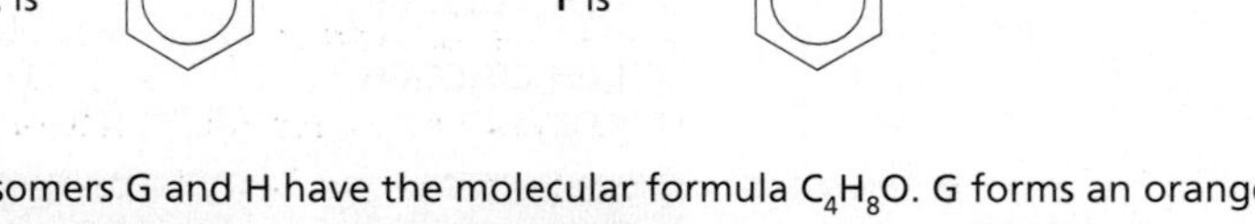

Reduction of Fehling's solution

The bright blue **Fehling's solution** is a solution of Cu^{2+} ions in an aqueous alkaline medium, complexed with salts of tartaric acid. When warmed with an aldehyde, the Cu^{2+} ions are reduced to Cu^{+} ions, which in the alkaline solution form a red precipitate of copper(I) oxide. The aldehyde is oxidised to the salt of the corresponding carboxylic acid:

$$\underset{\text{}}{CH_3CHO(aq)} + \underset{\text{blue solution}}{2Cu^{2+}(aq)} + 5OH^-(aq) \rightarrow CH_3CO_2^-(aq) + \underset{\text{red precipitate}}{Cu_2O(s)} + 3H_2O(l)$$

Formation of a silver mirror with Tollens' reagent

Tollens' reagent contains silver ions in aqueous alkaline solution (complexed with ammonia). These silver ions are readily reduced to silver metal on gentle warming with an aldehyde. The metal will often silver-plate the inside of the test tube. Once again, the aldehyde is oxidised to the salt of the corresponding carboxylic acid:

$$CH_3CHO + 2Ag^+ + 3OH^- \rightarrow CH_3CO_2^- + \underset{\text{silver mirror}}{2Ag} + 2H_2O$$

A dilute solution of the aldehyde sugar, glucose, was once used to make mirrors for domestic use by means of this reaction.

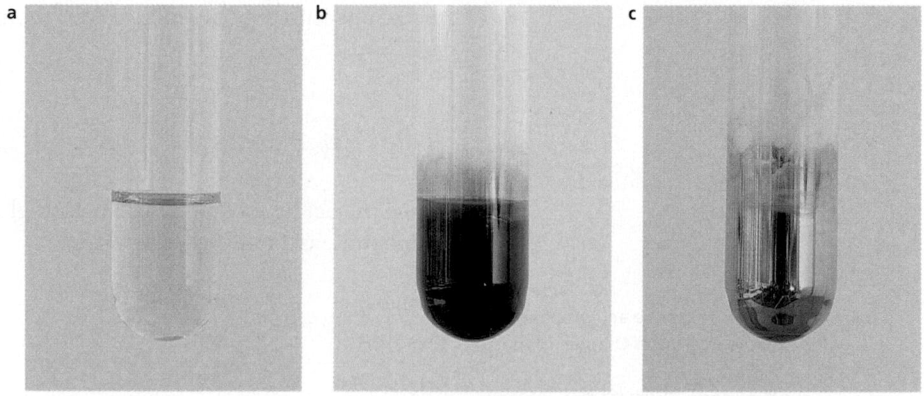

■ **Figure 28.6** An aldehyde produces a silver mirror with Tollens' reagent.

Aldehydes can be distinguished from ketones by one of the following tests:

1 they produce a red precipitate on warming with Fehling's solution
2 they produce a silver mirror on warming with Tollens' reagent.

Specific reactions

Ketones that contain the group —$COCH_3$ (that is, methyl ketones) undergo the triiodomethane (iodoform) reaction on treatment with aqueous alkaline solution of iodine. This reaction was introduced in Chapter 26, page 486, since secondary alcohols that can be oxidised to methyl ketones also undergo the reaction:

$$(CH_3)_2C{=}O + 3I_2 + 4OH^- \rightarrow CH_3CO_2^- + \underset{\text{pale yellow precipitate}}{CHI_3} + 3I^- + 3H_2O$$

There is one aldehyde, ethanal, that also undergoes the triiodomethane reaction:

$$CH_3{-}CHO \xrightarrow{I_2 + OH^-(aq)} HCO_2^- + CHI_3$$

The mechanism for the triiodomethane reaction is described in the panel overleaf.

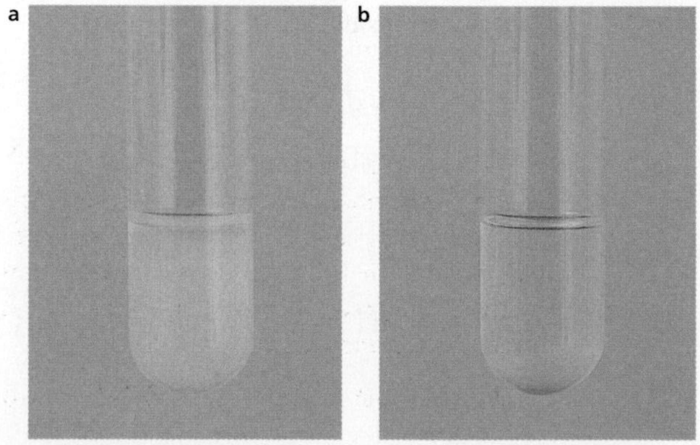

■ **Figure 28.7** The triiodomethane reaction **a** before and **b** after standing.

How it happens (3): the triiodomethane reaction

The highly polarised C=O bond in ketones and aldehydes withdraws electrons from the C—H bonds on adjacent atoms. This allows the molecule to act as a weak acid, reacting with bases as follows:

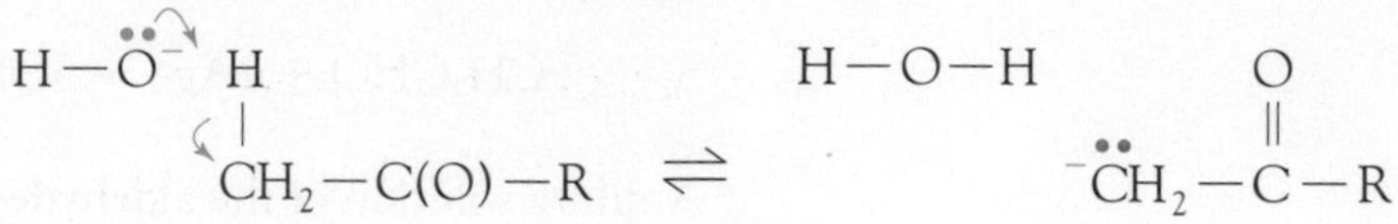

The negative charge on the carbon atom in the product (called an **enolate ion**) is delocalised onto the oxygen atom too (compare the carboxylate ion, see Chapter 3, page 78):

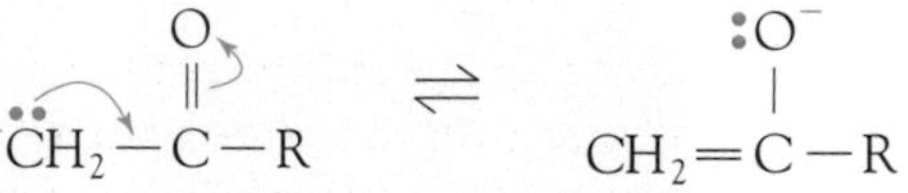

Hence the acidity of ketones is much higher than that of other C—H containing compounds, as Table 28.4 illustrates.

■ **Table 28.4**
The acidity of ethane and propanone. Remember from Chapter 12 that:

$$pK_a = -\log_{10} K_a \text{ where } K_a = \frac{[H^+][A^-]}{[HA]}$$

for acids undergoing the dissociation $HA \rightleftharpoons H^+ + A^-$.

Compound	pK_a
$CH_3—CH_3$	50
CH_3COCH_3	20

Iodine dissolved in aqueous sodium hydroxide undergoes the following successive equilibria:

$$I_2 + OH^- \rightleftharpoons HOI + I^-$$
$$HOI + OH^- \rightleftharpoons OI^- + H_2O$$

The first equilibrium lies well to the right, whereas the second lies to the left (HOI is a weak acid). The major species in solution is therefore HOI, and it is this that reacts with the enolate ion:

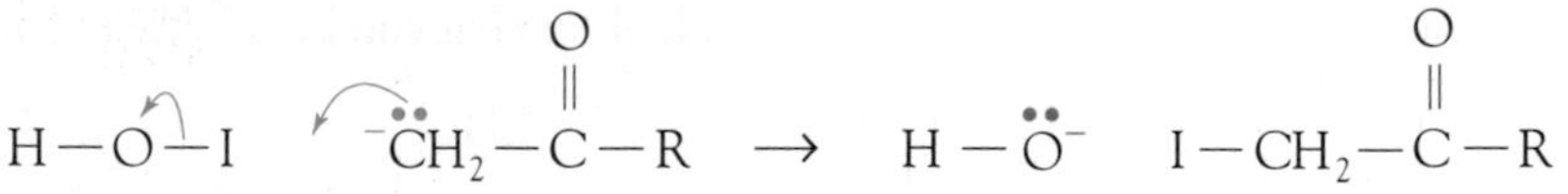

Further substitution occurs until all three hydrogen atoms are replaced by iodine.

$$I—CH_2—COR + OH^- \rightarrow I—\bar{C}H—COR + H_2O$$
$$I—\bar{C}H—COR + HOI \rightarrow I_2CH—COR + OH^- \rightarrow\rightarrow I_3C—COR$$

Substitution of hydrogen by electronegative atoms such as iodine withdraws electron density from the carbonyl $C^{\delta+}$ atom, allowing this to be attacked by an OH^- ion acting as a nucleophile. Subsequent C—C bond cleavage is followed by proton transfer to form the products.

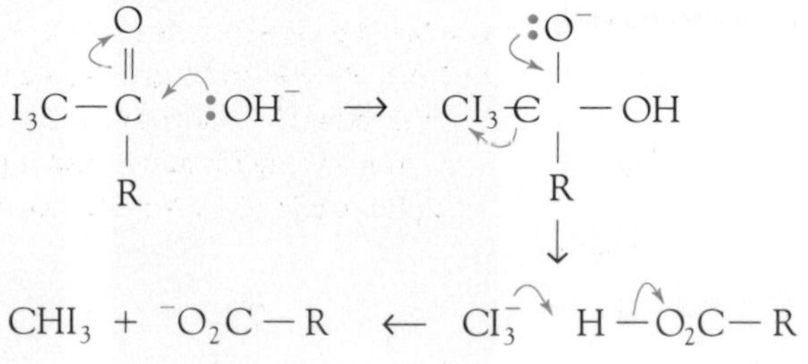

The carbonyl group as an electrophile

The driving force for the attack of nucleophiles and bases on carbonyl compounds is the ability of the oxygen atom to accommodate a negative charge.

attack by nucleophile X⁻:

$$\begin{array}{ccc} O & & :O^- \\ \| & & | \\ -C- & \longrightarrow & -C- \\ | & & | \\ X^- & & X \end{array}$$

attack by base Y⁻:

$$Y:^- \curvearrowright H-\overset{\overset{\displaystyle O}{\|}}{C}-C- \longrightarrow Y-H + \overset{:O^-}{C}=C-$$

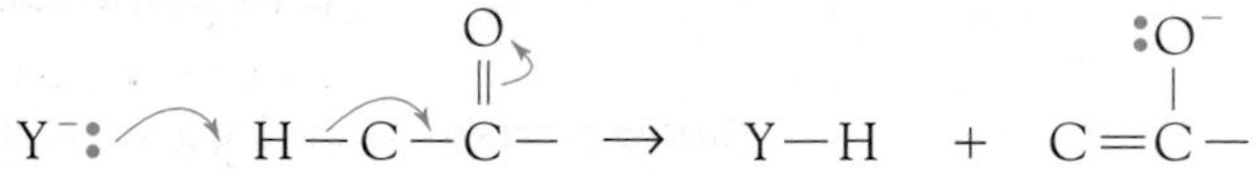

If the groups attached to the carbonyl group already spread a degree of negative charge over the oxygen atom, the ability of the carbonyl group to accommodate further electrons is hampered. So acids, esters and amides do not undergo condensation reactions or the iodoform reaction, and their reduction requires successively more powerful hydrogen donors.

In Table 28.5, the likelihood of attack by nucleophiles or bases decreases from top to bottom.

■ **Table 28.5**
Those carbonyl groups at the top of the table can spread negative charge to the oxygen and so are most easily attacked by nucleophiles or bases. Those at the bottom already have 'extra' negative charge on the carbonyl oxygen and so are more resistant to attack.

Compound	Comments
$$\overset{\overset{\displaystyle O}{\|}}{H-C-H}$$	Very ready addition: exists in water solution as $CH_2(OH)_2$.
$$\overset{\overset{\displaystyle O}{\|}}{CH_3-C-H}$$	Easily reduced by $NaBH_4$. Undergoes many reactions via $^-CH_2-CHO$.
$$\overset{\overset{\displaystyle O}{\|}}{CH_3-C-CH_3}$$	Easily reduced by $NaBH_4$. Undergoes several reactions via $^-CH_2-COCH_3$.
$$\overset{\overset{\displaystyle O}{\|}}{CH_3-C-OR}$$	Requires $LiBH_4$ (a stronger agent than $NaBH_4$) for reduction. Few reactions via carbanion.
$$\overset{\overset{\displaystyle O}{\|}}{CH_3-C-NH_2}$$	Requires $LiAlH_4$ (a very strong reducing agent) for reduction. No reactions via carbanion.
$$\overset{\overset{\displaystyle O}{\|}}{CH_3-C-O^-}$$	Requires $LiAlH_4$ (a very strong reducing agent) for reduction. No reactions via carbanion.

Example   Identify the following compounds.
a **A** has the molecular formula C_3H_6O. It reacts with hydrogen and nickel to give **B**, C_3H_8O. Both **A** and **B** give a pale yellow precipitate with aqueous alkaline iodine.
b **C** has the molecular formula C_4H_8O. It reacts with Fehling's solution. On treatment with sodium tetrahydridoborate(III) it gives **D**, which on warming with concentrated sulphuric acid gives 2-methylpropene.

Answer   a

$$\mathbf{A}\text{ is }\overset{\overset{\displaystyle O}{\|}}{\underset{CH_3 \quad CH_3}{C}} \qquad \mathbf{B}\text{ is }\overset{HO \quad H}{\underset{CH_3 \quad CH_3}{C}}$$

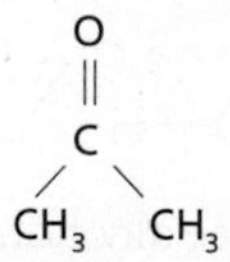

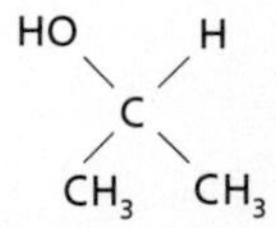

b A positive Fehling's solution test means that **C** is an aldehyde. Sodium tetrahydridoborate(III) reduces an aldehyde to a primary alcohol. Concentrated sulphuric acid converts this to the alkene.
Therefore:

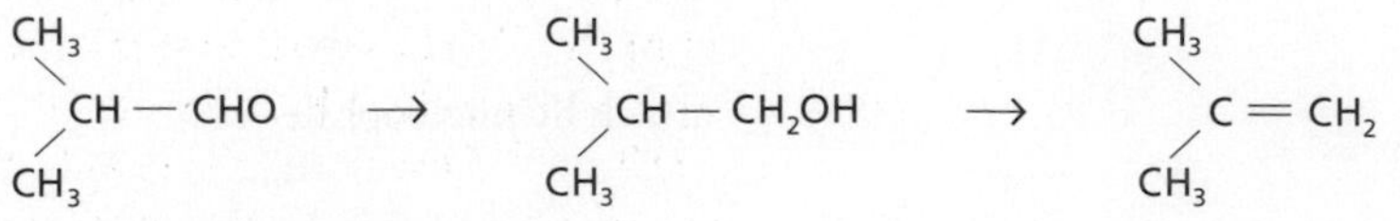

C 2-methylpropanal **D** 2-methylpropan-1-ol 2-methylpropene

Further practice Describe a test (a different one in each case) that you could use to distinguish between the following pairs of compounds:

1 CH_3CH_2CHO and CH_3COCH_3
2 $CH_3CH_2COCH_2CH_3$ and $CH_3CH_2CH_2COCH_3$
3 $CH_2{=}CH{-}CH_2OH$ and CH_3CH_2CHO.

28.4 Preparing carbonyl compounds

From alcohols, by oxidation

See Chapter 26, page 485, for details of this reaction. For example:

$$CH_3CH_2CH_2OH + [O] \xrightarrow[\text{heat}]{K_2Cr_2O_7 + H_2SO_4(aq)} CH_3CH_2CHO + H_2O$$

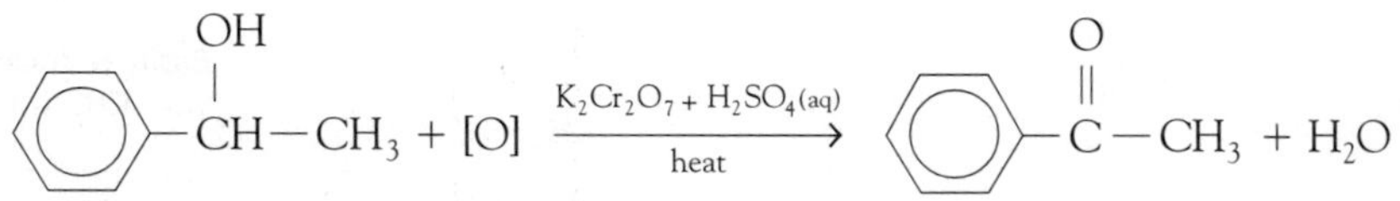

To prevent further oxidation to the acid, the oxidant is added slowly to an excess of the alcohol, and the aldehyde is distilled off as it is formed.

From arenes and acyl chlorides by the Friedel–Crafts reaction

See Chapter 27, page 506, for details of this reaction. For example:

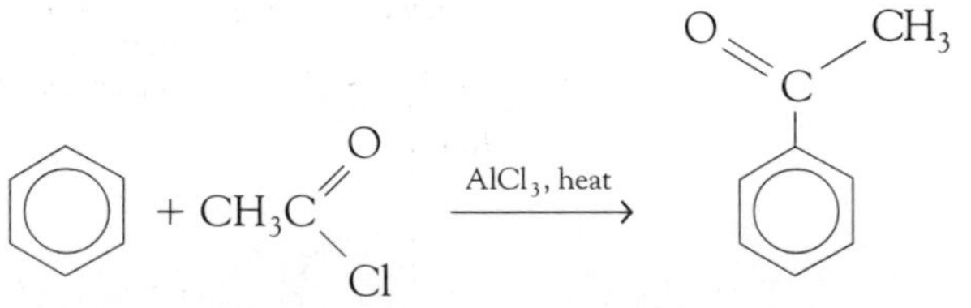

Note that it is not possible to prepare aldehydes from carboxylic acids using lithium tetrahydridoaluminate(III) (see Chapter 29, page 547). Aldehydes are not only more easily oxidised than alcohols (see Chapter 26, page 485), but are also more easily reduced than carboxylic acids. Therefore the reduction goes all the way to the alcohol, and it is not possible to stop at the half-way stage.

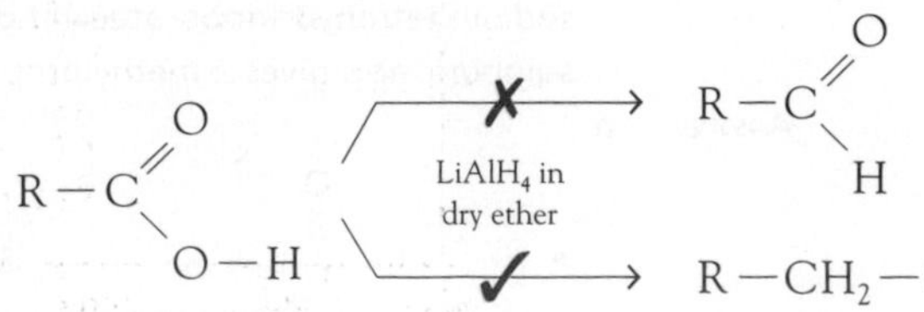

(To obtain the free alcohol, acid must be added after the $LiAlH_4$ in dry ether.)

28.5 The infrared spectra of carbonyl compounds

The polar $C^{\delta+}{=}O^{\delta-}$ bond is strongly infrared active. Because the bond is stronger than the C—O single bond, we expect the absorbance to be at a higher frequency, which indeed it is. Ketones absorb at 1715 cm^{-1}, and aldehydes a little higher, at 1730 cm^{-1}. Overlap of the C=O π bond with an adjacent aromatic ring causes the frequency to fall (see Table 28.6).

■ **Table 28.6**
IR absorption frequencies for some carbonyl compounds.

Compound	Formula	Frequency of absorption (wavenumber)/cm^{-1}
Ethanal	CH_3CHO	1730
Propanone	CH_3COCH_3	1715
Benzaldehyde	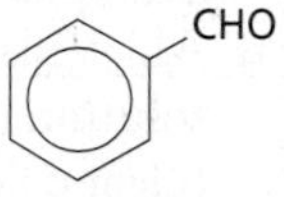	1705
Phenylethanone	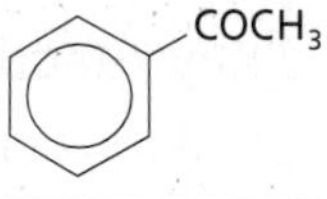	1695

■ **Figure 28.8**
IR spectra of **a** cyclohexane and **b** cyclohexanone.

a cyclohexane

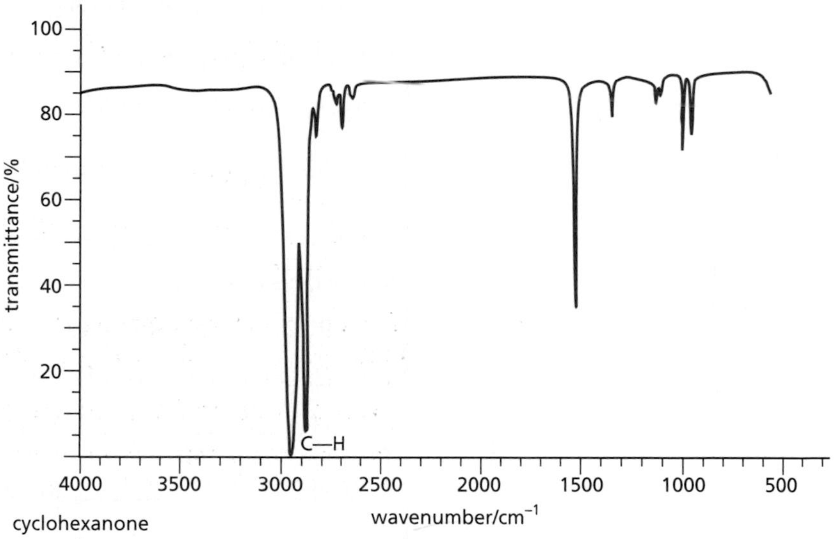

b cyclohexanone

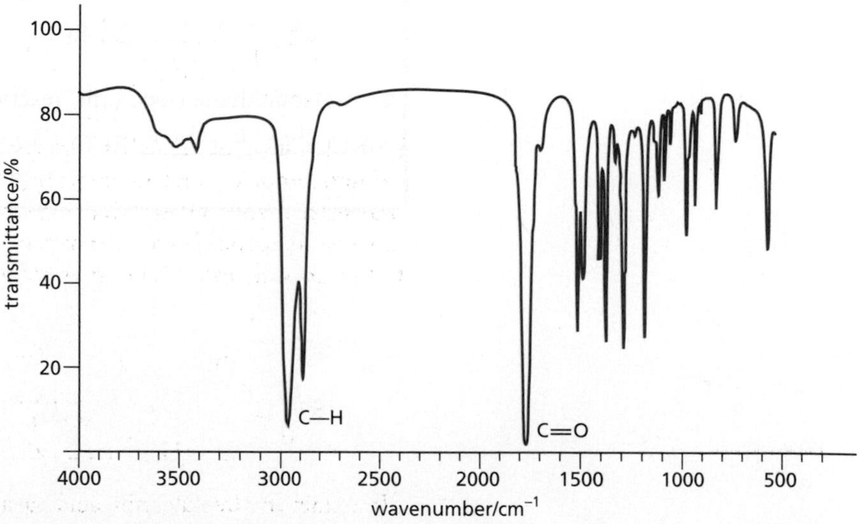

Summary

- The carbonyl group has the structure:

$$>C{=}O$$

- **Carbonyl compounds** comprise compounds in which the carbonyl group is not associated with other hetero-atoms in the same functional group. There are two types of carbonyl compound – **ketones** and **aldehydes**.
- Their characteristic reaction is nucleophilic addition.
- They can be reduced to alcohols, and aldehydes can be oxidised to carboxylic acids.
- They undergo condensation reactions.
- The condensation reaction with 2,4-dinitrophenylhydrazine, giving an orange precipitate, is a good test for the presence of a carbonyl compound in a sample.
- Aldehydes can be distinguished from ketones by their effects on **Fehling's solution** (deep blue solution→red-brown precipitate) and **Tollens' reagent** (clear colourless solution→silver mirror).

Key reactions you should know

(All reactions undergone by R_2CO are also undergone by RCHO.)

- Catalytic hydrogenation:

$$R_2CO + H_2 \xrightarrow{Ni} R_2CH(OH)$$

- Nucleophilic additions:

$$R_2CO \xrightarrow{NaBH_4} R_2CH(OH)$$

$$R_2CO + HCN \xrightarrow{NaCN} R_2C(OH)CN$$

$$R_2CO + CH_3MgBr \xrightarrow{\text{(followed by } H_3O^+\text{)}} R_2C(OH)CH_3$$

- Condensation reaction:

$$R_2CO + H_2N{-}R' \rightarrow R_2C{=}N{-}R' + H_2O \quad (H_2N{-}R' = \text{2,4-DNPH})$$

- Oxidation reactions (RCHO only):

$$RCHO \xrightarrow{Cr_2O_7^{2-} + H_3O^+/\text{heat}} RCO_2H$$

$$RCHO \xrightarrow{\text{Fehling's solution } (Cu^{2+})} RCO_2^- + \underset{\text{(red ppt)}}{Cu_2O}$$

$$RCHO \xrightarrow{\text{Tollens' reagent } (Ag^+)} RCO_2^- + \underset{\text{(silver mirror)}}{Ag}$$

- Triiodomethane (iodoform) reaction:

$$RCOCH_3 \xrightarrow{I_2 + OH^-(aq)} RCO_2^- + \underset{\text{(pale yellow precipitate)}}{CHI_3}$$

29 Carboxylic acids and their derivatives

Functional groups:

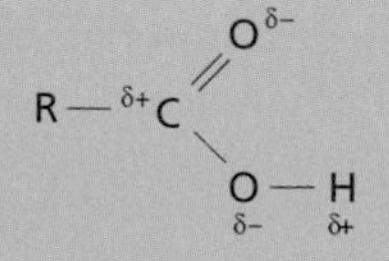

carboxylic acids

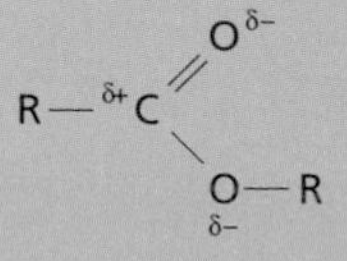

esters

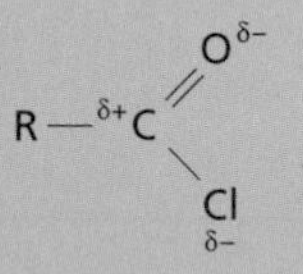

acyl chlorides

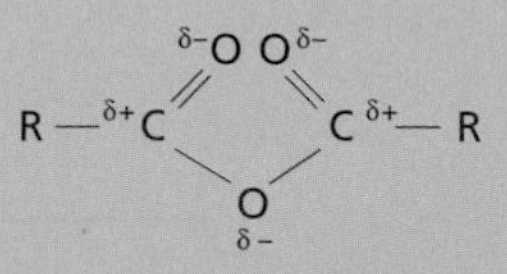

acid anhydrides

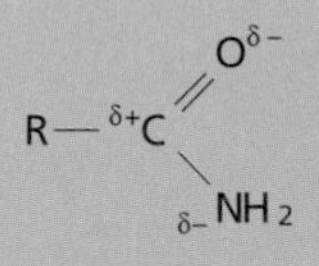

amides

In this chapter we look at the major class of organic acids, the carboxylic acids, and the compounds derived from them. Apart from acid–base reactions, their chemistry is dominated by nucleophilic substitution reactions, due to the δ+ carbon atom in the carbonyl group.

The chapter is divided into the following sections.

29.1 Introduction

When the carbonyl group is directly joined to an oxygen atom, the **carboxyl group** is formed. This occurs in **carboxylic acids**, **esters** and **acid anhydrides**.

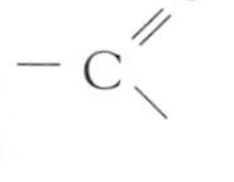

carbonyl group

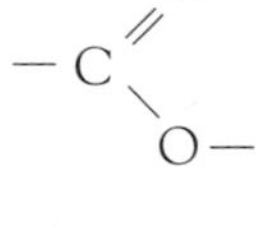

carboxyl group

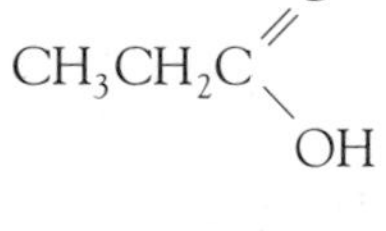

carboxylic acid:
propanoic acid

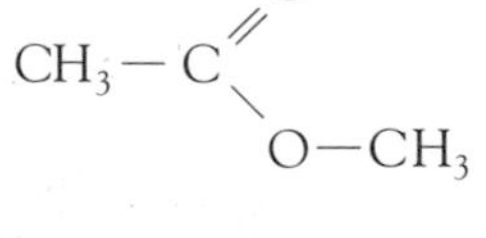

ester:
methyl ethanoate

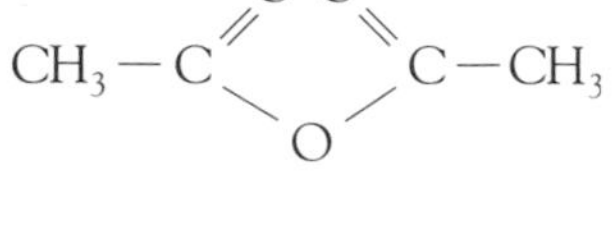

acid anhydride:
ethanoic anhydride

The reactions of the carbonyl group are drastically changed by the presence of the electronegative oxygen atom. These compounds have virtually none of the reactions of carbonyl compounds as described in Chapter 28. The same is true of two other classes of compound in which the carbonyl group is directly attached to an electronegative atom, namely the **acyl chlorides** and the **amides**:

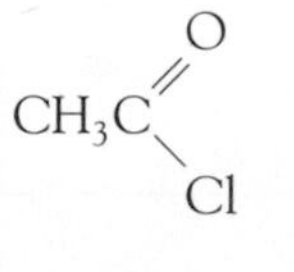

acyl chloride:
ethanoyl chloride

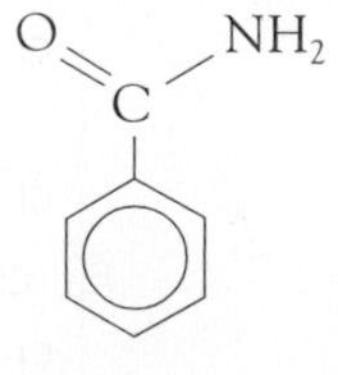

amide:
benzamide (benzenecarboxamide)

The predominant mode of reaction of the acid derivatives esters, acid anhydrides, acyl chlorides and amides is by nucleophilic substitution. For carboxylic acids, however, the reactivity is dominated by the tendency of the O—H bond to ionise to give hydrogen ions, hence the incorporation of the word 'acid' in their name. The extent of ionisation is small, however. For example, in a 1.0 mol dm^{-3} solution of ethanoic acid, about one molecule in 10^3 is ionised.

$$CH_3CO_2H \rightleftharpoons CH_3CO_2^- + H^+$$

Carboxylic acids are therefore classed as **weak acids** (see Chapter 6).

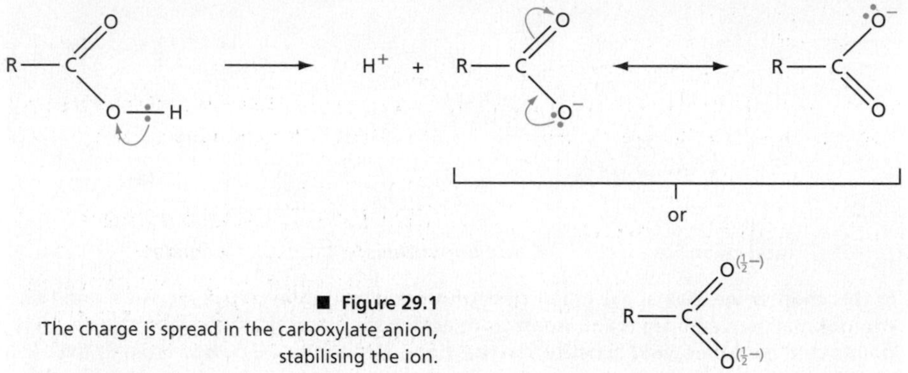

■ **Figure 29.1**
The charge is spread in the carboxylate anion, stabilising the ion.

The O—H bond breaks heterolytically far more easily in carboxylic acids than in alcohols or in phenols (see Chapter 27, page 517). This is because the anion that results from bond breaking is stabilised. The negative charge formed by heterolysis can be delocalised over two electronegative oxygen atoms. This spreading out of charge invariably leads to a stabilisation of the ion (see Figure 29.1).

Atoms or groups that draw electrons away from the $—CO_2^-$ group will help the anion to form, and this causes the acid to be more dissociated (that is, to become a stronger acid). On the other hand, groups that donate electrons will cause the acid to become weaker (see Table 29.1).

Electron-donating groups decrease the acid strength of carboxylic acids, whereas electron-withdrawing groups increase their acid strength.

■ **Table 29.1**
The acidity of some carboxylic acids. Remember from Chapter 12 that:

$$pK_a = -\log_{10} K_a \text{ where } K_a = \frac{[H^+][A^-]}{[HA]}$$

for acids undergoing the dissociation $HA \rightleftharpoons H^+ + A^-$.

Formula of acid	**pK_a**	**% dissociation in 1.0 mol dm^{-3} aqueous solution**
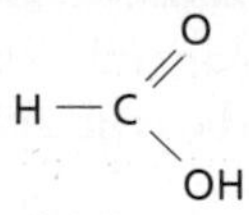	3.75	1.3%
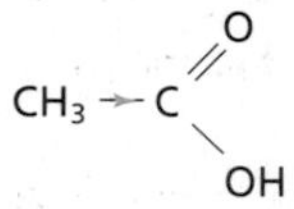	4.76	0.42%
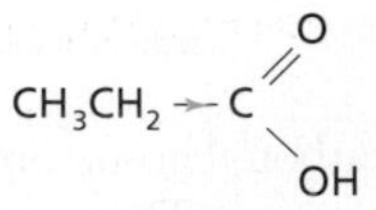	4.87	0.36%
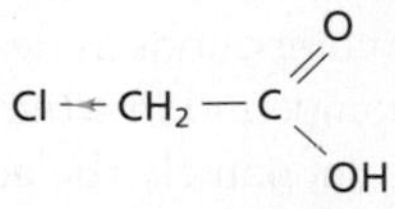	2.87	3.7%
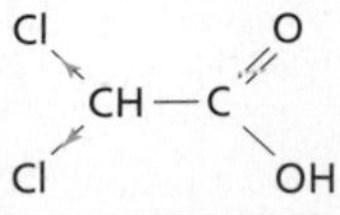	1.26	21%
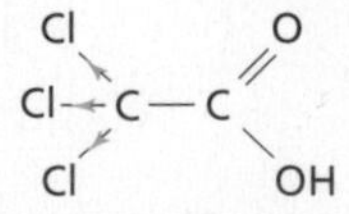	0.66	59%

Example The pK_a of fluoroethanoic acid is 2.57. Does this mean it is a stronger or a weaker acid than chloroethanoic acid? What pK_a would you expect of difluoroethanoic acid?

Answer The lower the pK_a, the larger is K_a. This means that the acid is more dissociated, and therefore stronger. Hence fluoroethanoic acid is a stronger acid than chloroethanoic acid. This is due to the greater electron-withdrawing ability of the highly electronegative fluorine atom. Difluoroethanoic acid would be expected to have a lower pK_a than dichloroethanoic acid – about 1.0.

Further practice

1 Use the data in Table 29.1 to predict the pK_a values of:
 a $(CH_3)_3C—CO_2H$
 b $Cl—CH_2—CH_2—CO_2H$.
 Explain your reasoning.

2 Explain why the pK_a values of 2-chlorobutanoic acid ($pK_a = 2.86$) and 4-chlorobutanoic acid ($pK_a = 4.53$) differ so much. Predict the pK_a value for 3-chlorobutanoic acid.

Formula of acid	pK_a	% dissociation
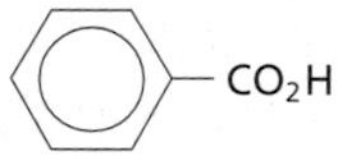	4.20	0.80%
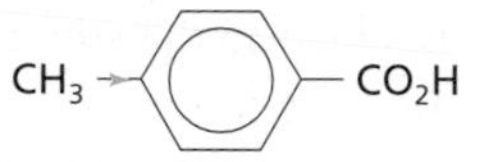	4.37	0.73%
	3.99	1.0%

■ **Table 29.2**
The effect on pK_a for benzoic acid of electron-donating and electron-withdrawing groups.

The effect of electron-donating or electron-withdrawing groups is seen even when such groups are situated on the opposite side of a benzene ring to the $—CO_2H$ group (see Table 29.2). The carboxylic acid group is a strongly hydrogen-bonded one, both to carboxylic acid groups on other carboxylic acid molecules and to solvent molecules such as water. This has an effect on two important properties – boiling points and solubilities. Table 29.3 compares these properties for carboxylic acids and the corresponding alcohols.

Number of carbon atoms	Alcohol			Carboxylic acid		
	Formula	Boiling point/°C	Solubility in 100 g of water/g	Formula	Boiling point/°C	Solubility in 100 g of water/g
1	CH_3OH	65	∞	HCO_2H	101	∞
2	CH_3CH_2OH	78	∞	CH_3CO_2H	118	∞
3	$CH_3CH_2CH_2OH$	97	∞	$CH_3CH_2CO_2H$	141	∞
4	$CH_3CH_2CH_2CH_2OH$	118	7.9	$CH_3CH_2CH_2CO_2H$	164	∞
5	$CH_3CH_2CH_2CH_2CH_2OH$	138	2.3	$CH_3CH_2CH_2CH_2CO_2H$	187	3.7

■ **Table 29.3**
Hydrogen bonding increases the boiling points and solubilities in carboxylic acids.

Both in the pure liquid state, and in solution in non-hydrogen-bonding solvents such as benzene, carboxylic acids can form hydrogen-bonded dimers (see Figure 29.2).

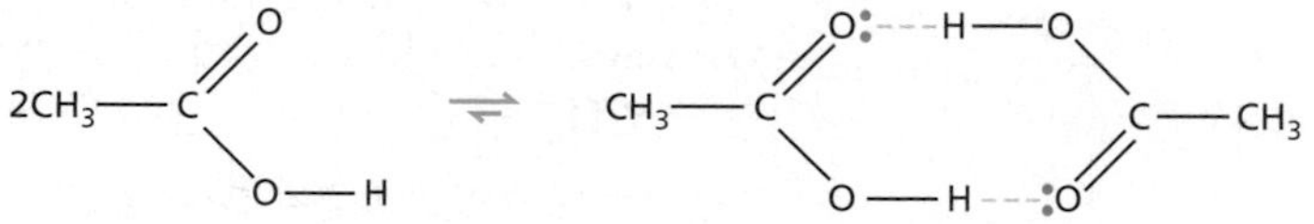

■ **Figure 29.2**
Carboxylic acids dimerise as two molecules hydrogen bond to each other.

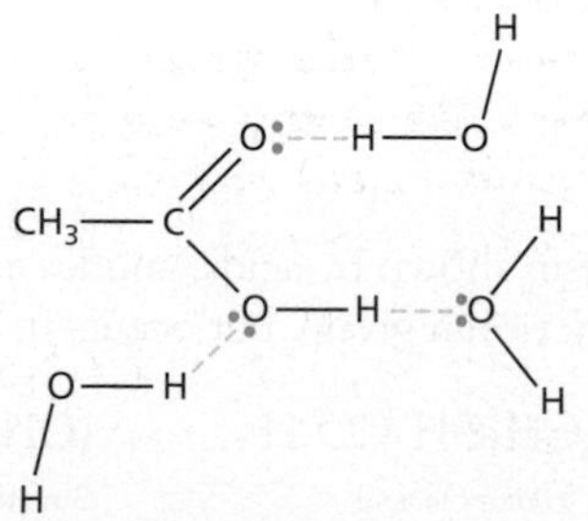

■ **Figure 29.3**
Carboxylic acids hydrogen bond to water molecules.

In water, both oxygen atoms and the —OH hydrogen atom can hydrogen bond with the solvent (see Figure 29.3).

29.2 Isomerism and nomenclature

Carboxylic acids are named by finding the longest carbon chain that contains the acid functional group, and adding the suffix '-oic acid' to the stem. Side groups off the chain are named in the usual way:

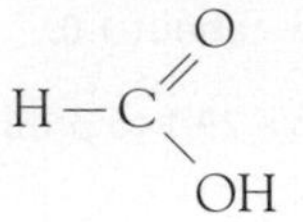

methanoic acid

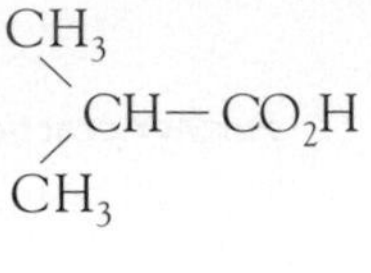

2-methylpropanoic acid

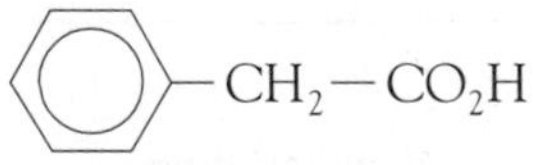

phenylethanoic acid

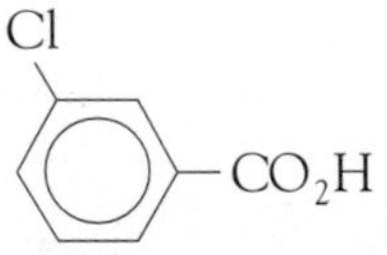

3-chlorobenzoic acid
(3-chlorobenzenecarboxylic acid)

Acyl chlorides are named similarly. The suffix '-oyl chloride' is added to the stem:

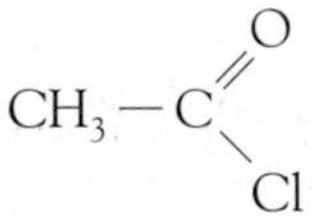

ethanoyl chloride

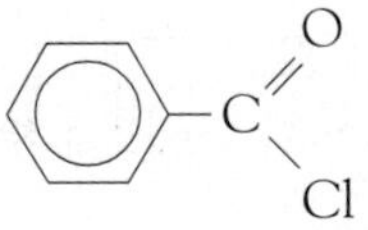

benzoyl chloride
(benzenecarbonyl chloride)

Acid amides are similarly named, adding the suffix 'amide' to the stem.

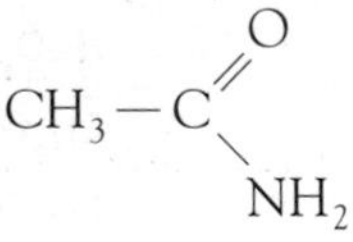

ethanamide

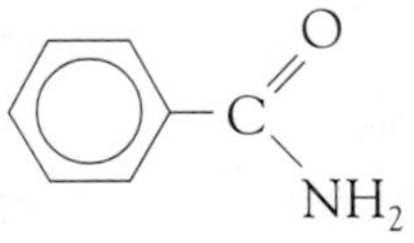

benzamide
(benzenecarboxamide)

Esters are named as alkyl derivatives of acids, similar to the naming of salts:

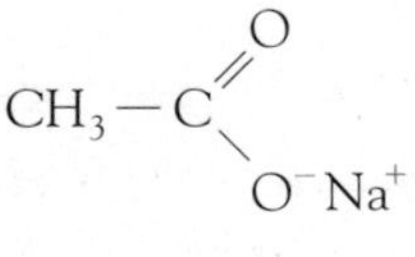

sodium ethanoate

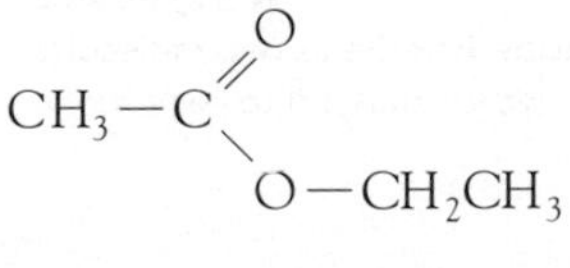

ethyl ethanoate

The isomerism shown by acids, amides and acid chlorides is not usually associated with the functional group, but occurs in the carbon chain to which it is attached.

$CH_3CH_2CH_2CO_2H$	$(CH_3)_2CHCO_2H$
butanoic acid	2-methylpropanoic acid

Esters, however, can show a particular form of isomerism, depending on the number of carbon atoms they have in the acid or alcohol part of the molecule. Two isomers of $C_4H_8O_2$ are shown below. Note that although the acid part is usually written first in a formula, by convention, it appears last in the name:

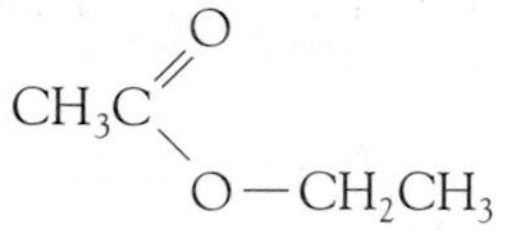

ethylethanoate

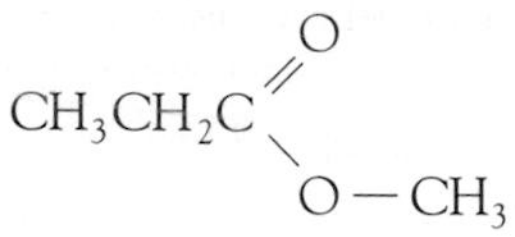

methyl propanoate

Example Draw the structures of the other two other isomeric esters with the formula $C_4H_8O_2$, and name them.

Answer

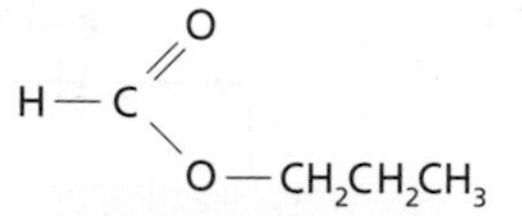

prop-1-yl methanoate

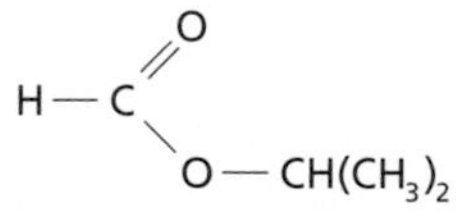

prop-2-yl methanoate

Further practice Draw and name possible ester isomers with the formula $C_5H_{10}O_2$. How many ester isomers are possible with this formula? Indicate in your answer which of the esters are chiral.

29.3 Reactions of carboxylic acids

Reactions of the O—H group

The acidity of carboxylic acids has already been mentioned. They show most of the typical reactions of acids.

With metals

Ethanoic acid reacts with sodium metal, liberating hydrogen gas (compare with alcohols, Chapter 26, page 481, and phenols, Chapter 27, page 517):

$$CH_3CO_2H + Na \rightarrow CH_3CO_2^-Na^+ + \tfrac{1}{2}H_2$$

They also react with other reactive metals, such as calcium and magnesium.

With sodium hydroxide

Like phenols, but unlike alcohols, carboxylic acids form salts with alkalis:

$$C_3CO_2H + NaOH \rightarrow CH_3CO_2^-Na^+ + H_2O$$

Salts are often much more soluble in water than the carboxylic acids from which they are derived. For example, soluble aspirin tablets contain the calcium salt of ethanoylsalicylic acid (see Chapter 27, page 521):

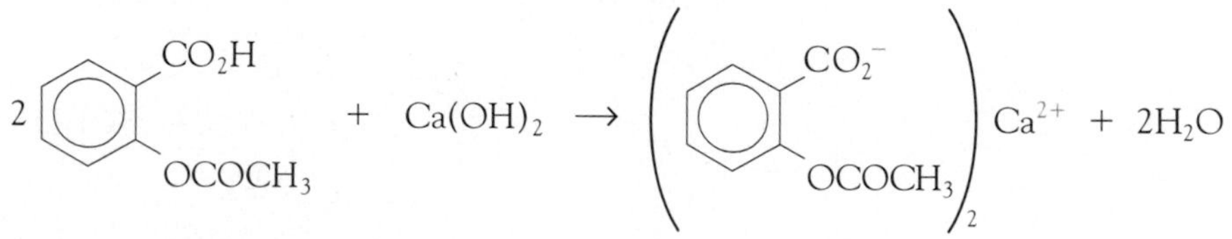

With sodium carbonate and sodium hydrogencarbonate

Like the usual bench acids, but unlike both alcohols and phenols, carboxylic acids will react with carbonates, liberating carbon dioxide gas:

$$2CH_3CO_2H + Na_2CO_3 \rightarrow 2CH_3CO_2^-Na^+ + CO_2 + H_2O$$

A similar reaction occurs with sodium hydrogencarbonate:

$$CH_3CO_2H + NaHCO_3 \rightarrow CH_3CO_2^-Na^+ + CO_2 + H_2O$$

These three reactions form the basis of a series of tests by which alcohols, phenols and carboxylic acids can be distinguished from each other, and from other functional groups (see Figure 29.4).

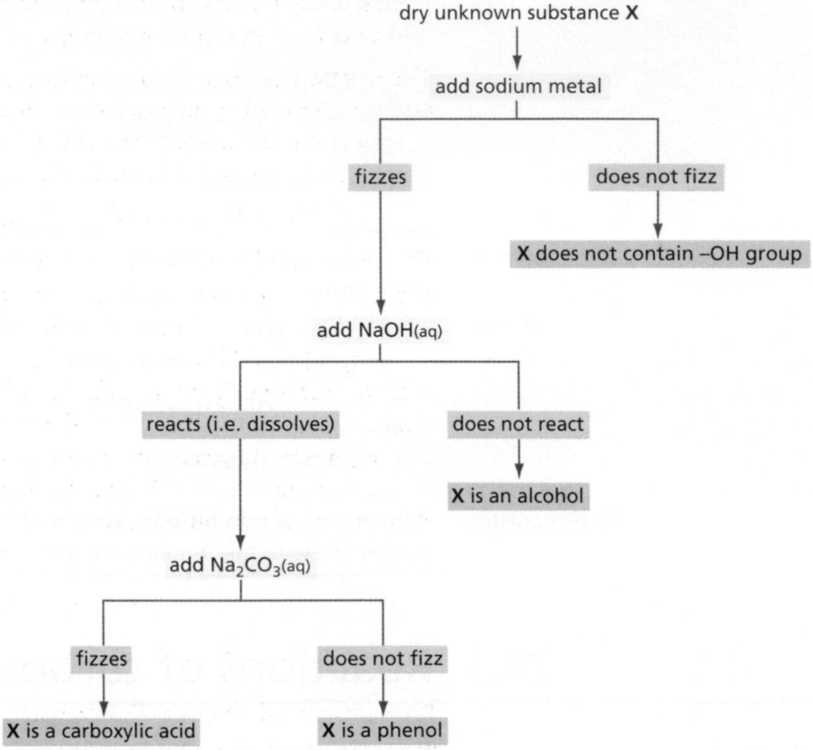

■ **Figure 29.4** How to distinguish between alcohols, phenols and carboxylic acids.

Acids, phenols and alcohols can be distinguished by their reactions with sodium metal, aqueous sodium hydroxide and aqueous sodium carbonate.

Reactions of the C—OH group

There are two main reactions which replace the —OH group in carboxylic acids with other atoms or groups:

With phosphorus(V) chloride or disulphur dichloride oxide

Carboxylic acids react with these reagents in exactly the same way as do alcohols (see Chapter 26, page 482), and according to the same mechanism. Acyl chlorides are produced:

$$CH_3C{-}C(=O)OH + PCl_5 \rightarrow CH_3C(=O)Cl + HCl + PCl_3O$$

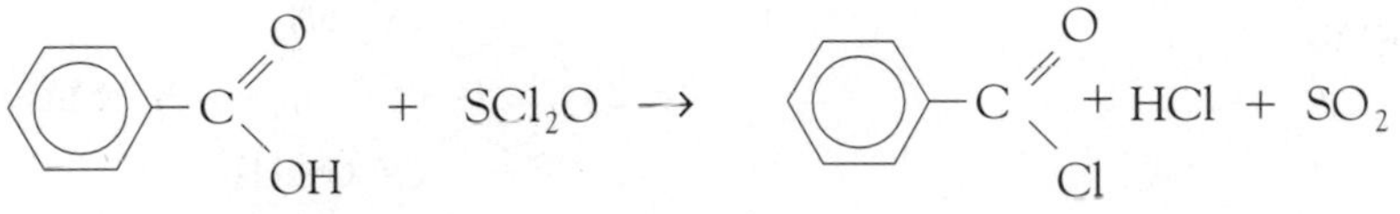

With alcohols in the presence of anhydrous acids

Carboxylic acids react with alcohols (but not with phenols) in an acid-catalysed equilibrium reaction to give esters:

$$CH_3C(=O)OH + CH_3CH_2OH \rightleftharpoons CH_3C(=O)O{-}CH_2CH_3 + H_2O \quad (1)$$

Typical reaction conditions are:

- heat under reflux an equimolar mixture of carboxylic acid and alcohol with 0.1 mole equivalent of concentrated sulphuric acid for 4 hours, or
- heat under reflux a mixture of carboxylic acid and alcohol, whilst passing through the mixture dry hydrogen chloride gas for 2 hours.

The reaction needs an acid catalyst, and like any equilibrium reaction it can be made to go in either direction depending on the conditions. To encourage esterification, an excess of one of the reagents (often the alcohol) is used, and the water is removed as it is formed (often by the use of concentrated sulphuric acid):

$$H_2SO_4 + H_2O \rightarrow H_3O^+ + HSO_4^-$$

This drives equilibrium (1) over to the right (see Chapter 9, page 177). To encourage the reverse reaction, the hydrolysis of the ester, the ester is heated under reflux with a dilute solution of sulphuric acid. This provides an excess of water to drive equilibrium (1) over to the left.

If an ester is made from an alcohol whose —OH group is labelled with an oxygen-18 atom, it is found that the ^{18}O is 100% incorporated into the ester. None of it appears in the water that is also produced. Therefore the water must come from combining the —OH from the carboxylic acid and the —H from the alcohol, rather than vice versa.

Reduction of the —CO_2H group

The powerful reducing agent lithium tetrahydridoaluminate(III) will reduce a carboxylic acid to an alcohol:

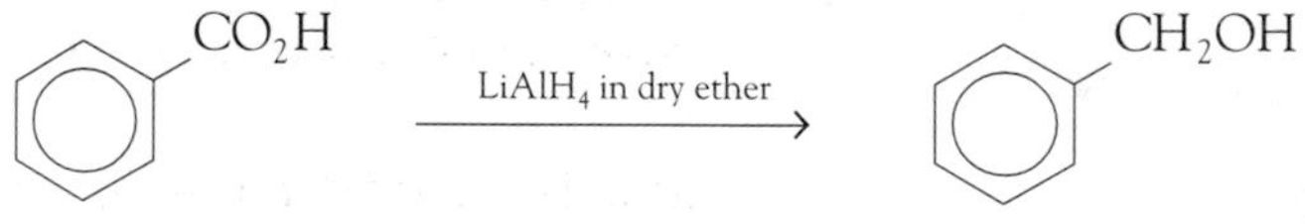

benzoic acid

benzyl alcohol (phenylmethanol)

Other common reducing agents, such as hydrogen and nickel, or sodium tetrahydridoborate(III) will not reduce carboxylic acids.

How it happens (1): the mechanism of esterification/hydrolysis

In a strongly acidic medium, both the alcohol and the carboxylic acid can undergo reversible protonation. However, only the protonated carboxylic acid undergoes further reaction.

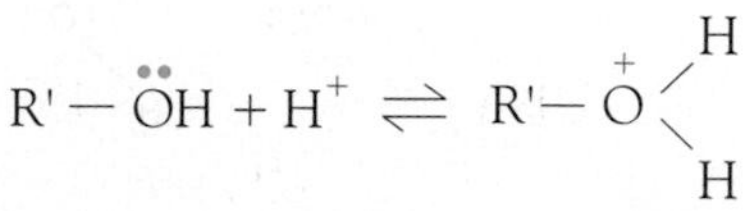

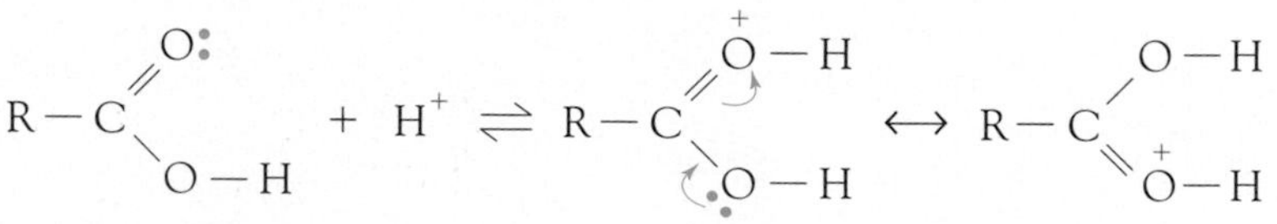

(I)

The protonation of the carboxylic acid in step (I) produces a highly δ+ carbon atom, which can undergo nucleophilic attack by a lone pair of electrons on the oxygen atom of an alcohol molecule:

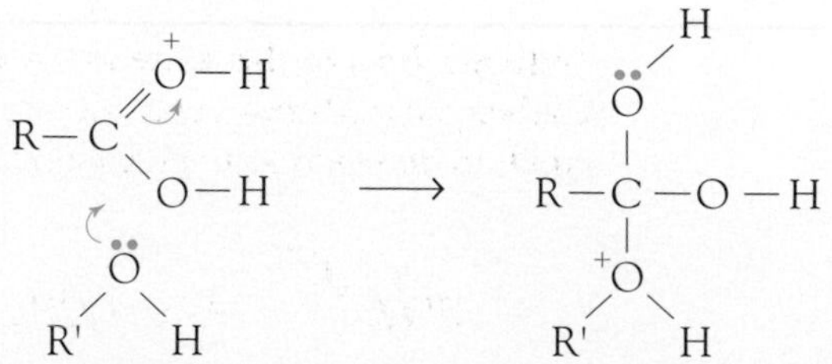

(II)

(continued)

The three oxygen atoms in this intermediate cation are very similar in basicity. Any one of them could be attached to the proton, and proton transfer from one to another can occur readily:

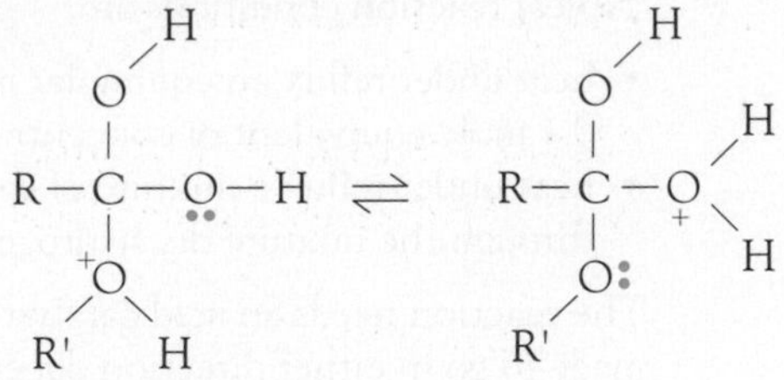

(III)

This new cation can undergo a carbon–oxygen bond breaking, which is essentially the reverse of reaction (II), to give a molecule of water, along with the protonated ester:

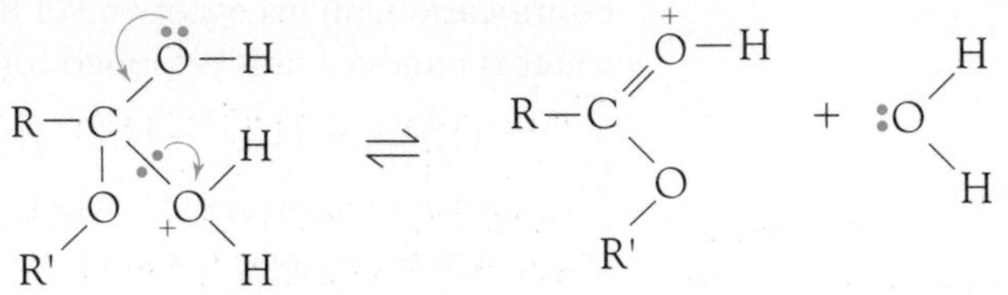

(IV)

Lastly, the cation loses a proton, to regenerate the H^+ catalyst, and form the ester product:

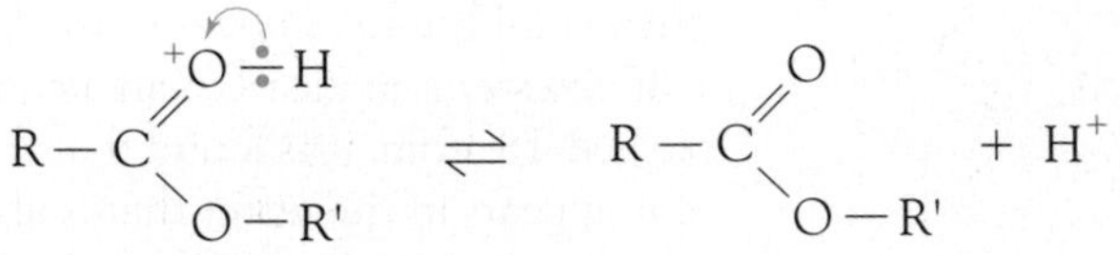

(V)

As mentioned above, the reaction is reversible. So is the mechanism. With an excess of water, esters undergo acid-catalysed hydrolysis.

Example Draw the mechanism for the acid-catalysed hydrolysis of methyl methanoate.

Answer

■ Figure 29.5

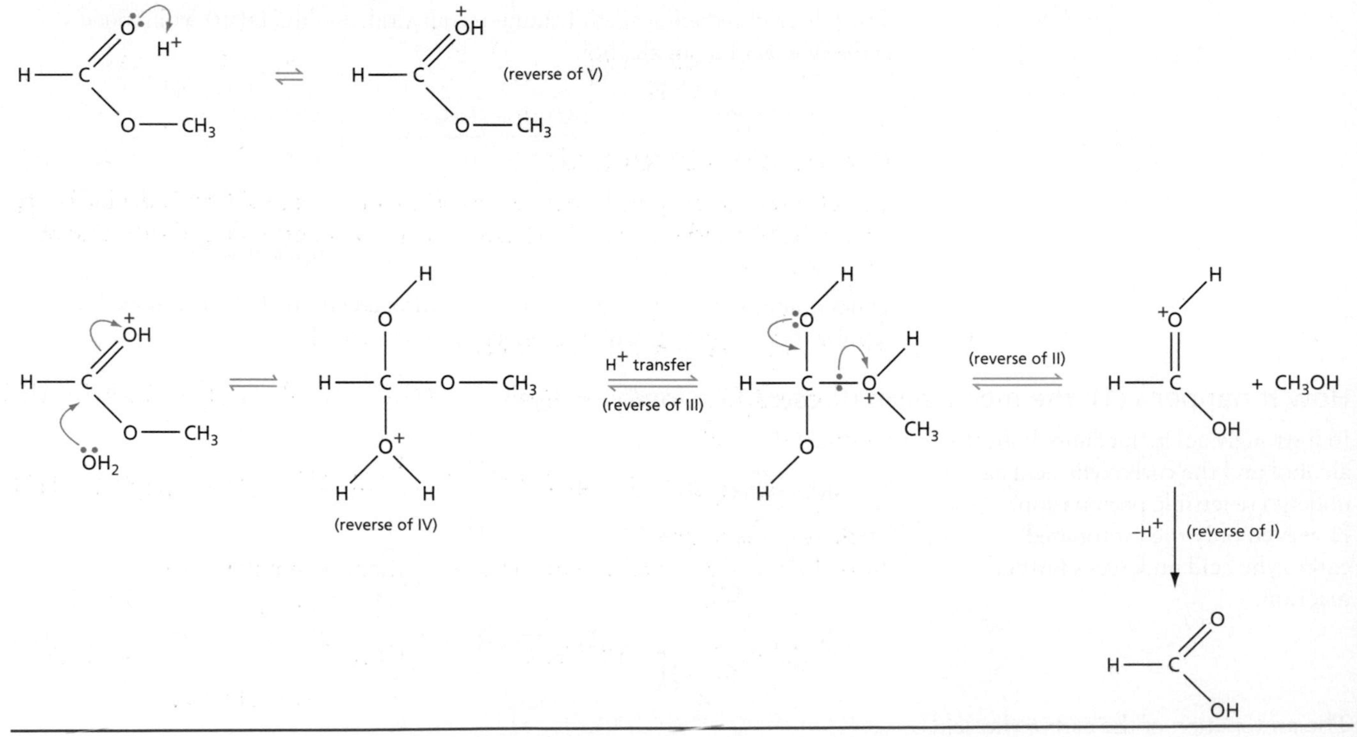

29.4 Acyl chlorides

Properties of acyl chlorides

Acyl chlorides have the functional group:

$$R — \overset{\delta+}{C} \overset{\overset{\displaystyle \overset{\delta-}{O}}{\diagup\!\diagdown}}{\underset{\underset{\delta-}{Cl}}{}}$$

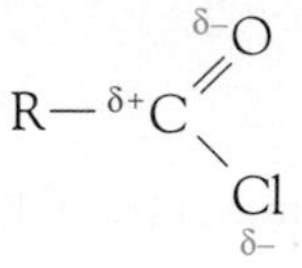

They are usually liquids which fume in moist air. They are immiscible with water, but react slowly with it and eventually dissolve (see below). They are not hydrogen bonded, their main intermolecular attractions being a combination of van der Waals forces and dipole–dipole forces (see Figure 29.6).

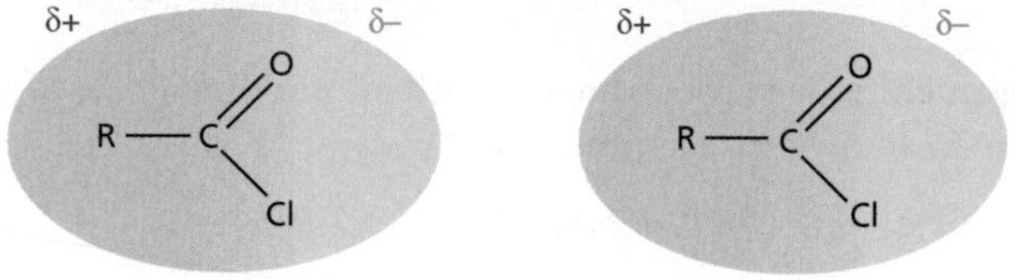

■ **Figure 29.6**
Dipole–dipole attractions in acyl chlorides.

The extra electron-withdrawing effect of the carbonyl group has the effect of increasing the boiling point by about 10–15 °C compared with the boiling point of the halogenoalkane with similar shape, due to extra dipole–dipole attractions (see Table 29.4).

■ **Table 29.4**
Strong dipole–dipole attractions increase the boiling points in acyl chlorides.

Chloroalkane	M_r	Boiling point °C	Acyl chloride	M_r	Boiling point/°C	Difference/ °C
[structure with Cl]	78.5	37	[structure with O and Cl]	78.5	52	15
[structure with Cl]	92.5	68	[structure with O and Cl]	92.5	80	12

Reactions of acyl chlorides

The electronegativity of the oxygen, and the easily polarised C=O double bond, have a dramatic effect on the reactivity of acyl chlorides compared with that of chloroalkanes.

With water

Acyl chlorides react readily with water:

$$CH_3 — COCl + H_2O \xrightarrow{\text{room temperature, complete within minutes}} CH_3 — COOH + HCl$$

Contrast this with:

$$CH_3 — CH_2Cl + H_2O \xrightarrow{\text{heat under pressure at 100 °C for 14 days}} CH_3 — CH_2OH + HCl$$

With alcohols or phenols

Acyl chlorides react readily with alcohols or phenols, forming esters:

$$CH_3C\overset{\displaystyle O}{\underset{Cl}{\diagup\!\diagdown}} + HOCH_2CH_3 \rightarrow CH_3C\overset{\displaystyle O}{\underset{O—CH_2CH_3}{\diagup\!\diagdown}} + HCl$$

$$CH_3C\overset{\displaystyle O}{\underset{Cl}{\diagup\!\diagdown}} + HO—\hexagon \rightarrow CH_3C\overset{\displaystyle O}{\underset{O—\hexagon}{\diagup\!\diagdown}} + HCl$$

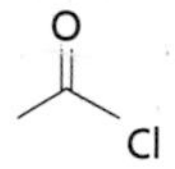

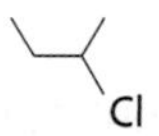

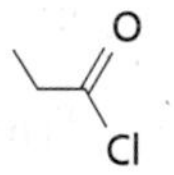

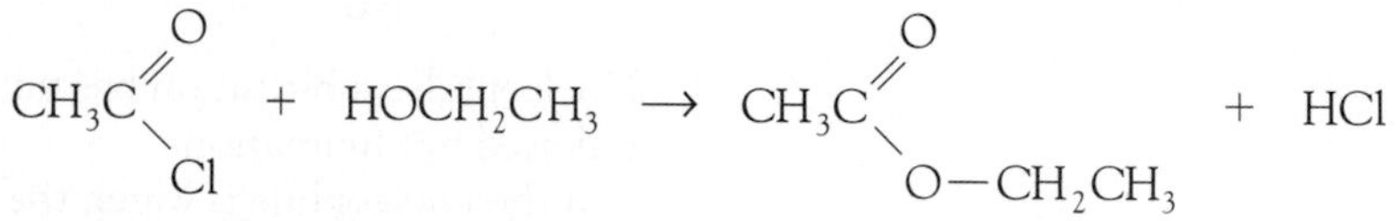

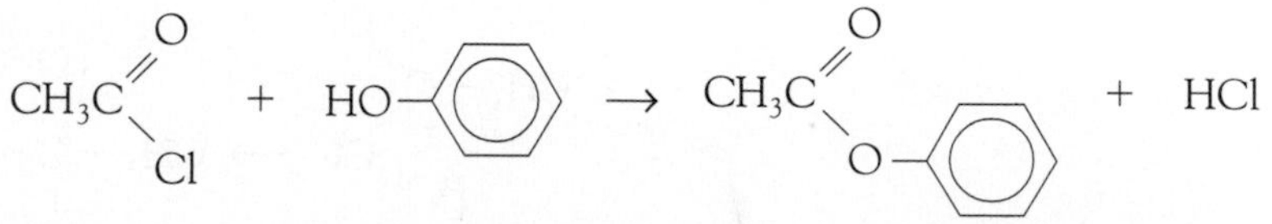

Because phenols do not react directly with carboxylic acids, this is the only method for preparing phenyl esters (see Chapter 27, page 518).

With ammonia and amines

Acyl chlorides react vigorously with ammonia, forming amides:

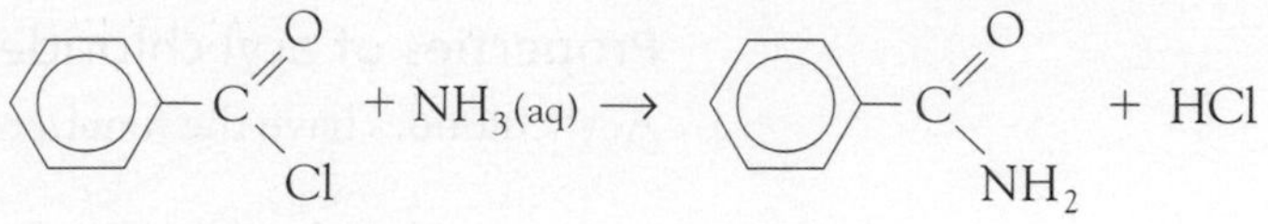

benzamide

and with primary amines (see Chapter 30), forming substituted amides:

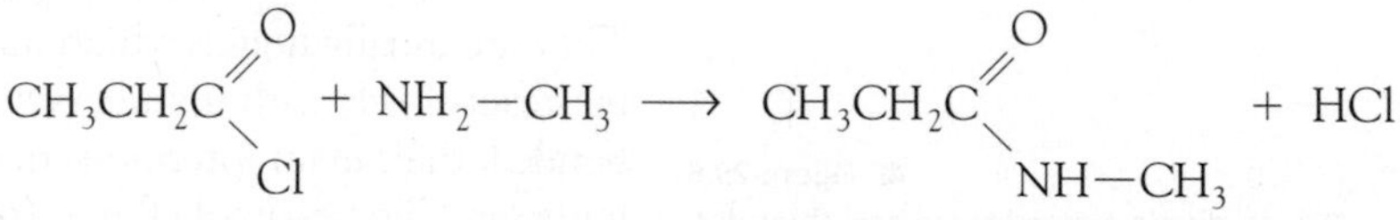

N-methylpropanamide

The hydrogen chloride produced reacts with another molecule of amine or ammonia in an acid–base reaction:

$$R{-}NH_2 + HCl \rightarrow R{-}NH_3^+Cl^-$$

Therefore an excess of amine or ammonia is used to ensure complete reaction.

With salts of carboxylic acids

Acyl chlorides react with salts of carboxylic acids forming acid anhydrides:

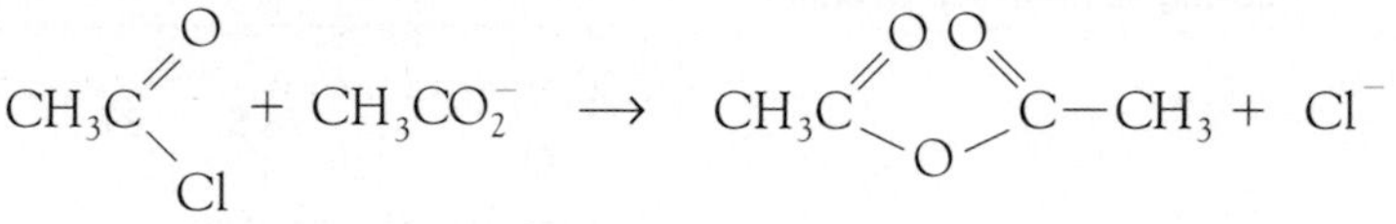

The mechanism of the reactions of acyl chlorides

The carbonyl group in acyl chlorides can undergo nucleophilic addition in a similar manner to carbonyl compounds:

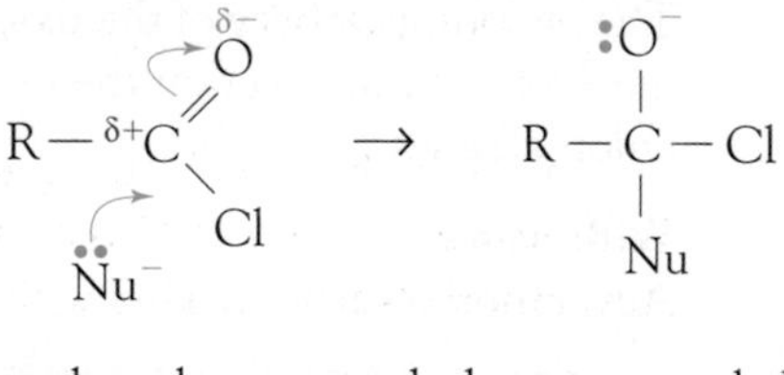

Unlike carbonyl compounds, however, acyl chlorides are provided with an easily removed leaving group, the chloride ion:

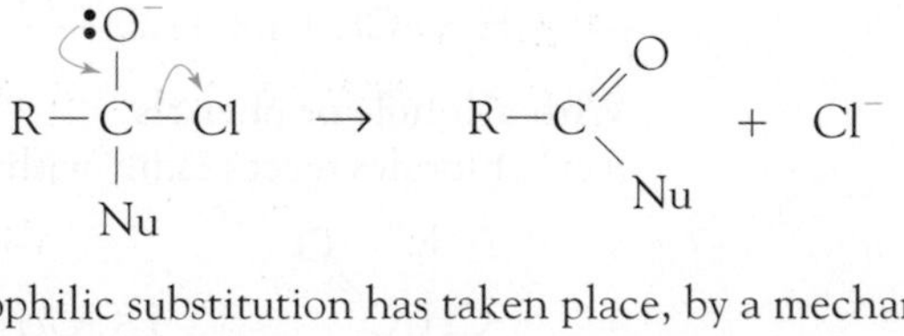

Nucleophilic substitution has taken place, by a mechanism involving addition, followed by elimination.

If the nucleophile is water, the carboxylic acid is formed:

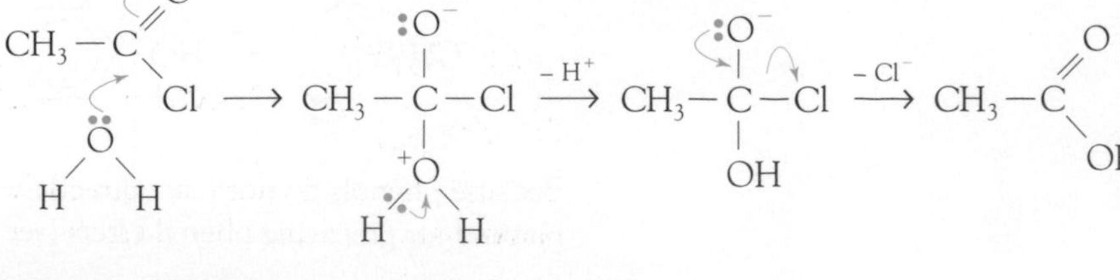

Example   Draw the mechanism for the reaction between ammonia and benzoyl chloride:

$$\text{C}_6\text{H}_5-\text{COCl} + 2\text{NH}_3 \longrightarrow \text{C}_6\text{H}_5-\text{CONH}_2 + \text{NH}_4\text{Cl}$$

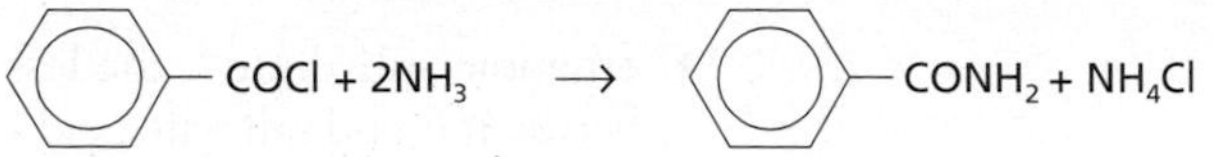

Answer

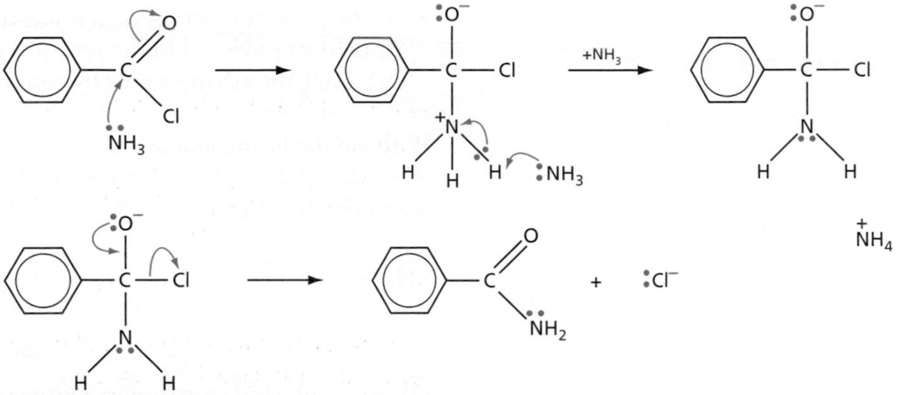

■ **Figure 29.7**

Further practice   Predict the products of, and suggest a mechanism for, the reaction between propan-1-ol and propanoyl chloride.

29.5 Acid anhydrides

Acid anhydrides have the functional group:

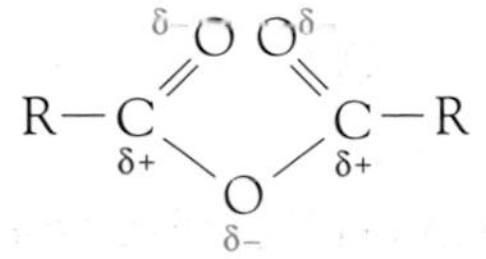

They are prepared from carboxylic acids by dehydration with phosphorus(V) oxide:

$$2\text{CH}_3\text{CO}_2\text{H} \xrightarrow[-\text{H}_2\text{O}]{\text{P}_2\text{O}_5} \text{CH}_3-\text{C}(\text{O})-\text{O}-\text{C}(\text{O})-\text{CH}_3$$

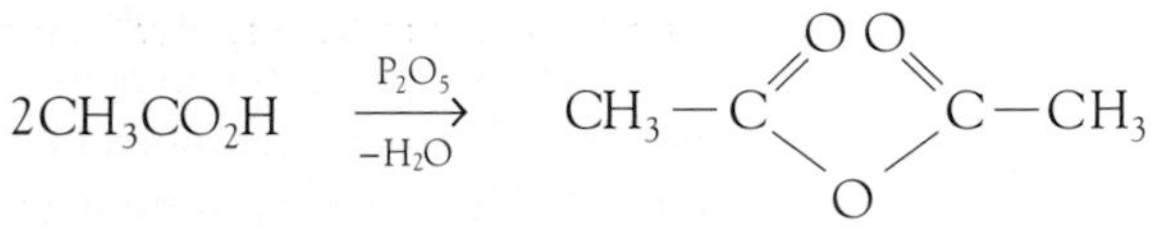

The water is absorbed by the phosphorus(V) oxide, forming phosphoric(V) acid.
   Acid anhydrides can also be made by reacting an acyl chloride with the salt of a carboxylic acid.

Reactions of acid anhydrides

They have similar reactions to those of acyl chlorides, but are less reactive.

With water
Acid anhydrides are slowly hydrolysed by water:

$$\text{CH}_3-\text{C}(\text{O})-\text{O}-\text{C}(\text{O})-\text{CH}_3 + \text{H}_2\text{O} \longrightarrow 2\text{CH}_3\text{CO}_2\text{H}$$

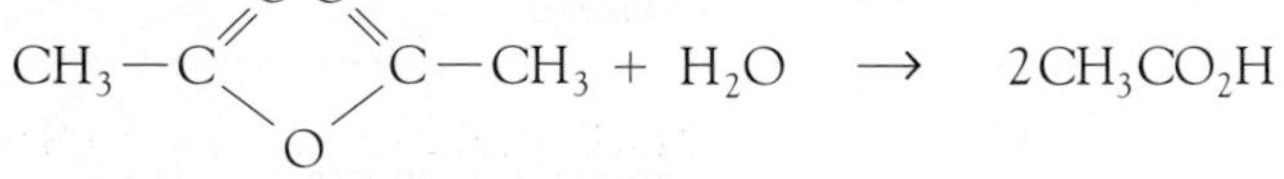

With alcohols or phenols
Acid anhydrides react with alcohols or phenols to form esters:

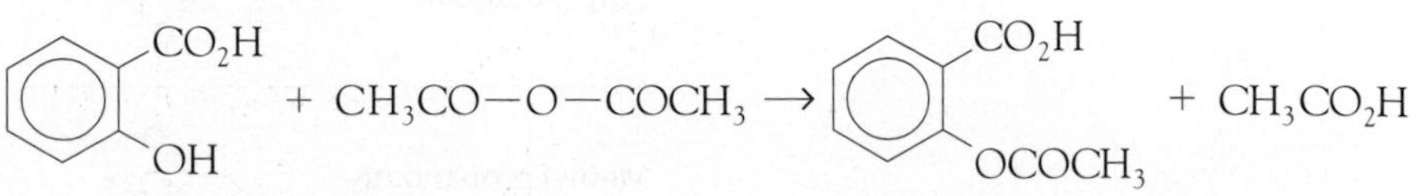

salicylic acid   aspirin

There are several advantages of using ethanoic anhydride over ethanoyl chloride in the manufacture of aspirin:

- ethanoic anhydride is the less reactive of the two, so the (exothermic) reaction between it and salicylic acid is more easily controlled on the large scale
- ethanoic anhydride is cheaper to make than ethanoyl chloride
- the by-product ethanoic acid is less hazardous than the hydrogen chloride by-product produced in the reaction using ethanoyl chloride. It is also possible to recycle the ethanoic acid by converting it into more ethanoic anhydride.

With ammonia and amines

Acid anhydrides react with ammonia to form primary amides, and with amines to form substituted amides:

$$HO-\text{C}_6\text{H}_4-NH_2 + CH_3CO-O-COCH_3 \rightarrow HO-\text{C}_6\text{H}_4-NH-COCH_3 + CH_3CO_2H$$

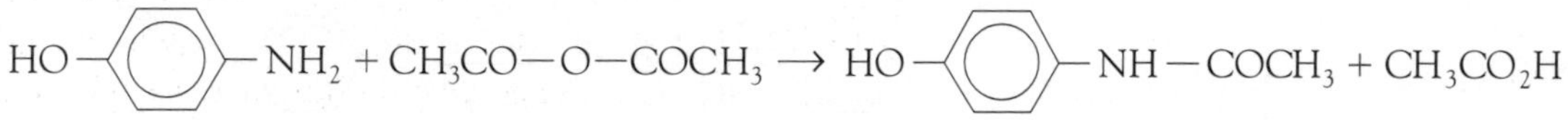

paracetamol

Ethanoic anhydride is therefore an important industrial compound, used in the manufacture of aspirin, paracetamol, and the ethanoyl esters of many alcohols.

29.6 Esters

Properties of esters

Esters have the functional group:

$$R-\overset{\delta+}{C}\begin{smallmatrix}\overset{\delta-}{O}\\\\O-R\end{smallmatrix}$$

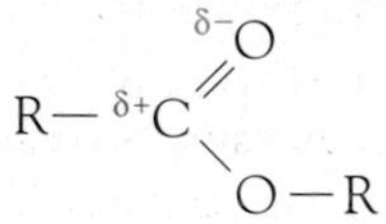

Many esters are liquids with sweet, fruity smells. They are immiscible with, and usually less dense than, water. Despite containing two oxygen atoms, they do not form strong hydrogen bonds with water molecules. Neither do they form hydrogen bonds with other ester molecules (because they do not contain $\delta+$ hydrogen atoms). Their major intermolecular bonding is van der Waals, supplemented by a small dipole–dipole contribution. Their boiling points are therefore a few degrees above those of the alkanes of similar molecular mass. The position of the carboxyl group along the chain has little effect on the strength of intermolecular bonding, and hence boiling point (see Table 29.5).

Esters find many uses as solvents and flavouring agents in fruit drinks and sweets. It is the ethyl esters formed from alcohol and the various acids in wines that account for much of their 'nose', or aroma, in the glass.

■ **Table 29.5**
Esters have boiling points a little higher than those of corresponding alkanes.

Compound		M_r	Boiling point/°C
Hexane		86	69
Propyl methanoate		88	81
Ethyl ethanoate		88	77
Methyl propanoate		88	79

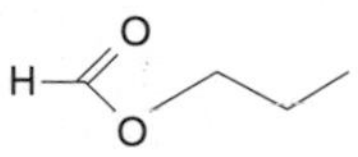

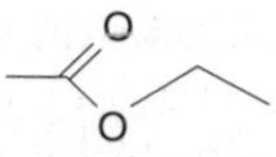

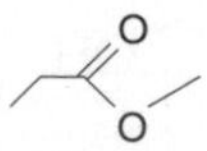

Reactions of esters

Like acyl chlorides, esters are susceptible to nucleophilic attack. They are less reactive than acyl chlorides, however.

With water

The hydrolysis of an ester is a slow process, taking several hours of heating under reflux with dilute aqueous acids:

$$CH_3-C(=O)-O-CH_2CH_3 + H_2O \xrightarrow{\text{heat with } H_2SO_4(aq)} CH_3-C(=O)-OH + C_2H_5OH$$

The mechanism of this acid-catalysed hydrolysis has been discussed in the panel on page 547–8 – it is the reverse of the acid-catalysed esterification described there.

Ester hydrolysis can also be carried out in alkaline solution. The reaction is quicker (OH^- is a stronger nucleophile than water) and does not reach equlibrium, but goes to completion. This is because the carboxylic acid produced reacts with an excess of base to form the carboxylate anion:

$$CH_3-C(=O)-O-CH_2CH_3 + OH^- \xrightarrow{\text{heat with NaOH(aq)}} CH_3-C(=O)-O^-Na^+ + C_2H_5OH$$

Further practice

Suggest a mechanism for this reaction, using OH^- as the initial nucleophile, and $C_2H_5O^-$ as the base in the last step.

The alkaline hydrolysis of esters has an important application in the manufacture of soap. Fats are glyceryl triesters of long-chain carboxylic acids. Heating fats with strong aqueous sodium hydroxide causes hydrolysis of the triesters. The sodium salts of the long-chain carboxylic acids are precipitated by adding salt to the mixture, and this solid is then washed and compressed into bars of soap (see Figure 29.8). Perfume and colour are also added.

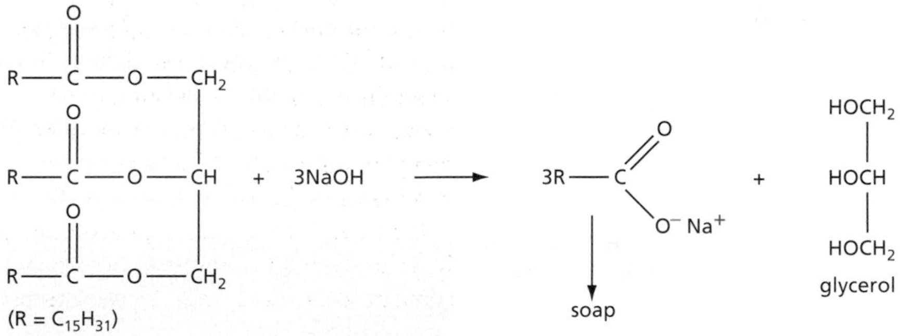

■ **Figure 29.8** Soap is made by the hydrolysis of the glyceryl triesters of fats.

With ammonia

Like acyl chlorides, esters react with ammonia, forming amides:

$$H-C(=O)-O-CH_2CH_3 + NH_3 \rightarrow H-C(=O)-NH_2 + CH_3CH_2OH$$

Reduction

As we saw for carboxylic acids on page 547, esters can be reduced by lithium tetrahydridoaluminate(III):

$$CH_3CH_2-C(=O)-O-CH_3 \xrightarrow{\text{LiAlH}_4 \text{ in dry ether}} CH_3CH_2CH_2OH + CH_3OH$$

Formation of polyesters

When a diol is esterified with a diacid (for example, by heating with hydrogen chloride gas), or reacted with a diacyl chloride, a polyester is produced (see Figure 29.9). For example:

$$n\text{HO}-\text{CH}_2\text{CH}_2-\text{OH} + n\text{HO}-\overset{\displaystyle O}{\underset{\displaystyle \|}{C}}-\bigcirc-\overset{\displaystyle O}{\underset{\displaystyle \|}{C}}-\text{OH} \xrightarrow{-n\text{H}_2\text{O}}$$

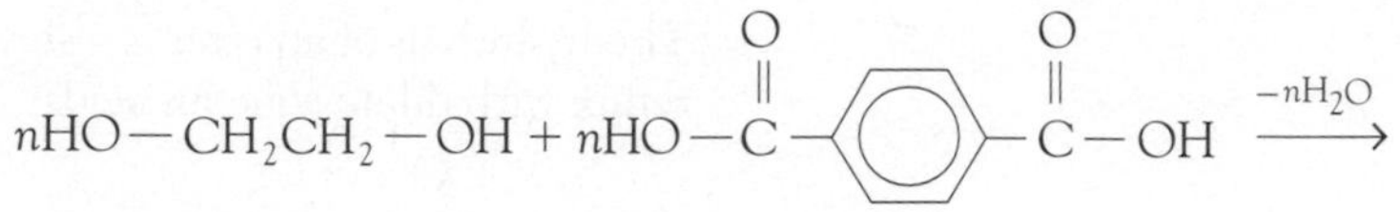

ethane-1,2-diol   benzene-1,4-dicarboxylic acid

$$\left[-\text{O}-\text{CH}_2\text{CH}_2-\text{O}-\overset{\displaystyle O}{\underset{\displaystyle \|}{C}}-\bigcirc-\overset{\displaystyle O}{\underset{\displaystyle \|}{C}}-\right]_n$$

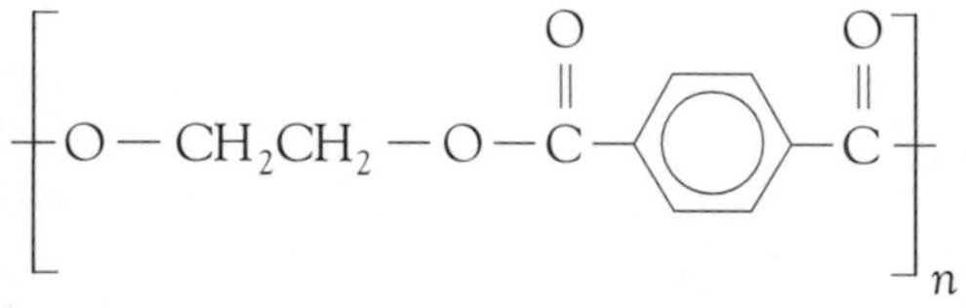

Terylene, a type of polyester

Polyesters are examples of a class of polymers known as **condensation polymers**.

> When monomers join together to form a **condensation polymer**, a small molecule such as H_2O, HCl or NH_3 is also produced.

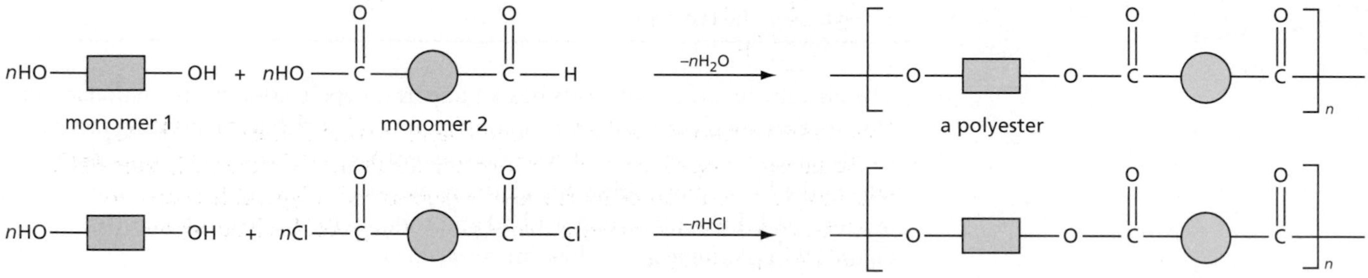

■ **Figure 29.9**
Formation of a polyester.

$$n\text{HO}-\boxed{}-\text{OH} + n\text{HO}-\overset{O}{\underset{\|}{C}}-\bigcirc-\overset{O}{\underset{\|}{C}}-\text{H} \xrightarrow{-n\text{H}_2\text{O}} \left[-\text{O}-\boxed{}-\text{O}-\overset{O}{\underset{\|}{C}}-\bigcirc-\overset{O}{\underset{\|}{C}}-\right]_n$$

monomer 1   monomer 2   a polyester

$$n\text{HO}-\boxed{}-\text{OH} + n\text{Cl}-\overset{O}{\underset{\|}{C}}-\bigcirc-\overset{O}{\underset{\|}{C}}-\text{Cl} \xrightarrow{-n\text{HCl}} \left[-\text{O}-\boxed{}-\text{O}-\overset{O}{\underset{\|}{C}}-\bigcirc-\overset{O}{\underset{\|}{C}}-\right]_n$$

The most commonly used polyester is Terylene, produced by polymerising ethane-1,2-diol and benzene 1,4-dicarboxylic acid, as shown above. The polymer can be drawn into fine fibres and spun into yarn. Jointly woven with cotton or wool, it is a component of many everyday textiles. Material made from Terylene is hard-wearing and strong, but under extreme alkaline or acidic conditions the ester bonds can be hydrolysed, causing the fabric to break up.

■ **Figure 29.10**
Polyester products.

29.7 Amides

Properties of amides

Amides have the functional group:

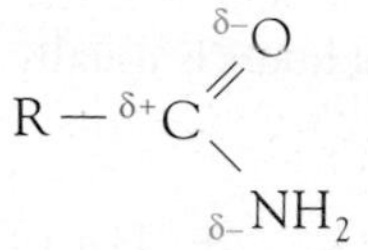

They are extensively hydrogen bonded, having both $H^{\delta+}$ atoms (on nitrogen) and lone pairs of electrons (on oxygen and nitrogen). Most amides are solids at room temperature, and are soluble in water.

Unlike amines (see Chapter 30), amides form neutral solutions in water, and can be protonated only by strong acids. The site of protonation is unusual, as we shall see, and sheds light on why amides are such weak bases.

The lone pair on the nitrogen atom in amides is in a p orbital, and can overlap with the π orbital of the adjacent carbonyl group (see Figure 29.11).

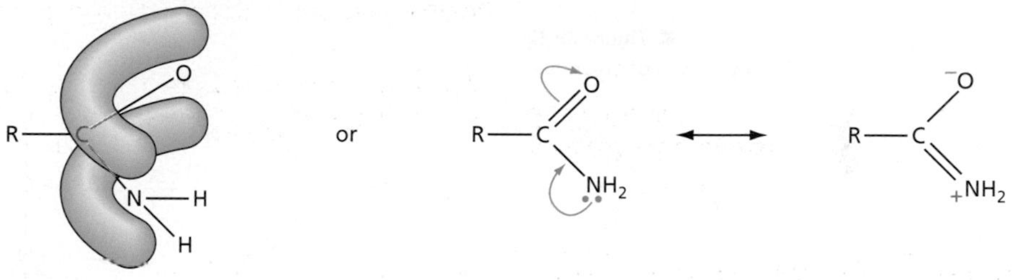

■ **Figure 29.11** Delocalisation of the nitrogen lone pair in the amide group.

This confers considerable double-bond characteristics to the C—N bond, including restricted rotation about it (compare alkenes, Chapter 24, page 441). This has great significance in the stereochemistry of polypeptide chains in proteins, which we shall meet in Chapter 30. Polypeptide chains contain secondary amide groups:

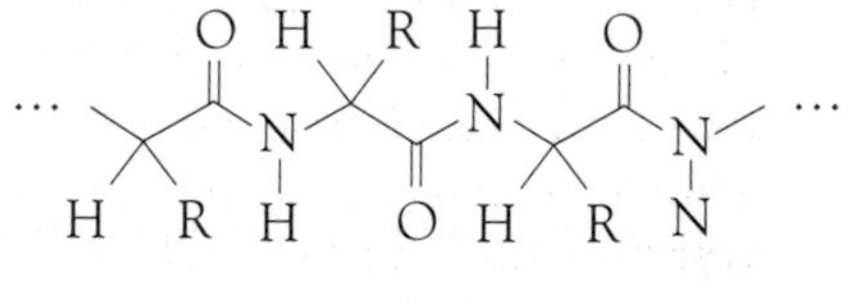

The properties of synthetic polyamides such as nylon are covered in Chapter 30.

When amides react with strong acids, it is the oxygen atom, rather than the nitrogen atom, that is protonated:

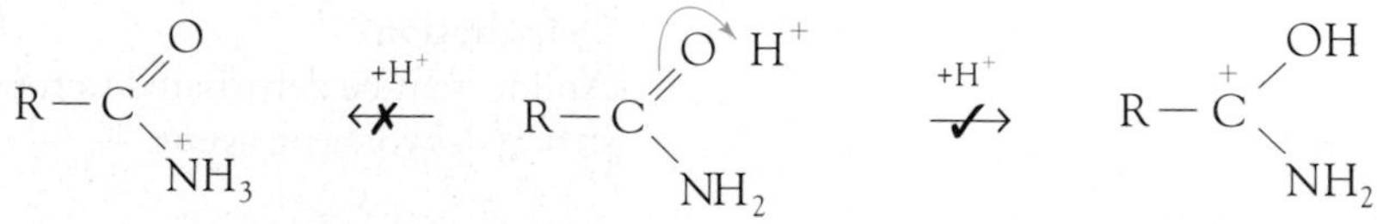

Protonation of the nitrogen atom would result in a positively charged nitrogen adjacent to the δ+ carbon of the carbonyl group – an unfavourable situation. On the other hand, protonation of the oxygen atom allows the positive charge on the cation to be delocalised over three atoms – an energetically favourable situation:

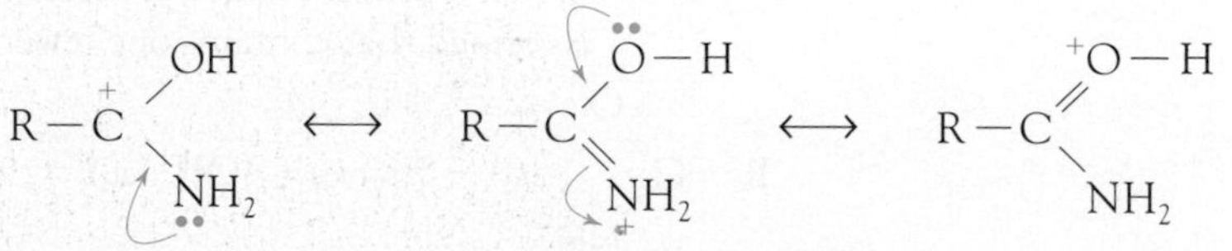

Reactions of amides

With water

Because of the high degree of positive charge on the carbon atom in protonated amides, they are susceptible to nucleophilic attack. Hydrolysis is usually carried out in dilute sulphuric acid. It is still quite a slow reaction, however – heating under reflux for several hours is usually required:

$$R-C{\overset{O}{\underset{NH_2}{}}} + H_2O + H^+ \rightarrow R-C{\overset{O}{\underset{OH}{}}} + NH_4^+$$

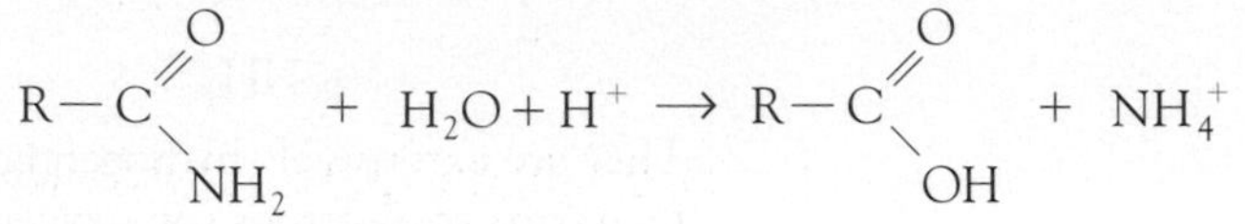

How it happens (2): the mechanism of the hydrolysis of amides

The initial reaction is the protonation of the amide by the mineral acid (this occurs on the oxygen atom, as mentioned above). The protonated amide is then attacked nucleophilically by a water molecule. Protons rearrange themselves amongst the oxygen and nitrogen atoms, and eventually the C—N bond cleaves, and a proton is lost (see Figure 29.12).

■ **Figure 29.12**
The hydrolysis of an amide.

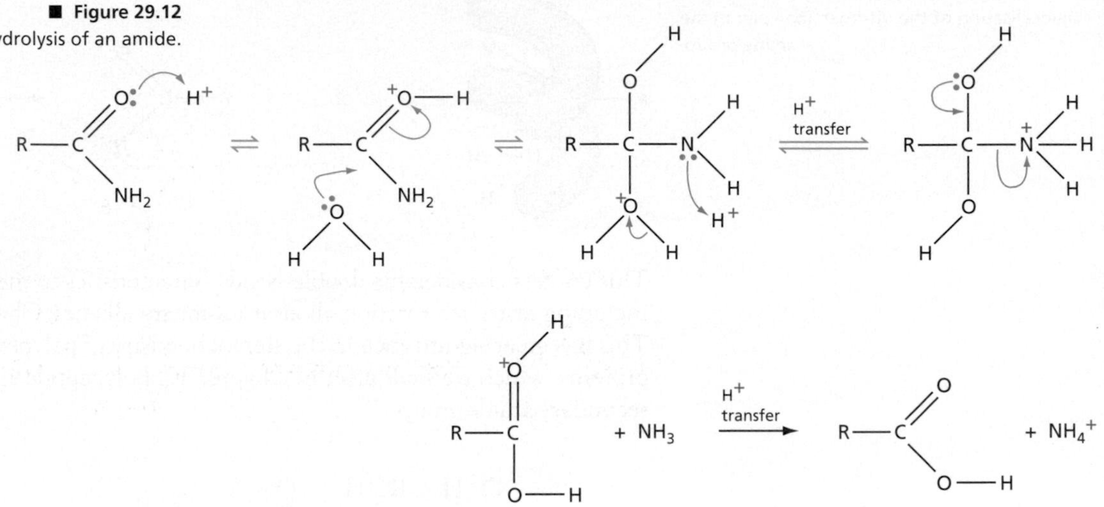

Amides are intermediates in the hydrolysis of nitriles to carboxylic acids:

$$R-C\equiv N + H_2O \rightarrow R-C{\overset{O}{\underset{NH_2}{}}} + H_2O \xrightarrow[\text{heat}]{H^+} R-C{\overset{O}{\underset{OH}{}}} + NH_3$$

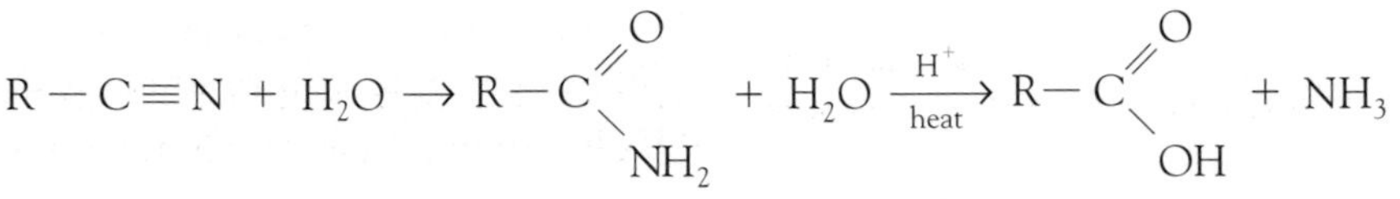

Dehydration

Amides can be dehydrated to nitriles by heating with phosphorus(V) oxide, a strong dehydrating agent.

$$R-C{\overset{O}{\underset{NH_2}{}}} - H_2O \xrightarrow{P_2O_5} R-C\equiv N$$

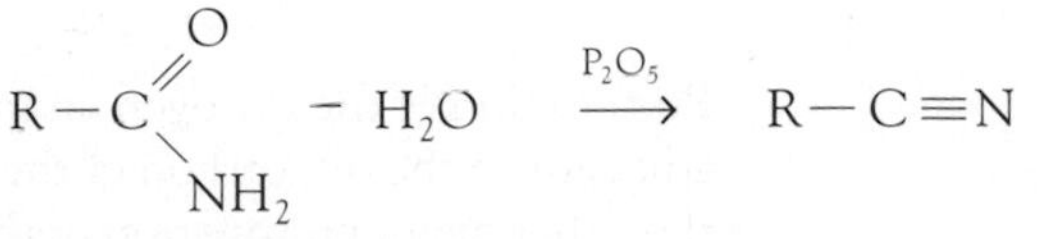

The Hofmann reaction

When a primary amide is treated with bromine in aqueous alkali, a primary amine is formed that contains one fewer carbon atom:

$$R-C{\overset{O}{\underset{NH_2}{}}} (aq) + Br_2(aq) + 4OH^-(aq) \xrightarrow{\text{warm}} R-NH_2(aq) + CO_3^{2-}(aq) + 2Br^-(aq) + 2H_2O(l)$$

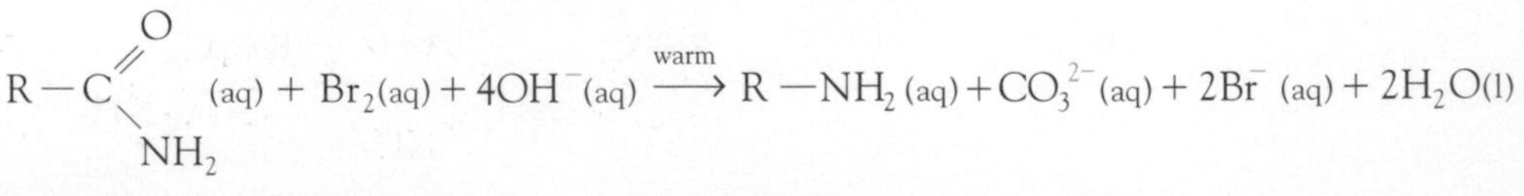

This reaction, discovered by August Hofmann in 1882, gives a good yield of the amine. It is particularly useful in the formation of primary amines at tertiary carbon atoms, which are difficult to make by other methods:

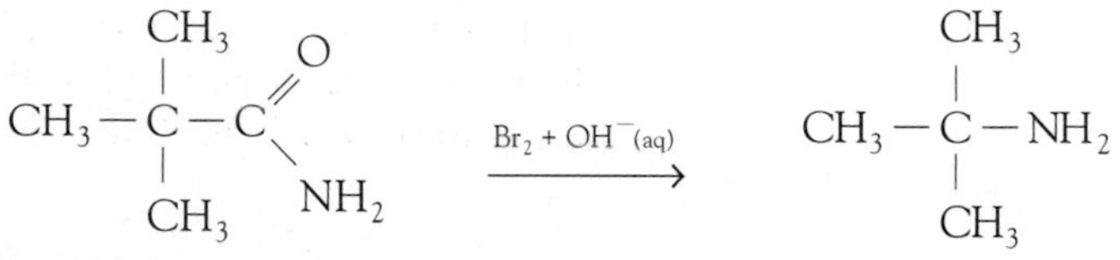

How it happens (3): the mechanism of the Hofmann reaction

The reaction involves an initial bromination of the nitrogen atom, followed by the deprotonation of the bromoamide by the aqueous base. There then follows an intramolecular rearrangement, in which an organic group moves sideways from the carbonyl carbon atom to the nitrogen atom (see Figure 29.13).

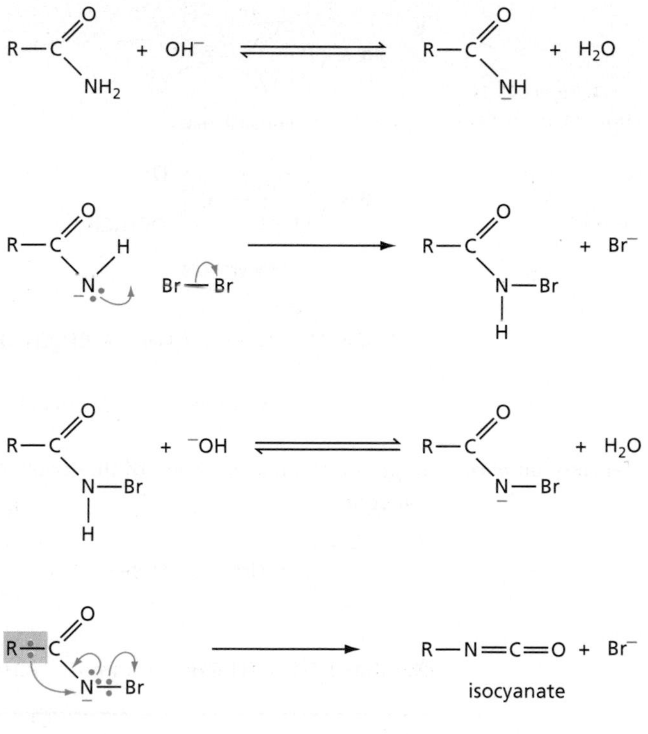

■ Figure 29.13
The formation of an isocyanate in the Hofmann reaction.

The resulting isocyanate then undergoes hydrolysis to an amide, and further hydrolysis to the target amine and carbonate ions (Figure 29.14).

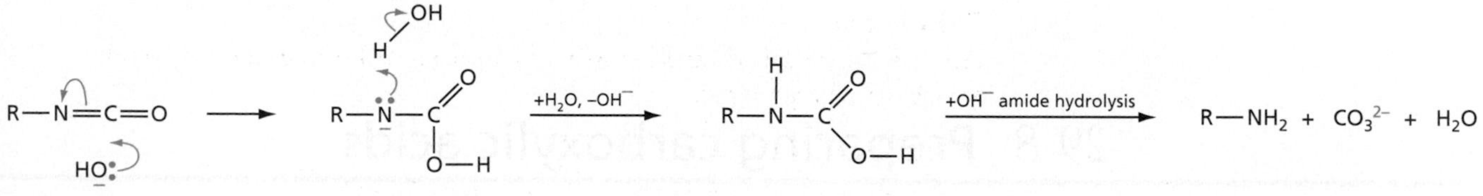

■ Figure 29.14
The hydrolysis of the isocyanate yields the amine.

Reduction to amines

Amides can be reduced to amines by lithium tetrahydridoaluminate(III).

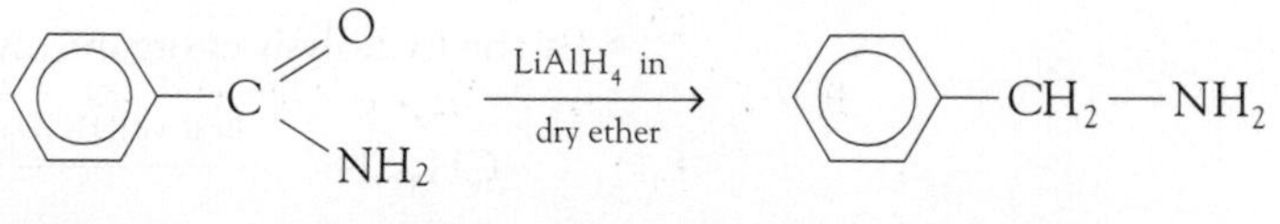

Example   Suggest products of the following reactions:

a

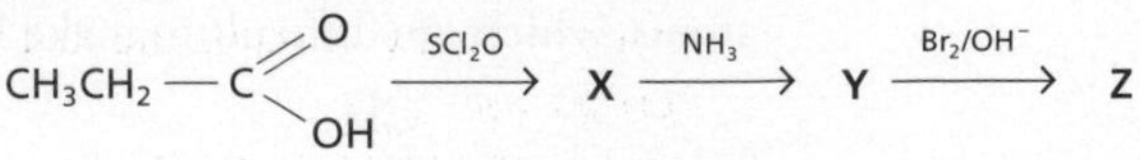

$$CH_3CH_2 - C \overset{O}{\underset{OH}{<}} \xrightarrow{SCl_2O} X \xrightarrow{NH_3} Y \xrightarrow{Br_2/OH^-} Z$$

b

$\xrightarrow{PCl_5}$ **A** $\xrightarrow{CH_3CH_2OH}$ **B** $\xrightarrow[\text{ether}]{LiAlH_4}$ **C**

Answer   **a**

X is $CH_3CH_2C \overset{O}{\underset{Cl}{<}}$ Y is $CH_3CH_2C \overset{O}{\underset{NH_2}{<}}$ Z is $CH_3CH_2NH_2$

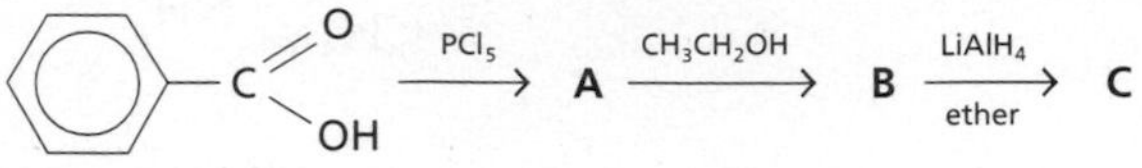

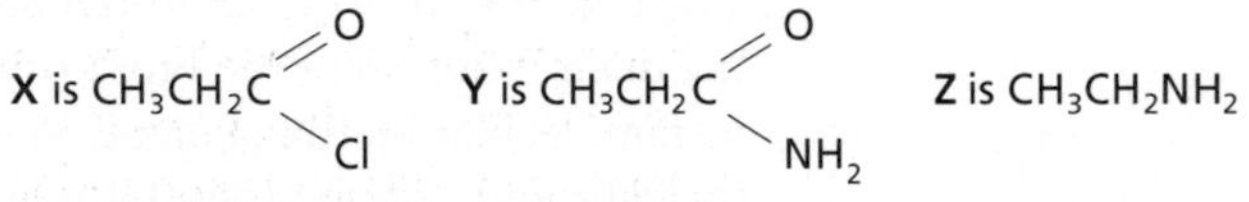

propanoyl chloride

propanamide

ethylamine
(See Chapter 30)

b

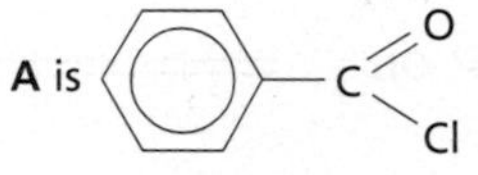

A is

benzoyl chloride

B is

ethyl benzoate

C is —CH₂OH (+ CH_3CH_2OH)

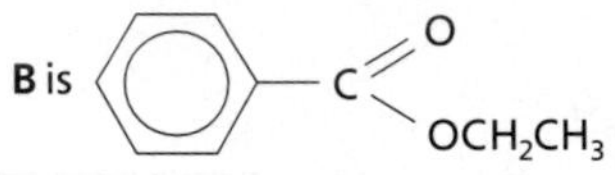

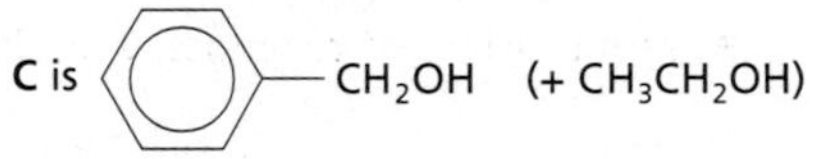

phenylmethanol   ethanol

Further practice   Suggest two-stage syntheses of the following compounds, starting from the stated compounds.

1   —NH₂   from   —COCl

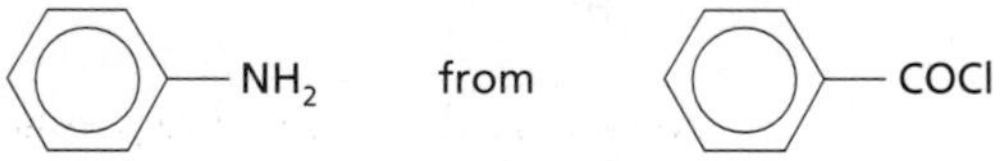

2 $(CH_3)_2CH - CH_2OH$ from $(CH_3)_2CH - CN$

29.8 Preparing carboxylic acids

Carboxylic acids may be prepared by the following reactions.

- By the oxidation of primary alcohols or aldehydes:

$$CH_3CH_2OH \xrightarrow[\text{heat}]{Na_2Cr_2O_7 + H_2SO_{4(aq)}} CH_3CO_2H$$

- By the hydrolysis of organic cyanides (nitriles):

$$CH_3CN \xrightarrow{\text{heat with } H_2SO_{4(aq)}} CH_3CO_2H \ (+ NH_4^+)$$

- By the hydrolysis of acid derivatives. These are not usually useful preparative reactions, although they proceed in high yields. Acid derivatives are usually made from the acids in the first place, so their hydrolysis forms no compounds of further use.
- By the oxidation of aryl side chains. When treated with hot alkaline potassium manganate(VII), aryl hydrocarbons produce benzoic acids by oxidation:

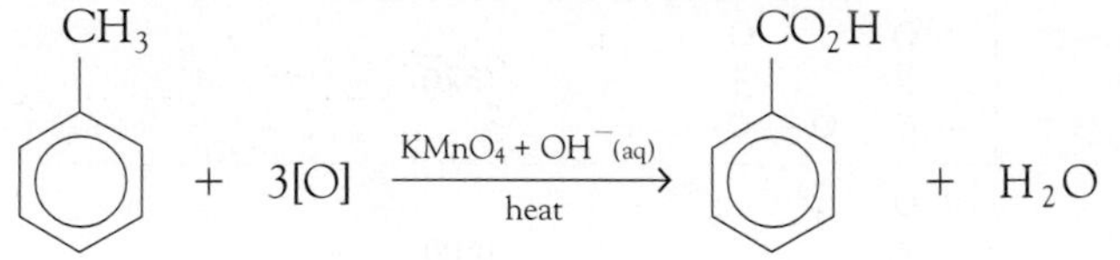

- From methyl ketones by the triiodomethane reaction. This reaction is normally used to test for the presence of the $-COCH_3$ group (see Chapter 28, page 535), but it can produce useful yields of acids:

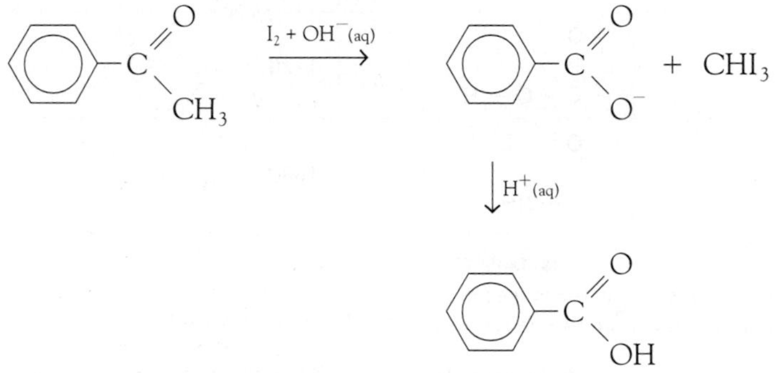

- From Grignard reagents (see Chapter 25, page 473):

$$CH_3CH_2Br \xrightarrow{\text{Mg in ether}} CH_3CH_2MgBr$$

$$\downarrow \text{dry } CO_2$$

$$CH_3CH_2CO_2MgBr \xrightarrow{H^+(aq)} CH_3CH_2CO_2H$$

Most acid derivatives are prepared from carboxylic acids by the reactions described in this chapter. Figure 29.15 shows a chart summarising the interrelationships between the various derivatives.

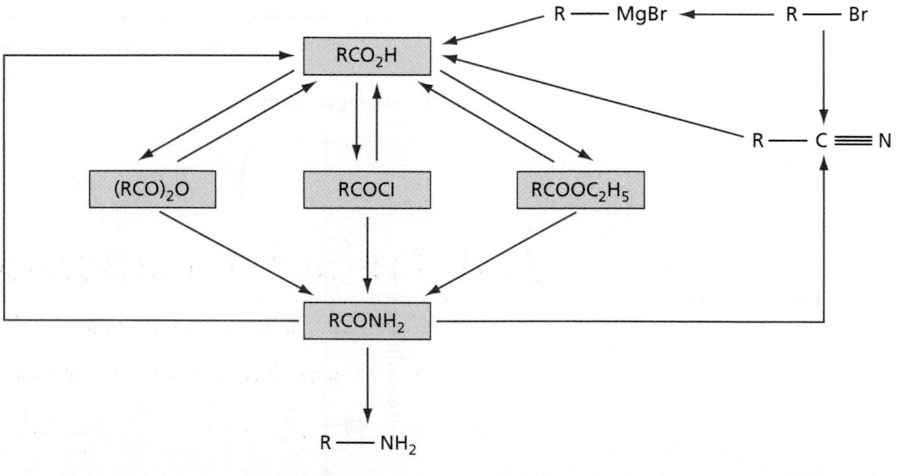

■ **Figure 29.15** Interconversions between carboxylic acids and their derivatives.

Further practice Copy the chart in Figure 29.15 and write on each arrow the correct reagents and conditions for each reaction.

29.9 The infrared spectra of carboxylic acids and their derivatives

Compound	Group	Frequency of absorption (wavenumber)/cm^{-1}
Acid anhydride	$-\overset{O}{\overset{\Vert}{C}}-O-\overset{O}{\overset{\Vert}{C}}-$	1820
Acyl chloride	$-\overset{O}{\overset{\Vert}{C}}-Cl$	1800
Ester	$-\overset{O}{\overset{\Vert}{C}}-O-R$	1740
Acid	$-\overset{O}{\overset{\Vert}{C}}-O-H$	1720
Amide	$-\overset{O}{\overset{\Vert}{C}}-NH_2$	1680

■ **Table 29.6**
C=O stretching frequencies in the IR spectra of carboxylic acids and their derivatives.

All the compounds described in this chapter contain the carbonyl group, which absorbs strongly in the region 1680–1820 cm^{-1}. The precise frequency depends on the environment (see Table 29.6).

Carboxylic acids, acid anhydrides and esters also show an absorbance due to the C—O bond, between 1100 and 1300 cm^{-1}.

Carboxylic acids and amides contain O—H and N—H bonds. The strongly hydrogen-bonded O—H group in carboxylic acid shows a characteristic very broad infrared absorption between 2500 and 3000 cm^{-1} (see Figure 29.16). The amide N—H bonds absorb at 3400 cm^{-1}. The C—Cl bond in acyl chlorides absorbs at its usual frequency of 600–800 cm^{-1}.

■ **Figure 29.16**
IR spectra of **a** ethanoic anhydride, **b** ethanoyl chloride, **c** ethyl ethanoate, **d** ethanoic acid and **e** ethanamide.

a ethanoic anhydride

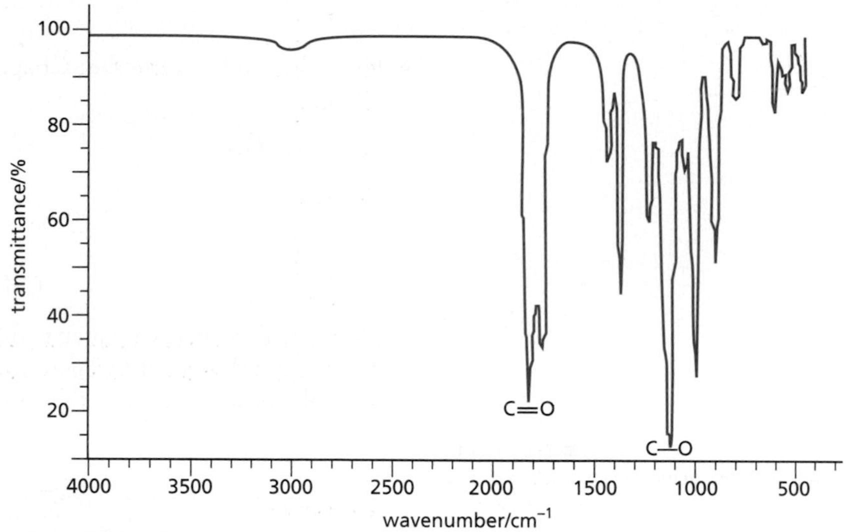

b ethanoyl chloride

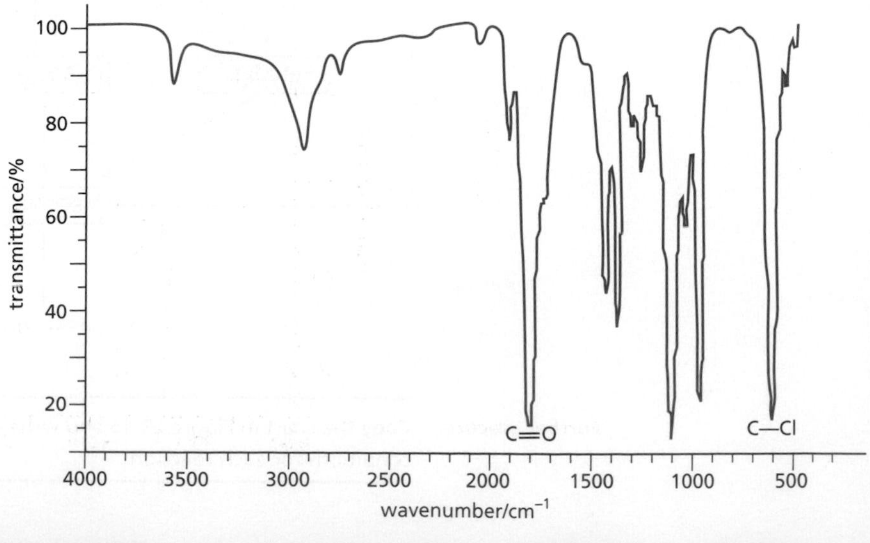

c ethyl ethanoate

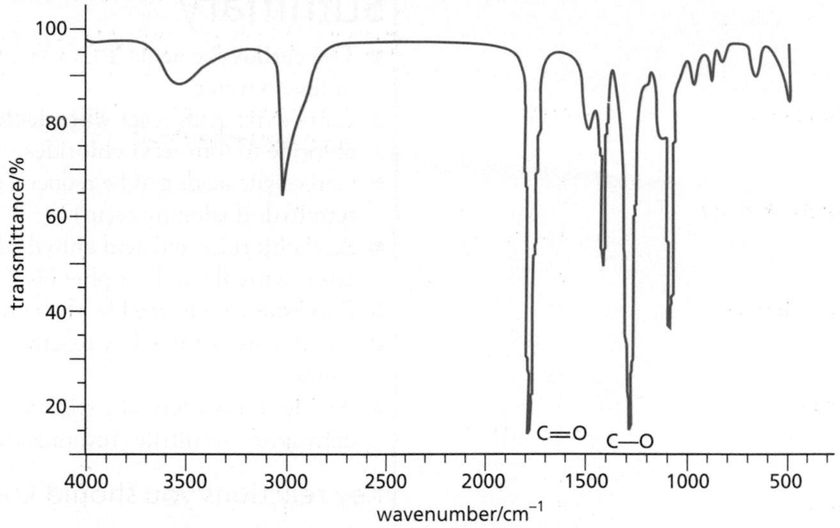

d ethanoic acid

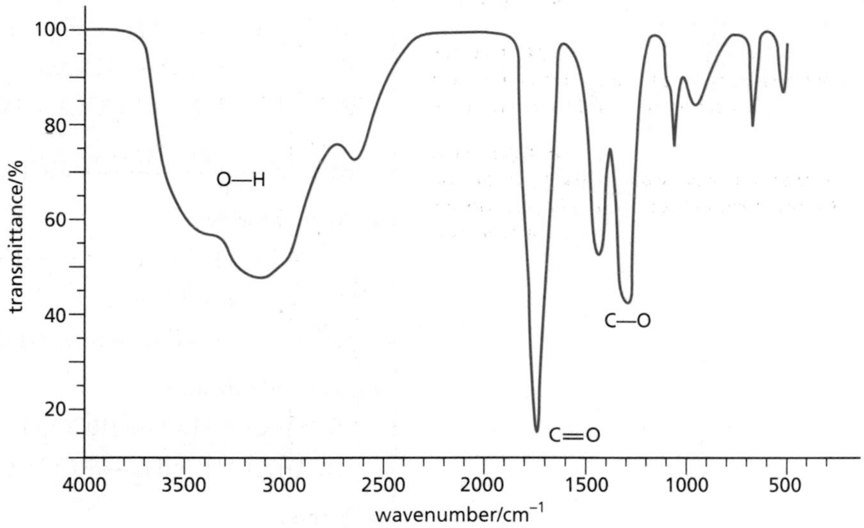

e ethanamide

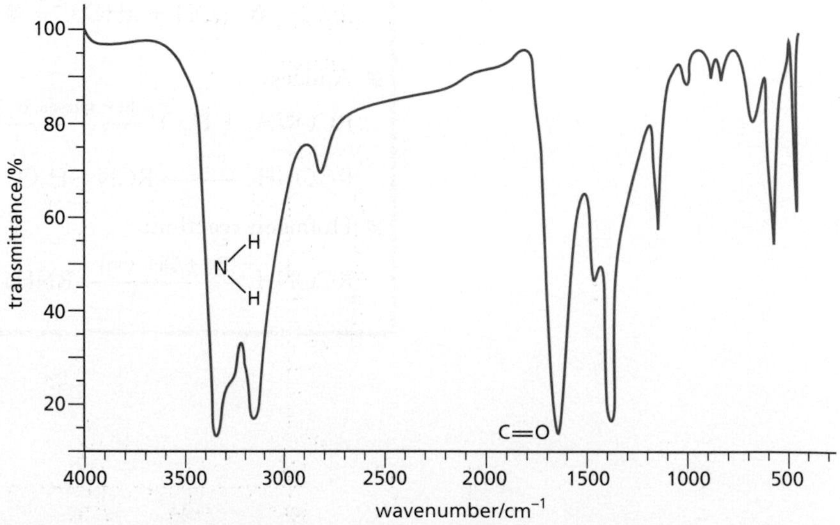

Summary

- The **carboxylic acids**, RCO_2H, are weak acids, ionising to the extent of 1% or less in water.
- Carboxylic acids react with alcohols to form **esters**, and with phosphorus(V) chloride to form **acyl chlorides**.
- Carboxylic acids can be reduced to alcohols with lithium tetrahydridoaluminate(III).
- Acyl chlorides and **acid anhydrides** are very useful intermediates, forming esters with alcohols or phenols, and amides with amines.
- **Polyesters** are formed by the condensation of a diol with a dicarboxylic acid.
- Amides are obtained by reacting acyl chlorides or esters with ammonia or amines.
- Amides form amines by reduction or by the **Hofmann reaction**, and can be dehydrated to nitriles (cyanides).

Key reactions you should know

- **Carboxylic acids:**

  $RCO_2H + Na \rightarrow RCO_2^-Na^+ + \frac{1}{2}H_2$

  $RCO_2H + NaOH \rightarrow RCO_2^-Na^+ + H_2O$

  $2RCO_2H + Na_2CO_3 \rightarrow 2RCO_2^-Na^+ + CO_2 + H_2O$

  $RCO_2H + SCl_2O \rightarrow RCOCl + SO_2 + HCl$

  $RCO_2H + PCl_5 \rightarrow RCOCl + PCl_3O + HCl$

  $RCO_2H \xrightarrow{\text{LiAlH}_4\text{ in dry ether}} RCH_2OH$

- **Acyl chlorides:**

  $RCOCl + H_2O \rightarrow RCO_2H + HCl$

  $RCOCl + R'OH \rightarrow RCO_2R' + HCl$ (R′ = alkyl or aryl)

  $RCOCl + 2R'NH_2 \rightarrow RCONHR' + R'NH_3Cl$

- **Acid anhydrides:**

  $(RCO)_2O + H_2O \rightarrow 2RCO_2H$

  $(RCO)_2O + R'NH_2 \rightarrow RCONHR' + RCO_2H$

- **Esters:**

  $RCO_2R' \xrightarrow{\text{boil with H}_3\text{O}^+\text{ or OH}^-\text{(aq)}} RCO_2H + R'OH$

- Condensation polymerisation:

  nHO—■—OH + nHO$_2$C—●—CO$_2$H →

  $\!\!+\!$O—■—OCO—●—CO$\!+_n$

- **Amides:**

  $RCONH_2 + H_3O^+ \xrightarrow{\text{heat with dil. H}_2\text{SO}_4} RCO_2H + NH_4^+$

  $RCONH_2 \xrightarrow{P_2O_5} RCN + H_2O$

- **Hofmann reaction:**

  $RCONH_2 \xrightarrow{Br_2 + OH^-(aq)} RNH_2 + CO_3^{2-}$

30 Amines and amino acids

Functional groups:

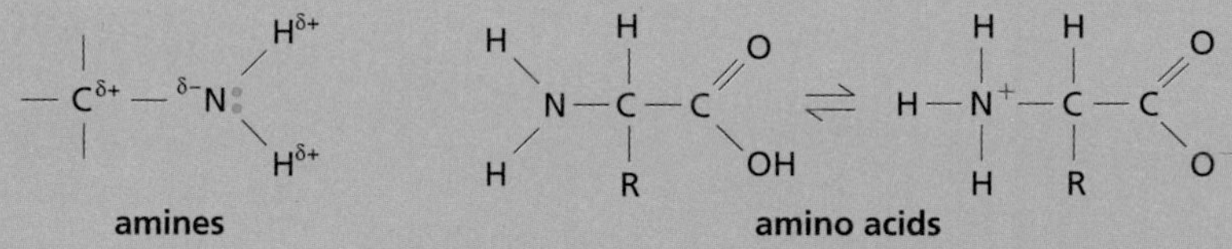

Amines are organic bases. They react with acids to form salts, and amines that are soluble in water form alkaline solutions. The reaction of aryl amines with nitrous acid is the first step in making dyes.

Amino acids are bi-functional molecules containing both the basic —NH_2 group and the acidic —CO_2H group. They are amphoteric, reacting with both acids and bases. They are the building blocks of the important class of biological polymer known as the polypeptides, or proteins. The chapter concludes with a look at synthetic polyamides, including the various types of nylon.

The chapter is divided into the following sections.

30.1 Introduction

In the early days of organic chemistry, when chemists were extracting, purifying and attempting to identify the hundreds of compounds that occur in natural organisms, they discovered a group of compounds that were insoluble in water, but dissolved in dilute acids. They called these the **alkaloids** ('alkali-like'). Many of the alkaloids extracted from plants were found to have strong physiological and psychological effects on humans and other animals. These included hypnotics such as morphine, narcotics such as cocaine, and cardiac poisons such as strychnine. All alkaloids contained at least one nitrogen atom, and all belonged to the class of compounds we now call **amines**. Figure 30.1 illustrates some of their structures.

■ **Figure 30.1** The structures of some naturally-occurring alkaloids.

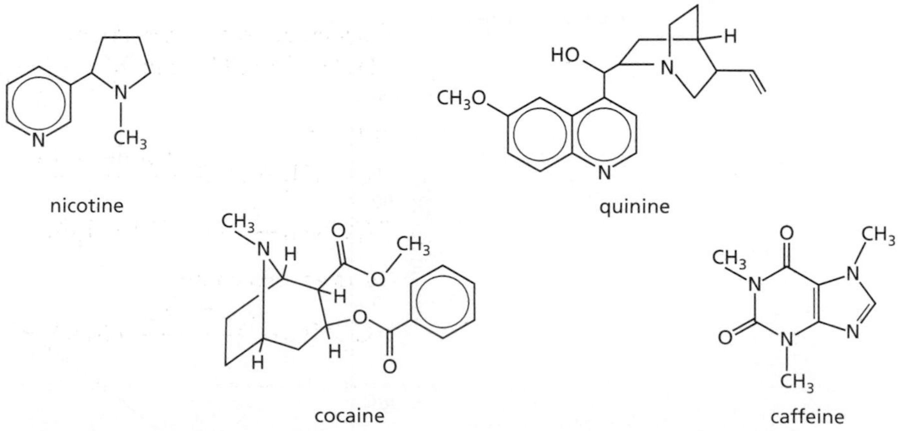

Most alkaloids have a bitter taste, but are too involatile to have an odour. Those amines that are volatile have distinctive smells. The lower members have an astringent ammonia-like smell, but this is rapidly replaced by a strong fishy odour in butylamine and higher amines. Aryl amines such as phenylamine have a more 'oily' smell.

The amines can be derived from ammonia, both conceptually and in the laboratory, by replacing one or more of the hydrogen atoms in NH_3 by organic groups:

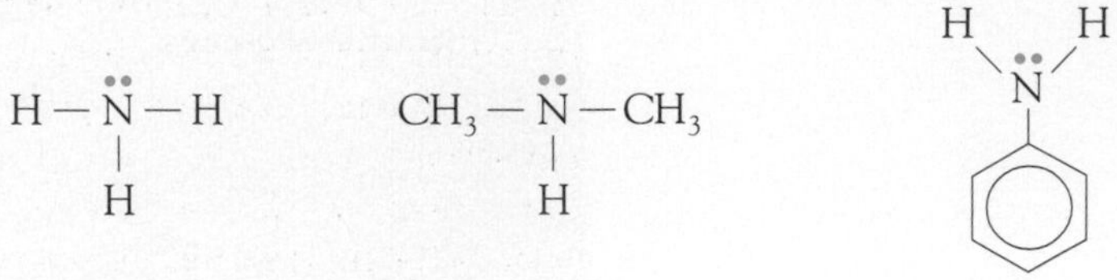

ammonia   dimethylamine   phenylamine

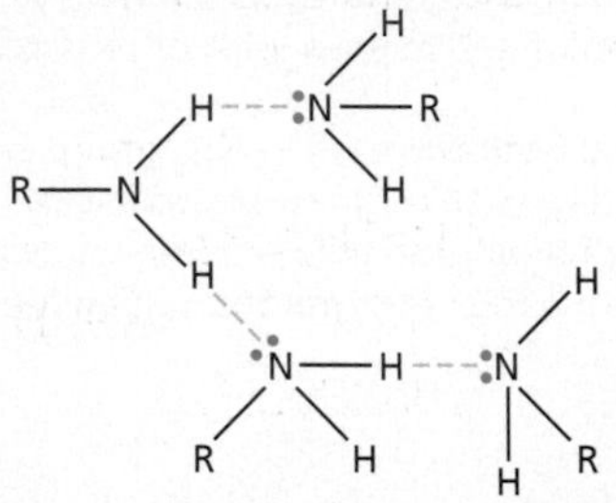

■ **Figure 30.2**
Intermolecular hydrogen bonding in amines.

The carbon–nitrogen bond, although quite strongly polarised, is fairly inert. It is not as easily broken as the C—O bond in alcohols (see Chapter 26, pages 481–5). There are few reactions in which the $C^{\delta+}$ atom takes part. The key point of reactivity in the amines is the lone pair of electrons on the nitrogen atom. This is nucleophilic and basic (that is, nucleophilic towards hydrogen).

Amines form strong hydrogen bonds to hydrogen-donor molecules such as water. Those amines that contain N—H bonds also form strong intermolecular hydrogen bonds (see Figure 30.2).

The lower members of the series are therefore water soluble and their boiling points are higher than those of the corresponding alkanes (see Table 30.1 and Figure 30.3). The effect of hydrogen bonding on boiling point is not as pronounced as it is in the alcohols, however (compare Table 30.1 with Table 26.1, page 478).

■ **Table 30.1**
Boiling points of some alkanes and corresponding amines.

Number of electrons in the molecule	Alkane		Amine		Enhancement of boiling point/°C
	Formula	Boiling point/°C	Formula	Boiling point/°C	
18	C_2H_6	−88	CH_3NH_2	−8	80
26	C_3H_8	−42	$C_2H_5NH_2$	17	59
34	C_4H_{10}	0	$C_3H_7NH_2$	49	49
42	C_5H_{12}	36	$C_4H_9NH_2$	78	42
50	C_6H_{14}	69	$C_5H_{11}NH_2$	104	35
58	C_7H_{16}	98	$C_6H_{13}NH_2$	130	32

■ **Figure 30.3**
Boiling points of some alkanes and corresponding amines.

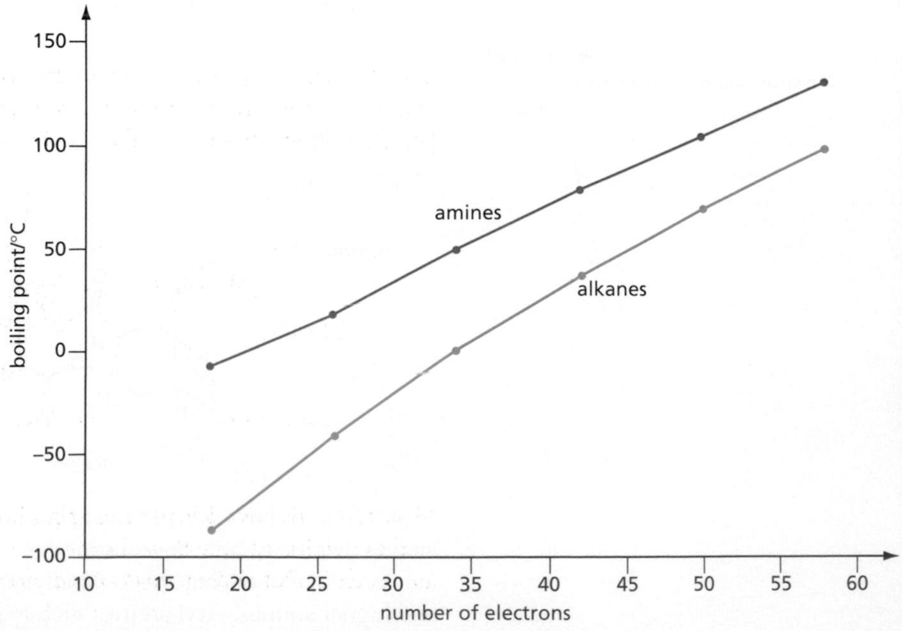

30.2 Isomerism and nomenclature

As we have mentioned, the most important feature of the amine molecule is the nitrogen atom. This is reflected in the nomenclature of amines. The simpler ones such as dimethylamine and phenylamine shown above are named as derivatives of ammonia.

Up to three hydrogen atoms in ammonia can be replaced by organic groups. Successive replacement forms **primary, secondary** and **tertiary amines** (see Table 30.2). Note that it is the branching that takes place at the nitrogen atom, rather than at the carbon atom attached to it, that determines whether an amine is primary, secondary or tertiary. This is not the same as in the case of the alcohols:

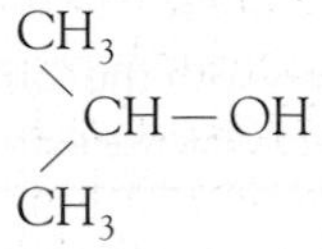

propan-2-ol
a secondary alcohol

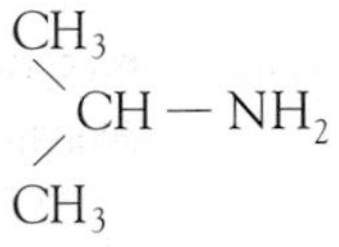

2-aminopropane
a primary amine

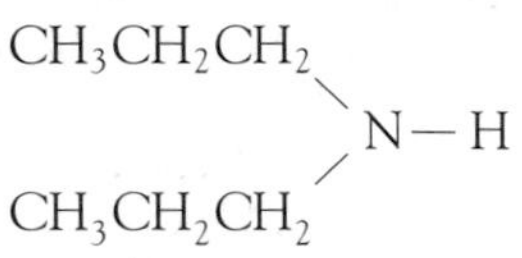

dipropylamine
a secondary amine

■ Table 30.2
Some primary, secondary and tertiary amines.

Primary amines	**Secondary amines**	**Tertiary amines**
$CH_3CH_2NH_2$ Ethylamine	$(CH_3CH_2)_2NH$ Diethylamine	$(CH_3CH_2)_3N$ Triethylamine
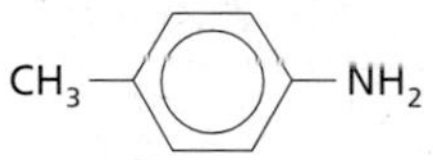 4-Methylphenylamine	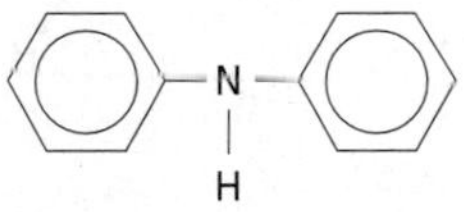 Diphenylamine	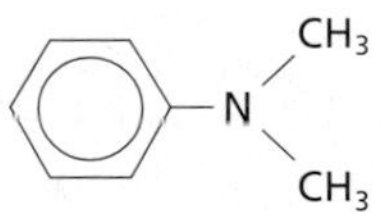 *N,N*-Dimethylphenylamine
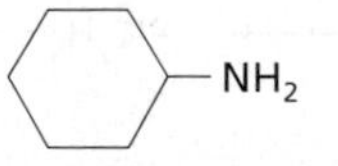 Cyclohexylamine	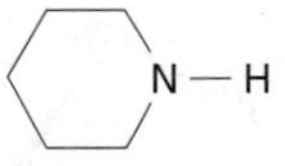 Piperidine	

An alternative way of naming amines is used when the chain becomes more branched or where other functional groups are present. This views the $—NH_2$ group as a substituent, called an **amino group**:

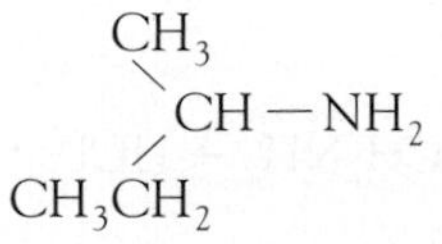

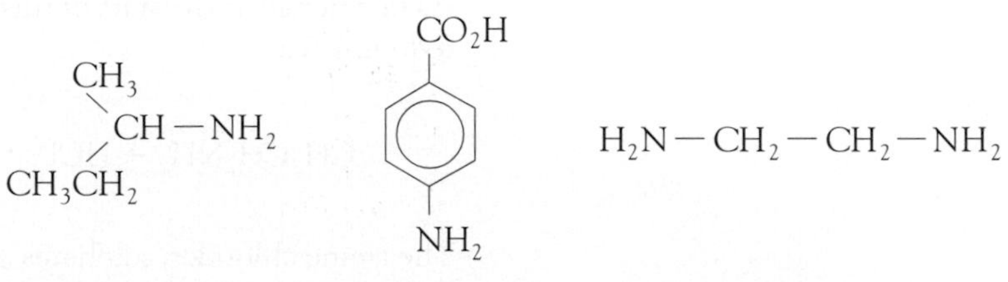

2-aminobutane

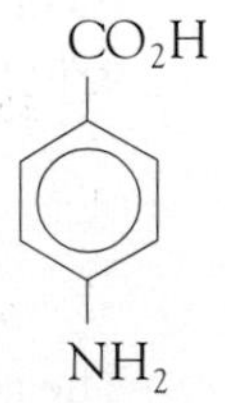

4-aminobenzoic acid

$H_2N-CH_2-CH_2-NH_2$

1,2-diaminoethane
(abbreviated 'en')

Apart from the usual structural or positional isomerism that can occur in the carbon chain, amines also demonstrate structural isomerism around the nitrogen atom, through the formation of secondary and tertiary amines as well as primary amines.

Example How many isomers are there with the formula C_3H_9N?

Answer There are four isomers. Structural isomerism in the propyl group allows two primary amines:

$CH_3CH_2CH_2NH_2$	$(CH_3)_2CH—NH_2$
propylamine (1-aminopropane)	2-aminopropane

In addition to these, we can split the three-carbon chain into two or three chains, to form a secondary amine:

$CH_3—CH_2—NH—CH_3$

ethylmethylamine

and a tertiary amine:

$(CH_3)_3N$

trimethylamine

Further practice There are 11 isomers with the formula $C_8H_{11}N$, all containing a benzene ring. Draw their structures, and state which are primary, secondary and tertiary amines.

The nitrogen atom in amines is pyramidal, as it is in ammonia. In theory, the secondary amine ethylmethylamine could exist as a mirror-image pair of compounds (see Figure 30.4). But in practice the nitrogen atom undergoes a rapid inversion, like an umbrella turning inside out, causing a 50 : 50 mixture of the two forms to exist at room temperature (see Figure 30.5).

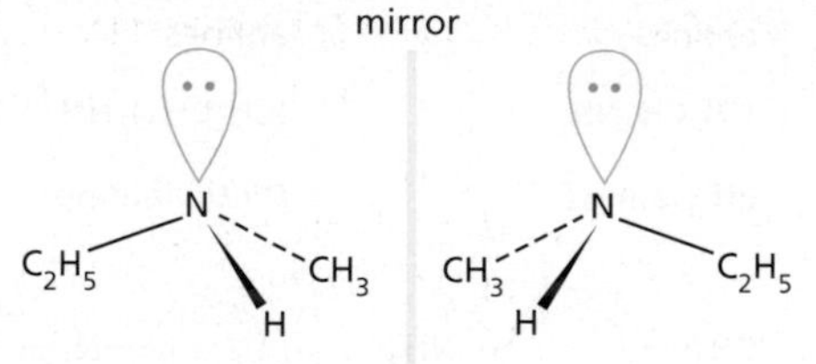

■ **Figure 30.4** Theoretical enantiomers of ethylmethylamine.

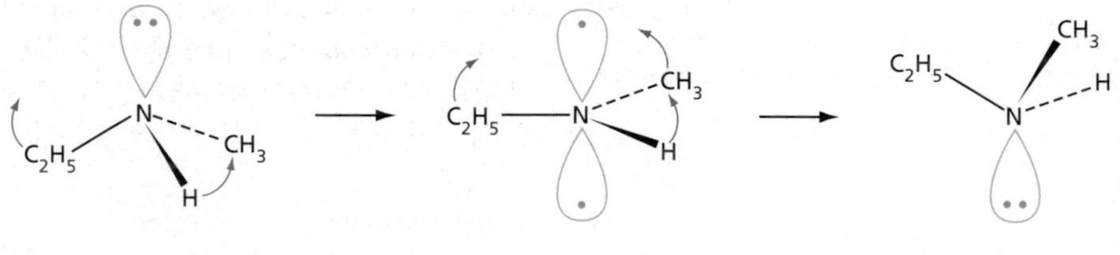

■ **Figure 30.5** Inversion at the nitrogen atom produces a racemic mixture.

30.3 Reactions of amines

Basicity

As mentioned at the start of this chapter, amines are basic. They react with acids to form salts:

$$CH_3CH_2\ddot{N}H_2 + HCl \rightarrow CH_3CH_2—\overset{+}{N}H_3 + :Cl^-$$

The amine chlorides, sulphates and nitrates are white crystalline solids, soluble in water, but insoluble in organic solvents.

Those amines that are soluble in water form weakly alkaline solutions, just as ammonia does, due to partial reaction with the solvent, producing OH^- ions:

$$CH_3CH_2\ddot{N}H_2 + H_2O \rightleftharpoons CH_3CH_2NH_3^+ + OH^-$$

Table 30.3 lists some amines, with their K_b and pK_b values. Ammonia is included for comparison.

■ **Table 30.3**
The basicity of some amines.

$pK_b = -\log_{10} K_b$ where $K_b = \dfrac{[RNH_3^+][OH^-]}{[RNH_2]}$

Amine	Formula	K_b/mol dm^{-3}	pK_b
4-Nitrophenylamine	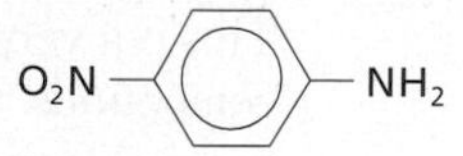	1.0×10^{-13}	13.0
Phenylamine	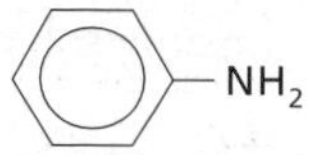	4.2×10^{-10}	9.38
Dimethylphenylamine	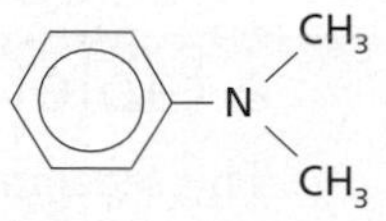	7.1×10^{-10}	9.15
4-Methylphenylamine	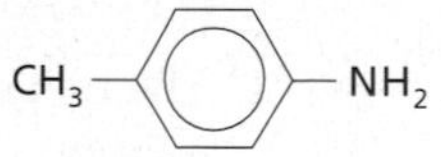	1.2×10^{-9}	8.92
Ammonia	NH_3	1.8×10^{-5}	4.75
Ethylamine	$CH_3CH_2NH_2$	5.1×10^{-4}	3.29
Diethylamine	$(CH_3CH_2)_2NH$	1.0×10^{-3}	3.00

The larger K_b is (and hence the smaller pK_b is), the stronger is the base. From Table 30.3 we can see that electron-donating alkyl groups attached to the nitrogen atom or to the benzene ring increase the basicity of amines, whereas the electron-withdrawing nitro group decreases the basicity. This is as we would expect, since the basicity depends on the availability of the lone pair of electrons on nitrogen to form a dative bond with a proton (see Figure 30.6). Electron donation from an alkyl group will encourage dative bond formation.

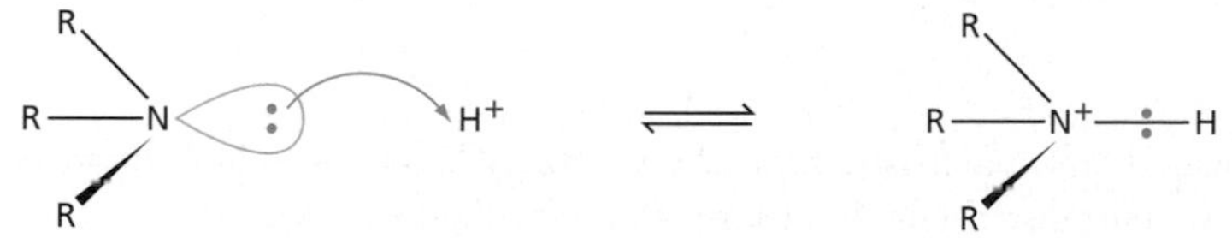

■ **Figure 30.6**
When an amine acts as a base, the nitrogen lone pair forms a dative bond with a proton.

The most dramatic difference in basicities to be seen in Table 30.3 is between that of the aryl amines (pK_b range 8.9–13.0) and the alkyl amines (pK_b range 3.0–3.3). Taking two compounds of about the same relative molecular mass and shape, we see that phenylamine is about a million times less basic than cyclohexylamine:

phenylamine
$K_b = 4.2 \times 10^{-10}$

cyclohexylamine
$K_b = 3.3 \times 10^{-4}$

This is because in phenylamine, the lone pair of electrons on the nitrogen atom is delocalised over the benzene ring. The nitrogen atom is planar, with its lone pair in a p orbital. This can overlap with the delocalised π bond of the benzene ring (see Figure 30.7). This overlap, causing a drift of electron density from nitrogen to the ring, has two effects on the reactivity of phenylamine:

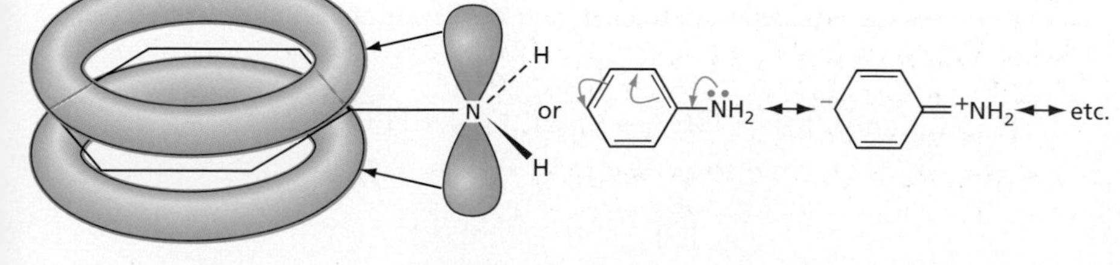

■ **Figure 30.7**
Delocalisation of the nitrogen lone pair in phenylamine.

- It causes the lone pair to be much less basic (see above) and also much less nucleophilic.
- It causes the ring to be more electron rich, and so to undergo electrophilic substitution reactions much more readily than benzene. The enhanced reactivity of phenylamine in this regard is on a par with that of phenol (see Chapter 27, page 518).

Only the lower amines (with five or fewer carbon atoms) are soluble in water. But the ionic nature of their salts allows all amines to dissolve in dilute aqueous acids. This is a very useful way of purifying amines from a mixture with non-basic compounds. The amines can be regenerated by adding an excess of aqueous sodium hydroxide to the solution, after the non-basic impurity has been extracted with an organic solvent (see the experiment below).

■ *Experiment* The separation of codeine and paracetamol from a tablet

This separation makes use of the fact that codeine is basic, forming water-soluble salts with acids, but paracetamol is slightly acidic, and is insoluble in water.

1 The tablet is ground up with an organic solvent such as dichloromethane. The solution is placed in a separating funnel along with dilute hydrochloric acid and shaken (see Figure 30.8).

$$\underset{\text{codeine}}{C_{18}H_{21}O_3N} + HCl \rightarrow \underset{\text{codeine hydrochloride}}{[C_{18}H_{21}O_3NH^+]Cl^-}$$

■ **Figure 30.8**
The first separation removes the paracetamol in the organic layer.

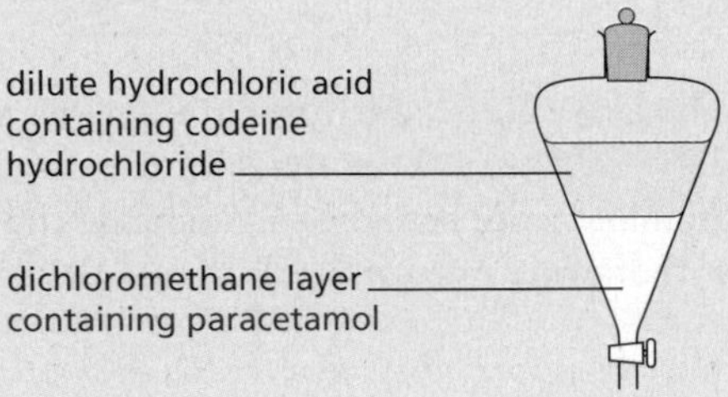

2 The lower dichloromethane layer is run off, and aqueous sodium hydroxide added to neutralise the hydrochloric acid (see Figure 30.9).

■ **Figure 30.9**
Sodium hydroxide neutralises the aqueous layer.

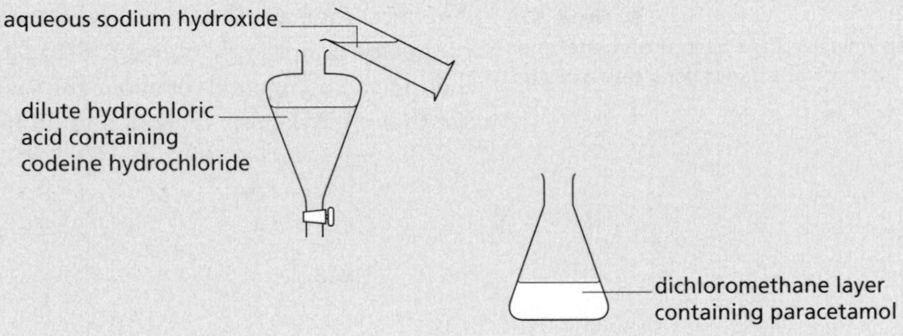

3 The codeine is liberated from its salt by the aqueous sodium hydroxide. More dichloromethane is added to extract it, and the layers are again separated (see Figure 30.10).

$$[C_{18}H_{21}O_3NH^+]Cl^- + OH^- \rightarrow C_{18}H_{21}O_3N + H_2O + Cl^-$$

■ **Figure 30.10**
The neutral codeine is re-formed and dissolves in the organic layer.

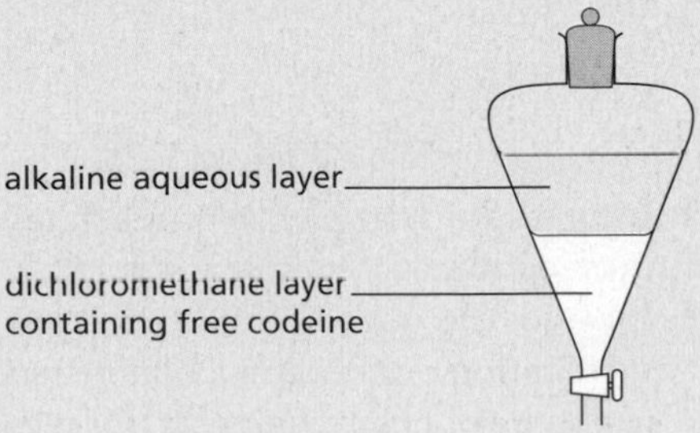

Reactions as nucleophiles

Another similarity between amines and ammonia is their nucleophilicity. The lone pair on the nitrogen atom of amines makes them good nucleophiles. They react with alkyl and acyl halides, with acid anhydrides and with esters (see Chapter 25, page 468, and Chapter 29, pages 550, 552 and 553).

$$CH_3CH_2NH_2 + CH_3Br \rightarrow CH_3CH_2NHCH_3 + HBr$$

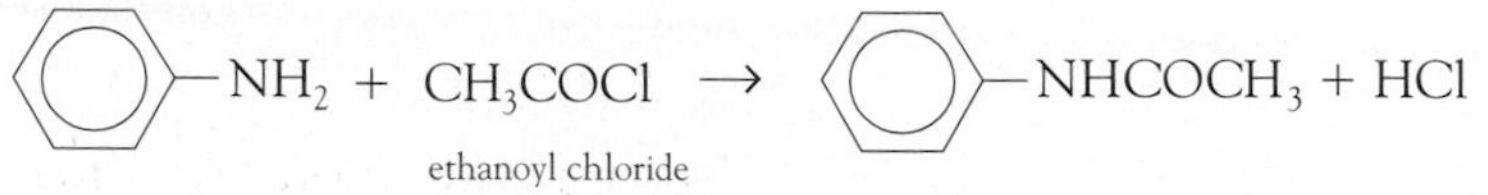

$$\langle\bigcirc\rangle\!-\!NH_2 + \underset{\text{ethanoyl chloride}}{CH_3COCl} \rightarrow \langle\bigcirc\rangle\!-\!NHCOCH_3 + HCl$$

$$CH_3NH_2 + \underset{\text{ethanoic anhydride}}{CH_3CO\!-\!O\!-\!COCH_3} \rightarrow CH_3CO\!-\!NHCH_3 + CH_3CO_2H$$

In the presence of an excess of bromoalkane, amines can be successively alkylated, first to secondary and then to tertiary amines. These are still strongly nucleophilic, and are readily alkylated to quaternary ammonium salts:

$$CH_3CH_2NH_2 + CH_3CH_2Br \rightarrow (CH_3CH_2)_2NH + HBr$$
$$(CH_3CH_2)_2NH + CH_3CH_2Br \rightarrow (CH_3CH_2)_3N + HBr$$
$$(CH_3CH_2)_3N + CH_3CH_2Br \rightarrow (CH_3CH_2)_4N^+Br^-$$

Quarternary ammonium salts are water-soluble solids, with no basic character at all, because there is no lone pair of electrons on the nitrogen atom. An important naturally occurring ammonium salt is choline. Phosphatidylcholine is a key phospholipid component of cell membranes. Acetylcholine is an important neurotransmitter, allowing a nerve impulse to pass from the end of one nerve to the start of the next one (see Figures 30.11 and 30.12).

■ **Figure 30.11**
Some important naturally occurring quaternary ammonium salts.

$$C_{15}H_{31}\!-\!\overset{\overset{\displaystyle O}{\|}}{C}\!-\!O\!-\!CH_2$$

$$C_{17}H_{33}\!-\!\overset{\overset{\displaystyle O}{\|}}{C}\!-\!O\!-\!CH$$

phosphatidylcholine

$$HO\!-\!CH_2\!-\!CH_2\!-\!{}^+N(CH_3)_3Cl^-$$
choline chloride

$$CH_3\!-\!\overset{\overset{\displaystyle O}{\|}}{C}\!-\!O\!-\!CH_2\!-\!CH_2\!-\!{}^+N(CH_3)_3Cl^-$$
acetylcholine chloride

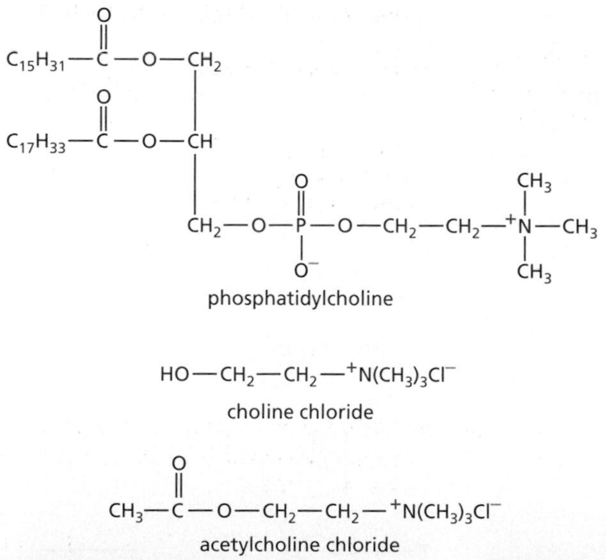

■ **Figure 30.12**
a Acetylcholine transmits a nerve impulse from one end of a nerve to another at a synapse or to a muscle at a motor end plate. **b** Curare, a deadly poison used by South American Indians on their arrow tips, blocks the action of acetylcholine.

a**b**

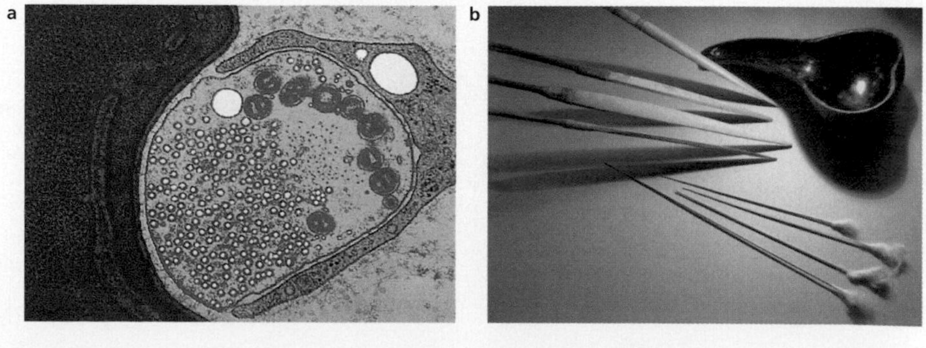

Quaternary ammonium salts containing long carbon chains (see Figure 30.13) are important cationic surfactants. They act in a similar way to anionic detergents (see Chapter 27, page 508), reducing the surface tension, and forming micelles with oil and grease particles.

■ **Figure 30.13**
A cationic surfactant molecule acts in a similar way to an anionic detergent.

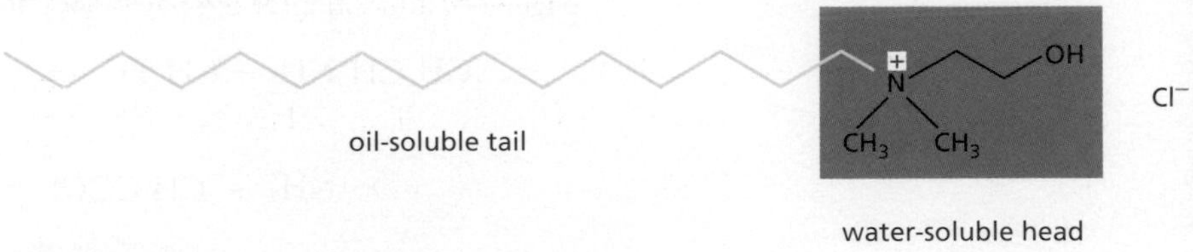

Reaction with nitrous acid (nitric(III) acid)

Primary amines react with nitrous acid (nitric(III) acid) in warm aqueous acidic solution to give nitrogen gas:

$$R{-}NH_2 + HNO_2 \xrightarrow[\text{in water}]{30\text{–}50\,°C} R{-}OH + N_2 + H_2O$$

Nitrous acid is unstable, and has to be made as required by reacting together sodium nitrite and hydrochloric acid:

$$NaNO_2 + HCl \rightarrow NaCl + HNO_2$$

How it happens: the reaction of amines with nitrous acid

The mechanism of the reaction is as follows. The lone pair of electrons on the nitrogen atom of the amine attacks the nitrogen atom of the covalent nitric(III) acid molecule. Proton transfers then take place, eventually breaking an N—O bond to form a **diazonium salt**. Because of the great stability of the nitrogen molecule (bond strength = 945 kJ mol^{-1}), the C—N bond breaks, forming a carbocation. This then reacts with the solvent, water, to form the alcohol (see Figure 30.14).

■ **Figure 30.14**
The reaction of an amine with nitrous acid forms a diazonium salt, which breaks down to form the alcohol.

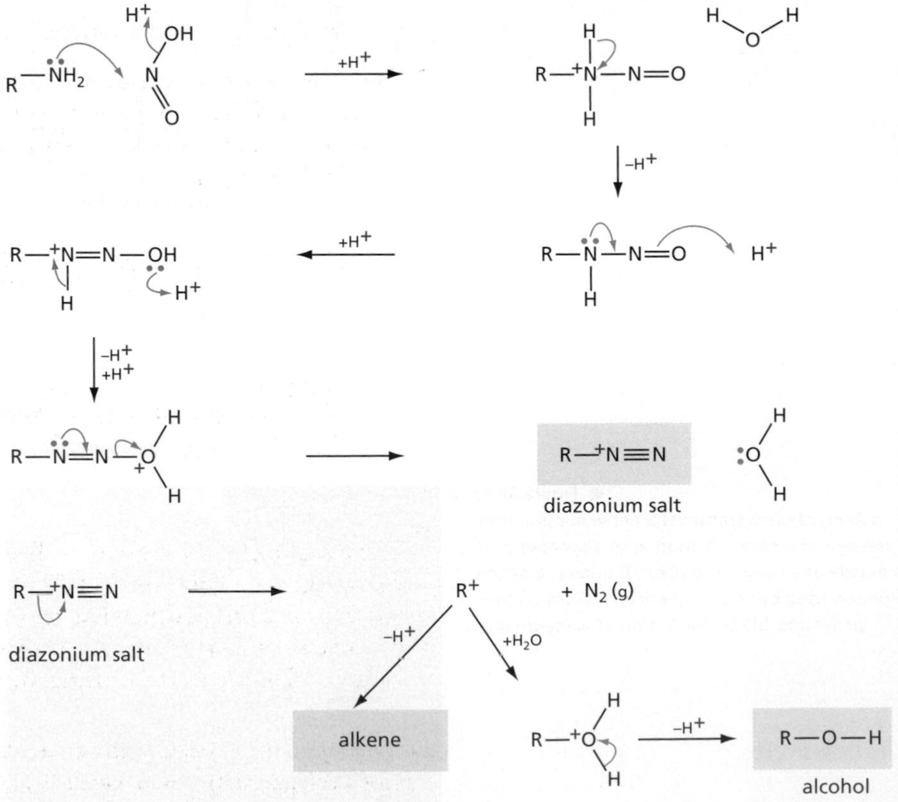

The panel opposite shows that a **diazonium salt** is an intermediate in the reaction between nitrous acid and a primary amine. The diazonium salts derived from alkyl amines are unstable under all conditions, and cannot be isolated. Aryl amines, however, form fairly stable diazonium salts at low temperatures. Under these conditions nitrogen is not evolved, but a solution of the diazonium salt is formed:

$$\text{C}_6\text{H}_5-\text{NH}_2 + \text{HNO}_2 + \text{H}^+ \xrightarrow[\text{at } T < 5\,^\circ\text{C}]{\text{NaNO}_2 + \text{HCl}} \text{C}_6\text{H}_5-\overset{+}{\text{N}}\equiv\text{N}\overset{-}{\text{Cl}} + 2\text{H}_2\text{O}$$

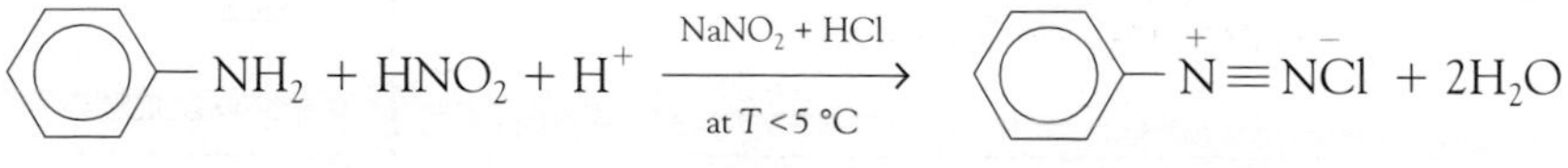

phenyldiazonium chloride

Aryl diazonium salts are unstable and explosive when dry, but can be kept for several days in solution in a refrigerator.

Reactions of diazonium salts

1 Diazonium salts are important intermediates in the synthesis of other substituted aromatic compounds:

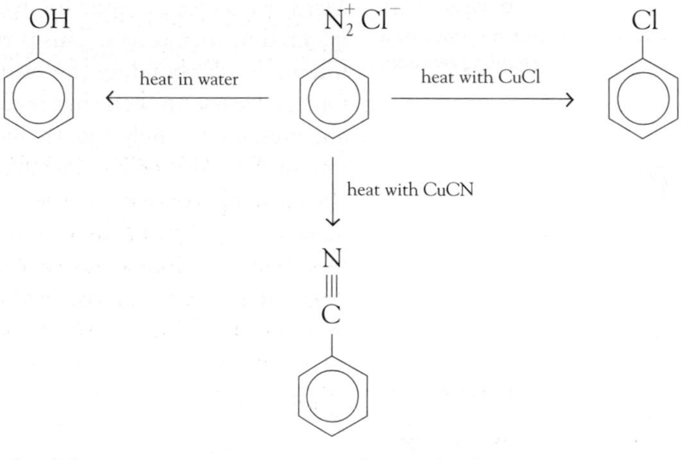

In each of these reactions nitrogen is also evolved.

2 The most important reactions of diazonium salts are their use in the formation of **azo dyes**. When a solution of a diazonium salt is added to an alkaline solution of a phenol, an electrophilic substitution reaction (known as a **coupling reaction**) takes place:

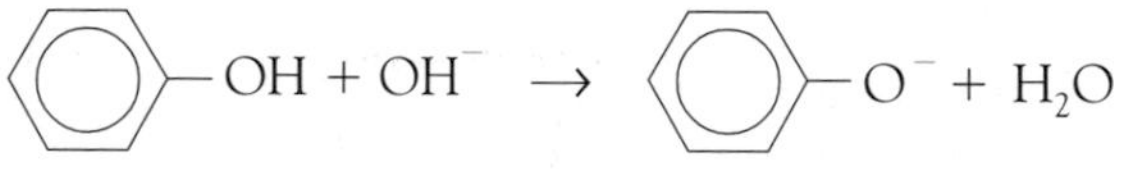

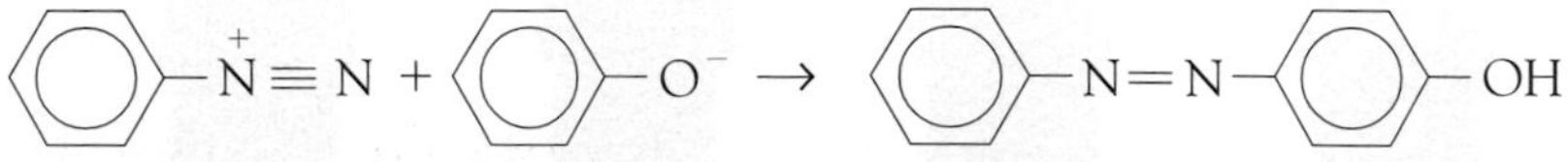

an azo compound

The azo group, —N=N— is called a **chromophore**. Compounds containing this group are highly coloured. Their colours range from yellow and orange to red, blue and green, depending on what other groups are attached to the benzene rings. The common acid–base indicator methyl orange is an azo compound, made by the coupling reaction shown in Figure 30.15 (overleaf).

A **chromophore** is a group which, especially when joined to other unsaturated groups, causes a compound to absorb visible light, and so become coloured.

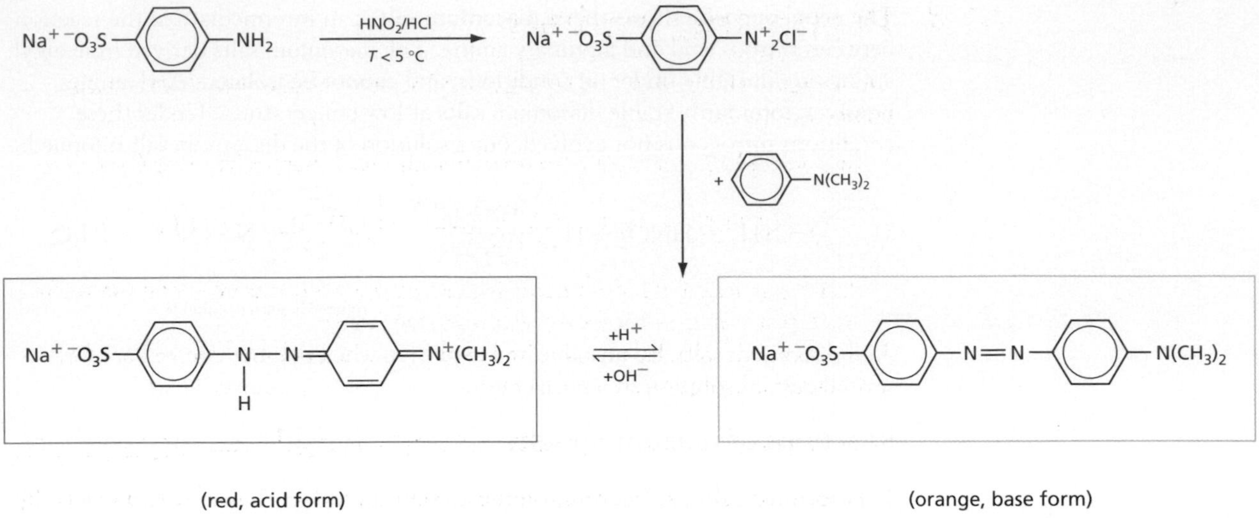

■ **Figure 30.15**
Methyl orange is an azo compound formed by a coupling reaction.

Many dye molecules used for dyeing clothes do not easily stick to the fibres of the material by themselves. This is especially the case if their predominant method of intermolecular bonding (for example, van der Waals, hydrogen bonding, ionic forces) does not match that of the molecules that make up the material. Mordants are often used to help the dye molecules stick. A **mordant** is a polyvalent metal ion, such as Al^{3+} or Fe^{3+}, which can form coordination complexes with both the dye molecule and with —OH, —CO or —NH groups on the molecules that make up the fibres of the material. By this means the dye molecule and the fibre molecule are permanently held together by the metal ion.

A great number of dyes for clothes, colour printing and food colouring are azo dyes. Figure 30.16 shows two examples.

■ **Figure 30.16**
Some azo dyes used commercially.

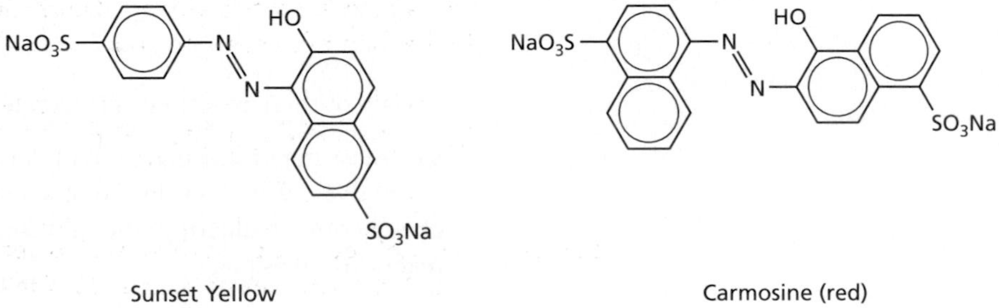

■ **Figure 30.17**
Food colours often contain azo dyes.

30.4 Preparing amines

There are two main ways of preparing alkyl amines, both of which start with halogenoalkanes.

- By nucleophilic substitution with ammonia (see Chapter 25, page 468):

$$CH_3CH_2Br + 2NH_3 \xrightarrow[\text{under pressure}]{\text{heat in ethanol}} CH_3CH_2NH_2 + NH_4Br$$

  As long as an excess of ammonia is used, further substitutions giving secondary and tertiary amines can be avoided (see page 569).
- By nucleophilic substitution with sodium cyanide, followed by reduction (see Chapter 25, page 469):

$$CH_3CH_2Br + NaCN \xrightarrow{\text{heat in ethanol}} CH_3CH_2CN \xrightarrow{H_2 + Ni} CH_3CH_2CH_2NH_2$$

  Notice that during this reaction the carbon chain length has been extended by one carbon atom.

Aryl amines are most commonly prepared by the reduction of aromatic nitro compounds (see Chapter 27, page 505):

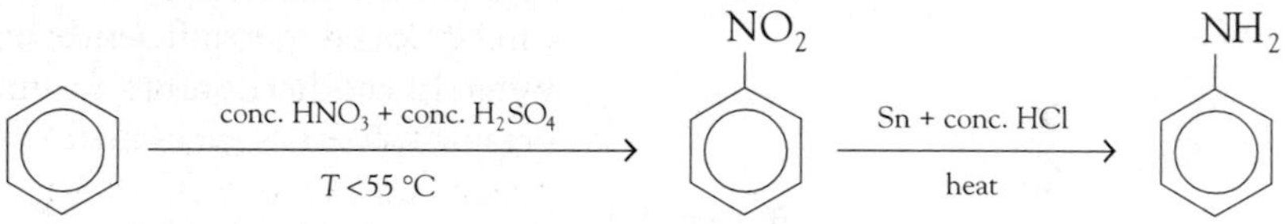

To produce the free amine, excess sodium hydroxide must be added after the reduction is completed.

An unusual way of making amines, which is applicable to both alkyl and aryl amines, is the Hofmann reaction (see Chapter 29, page 557). In this reaction an acid amide is reacted with an alkaline solution of bromine. A carbon atom is lost, in the form of a carbonate ion:

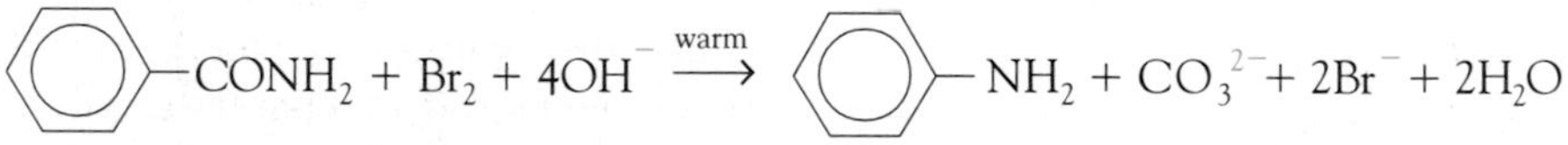

Example The 'feel good factor' in chocolate has been identified as 2-phenylethylamine, $C_6H_5CH_2CH_2NH_2$. Suggest a synthesis of this compound from chloromethylbenzene, $C_6H_5CH_2Cl$.

Answer The target amine has one more carbon atom than the suggested starting material, so the cyanide route is required:

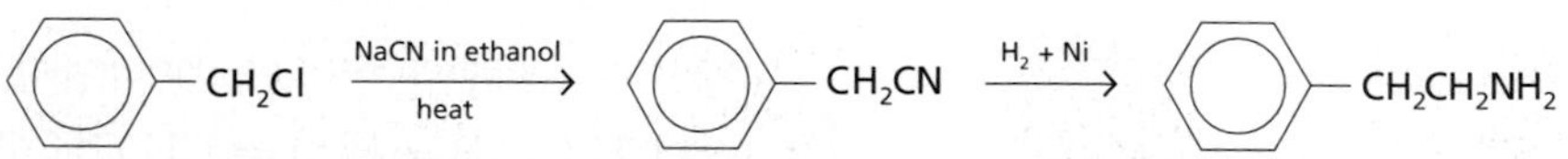

Further practice Suggest three ways of making butylamine ($CH_3CH_2CH_2CH_2NH_2$), each method starting from a compound containing a different number of carbon atoms.

30.5 Amino acids

Amino acids have two functional groups:

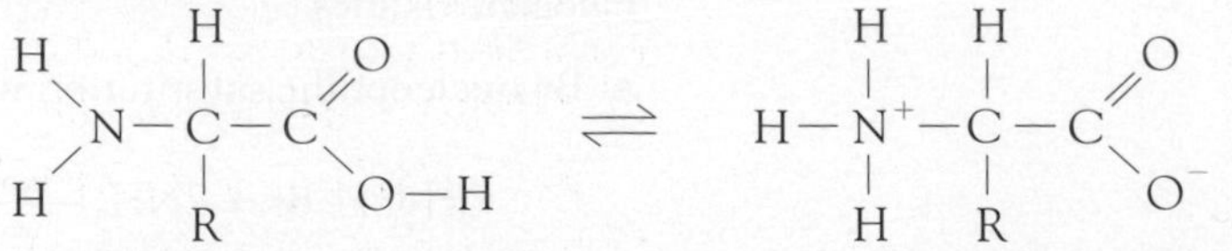

The 2-amino acids (α-amino acids) form an interesting class of compounds, apart from their great importance as the building blocks of proteins. In many compounds containing two or more functional groups, the reactions of one group can be considered independently of those of the other group. They are often widely separated, and dissimilar to each other.

In amino acids, however, the two groups are adjacent to each other. What is more, they are of opposite chemical type: the $-NH_2$ group is basic, whilst the $-CO_2H$ group is acidic. Interaction between the two is virtually inevitable.

Physical and chemical properties

Whilst alkyl amines and the lower aliphatic carboxylic acids are liquids, the amino acids are all solids. They have high melting points (often decomposing before they can be heated to a sufficiently high temperature to melt them), and are soluble in water. In another contrast to amines and carboxylic acids, they are insoluble in organic solvents such as methylbenzene.

Table 30.4 A comparison of some properties of the amino acids glycine and alanine with those of amines and carboxylic acids.

Compound	Formula	Melting point/°C	K_a/mol dm^{-3}	K_b/mol dm^{-3}
Glycine	$H_2NCH_2CO_2H$	233	1.4×10^{-10}	2.4×10^{-12}
Alanine	$H_2NCH(CH_3)CO_2H$	297	9.8×10^{-11}	2.2×10^{-12}
Ethylamine	$H_2NCH_2CH_3$	−81	–	5.1×10^{-4}
Propanoic acid	$CH_3CH_2CO_2H$	−21	1.3×10^{-5}	–

Table 30.4 compares the properties of the two simplest amino acids with those of a simple amine and a simple carboxylic acid. Apart from their extremely high melting points, another clear difference between the amino acids and their mono-functional counterparts is their much reduced basicity and acidity. Compared with ethylamine,

$$CH_3CH_2NH_2 + H_2O \rightleftharpoons CH_3CH_2NH_3^+ + OH^- \qquad K_b = 5.1 \times 10^{-4}\,\text{mol dm}^{-3}$$

glycine is 200 million times less basic:

$$H_2NCH_2CO_2H + H_2O \rightleftharpoons H_3N^+CH_2CO_2H + OH^- \qquad K_b = 2.4 \times 10^{-12}\,\text{mol dm}^{-3}$$

Likewise, compared with propanoic acid:

$$CH_3CH_2CO_2H + H_2O \rightleftharpoons CH_3CH_2CO_2^- + H_3O^+ \qquad K_a = 1.3 \times 10^{-5}\,\text{mol dm}^{-3}$$

alanine is over 100 000 times less acidic:

$$H_2NCH(CH_3)CO_2H + H_2O \rightleftharpoons H_2NCH(CH_3)CO_2^- + H_3O^+ \qquad K_a = 9.8 \times 10^{-11}\,\text{mol dm}^{-3}$$

This large reduction in basicity and acidity is readily explained by the idea that the acidic and basic groups with an amino acid molecule have already reacted with each other:

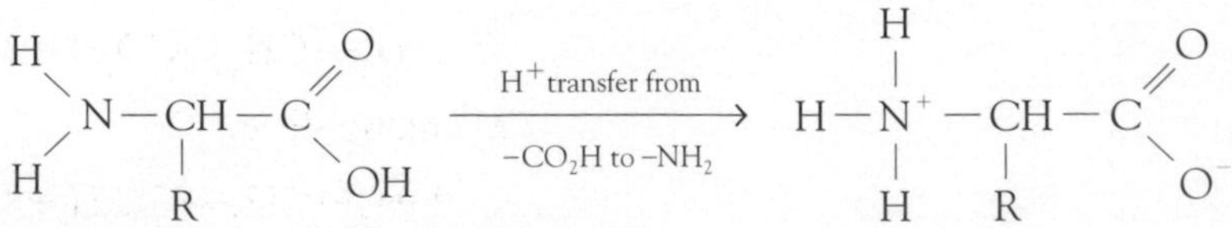

The product is an 'internal salt', called a **zwitterion** (a name first coined by Kuster in 1897). The salt-like nature of zwitterions also explains the high melting points and the solubility characteristics of amino acids. It is thought that all amino acids exist as zwitterions (or dipolar ions) in solution and in the solid lattice. Their low acidity is therefore due to proton donation from the ammonium group rather than from the carboxylic acid group:

$$H_3N^+CH_2CO_2^- + H_2O \rightleftharpoons NH_2CH_2CO_2^- + H_3O^+ \quad K_a = 1.4 \times 10^{-10} \, \text{mol dm}^{-3}$$

This value is directly comparable with the ionisation of other ammonium salts:

$$NH_4^+ + H_2O \rightleftharpoons NH_3 + H_3O^+ \quad\quad\quad\quad K_a = 5.6 \times 10^{-10} \, \text{mol dm}^{-3}$$

> A **zwitterion** is a molecule that contains both a cationic group and an anionic group.

One further characteristic property of 2-amino acids is their optical activity. All except glycine possess a chiral carbon atom, having four different atoms or groups attached to it. There are two distinct and different ways in which these four groups can arrange themselves around the central carbon atom, shown in Figure 30.18 (see Chapter 22, pages 407–10). All amino acids derived from the hydrolysis of proteins have the *l* configuration.

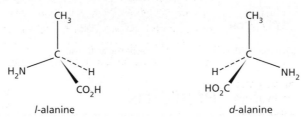

■ **Figure 30.18**
l- and *d*-alanine.

l-alanine   *d*-alanine

Reactions of amino acids

Because the dipolar form of amino acids is in equilibrium with a small amount of the un-ionised form, amino acids show many of the typical reactions of amines and carboxylic acids.

1 The amino group can be acylated (see page 569) and reacts with nitrous acid (see page 570):

$$NH_2 - CH(R) - CO_2H \xrightarrow{CH_3COCl} CH_3\overset{O}{\overset{\|}{C}} - NH - CH(R) - CO_2H$$

$$NH_2 - CH(R) - CO_2H \xrightarrow{HNO_2} HO - CH(R) - CO_2H + N_2(g)$$

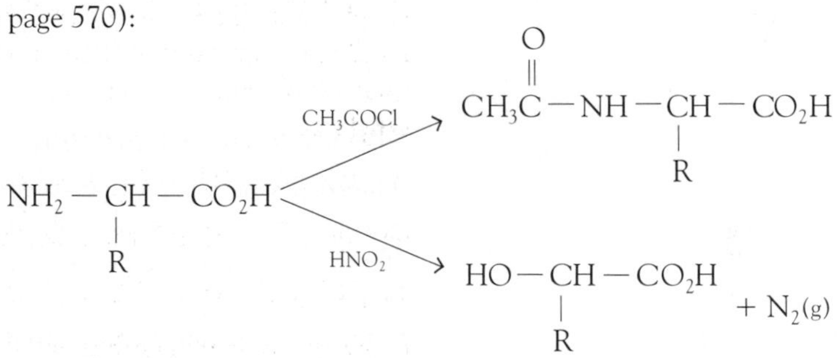

2 The carboxylic acid group can be esterified:

$$NH_2 - CH(R) - C\overset{O}{\underset{OH}{}} \xrightarrow[\text{heat}]{CH_3OH + \text{conc. } H_2SO_4} NH_2 - CH(R) - C\overset{O}{\underset{OCH_3}{}}$$

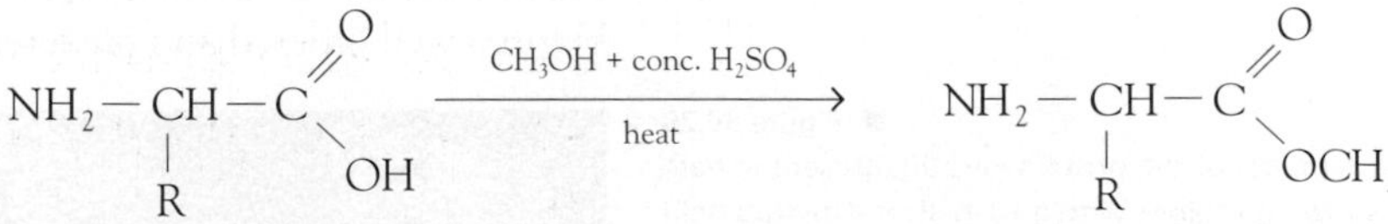

Amino acids act as buffers, stabilising the pH of a solution if excess acid or alkali is added (see Chapter 12, page 247).

On adding acid:

$$NH_2-CH_2-CO_2H + H^+ \rightarrow {}^+NH_3-CH_2-CO_2H$$

On adding alkali:

$$NH_2-CH_2-CO_2H + OH^- \rightarrow NH_2-CH_2-CO_2^- + H_2O$$

In addition to the reactions of the —NH_2 and the —CO_2H groups, amino acids with side chains that contain functional groups also show the reactions of that group. Figure 30.19 shows the side chains of some amino acids. For example, serine reacts with phosphorus(V) chloride to form a chloroalkyl side chain, and tyrosine reacts with bromine water just as does phenol (see Chapter 27, page 518).

■ **Figure 30.19** A selection of amino acids.

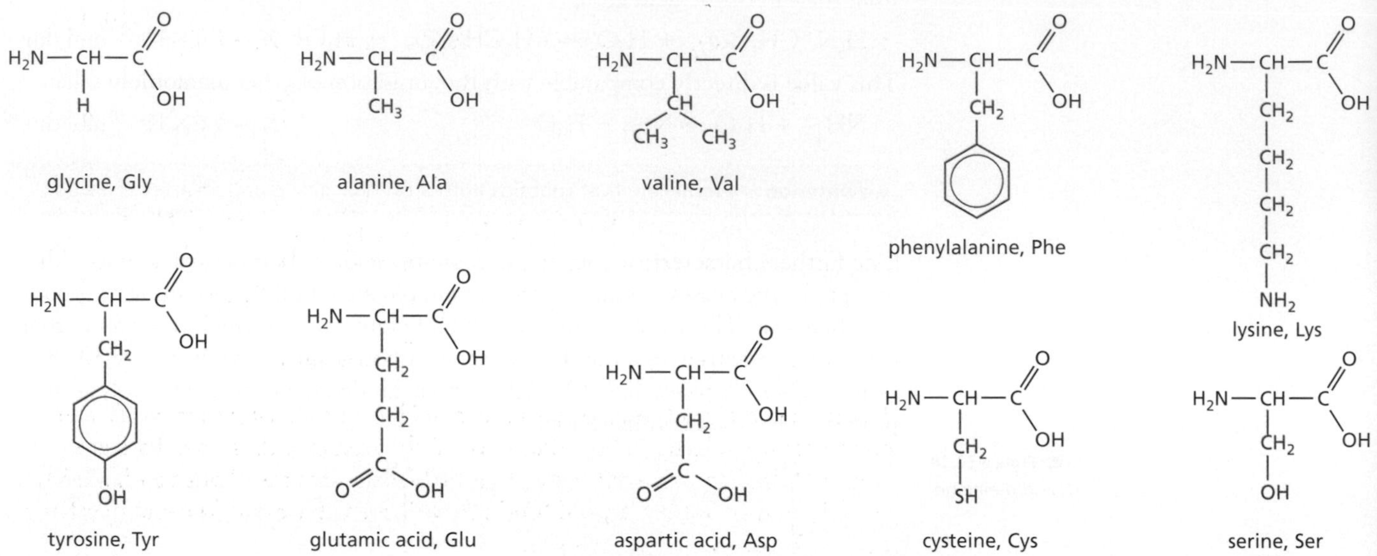

30.6 Proteins

Proteins are one of several types of polymer found in nature.

- They are important structurally, in skin, muscles and tendons.
- They are essential in metabolism – all enzymes are proteins.
- Many proteins have a function in transport – oxygen is carried in the blood by haemoglobin, the major part of which is a protein, and ionic compounds are transported through cell walls by specific complexing proteins.
- Proteins are also involved in communication – the major part of a nerve is made from protein, and many hormones (for example, insulin) are small proteins or peptides.

The structure of proteins

The backbone of a protein is one or more chains made up of polymers of amino acids. Some proteins also have non-amino-acid groups called **prosthetic groups** (for example, the haem group in haemoglobin). The shape and physical and biochemical properties of a protein are the direct result of the number, type and sequence of the amino acids that make it up. This sequence is determined by the **gene**, the sequence of organic bases in the length of DNA that codes for the protein. Amino acids do not join together to form proteins in the body randomly, but in a well ordered and predetermined sequence.

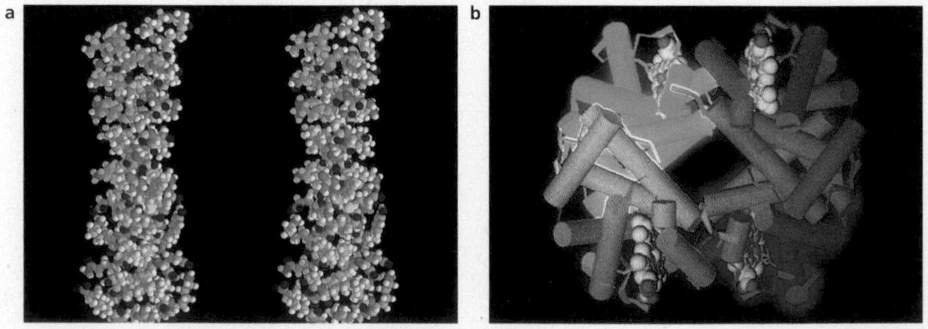

■ **Figure 30.20** a A model of the protein keratin, present in hair, showing regions with an α-helical structure held together by hydrogen bonding between the amino and carboxyl groups of amino acids along the protein chain. **b** Haemoglobin is a protein that contains four iron-containing haem groups.

The amino acids are joined by amide bonds between the amino group of one acid and the carboxyl group of another. So proteins are polyamides, and are condensation polymers (see Chapter 29, page 554). When an amide bond is formed between two amino acids, it is called a **peptide bond**, and proteins are also known as **polypeptides**:

$$H_2N-CH-CO_2H + H_2N-CH-CO_2H + H_2N-CH-CO_2H \xrightarrow{-2H_2O}$$
$$\qquad\quad | \qquad\qquad\qquad | \qquad\qquad\qquad |$$
$$\qquad\quad R \qquad\qquad\qquad R \qquad\qquad\qquad R$$

peptide group

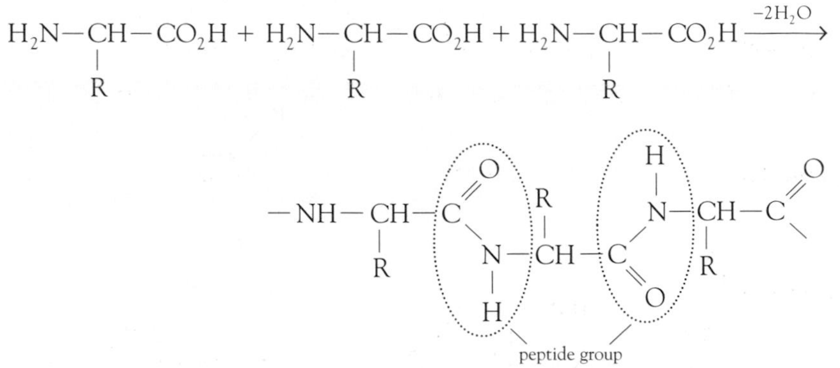

The formulae of peptides are often described by the following shorthand. The customary three-letter abbreviations for all the amino acids in the chain are written down, starting with the **N terminal** (the end of the chain that has a free —NH$_2$ group) on the left, and finishing with the **C terminal** (the end that has a free —CO$_2$H group) on the right.

Example Convert the following structural formula into the correct shorthand:

$$NH_2-CH_2-CO-NH-CH(CH_3)-CO_2H$$

Answer This structure contains one peptide bond in the middle, so it is a **dipeptide**, containing two amino acid residues. The residue to the left of the peptide bond is glycine (Gly) and that to the right is alanine (Ala) (see Figure 30.19). The shorthand formula is therefore:

Gly–Ala

Note that this is a different compound from Ala–Gly, because of the left-to-right direction of the peptide bond. The structural formula of Ala–Gly is:

$$NH_2-CH(CH_3)-CO-NH-CH_2-CO_2H$$

Further practice 1 Draw the structural formulae of the following two tripeptides (use Figure 30.19 to help you):
   a Val–Ser–Tyr
   b Tyr–Ala–Gly.

2 Use the customary three-letter abbreviations to describe the following tetrapeptide:

$$NH_2-CH-CO-NH-CH-CO-NH-CH_2-CO-NH-CH-CO_2H$$
$$\qquad\quad | \qquad\qquad\qquad | \qquad\qquad\qquad\qquad\qquad\qquad |$$
$$\qquad CH_2OH \qquad\qquad CH_2CO_2H \qquad\qquad\qquad\qquad CH(CH_3)_2$$

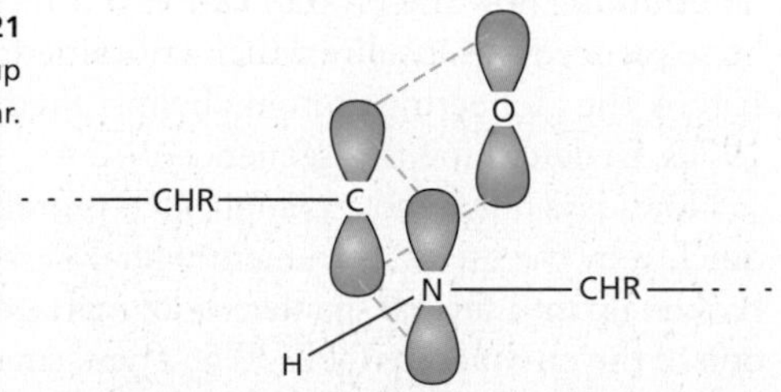

■ Figure 30.21
Overlap of p orbitals makes the peptide group planar.

Because of the overlap between the lone pair of electrons on the nitrogen atom (in a p orbital) and the carbonyl group next to it, the peptide group is rigid and planar. It exists in the *trans* arrangement (see Figure 30.21).

This inflexibility of the peptide group allows many proteins to take up an **α-helical arrangement** (see Figure 30.22, overleaf). This lends strength and rigidity to some protein structures. Strong hydrogen bonds can form between the N—H of one peptide group and the C=O of another one.

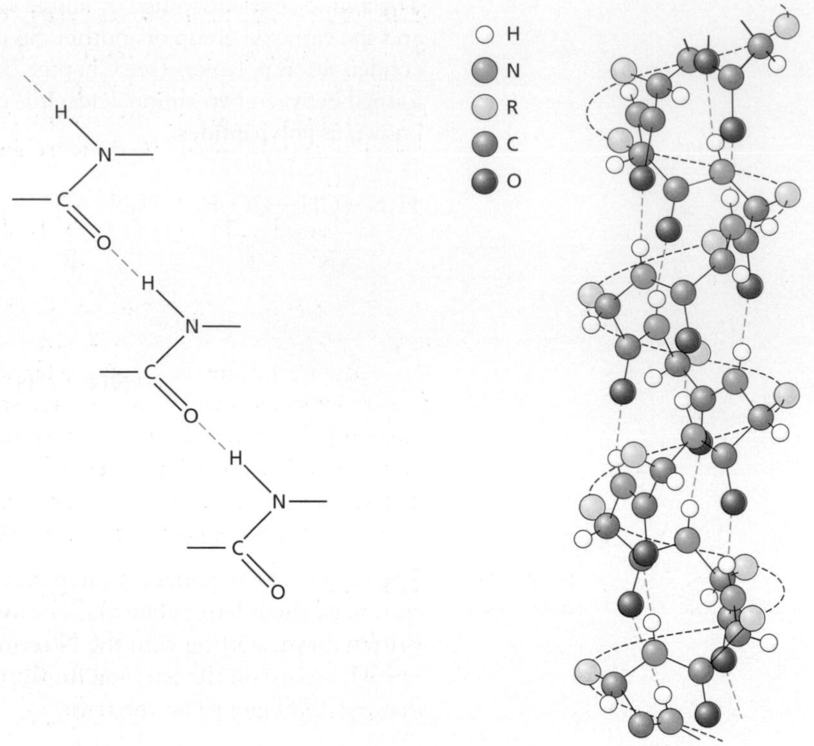

■ **Figure 30.22** The α helix in proteins.

Like all amides, proteins are fairly resistant to hydrolysis by chemical means. Heating under reflux for 8 hours with concentrated hydrochloric acid is often sufficient to break up the polypeptide chains, however.

$$\underset{\text{protein}}{-NH-CHR-CO-NH-CHR'-CO-} \xrightarrow{\text{conc. HCl for 8 hours at 110 °C}}$$

$$\underset{\text{amino acids}}{NH_2-CHR-CO_2H + NH_2-CHR'-CO_2H}$$

Enzymes can carry out this hydrolysis under much milder conditions (35 °C, pH 3–7). Some of them cut the peptide chain specifically at certain amino acids, whilst others are more general (see Table 30.5).

■ **Table 30.5** The specificity of some protein-hydrolysing enzymes.

Enzyme	Origin	Specificity
Trypsin	Pancreas	Breaks the chain on the —CO— (right-hand) side of basic amino acids such as lysine
Chymotrypsin	Pancreas	Breaks the chain on the —CO— (right-hand) side of amino acids containing aromatic rings, such as phenylalanine and tyrosine
Carboxypeptidase	Pancreas	Breaks off one amino acid at a time, starting at the CO_2H end of the chain

The analysis of proteins

As mentioned above, the amino-acid sequence of a protein is all-important, determining how the protein carries out its biochemical role. The first protein to be sequenced was insulin, which contained 51 amino acid residues in two chains. It took the pioneering protein chemist Fred Sanger several years, in the early 1950s, to determine this sequence.

Nowadays much more complicated proteins can be sequenced considerably more quickly, by the use of instrumental analysis and automation. Firstly, the protein is broken up into several smaller, more manageable chunks by partial hydrolysis using one of the enzymes in Table 30.5. These smaller chunks are separated and each is hydrolysed further, often using an enzyme of different specificity to the first one. The process is repeated until chunks containing just a few (5–10) amino acid monomers are formed. These can then be identified by looking at the fragmentation pattern in their mass spectra. The various pieces of the jigsaw can then be fitted together, to arrive at the sequence in the original protein.

Isoelectric points and electrophoresis

If we have a solution of an amino acid in water, the average charge on the many molecules of amino acid in the solution depends on the pH of the solution. This is because of the amphoteric properties of amino acids. The pH at which the net overall charge is zero is called the **isoelectric point** of that amino acid. For glycine, this is 6.07.

$$\underset{\text{at pH}>6.07}{NH_2CH_2CO_2^-} \xleftarrow{-H^+} \underset{\text{at pH}=6.07}{^+NH_3CH_2CO_2^-} \xrightarrow{+H^+} \underset{\text{at pH}<6.07}{^+NH_3CH_2CO_2H}$$

In solution at pH > 6.07, the average charge on glycine molecules becomes negative; at pH < 6.07, the molecules become positively charged.

Depending on the side groups, different amino acids have different tendencies to form cations or anions in solution, and so their isoelectric points will occur at different pH values. The more basic ones such as lysine, which have a tendency to form cations if dissolved in water, will have their isoelectric points at a more alkaline pH than most. On the other hand, the acidic amino acids such as aspartic acid will have their isoelectric points at a more acidic pH than most (see Table 30.6).

■ **Table 30.6**
The isoelectric points of some amino acids.

Amino acid	R group	Isoelectric point (IEP)
Glycine	—H	6.07
Lysine	$—(CH_2)_4NH_2$	9.74
Aspartic acid	$—CH_2CO_2H$	2.98

Conversely, if all three amino acids were dissolved together in the same buffer solution kept at pH 6.07, on average, the molecules of glycine would be electrically neutral:

$$NH_3^+—CH_2—CO_2^-$$

the molecules of lysine would be positively charged:

$$NH_3^+—\underset{\underset{CO_2^-}{|}}{CH}—(CH_2)_4NH_3^+$$

and the molecules of aspartic acid would be negatively charged:

$$NH_3^+—\underset{\underset{CO_2^-}{|}}{CH}—CH_2—CO_2^-$$

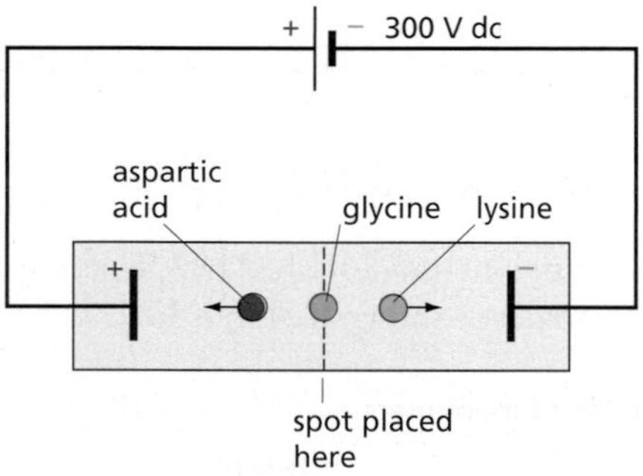

■ **Figure 30.23**
Electrophoresis separates lysine, aspartic acid and glycine.

If a drop of this buffer solution at pH 6.07 containing these three amino acids was spotted onto a gel-coated plate, immersed in a conducting liquid, and a potential difference applied, the lysine would move to the cathode, the aspartic acid would move to the anode, and the glycine, having no overall electrical charge, would not move at all.

This separation of a mixture in an electric field is called **electrophoresis**. The electrophoretic mobility depends not only on average charge but also on the size and shape of a molecule. Large, spiky molecules travel slower than small, spherical, ones:

$$v = \frac{EZ}{F}$$

where v = velocity, E = electric field, Z = average charge, F = frictional resistance

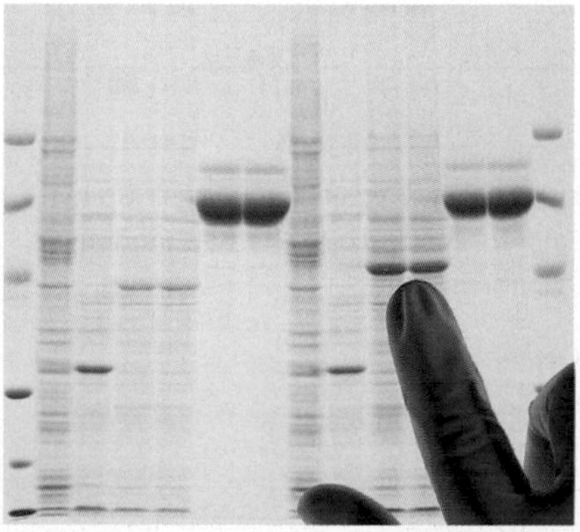

■ **Figure 30.24**
Electrophoresis of proteins.

Electrophoresis is used regularly to separate and identify not only mixtures of amino acids, but also mixtures of peptides obtained from proteins, and even mixtures of proteins themselves.

Example Partial hydrolysis of a pentapeptide produced the following fragments. What is the sequence of the five amino acids in the peptide?

Ala–Gly   Tyr–Lys   Ser–Ala   Gly–Tyr

Answer Line up the dipeptides, with the same amino acid residues above each other:

Ala — Gly

Ser — Ala

Gly — Tyr

Tyr — Lys

Hence the sequence is:

Ser–Ala–Gly–Tyr–Lys

Further practice Use Table 30.5 to determine what fragments will be formed as a result of the hydrolysis of the peptide

Ala–Tyr–Lys–Gly–Ser

by: **1** the enzyme trypsin
     2 the enzyme chymotrypsin.

30.7 Synthetic polyamides (nylons)

Synthetic polyamides (nylons) are an important class of manufactured polymer. They are condensation polymers (see Chapter 29, page 554) formed by the reaction between a diamine and a dicarboxylic acid:

$$H_2N-(CH_2)_n-NH_2 + HO_2C-(CH_2)_m-CO_2H \rightarrow$$

$$H_2N-(CH_2)_n-NH-CO-(CH_2)_m-CO_2H + H_2O$$

$$\downarrow$$

$$-HN-(CH_2)_n-NH-CO-(CH_2)_m-CO-$$

Values of n and m vary from 4 to 10. Different nylons are described by specifying the carbon atoms contained within each monomer as shown in Table 30.7.

■ **Table 30.7**
The monomers of some nylons.

Name	Formulae of monomers	
	Diamine	**Diacid**
Nylon-6,6	$NH_2-(CH_2)_6-NH_2$ (6 carbons)	$HO_2C-(CH_2)_4-CO_2H$ (6 carbons)
Nylon-6,10	$NH_2-(CH_2)_6-NH_2$ (6 carbons)	$HO_2C-(CH_2)_8-CO_2H$ (10 carbons)
Nylon-6	$NH_2-(CH_2)_5-CO_2H$ (only one monomer of 6 carbons)	

Example Draw the repeat unit of the chains of:
a nylon-6,10   **b** nylon-6.

Answer The repeat unit is the smallest unit from which the polymer chain can be built up by repetition. That of nylon-6,10 will include one molecule of each of the two monomers, while that of nylon-6 will contain just the one monomer unit:

a

$$\left[NH-(CH_2)_6-NH-\overset{O}{\overset{\|}{C}}-(CH_2)_8-\overset{O}{\overset{\|}{C}}\right]_n \quad \text{nylon-6,10}$$

b

$$\left[NH-(CH_2)_5-\overset{O}{\overset{\|}{C}}\right]_n \quad \text{nylon-6}$$

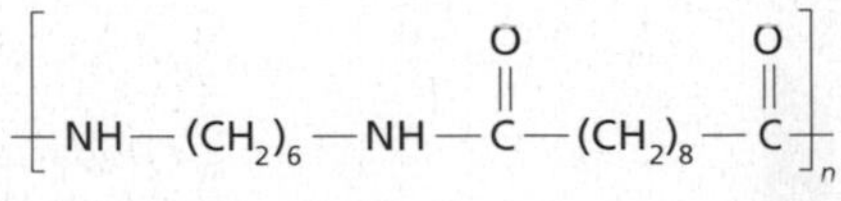

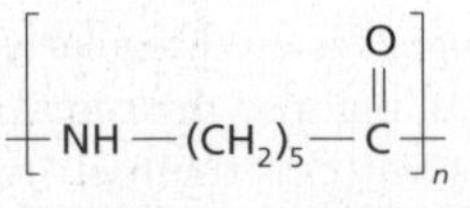

Both in proteins and in nylon-6, the amide bonds are all connected in the same direction along the chain:

$$\text{protein:} \quad -NH-\underset{\underset{R}{|}}{CH}-\overset{\overset{O}{||}}{C}-NH-\underset{\underset{R}{|}}{CH}-\overset{\overset{O}{||}}{C}-NH-$$

$$C \rightarrow N \qquad\qquad C \rightarrow N$$

$$\text{nylon-6:} \quad -NH-(CH_2)_5-\overset{\overset{O}{||}}{C}-NH-(CH_2)_5-\overset{\overset{O}{||}}{C}-NH-$$

$$C \rightarrow N \qquad\qquad C \rightarrow N$$

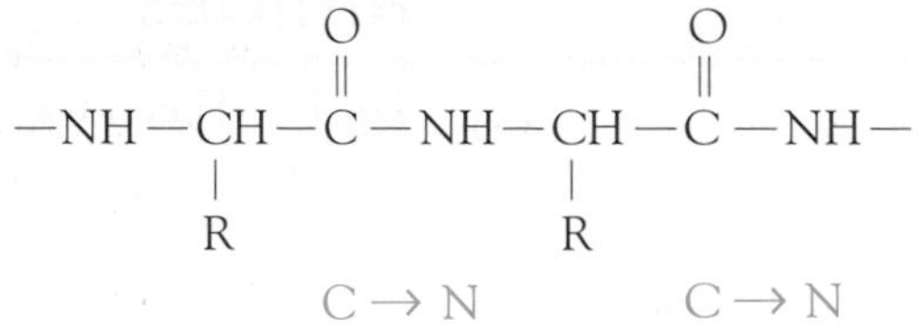

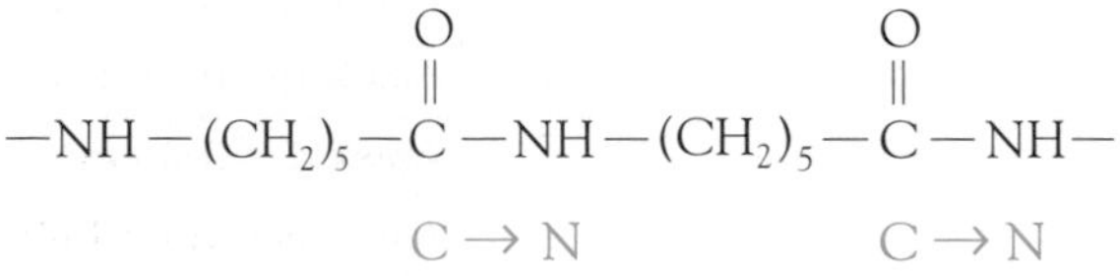

but in nylon–6,6, the direction of the amide linkage alternates:

$$-\overset{\overset{O}{||}}{C}-NH-(CH_2)_6-NH-\overset{\overset{O}{||}}{C}-(CH_2)_4-\overset{\overset{O}{||}}{C}-NH-$$

$$C \rightarrow N \qquad\qquad N \rightarrow C \qquad\qquad C \rightarrow N$$

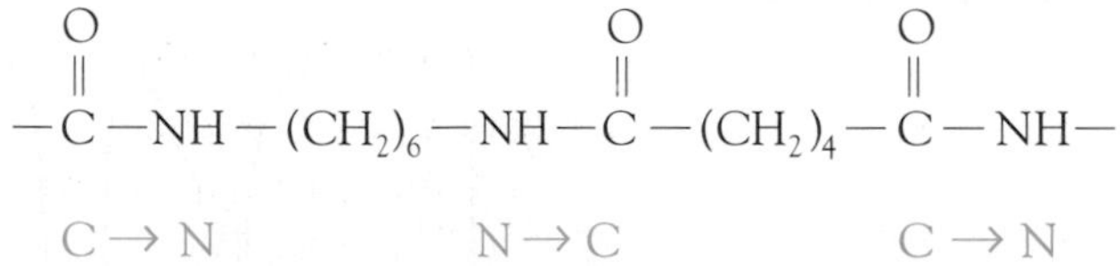

Like simple amides, polyamides can be hydrolysed by strong acids or alkalis. Many common fibres used for clothing, such as wool and silk (both are proteins), and nylon, may be dissolved accidentally by careless splashes of laboratory acids or alkalis.

■ **Figure 30.25**
Some uses of nylon.

30.8 The infrared spectra of amines and amides

The N—H bond is polar, and absorbs fairly strongly in the higher-frequency end of the infrared spectrum. The C—N bond is less polarised, and absorbs only weakly. Its absorbance is of no diagnostic value. Table 30.8 lists useful absorbances.

Table 30.8 IR absorption frequencies for amines and amides.

Bond and vibration mode	Frequency of absorption (wavenumber)/cm^{-1}
Amine N—H stretch	3400
Amide N—H stretch	3100
Amide C=O stretch	1690
Amine and amide N—H bend	1600

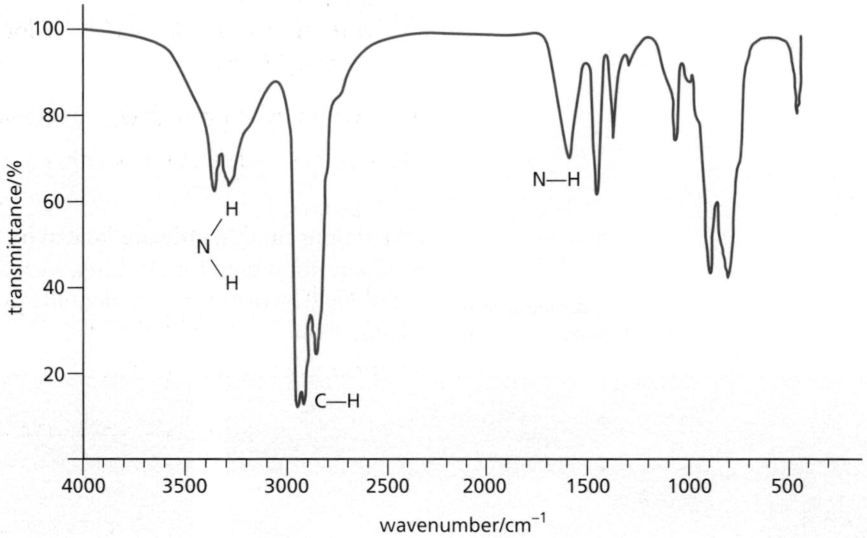

Figure 30.26 IR spectrum of propylamine.

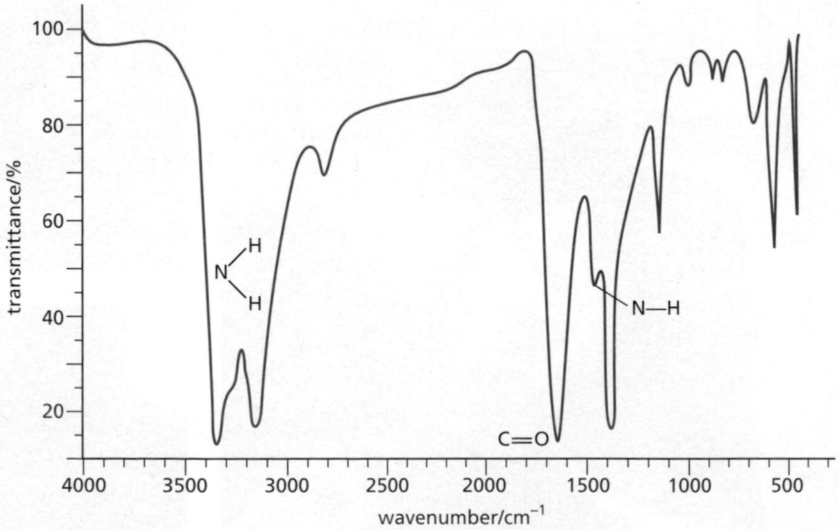

Figure 30.27 IR spectrum of ethanamide.

Summary

- **Amines** are weak organic bases, reacting with acids to form salts, and dissolving in water to give alkaline solutions.
- There are three types of amine – **primary**, RNH_2, **secondary**, R_2NH, and **tertiary**, R_3N.
- Aryl amines react with nitric(III) acid to form **diazonium salts**, from which many useful **azo dyes** are manufactured.
- **Amino acids** contain the $—NH_2$ and $—CO_2H$ functional groups adjacent to each other. In solution and in the solid they exist as **zwitterions**, $NH_3^+CHRCO_2^-$.
- Most amino acids are chiral. Those isolated from natural proteins have the *l* configuration.
- **Proteins** are polymers made up from hundreds of amino acids, joined by **peptide bonds**.
- **Synthetic polyamides** such as nylon are made by condensing diamines with dicarboxylic acids.

Key reactions you should know

(R = alkyl or aryl (Ar) unless otherwise stated)

- Alkylation:
  $R—NH_2 + CH_3Br \rightarrow R—NHCH_3 + HBr$
- Acylation:
  $R—NH_2 + CH_3COCl \rightarrow R—NHCOCH_3 + HCl$
- Nitrous acid (nitric (III) acid):

  $$R—NH_2 + HNO_2 \xrightarrow{\text{at room temperature}} R—OH + N_2(g) + H_2O$$
- Diazotisation:

  $$Ar—NH_2 + HNO_2 \xrightarrow{NaNO_2 + HCl \text{ at } 0\,°C} Ar—\overset{+}{N}{\equiv}N\ Cl^-$$

  $$Ar—\overset{+}{N}{\equiv}N\ Cl^- \xrightarrow{\text{hot } H_2O} Ar—OH$$

  $$Ar—\overset{+}{N}{\equiv}N\ Cl^- \xrightarrow{+\text{ phenols}/OH^-} \text{dyes}$$
- Polymerisation:
  $n H_2N—●—NH_2 + n H_2OC—■—CO_2H \rightarrow$
  $\left[NH—●—NHCO—■—CO\right]_n$

31 Organic analysis, spectra and synthesis

The previous chapters in the organic chemistry section of this book have looked at the chemistry of the various functional groups that organic molecules contain. This chapter provides an overview of these reactions, and explains how they can be coupled together to provide routes for the synthesis of more complex and useful organic compounds from simpler precursors.

We start by describing how the structure of an organic compound can be determined, using chemical analysis and spectroscopy.

The chapter is divided into the following sections.

31.1 Introduction to organic analysis

The structural formula of an organic compound is a vital part of our knowledge of the compound's properties and reactions. Many organic compounds, especially those of industrial or medical importance, have quite complicated structures. Modern analytical chemists have many techniques available to them for determining the structure of an organic compound. These include both chemical tests and spectroscopic data.

Determining the structure of natural products, whether they be of plant, animal or bacterial origin, is an essential part of increasing our understanding of the world around us. Many natural products have important pharmaceutical uses. The determination of their structures is an important step in explaining their physiological activity. Many pharmaceutical drugs obtain their effects by interacting with the active sites of enzymes. The size, shape, surrounding electronic field and functional groups of a molecule are the factors that determine how it binds to an enzyme. All these factors can be predicted once the structural formula has been determined.

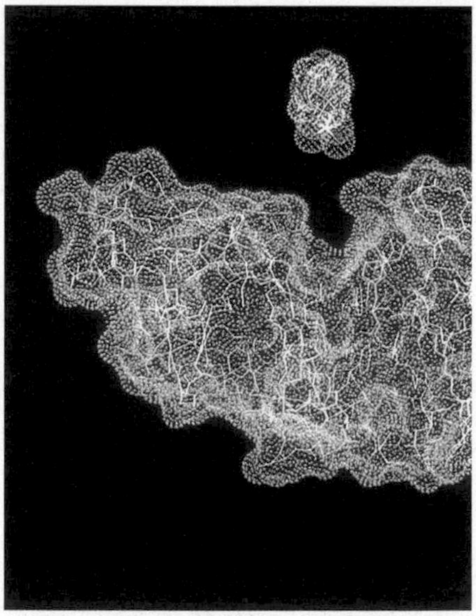

■ **Figure 31.1**
Computer models of enzyme molecules allow chemists to see which molecules may bind to the active site.

Knowledge of the geometry of, and the chemical groups present in, the active site of an enzyme allows the pharmaceutical chemist to design new molecules that have specific interactions with the active site. These molecules often bear structural similarities to natural products that are known to have a pharmaceutical use. Synthetic organic chemists can then use methods similar to the ones described in later sections of this chapter to make these new compounds. Many thousands of new organic substances are synthesised each year, and only a few of them are found to be useful drugs. But the search is well worthwhile. The handful of synthetic antibiotics, anti-cancer and heart-regulating drugs in use today have saved countless millions of lives throughout the world.

Figure 31.2 outlines the stages involved in determining the structure of an organic compound, together with the techniques that can be employed at each stage.

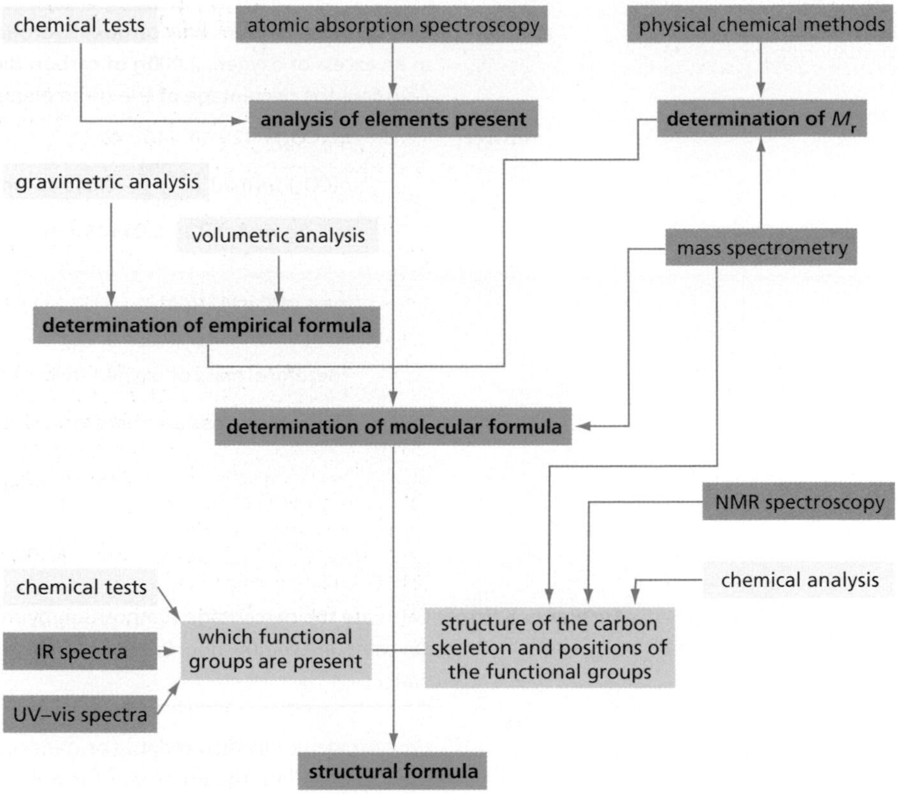

■ Figure 31.2
Outline of methods to find the structure of an organic compound.

31.2 The determination of empirical formulae

The **empirical formula** is the simplest formula that shows the relative number of atoms of each element present in a compound.

We saw in Chapter 1, page 11, how the empirical formula of a compound can be determined from its percentage composition by mass. For compounds that contain only carbon, hydrogen and oxygen, this can often be found by a quantitative combustion experiment in an excess of oxygen.

$$C_lH_mO_n \xrightarrow{O_2} lCO_2 + \tfrac{m}{2}H_2O$$

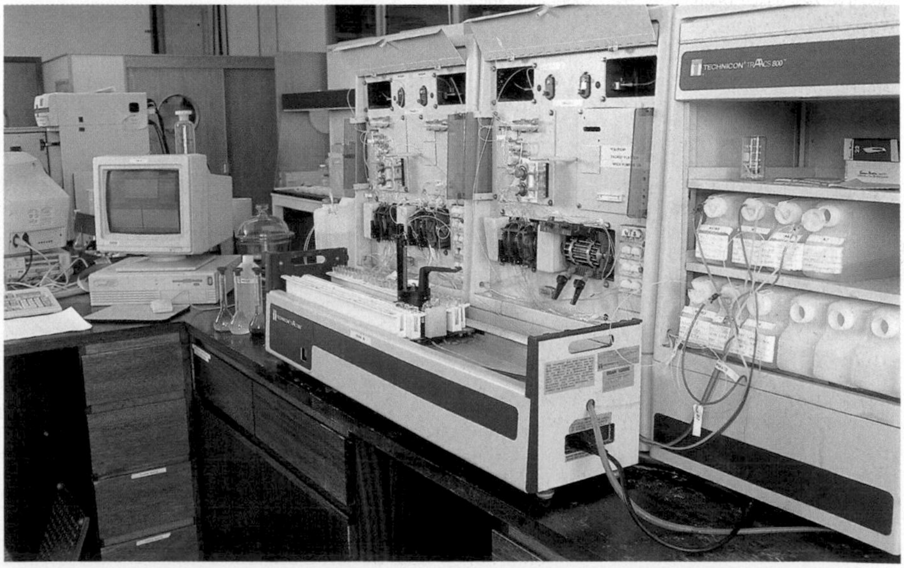

■ Figure 31.3
An autoanalyser can be used to find the empirical formula of a compound.

Example Compound **A** contains only carbon, hydrogen and oxygen. When 1.000 g of **A** was burned in an excess of oxygen, 2.000 g of carbon dioxide and 0.818 g of water were produced. Calculate the percentage of the three elements present in **A**.

Answer

$M_r(CO_2) = 12 + 16 + 16 = 44$ $M_r(H_2O) = 1 + 1 + 16 = 18$

$n(CO_2)$ formed $= \frac{2.000}{44.0} = \mathbf{0.04545\,mol}$ $n(H_2O)$ formed $= \frac{0.818}{18.0} = \mathbf{0.04544\,mol}$

in 1.0 g of **A**: $n(C) = \mathbf{0.04545\,mol}$ $n(H) = 2 \times 0.04544 = \mathbf{0.0909\,mol}$

(because 1 mol of H_2O contains 2 mol of H)

mass of **C** in 1.0 g of **A** $= 0.04545 \times 12 = 0.545$ g mass of H in 1.0 g of **A** $= 0.0909 \times 1 = 0.091$ g

Therefore: mass of oxygen in **A** $= 1.000 - 0.545 - 0.091 = 0.364$ g

The percentages are therefore: carbon $= \frac{0.545}{1.00} \times 100 = 54.5\%$

hydrogen $= \frac{0.091}{1.00} \times 100 = 9.1\%$

oxygen $= \frac{0.364}{1.00} \times 100 = 36.4\%$

Further practice Calculate the percentage composition by mass of compound **B** from the following data: On complete combustion, 0.550 g of **B** produced 0.834 g of carbon dioxide and 0.171 g of water.

Percentage composition data (or data on the masses of individual elements) can be used to calculate the empirical formulae of compounds. The steps in the calculation were explained in Chapter 1, page 11. An example will revise the method.

Example Use the percentage composition data calculated for **A** in the example above to work out the empirical formula for **A**.

Answer Step **1**

C: $\frac{54.5}{12} = 4.54$

H: $\frac{9.1}{1} = 9.1$

O: $\frac{36.4}{16} = 2.28$

Step **2**

C: $\frac{4.54}{2.28} = 1.99$ (2)

H: $\frac{9.1}{2.28} = 3.99$ (4)

O: $\frac{2.28}{2.28} = 1$ (1)

(Step **3** is not needed here.) The empirical formula is $\mathbf{C_2H_4O}$.

Further practice

1 Use the percentage composition data you have calculated for compound **B** above to deduce the empirical formula of **B**.

2 Calculate the empirical formulae of compounds **C**, **D** and **E**, given the mass data in Table 31.1.

■ **Table 31.1**

Compound	Mass of carbon/g	Mass of hydrogen/g	Mass of oxygen/g	Mass of nitrogen/g
C	0.200	0.033	0.267	–
D	1.8	0.45	–	0.70
E	35.3	5.9	31.4	27.4

31.3 The determination of molecular formulae

The **molecular formula** shows the number of atoms of each element contained in one molecule of a compound.

For very simple molecules (or for very complicated molecules!) the empirical and molecular formulae may be identical. For example:

- both the empirical and molecular formulae of methanal are CH_2O
- for ethanol they are both C_2H_6O
- for cocaine they are both $C_{17}H_{21}NO_4$
- for glycerol they are both $C_3H_8O_3$.

For many molecules the molecular formula is a multiple of the empirical formula, as shown in Table 31.2.

■ **Table 31.2** The molecular formula is a multiple of the empirical formula for these compounds.

Compound	Empirical formula	Molecular formula
Caffeine	$C_4H_5N_2O$	$C_8H_{10}N_4O_2$
Quinine	$C_{10}H_{12}NO$	$C_{20}H_{24}N_2O_2$
Trinitrobenzene	C_2HNO_2	$C_6H_3N_3O_6$

To derive the molecular formula from the empirical fomula, we need to know either the approximate relative molecular mass or the number of carbon atoms each molecule possesses. The relative molecular mass can most easily be measured by the use of low resolution mass spectrometry (see Chapter 2, pages 24–5).

Example The mass spectrum of compound **A** (empirical formula C_2H_4O) has peaks at mass numbers 15, 29, 43 and 44. No peaks of higher mass number are observed. What is the molecular formula of **A**?

Answer The relative formula mass of the empirical formula is equal to $2\times12+4\times1+16=44$. This equals the highest mass number peak in the mass spectrum. Therefore the molecular formula is identical to the empirical formula, $\mathbf{C_2H_4O}$.

The number of carbon atoms contained in a molecule can be calculated if the volume of carbon dioxide produced by combusting a known volume of gaseous compound can be measured.

Example When 10 cm³ of gaseous hydrocarbon **F** (empirical formula CH_2) is burned in an excess of oxygen, 30 cm³ of carbon dioxide are produced (all volumes measured at 25 °C and 1.0 atm, where 1 mol of gas has a volume of 24 dm³). Calculate the molecular formula of **F**.

Answer

$$n(\mathbf{F})=\frac{10}{24\,000}=4.17\times10^{-4}\,\text{mol}$$

$$n(CO_2)=\frac{30}{24\,000}=1.25\times10^{-3}\,\text{mol}$$

$$\text{Therefore: } n(CO_2)\text{ per mole of }\mathbf{F}=\frac{1.25\times10^{-3}}{4.17\times10^{-4}}=3.0$$

An alternative method uses Avogadro's Law to calculate the ratio (moles of CO_2):(moles of **F**) directly.

Avogadro's Law
Equal volumes of gases (at the same temperature and pressure) contain equal numbers of molecules.

If: $10\,cm^3$ of **F** give $30\,cm^3$ of CO_2
then: 1 volume of **F** would give 3 volumes of CO_2 (1 vol = $10\,cm^3$)
so: 1 mol of **F** would give 3 mol of CO_2.

There are therefore 3 atoms of carbon per molecule of **F**, and the molecular formula is $3 \times CH_2 = \mathbf{C_3H_6}$.

Further practice

Calculate the molecular formulae of the following compounds:

1 compound **G** (empirical formula C_2H_4O, M_r about 90)
2 compound **H** (empirical formula $C_2H_2O_2$, M_r about 120)
3 compound **I** (empirical formula C_6H_5N, M_r about 180).

31.4 Chemical tests for functional groups

Here we bring together the tests that have been described in the various chapters of the organic section of this book.

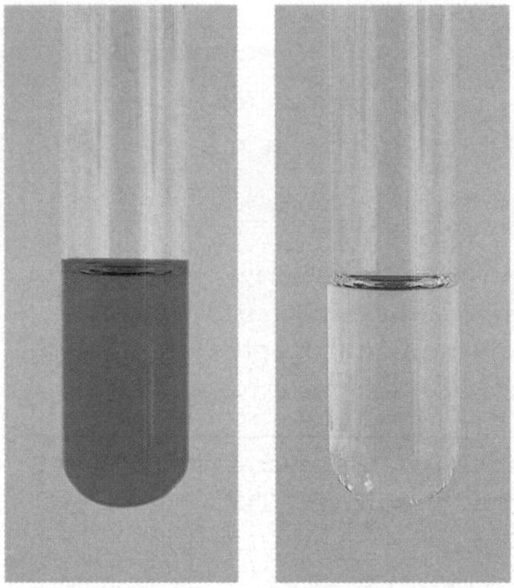

■ **Figure 31.4** (above) Alkenes decolorise bromine water.

Alkanes (Chapter 23)

These burn with a non-smoky flame, are immiscible with water, and less dense than water. They do not decolorise bromine water or dilute potassium manganate(VII).

Alkenes (Chapter 24)

These burn with a slightly smoky flame, are immiscible with water, and less dense than water. They decolorise bromine water and dilute potassium manganate(VII) at room temperature.

Alcohols (Chapter 26)

These are neutral to litmus solution and (if insoluble in water) do not dissolve in aqueous sodium hydroxide or sodium carbonate. They effervesce with sodium metal (giving off hydrogen) and with phosphorus(V) chloride (giving off hydrogen chloride). Primary, secondary and tertiary alcohols are most easily distinguished from each other by attempted oxidation with acidified dichromate(VI), followed by distillation and performing tests on the products (see Figure 31.5). They can also be distinguished by using the Lucas test (see page 483).

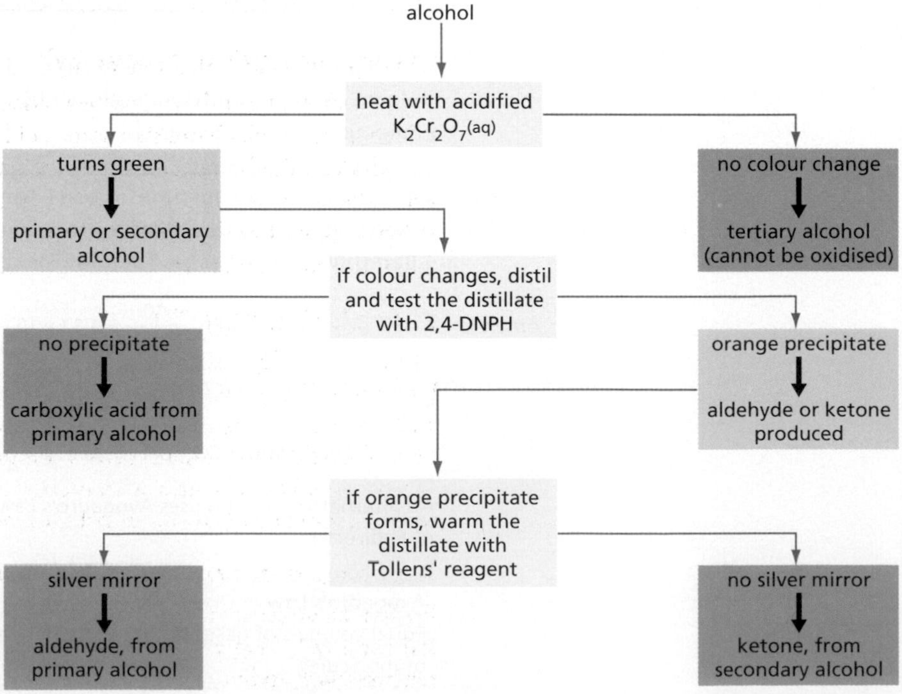

■ **Figure 31.5** (right) How to find whether an alcohol is primary, secondary or tertiary.

Carbonyl compounds (Chapter 28)

Both aldehydes and ketones produce orange precipitates with 2,4-dinitrophenylhydrazine. They may be distinguished from each other by warming with Tollens' reagent (ammoniacal silver nitrate solution) or Fehling's solution (alkaline copper(II) solution). Aldehydes reduce these reagents (to a silver mirror, or a red precipitate of copper(I) oxide, respectively), whereas ketones have no effect upon them.

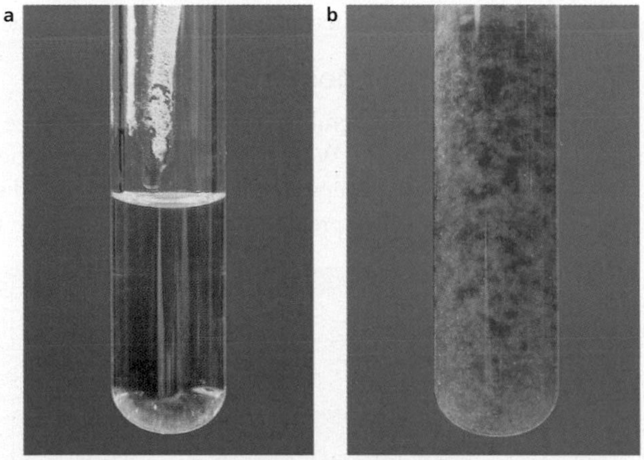

■ **Figure 31.6** Aldehydes and ketones give a precipitate with 2,4-DNPH. **a** Ethanal reacting and **b** forming an orange precipitate.

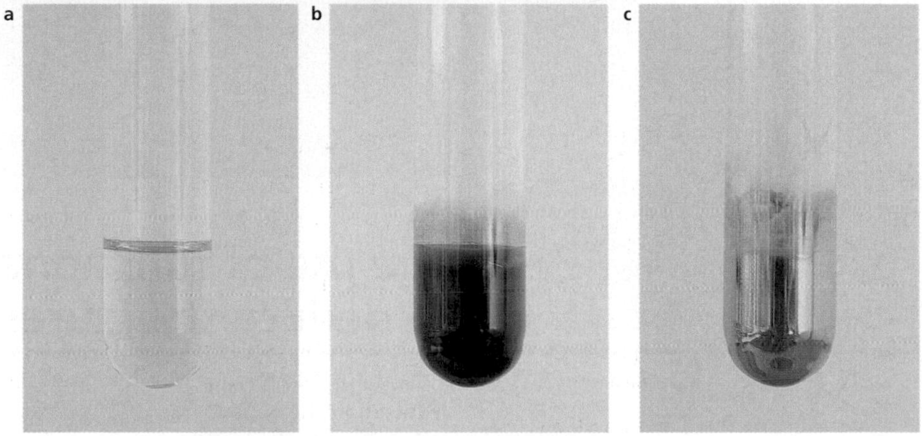

■ **Figure 31.7** Aldehydes give a silver mirror with Tollens' reagent.

Carboxylic acids (Chapter 29)

When a drop of a carboxylic acid is added to a solution of universal indicator, the solution turns red. (Note that some acid derivatives such as acyl chlorides and acid anhydrides will also turn universal indicator red due to their hydrolysis.) Carboxylic acids effervesce with sodium (giving off hydrogen) and with phosphorus(V) chloride (giving off hydrogen chloride). They also dissolve in aqueous sodium hydroxide and sodium carbonate (giving off carbon dioxide in the process).

Phenols (Chapter 27)

These are usually insoluble in water, but will dissolve in aqueous sodium hydroxide. They do not dissolve in aqueous sodium carbonate. They effervesce with sodium (giving off hydrogen), and with phosphorus(V) chloride. When a phenol is added to a dilute neutral solution of iron(III) chloride, a violet coloration is produced. Phenols react with bromine water, decolorising it, and producing a white precipitate of the di- or tribromophenol.

Acyl chlorides and acid anhydrides (Chapter 29)

Both of these react with water to produce acidic solutions which turn universal indicator red. They may be distinguished from carboxylic acids by their not reacting with sodium metal or phosphorus(V) chloride. They may be distinguished from each other by adding aqueous silver nitrate to the solution resulting from their reactions with water. Acyl chlorides produce a white precipitate (of silver chloride) whereas acid anhydrides do not.

Esters (Chapter 29)

These are neutral, sweet-smelling liquids that are immiscible with water. They do not react with sodium metal or phosphorus(V) chloride. On heating with dilute acids or alkalis, however, they are hydrolysed to alcohols and carboxylic acids. These products can be tested for in the usual way.

The triiodomethane reaction

This is a useful test for the presence of the groups $CH_3CH(OH)-$ or CH_3CO- in a molecule. When alkaline aqueous iodine is added to a compound containing one of these groups (ethanol, ethanal, methyl secondary alcohols or methyl ketones), a pale yellow precipitate of triiodomethane (iodoform) is formed.

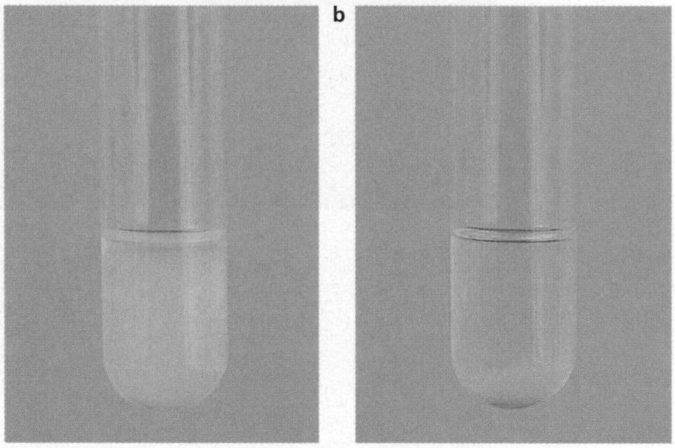

■ **Figure 31.8** a b The triiodomethane reaction **a** before and **b** after standing.

31.5 The use of functional-group tests to deduce structures

If the molecular formula of a compound is known, its structure can often be deduced on the basis of the results of various functional-group tests. An example will make this clear.

Example Three compounds, **J**, **K** and **L** are isomers with the molecular formula $C_4H_8O_2$.

a Compound **J** is unaffected by hot dilute sulphuric acid, but reacts with sodium metal, Fehling's solution and alkaline aqueous iodine.

b Both **K** and **L** react with hot dilute sulphuric acid. Under these conditions compound **K** gives **M** (CH_2O_2) and **N** (C_3H_8O) and compound **L** gives **P** ($C_2H_4O_2$) and **Q** (C_2H_6O). Both **M** and **P** effervesce with sodium carbonate solution. **N** and **Q** both react with sodium metal, and are both oxidised by acidified potassium dichromate(VI). The oxidation product from **N** gives an orange precipitate with 2,4-dinitrophenylhydrazine, but does not produce a silver mirror when warmed with ammoniacal silver nitrate. The oxidation product from **Q** is identical to compound **P**. Identify the compounds **J–Q**, and write equations for all reactions that occur.

Answer **a** Compound **J** is not an ester (because it does not react with hot dilute sulphuric acid) but it could be an alcohol or a carboxylic acid (reaction with sodium). Reaction with Fehling's solution suggests an aldehyde, and reaction with alkaline aqueous iodine suggests the group $CH_3CH(OH)-$ or CH_3CO-. We now consider the second oxygen atom in the formula – since at least one of the two oxygen atoms is taken up by the aldehyde group, **J** cannot be a carboxylic acid. It must therefore be an alcohol (reaction with sodium). So **J** contains the groups $CH_3CH(OH)-$ and $-CHO$, and must therefore be $CH_3CH(OH)CH_2CHO$.

Equations:

$$CH_3CH(OH)CH_2CHO + Na \rightarrow CH_3CH(ONa)CH_2CHO + \frac{1}{2}H_2$$

$$CH_3CH(OH)CH_2CHO + [O] \xrightarrow{\text{(from Fehling's solution)}} CH_3CH(OH)CH_2CO_2H$$

$$CH_3CH(OH)CH_2CHO + 4I_2 + 6OH^- \rightarrow CHI_3 + {}^-O_2CCH_2CHO + 5I^- + 5H_2O$$

b Both compounds **K** and **L** are esters (reaction with hot dilute sulphuric acid), being hydrolysed to acids (**M** and **P**) and alcohols (**N** and **Q**). **M** and **P** have unique structures for their molecular formulae. **M** is methanoic acid, HCO_2H, and **P** is ethanoic acid, CH_3CO_2H. Both acids will effervesce with aqueous sodium carbonate, giving off carbon dioxide. The alcohols **N** and **Q** are both oxidised by acidified potassium dichromate(VI) solution, so neither is a tertiary alcohol. **N** produces a ketone on oxidation (orange precipitate with 2,4-DNPH, but no silver mirror with Tollens' reagent). So **N** is the secondary alcohol, propan-2-ol. **Q** must be ethanol, which is oxidised to ethanoic acid (**P**).

Piecing all this together means that **K** must be prop-2-yl methanoate, and **L** must be ethyl ethanoate.

$$\underset{\textbf{K}}{HCO_2CH(CH_3)_2} \xrightarrow{H_2SO_4(aq)} \underset{\textbf{M}}{HCO_2H} + \underset{\textbf{N}}{HOCH(CH_3)_2}$$

$$\textbf{N} \xrightarrow{[O]} O{=}C(CH_3)_2$$

$$\underset{\textbf{L}}{CH_3CO_2CH_2CH_3} \xrightarrow{H_2SO_4(aq)} \underset{\textbf{P}}{CH_3CO_2H} + \underset{\textbf{Q}}{HOCH_2CH_3}$$

$$\textbf{Q} \xrightarrow{[O]} \textbf{P}$$

Further practice Deduce the structures of the following compounds, explaining your reasoning:

1 Compound **R** has the molecular formula C_8H_8O. It effervesces with sodium metal, but not with phosphorus(V) chloride. It decolorises bromine water, giving a white precipitate. It also decolorises dilute aqueous potassium manganate(VII) solution.

2 Compound **S** has the molecular formula $C_6H_{12}O_2$. It is unaffected by hot dilute sulphuric acid and also by hot acidified dichromate, but reacts with both sodium metal and alkaline aqueous iodine. It forms an orange precipitate with 2,4-dinitrophenylhydrazine, but does not react with Fehling's solution.

Functional-group tests are also useful for distinguishing between isomers. The following examples illustrate this.

Example For each of the following pairs of isomers, suggest a test that will distinguish between the two compounds.

a $CH_3CH_2CH_2CHO$ and $CH_3CH_2COCH_3$

b $(CH_3)_3COH$ and $(CH_3)_2CHCH_2OH$

c $CH_2{=}CHCH_2OH$ and CH_3CH_2CHO

Answer **a** One of the pair is an aldehyde, and the other is a methyl ketone. We could use either Fehling's solution (red precipitate with the aldehyde, no change with the ketone) or alkaline aqueous iodine (yellow precipitate with the methyl ketone, no change with the aldehyde). Note that 2,4-DNPH would not distinguish – both will give orange precipitates.

b Both compounds are alcohols, so sodium metal or phosphorus(V) chloride would not distinguish between them. One is a tertiary alcohol, so would not be affected by warming with acidified potassium dichromate(VI). The other is a primary alcohol, which would turn acidified dichromate(VI) from orange to green.

c Several tests could be used for this pair. The first compound is an alkene alcohol, so it would react with aqueous bromine, cold potassium manganate(VII) solution, sodium metal or phosphorus(V) chloride. None of these reagents would react with the second compound. Being an aldehyde, however, this second compound would react with 2,4-DNPH, Fehling's solution or Tollens' reagent. Note that both compounds would be oxidised by warm acidified dichromate(VI), so this reagent would not distinguish between them.

Further practice Suggest tests that could be carried out on each of the following pairs of isomers that would distinguish between them.

1 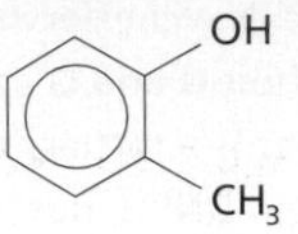

and 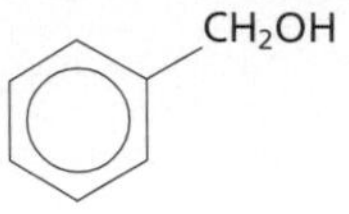

2 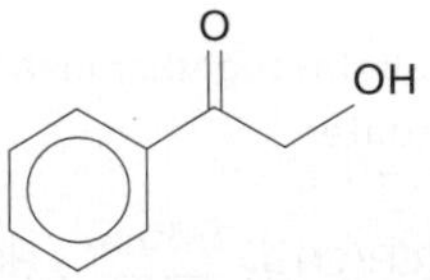

and 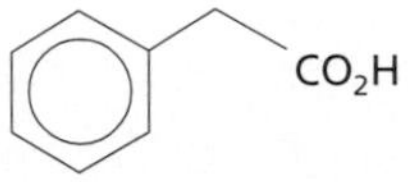

3 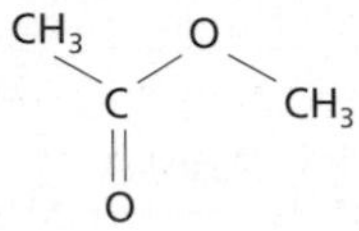

and 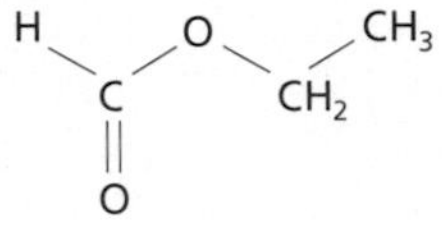

4 $CH_3—CH_2—COCl$ and $CH_3—CHCl—CHO$

31.6 Ultraviolet and visible spectroscopy

We saw in Chapter 14, page 274, that organic compounds containing double bonds and benzene rings absorb ultraviolet light, due to electrons within the molecules undergoing transitions to higher energy levels.

Many organic compounds, even those containing double bonds, do not absorb in easily accessible regions of the spectrum (that is, at a reasonably high wavelength, near the visible region). But if a compound contains several conjugated double bonds, or one of several **chromophoric groups**, the absorption occurs at higher wavelengths, and can even occur in the visible region, causing the compound to be coloured.

Ultraviolet/visible spectroscopy is used in two ways:

- as evidence for conjugated unsaturated systems
- as a tool for the quantitative estimation of amounts of absorbing compounds.

The latter application makes use of the Beer–Lambert Law (see Chapter 14, page 269):

$$\text{absorbance} = \varepsilon c l$$

If ε and l are constant (which they will be if we are studying just one compound, and using an identical cell for each measurement), the absorbance is proportional to the concentration. Hence the concentration of the compound can be measured. Depending on how strongly a compound absorbs, concentrations as low as $1 \times 10^{-6}\,\text{mol}\,\text{dm}^{-3}$ can be measured by UV spectroscopy.

The absorption characteristic of some electron transitions

Absorption of UV light occurs when an electron in a molecule moves from a lower to a higher energy level. The lowest energy transition that can occur in molecules that do not have double bonds or lone pairs of electrons is that from a σ orbital to a σ* orbital (see the panel on page 54). This large energy gap corresponds to a short wavelength (see Figure 31.9).

Compounds containing lone pairs of electrons in non-bonding orbitals can absorb at longer wavelengths (see Figure 31.10).

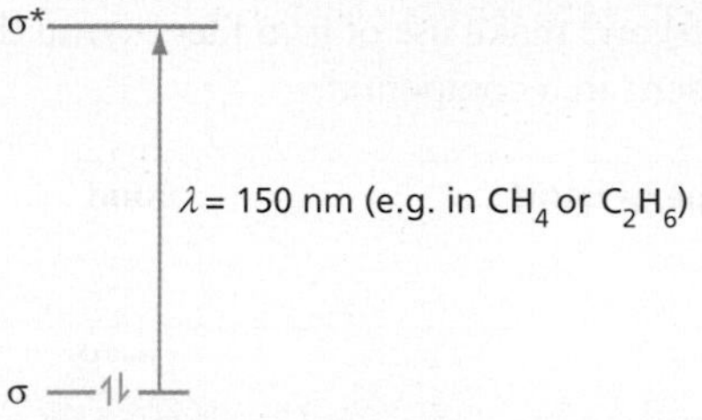

■ Figure 31.9
Absorption by bonding pairs.

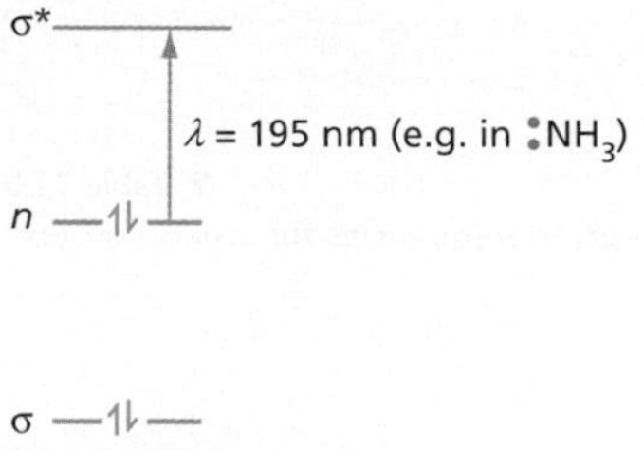

■ Figure 31.10
Absorption by a lone pair.

Compounds containing C=C double bonds, but no lone pairs of electrons, absorb at about the same wavelength as those with lone pairs but no double bonds (see Figure 31.11).

Compounds containing both double bonds and a lone pair of electrons on a doubly bonded atom absorb at the longest wavelength (see Figure 31.12).

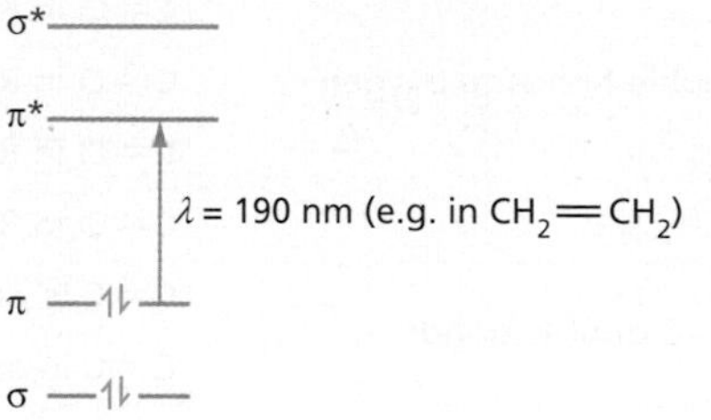

■ Figure 31.11
Absorption by an electron pair in a π bond.

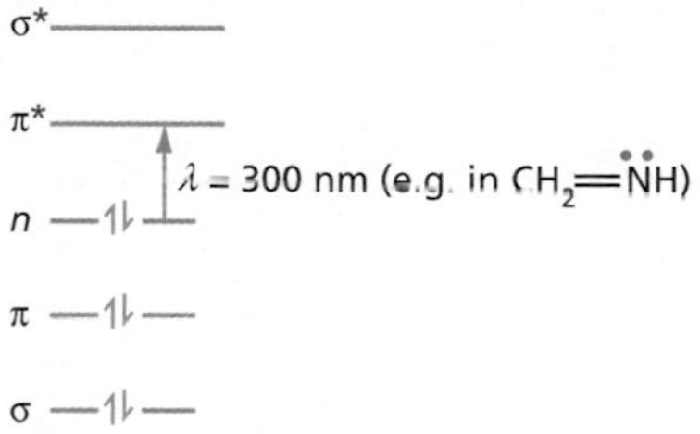

■ Figure 31.12
Absorption by a lone pair in a molecule containing a π bond.

Conjugation of π bonds reduces the energy gap between the π and the π* orbitals, as shown in Table 31.3.

■ Table 31.3
The energy gap between π and π* orbitals reduces in conjugated systems.

Group	π → π* energy gap/kJ mol^{-1}	Absorption wavelength (λ)/nm
R—CH=CH—R	690	190
RCH=CH—CH=CH—R	520	230
R—CH=CH—CH=CH—CH=CH—R	435	275

The incorporation of an atom with a lone pair of electrons can cause a major shift of absorbance to longer wavelengths (see Table 31.4).

■ Table 31.4
A lone pair shifts absorption to a longer wavelength.

Group	Absorption wavelength (λ)/nm
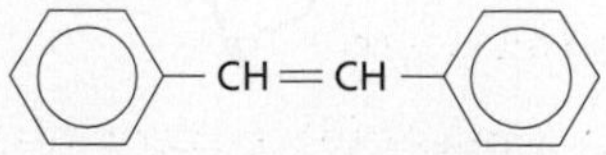	296 (colourless)
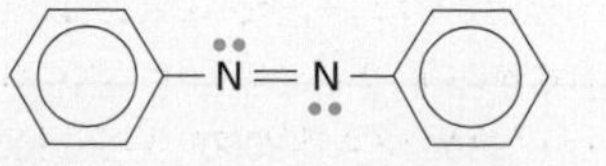	400 (orange, due to absorption of blue/violet)

31.7 Infrared spectroscopy

In Chapter 14, page 278, we discussed how the vibrations of bonds within a molecule give rise to absorption in the infrared region of the spectrum. The frequency of absorption is dependent on the stiffness of the bond and the masses of the atoms at each end.

Throughout the chapters dealing with the functional groups in the organic chemistry section of this book, characteristic infrared frequencies associated with each group have been described. Here we collect together all these data (see Table 31.5) and make use of it in identifying the functional groups that might possibly be present in a compound.

Table 31.5 IR absorption frequencies for organic groups.

Type of bond	Bond	Frequency of absorption (wavenumber)/cm^{-1}
Bonds to hydrogen	O—H	3600
	O—H (hydrogen bonded)	3200–3500
	N—H	3400
	O—H in RCO_2H (strongly hydrogen bonded)	2500–3300
	C—H	2800–3100
Triple bonds	—C≡C— or —C≡N	2200
Double bonds to oxygen	C=O in RCOCl	1800
	C=O in RCO_2R	1740
	C=O in RCHO	1730
	C=O in RCO_2H	1720
	C=O in R_2CO	1715
C—C double bonds	C=C in alkenes	1650
	C=C in arenes	1600 and 1500
Single bonds	C—O	1100–1250
	C—F	1200
	C—Cl	700

Example Compounds **T** and **U** are isomers with the molecular formula $C_3H_6O_2$. Suggest their structures based on the spectra shown in Figures 31.13 and 31.14.

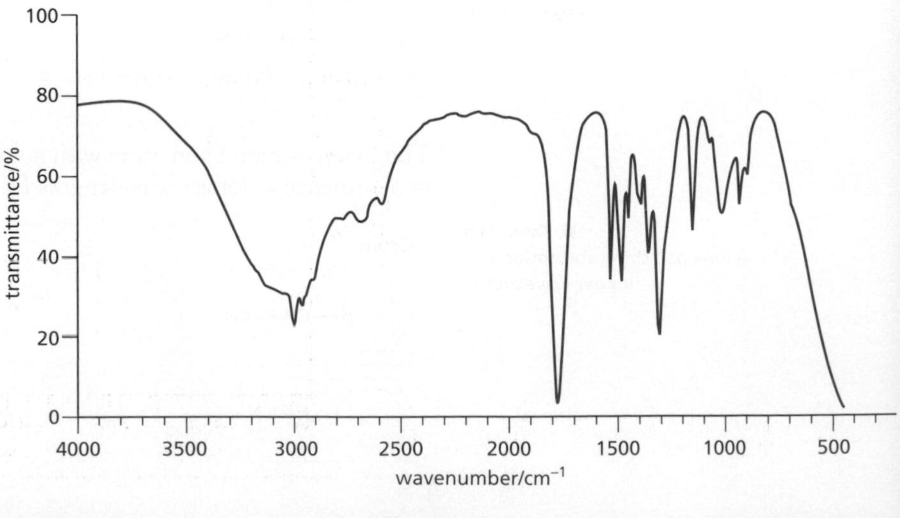

Figure 31.13 IR spectrum of **T**.

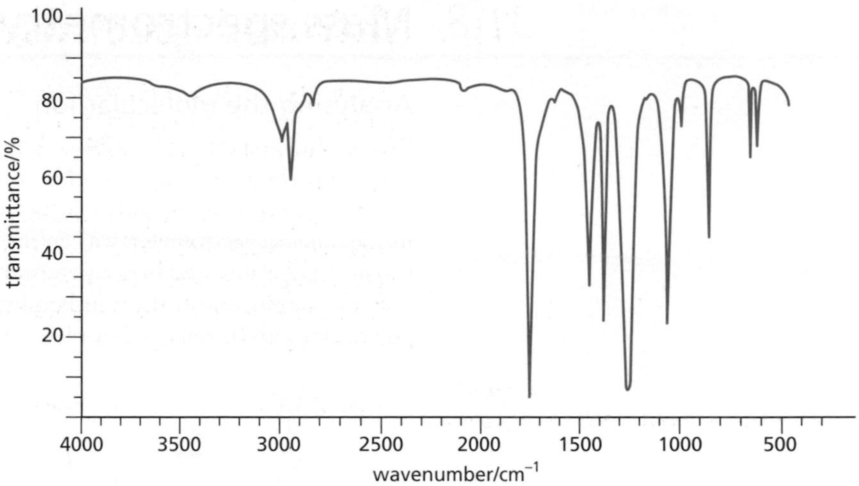

■ **Figure 31.14** IR spectrum of **U**.

Answer Both **T** and **U** show a C=O absorption in their spectrum at about 1700 cm^{-1}, and a C—O absorption at about 1250 cm^{-1}. **T** shows a broad hydrogen-bonded O—H band from 3300 to 2500 cm^3, whilst **U** shows no O—H band at all.

So **T** is **$CH_3CH_2CO_2H$** (propanoic acid) and **U** could be either the ester **$CH_3CO_2CH_3$** (methyl ethanoate) or the ester **$HCO_2CH_2CH_3$** (ethyl methanoate).

Further practice Compound **V** (C_3H_6O) gives a silver mirror when warmed with Tollens' reagent. It can be converted to compound **W** by reagent **X**. Use the spectra in Figures 31.15 and 31.16 to identify the functional groups present in **V** and **W**, and suggest the identity of reagent **X**.

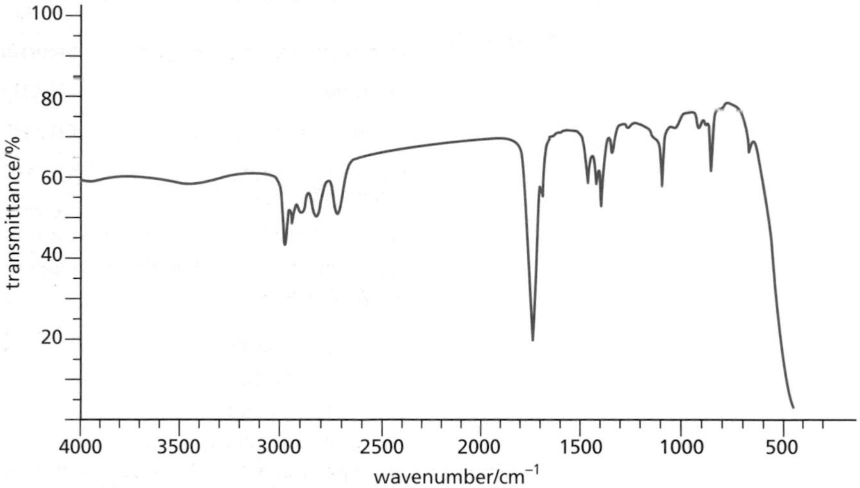

■ **Figure 31.15** IR spectrum of **V**.

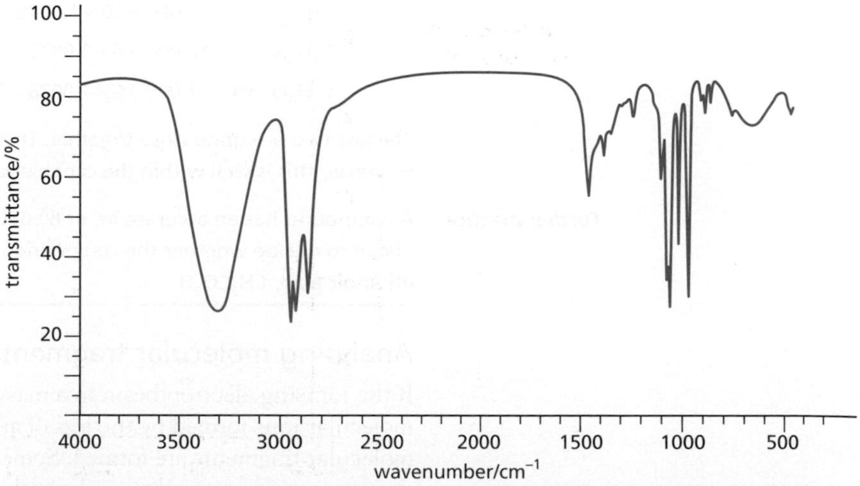

■ **Figure 31.16** IR spectrum of **W**.

31.8 Mass spectrometry

Analysing the molecular ion

We saw in Chapter 2, pages 24–5, how a mass spectrometer can be used to determine the isotopic masses of individual atoms.

If we vaporise an organic molecule and subject it to the ionising conditions inside a mass spectrometer, we can measure the mass/charge ratio (*m*/*e* ratio) for the molecular ion, and hence determine the relative molecular mass.

For example, one of the non-bonding electrons on the oxygen atom of propanone can be removed by electron bombardment, to give an ionised molecule:

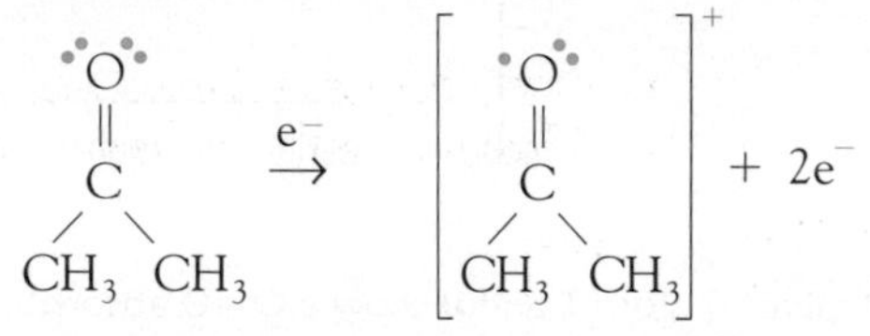

The *m*/*e* ratio for the resulting molecular ion is $(3\times12+6\times1+16):1$, which is 58.

Using **very high resolution mass spectrometry**, we can measure *m*/*e* ratios to an accuracy of five significant figures (1 part in 100 000). By this means, it is not only possible to measure the M_r value of a compound, but also to determine its molecular formula. We can do this because the accurate relative atomic masses of individual atoms are not exact whole numbers.

Example The three compounds in Table 31.6 all have an approximate M_r of 70.

■ Table 31.6

Name	Structure	Molecular formula
Pentene	$CH_3CH_2CH_2CH{=}CH_2$	C_5H_{10}
2-Aminopropanenitrile	$CH_3CH(NH_2)CN$	$C_3H_6N_2$
But-1-ene-3-one	$CH_2{=}CHCOCH_3$	C_4H_6O

Use the following accurate relative atomic masses to calculate their accurate M_r values, and decide how sensitive the mass spectrometer needs to be in order to distinguish between them:

H = 1.0078
C = 12.000
N = 14.003
O = 15.995

Answer The accurate M_r values are as follows:

$C_5H_{10} = 5\times12.000 + 10\times1.0078 = 70.078$

$C_3H_6N_2 = 3\times12.000 + 6\times1.0078 + 2\times14.003 = 70.053$

$C_4H_6O = 4\times12.000 + 6\times1.0078 + 15.995 = 70.042$

The last two are quite close together. They differ by 11 parts in 70 000, or about 0.16%. However, this is well within the capabilities of a high resolution mass spectrometer.

Further practice A compound has an accurate M_r of 60.068. Use the accurate relative atomic masses given above to decide whether the compound is 1,2-diaminoethane, $H_2NCH_2CH_2NH_2$, or ethanoic acid, CH_3CO_2H.

Analysing molecular fragments

If the ionising electron beam in a mass spectrometer has enough energy, the molecular ions formed by the loss of an electron can undergo bond fission, and molecular fragments are formed. Some of these will carry the positive charge, and therefore appear as further peaks in the mass spectrometer (see Figure 31.17).

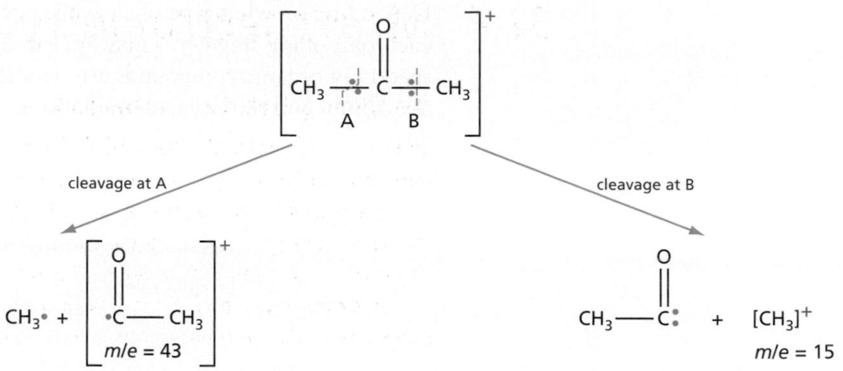

■ **Figure 31.17**
Ionic fragments formed from propanone.

We therefore expect the mass spectrum of propanone to contain peaks at *m*/*e* = 15 and 43, as well as the molecular ion peak at 58 (see Figure 31.18).

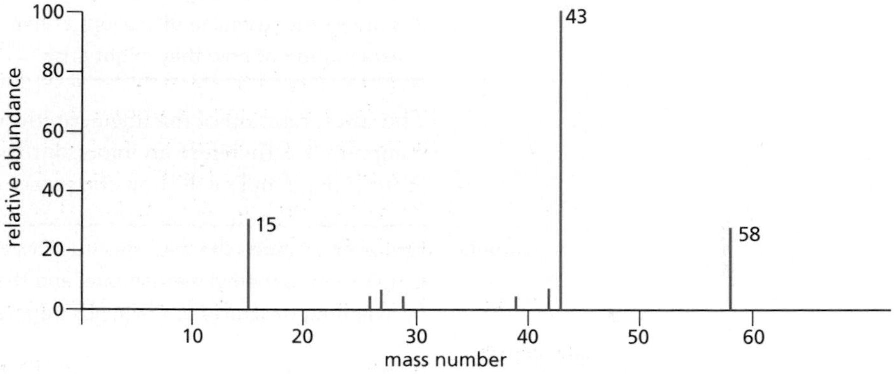

■ **Figure 31.18**
Mass spectrum of propanone.

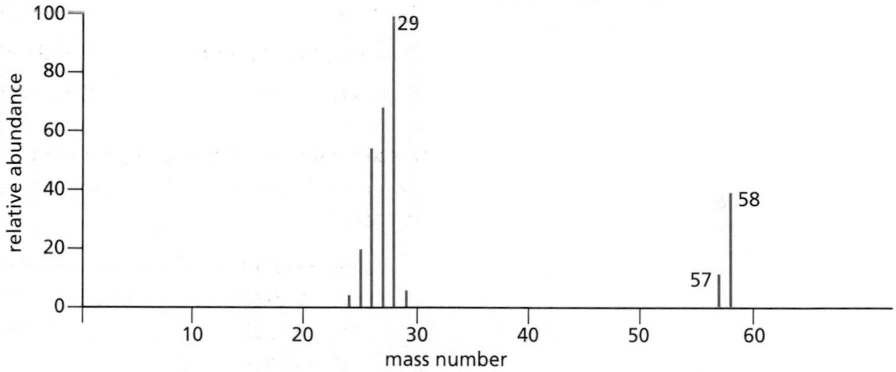

■ **Figure 31.19**
Mass spectrum of propanal.

The fragmentation pattern can readily distinguish between isomers. Compare Figure 31.18 with Figure 31.19 which shows the mass spectrum of propanal. Here there is no peak at *m*/*e* = 15, nor one at *m*/*e* = 43. Instead, there are peaks at *m*/*e* = 57 and several from *m*/*e* = 26 to 29. This is readily explained by the fragmentations shown in Figure 31.20.

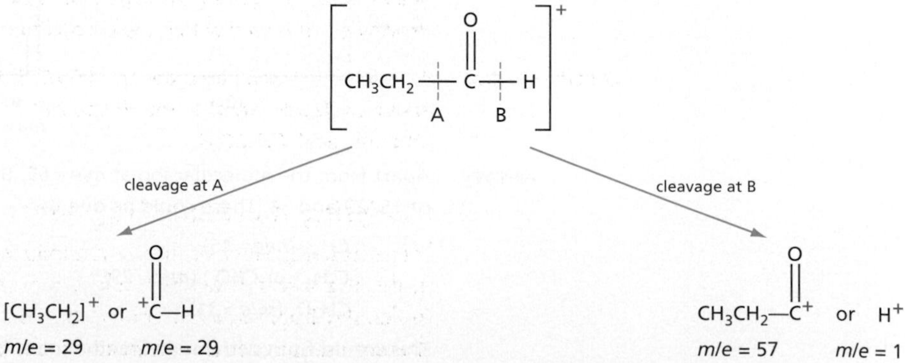

■ **Figure 31.20**
Ionic fragments formed from propanal.

Depending on what type of cleavage occurs at A and B, one or other or both of each pair of ion fragments may appear. The peaks of highest abundance in the mass spectra of organic compounds are associated with particularly stable cations, such as acylium and tertiary carbonium ions:

$$R-\overset{+}{C}=O \longleftrightarrow R-C\equiv\overset{+}{O}$$

the acylium ion

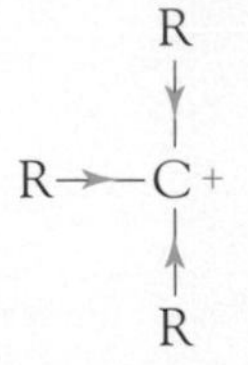

a tertiary carbonium ion

Further practice

1 High resolution mass spectrometry could decide whether $CH_3CH_2^+$ or CHO^+ was responsible for the peak at $m/e = 29$. Calculate the accurate M_r values of these two ions, using the data on page 596.

2 Suggest the formulae of the ions at m/e values 26, 27 and 28, and suggest an explanation of how they might arise.

The interpretation of the fragmentation pattern in the mass spectra of organic compounds is therefore an important tool in the elucidation of their structures. A further example will show the power of the technique.

Example

Figure 31.21 shows the mass spectra of two compounds with the molecular formula $C_2H_4O_2$. One is methyl methanoate, and the other is ethanoic acid. Decide which is which by assigning structures to the major fragments.

■ **Figure 31.21** Mass spectra of methyl methanoate and ethanoic acid. Which is which?

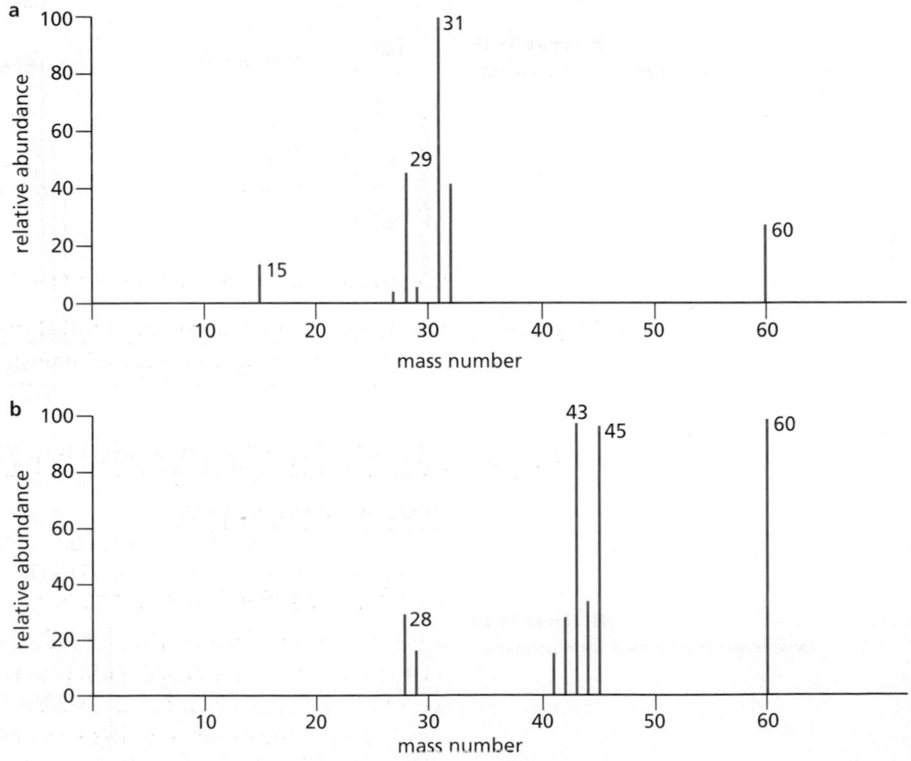

Answer

Apart from the molecular ion at $m/e = 60$, the major peaks in spectrum **a** are at m/e values of 15, 29 and 31. These could be due to:

CH_3^+ ($m/e = 15$)
$C_2H_5^+$ or CHO^+ ($m/e = 29$)
CH_3O^+ ($m/e = 31$)

This would fit in with methyl methanoate (see Figure 31.22).

■ **Figure 31.22**
Ionic fragments formed from methyl methanoate.

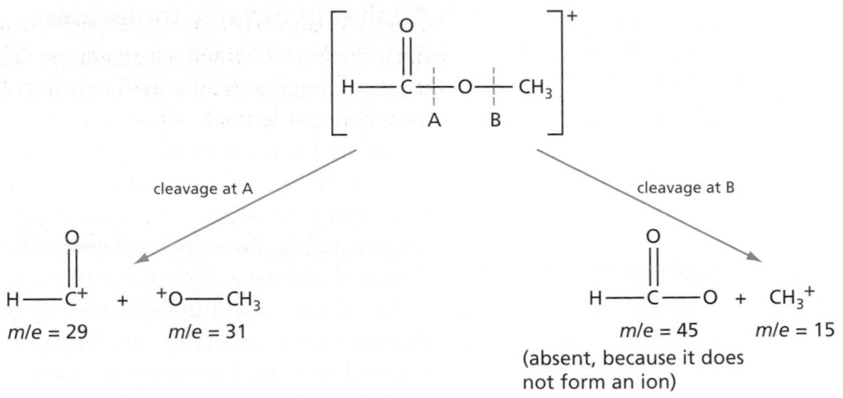

The peak at $m/e = 31$ can come only from methyl methanoate, and not from ethanoic acid. The major peaks in Figure 31.21**b**, apart from the molecular ion at $m/e = 60$, are at m/e values of 28, 43 and 45. These could be due to:

CO^+ ($m/e = 28$)
CH_3CO^+ ($m/e = 43$)
CO_2H^+ ($m/e = 45$)

These could arise from the fragmentations shown in Figure 31.23.

■ **Figure 31.23**
Ionic fragments formed from ethanoic acid.

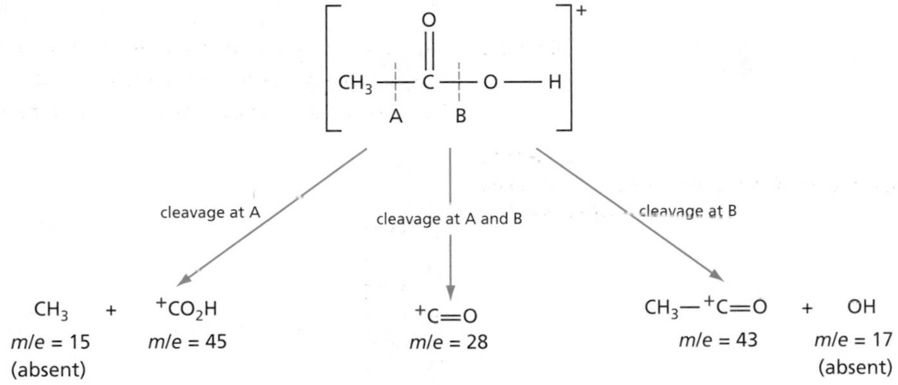

The peak at $m/e = 43$ can come only from ethanoic acid, and not from methyl methanoate.

Further practice A compound of molecular formula $C_3H_6O_2$ has major peaks at $m/e = 27$, 28, 29, 45, 57, 73 and 74. Suggest formulae for these fragments, and a structure for the compound.

31.9 Nuclear magnetic resonance (NMR) spectroscopy

The basis of NMR spectroscopy

We cannot see molecules with our naked eyes, but some of the most direct evidence for their structures and shapes comes to us from NMR spectroscopy. Every aspect of an NMR spectrum demands an exact interpretation, and there is usually a unique molecular structure that gives rise to a particular spectrum. NMR is potentially the most powerful technique at the disposal of the structural organic chemist.

We saw in Chapter 14, page 281, that nucleons have spin. If an atom has an odd number of nucleons, the atoms with nucleons in the two different spin states are split into different energy groups when placed in a strong magnetic field. Nucleons can be persuaded to flip from the lower energy spin state to the higher (to **resonate**) by irradiating a sample with electromagnetic radiation of the right frequency. The absorption of this frequency is detected by the NMR spectrometer.

Analysing organic molecules

Both ^{1}H and ^{13}C NMR spectroscopy find an important use in the analysis of organic compounds. We shall confine our description here to ^{1}H NMR, as it is the most commonly used.

The frequency at which a proton, ^{1}H, absorbs radiation depends on the strength of the magnetic field around it. However, even in a constant external field, protons in different chemical environments within a molecule absorb at different frequencies, because the local magnetic field they experience depends on the electrical and magnetic environment around them.

The electrons within molecules are usually 'paired' (that is, they occur as pairs of electrons spinning in opposite directions). When a molecule is placed in an external field, the electron pairs rotate in their orbits in such a way that they produce a magnetic field which opposes the external field. This phenomenon is called **diamagnetism** (see Figure 31.24). The effect is to shield nearby protons from the external field. This in turn reduces the frequency at which they absorb energy when they flip from their lower to their higher energy state.

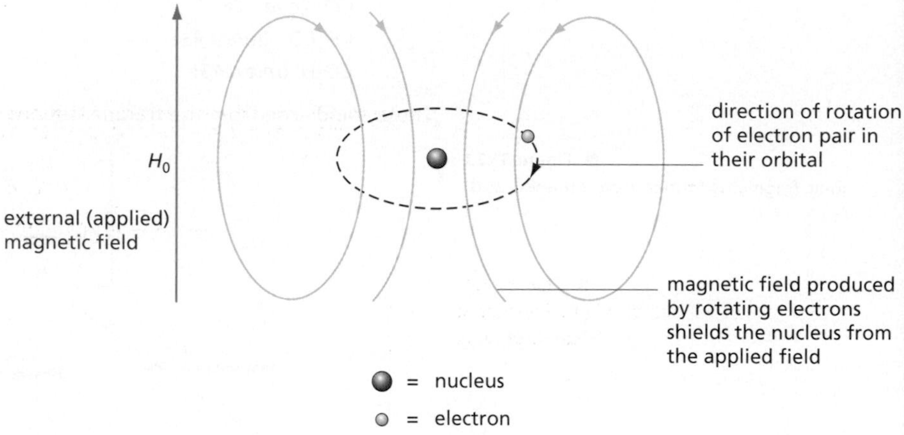

■ **Figure 31.24** In an external magnetic field, the electron pairs rotate in such a way that they produce an opposing magnetic field.

When, however, a proton is near an electronegative atom (or group) within a molecule, the bonding electrons are drawn away from the proton to the electronegative atom. The proton is less shielded from the external magnetic field, and hence it absorbs radiation at a higher frequency. The effect is very pronounced if the proton is attached to a benzene ring. In this situation the mobile delocalised π electrons in the ring can create a strong diamagnetic effect, opposing the external field. This has the effect of strengthening the magnetic field within the vicinity of the protons (see Figure 31.25).

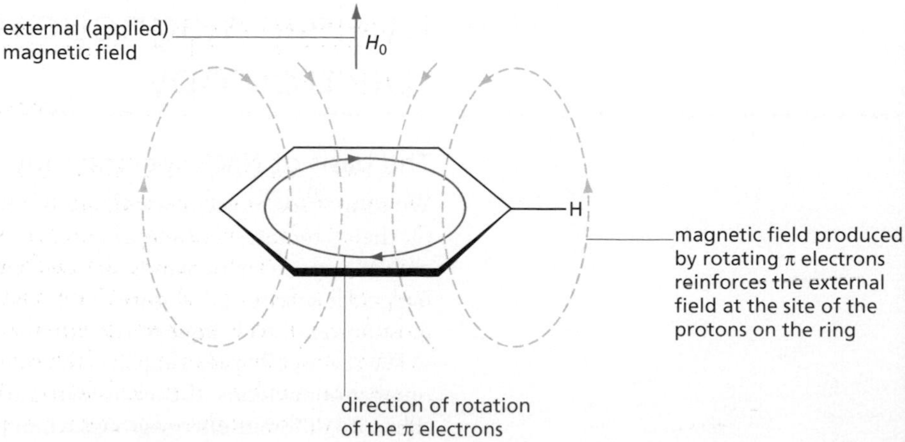

■ **Figure 31.25** In benzene, the field created by the rotation of the π electrons reinforces the applied field.

The chemical shifts (δ, see Chapter 14, page 282) of protons in various environments are listed in Table 14.6, page 282.

Low and high resolution NMR spectroscopy

The low resolution NMR spectrum of ethanol was discussed in Chapter 14, page 283. When the NMR spectrum of ethyl ethanoate is scanned at low resolution, three peaks are observed, corresponding to the three different chemical environments of the protons in the molecule (see Figure 31.26).

■ Figure 31.26 NMR spectrum of ethyl ethanoate at low resolution. Note that the δ scale, by convention, has its zero on the right of the *x*-axis.

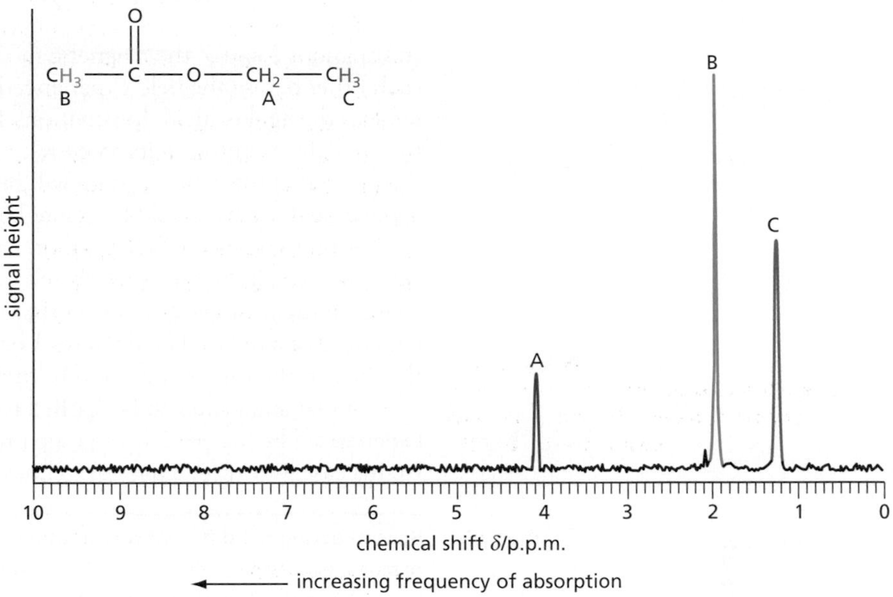

■ Figure 31.27 NMR spectrum of ethyl ethanoate at high resolution.

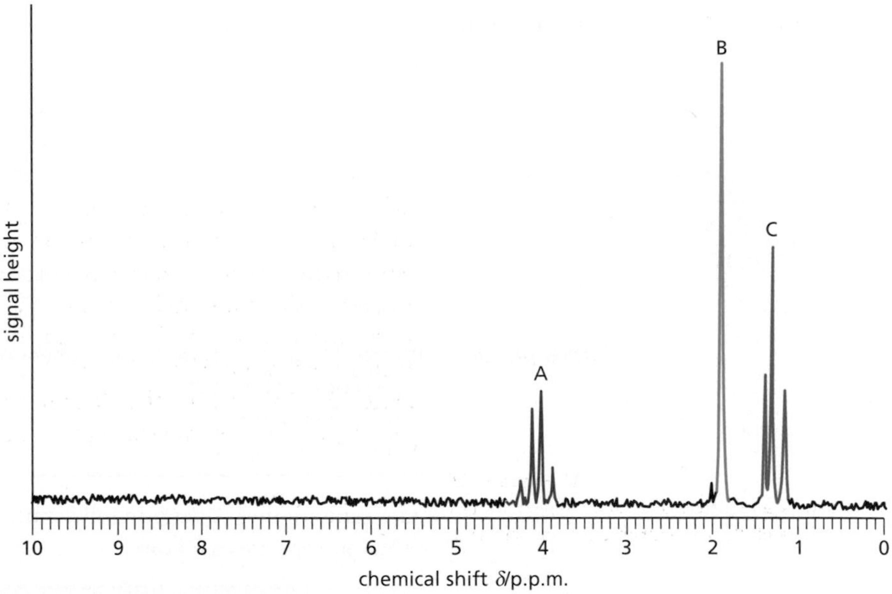

The areas under the three peaks are in the ratio 2 : 1 : 3, being proportional to the number of protons at each chemical environment. At higher resolution (see Figure 31.27), peaks A and C are seen to be multiple peaks, although B remains a single peak. This is because the nuclear spins of the protons in the ethyl group, responsible for peaks A and C, interact with each other. This is called **spin–spin coupling**, and it is a general phenomenon observed whenever protons on adjacent carbon atoms are in different chemical environments. The splitting of the peak arises because the magnetic field experienced by a proton is slightly altered due to the orientation of the magnetic moments (the spin states) of the protons on the adjacent carbon atom. Consider the protons in the CH_3 group of the ethyl chain. The field they experience will depend on the orientation of the magnetic moments of the —CH_2— protons, as shown in Figure 31.28 (overleaf).

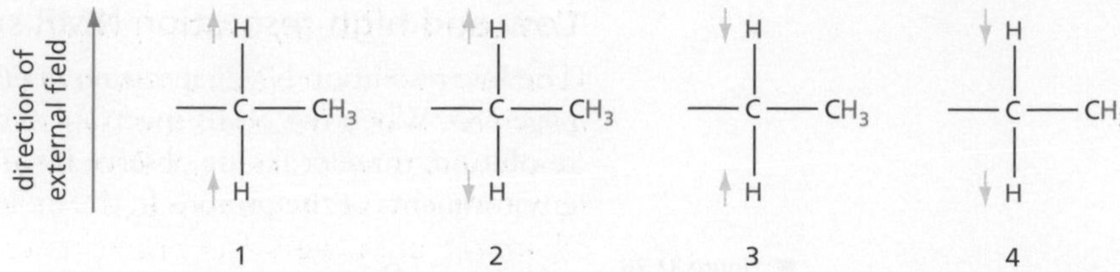

■ **Figure 31.28** The directions of the magnetic moments of the $—CH_2—$ protons have an effect on the $—CH_3$ protons.

In situations 2 and 3, the magnetic moments of the two $—CH_2—$ protons cancel each other out, so the field experienced by the $—CH_3$ group protons will be the same as the applied field. In situations 1 and 4, however, the magnetic moments of the $—CH_2—$ protons reinforce each other. Consequently the field experienced by the protons of the $—CH_3$ group will be, respectively, higher and lower than the applied field. There should therefore be a total of three frequencies at which these $—CH_3$ protons absorb. What is more, the probabilities of the four states 1 to 4 are equal, so overall there is twice the chance of the $—CH_3$ protons experiencing no change in field (states 2 and 3) as there is for the protons to experience either an enhanced or a reduced field (states 1 or 4). We therefore expect the intensities of the lines in the triplet of lines to be in the ratio 1 : 2 : 1.

A similar argument can be applied to the modification of the magnetic field experienced by the $—CH_2—$ group protons, by the protons in the $—CH_3$ group. In this case we expect a quartet of lines, in the ratio of 1 : 3 : 3 : 1.

Example By considering the different combinations of ↑ and ↓ magnetic moments of the $—CH_3$ protons, explain how the ratio 1:3:3:1 arises.

Answer The possible combinations of the three $—CH_3$ protons are shown in Figure 31.29.

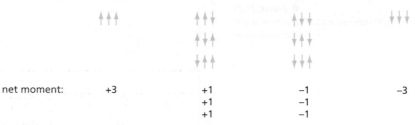

■ **Figure 31.29**

net moment:	+3	+1	−1	−3
		+1	−1	
		+1	−1	

There are three times as many combinations giving a net magnetic moment of +1 or −1, compared with +3 or −3.

Further practice Predict the splitting pattern (the number of lines and the relative intensities of the lines) for a proton adjacent to:

1 one other proton

2 four other protons.

The general rules concerning the splitting of the resonance peak of a proton by other protons are as follows.

- Protons in identical chemical environments do not split each other's peaks.
- The peak of a proton adjacent to n protons in a different environment is split into $(n + 1)$ lines.
- The relative intensities of the $(n + 1)$ lines are in the pattern shown in Table 31.7.

■ **Table 31.7** The splitting of a peak for a proton next to protons in a different chemical environment.

Number of protons adjacent to resonating proton	Number of lines in multiplet	Relative intensities
1	2	1:1
2	3	1:2:1
3	4	1:3:3:1
4	5	1:4:6:4:1

The use of 'heavy water', D_2O

Protons directly attached to oxygen or nitrogen atoms can appear almost anywhere in an NMR spectrum. The field strength at which they resonate depends on the acidity of the solution. Because of easy proton exchange with other O—H or N—H protons in the sample, these protons often do not cause the splitting of adjacent protons' peaks. They can, however, be identified by **deuterium exchange**. If the compound containing them is dissolved in D_2O ('heavy water', D=^{2}H), the protons are exchanged with deuterium atoms in the water. The peaks due to the —OH or —NH_2 protons disappear (deuterium atoms, having an even number of nucleons, do not resonate):

$$CH_3CH_2{-}OH + D_2O \rightleftharpoons CH_3CH_2{-}OD + HDO$$

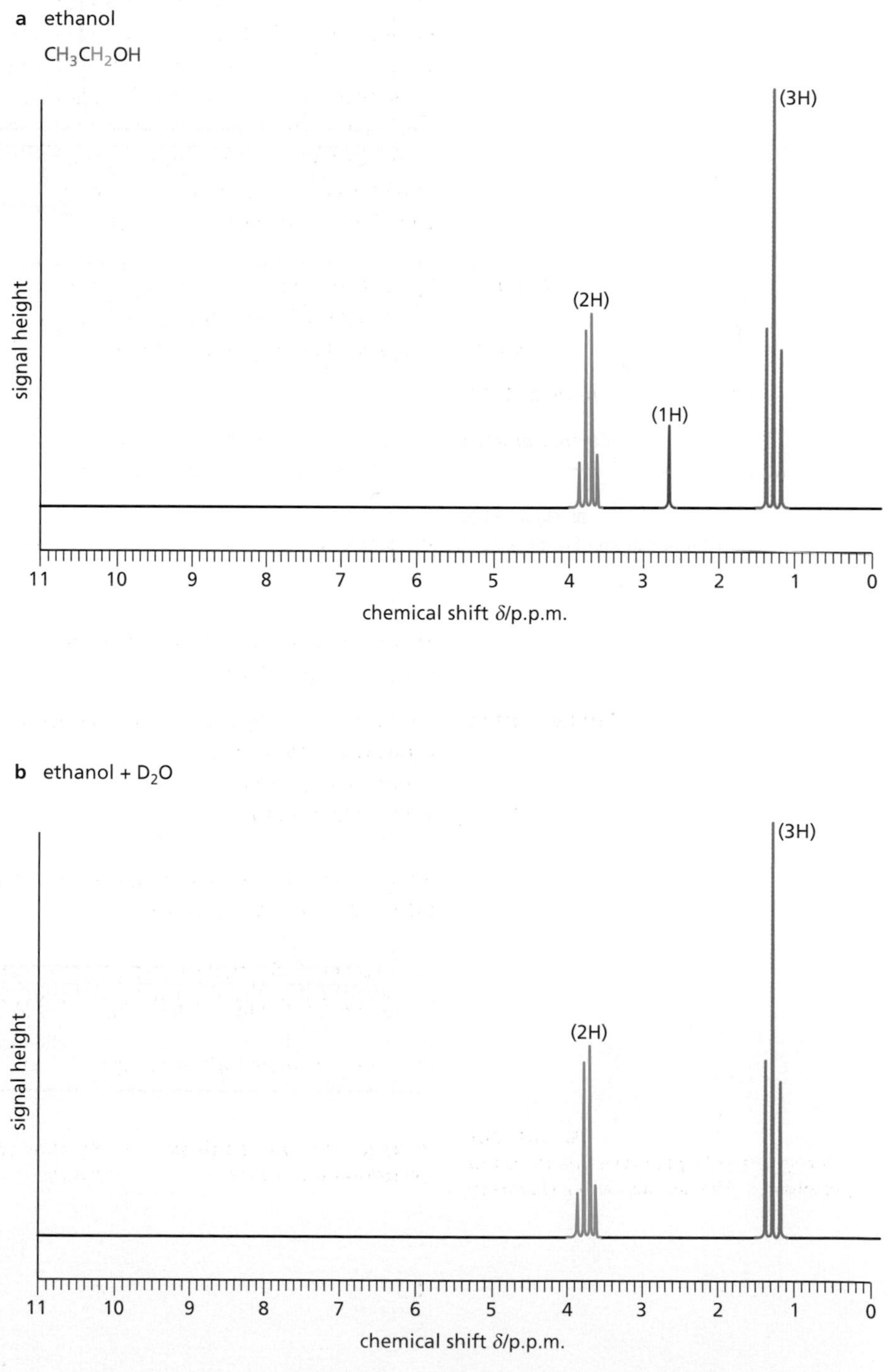

■ Figure 31.30 NMR spectra of ethanol **a** showing the —OH peak and **b** with D_2O added.

Example Figure 31.31 shows the NMR spectrum of an acid with the molecular formula $C_8H_8O_2$. Work out its structure. (Use the δ values in Table 14.6, page 282 to help you.)

■ **Figure 31.31**

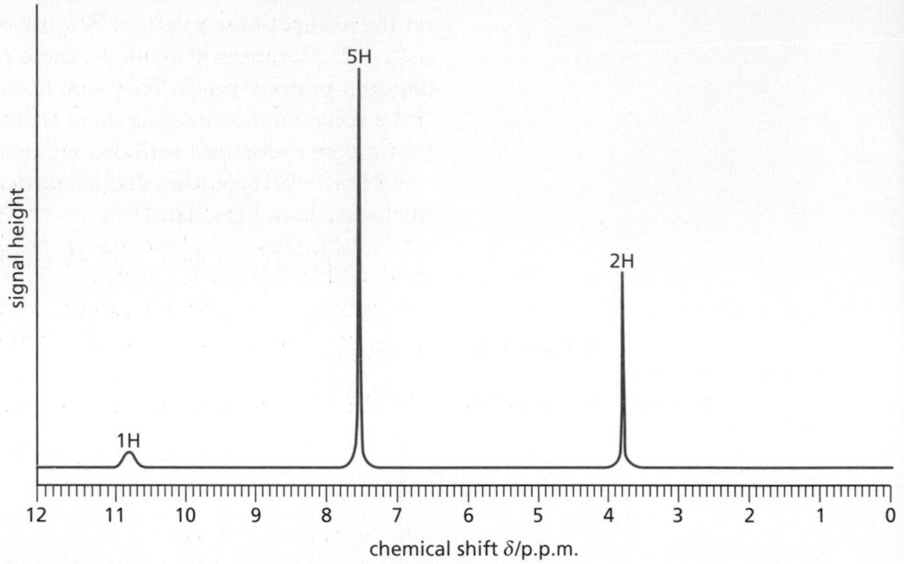

Answer The high C : H ratio in the molecular formula suggests the presence of a benzene ring, and this is confirmed by the peak at $\delta = 7.6$. The broad peak at $\delta = 10.8$ is typical of the O—H hydrogen of a carboxylic acid. The two-proton single peak at $\delta = 3.7$ is a CH_2 group flanked by both an aryl ring and a CO_2H group, both of which would cause a high-field shift in resonance (by about 1 δ unit each). The structure is therefore $C_6H_5—CH_2—CO_2H$.

Further practice Figure 31.32 shows the NMR spectrum of compound **Y**, whose molecular formula is $C_3H_6O_2$. Suggest a possible structure for **Y**, with reasons.

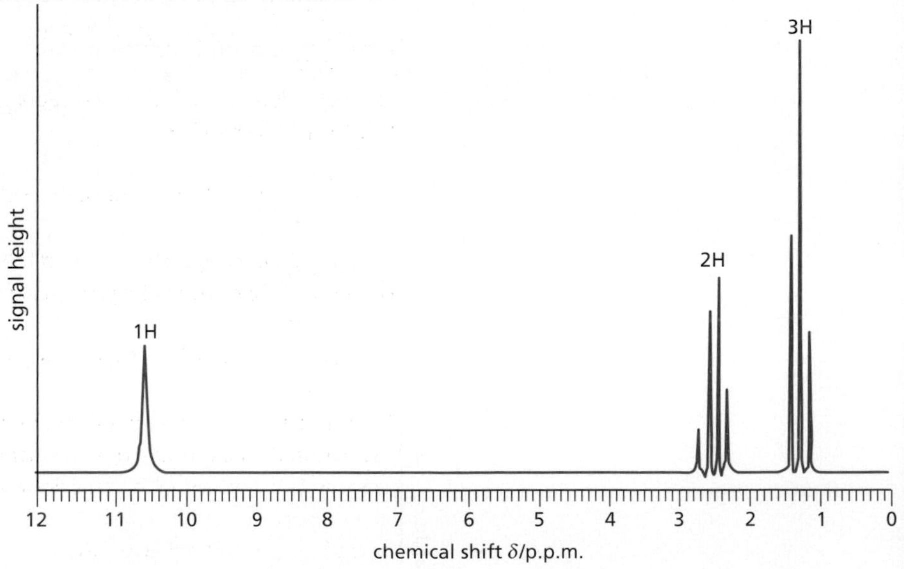

■ **Figure 31.32**
NMR spectrum of compound **Y**.

31.10 Multifunctional compounds

Most of the organic compounds that we have looked at in this book have contained only one functional group. But in general, many organic compounds are likely to have two or more functional groups. In many situations the reactions of the functional groups can be considered separately, but in some cases the functional groups react with each other. The amino acids of Chapter 30 are a good example of this interaction.

Table 31.8 on page 611 lists many of the reagents you have come across in this book, and the seven charts on pages 612–13 show in diagrammatic form the ways of preparing, and the transformations of, the various functional groups. In addition to these, you should recall the simple acid–base reactions of alcohols, phenols, carboxylic acids and amines with sodium, aqueous sodium hydroxide, aqueous sodium carbonate and hydrochloric acid. We can use Table 31.8 and the charts to predict what might occur when a multifunctional compound is reacted with a particular reagent.

Example

How might the compound cinnamaldehyde:

$$\text{C}_6\text{H}_5 - \text{CH} = \text{CH} - \text{CHO}$$

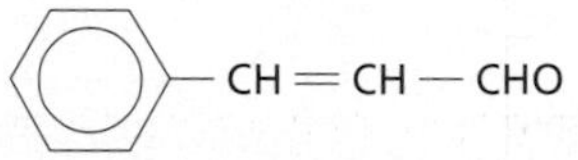

react with the following?
a hydrogen with nickel at room temperature
b cold aqueous potassium manganate(VII)
c hot potassium manganate(VII) and dilute sulphuric acid
d hot potassium dichromate(VI) and dilute sulphuric acid

Answer

Cinnamaldehyde contains three functional groups – the aromatic ring, the C=C double bond and the aldehyde group.

a Hydrogen reacts with C=C double bonds (chart A reaction A5) *and* with C=O double bonds (reaction D3). (Hydrogen reacts with benzene rings only at high temperatures – see Chapter 27, page 509.) The product will therefore be:

$$\text{C}_6\text{H}_5 - \text{CH}_2 - \text{CH}_2 - \text{CH}_2\text{OH}$$

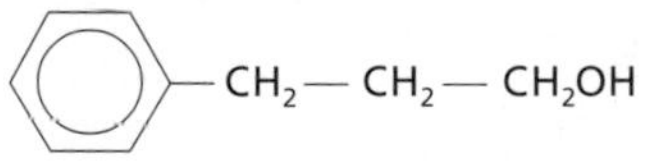

b Cold potassium manganate(VII) solution reacts only with C=C double bonds (reaction A6). The product will be:

$$\text{C}_6\text{H}_5 - \text{CH(OH)} - \text{CH(OH)} - \text{CHO}$$

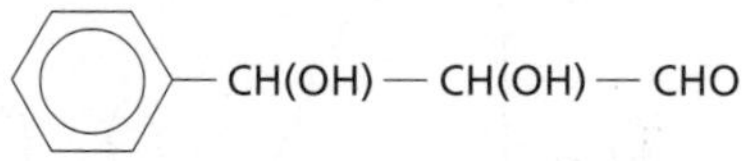

c Hot potassium manganate(VII) and dilute sulphuric acid react with C=C double bonds (reaction A7) *and* with aldehydes (reaction D4). The products are:

$$\text{C}_6\text{H}_5 - \text{CO}_2\text{H} + \text{HO}_2\text{C} - \text{CO}_2\text{H}$$

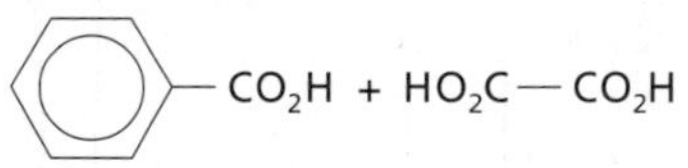

(Note that ethanedioic acid is further oxidised by manganate(VII) to carbon dioxide.)
d Hot potassium dichromate(VI) and dilute sulphuric acid react only with aldehydes (reaction D4), and not with C=C double bonds. The product will therefore be:

$$\text{C}_6\text{H}_5 - \text{CH} = \text{CH} - \text{CO}_2\text{H}$$

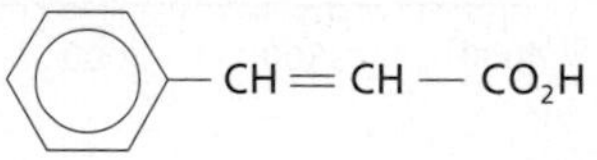

Further practice

Compound **X** is:
$$\text{Cl} - \text{CH}_2 - \text{CH}_2 - \overset{\overset{\displaystyle O}{\|}}{\text{C}} - \text{CH}_3$$

Predict the products of the reactions of compound **X** with:
1 aqueous ammonia
2 hydrogen cyanide with a trace of sodium cyanide
3 aqueous alkaline iodine
4 aqueous hydroxide ions and heat.

As we have seen, the infrared spectrum of a compound can often be used to identify the functional groups present. The absorptions in the spectrum of a multifunctional compound are the sum of the absorptions of each functional group.

Example Compound **Y** has the molecular formula C_2H_3N, and compound **Z** has the molecular formula C_2H_7N. Figure 31.33 shows their IR spectra. Deduce the functional groups in the two molecules, and suggest their structures.

a Predict the product of the reaction of **Y** with hot dilute sulphuric acid.

b Predict the product of the reaction of **Z** with cold dilute sulphuric acid.

■ **Figure 31.33** IR spectra of **a Y** and **b Z**.

a Y

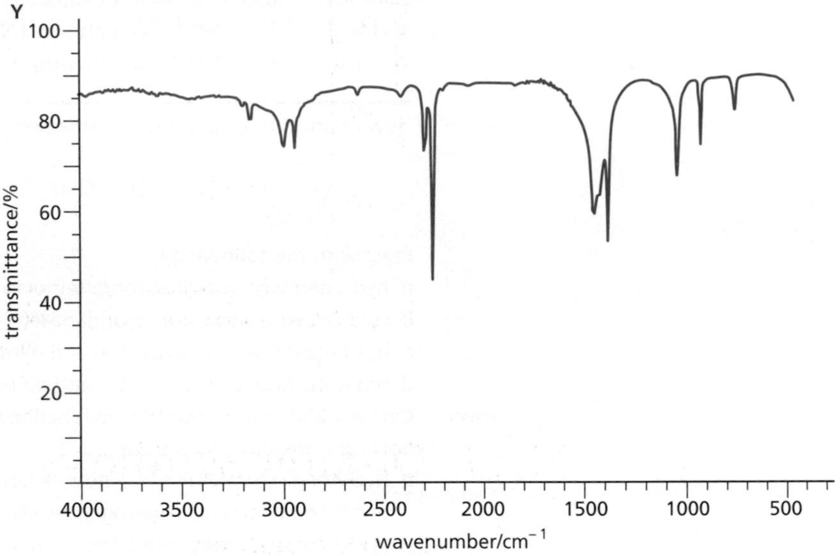

b Z

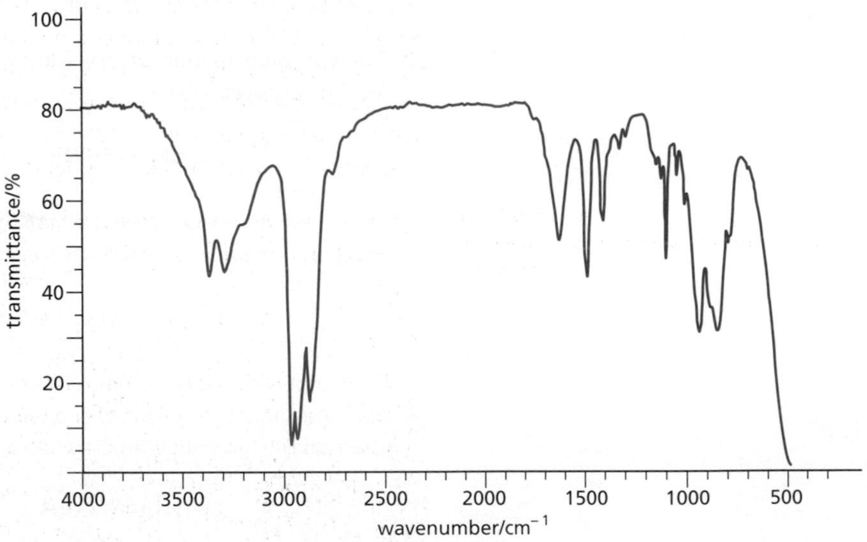

Answer The spectrum of **Y** is a simple one, suggesting a small molecule with few bonds to vibrate. The absorption at 2200 cm^{-1} suggests a cyanide group. The simple weak absorptions in the C—H region around 3000 cm^{-1} suggest a CH_3 group. Therefore **Y** is CH_3—C≡N.

The double absorption at 3300 cm^{-1} in the spectrum of **Z** suggests an —NH_2 group. The C—H absorptions in this spectrum are much stronger. **Z** is ethylamine, $CH_3CH_2NH_2$.

The products of the two reactions are:

a CH_3CO_2H

b $(CH_3CH_2NH_3^{+})_2SO_4^{2-}$

Further practice Compound **P** has the molecular formula $C_8H_8O_3$. Its IR spectrum is shown in Figure 31.34. Deduce the functional groups present in **P**, and suggest a structure for it. Predict the structural formulae of the products formed when **P** is reacted with:

1 potassium dichromate(VI) + dilute sulphuric acid + heat

2 ethanol + concentrated sulphuric acid + heat.

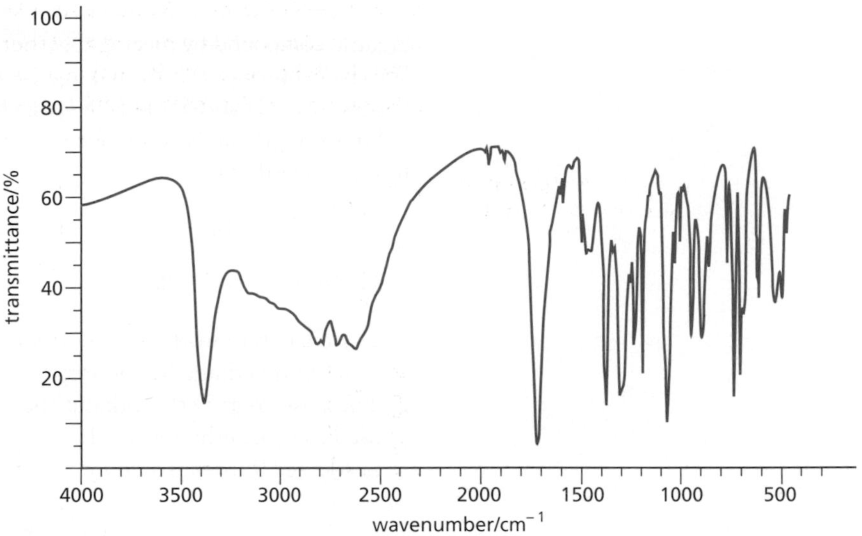

■ **Figure 31.34** IR spectrum of **P**.

31.11 Organic synthesis

Simple molecules

As was mentioned at the start of this chapter, many organic compounds find important uses as pharmaceuticals, pesticides, perfumes and dyes. These compounds often have quite complicated structures, and most of them are manufactured by organic chemists from much simpler starting materials. The science of organic synthesis is akin to molecular 'Lego'. A structure is built up by combining a sequence of reactions – from two to perhaps 20 or so – which start from known, readily available, chemicals and end up with the target molecule.

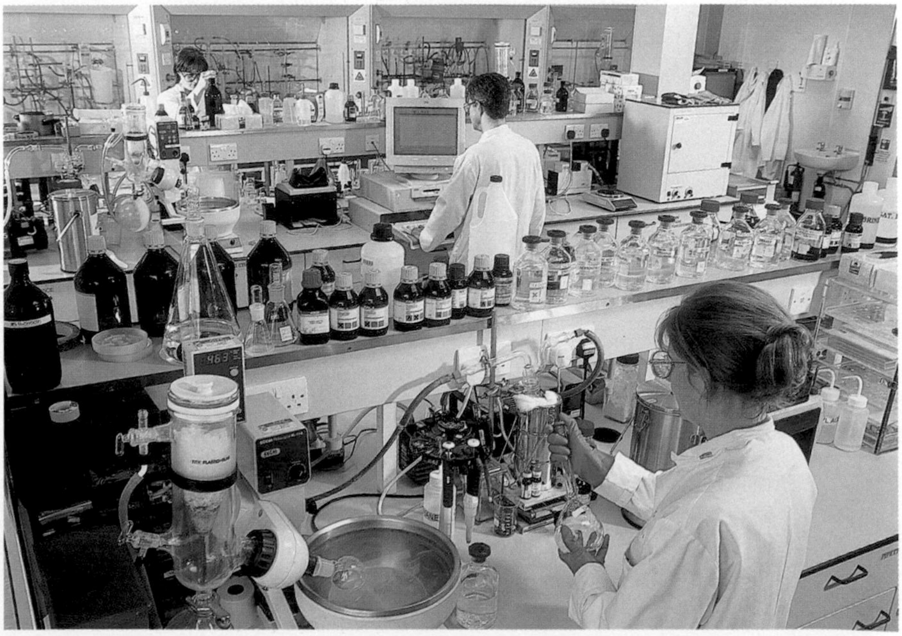

■ **Figure 31.35** A laboratory where new organic compounds are synthesised.

As well as science, there is much art and artisanship associated with organic synthesis. Students who have attempted organic preparations in the laboratory will know that the yield and purity of their product will often vary from those given in the 'recipe'. It takes skill and practice to perfect practical techniques. The reagents and conditions that give an excellent yield with one compound may not be effective for another compound having the identical functional group.

Nevertheless, it is possible to devise a multistep method for synthesising a given organic compound by piecing together successive standard organic transformations. We shall illustrate this by making use of the reactions summarised in charts A–G (pages 612–13) and Table 31.8 (page 611).

For example, suppose we need to devise a synthesis of ethanoic acid, starting from bromoethane:

$$CH_3CH_2Br ----- \rightarrow CH_3CO_2H$$

The strategy is as follows:

1 Use chart B to work out the structures of all the compounds that can be made from bromoethane in one step.
2 Then use chart D to work out the structures of all the compounds that could be used to make ethanoic acid.
3 Lastly, see if there is a compound that is common between both charts.

$$CH_3CH_2Br \Bigg\langle \begin{matrix} \nearrow CH_3CH_2OH \\ \rightarrow CH_3CH_2CN \\ \rightarrow CH_3CH_2NH_2 \\ \searrow CH_2{=}CH_2 \end{matrix}$$

$$\begin{matrix} CH_3CH_2OH \searrow \\ CH_3CN \longrightarrow CH_3CO_2H \\ CH_3MgBr \nearrow \end{matrix}$$

        chart B   chart D

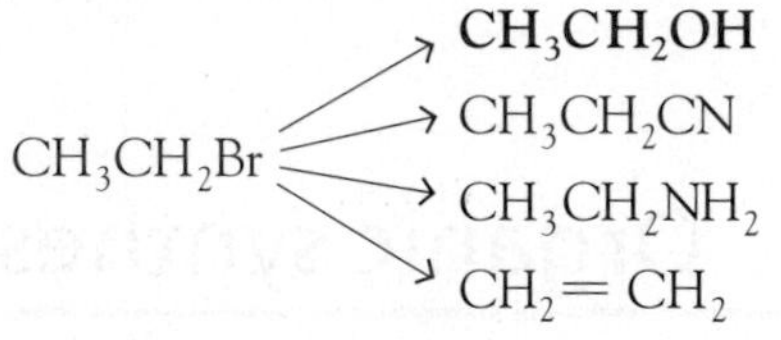

The compound ethanol is common to both, so a viable synthetic route would have ethanol as an intermediate:

$$CH_3CH_2Br \rightarrow CH_3CH_2OH \rightarrow CH_3CO_2H$$

Finally, find the reagents and conditions required to carry out these two transformations:

$$CH_3CH_2Br \xrightarrow{\text{heat with NaOH}_{(aq)}} CH_3CH_2OH \xrightarrow[\text{Na}_2\text{Cr}_2\text{O}_7 + \text{H}_2\text{SO}_{4(aq)}]{\text{heat with an excess of}} CH_3CO_2H$$

Steps **1** and **2** above using charts A–G may need to be repeated if a common compound is not found.

Example Devise a synthesis of propylamine, CH$_3$CH$_2$CH$_2$NH$_2$, from ethene, C$_2$H$_4$.
Answer Using chart A, there are four reactions of ethene that may be useful:

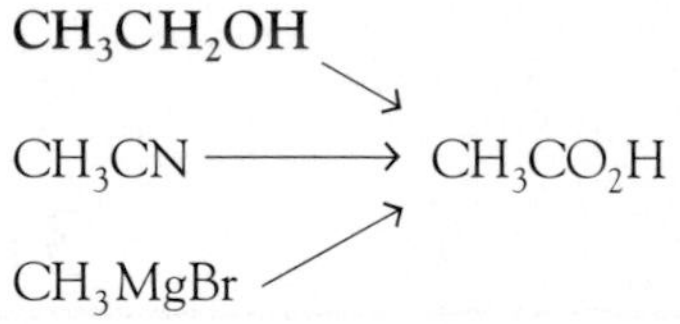

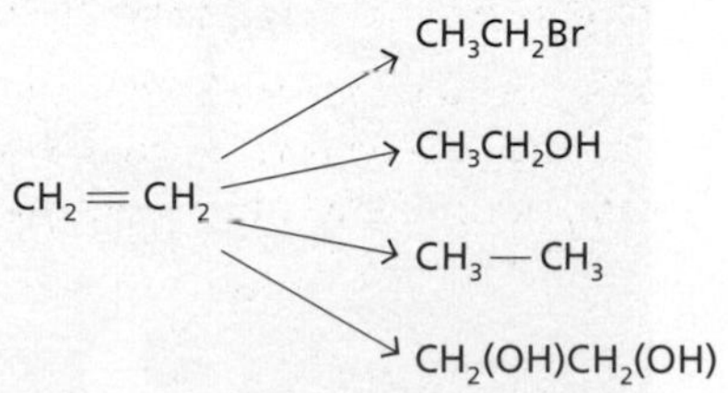

chart A

Starting from the target molecule there are just three methods of making amines (chart F):

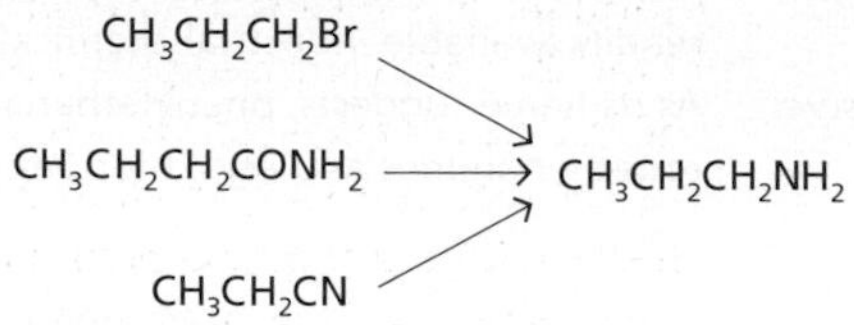

chart F

None of these three starting materials for making propylamine is common to the four products from ethene, so another step in the synthesis is needed. We therefore take each of the four products derived from ethene, and carry out the same analysis on them – what compounds can we make from each one in turn?

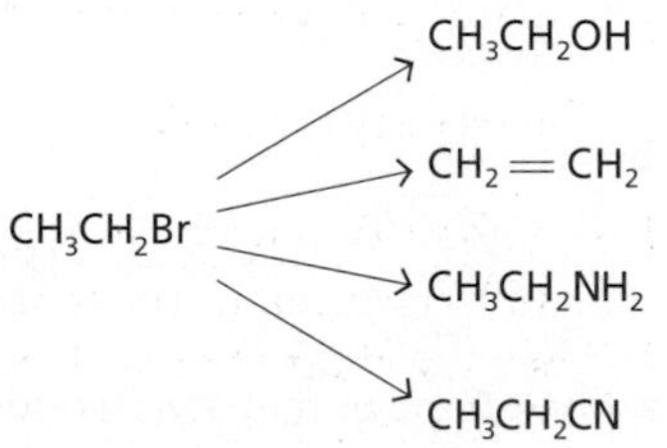

chart B

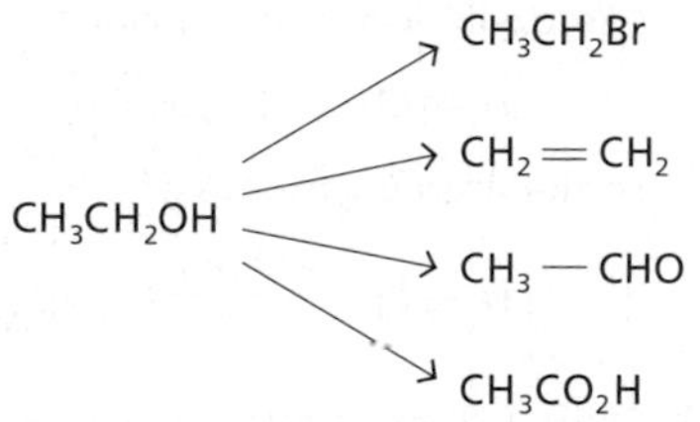

chart C

We'll stop there. We can see immediately that there is a compound in common between chart B and chart F, which is propanenitrile, CH_3CH_2CN. The synthetic route therefore includes two intermediates:

$$CH_2{=}CH_2 \rightarrow CH_3CH_2Br \rightarrow CH_3CH_2CN \rightarrow CH_3CH_2CH_2NH_2$$

Adding the reagents and conditions, the whole synthesis can be described as follows:

$$CH_2{=}CH_2 \xrightarrow{HBr(g)} CH_3CH_2Br \xrightarrow[\text{in ethanol}]{\text{heat with NaCN}} CH_3CH_2CN \xrightarrow{H_2+Ni} CH_3CH_2CH_2NH_2$$

Note that a 'shortcut' to the process here is given by the fact that a product containing one more carbon atom in the chain than the starting material must have been synthesised using a step involving a nitrile.

Further practice

Devise syntheses for the following compounds, starting with the specified compounds:

1 CH_3CH_2Br from CH_3CO_2H (two steps)
2 $CH_3CH(OH)CN$ from CH_3CH_2OH (three steps)
3 $CH_3CO_2CH_3$ from CH_3CN (two steps)
4 $C_6H_5{-}CH_2CO_2H$ from C_6H_6 (benzene) (four steps).

The synthesis of more complicated organic molecules

Many compounds of pharmaceutical interest are derived from joining together two or more organic parts, each of which needs its own synthesis. For example, the compound phenylethanamide has been used as the drug antifebrin:

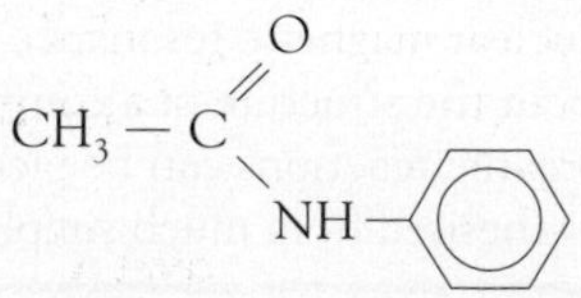

Example Devise a synthesis of phenylethanamide, starting from ethene and benzene (both are readily available industrial chemicals which are derived from petroleum).

Answer As its name suggests, phenylethanamide is an amide. Amides are formed by reacting together amines and acyl chlorides. Phenylethanamide can therefore be made as follows:

$$CH_3COCl + NH_2 - \bigcirc \longrightarrow CH_3 - C \overset{O}{\underset{NH-\bigcirc}{\Big<}}$$

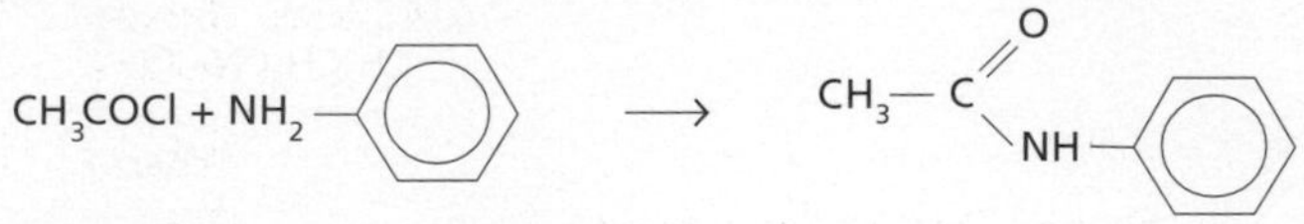

The synthesis is therefore in three parts:

a

$$CH_2{=}CH_2 \rightarrow \rightarrow \rightarrow CH_3COCl$$

b

$$\bigcirc \rightarrow \rightarrow \rightarrow H_2N-\bigcirc$$

$$\longrightarrow CH_3CONH-\bigcirc$$

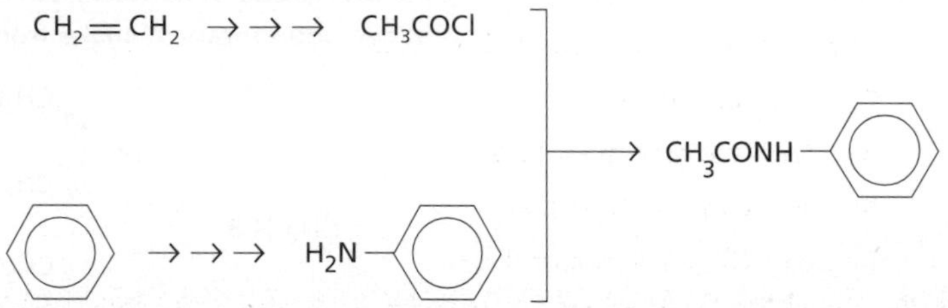

a Chart D shows that acyl chlorides are made from carboxylic acids:

$$CH_3CO_2H \rightarrow CH_3COCl$$

Carboxylic acids can be made from alcohols, which in turn can be made from alkenes:

$$CH_2{=}CH_2 \rightarrow CH_3CH_2OH \rightarrow CH_3CO_2H$$

So the overall synthesis of ethanoyl chloride from ethene is as follows:

$$CH_2{=}CH_2 \xrightarrow[\text{heat}]{\text{dil. H}_2\text{SO}_4} CH_3CH_2OH \xrightarrow[\text{Na}_2\text{Cr}_2\text{O}_7+\text{H}^+\text{(aq)}]{\text{heat with excess}} CH_3CO_2H \xrightarrow[\text{heat}]{\text{PCl}_5} CH_3COCl$$

b Chart G shows a two-step synthesis of phenylamine from benzene:

$$\bigcirc \xrightarrow[\text{conc. H}_2\text{SO}_4 \text{ at } T < 55\,°\text{C}]{\text{conc. HNO}_3 +} \overset{NO_2}{\bigcirc} \xrightarrow[\text{heat}]{\text{Sn + conc. HCl}} \overset{NH_2}{\bigcirc}$$

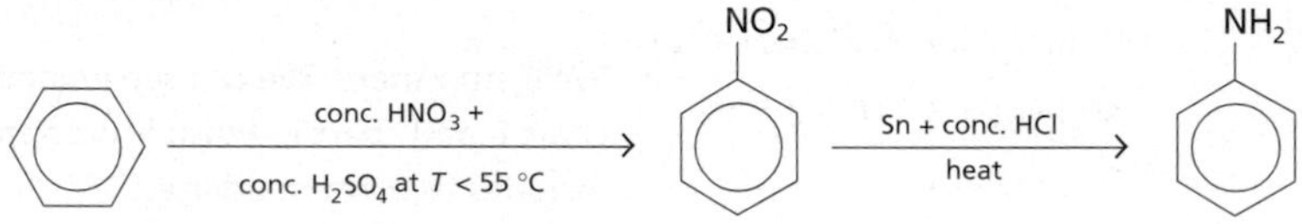

The complete synthesis therefore comprises six steps:

$$CH_2{=}CH_2 \longrightarrow CH_3CH_2OH \longrightarrow CH_3CO_2H \longrightarrow CH_3COCl$$

$$\bigcirc \rightarrow \bigcirc\!-NO_2 \rightarrow \bigcirc\!-NH_2$$

$$\longrightarrow CH_3CONH-\bigcirc$$

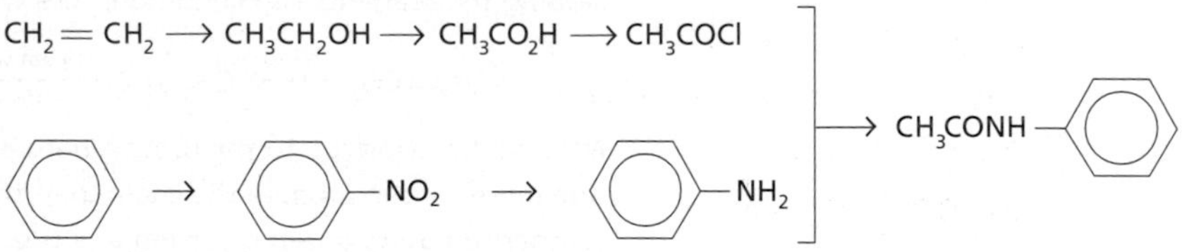

Further practice Devise a synthesis of the ester prop-2-yl phenylethanoate, $C_6H_5CH_2CO_2CH(CH_3)_2$, from methylbenzene and propene (five steps in all).

Summary

- Mass spectrometry can be used to determine the molecular formula of a compound, and often its structure too.
- Structures can also be deduced from molecular formulae by performing functional group tests on a compound.
- Functional groups can be identified by the use of infrared and UV/visible spectroscopy.
- Nuclear magnetic resonance spectroscopy provides very detailed evidence about the structure of a compound.
- Organic reactions can be pieced together to allow complex molecules to be synthesised from much simpler starting materials.

■ **Table 31.8** Summary of reagents used for organic reactions.

Reference (see key below)	Reagent and conditions	Reaction number on charts A–G (pages 612–13)						
		A	B	C	D	E	F	G
		Alkenes	Bromo-alkanes	Alcohols	Aldehydes and acids	Methyl ketones	Amines	Benzene
n1	OH^-(aq) + heat		B4	C1				
n2	NH_3 in ethanol + heat under pressure		B6		D10		F1	
n3	NaCN in ethanol + heat		B7					
n4	HCN + NaCN in ethanol				D2	E5		
n5	HBr(conc.) *or* NaBr + conc. H_2SO_4 + heat		B1	C6				
n6	$SOCl_2$ *or* PCl_5 + heat				D8			
n7	R—OH at room temperature				D12			
n8	R—OH + conc. H_2SO_4 + heat				D9			
n9	phenol + OH^-(aq) in the cold				D11			G13
e1	OH^- in ethanol + heat	A1	B5					
e2	Al_2O_3 + heat	A2		C7				
e3	conc. H_2SO_4 + heat	A2		C7				
r1	H_2(g) + Ni catalyst at room temperature	A5					F3	
r2	$NaBH_4$ in aqueous methanol + heat			C3	D3	E4		
r3	$LiAlH_4$ in dry ether			C3, C4	D3, D7	E4	F3	
r4	Sn + conc. HCl(aq) + heat							G10
o1	$Na_2Cr_2O_7$(aq) + H_2SO_4(aq) + heat			C8, C9	D1, D4	E3		
o2	$KMnO_4$(aq) (cold)	A6						
o3	$KMnO_4$(aq) + H_2SO_4(aq) + heat	A7		C8, C9	D1, D4	E3		
o4	$KMnO_4$(aq) + OH^-(aq) + heat							G7, G8
h1	Cl_2(g) + light							G9
h2	Br_2(g) + light		B2					
h3	Br_2 + $AlCl_3$ + heat							G2
h4	Br_2(aq) + OH^-(aq) (warm)						F2	
h5	I_2(aq) + OH^-(aq) at room temperature					E6		G7
fc1	CH_3Cl + $AlCl_3$ + heat							G4
fc2	CH_3COCl + $AlCl_3$ + heat							G3
m1	H_2O at room temperature		B9					G12
m2	HBr(g) at room temperature	A3	B3					
m3	H_2SO_4(aq) at room temperature	A4		C2		E1		
m4	H_2SO_4(aq) + heat				D5			
m5	conc. HNO_3 + conc. H_2SO_4 at $<55\,°C$							G1
m6	HNO_2, i.e. $NaNO_2$ + HCl(aq) at $<5\,°C$						F5	G11
m7	R″COCl at room temperature						F4	
m8	Mg in dry ether at room temperature		B8					G5
m9	R_2CO(l), then H^+(aq) at room temperature		B10	C5		E2		
m10	CO_2(g), then H^+(aq) at room temperature		B11		D6			G6

Key to reagent types:
- n — nucleophilic reagents
- e — reagents that cause elimination reactions
- r — reducing agents
- o — oxidising reagents
- h — halogenation agents
- fc — Friedel–Crafts reagents
- m — miscellaneous

Chart A: Synthetic routes involving alkenes

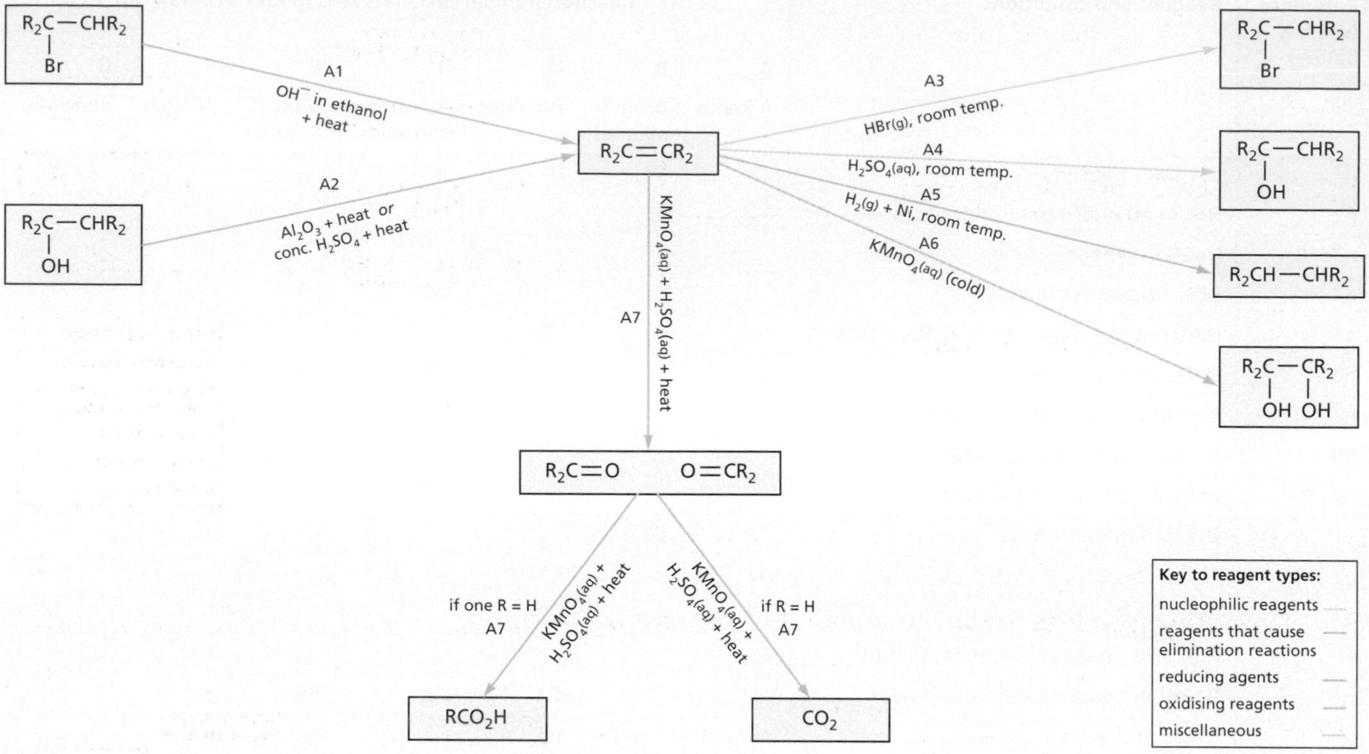

Chart B: Synthetic routes involving bromoalkanes

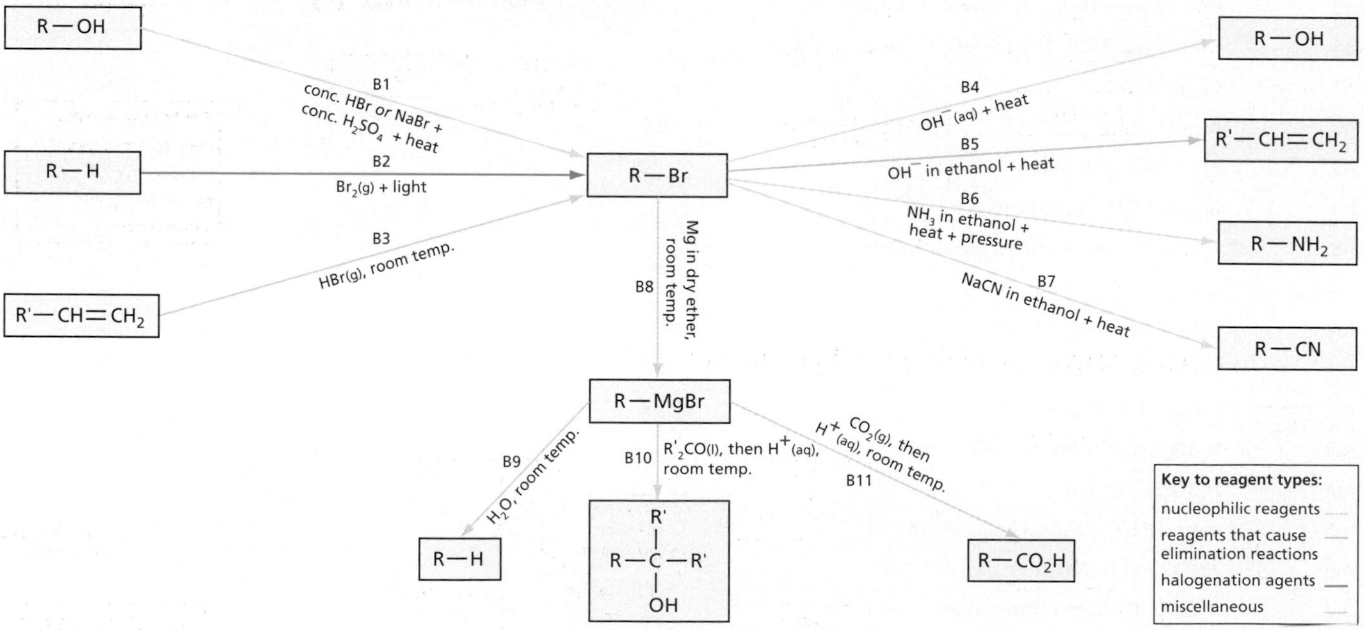

Chart C: Synthetic routes involving alcohols

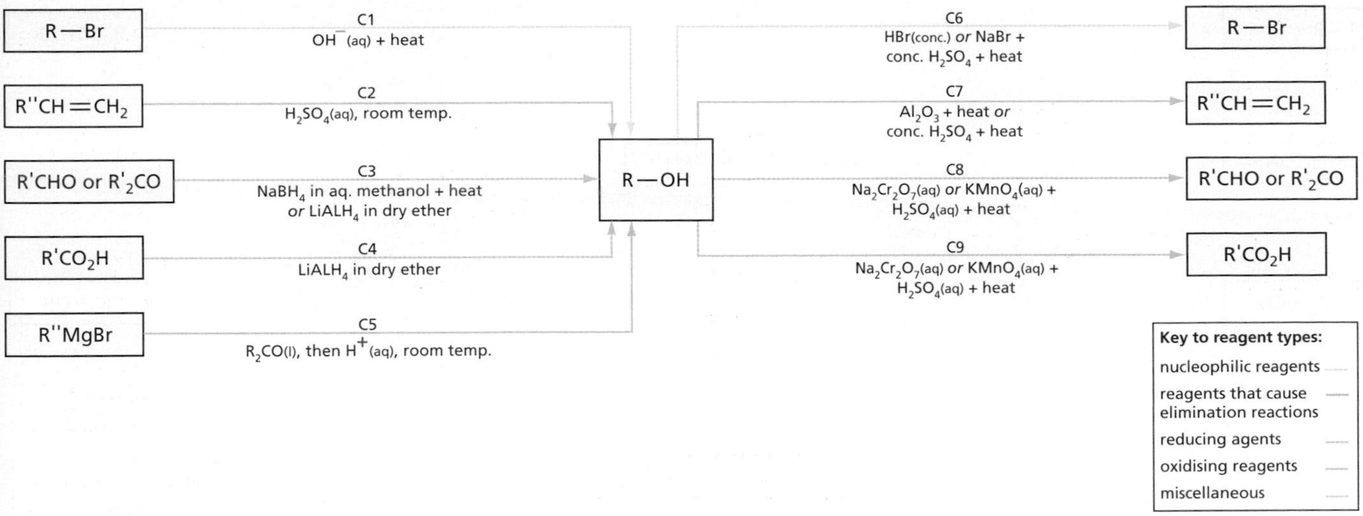

Chart D: Synthetic routes involving aldehydes and carboxylic acids

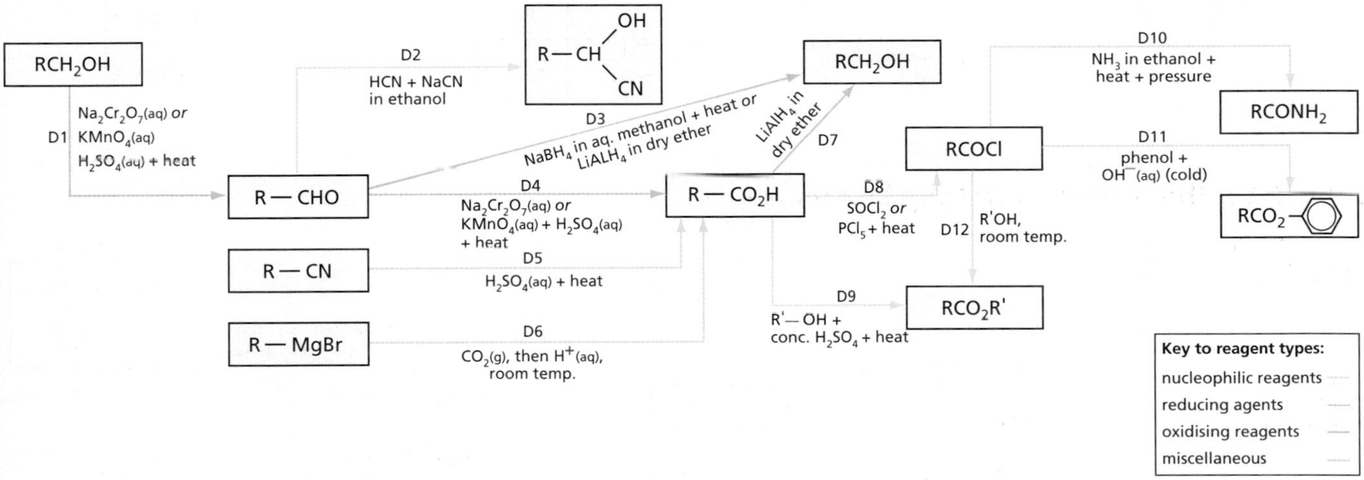

Chart E: Synthetic routes involving methyl ketones

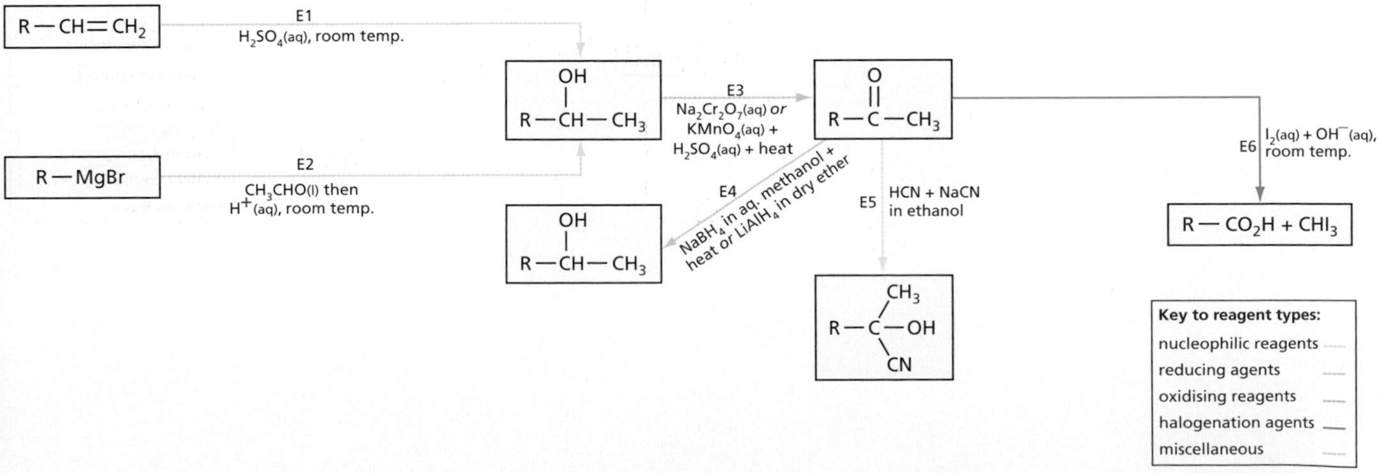

Chart F: Synthetic routes involving amines

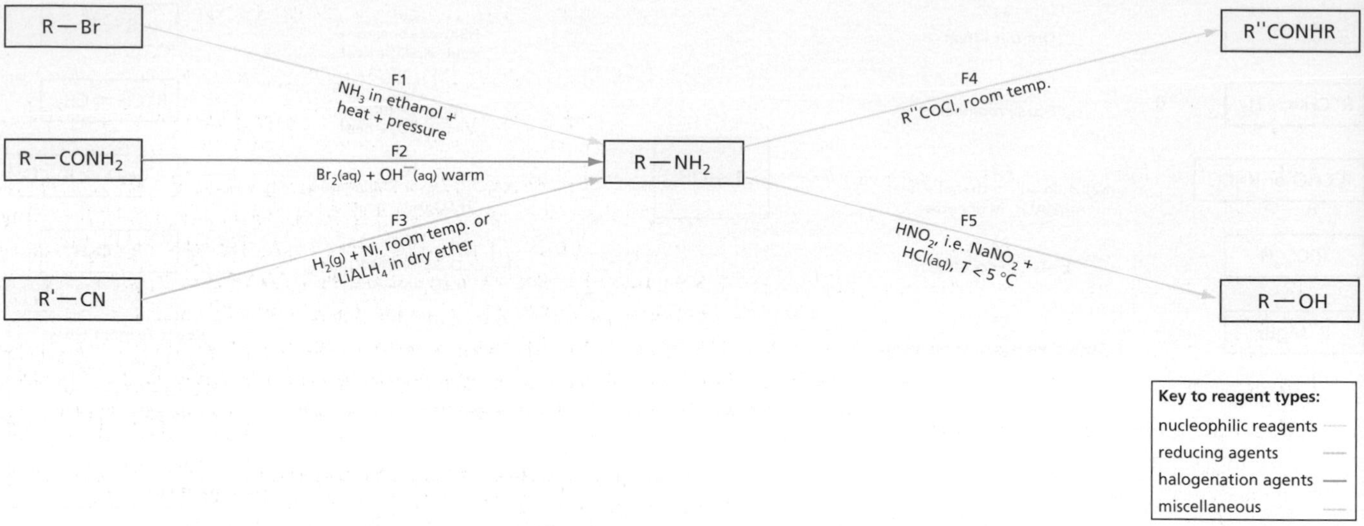

Chart G: Synthetic routes involving benzene

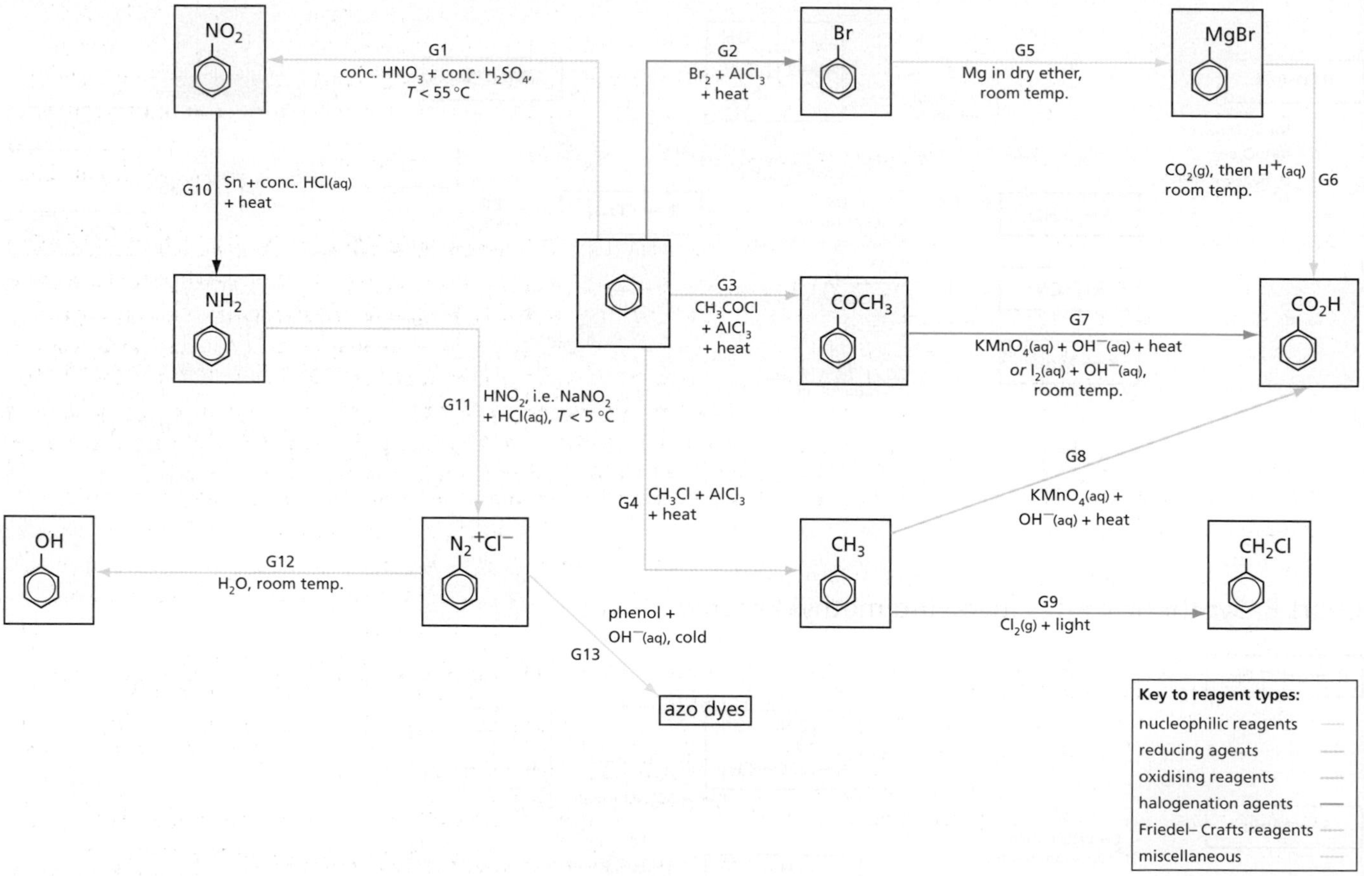

Examination questions

Chapter 1

1 Calculate the molar masses of the following:
a $Al_2(SO_4)_3$
b $K_4Fe(CN)_6$ (4)

2 How many moles of substance are present in the following?
a 0.250 g of calcium carbonate
b 5.72 g of sodium carbonate crystals, $Na_2CO_3.10H_2O$. (4)

3 Calculate the percentage by mass of:
a Mg in Mg_3N_2
b Br in $CaBr_2$ (4)

4 Calculate the empirical formula of the compounds for which the following analysis results were obtained:
a 27.3% C 72.7% O
b 29.1% Na 40.5% S 30.4% O (4)

5 A weighed sample of $CaSO_4.xH_2O$ was heated until the residue reached a constant mass, to drive off all the water of crystallisation. The following results were obtained:

1.124 g of $CaSO_4.xH_2O$ gave 0.889 g of anhydrous residue.

Calculate the value of *x*. (3)

6 The sulphur present in 0.100 g of an organic compound is converted into barium sulphate. A precipitate of 0.1852 g of dry $BaSO_4$ is obtained. Calculate the percentage by mass of sulphur in the organic compound. (4)

7 What mass of sodium carbonate, Na_2CO_3 must be reacted with hydrochloric acid to produce 11.70 g of sodium chloride? (2)

8 What volume of hydrogen (at s.t.p.) is formed when 3.00 g of magnesium react with an excess of dilute sulphuric acid? (2)

9 A solution is made by dissolving 5.00 g of impure sodium hydroxide in water and making it up to 1.00 dm^3 of solution. 25.0 cm^3 of this solution is neutralised by 30.3 cm^3 of hydrochloric acid, of concentration 0.102 mol dm^{-3}. Calculate the percentage purity of the sodium hydroxide. (2)

10 a What experimental data are required in order to calculate the empirical formula of a compound? (1)
b Give the meaning of the term *molecular formula*. (1)
c Define the term *relative atomic mass* of an element. (2)
d When barium nitrate is heated it decomposes as follows:

$$Ba(NO_3)_2(s) \rightarrow BaO(s) + 2NO_2(g) + \tfrac{1}{2}O_2(g)$$

i Calculate the total volume, measured at 298 K and 100 kPa, of gas which is produced by decomposing 5.00 g of barium nitrate.
ii Calculate the volume of 1.20 M hydrochloric acid which is required to neutralise exactly the barium oxide formed by decomposition of 5.00 g of barium nitrate. Barium oxide reacts with hydrochloric acid as follows.

$$BaO(s) + 2HCl(aq) \rightarrow BaCl_2(aq) + H_2O(l)$$ (7)

AQA, CH01, June 2000

11 The female of the American cockroach (*Periplaneta americana*) secretes a chemical (pheromone) of molecular formula $C_{11}H_{18}O_2$ to which the male of the species is attracted. It is reported that the male will respond to as few as 60 molecules of the pheromone.
What is the mass, in grams, of these 60 molecules?
Use the following headings to assist you in your calculation: relative molecular mass; mass of one mole; mass of one molecule; mass of sixty molecules in grams. (4)

CCEA, AS1, Module 1, June 2001

12 Calcium carbonate is added to an excess of hydrochloric acid.

$$CaCO_3(s) + 2HCl(aq) \rightarrow CaCl_2(aq) + CO_2(g) + H_2O(l)$$

a Deduce *two* observations that you would expect to see during this reaction. (2)
b In this experiment, 0.040 g $CaCO_3$ is added to 25 cm^3 of 0.050 mol dm^{-3} HCl. (2)
i Explain what is meant by 0.050 mol dm^{-3} HCl. (2)
ii Calculate how many moles of $CaCO_3$ were used in this experiment. (2)
iii Calculate how many moles of HCl are required to react with this amount of $CaCO_3$. (1)
iv Hence show that the HCl is in excess. (1)
c State *one* large-scale use of a named Group 2 compound that is being used to reduce acidity. (1)

OCR, AS, 2811, Jan 2001

13 Well over 2 000 000 tonnes of sulphuric acid, H_2SO_4, are produced in the UK each year. This is used in the manufature of many important materials such as paints, fertilisers, detergents, plastics, dyestuffs and fibres.
The sulphuric acid is prepared from sulphur in a 3-stage process.

Stage 1: The sulphur is burnt in oxygen to produce sulphur dioxide.

$$S + O_2 \rightarrow SO_2$$

Stage 2: The sulphur dioxide reacts with more oxygen using a catalyst to form sulphur trioxide.

$$2SO_2 + O_2 \rightarrow 2SO_3$$

Stage 3: The sulphur trioxide is dissolved in concentrated sulphuric acid to form 'oleum', $H_2S_2O_7$, which is then diluted in water to produce sulphuric acid.

a 100 tonnes of sulphur dioxide were reacted with oxygen in stage 2.
Assuming that the reaction was complete, calculate
i how many moles of sulphur dioxide were reacted.
M_r: SO_2, 64.1. (1 tonne = 1×10^6 g) (1)
ii the mass of sulphur trioxide that formed.
M_r: SO_3, 80.1 (1)

b Construct a balanced equation for the formation of sulphuric acid from oleum. (1)

c The concentration of the sulphuric acid can be checked by titration. A sample of the sulphuric acid was analysed as follows.

- 10.0 cm³ of sulphuric acid was diluted with water to make 1.00 dm³ of solution.
- The diluted sulphuric acid was then titrated with aqueous sodium hydroxide, NaOH.

$$H_2SO_4(aq) + 2NaOH(aq) \rightarrow Na_2SO_4(aq) + 2H_2O(l)$$

- In the titration, 25.0 cm³ of 0.100 mol dm^{-3} aqueous sodium hydroxide required 20.0 cm³ of the *diluted* sulphuric acid for neutralisation.

i Calculate how many moles of NaOH were used. (1)
ii Calculate the concentration, in mol dm^{-3}, of the *diluted* sulphuric acid, H_2SO_4. (2)
iii Calculate the concentration, in mol dm^{-3}, of the original sulphuric acid sent for analysis. (1)

OCR, AS, 2811, Jan 2001

14 Lead compounds are extensively used to provide the colour in paints and pigments.

a 'White lead', used for over 2000 years as a white pigment, is based upon lead carbonate. Analysis shows that lead carbonate has the following percentage composition by mass: Pb, 77.5%; C, 4.5%; O, 18.0%.
Calculate the empirical formula of lead carbonate.
(A_r: C, 12.0; O, 16.0; Pb, 207.0.) (3)

b 'Red lead' is the pigment in paint used as a protective coating for structural iron and steel. It is based upon lead oxide, Pb_3O_4, a scarlet powder formed by oxidising lead(II) oxide with oxygen.
i Balance the equation for the oxidation of PbO.

$$PbO\,(s) + O_2\,(g) \rightarrow Pb_3O_4\,(s)$$

ii What is the molar mass of Pb_3O_4? (A_r: O, 16.0; Pb, 207.0.)
iii Calculate the mass of Pb_3O_4 that could be formed from 0.300 mol of PbO. (4)

OCR, AS, Specimen 2811, 2001/2

Chapter 2

1 a Give the relative mass and relative charge of a neutron. (2)

b In terms of the number of their fundamental particles, what do two isotopes of an element have in common and how do they differ? (2)

c Give the complete atomic symbol, including mass number and atomic number, for an atom of the isotope with 22 neutrons and 19 electrons. (2)

d In a mass spectrometer the isotopes of an element are separated and two measurements are made for each isotope.
i Which two measurements are made for each isotope?
ii State how the detector in a mass spectrometer works.
iii Why is a mass spectrometer incapable of distinguishing between the ions $^{14}N^+$ and $^{14}N_2^{2+}$? (5)

e Using arrows ↑ and ↓ to represent electrons, copy and complete the energy-level diagram below to show the electronic arrangement in an atom of carbon.

— — — 2p

— 2s

— 1s (2)

f In terms of sub-levels, give the electronic configuration of the carbon ion C^{2+}. (1)

AQA, AS, Unit 1, Jan 2001

2 a Define the terms *mass number* and *atomic number* of an atom. (2)

b Give the symbol, including the mass number and the atomic number, for the atom which has 3 fewer neutrons and 2 fewer protons than $^{14}_{7}N$. (2)

c In terms of sub-levels, give the complete electronic configuration of the nitrogen atom, N, and of the nitride ion, N^{3-}. (2)

d Define the term *relative atomic mass* of an element. (2)

e When a pure, gaseous sample of element X is introduced into a mass spectrometer, four mononuclear, singly-charged ions are detected, as shown in the spectrum below.
i Describe the process by which the gaseous sample of X is converted into ions in a mass spectrometer.
ii What adjustment is made to the operating conditions in order to direct the different ions, in turn, onto the detector of a mass spectrometer?
iii Use data from the spectrum below to calculate the relative atomic mass of X.
iv Identify the element X. (7)

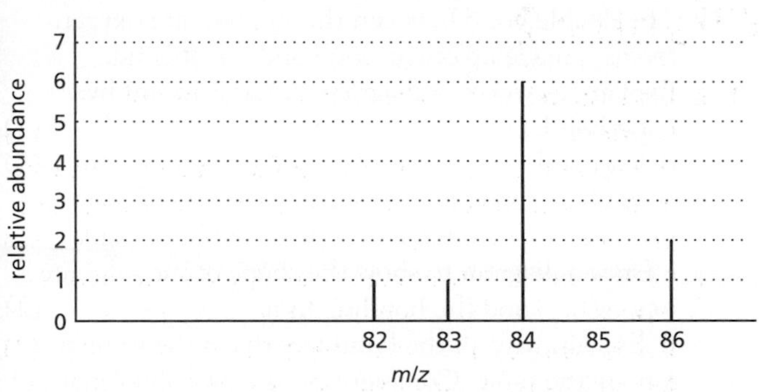

AQA, AS, Unit 1, June 2001

3 a When a sample of copper is analysed using a mass spectrometer, its atoms are ionised and then accelerated.
i Explain how the atoms of the sample are ionised. (2)
ii State how the resulting ions are then accelerated. (1)

b For a particular sample of copper two peaks were obtained in the mass spectrum.

Peak at *m/e*	Relative abundance
63	69.1
65	30.9

i Give the formula of the species responsible for the peak at $m/e = 65$. (1)
ii State why **two** peaks, at *m/e* values of 63 and 65, were obtained in the mass spectrum. (1)
iii Calculate the relative atomic mass of this sample of copper, using the table of results above. (2)

Edexcel, AS, Unit 1, June 2001

4 In the periodic table, where elements are arranged by atomic number, chlorine is a p-block element whereas manganese, a transition element, is in the d-block.

a i Define the term *atomic number*. (1)
ii Define the term *d-block element*. (1)
iii Define the term *transition element*. (1)

b The electron configuration of chlorine is $1s^2 2s^2 2p^6 3s^2 3p^5$. Write the electron configuration for manganese in a similar manner. (1)

c i Define the term *first ionisation energy of chlorine*. (2)
ii Sketch the *pattern* you would expect to see in a plot of successive ionisation energies of chlorine against the number of electrons removed. (3)

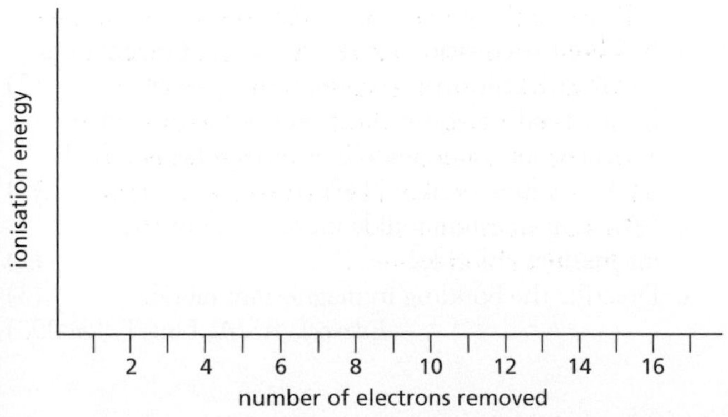

d Manganese(IV) oxide, MnO_2, reacts with concentrated hydrochloric acid to produce chlorine, water and a salt. The salt has a composition of 43.7% manganese and 56.3% chlorine by mass.
i Determine the empirical formula of the salt. (3)
ii The molecular formula of the salt is the same as its empirical formula.
Write the balanced equation for the reaction between manganese(IV) oxide and concentrated hydrochloric acid. (2)

Edexcel, A, Module 1, June 2001

5 The first ionisation energies of Group I elements are low and decrease as the Group is descended.

a A plot of the successive ionisation energies for sodium is shown below.

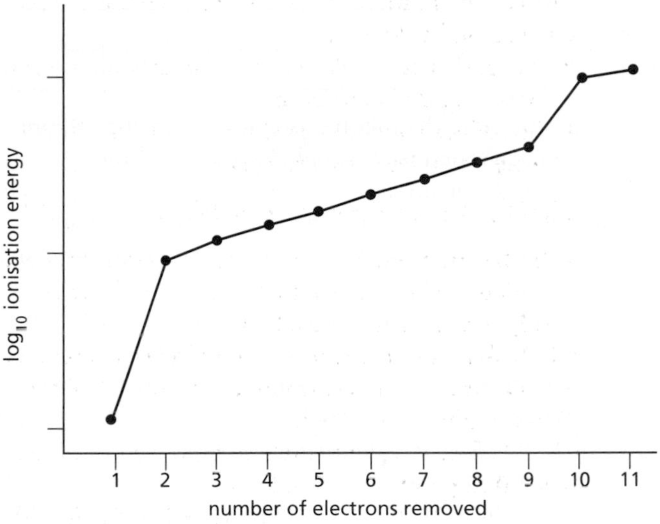

i Define the term *first ionisation energy*. (2)
ii Write an equation, including state symbols, for the first ionisation energy of sodium. (1)
iii Explain the sharp rise in the energy required to remove the second electron. (1)
iv Explain why the first ionisation energy of Group I elements decrease as the Group is descended. (2)

b Explain why the first ionisation energy of magnesium is greater than that of sodium. (3)

CCEA, A, Module 1, June 2001

6 Lithium was discovered in 1817 by the Swedish chemist Arfvedson. Lithium exists naturally as a mixture of isotopes.

a Explain the term *isotopes*. (1)

b Which isotope is used as the standard against which relative atomic masses are measured? (1)

c The mass spectrum overleaf shows the isotopes present in a sample of lithium.

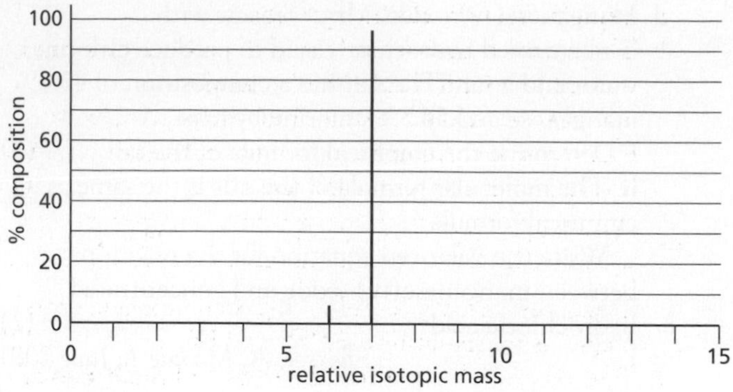

i Use this mass spectrum to help you complete the table below for each lithium isotope in the sample. (3)

Isotope	Percentage composition	Number of	
		Protons	Neutrons
^{6}Li			
^{7}Li			

ii Calculate the relative atomic mass of this lithium sample. Your answer should be given to three significant figures. (2)

d The species responsible for the peaks in this mass spectrum are lithium ions, produced and separated in a mass spectrometer.
i How are the electrons removed from lithium atoms to form lithium ions in a mass spectrometer? (1)
ii How does a mass spectrometer separate the ions?(1)

e The first ionisation energy of lithium is $+520 \text{ kJ mol}^{-1}$.
i Define the term *first ionisation energy*. (3)
ii The first ionisation energy of sodium is $+496 \text{ kJ mol}^{-1}$.
Explain why the first ionisation energy of sodium is less than that of lithium. Your answer should compare the atomic structures of each element. (3)

OCR, AS, 2811, Jan 2001

Chapter 3

1 a i State what is meant by the term *polar bond*. (1)
ii Sulphuric acid is a liquid that can be represented by the formula drawn below.

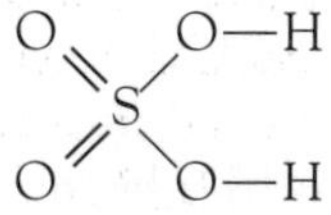

Given that the electronegativity values for hydrogen, sulphur and oxygen are 2.1, 2.5 and 3.5, respectively, clearly indicate the polarity of each bond present in the formula given. (2)
iii Suggest the strongest type of intermolecular force present in pure sulphuric acid. Briefly explain how this type of intermolecular force arises. (2)

b The double bond between the sulphur and oxygen atoms is made up one σ bond and one π bond. Explain, in words or diagrams, what is meant by:
i a σ bond. (1)
ii a π bond. (1)

c A sulphate ion can be formed by the loss of two hydrogen ions from a molecule of sulphuric acid.
i Draw a diagram to show the shape of the sulphate ion, SO_4^{2-}, and the bonding in it. (2)
ii Explain why all the bond lengths in the sulphate ion are the same. Comment on the probable length of these bonds compared to the lengths of the bonds between sulphur and oxygen in sulphuric acid. (2)

AQA, A, Module 8, June 2000

2 a Name the type of force that holds the particles together in an ionic crystal. (1)
b What is a covalent bond? (1)
c State how a co-ordinate bond is formed. (2)
d Describe the bonding in a metal. (2)
e A molecule of hydrogen chloride has a dipole and molecules of hydrogen chloride attract each other by permanent dipole–dipole forces. Molecules of chlorine are non-polar.
i What is a permanent dipole?
ii Explain why a molecule of hydrogen chloride is polar.
iii Name the type of force which exists between molecules of chlorine. (5)
f Show, by means of a diagram, how two molecules of hydrogen fluoride are attracted to each other by hydrogen bonding; include all lone-pair electrons and partial charges in your diagram. (3)
g Why is there no hydrogen bonding between molecules of hydrogen bromide? (1)

AQA, AS, Unit 1, June 2001

3 Deduce and draw the shapes of the following molecules or ions. Suggest a value for the bond angle in each case. Give a brief explanation of why each has the shape you give.
a SF_6 (3)
b PH_3 (3)
c PF_4^{+} (3)

Edexcel, AS, Unit 1, June 2001

4 a i Write the equation for the reaction of magnesium metal with chlorine, showing state symbols. (2)
ii The product of this reaction is ionic. Use this information to explain why it has a relatively high melting temperature. (714 °C) (2)
b Why is magnesium iodide more covalent than magnesium chloride? (2)
c Describe the bonding in magnesium metal. (3)

Edexcel, AS/A, Unit 1, Jan 2001

5 The shape and polarity of the water molecule have a pronounced effect on its physical and chemical properties.

a i Draw a dot and cross diagram to show the bonding present in a water molecule. (2)

ii State and explain the shape of a water molecule. (3)

b The polarity of the O—H bond in water may be deduced using the electronegativity values below:

hydrogen 2.5
oxygen 3.5

i Explain the term *electronegativity*. (2)

ii Copy the diagram below and label it to show the polarity of the O—H bond in water.

O—H (1)

c Acids dissolve in water to form hydroxonium (oxonium) ions, H_3O^+.

i Name the type of bond formed between a water molecule and a hydrogen ion, H^+, in the hydroxonium ion. (1)

ii Suggest the shape of the hydroxonium ion. (1)

CCEA, A, Module 1, June 2001

6 a Showing outer electron shells only, draw 'dot-and-cross' diagrams to show the bonding in ammonia and water. (2)

b Draw diagrams to illustrate the shape of a molecule of each of the compounds, NH_3 and H_2O. State the size of the bond angles on each diagram and name each shape. (6)

c On mixing with water, ammonia forms an alkaline solution containing the ammonium ion, NH_4^+:

$$NH_3(g) + H_2O(l) \rightarrow NH_4^+(aq) + OH^-(aq)$$

i The ammonium ion shows *dative covalent (co-ordinate)* bonding. Explain what is meant by this term.

ii Draw a 'dot-and-cross' diagram of the ammonium ion. Label on your diagram a dative covalent bond. (5)

OCR, AS, Specimen 2811, 2001/2

7 The compounds NH_3, BF_3 and HI all have covalent bonding and simple molecular structures. The Pauling electronegativity values shown in the table below can be used to predict polarity in these compounds.

			H 2.1			
Li 1.0	Be 1.5	B 2.0	C 2.5	N 3.0	O 3.5	F 4.0
Na 0.9						Cl 3.0
K 0.8						Br 2.8
						I 2.5

a Explain the term *electronegativity*. (2)

b The electronegativity values in the table above can be used to predict the polarity of a bond.

Copy the boxes below, and show the polarity of each bond by adding δ+ or δ− to each bond. The first box has been completed for you. (2)

$^{\delta-}$O—H$^{\delta+}$	H—N	F—B	H—I

c Using outer electron shells only, draw 'dot-and-cross' diagrams for molecules of NH_3 and BF_3. (2)

d The diagrams below show the shapes of molecules of NH_3 and BF_3.

For each diagram below, state the bond angle in each molecule and state the name of each shape. (4)

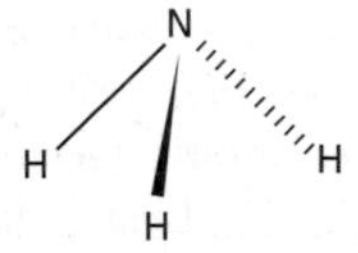

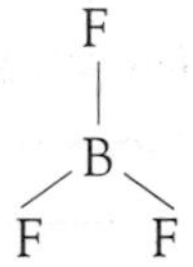

e Explain why NH_3 has polar molecules whereas molecules of BF_3 are non-polar. (2)

f Polar molecules of NH_3 form hydrogen bonds. Draw a diagram to show this hydrogen bonding. (1)

g NH_3 reacts with HI to form the ionic compound NH_4I, made up of NH_4^+ and I^- ions.

$$NH_3 + HI \rightarrow NH_4I$$

i Explain why the H—N—H bond angle in NH_3 is less than that in NH_4^+. (2)

ii Describe a simple test to confirm the presence of I^- ions in an acidified solution of NH_4I. (2)

OCR, AS, 2811, Jan 2002

Chapter 4

1 a With the aid of diagrams, describe the structure of, and bonding in, crystals of sodium chloride, graphite and magnesium. In each case, explain how the melting point and the ability to conduct electricity of these substances can be understood by a consideration of the structure and bonding involved. (23)

b Explain how the electron-pair repulsion theory can be used to predict the shapes of the molecules H_2O and PF_5. Illustrate your answer with diagrams of the molecules on which the bond angles are shown. (7)

AQA, AS, Unit 1, Jan 2001

2 a What is the name given to the number of molecules in one mole of carbon dioxide? (1)

b i State the ideal gas equation.

ii Calculate the volume of 1.00 mol of carbon dioxide gas at 298 K and 100 kPa.
(The gas constant $R = 8.31\,J\,mol^{-1}\,K^{-1}$)

iii Calculate the mass of carbon dioxide gas at 273 K and 500 kPa contained in a cylinder of volume $0.00500\,m^3$. (7)

c Hydrogen can be made by the reaction of hydrochloric acid with magnesium according to the equation

$2HCl + Mg \rightarrow MgCl_2 + H_2$

What mass of hydrogen is formed when 100 cm^3 of hydrochloric acid of concentration 5.0 mol dm^{-3} reacts with an excess of magnesium? (3)

d A compound of iron contains 38.9% by mass of iron and 16.7% by mass of carbon, the remainder being oxygen.

i Determine the empirical formula of the iron compound.

ii When one mole of this iron compound is heated, it decomposes to give one mole of iron(II) oxide, FeO, one mole of carbon dioxide and one mole of another gas. Identify this other gas. (The molecular formula of the iron compound is the same as its empirical formula.) (4)

AQA, AS, Unit 1, June 2001

3 a Explain the following observations. Include details of the *bonding* in and the *structure* of each substance.

i The melting temperature of diamond is much higher than that of iodine. (5)

ii Sodium chloride has a high melting temperature (approximately 800 °C). (3)

b Explain why aluminium metal is a good conductor of electricity. (3)

Edexcel, AS, Unit 1, June 2001

4 Sodium chloride is an example of an ionic crystal.

a What is an ion? (1)

b What is the formula of sodium chloride? (1)

c Write the electronic configuration of the sodium ion. (1)

d Write the electronic configuration of the chloride ion. (1)

e Solid sodium chloride dissolves in water. The enthalpy change is slightly endothermic.

i Write the equation for the dissolving of sodium chloride in water using state symbols. (2)

ii Explain the meaning of the term *endothermic*. (1)

CCEA, AS, AS1 Module 1, June 2001

5 0.145 cm^3 of an aliphatic amine of density 0.689 g cm^{-3} was vaporised and 60.0 cm^3 of the vapour was obtained at 150 °C and one atmosphere pressure (1.0×10^5 Pa).

a Write the ideal gas equation. (1)

b Use the ideal gas equation to determine the relative molecular mass of the amine. (3)

c Suggest a molecular formula for the amine and draw a possible structure. (2)

CCEA, A, Module 3, June 2001

6 a Describe the nature of the chemical and intermolecular bonding in

i ice (2)

ii iodine. (2)

b Describe the nature of the bonding in copper metal. (2)

c Explain why water and ethanol, C_2H_5OH, are totally miscible (mix in all proportions) but a hydrocarbon such as paraffin oil is barely soluble in water, forming an immiscible layer on top of the water. (3)

d Using the Valence Shell Electron Pair Repulsion (VSEPR) theory, describe and explain the shape of a molecule of gaseous boron trifluoride, BF_3. (2)

WJEC, AS/A, C1, June 2000

7 The table below shows the boiling points of the elements sodium to chlorine in Period 3 of the periodic table.

Element	Na	Mg	Al	Si	P	S	Cl
Boiling point/°C	883	1107	2467	2355	280	445	−35
Bonding							
Structure							

a i Copy and complete the *bonding* row of the table using
- M for *metallic bonding*
- C for *covalent bonding*. (1)

ii Copy and complete the *structure* row of the table using
- S for a *simple molecular structure*
- G for a *giant structure*. (1)

b State what is meant by *metallic bonding*. You should draw a diagram as part of your answer. (3)

c Explain, in terms of their structure and bonding, why the boiling point of

i phosphorus is much *lower* than that of silicon (2)

ii aluminium is much *higher* than that of magnesium. (2)

OCR, AS, 2811, June 2001

In the following question, two marks are available for the quality of written communication.

You should use diagrams to illustrate your answer.

8 Sodium reacts with chlorine forming sodium chloride.

a Describe the bonding in Na, Cl_2 and NaCl. (8)

b Relate the *physical* properties of Cl_2 and NaCl to their structure and bonding. (8)

OCR, AS, 2811, Jan 2001

Chapter 5

1 a Define the terms *standard enthalpy of formation* and *standard enthalpy of combustion*. (6)

b Use the standard enthalpies of formation, $\Delta H_f^{\ominus}$, given below

Compound	**$\Delta H_f^{\ominus}$/kJ mol^{-1}**
$CO_2(g)$	−394
$C_3H_7OH(l)$	−304
$H_2O(l)$	−286

to calculate the standard enthalpy of combustion of an alcohol C_3H_7OH, as shown by the equation:

$$C_3H_7OH(l) + 4\tfrac{1}{2}O_2(g) \rightarrow 3CO_2(g) + 4H_2O(l) \qquad (3)$$

c A value for the enthalpy of combustion of the alcohol C_3H_7OH was determined in the laboratory using the apparatus shown below.

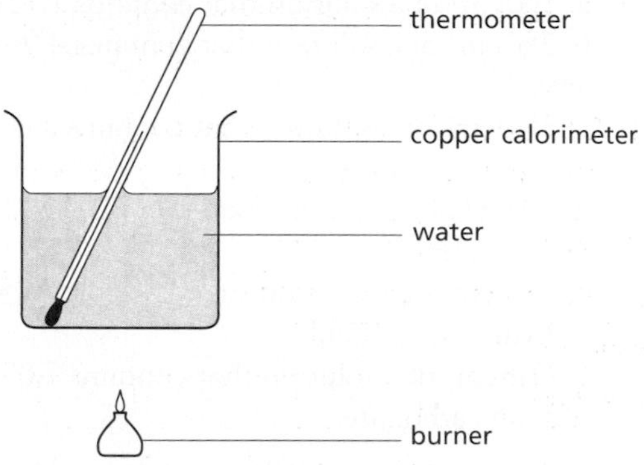

The following results were obtained.
Mass of water in the calorimeter = 200 g
Initial temperature of water = 15 °C
Final temperature of water = 30 °C
Mass of alcohol burned = 0.90 g

i Calculate the heat energy required to raise the temperature of the water from 15 °C to 30 °C. The specific heat capacity of water is $4.2\,J\,g^{-1}K^{-1}$.
ii Calculate the number of moles of the alcohol, C_3H_7OH, burned.
iii Hence, calculate a value for the enthalpy of combustion of 1.0 mol of the alcohol.
iv Give *two* reasons why you would expect your answer to part **c iii** to differ from that in part **b**. (8)

AQA, AS, Unit 2, Jan 2001

2 a Define the term *standard enthalpy of combustion*. (3)
b Using the data given below, calculate the standard enthalpy change for the following reaction.

$$CH_4(g) + 2O_2(g) \rightarrow CO_2(g) + 2H_2O(l)$$

$\Delta H_f^\ominus\ CO_2(g) = -394\,kJ\,mol^{-1}$

$\Delta H_f^\ominus\ H_2O(l) = -286\,kJ\,mol^{-1}$

$\Delta H_f^\ominus\ CH_4(g) = -75\,kJ\,mol^{-1}$ (3)

c i State what is meant by the term *mean bond enthalpy*.
ii Using the standard enthalpy of formation of methane given in part **b** and the data given below, calculate the mean bond enthalpy of the C—H bond in methane.

$C(s) \rightarrow C(g) \qquad \Delta H^\ominus = +715\,kJ\,mol^{-1}$

$H_2(g) \rightarrow 2H(g) \qquad \Delta H^\ominus = +436\,kJ\,mol^{-1}$

iii Using the C—H bond enthalpy calculated in part **ii** and the standard enthalpy change for the reaction given below, calculate the mean bond enthalpy of the C—C bond in propane.

Note: If you failed to complete part **c ii**, you may assume that the mean bond enthalpy of the C—H bond is $+390\,kJ\,mol^{-1}$. (This is not the correct value.)

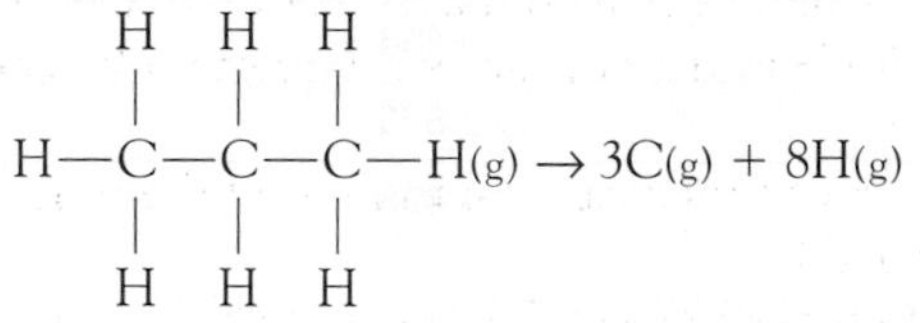

$\Delta H^\ominus = +4020\,kJ\,mol^{-1}$

(7)

AQA, AS, Unit 2, June 2001

3 a The formation of compounds is accompanied by enthalpy changes.
i Explain the term *standard enthalpy change of formation*. (2)
ii State the conditions under which standard enthalpy changes are measured. (2)
b Oxidation reactions are normally exothermic.
i What do you understand by the term *exothermic*? (1)
ii State *one* example of an exothermic oxidation reaction that is important in industry or everyday life. (1)
c The oxidation of hydrazine, N_2H_4, by dinitrogen tetroxide, N_2O_4, has been used in rocket propulsion.

$$2N_2H_4(l) + N_2O_4(l) \rightarrow 3N_2(g) + 4H_2O(g)$$

i Use the following standard enthalpy changes of formation to calculate the enthalpy change for this reaction.

Compound	$\Delta H_f^\ominus/kJ\,mol^{-1}$
$CO_2(g)$	−394
$C_3H_7OH(l)$	−304
$H_2O(l)$	−286

(3)

ii Suggest what feature, other than the value of ΔH, makes this reaction suitable for propelling a rocket. (1)

OCR, AS, 2813, Jan 2002

4 The question refers to the enthalpy changes of some reactions of hydrocarbons. Table 1 below lists some average bond enthalpies.

Table 1

Bond	Bond enthalpy/kJ mol^{-1}
C—H	+413
C—C	+347
O—H	+464
C=O	+805
O=O	+498

a i Explain the term *average bond enthalpy*. (2)
ii Write an equation, including state symbols, to represent the average C—H bond enthalpy in methane, CH_4. (2)

b Using the information in Table 1, calculate the enthalpy change of combustion of propane.

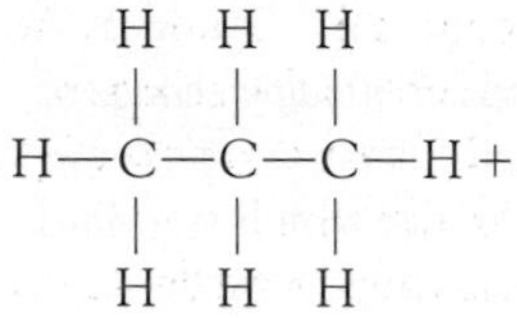

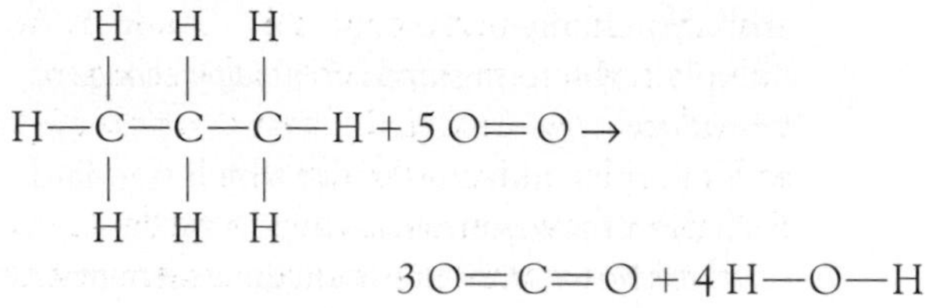

3 O=C=O + 4 H—O—H (3)

c Table 2 shows the *actual* standard enthalpy changes of combustion of some alkanes.

Table 2

Compound	Molecular formula	$\Delta H_c^\ominus$/kJ mol^{-1}
Propane	C_3H_8	−2220
Butane	C_4H_{10}	−2870
Pentane	C_5H_{12}	
Hexane	C_6H_{14}	−4160

i Using average bond energies, the enthalpy change of combustion of butane was calculated as −2672 kJ mol^{-1}. The *actual* value for butane is shown in Table 2. Suggest a reason for the difference. (1)
ii Copy the axes below and plot a graph below using the data in Table 2. (1)

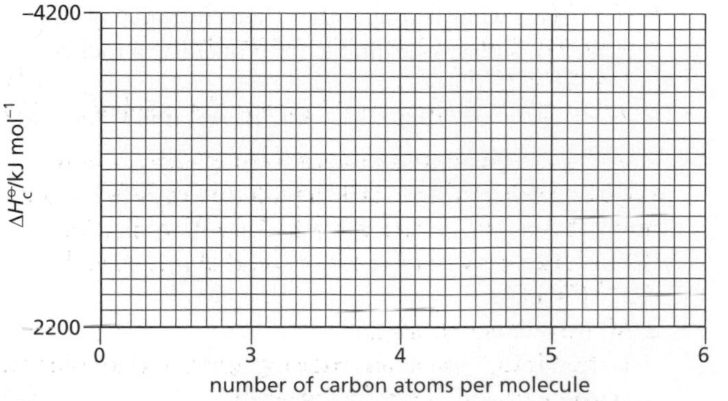

iii Use your graph to determine a value for the standard enthalpy change of combustion of pentane. (1)

iv Suggest why there is a regular trend in $\Delta H_c^\ominus$ for the alkanes. (1)

OCR, AS, 2813, Jan 2001

Chapter 6

1 Calculate the concentration, in mol dm^{-3}, of each of the following solutions.
a 100 cm^3 of a solution that contains 1.12 g of KOH. (1)
b 250 cm^3 of a solution that contains 4.90 g of sulphuric acid. (1)
c 25.0 cm^3 of a solution that contains 2.00 g of $KHCO_3$. (1)
d 20 cm^3 of a solution that contains 2.720 g of $H_2C_2O_4.2H_2O$. (1)
e 13.8 cm^3 of a solution that contains 3.05×10^{-3} mol of hydrochloric acid. (1)
f 27.8 cm^3 of a solution that contains 7.05×10^{-4} mol of sodium carbonate. (1)

2 Calculate the amount (in moles) in each of the following.
a 26.0 cm^3 of 0.0125 mol dm^{-3} sodium hydroxide. (1)
b 17.6 cm^3 of 1.38 mol dm^{-3} nitric acid. (1)

3 A 5.30 g sample of Na_2CO_3 was made up to 250 cm^3 in a standard flask. To 25.0 cm^3 of this solution, three drops of phenolphthalein indicator were added. The indicator changed from pink to colourless when 20.0 cm^3 of 0.025 mol dm^{-3} HCl had been added.
a Calculate the number of moles of Na_2CO_3 weighed out. (1)
b Calculate the number of moles of HCl used in the titration. (1)
c Calculate how many moles of HCl react with one mole of Na_2CO_3. (1)
d Hence suggest a balanced equation for the reaction. (1)

4 A 3.35 g sample of a Group 1 hydroxide, of formula MOH, was made up to 250 cm^3 in a standard flask. A 25.0 cm^3 sample of this solution required 24.2 cm^3 of 0.135 mol dm^{-3} HCl for neutralisation.
a Calculate the number of moles of HCl used. (1)
b Write a balanced equation for the reaction of MOH with HCl. (1)
c Calculate the number of moles of MOH in the 250 cm^3 sample. (1)
d Hence calculate the number of moles of MOH in the 250 cm^3 standard flask. (1)
e Use the mass of MOH to calculate the relative molecular mass of MOH. (1)
f Hence calculate the relative atomic mass of M and suggest its identity. (2)

5 A student added 50.0 cm^3 of hydrochloric acid to 50.0 cm^3 of sodium hydroxide solution in a polystyrene cup. The temperature rose by 6.5 °C. The initial concentration of each solution was 1.00 mol dm^{-3}.
a Write an ionic equation for the reaction occurring. (1)

b Calculate the number of moles of acid used in the reaction. (1)
c Calculate the heat energy evolved in the reaction. (Assume that the final solution has a specific heat capacity of $4.18\,J\,g^{-1}\,K^{-1}$ and a density of $1.00\,g\,cm^{-3}$.) (2)
d Calculate the molar enthalpy change for the reaction. (2)

AQA, AS, Unit 2, June 2001

6 When gaseous ammonia, NH_3, is passed into dilute sulphuric acid it reacts to form a solution of ammonium sulphate, $(NH_4)_2SO_4$, which may then be crystallised.
a Write the equation for the neutralisation of sulphuric acid by ammonia. (1)
b $2.50\,dm^3$ of *impure* ammonia gas was passed into excess sulphuric acid. 6.23 g of crystals of ammonium sulphate was produced.
i Calculate the amount (in moles) of ammonium sulphate in 6.23 g of ammonium sulphate. (M_r = 132.) (1)
ii Calculate the amount (in moles) of ammonia required to produce this mass of ammonium sulphate. (1)
iii What is the number of moles of pure ammonia in the $2.50\,dm^3$ sample of impure ammonia? (1)
iv Calculate the percentage purity of the ammonia used in this experiment. (Molar volume of a gas at the temperature and pressure of the experiment = $24.0\,dm^3$.) (2)
v Calculate the volume of $1.30\,mol\,dm^{-3}$ sulphuric acid needed to make 6.23 g of ammonium sulphate. (2)

Edexcel, A, Module 1, June 2001

7 Sodium hydroxide may be titrated with hydrochloric acid. The apparatus shown below is required for the experiment.

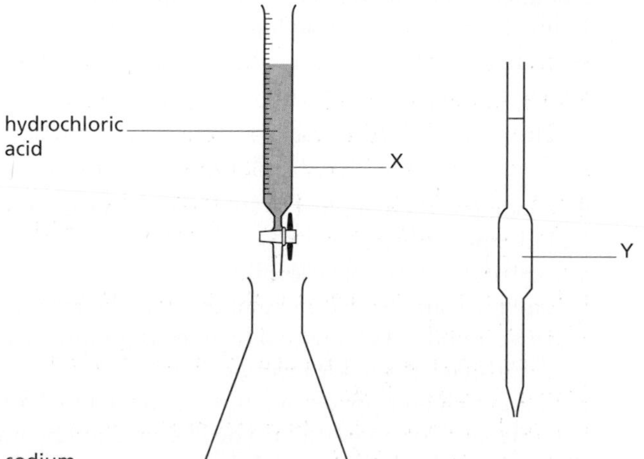

The acid is added to $25.0\,cm^3$ of the alkali which has a concentration of 0.1 M. The alkali contains a few drops of phenolphthalein indicator. The end point occurs when $24.0\,cm^3$ of the acid is added.

a Name the pieces of apparatus X and Y. (2)
b Write the equation for the reaction taking place between the hydrochloric acid and the sodium hydroxide. (1)
c Explain what is meant by the *end point*. (1)
d State the colour of the phenolphthalein in acid and in alkaline solution. (2)
e i Calculate the number of moles of NaOH present in $25.0\,cm^3$ of 0.1 M sodium hydroxide solution. (2)
ii If one mole of NaOH reacts with one mole of HCl deduce the number of moles of HCl present in $24.0\,cm^3$ of solution. (1)
iii Use your answer from part ii to determine how many moles of HCl there are in $1\,cm^3$ of the hydrochloric acid. (1)
iv Calculate how many moles of HCl there are in $1\,dm^3$ ($1000\,cm^3$) of the hydrochloric acid. (1)
f Sodium chloride solution, together with indicator, is left in the conical flask at the end of the titration. Addition of powdered charcoal followed by heating absorbs any indicator present in the solution.
i Outline how you could obtain a sample of pure solid sodium chloride from the solution. (2)
ii Explain how you could carry out a flame test on the solid to show that it was a sodium salt. State the result expected.
(Up to two marks may be obtained for the quality of written communication in this part.) (5)
iii State the type of bonding found in sodium chloride crystals. (1)

CCEA, AS, AS1 Module 1, Jan 2001

Chapter 7

1 Calculate the concentration, in $mol\,dm^{-3}$, of each of the following solutions.
a $250\,cm^3$ of a solution that contains 0.395 g of potassium manganate(VII). (1)
b $25.0\,cm^3$ of a solution that contains 2.00 g of $FeSO_4 \cdot 7H_2O$. (1)

2 Calculate the amount (in moles) in each of the following.
a $16.2\,cm^3$ of $0.0125\,mol\,dm^{-3}$ potassium manganate(VII). (1)
b $17.8\,cm^3$ of $0.126\,mol\,dm^{-3}$ sodium thiosulphate. (1)

3 A 1.50 g sample of iron was dissolved in excess dilute sulphuric acid and the resulting solution of $FeSO_4$ made up to $250\,cm^3$ in a standard flask. A $25.0\,cm^3$ portion of this solution was oxidised by $23.5\,cm^3$ of $0.0200\,mol\,dm^{-3}$ $KMnO_4$.
a Write half equations for
i the MnO_4^- ion as an oxidising agent in acid solution
ii the Fe^{2+} ion as a reducing agent. (2)
b Hence write a balanced equation for the oxidation of Fe^{2+} ions by MnO_4^- ions. (1)

c Calculate the amount (in moles) of MnO_4^- ions used in the titration. (1)
d Calculate the amount (in moles) of Fe^{2+} ions used in the titration. (1)
e Calculate the mass of iron in the 250 cm^3 standard flask. (1)
f Calculate the percentage purity of the sample of iron. (1)

4 Household bleach contains the ion ClO^-. The concentration of ClO^- ions can be determined by using their ability, under acid conditions, to oxidise iodide ions to iodine. The liberated iodine can then be estimated by titration with a standard solution of sodium thiosulphate, $Na_2S_2O_3$.
a i Define oxidation in terms of electrons.
ii Give the oxidation state of chlorine in ClO^-. (2)
b Write an equation for the reaction between thiosulphate ions and iodine. (2)
c An excess of solid potassium iodide was added to 25.0 cm^3 of a bleach solution and the resulting mixture was acidified with dilute sulphuric acid. The liberated iodine was then titrated with a sodium thiosulphate solution of concentration 0.100 $mol\,dm^{-3}$ and 24.4 cm^3 were required for complete reaction.
i Calculate the number of moles of $S_2O_3^{2-}$ present in 24.4 cm^3 of sodium thiosulphate solution.
ii Calculate the number of moles of I_2 liberated in this reaction.
iii Write an ionic equation for the reaction between ClO^- ions and I^- ions, in the presence of acid, to form I_2 and Cl^- ions.
iv Calculate the concentration, in $mol\,dm^{-3}$, of ClO^- in the bleach solution.
v Name the indicator used in this titration and state the colour change at the end-point. (9)

AQA, AS, Unit 2, Jan 2001

5 a Define *reduction* in terms of electrons. (1)
b The oxide of nitrogen formed when copper reacts with nitric acid depends upon the concentration and the temperature of the acid. The reaction of copper with cold, dilute acid produces NO as indicated by the following equation.

$$3Cu + 8H^+ + 2NO_3^- \rightarrow 3Cu^{2+} + 4H_2O + 2NO$$

In warm, concentrated acid, NO_2 is formed.
i Give the oxidation states of nitrogen in NO_3^-, NO_2 and NO.
ii Identify, as oxidation or reduction, the formation of NO from NO_3^- in the presence of H^+ and deduce the half-equation for the reaction.
iii Deduce the half-equation for the formation of NO_2 from NO_3^- in the presence of H^+.
iv Deduce the overall equation for the reaction of copper with NO_3^- and H^+ to produce Cu^{2+} ions, NO_2 and water. (8)

AQA, AS, Unit 2, June 2001

6 Wines often contain a small amount of sulphur dioxide that is added as a preservative. The amount of sulphur dioxide added needs to be carefully calculated: too little and the wine readily goes bad, too much and the wine tastes of sulphur dioxide.

The sulphur dioxide content of a wine can be found using its reaction with aqueous iodine.

$$SO_2(aq) + I_2(aq) + 2H_2O(l) \rightarrow SO_4^{-2}(aq) + 2I^-(aq) + 4H^+(aq)$$

a i State the oxidation number of sulphur in SO_2 and in SO_4^{2-}.
ii State, with a reason, whether sulphur is oxidised or reduced in the conversion of SO_2 into SO_4^{2-}. (3)
b The sulphur dioxide content of a wine can be found by titration. An analyst found that the sulphur dioxide in 50.0 cm^3 of a sample of white wine reacted with exactly 16.4 cm^3 of 0.0100 $mol\,dm^{-3}$ aqueous iodine.
i How many moles of iodine, I_2, did the analyst use in the titration?
ii How many moles of sulphur dioxide were in the 50.0 cm^3 of wine?
iii What was the concentration, in $mol\,dm^{-3}$, of sulphur dioxide in the wine?
iv What was the concentration, in $g\,dm^{-3}$, of sulphur dioxide in the wine? (5)
c The generally accepted maximum concentration of sulphur dioxide in wine is 0.25 $g\,dm^{-3}$.
A concentration of less than 0.01 $g\,dm^{-3}$ is insufficient to preserve the wine.
Comment on the effectiveness of the sulphur dioxide in the wine analysed in **b**. (1)

OCR, AS/A, 4820, June 1999

7 Most of the world's iodine is obtained from small amounts of sodium iodate(V) present in Chile saltpetre. Iodine is precipitated by adding sodium hydrogensulphite to a solution of the saltpetre according to the equation:

$$2IO_3^-(aq) + 5HSO_3^-(aq) \rightarrow 2SO_4^{2-}(aq) + 3HSO_4^-(aq) + H_2O(l) + I_2(s)$$

The iodine is then purified by sublimination which is an endothermic process.
a Explain the term *endothermic*. (1)
b Explain, using oxidation numbers, why the reaction between iodate(V) ions and hydrogensulphite ions can be classified as a redox reaction. (3)
c Excess sodium hydrogensulphite was added to 2.5 g of Chile saltpetre in solution. The liberated iodine was dissolved in potassium iodide solution and reacted with exactly 12.4 cm^3 of 0.025 M sodium thiosulphate solution. Calculate the percentage of sodium iodate(V) in Chile saltpetre. (4)

d Copy the diagrams of the iodine molecules below. Indicate and label the covalent and van der Waals radii.

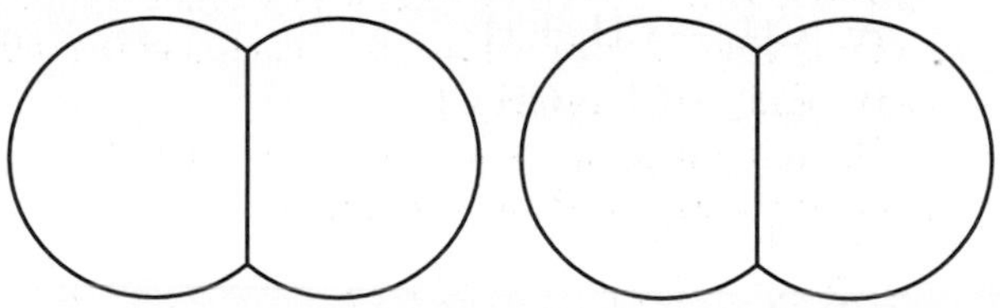

(2)

CCEA, A, Module 1, Feb 2001

8 The formation of magnesium oxide, MgO, from its elements involves both oxidation and reduction in a redox reaction.

a i What is meant by the terms *oxidation* and *reduction*? (2)

ii Write a full equation, including state symbols, for the formation of MgO from its elements. (2)

iii Write half equations for the oxidation and reduction processes that take place in this reaction. (2)

b MgO reacts when heated with acids such as nitric acid, HNO_3.

$$MgO(s) + 2HNO_3(aq) \rightarrow Mg(NO_3)_2(aq) + H_2O(l)$$

A student added MgO to 25.0 cm^3 of a warm solution of 2.00 mol dm^{-3} HNO_3 until all the acid had reacted.

i How would the student have known that the reaction was complete? (1)

ii Calculate how many moles of HNO_3 were used. (1)

iii Deduce how many moles of MgO reacted with this amount of HNO_3.

iv Calculate what mass of MgO reacted with this amount of HNO_3. (A_r: Mg, 24.3; O, 16.0.)
Give your answer to three significant figures. (3)

iv Using oxidation numbers, explain whether the reaction between MgO and HNO_3 is a redox reaction. (2)

c MgO has a very high melting point. Explain this property of MgO. (2)

OCR, AS, 2811, Jan 2002

Chapter 8

1 a State what is meant by the terms *rate of reaction* and *activation energy*. (4)

b The diagram below shows the Maxwell–Boltzmann energy distribution curve for a sample of gas at a fixed temperature. E_a is the activation energy for the decomposition of this gas.

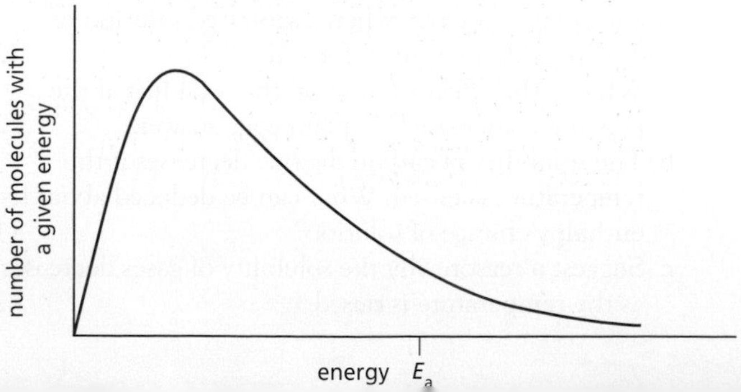

i Copy this diagram and sketch the distribution curve for the same sample of gas at a higher temperature.

ii What is the effect of an increase in temperature on the rate of a chemical reaction? Explain your answer with reference to the Maxwell–Boltzmann distribution.

iii What is the effect of the addition of a catalyst on the rate of a chemical reaction? Explain your answer with reference to the Maxwell–Boltzmann distribution. (9)

AQA, AS, Unit 2, June 2001

2 The rate of any chemical reaction is increased if the temperature is increased.

a Copy these axes and represent the Maxwell–Boltzmann distribution of molecular energies at a temperature T_1 and at a higher temperature T_2.

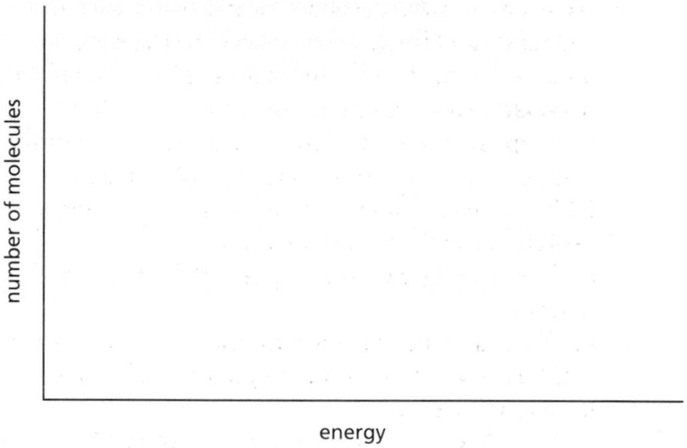

(3)

b Use your diagram and the idea of activation energy to explain why the rate of a chemical reaction increases with increasing temperature. (4)

Edexcel, AS, Unit 2, June 2001

3 Heterogeneous catalysis is used in the fight against air pollution. Harmful gases produced in a car engine are passed through a catalytic converter to reduce environmental damage.

a i Explain the term *heterogeneous* as applied to a catalytic converter. (2)

ii Name the products formed when the following compounds are passed through a catalytic converter.

carbon monoxide (1)
hydrocarbons (1)

iii Explain why leaded petrol must not be used in cars fitted with a catalytic converter. (1)

b In the catalytic converter oxides of nitrogen, NO_x, are converted into nitrogen in an exothermic reaction.

i Copy these axes (overleaf) and draw a reaction profile for the conversion of NO_x to nitrogen, labelling the activation energy. (2)

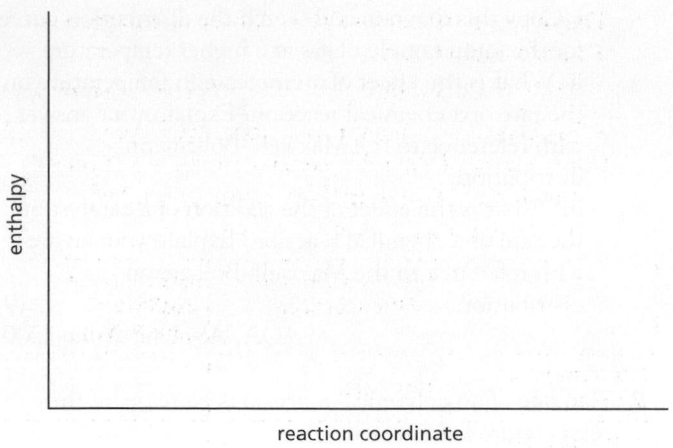

ii Explain, with reference to activation energy, how a catalytic converter increases the rate of conversion of NO_x to nitrogen. (1)

iii The conversion of NO_x to nitrogen is extremely *slow* at room temperature as well as being *exothermic*.

Explain why NO_x, with respect to nitrogen, may be described as kinetically stable and thermodynamically unstable. (2)

CCEA, AS, AS2, Module 2, June 2001

4 a i On a copy of the following axes, sketch the Boltzmann distribution of molecular energies for a fixed amount of gas at a temperature labelled as T_1.

E_{act} represents the activation energy of the reaction.

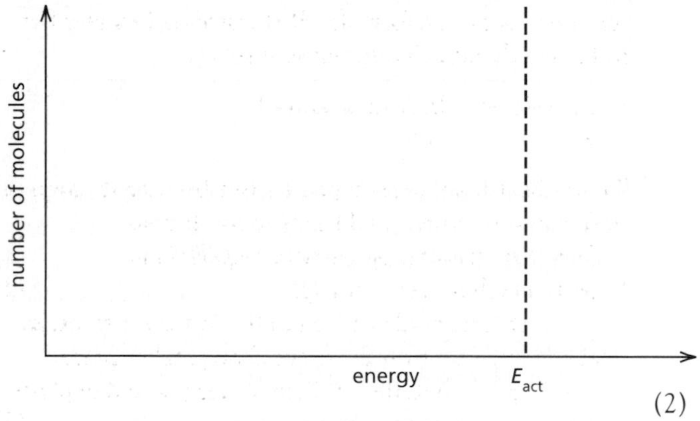

(2)

ii On the same axes, sketch another distribution for the same amount of gas, at a higher temperature, labelled as T_2.

b What do you understand by the term *activation energy*, E_{act}? (1)

c Using your answers to **a** and **b**, explain why the rate of a chemical reaction is affected by changes in temperature. (3)

d The table below lists some average bond enthalpies.

Bond	Bond enthalpy/kJ mol^{-1}
C—H	+413
C—Cl	+327
Cl—Cl	+243
Br—Br	+193

i Use the values in the table above to suggest the order of *increasing* E_{act} values for the following four reactions.

A $CH_4 \rightarrow CH_3 + H$

B $CH_3 + Cl \rightarrow CH_3Cl$

C $Br_2 \rightarrow 2Br$

D $Cl_2 \rightarrow 2Cl$ (2)

ii Explain your choice of order in **i**. (2)

OCR, AS, 2813, Jan 2002

5 Benzene, C_6H_6, can be manufactured by passing the gaseous hydrocarbon ethyne, C_2H_2, over finely divided nickel. The equation is shown below.

$$3C_2H_2(g) \xrightarrow{\text{nickel}} C_6H_6(l)$$

a Write a chemical equation, including state symbols, to represent the standard enthalpy change of combustion of benzene, $C_6H_6(l)$. (2)

b Use the following standard enthalpy changes of combustion to calculate the standard enthalpy change for the above reaction.

Compound	$\Delta H_c^{\ominus}$/kJ mol^{-1}
C_2H_2	−1301
C_6H_6	−3267

(3)

c Suggest, with explanations, how the *rate* of the above reaction might be affected by

i an increase in temperature (3)

ii an increase in pressure. (2)

d Describe and explain the purpose of the nickel in the above reaction. (2)

OCR, AS, 2813, June 2001

Chapter 9

1 Write an equation (with state symbols) to show the equilibrium that exists when a saturated solution of potassium nitrate in water is formed. (2)

a What is the effect (if any) on the solubility if the pressure is increased? Explain your answer. (2)

b The solubility of potassium nitrate increases if the temperature is raised. What can be deduced about its enthalpy change of solution? (1)

2 Write an equation (with state symbols) to show the equilibrium that exists when a saturated solution of carbon dioxide in water is formed. (2)

a What is the effect (if any) on the solubility if the pressure is increased? Explain your answer. (2)

b The solubility of carbon dioxide decreases if the temperature is raised. What can be deduced about its enthalpy change of solution? (1)

c Suggest a reason why the solubility of gases decreases as the temperature is raised. (2)

3 When carbon monoxide reacts with steam at 670 K, the following homogeneous dynamic equilibrium is established.

$$CO(g) + H_2O(g) \rightleftharpoons CO_2(g) + H_2(g) \qquad \Delta H^\ominus = -42\,kJ\,mol^{-1}$$

a i Explain why this reaction is described as a *homogeneous* reaction.
ii State what is meant by the term *dynamic equilibrium*. (3)

b State the effect, if any, of the following changes on the concentration of hydrogen in the equilibrium mixture. In each case, explain your answer.
i An increase in the concentration of steam.
ii A decrease in the temperature. (6)

c State and explain the effect, if any, of a catalyst on the position of this equilibrium. (3)

AQA, AS, Unit 2, June 2001

4 The equilibrium yield of product in a gas-phase reaction varies with changes in temperature and pressure as shown below.

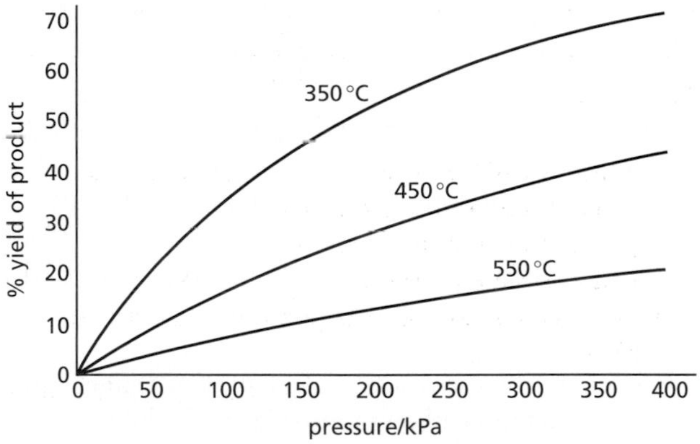

a Use the information given above to deduce whether the forward reaction involves an increase, a decrease, or no change in the number of moles present. Explain your deduction. (4)

b Use the information given above to deduce whether the forward reaction is exothermic or endothermic. Explain your answer. (3)

c i Estimate the percentage yield of product which would be obtained at 350 °C and a pressure of 250 kPa.
ii The reaction is an example of a dynamic equilibrium. State what is meant by the term *dynamic equilibrium*.
iii State what effect, if any, a catalyst has on the position of the equilibrium. Explain your answer. (6)

d A 70% equilibrium yield of product is obtained at a temperature of 350 °C and a pressure of 400 kPa. Explain why an industrialist may choose to operate the plant at
i a temperature higher than 350 °C
ii a pressure lower than 400 kPa. (2)

AQA, AS, Specimen Unit 2, 2001/2

5 The reaction between sulphur dioxide and oxygen is a dynamic equilibrium.

$$2SO_2 + O_2 \rightleftharpoons 2SO_3 \qquad \Delta H = -196\,kJ\,mol^{-1}$$

a Explain what is meant by *dynamic equilibrium*. (2)

b On a copy of the table below state the effect on this reaction of increasing the temperature and of increasing the pressure. (3)

	Effect on the rate of the reaction	**Effect on the position of equilibrium**
Increasing the temperature	Increases	
Increasing the pressure		

c This reaction is one of the steps in the industrial production of sulphuric acid. The normal operating conditions are a temperature of 450 °C, a pressure of 2 atmospheres and the use of a catalyst.
Justify the use of these conditions.
i A temperature of 450 °C. (3)
ii A pressure of 2 atmospheres. (2)
iii A catalyst. (1)

d Give the name of the catalyst used. (1)

e Give one large-scale use of sulphuric acid. (1)

Edexcel, AS, Unit 2, June 2001

6 When the indicator bromothymol blue (which can be represented by the formula HIn) dissolves in water the following dynamic equilibrium is set up.

$$\underset{\text{yellow}}{HIn(aq)} \rightleftharpoons \underset{\text{blue}}{In^-(aq)} + H^+(aq)$$

The indicator solution appears green because it contains both the yellow HIn and the blue In^- forms.

a State two features of a *dynamic equilibrium*. (2)

b State le Chatelier's principle. (2)

c Hydrochloric acid is added to the indicator solution above until no further colour change takes place.
Using le Chatelier's principle, suggest and explain what colour change you might see. (2)

d Aqueous sodium hydroxide is gradually added to the resulting solution in **c** until no further colour changes take place.
Suggest *all* the colour changes you might see.
Explain your answer. (4)

OCR, AS, 2813, June 2001

7 Ammonia, NH_3, is made industrially from its elements by the Haber process. This is an exothermic equilibrium reaction.

$$N_2(g) + 3H_2(g) \rightleftharpoons 2NH_3(g) \qquad \Delta H = -92\,kJ\,mol^{-1}$$

a State *three* reaction conditions that are used in the Haber process. (3)

b Describe and explain the effect of increasing the pressure on the *rate* of this reaction. (2)

c Describe and explain how the *equilibrium position* of this reaction is affected by
i increasing the temperature (2)
ii increasing the pressure. (2)
d Why is the temperature used described as a compromise? (2)
e Some of the ammonia from the Haber process reacts with carbon dioxide to make the fertiliser urea.

$$\_NH_3(g) + \_CO_2(g) \rightarrow \_\underset{\text{urea}}{NH_2CONH_2(s)} + \_H_2O(l)$$

i Copy the above equation and balance. (1)
ii Calculate the maximum mass of urea that could be obtained from 1.00 kg of ammonia. (2)

OCR, AS, 2813, Jan 2002

Chapter 10

1 An equation for the combustion of cyclopropane, C_3H_6, is shown below.

$$C_3H_6(g) + 4\tfrac{1}{2}O_2(g) \rightarrow 3CO_2(g) + 3H_2O(g)$$

The standard enthalpy of combustion of cyclopropane can be calculated either from standard enthalpies of formation or by using mean bond enthalpies.

a Use the standard enthalpies of formation given below to calculate the standard enthalpy of combustion of cyclopropane.

$\Delta H_f^{\ominus}\ C_3H_6(g) = +53\,kJ\,mol^{-1}$
$\Delta H_f^{\ominus}\ CO_2(g) = -393\,kJ\,mol^{-1}$
$\Delta H_f^{\ominus}\ H_2O(g) = -242\,kJ\,mol^{-1}$ (3)

b State what is meant by the term *mean bond enthalpy*. (2)

c The following mean bond enthalpies have been obtained from a data book. Use these values to calculate the standard enthalpy of combustion of cyclopropane.

C—C 347 kJ mol^{-1}
C—H 413 kJ mol^{-1}
C=O 805 kJ mol^{-1}
O=O 498 kJ mol^{-1}
O—H 464 kJ mol^{-1} (3)

d Explain, by reference to the structure of cyclopropane, why the enthalpy of combustion calculated in part **a** is more exothermic than that calculated in part **c**. (2)

AQA, AS/A, CH04, June 2000

2 a Construct a Born–Haber cycle for the formation of sodium oxide, Na_2O, and use the data given below to calculate the second electron affinity of oxygen.

$Na(s) \rightarrow Na(g)$ $\Delta H^{\ominus} = +107\,kJ\,mol^{-1}$
$Na(g) \rightarrow Na^+(g) + e^-$ $\Delta H^{\ominus} = +496\,kJ\,mol^{-1}$
$O_2(g) \rightarrow 2O(g)$ $\Delta H^{\ominus} = +249\,kJ\,mol^{-1}$
$O(g) + e^- \rightarrow O^-(g)$ $\Delta H^{\ominus} = -141\,kJ\,mol^{-1}$
$O^-(g) + e^- \rightarrow O^{2-}(g)$ $\Delta H^{\ominus}$ to be calculated
$2Na^+(g) + O_2^-(g) \rightarrow Na_2O(s)$ $\Delta H^{\ominus} = -2478\,kJ\,mol^{-1}$
$2Na(s) + \tfrac{1}{2}O_2(g) \rightarrow Na_2O(s)$ $\Delta H^{\ominus} = -414\,kJ\,mol^{-1}$ (9)

b Explain why the second electron affinity of oxygen has a large positive value. (2)

c Explain, by reference to steps from relevant Born–Haber cycles, why sodium forms a stable oxide consisting of Na^+ and O^{2-} ions but not oxides consisting of Na^+ and O^- or Na^{2+} and O^{2-} ions. (4)

AQA, AS/A, CH04, June 2000

3 Use the entropy data in the table below to answer the following questions.

Species	$S^{\ominus}$/J K^{-1} mol^{-1}	Species	$S^{\ominus}$/J K^{-1} mol^{-1}
C(graphite)	6	$H_2O(g)$	189
C(diamond)	3	$H_2O(l)$	70
$H_2(g)$	131	$CH_4(g)$	186
CO(g)	198	CaO(s)	40
$CO_2(g)$	214	$CaCO_3(s)$	90

Give chemical equations and calculate numerical values of ΔS wherever possible.

a At all temperatures below 100 °C, steam at atmospheric pressure condenses spontaneously to form water. Explain this observation in terms of ΔG and calculate the enthalpy of vaporisation of water at 100 °C. (4)

b Explain why the reaction of 1 mol of methane with steam to form carbon monoxide and hydrogen ($\Delta H^{\ominus} = +210\,kJ\,mol^{-1}$) is spontaneous only at high temperatures. (6)

c Explain why the change of 1 mol of diamond to graphite ($\Delta H^{\ominus} = -2\,kJ\,mol^{-1}$) is feasible at all temperatures yet does not occur at room temperature. (3)

d The reaction between 1 mol of calcium oxide and carbon dioxide to form calcium carbonate ($\Delta H^{\ominus} = -178\,kJ\,mol^{-1}$) ceases to be feasible above a certain temperature, T_s. Determine the value of T_s. (2)

AQA, A, Specimen Unit 5, 2001/2

4 a The Born–Haber cycle for the formation of sodium chloride is shown below.

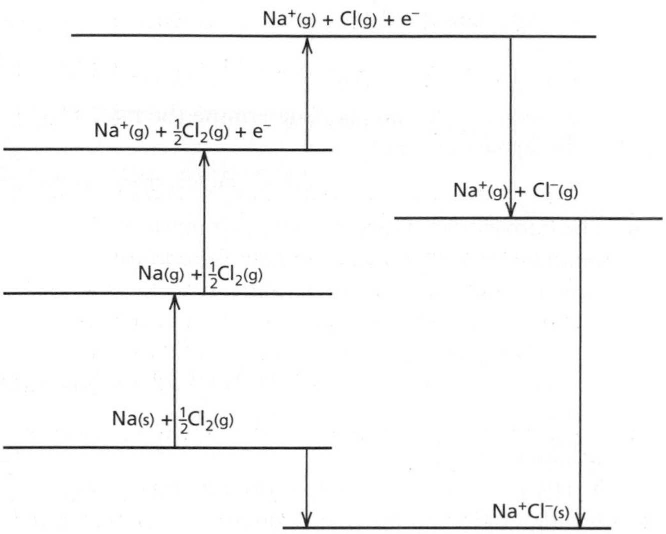

Use the data below to calculate the lattice enthalpy of sodium chloride.

Enthalpy change	Value of the enthalpy change /kJ mol^{-1}
Enthalpy of atomisation of sodium	+109
1st ionisation energy of sodium	+494
Enthalpy of formation of sodium chloride	−411
Enthalpy of atomisation of chlorine	+121
Electron affinity of chlorine	−364

(2)

b Sodium chloride and magnesium oxide have very similar crystal lattices. Suggest why the lattice enthalpy of magnesium oxide is very much larger than that of sodium chloride. (2)

c The lattice enthalpy of silver iodide can be calculated but the experimental value does not match the calculated value as well as those for sodium chloride match each other.

Explain why the calculated and experimental values for silver iodide are different. (2)

Edexcel, A, Module 3, June 2001

5 a i Draw a Born–Haber cycle for the formation of magnesium chloride, $MgCl_2$. Use the values below to calculate the lattice enthalpy of magnesium chloride.

	$\Delta H^\ominus$/kJ mol^{-1}
1st electron affinity of chlorine	−364
1st ionisation energy of magnesium	+736
2nd ionisation energy of magnesium	+1450
Enthalpy of atomisation of chlorine	+121
Enthalpy of atomisation of magnesium	+150
Enthalpy of formation of $MgCl_2$(s)	−642

(5)

ii The value of the lattice enthalpy of magnesium chloride calculated from a purely ionic model is −2326 kJ mol^{-1}.

Explain why this differs from the value determined from a Born–Haber cycle. (2)

b Use the following data to answer the questions in this section.

	$\Delta H^\ominus$/kJ mol^{-1}
$\Delta H_{hydration}$ of Sr^{2+}	−1480
$\Delta H_{hydration}$ of Ba^{2+}	−1360
$\Delta H_{hydration}$ of OH^-	−460
Lattice enthalpy of $Sr(OH)_2$	−1894
Lattice enthalpy of $Ba(OH)_2$	−1768

i Explain why the lattice enthalpy of strontium hydroxide is different from that of barium hydroxide. (2)

ii Explain why the hydration enthalpy of a cation is exothermic. (2)

iii Use the lattice enthalpy and hydration enthalpy values to explain why barium hydroxide is more soluble in water than strontium hydroxide. (4)

Edexcel, A, Module 3, Jan 2001

Chapter 11

1 a A chemical reaction is first order with respect to compound X and second order with respect to compound Y.

i Write the rate equation for this reaction.

ii What is the overall order of this reaction?

iii By what factor will the rate increase if the concentrations of X and Y are *both* doubled? (4)

b The table below shows the initial concentrations of two compounds, A and B, and also the initial rate of the reaction that takes place between them at constant temperature.

Experiment	[A]/mol dm^{-3}	[B]/mol dm^{-3}	Initial rate/ mol dm^{-3} s^{-1}
1	0.2	0.2	3.5×10^{-4}
2	0.4	0.4	1.4×10^{-3}
3	0.8	0.4	5.6×10^{-3}

i Determine the overall order of the reaction between A and B. Explain how you reached your conclusion.
ii Determine the order of reaction with respect to compound B. Explain how you reached your conclusion.
iii Write the rate equation for the overall reaction.
iv Calculate the value of the rate constant, stating its units. (7)

AQA, A, Specimen Unit 4, 2001/2

2 a The kinetics of the hydrolysis of the halogenoalkane RCH_2Cl with aqueous sodium hydroxide (where R is an alkyl group) was studied at 50 °C. The following results were obtained:

Experiment	$[RCH_2Cl]$	$[OH^-]$	Initial rate/mol dm^{-3} s^{-1}
1	0.050	0.10	4.0×10^{-4}
2	0.15	0.10	1.2×10^{-3}
3	0.10	0.20	1.6×10^{-3}

i Deduce the order of reaction with respect to the halogenoalkane, RCH_2Cl, and with respect to the hydroxide ion, OH^-, giving reasons for your answers. (4)
ii Hence write the rate equation for the reaction. (1)
iii Calculate the value of the rate constant with its units for this reaction at 50 °C. (2)
iv Using your answer to part **ii**, write the mechanism for this reaction. (3)

b i Write the equation for the reaction of concentrated sulphuric acid with solid sodium chloride. (1)
ii When concentrated sulphuric acid is added to solid sodium bromide a different type of reaction occurs. Explain why the reactions are different. Identify the gases produced with sodium bromide and write an equation to show the formation of these gases. (4)

Edexcel, A, Module 6 (Synoptic), Jan 2001

3 Oxygen reacts with nitrogen monoxide to make nitrogen dioxide.

$$O_2(g) + 2NO(g) \rightleftharpoons 2NO_2(g)$$

In an experiment to investigate the effects of changing concentrations on the rate of reaction, the following results were obtained.

Experiment number	Initial concentration O_2/10^{-2} mol dm^{-3}	Initial concentration NO/10^{-2} mol dm^{-3}	Initial rate of disappearance of NO/10^{-4} mol dm^{-3} s^{-1}
1	1.0	1.0	0.7
2	1.0	2.0	2.8
3	1.0	3.0	6.3
4	2.0	2.0	5.6
5	3.0	3.0	18.9

a Deduce the order of reaction with respect to O_2 and NO, explaining your reasoning. (4)
b i Write the rate equation for this reaction.
ii Calculate the value of the rate constant, *k*. State its units. (3)
c When the rate of disappearance of NO is 2.8×10^{-4} mol dm^{-3} s^{-1}, determine the rate of disappearance of O_2. (1)

OCR, AS/A, 4826, June 1999

4 The conversion of succinic acid to fumaric acid is catalysed by the enzyme succinate dehydrogenase. The enzyme is specific to the reaction.

$$\underset{\text{succinic acid}}{HOOCCH_2CH_2COOH} \xrightarrow{\text{succinate dehydrogenase}} \underset{\text{fumaric acid}}{HOOCCH{=}CHCOOH}$$

a Explain the term *specific*. (2)
b The initial rate of reaction was investigated using different succinic acid concentrations and different temperatures. The following results were obtained:

Succinic acid concentration $\times 10^3$/mol dm^{-3}	10.20	5.10	2.60	1.30	0.60
Initial rate at 25 °C/ mol dm^{-3} s^{-1}	1.20	0.66	0.32	0.15	0.07
Initial rate at 30 °C/ mol dm^{-3} s^{-1}	1.80	0.88	0.45	0.19	0.10

(The order with respect to the catalyst is zero.)
i Explain why the rate of the reaction increases with temperature. (2)
ii Use the data to deduce the order with respect to succinic acid for the reaction at 25 °C, and write the rate equation. (2)
iii Calculate the value of the rate constant for the reaction at 25 °C and state its units. (3)

c i Succinic acid reacts with ethanol to form a diester. Write an equation for this reaction. (2)
ii Explain why succinic acid has two different pK_a values. (2)

d One mole of fumaric acid reacts with two moles of ammonia to form a salt. Write an equation for this reaction. (2)

CCEA, A, Module 2, June 2001

5 In concentrated ethanoic acid, bromine adds to propenyl ethanoate, $CH_2{=}CHCH_2OCOCH_3$, slowly enough for the reaction to be followed by usual laboratory techniques. The following results were obtained at 25 °C.

Experiment number	[Ester]/mol dm^{-3}	$[Br_2]$/mol dm^{-3}	Rate/mol dm^{-3} s^{-1}
1	2.0×10^{-2}	2.0×10^{-2}	6.51×10^{-4}
2	4.0×10^{-2}	2.0×10^{-2}	1.29×10^{-3}
3	6.0×10^{-2}	2.0×10^{-2}	1.94×10^{-3}
4	2.0×10^{-2}	4.0×10^{-2}	2.55×10^{-3}
5	2.0×10^{-2}	6.0×10^{-2}	5.85×10^{-3}

a i Write the equation for the addition of bromine to propenyl ethanoate.
ii Suggest *two* reasons why concentrated ethanoic acid was used as the solvent rather than water. (3)

b i Deduce, showing your working in each case, the order of the reaction with respect to

propenyl ethanoate
bromine

ii Write the rate equation for the reaction.
iii Calculate the rate constant for the reaction.
iv Suggest a possible mechanism for the rate-determining step. (6)

c i Estimate how long it would take for 1.0% of the ester to react in experiment number 3.
ii State if the time for 1.0% of the ester to react would be exactly the same in experiment number 1 as in experiment number 3. Explain your reasoning. (4)

d Suggest an outline experimental technique for measuring the rate of the reaction using
i a physical method
ii a chemical method. (5)

e Experiment number 1 was repeated with 2.0×10^{-2} mol dm^{-3} bromine, under the same experimental conditions, but using 2.0×10^{-2} mol dm^{-3} propenyl chloroethanoate, $CH_2{=}CHCH_2OCOCH_2Cl$, instead of propenyl ethanoate. The rate of the reaction was 2.80×10^{-4} mol dm^{-3} s^{-1}. Suggest a reason for this change in rate. (2)

UCLES, S, 9254/0, Nov 1998

6 Bromine oxidises methanoic acid to carbon dioxide according to the following equation:

$$HCO_2H(aq) + Br_2(aq) \rightarrow CO_2(g) + 2HBr(aq)$$

The rate of the reaction is found to depend upon the concentrations of bromine, methanoic acid and H^+ ions.

i.e. rate $= k[Br_2]^x[HCO_2H]^y[H^+]^z$

where x, y and z are integers.
In an experiment to determine x, the order with respect to bromine, the initial concentrations of H^+ and HCO_2H were made much greater than the initial concentration of the bromine. The rate law may then be written:

rate $= k_{obs}[Br_2]^x$

where $k_{obs} = k[HCO_2H]^y[H^+]^z$.

a Suggest how the conditions used help to simplify an analysis of the kinetics of the reaction. (1)

b In order to determine the concentration of bromine, samples of the reaction mix were removed at regular intervals and mixed with an excess of potassium iodide solution. The resultant mixture was then titrated with sodium thiosulphate solution of known concentration.
i Write an equation for the reaction between bromine and potassium iodide solution. (1)
ii Write an equation for the reaction involving the sodium thiosulphate solution. (1)

c The quantity of thiosulphate needed in the titration is directly proportional to the concentration of bromine present. The graph below shows how the quantity of this titre varies with time.

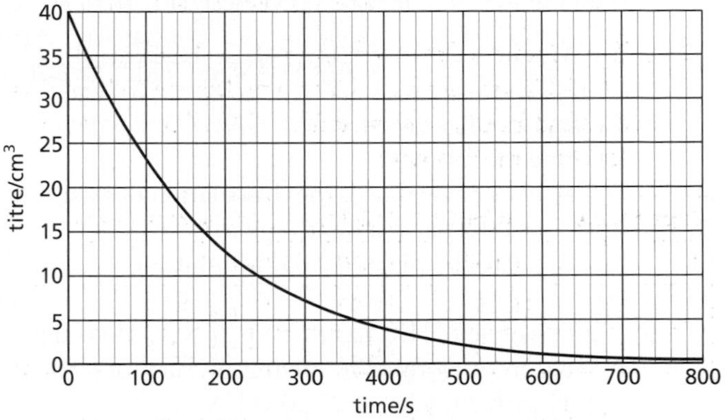

Explain how this graph tells us that the reaction is first order with respect to bromine. (2)

d The concentration of methanoic acid was then varied, keeping the concentration of hydrogen ions constant.

$[HCO_2H]$/mol dm^{-3}	0.1409	0.2113	0.2818	0.4227
k_{obs}/s^{-1}	0.096	0.146	0.192	0.284

e i What is y, the order of reaction with respect to $[HCO_2H]$? Explain your reasoning. (2)
ii The concentration of hydrogen ions was then varied, keeping the concentration of methanoic acid constant.

$[H^+]$/mol dm^{-3}	0.121	0.242	0.362
k_{obs}/s^{-1}	0.353	0.176	0.117

What is z, the order of reaction with respect to $[H^+]$? Explain your reasoning. (2)

f i Sketch the form of the graph showing how the log of the rate of reaction varies with pH. (2)
ii Write the rate law for the overall equation. What are the units for the rate constant, k? (2)

g i Write an expression for the concentration of methanoate ions in terms of the concentrations of methanoic acid and hydrogen ions, and the K_a for methanoic acid. (2)
ii The rate determining step for the reaction is thought to be between a molecule of bromine and the methanoate ion. Show how your expression for the rate law in part **f ii** is consistent with this hypothesis. (2)
iii Draw an energy profile diagram for the reaction marking clearly any intermediates and the overall activation energy for the reaction. (3)

OCR, A, 9485 (STEP), 2001

Chapter 12

1 a At 25 °C, the constant K_w has the value $1.00 \times 10^{-14}\,mol^2\,dm^{-6}$. Define the term K_w. (1)
b Define the term *pH*. (1)
c Calculate the pH at 25 °C of 2.00 M HCl. (1)
d Calculate the pH at 25 °C of 2.50 M NaOH. (2)
e Calculate the pH at 25 °C of the solution that results from mixing 19.0 cm^3 of 2.00 M HCl with 16.0 cm^3 of 2.50 M NaOH. (6)
AQA, AS/A, CH02, June 2000

2 a An acid HA has $pK_a = 4.20$.
i Define the term pK_a.
ii Calculate the value of the dissociation constant, K_a, for the acid HA and state its units.
iii Calculate the pH of a 0.830 M solution of the acid HA. (7)
b A different acid, HX, has $K_a = 5.25 \times 10^{-5}\,mol\,dm^{-3}$. A solution was formed by mixing 10.5 cm^3 of 0.800 M NaOH with 25.0 cm^3 of 0.920 M HX.
i Calculate the number of moles of X^- ions present in the solution formed. (Ignore any X^- ions formed by dissociation of the excess acid HX).
ii Calculate the number of moles of HX which remain unreacted.
iii Calculate the concentrations of both X^- and HX and use these to determine the pH of the solution formed. (9)
c State qualitatively how the pH of the solution formed in part **b** changes when a small volume of dilute hydrochloric acid is added. Use appropriate equations to explain your answer. (3)
AQA, A, CH04, June 2000

3 When heated, dinitrogen tetroxide, N_2O_4, dissociates to form nitrogen dioxide, NO_2, as shown by the equation

$$N_2O_4(g) \rightleftharpoons 2NO_2(g)$$

a Write an expression for the equilibrium constant, K_p, for this reaction. (1)
b In an experiment, 1.00 mol of gaseous dinitrogen tetroxide was sealed in a flask and heated to 360 K. When equilibrium had been established, the total pressure in the flask was 150 kPa and the gaseous mixture contained 0.91 mol of dinitrogen tetroxide.
i Calculate the total number of moles of gas present at equilibrium.
ii Calculate the mole fraction of each of the gases in the equilibrium mixture.
iii Calculate the partial pressure of each of the gases in the equilibrium mixture.
iv Hence, calculate the equilibrium constant, K_p, for this equilibrium at 360 K and state its units. (8)
c An inert gas was introduced into the equilibrium mixture which was held in a vessel of constant volume. State the change, if any, in
i the partial pressures of $NO_2(g)$ and $N_2O_4(g)$
ii the equilibrium position. (2)
AQA, A, CH04, June 2000

4 a i Write an expression for the dissociation constant, K_a, of propanoic acid, CH_3CH_2COOH.
ii Write an expression for pK_a in terms of K_a.
iii Calculate the pH of a 0.10 M solution of propanoic acid, given that $K_a = 1.35 \times 10^{-5}\,mol\,dm^{-3}$ for this acid at 25 °C. (6)
b Explain why an aqueous solution containing propanoic acid and its sodium salt constitutes a buffer system able to minimise the effect of added hydrogen ions. (3)
AQA, A, Specimen Unit 4, 2001/2

5 a Explain why two different liquids usually have different vapour pressures at the same temperature. (2)
b Explain why the boiling point of a mixture of two miscible volatile liquids varies with the composition of the mixture. (2)
c Describe how the components of air may be separated from one another on an industrial scale. (3)
Edexcel, A, Module 6, June 2001

6 a i Calculate the concentration in $mol\,dm^{-3}$, of a solution of hydrochloric acid, HCl, which has a pH of 1.13. (1)
ii Calculate the concentration, in $mol\,dm^{-3}$, of a solution of chloric(I) acid, HOCl, which has a pH of 4.23.
Chloric(I) acid is a weak acid with $K_a = 3.72 \times 10^{-8}\,mol\,dm^{-3}$. (4)
b The pH of 0.100 $mol\,dm^{-3}$ sulphuric acid is 0.98.
i Calculate the concentration of hydrogen ions, H^+, in this solution. (1)
ii Write equations to show the two successive ionisations of sulphuric acid, H_2SO_4, in water. (2)
iii Suggest why the concentration of hydrogen ions is not 0.20 $mol\,dm^{-3}$ in 0.100 $mol\,dm^{-3}$ sulphuric acid. (1)
c Many industrial organic reactions produce hydrogen chloride as an additional product. This can be oxidised to chlorine by the Deacon process:

$$4HCl(g) + O_2(g) \rightleftharpoons 2Cl_2(g) + 2H_2O(g) \qquad \Delta H = -115\,kJ\,mol^{-1}$$

0.800 mol of hydrogen chloride was mixed with 0.200 mol of oxygen in a vessel of volume 10.0 dm^3 in the presence of a copper(I) chloride catalyst at 400 °C. At equilibrium it was found that the mixture contained 0.200 mol of hydrogen chloride.
i Write an expression for the equilibrium constant, K_c. (1)
ii Calculate the value of K_c at 400 °C. (4)

d State and explain the effect, if any, on the *position of equilibrium* in **c** of:
i decreasing the temperature (2)
ii decreasing the volume (2)
iii removing the catalyst. (2)
Edexcel, A, Module 2, Jan 2001

7 Barium sulphate, $BaSO_4$, is the basis for 'barium meals', which are used in the investigation of digestive problems because barium ions can be detected by X-rays.
$Ba^{2+}(aq)$ ions are poisonous to humans.
a Write the equilibrium equations for what happens when barium sulphate is added to water. Include state symbols. (1)
b Write the expression for the solubility product, K_{sp}, for barium sulphate. (4)
c At 25 °C the solubility product of barium sulphate is 1.00×10^{-10} mol^2 dm^{-6} and that of barium carbonate is 5.50×10^{-10} mol^2 dm^{-6}.
Calculate the solubility of barium sulphate in
i mol dm^{-3}
ii g dm^{-3}. (2)
d Suggest why barium sulphate, but not barium carbonate, can be used as a barium meal. (1)
OCR, AS/A, 4826, Mar 2001

Chapter 13

1 The table below shows some values for standard electrode potentials. These data should be used, where appropriate, to answer the questions that follow concerning the chemistry of copper and iron.

Electrode reaction	$E^\ominus$/V
$Fe^{2+}(aq) + 2e^- \rightleftharpoons Fe(s)$	−0.44
$2H^+(aq) + 2e^- \rightleftharpoons H_2(g)$	0.00
$Cu^{2+}(aq) + 2e^- \rightleftharpoons Cu(s)$	+0.34
$O_2(g) + 2H_2O(l) + 4e^- \rightleftharpoons 4OH^-(aq)$	+0.40
$NO_3^-(aq) + 4H^+(aq) + 3e^- \rightleftharpoons NO(g) + 2H_2O(l)$	+0.96

a Write an equation to show the reaction that occurs when iron is added to a solution of a copper(II) salt. (1)
b A similar overall reaction to that shown in **a** would occur if an electrochemical cell was set up between copper and iron electrodes.
i Write down the cell diagram to represent the overall reaction in the cell. (2)
ii Calculate the e.m.f. of the cell. (1)
iii Calculate the standard free energy change, $\Delta G^\ominus$, for the reaction occurring in the cell. Faraday constant = 96 500 C mol^{-1}. (3)
c i Use the standard electrode potential data given to explain why copper reacts with dilute nitric acid but has no reaction with dilute hydrochloric acid. (3)
ii Write an equation for the reaction between copper and dilute nitric acid. (2)
d Although iron is a widely used metal, it has a major disadvantage in that it readily corrodes in the presence of oxygen and water. The corrosion is an electrochemical process which occurs on the surface of the iron.
i Use the standard electrode potential data given to write an equation for the overall reaction that occurs in the electrochemical cell set up between iron, oxygen and water. (1)
ii State, with a reason, whether the iron acts as the anode or cathode of the cell. (2)

Electrode reaction	$E^\ominus$/V
$\frac{1}{2}F_2 + e^- \rightleftharpoons F^-$	+2.87
$\frac{1}{2}Cl_2 + e^- \rightleftharpoons Cl^-$	+1.36
$\frac{1}{2}Br_2 + e^- + \rightleftharpoons Br^-$	+1.07
$\frac{1}{2}I_2 + e^- \rightleftharpoons I^-$	+0.54

iii Predict and explain whether or not you would expect a similar corrosion reaction to occur with copper in the presence of oxygen and water. (2)
AQA, A, Module 9, June 2000

2 This question is about Group 7 of the periodic table – the halogens. The standard electrode potentials for these elements are given below.
a i Define the term *standard electrode potential.* (2)
ii State which element or ion in the table above is the strongest oxidising agent. (1)
b The standard electrode potentials for chromium(III) changing to chromium(II) and for chromium(VI) changing to chromium(III) are given below.

$Cr^{3+} + e^- \rightleftharpoons Cr^{2+}$ $E^\ominus = -0.41$ V

$\frac{1}{2}Cr_2O_7^{2-} + 7H^+ + 3e^- \rightleftharpoons Cr^{3+} + \frac{7}{2}H_2O$ $E^\ominus = +1.33$ V

i On the basis of the data provided, list those halogens which will oxidise chromium(II) to chromium(III). (1)
ii On the basis of the data provided, list those halogens which will oxidise chromium(II) to chromium(III) but not to chromium(VI). (1)
iii Chromium(II) in aqueous solution is sky blue while aqueous chromium(III) solution is dark green. Describe how you would show that your prediction in part **ii** actually worked in practice. (2)
Edexcel, A, Module 3, June 2001

3 a Complete, and label, the diagram below to show how you would measure the standard electrode (reduction) potential of the Fe^{3+}/Fe^{2+} system.

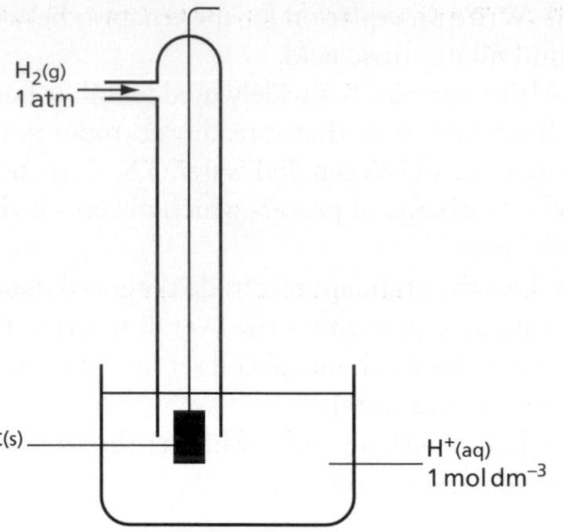

(5)

b A standard Fe^{3+}(aq)/Fe^{2+}(aq) electrode is connected to a standard gold electrode, Au^{3+}(aq)/Au(s), at 25 °C. The conventional cell diagram for this is:

Pt(s) | Fe^{2+}(aq), Fe^{3+}(aq) || Au^{3+}(aq) | Au(s)

i Write the half equations for the reactions that take place at each electrode when an electric current is drawn from this cell. (2)
ii State which would be the positive electrode. (1)
iii The potential of this cell is +0.73 V and the standard electrode potential for a Fe^{3+}(aq)/Fe^{2+}(aq) electrode is +0.77 V.
Calculate the standard electrode potential of the gold electrode. (2)

c Thallium, atomic number 81, exists in more than one oxidation state. Thallium(III) is strongly oxidising and it will oxidise iodide ions to iodine.
25.0 cm^3 of a 0.0480 mol dm^{-3} solution of Tl^{3+} ions was added to excess potassium iodide solution and the liberated iodine titrated against standard 0.106 mol dm^{-3} sodium thiosulphate solution. 22.6 cm^3 of the sodium thiosulphate solution was required to react with the iodine.

$$I_2 + 2S_2O_3^{2-} \rightarrow 2I^- + S_4O_6^{2-}$$

i Calculate the amount, in moles, of Tl^{3+} used. (1)
ii Calculate the amount, in moles, of sodium thiosulphate used in the titration, and hence the amount of iodide ions, I^-, oxidised by the Tl^{3+} ions. (2)
iii Hence deduce the change in oxidation number of the thallium. (2)

Edexcel, A, Module 3, Jan 2001

4 Cadmium is an environmental poison, so its concentration in rivers and other aqueous solutions is frequently measured. To do this, the standard electrode potential for the cadmium half-cell is needed.
A student determined this half-cell potential by setting up a standard cell in which the reactions are represented by the following half-equations.

$$Cd(s) \rightarrow Cd^{2+}(aq) + 2e^-$$
$$Ag^+(aq) + e^- \rightarrow Ag(s)$$

a i Construct the overall redox equation representing this cell, which the student labelled *cell* A.
ii Sketch a labelled diagram of *cell* A. (4)

b The student measured the standard cell potential of *cell* A as +1.2 V.
i Calculate the standard electrode potential of the cadmium half-cell.
ii Explain why cadmium is the negative electrode of *cell* A. (3)

c The student made *cell B* from the standard cadmium half-cell and a standard chromium half-cell. The chromium half-cell had a standard electrode potential of −0.74 V.
Calculate the standard electrode potential of *cell B*. (1)

OCR, A, 4822, Mar 2001

5 An electrochemical cell was set up using a nickel half-cell and a silver half-cell. The cell potential was found to be 1.08 V.
a Draw a labelled diagram of this cell. (2)
b i Use a data booklet to calculate the standard cell potential for a nickel–silver cell.
ii Give *one* reason why there is a difference between the actual cell potential and that calculated in **i**. (3)
c Write an equation showing the overall cell reaction. (1)
d i From which half-cell are electrons released into the external circuit?
ii If the silver half-cell is replaced with a magnesium half-cell, the direction of electron flow in the external circuit changes. Explain why. (2)

OCR, A, 4822, June 1999

Chapter 14

1 a Spectroscopic techniques which rely on different atomic or molecular processes absorb energy in different parts of the electromagnetic spectrum.
Copy and complete the table below.

Process	**Region in electromagnetic spectrum**
Electron promotion	
	Radio waves
Molecular vibrations	

(3)

b Part of the absorption spectrum of hydrogen is shown below.

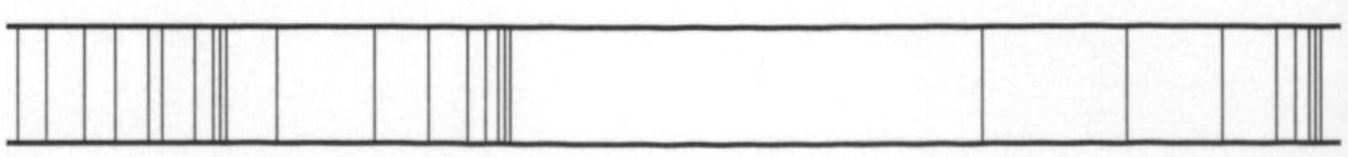

i Why does the spectrum consist of lines?
ii Why are there several series of lines?
iii Why does each of the series of lines converge? (3)
c Organic molecules such as ethanal, CH_3CHO, absorb energy in the uv/visible range. What electronic transitions are responsible for such absorptions? (2)
OCR, A, 4825, Mar 2001

2 Astronomers can detect the presense of hydrogen in stars using spectroscopy. A diagram, representing the line emission spectrum of atomic hydrogen in the ultraviolet region, is shown below.

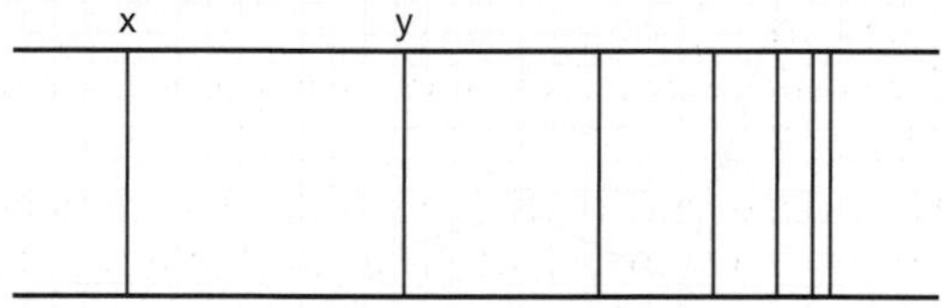

a Draw a labelled diagram to show the electronic transitions which would give rise to the lines x and y in the spectrum. (3)
b Explain the following features of the spectrum:
i The spectrum is composed of discrete lines. (1)
ii The spaces between the lines decrease as the frequency increases. (1)
c Explain how the spectrum may be used to determine the ionisation energy of atomic hydrogen. (3)
d Electronic transitions in metal ions produce characteristic flame colours. Copy and complete the table below.

Metal ion	Flame colour
	Yellow
Calcium	
	Green

(3)
CCEA, A, Module 1, Feb 2001

3 The equation $E = h\upsilon$ is used in spectroscopy. State what each of these symbols represents and the units used for their measurement. (3)
CCEA, AS, AS1 Module 1, Jan 2001

4 The electronic energy levels of atomic hydrogen are shown below. Draw an arrow on a copy of the diagram which represents the energy change associated with the lowest frequency line in the ultraviolet *emission* spectrum.

(3)
CCEA, AS, AS1 Module 1, June 2001

5 Transition metal complexes, such as $[Cu(H_2O)_6]^{2+}$, and some organic molecules, such as nitrobenzene, $C_6H_5NO_2$, absorb strongly in the u.v./visible range of the spectrum.
Using the above examples, explain:
a the reasons for such absorptions (4)
b the factors which influence the positions of the absorptions and hence the colour of the compound. (6)
UCLES, A, 9254, June 1995

6 Different forms of spectroscopy are used to assist in various aspects of diagnosis in medicine.
a Outline the use of atomic absorption spectroscopy in the determination of sodium ions in blood serum. (4)
b Suggest why n.m.r. spectroscopy may be used as an important non-invasive diagnostic technique. (3)
c Suggest why i.r. spectroscopy is not a useful technique for investigating biological molecules in living tissue. (3)
UCLES, A, 9254, June 1996

7 In 1997, Comet Hale-Bopp passed relatively close to the Earth, enabling scientists to study the substances which were contained in it. As the comet came closer to the Sun, material was 'boiled off' producing distinct 'tails' of debris.
a The presence of sodium was detected in the tail of the comet.
i State which branch of spectroscopy could have been used to detect sodium.
ii Outline a process which makes the detection of sodium by this method possible. (3)
b The solid present in the 'nucleus' of the comet is composed mainly of water, methane and ammonia, which vaporise as the comet approaches the Sun.

i In what region of the spectrum do all three of these molecules absorb energy?
ii Explain why each molecule can be detected in the presence of the others.

c Scientists believe they detected in the tail of the comet a molecule with the structure shown below.

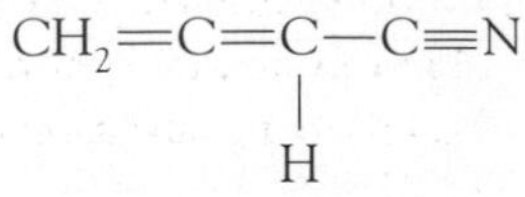

i State what electronic transitions this molecule would show in order to give u.v./visible absorptions.
ii Explain which branch of spectroscopy you could use to distinguish the structure shown above from the following compound.

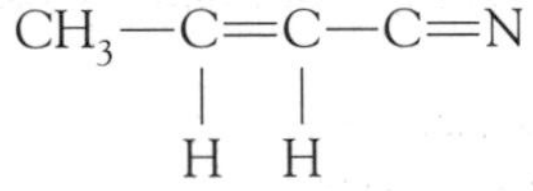

(4)

OCR, A, 9254, June 1999

8 a The diagram below shows the visible spectrum of $[M(H_2O)_6]^{2+}$, where M is a metal.

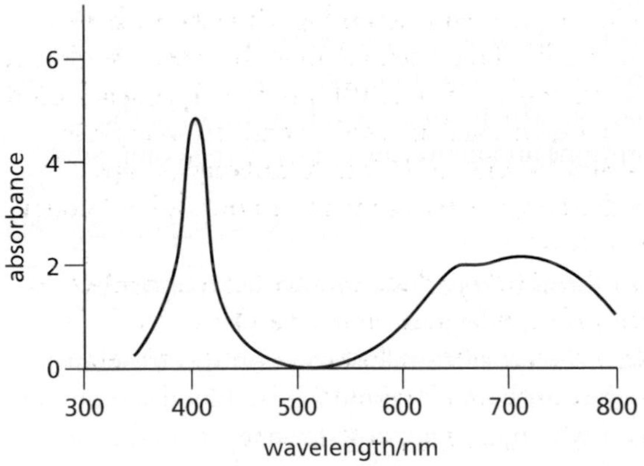

i Predict, with reasoning, the colour of the solution.
ii The spectrum was recorded using a cell of path length 2.0 cm. At the maximum absorbance, the extinction coefficient, ε, was $165\,dm^3\,mol^{-1}\,cm^{-1}$. Using these data, calculate the concentration of the solution. (4)

b Two compounds J and K have the same formula C_6H_{12}.
i Compound J absorbs in the ultraviolet region of the spectrum, whereas compound K does not. Suggest structures for J and K and explain the difference in their behaviour.
ii Compound J reacts with bromine to form $C_6H_{12}Br_2$. The original u.v. absorption disappears, and a new absorption appears.
Suggest reasons for these three observations. (6)

OCR, A, 9254, Nov 1999

Chapter 15

1 The diagram below shows the trend in the first ionisation energies of the elements from neon to aluminium.

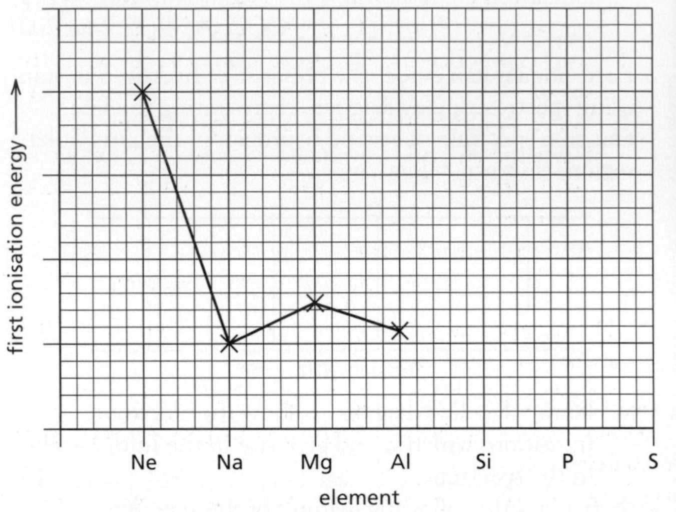

a Draw crosses on a copy of the graph to show the first ionisation energies of silicon, phosphorus and sulphur. (3)
b Write an equation to illustrate the process which occurs during the first ionisation of neon. (1)
c Explain why the first ionisation energy of neon and that of magnesium are both higher than that of sodium. (4)
d Explain why the first ionisation energy of aluminium is lower than that of magnesium. (2)
e State which one of the elements neon, sodium, magnesium, aluminium and silicon has the lowest melting point and explain your answer in terms of the structure and bonding present in that element. (3)
f State which one of the elements neon, sodium, magnesium, aluminium and silicon has the highest melting point and explain your answer in terms of the structure and bonding present in that element. (3)

AQA, AS, Unit 1, Jan 2001

2 a In terms of structure and bonding, describe and explain fully the difference between the melting points of the Period 3 elements aluminium, silicon and phosphorus. (12)
b Describe and explain the difference between the electrical conductivities of the elements aluminium, silicon and phosphorus. (4)
c State appropriate conditions under which magnesium and calcium react with water. Give equations for the reactions and describe what you would observe. (7)
d 'Beryllium is an atypical element in Group II.' Justify this statement by comparing the reactions of beryllium hydroxide and magnesium hydroxide with hydrochloric acid and also with sodium hydroxide. Write equations to illustrate your answer. (7)

AQA, AS, Unit 1, June 2001

3 a Write equations to show what happens when the following oxides are added to water and predict approximate values for the pH of the resulting solutions:
i sodium oxide
ii sulphur dioxide. (4)
b What is the relationship between bond type in the oxides of the Period 3 elements and the pH of the solutions which result from addition of the oxides to water? (2)
c Write equations to show what happens when the following chlorides are added to water and predict approximate values for the pH of the resulting solutions:
i magnesium chloride
ii silicon tetrachloride. (4)
AQA, A, Specimen Unit 5, 2001/2

4 The atomic radii of the elements Li to F and Na to Cl are shown in the table below.

Element	Li	Be	B	C	N	O	F
Atomic radius/nm	0.134	0.125	0.090	0.077	0.075	0.073	0.071
Element	Na	Mg	Al	Si	P	S	Cl
Atomic radius/nm	0.154	0.145	0.130	0.118	0.110	0.102	0.099

a Using *only* the elements in this table, select
i an element with *both* metallic and non-metallic properties, (1)
ii the element with the largest first ionisation energy, (1)
iii an element with a giant molecular structure. (1)
b Explain what causes the general *decrease* in atomic radii across each period? (3)
c Predict and explain whether a sodium **ion** is *larger*, *smaller* or the *same size* as a sodium **atom**. (3)
OCR, AS, 2811, Jan 2001

5 The table below relates to oxides of Period 3 in the periodic table.

Oxide	Na_2O	MgO	Al_2O_3	SiO_2	P_4O_{10}	SO_3
Melting point/°C	1275	2827	2017	1607	580	33
Bonding						
Structure						

a Copy and complete the table using the following guidelines.
i Complete the '*bonding*' row using only the words: *ionic* or *covalent*.
ii Complete the '*structure*' row using only the words: *simple molecular* or *giant*.
iii Explain, in terms of forces, the difference between the melting points of MgO and SO. (5)
b The oxides Na_2O and SO_3 were each added separately to water. For each oxide, construct a balanced equation for its reaction with water.
i SO_3 reaction with water
ii Na_2O reaction with water (2)
OCR, A, Specimen 2815, 2001/2

6 As well as showing the usual Group trends, the first three elements of the second Period also show diagonal relationships to elements of the third Period in the next Group. Thus, some of the properties of lithium are similar to those of magnesium; the elements in the pairs beryllium/aluminium and boron/silicon also show similar properties.
a Use these relationships and your knowledge of the reactions of magnesium, aluminium and silicon to make predictions about the following situations:
i the action of heat on Li_2CO_3
ii the molecular state of $BeCl_2$ vapour and the shape of the resulting molecule
iii the pH of an aqueous solution of $BeCl_2$ and the effect of adding NaOH(aq) to it
iv the reaction of BCl_3 with water
v the reaction between boron oxide and the oxides Na_2O and CaO. (12)
In your answers, you should:
• write equations for all reactions
• contrast, where appropriate, the diagonal relationship property with the one expected from the usual Group trend.
b Suggest explanations as to *why* these pairs of elements show diagonal relationships. (2)
c Despite its diagonal relationship with silicon, boron still shows similarities to aluminium in the reactions of some of its compounds. Thus, $LiBH_4$ and $LiAl_4$ are each good reducing agents and are used extensively in organic chemistry for adding H^- ions to carbonyl groups.
i Of the two, $LiAlH_4$ is the more powerful H^- donor. Why is this?
ii Unlike $LiAlH_4$, which catches fire when added to water, $LiBH_4$ is soluble in water and is fairly unreactive in neutral solutions. In acidic solutions, however, hydrogen is evolved.

$$LiBH_4 + HCl \rightarrow LiCl + BH_3 + H_2$$

Two mechanisms have been proposed for this reaction. The first involves a single-step transfer of a H^- ion to a proton.

$$H^+ + H{:}BH_3^- \rightarrow H_2 + BH_3$$

The second mechanism involves the formation of a symmetrical neutral boron hydride, which then decomposes.

$$H^+ + BH_4^- \rightarrow [BH_5] \rightarrow H_2 + BH_3$$

When a sample of $LiBD_4$ ($D = {}^2H$) is treated with HCl(aq) and the resulting hydrogen analysed by mass spectrometry, the H_2:HD:D_2 ratios are found to be 0:4:6.
Which of the above mechanisms does this result support? Explain how the observed ratios arise. (6)
OCR, S, 9434/0, June 2000

Chapter 16

1 a Define the term *electronegativity*. (2)
b State and explain the trend in electronegativity down Group II. (3)
c Write an equation for the reaction of strontium with water, and suggest an approximate value for the pH of the resulting solution (2)
d Describe what is seen when an aqueous solution of barium chloride is added to dilute sulphuric acid. Write an equation for the reaction which occurs. (2)
e Give two examples which illustrate the atypical properties of beryllium compounds in Group II. (2)
f Give one feature of the beryllium ion which causes the atypical properties of beryllium compounds. (1)
AQA, AS, Unit 1, Jan 2001

2 a When the Group 2 element calcium is added to water, calcium hydroxide and hydrogen are produced. Write an equation for the reaction. (1)
b State the trend in solubility of the hydroxides of the Group 2 elements as the atomic mass of the metal increases. (1)
c i Define the term *first ionisation energy*, and write an equation to represent the change occurring when the first ionisation energy of calcium is measured. (4)
ii State and explain the trend in the first ionisation energy of the Group 2 elements. (3)
Edexcel, AS, Unit 1, June 2001

3 The elements lithium, sodium and potassium of Group I in the periodic table were each separately reacted with water. The increase in reactivity from lithium to potassium depends in part on their first ionisation energies.
a i Define the term *first ionisation energy*.
ii Explain, in terms of first ionisation energies, why the Group I elements become more reactive as the group is descended.
iii Explain why the second ionisation energy of potassium differs significantly from its first ionisation energy. (7)
b 7.82 g of potassium were reacted with an excess of water producing hydrogen gas and 250 cm³ of aqueous potassium hydroxide.

$$2K(s) + 2H_2O(l) \rightarrow 2KOH(aq) + H_2(g)$$

i How many moles of K(s) reacted?
ii Calculate the concentration, in mol dm^{-3}, of the aqueous potassium hydroxide formed. (2)
OCR, AS/A, 4820, Mar 2001

4 Group II metals show a steady change in both their physical and chemical properties as the Group is descended.
a Reaction of each metal with cold water produces the hydroxide.
i State *two* observations which may be made when barium metal is added to water. (2)
ii Write an equation for the reaction of barium with water. (2)
iii State and explain the difference in solubility between magnesium hydroxide and barium hydroxide. (3)
iv Explain, without practical details, how solid samples of magnesium hydroxide and barium hydroxide may be distinguished using sulphuric acid. (2)
b Calcium carbonate occurs naturally as marble, limestone, chalk and shells. The percentage of calcium carbonate in mussel shells may be determined by a back titration method. The shells are treated with excess dilute hydrochloric acid.

$$CaCO_3 + 2HCl \rightarrow CaCl_2 + H_2O + CO_2$$

The unreacted acid is then determined by titration with standard sodium hydroxide solution.

$$HCl + NaOH \rightarrow NaCl + H_2O$$

i Explain the term *standard* solution. (1)
ii Name a suitable indicator for the titration and state the colour change occurring at the end point. (3)
iii A student weighed out 1.05 g of mussel shell fragments and reacted them with 40.0 cm³ of 1 M hydrochloric acid. The resultant solution was transferred to a 250 cm³ volumetric flask and made up to the mark with distilled water. 25.0 cm³ portions of the solution were titrated with 0.1 M sodium hydroxide solution. The average titre was found to be 20.2 cm³. Calculate the percentage of calcium carbonate in the mussel shells. (4)
CCEA, AS, Module 2, June 2001

5 In the UK, over 60 million tonnes of limestone are quarried each year. Much of this limestone is used to produce cement. The main chemical in limestone is calcium carbonate, $CaCO_3$.
a Copy and complete the flow-chart below for reactions starting from calcium carbonate. You should identify each of the substances A–F by name or formula.

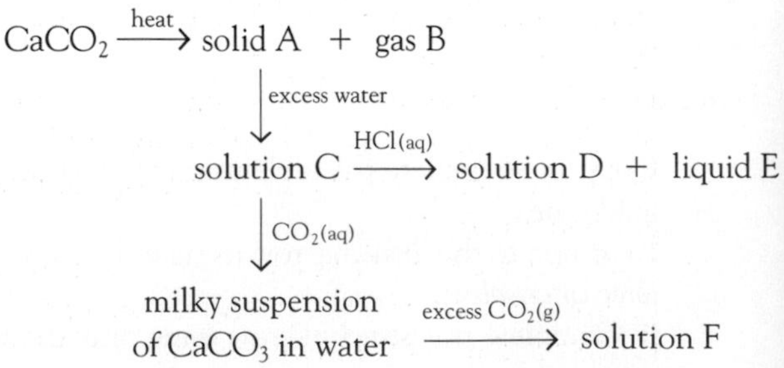

b Cement is a mixture of calcium and aluminium silicates, formed by heating limestone with clay.

$$4CaCO_3(s) + \underset{\text{clay}}{Al_2Si_2O_7(s)} \longrightarrow \underbrace{2CaSiO_3(s) + Ca_2Al_2O_5(s)}_{\text{cement}} + 4CO_2(g)$$

A typical bag of cement has a mass of 25 kg. Calculate the mass of limestone (taken as calcium carbonate) required to make 25 kg of cement.

The molar mass of cement, taken as ($2CaSiO_3$ + $Ca_2Al_2O_5$), is 446.6 g mol^{-1}. [A_r: Al, 27.0; C, 12.0; Ca, 40.1; O, 16.0; Si, 28.1.] (3)

c Lime mortar is a thick paste made by adding water to a mixture of slaked lime, $Ca(OH)_2$, and sand. As mortar dries out the slaked lime reacts with carbon dioxide in the air, forming calcium carbonate which causes the mortar to harden.

i Write an equation to represent the hardening of mortar. Assume the sand does not react. (1)

ii In time, lime mortar crumbles and needs to be replaced. Suggest why this happens more quickly when the mortar is exposed to air contaminated with acidic pollution. (1)

OCR, AS, 2811, Jan 2002

6 This question is about the metals of Group I and Group II of the periodic table.

a i Suggest how the lattice enthalpies (ΔH_{latt}) of the sulphates, MSO_4, and carbonates, MCO_3, vary down Group II. (1)

ii The hydration enthalpy, ΔH_{hyd}, for an ion M^{n+} is the enthalpy change when one mole of gaseous ions is dissolved in water. The hydration enthalpy of the ions M^{2+} decreases down Group II. Suggest an explanation for this observation. (2)

iii Construct an enthalpy cycle for the dissolution of metal salts in aqueous solution. Use your cycle to justify the observed decrease in solubilities of the salts MCO_3 and MSO_4 on descending Group II. (2)

b Using any necessary mineral acids, suggest the step(s) you might use to prepare:

i anhydrous $MgSO_4$ from $MgCO_3$ (2)

ii $BaSO_4$ from $BaCO_3$. (3)

c Conduction involves the transport of charge by electrons or ions between electrodes.

i Explain why most chloride salts of Groups I and II elements are insulators in the solid state, but conduct in the molten form. (1)

ii $BeCl_2$ is an insulator in the liquid state and can be dissolved in organic solvents. Describe the bonding in $BeCl_2$ and rationalise these observations in terms of the bonding in $BeCl_2$. (3)

d The ionic radii of Li^+ and Mg^{2+} are very similar yet their hydration enthalpies are very different.

i Explain the observed similarity in the ionic radii of Li^+ and Mg^{2+}. (2)

ii Suggest, with reasons, which ion has the larger hydration enthalpy. (2)

iii Li reacts with N_2 to form compound A, containing 59.8% Li. Identify A and suggest a formula for the product from the reaction of Mg with N_2.
[A_r: Li = 7; N = 14.] (2)

OCR, A, 9485 (STEP), 2001

Chapter 17

1 a State the trend in oxidising power of the halogens chlorine, bromine and iodine. (1)

b State what would be observed if aqueous bromine were to be added separately to samples of aqueous potassium chloride and aqueous potassium iodide. Write an ionic equation for any reaction occurring. (3)

c When chlorine is dissolved in cold water a pale-green solution, chlorine water, is formed. A piece of universal indicator paper, dipped into chlorine water, first turns red and then becomes white.

i Give the formula of the species responsible for the green colour of chlorine water.

ii Write an equation for the reaction between chlorine and cold water.

iii Explain the colour changes observed when universal indicator paper is dipped into chlorine water. (4)

d Write an equation for the reaction that occurs when chlorine is bubbled into cold, dilute aqueous sodium hydroxide. (1)

AQA, AS, Unit 2, Jan 2001

2 a State and explain the trend in electronegativity of the halogens down Group VII. (4)

b State and explain the trend in boiling points of the halogens down Group VII. (3)

AQA, AS, Unit 2, June 2001

3 a Define the term 'oxidation number'. (2)

b The equation below shows the disproportionation of chlorine.

$$Cl_2(g) + H_2O(l) \rightarrow HClO(aq) + HCl(aq)$$

i For each of the chlorine containing species write the oxidation number of chlorine in each case. (1)

ii Use these oxidation numbers to explain the term disproportionation. (2)

c Explain why hydrogen chloride dissolves in water to form an acidic solution. (2)

d Outline how aqueous silver nitrate followed by aqueous ammonia may be used in the identification of chloride, bromide and iodide ions in aqueous solution. (6)

Edexcel, A/AS, Unit 1, Jan 2001

4 a Seawater contains aqueous bromide ions. During the manufacture of bromine, seawater is treated with chlorine gas and the following reaction occurs:

$$2Br^- + Cl_2 \rightarrow Br_2 + 2Cl^-$$

i Explain the term *oxidation* in terms of electron transfer. (1)

ii Explain the term *oxidising agent* in terms of electron transfer. (1)

iii State which of the elements chlorine or bromine is the stronger oxidising agent and explain the importance of this in the extraction of bromine from seawater, as represented in the equation above. (2)

b When sodium chlorate(I), NaClO, is heated, sodium chlorate(V) and sodium chloride are formed.
i Write the *ionic* equation for this reaction. (2)
ii What type of reaction is this? (1)

c During one process for the manufacture of iodine the following reaction occurs:

$$2IO_3^- + 5SO_2 + 4H_2O \rightarrow I_2 + 8H^+ + 5SO_4^{2-}$$

i Deduce the oxidation number of sulphur in SO_2 and SO_4^{2-}. (2)
ii Use your answers to part **c i** to explain whether SO_2 has been oxidised or reduced in the above reaction. (1)
iii Name a reagent that could be used to confirm that a solution contains iodine, and state what would be *seen*. (2)

Edexcel, AS, Unit 1, June 2001

5 Hydrogen chloride, HCl, is a colourless gas which dissolves very readily in water forming hydrochloric acid. [1 mol of gas molecules occupy $24.0\,dm^3$ at room temperature and pressure, r.t.p.]

a At r.t.p, $1.00\,dm^3$ of water dissolved $432\,dm^3$ of hydrogen chloride gas.
i How many moles of hydrogen chloride dissolved in the water? (1)
ii The hydrochloric acid formed has a volume of $1.40\,dm^3$. What is the concentration, in $mol\,dm^{-3}$, of the hydrochloric acid? (1)

b In solution, the molecules of hydrogen chloride ionise.

$$HCl(aq) \rightarrow H^+(aq) + Cl^-(aq)$$

Describe a simple test to confirm the presence of chloride ions. (2)

c Hydrochloric acid reacts with magnesium oxide, MgO, and magnesium carbonate, $MgCO_3$. For each reaction, state what you would see and write a balanced equation.
i MgO (2)
ii $MgCO_3$ (2)

OCR, AS, 2811, June 2001

6 The halogens have different reactivities.

a Describe how displacement reactions can be used to show the different reactivities of chlorine, bromine and iodine.
In your answer, you should include equations and observations. (4)

b Explain the trend in reactivity of the halogens. (4)

OCR, AS, 2811, Jan 2002

7 a The following reactions of iodine and fluorine and their compounds occur under the conditions stated.

- Reaction I
Iodine reacts with fluorine at −78 °C to form iodine trifluoride, IF_3.
- Reaction II
When the temperature of the IF_3 is allowed to rise to 25 °C, two products are formed, one of which is iodine pentafluoride, IF_5.
- Reaction III
When iodine pentafluoride and fluorine are passed through a platinum tube at 300 °C, iodine heptafluoride, IF_7, is formed.

IF_5 is a colourless liquid which freezes at 8.5 °C and boils at 97 °C.
IF_7 is also colourless with melting point 4.5 °C and boiling point 5.5 °C.
i Write an equation for Reaction II. What type of reaction is this?
ii Draw 'dot and cross' diagrams to show the bonding in IF_5 and in IF_7. You need only show the electrons from the fluorine atoms which are actually involved in bonding.
iii By references to your diagrams in **ii**, deduce and draw the shapes of the IF_5 and IF_7 molecules.
iv Suggest why IF_5 has a much higher boiling point than IF_7. (8)

b
- Under suitable conditions, iodine reacts with chlorine to form a yellow solid, A, which contains 54.4% by mass of iodine.
- When A is heated, a red solid, B, and a greenish gas, C, are formed as the only products.
- The solid A is soluble in concentrated hydrochloric acid forming a yellow solution, D. When aqueous ammonia is added to D, a yellow solid, E, is formed. E is ionic, consisting of one anion and one cation, and contains 44.3% by mass of iodine and 49.5% by mass of chlorine.
- When solid B is added to aqueous potassium iodide, a brown solution is obtained. When gas C is bubbled through aqueous potassium iodide, the same result is observed. In each case, the brown colour is removed when aqueous sodium thiosulphate is added.
- When solid B is melted, the liquid conducts electricity.

Account for all these observations and identify compounds A to E, giving chemical equations where possible. (12)

OCR, S, 9434/0, June 2000

Chapter 18

1 This question is about compounds of the Group IV elements, carbon to lead.

a Explain why the +2 compounds become more stable going from silicon to lead. (2)

b Carbon dioxide, CO_2, is an acidic oxide. Write an equation which illustrates this property for CO_2. (1)

c Tin(II) oxide, SnO, can act as an acid or a base.
i Write an equation to show how SnO reacts with aqueous hydrochloric acid.

ii Suggest the formula of a tin-containing product formed when SnO reacts with aqueous sodium hydroxide. (3)

d Group IV elements form tetrachlorides when they react with chlorine.

i State and explain the trend in thermal stability of these tetrachlorides.

ii Predict the polarity of the Si—Cl bonds in $SiCl_4$.

iii Explain how a water molecule attacks this bond.

iv Explain why a water molecule does not react with a CCl_4 molecule. (5)

OCR, A, 4822, March 2001

2 a Silicon tetrachloride, $SiCl_4$, reacts vigorously with water.

i Write an equation for this reaction.

ii Explain this reaction in terms of structure and bonding. (You may find a diagram helpful.)

iii Suggest why CCl_4 does not react with water. (4)

b Describe the shape of the CCl_4 molecule. (1)

c Tetrachloromethane, CCl_4, is a volatile liquid which was used as a dry-cleaning fluid. Suggest which property made it suitable for this use. (1)

d Suggest why $PbCl_4$ is *not* thermally stable whereas CCl_4 is. (1)

OCR, A, 4822, June 1999

3 a An important use of lead is in car batteries. Lead(IV) oxide acts as the positive plate of the battery, the negative plate is lead metal, and the electrolyte is aqueous sulphuric acid.

The standard electrode potentials for the process are

$$PbO_2(s) + 4H^+(aq) + 2e^- \rightleftharpoons Pb^{2+}(aq) + 2H_2O(l) \qquad E^\ominus = +1.46\,V$$

$$Pb^{2+}(aq) + 2e^- \rightleftharpoons Pb(s) \qquad E^\ominus = +0.13\,V$$

i Calculate the e.m.f. (standard potential) of the cell. (1)

ii A single cell in a car battery has an e.m.f. of 2 V. State why this value is different from the value obtained in **i**. (1)

iii When the battery becomes discharged lead(II) sulphate is formed. State what would be *seen* when this occurs. (1)

iv Explain why the concentration of the sulphuric acid falls as the battery becomes discharged. (1)

b Lead(IV) oxide is a powerful oxidising agent.

i Write the balanced equation for the reaction of the lead(IV) oxide with warm concentrated hydrochloric acid. (1)

ii Explain why many reactions of lead(IV) oxide produce a lead(II) compound as one of the products but the reactions of carbon dioxide do not generally give carbon monoxide. (1)

c Lead(II) hydroxide is described as an *amphoteric* compound.

Describe how lead(II) hydroxide would react with an acid such as nitric acid and with an alkali such as sodium hydroxide.

Give a balanced equation for the *latter* reaction. (2)

d Aqueous tin(II) ions react with an aqueous solution of iodine to form a mixture of products in solution.

Explain why a yellow precipitate is seen when aqueous silver ions (Ag^+) are then added to the mixture. (2)

WJEC, A, C2, June 2000

4 a Describe and explain the variation in electrical conductivity of the elements in Group IV. (6)

b One of the three chlorides CCl_4, $SiCl_4$ and $GeCl_4$ does not react with water but the other two do. Write an equation for one of the reactions, and suggest an explanation for why the third chloride does not react. (3)

c 'Red lead' is an oxide of lead used extensively as a surface coating to prevent corrosion of iron and steel. It contains 90.66% by mass of lead, which is present in both the +II and +IV oxidation states.

Calculate the empirical formula of red lead, and predict the reaction it would undergo on being heated strongly in air. Write an equation for the reaction. (3)

OCR, S, 9254/01, June 2001

5 Suggest reasons for the following, writing equations where appropriate. Your answers should include the structure and bonding of the substances under discussion.

a Silicon exists in only one form which is a poor conductor of electricity. Diamond and graphite are allotropes of carbon. Graphite conducts electricity but diamond does not.

It is calculated that to convert graphite to diamond at a normal laboratory temperature requires a pressure of 15000 atmospheres, but even at this pressure there is little conversion apparent until the temperature is also considerably raised. (8)

b Silicon itself is not nearly as hard as diamond. Silica can be reduced with coke in an electric furnace to produce carborundum, which has the formula SiC. Carborundum is used industrially to form an abrasive powder. (2)

c $SiCl_4$ is a liquid but SiS_2 and $BeCl_2$ are solids with high melting points and chain structures. (4)

d When the acid $CH_2(CO_2H)_2$ is heated with P_4O_{10}, an oxide of carbon which has a boiling point of 279 K and a relative molecular mass of 68 is produced. (3)

e $SiO_2(s)$ is normally insoluble in liquid argon. However, a solution of silicon dioxide in liquid argon can be obtained as follows.

Firstly, SiO_2 is heated to 2000 K and SiO gas is formed.

Then, SiO and O_3 react together using liquid argon as the solvent forming a solution of silicon dioxide. (3)

OCR, S, 9234/0, June 1999

6 **a** Describe and explain the variation in boiling points of the Group IV tetrachlorides from carbon to lead. (3)

b How and why do CCl_4 and $SiCl_4$ differ in their reactions with water? Include in your answer an equation for any reaction that occurs. (3)

c Suggest a use for each of the following elements or compounds, and explain how each use relates to its chemical or physical properties and structure.

i silicon

ii silicon(IV) oxide

iii carbon dioxide (3)

d Lead white, a white pigment used in old paintings, contains basic lead(II) carbonate. It darkens when exposed to air containing traces of hydrogen sulphide, due to the formation of black lead(II) sulphide, PbS. The white colour can be restored by treating the painting with aqueous hydrogen peroxide, H_2O_2, which converts the lead(II) sulphide into lead(II) sulphate.

Write an equation for this reaction and use it to calculate the volume of 0.10 $mol\,dm^{-3}$ hydrogen peroxide required to react with 0.25 g of lead(II) sulphide. (3)

UCLES, A, 9254, Nov 1998

Chapter 19

1 **a** When aqueous cobalt(II) chloride is treated with aqueous ammonia, a precipitate forms.

i Give the formula of this precipitate and write an equation, or equations, to show how it is formed.

ii This precipitate dissolves when an excess of aqueous ammonia is added and a pale brown solution is formed. Give the formula of the cobalt species present in the pale brown solution and write an equation to show how it is formed from the precipitate.

iii State what is observed when this pale brown solution is allowed to stand in air and give the formula of the new cobalt species formed. (8)

b In order to determine the concentration of a solution of cobalt(II) chloride, a 25.0 cm^3 sample was titrated with a 0.0168 M solution of $EDTA^{4-}$; 36.2 cm^3 were required to reach the end-point. The reaction occurring in the titration is

$$[Co(H_2O)_6]^{2+} + EDTA^{4-} \rightarrow [Co(EDTA)]^{2-} + 6H_2O$$

i What type of ligand is $EDTA^{4-}$?

ii Calculate the molar concentration of the cobalt(II) chloride solution.

iii Suggest an alternative analytical method for determining the concentration of a solution which contains only cobalt(II) chloride. (7)

AQA, A, CH05, June 2000

2 **a** Vanadium is a transition element. State *three* characteristic features of the chemistry of vanadium and its compounds. (3)

b Vanadium(IV) chloride is a Lewis acid.

i Define the term *Lewis acid*.

ii Predict, with a reason in *each* case, whether or not vanadium(IV) chloride would react with hexane or with ethanol. (5)

c When an aqueous solution of vanadium(III) chloride is treated with sodium carbonate, effervescence occurs and a precipitate forms. Deduce the formula of the gas and of the precipitate. (2)

AQA, A, Specimen Unit 5, 2001/2

3 **a** Small amounts of manganese are used in the production of specialist steels that are used in the construction of nuclear reactors.

While in the nuclear reactor the steel becomes irradiated with neutrons and some of the manganese in the steel is converted into the radioisotope $^{56}_{25}Mn$. This isotope decays by β-emission with a half-life of 2.6 hours.

i Write the nuclear equation for the decay of $^{56}_{25}Mn$. (2)

ii A sample of the steel removed from a reactor was found to contain 0.80 g of $^{56}_{25}Mn$. Calculate the mass of $^{56}_{25}Mn$ that will remain after 13 hours. (2)

b Copy this and write the electronic structure of a manganese atom and a Mn^{2+} ion.

	3d					4s
Mn [Ar]						
Mn^{2+} [Ar]						

(2)

c Solutions of manganese(II) sulphate contain the hydrated manganese(II) ion.

i Write the formula of this ion. (1)

ii When aqueous ammonia is added to a solution of manganese(II) sulphate, a buff coloured precipitate is obtained. Write an *ionic equation* for this reaction and state the type of reaction taking place. (3)

iii The precipitate produced slowly darkens on exposure to air. Suggest a reason for this and state *two* characteristic properties of transition elements that are being shown by manganese. (4)

d Potassium manganate(VII), $KMnO_4$, reacts with sulphite ions, SO_3^{2-}, in acidic solution according to the equation

$$2MnO_4^- + 5SO_3^{2-} + 6H^+ \rightarrow 2Mn^{2+} + 5SO_4^{2-} + 3H_2O$$

Sodium sulphite, Na_2SO_3, is slowly oxidised in air to sodium sulphate, Na_2SO_4, and hence it is very difficult to keep it pure.

1.75 g of an impure sample of sodium sulphite was dissolved in water and made up to 250 cm^3 with distilled water. 25.0 cm^3 of this solution required 22.8 cm^3 of 0.0216 mol dm^{-3} potassium manganate(VII) solution for complete oxidation.

i Calculate the change in oxidation number of sulphur in the reaction of sulphite ions with manganate(VII) ions. (1)

ii Calculate the amount (in moles) of manganate(VII) ions used in the titration. (1)

iii Calculate the amount (in moles) of sodium sulphite present in 25.0 cm^3 of the solution. (1)

iv Calculate the total mass of pure sodium sulphite in 250 cm^3 of the solution. (2)

v Calculate the percentage purity of the sample of sodium sulphite. (1)

Edexcel, A, Module 1, Jan 2001

4 Copper and vanadium are transition metals. Vanadium was named after Vanadis, the Scandinavian goddess of beauty and it forms many coloured compounds. Copper was named after the island of Cyprus which was once rich in copper ores.

a What, in terms of electronic configuration, do the positive ions of vanadium and copper have in common? (1)

b Vanadium was first extracted in 1867 by Henry Roscoe by the reduction of vanadium(III) chloride.

i Write an equation for the reduction of vanadium(III) chloride to metallic vanadium using hydrogen. (2)

ii Using the following standard redox potentials, choose a reagent which will convert vanadium(V) to vanadium(IV) but not to vanadium(III). Write an equation for the conversion.

$VO_2^+ + 2H^+ + e^- \rightleftharpoons VO^{2+} + H_2O$ $E^\ominus = +1.00\,V$

$SO_4^{2-} + 2H^+ + 2e^- \rightleftharpoons SO_3^{2-} + H_2O$ $E^\ominus = +0.93\,V$

$VO^{2+} + 2H^+ + e^- \rightleftharpoons V^{3+} + H_2O$ $E^\ominus = +0.32\,V$

$V^{3+} + e^- \rightleftharpoons V^{2+}$ $E^\ominus = -0.26\,V$

$Zn^{2+} + 2e^- \rightleftharpoons Zn$ $E^\ominus = -0.76\,V$

(2)

iii Describe an experiment in which the oxidation state of vanadium is successfully reduced from +5 to +2. Give practical details, and state and explain any colour changes you would observe.
(Up to two marks may be obtained for the quality of language in this part.) (6)

c Copper(II) ions in aqueous solution react with excess chloride ions to form a complex ion according to the equation:

$[Cu(H_2O)_6]^{2+} + 4Cl^- \rightleftharpoons [CuCl_4]^{2-} + 6H_2O$

i Write an expression for the overall stability constant for the $[CuCl_4]^{2-}$ complex. (2)

ii The numerical value for this stability constant is 5.6, and that for the formation of $[Cu(NH_3)_4(H_2O)_2]^{2+}$ from the aqueous cation is 13.1.
Suggest and explain what will happen if excess ammonia solution is added to the $[CuCl_4]^{2-}$ solution. (2)

iii State the colours of $[Cu(NH_3)_4(H_2O)_2]^{2+}$ and $[CuCl_4]^{2-}$ in aqueous solution. (2)

iv State the coordination number and oxidation number of copper in the complex $[CuCl_4]^{2-}$. (2)

CCEA, A, Module 2, Feb 2001

5 a i The ground state electronic configuration of scandium may be written as $1s^2 2s^2 2p^6 3s^2 3p^6 3d^1 4s^2$.
Write down the ground state electronic configuration of nickel in the same way. (1)

ii State *three* general *chemical* properties of nickel which are characteristic of the d-block elements. (3)

b A *particular* sample of the element nickel contains three isotopes with abundances as shown below.

Relative isotopic mass	Percentage abundance
58	60.0
60	30.0
61	10.0

Calculate the relative atomic mass of nickel for *this* sample. (2)

c In the manufacture of nickel, the impure element is converted into the volatile compound nickel tetracarbonyl.

$4CO(g) + Ni(s) \xrightarrow{60\,°C} Ni(CO)_4(g)$

On heating to around 200 °C the nickel tetracarbonyl decomposes to leave pure nickel.

$Ni(CO)_4(g) \rightarrow 4CO(g) + Ni(s)$

Calculate the minimum volume, in cm^3, of carbon monoxide, measured at 0 °C and 101 kPa, which is required to produce 500 kg of pure nickel.
[A_r (Ni) = 58.71; the molar gas volume at 0 °C and 101 kPa is 2.241×10^4 cm^3.] (3)

WJEC, AS/A, C1, June 2000

6 1,2-Diaminoethane, $NH_2CH_2CH_2NH_2$, is a *bidentate ligand*.

a Explain the term *bidentate ligand*. (2)

b There are three isomeric complexes with the formula $[Cr(NH_2CH_2CH_2NH_2)_2Cl_2]^+$, all having the same basic shape.

i State the shape of these complexes.

ii Draw structures of these three complexes, I, II and III, to show the differences between them.

iii Which of the complexes you have drawn above will have a dipole? (5)

OCR, A, 2815, 2001/2

7 This question relates to the oxides of iron.
a When steam is passed over heated iron, Fe_3O_4 is formed, together with a flammable gas. Heating iron(III) oxide in a vacuum also forms Fe_3O_4, but in the presence of air, no reaction is observed. The crystal structure of Fe_3O_4 reveals iron atoms with two different ionic radii; 0.078 and 0.092 nm.
i Write a balanced equation for the formation of Fe_3O_4 from iron and steam. (1)
ii Write an equation for the reaction for the formation of Fe_3O_4 from iron(III) oxide. Suggest why the reaction occurs in a vacuum but not in the presence of air. (3)
iii What is the average oxidation state of iron in Fe_3O_4? (1)
iv Suggest why there are two different ionic radii for the iron atoms in Fe_3O_4. In the light of your answer, suggest how the average oxidation state of iron may be rationalized in Fe_3O_4. (3)
b On heating iron in oxygen, iron(III) oxide is formed. However, if iron is heated in a limited supply of oxygen, an unstable black powder, A, is obtained. A may also be obtained through the thermal decomposition of iron(II) ethanedioate, $Fe(C_2O_4)$. The crystal structure of A is found to be the same as that of sodium chloride.
i Deduce the formula of A and hence write a balanced equation for its formation from iron and oxygen. (2)
ii Write a balanced equation for the decomposition of iron(II) ethanedioate. (1)
On standing, 2.400 g of A forms 0.467 g of iron and compound B. [A_r: Fe = 56; O = 16.]
iii Determine the formula of B and write a balanced equation for its formation. (3)
c In water, Fe^{2+} and Fe^{3+} ions each have six water molecules coordinated to them, for example, Fe^{2+}(aq) exists as the complex $[Fe(H_2O)_6]^{2+}$. These complexes are acidic, i.e. they readily lose protons.
i Draw the structure of the complex $[Fe(H_2O)_6]^{2+}$. (1)
ii Write an equation for $[Fe(H_2O)_6]^{2+}$ acting as a monoprotic (monobasic) acid. (1)
iii Suggest a mechanism by which hydroxide ions form a precipitate of $Fe(OH)_2.xH_2O$ with Fe^{2+} ions in aqueous solution. (2)
iv $Fe(OH)_2.xH_2O$ decomposes on heating to give two oxides of iron, depending on whether oxygen is present or not. Write balanced equations for the decomposition of $Fe(OH)_2.xH_2O$ in both the presence and absence of oxygen. (2)

OCR, A, 9485 (STEP), 2001

Chapter 20

1 a Iron is extracted from the oxide Fe_2O_3 by reduction with carbon and carbon monoxide in a blast furnace.
i In the blast furnace, carbon monoxide is produced from carbon in a two-stage process. Write equations to show these two stages.
ii Write an equation for the reduction of Fe_2O_3 by carbon and an equation for the reduction of Fe_2O_3 by carbon monoxide.
iii Limestone is added to the blast furnace to remove impurities present in the iron ore. Identify the main impurity removed by limestone. Write an equation or equations to show how limestone acts to remove this impurity. (7)
b State the main impurity in iron obtained from the blast furnace. Explain how this impurity is removed in the conversion of impure iron into steel. (3)
c Identify a gas which is released from the blast furnace which leads to environmental problems. State the environmental problem. (2)
d Identify another gas which leads to environmental problems when sulphide ores are used in the extraction of metals other than iron. State the environmental problem. (2)

AQA, AS, Unit 2, Jan 2001

2 Aluminium and titanium are extracted from their purified oxides by different methods.
a Discuss, with the aid of chemical equations, the method used for each metal. (10)
b Explain why each method is chosen. (4)
c Explain why aluminium is recycled although aluminium oxide is in plentiful supply. (3)

AQA, AS, Unit 2, June 2001

3 The manufacture of ammonia is an important industrial process based on the equilibrium.

$$N_2 + 3H_2 \rightleftharpoons 2NH_3 \qquad \Delta H = -92.4\,\text{kJ mol}^{-1}$$

a Explain the meaning of the term *dynamic equilibrium*. (2)
b Explain why raising the equilibrium temperature results in *less* ammonia being produced. (1)
c State why, despite the lower yield of ammonia, the industrial process operates at about 450 °C. (1)
d How is ammonia removed from the mixture of gases? (1)

e **i** Give the meaning of the term *catalyst*. (2)
ii Draw an energy level diagram (energy profile) for the reaction with and without a catalyst and use it to explain how a catalyst works.

energy

(4)

iii State the effect of the presence of a catalyst on the yield of ammonia. (1)

f Give one large scale use of ammonia. (1)

g The first stage in the conversion of ammonia to nitric acid involves mixing it with hot air and passing the mixture over a catalyst.

Copy and complete the equation for the first stage.

$\_NH_3 + \_O_2 \rightarrow \_NO + \_H_2O$ (1)

h The next stage involves the conversion of NO to nitrogen dioxide, NO_2, by adding cold air.
i Write an equation for this conversion. (1)
ii Suggest why the temperature must be lowered at this point. (1)

Edexcel, A, Module 2, June 2001

4 The diagram below represents a diaphragm cell for the production of sodium hydroxide from brine as an electrolyte.

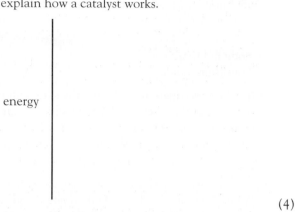

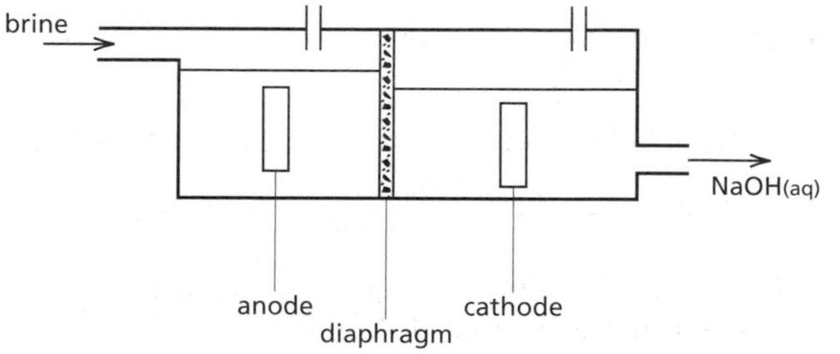

a What is brine? (1)

b **i** Identify the gases produced at the anode and at the cathode and give one use for each on a commercial scale. Describe a test for the gas produced at the anode and state the result you would expect for a positive test.

	Anode	Cathode
Gas		
Use		
Test		

(5)

ii State the electrode at which oxidation takes place, giving your reasoning. (1)

c **i** Explain how the process results in the formation of sodium hydroxide in the cathode compartment. (2)
ii Suggest why the brine must be purified to remove calcium and magnesium ions. (1)

d In the more modern processes the diaphragm is replaced by a membrane which is ion-selective. It will allow sodium ions to pass through but will not allow chloride ions through.

Suggest *ONE* advantage of using such a process. (1)

Edexcel, A, Module 3, June 2001

5 The economic importance of sulphuric acid was recognised over 100 years ago by Baron Justus von Liebig who stated that:

'The commercial prosperity of a nation can be measured by the amount of sulphuric acid it consumes.'

The Contact Process for the production of sulphuric acid involves the following equilibrium:

$2SO_2(g) + O_2(g) \rightleftharpoons 2SO_3(g) \qquad \Delta H = -192\,kJ\,mol^{-1}$

a State and explain the effect of increasing the total pressure on the yield of sulphur trioxide. (3)

b State, in terms of equilibrium and kinetic considerations, why a moderate temperature of 450 °C is used for the production of sulphur trioxide. (2)

c A catalyst is used in the Contact Process.
i Name the catalyst. (1)
ii State the effect, if any, of the catalyst on the equilibrium yield of sulphur trioxide. (1)

d Sulphuric acid is used in the manufacture of the fertiliser ammonium sulphate.

$H_2SO_4 + 2NH_3 \rightarrow (NH_4)_2SO_4$

i State a chemical test for ammonia gas. (2)
ii Calculate the maximum mass of ammonium sulphate that can be manufactured from 3.00 tonnes of sulphuric acid and 0.85 tonne of ammonia using the following headings.

number of moles of sulphuric acid
number of moles of ammonia
name of reactant present in excess
number of moles of ammonium sulphate formed
mass of ammonium sulphate formed (4)

CCEA, AS, AS2 Module 2, June 2001

6 Aluminium metal is manufactured by a process in which purified bauxite, dissolved in molten cryolite, is electrolysed at 800 °C. Graphite electrodes and a current of about 120 000 amperes are used.

a **i** Give the ionic equations for the reactions taking place at the anode and the cathode. (2)
ii State which of these reactions is an oxidation process. (1)

iii Explain why the anodes need to be replaced frequently. (2)
iv Explain why an electrolyte of pure molten bauxite is not used. (2)

b The production of aluminium is expensive.
i Explain why, despite this high cost, aluminium is manufactured in large quantities. (2)
ii Explain why it is worthwhile to recycle aluminium. (2)

Edexcel, AS, Unit 2, June 2001

Chapter 21

1 a State what you would observe on adding aqueous chlorine to separate aqueous solutions of sodium bromide and sodium iodide. Write equations for the reactions occurring. (4)

b State what you would observe on adding concentrated sulphuric acid to separate solid samples of sodium bromide and sodium iodide. In each case, identify all the reduction products. Using half-equations, construct an overall ionic equation for the oxidation of bromide ions by concentrated sulphuric acid. (9)

AQA, AS, Unit 2, June 2001

2 Read the passage below. Identify each of A, B, C, D, E, F, G, H and I, and write equations for all the reactions occurring.

A is a black solid which dissolves in water to form a blue solution which contains a cation B and an anion C.

The addition of aqueous ammonia to the blue solution gives initially a blue precipitate D which dissolves when an excess of aqueous ammonia is added giving a deep blue solution containing species E.

The addition of concentrated hydrochloric acid to the blue solution of A gives a yellow–green solution containing species F.

The addition of aqueous silver nitrate to the blue solution of A gives a cream precipitate G. Precipitate G is insoluble in dilute aqueous ammonia but dissolves forming a colourless solution containing species H when concentrated aqueous ammonia is added. Precipitate G also dissolves when an excess of an aqueous solution of sodium thiosulphate is added giving a colourless solution containing species I. (15)

AQA, A, CH05, June 2000

3 a The compounds lithium chloride, sodium bromide and potassium iodide can be distinguished from one another by the use of flame tests.
i Copy and complete the following table.

Compound	Flame colour
Lithium chloride	
Sodium bromide	
Potassium iodide	

(3)

ii Explain the origin of the colours in flame tests. (2)

b These compounds can also be distinguished from one another by the use of concentrated sulphuric acid.
i State what would be seen when concentrated sulphuric acid is added to separate solid samples of each of these compounds.

Lithium chloride
Sodium bromide
Potassium iodide (4)

ii Write an equation, including the state symbols, for the reaction between solid lithium chloride and concentrated sulphuric acid. (2)

Edexcel, AS, Unit 1, June 2001

4 Suggest reasons for the following and put forward formulae for A, B, C and D. [All gas volumes are measured at s.t.p.]

a When 0.1 mol dm^{-3} lead(II) ethanoate is added to an equal volume of 0.2 mol dm^{-3} hydrochloric acid, a white precipitate forms; when aqueous potassium iodide is added to the mixture, the precipitate turns yellow. (3)

b When aqueous ammonia is added to aqueous magnesium chloride, a white precipitate forms; when aqueous ammonium chloride is stirred into the mixture, the precipitate dissolves. (4)

c When aqueous solutions of copper(II) sulphate and sodium carbonate are mixed, a green solid, A, separates. Solid A contains 53.0% of copper by mass. The action of heat on a 1.00 g sample of A produces 0.660 g of a black powder, B, water vapour, and 94.0 cm^3 of a colourless gas. Solid B contains 79.9% of copper by mass. (6)

d When aqueous solutions of copper(II) sulphate, sodium carbonate and sodium hydrogencarbonate are mixed, blue crystals, C, can be obtained. Solid C contains 22.4% of copper by mass.

The action of heat on a 1.00 g sample of C produces a black powder, D, water vapour and 79.0 cm^3 of a colourless gas.

When C and D are each treated with dilute hydrochloric acid, carbon dioxide is produced.

When a large potential difference is applied across a solution of C, the blue colour is attracted to the positive electrode. Suggest a structure for C. (7)

UCLES, S, 9254/0, Nov 1998

Chapter 22

1 a The mass of one atom of ^{12}C is 1.99×10^{-23} g. Use this information to calculate a value for the Avogadro constant. Show your working. (2)

b Give the meaning of the term *empirical formula*. (1)

c Define the term *relative molecular mass*. (2)

d The empirical formula of a compound is CHO and its relative molecular mass has the value 174. Determine the molecular formula of this compound and show your working. (2)

e A compound with molecular formula CH_4O burns in air to form carbon dioxide and water. Write a balanced equation for this reaction. (1)

AQA, AS, Unit 1, Jan 2001

2 a Compound A, consisting of carbon and hydrogen only, was found to contain 80.0% carbon by mass.
i Calculate the empirical formula of compound A, using the data above and the periodic table. (3)
ii The relative molecular mass of compound A was found to be 30. Use this information to deduce the molecular formula of compound A. (1)

b Propane has the molecular formula C_3H_8. Propane burns completely in oxygen to form carbon dioxide and water as shown in the equation.

$$C_3H_8(g) + 5O_2(g) \rightarrow 3CO_2(g) + 4H_2O(g)$$

i Calculate the mass of water produced when 110 g of propane burns completely in oxygen. (3)
ii Calculate the volume of oxygen required to completely burn 110 g of propane. (1 mole of gas has a volume of 24 dm^3 under the conditions of the experiment.) (2)

Edexcel, AS, Unit 1, June 2001

3 a But-2-ene, $CH_3CH{=}CHCH_3$, exists as geometric isomers.
i Draw the geometric isomers of but-2-ene. (2)
ii Explain how geometric isomerism arises. (1)

b i Draw the structural formula of a compound which is an isomer of but-2-ene but which does not show geometric isomerism. (1)
ii Explain why the isomer drawn in **i** does not show geometric isomerism. (1)

Edexcel, AS, Unit 2, June 2001

4 The table below shows information about some alcohols which form part of an homologous series.

Name	Formula	Boiling point/°C	Relative molecular mass
Methanol	CH_3OH	65	32
Ethanol	C_2H_5OH	78	46
Propan-1-ol	C_3H_7OH	97	60
Butan-1-ol	C_4H_9OH		74
Pentan-1-ol	$C_5H_{11}OH$	138	
Hexan-1-ol	$C_6H_{13}OH$	158	102

a i Identify the functional group common to all alcohols. (1)
ii What is the general formula for these alcohols? (1)
iii What is the formula of the next alcohol in the series? (1)

b Calculate the relative molecular mass of pentan-1-ol. (1)

c i Copy the axes below and plot a graph of boiling point against number of carbon atoms in a molecule of the alcohol. (2)

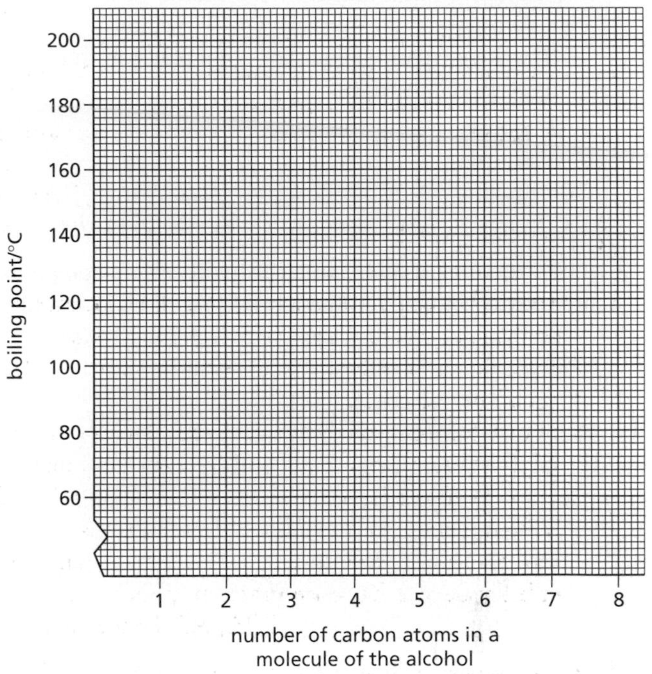

Use the graph to estimate the boiling points of butan-1-ol and $C_8H_{17}OH$. (2)
ii State the connection between boiling point and the relative molecular mass of these alcohols. (1)

OCR, AS, 2812, Jan 2001

5 Myrcene is a naturally occurring oil present in bay leaves. The structure of myrcene is shown below.

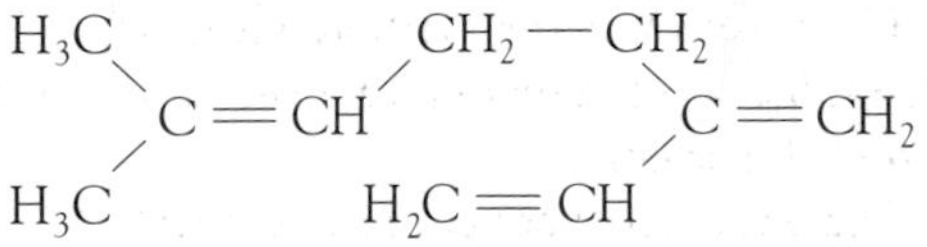

a State the molecular formula of myrcene. (1)

b Reaction of a 0.100 mol sample of myrcene with hydrogen produced a saturated alkane A.
i Explain what is meant by the term *saturated* alkane.
ii Determine the molecular formula of the saturated alkane A.
iii Construct a balanced equation for this reaction.
iv Calculate the volume of hydrogen, measured at room temperature and pressure (r.t.p.) that reacted with the sample of myrcene. [1 mole of gas molecules occupy 24.0 dm^3 at r.t.p.] (5)

c Squalene is a naturally occurring oil present in shark liver oil. A 0.100 mol sample of squalene reacted with 14.4 dm^3 of hydrogen, measured at r.t.p., to form a saturated hydrocarbon $C_{30}H_{62}$.
i Calculate how many double bonds there are in each molecule of squalene.
ii Suggest the molecular formula of squalene. (3)

OCR, AS, 2812, 2001/2

6 Reagents used in organic chemistry may contain electrophiles, nucleophiles or free radicals. For each of the terms **a**, **b** and **c**

i explain what is meant by each term (1)

ii give an example of each type of species (1)

iii write a balanced equation for a reaction that involves each species. (1)

a *electrophile*

b *nucleophile*

c *free radical*

OCR, AS, 2812, June 2001

7 The structures of some saturated hydrocarbon compounds are shown below. They are labelled A, B, C, D, E and F.

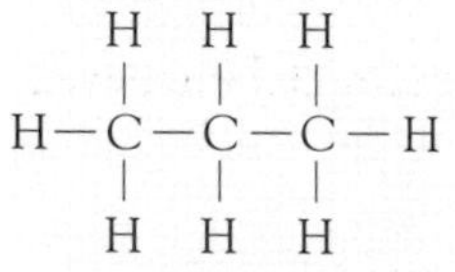

A

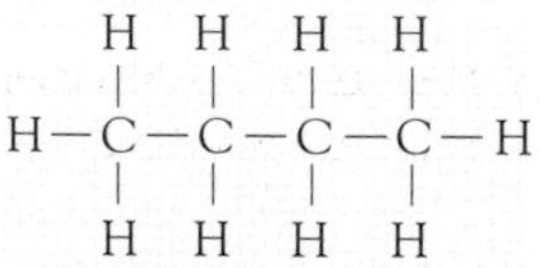

B

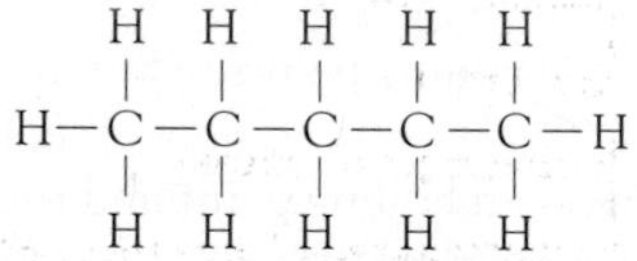

C

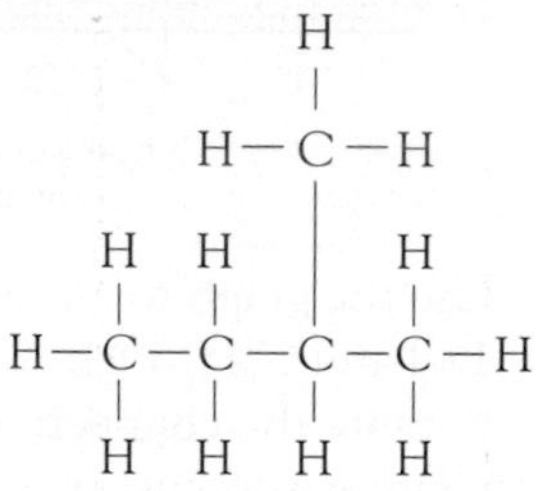

D

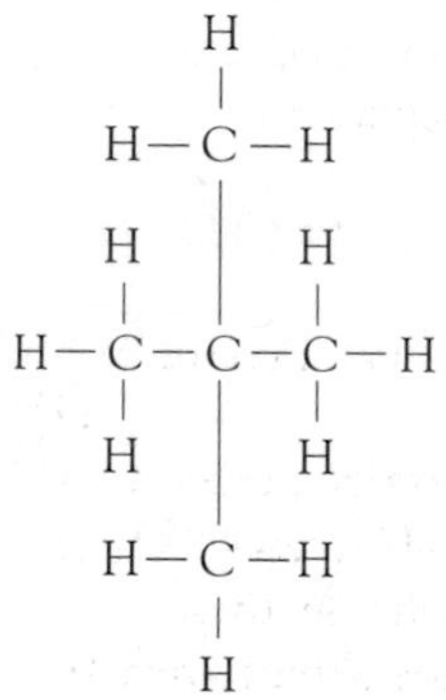

E

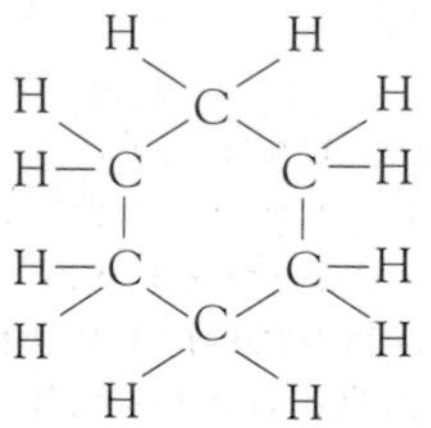

F

a The general formula of an alkane is C_nH_{2n+2}.

i Which of the compounds, A to F, has a formula that does *not* fit with this general formula? (1)

ii A, B and C are successive members of the alkane series.

What is the *molecular* formula of the next member in the series? (1)

iii What is the *empirical* formula of compound F? (1)

b Three of compounds A to F are structural isomers of each other.

i Identify, by letter, which of the compounds are the three structural isomers. (1)

ii Explain what is meant by the term *structural isomer*. (1)

c i Which of the compounds, A, B or C, would you expect to have the highest boiling point?

ii Which of the compounds, C, D or E, would you expect to have the highest boiling point?

iii Explain your answers to **c i** and **c ii** in terms of intermolecular forces. (3)

OCR, AS, 2812, June 2002

Chapter 23

1 Petrol is a fuel sold by the way it behaves rather than by composition. Most petrol has a Relative Octane Number of 95. This means that it behaves in the same way as a mixture of 95% 2,2,4-trimethylpentane and 5% heptane.

Octane is a straight chain hydrocarbon C_8H_{18}.

a i Draw the full structural formula of 2,2,4-trimethylpentane. (2)

ii State the relationship between 2,2,4-trimethylpentane and octane. (1)

iii Write the equation for octane burning completely in oxygen. (2)

iv Octane has to be vaporised before burning in an engine. Determine the fuel to air ratio by volume for the complete combustion of gaseous octane. (1)

b State one advantage of using liquid fuels rather than gaseous fuels for internal combustion engines. (1)

c Lead tetraethyl used to be added to petrol to boost its Relative Octane Number but this has now been replaced by compounds such as benzene or methyl tertiary butyl ether (MTBE). The latter compounds are not as effective as the lead compound and so need to be added in larger quantities and this causes solubility problems.

i Why has the addition of lead tetraethyl to petrol been stopped in the UK? (1)

ii How might the difficulty in keeping MTBE in solution in the petrol cause a problem in the running of the car? (1)

iii Apart from solubility, state one problem associated with the use of benzene as an additive to petrol. (1)

d The exhaust gases from a petrol engine were found to contain a compound E. On analysis E was found to have a composition of 54.6% C, 36.4% O and 9.1% H by mass.

The mass spectrum and infra-red spectrum of E are given below.

i Calculate the empirical formula of E. (2)
ii Use the spectra and the infra-red data below to identify E. Give your reasoning.

mass spectrum of compound E

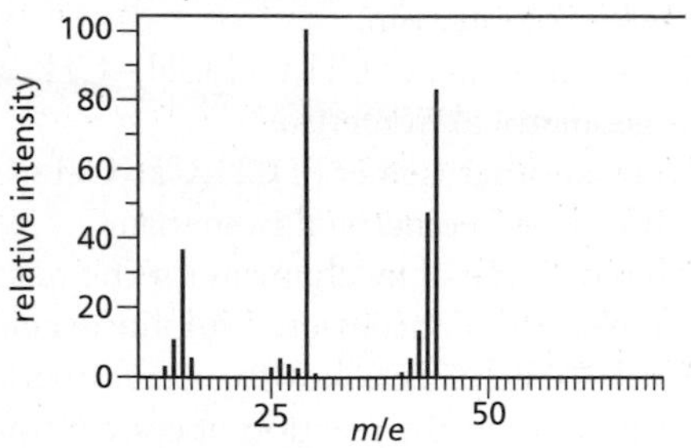

infra-red spectrum of compound E

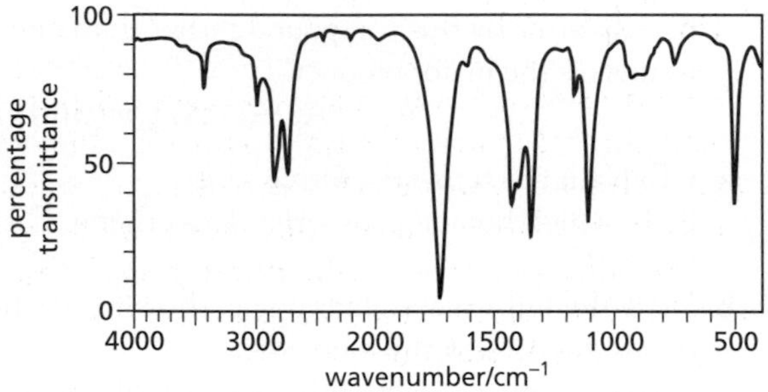

Bond	Wavenumber/cm^{-1}	Bond	Wavenumber/cm^{-1}
C=C	1680–1610	O—H	3550–3230
C=O	1750–1680	C—H	3000–2850

(3)
Edexcel, A, Module 4, June 2001

2 Three types of formulae commonly used in chemistry are empirical, molecular and structural.
a 4.3 g (0.05 mole) of a hydrocarbon were analysed and found to contain 3.6 g of carbon. Calculate its empirical formula and deduce its molecular formula. (4)
b Draw the structures and give the systematic names of the two branched chain isomers with the molecular formula C_5H_{12}. (4)
c Structural formulae may be used to show *cis–trans* isomerism in alkenes.
i Explain why pent-2-ene exhibits *cis–trans* isomerism. (2)
ii Draw and label the structures of the *cis* and *trans* isomers of pent-2-ene. (2)
d Pent-2-ene undergoes catalytic hydrogenation forming pentane.
i Name a suitable catalyst for this reaction. (1)
ii Calculate the volume of hydrogen, measured at 20 °C and one atmosphere pressure, required to saturate 5.0 g of pent-2-ene. (2)
CCEA, A, Module 1, Feb 2001

3 Petroleum is a mixture of hydrocarbons which are separated into fractions by fractional distillation.
a Explain what is meant by the terms, *hydrocarbon* and *fraction*. (3)
b Long chain hydrocarbons may be converted into shorter ones by cracking. Write an equation for the cracking of dodecane, $C_{12}H_{26}$, to form propene as one of the products. (2)
c 20 cm^3 of a saturated hydrocarbon required 160 cm^3 of oxygen for complete combustion; 100 cm^3 of carbon dioxide and 120 cm^3 of water vapour were produced, all measurements being made at the same temperature and pressure. Deduce the formula of the hydrocarbon. (3)
CCEA, AS, AS2 Module 2, June 2001

4 **a** The diagram below represents the industrial fractional distillation of crude oil.

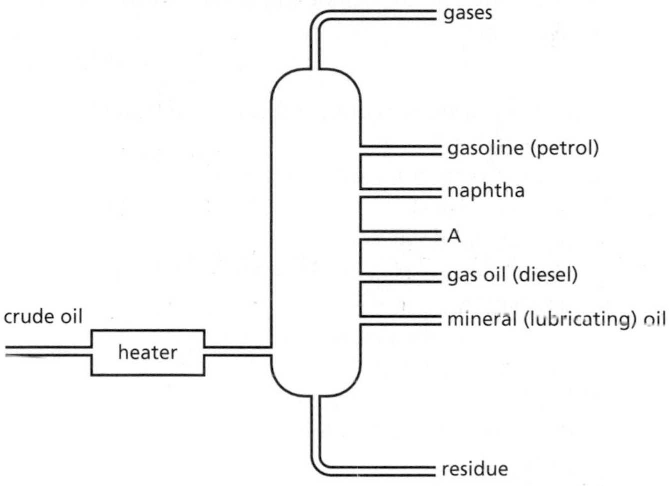

i Identify fraction A.
ii What property of the fractions allows them to be separated in the column? (2)
b A gas oil fraction from the distillation of crude oil contains hydrocarbons in the C_{15} to C_{19} range. These hydrocarbons can be cracked by strong heating.
i Write the molecular formula for the alkane with 19 carbon atoms.
ii Name the type of reaction involved in cracking.
iii Write an equation for one possible cracking reaction of the alkane $C_{16}H_{34}$ when the products include ethene and propene in the molar ratio 2 : 1 and only one other compound. (4)
AQA, AS, Unit 3(a), 2001/2

5 The enthalpy change of combustion of two fuels is listed in the table below.

Fuel	Enthalpy of combustion/kJ mol^{-1}
Hydrogen, H_2	−280
Octane, C_8H_{18}	−5510

a Calculate the enthalpy change per unit mass for each of the fuels, hydrogen and octane. (3)
b Suggest, giving two reasons, which substance is the more useful as a fuel for motor cars and give your reasoning. (2)
c Suggest one disadvantage of using the fuel chosen in **b**. (1)

Edexcel, AS, Unit 2, June 2001

6 Describe the reaction of a named alkane with bromine. Your answer should include full details of the reaction mechanism. (8)

OCR, AS, 2812, 2001/2

7 The hydrocarbons in crude oil can be separated by fractional distillation.
a Explain what is meant by the terms
i *hydrocarbons* (1)
ii *fractional distillation* (1)
b Undecane, $C_{11}H_{24}$, can be isolated by fractional distillation.
Calculate the percentage composition by mass of carbon in undecane. (3)
c Undecane can be cracked into nonane and compound A. One molecule of nonane contains nine carbon atoms.
i Write a balanced equation for this reaction. (2)
ii Name compound A. (1)
d Hydrocarbons of formula C_5H_{12} can also be isolated from crude oil.
i Draw the three structural isomers, B, C and D, of C_5H_{12}. (3)
ii Isomers B, C and D can be separated by fractional distillation. State the order, lowest boiling point first, in which they would distil. (1)
iii Justify the order stated in **d ii**. (1)
iv Write a balanced equation for the *complete* combustion of pentane, C_5H_{12}. (2)
v Why do oil companies isomerise alkanes such as pentane? (1)

OCR, AS, 2812, Jan 2001

Chapter 24

1 a i Name the alkene $CH_3CH_2CH{=}CH_2$.
ii Explain why $CH_3CH_2CH{=}CH_2$ does not show geometrical isomerism.
iii Draw an isomer of $CH_3CH_2CH{=}CH_2$ which does show geometrical isomerism.
iv Draw another isomer of $CH_3CH_2CH{=}CH_2$ which does not show geometrical isomerism. (4)
b i Name the type of mechanism for the reaction shown by alkenes with concentrated sulphuric acid.
ii Write a mechanism showing the formation of the major product in the reaction of concentrated sulphuric acid with $CH_3CH_2CH{=}CH_2$.
iii Explain why this compound rather than one of its isomers is the major product. (6)

AQA, AS, Unit 3(a), 2001/2

2 a i Explain the term *homologous series*. (2)
ii To which homologous series does ethene, C_2H_4, belong? (1)
b Draw the full structural formulae, showing all the bonds, for each of the following.
i The organic product of the reaction of ethene, C_2H_4, with aqueous potassium manganate(VII) and sulphuric acid. (2)
ii 3,4-dimethylhex-2-ene. (2)
iii A repeating unit of poly(propene). (2)
c Ethene reacts with hydrogen chloride gas to form C_2H_5Cl.
i What type of reaction is this? (2)
ii Give the systematic name for C_2H_5Cl. (1)

Edexcel, AS, Unit 2, June 2001

3 a Using alkanes and alkenes as examples, explain what is meant by the following terms:
i homologous series
ii structural and *cis–trans* isomerism. (7)
b Including a mechanism, describe the reaction of bromine *either* with an alkane *or* with an alkene. (7)
c Compound E is an unbranched alkene. Complete combustion of 0.441 g of E was carried out with minimal heat loss. This produced sufficient heat to raise the temperature of 100 g of water by 50 °C.
- The specific heat capacity of water = 4.2 J $g^{-1}K^{-1}$.
- The enthalpy change of combustion of E is −4000 kJ mol^{-1}.

Using the information above and showing each step in your working clearly,
i calculate how much heat was evolved
ii calculate how many moles of E were burnt
iii calculate the relative molecular mass of E
iv deduce the molecular formula of E
v suggest *two* possible structures for E which have *cis–trans* isomers. (7)

OCR, AS/A, 4820, Mar 2001

4 The alkynes are unsaturated hydrocarbons containing a triple bond. Reactions of the alkynes are very similar to those of the alkenes. The first member of the series is ethyne (acetylene), C_2H_2, which has the following structure:

$$H-C\equiv C-H$$

a The triple bond in the ethyne molecule is composed of a sigma (σ) bond and two pi (π) bonds.
i Using a diagram explain the formation of a π bond. (2)
ii Suggest how the length and strength of the triple bond in ethyne compares with the double bond in ethene. (2)
b Ethyne may be hydrogenated to form ethane.
i Write an equation for the complete hydrogenation of ethyne. (1)
ii Standard enthalpies of combustion may be used to determine ΔH for this reaction. Define the term *standard enthalpy change of combustion*. (3)
iii Use the following standard enthalpy changes of combustion to calculate ΔH for the complete hydrogenation of ethyne.

	$\Delta H_{combustion}$/kJ mol^{-1}
H_2	−286
C_2H_2	1301
C_2H_6	−1560

(3)

c But-2-yne, C_4H_6, is a liquid alkyne.
i Draw the structural formula of but-2-yne. (1)
ii But-2-yne reacts in a similar way to but-2-ene. Suggest what would be observed when bromine reacts with excess but-2-yne. (2)

CCEA, A, Module 1, June 2001

5 a Propene, C_3H_6, readily undergoes electrophilic addition reactions. Copy the equation and show, with the aid of curly arrows, the mechanism of the electrophilic addition reaction of propene with bromine.

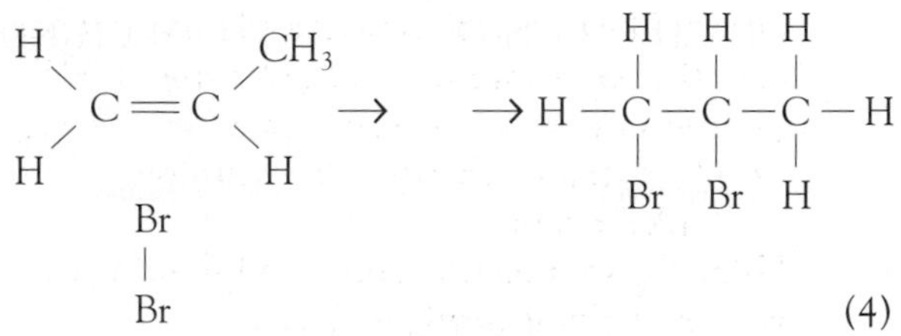

(4)

b Propene also reacts as shown below:

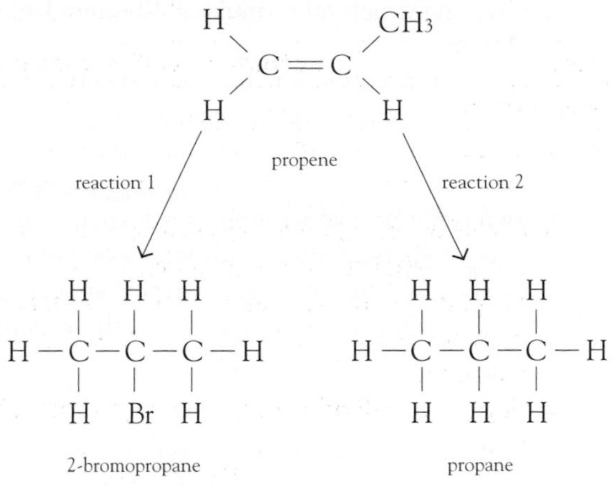

i State a suitable reagent for reaction 1. (1)
ii State a suitable reagent and conditions for reaction 2. (2)
iii In the presence of an acid catalyst, propene can react with steam to form a mixture of two alcohols. Draw the structures of the two alcohols. (2)
c The scientists Ziegler and Natta were awarded a Nobel Prize for chemistry in 1963 for their work on polymerisation. Part of this work involved the polymerisation of propene into poly(propene).
i What type of polymerisation forms poly(propene)? (1)
ii Draw a section of poly(propene) to show *two* repeat units. (1)
iii State two difficulties in the disposal of poly(propene). (2)

OCR, AS, 2812, Jan 2001

6 a Describe the differences between addition polymerisation and condensation polymerisation.
For an example of each type of polymerisation:
i give the structural formulae of the starting material(s)
ii draw a section of the polymer chain, showing at least one repeat unit. (8)
b The commercial plastic ABS is rigid and tough. It is used to make suitcases and car body panels. It is made from three monomers, known in industry as Acrylonitrile, Butadiene and Styrene.

$CH_2{=}CH{-}CN$ Acrylonitrile

$CH_2{=}CH-CH{=}CH_2$ Butadiene

$C_6H_5CH{=}CH_2$ Styrene

i Assuming that the monomers join together in a 1:1:1 ratio, suggest a repeat unit for ABS.
ii What type of polymerisation is this an example of?
iii The repeat unit still contains a double bond. Suggest why this could help make ABS the rigid plastic it is. (4)

OCR, A, 9254, Nov 1999

Chapter 25

1 a Give the structural formula of 2-bromo-3-methyl butane. (1)

  b Write an equation for the reaction between 2-bromo-3-methylbutane and dilute aqueous sodium hydroxide. Name the type of reaction taking place and outline a mechanism. (4)

  c Two isomeric alkenes are formed when 2-bromo-3-methylbutane reacts with ethanolic potassium hydroxide. Name the type of reaction occurring and state the role of the reagent. Give the structural formulae of the two alkenes. (4)

AQA, AS, Unit 3(a), 2001/2

2 a Ethene reacts with bromine as follows:

$$C_2H_4 + Br_2 \rightarrow C_2H_4Br_2$$

    i State the conditions necessary for ethene to react with bromine. (1)

    ii Give the name of the product. (1)

    iii Using the bond enthalpy data below, calculate the enthalpy change for the reaction.

Bond	Bond enthalpy/kJ mol^{-1}
C—C	+348
C=C	+612
C—H	+412
C—Br	+276
Br—Br	+193

(3)

    iv The enthalpy change for this reaction found using the enthalpies of formation of ethene and the product is $-90\,kJ\,mol^{-1}$. Suggest which value is more likely to be accurate and explain your answer. (2)

  b Give the mechanism for this reaction. (3)

  c The product, $C_2H_4Br_2$, is a typical bromoalkane. Suggest the structural formulae of each of the products of the reaction of $C_2H_4Br_2$ with the reagents given below and identify the *type* of reaction involved.

    i Aqueous sodium hydroxide. (2)

    ii Sodium hydroxide in ethanol (heated under reflux). (2)

  d Suggest, giving the reagents and conditions, how compound A could be converted in two steps into compound B. (4)

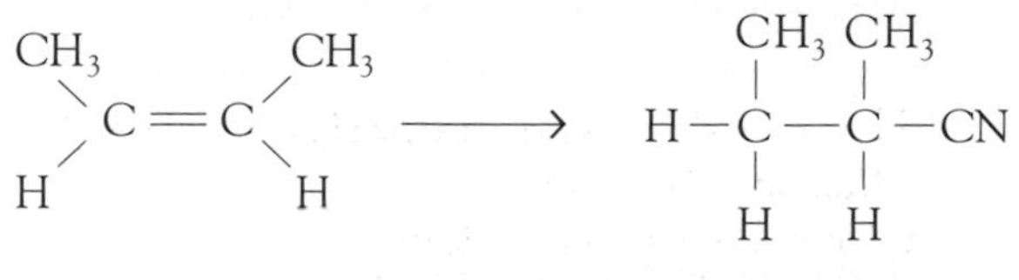

A   B

  e Molecules of B are chiral.

    i Explain the term *chiral molecule*. (1)

    ii Draw the optical isomers of B. (2)

Edexcel, A, Module 2, June 2001

3 The halogenation of hydrocarbons leads to the formation of compounds which have a wide variety of uses.

  a Chloroform (trichloromethane) was used as an early anaesthetic. It may be formed, under suitable conditions, by the chlorination of methane.

    i State *one* reaction condition necessary for the chlorination of methane. (1)

    ii Write *one* overall equation for the formation of chloroform (trichloromethane) from methane and chlorine. (2)

    iii The monochlorination of ethane (C_2H_6) has a mechanism similar to that of methane. Suggest steps (under the headings below) for the mechanism of this reaction using the formula C_2H_6.

    Initiation
    Propagation
    Termination (4)

CCEA, A, Module 1, Feb 2001

4 Halothane is used as an anaesthetic and has the structure shown below. The molecule is chiral.

$$\begin{array}{ccc} & F & H \\ & | & | \\ F - & C - & C - Cl \\ & | & | \\ & F & Br \end{array}$$

  a i Copy the structure and mark the asymmetric (chiral) centre in the molecule with an asterisk (*). (1)

    ii Explain the term *chiral*. (1)

  b Chloroethane was used as an anaesthetic before it was found to be toxic.

    i Write the equation for the reaction between chloroethane and hydroxide ions in ethanol. (2)

    ii What type of reaction is this? (1)

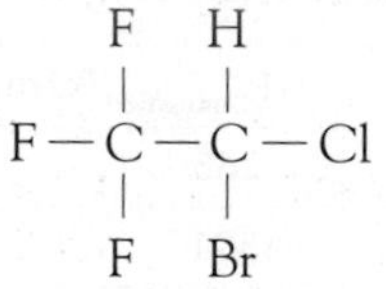

CCEA, A, Module 2, June 2001

5 Some reactions of 1-bromobutane, $CH_3CH_2CH_2CH_2Br$, are shown below.

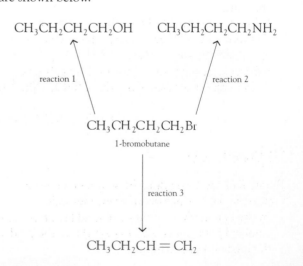

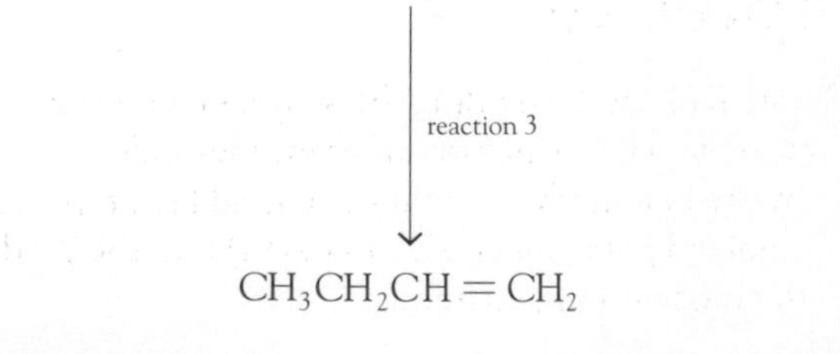

a For each of the reactions 1, 2 and 3, name the reagent and the solvent used. (6)

b Under the same conditions, 1-chlorobutane was used in reaction 1 in place of 1-bromobutane.

What difference (if any) would you expect in the rate of reaction? Explain your answer. (2)

c Dichlorodifluoromethane, CCl_2F_2, is an example of a chlorofluorocarbon, CFC, that was commonly used as a propellant in aerosols. Nowadays CFCs have limited use because of the damage caused to the ozone layer.

i Draw a diagram to show the shape of a molecule of CCl_2F_2. (1)

ii Predict an approximate value for the bond angles in a molecule of CCl_2F_2. (1)

iii Suggest a property that makes CCl_2F_2 suitable as a propellant in an aerosol. (1)

iv When CFCs are exposed to strong ultraviolet radiation in the upper atmosphere, homolytic fission takes place to produce free radicals.

Explain what is meant by the term *homolytic fission*. (1)

v Suggest which bond is most likely to be broken when CCl_2F_2 is exposed to ultraviolet light. Explain your answer. (2)

vi Identify *two* free radicals most likely to be formed when CCl_2F_2 is exposed to ultraviolet light. (2)

OCR, AS, 2812, June 2001

6 a Describe the mechanism of the reaction of chloroethane with ammonia. Give the name of the type of reaction undergone and the structural formula of the organic product. (3)

b State and explain how the rate of this reaction changes when chloroethane is replaced by bromoethane or iodoethane. (2)

c Draw structural formulae of compounds P, Q and R in the following scheme.

$$CH_3CH_2CH_2Br \xrightarrow{NaCN} P \xrightarrow{H_3O^+/heat}$$

$$Q \xrightarrow{C_2H_5OH/H^+/heat} R,\ C_6H_{12}O_2$$

(3)

d When hydrogen chloride is eliminated from 2-chlorobutane, three isomeric alkenes with the formula C_4H_8 are produced.

i Suggest reagents and conditions for this reaction.

ii Draw the structures of the three butenes produced. (4)

OCR, S, 9254, Nov 1999

Chapter 26

1 Butan-1-ol can be oxidised by acidified potassium dichromate(VI) using two different methods.

a In the first method, butan-1-ol is added dropwise to acidified potassium dichromate(VI) and the product is distilled off immediately.

i Using the symbol [O] for the oxidising agent, write an equation for this oxidation of butan-1-ol, showing clearly the structure of the product.

State what colour change you would observe.

ii Butan-1-ol and butan-2-ol give different products on oxidation by this first method. By stating a reagent and the observation with each compound, give a simple test to distinguish between these two oxidation products. (6)

b In a second method, the mixture of butan-1-ol and acidified potassium dichromate(VI) is heated under reflux. Identify the product which is obtained by this reaction. (1)

c Give the structures and names of two branched chain alcohols which are both isomers of butan-1-ol. Only isomer 1 is oxidised when warmed with acidified potassium dichromate(VI). (4)

AQA, AS, Unit 3(a), 2001/2

2 a i Draw the structural formula of the tertiary alcohol $C_5H_{12}O$. (1)

ii For this tertiary alcohol suggest reagents and conditions which would enable its preparation via a reaction involving a Grignard reagent.
(You are not expected to describe how to prepare a Grignard reagent.) (3)

b i Draw the structural formula of the secondary alcohol, $C_5H_{12}O$, which does *not* exist as optical isomers. (1)

ii X is obtained by oxidising this secondary alcohol with potassium dichromate(VI) acidified with dilute sulphuric acid.

Draw the structural formula of X. (1)

iii To which class of organic compounds does X belong? (1)

iv Describe a test you would do on X, and the results you would expect, to show that your classification was correct. (2)

v X does not give a yellow precipitate when treated with iodine in the presence of sodium hydroxide solution. Explain why not. (1)

Edexcel, A, Module 4, June 2001

3 One of the isomers of $C_4H_{10}O$ is the alcohol 2-methylpropan-2-ol which has the structural formula

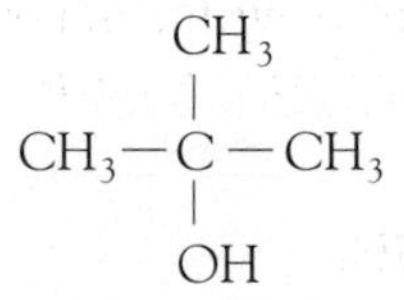

a There are three other *structural* isomers of $C_4H_{10}O$ which are also alcohols.

i Draw their structural formulae. (3)

ii One of these isomers exhibits stereoisomerism. Name the type of isomerism shown and draw diagrams showing clearly how these stereoisomers differ from one another. (3)

iii Describe a test to show that each of the isomers in **i** contains an OH group. (2)

b Draw the structural formula of the final organic product of the reaction when each of the three alcohols in **a i** is heated under reflux with a solution of potassium dichromate(VI) in dilute sulphuric acid. (3)

c 2-methylpropan-2-ol can be prepared by the reaction of 2-bromo-2-methylpropane with dilute aqueous potassium hydroxide.

i Give the mechanism for this reaction. (3)

ii If a concentrated solution of potassium hydroxide in ethanol is used instead of dilute aqueous potassium hydroxide, a different organic product is obtained.

Draw the structural formula of this product. (1)

Edexcel, A, Module 2, Jan 2001

4 a Primary alcohols can be oxidised to aldehydes. The following are the instructions for the oxidation of ethanol (boiling temperature 78 °C) to ethanal (boiling temperature 21 °C).

- Place 50 cm^3 of water in a round bottom flask and add, carefully with stirring, 17 cm^3 of concentrated sulphuric acid. Arrange the flask, with a tap funnel attached, in a distillation apparatus.
- Dissolve 50 g of sodium dichromate(VI) in 50 cm^3 of water and add 40 cm^3 of ethanol, mix and place in the tap funnel.
- Heat the dilute acid in the flask until it is boiling and then remove the flame.
- Slowly run the solution of sodium dichromate(VI) into the flask. A vigorous reaction takes place.
- Collect the mixture of ethanal and water, which distills off, in a conical flask, which is surrounded by iced water.

i Draw a diagram of the apparatus which could be used for this reaction. (3)

ii Explain why the ethanol and the oxidising agent are not initially placed in the round-bottomed flask with the sulphuric acid. (1)

iii Explain why the heat is removed before the solution of ethanol and sodium dichromate(VI) is added. (1)

iv Suggest why the flask in which the product is collected is surrounded by iced water. (1)

b Consider the following reaction scheme.

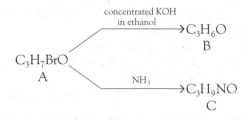

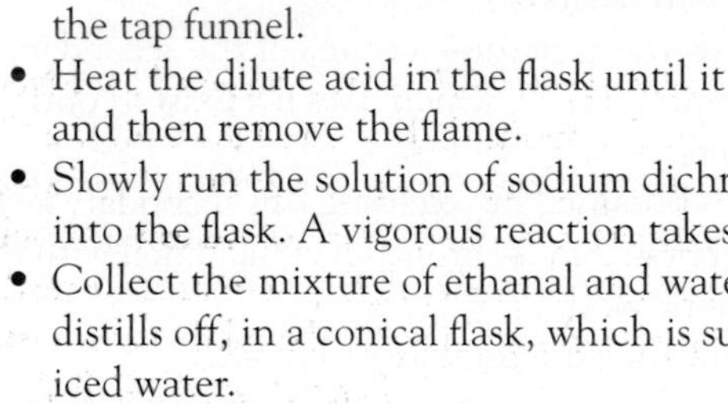

- A produces steamy fumes when PCl$_5$ is added.
- A can be oxidised to an aldehyde.
- A is chiral.
- B will decolourise bromine water.

i Identify the functional groups present in A. (2)

ii Draw the structural formulae of A, B and C. (3)

iii State the conditions needed for the conversion of A into C. (2)

c B can be polymerised.

i Suggest a structural formula for this polymer, showing clearly at least one repeating unit. (2)

ii State the type of polymerisation which B undergoes. (1)

d C reacts with ethanoyl chloride to form a solid product. Describe how an impure sample of this solid could be purified by recrystallisation. (4)

Edexcel, A, Module 4, Jan 2001

5 The original breathalysers used to detect alcohol (ethanol) vapour in exhaled breath contained acidified potassium dichromate(VI).

a State the colour change noted for a positive result. (2)

b Name the major organ in the body which breaks down alcohol and may be damaged by an excess. (1)

CCEA, AS, AS2 Module 2, June 2001

6 a Predict the structural formula of the organic product from the reaction of 1-bromopropane, $CH_3CH_2CH_2Br$, with:

i aqueous potassium cyanide solution (1)

ii ammonia gas. (1)

b Give details of a chemical test you could do to distinguish between 2-chlorobutane and butan-2-ol, including the expected observations with each compound. (2)

c i Draw the full structural formula showing all the bonds for the isomer of butan-2-ol that is a tertiary alcohol. (1)

ii Give details of a chemical test you could do to distinguish between butan-2-ol and its isomer drawn in **i** and the observations you would expect to make. (4)

iii Explain the chemistry involved in the test you described in part **ii**. (2)

Edexcel, AS, Unit 2, June 2001

7 Compounds J and K contribute to the 'leafy' odour of violet oil.

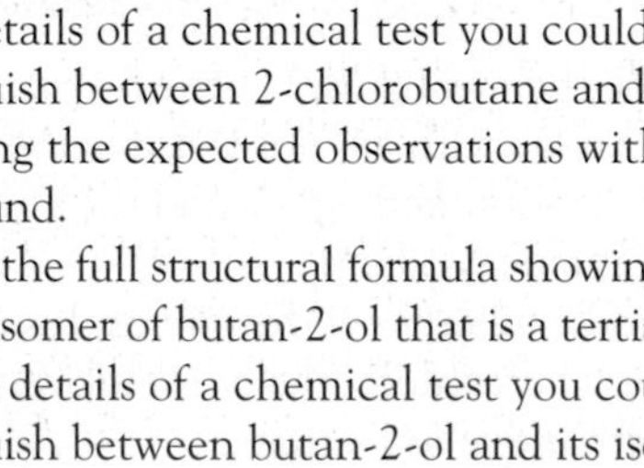

J

a Name the functional groups present in compound J. (2)

b What is the molecular formula of compound J? (1)

c Draw the structure of the organic product formed by the reaction of compound J with

i Br_2

ii CH_3COOH in the presence of an acid catalyst. (2)

d A chemist reacted compound J with HBr. He separated 2 structural isomers K and L with the molecular formula $C_5H_{10}Br_2$. Draw structures for K and L. (2)

e Compound M below can be prepared from compound J.

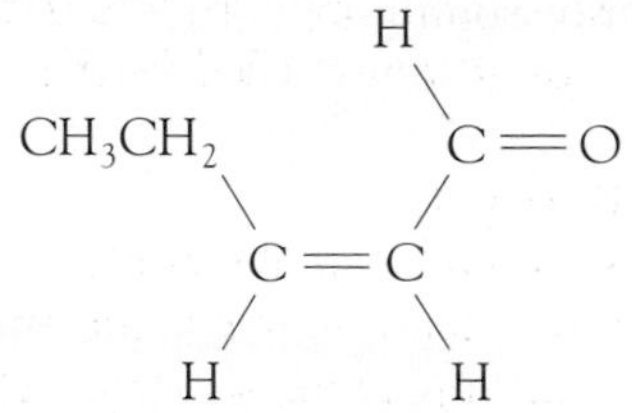

M

i Suggest reagent(s) for the conversion of J into M.
ii Draw the structure of a possible organic impurity (other than J) which might contaminate the product. Explain your choice. (3)

OCR, AS, 2812, 2001/2

8 There are two different industrial methods for the production of ethanol. Describe and explain, with the aid of equations, the industrial production of ethanol from glucose and from ethene.

(In this question, 1 mark is available for the quality of written communication.)

OCR, AS, 2812, Jan 2001

Chapter 27

1 Benzene, C_6H_6, reacts with ethanoyl chloride, CH_3COCl, by an electrophilic substitution reaction in the presence of aluminium chloride as a catalyst.

a Identify the electrophile involved in this reaction and write an equation to show its formation. (2)

b Draw the mechanism for the electrophilic substitution of benzene by ethanoyl chloride. (3)

c Suggest a reaction scheme, stating reagents and conditions, to convert the product of the above reaction into

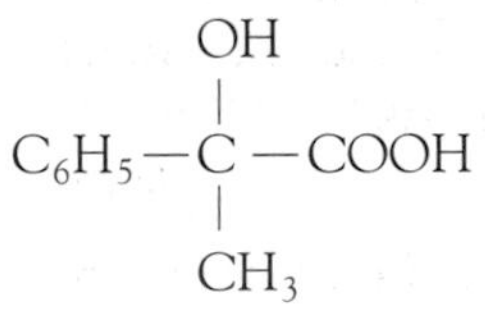

(5)

Edexcel, A, Module 6, June 2001

2 a Define:

i the standard enthalpy of formation of benzene, $C_6H_6(l)$ (2)
ii the standard enthalpy of combustion of benzene, $C_6H_6(l)$. (2)

b Calculate the standard enthalpy of formation of benzene, $C_6H_6(l)$, using the following enthalpy of combustion data:

Substance	$\Delta H^\ominus_c$/kJ mol^{-1}
$C_6H_6(l)$	−3273
$H_2(g)$	−286
C(s)	−394

(3)

c If the standard enthalpy of formation is calculated from average bond enthalpy data assuming that benzene has three C=C and three C—C bonds, its value is found to be +215 kJ mol^{-1}.

Explain, with reference to the structure and stability of benzene, why this value differs from that calculated in **b**. Use an enthalpy level diagram to illustrate your answer. (4)

d Benzene reacts with bromine when gently warmed in the presence of a catalyst of anhydrous iron(III) bromide.

i The reaction is first order with respect to benzene and first order with respect to bromine. Write the rate equation for the reaction. (1)

ii The mechanism of this reaction involves an attack by Br^+ followed by loss of H^+.

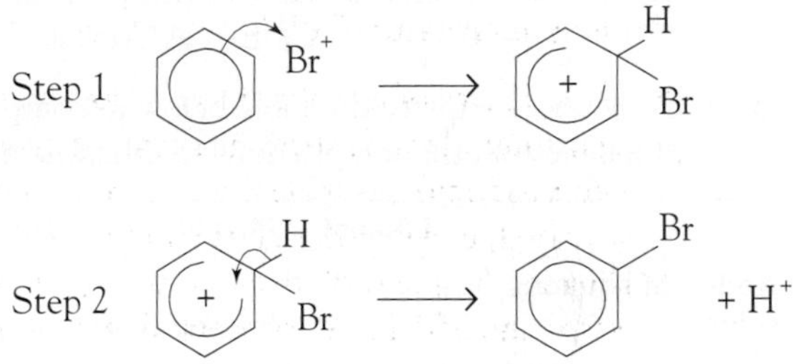

Deuterium, symbol D, is an isotope of hydrogen, and the C—D bond is slightly stronger than the C—H bond. If step 2 were the rate-determining (slower) step, suggest how the rate of this reaction would alter if deuterated benzene, C_6D_6, were used instead of ordinary benzene, C_6H_6, and explain your answer. (2)

Edexcel, A, Module 2, Jan 2001

3 Nitration of the benzene ring may be safely demonstrated by reacting methyl benzoate with a nitrating mixture to form methyl 3-nitrobenzoate.

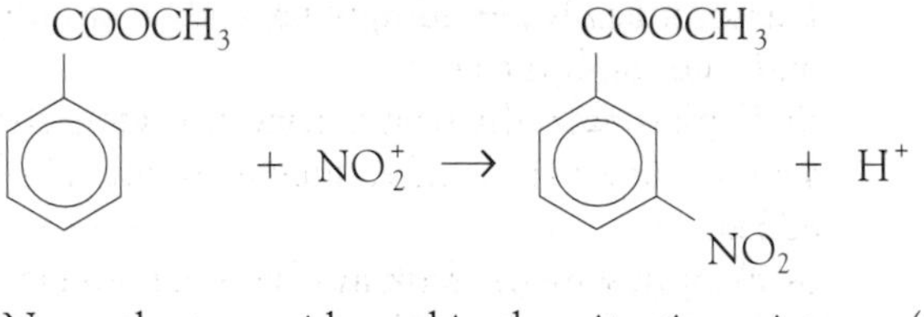

a Name the two acids used in the nitrating mixture. (2)

b Describe how the product may be purified. (3)

c A student reacted 9.52 g of methyl benzoate with excess nitrating mixture and obtained 8.05 g of pure methyl 3-nitrobenzoate.

Calculate the percentage yield. (3)

d Use a flow scheme to show the mechanism for the nitration of benzene. (3)

CCEA, A, Module 1, June 2001

4 Phenol is an antiseptic although its use is limited by the toxicity of its vapour and the corrosive nature of the solid.

a The major source of phenol is the petrochemical industry in which benzene is converted to phenol via cumene.

i Write an equation for the formation of cumene from benzene, indicating the conditions used. (4)
ii The cumene is oxidised by air. Draw the structure of the product formed. (1)
iii Further reaction of this product with sulphuric acid produces phenol and propanone. State how these products are separated. (1)

b A substituted phenol, which is used as an antiseptic, is shown below:

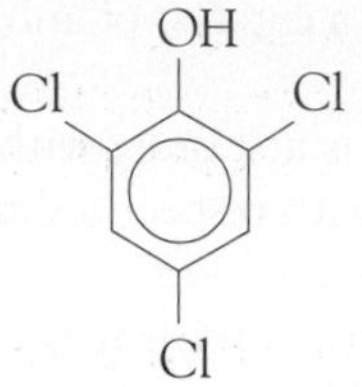

Suggest the systematic name of this phenol. (2)

CCEA, A, Module 3, Feb 2001

5 a Copy and complete the table below to compare the properties of ethanol, phenol and ethanoic acid.

	Ethanol	Phenol	Ethanoic acid
Acidity of aqueous solution		Weakly acidic	
Reaction of aqueous solution with sodium hydrogencarbonate		No reaction	
Reaction with ethanoyl chloride			Formation of ethanoic anhydride

(3)

b i Both propan-1-ol and propan-2-ol are oxidised by an aqueous mixture of sodium dichromate(VI) and sulphuric acid.

Depending upon the reaction conditions a total of three products, containing three carbon atoms per molecule, is possible.

Draw the full structural (graphic) formulae for these *three* products. (3)

ii State how the *two isomeric products* in **b i** may be distinguished by a *chemical* method. (2)

iii Describe *one* test-tube reaction to distinguish between propan-1-ol and propan-2-ol, other than by oxidation by dichromate(VI) ions in aqueous sulphuric acid. Your answer should give the names of the reagent(s) and the *observations*, if any, with *each* compound. (2)

WJEC, A, C3, June 2000

6 This question is about the structure and reactivities of aromatic compounds.

a In general, aromatic compounds are less reactive than alkenes. Describe the structure of benzene and explain why it is resistant to addition reactions. (4)

b Cyclic molecules show aromatic character if the number of delocalised electrons is equal to $4n + 2$ where n is a whole number.

i Deduce which of the compounds A–C are aromatic, showing your reasoning. (3)

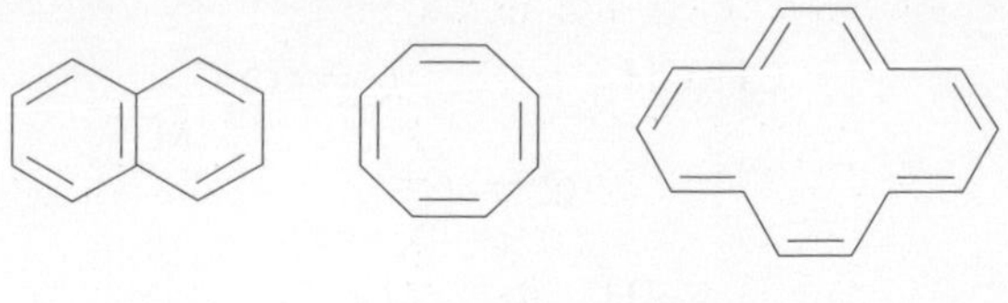

A B C

ii Suggest why the molecule D is considered to show aromatic character. (3)

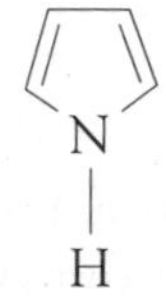

D

c Explain what is meant by the term *electrophilic substitution* when applied to aromatic molecules. Use the following reactions to illustrate your answer and give a mechanism for one of them:
i the reaction of benzene with a mixture of concentrated nitric and sulphuric acids;
ii the reaction of benzene with bromine in the presence of iron filings. (5)

d i When nitrobenzene is prepared from benzene there is only a small contamination by disubstituted products. Draw structures to show all the possible disubstituted products. (1)
ii In practice, on nitrating benzene, only a small amount of one disubstituted product is formed. Suggest an explanation for this observation. (4)

OCR, A, 9485 (STEP), 2001

Chapter 28

1 a Consider the following reaction scheme.

$$CH_3CH_2OH \xrightarrow{\text{step 2}} CH_3CH_2Br \xrightarrow{\text{step 3}} CH_3CH_2CN$$

$$\uparrow \text{step 1}$$

$$\underset{\text{ethanal}}{CH_3CHO}$$

$$\downarrow \text{step 4}$$

$$CH_3CH(OH)CN$$

i State the reagents and conditions for steps 1, 2 and 3. (7)
ii Give the mechanism in step 4 which is the reaction between ethanal and hydrogen cyanide. (3)
iii What type of mechanism is this? (1)
iv State and explain the conditions necessary for step 4. (2)

b The nitrile group, —C≡N, can also be introduced into a molecule by dehydration of an amide.

Outline a reaction scheme, giving names or formulae for the reagents, for the preparation of ethanonitrile, CH_3CN, from ethanal, CH_3CHO. (7)

Edexcel, A, Module 4, Jan 2001

2 Consider the following two reactions of cinnamaldehyde, one of the main components of the food flavouring, cinnamon.

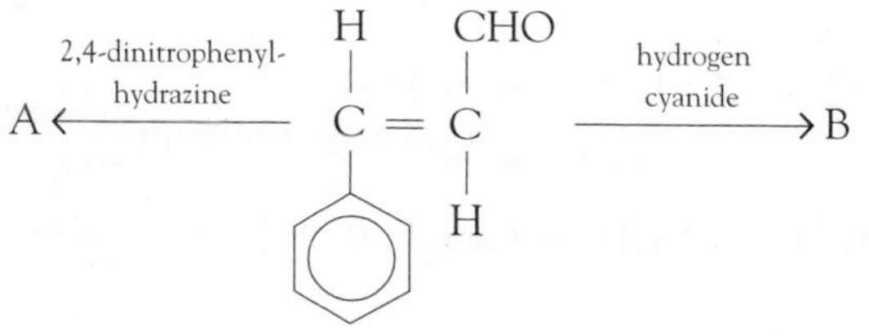

a i Draw the structure of A and state its colour. (3)
ii Write an equation for the formation of B, showing its structure. (2)
iii Name the mechanism used to describe the reaction between hydrogen cyanide and cinnamaldehyde. (1)
b Organic acids give the sharp flavour to many fruits. They can be neutralised by alkalis such as sodium hydroxide.
i Copy the axes below and draw the graph you would expect if $25.0\,cm^3$ of a $0.1\,mol\,dm^{-3}$ monobasic acid solution (pH = 3.8) is titrated against $0.1\,mol\,dm^{-3}$ sodium hydroxide solution.

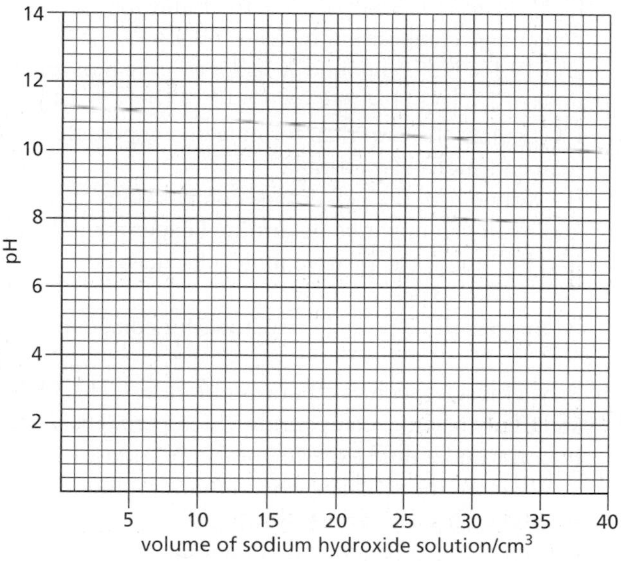

(3)
ii Name a suitable indicator for this titration. (1)
iii What colour change would be observed at the end point? (2)

CCEA, A, Module 2, Feb 2001

3 a Describe the mechanism of the reaction between ethanal and hydrogen cyanide. Name the type of reaction undergone, state any other reagents needed, and include the structural formula of any intermediate, as well as of the product. (4)
b The above reaction produces two isomeric products in equal amounts. Name the type of isomerism involved, and draw displayed formulae to illustrate it. (3)
c As a consequence of starvation or diabetes, the blood plasma and urine of patients can contain large amounts of 'ketone bodies'. These include propanone, 3-oxobutanoic acid, and 3-hydroxybutanoic acid.

CH_3COCH_3
propanone

$CH_3COCH_2CO_2H$
3-oxobutanoic acid

$CH_3CH(OH)CH_2CO_2H$
3-hydroxybutanoic acid

Describe a separate simple chemical test in each case to distinguish 3-oxobutanoic acid from
i propanone
ii 3-hydroxybutanoic acid.
For each test, give reagents and conditions, and state what would be seen with each compound. (5)

OCR, A, 9254, June 1999

4 This question is about addition reactions to alkenes and simple carbonyl compounds (aldehydes and ketones).
a Outline the mechanism of the reaction between but-2-ene and hydrogen bromide, showing clearly the structure of the product. Use your mechanism to suggest how the rate of reaction might vary with an increase in solvent polarity. (4)
b Alkenes will rapidly decolourise aqueous bromine solutions. The mechanism for this reaction is similar to that with hydrogen bromide. When hex-1-ene is reacted with bromine dissolved in aqueous sodium chloride, a mixture of products is isolated. Suggest a mechanistic explanation for the reactions and the structures of possible products. (4)
c i Outline the mechanism of the reaction between benzaldehyde, C_6H_5CHO, and hydrogen cyanide, HCN, showing the structure of the product. (3)
ii The product from i exhibits stereoisomerism. Draw the structures of the isomers. (2)
d Sodium tetrahydridoborate, $NaBH_4$, acts as a source of H^-. The Grignard reagent, CH_3CH_2MgBr, acts as a source of $CH_3CH_2^-$.
i For the reactions shown below, suggest structures for products, A and B, giving a clear explanation for their formation.

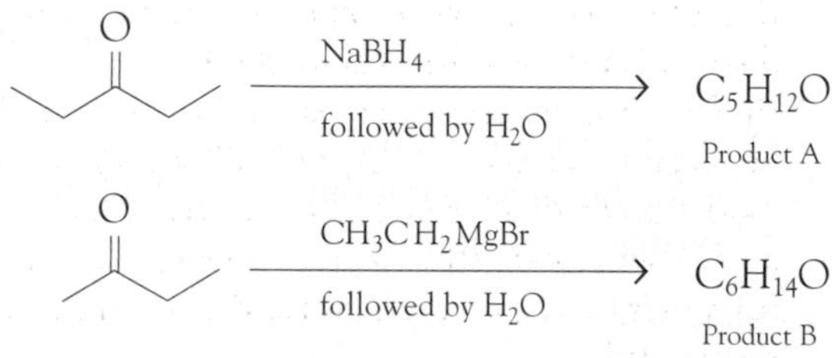

(5)
ii Suggest why CH_3CH_2MgBr acts in this manner. (2)

OCR, A, 9485 (STEP), 1999

Chapter 29

1 **a** Write an equation for the formation of ethyl ethanoate from ethanoyl chloride and ethanol. Name and outline the mechanism for the reaction taking place. (6)

b Explain why dilute sodium hydroxide will cause holes to appear in clothing made from polymers such as Terylene but a poly(phenylethene) container can be used to store sodium hydroxide. (2)

AQA, A, Specimen Unit 4, 2001/2

2 **a** State the reagent(s) used for the conversion of ethanol to:

i iodoethane (1)

ii ethene. (1)

b Suggest a series of reactions by which ethanol can be converted to 2-hydroxypropanoic acid, $CH_3CH(OH)COOH$. For each reaction specify the reagents and conditions necessary. (6)

c Explain whether a solution of 2-hydroxypropanoic acid, *made in this way*, would have any effect on a beam of plane polarised monochromatic light. (2)

d 2-hydroxypropanoic acid reacts with lithium tetrahydridoaluminate(III) (lithium aluminium hydride).

State the conditions necessary for this reaction and give the structural formula of the organic product. (2)

Edexcel, A, Module 4, June 2000

3 Scientists have recently investigated an ester called ethyl oleate, the structure of which is shown below.

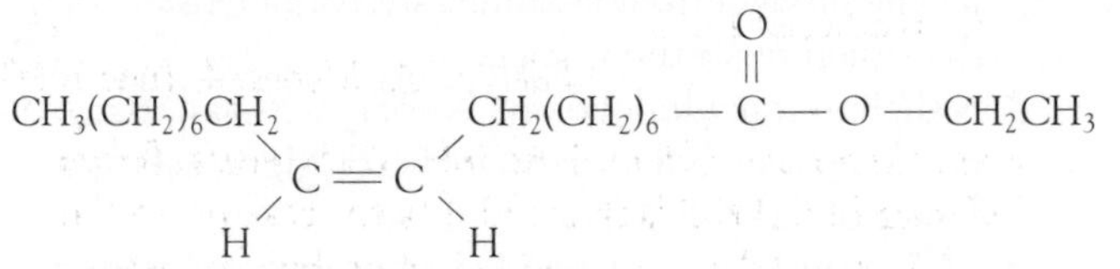

This ester may affect the release of neurotransmitters in the brain.

It is possible that the body could make this ester from the ethanol present in alcoholic drinks and the natural fatty acid, oleic acid.

a Write the structural formula of oleic acid. (1)

b **i** Write the molecular formula of each of the following:

oleic acid
ethanol
ethyl.

ii Construct a balanced equation for the formation of ethyl oleate from ethanol and oleic acid. (4)

c Suggest how oleic acid could be obtained from the glyceryl ester shown below.

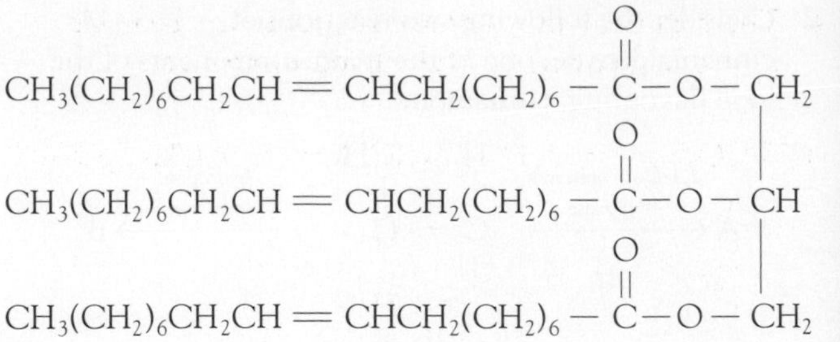

(1)

OCR, A, 4821, June 1999

4 Esters are derivatives of carboxylic acids and occur widely in nature.

a Draw the structure of the ester formed when ethanol reacts with ethanoic acid. (2)

b Name an alternative reagent which may be used in place of ethanoic acid to produce the ester. (1)

CCEA, AS, AS2 Module 2, June 2001

5 This question is about the reactions of carboxylic acids and their derivatives.

a A carboxylic acid derivative X was found to contain C, H, N and O. Analysis gave the following percentage composition by mass: 49.4% C, 9.6% H and 19.1% N. Compound X had a relative molecular mass of 73. [A_r: H = 1; C = 12; O = 16; N = 14.]

i Calculate the empirical formula of compound X and hence its molecular formula. (3)

ii Outline how the percentage compositions of carbon and hydrogen might be determined experimentally. (4)

iii How might the relative molecular mass be determined experimentally? (1)

iv Suggest three possible structures for compound X. (3)

b Acyl chlorides such as ethanoyl chloride, CH_3COCl, are highly reactive compounds which permit reactions to occur under mild experimental conditions.

What are the products when ethanoyl chloride is reacted with each of the following?

i water **ii** 2-propanol **iii** ammonia
iv hydrogen sulphide (4)

c R and S are two isomeric amides which can be hydrolysed by aqueous acid.

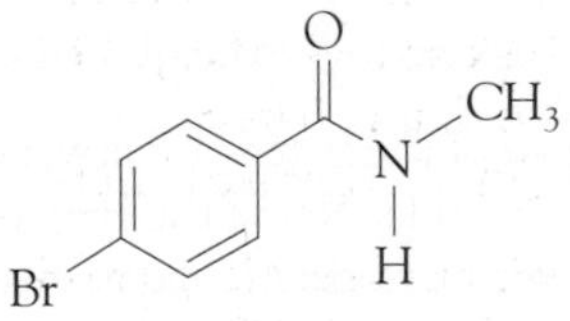

R

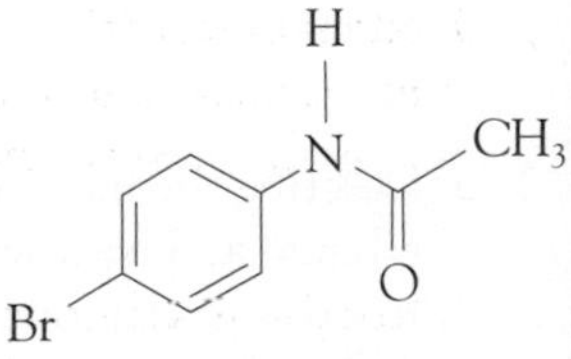

S

i Draw the structures of the products formed on hydrolysing R and S. (2)

ii The products formed by hydrolysing S are soluble in aqueous acid whereas a precipitate remains on hydrolysing R. Suggest an explanation for these observations. (3)

OCR, A, 9485 (STEP), 2001

6 Three chlorine-containing compounds can be made from ethanoic acid (A) by the following routes:

$$\underset{\text{A}}{CH_3CO_2H} \xrightarrow{\text{I}} \underset{\text{B}}{CH_2ClCO_2H} \longrightarrow \underset{\text{C}}{CHCl_2CO_2H}$$

$$\underset{\text{A}}{CH_3CO_2H} \xrightarrow{\text{II}} \underset{\text{D}}{CH_3COCl}$$

a Suggest reagents and conditions for reactions I and II. (2)

b When D reacts with ethanol, steamy fumes are evolved and a fruity-smelling liquid is produced.

Write a balanced equation for this reaction, drawing the structural formula of the organic product. Name the functional group that it contains. (3)

c When compounds A, B and D (not necessarily in that order) are added to separate portions of water, solutions are formed with pH values of 0.5, 2.5 and 3.0. When aqueous silver nitrate is added to these three solutions, two show no reaction but the third one produces a thick white precipitate.

Suggest, with explanations, which pH value is associated with each of A, B and D. Explain the formation of the white precipitate. (5)

d Predict, with a reason, the likely pH value of an aqueous solution of compound C. (2)

OCR, A, 9254, Nov 2001

Chapter 30

1 a Explain why ethylamine is a Brønsted–Lowry base. (2)

b Why is phenylamine a weaker base than ethylamine? (2)

c Ethylamine can be prepared from the reaction between bromoethane and ammonia

i Name the type of reaction taking place.

ii Give the structures of *three* other organic substitution products which can be obtained from the reaction between bromoethane and ammonia. (4)

d Write an equation for the conversion of ethanenitrile into ethylamine and give one reason why this method of synthesis is superior to that in part c. (2)

AQA, A, Specimen Unit 4, 2001/2

2 Adrenalin is a hormone which raises blood pressure, increases the depth of breathing and delays fatigue in muscles, thus allowing people to show great strength under stress.

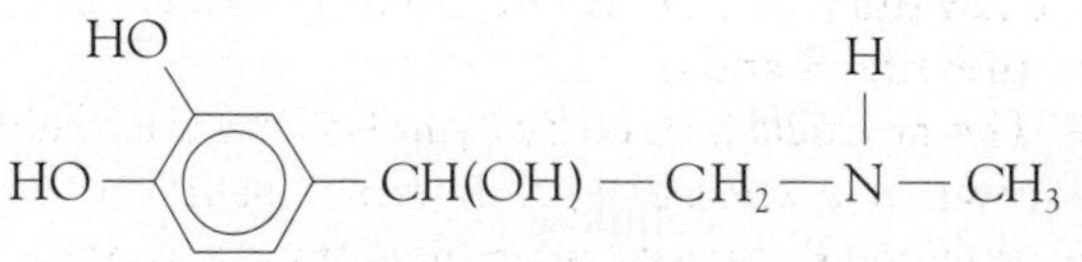

Adrenalin

Benzedrine is a pharmaceutical which stimulates the central nervous system in a similar manner to adrenalin.

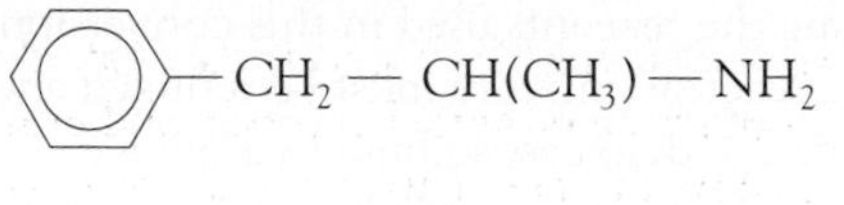

Benzedrine

a i Copy the structure for benzedrine and mark with a * any asymmetric carbon atom that causes chirality. (1)

ii Suggest why adrenalin is more soluble in water than is benzedrine. (2)

b Give the structural formulae of the organic products obtained when benzedrine reacts with:

i an aqueous acid such as dilute hydrochloric acid (1)

ii ethanoyl chloride in the absence of a catalyst (1)

iii excess ethanoyl chloride in the pressence of the catalyst anhydrous aluminium chloride. (2)

c State the two reagents needed to convert

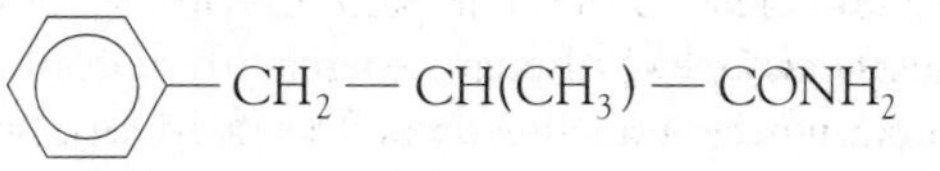

into benzedrine. (2)

d It is possible to eliminate a molecule of water from adrenalin which for the purpose of this question may be represented as R—CH(OH)—CH_2—NH—CH_3. Draw the structural formulae of the two stereoisomers produced. (2)

e The mass spectra of both benzedrine and adrenalin have a peak at a mass/charge ratio of 44. Draw the structure of the species which give these peaks:

i in benzedrine (1)

ii in adrenalin. (1)

Edexcel, A, Module 4, June 2000

3 Phenylketonuria (PKU) is an inherited disease. Babies suffering from PKU lack the ability to break down the surplus phenylalanine present in their diet. Excess phenylalanine is toxic and may cause mental retardation. Phenylalanine is an α-amino acid with the structural formula:

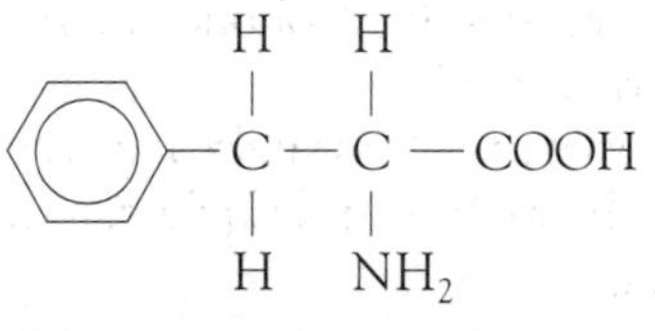

a i Explain the term *α-amino acid.* (2)

ii Draw the structural formula of the dipolar ion present in an aqueous solution of phenylalanine. (1)

iii Phenylalanine acts as a buffer. Explain why the addition of small quantities of acid to an aqueous solution of phenylalanine causes little change in the pH. (2)

b Alanine reacts with phenylalanine to form two different dipeptides. Draw the structural formula for *one* of these dipeptides and circle the peptide link. (3)

CCEA, A, Module 3, Feb 2001

4 Phenylamine is prepared in the laboratory from nitrobenzene by reduction.

a Name the reagents used in this conversion. (2)

b The reaction mixture must be refluxed and then treated with excess sodium hydroxide.

i Explain what is meant by the term *reflux*. (2)

ii With the aid of an equation explain the function of the sodium hydroxide. (2)

c Phenylamine may be obtained from the final mixture by steam distillation followed by extraction with ether, a volatile organic solvent.

i If 75 cm^3 of aqueous distillate containing 5.5 g of phenylamine is shaken with 100 cm^3 of ether, 5.12 g is extracted into the organic layer. Calculate the partition coefficient, K_d, of phenylamine between ether and water. (3)

ii Explain how 100 cm^3 of ether could be used to extract a higher proportion of phenylamine from the aqueous layer. (1)

d Phenylamine, like phenol, reacts with diazonium compounds forming azo-dyes. The dye Disperse Orange is prepared in a similar way to other azo-dyes.

i The first step involves conversion of 4-nitrophenylamine to a diazonium salt as shown:

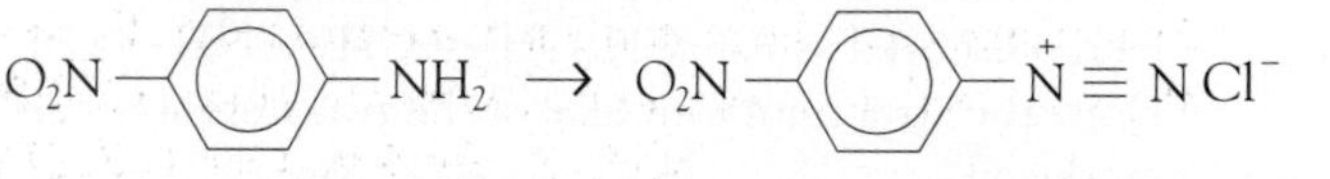

Name the reagent(s) required for this step. (1)

ii Reaction of this diazonium salt with phenylamine produces Disperse Orange. Suggest the structure of Disperse Orange and state the type of reaction involved. (3)

iii State *one* other use of azo-compounds apart from making azo dyes. (1)

CCEA, A, Module 3, Feb 2001

5 a Diazo dyes can be made by coupling aromatic amines with other compounds. Study the following scheme showing the synthesis of dye C and answer the questions below it.

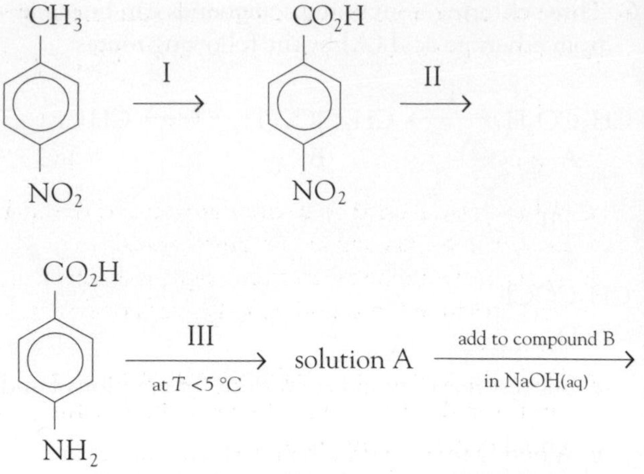

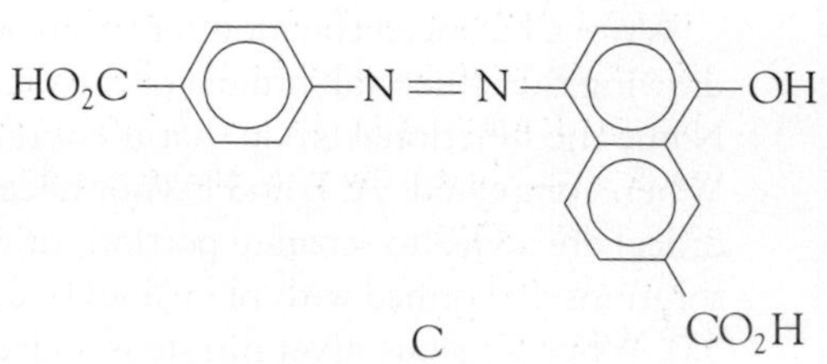

C

i Suggest reagents for steps I, II and III.

ii What types of reaction are steps I and II?

iii Draw the structural formula of compound B and of the compound in solution A. (7)

b Draw the displayed formula of the repeat unit of Terylene and state the functional group present. (2)

c How well a dye colours a material depends on how strongly the dye molecules are attracted to the molecules of the cloth. By describing the types of intermolecular interactions involved, suggest, with reasons, which dye molecule C (above) or D (below) is more strongly attracted to

i Terylene,

ii the cellulose in cotton.

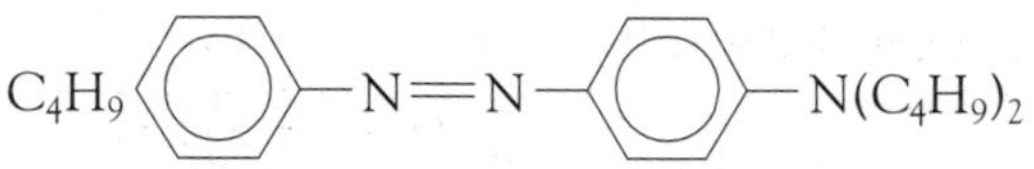

D

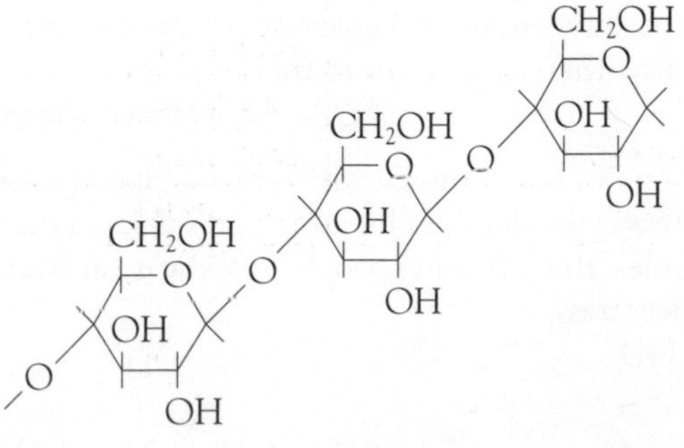

cellulose

(3)

OCR, S, 9254, Nov 2001

6 The following structure shows part of a protein molecule.

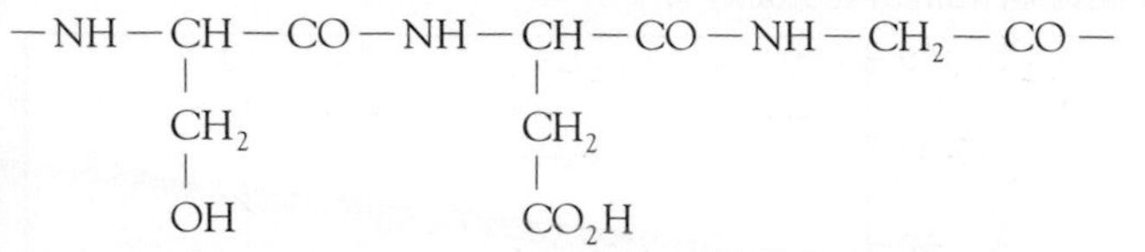

a What reagents and conditions are needed to break the protein into its constituent amino acids? (2)

b What is the name given to the type of reaction you have described in **a**, and what type of bond is broken during it? (2)

c Draw the structural formula of each of the three amino acids produced by the reaction, labelling any chiral centres in the molecules. (5)

d In solution, amino acids exist as *zwitterions*. Choose *one* of the amino acids you have drawn in **c** to illustrate what is meant by this term. (1)

e Amino acids act as *buffers* in solution. By means of equations, show how your chosen amino acid can act as a buffer when

i dilute HCl

ii dilute NaOH

is added to its solution. (2)

OCR, A, 9254, Nov 1999

Chapter 31

1 Compound A, $C_5H_{10}O$, reacts with $NaBH_4$ to give B, $C_5H_{12}O$. Treatment of B with concentrated sulphuric acid yields compound C, C_5H_{10}. Acid-catalysed hydration of C gives a mixture of isomers, B and D.

Fragmentation of the molecular ion of A, $[C_5H_{10}O]^+$, leads to a mass spectrum with a major peak at m/z 57. The infra-red spectrum of compound A has a strong band at $1715\,cm^{-1}$ and the infra-red spectrum of compound B has a broad absorption at $3350\,cm^{-1}$ (see table below). The proton n.m.r. spectrum of A has two signals at δ 1.06 (triplet) and 2.42 (quartet), respectively (see figure below).

Bond	Wavenumber/cm^{-1}
C—H	2850–3300
C—C	750–1100
C=C	1620–1680
C=O	1680–1750
C—O	1000–1300
O—H (alcohols)	3230–3550
O—H (acids)	2500–3000

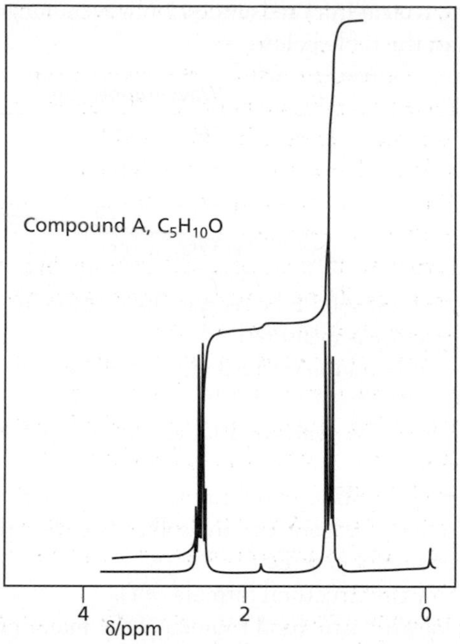

Use the analytical and chemical information provided to deduce structures for compounds A, B, C and D, respectively. Include in your answer an equation for the fragmentation of the molecular ion of A and account for the appearance of the proton n.m.r. spectrum of A. Explain why isomers B and D are formed from compound C. (20)

AQA, A, Specimen Unit 4, 2001/2

2 This question concerns the compounds linked by the reaction scheme

$$\underset{A}{C_4H_8O} \rightarrow \underset{B}{C_4H_{10}O} \rightarrow \underset{C}{C_4H_9Br} \rightarrow \underset{D}{C_4H_8}$$

A reacts with 2,4-dinitrophenylhydrazine to give a solid E which, when recrystallised, has a melting temperature of 126 °C. The melting temperatures of some 2,4-dinitrophenylhydrazine derivatives are listed below:

Compound	Melting temperature of 2,4-dinitrophenylhydrazine derivative/°C
Propanone	126
Butanone	116
Propanal	155
Methyl propanal	187
Butanal	126

The infra red spectrum of A has a peak at $1720\,cm^{-1}$, but none at about $3500\,cm^{-1}$ or $1650\,cm^{-1}$. The spectrum of B has a very broad peak at $3500\,cm^{-1}$, but none at about $1720\,cm^{-1}$ or $1650\,cm^{-1}$.

Some typical infra red absorption wavenumbers are shown in the table below:

Bond	Wavenumber/cm^{-1}
O—H	3600 to 3300
C—O	1200 to 1150
C=C	1680 to 1620
C=O	1750 to 1680

a **i** Why must solid E be recrystallised before its melting temperature is measured? (1)
ii What bond in A is responsible for the peak at 1720 cm^{-1}? (1)
iii Why is the peak at 3500 cm^{-1} in the spectrum of B very broad? (2)
iv Draw the structural formula of A and of B. (2)

b **i** Name the reagent and the solvent used for the conversion of C to D. (2)
ii Draw the structural formula of D. (1)
iii Draw the structural formula of the major product of the addition of HBr to D. (1)
iv Suggest why the reaction in **iii** does not produce C as the major product. (1)

Edexcel, A, Module 4, June 2000

3 This question concerns the three isomers A, B and C, each of which has a relative molecular mass of 134.

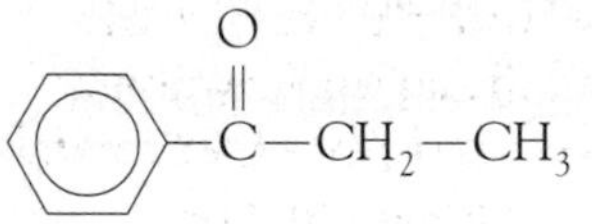

A

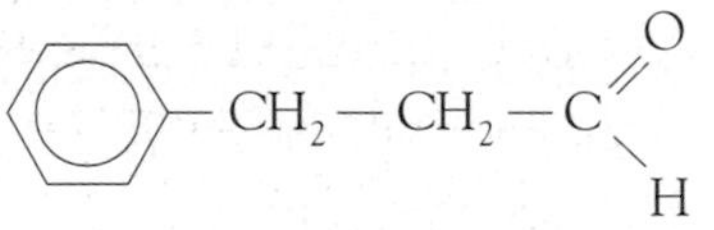

B

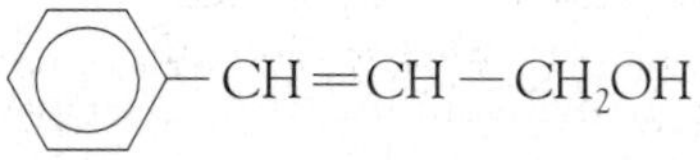

C

a The mass spectrum of substance A is shown below. Identify the species responsible for the peaks labelled 1, 2 and 3. (3)

b The infra-red spectra of two of these substances were also measured.
i Use the table below and the spectra below to identify which spectrum is that of substance C.
ii Give *one* reason for your choice. (1)
iii Give one *other* reason why the other spectrum could *not* be that of substance C. (1)

Bond	Wavenumber/ cm^{-1}	Bond	Wavenumber/ cm^{-1}
C—H (arenes)	3000–3100	O—H (hydrogen bonded)	3200–3570
C—H (alkanes)	2850–3000	O—H (not hydrogen bonded)	3580–3650
C=O	1680–1750	C=C (arenes)	1450–1600

mass spectrum of substance A

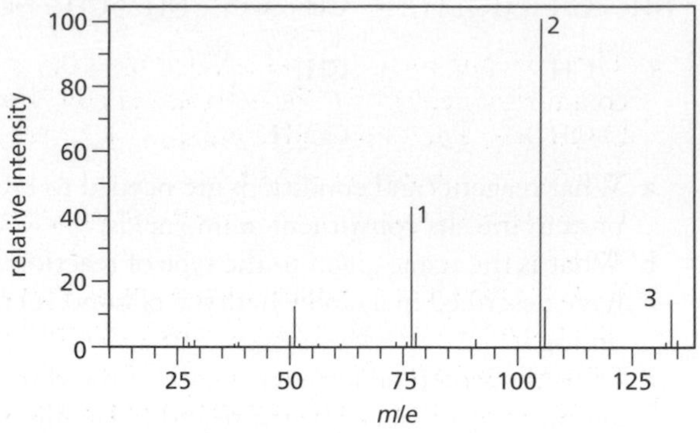

infra-red spectrum number 1

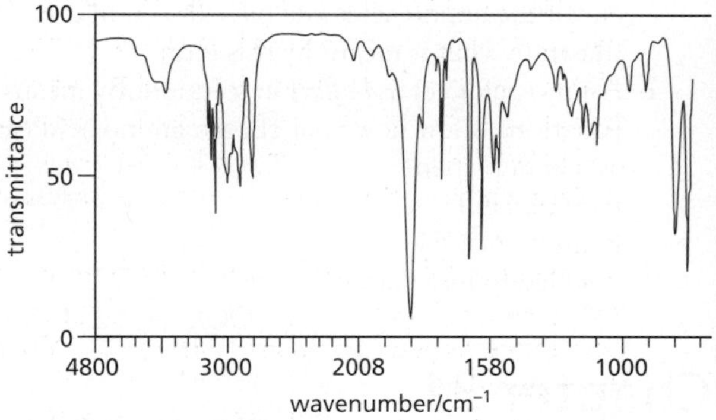

infra-red spectrum number 2

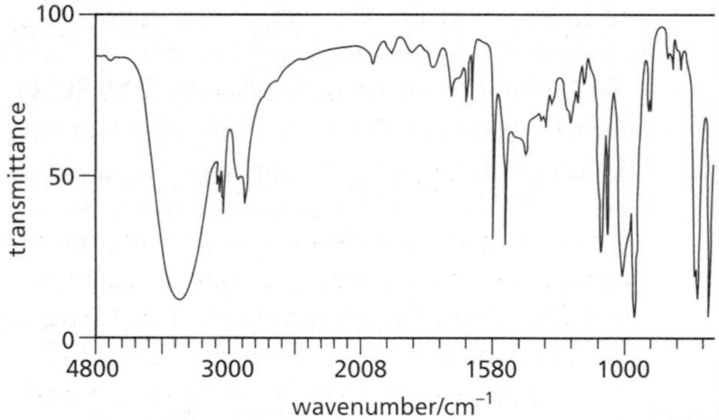

c State which of the substances A, B and C will react with the following reagents and state what would be observed:
i bromine dissolved in hexane (2)
ii a warm ammoniacal solution of silver nitrate (2)
iii 2,4-dinitrophenylhydrazine solution. (3)
iv Give the structural formula of the organic product(s) obtained in **c i**. (1)
v Give the structural formula of the organic product(s) obtained in **c ii**. (1)

d **i** Give the reagents and the conditions necessary for the preparation of A from benzene. (3)
ii This is an electrophilic substitution reaction. Write an equation to show the formation of the electrophile. (2)

Edexcel, A, Module 4, Jan 2001

Synoptic questions

1 a The covalent compound urea, $(NH_2)_2C{=}O$, is commonly used as a fertiliser in most of the European Union whereas in the UK the most popular fertiliser is ionic ammonium nitrate, NH_4NO_3.

i Calculate the percentage of available nitrogen in urea. (2)

ii Apart from the nitrogen content, suggest *two* advantages of using urea as a fertiliser compared with using ammonium nitrate. (2)

b Some organic nitrogen compounds are used to manufacture polyamides by condensation polymerisation.

With the aid of diagrams, define the terms *condensation polymerisation* and *polyamide*. (4)

c The ammonium ion in water has an acid dissociation constant, $K_a = 5.62 \times 10^{-10}\,\text{mol dm}^{-3}$. The conjugate acid of urea has $K_a = 0.66\,\text{mol dm}^{-3}$. Use this data to explain which of ammonia or urea is the stronger base. (2)

d Ethanamide, CH_3CONH_2, can be converted into methylamine, CH_3NH_2.

i State the reagents and conditions for carrying out the conversion. (3)

ii Suggest the formula of the likely product if urea were used instead of ethanamide in this conversion. (1)

e Ammonium nitrate can explode when heated strongly.

$$NH_4NO_3(l) \rightarrow N_2O(g) + 2H_2O(g) \quad \Delta H = -23\,\text{kJ mol}^{-1}$$

With moderate heating the ammonium nitrate volatilises reversibly.

$$NH_4NO_3(s) \rightleftharpoons NH_3(g) + HNO_3(g) \quad \Delta H = +171\,\text{kJ mol}^{-1}$$

i State why the expression for K_p for the reversible change does not include ammonium nitrate. (1)

ii 6.00 g of ammonium nitrate was gently heated in a sealed vessel until equilibrium was reached. The equilibrium constant was found to be $15.7\,\text{atm}^2$ under these conditions. Calculate the partial pressure of ammonia present at equilibrium and, hence, the percentage of the ammonium nitrate which has dissociated. (One mole of gas under these conditions exerts a pressure of 50 atm.) (5)

iii Explain the concepts of *thermodynamic* and *kinetic stability* with reference to these two reactions. (5)

Edexcel, A, Module 6, June 2001

2 Many of the natural compounds which are responsible for aromas in perfumes were first extracted from plants. Geraniol is found in roses and citral in lemon grass.

$$(CH_3)_2C{=}CHCH_2CH_2\underset{\displaystyle CH_3}{\underset{|}{C}}{=}CHCH_2OH$$

geraniol

$$(CH_3)_2C{=}CHCH_2CH_2\underset{\displaystyle CH_3}{\underset{|}{C}}{=}CH\underset{\displaystyle H}{\underset{|}{C}}{=}O$$

citral

Both of these are attractants for honey bees whereas

$$(CH_3)_2CHCH_2CH_2O\underset{\displaystyle CH_3}{\underset{|}{C}}{=}O$$

3-methylbutyl ethanoate

is a bee alarm pheromone and signifies danger to a honey bee but is also the principal cause of the smell of bananas.

a Describe how infra-red spectroscopy could be used to distinguish between geraniol and citral. (3)

b i Suggest the reagents and conditions necessary to convert geraniol to citral in the laboratory. (3)

ii Describe a chemical test you could do to check that geraniol has been converted to citral. (2)

c i What type of compound is 3-methylbutyl ethanoate? (1)

ii If 3-methylbutyl ethanoate is warmed with aqueous sodium hydroxide solution a slow reaction takes place to produce sodium ethanoate and another product, A. Suggest the name of the other product A. (1)

iii This reaction is first order with respect to both 3-methylbutyl ethanoate and the aqueous hydroxide ion. Explain the term *first order* and give experimental details showing how this information could be obtained. (8)

iv Suggest the identity of the organic product if A is treated with phosphorus pentachloride. Give the equation for the reaction. (2)

v If A is oxidised it produces a carboxylic acid $(CH_3)_2CHCH_2CO_2H$. The acid has a boiling temperature of 176 °C. 3-methylbutyl ethanoate has a boiling temperature of 142 °C. Comment on the intermolecular forces involved in the two liquids and hence account for the relative values of their boiling temperatures. (5)

Edexcel, A, Module 6, June 2001

3 a Give the principles which enable you to predict the shape of a particular molecule, given its formula. Draw dot-and-cross diagrams for $SiCl_4$ and XeF_4, and hence draw the shapes of each of these molecules. (6)

b When 2-bromopropanoic acid, $CH_3CH(Br)COOH$, is reacted with sodium hydroxide in aqueous solution, 2-hydroxypropanoic acid is formed. The reaction takes place by an S_n1 mechanism.

i Write an equation for the reaction between 2-bromopropanoic acid and hydroxide ions. (1)

ii Explain why 2-bromopropanoic acid can show optical isomerism. State how this could be detected. (2)

iii If a single optical isomer of 2-bromopropanoic acid is reacted with sodium hydroxide, the resulting 2-hydroxypropanoic acid mixture is not optically active. Write a reaction mechanism for the reaction, and explain why optical activity is lost. (5)

c Optical activity is not confined to organic compounds; chiralty is also seen in some chromium compounds. 1,2-diaminoethane, $H_2NCH_2CH_2NH_2$, forms an octahedral complex with the chromium(III) ion. This complex ion $[Cr(H_2NCH_2CH_2NH_2)_3]^{3+}$ is chiral, and the structure of one isomer is shown below.

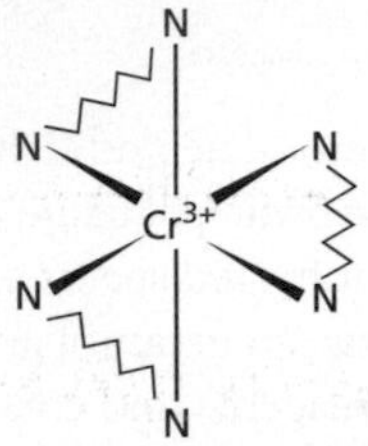

where 1,2-diaminoethane is shown as N〰N.

Sketch the structure of the other isomer and say why the complex ion is chiral. (2)

d A student was given a chromium compound Z to analyse. The following information was obtained:

- Z was orange
- a solution of Z acidified with sulphuric acid converted propan 2-ol to propanone
- a solution of Z acidified with sulphuric acid, reacted with a solution of iron(II) ions in a 1:6 molar ratio Z: Fe^{2+} to give a green solution
- on heating an alkaline solution of Z, ammonia was evolved.

Account as fully as you can for the above observations and identify Z. (9)

Edexcel, A, Module 6, June 2000

4 This question explores the differences and similarities between manganese, a transition metal, and aluminium, a main group metal.

a The following information will be required for this question.

	$E^\ominus$/V
$2H_2O + O_2 + 4e^- \rightleftharpoons 4OH^-$	+0.40
$MnO_4^- + e^- \rightleftharpoons MnO_4^{2-}$	+0.56
$4H^+ + MnO_4^{2-} + 2e^- \rightleftharpoons MnO_2 + 2H_2O$	+2.26

Manganate(VI), MnO_4^{2-}, is a green ion of manganese.

i One way of preparing manganate(VI) ions might be to oxidise manganese(IV) oxide with manganate(VII) ions in aqueous acidic solution. Derive the overall equation for this reaction, and show whether it is feasible under standard conditions. (3)

ii What effect would an increase of pH of the reaction mixture have on the feasibility of the reaction? Explain your answer. (2)

iii If potassium manganate(VII) solution on its own is made very alkaline, it turns green.

Write the equation for the reaction which is occurring. State, with reasons, what is being oxidised. (4)

b The reactions of manganese considered so far arise from its ability to show a variety of oxidation states. Unlike manganese, aluminium does not have this property.

Explain why manganese has several oxidation states whereas aluminium does not. (4)

c Aluminium forms a hexaquo ion, $[Al(H_2O)_6]^{3+}$, similarly to the transition metals.

i Explain why aqueous solutions of aluminium salts are acidic. (2)

ii If sodium hydroxide solution is added to an aqueous solution of an aluminium salt, a white precipitate appears which disappears as more sodium hydroxide is added. Write a series of equations to explain this behaviour. (3)

d i What is the significance of the reactions in **c ii** with respect to the purification of bauxite in the extraction of aluminium? (2)

ii Outline the manufacture of aluminium metal from purified bauxite. (4)

e The standard electrode potential for the aluminium electrode is such that the metal could, in principle, be oxidised by an acidic solution of manganate(VII) ions. Why in practice does this reaction not occur? (1)

Edexcel, A, Module 6, June 2000

5 a Using a Boltzmann distribution diagram, explain how altering the temperature affects the rate of a chemical reaction. (7)

b Chlorate(I) ions can be converted into chlorate(V) ions according to the stoichiometric equation below.

$$3ClO^-(aq) \rightarrow ClO_3^-(aq) + 2Cl^-(aq)$$

chlorate(I) chlorate(X)

When the rate of this reaction was investigated, it was found to be second order overall and second order with respect to ClO^-.

i Using this reaction as an example, state the meaning of the terms

- stoichiometric equation
- second order reaction.

ii Deduce a rate equation for this reaction.

iii Suggest a mechanism for this reaction, indicating the relative rates of any stages. (6)

c Chloric(I) acid, HClO, is a weak acid with an acid dissociation constant of 3.7×10^{-8} mol dm^{-3} at 25 °C.

i Define the terms

- weak acid
- acid dissociation constant.

ii Calculate the pH of 0.010 mol dm^{-3} chloric(I) acid at 25 °C.

iii Calculate the OH^- concentration in this 0.010 mol dm^{-3} chloric(I) acid solution. (8)

OCR, AS/A, 4826, Mar 2001

6 Hydrogen cyanide has been used in America to execute convicted murderers. Potassium cyanide capsules were dropped into hydrochloric acid to form lethal hydrogen cyanide gas.

a Write the equation for the reaction between potassium cyanide and hydrochloric acid. (1)

b Hydrogen cyanide has also been used in chemical warfare.

Concentration/mg m^{-3}	Effect
300	Immediately lethal
150	Lethal after 30 minutes
30	Light symptoms of dizziness/nausea after several hours.

A ketone-based antidote has been developed.

i Name the organic product formed in the reaction between propanone and hydrogen cyanide. (2)

ii Draw a flow scheme to illustrate the mechanism of this reaction. (3)

iii What name is given to this mechanism? (1)

c Cyanide ions act as ligands and bond to Fe^{2+} ions to form $[Fe(CN)_6]^{4-}$. What is meant by the term *ligand?* (2)

d Cyanide ions can act as nucleophiles when reacting with 1-iodobutane.

i Write the equation for this reaction. (2)

ii What is meant by the term *nucleophile?* (2)

e Aqueous hydroxide ions can hydrolyse halogenoalkanes. Without practical details, state and explain the ease of hydrolysis of 1-chlorobutane, 1-bromobutane and 1-iodobutane in terms of bond enthalpy and bond polarity.
(Up to two marks may be obtained for the quality of language in this part.) (6)

CCEA, A, *Module 2, June 2001*

7 Potassium iodide is a typical salt and is completely ionised under all conditions.

a i Use the Born–Haber cycle below to determine the lattice enthalpy for potassium iodide.

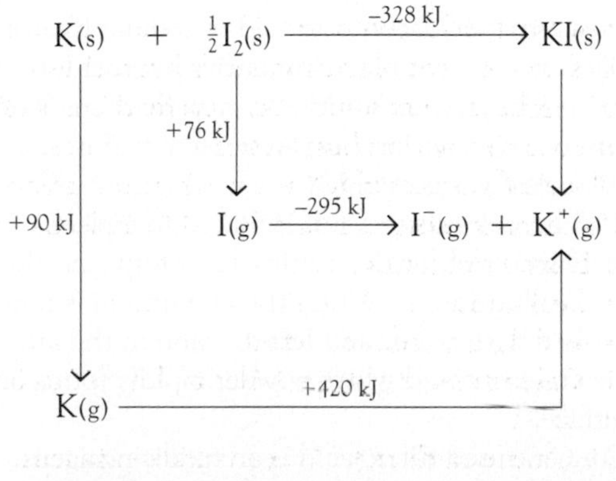

(3)

ii The lattice enthalpy can also be obtained from the enthalpy of solvation and the enthalpy of solution. Name the enthalpy changes represented by X, Y and Z in the diagram below.

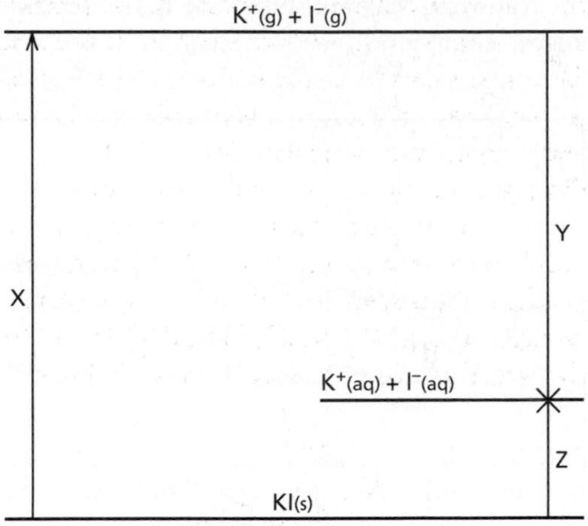

(2)

iii Use the following values to determine the lattice enthalpy.

enthalpy of solvation of potassium iodide = $-10\,kJ\,mol^{-1}$
enthalpy of solution of potassium iodide = $-22\,kJ\,mol^{-1}$

iv The lattice enthalpies of the other iodides in Group 1 are:

LiI	NaI	RbI
753	743	624

Explain this trend. (2)

b Potassium iodide is very soluble in water; 100 g of water dissolves 144 g of potassium iodide at 20 °C.

i Calculate the molarity of the solution. (2)

ii Explain why a solution of potassium iodide does *not* undergo salt hydrolysis. (2)

iii What colour will be produced when potassium iodide solution is sprayed into a Bunsen flame? (1)

c State what is observed when solid potassium iodide is reacted with concentrated sulphuric acid. (3)

d Iodobenzene is not made by the direct iodination of benzene as the reaction is reversible.

$$C_6H_6 + I_2 \rightleftharpoons C_6H_5I + HI$$

Potassium iodide and phenylamine may be used to prepare iodobenzene via a diazonium salt. Use a flow scheme to explain how this is done. (3)

CCEA, A, *Module 3, June 2001*

8 a Design a *two stage* synthesis of benzocaine, a modern local anaesthetic, from 4-nitrobenzenecarboxylic acid. State the names of the functional groups formed and the reagent(s) and condition(s) required at each stage.

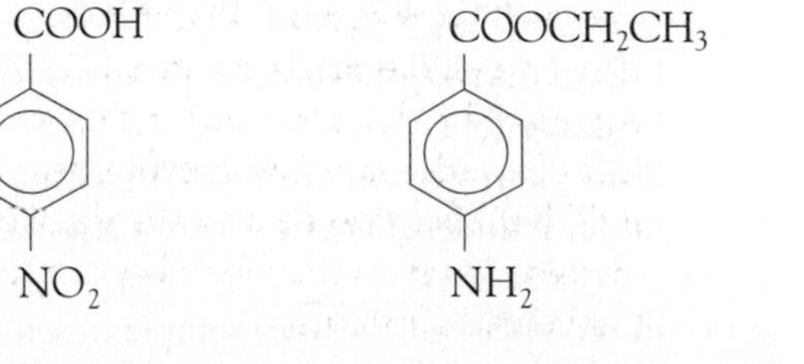

4-nitrobenzenecarboxylic acid   benzocaine

(4)

b The following diagram shows the mass spectrum of an organic compound.

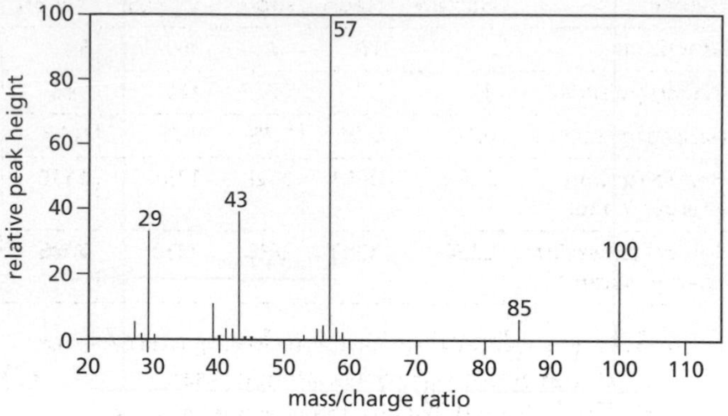

Information

The molecular ion is present.

The compound:

- reacts with a warm solution of sodium hydroxide and iodine to give a yellow crystalline precipitate.
- forms a secondary alcohol on reduction by sodium tetrahydridoborate(III) (sodium borohydride), $NaBH_4$.
- contains only one functional group.

i Draw a possible structure for the fragment ion of mass/charge ratio 57, which contains 4 carbon atoms in a branched chain. (1)

ii Draw the full structural (graphic) formula of the organic compound which contains the group in **b i**. Give a brief reasoning for your choice. (3)

c i Experimentally determined values of the standard molar enthalpy changes of combustion, $\Delta H_c^\ominus$, for benzene, carbon and hydrogen were found to be −3280, −393 and −286 kJ mol^{-1} respectively. Using these values, calculate the standard molar enthalpy change of formation, $\Delta H_f^\ominus$, for benzene. *Show your working.* (3)

ii Using average bond energy data, the theoretical value for the enthalpy change of combustion for benzene is −3429 kJ mol^{-1}. The experimentally determined value is −3280 kJ mol^{-1}. Explain the difference between these two values. (2)

WJEC, A, A2, June 2000

9 a Bromine was isolated from sea water, from which it is still obtained, by the French chemist, Balard, in 1826. In the first stage of extraction chlorine is passed through sea water to oxidise bromide ions.

$$Cl_2 + 2Br^- \rightarrow 2Cl^- + Br_2$$

i Explain why this process involves the *oxidation* of bromide ions. (1)

ii Explain why, in terms of its position in the periodic table, iodine *cannot* be used for this oxidation process. (1)

iii A 1000 kg (1 tonne) sample of sea water contains 64 g of bromide ions.

I Calculate how many tonnes of sea water need to be processed to produce 1 tonne of bromine. (1)

II Calculate the minimum mass of chlorine required to convert all the bromide ions to bromine in 1000 kg of sea water. (1)

b Calcium bromide is a relatively safe compound of bromine and could be used to avoid transporting corrosive bromine.

i When sodium carbonate solution is added to calcium bromide solution a precipitate of calcium carbonate occurs. Explain *briefly*, in terms of lattice enthalpies and the hydration enthalpies of the ions, why calcium carbonate is largely insoluble in water. (1)

ii State what would be *observed* and give an *ionic* equation for the reaction between aqueous bromide ions and aqueous silver ions. (1)

c The agricultural chemical 1,2-dibromoethane is made by reacting ethene with bromine.

$$\begin{array}{c} H \\ \diagdown \\ \diagup \\ H \end{array} C = C \begin{array}{c} H \\ \diagup \\ \diagdown \\ H \end{array} + Br_2 \longrightarrow H - \begin{array}{c} H \\ | \\ C \\ | \\ Br \end{array} - \begin{array}{c} H \\ | \\ C \\ | \\ Br \end{array} - H$$

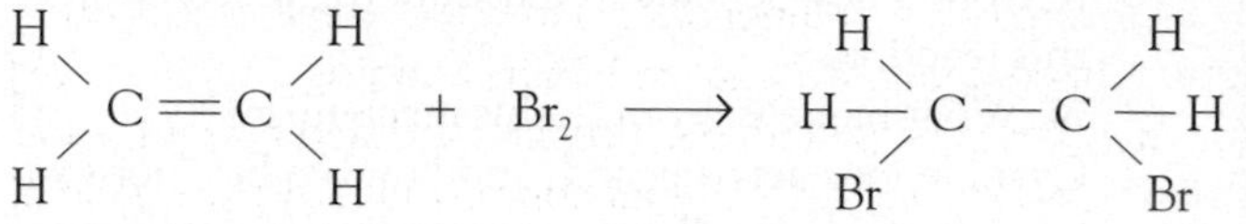

Given the bond energies below calculate the enthalpy change, in kJ mol^{-1}, for the reaction assuming that the products and reactants are gaseous.

Bond	Energy/kJ mol^{-1}
C—H	413
C=C	612
Br—Br	193
C—C	347
C—Br	290

(3)

WJEC, A, C4, June 2000

10 a Suggest an explanation for the following observations and write equations for the reactions occurring.

i When left in air, pure aluminium forms a protective coating. This coating can be removed by treatment with either hydrochloric acid or sodium hydroxide solutions. When placed in either hydrochloric acid or sodium hydroxide solutions, initially there is no observable reaction but, after a few minutes, a colourless gas is evolved. (10)

ii When aluminium foil is placed in aqueous mercury(II) chloride, a shiny layer forms on the surface of the aluminium. When the aluminium is removed, washed with water and left to stand in the air, it becomes hot and white powder rapidly forms on its surface. (5)

b Concentrated nitric acid is an oxidising agent. Consider the following reaction scheme and answer the questions.

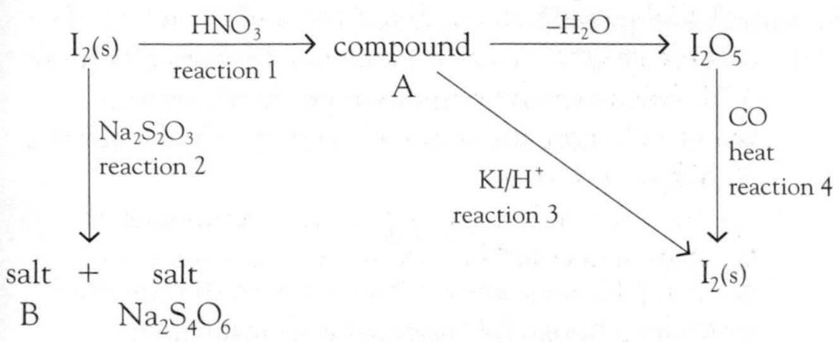

i Identify compound A and salt B. (2)
ii Write a half equation for the reduction of concentrated nitric acid to NO_2 and write equations for reactions 1 to 4. (8)

c When boron trichloride (BCl_3) and iodine monochloride (ICl) react, a single product is formed. This product, compound C, contains one anion and one cation. The cation acts as an electrophile in the preparation of iodobenzene from benzene. Suggest a formula for compound C. Draw the shape of the anion in compound C and explain its bonding. (5)

d i Four compounds each contain a different octahedral Co(III) complex. All four compounds have the general empirical formula $[CoCl_x(NH_3)_y]Cl_z$. Two of the complexes have the same formula but have the ligands arranged differently in space and so are isomers of each other.

Treatment of an aqueous solution of each compound with $AgNO_3$(aq) will give a white precipitate of AgCl. The table below contains details of the colours of each complex and of the number of moles of $AgNO_3$ which would react with 1.0 mol of each compound.

Number of complex	1	2	3	4
Colour of solid complex	Golden brown	Purple	Green	Violet
Amount of $AgNO_3$ required per mole of compound/mol	3.0	2.0	1.0	1.0

Suggest the structures of the four complexes. Explain your reasoning. Draw the octahedral structures of the two isomers.

ii The ligand, 1,2-diaminoethane ($H_2NCH_2CH_2NH_2$), can be represented as 'en'.

The ion Co^{3+} reacts with 'en' to give $[Co(en)_3]^{3+}$, which is an octahedral complex ion. The complex can exist in two different forms, which are mirror images of each other.

Identify the feature of 'en' that enables it to form two co-ordinate bonds with a transition metal ion. Explain why only three molecules of 1,2-diaminoethane are required to form $[Co(en)_3]^{3+}$.

Draw a diagram of $[Co(en)_3]^{3+}$ which clearly shows the three-dimensional structure of the complex ion. Draw a second diagram to show the structure of its mirror image. (5)

AQA, A, Extension, 2001/2

11 The data presented below concern the first five elements which are found in Group VIII of the periodic table.

Element	**Helium**	**Neon**	**Argon**	**Krypton**	**Xenon**
Atomic no.	2	10	18	36	54
Boiling point/K	4	27	87	121	166
Density/g cm^{-3}	0.18	0.90	1.78	3.75	5.90
First ionisation energy/ kJ mol^{-1}	2372	2081	1520	1351	1170
Second ionisation energy/ kJ mol^{-1}	5250	3952	2665	2350	2046

a Suggest explanations for the following observations.
i All the elements are monatomic gases. (1)
ii The boiling points increase as the group is descended. (2)
iii Helium, neon and argon can truly be described as 'inert gases', but the later elements in the group can be oxidized under extreme conditions to form a number of compounds. (2)
iv The ionisation energies decrease down the group, but the differences between helium and neon or between neon and argon, are much greater than those between argon and krypton, or between krypton and xenon. (2)
v The difference between the first and second ionisation energies decreases as the group is descended. (2)

b By considering the number of electron pairs, draw the structures of the Xe compounds below. You should include all lone pairs on xenon and indicate approximate bond angles.

XeF_2 $XeOF_2$ XeO_4 XeF_4 $XeOF_4$ (8)

c The fluoride XeF_4 can be prepared by direct fluorination of Xe gas. The oxyfluoride $XeOF_2$ can be prepared by hydrolysis of XeF_4. $XeOF_2$ disproportionates above −20 °C to form XeO_2F_2 and XeF_2.
i Write balanced equations for the preparation of $XeOF_2$. (1)
ii Suggest what is meant by the term *disproportionate*. (1)
iii Write a balanced equation for the disproportionation of $XeOF_2$. (1)

OCR, A, 9485 (STEP), 1999

12 a i Write out the electronic configurations of fluorine, F, and selenium, Se, in terms of s, p and d electrons. (2)
ii Define the term *ionisation energy*. (1)
iii Which will be greater – the first ionisation energy of fluorine or the first ionisation energy of selenium? Explain your answer. (2)
iv Sketch a graph to show the energies needed to remove in succession the first seven electrons from an atom of selenium. Label clearly any jumps you predict and explain the form of your graph. (4)

b Natural fluorine exists as a single isotope with relative atomic mass 18.9984.

Using fluorine as an example, define the term *isotope*. (2)

c Selenium tetrafluoride, SeF_4, can be prepared by the controlled reaction of selenium and fluorine at 0 °C. When SeF_4 is analysed using mass spectrometry, six peaks between 149 and 158 are observed. The accurate masses of the peaks and their relative intensities are given in the table below.

Mass of peak	Relative intensity/%
149.9161	0.89
151.9128	9.36
152.9135	7.63
153.9109	23.78
155.9101	49.61
157.9103	8.73

i Predict the shape of the SeF_4 molecule. (2)

ii Suggest why there are six peaks in the mass spectrum of SeF_4. (2)

iii Calculate the relative atomic mass of naturally occurring selenium. (3)

iv Suggest why the relative atomic mass of selenium is only usually quoted to two decimal places whereas that of fluorine can be quoted to seven decimal places. (2)

OCR, A, 9485 (STEP), 2001

13 a The pK_w for water at 25 °C is 14.000, but at 100 °C it is 12.322.

i Showing your working, calculate the value of K_w at 25 °C and at 100 °C. (2)

ii What is the sign of $\Delta H^\ominus$ for the dissociation of water? Justify your answer. (2)

iii Calculate the pH of water at 100 °C. (2)

iv Showing your working, calculate the pH of 0.01 mol dm^{-3} sodium hydroxide at 25 °C and at 100°C. (3)

b Hydrogen exists as a mixture of three isotopes: normal hydrogen, deuterium (which can be represented by the symbol D), and tritium, T. Relative atomic masses: D = 2; T = 3.

i Explain what the term *isotope* means with reference to the isotopes of hydrogen. (2)

ii Showing your working, calculate the density of liquid T_2O, stating any assumptions you make. (3)

c Consider the following thermodynamic data.

Enthalpy change of formation of HCl(g) = −92 kJ mol^{-1}

Enthalpy change of dissociation of Cl_2(g), $Cl_2(g) \rightarrow 2Cl(g)$ = +242 kJ mol^{-1}

Enthalpy change of dissociation of H_2(g), $H_2(g) \rightarrow 2Cl(g)$ = +435 kJ mol^{-1}

Electron affinity of Cl(g), $Cl(g) + e^- \rightarrow Cl^-(g)$ = −364 kJ mol^{-1}

Ionisation energy of H(g), $H(g) \rightarrow H^+(g) + e^-$ = +569 kJ mol^{-1}

i By constructing an appropriate thermodynamic cycle, calculate the standard enthalpy change for the following reaction.

$HCl(g) \rightarrow H^+(g) + Cl^-(g)$ (4)

ii Compare your answer from **c i** with the standard enthalpy change for the process shown below.

$HCl(g) + aq \rightarrow H^+(aq) + Cl^-(aq)$

$\Delta H^\ominus = -75\ kJ\ mol^{-1}$. (2)

OCR, A, 9485 (STEP), 1998

14 a The data below refer to three metallic elements of the third row of the periodic table: potassium, K; calcium, Ca; and copper, Cu.

Element	Potassium	Calcium	Copper
Proton (atomic) number	19	20	29
Metallic radius/nm	0.231	0.197	0.128
Density/g cm^{-3}	0.86	1.54	8.92
Electronic configuration	[Ar]4s^1	[Ar]4s^2	[Ar]3d^{10}4s^1
First ionisation energy/ kJ mol^{-1}	418	590	745
Second ionisation energy/ kJ mol^{-1}	3070	1150	1960

Explain the following observations, using the data as necessary.

i The metallic radii decrease as the proton number increases. (1)

ii Copper has a much higher density than either potassium or calcium. (1)

iii Copper can form both Cu^+ and Cu^{2+} ions whereas potassium and calcium favour K^+ and Ca^{2+} ions respectively. (2)

iv The chlorides KCl, $CaCl_2$ and CuCl are typically white whereas $CuCl_2.2H_2O$ is green. (2)

b The following question refers to the *Solvay Process*, used in the industrial preparation of Na_2CO_3 from limestone ($CaCO_3$) and sodium chloride.

i During the Solvay Process, limestone is strongly heated to form a gaseous product. Identify the gas and write a balanced equation for this reaction. (1)

ii This gaseous product is then reacted with aqueous ammonia and aqueous sodium chloride to form sodium hydrogencarbonate. Write a balanced equation for this reaction. (1)

iii The sodium hydrogencarbonate decomposes on heating to form Na_2CO_3. Write a balanced equation for this reaction. (1)

iv Write a balanced equation for the overall reaction. (1)

v The direct reaction between limestone and sodium chloride alone does not yield Na_2CO_3. Discuss why this might be so, and comment on the role of the aqueous ammonia in the reaction. (2)

c Copper reacts with dilute nitric acid to produce a colourless, diatomic gas (A) and a blue-green solution, which on evaporation yields a blue solid (B) containing 21.5% Cu, by mass. Exposure of A to air rapidly leads to the evolution of a brown gas (C) containing 30.4% N, by mass. On gentle heating, B evolves a colourless gas (D) which readily condenses on a cold watch glass. On stronger heating, more brown gas (C) forms and a black residue (E) remains which contains 79.9% Cu, by mass. [Relative atomic masses: H = 1; N = 14; O = 16; Cu = 63.5.]

i Identify compounds A to E. (4)

ii Write balanced equations for the chemical processes involved. (4)

OCR, A, 9485 (STEP), 1998

Periodic table

p block

d block

Period	Group 1	2											3	4	5	6	7	0
1							1 **H** hydrogen 1											4 **He** helium 2
2	7 **Li** lithium 3	9 **Be** beryllium 4											11 **B** boron 5	12 **C** carbon 6	14 **N** nitrogen 7	16 **O** oxygen 8	19 **F** fluorine 9	20 **Ne** neon 10
3	23 **Na** sodium 11	24 **Mg** magnesium 12											27 **Al** aluminium 13	28 **Si** silicon 14	31 **P** phosphorus 15	32 **S** sulphur 16	35.5 **Cl** chlorine 17	40 **Ar** argon 18
4	39 **K** potassium 19	40 **Ca** calcium 20	45 **Sc** scandium 21	48 **Ti** titanium 22	51 **V** vanadium 23	52 **Cr** chromium 24	55 **Mn** manganese 25	56 **Fe** iron 26	59 **Co** cobalt 27	59 **Ni** nickel 28	63.5 **Cu** copper 29	65 **Zn** zinc 30	70 **Ga** gallium 31	73 **Ge** germanium 32	75 **As** arsenic 33	79 **Se** selenium 34	80 **Br** bromine 35	84 **Kr** krypton 36
5	85 **Rb** rubidium 37	88 **Sr** strontium 38	89 **Y** yttrium 39	91 **Zr** zirconium 40	93 **Nb** niobium 41	96 **Mo** molybdenum 42	99 **Tc** technetium 43	101 **Ru** ruthenium 44	103 **Rh** rhodium 45	106 **Pd** palladium 46	108 **Ag** silver 47	112 **Cd** cadmium 48	115 **In** indium 49	119 **Sn** tin 50	122 **Sb** antimony 51	128 **Te** tellurium 52	127 **I** iodine 53	131 **Xe** xenon 54
6	133 **Cs** caesium 55	137 **Ba** barium 56		178.5 **Hf** hafnium 72	181 **Ta** tantalum 73	184 **W** tungsten 74	186 **Re** rhenium 75	190 **Os** osmium 76	192 **Ir** iridium 77	195 **Pt** platinum 78	197 **Au** gold 79	201 **Hg** mercury 80	204 **Tl** thallium 81	207 **Pb** lead 82	209 **Bi** bismuth 83	209 **Po** polonium 84	210 **At** astatine 85	222 **Rn** radon 86
7	223 **Fr** francium 87	226 **Ra** radium 88		261 **Rf** rutherfordium 104	262 **Db** dubnium 105	263 **Sg** seaborgium 106	262 **Bh** bohrium 107	265 **Hs** hassium 108	266 **Mt** meitnerium 109	269 **Uun** ununnillium 110	272 **Uuu** unununium 111	277 **Uub** ununbium 112						

s block

139 **La** lanthanum 57	140 **Ce** cerium 58	141 **Pr** praseodymium 59	144 **Nd** neodymium 60	145 **Pm** promethium 61	150 **Sm** samarium 62	152 **Eu** europium 63	157 **Gd** gadolinium 64	159 **Tb** terbium 65	162 **Dy** dysprosium 66	165 **Ho** holmium 67	167 **Er** erbium 68	169 **Tm** thulium 69	173 **Yb** ytterbium 70	175 **Lu** lutetium 71
227 **Ac** actinium 89	232 **Th** thorium 90	231 **Pa** protactinium 91	238 **U** uranium 92	237 **Np** neptunium 93	244 **Pu** plutonium 94	243 **Am** americium 95	247 **Cm** curium 96	247 **Bk** berkelium 97	251 **Cf** californium 98	252 **Es** einsteinium 99	257 **Fm** fermium 100	258 **Md** mendelevium 101	259 **No** nobelium 102	260 **Lr** lawrencium 103

f block

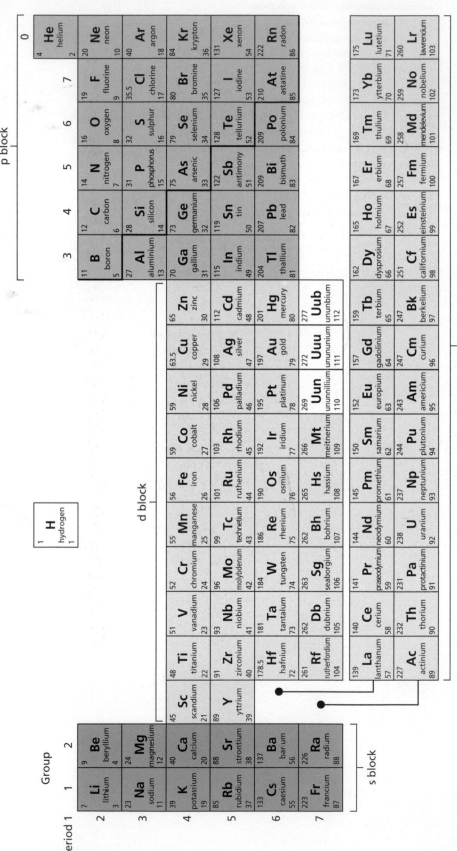

Relative atomic masses of the elements

Hydrogen	1
Helium	4
Lithium	7
Beryllium	9
Boron	11
Carbon	12
Nitrogen	14
Oxygen	16
Fluorine	19
Neon	20
Sodium	23
Magnesium	24
Aluminium	27
Silicon	28
Phosphorus	31
Sulphur	32
Chlorine	35.5
Argon	40
Potassium	39
Calcium	40
Scandium	45
Titanium	48
Vanadium	51
Chromium	52
Manganese	55
Iron	56
Cobalt	59
Nickel	59
Copper	63.5
Zinc	65
Gallium	70
Germanium	73
Arsenic	75
Selenium	79
Bromine	80
Krypton	84
Rubidium	85
Strontium	88
Yttrium	89
Zirconium	91
Niobium	93
Molybdenum	96
Technetium	99
Ruthenium	101
Rhodium	103
Palladium	106
Silver	108
Cadmium	112
Indium	115
Tin	119
Antimony	122
Tellurium	128
Iodine	127
Xenon	131
Caesium	133
Barium	137
Lanthanum	139
Cerium	140
Praseodymium	141
Neodymium	144
Promethium	145
Samarium	150
Europium	152
Gadolinium	157
Terbium	159
Dysprosium	162
Holmium	165
Erbium	167
Thulium	169
Ytterbium	173
Lutetium	175
Hafnium	178.5
Tantalum	181
Tungsten	184
Rhenium	186
Osmium	190
Iridium	192
Platinum	195
Gold	197
Mercury	201
Thallium	204
Lead	207
Bismuth	209
Polonium	209
Astatine	210
Radon	222
Francium	223
Radium	226
Actinium	227
Thorium	232
Protactinium	231
Uranium	238
Neptunium	237
Plutonium	244
Americium	243
Curium	247
Berkelium	247
Californium	251
Einsteinium	252
Fermium	257
Mendelevium	258
Nobelium	259
Lawrencium	260
Rutherfordium	261
Dubnium	262
Seaborgium	263
Bohrium	262
Hassium	265
Meitnerium	266
Ununnilium	(269)
Unununium	(272)
Ununbium	(277)

Numerical answers to further practice questions

Chapter 1

Page 8
1 1.5 mol
2 0.75 mol
3 0.67 mol

Page 9
1 9
2 11
3 12
4 24
5 22

Page 11
1 **a** 151.9 **b** 162.1 **c** 60 **d** 132.1 **e** 388.8
2 **a** 0.50 mol **b** 2.50 mol **c** 0.60 mol **d** 2.00 mol **e** 0.38 mol
3 **a** 180.6 g **b** 44.46 g

Page 12
1 Cu_2S
2 C_3H_8
3 Fe_2CaO_4

Page 16
1 17.0 g
2 6.07 g

Pages 18–19
1 **a** 2.67 mol dm^{-3} **b** 0.025 mol dm^{-3} **c** 0.833 mol dm^{-3}
2 **a** 0.75 mol **b** 0.044 mol **c** 0.01 mol

Page 19
1 1.12 dm^3
2 12 dm^3
3 0.411 dm^3 of H_2SO_4, 4.9 dm^3 of H_2
4 0.64 g
5 1/27 mol dm^{-3}

Chapter 2

Page 26
1 52.06
2 B = 10.8; Ne = 20.2; Mg = 24.3
3 ^{193}Ir : ^{192}Ir = 1.6 : 1

Page 28
6.4×10^{13} tonnes

Chapter 4

Page 106
1 0.12 mol
2 3.7 mol
3 2.9×10^{-3} mol

Chapter 5

Page 113
1 **b** +1800 J **c** +18 kJ mol^{-1}
2 +26 kJ mol^{-1}

Page 114
1 **a** +1380 J **b** 0.025 mol **c** –55 kJ mol^{-1}
2 +52 kJ mol^{-1}
3 **b** –640 kJ mol^{-1}

Page 119
2 –900 kJ mol^{-1}

Page 124
2 –265 kJ mol^{-1}
4 **a** –43.6 kJ mol^{-1} **b** –60.5 kJ mol^{-1} **c** +330.2 kJ mol^{-1} **d** –479.4 kJ mol^{-1} **e** –132.7 kJ mol^{-1}

Chapter 6

Page 140
1 97.3%
2 $x = 2$

Page 143
1 98.6%
2 $x = 1$
3 24.8%

Chapter 7

Page 154
1 **a** 0.0393 mol dm^{-3} **b** 0.550 g **c** 22.3%
2 **a** 2.5
3 **a** 2.85×10^{-4} mol **b** 7.125×10^{-3} mol **c** 184 **d** $x = 1$

Page 156
1 **a** 1.05×10^{-3} mol **b** 5.25×10^{-4} mol
2 67.0%

Chapter 8

Page 162
2 **b i** 0.0020 s^{-1} **ii** 0.0016 s^{-1}
3 **c** 0.26 mol dm^{-3}

Chapter 10

Page 185
1 366.1 kJ mol^{-1}
2 +463.5 kJ mol^{-1}
3 +313.2 kJ mol^{-1}

Page 186
1 **a** +1722 kJ mol^{-1} **c** +158.0 kJ mol^{-1} **d** –92 kJ mol^{-1}
2 +345.5 and 413.8 kJ mol^{-1}

Page 189
1 a 7039.3 kJ mol^{-1} b +347.2 kJ mol^{-1}
2 +391 kJ mol^{-1}

Page 193
1 b –3142 kJ mol^{-1}

Page 195
2 –6286 kJ mol^{-1}
4 +519 kJ mol^{-1}

Chapter 11

Page 216
3 0.016 mol^{-1} dm^3 s^{-1}

Page 221
1 a 54 kJ mol^{-1} b 580 kJ mol^{-1}
2 a 88 kJ mol^{-1} b 1×10^{15} mol^{-1} dm^3 s^{-1}

Page 223
1 a 3.6×10^{-18} b 5.8×10^{-6}
2 1.6×10^{12}

Chapter 12

Page 233
1 b 3.0
2 a H_2 = 0.5 mol, N_2 = 1.5 mol, NH_3 = 1.0 mol
b 2.1×10^3 mol^{-2} dm^6

Page 234
NO_2 = 0.41 mol, N_2O_4 = 0.29 mol (reject –ve solution)

Page 237
1 p_{SO_2} = 0.80 atm, p_{O_2} = 0.40 atm, p_{SO_3} = 0.12 atm
2 5.6 atm^{-1}

Page 239
0.69

Page 243
1 a 7.6×10^{-4} mol dm^{-3} b 2.5×10^{-5} mol dm^{-3}
2 a 1.6×10^{-3} mol dm^{-3} b 1.1×10^{-3} mol dm^{-3}

Page 245
1 3.7
2 11.5
3 6.8

Page 252
+54.3 kJ mol^{-1}

Chapter 13

Page 261
1 0.089 V more positive
2 zero
3 0.37 V less positive

Page 264
1 a 448 g b 37 cm^3

Chapter 22

Page 418
1 69%
2 61%

Chapter 23

Page 432
55%
C_5H_8

Page 434
ΔH_3 = +47 kJ mol^{-1}
ΔH_4 = +137 kJ mol^{-1}

Chapter 31

Page 586
C = 41.3%, H = 3.4%; O = 55.3%
1 **B** = CHO
2 **C** = CH_2O **D** = C_3H_9N **E** = $C_3H_6O_2N$

Index